OF THE ELEMENTS

10	11 IB	12 IIB	13 IIIA	14 IVA	15 VA	16 VIA	17 VIIA	
								Helium **He** 4.00
			5 Boron **B** 10.81	6 Carbon **C** 12.01	7 Nitrogen **N** 14.01	8 Oxygen **O** 16.00	9 Fluorine **F** 19.00	10 Neon **Ne** 20.18
			13 Aluminum **Al** 26.98	14 Silicon **Si** 28.09	15 Phos-phorus **P** 30.97	16 Sulfur **S** 32.06	17 Chlorine **Cl** 35.45	18 Argon **Ar** 39.95
28 Nickel **Ni** 58.69	29 Copper **Cu** 63.54	30 Zinc **Zn** 65.37	31 Gallium **Ga** 69.72	32 Germa-nium **Ge** 72.61	33 Arsenic **As** 74.92	34 Selenium **Se** 78.96	35 Bromine **Br** 79.91	36 Krypton **Kr** 83.80
46 Palladium **Pd** 106.42	47 Silver **Ag** 107.87	48 Cadmium **Cd** 112.41	49 Indium **In** 114.82	50 Tin **Sn** 118.69	51 Antimony **Sb** 121.75	52 Tellurium **Te** 127.60	53 Iodine **I** 126.90	54 Xenon **Xe** 131.29
78 Platinum **Pt** 195.08	79 Gold **Au** 196.97	80 Mercury **Hg** 200.59	81 Thallium **Tl** 204.37	82 Lead **Pb** 207.19	83 Bismuth **Bi** 208.98	84 Polonium **Po** (209)	85 Astatine **At** (210)	86 Radon **Rn** (222)

65 Terbium **Tb** 158.92	66 Dys-prosium **Dy** 162.50	67 Holmium **Ho** 164.93	68 Erbium **Er** 167.26	69 Thulium **Tm** 168.93	70 Ytterbium **Yb** 173.04	71 Lutetium **Lu** 174.97
97 Berkelium **Bk** (247)	98 Califor-nium **Cf** (251)	99 Einstei-nium **Es** (252)	100 Fermium **Fm** (257)	101 Mende-levium **Md** (258)	102 Nobelium **No** (259)	103 Lawren-cium **Lr** (260)

PRINCIPLES & APPLICATIONS OF Organic & Biological Chemistry

PRINCIPLES & APPLICATIONS OF

Organic & Biological Chemistry

ROBERT L. CARET

Towson State University

KATHERINE J. DENNISTON

Towson State University

JOSEPH J. TOPPING

Towson State University

Wm. C. Brown Publishers
A Division of Wm. C. Brown Communications, Inc.

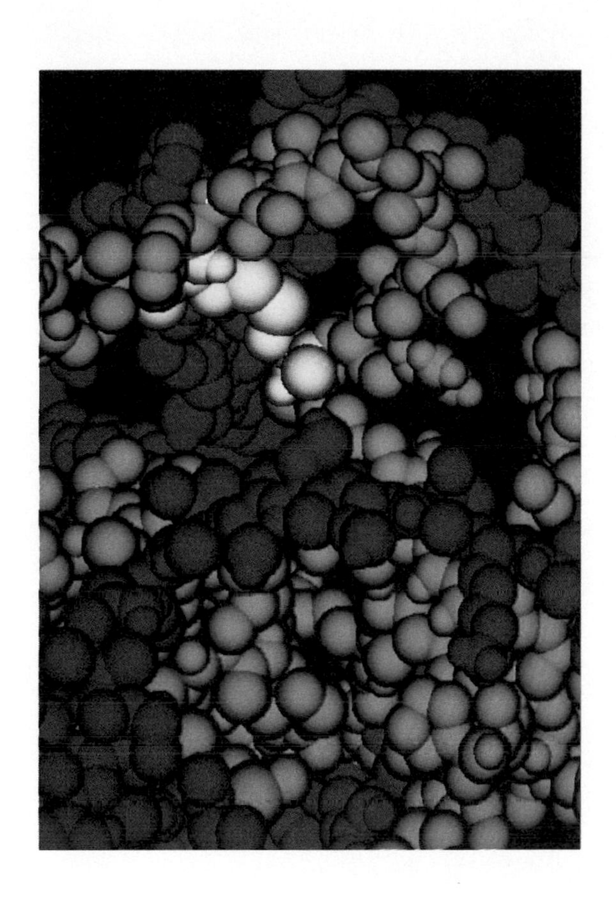

To all our children.

RLC
KJD
JJT

Book Team

Editor Craig Marty
Developmental Editor Elizabeth M. Sievers
Production Editorial Manager Julie A. Kennedy
Visuals/Design Freelance Specialist Barbara J. Hodgson

Wm. C. Brown Publishers
A Division of Wm. C. Brown Communications, Inc.

Vice President and General Manager *George Bergquist*
National Sales Manager *Vincent R. Di Blasi*
Assistant Vice President, Editor-in-Chief *Edward G. Jaffe*
Marketing Manager *John W. Calhoun*
Advertising Manager *Amy Schmitz*
Managing Editor, Production *Colleen A. Yonda*
Manager of Visuals and Design *Faye M. Schilling*
Publishing Services Manager *Karen J. Slaght*
Permissions/Records Manager *Connie Allendorf*

Wm. C. Brown Communications, Inc.

Chairman Emeritus *Wm. C. Brown*
Chairman and Chief Executive Officer *Mark C. Falb*
President and Chief Operating Officer *G. Franklin Lewis*
Corporate Vice President, Operations *Beverly Kolz*
Corporate Vice President, President of WCB Manufacturing *Roger Meyer*

Front/back cover: © John Colletti/THE PICTURE CUBE

Copyediting, permissions, and production by York Production Services, Inc.

Cover and interior design by York Production Services, Inc.

The credits section for this book begins on page 543 and is considered an extension of the copyright page.

Library of Congress Catalog Card Number: 91–76894

ISBN 0–697–13742–2

Printed in the United States of America by Wm. C. Brown Communications, Inc., 2460 Kerper Boulevard, Dubuque, IA 52001

10 9 8 7 6 5 4 3 2 1

Contents

Brief Contents

Expanded Contents

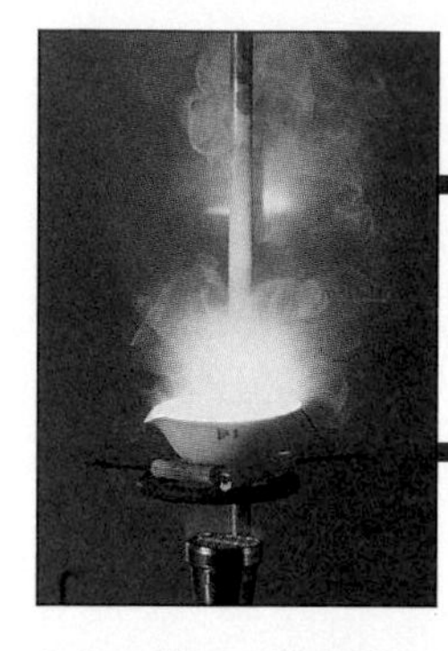

CHAPTER 2

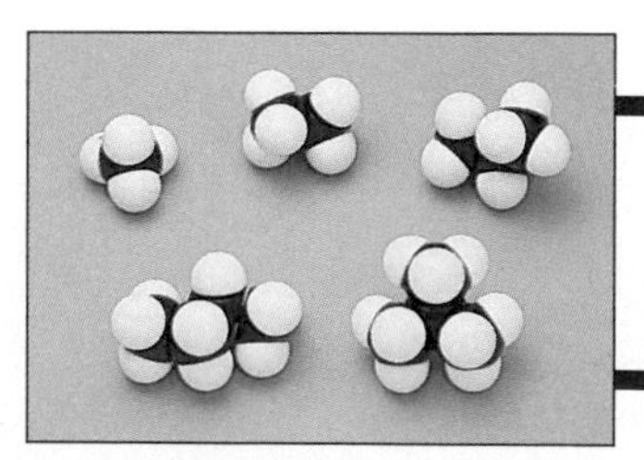

CHAPTER 3

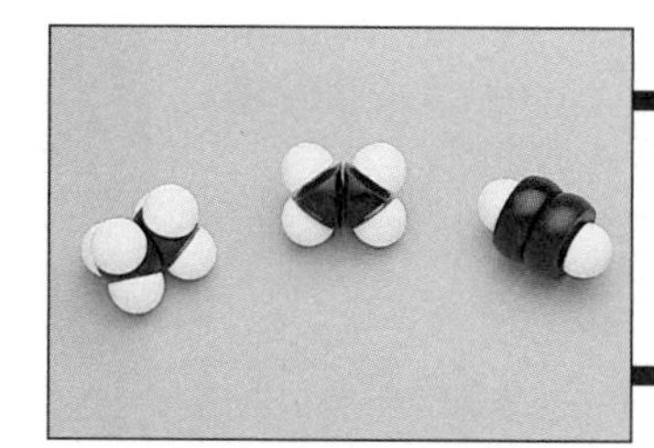

CHAPTER 4

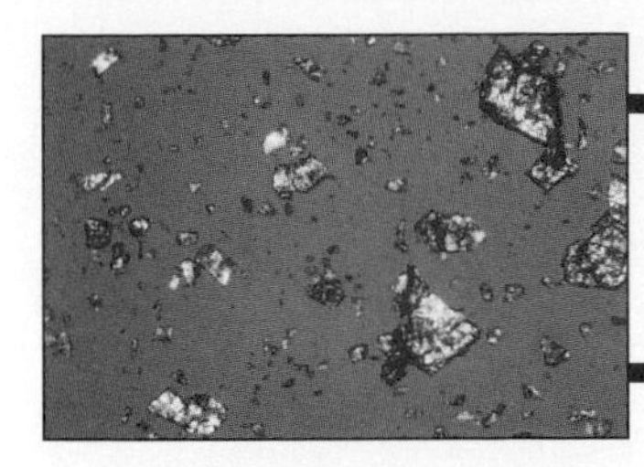

CHAPTER 5

Alcohols, Phenols, Thiols, and Ethers 102

CHAPTER 6

Aldehydes and Ketones 128

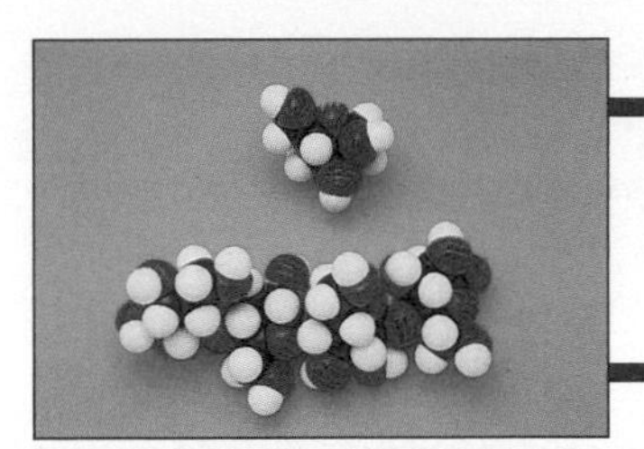

CHAPTER 7

Carbohydrates 151

CHAPTER 8

Carboxylic Acids and Carboxylic Acid Derivatives 177

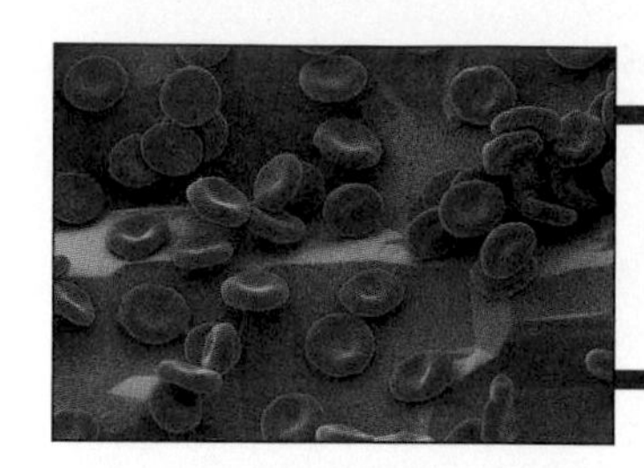

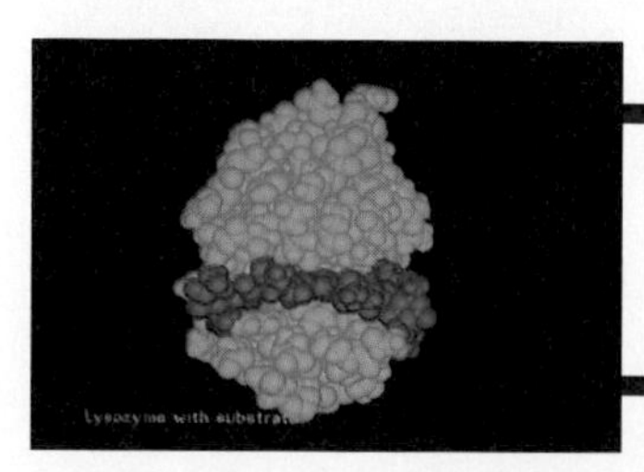

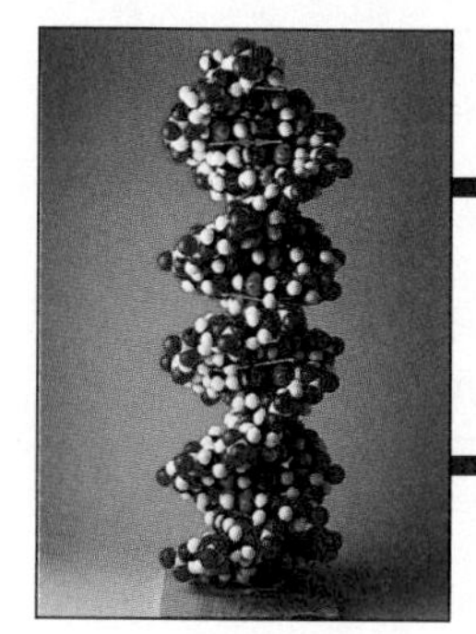

CHAPTER 16

Introduction to Molecular Genetics 401

CHAPTER 17

Nutrition 433

The Learning System

Detailed Chapter Outline

A detailed table of contents is provided for each chapter. Topics are presented according to section number and are further sub-divided in outline form to help students organize the material.

Chapter Objectives

A set of chapter objectives is presented at the beginning of each chapter to alert the students to the concepts that will be covered in the chapter. Using the objectives, the students have the opportunity to preview the material and to establish goals for comprehending concepts. The objectives are also a valuable study tool when reviewing chapter material for examinations.

C H A P T E R 1 5

Fatty Acid Metabolism

O B J E C T I V E S

- Recognize the pivotal role of acetyl CoA in fatty acid and lipid metabolism.
- Summarize the role of bile salts and lipases in the digestion of lipids.
- Understand the importance of fatty acid metabolism in energy production in the cell.
- Describe the degradation of fatty acids, β-oxidation.
- Understand the role of "ketone body" production in β-oxidation.
- Describe the major differences between β-oxidation and fatty acid biosynthesis.
- Describe the regulation of lipid and carbohydrate metabolism in relation to the liver, adipose tissue, muscle tissue, and the brain.
- Summarize the antagonistic effects of glucagon and insulin.
- Relate insulin and insulin production to the disease diabetes mellitus.

O U T L I N E

10.3 Heterocyclic Amines

Question 10.7 Draw complete structural formulas for each of the following compounds:

a. methyldiethylamine
b. tri-*sec*-butylamine
c. 4-aminopentanal
d. isopentylamine
e. *N*-methylaniline
f. *N*-ethyl-1-aminohexane

10.3 HETEROCYCLIC AMINES

Heterocyclic compounds were introduced in Section 5.7. We will now discuss nitrogen heterocycles, the most common and important of all the heterocyclic compounds. They are found throughout nature: in DNA, RNA, many alkaloids (cocaine, nicotine, quinine, morphine, heroin, and LSD for example), and in many enzymes and coenzymes (NAD^+ and NADPH, for example). They also play important roles in the manufacture of polymers, medicinals, pharmaceuticals, and food additives; some examples of products are Vitamin B_6, nylon, and saccharin. The common names for several of the more common nitrogen heterocycles are presented here:

Sections 5.7 and 5.8

Pyrrole Imidazole Pyridine Pyrimidine

Metal atom

Purine Indole Porphyrin

The pyrimidine and purine rings are found in the nucleic acids (DNA and RNA). The porphyrin ring is found in hemoglobin (a blood protein), myoglobin (a protein found in muscle tissue), and chlorophyll (a plant pigment), while imidazole is found in amino acids, peptides, and proteins. The indole and pyridine rings are found in many alkaloids such as strychnine and nicotine. Several examples are provided in Figure 10.3.

Section 16.1

Chapter 11

Question 10.8 The general structure of a porphyrin ring system is provided in the previous section. Review this structure and refer to a reference text to find the complete structure of chlorophyll. Explain in your own words the bonding that is present between the porphyrin ring and the magnesium metal ion in chlorophyll.

Cross-Referencing

Often a discussion in the text demands an understanding of key principles and concepts described previously or the topic under discussion may be very important to the comprehension of an upcoming concept. Marginal cross-references are used to help students locate pertinent background material or alert them to important upcoming discussions.

Chapter Summary

At the end of each chapter, there is a summary of the major concepts covered. Major topics are reviewed in paragraph form. The summary, along with the chapter outline, provides a complete overview of the chapter material.

Glossary of Key Terms

All bold-faced terms in the chapter are defined at the end of each chapter along with a reference to the section in which the term appears. These terms and references also appear in a complete glossary at the end of the textbook. The dual placement of glossary terms make the definitions readily available to the students whether they are reviewing a specific chapter or reviewing material in general.

Glossary of Key Terms 375

The reaction catalyzed by pyruvate carboxylase is called an **anaplerotic reaction.** The term "anaplerotic" means "to fill up." Indeed, this critical enzyme must constantly replenish the oxaloacetate and thus indirectly all the citric acid cycle intermediates that are withdrawn as biosynthetic precursors for the reactions summarized in Figure 14.12.

Question 14.15 What is the function of an anaplerotic reaction?

SUMMARY

The complete oxidation of glucose, which occurs under aerobic conditions, provides most of the energy for the cell. These reactions occur in the mitochondria.

The mitochondrion is an aerobic cell organelle that is responsible for most energy production in eukaryotic cells. It is enclosed by a double membrane. The outer membrane permits low molecular weight molecules to pass through. The inner mitochondrial membrane, by contrast, is almost completely impermeable to most molecules. The inner mitochondrial membrane is the site where oxidative phosphorylation occurs. The enzymes of the citric acid cycle, of amino acid catabolism, and of fatty acid oxidation are located in the matrix space of the mitochondrion.

The citric acid cycle is the final common pathway for the degradation of carbohydrates, amino acids, and fatty acids. The citric acid cycle occurs in the matrix of the mitochondrion. It is a cyclic series of biochemical reactions that accomplishes the complete oxidation of the carbon skeletons of food molecules.

Oxidative phosphorylation, or respiration, is the process by which NADH and $FADH_2$ are oxidized, with concomitant production of ATP. Two molecules of ATP are produced when $FADH_2$ is oxidized, and three molecules of ATP are produced when NADH is oxidized. The complete oxidation of one glucose molecule by the citric acid cycle and oxidative phosphorylation yields 36 molecules of ATP, versus two molecules of ATP for anaerobic degradation of glucose by glycolysis.

Because the rate of energy production by the cell must vary with the amount of available oxygen and the energy requirements of the body at any particular time, the citric acid cycle is regulated at several steps. This allows the cell to generate more energy when needed, as for exercise, and less energy when the body is at rest.

Amino acids are oxidized in the mitochondria. The first step of amino acid catabolism is deamination. The carbon skeletons of amino acids are converted to citric acid cycle intermediates. In the urea cycle the toxic ammonium ions that are released by deamination of amino acids are incorporated in urea and excreted.

In addition to its role in catabolism, the citric acid cycle also plays an important role in cellular anabolism, or biosynthetic reactions. Many of the citric acid cycle intermediates are precursors for the synthesis of amino acids and other macromolecules that are required by the cell. A pathway that functions in both catabolic and anabolic reactions is called an amphibolic pathway.

GLOSSARY OF KEY TERMS

aconitase (14.2) the enzyme that catalyzes the isomerization of citrate to isocitrate. This is a two-step reaction in which citrate is dehydrated to produce *cis*-aconitate, which is then rehydrated to produce isocitrate.
alanine aminotransferase (14.5) the enzyme that catalyzes the transfer of the α-amino group of alanine to α-ketoglutarate, producing pyruvate and glutamate.
aminotransferase (14.5) an enzyme that catalyzes the transfer of an amino group from one molecule to another. (Also called transaminase.)
amphibolic pathways (14.7) metabolic pathways that function in both anabolism and catabolism.
anabolism (14.7) energy-requiring biosynthetic pathways.
anaplerotic reactions (14.7) a reaction that replenishes a substrate needed for a biochemical pathway.
aspartate aminotransferase (14.5) an enzyme that mediates the transfer of the α-amino group of aspartate to α-ketoglutarate, producing oxaloacetate and glutamate.
ATP synthase (14.3) a multiprotein complex within the inner mitochondrial membrane that uses the energy of the proton gradient to produce ATP. (Also called F_0F_1 complex.)
brown fat (14.3) specialized tissue that burns lipids and produces heat.
catabolism (14.7) energy-generating degradative pathways.
chemiosmotic theory (14.3) the theory that explains the way in which the oxidation of glucose is coupled to the phosphorylation of ADP to generate ATP.
citrate synthase (14.2) the enzyme that catalyzes the condensation of the acetyl group of acetyl CoA with oxaloacetate in the first reaction of the citric acid cycle.
citric acid cycle (14.2) a cyclic biochemical pathway that is the final stage of degradation of carbohydrates, fats, and amino acids. It results in the complete oxidation of acetyl groups derived from these dietary fuels. Also called the Krebs Cycle.
creatine phosphate (14.1) a molecule in muscle tissue that serves as a storage reservoir for high-energy phosphate groups.
cristae (14.1) the highly folded inner membrane of the mitochondria.
fast twitch muscle fibers (14.1) muscle fibers that are specialized to provide ATP energy principally by anaerobic glycolysis.
F_0F_1 complex (14.3) the multiprotein complex in the inner mitochondrial membrane that uses the energy of the proton gradient to produce ATP. (Also called ATP synthase.)

In-Chapter Examples with Solutions

Throughout the text there are a number of examples that show the student the step-by-step solution to the problem. The examples show the student how the textual information presented is used to solve problems.

1.1 Review of Atomic Structure 9

TABLE 1.1 Selected Properties of the Three Basic Subatomic Particles

Name	Charge	Mass (amu)	Mass (grams)
Electron (e)	−1	5.4×10^{-4}	9.11×10^{-28}
Proton (p)	+1	1.00	1.67×10^{-24}
Neutron (n)	0	1.00	1.67×10^{-24}

ement may be symbolically represented as follows:

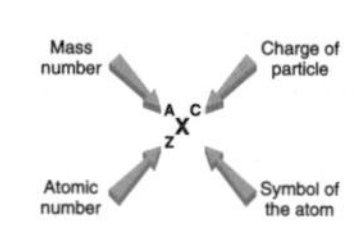

The **atomic number** (Z) is equal to the number of protons in the atom, and the **mass number** (A) is equal to the *sum* of the protons and neutrons (the mass of the electrons is so small as to be insignificant). If

number of protons + number of neutrons = mass number

then, by subtracting the number of protons from each side,

number of neutrons = mass number − number of protons

or, since the number of protons equals the atomic number,

number of neutrons = mass number − atomic number

For a neutral atom, an atom in which positive and negative charges cancel, the number of protons and electrons must be equal and identical to the atomic number.

Example 1.1

Calculate the number of protons, neutrons, and electrons in a neutral atom of fluorine.

The atomic symbol for the fluorine atom is

$$^{19}_{9}F$$

We know that the mass number, 19, is telling us that the total number of protons + neutrons is 19. The atomic number, 9, represents the number of protons. The difference, 19 − 9, or 10, is the number of neutrons. The number of electrons must be the same as the number of protons (hence 9) for a neutral atom.

In-Chapter Questions

In-chapter examples are occasionally followed by similar in-chapter questions. The questions provide students with an opportunity to solve a problem following the steps presented in the example. This builds students' confidence in solving problems.

Question 1.1 Calculate the number of protons, neutrons, and electrons in each of the following neutral atoms:

a. $^{32}_{16}S$ d. $^{244}_{94}Pu$
b. $^{23}_{11}Na$ e. $^{40}_{18}Ar$
c. $^{1}_{1}H$

Perspectives

"Perspectives" are included in each chapter to relate the subject matter to the real world. These special topics take an in-depth look at timely and interesting problems that apply chemical principles to clinical, medical, and ecological problems.

A HUMAN PERSPECTIVE

Brown Fat: The Fat That Makes You Thin?

Humans have two types of fat, or adipose, tissue. *White fat* is distributed throughout the body and is composed of aggregations of cells having membranous vacuoles containing stored triglycerides. The size and number of these storage vacuoles determines whether a person is overweight or not. The other type of fat is **brown fat.** Brown fat is a specialized tissue for heat production, called **nonshivering thermogenesis.** As the name suggests, this is a means of generating heat in the absence of the shivering response. The cells of brown fat look nothing like those of white fat. They do contain small fat vacuoles; however, the distinguishing feature of brown fat is the huge number of mitochondria within the cytoplasm. In addition, brown fat tissue contains a great many blood vessels. These provide oxygen for the thermogenic metabolic reactions.

(a)

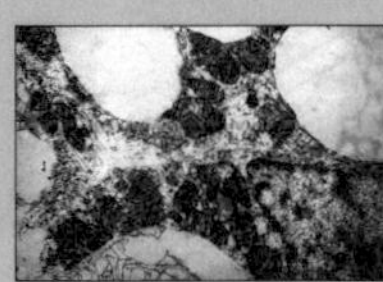

(b)

(a) A light micrograph of white fat cells. (b) An electron micrograph of brown fat cells. The cytoplasm contains few lipid storage vacuoles and a large number of mitochondria.

Brown fat is most pronounced in newborns, cold-adapted mammals, and hibernators. One major difficulty faced by a newborn is temperature regulation. The baby leaves an environment in which he or she was bathed in fluid of a constant 37°C, body temperature. Suddenly the child is thrust into a world that is much colder and in which he or she must generate his or her own warmth internally. By having a good reserve of active brown fat to generate that heat, the newborn is protected against cold shock at the time of birth. However, this thermogenesis literally burns up most of the brown fat tissue, and adults typically have so little brown fat that it can be found only by using a special technique called thermography, which detects temperature differences throughout a body. However, in some individuals, brown fat is very highly developed. For instance, the Korean diving women who spend 6–7 hours every day diving for pearls in cold water have a massive amount of brown fat to warm them by nonshivering thermogenesis. Thus, development of brown fat is a mechanism of cold adaptation.

When it was noticed that such cold-adapted individuals were seldom overweight, a correlation was made between the amount of brown fat in the body and the tendency to become overweight. Studies done with rats suggest that, to some degree, fatness is genetically determined. In other words, you are as lean as your genes allow you to be. In these studies, cold-adapted and non-cold-adapted rats were fed cafeteria food—as much as they wanted—and their weight gain was monitored. In every case the cold-adapted rats, with their greater quantity of brown fat, gained significantly less weight than their non-cold-adapted counterparts, despite the fact that they ate as much as the non-cold-adapted rats. This and other studies led researchers to conclude that brown fat burns excess fat in a highly caloric diet, a phenomenon called **thermogenic hyperphagia.**

How does brown fat generate heat and burn excess calories? For the answer we must turn to the mitochondrion. In addition to the ATP synthase and the electron transport system proteins that are found in all mitochondria, there is a protein in the inner mitochondrial membrane of brown fat tissue called **thermogenin.** This protein has a channel in the center through which the protons (H^+) of the intermembrane space could pass back into the mitochondrial matrix. Under normal conditions this channel is plugged by a GDP molecule so that it remains closed and the proton gradient can continue to drive ATP synthesis by oxidative phosphorylation.

When brown fat is turned on, by cold exposure or in response to certain hormones, there is an immediate increase in the rate of glycolysis and β-oxidation of the stored fat (Chapter 15). These reactions produce acetyl CoA, which then fuels the citric acid cycle. The citric acid cycle, of course, produces NADH and $FADH_2$, which carry electrons to the electron transport system. Finally, the electron transport system pumps protons into the intermembrane space. Under usual conditions the energy of the proton gradient would be used to synthesize ATP. However, when brown fat is stimulated, the GDP that had plugged the pore in thermogenin is lost. Now protons pass freely back into the matrix space, and the proton gradient is dissipated. The energy of the gradient, no longer useful for generating ATP, is released as *heat,* the heat that warms and protects newborns and cold-adapted individuals.

Brown fat is just one of the body's many systems for maintaining a constant internal environment regardless of the conditions in the external environment. Such mechanisms, called **homeostatic mechanisms,** are absolutely essential to allow the body to adapt to and survive in an ever-changing environment.

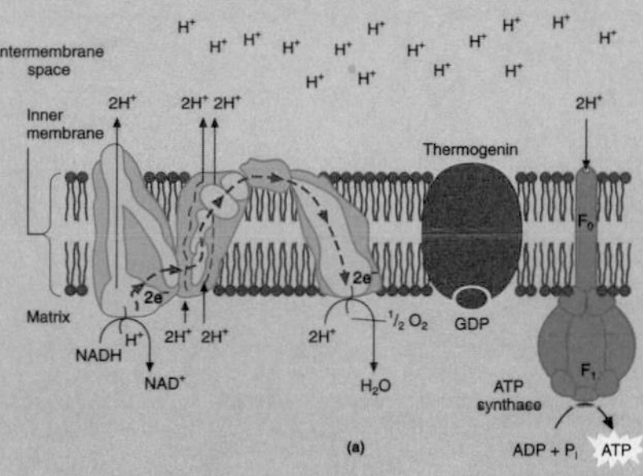

(a)

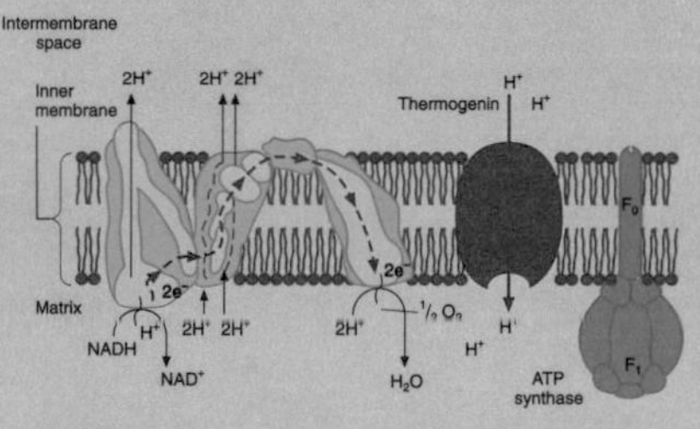

(b)

(a) The inner membrane of brown fat mitochondria contains thermogenin. In the normal state the pore in the center of thermogenin is plugged by a GDP molecule. (b) When brown fat is activated for thermogenesis, the GDP molecule is removed from the pore, and the protons from the H^+ reservoir are free to flow back into the matrix of the mitochondrion. As the gradient dissipates, the energy is used to generate heat.

12.5 Enzyme-Catalyzed Reactions: Specificity of the Enzyme-Substrate Complex 289

Enzyme + Substrate ⇌ Enzyme – substrate complex

FIGURE 12.5
The induced fit model of enzyme substrate binding. As it binds to the substrate, the shape of the active site conforms precisely to the shape of the substrate. The shape of the substrate may also change.

substrate cause a change in the three-dimensional structure of the active site, resulting in an active site that conforms to the surface of the substrate (Figure 12.5).

For this enzyme-substrate interaction to occur, the surface characteristics of the enzyme and substrate must still be complementary to one another. It is this critical requirement for specific complementarity that will determine whether an enzyme will bind to a particular substrate and mediate a chemical reaction.

Question 12.5 Compare the lock-and-key and induced fit models of enzyme-substrate binding.

To illustrate the specificity of enzymes, consider the following reactions:

The enzyme urease catalyzes the decomposition of urea to carbon dioxide and ammonia as follows:

$$H_2N{-}\overset{O}{\overset{\|}{C}}{-}NH_2 + H_2O \xrightarrow{\text{urease}} CO_2 + 2NH_3$$

Urea — Carbon dioxide — Ammonia

Methylurea, in contrast, though structurally similar to urea, is catalytically unaffected by urease:

$$H_2N{-}\overset{O}{\overset{\|}{C}}{-}NHCH_3 + H_2O \xrightarrow{\text{urease}} \text{NO REACTION}$$

Methylurea

Not all enzymes exhibit the same degree of specificity. For convenience we generally classify enzymes into one of four groups:

1. *Absolute specificity*. An enzyme that catalyzes the reaction of only one substrate is **absolutely specific.** Aminoacyl-tRNA synthetases exhibit absolute specificity. Each must attach the correct amino acid to the appropriate transfer RNA molecule. If the wrong amino acid were attached to the transfer RNA, it would be mistakenly added to a peptide chain. Our study of sickle cell anemia reveals how serious the consequences of the alteration of a single amino acid can be. *Section 16.11*
2. *Group specificity*. An enzyme that catalyzes processes involving similar molecules containing the same functional group is **group-specific.** One such enzyme is hexokinase, which catalyzes the phosphorylation of the hexose sugar glucose in the first step of glycolysis. Hexokinase also catalyzes the phosphorylation of several other six-carbon sugars. *Section 16.3*

In-Chapter Questions

There are many questions provided throughout the chapters that allow students to review information just covered and test their comprehension.

268 Protein Conformation and Function

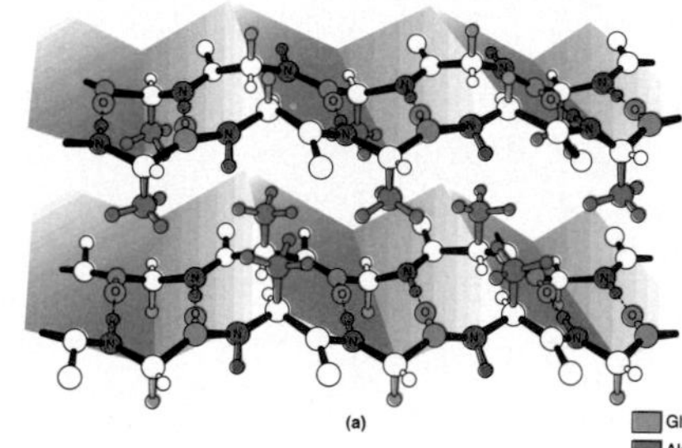

FIGURE 11.14
The structure of silk fibroin is almost entirely antiparallel β-pleated sheet. (a) The molecular structure of a portion of the silk fibroin protein. (b) A schematic representation of the antiparallel β-pleated sheet with the nestled R groups.

Illustrations

Because a picture is worth a thousand words, each chapter is amply illustrated with figures, tables, and chemical structures and formulas. Difficult concepts are more easily understood when presented both textually and visually.

The structure of collagen

Collagen is made from triple-stranded fibers. Two hydroxylated amino acid residues, *4-hydroxyproline* (Hyp) and *5-hydroxylysine* (Hyl), account for nearly one fourth of the amino acid composition of collagen (Figure 11.15), and glycine makes up about a third of the amino acid composition. The individual peptide chains of tropocollagen are left-handed helices. These left-handed strands are wrapped around one another in a right-handed sense. The triple-stranded helical fiber of collagen is known as **tropocollagen** (Figure 11.16).

FIGURE 11.15
Structures of 4-hydroxyproline and 5-hydroxylysine, two amino acids found only in collagen.

4-Hydroxyproline

5-Hydroxylysine

The role of vitamin c in collagen metabolism

The major known physiological function of **vitamin C** is in the hydroxylation of collagen. When the collagen protein is made, the amino acids proline and lysine are incorporated into the chain of amino acids. These are later modified by two enzymes to form 4-hydroxyproline and 5-hydroxylysine. These enzymes, proline hydroxylase and lysine hydroxylase, require vitamin C to carry out these important hydroxylation reactions. People who are deprived of vitamin C, as were sailors on long voyages prior to the eighteenth century, develop **scurvy,** a disease of collagen metabolism. The symptoms of scurvy include skin lesions, fragile blood vessels, and bleeding gums.

Ascorbic acid

The British Navy provided the antidote to scurvy by including limes, a fruit that is rich in vitamin C (as are all citrus fruits) in the diets of its sailors. The epithet "limey," a slang term for "British," entered the English language as a result.

Preface

AUDIENCE

This textbook is intended primarily to serve the undergraduate health-related majors, generally in their first or second year of college. However, it is not intended to be limited to that audience. The book will also be useful to students in all majors that require a one- or two-semester introduction to chemistry. Such introductory courses necessarily emphasize a more application-oriented, less theoretical approach to chemistry.

These students need the same kinds of information that are taught in a course for science majors but with less detail. At the same time students should be exposed to the practical aspects of chemistry that relate to health sciences and the environment. We have attempted to reach a balance between these two approaches while emphasizing material that is unique to health-related studies.

The level of the text is intended for students whose professional goals require an education that falls between purely technical training, on the one hand, and a very theoretical treatment, on the other. Thus the level of the text is intended for the average student, one whose goals in life do not include a mastery of chemistry but for whom an understanding of the principles of chemistry and their practical ramifications is a necessity.

PHILOSOPHY

Chemistry is a unified discipline. In this text we do, of course, treat the individual disciplines of inorganic, organic, and biological chemistry. However, the emphasis is on the integration of these areas to help the student understand the interrelatedness of these three disciplines. We believe that such an approach is both desirable and necessary. It is intended to provide a sound foundation in the principles of chemistry. At the same time the students will learn that life is not a magical property but rather is the result of a set of chemical reactions that obey the laws of chemistry (and physics). We want the students to develop a "feel" for the importance of chemistry in the normal biochemical function of the cell and the body as a whole, the medical ramifications that result when such reactions go awry, the action of drugs in the body, and the way in which this chemistry can be applied to alleviate disease and suffering. Indeed, these are the aspects of chemistry that are most critical in health-related occupations. Thus it is the goal of this text to provide both a sound theoretical background in chemistry and an understanding of the relationship of chemistry to society in general and the health-related professions in particular.

The approach to writing the book has been, very simply, to provide an interesting and logical sequence that will develop, for the non-chemistry major, the very basic principles of inorganic chemistry and an in-depth understanding of the basic biological molecules that make up the cell and the biochemical reactions that allow it to function. We hope the students will attain an appreciation of, if not complete facility with, the chemistry that allows the biological world to function. No attempt is made to teach "a little of everything" or to present material for its own sake. Topics have been chosen or omitted according to the philosophy stated above.

ORGANIZATION AND CONTENT

We have attempted to provide a comfortable flow of material and have maintained a rather traditional organization. The first eight chapters supply the type of background that is generally found in a two-semester science major's course, but the material is intended to be covered in half the time. Chapter 9 serves as a bridge between two topics that are often treated as unrelated: inorganic and organic chemistry. In this pivotal chapter the inorganic topics that are most essential to developing an understanding of organic chemistry are reviewed and applied to the chemistry of carbon. The chapters that follow (Chapters 10–13) present the basics of nomenclature, hydrocarbon structure and chemistry, and common functional groups. Immediately thereafter, we introduce the more biochemistry-oriented organic chemistry. Specifically, carbohydrate chemistry (Chapter 14) immediately follows the discussion of ketones and aldehydes (Chapter 13). This allows the students to apply what they have just learned about ketones and aldehydes to the important group of biological molecules that *are* ketones and aldehydes. A similar approach has been taken with Chapter 15 (Carboxylic Acids and Carboxylic Acid Derivatives), Chapter 16 (Lipids and Their Functions in Biochemical Systems), Chapter 17 (Amines and Amides), and

Chapter 18 (Protein Conformation and Function). This intermingling of organic chemistry and biochemistry softens the boundary between the two areas. Furthermore, this approach allows the instructor to begin coverage of biochemistry earlier in the course and provides a solution to one of the common problems encountered in teaching this type of course, namely, that there is so much material in such a course that there is insufficient time to adequately cover the focal point of the course: the biochemistry.

The book is designed to allow the instructor a great deal of flexibility in the organization of the course. For instance, an overview of all the major topics may be undertaken in one semester. Alternatively, following coverage of the first 12 chapters, the time spent on organic chemistry can be decreased and major emphasis can be applied to the final nine chapters. One could also design a course that would compress, to some extent, the study of those chapters dealing with stoichiometry, energy, and solutions and still present the basic concepts that are required for an understanding of organic and biological chemistry.

PEDAGOGICAL DEVICES

The text includes a variety of pedagogical devices. In designing these we asked ourselves the question, "If we were students, what would help us organize and understand the material covered in this chapter?" We chose the following pedagogical supports:

1. **Detailed chapter contents:** A detailed table of contents is provided for each chapter. Topics are divided and subdivided in outline form to help the students organize the material in their own minds.
2. **Complete chapter objectives:** A set of chapter objectives at the beginning of each chapter alerts the students to the concepts that will be covered in the chapter. The students have the opportunity to preview the material and become aware of the information and ideas that they are expected to learn. These chapter objectives are also useful as a self-test after the chapter has been read.
3. **In-chapter examples and problems:** Throughout the text of chapters that require a great deal of problem solving, there are a number of examples that show the student, step by step, precisely how to properly determine the correct answer. In addition, there are many problems within the text that allow the students to test their mastery of the information just covered.
4. **Glossaries of key terms:** All key terms are in boldface, and each term is defined at the end of the chapter. These terms also appear in an alphabetized glossary at the end of the textbook. Although on first examination this approach might appear redundant, our intent is to make the definitions readily available either at the time the student is involved in the study of a particular chapter (glossaries at the end of the chapter) or when the student is reviewing material in general and might not remember which chapter was appropriate (glossary at the end of the book). This means that some terms are defined in more than one chapter. Sometimes, the definitions are stated differently in different chapters; the concept is the same but it is the orientation that changes. A valuable lesson that students may learn from this is that a definition is not a series of words set in stone that must be memorized, but rather a functional description of a term or a concept.
5. **Chapter cross-references:** Often a discussion in the text demands an understanding of key principles and concepts described previously. To help the students locate the pertinent background material, references to previous chapters, sections, and perspectives are noted in the margins of the text. These marginal cross-references are also used to alert the students to upcoming topics that require an understanding of the information that is currently being studied.
6. **Chapter summary:** At the end of each chapter there is a summary of the major concepts covered. Each of the major topics of the chapter is briefly reviewed in paragraph form.
7. **End-of-chapter problem sets:** There are at least 50 questions and problems at the end of each chapter. These are arranged and labeled by topic. There are additional problems in most chapters that are more challenging. These are labeled "Further Problems."
8. **Answers to in-chapter and end-of-chapter problems:** The answers to all of the odd-numbered problems in the textbook are provided at the end of the book. Each answer has been verified by at least two reviewers to ensure the highest level of accuracy.
9. **Perspectives:** Special sections, entitled perspectives, are included in each chapter to relate the subject matter to the real world. These special topics take an in-depth look at timely and interesting problems that apply chemical principles to clinical, medical, and ecological problems. In addition to helping the students apply the chemistry, the goal is to make the students aware of current important problems and potential solutions.
10. **Illustrations:** Each chapter is amply illustrated with figures, tables, and chemical formulas. All of these illustrations are carefully annotated for clarity.

SUPPLEMENTARY MATERIALS

An extensive supplemental package has been designed to support this text. It includes the following elements.

1. **Instructor's Manual.** The Instructor's Manual contains the printed test item file, solutions to half of the text's problems and exercises, and a list of the transparencies.
2. **Instructor's Resource Guide.** Written by the authors, this unique ancillary contains suggestions for organizing lectures, additional "Perspectives," and a list of each chapter's key problems and concepts.
3. **Student Solutions Manual.** A separate Student Solutions Manual is available for the students. It contains the answers and solutions for the half of the chapter problems not found in the Instructor's Manual.
4. **Student Study Guide.** Written by the authors, and Larry C. Byrd, of Western Kentucky University, the study guide offers students a variety of exercises and keys for testing their comprehension of basic, as well as difficult, concepts.
5. **Transparencies.** A set of 60 transparencies is available to help the instructor coordinate the lecture with the key illustrations from the text.
6. **Customized Transparency Service.** For those adopters interested in receiving acetates of text figures not included in the standard transparency package, a select number of acetates will be custom made upon request. Contact your local Wm. C. Brown Publishers sales representative for more information.
7. **TestPak.** This computerized classroom management system/service includes a database of text questions, reproducible student self-quizzes, and a grade-recording program. Disks are available for IBM, Apple and Macintosh computers, and require no programming experience. If a computer is not available, instructors can choose questions from the Test Item File, and phone or FAX in their request for a printed exam, which will be returned within 48 hours.
8. **Laboratory Manual.** Written by Charles H. Henrickson, Larry C. Byrd, and Norman W. Hunter, all of Western Kentucky University, *Experiments in Inorganic, Organic and Biochemistry* strongly and safely guides students through the process of scientific inquiry. The manual features 26 self-contained experiments that can easily be reorganized to suit individual course needs. In addition, you can delete experiments and/or add your own materials to create a custom manual to fit your students' precise needs.
9. **Laboratory Resource Guide.** This helpful prep guide contains the hints that the authors have learned over the years to ensure student success.
10. **Chemistry Study Cards.** Written by Kyle Van De Graaff and Kent Van De Graaff, of Brigham Young University, and Paul Fore of Snow College, this boxed set of 300, two-sided study cards provides a quick, yet thorough visual synopsis of the key chemistry terms and concepts covered in a freshman chemistry curriculum. Each card features a masterful illustration, pronunciation guide, definition and description in context.
11. **Is Your Math Ready for Chemistry?** Developed by Walter Gleason of Bridgewater State College, this unique booklet provides a diagnostic test that measures your students' math ability. Part II of the booklet provides helpful hints on the necessary math skills needed to successfully complete a chemistry course.
12. **Problem Solving Guide to General Chemistry.** Written by Ronald DeLorenzo of Middle Georgia College, this exceptional supplement provides your students with over 2,500 problems and questions. The guide holds students' interests by integrating the solution of chemistry problems with real-life applications, analogies, and anecdotes.
13. **How to Study Science.** Written by Fred Drewes of Suffolk County Community College, this excellent workbook offers students helpful suggestions for meeting the considerable challenges of a science course. It offers tips on how to take notes and how to get the most out of laboratories, as well as on how to overcome science anxiety. The book's unique design helps to stir critical thinking skills, while facilitating careful note taking on the part of the student.
14. **Video Tapes.** Narrated by Ken Hughes of the University of Wisconsin-Oshkosh, the tapes provide six hours of laboratory demonstrations. Many of the demonstrations are of high interest experiments, too expensive or dangerous to be performed in the typical freshmen laboratory. Contact your local Wm. C. Brown Publishers sales representative for more details.
15. **Doing Chemistry Videodisc.** This critically acclaimed image database contains 136 experiments and demonstrations. It can be used as a pre-lab demonstration of equipment setup, laboratory techniques, and safety precautions. It may also be used as a substitute for lab experiences for which time or equipment is not available. Contact your local Wm. C. Brown Publishers sales representative for more details.

Acknowledgments/Reviewers

The preparation of a textbook is a family effort, and the quality of the final product is a reflection of the dedication of all the family members. We are profoundly grateful to our own families, whose patience and support made it possible for us to undertake this project. We also thank the members of our extended family, our colleagues at Wm. C. Brown Publishers, especially our Developmental Editor, Elizabeth Sievers, and our Acquisitions Editor, Craig Marty. We also thank Darlene Schueller, Senior Secretary, Barbara Hodgson, Visuals Design Specialist, and Julie Kennedy, Production Editorial Manager, who have worked quietly behind the scene to ensure that the book is both accurate and elegant. Finally we extend our sincere gratitude and appreciation to Mary Jo Gregory of York Production Services. They have guided us patiently through the winding paths of textbook publication; they have prodded us gently to meet our deadlines; and, above all, they have always insisted on excellence in the quality of the science, the writing, the artwork, and the overall process of publication.

Sister Helen Burke
Chestnut Hill College

Sharmaine S. Cady
East Stroudsburg University

Donald R. Evers
Iowa Central Community College

Lidija Kampa
Kean College of New Jersey

Judith Kasperek
Pitt Community College

William Moeglein
Northland Community College

John A. Paparelli
San Antonio College

Introduction

When you awoke this morning, a flood of chemicals called neurotransmitters was sent from cell to cell in your nervous system. As these chemical signals accumulated, you gradually become aware of your surroundings. Chemical signals from your nerves to your muscles propelled you from your warm bed to prepare for your day.

For breakfast you had a glass of milk, two eggs, and buttered toast, thus providing your body with needed molecules in the form of carbohydrates, protein, lipids, vitamins, and minerals. As you ran out the door, enzymes of your digestive tract were dismantling the macromolecules of your breakfast. Other enzymes in your cells were busy converting the chemical energy of food molecules into ATP (adenosine triphosphate), the universal energy currency of all cells.

As you continue through your day, thousands of biochemical reactions will keep your cells functioning optimally. Hormones and other chemical signals will regulate the conditions within your body. They will let you know if you are hungry or thirsty. If you injure yourself or come into contact with a disease-causing microorganism, chemicals in your body will signal cells to begin the necessary repair or defense processes.

Life is an organized array of large, carbon-based molecules that are maintained by biochemical reactions. To understand and appreciate the nature of a living being, you must understand the principles of organic chemistry as they are applied to biological molecules.

You have already studied the principles and concepts of *inorganic chemistry,* which focus primarily on elements other than carbon. You have studied the scientific method and experimentation, the structure of the atom, chemical bonding, physical and chemical properties and the periodic table, and the states of matter, as well as equilibrium and energy changes.

In the following chapters we shall use many of these concepts to develop an understanding of the chemistry of compounds that principally contain carbon, *organic chemistry*. The significance of organic chemistry to the study of life processes is revealed by the name: the word ''organic'' means derived from or having characteristics of life. Although the present-day meaning of organic chemistry is significantly broader, our study of organic chemistry will provide the foundation for our understanding of *biochemistry,* the chemistry of living systems.

THE ORIGIN OF MODERN ORGANIC CHEMISTRY

Originally, all organic compounds were believed to be exclusively derived from natural sources (living or once living matter). All early attempts to synthesize organic compounds (such as starch, protein, or components of petroleum) in the laboratory failed, and it was proposed that a *vital force* was necessary for their formation. In 1828 a German chemist, Friedrich Wöhler, converted ammonium cyanate (an inorganic salt) to urea (an organic chemical). Urea is found in urine and is a common waste product of the metabolism of proteins. The following equation represents the reaction carried out by Wöhler:

$$NH_4OCN \xrightarrow{\text{heat}} H_2N-\overset{\overset{\displaystyle O}{\|}}{C}-NH_2$$

Ammonium cyanate (inorganic salt) — **Urea** (organic compound)

Soon after the discovery, many other organic compounds were successfully synthesized, and the ''vital force'' theory was laid to rest. Wöhler's experiment is now widely recognized as the beginning of modern organic chemistry. Today, the study of organic chemistry is the study of all carbon-containing compounds, regardless of their source.

THE SCOPE OF ORGANIC CHEMISTRY

In light of the fact that over 100 different elements have been isolated and identified, it might seem surprising that such a large part of our introductory study of chemistry is devoted to the study of compounds that contain only one common element, carbon. This emphasis reflects the fact that the majority of compounds, from both living systems and industrial production, are organic compounds.

Why are there so many organic compounds? In the first place, carbon atoms are able to form *stable, covalent* carbon-to-carbon bonds with other carbon atoms. This process of forming ''chains'' of carbon-to-carbon bonds is called *catenation*. As a result of this property, elemental (pure) carbon may exist in three *allotropic forms:* amorphous carbon (without a defined structure), graphite, and diamond.

The existence of an element in two or more forms is called allotropy.

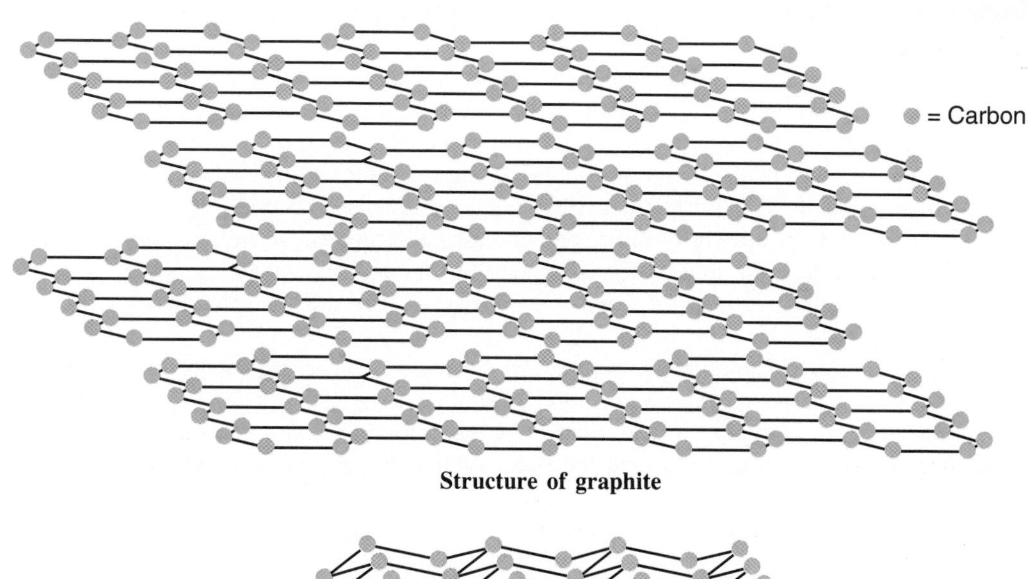

Structure of graphite

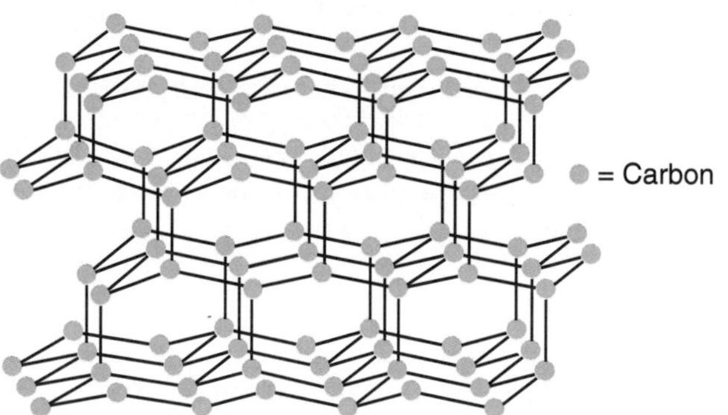

Structure of diamond

Although both graphite and diamond are composed of pure carbon, each is a different entity because of structural differences that cause very different chemical and physical properties. Graphite, a planar unit composed of numerous carbon-carbon bonds extending in two dimensions, is used as a solid lubricant. The planar structure allows the units to slide over one another, facilitating motion. Diamond is composed of a large three-dimensional covalent unit. This structure results in the characteristic properties of diamond: hardness and stability.

The majority of carbon-containing compounds also exhibit this ability to form carbon chains. Consider the first four members of the family of alkanes:

Chapter 3

```
   H            H  H             H  H  H              H  H  H  H
   |            |  |             |  |  |              |  |  |  |
H—C—H       H—C—C—H        H—C—C—C—H         H—C—C—C—C—H
   |            |  |             |  |  |              |  |  |  |
   H            H  H             H  H  H              H  H  H  H

Methane       Ethane          Propane              Butane
```

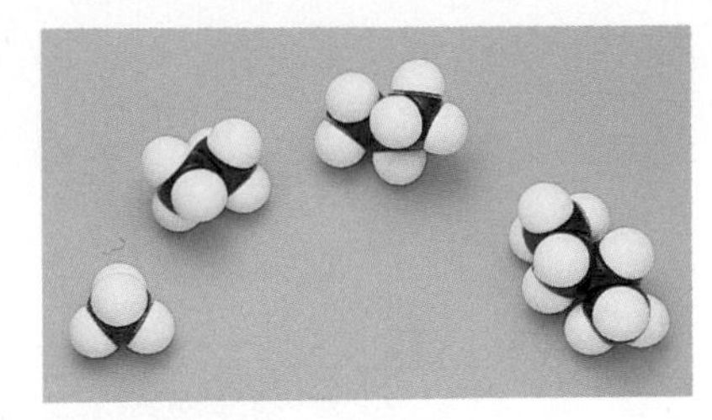

FIGURE 1
Space-filling models of (left to right) methane (CH_4), ethane (C_2H_6), propane (C_3H_8), and butane (C_4H_{10}).

Figure 1 depicts these four molecules, using space-filling models. These models are often used in both organic chemistry and biochemistry to realistically depict the shape, size, and geometry of molecules.

Each alkane molecule contains *only* C and H, but each is a different chemical substance with unique chemical and physical properties. The number of possible alkanes is infinite; alkanes with hundreds, indeed thousands, of carbon and hydrogen atoms may be synthesized easily.

A second reason for the huge number of organic compounds is that, in addition to C—C and C—H bonds, carbon is capable of forming stable bonds with other elements, for example, nitrogen, oxygen, sulfur, and the halogens.

Finally, the number of ways in which these elements may combine to form unique structures is practically limitless. Consider the following examples:

```
                                  H
                                  |
                                H—C—H
     H  H  H  H               H   |   H
     |  |  |  |               |   |   |
   H—C——C——C——C—H           H—C———C———C—H
     |  |  |  |               |   |   |
     H  H  H  H               H   H   H

      n-Butane                 iso-Butane
```

Each compound has four carbon atoms and ten hydrogen atoms. Each may be represented by the molecular formula C_4H_{10}. However, each is a different compound with different properties, and each can be isolated as a separate entity. Compounds like these, having the same molecular formula but different structures (hence different properties), are termed *isomers*. A photograph of the space-filling models of each of these compounds is seen in Figure 2.

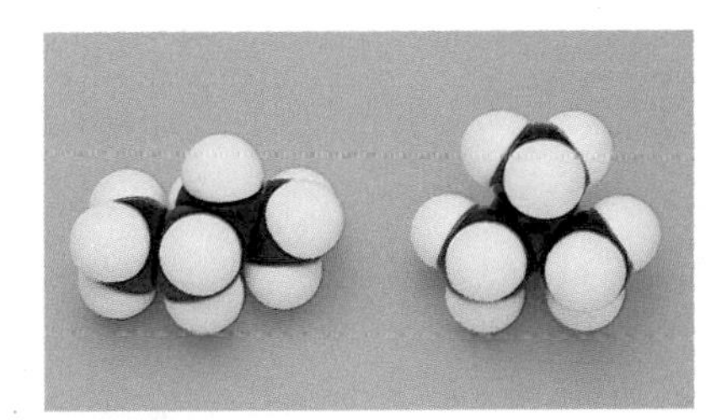

FIGURE 2
Models of *n*-butane (left) and isobutane (right), the isomeric forms of C_4H_{10}.

CLASSIFICATION OF ORGANIC COMPOUNDS

Because of the sheer number of different compounds and the wide variety of structures, organic chemistry is a complex subject. To simplify the study of organic chemistry, compounds are classified according to groups, or *families,* just as the elements are classified into families of the periodic table. The two most general classifications are hydrocarbons and substituted hydrocarbons.

Hydrocarbons

Hydrocarbons contain only carbon and hydrogen and are subdivided into two principal classes: *aliphatic* and *aromatic hydrocarbons*. Aliphatic hydrocarbons are further subdivided into three families: *alkanes, alkenes,* and *alkynes* (Figure 3). The classifications are based on similarities in structure and properties.

Alkanes and cycloalkanes are also called *saturated hydrocarbons* and are composed solely of C—C and C—H *single bonds. Unsaturated hydrocarbons* contain at least one C═C (carbon-carbon double bond, alkenes) or C≡C (carbon-carbon triple bond, alkynes). Aromatic compounds are cyclic and have a unique carbon-carbon bonding arrangement.

FIGURE 3
Molecular models of (left to right) an alkane, ethane, C_2H_6; an alkene, ethene, C_2H_4; and an alkyne, ethyne (commonly, acetylene) C_2H_2.

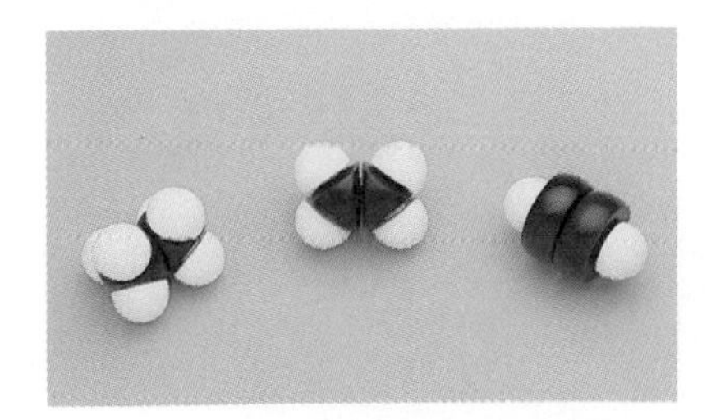

Substituted hydrocarbons

The chemistry of the hydrocarbons is quite predictable; for instance, there exist no great differences in chemical and physical properties between a hydrocarbon containing five carbons and one containing six carbon atoms. Furthermore, the hydrocarbons do not possess significant biochemical properties (except for toxicity!). However, the substitution of one or more functional groups for hydrogens in a hydrocarbon brings about huge

FIGURE 4
Molecular models of carbohydrates: glucose (top), a portion of the extended structure of cellulose (bottom).

changes in properties. A *functional group* is an atom or group of atoms in a molecule that is principally responsible for the chemical and physical properties of that molecule. All compounds that contain a particular functional group (for example, the —OH or hydroxyl group) are classified within the same family (the —OH group is the functional group of the alcohols). The following compounds are a simple alcohol and a complex alcohol; both contain the —OH group.

CH_3—CH_2—**OH**

Ethanol
(ethyl alcohol)
alcohol in fermented beverages

CH_3
$CH_3CHCH_2CH_2CH_2CHCH_3$
CH_3
CH_3
HO

Cholesterol

alcohol found in eggs, dairy products, and red meats

The functional group is critical to a study of organic and biological chemistry. Common functional groups that will be encountered in future chapters are found in Table 1. In most cases, the chemistry of these compounds is controlled by the functional group or groups that are present in the molecule. It is impossible to learn the chemistry of each new compound, but it is quite possible to learn the chemistry associated with each functional group and to use this as the basis for understanding the reactivity of virtually all organic and biological molecules. This functional group approach will be used throughout this book as we study the properties and chemical reactions of the families of organic and biological molecules.

THE SCOPE OF BIOCHEMISTRY

As the first few paragraphs of this introduction suggested, all aspects of life are a function of the structure of biological molecules and the reactions in which they participate. Biochemistry involves the study of the families of compounds found in living tissues, including carbohydrates, proteins, lipids, and nucleic acids. It is vital that we understand, as completely as possible, the chemistry of the normal pathways for the synthesis, utilization, regulation, and elimination of bio-organic compounds in living systems. Only then can medical science understand and devise treatment for the pathological conditions that arise when one of the normal pathways goes awry.

The major families of biological molecules that we will consider are the carbohydrates, lipids, proteins, and nucleic acids.

The family of *carbohydrates* (Figure 4) includes small, simple sugars like glucose, fructose, and sucrose and long polymers of simple sugars such as starch and cellulose. The carbohydrates are our major dietary source of energy. Therefore we will study not only their structure and properties, but also the biochemical reactions by which they are broken down to produce energy in a form that can be used for cellular work.

The family of *lipids* (Figure 5) includes a variety of fats, oils, and waxes. Lipids are also a dietary energy source. In addition, they serve as the largest reserve of stored chemical energy in the human body (adipose tissue), are important hormones (the sex hormones and prostaglandins), are the major structural components of our cell membranes (phospholipids and cholesterol), and aid in digestion (bile salts). We will study the structure of the lipids as well as the biochemical pathways by which they are broken down to generate cellular energy and the pathways by which they are synthesized and stored.

FIGURE 5
Molecular models of two lipids: a triglyceride (top) and a monoglyceride (bottom).

TABLE 1 Common Functional Groups

Functional Group	Name	Family of Organic Compounds
$H_2C{=}CH_2$	Carbon-carbon double bond	Alkene
H—C≡C—H	Carbon-carbon triple bond	Alkyne
C_6H_6 (or) benzene ring symbol	Benzene ring	Aromatic
—C—X (X = F, Cl, Br, I)	Halogen atom	Alkyl halide
—C—OH	Hydroxyl group	Alcohol
—C—O—R*	Alkoxy group	Ether
>C=O	Carbonyl group	Aldehyde or ketone
—C(=O)—OH	Carboxyl group	Carboxylic acid
—C(=O)—G (G = Cl, OR*, and others)	Acyl group	Carboxylic acid derivatives
—C—N—	Amino group	Amine

*R is an abbreviation for any alkyl or aryl group; aryl is used for aromatic compounds in the same way that alkyl is used for aliphatic compounds (for example, methyl, ethyl, isopropyl). An aryl group is an aromatic compound with one hydrogen removed (for example, phenyl—the phenyl group is benzene with one hydrogen removed.)

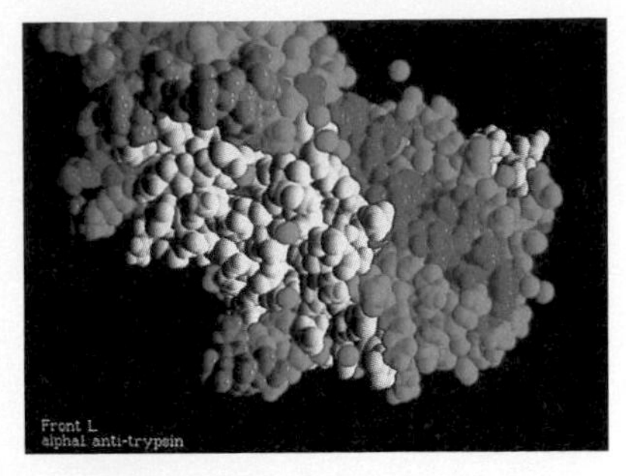

FIGURE 6
A protein model, illustrating its unique three-dimensional structure.

FIGURE 7
A space-filling model of the double helix of DNA.

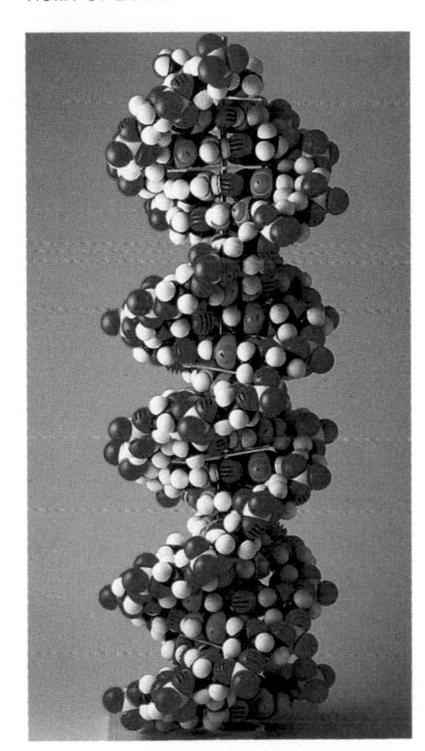

Our study of *proteins* (Figure 6) will touch on every facet of life processes. Proteins are essential structural elements of the cell, but they also catalyze all the biochemical reactions of the cell (enzymes), transport materials into and out of the cell, defend the body against foreign invaders (antibodies), function as hormones (endorphins), and much, much more. Because life could not exist without the activity of enzymes, we will devote an entire chapter to the study of these fascinating proteins.

Our discussion of the family of *nucleic acids* (Figure 7) will lead us into the new and rapidly developing field of molecular biology. The nucleic acids include deoxyribonucleic acid (DNA), which carries the genetic information of every cell, and ribonucleic acid (RNA), which carries out the interpretation of the genetic information and assists in the synthesis of all the proteins that are required for the maintenance of life. We will take a brief look at the newly emerging field of *genetic engineering,* concentrating particularly on the products of genetic engineering that are making it possible to alleviate suffering caused by a variety of human genetic diseases and infectious diseases.

The first two chapters and the ten appendixes of this book provide an opportunity for review and reinforcement of several of the key concepts of inorganic chemistry. These chapters and appendixes will serve as our point of departure for the study of organic chemistry and biochemistry. Such topics as atomic structure, the periodic table, chemical bonding, including an emphasis on the important differences between covalent and ionic compounds, are discussed in Chapter 1. Chapter 2 includes a discussion of thermodynamics and kinetics of chemical reactions and a review of chemical change, including the classification of chemical reactions, acid-base chemistry, and oxidation-reduction processes.

CHAPTER 1

A Review of Structure and Properties of Atoms, Ions, and Molecules

OBJECTIVES

Chapter 1 is a review of the structures and properties of elements and compounds and of the ways in which atoms (the smallest unit of an element) interact with one another to produce molecules (the smallest unit of a compound).

After reading Chapter 1, you should have fulfilled the following objectives, which will enhance your study of organic chemistry and biochemistry in Chapters 3–17.

- Describe the structure of an atom.
- Be able to obtain information about an element from the periodic table.
- Understand the meaning of, and utility of, ionization energies, electron affinities, and electronegativities in predicting bond formation.
- Use the octet rule to predict the charge of common cations and anions.
- Write Lewis symbols for the common atoms and ions.
- Distinguish between ionic and covalent bonding.
- Draw Lewis structures for covalent compounds and complex inorganic ions.
- Draw resonance structures for simple chemical compounds.
- Know the relationships among stability, bond energy, and resonance.
- Predict the geometry of molecules and ions using the octet rule and Lewis structures.
- Understand the role that molecular geometry plays in determining the solubility and melting and boiling points of compounds.

OUTLINE

The properties of matter are a direct result of the structure of the individual units, the elements or compounds, that make up the bulk material. Individual units of elements, **atoms,** can be combined in chemical reactions to produce compounds. The properties of a bulk (macroscopic) material are determined by the structure of the individual units of the compound making up the bulk material. These units are called **molecules.**

$$\text{atoms} \xrightarrow[\text{reaction}]{\text{chemical}} \text{compounds} \xrightarrow[\text{possess}]{\text{which}} \begin{array}{c}\text{properites}\\ \text{and}\\ \text{function}\end{array}$$

Consider the compound water. The utility of water as a solvent and as a biological fluid, its boiling and melting temperature, indeed, all of its properties result from the structure of its individual units and, ultimately, from the structure of its component atoms, hydrogen and oxygen, and the way in which they are joined.

An understanding of the function of compounds of biological interest is one of the primary goals of this course. From your study of inorganic chemistry you know that an understanding of atomic structure is the basis for prediction of structures and properties of all chemical compounds, including antibiotics, therapeutic drugs, and other compounds that we encounter in our daily lives.

1.1 REVIEW OF ATOMIC STRUCTURE

Over the last 100 years the theory of atomic structure has developed rapidly from the initial work of John Dalton to its present level of sophistication. A brief summary of our present understanding of the composition of the atom is presented below. Before proceeding, we inject a note of caution. One must not think of the present picture of the atom as final. Scientific inquiry continues, and we should view the present theory as a step in an evolutionary process. *Theories are subject to constant refinement.*

Modern view of atomic structure

The basic structural unit of an element is the atom, which is the smallest unit of an element that retains the chemical properties of that element. A tiny sample of an element such as carbon, too small to be seen by the naked eye, is, in reality, composed of billions of carbon atoms arranged in some orderly fashion. Each atom is incredibly small, too small to be seen in any detail, even with the most sophisticated of modern instruments.

The atom is composed of three primary particles: the *electron,* the *proton,* and the *neutron.* While other ''subatomic fragments'' with unusual names (neutrinos, gluons, quarks, and so forth) have been discovered recently, we shall concern ourselves only with the primary particles—protons, neutrons, and electrons.

The atom is comprised of two distinct regions:

1. The **nucleus** is a small, dense, positively charged region in the center of the atom. The nucleus is composed of positively charged **protons** and uncharged **neutrons.**
2. Surrounding the nucleus is a diffuse region of negative charge populated by **electrons,** the source of the negative charge. Electrons are tiny in comparison to the protons and neutrons. The properties of these particles are summarized in Table 1.1.

Atoms of various types differ in their number of protons, neutrons, and electrons. The number of protons determines the identity of the atom. As such, the number of protons is *characteristic* of the element. When the number of protons is equal to the number of electrons, the atom is neutral; the charges are balanced and effectively cancel.

TABLE 1.1 Selected Properties of the Three Basic Subatomic Particles

Name	Charge	Mass (amu)	Mass (grams)
Electron (e)	−1	5.4×10^{-4}	9.11×10^{-28}
Proton (p)	+1	1.00	1.67×10^{-24}
Neutron (n)	0	1.00	1.67×10^{-24}

An element may be symbolically represented as follows:

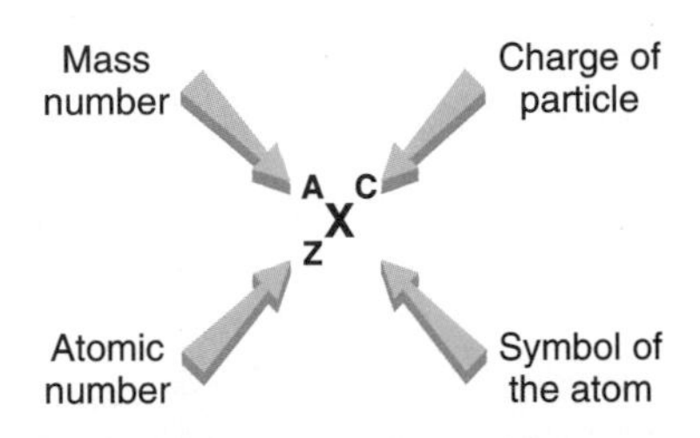

The **atomic number** (Z) is equal to the number of protons in the atom, and the **mass number** (A) is equal to the *sum* of the protons and neutrons (the mass of the electrons is so small as to be insignificant). If

$$\text{number of protons} + \text{number of neutrons} = \text{mass number}$$

then, by subtracting the number of protons from each side,

$$\text{number of neutrons} = \text{mass number} - \text{number of protons}$$

or, since the number of protons equals the atomic number,

$$\text{number of neutrons} = \text{mass number} - \text{atomic number}$$

For a neutral atom, an atom in which positive and negative charges cancel, the number of protons and electrons must be equal and identical to the atomic number.

Example 1.1

Calculate the number of protons, neutrons, and electrons in a neutral atom of fluorine.

The atomic symbol for the fluorine atom is

$$^{19}_{9}\text{F}$$

We know that the mass number, 19, is telling us that the total number of protons + neutrons is 19. The atomic number, 9, represents the number of protons. The difference, 19 − 9, or 10, is the number of neutrons. The number of electrons must be the same as the number of protons (hence 9) for a neutral atom.

Question 1.1 Calculate the number of protons, neutrons, and electrons in each of the following neutral atoms:

a. $^{32}_{16}\text{S}$
b. $^{23}_{11}\text{Na}$
c. $^{1}_{1}\text{H}$
d. $^{244}_{94}\text{Pu}$
e. $^{40}_{18}\text{Ar}$

Isotopes

Isotopes are atoms of the same element having different atomic masses *due to different numbers of neutrons*. For example, each of the following is an isotope of carbon:

$$^{12}_{6}C \quad ^{13}_{6}C \quad ^{14}_{6}C$$

The existence of isotopes explains why the atomic masses [measured in atomic mass units (amu)] of the various elements are not whole numbers, as would be expected from proton and neutron masses, which are whole numbers. Consider, for example, the mass of one chlorine atom, containing 17 protons (atomic number) and 18 neutrons to be

$$17 \text{ protons} \times 1.00 \text{ amu/proton} = 17.00 \text{ amu}$$

$$18 \text{ neutrons} \times 1.00 \text{ amu/neutron} = 18.00 \text{ amu}$$

$$17.00 \text{ amu} + 18.00 \text{ amu} = 35.00 \text{ amu (mass of chlorine atom)}$$

The mass of electrons is considered insignificant in the calculation of atomic mass to three or four significant figures.

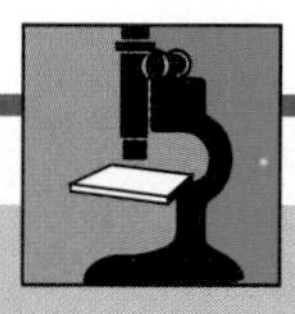

A CLINICAL PERSPECTIVE

Isotopes in Medicine

Fear of nuclear proliferation and concerns about the safety of nuclear reactors for electric power generation have overshadowed developments in nuclear medicine that have resulted in improvements in the quality of life for countless individuals. Nuclear medicine involves the application of radioisotopes in two quite different but complementary ways: diagnosis and therapy.

Diagnosis of internal problems was often possible only with surgery. Any surgical procedure entails some risk; infection, internal bleeding, even death may occur. The use of radioactive isotopes as tracers is a noninvasive procedure. It is certainly not risk-free; few human activities are. The principal risk is associated with the exposure of internal organs to radioactive emissions from the diagnostic isotopes.

The success of such procedures is directly attributable to the fact that the *chemical behavior* of a radioactive isotope is virtually identical to that of the nonradioactive isotopes of the same element. The radioactive isotope, introduced into the human body, will perform the same function and be transported to the same organs as its nonradioactive isotopes. But it performs one additional function: it behaves as a tiny signal transmitter. The emitted radiation can be detected by using sophisticated electronics to produce a remarkable image of the organ of concern. Internal damage, swelling, tumors, and other irregularities become apparent without the need for surgical investigation.

The isotope ^{123}I is concentrated in the thyroid gland and can provide information about the activity of this gland. ^{201}Tl is preferentially accumulated in the heart muscle, and one of the isotopes of the element technetium is particularly useful for studies of the brain and kidneys, as well as the thyroid.

Selection of the proper isotope involves more than a consideration of the organ or organs in which it may accumulate. The need to quickly remove the radioisotope from the body dictates the use of an isotope that is easily and swiftly eliminated by the excretory system. Additionally, a short half-life ensures rapid conversion to a harmless, nonradioactive daughter product.

Therapeutic uses for radioactive isotopes center on the treatment for cancer. Studies indicate that rapidly dividing tumor cells are more susceptible to radiation than are normal cells. ^{60}Co, a cobalt isotope that produces intense gamma radiation, is very commonly used. The beam of gamma radiation is focused on a small area of the body where the tumor is located. Time of exposure and total radiation dosage must be closely monitored to avoid undue damage to noncancerous regions of the body that also lie in the path of the radiation.

Current studies involve the development of selective ways of delivering the radiation to concentrate a very high percentage of emitted radiation directly at the site of the tumor; this maximizes the therapy and minimizes harmful effects. This type of approach is illustrated by the use of ^{131}I in the treatment of thyroid cancer. The isotope, as a solution of potassium iodide, is ingested and transported to the thyroid, where it accumulates, and beta emission from the ^{131}I destroys the cancerous cells. Normal cell damage occurs as well, but far less than that experienced when ^{60}Co is used.

Inspection of the periodic table reveals that the atomic mass of chlorine is actually 35.45 amu, *not* 35.00 amu. The existence of isotopes accounts for this difference. A naturally occurring sample of chlorine is principally composed of two isotopes, $^{35}_{17}Cl$ and $^{37}_{17}Cl$, in approximately a 3:1 ratio, and the tabulated atomic mass is the *weighted average* of the two isotopes.

Certain kinds of atoms can "split" into smaller particles and release large amounts of energy. These *radioisotopes* emit particles and energy, **radioactivity,** that are useful in tracing the behavior of biochemical systems. These isotopes have identical chemical behavior to that of any other isotope of the same element; it is their nuclear behavior that is unique. As a result, one of these isotopes can be substituted for the "nonradioactive" isotope, and its biochemical activity can be followed by monitoring the particles of energy emitted by the isotope as it progresses through the human body.

Ions

Ions are charged atoms or groups of atoms. They result from a gain of one or more electrons by the parent atom (forming negative ions, or **anions**) or a loss of one or more electrons from the parent atom (forming positive ions, or **cations**).

Formation of an anion may occur as follows:

$$^{1}_{1}H + 1e^{-} \longrightarrow {}^{1}_{1}H^{-}$$

Neutral atom gains electron — The hydride ion is formed

Formation of a cation of hydrogen may proceed as follows:

$$^{1}_{1}H \longrightarrow 1e^{-} + {}^{1}_{1}H^{+}$$

Neutral atom loses electron — Producing the hydrogen cation (also called the proton)

Note that the electrons gained are written to the left of the reaction arrow (they are reactants), while electrons lost are written as products to the right of the reaction arrow. It is cumbersome to include the atomic and mass numbers in chemical equations. For simplification, the atomic and mass numbers are often omitted. For example, the hydrogen cation would be written as H^{+} and the anion as H^{-}.

$$H \longrightarrow 1e^{-} + H^{+}$$

$$H + 1e^{-} \longrightarrow H^{-}$$

Formal charge

The concept of **formal charge** is often used to represent the charge carried by the various atoms in more complex molecules. The formal charge is simply the overall charge on an atom or group of atoms in a molecule. One can view an atom (or group of atoms) as a combination of protons, neutrons, and electrons. The neutrons have no charge and so have no bearing on the formal charge of the atom (or group of atoms). A neutral, uncharged atom has an equal number of protons (positive charges) and electrons (negative charges). The atom is electronically neutral. If an electron is removed from the atom (or group of atoms), we are left with one more proton than electron, and the atom (or group of atoms) takes on a positive formal charge of 1 (1+). If two electrons are removed, we are left with two protons in excess, and the atom (or group of atoms) takes on a positive charge of 2 (2+). This process could theoretically continue until all of the electrons are stripped from the atom. In reality this does not occur because the energy required to remove electrons from an atom that is already positively charged becomes prohibitive. Most atoms (or groups of atoms) bear a formal positive charge between 1+ and 3+. In contrast, a negative charge develops when the atom (or group of atoms) has more electrons than protons, and we generally find negatively charged atoms (or groups of atoms) with charges that range from 1− to 3−.

These processes are shown schematically below.

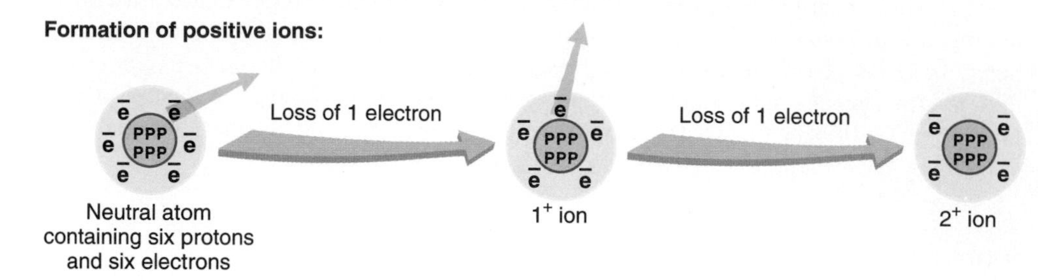

Formation of negative ions:

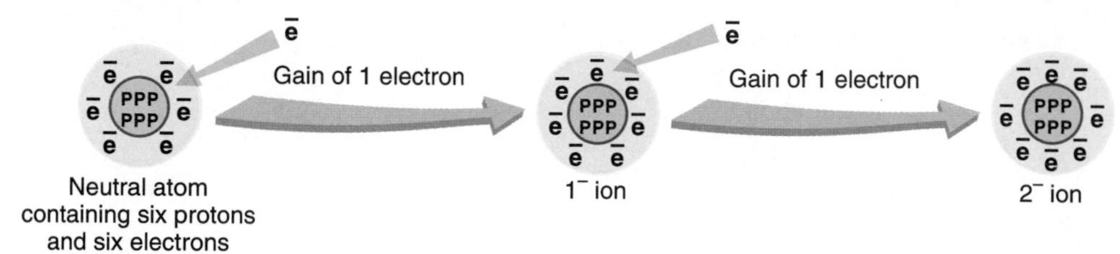

Example 1.2

The neutral hydrogen atom is shown below, as is the neutral carbon atom. The formation of both the positive ion (cation) and negative ion (anion) of each of these atoms is also shown. Note that the process of electron gain or loss is the same as that shown in the generic sense in the preceding section. In actuality the electron can be gained or lost through a variety of pathways. Examples include electron transfer through oxidation-reduction and electron loss through radioactive bombardment or spontaneous decay.

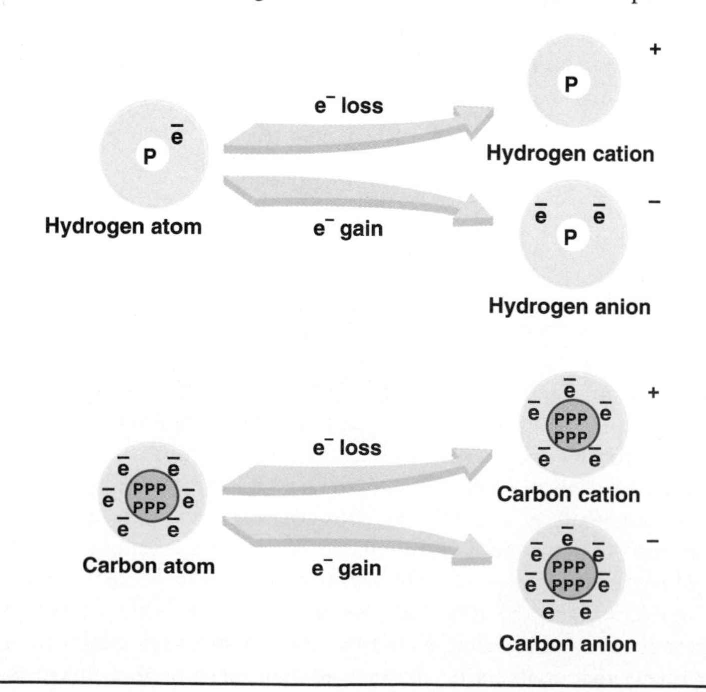

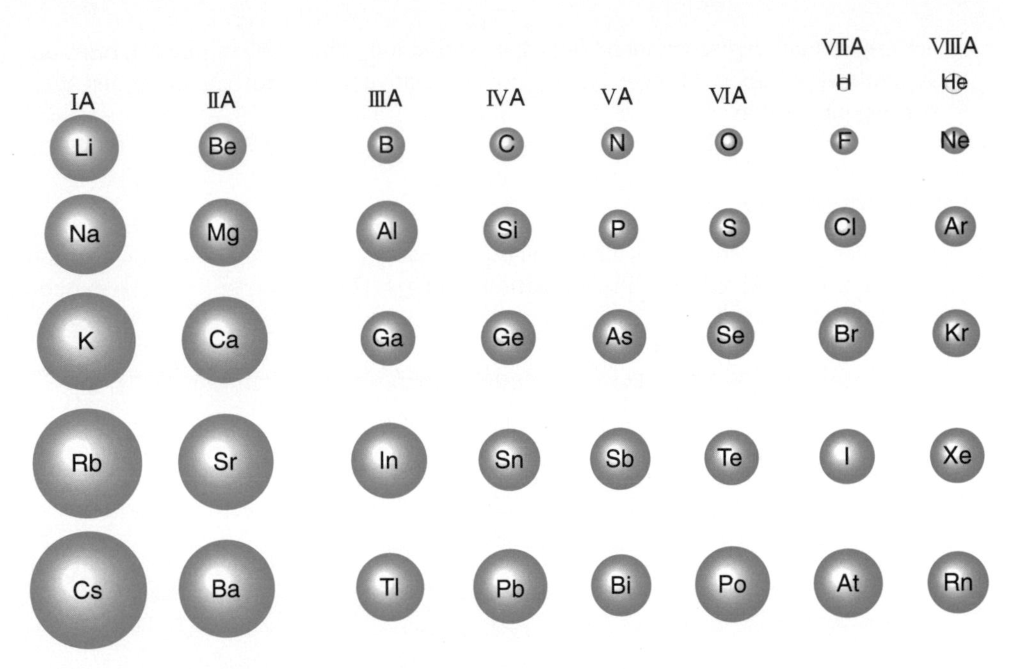

FIGURE 1.1
Variation in the size of atoms as a function of their position in the periodic table.

1.2 TRENDS IN THE PERIODIC TABLE

Atomic size

If the atom is depicted as a tiny sphere whose radius is determined by the distance between the center of the nucleus and the boundary of the region where the outermost electrons are located, the size of the atom will be determined principally by two factors:

1. The energy level (n-level) in which the outermost electron or electrons are located increases as we go *down* a group (recall that the outermost n-level correlates with period number). Thus the size of atoms should increase from top to bottom of the periodic table as successive electrons are added to the atom (see Figure 1.1).
2. As the magnitude of the positive charge of the nucleus increases, its ''pull'' on all of the electrons increases, and the electrons are drawn closer to the nucleus. This results in a contraction of the atomic radius and therefore a decrease in atomic size. This effect is apparent as we read *across* the periodic table within a period. Atomic size decreases from left to right in the periodic table (see Figure 1.1).

Ion size

Three generalizations can be made about the size of ions:

1. Positive ions (cations) are smaller than the parent atom. The cation has more protons than electrons (more positive than negative charge). The excess positive nuclear charge pulls the remaining electrons closer to the nucleus. This is analogous to the situation that results in a decrease in the size of atoms from left to right across a row in the periodic table.
2. Negative ions (anions) are larger than the parent atom. The anion has more electrons than protons (a net negative charge). Because of this excess negative charge, the effective nuclear ''pull'' on each individual electron is reduced. The electrons are held less tightly, and the increased repulsion of the identically charged electrons results in an increase in the radius of the ion in comparison with the neutral atom.
3. Ions with multiple positive charges (such as Cu^{2+}) are even *smaller* than their corresponding monopositive ion (Cu^{1+}); ions with multiple negative charge (such as O^{2-})

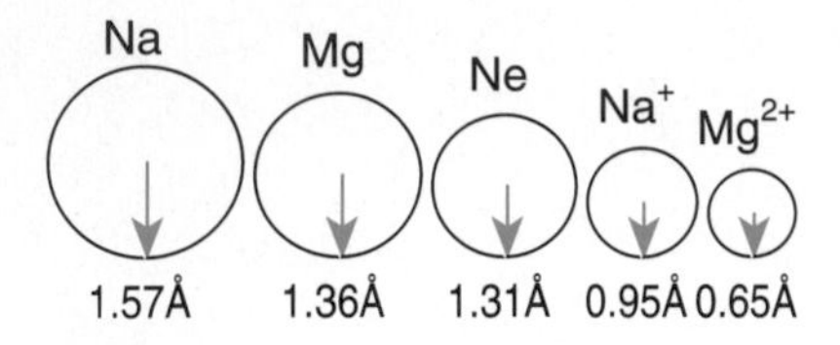

FIGURE 1.2
Relative sizes of ions and their parent atoms.

are *larger* than their corresponding less negative ion. This follows directly from the generalizations above. Figure 1.2 depicts the relative size of certain atoms and their corresponding ions.

Ionization energy

If we consider an isolated atom in the gas phase, the energy required to remove an electron from that atom is referred to as the **ionization energy.** The process for sodium may be represented as follows:

$$\underset{(gas)}{Na} + \underset{energy}{\overset{ionization}{}} \longrightarrow \underset{(gas)}{Na^{1+}} + e^-$$

The magnitude of the ionization energy should correlate with the strength of the attractive force between the nucleus and the outermost electron. Consider the following:

- Reading *down* a group, the ionization energy decreases, since the atom's size is increasing. The outermost electron is progressively farther from the nuclear charge and hence easier to remove.
- Reading across a period, atomic size decreases, as the outermost electrons are closer to the nucleus, more tightly held, and more difficult to remove. Therefore the ionization energy must increase.

A correlation does indeed exist between trends in atomic size and ionization energy. Atomic size *decreases* from bottom to top of a group and from left to right in a period. Ionization energies *increase* in the same periodic way. Note also that ionization energies are highest for the inert gases (see Figure 1.3). A high value for ionization energy means that it is difficult to remove electrons from the atom, and this in part accounts for the extreme stability and nonreactivity of the inert gases.

Electron affinity

The energy that is released when a single electron is added to a neutral atom in the gaseous state is known as the **electron affinity.** If we consider ionization energy in relation to the formation of positive ions (remember that the magnitude of the ionization energy tells us the ease of *removal* of an electron, or the ease of forming positive ions), then electron affinity gives us a measure of the ease of forming negative ions. A large value of electron affinity (energy released) indicates that the atom becomes more stable as it becomes a

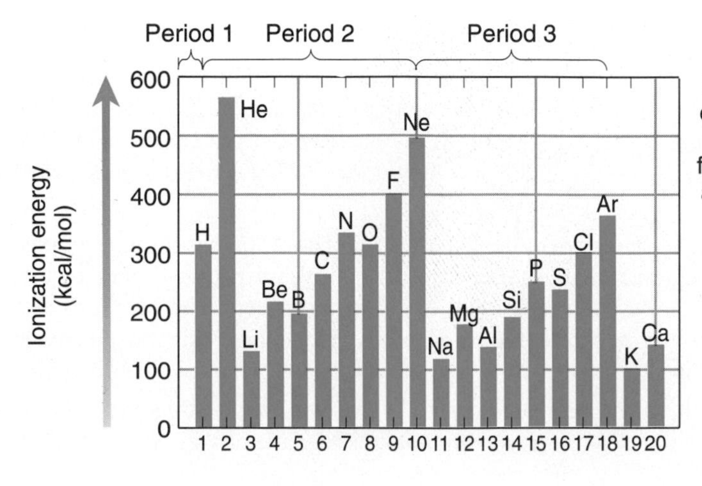

FIGURE 1.3
The ionization energies of the first 20 elements versus their atomic numbers.

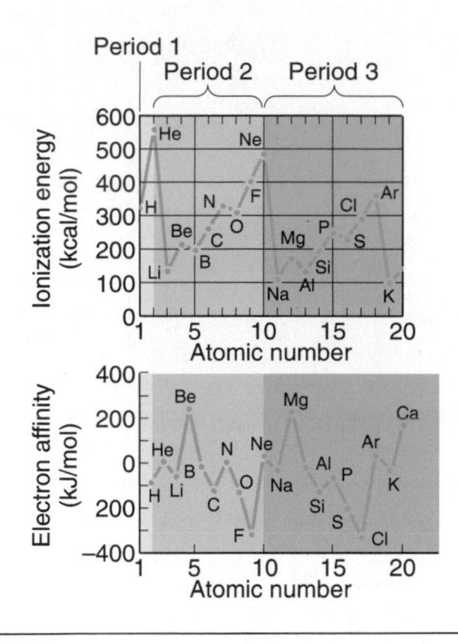

A high ionization energy indicates the difficulty of forming a positive ion. Note the high values for the inert gases and the halogens.

A negative sign for electron affinity indicates a large amount of energy released when an electron is added. F and Cl easily form negative ions, Be, Mg, and Ca do not under any circumstances.

FIGURE 1.4
The periodic variation of ionization energy and electron affinity.

negative ion (through gaining an electron). Consider the gain of an electron by a gaseous bromine atom to produce the bromide ion:

$$\underset{\text{(gas)}}{\mathrm{Br}} + e^- \longrightarrow \underset{\text{(gas)}}{\mathrm{Br}^-} + \underset{\textit{Electron affinity}}{\text{energy}}$$

Periodic trends for electron affinity are as follows:

- Electron affinities generally decrease as we go down a group.
- Electron affinities generally increase as we go across a period.

These trends (see Figure 1.4) are not easily explained, but one should recognize two factors that affect electron affinity. First, the magnitude of electron affinities is a complex function of nuclear charge, atomic size, and electronic configuration. Second, there are exceptions to these generalizations. Perhaps the most notable exception involves the inert gases. From the second generalization above, we would predict a larger electron affinity for the inert gas atoms than for the corresponding halogen in the same period. Such is *not the case;* electron affinities for the inert elements are very low, implying that they do not gain extra stability by forming negative ions. This is another indication of the stability of nonreactivity of the inert gases.

Electronegativity

Electronegativity is the ability of an atom in a molecule to attract electrons to itself. Electronegativity is represented by a scale derived from the measurement of energies of chemical bonds. The scale was developed by Linus Pauling, and values range from 4.0 (most electronegative element) to 0.7 (least electronegative element). The periodic trends for electronegativity, which decreases from top to bottom and increases from left to right, are similar to those of both ionization energy and electron affinity and are summarized in Figure 1.5.

All the trends discussed in Section 1.2 are summarized in Figure 1.6. A knowledge of the meaning and relative magnitude of the various properties discussed above is essential to an understanding of our next topic: the formation of chemical bonds.

Question 1.2 Rank the elements B, N, O, Be, and Ne in order of increasing

a. Atomic size
b. Ionization energy
c. Electron affinity
d. Electronegativity

Nonmetals

H 2.1																H 2.1	He ...
Li 1.0	Be 1.5											B 2.0	C 2.5	N 3.0	O 3.5	F 4.1	Ne ...
Na 1.0	Mg 1.2											Al 1.5	Si 1.7	P 2.1	S 2.4	Cl 2.8	Ar ...
K 0.9	Ca 1.0	Sc 1.2	Ti 1.3	V 1.4	Cr 1.6	Mn 1.6	Fe 1.6	Co 1.7	Ni 1.8	Cu 1.8	Zn 1.7	Ga 1.8	Ge 2.0	As 2.2	Se 2.5	Br 2.7	Kr ...
Rb 0.9	Sr 1.0	Y 1.1	Zr 1.2	Nb 1.2	Mo 1.3	Tc 1.4	Ru 1.4	Rh 1.4	Pd 1.4	Ag 1.4	Cd 1.5	In 1.5	Sn 1.7	Sb 1.8	Te 2.0	I 2.2	Xe ...
Cs 0.9	Ba 1.0	La-Lu 1.1-1.2	Hf 1.2	Ta 1.3	W 1.4	Re 1.5	Os 1.5	Ir 1.6	Pt 1.4	Au 1.4	Hg 1.4	Tl 1.4	Pb 1.6	Bi 1.7	Po 1.8	At 2.1	Rn ...
Fr 0.9	Ra 1.0	Ac-Lr 1.1-	Unq ...	Unp ...	Unh ...	Uns ...	Uno ...	Une ...									

Metals

Transition metals

Increasing values (→ across; ↑ up)

(Note the existence of several exceptions to the general trends)

FIGURE 1.5
Electronegativity values for some common elements.

Question 1.3 Rank the elements Cl, Br, I, and F in order of increasing

a. Atomic size
b. Ionization energy
c. Electron affinity
d. Electronegativity

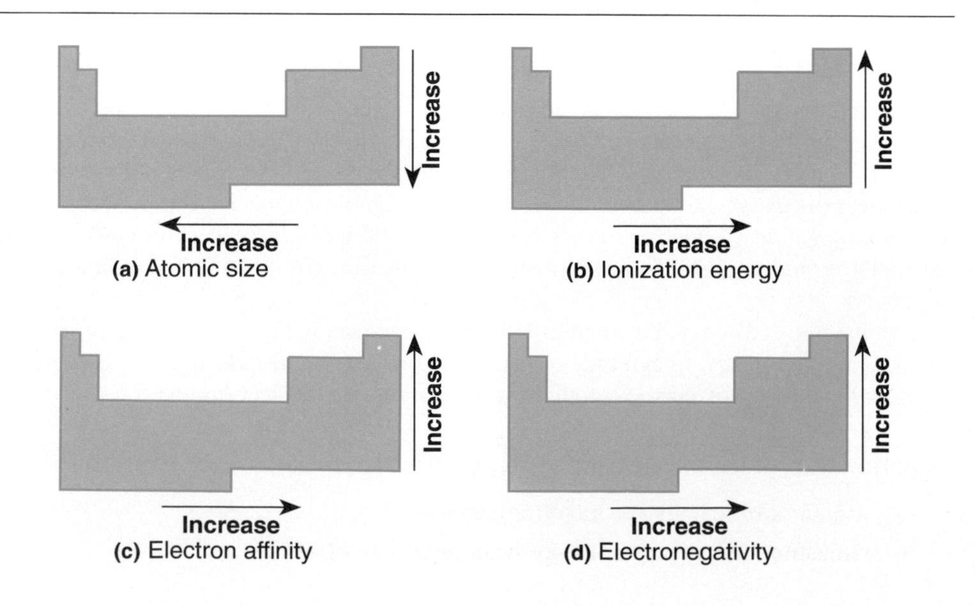

FIGURE 1.6
Periodic trends for atomic size, ionization energy, electron affinity, and electronegativity.

1.3 CHEMICAL BONDING

When two atoms are joined together to make a chemical compound, the force of attraction between the two species is referred to as a **chemical bond.** The attraction is the force that prevents the positive nuclei from coming apart.

Interactions involving **valence electrons** are responsible for the chemical bond. We shall focus our attention on these electrons and the electron configuration of atoms before and after bond formation.

Valence electrons are the outermost electrons in an atom. The number of valence electrons in an atom of a representative element is indicated by the group number.

Lewis symbols

The **Lewis symbol,** developed by G. N. Lewis early in this century, is a convenient way of representing atoms singly or in combination. Its principal advantage is that *only* valence electrons (those that may participate in bonding) are shown. This results in simple structures and great clarity. To draw Lewis structures, the chemical symbol of the atom is written; this symbol represents the nucleus and all of the lower-energy nonvalence electrons, which do not directly participate in bonding. The valence electrons are indicated by dots (·) or crosses (×) arranged around the atomic symbol as illustrated below:

Recall that the number of valence electrons is equal to the group number for representative elements.

H·	He:	Li·	Be:	·Ḃ·
Hydrogen	Helium	Lithium	Beryllium	Boron
·Ċ·	·N̈: (·N:)	:Ö:	:F:	:Ne:
Carbon	Nitrogen	Oxygen	Fluorine	Neon

Note particularly that the number of dots corresponds to the number of valence electrons in the atoms.

Types of chemical bonds: ionic and covalent

Two principal classes of chemical bonds exist: ionic and covalent.

Ionic bonding is characterized by an electron transfer process occurring before bond formation. In **covalent bonding,** electrons are *shared* between atoms in the bonding process. Before discussing each type, we should recognize that the distinction between ionic and covalent bonding is not always clear-cut. Some compounds are clearly ionic, some are clearly covalent; many others possess both ionic and covalent character.

Ionic bonding

Consider the reaction of a sodium atom and a chlorine atom to produce sodium chloride:

$$\mathrm{Na + Cl \longrightarrow NaCl}$$

Using Lewis symbols, we may rewrite the equation as follows:

$$\mathrm{Na^{\times} + \cdot\ddot{\underset{\cdot\cdot}{Cl}}: \longrightarrow Na^{+}Cl^{-}}$$

We have, for convenience, chosen to represent the valence electron of sodium with a "×" and the valence electrons of chlorine with a "·"; this is not meant to imply any difference between electrons, since all electrons are identical; it merely simplifies our "bookkeeping" procedure. Recall that the sodium atom has the following properties:

- a low ionization energy (it wants to lose an e^-),
- a low electron affinity (it does not want any more electrons).

Therefore if sodium loses an electron, it will become **isoelectronic** (same number of electrons) with neon, an indication that the sodium atom would be a good electron donor, and form the sodium cation:

$$Na \longrightarrow Na^+ + 1e^-$$

Recall that eight valence electrons, as found in the inert gas elements, is a particularly stable arrangement: the **octet rule.**

Chlorine, on the other hand, ''wishes'' to gain one more electron. By doing so, it will complete an octet of electrons and be isoelectronic with argon, a stable inert gas. Therefore chlorine should behave as a willing electron acceptor, forming an anion:

$$Cl + 1e^- \longrightarrow Cl^-$$

The electron released by sodium (*electron donor*) is the electron received by chlorine (*electron acceptor*):

$$Na \longrightarrow Na^+ + 1e^-$$

$$1e^- + Cl \longrightarrow Cl^-$$

The resulting ions of opposite charge, Na^+ and Cl^-, are attracted to each other (opposite charges attract) and held together by this **electrostatic force** as an **ion pair:** Na^+Cl^-. The electrostatic force, which is quite strong and holds the ion pair together, is called the ionic bond.

The essential features of ionic bonding are as follows:

- Elements with low ionization energy and low electron affinity tend to form positive ions.
- Elements with high ionization energy and high electron affinity tend to form negative ions.
- Ion formation results from electron transfer.
- The positive and negative ions are held together by the electrostatic force between ions of opposite charge in an ionic bond.
- Reactions between metals and nonmetals tend to result in ionic bonds.

Metals are elements far to the left and nonmetals are elements far to the right, respectively, on the periodic table.

Covalent bonding

Consider the bond between two hydrogen atoms to give the ''normal'' form of hydrogen: H_2. Individual hydrogen atoms are not stable but adopt a more stable configuration when two hydrogen atoms combine to produce diatomic hydrogen:

$$H + H \longrightarrow H_2$$

If a hydrogen atom, H ·, were to gain a second electron, it would be isoelectronic with the stable electron configuration of helium. However, since two identical hydrogen atoms have an equal tendency to gain or lose electrons, an electron transfer from one atom to the other does not occur. Each atom may attain an inert gas structure by *sharing* its electron with the other. This is shown below, using Lewis symbols:

$$H\cdot + \cdot H \longrightarrow H:H$$

Bohr's representation of the atom involved a series of concentric orbits around the nucleus. The radius of each orbit is proportional to the energy of the electrons contained in that orbit.

The formation of a covalent bond using Bohr's representation for the hydrogen molecule is shown below:

$$H\ 1e^- + 1e^-\ H \longrightarrow H{-}H$$

When electrons are shared rather than transferred, the *shared electron pair* is referred to as a *covalent bond*. Compounds characterized by covalent bonding are called *covalent compounds*. Covalent bonds tend to form between atoms with similar tendencies to gain or lose electrons. The most obvious examples are the diatomic molecules: H_2, N_2, O_2, F_2, Cl_2, I_2, and Br_2. Bonding in these molecules is *totally covalent* because there is no net tendency for electron transfer between identical atoms.

The formation of F_2, for example, may be represented as

$$:\ddot{\underset{..}{F}}\cdot + \cdot\ddot{\underset{..}{F}}: \longrightarrow :\ddot{\underset{..}{F}}:\ddot{\underset{..}{F}}:$$

$$F\ 7\bar{e} + 7\bar{e}\ F \longrightarrow F_2$$

As in H_2, a single covalent bond is formed. The bonding electron pair is said to be *localized,* or largely confined to the region between the two fluorine nuclei.

Two atoms do not have to be identical to form a covalent bond. Consider compounds such as the ones shown below:

$H:\ddot{\underset{..}{F}}:$	$H:\ddot{\underset{..}{O}}:H$	$H:\overset{H}{\underset{H}{\ddot{\underset{..}{C}}}}:H$	$H:\overset{H}{\ddot{\underset{..}{N}}}:H$
Hydrogen fluoride	Water	Methane	Ammonia
$7e^-$ from F	$6e^-$ from O	$4e^-$ from C	$5e^-$ from N
$1e^-$ from H	$2e^-$ from 2H	$4e^-$ from 4H	$3e^-$ from 3H
$8e^-$ around F	$8e^-$ around O	$8e^-$ around C	$8e^-$ around N

In each of these cases, bond formation satisfies the octet rule. There are a total of eight electrons around each atom other than hydrogen. Hydrogen "sees" only two electrons.

Distinction between ionic and covalent compounds

In the four examples given above, the atoms joined differ in their electron donating and withdrawing characteristics. A convenient measure of this is the *electronegativity* (EN) or, preferably, the *difference in electronegativity* (ΔEN) between the atoms forming the bond. A series of representative compounds, correlating electronegativity differences and bond type, is shown in Table 1.2.

The following general statements, based on the concept of electronegativity, apply here. Two atoms with greatly different electronegativities, such as sodium and chlorine, form ionic bonds. Pairs of atoms with no electronegativity difference, such as two H or two F atoms, form pure covalent bonds. Between these extremes exist bonds that have both ionic and covalent character. An approximate "boundary" (defined by electronegativity) exists between ionic and covalent compounds, which is an electronegativity differ-

TABLE 1.2 Correlation of Electronegativity Difference (ΔEN) with Bond Type for a Series of Compounds

Compound	EN (atom 1)	EN (atom 2)	ΔEN (2 – 1)	Bond Type
NaCl	0.9 (Na)	3.0 (Cl)	2.1	I
HF	2.1 (H)	4.0 (F)	1.9	PC
H_2O	2.1 (H)	3.5 (O)	1.4	PC
NH_3	2.1 (H)	3.0 (N)	0.9	PC
CH_4	2.1 (H)	2.5 (C)	0.4	PC
F_2	4.0 (F)	4.0 (F)	0.0	C
H_2	2.1 (H)	2.1 (H)	0.0	C

I = ionic, PC = polar covalent, C = covalent.

ence of 1.7. Compounds with an electronegativity difference $\Delta EN \geq 1.7$ are generally classified as ionic; those with $\Delta EN < 1.7$ are considered to be covalent.

The majority of covalently bonded compounds contain one or more bonds involving the sharing of electrons between nonidentical atoms, resulting in a nonzero electronegativity difference. This type of covalent bonding is distinguished from "pure" covalent bonding (a zero electronegativity difference) and is described as *polar covalent*. The bonds are polar; that is, the electron density is not equally distributed throughout the bond. Rather, the atom with the greater electronegativity has a greater negative character, leaving the other bonded atom with an apparent positive charge. A molecule containing several bonds, some or all of which are polar bonds, may itself be either polar or nonpolar, depending upon the arrangement of the bonds in the molecule. If the bonds are arranged in such a way as to cancel the resulting dipoles of the various bonds, the molecule will be nonpolar. Otherwise, a polar molecule will result.

Section 1.6

1.4 FORMULAS OF ORGANIC COMPOUNDS

The molecular formula describes the atomic composition of a molecule, that is, the number of each type of atom in the molecule. However, molecular formulas have a limitation that is particularly serious in organic chemistry; they provide no information about the structure of the molecule. This is less a problem in inorganic chemistry, since, often, only one obvious structure is possible. For example, HCl must be an acid, NaOH must be a base, and NH_4Cl must be a salt. The most important characteristics of the compound are readily deduced from the formula.

In contrast, organic compounds may be made up of many atoms in a variety of arrangements. Here, knowing the structure of the molecule is necessary to avoid any ambiguity.

Carbon-containing organic compounds often exist in a variety of different structures, each of which is represented by the same molecular formula. An example is the compound butane, C_4H_{10}, which was described in the introduction of this textbook. This formula represents two different compounds with different structures, referred to as **isomers.** Although butane has only two isomers, as the number of carbon atoms in the formula increases, the number of possible isomers increases dramatically.

The molecular formula for both butane isomers is identical; therefore a different representation, a **structural formula,** is required to unambiguously describe these isomers. However, structural formulas are somewhat cumbersome for routine use. A suitable compromise between the convenience of molecular formulas and the detail of the structural formula is the condensed structural formula. The condensed structural formula for pentane (n-pentane) is

$$CH_3CH_2CH_2CH_2CH_3, \text{ representing } \begin{array}{ccccccccccccc} & & H & & H & & H & & H & & H & & \\ & & | & & | & & | & & | & & | & & \\ H & - & C & - & C & - & C & - & C & - & C & - & H \\ & & | & & | & & | & & | & & | & & \\ & & H & & H & & H & & H & & H & & \end{array}$$

that for 2-methyl butane (isopentane) is

$$CH_3CH(CH_3)CH_2CH_3, \text{ representing } \begin{array}{ccccccccccc} & & & & H & & & & & & \\ & & & & | & & & & & & \\ & & & H- & C & -H & & & & & \\ & & H & & | & & H & & H & & \\ & & | & & | & & | & & | & & \\ H & - & C & - & C & - & C & - & C & - & H \\ & & | & & | & & | & & | & & \\ & & H & & H & & H & & H & & \end{array}$$

and that for 2,2-dimethylpropane (neopentane) is

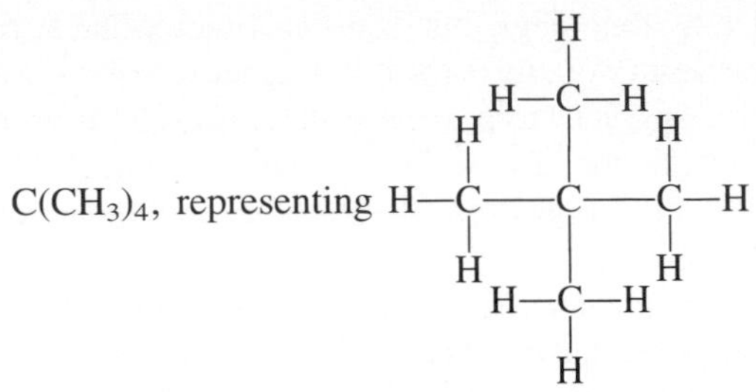

```
                                   H
                                   |
                                 H-C-H
                                H  |  H
                                |  |  |
$C(CH_3)_4$, representing   H—C——C——C—H
                                |  |  |
                                H  |  H
                                 H-C-H
                                   |
                                   H
```

Question 1.4 Write the structural formula corresponding to each of the following condensed structural formulas:

a. $(CH_3)_3COH$
b. $CH_3CH_2C(CH_3)_3$
c. $(CH_3)_3CC(CH_3)_3$

Question 1.5 Write the structural formula corresponding to each of the following condensed structural formulas:

a. $CH_3CH_2CH[CH(CH_3)_2]CH_3$
b. $CH_3C(CH_3)BrCH_2CH_3$
c. $CH_3C[CH(CH_3)_2]_2CH_3$

1.5 SHAPES OF ORGANIC MOLECULES

Many chemical and physical properties of organic compounds depend upon their shape. To a large extent, the molecular shape is determined by the *bond order* of each carbon atom in the molecule.

A single bond = bond order of 1

A double bond = bond order of 2

A triple bond = bond order of 3

This is true because the bond order is related to the **bond angle** associated with the carbon atom. Consider three cases.

Four single bonds

When carbon is bonded by four single bonds such as in methane, the bond angle (the angle formed by any two of the bonds joined at the carbon atom) is ideally 109.5° as predicted by the **valence shell electron pair repulsion (VSEPR) theory.**

VSEPR theory describes the geometry of a molecule on the basis of the repulsion of electron pairs (lone pairs) and other atoms or groups of atoms attached to a central atom.

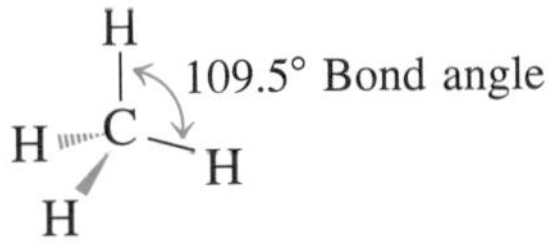

```
    H
    |  109.5° Bond angle
 H--C--
   /   H
  H
```

Methane

Chemists have a variety of experimental techniques for experimentally determining bond angles. Factors such as attraction and repulsion between atoms may cause this angle to vary (but only slightly) from compound to compound.

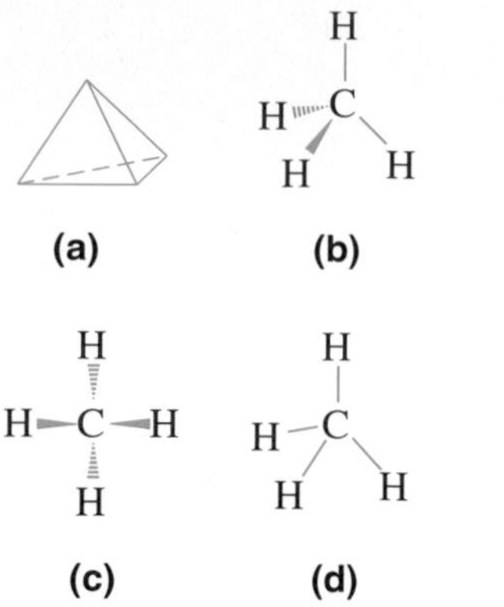

(a) (b)

(c) (d)

FIGURE 1.7
The tetrahedral carbon atom: (a) a tetrahedron; (b) the tetrahedral carbon drawn with dashes and wedges; (c) a different orientation from that in part (b); (d) the stick drawing of the tetrahedral carbon atom.

Carbon surrounded by four single bonds is described as the tetrahedral carbon atom. The tetrahedral carbon atom is named for the geometric solid—a triangular solid contained by four planar, triangular faces. This structure results from the outline of carbon surrounded by four single bonds. The solid is shown in Figure 1.7. Note the relationship between the two forms—the tetrahedron (Figure 1.7a) and the tetrahedral carbon (Figure 1.7b).

Chemists use several methods to draw the tetrahedral carbon in a way that will illustrate its three-dimensional nature. These are illustrated in Figure 1.7. In Figure 1.7b, dashes and wedges are used; dashes go back into the page, away from you, and wedges come out of the page, toward you. The other bonds are in the plane of the page. In Figure 1.7c, dashes and wedges are also used. In this method, however, two of the bonds are oriented away from the viewer (behind the page, dashes) and two of the bonds are oriented toward the viewer (in front of the page, wedges). None of the bonds are in the plane of the page. In Figure 1.7d, lines are used in an attempt to show three-dimensionality. Figure 1.7d is equivalent to Figure 1.7c, yet it leaves a great deal more to the imagination of the viewer. Dash/wedge structures are preferred and are used almost exclusively throughout this text in depicting the three-dimensional arrangement of atoms in space.

Question 1.6 Using a molecular model kit (gum drops and toothpicks will also do), build the structure for each compound in Question 1.4.

One double bond and two single bonds

When carbon is bonded by one double and two single bonds such as in the compound ethene (ethylene), CH_2CH_2:

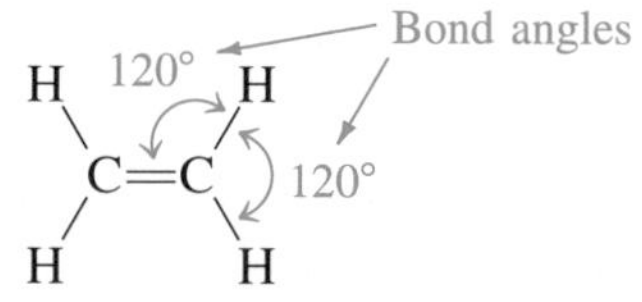

Bond angles
120°
120°

Ethene
(ethylene)

the molecule is *planar* (all atoms involved in the double bond lie in a common plane), and *each* bond angle is approximately 120° as predicted by VSEPR theory.

A triple bond

When two carbons are bonded by a triple bond such as in the compound ethyne (acetylene), HCCH:

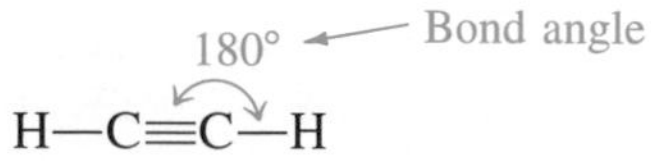

180° Bond angle

H—C≡C—H

Ethyne
(acetylene)

the molecule is *linear* (all the atoms are positioned in a straight line), and *each* bond angle is 180°.

The three cases discussed above give rise to the three basic geometries:

- tetrahedral (a three-dimensional pyramid)
- planar (all atoms in one plane)
- linear (all atoms in one line)

In dealing with larger, more complex molecules containing atoms other than H and C, prediction of molecular geometry becomes very difficult. In fact the elucidation of the shape and structure of one complex biological molecule, DNA (deoxyribonucleic acid), the molecule that is intimately linked to our genetic heritage, resulted in a Nobel Prize for James Watson and Francis Crick. Certainly, as you become adept as assessing molecular geometry, understanding the properties and reactions of organic and bio-organic compounds will become less difficult.

see Figure 7

Question 1.7 Predict approximate bond angles for the indicated bond in each of the following compounds. Structures are not necessarily drawn with correct bond angles. Use your knowledge of structure to determine the bond angles; don't rely on the drawings.

a. $CH_3—Cl$
b. $CH_3—CH_3$
c. $CH_3—C{\equiv}C—H$
d. $CH_2{=}C(Br)—CH_3$ (Br on the central C)
e. $H—CH_2—CH_2—O—H$
f. $H—CH_2—C{\equiv}C—C(H)(H)—CH_3$ (two H on the fourth C, drawn above and below)

Question 1.8 What is the bond order of each of the bonds indicated in Question 1.7? For each compound in Question 1.7, what is the highest bond order present in the molecule? Which bond has this bond order?

The shapes of organic molecules are vital to the reactivity of the molecules. Very often, a chemist will find that one molecule will undergo a reaction but another, similar molecule with a slightly different shape will not react. In biochemistry this observation is common in studying the activity of enzymes.

Section 12.2

Enzymes are molecules that behave as catalysts in biochemical systems.

Enzymes generally bind a specific compound and facilitate a reaction. Only this "specific" compound can orient itself appropriately within a pocket on the surface of the enzyme. This is called the lock-and-key model of enzyme reactivity. The enzyme acts as the lock, and the reactive compound with the correct shape acts as the key. Only certain keys fit the lock.

Section 12.5

Bond length (the distance between the nuclei of two atoms held together by chemical bonds) and **bond strength** (the energy necessary to break a bond between two atoms) are also important in discussing the shape of molecules and their reactivity. For example, single bonds are longer than double bonds, and double bonds are longer than triple bonds. Triple bonds are stronger than double bonds, and double bonds are stronger than single bonds. Trends such as these are useful in predicting reactivity and designing reactions. They are also important in predicting and understanding the shapes of molecules. All of these concepts are applied and expanded upon in upcoming chapters.

1.6 IMPORTANT DIFFERENCES BETWEEN COVALENT AND IONIC COMPOUNDS

Organic compounds are characterized by covalent bonding, whereas many nonorganic (inorganic) compounds exist as ionic substances. Therefore the overall physical and chemical characteristics of organic compounds are quite different from those of inorganic compounds. The major differences are summarized in Table 1.3.

Methane is an example of a typical, simple organic molecule, and sodium chloride has properties that are generally representative of inorganic substances. Remember that there are exceptions in any generalization of this type. However, this comparison does provide you with some feeling for the huge difference in the properties and reactivities of these two classes.

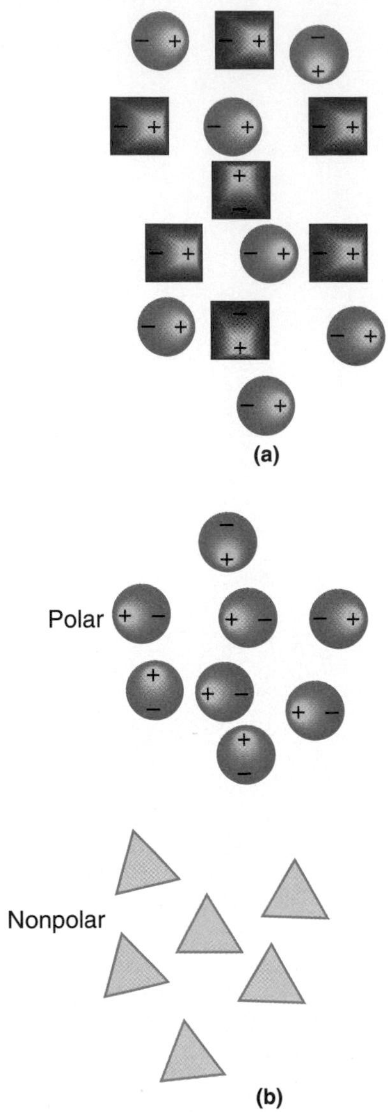

FIGURE 1.8
The polarity of molecules: "like dissolves like." (a) Here, two different polar molecules (circles and squares) are completely intermingling. Complete solubility results. (b) Here, two sets of molecules, polar and nonpolar, do not interact with each other. There is no intermixing, and the two are mutually insoluble.

TABLE 1.3 Comparison of the Major Properties of Typical Organic and Inorganic Compounds: Methane Versus Sodium Chloride

Organic (Methane)	Inorganic (Sodium Chloride)
Covalent bonding: CH_4	Ionic bonding: Na^+Cl^-
Low boiling point: −164°C	High boiling point: 1433°C
Low melting point: −182°C	High melting point: 801°C
Insoluble in water: trace	Soluble in water: 36 g/100 mL
Flammable	Nonflammable
Chemical reactivity: often slow	Chemical reactivity: often fast
Gas at room temperature	Solid at room temperature
Nonconductor of electricity in solution	Conductor of electricity in solution

Melting and boiling points

Ionic compounds do not exist as individual units—molecules—but rather as large three-dimensional crystals. The strong attractive forces between positive and negative ions make it difficult to break apart the crystal. Large amounts of energy are required to accomplish this. Hence the melting points of this class of compounds are generally very high. Strong attractive forces are also responsible for the very high boiling points of these compounds.

The situation with respect to covalently bonded organic compounds is quite different. Here the individual unit is a single molecule, and the attractive forces *between* molecules (Van der Waals and London forces) are relatively weak. These intermolecular forces are overcome at lower temperatures; hence we expect and find lower melting and boiling points for organic compounds.

Solubility

Recall the rule ''like dissolves like.'' Ionic compounds, composed of individual charged units, if soluble at all, tend to be much more soluble in water than less polar solvents. Organic molecules, in contrast, generally are nonpolar or, at best, moderately polar; consequently, they have limited water solubility. The degree of polarity—hence solubility—depends upon the number of **heteroatoms** (atoms other than C and H, such as N, O, S, X—the symbol X will be used to represent the halogens) as well as on their arrangement in the molecule.

Viewing polar molecules as tiny magnets helps one to understand the differences in physical properties between polar and nonpolar organic compounds. Behaving analogously to small magnets, polar molecules are attracted to one another and to other polar molecules quite strongly. Nonpolar molecules do not exhibit these (relatively) strong interactions.

When two polar liquids are mixed, the molecules attract each other, and complete intermixing and mutual solubility (miscibility) result. When a polar and a nonpolar liquid are mixed, the polar molecules are attracted to each other, but there is no attraction between the polar and nonpolar molecules. No mixing occurs—the two liquids are immiscible (see Figure 1.8).

Van der Waals and London forces are responsible for intermolecular attractive forces between molecules. These forces result from either permanent or induced dipole interaction between covalent or polar covalent molecules.

Using similar logic, one would predict that polar molecules would have a much higher boiling point than nonpolar molecules, and this prediction is generally correct. On the molecular level, when a liquid is converted to the gas phase, the molecules move far

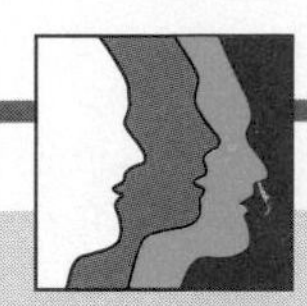

A HUMAN PERSPECTIVE

An Extraordinary Molecule

Think for a moment. What is the only common molecule to exist in all three physical states of matter (solid, liquid, and gas) under natural conditions on the earth? This molecule is absolutely essential for life; in fact, life probably arose in this substance. It is the most abundant molecule in the cells of living organisms (70–95%) and covers 75% of the earth's surface. Without it, cells quickly die, and without it the earth would not be a fit environment in which to live. By now you have guessed that we are talking about the water molecule. It is so abundant on the earth that we take this deceptively simple molecule for granted.

What are some of the properties of water that cause it to be essential to life as we know it? Water has the ability to stabilize temperatures on the earth and in the body. This ability is due in part to the energy changes that occur when water changes physical state; but ultimately, this ability is due to the polar nature of the water molecule.

Life can exist only within a fairly narrow range of temperatures. Above or below that range, the chemical reactions necessary for life, and thus life itself, will cease. Water can moderate temperature fluctuation and maintain the range necessary for life, and one property that allows it to do so is its unusually high specific heat, 1 calorie/gram °C. This means that water can absorb or lose more heat energy than other substances without undergoing a significant temperature change. This is because in the liquid state, every water molecule is hydrogen bonded to other water molecules. Since a temperature increase is really just a measure of increased molecular motion, we must get the water molecules moving more rapidly, independent of one another, to register a temperature increase. Before we can achieve this independent, increased activity, the hydrogen bonds between molecules must be broken. Much of the heat energy absorbed by water is involved in breaking hydrogen bonds and is *not* used to increase molecular movement. Thus a great deal of heat is needed to raise the temperature of water even a little bit.

Water also has a very high heat of vaporization. It takes 540 calories to change one gram of liquid water at 100°C to a gas and even more, 603 cal/g, when the water is at 37°C, human body temperature. That is about twice the heat of vaporization of alcohol. As water molecules evaporate, the surface of the liquid cools because only the highest-energy (or "hottest") molecules leave as a gas. Only the "hottest" molecules have enough energy to break the hydrogen bonds that bind them to other water molecules. Indeed, evaporation of water molecules from the surfaces of lakes and oceans helps to maintain stable temperatures in those bodies of water. Similarly, evaporation of perspiration from body surfaces helps to prevent overheating on a hot day or during strenuous exercise.

Even the process of freezing helps to stabilize and moderate temperatures; this is especially true in the fall. Water releases heat when hydrogen bonds are formed. This is an example of an exothermic process; thus when water freezes, solidifying into ice, additional hydrogen bonds are formed, and heat is released into the environment. As a result, the temperature change between summer and winter is more gradual, allowing organisms to adjust gradually to the change.

One last feature that we take for granted is the fact that when we put ice in our iced tea on a hot summer day, it will float. This means that the solid state of water is actually *less* dense than the liquid state! In fact, it is about 10% less dense, having an open lattice structure with each molecule hydrogen bonded to a maximum of four other water molecules. What would happen if ice did sink? Eventually, all bodies of water, including the mighty oceans, would freeze solid, killing all aquatic and marine plant and animal life. Even in the heat of summer, only a few inches of ice at the surface would thaw. Instead, the ice forms at the surface and provides a layer of insulation that prevents the water below from freezing.

As we continue our study of chemistry we will refer again and again to this amazing molecule. In other Human Perspective features, we will examine other properties of water that make it essential to life.

apart. More energy is required to separate polar molecules from one another than is required to separate nonpolar molecules. Since the average kinetic energy increases as the Kelvin temperature increases, the magnitude of the boiling temperature depends upon the strength of the attractive forces (intermolecular forces) between the molecules.

This simple model explains many of the physical properties of compounds that will be studied in the chapters ahead. There are numerous biochemical implications. For example, the polar/nonpolar nature of molecules plays a vital role in the formation and function of cell membranes and in nutrient and waste transport in the blood.

Section 9.6, Section 9.5

Lewis structure and resonance

In some cases we find that it is possible to write more than one Lewis structure that satisfies the octet rule for a particular compound. Consider sulfur dioxide, SO_2. Its skeletal structure is

O—S—O

The total number of valence electrons may be calculated as follows:

$$\begin{array}{rl} 1 \text{ sulfur atom } \times 6 \text{ valence e/atom} = & 6e^- \\ + \ 2 \text{ oxygen atom} \times 6 \text{ valence e/atom} = & 12e^- \\ \hline & 18e^- \text{ total} \end{array}$$

The resulting Lewis structures are

O: :S:O: and :O:S: :O

Both satisfy the octet rule. However, experimental evidence shows the absence of a double bond in SO_2. The two sulfur-oxygen bonds are equivalent. Apparently, neither structure accurately represents the structure of SO_2, and neither actually exists. The actual structure is said to be an *average* or *hybrid* of these two Lewis structures. When a compound has two or more Lewis structures that contribute to the real structure, we say that the compound displays the property of *resonance*. The contributing Lewis structures are *resonance forms*. The true structure, a hybrid of the resonance forms, is known as a *resonance hybrid* and is represented as:

O: :S:O: ⟺ :O:S: :O

A common analogy might help to clarify this concept. Two fruit trees, an orange and a tangerine, may be cross-bred to produce a hybrid, the tangelo. The tangelo doesn't look or taste exactly like either an orange or a tangerine, yet it has attributes of both. The resonance hybrid of a molecule has properties of each resonance form but is not identical to any one form.

The existence of resonance enhances molecular stability. This concept is important in understanding the chemical reactions of many complex organic molecules and will be used extensively in organic chemistry.

Question 1.9 SeO_2, like SO_2, has two resonance forms.

a. Draw their Lewis structures.

b. Explain any similarities between these structures and those for SO_2 in light of periodic relationships.

SUMMARY

Our modern view of the atom, the basic unit of matter, is based upon a gradual evolution of theory, spurred on by a series of significant experimental discoveries. John Dalton's atomic theory was the first scientific attempt to explain the matter around us. Discovery of the electron, proton, and neutron, the nucleus, and radioactivity paved the way for the development of the modern theory of atomic structure. The periodic table provides the basis for prediction of such properties as relative atomic and ionic size, ionization energy, electron affinity, and electronegativity, as well as metallic or nonmetallic character and reactivity.

The importance of valence electrons and the octet rule in bonding has been presented. The periodic table provides an easy way of determining the number of valence electrons in atoms. In bonding, the combination of atoms to form bonds generally requires that the resulting valence electron configuration satisfy the octet rule. In doing so, atoms may form ionic, covalent, or polar covalent bonds. The type of bond that results is directly related to the electronegativity differences (ΔEN) between the bonding atoms. The fundamental differences between the ionic bond, which is characteristic of most inorganic substances, and the covalent bond in organic compounds are responsible for the great contrast in properties and reactivity between organic and inorganic compounds. For example, organic compounds generally have lower melting and boiling points and lower water solubility.

The molecular formula is insufficient to uniquely represent most organic compounds because of the existence of isomers. The expanded structural formula and the condensed structural formula are preferred.

The shape of an organic molecule and the presence or absence of functional groups determine both properties and the kinds of chemical reactions that the molecule will undergo. Shape is determined in part by the bond order and bond arrangement around carbon. Resonance, when present in a molecule, may have a significant effect on the bond order, shape, and reactivity of a molecule.

GLOSSARY OF KEY TERMS

anion (1.1) a negatively charged atom.
atom (Introduction) the smallest unit of an element that retains the properties of that element.
atomic number (1.1) the number of protons in the nucleus of an atom. It is a characteristic identifier of an element.
bond angle (1.5) the angle (in degrees) formed by two covalent bonds involving a common atom.
bond length (1.5) the distance between the nuclei of two atoms joined by a chemical bond.
bond strength (1.5) the energy necessary to break a bond between two atoms.
cation (1.1) a positively charged atom.
chemical bond (1.3) the attractive electrical force holding two atomic nuclei together in a chemical compound.
covalent bond (1.3) a pair of electrons shared between two atoms.
electron (1.1) a negatively charged particle outside of the nucleus of an atom.
electron affinity (1.2) the energy released when an electron is added to an isolated atom.
electronegativity (1.2) a measure of the tendency of an atom in a molecule to attract shared electrons.
electrostatic force (1.3) the attractive force between two oppositely charged particles.
formal charge (1.1) the charge on each atom in a molecule.
heteroatoms (1.6) atoms other than carbon and hydrogen contained in an organic molecule.
ion (1.1) a charged atom.
ion pair (1.3) the empirical formula unit for an ionic compound.
ionic bond (1.3) an electrostatic attractive force between ions resulting from electron transfer.
ionization energy (1.2) the energy needed to remove an electron from an atom in the gas phase.
isoelectronic (1.3) atoms and ions containing the same number of electrons.
isomers (1.4) two or more substances that have the same molecular formula but different structures and hence different properties.
isotopes (1.1) atoms of the same element that differ in mass because they contain different numbers of neutrons.
Lewis symbol (1.3) representation of an atom or ion using the atomic symbol (for the nucleus and core electrons) and dots to represent valence electrons.
mass number (1.1) the sum of the number of protons and neutrons in an atom.
molecule (Introduction) a unit of two or more atoms held together by chemical bonds.
neutron (1.1) an uncharged particle, with the same mass as the proton, in the nucleus of an atom.
nucleus (1.1) the small, dense center of positive charge in the atoms.
octet rule (1.3) a rule predicting that atoms of nonmetals form the most stable molecules or ions when they are surrounded by eight electrons in their highest occupied energy level.
proton (1.1) a positively charged particle in the nucleus of an atom.
radioactivity (1.1) the process by which atoms emit high-energy particles or rays; the spontaneous decomposition of a nucleus to produce a different nucleus.
structural formula (1.4) a formula showing all of the atoms in a molecule and exhibiting all bonds as lines.
valence electrons (1.3) electrons in the outermost shell (principal quantum level) of an atom.
valence shell electron pair repulsion (VSEPR) theory (1.5) a model accounting for the geometrical arrangement of shared and unshared electron pairs around a central atom in terms of electron pair repulsion.

QUESTIONS AND PROBLEMS

The following problem set provides an opportunity for a review of the key concepts of inorganic chemistry that have particular applicability to the study of organic chemistry and biochemistry. These concepts have been discussed in this chapter and are supplemented by material contained in the appendices at the end of this textbook. These problems presume some basic background in the principles of inorganic chemistry.

The Periodic Table

1.10 Define each of the following terms:
a. periodic law
b. period
c. group
d. ion
e. electron configuration
f. octet rule
g. ionization energy
h. electronegativity

1.11 Label each of the following statements as true or false:
a. Elements of the same group have similar properties.
b. Atomic size decreases from left to right across a period.
c. Ionization energy increases from top to bottom within a group.
d. Electronegativity increases from left to right across a period.

1.12 For each of the elements Na, Ni, Al, P, Cl, and Ar, provide the following information:
a. Which are metals?
b. Which are representative metals?
c. Which tend to form positive ions?
d. Which are inert?
e. Which are in group III?
f. Which are in period 3?

1.13 Provide the name of the element represented by each of the following symbols:
a. Na d. Ca
b. K e. Cu
c. Mg f. Co

1.14 Give three examples of elements that are:
a. metals
b. metalloids
c. nonmetals

1.15 Which group of the periodic table is known as the alkali metals? List them.

1.16 Which group of the periodic table is known as the alkaline earth metals? List them.

1.17 Which group of the periodic table is known as the halogens? List them.

1.18 Which group of the periodic table is known as the inert gases? List them.

1.19 How many valence electrons are in each of the following elements?
a. H d. F
b. Na e. Ne
c. B f. He

1.20 Give the most probable ion formed from each of the following elements:
a. Li d. Br
b. O e. S
c. Ca f. Al

1.21 Which of the following pairs of species are isoelectronic with one another?
a. O^{2-}, Ne c. F^-, Cl^-
b. S^{2-}, Cl^- d. K^+, Ar

1.22 Which species in each of the following groups would you expect to be most stable?
a. Na, Na^+, Na^-
b. S^{2-}, S^-, S^+
c. Cl, Cl^-, Cl^+

1.23 Why are inert gases so nonreactive?

1.24 Write the Lewis structure for each of the following elements:
a. C d. H
b. N e. S
c. O

1.25 Write Lewis structures for each of the following elements:
a. Si d. He
b. P e. Se
c. Cl

1.26 Why do we make a distinction between the number of valence electrons and the total number of electrons in an atom?

Electron Configuration

1.27 Using only the periodic table or list of elements, write the electron configuration of each of the following atoms:
a. B d. V
b. S e. Cd
c. Ar f. Re

1.28 Using only the periodic table or list of elements, write the electron configuration of each of the following ions:
a. I^- c. Se^{2-}
b. Ba^{2+} d. Al^{3+}

1.29 What is the common feature of the electron configuration of elements in Group IA?

1.30 What is the common feature of the electron configurations of elements in group VIIIA?

Atomic Properties

1.31 Arrange each of the following lists of elements in order of increasing electronegativity:
a. N, O, F
b. Li, K, Cs
c. Cl, Br, F

1.32 Arrange each of the following lists of elements in order of increasing ionization energy:
a. N, O, F
b. Li, K, Cs
c. Cl, Br, I

1.33 Arrange each of the following lists of elements in order of increasing atomic size:
a. N, O, F
b. Li, K, Cs
c. Cl, Br, I

1.34 Arrange each of the following lists of elements in order of decreasing electron affinity:
a. Na, Li, K
b. Br, F, Cl
c. S, O, Se

1.35 a. Which of the elements has the highest electronegativity?
b. Which of the elements has the highest electron affinity?
c. Which of the elements has the highest ionization energy?

1.36 Arrange each of the following lists of elements in order of increasing size:
a. Al, Si, P, Cl, S
b. In, Ga, Al, B, Tl
c. Sr, Ca, Ba, Mg, Be
d. P, N, Sb, Bi, As

1.37 Explain why the size of a positive ion is always less than the size of its parent ion.

1.38 Explain why the fluoride ion (F^-) is more stable than the fluoride atom (F).

Chemical Bonding

1.39 Classify each of the following compounds as ionic or covalent:
a. $MgCl_2$ d. H_2S
b. CO_2 e. NO_2
c. NaCl f. ICl

1.40 Using Lewis symbols, write an equation predicting the product of the reaction of each of the following pairs of atoms:
a. H + Cl d. Na + O
b. Li + Br e. Na + S
c. Mg + Cl

1.41 Discuss the principal difference between ionic and covalent bonding.

1.42 Give the Lewis structure for each of the following compounds:
a. NCl_3 d. CH_3OH
b. CS_2 e. HNO_3
c. CCl_4 f. PBr_3

1.43 Give the Lewis structure for each of the following:
a. CO_3^{2-} d. SO_3^{2-}
b. NH_4^+ e. HNO_3
c. CF_4 f. Br_2

1.44 Which element has six protons and an isotope with a mass of 12? Other isotopes have masses of 13 and 14. Give the chemical symbol of each isotope.

1.45 Fill in the blanks:

Atomic Symbol	No. Protons	No. Neutrons	No. Electrons
Example:			
$^{40}_{20}Ca$	20	20	20
$^{23}_{11}Na$	11	____	11
$^{32}_{16}S^{2-}$	16	16	____
____	8	8	8
$^{24}_{12}Mg^{2+}$	____	12	____
____	19	20	18

1.46 Fill in the blanks:
a. An isotope of an element differs in mass from the atom because the atom has as a different number of ____.
b. The atomic number gives the number of ____ in the nucleus.
c. The mass number of an atom is due to the number of ____ and ____ in the nucleus.
d. A charged atom is called a(n) ____.
e. Electrons surround the ____ and have a ____ charge.

Properties of Ionic and Covalent Compounds

1.47 Contrast ionic and covalent compounds with respect to the nature of the solid state.

1.48 Contrast ionic and covalent compounds with respect to their relative melting points.

1.49 Contrast ionic and covalent compounds with respect to their relative boiling points.

1.50 Draw the appropriate Lewis structure for each of the following atoms:
a. H f. C
b. He g. N
c. Li h. O
d. Be i. F
e. B j. S

1.51 Draw the appropriate Lewis structure for each of the following ions:
a. Li^+ f. Cl^-
b. Mg^{2+} g. P^{3-}
c. Be^{2+} h. O^{2-}
d. Al^{3+} i. S^{2-}
e. NH_4^+ j. $C_3H_3O_2^-$

1.52 Predict whether the bond formed between each of the following pairs of atoms would be ionic, nonpolar, or polar covalent:
a. Cl and Cl f. C and N
b. H and H g. S and O
c. C and H h. Na and Cl
d. Li and F i. Na and O
e. O and O j. Ca and Br

1.53 Draw an appropriate covalent Lewis structure formed by the simplest combination of atoms in Question 1.52.

1.54 Discuss the relationship between bond order and stability.

1.55 Discuss the concept of resonance, being certain to define resonance, resonance form, and resonance hybrid.

1.56 Why is resonance an important concept in bonding?

1.57 What is the relationship between the polarity of a bond and the polarity of the molecule?

1.58 What effect does polarity have on the solubility of a compound in water?

1.59 What effect does polarity have on the melting point of a pure compound?

1.60 What effect does polarity have on the boiling point of a pure compound?

1.61 Which of the following molecules are polar? Which are nonpolar?
a. CH_4
b. NH_3
c. CO_3^{2-} (Hint: Remember resonance!)
d. CH_3OH
e. HCOOH (formic acid, which causes the "sting" of an ant bite)
f.
```
          H
          |
   H     C
     \ //  \
      C      C—H
      |      ||
      C      C—H
     / \\  /
   H     C
          |
          H
```
g. NH_4^+

Differences Between Covalent and Ionic Compounds

1.62 Comparing organic and inorganic compounds:
a. Which compounds make good electrolytes?
b. Which compounds exhibit ionic bonding?
c. Which compounds have lower melting points?
d. Which compounds are soluble in water?
e. Which compounds are flammable?

1.63 Using the octet rule, explain why carbon forms four bonds in a stable compound.

1.64 Using the octet rule, explain why oxygen forms two bonds in a stable compound.

CHAPTER 2

Chemical Change

OBJECTIVES

Chapter 2 is a review of the chemical reaction, the various classes of chemical change, acids and bases, and oxidation-reduction; additionally, the fundamentals of chemical change, kinetics, and thermodynamics are discussed.

After reading this chapter, you should be able to achieve the following objectives, which are needed for your study of organic chemistry and biochemistry in Chapters 3–17.

- Be able to classify chemical reactions by type.
- Know how to recognize acids and bases and describe the role of the solvent in acid-base reactions.
- Calculate hydronium and/or hydroxide ion concentration from pH data and vice versa.
- Know what is meant by a neutralization reaction.
- Know the meaning of the term ''buffer'' and their application to chemical and biochemical systems.
- Define oxidation and reduction.
- Calculate oxidation numbers of atoms in ions and molecules and use them to predict redox processes.
- State the first and second laws of thermodynamics, and know their implications.
- Know what is meant by the terms ''activation energy'' and the ''activated complex.''
- Develop a feel for the way in which reactant structure, concentration, and temperature affect the rate of a chemical reaction.
- Know the role of a catalyst in a chemical reaction.

OUTLINE

Chemistry may be described as the study of matter and the changes that matter undergoes. Changes in matter are important not only to the professional chemist; these changes affect all of us. Literally thousands of examples of chemical change are known to take place in the human body alone. The fire that warms us, the explosive reactions in an automobile engine, and subtle changes in skin pigmentation after a day at the beach are a few examples of chemical change that are directly applicable to our lives.

This chapter will serve as a review of the major classes of chemical reactions, which describe chemical change, followed by an overview of the most important types of chemical reactions, namely, those of acids and bases as well as oxidation and reduction. Both have widespread applications in organic and biological chemistry.

Why does chemical change take place? All chemical processes either gain or lose energy as they occur. Energy—or, more properly, energy change—is the driving force for chemical change; thermodynamics is the study of the energetics of change, and this topic will be reviewed in Section 2.4.

How quickly do chemical reactions occur? Some reactions, such as those involved in respiration and nerve transmission, must be lightning fast. However, many of these biochemical reactions proceed extraordinarily slowly under laboratory conditions. In fact, life processes would not be possible if these reactions took place at their normal rate. However, biological catalysts, enzymes, increase the rate of these reactions by factors of 10^3 to 10^6. The study of rates of chemical reactions is termed kinetics and is reviewed in conjunction with the discussion of thermodynamics.

2.1 CLASSIFICATION OF CHEMICAL REACTIONS

Spontaneous chemical reactions occur for a variety of reasons, linked by the tendency to achieve the lowest (most stable) electronic energy state. Strong electrolytes will react to form *weak* (less dissociated) *electrolytes* when possible:

$$HCl(aq) + NaOH(aq) \longrightarrow NaCl(aq) + H_2O(l)$$

The strong electrolytes HCl and NaOH react completely to produce H_2O, a weak electrolyte. In contrast, the reaction

$$NaCl(aq) + KNO_3(aq) \longrightarrow NaNO_3(aq) + KCl(aq)$$

does not occur; both reactants and products are strong electrolytes. In actuality, such a mixture would produce an aqueous solution containing Na^+, K^+, Cl^-, and NO_3^- ions.

Reactions forming *gaseous products* are favored:

$$2HCl(aq) + Na_2S(aq) \longrightarrow H_2S(g) + 2NaCl(aq)$$

The gas may be nondissociating or, in the case of H_2S, weakly dissociating in solution.

In addition, formation of *insoluble solid products* favors a chemical reaction.

$$AgNO_3(aq) + NaCl(aq) \longrightarrow AgCl(s) + NaNO_3(aq)$$

There are thousands of different chemical reactions, and it is necessary and convenient to categorize these reactions into reaction types. One approach involves describing chemical reactions in terms of the reactants and the changes that the reactants undergo. In a broad sense, chemical reactions involve the *combination* of reactants to produce products, the *decomposition* of reactant(s) into products, or the *replacement* of one or more elements in a compound to yield products. Each of the major reaction types is discussed below.

Combination reactions

Combination reactions involve the joining of two or more atoms or compounds, producing a product of different composition. The general form of a combination reaction is

$$A + B \longrightarrow AB$$

where A and B represent reactant atoms or compounds and AB is the product. Examples include:

1. The combination of nitrogen atoms to produce the nitrogen molecule:

$$N(g) + N(g) \longrightarrow N_2(g)$$

2. The combination of hydrogen and chlorine molecules to produce hydrogen chloride:

$$H_2(g) + Cl_2(g) \longrightarrow 2HCl(g)$$

3. Formation of water from hydrogen and oxygen molecules:

$$2H_2(g) + O_2(g) \longrightarrow 2H_2O(g)$$

4. The reaction of magnesium oxide and carbon dioxide to produce magnesium carbonate:

$$MgO(s) + CO_2(g) \longrightarrow MgCO_3(s)$$

Decomposition reactions

Decomposition reactions produce two or more products from a single reactant. The general form of these reactions is the reverse of a combination reaction:

$$AB \longrightarrow A + B$$

Some examples are:

1. The heating of calcium carbonate to produce calcium oxide and carbon dioxide:

$$CaCO_3(s) \longrightarrow CaO(s) + CO_2(g)$$

A *hydrate* is a substance that has water molecules incorporated in its crystalline or solid structure.

2. The removal of water from a hydrated material:

$$CuSO_4 \cdot 5H_2O(s) \longrightarrow CuSO_4(s) + 5H_2O(g)$$

3. The dissociation of an ionic substance:

$$NaCl(aq) \longrightarrow Na^+(aq) + Cl^-(aq)$$

or

$$HCl(aq) \longrightarrow H^+(aq) + Cl^-(aq)$$

or

$$NaOH(aq) \longrightarrow Na^+(aq) + OH^-(aq)$$

Replacement reactions

Replacement reactions are subcategorized as either *single-replacement* or *double-replacement*. **Single-replacement reactions** occur between *atoms and compounds;* the atom replaces another atom in the compound producing a new compound and the replaced atom:

$$A + BC \longrightarrow AC + B$$

Examples include:

1. The replacement of copper by zinc in copper sulfate:

$$Zn(s) + CuSO_4(aq) \longrightarrow ZnSO_4(aq) + Cu(s)$$

2. The formation of solid silver metal through the reaction of silver ion with metallic copper:

$$Cu(s) + 2Ag^+(aq) \longrightarrow Cu^{2+}(aq) + 2Ag(s)$$

Double-replacement reactions, on the other hand, involve *two compounds* undergoing a "change of partners." Two compounds react by exchanging atoms to produce two new compounds:

$$AB + CD \longrightarrow AD + CB$$

Examples include:

1. The reaction of an acid (hydrochloric acid) and a base (sodium hydroxide) to produce water and salt:

$$HCl(aq) + NaOH(aq) \longrightarrow H_2O(l) + NaCl(aq)$$

2. The formation of solid barium sulfate and potassium chloride from barium chloride and potassium sulfate:

$$BaCl_2(aq) + K_2SO_4(aq) \longrightarrow BaSO_4(s) + 2KCl(aq)$$

Question 2.1 Classify each of the following reactions as decomposition (D), combination (C), single-replacement (SR), or double-replacement (DR):

a. $HNO_3(aq) + KOH(aq) \longrightarrow KNO_3(aq) + H_2O(l)$
b. $Al(s) + 3NiNO_3(aq) \longrightarrow Al(NO_3)_3(aq) + 3Ni(s)$
c. $KCN(aq) + HCl(aq) \longrightarrow HCN(g) + KCl(aq)$
d. $Na^+(aq) + Cl^-(aq) \longrightarrow NaCl(aq)$
e. $2Al(OH)_3(s) \xrightarrow{\Delta} Al_2O_3(s) + 3H_2O(g)$
f. $Fe_2S_3(s) \longrightarrow 2Fe(s) + 3S(s)$
g. $Na_2CO_3(aq) + BaCl_2(aq) \longrightarrow BaCO_3(s) + 2NaCl(aq)$
h. $C(s) + O_2(g) \longrightarrow CO_2(g)$

Another approach to the classification of chemical reactions is based upon a consideration of *charge transfer*. **Acid-base reactions** involve the transfer of a *hydrogen ion,* H^+, from one reactant to another. Another important reaction type, **oxidation-reduction,** takes place because of the transfer of negative charge, one or more *electrons,* from one reactant to another.

2.2 ACIDS AND BASES

Acids and bases include some of the most important compounds in nature. Historically, it was recognized that certain compounds—acids—had a sour taste, were able to dissolve metals, and caused vegetable dyes to change color. We now recognize that digestion of proteins is promoted by stomach acid (hydrochloric acid), and many biochemical processes such as enzyme catalysis depend upon the proper level of acidity. Indeed, a wide variety of chemical reactions critically depend on the acid-base composition of the solution (see Figure 2.1). For example, the regulation of acidity in living organisms is a chemical process controlled by substances we call *buffers*.

Section 12.8

Bases, in contrast, have long been recognized by their bitter taste, slippery feel, and corrosive nature. Bases react strongly with acids and cause many metal ions in solution to form a solid precipitate.

FIGURE 2.1
The dependence of chemical species on the acidity of the solvent. Both aqueous solutions initially contained Na_2CrO_4 (sodium chromate). The solution on the left was made acidic, and the yellow color of CrO_4^{-2} is apparent. The solution on the right was made basic, and the yellow CrO_4^{-2} (chromate ion) was converted to the reddish-brown $Cr_2O_7^{-2}$ (dichromate ion).

Acid-base theories

When we use acids and bases in the laboratory, they are generally present as solutions of some specified concentration, not as pure substances. Acids and bases in foods, biological fluids, and industrial products are generally in solution as well. For this reason we shall deal with properties and reactions of solutions of acids and bases.

The earliest definition of acids and bases is the **Arrhenius Theory.** According to this theory, an acid dissociates to form *hydrogen ions,* H^+, and a base dissociates to form *hydroxide ions,* OH^-. For example, hydrochloric acid dissociates in solution according to the reaction

$$HCl(aq) \longrightarrow H^+(aq) + Cl^-(aq)$$

Sodium hydroxide, a base, produces hydroxide ions in solution:

$$NaOH(aq) \longrightarrow Na^+(aq) + OH^-(aq)$$

The Arrhenius theory satisfactorily explains the behavior of many acids and bases. However, a substance such as ammonia, NH_3, has basic properties but cannot be an Arrhenius base, since it contains no OH^-. The **Brönsted-Lowry Theory** adequately explains this enigma and provides a broader view of acid-base theory through a consideration of the role of the solvent in the dissociation process.

The Brönsted-Lowry Theory defines an acid as a proton (H^+) donor and a base as a proton acceptor.

Hydrochloric acid in solution *donates* a proton to the solvent water and behaves as a Brönsted acid:

$$HCl(aq) + H_2O(l) \longrightarrow H_3O^+(aq) + Cl^-(aq)$$

H_3O^+, a hydrated proton, is referred to as the **hydronium ion.**

The basic properties of ammonia are clearly accounted for by the Brönsted-Lowry Theory. Ammonia *accepts* a proton from the solvent water, producing OH^-:

$$H_2O(l) + NH_3(g) \longrightarrow OH^-(aq) + NH_4^+(aq)$$

Acid-base strength

The terms *acid or base strength* and *acid or base concentration* are easily confused. *Strength* is a measure of the *degree of dissociation* of an acid or base in solution, independent of its concentration. Concentration, as we have learned, refers to the amount of acid or base per quantity of solution.

The strength of acids and bases in water depends upon the extent to which they react with the solvent water. Acids and bases are classified as *strong* when the reaction with water is virtually 100% complete and as *weak* when the reaction with water is much less than 100% complete. Important strong acids include:

Hydrochloric acid: $HCl(aq) + H_2O(l) \longrightarrow H_3O^+(aq) + Cl^-(aq)$

Nitric acid: $HNO_3(aq) + H_2O(l) \longrightarrow H_3O^+(aq) + NO_3^-(aq)$

Sulfuric acid: $H_2SO_4(aq) + H_2O(l) \longrightarrow H_3O^+(aq) + HSO_4^-(aq)$

Note that the equation for the dissociation of each of the above acids is written with a single arrow. This indicates that the reaction has little or no tendency to proceed in the reverse direction to establish equilibrium. All of the molecules of acid are converted to ions.

All *strong bases* are *metal hydroxides*. Strong bases completely dissociate, or ionize, in aqueous solution to produce hydroxide ions and metal cations. Of the common metal hydroxides, only NaOH and KOH are soluble in water and are the only readily usable strong bases:

Sodium hydroxide: $NaOH(aq) \longrightarrow Na^+(aq) + OH^-(aq)$

Potassium hydroxide: $KOH(aq) \longrightarrow K^+(aq) + OH^-(aq)$

Weak acids and *weak bases* dissolve in water principally in the molecular form. Only a small percent of the molecules dissociate to form the hydronium or hydroxide ion. The most important weak acids are:

Acetic acid: $CH_3COOH(aq) + H_2O(l) \rightleftharpoons H_3O^+(aq) + CH_3COO^-(aq)$

Carbonic acid: $H_2CO_3(aq) + H_2O(l) \rightleftharpoons H_3O^+(aq) + HCO_3^-(aq)$

The double arrow denotes an equilibrium between dissociated and undissociated species.

We have already mentioned the most common weak base, ammonia. Many organic compounds function as weak bases. Several examples follow:

Pyridine: $C_5H_5N(aq) + H_2O(l) \rightleftharpoons C_5H_5NH^+(aq) + OH^-(aq)$

Anilinc: $C_6H_5NH_2(aq) + H_2O(l) \rightleftharpoons C_6H_5NH_3^+(aq) + OH^-(aq)$

Methylamine: $CH_3NH_2(aq) + H_2O(l) \rightleftharpoons CH_3NH_3^+(aq) + OH^-(aq)$

The most fundamental chemical difference between strong and weak acids and bases is their equilibrium situation.

A strong acid, such as HCl, does not, in aqueous solution, exist to any measurable degree in equilibrium with its ions, H_3O^+ and Cl^-. On the other hand, a weak acid, such as acetic acid, establishes a dynamic equilibrium with its ions, H_3O^+ and $C_2H_3O_2^-$. This equilibrium may be represented as the ratio of concentration ([ ] indicates molar concentration) of products and reactants (excluding the solvent):

$$\text{ratio of the concentration of products to reactants} = \frac{[H_3O^+][CH_3COO^-]}{[CH_3COOH]}$$

At a fixed temperature and in a specified solvent, this ratio is a *constant value,* K_a

$$K_a = \frac{[H_3O^+][CH_3COO^-]}{[CH_3COOH]}$$

where K_a is the **equilibrium constant** or acid dissociation constant for acetic acid. The *magnitude* of this constant denotes the strength of the acid. The larger the dissociation constant (larger numerator, smaller denominator), the more dissociated (hence stronger) the acid will be.

A more complete discussion of equilibrium and equilibrium constants may be found in Appendix IX.

The situation for bases is analogous. The equilibrium expression for the weak base pyridine in water is

$$\underset{\text{Pyridine}}{C_5H_5N(aq)} + H_2O(l) \rightleftharpoons \underset{\text{Pyridinium ion}}{C_5H_5NH^+(aq)} + OH^-(aq)$$

and the equilibrium constant for this weak base is:

$$K_b = \frac{[C_5H_5NH^+][OH^-]}{[C_5H_5N]}$$

Known, tabulated values of acid and base dissociation constants are useful in performing numerous calculations related to acid-base solution composition. Calculations of this type are essential to many areas of chemistry.

Many organic compounds have acid or base properties. The chemistry of organic acids and bases will be discussed in Chapters 8 and 10.

Solutions of acids and bases that are used in the laboratory must be handled with care. Acids produce burns because of their exothermic reaction with water present in and on the skin. Acids and bases also react with and denature (destroy) proteins, principal components of the skin and eyes.

Such solutions are more hazardous if they are strong or concentrated. A strong acid or base produces more H_3O^+ or OH^- than the corresponding weak substances. The more concentrated acid or base contains more H_3O^+ or OH^- than a less concentrated solution of the same strength.

The dissociation of water

Aqueous solutions of acids and bases are electrolytes; the dissociation of the acid or base produces ions that are capable of conducting an electrical current. As a result of the differences in the degree of dissociation, *strong acids and bases are strong electrolytes; weak acids and bases are weak electrolytes*. The conductivity of these solutions is due to the solute and not the solvent, water, which is nondissociated and nonconducting.

Although pure water is virtually 100% molecular, a small number of water molecules do ionize. This process occurs by the transfer of a proton from one water molecule to another, producing a hydronium and a hydroxide ion. This equilibrium is shown below:

$$H_2O(l) + H_2O(l) \rightleftharpoons H_3O^+(aq) + OH^-(aq)$$

This process is the **autoionization,** or self-ionization, of water. Water is therefore a *very* weak electrolyte and a poor conductor of electricity. Water has *both* acid and base properties; the dissociation produces both the hydronium and hydroxide ion.

Pure water at room temperature has a hydronium ion concentration of 1.0×10^{-7} *M*. One hydroxide ion is produced for each hydronium ion; therefore the hydroxide ion concentration is also 1.0×10^{-7} *M*:

$$[H_3O^+] = 1.0 \times 10^{-7}$$
$$[OH^-] = 1.0 \times 10^{-7}$$

The product of hydronium and hydroxide ion concentration is referred to as the **autoionization constant** of water, K_w, or

$$K_w = [H_3O^+][OH^-]$$
$$K_w = [1.0 \times 10^{-7}][1.0 \times 10^{-7}]$$
$$K_w = 1.0 \times 10^{-14}$$

K_w is a constant because its value does not depend on the nature or concentration of the solute, as long as the temperature does not change. K_w is a temperature-dependent quantity.

The nature and concentration of the solutes added to water do alter the relative concentrations of H_3O^+ and OH^- present, but the product, $[H_3O^+][OH^-]$, always equals 1.0×10^{-14}. This relationship is the basis for a scale that is useful in measuring the level of acidity or basicity of solutions. This scale, the pH scale, was devised by the Danish chemist Sören Sörensen. The designation pH is an abbreviation for *pouvoir hydrogène,* French for the power of the hydrogen ion.

The pH scale

The **pH scale** correlates the hydronium ion concentration with a number, the pH, which serves as a useful indicator of the degree of acidity or basicity of a solution. This relationship is represented as

$$pH = -\log [H_3O^+]$$

The pH scale is somewhat analogous to the temperature scale used for assignment of relative levels of ''hot'' or ''cold.'' The temperature scale was developed to allow us to assess ''how cold'' or ''how hot'' an object is. The pH scale specifies ''how acidic'' or ''how basic'' a solution is (see Figure 2.2).

To help develop a concept of pH, consider the following:

1. Addition of an acid (proton donor) to water *increases* the $[H_3O^+]$ and decreases the $[OH^-]$.
2. Addition of a base (proton acceptor) to water *decreases* the $[H_3O^+]$ and increases the $[OH^-]$.
3. $[H_3O^+] = [OH^-]$ when *equal* amounts of acid and base are present.
4. In all of the above cases, $[H_3O^+][OH^-] = 1.0 \times 10^{-14} = K_w$.

(a)

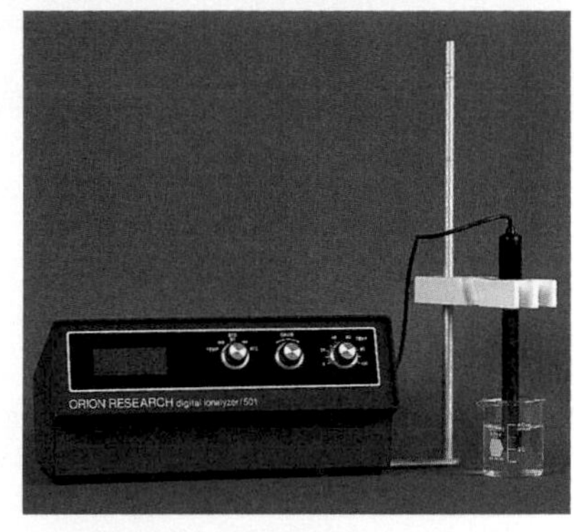

(b)

FIGURE 2.2
The measurement of pH. (a) A strip of test paper impregnated with indicator (a material that changes color as the acidity of the surroundings changes) is put in contact with the solution of interest. The resulting color is matched with a standard color chart (colors shown as a function of pH) to obtain the approximate pH. (b) A pH meter uses a sensor (a pH electrode), which develops an electrical potential that is proportional to the pH of the solution.

The importance of pH and pH control

Solution pH and pH control play a major role in many facets of our everyday lives. Consider a few examples:

1. *Agriculture:* Crops grow best in a soil of proper pH. Proper fertilization involves the maintenance of a suitable pH.
2. *Biochemistry:* If the pH of our blood were to shift by one unit, we would die. Many biochemical reactions in living organisms are extremely pH-dependent.
3. *Organic chemistry:* We shall see that the yield and rate of many organic reactions is pH-dependent.
4. *Industry:* From manufacture of processed foods to the plating of chrome auto bumpers, industrial processes often require rigorous pH control.
5. *Municipal services:* Purification of drinking water and treatment of sewage must be undertaken at their optimum pH.
6. *Acid rain:* Nitric acid and sulfuric acid, resulting from the reaction of industrial emission (nitrogen and sulfur oxides) with water, are carried down by precipitation and enter natural aquatic systems (lakes and streams), lowering the pH of the water. A less-than-optimum pH poses serious problems for native animal and plant populations.

The list could continue on for many pages. However, in summary, any change that takes place in aqueous solution generally has at least some pH dependence.

Neutralization

The reaction of an acid with a base to produce a salt and water is referred to as **neutralization.** In the strictest sense, neutralization requires equal numbers of moles of H_3O^+ and OH^- to produce a *neutral solution* (no excess acid or base).

Consider the reaction of hydrochloric acid and sodium hydroxide:

$$\underset{\text{Acid}}{HCl(aq)} + \underset{\text{Base}}{NaOH(aq)} \longrightarrow \underset{\text{Salt}}{NaCl(aq)} + \underset{\text{Water}}{H_2O(l)}$$

Our objective is to make the balanced equation represent the process that actually occurs.

AN ENVIRONMENTAL PERSPECTIVE

Acid Rain

Acid rain is a global environment problem that has raised public awareness of the chemicals that are polluting the air through the activities of our industrial society. Normal rain has a pH of about 5.6 as a result of the chemical reaction between carbon dioxide gas and water in the atmosphere. The following equation shows this reaction:

$$CO_2(g) + H_2O(l) \longrightarrow H_2CO_3(aq)$$

Carbon dioxide Water Carbonic acid

Acid rain refers to conditions that are much more acid than this. In upstate New York the rain has as much as 25 times the acidity of normal rainfall. One rainstorm recorded in West Virginia produced rainfall that measured 1.5 on the pH scale. This is approximately the pH of stomach acid, or about 10,000 times more acidic than "normal rain" (remember that the pH scale is logarithmic).

Acid rain is destroying life in streams and lakes. More than half the highland lakes in the western Adirondack Mountains have no native gamefish. In addition to these 300 lakes, there are 140 lakes in Ontario that have suffered a similar fate. It is estimated that 48,000 other lakes in Ontario and countless others in the northeastern and central United States are threatened. Our forests are also endangered. The acid rain decreases soil pH, which, in turn, alters the solubility of minerals needed by plants. Studies have shown that about 40% of the red spruce and maple trees in New England have died. Increased acidity of rainfall appears to be the major culprit.

What is the cause of this acid rain? The combustion of fossil fuels (gas, oil, and coal) by power plants produces nitrogen and sulfur oxides. These react with water, as does the CO_2 in normal rain, but the products are strong acids: sulfuric and nitric acid. Let us look at the equations for these processes.

In the atmosphere, nitric oxide (NO) can react with oxygen to produce nitrogen dioxide as shown in the following equations:

$$2NO(g) + O_2(g) \longrightarrow 2NO_2(g)$$

Nitric oxide Oxygen Nitrogen dioxide

Nitrogen dioxide (which is reddish brown and causes the brown air color of smog) then reacts with water to form nitric acid:

$$3NO_2(g) + H_2O(l) \longrightarrow 2HNO_3(aq) + NO(g)$$

A similar chemistry is seen with the sulfur oxides. Coal may contain as much as 3% sulfur. When the coal is burned, the sulfur also burns; this produces the choking, acrid sulfur dioxide gas:

$$S(s) + O_2(g) \longrightarrow SO_2(g)$$

By itself, sulfur dioxide can cause serious respiratory problems for people with asthma or other lung diseases, but matters are worsened by the reaction of SO_2 with atmospheric oxygen:

$$2SO_2(g) + O_2(g) \longrightarrow 2SO_3(g)$$

SO_3 readily reacts with water:

$$SO_3(g) + H_2O(l) \longrightarrow H_2SO_4(aq)$$

The product, sulfuric acid, is even more irritating to the respiratory tract. When the acid rain created by the reactions shown above falls to earth, the impact is significant.

It is easy to balance these chemical equations, but it could require decades to balance the ecological system that we have disrupted by our massive consumption of fossil fuels. A sudden decrease of even 25% in the use of fossil fuels would lead to worldwide financial chaos. It is hoped that development of alternative fuel sources, such as solar energy and safe nuclear power, will reduce our dependence on fossil fuels and help us to balance the global equation.

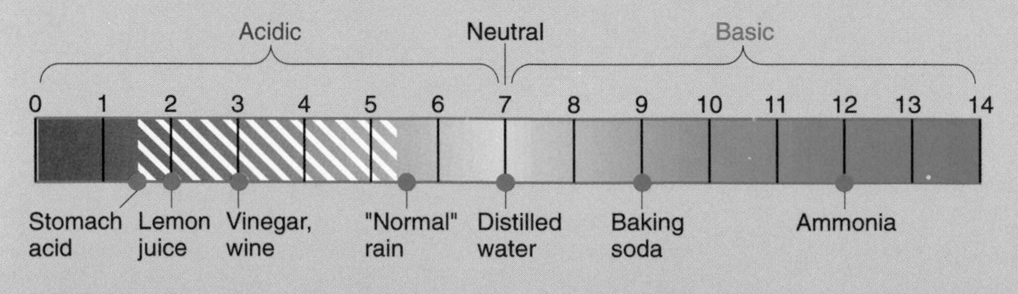

pH values for a variety of substances compared with the pH of acid rain

We recognize that HCl, NaOH, and NaCl are dissociated in solutions:

$$H^+(aq) + Cl^-(aq) + Na^+(aq) + OH^-(aq) \longrightarrow Na^+(aq) + Cl^-(aq) + H_2O(l)$$

We know further that Na^+ and Cl^- are unchanged in the reaction; they are referred to as "spectator" ions. If we write only the components that actually change, we produce a *net, balanced ionic equation* as shown below:

$$H^+(aq) + OH^-(aq) \longrightarrow H_2O(l)$$

If we realize that the H^+ occurs in aqueous solution as the hydronium ion, H_3O^+, the most correct representation would be

$$H_3O^+(aq) + OH^-(aq) \longrightarrow 2H_2O(l)$$

The equation for any strong acid/strong base neutralization reaction is the same as the equation shown above.

A neutralization reaction may be used to determine the concentration of an unknown acid or base solution. The technique of **titration** involves the addition of measured amounts of a **standard solution** (one whose concentration is known) to neutralize the second, unknown solution. From the volumes of the two solutions and the concentration of the standard solution, the concentration of the unknown solution may be determined.

Buffers

A **buffer solution** contains components that enable the solution to resist large changes in pH when acids or bases are added. Buffer solutions may be prepared in the laboratory to maintain optimum conditions for a chemical reaction. Buffers are routinely used in commercial products to maintain optimum conditions for product behavior (see Figure 2.3).

Buffer solutions also occur naturally. Blood, for example, is naturally buffered, maintaining a pH of approximately 7.4, which is optimum for the life processes of the cells of the body. The buffering agent in blood is a result of the unique interaction of carbonic acid (H_2CO_3) and bicarbonate ions (HCO_3^-).

The buffer process

The basis of buffer action is the establishment of an equilibrium between either

a weak acid and its salt (the salt is a conjugate base of the weak acid) or

a weak base and its salt (the salt is a conjugate acid of the weak base).

Consider the case of a weak acid and its salt: A common buffer solution may be prepared from the weak acid acetic acid (CH_3COOH) and the salt sodium acetate (CH_3COONa), which behaves as a base. An *equilibrium* is established in solution between the weak acid and the salt anion:

$$\underset{\text{Acetic acid (weak acid)}}{CH_3COOH(aq)} + \underset{\text{Water}}{H_2O(l)} \rightleftharpoons \underset{\text{Hydronium ion}}{H_3O^+(aq)} + \underset{\text{Acetate ion (salt)}}{CH_3COO^-(aq)}$$

The properties of a buffer solution are a direct result of **LeChatelier's Principle,** which states that an equilibrium system, when stressed, will shift its equilibrium to alleviate that stress. This principle may be best illustrated by the following examples.

Addition of base (OH^-) to a buffer solution: Addition of a basic substance to a buffer solution causes the following changes:

1. OH^- from the base reacts with H_3O^+ producing water.

FIGURE 2.3
Commercial products that claim improved function due to their ability to control pH.

2. Molecular acetic acid *dissociates* to replenish the H_3O^+ consumed by the base, maintaining the pH close to the initial level.

This is an example of LeChatelier's Principle, since the loss of H_3O^+ (the *stress*) is compensated by the dissociation of acetic acid to produce more H_3O^+.

Addition of acid (H_3O^+) to a buffer solution: Addition of an acidic substance to a buffer results in:

1. H_3O^+ from the acid increases the overall $[H_3O^+]$.
2. The system reacts to this stress, in accordance with LeChatelier's Principle, to form more molecular acetic acid; the acetate ion combines with H_3O^+

These effects may be summarized as follows:

$$CH_3COOH(aq) + H_2O(l) \rightleftharpoons H_3O^+(aq) + CH_3COO^-(aq)$$

OH^- added, equilibrium shifts to the right $\longrightarrow$

H_3O^+ added, equilibrium shifts to the left $\longleftarrow$

Note that buffering against base is a function of the concentration of the weak acid (in this case, CH_3COOH). Buffering against acid depends upon the concentration of the anion of the salt (CH_3COO^- here).

Control of blood pH

A pH of 7.4 is maintained in blood partly by a carbonic acid–bicarbonate buffer system based on the following equilibrium:

$$\underset{\text{(weak acid)}}{\underset{\text{Carbonic acid}}{H_2CO_3(aq)}} + H_2O(l) \rightleftharpoons H_3O^+(aq) + \underset{\text{(salt)}}{\underset{\text{Bicarbonate ion}}{HCO_3^-(aq)}}$$

This regulation process, based upon LeChatelier's Principle, is one of the major mechanisms by which blood pH is controlled.

Blood transports O_2, bound to hemoglobin in the red blood cells, from the lungs to all the cells of the body. Similarly, blood transports CO_2, a waste product of cellular metabolism, back to the lungs for removal from the body. Of the CO_2 entering the blood, 7% remains dissolved in the liquid portion of the blood (plasma) and 93% enters the red blood cells. Of the 93% that enters the red blood cells, 23% is bound to hemoglobin molecules and 70% remains dissolved in the cytoplasm of the cells.

The 70% of the CO_2 that is dissolved in the cytoplasm of the red blood cells participates in a buffer equilibrium that helps to maintain the pH of the blood near 7.4. Carbon dioxide reacts with water to form carbonic acid. It should be noted that this reaction occurs very efficiently because it is speeded up by an enzyme called carbonic anhydrase. The same reaction involving CO_2 dissolved in the plasma is much less efficient:

$$CO_2(aq) + H_2O(l) \rightleftharpoons H_2CO_3(aq)$$

At this point the buffer equilibrium becomes more complex:

$$CO_2(aq) + 2H_2O(l) \rightleftharpoons H_2CO_3(aq) + H_2O(l) \rightleftharpoons H_3O^+(aq) + HCO_3^-(aq)$$

The bicarbonate ion is actively pumped out of the red blood cells into the plasma, where it functions as a buffer to control the pH of the blood. Through this sequence of events, the concentration of CO_2 in the blood buffers blood pH.

Higher-than-normal CO_2 levels shift the above equilibrium to the right (LeChatelier's Principle), increasing $[H_3O^+]$ and lowering the pH. The blood becomes too acidic, leading to numerous medical problems. A situation of high blood CO_2 levels and low pH is called *acidosis*. Acidosis results from various diseases (emphysema, pneumonia) that restrict the respiration process, allowing the buildup of waste CO_2 in the blood.

Lower-than-normal CO_2 levels, on the other hand, shift the equilibrium to the left, decreasing $[H_3O^+]$ and making the pH more basic; this condition is termed *alkalosis* (from alkali, implying basic in nature). Hyperventilation, or rapid breathing, is a common cause of alkalosis.

Question 2.2 Explain how the molar concentration of H_2CO_3 in the blood would change if:

a. The partial pressure of CO_2 in the lungs were to increase.
b. The partial pressure of CO_2 in the lungs were to decrease.

Question 2.3 Explain how the molar concentration of hydronium ion in the blood would change under each of the conditions described in Question 2.2.

Question 2.4 Explain how the pH of blood would change under each of the conditions described in Question 2.2.

2.3 OXIDATION-REDUCTION PROCESSES

Oxidation-reduction processes are the basis for many types of chemical change. Corrosion, the operation of a battery, and biochemical processes of cellular respiration are but a few examples. In this section we shall explore the basic concepts underlying this class of chemical reactions.

Oxidation and reduction

Oxidation is defined as a loss of electrons. *Sodium metal* is, for example, oxidized to a *sodium ion,* losing one electron:

$$Na \longrightarrow Na^+ + e^-$$

Reduction is defined as a gain of electrons. A *chlorine atom* is reduced to a *chloride ion* by gaining one electron:

$$Cl + e^- \longrightarrow Cl^-$$

Oxidation and reduction are complementary processes. The *oxidation half-reaction* produces an electron that is the reactant for the *reduction half-reaction*. Oxidation and reduction go hand in hand. The combination of two half reactions, one oxidation and one reduction, produces the complete reaction:

Oxidation 1/2-reaction:	$Na \longrightarrow Na^+ + e^-$
Reduction 1/2-reaction:	$Cl + e^- \longrightarrow Cl^-$
Complete reaction:	$Na + Cl \longrightarrow Na^+ + Cl^-$

Note that the electrons cancel; in the electron transfer process, no free electrons remain.

In the reaction described above, sodium metal is the **reducing agent;** it releases electrons for the reduction of chlorine. Chlorine is the **oxidizing agent;** it accepts electrons from the sodium, which is oxidized.

The characteristics of oxidizing and reducing agents are summarized below:

Oxidizing Agent	Reducing Agent
Is reduced	Is oxidized
Gains electrons	Loses electrons
Causes oxidation	Causes reduction

Oxidation state

Identification of an oxidation-reduction (or *redox*) reaction and the oxidizing and reducing agents themselves is facilitated by an electron bookkeeping device known as the oxidation number or **oxidation state.** The following rules for assignment of oxidation states for isolated atoms, atoms in ions, and atoms in molecules illustrate the concept.

1. *The oxidation state of an element in its pure form is zero.*

Element	Oxidation State
Cu	0
Fe	0
He	0
H_2	0
P_4	0

2. *The oxidation state of an element in a monatomic ion is equal* to the charge on the ion.

Ion	Oxidation State
Na^+	+1
Cl^-	−1
Fe^{+3}	+3

3. *Certain elements have the same oxidation state in (almost) all compounds.*

Elements	Oxidation State
Group IA metals (e.g., Na)	+1
Group IIA metals (e.g., Mg)	+2
Halogens (e.g., F)	−1
Oxygen	−2
H (bonded to a metal)	−1
H (bonded to a nonmetal)	+1

4. The sum of the oxidation numbers of all the atoms in the formula of a neutral compound is 0, and the sum of the oxidation numbers of all the atoms in the formula of a polyatomic ion must equal the charge of the ion.

2.4 THE BASIS FOR CHEMICAL CHANGE

Thermodynamics

In very general terms, **thermodynamics** deals with the relationships between energies of systems, work, and heat flow. Thermodynamics may be applied to chemical change, such as the calculation of the quantity of heat that may be obtained from the combustion of

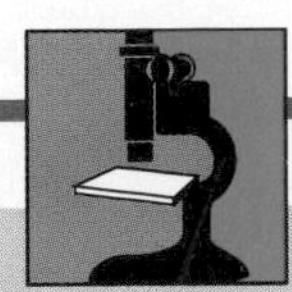

A CLINICAL PERSPECTIVE

Oxidizing Agents for Chemical Control of Microbes

Prior to the twentieth century, hospitals were not particularly sanitary establishments. Refuse, including human waste, was disposed of on hospital grounds. Because many hospitals had no running water, physicians often cleaned their hands and instruments by wiping them on their lab coats and then proceeded to treat the next patient! As you can imagine, many patients died of infections in hospitals.

By the late nineteenth century a few physicians and microbiologists had begun to realize that infectious diseases are transmitted by microbes, including bacteria and viruses. To decrease the number of hospital-acquired infections, physicians like Joseph Lister and Ignatz Semmelweis experimented with chemicals and procedures that were designed to eliminate the pathogens from environmental surfaces and from wounds.

Many of the common disinfectants and antiseptics are oxidizing agents. A disinfectant is a chemical that is used to kill or inhibit the growth of pathogens, disease-causing microorganisms, on environmental surfaces. An antiseptic is a milder chemical that is used to destroy pathogens associated with living tissue.

Hydrogen peroxide is an effective antiseptic that is commonly used to cleanse cuts and abrasions. We are all familiar with the furious bubbling that occurs as the enzyme catalase from our body cells catalyzed the breakdown of H_2O_2:

$$2H_2O_2(aq) \longrightarrow 2H_2O(l) + O_2(g)$$

A highly reactive and deadly form of oxygen, the superoxide radical (O_2^-), is produced during this reaction. This molecule inactivates proteins, especially critical enzyme systems.

At higher concentrations (3–6%), H_2O_2 is used as a disinfectant. It is particularly useful for disinfection of soft contact lenses, utensils, and surgical implants because there is no residual toxicity. Concentrations of 6–25% are even used for complete sterilization of environmental surfaces.

Benzoyl peroxide is another powerful oxidizing agent. Ointments containing 5–10% benzoyl peroxide have been used as antibacterial agents to treat acne. It is currently found in over-the-counter facial scrubs because it is also an exfoliant, causing sloughing of old skin and replacement with smoother-looking skin. A word of caution is in order; in sensitive individuals, benzoyl peroxide can cause swelling and blistering of tender facial skin.

Chlorine is a very widely used disinfectant and antiseptic. Calcium hypochlorite [$Ca(OCl)_2$] was first used in hospital maternity wards in 1847 by the pioneering Hungarian physician Ignatz Semmelweis. Semmelweis insisted that hospital workers cleanse their hands in a $Ca(OCl)_2$ solution and dramatically reduced the incidence of infection. Today, calcium hypochlorite is more commonly used to disinfect bedding, clothing, restaurant eating utensils, slaughterhouses, barns, and dairies.

Sodium hypochlorite (NaOCl), Clorox®, is used as a household disinfectant and deodorant but is also used to disinfect swimming pools, dairies, food-processing equipment, and kidney dialysis units. It can be used to treat drinking water of questionable quality. Addition of 1/2 teaspoon of household bleach (5.25% NaOCl) to 2 gallons of clear water renders it drinkable after 1/2 hour. The Centers for Disease Control even recommend a 1:10 dilution of bleach as an effective disinfectant against human immunodeficiency virus, the virus that causes Acquired Immune Deficiency Syndrome (AIDS).

Chlorine gas (Cl_2) is used to disinfect swimming pool water, sewage, and municipal water supplies. This treatment has successfully eliminated epidemics of water-borne diseases; however, chlorine is inactivated in the presence of some organic materials and, in some cases, may form toxic chlorinated organic compounds. For these reasons, many cities are considering the use of ozone (O_3) rather than chlorine.

Ozone is produced from O_2 by high-voltage electric discharges. (That fresh smell in the air after an electrical storm is ozone.) Several European cities use ozone to disinfect drinking water. It is a more effective killing agent than chlorine, especially with some viruses; less ozone is required for disinfection; there is no unpleasant residual odor or flavor; and there appear to be fewer toxic byproducts. However, ozone is more expensive than chlorine, and maintaining the required concentration in the water is more difficult. Nonetheless, the benefits seem to outweigh the drawbacks, and many U.S. cities may soon follow the example of the European cities and convert to the use of ozone for water treatment.

some quantity of fuel oil; similarly, the energy produced or consumed in physical change, such as the boiling or freezing of water, may be calculated.

There are three basic laws of thermodynamics; only the first two will be of concern here. Both laws will help us to understand why certain chemical reactions occur readily, while many others do not. For instance, mixing concentrated solutions of hydrochloric acid and sodium hydroxide produces a violent reaction that liberates a large quantity of

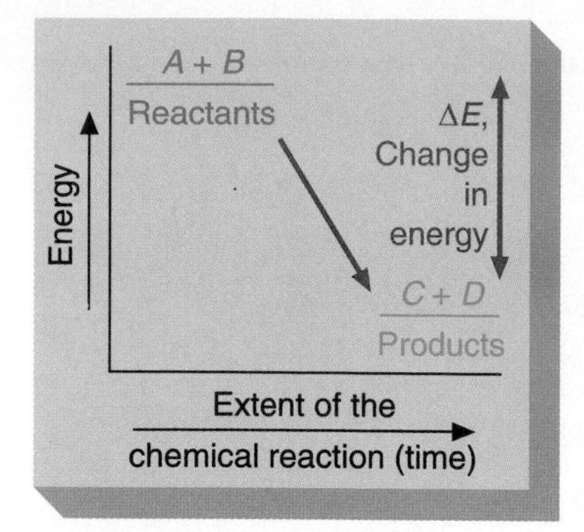

FIGURE 2.4
A potential energy diagram. ΔE represents the energy released during the progress of the exothermic reaction $A + B \rightarrow C + D + \Delta E$.

heat. On the other hand, nitrogen and oxygen have coexisted in the atmosphere for thousands of years without significant interaction.

The first law

The **first law of thermodynamics,** also known as *the law of conservation of energy,* states that energy cannot be created or destroyed in chemical systems, only converted from one form to another or transferred from one component of the system to another. Consider the combustion of methane in an oxygen atmosphere:

$$CH_4(g) + 2O_2(g) \longrightarrow CO_2(g) + 2H_2O(l) + 213 \text{ kcal}$$

We see that 213 kilocalories of heat energy have been released by the burning of one mole of methane in oxygen. Reactions of this type, which produce heat energy, are termed **exothermic** (*L.: exo,* out; *L.: therm,* heat). The energy released in this reaction was not created; it was present as potential (stored) energy in the chemical bonds of the reacting molecules (bond energy). The conversion of this stored energy or **chemical energy** into **heat energy** may be represented in a **potential energy diagram** of the reaction (Figure 2.4).

Note that the products of the reaction are lower in energy than the reactants. This very fact may be used as an argument for the ease with which the reaction takes place. A driving force for a chemical reaction is the necessity that the components of a system achieve the lowest possible potential *energy,* or greatest *stability.*

The *system* is simply a portion of the universe with which we are dealing. In this case, the system is made up of the products and reactants as well as the container. Systems may be closed, open, or isolated. A **closed system** is one in which the total mass and energy of the system does not change. An **open system,** on the other hand, may exchange matter and energy with the surroundings. If carbon dioxide and water were released to the atmosphere, we would have an open system; were they retained in a sealed vessel, the system would be closed. An **isolated system** cannot interact with its surroundings. It cannot lose or gain matter *or* energy, as is true of a closed system. Additionally, it cannot do or receive work. A Thermos bottle containing hot coffee represents an attempt to create an isolated system. The coffee cannot spill (lose matter), and the design of the Thermos prevents heat loss to the surroundings. If iced tea were in the Thermos, heat gain by the tea would also be prevented.

In contrast to exothermic reactions, chemical reactions that require energy in order to take place are termed **endothermic** (*L.: endo,* to take on; *L.: therm,* heat). An example of an endothermic reaction is the decomposition of ammonia:

$$22 \text{ kcal} + 2NH_3(g) \longrightarrow N_2(g) + 3H_2(g)$$

22 kilocalories of energy are required to make this reaction occur. The potential energy diagram for the reaction is shown in Figure 2.5. It should be noted that in this reaction the products are higher in energy than the reactants. The heat energy absorbed during the course of the reaction is stored as chemical potential energy in the products.

Experimental measurement of the quantity and direction of heat flow in a large number of chemical reactions has revealed some important correlations. *Most, but not all,* exothermic reactions are spontaneous; likewise, *most, but not all,* endothermic reactions are not spontaneous. We must conclude from these imperfect correlations that some other factor, in addition to the drive toward lower potential energy, is responsible for chemical reactions. This other factor is explained in the second law of thermodynamics.

FIGURE 2.5
A potential energy diagram for the decomposition of ammonia. 22 kcal of energy are required for the reaction 22 kcal + $2NH_3 \rightarrow N_2 + 3H_2$.

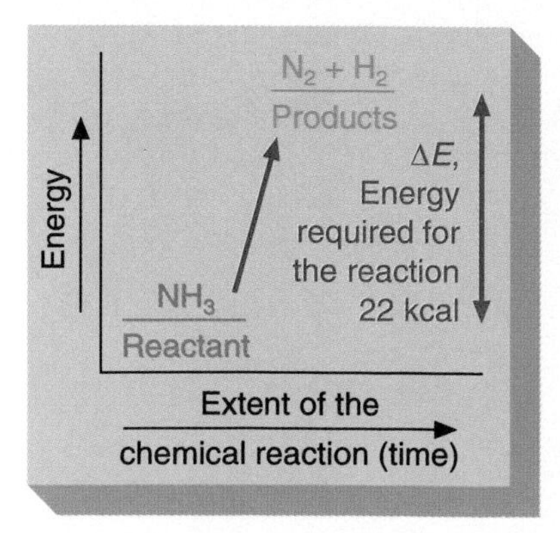

The second law

The **second law of thermodynamics** states that a *system and its surroundings* spontaneously tend toward increasing disorder (randomness). A measure of the randomness of a chemical system is referred to as *entropy.* A random or disordered system is characterized by *high entropy;* a well-ordered system is said to have *low entropy*. This drive toward increased entropy, in conjunction with the tendency to achieve a lower potential energy, is responsible for spontaneous chemical reactions.

As a general rule of thumb, reactions that are exothermic and whose products are more disordered (higher in entropy) will occur spontaneously, while endothermic reactions producing products of lower entropy will not be spontaneous. If they occur at all, they will require external energy input.

What do we mean by disorder in chemical systems? Disorder or randomness increases as we proceed from the solid to liquid to the vapor state. Solids often have an ordered crystalline structure, and liquids have, at best, a loose structure, while gas particles are virtually random in their distribution. A gas is an example of a substance with high entropy. A crystalline solid has a very low entropy.

Free energy and spontaneity of chemical change

In chemical change, the maximum amount of energy that can be converted to a useful form is called *free energy* (G). Free energy incorporates two factors, the energy factor and the entropy factor, into a single expression for predicting the spontaneity of chemical change.

If we operate at constant pressure, a common experimental condition, the heat absorbed or liberated by the reaction of interest is termed the *enthalpy* (H). Entropy is represented by the symbol S. Of main interest to us is the *change* in the value of these quantities as the reaction proceeds from reactants to products.

Change in enthalpy: $\Delta H = H_{\text{products}} - H_{\text{reactants}}$
Change in entropy: $\Delta S = S_{\text{products}} - S_{\text{reactants}}$
Change in free energy: $\Delta G = G_{\text{products}} - G_{\text{reactants}}$

The relationship among these quantities is

$$\Delta G = \Delta H - T\Delta S$$

We find that a spontaneous reaction will have a *negative* ΔG. A *positive* ΔG indicates a nonspontaneous reaction. In fact, a change for which ΔG is positive will occur *only if* energy is added to the system, usually by heating the reaction.

Entropy is a characteristic of all of the universe. Humans also obey the law of increasing disorder. Chaos in our room or workplace comes almost effortlessly. Reversal of this process requires work and energy, just as is the case at the molecular level. We may work tirelessly to trim hedges and mow the grass in our yard; this must be done many times each summer, year after year, just to maintain the status quo. Leave it unattended for just one summer (no energy expended on our part) and see what happens!

Kinetics

The first two laws of thermodynamics provide the basis for deciding upon whether a chemical reaction will take place. Knowing that a reaction is energetically and entropically favored, however, tells us nothing about the time required for the reaction to go to completion. The study of chemical **kinetics** provides the basis for the measurement, control, and possibly the prediction of the **rate** (speed) of a chemical reaction. Kinetics may also provide information about the *mechanism* of a reaction, a step-by-step description of how reactants become products of a reaction.

The reaction

Consider the exothermic reaction

$$CH_4(g) + 2O_2(g) \longrightarrow CO_2(g) + 2H_2O(l) + 213 \text{ kcal}$$

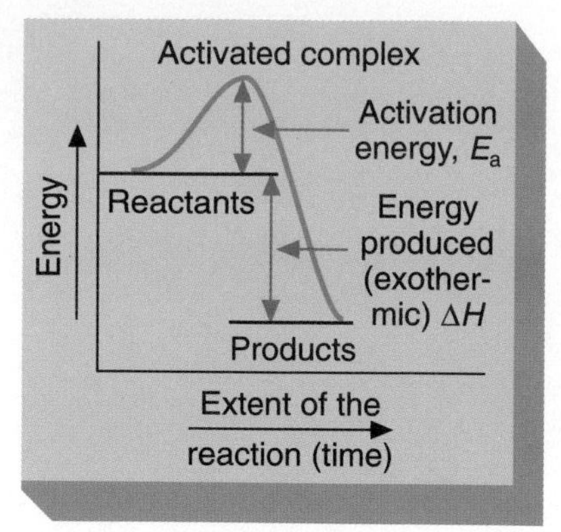

FIGURE 2.6
The change in potential energy as a function of reaction time for an exothermic chemical reaction. Note particularly the energy barrier associated with the formation of the activated complex. This energy barrier *is* the activation energy.

In order for the reaction to proceed, C—H and O—O bonds must be broken, and C—O and H—O bonds must be formed from the ''fragments'' of methane and oxygen molecules. Sufficient energy must be available to cause bond rupture if the reaction is to proceed. This energy is provided by the collision of two or more molecules. If the energy made available by the collision exceeds the bond energies, the bonds will break, and the resulting atoms will recombine in a lower energy configuration—in this case, as carbon dioxide and water. A collision that meets the above conditions and produces one or more product molecules is referred to as an *effective collision*. Only effective collisions lead to chemical reaction.

Activation energy and the activated complex

The minimum amount of energy required to produce a chemical reaction is called the **activation energy** for the reaction. As implied above, a large component of the activation energy is the bond energy of the reacting molecules.

We may view the chemical reaction in terms of the changes in potential energy that occur as a function of the time of the reaction. This is graphically represented in Figure 2.6.

Several important characteristics of this graph should be recognized:

1. The reaction proceeds from reactants to products through an extremely unstable intermediate state that we term the **activated complex.** The activated complex cannot be isolated from the reaction mixture but may be thought of as a short-lived grouping of atoms structured in such a way that it quickly and easily breaks apart into the products of the reaction.
2. Formation of the activated complex requires energy. The difference between the energy of reactants and activated complex is the activation energy.
3. Because this is an exothermic reaction, the overall energy change must be a *net* release of energy. The *net* release of energy is the difference in energy between products and reactants.

For an endothermic reaction, such as the decomposition of water,

$$\text{energy} + 2H_2O(l) \longrightarrow 2H_2(g) + O_2(g)$$

the variation of potential energy with reaction time is shown in Figure 2.7.

We would predict that liquid water is *stable* because the products are less stable (higher energy) than the reactants and that the conversion would proceed slowly because of the large activation energy required for conversion of water into the elements, hydrogen and oxygen. In experimental fact, this is true.

An understanding of the concepts of activation energy and the activated complex allows us to look at the term ''effective collision'' in a different light. An effective collision is one in which there is sufficient kinetic energy for the formation of the activated complex. Once the activated complex has formed, there is an equal probability that it will fall *forward* ''down the hill'' to make product(s) or fall *backward* ''down the hill'' to reform the reactant(s). So an effective collision not only must have the required energy, but must ''fall the right way'' as well.

FIGURE 2.7
The change in potential energy as a function of reaction time for an endothermic chemical reaction.

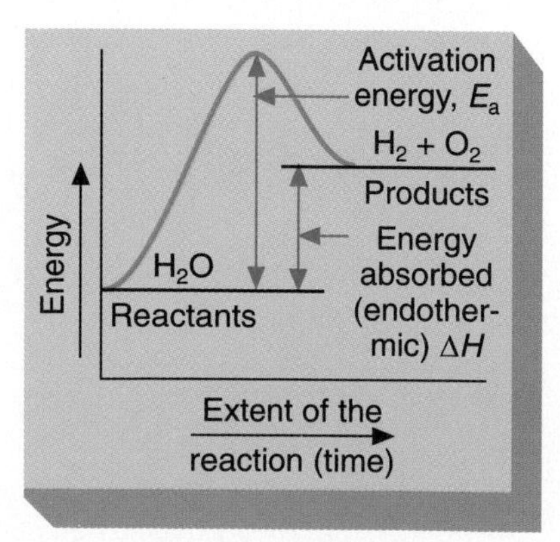

Experimental factors affecting the rate of reaction

Five major experimental conditions influence the rate of a chemical reaction.

Structure of the reacting species. As a result of electrostatic attraction, oppositely charged species generally react more rapidly than neutral species. Ions with the same charge are unreactive because of the repulsion of like charges. In contrast, oppositely charged ions attract and are often reactive. If the reactants are molecules, are they polar or nonpolar? Since reactants must *collide* to react, the attractive forces between polar molecules would be expected to favor a faster reaction when compared to nonpolar reactants. The bond strength certainly plays a role in determining reaction rates as well, since the magnitude of the activation energy, or energy barrier, is related to the bond strength. As will be especially evident in organic reactions, the sizes and shapes of reactant molecules are also instrumental in dictating the rate of the reaction. Large molecules, containing bulky groups of atoms, may ''hide'' the reactive part of the molecule from another reactive substance, thus slowing the reaction.

The concentration of reactants. The rate of a chemical reaction is often a complex function of the concentration of one or more of the reacting substances. The rate will generally *increase* as concentration *increases* simply because a higher concentration means more reactant molecules in a given volume and therefore a larger number of collisions per unit time. Assuming that other variables are held constant, a larger number of collisions means a larger number of effective collisions. The magnitude of the effect of concentration on rate is related to the particular type of reaction being conducted. The explosion (very fast exothermic reaction) of gunpowder is a dramatic example of a rapid rate at high reactant concentration.

The temperature of reactants. The rate of the reaction *increases* as the temperature increases, since the kinetic energy of particles is directly proportional to the absolute temperature. The more rapid motion of particles increases the probability of collision. The higher kinetic energy means that a higher percentage of these collisions will result in effective collisions. As a general rule of thumb, a 10°C rise in temperature doubles the reaction rate.

The physical state of reactants. The rate of a reaction is a function of the physical state of the reactants: solid, liquid, or vapor. Distance of separation between particles, strength of attractive forces and, particle mobility are largely related to the particular physical state of the reactants, and these factors all affect reaction rates.

The presence of a catalyst. A **catalyst** is a substance that *increases* the rate of a reaction. If added to a reaction mixture, the catalytic substance undergoes no net change, nor does it alter the outcome of the reaction. However, the catalyst interacts with the reacting substances to produce an alternative ''pathway'' for production of products. This alternative route or path has a lower activation energy, facilitating the reaction and increasing the rate. This effect is illustrated in Figure 2.8.

SUMMARY

Chemical reactions may be classified as involving combination, decomposition, or replacement. Another approach to classification depends upon the type of charge-transfer process occurring between reactants. Acid-base reactions involve proton transfer and oxidation-reduction reactions involve electron transfer.

Acids and bases may be described according to either the Arrhenius Theory or the Brönsted-Lowry Theory. Acids increase $[H_3O^+]$ of aqueous solutions and bases increase $[OH^-]$. However, the product of $[H_3O^+]$ $[OH^-]$ is a constant, 1.0×10^{-14}, the autoionization constant of water. The acidity or basicity of a solution is best represented by using the pH scale; pH is defined as $-\log [H_3O^+]$.

Neutralization involves the reaction of an acid and a base, resulting in a neutral aqueous salt solution. A titration is a special application of a neutralization reaction, useful in determining the concentration of acid or base solution.

Buffer solutions, consisting of a mixture of a weak acid and its salt or a weak base and its salt, are resistant to pH change upon addition of acidic or basic substances. Buffer solutions function in accordance with LeChatelier's Principle. An important application of buffers involves blood pH regulation in the human body.

Oxidation-reduction reactions are a second major class of chemical reactions. Oxidation involves a loss of one or more electrons, resulting in an increase in oxidation state; reduction is a gain of one or more electrons, resulting in a decrease in oxidation state. Electron-transfer processes may be viewed as analogous to proton transfer (acid-base) reactions.

Two laws of thermodynamics, the science involved with energy flow in physical and chemical change, are of particular importance. The first law, the law of conservation of energy,

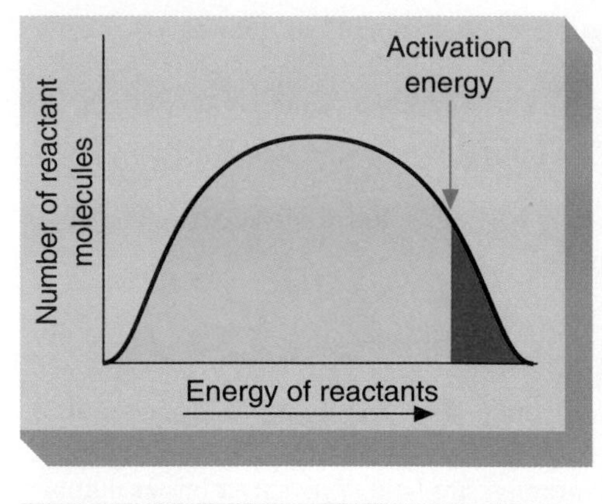

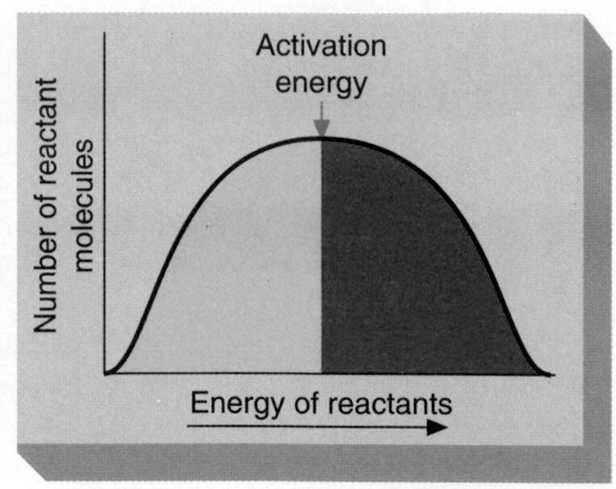

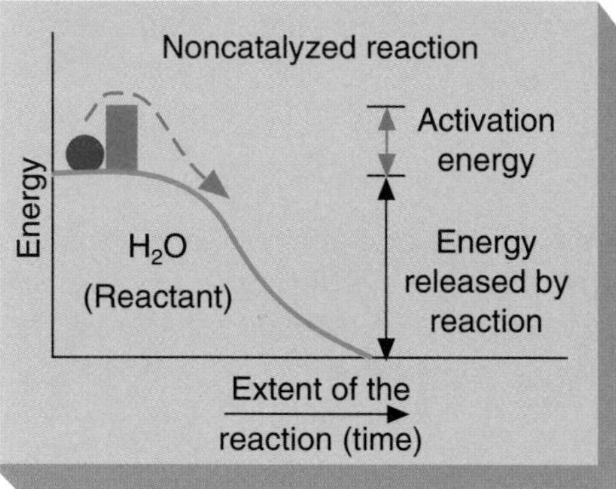

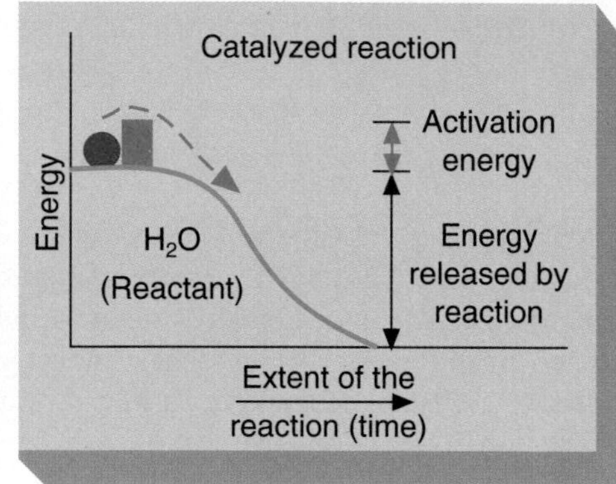

FIGURE 2.8
The effect of a catalyst on the magnitude of the activation energy of a chemical reaction. Note that the presence of a catalyst decreased the activation energy, thus increasing the rate of the reaction.

and the second law, the law of entropy or disorder, can provide basic information regarding the spontaneity of chemical or physical processes as well as the amount of energy absorbed or released by the reaction.

Kinetics deals with the rate and mechanism of a chemical reaction. Recognizing that chemical reactions result from effective collisions between potentially reacting particles, we saw that factors such as the structure of the reactant, its concentration, reaction temperature, and the presence or absence of a catalyst influence reaction rates. The role of the activation energy in chemical processes determines reaction rates; the formation of the activated complex is an energy barrier to be overcome.

GLOSSARY OF KEY TERMS

acid-base reaction (2.1) the transfer of a hydrogen ion from one reactant to another.

activated complex (2.4) the arrangement of atoms at the top of the potential energy barrier as the reaction proceeds.

activation energy (2.4) the threshold energy that must be overcome to produce a chemical reaction.

Arrhenius Theory (2.2) a theory that defines an acid as a substance that dissociates to produce H^+ and a base as a substance that dissociates to produce OH^-.

autoionization (2.2) or self-ionization, the reaction of a substance, such as water, with itself to produce a positive and a negative ion.

autoionization constant (2.3) the equilibrium constant that mathematically describes the self-ionization process.

Brönsted-Lowry Theory (2.2) a theory that describes an acid as a proton donor and a base as a proton acceptor.

buffer solution (2.2) a solution with a conjugate acid-base pair that is resistant to large changes in pH upon addition of strong acids or bases.

catalyst (2.4) any substance that increases the rate of a chemical reaction (by lowering the activation energy of the reaction) and that is not destroyed in the course of the reaction.

chemical energy (2.4) energy stored in substances that can be released during a chemical reaction.

closed system (2.4) a system in which the total mass and energy of the system does not change.

combination reaction (2.1) the joining of two or more atoms or compounds to produce a product of different composition.

decomposition reaction (2.1) a reaction that involves the breakdown of a substance into two or more substances.

double-replacement reaction (2.1) a reaction that involves a chemical change in which cations and anions ''exchange partners.''

endothermic process (2.4) a process that absorbs energy in a chemical change.

equilibrium constant (2.2) a number equal to the ratio of the equilibrium concentrations of products to the equilibrium concentrations of reactants, each raised to a power equal to its stoichiometric coefficient.

exothermic process (2.4) a process in which energy is released in a chemical change.
first law of thermodynamics (2.4) the law of conservation of energy; in chemical change, energy cannot be created or destroyed.
heat energy (2.4) random kinetic energy absorbed or released by a substance or substances.
hydronium ion (2.2) a hydrated proton.
isolated system (2.4) a system that cannot interact with its surroundings.
kinetics (2.4) the study of rates of chemical reactions.
LeChatelier's Principle (2.2) a law that states that when a system in equilibrium is disturbed, the equilibrium shifts in the direction that minimizes the disturbance.
neutralization reaction (2.2) the reaction between an acid and a base.
open system (2.4) a system that can exchange matter and energy with its surroundings.
oxidation (2.3) a loss of electrons or increase in oxidation state.
oxidation-reduction reaction (2.1) or redox reaction, a reaction involving the transfer of one or more electrons from one reactant to another.
oxidation state (2.3) the apparent charge of an atom in a molecule or the charge of a monatomic ion.
oxidizing agent (2.3) a substance that oxidizes, or removes electrons from, another substance; the oxidizing agent is reduced in the process.
pH scale (2.2) a numerical representation of acidity or basicity of a solution (pH = $-\log [H^+]$).
potential energy diagram (2.4) a diagram that represents stored energy as a function of distance or time (as in the progress of a chemical reaction).
rate of a reaction (2.4) the change in concentration of a reactant or product per unit time.
reducing agent (2.3) a substance that reduces, or donates electrons to, another substance; the reducing agent is itself oxidized in the process.
reduction (2.3) the gain of electrons or decrease in oxidation state.
second law of thermodynamics (2.4) a law that states that an increase in entropy (disorder) accompanies any spontaneous process.
single-replacement reaction (2.1) or substitution reaction, one in which one atom in a molecule is displaced by another.
standard solution (2.2) a solution whose concentration is accurately known.
thermodynamics (2.4) the study of energy and its interconversions.
titration (2.2) the process of adding a solution from a buret to a sample until a reaction is complete, at which time the volume is accurately measured.

QUESTIONS AND PROBLEMS

The following problem set provides an opportunity for a review of the key concepts of chemical change that have particular applicability to the study of organic chemistry and biochemistry. These concepts have been discussed in this chapter and are supplemented by material contained in the appendices at the end of this textbook. These problems presume some basic background in the principles of inorganic chemistry.

Classification of Chemical Reactions

2.5 Give at least three examples of a decomposition reaction.

2.6 Give at least three examples of a combination reaction.

2.7 Give at least three examples of a replacement reaction.

2.8 Balance each of the following equations:
a. $FeCl_2(s) + Cl_2(g) \longrightarrow FeCl_3(s)$
b. $Ca(s) + H_2O(l) \longrightarrow Ca(OH)_2(s) + H_2(g)$
c. $Mg(s) + H_2O(l) \longrightarrow MgO(s) + H_2(g)$

2.9 Balance each of the following equations:
a. $CH_4(g) + O_2(g) \longrightarrow H_2O(l) + CO_2(g)$
b. $C_3H_8(g) + O_2(g) \longrightarrow H_2O(l) + CO_2(g)$
c. $NH_3(g) + O_2(g) \longrightarrow N_2(g) + H_2O(l)$

2.10 Write a balanced chemical equation representing:
a. the combustion of octane (C_8H_{18})
b. the rusting of elemental iron (rusting is oxide formation)

2.11 Write at least three chemical reactions representing the dissociation of a strong electrolyte.

2.12 Write at least three chemical reactions representing the interaction of two strong electrolytes to form a weak electrolyte.

2.13 Write at least three chemical reactions representing the formation of a gaseous product.

2.14 Write at least three chemical reactions representing the formation of an insoluble solid product.

Acids and Bases

2.15 a. Define an acid according to Arrhenius.
b. Define a base according to Arrhenius.

2.16 a. Define an acid according to Brönsted.
b. Define a base according to Brönsted.

2.17 Methylamine reacts with water according to the following reaction:

$$CH_3NH_3(aq) + H_2O(l) \longrightarrow CH_3NH_4^+(aq) + OH^-(aq)$$

a. Is methylamine best described as an Arrhenius base or a Brönsted base?
b. Why?

2.18 Propionic acid reacts with water according to the following reaction:

$$CH_3CH_2COOH(aq) + H_2O(l) \longrightarrow H_3O^+(aq) + CH_3CH_2COO^-(aq)$$

a. Is propionic acid best described as an Arrhenius acid or a Brönsted acid?
b. Why?

2.19 What is a neutralization reaction?

2.20 Describe the purpose of a titration.

2.21 Describe the steps involved in the titration of an unknown acid with a standard solution of base in order to determine the molarity of the acid.

2.22 Which of the following pairs are capable of behaving as buffer solutions?
a. NH_3 and NH_4Cl
b. HNO_3 and KNO_3
c. HBr and $NaHCO_3$
d. H_2CO_3 and $NaHCO_3$

2.23 Label each of the following statements as either true or false:
a. Buffer solutions resist pH change.
b. An acidic buffer consists of a weak acid and its salt.
c. An alkaline buffer consists of a strong base and its salt.
d. The metabolism of the human body can tolerate only a narrow range in pH, and therefore buffers play an important role in body chemistry.

Oxidation-Reduction Processes

2.24 Define each of the following terms:
a. oxidation c. oxidizing agent
b. reduction d. reducing agent

2.25 Give the oxidation number of each of the elements in the following compounds:
a. S_8 d. SO_2
b. H_2SO_3 e. SO_3
c. H_2SO_4

2.26 List three examples of oxidation-reduction processes that occur in the course of your everyday life.

2.27 For each of the following chemical changes, state whether the species was oxidized or reduced:
a. NO_2^- to NO_3^-
b. MnO_4^- to Mn^{2+}
c. Cl^- to Cl_2
d. Cr^{3+} to CrO_4^{2-}
e. H_2O_2 to OH^-

2.28 Fill in the blanks:
a. During an oxidation process, the species oxidized ________ electrons.
b. During an oxidation-reduction reaction the species ________ is the oxidizing agent, and the species ________ is the reducing agent.
c. Metals tend to be good ________ agents.

2.29 In each of the following reactions, identify the oxidized species, reduced species, oxidizing agent, and reducing agent:
a. $Zn(s) + 4HNO_3(aq) \longrightarrow Zn(NO_3)_2(aq) + 2NO_2(g) + 2H_2O(l)$
b. $Cl_2(g) + 2KI(s) \longrightarrow 2KCl(s) + I_2(g)$
c. $14H^+(aq) + Cr_2O_7^{2-}(aq) + 6Fe^{2+}(aq) \longrightarrow 2Cr^{3+}(aq) + 6Fe^{3+}(aq) + 7H_2O(l)$

2.30 Silver metal tarnishes in the presence of hydrogen sulfide (perhaps from the eggs you had for breakfast) in two steps:

$$Ag(s) \longrightarrow Ag^+(aq) + 1e^-$$

$$2Ag^+(aq) + H_2S(aq) \longrightarrow Ag_2S(s) + 2H^+(aq)$$

Which reaction classification describes each step?

2.31 What is the oxidation number of the boldfaced element in each of the following?
a. $\mathbf{S}O_4^{2-}$ d. $\mathbf{Cl}_2$
b. $\mathbf{S}O_3^{2-}$ e. $\mathbf{O}^{2-}$
c. **Mg**

2.32 What is the oxidation number of the boldfaced element in each of the following?
a. $\mathbf{N}O_3^-$ d. $\mathbf{Cl}O^-$
b. $\mathbf{Br}^-$ e. $\mathbf{O}H^-$
c. $\mathbf{Mn}O_4^-$

Thermodynamics

2.33 Explain what is meant by the first law of thermodynamics.

2.34 Explain what is meant by the second law of thermodynamics.

2.35 Define or explain each of the following terms:
a. exothermic reaction
b. endothermic reaction
c. closed system
d. open system
e. isolated system

2.36 Define or explain each of the following terms:
a. entropy
b. specific heat
c. fuel value
d. calorimeter
e. driving force

2.37 Predict whether each of the following processes increases or decreases entropy, and explain your reasoning:
a. melting of a solid metal
b. boiling of water
c. burning a log in a fireplace
d. combustion of gasoline
e. condensation of water vapor on a cold surface

2.38 Explain why an exothermic reaction produces products that are more stable than the reactants.

2.39 Provide an example of entropy from your own life experience.

2.40 Isopropyl alcohol, commonly known as rubbing alcohol, feels cool when applied to the skin. Explain.

2.41 Energy is required to break chemical bonds during the course of a reaction. When is energy released?

2.42 Explain the terms "endothermic" and "exothermic" in relation to the amount of energy absorbed when bonds are broken and energy released when bonds are formed during the course of a reaction.

2.43 A fundamental law of thermodynamics states that heat will always travel from a hotter substance to a colder substance. With this statement in mind, why does a glass of iced tea feel cold? Why does a cup of hot chocolate feel hot?

Kinetics

2.44 Define or explain each of the following terms:
a. rate
b. mechanism
c. activation energy
d. catalyst

2.45 Fill in the blank:

a. During the course of a reaction the highest energy state reached is called the ________.

b. The condition at which the rate of the forward reaction equals the rate of the reverse reaction is called ________.

c. The steps that show the path a reaction takes is called a ________.

d. The rate-determining step in a mechanism is the ________ step of the mechanism.

2.46 List the factors that change the rate of a chemical reaction.

2.47 Explain the effect of temperature on the rate of a chemical reaction.

2.48 Explain the effect of concentration on the rate of a chemical reaction.

2.49 Describe the way in which a chemical catalyst alters the rate of a reaction.

2.50 What is the difference between a collision and an *effective* collision?

CHAPTER 3

Saturated Hydrocarbons: Alkanes and Alkyl Halides

OBJECTIVES

- Draw and name the simple alkanes, using the I.U.P.A.C. Nomenclature System.
- Draw and name the simple alkyl halides, using the I.U.P.A.C. Nomenclature System.
- Draw and name the simple cycloalkanes and substituted cycloalkanes, using the I.U.P.A.C. Nomenclature System.
- Predict and explain the general trends in the physical properties of the alkanes and simple alkyl halides.
- Be able to work with (draw and name) structural isomers.
- Write and work with equations for the halogenation and combustion of alkanes.
- Write and work with equations for the nucleophilic substitution of primary and secondary alkyl halides.
- Be aware of the relevant (commercial, medicinal, environmental, and so forth) roles of the simple alkanes, cycloalkanes, and alkyl halides.
- Be familiar with the petroleum industry and its intimate role in our day-to-day lives.

OUTLINE

We begin our study of organic chemistry with an introduction to the principal class of organic compounds: the hydrocarbons. Alkanes are presented in Chapter 3, and alkenes, alkynes, and aromatic hydrocarbons in Chapter 4. Nomenclature systems, molecular structure, and the physical and chemical properties of each of these families are discussed. The cycloalkanes and alkyl halides, two additional families of organic compounds, are also introduced in this chapter.

3.1 ALKANES: STRUCTURE AND PROPERTIES

Structure

Alkanes are hydrocarbons that contain only carbon and hydrogen and that are bonded together through carbon-hydrogen and carbon-carbon single bonds (see Table 3.1). The structures of several simple alkanes are provided here. Additional examples are provided throughout the chapter and in Table 3.3.

Structural formula

The **structural formula** shows all of the atoms in a molecule and exhibits all bonds as lines. For example,

```
    H  H
    |  |
 H—C—C—H
    |  |
    H  H
```

Ethane
(ethane)

Molecular formula

The **molecular formula** provides the atoms and number of each type of atom in a molecule but gives no information regarding the bonding pattern involved. For example,

$$C_2H_6$$

Condensed structural formula

The **condensed structural formula** shows all of the atoms in a molecule and places them in a sequential arrangement that details which atoms are bonded to each other. For example,

$$CH_3CH_3$$

TABLE 3.1 Examples of Alkanes

General Formula for Alkanes	Typical Examples				
C_nH_{2n+2}	Methane	$n = 1$	C_1H_4	or	CH_4
	Ethane	$n = 2$	C_2H_6	or	$CH_3—CH_3$
	Propane	$n = 3$	C_3H_8	or	$CH_3—CH_2—CH_3$
	Butane	$n = 4$	C_4H_{10}	or	$CH_3—CH_2—CH_2—CH_3$

TABLE 3.2 Physical Properties of Selected Alkanes

Name	Formula	b.p. (°C)	m.p. (°C)	Density*
Methane	CH_4	−161.7	−182.6	0.424
Ethane	C_2H_6	−88.6	−172.0	0.546
Propane	C_3H_8	−42.2	−187.1	0.582
Butane	C_4H_{10}	−0.5	−135.0	0.579
Pentane	C_5H_{12}	36.1	−129.7	0.626
Hexane	C_6H_{14}	68.7	−94.0	0.659
Heptane	C_7H_{16}	98.4	−90.5	0.684
Octane	C_8H_{18}	125.6	−56.8	0.703
Nonane	C_9H_{20}	150.7	−53.7	0.718
Decane	$C_{10}H_{22}$	174.0	−29.7	0.730

*At 20°C.

Three-dimensional drawings of several simple members of this family are provided in Figure 3.1. Note that each carbon atom is surrounded by four bonds, while each hydrogen has only one bond. Although, as we will see, there may be differences in the types of bonding (single, double, and triple), the total number of bonds (and therefore the number of electrons involved in bonding) remains constant. The bonding of carbon and hydrogen to form alkanes produces carbon atoms that have a tetrahedral geometry. That is, the four bonds to the central carbon in alkanes form a tetrahedron structure (see Figure 3.1).

A series of compounds in which the structural formula of each differs from the immediately preceding member of the family and the following member of the family by a constant number of atoms (usually CH_2) and in no other way is called a **homologous series.** Any compound in the series is called a **homolog** of the others. The family of alkanes is an example of a homologous series.

FIGURE 3.1
(a) Three-dimensional molecular model drawing of ethane; bond angles, bond lengths, and bond strengths are provided. (b) Molecular model drawings of a more complex long-chain hydrocarbon, butane.

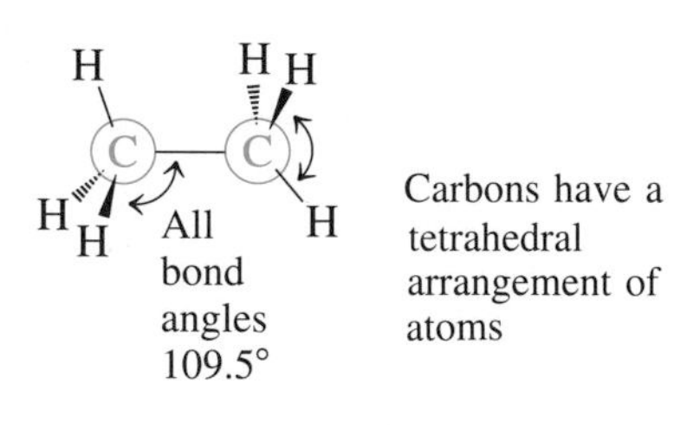

(a) Ethane

Bond length:

C—C 1.54 Å

C—H 1.10 Å

Bond strength:

C—C 88 kcal/mole

C—H 98 kcal/mole

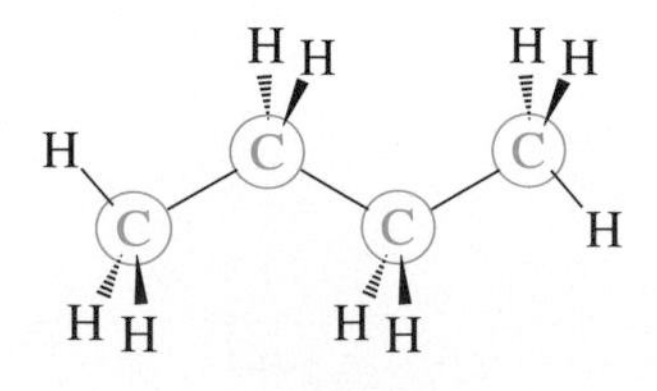

(b) Butane

Physical properties

Since all of the hydrocarbons are composed of nonpolar carbon-carbon and carbon-hydrogen bonds, hydrocarbons are nonpolar molecules. As a result, they are not water-soluble but are readily soluble in nonpolar solvents (for example, other hydrocarbons and common organic solvents such as ether and carbon tetrachloride). The rule of thumb ''like dissolves like'' holds true.

This simple rule has far-reaching implications. We find, for example, that both the structure of cell membranes and nutrient transport through the membranes depend on the polarity of the molecules involved. Similarly, many medicinal compounds are designed with functional groups (polar or nonpolar), allowing them to be administered to patients, transported through the body, or dissolved at a particular site in the body, all with the aim of making them more effective. Aspirin, for example, has several undesirable side effects, including gastric bleeding (bleeding in the stomach). This bleeding appears to be due to a particular functional group. When the acetyl group of aspirin is replaced with a hydroxyl group, many of the desirable aspirinlike properties are retained, and the undesirable side effect disappears.

As a result of their nonpolar nature, virtually all of the hydrocarbons are less dense than water and have relatively low melting points and boiling points (see Table 3.2). The smaller members of the series are gases at room temperature. A list of the simpler members of the alkane family and their main physical properties is provided in Table 3.2.

3.2 ALKANES: NOMENCLATURE SYSTEM

The I.U.P.A.C. System

The International Union of Pure and Applied Chemistry (I.U.P.A.C.) is the organization charged with the responsibility of maintaining a standard, universal system for the naming

I.U.P.A.C. names are provided in parentheses. If both common and I.U.P.A.C. names are given, the common name is in italics.

TABLE 3.3 I.U.P.A.C. Names of Alkyl Substituents

Alkane R—H	Alkyl Substituent R—	Alkane R—H	Alkyl Substituent R—
$CH_3—H$ (Methane)	$CH_3—$ (Methyl)	CH_3 | $CH_3CHCH_2—H$ (2-Methylpropane; *isobutane*)	CH_3 | $CH_3CHCH_2—$ (*isobutyl*)
$CH_3CH_2—H$ (Ethane)	$CH_3CH_2—$ (Ethyl)		
$CH_3CH_2CH_2—H$ (Propane)	$CH_3CH_2CH_2—$ (Propyl or *n-propyl*)	CH_3 | CH_3CCH_3 | H (2-Methylpropane; *isobutane*)	CH_3 | CH_3CCH_3 | (*tert-Butyl*)
$CH_3CH—H$ | CH_3 (Propane)	$CH_3CH—$ | CH_3 (Isopropyl)		
$CH_3CH_2CH_2CH_2—H$ (Butane)	$CH_3CH_2CH_2CH_2—$ (Butyl or *n-butyl*)	CH_3 | CH_3CCH_3 | $CH_2—H$ (2,2-Dimethylpropane; *neopentane*)	CH_3 | CH_3CCH_3 | $CH_2—$ (*Neopentyl-*)
$CH_3CH_2CH—H$ | CH_3 (Butane)	$CH_3CH_2CH—$ | CH_3 (*sec-Butyl*)		

of organic compounds. The basic rules used in the **I.U.P.A.C. Nomenclature System** are as follows:

1. The name of the compound is defined by the longest continuous carbon chain containing the functional group in the compound. This chain is the **parent compound.** For example:

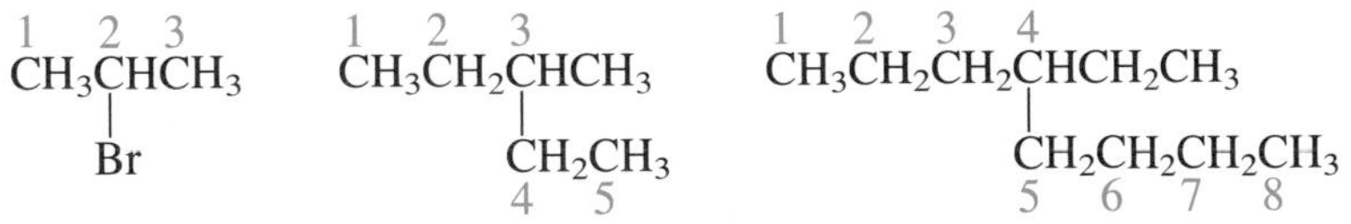

1 2 3
CH_3CHCH_3
|
Br

Parent name:
Propane

1 2 3
$CH_3CH_2CHCH_3$
|
CH_2CH_3
4 5

Pentane

1 2 3 4
$CH_3CH_2CH_2CHCH_2CH_3$
|
$CH_2CH_2CH_2CH_3$
5 6 7 8

Octane

Refer to Table 3.2 for the names of the simple alkanes.

2. Each substituent attached to the main, parent compound is given a name and a number. The number designates the position of the substituent on the main chain, and the name tells what type of substituent is present at that position. For example:
   a. **The halogens:**
      —F fluoro-
      —Cl chloro-
      —Br bromo-
      —I iodo-
   b. **Alkyl groups:** The name of a hydrocarbon substituent is derived from the name of the parent alkane; that is, the alkane of the same number of carbon atoms. The alkyl group results from the removal of one hydrogen from the original alkane (e.g., methyl, $CH_3—$; ethyl, $C_2H_5—$). The *-ane* is replaced by the *-yl* ending of an alkyl substituent (see Table 3.3).

3. The chain must be numbered from one end to the other to provide the lowest position number for each substituent. If more than one substituent is present, number from the end that gives the lowest number to the first substituent encountered, regardless of the numbers that result on the other substituents. For example:

$$\underset{\underset{Br}{|}}{\overset{1\ \ 2\ \ 3}{CH_3CHCH_3}} \qquad \overset{3\ \ 2\ \ 1}{CH_3CHCH_2CH_3} \text{ with } \underset{4\ \ 5}{CH_2CH_3} \text{ on C-3} \qquad \overset{1\ \ 2\ \ 3\ \ 4}{CH_3CH_2CH_2CHCH_2CH_2CH_3} \text{ with } \underset{5\ \ 6\ \ 7\ \ 8}{CH_2CH_2CH_2CH_3} \text{ on C-4}$$

Substituent:
2-Bromo 3-Methyl 4-Propyl (4-*n*-propyl)

4. If the same substituent occurs more than once in the compound, a separate position number is supplied for each substituent, and the prefixes di-, tri-, tetra-, penta-, hexa-, hepta-, etc. are used. Examples with multiple substituents are the following:

$$\underset{1\ \ 2\ \ 3\ \ 4\ \ 5\ \ 6}{CH_3\overset{Br}{\overset{|}{C}}HCH_2CH_2\overset{Br}{\overset{|}{C}}HCH_3}$$

2,5-Dibromo

$$\underset{\substack{1\ \ 2\ \ 3\ \ 4\ \ 5\ \ 6\ \ 7\ \ 8\ \ 9\ \ 10 \\ 10\ \ 9\ \ 8\ \ 7\ \ 6\ \ 5\ \ 4\ \ 3\ \ 2\ \ 1}}{CH_3CH_2\overset{CH_3}{\overset{|}{C}}HCH_2\overset{CH_3}{\overset{|}{C}}HCH_2\overset{CH_3}{\overset{|}{C}}HCH_2CH_2CH_3}$$

3,5,7-Trimethyl
NOT 4,6,8-Trimethyl

The alphabetical ordering is less subjective than the ordering based on the complexity of the substituent.

5. Place the names of the substituents in alphabetical order (or order of increasing complexity of the substituent) before the name of the parent compound. Numbers are separated from each other by commas, and numbers are separated from names by hyphens. Note that by convention, halogen substituents are generally placed before alkyl substituents in this priority sequence, regardless of the alphabetization.

Example 3.1

Naming substituted alkanes using the I.U.P.A.C. System.

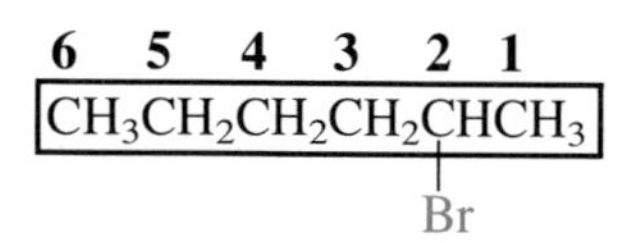

Parent chain: hexane

Substituent: 2-bromo (not 5-bromo)
Name: 2-bromohexane

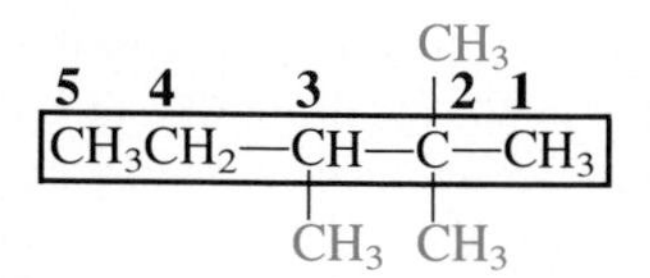

Parent chain: pentane

Substituents: 2,2,3-trimethyl (not 3,4,4-trimethyl)
Name: 2,2,3-trimethylpentane (not 2-*tert*-butylbutane)

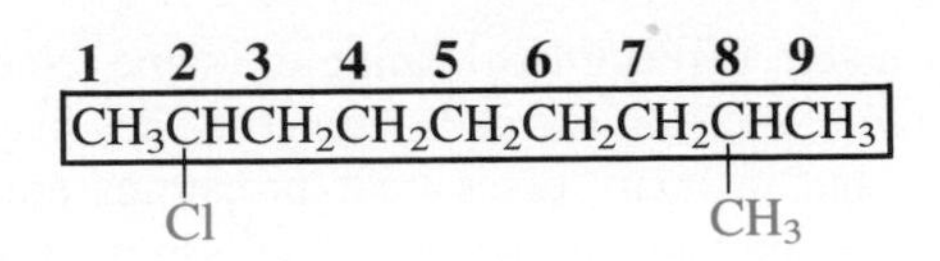

Parent name: nonane

Substituents: 2-chloro and 8-methyl
Name: 2-chloro-8-methylnonane

Question 3.1 Name the following compounds, using the I.U.P.A.C. Nomenclature System:

a. CCl_4

b. 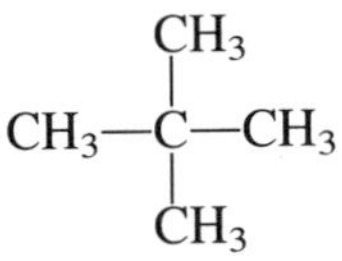 $CH_3—C(CH_3)_2—CH_3$ (central C bearing CH_3 above and CH_3 below)

c. $CH_3—C—CH_3$ with $CH_2CH_2CH_3$ above and CH_3 below the central C

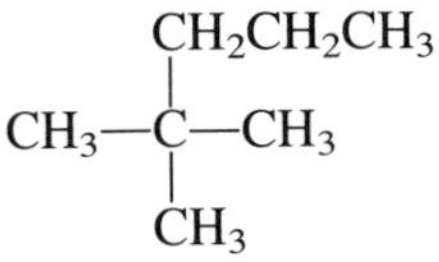

d. 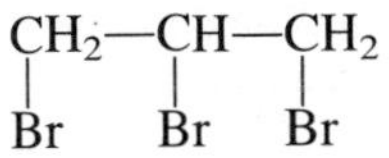 $CH_2—CH—CH_2$ with Br on each carbon

e. 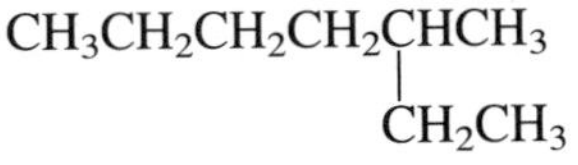 $CH_3CH_2CH_2CH_2CHCH_3$ with CH_2CH_3 on the fifth carbon

f. 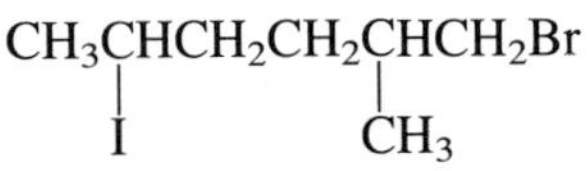 $CH_3CHCH_2CH_2CHCH_2Br$ with I on C-2 and CH_3 on C-5

g. 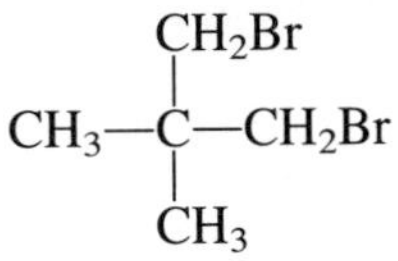 $CH_3—C—CH_2Br$ with CH_2Br above and CH_3 below the central C

h. 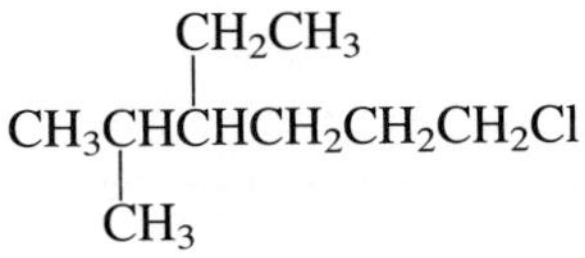 $CH_3CHCHCH_2CH_2CH_2Cl$ with CH_3 on C-2 and CH_2CH_3 on C-3

The common system of nomenclature

The I.U.P.A.C. System will be the preferred system of nomenclature in this text. However, the older, well-established system of nomenclature, called the **common system,** is also widely used. A few examples are listed below, and others are listed in Table 3.3.

I.U.P.A.C.	*Common*
Methane	Methane
Ethane	Ethane
Propane	*n*-Propane
Butane	*n*-Butane
Pentane	*n*-Pentane

In future discussions we will provide the common name, as well as the I.U.P.A.C. name, where appropriate.

3.3 STRUCTURAL ISOMERS

Two molecules that have the same molecular formula but different structures are called **isomers.** For example:

$CH_3—CH_2—CH_2—CH_3$

Butane
(*n*-butane)

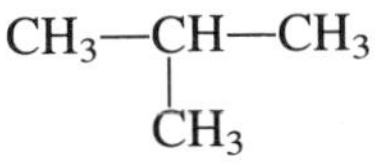

$CH_3—CH(CH_3)—CH_3$

2-Methylpropane
(isobutane)

As we have already noted, two isomers bearing the same molecular formula are two unique compounds because of their structural differences. They may have similar physical and chemical properties, but in many cases their properties are quite dissimilar. For example, butane and 2-methylpropane (drawn above) are isomeric alkanes. Their physical and chemical properties are very similar:

	Butane	**2-Methylpropane**
b.p. (°C)	−0.5	−12
	Nonpolar	Nonpolar

The situation is altogether different in the case of ethanol and dimethyl ether. Here the structures are radically different. Ethanol, for example, has a hydroxyl group (—O—H), but dimethyl ether does not. Consequently, their physical and chemical properties are predicted to be different. We find that they are.

$$CH_3—CH_2—O—H \qquad CH_3—O—CH_3$$
$$(C_2H_6O) \qquad (C_2H_6O)$$

	Ethanol	**Dimethyl Ether**
	An alcohol	An ether
b.p. (°C)	78.3	−24.9
	Very polar	Slightly polar

The two pairs of structural isomers (butane and 2-methylpropane; ethanol and dimethyl ether) shown above do indeed have a different bonding framework. As with all structural isomers, they differ in the manner in which their atoms are bonded together, and their physical and chemical properties reflect these differences.

Example 3.2

Drawing structural isomers

Write all the structural isomers corresponding to the molecular formula C_6H_{14}.

1. Begin with the continuous-chain structure:

$$CH_3—CH_2—CH_2—CH_2—CH_2—CH_3 \tag{3.1}$$

2. Next consider the five-carbon chain structure in which a methyl group is attached to one of the nonterminal carbons of the chain:

$$\begin{array}{l} CH_3—CH—CH_2—CH_2—CH_3 \\ \qquad\quad | \\ \qquad\ \ CH_3 \end{array} \tag{3.2}$$

and

$$\begin{array}{l} CH_3—CH_2—CH—CH_2—CH_3 \\ \qquad\qquad\qquad | \\ \qquad\qquad\quad CH_3 \end{array} \tag{3.3}$$

3. Next consider the four-carbon structure in which the two remaining carbons may be attached as either two methyl ($—CH_3$) groups or one ethyl ($—CH_2CH_3$) group:

$$\begin{array}{l} CH_3—CH—CH—CH_3 \\ \qquad\quad |\qquad\ | \\ \qquad\ \ CH_3\ CH_3 \end{array} \tag{3.4}$$

and

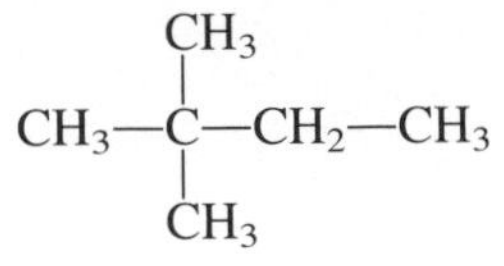

$$CH_3-\overset{\displaystyle CH_3}{\underset{\displaystyle CH_3}{C}}-CH_2-CH_3 \qquad 3.5$$

These are the five possible structural isomers of hexane. At first it might appear that there are other isomers as well, but careful comparison will show that they are duplicates of those already constructed. For example:

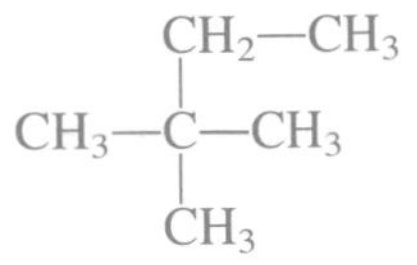

$$CH_3-\overset{\displaystyle CH_2-CH_3}{\underset{\displaystyle CH_3}{C}}-CH_3$$

is the same as structure 3.5, and

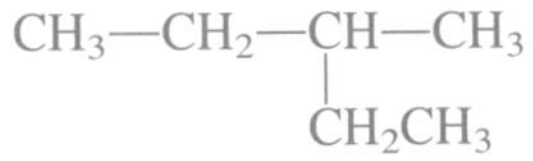

$$CH_3-CH_2-\underset{\displaystyle CH_2CH_3}{CH}-CH_3$$

is the same as structure 3.3.

A good check, to be sure that you have not inadvertently constructed duplicate isomers, makes use of the I.U.P.A.C. Nomenclature System. If you apply the nomenclature rules correctly, all isomers must have different I.U.P.A.C. names. If two names are identical, the structures are identical. In the example provided above, compare isomers 3.1–3.5 and the two duplicate isomers provided at the end of the section. Name them, using the I.U.P.A.C. System. It should be obvious that the last two structures are indeed duplicates of structures 3.3 and 3.5, as stated.

Question 3.2 Draw complete structural formulas of each of the structural isomers of the following molecular formulas:

a. C_4H_9Br b. $C_4H_8Br_2$ c. $C_5H_{11}I$

3.4 CYCLOALKANES

The **cycloalkanes** are another family of hydrocarbons that are closely related to the alkanes. Cycloalkanes have the general molecular formula C_nH_{2n}. Note that they contain two fewer hydrogens than the corresponding alkane. The relationship that exists between an alkane and a cycloalkane is shown below for hexane and cyclohexane:

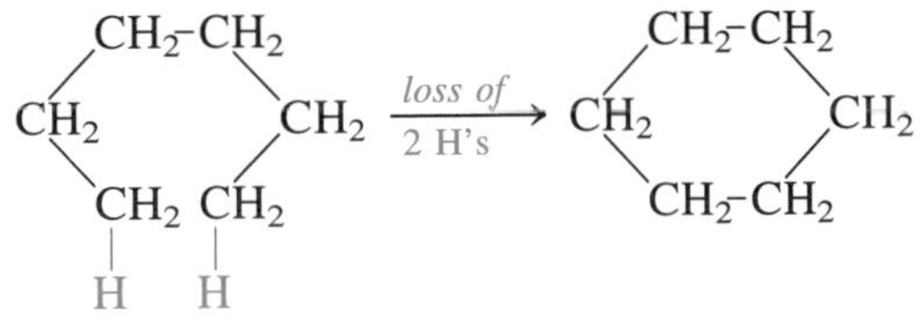

Hexane (CH$_2$–CH$_2$, CH$_2$, CH$_2$, CH$_2$ CH$_2$ with terminal H H)	$\xrightarrow[\text{2 H's}]{\text{loss of}}$	Cyclohexane (ring of six CH$_2$)
Hexane		Cyclohexane
Alkane		*Cycloalkane*
C_nH_{2n+2}		C_nH_{2n}

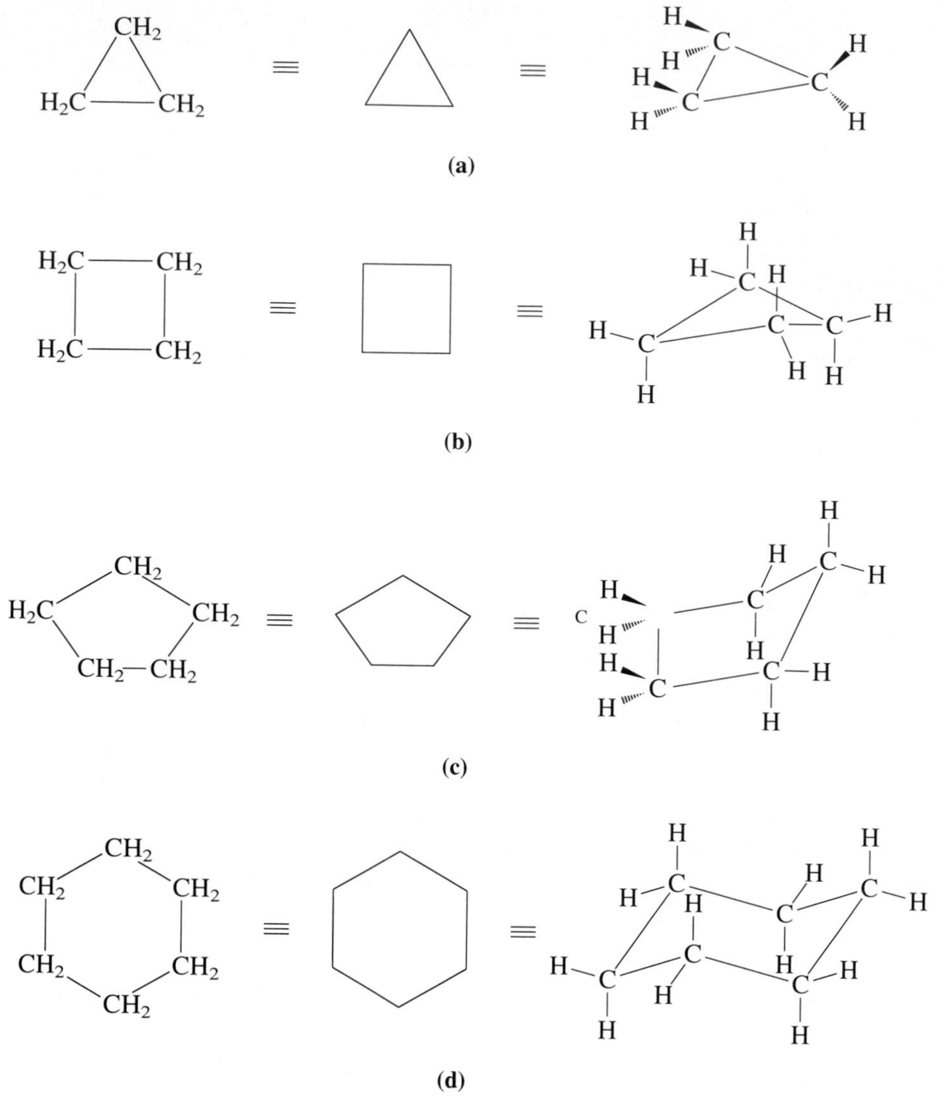

(a)

(b)

(c)

(d)

FIGURE 3.2
Cycloalkanes: (a) cyclopropane; (b) cyclobutane; (c) cyclopentane; (d) cyclohexane. All of the cycloalkanes are shown in their condensed structural forms (left column), abbreviated planar structures (center column), and three-dimensional structures (right column).

The structures and names of the simpler cycloalkanes are presented in Figure 3.2.

Cycloalkanes derive their names from the alkane with the same number of carbon atoms, and the prefix *cyclo-* is added. For example, cyclopentane is a cyclic alkane that has five carbon atoms. Substituted cycloalkanes are named by placing the name of the substituent before the name of the cycloalkane. No number is needed if only a single substituent is present. If more than one substituent is present, then the numbers that result in the *lowest possible position numbers* for the substituents are used. This practice is illustrated in the following example:

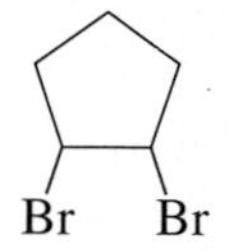

1,2-Dibromocyclopentane

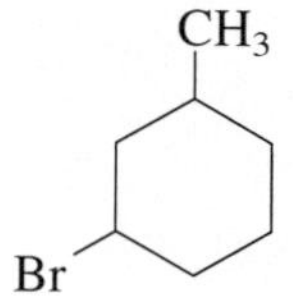

3-Bromo-1-methylcyclohexane
(not 5-Bromo-1-methylcyclohexane)

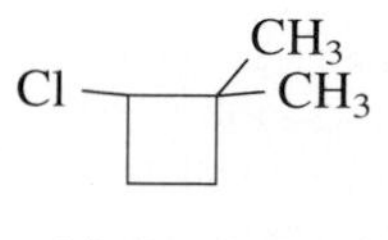

2-Chloro-1,1-dimethylcyclobutane

The cycloalkanes and alkanes, as would be predicted, have very similar physical and chemical properties.

Question 3.3 There are actually three isomers of dichlorocyclopropane. Use a set of molecular models to construct the isomers and to contrast their differences. Draw the three isomers.

Question 3.4 How many isomers of dibromocyclobutane can you construct? As in Question 3.3, use a set of molecular models to construct the isomers and then draw them.

3.5 CONFORMATIONAL ANALYSIS

As we have seen, there is free rotation around a carbon-carbon single bond (see Figure 3.3), and at room temperature, rotations occur on the order of millions of times per second. It is important to realize that these molecules are not static, but rather exist in an infinite variety of arrangements in space that are rapidly interconverting. These varying arrangements, which are capable of interconversion by simple rotation about a bond, are called conformations or **conformers;** the molecules are also called *conformational isomers*.

Conformational isomers are discrete, distinct isomeric structures that may be interconverted, one to the other, by rotation about the bonds in the molecule. Because these conversions (of one conformer to the others) occur so rapidly, the various conformers cannot generally be isolated from one another. However, as we will see in the chapter on proteins, an understanding of the various possible conformers for a given molecule can help us to understand the chemical and (where applicable) biochemical reactivity of that species.

Chapter 11

Cycloalkanes are also composed of carbon-carbon single bonds, and here too free rotation results in a variety of different conformational forms (see Figure 3.4).

The shapes of these conformational isomers of cyclohexane suggest the names used to describe them. One form is called the chair form because of its resemblance to a lawn chair. Another is called the boat form because it resembles a row boat.

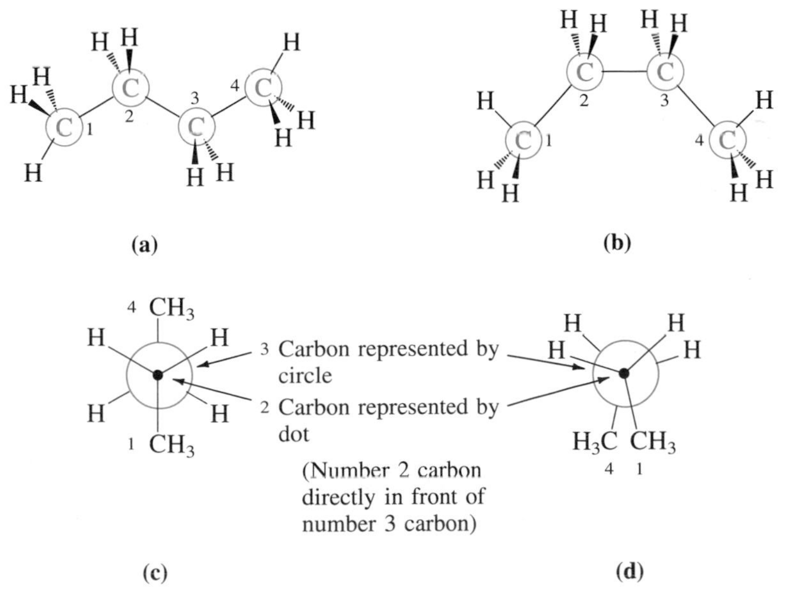

(a) (b)

3 Carbon represented by circle

2 Carbon represented by dot

(Number 2 carbon directly in front of number 3 carbon)

(c) (d)

FIGURE 3.3
Conformational isomers of butane. Note the extreme crowding of atoms in the form depicted in (b) compared with the form shown in (a). The form shown in (a) is energetically favored.

A HUMAN PERSPECTIVE

The Petroleum and Coal Industries

Virtually all organic chemicals have their origin in the petroleum and coal industries, and were defined as having been derived from what was once living matter. Petroleum (L.: *petra,* rock; *oleum,* oil) and coal result from the anaerobic (without air) decay of the remains of plants, animals, and microorganisms over millions of years.

Petroleum consists primarily of alkanes and small amounts of alkenes and aromatic hydrocarbons. (Substituted hydrocarbons, for example, phenols, are also present in very small quantities.) We also find that petroleum composition varies with the source of the petroleum (for example, the United States, South America, the Persian Gulf, and so forth). The mixture of hydrocarbons in petroleum is separated into its component parts on the basis of differences in the boiling points of the various hydrocarbons. The process of separating materials on the basis of differences in boiling points is called **distillation.** Often, several successive distillations of various fractions of the original mixture are required to effect complete separation of the desired materials. In the first distillation the petroleum is separated, as a function of boiling point, into several fractions, which are themselves mixtures of several hydrocarbons. Each of these fractions can be further purified by successive distillations. These distillations are carried out on a very large industrial scale, and the distillation columns used may be hundreds of feet high, in contrast to the few inches of a laboratory scale column.

The purification of a given compound is a formidable and expensive proposition. To minimize cost, mixtures of these chemicals are used whenever possible. For example, gasoline is a mixture of alkanes and cycloalkanes containing six to twelve carbon atoms (other additives, such as tetraethyl lead, are included to increase the efficiency of the fuel). The following table provides a breakdown of the many possible fractions obtained from the petroleum distillation process.

Fraction	Number of C Atoms	b.p. (°C)
Natural gas	C_1–C_4	<20
Petroleum ether	C_5–C_6	20–60
Gasoline	C_6–C_{12}	50–200
Kerosene	C_{12}–C_{18}	175–275
Heating fuel	$> C_{18}$	>275
Lubricating oils	C_{20}–C_{30}	>350
Paraffins	C_{23}–C_{40}	Nonvolatile
Residue	C_{25}–C_{50}	Nonvolatile

Many other hydrocarbons are also obtained indirectly from petroleum by various industrial, synthetic processes. Among these are *catalytic cracking, catalytic reforming, hydroforming,* and *platforming.* In general, all of these processes use specific conditions (for example, high temperature, catalysts, and/or the addition of inexpensive reagents such as water or hydrogen) to convert readily available, naturally occurring hydrocarbons into other desired organic chemicals. The reactions often involve fragmenta-

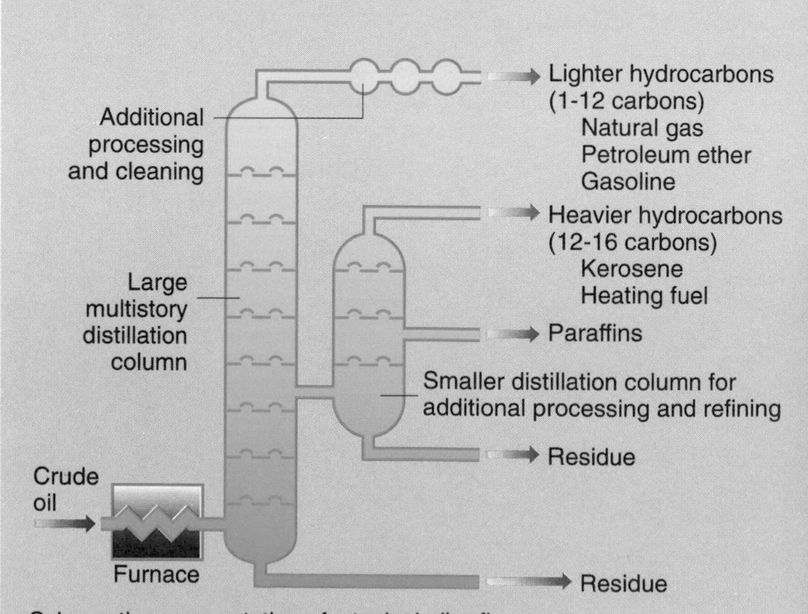

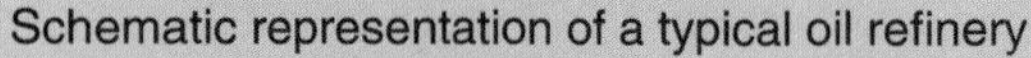
Schematic representation of a typical oil refinery

An oil refinery with the principal components labeled: *left*, schematic representation; *right*, oil refinery at Carson, California.

tion of the reactant into one or more products, a process known as **catalytic cracking,** or the rearrangement of the reactant into a more useful product, a process known as **catalytic reforming.** Examples of the two processes are shown here:

Cracking:

$$CH_3CH_2CH_2CH_3 \xrightarrow[\text{high pressure}]{\text{heat (500°C),}} CH_2{=}CH_2 + CH_3CH{=}CH_2 + CH_3CH{=}CHCH_3 + CH_3CH_2CH{=}CH_2 \quad 3.6$$

Reforming:

$$CH_3(CH_2)_4CH_3 \xrightarrow[\text{Al}_2\text{O}_3\text{ (catalyst)}]{\text{heat, pressure,}} \text{Benzene} \quad 3.7$$

Hexane   Benzene

$$CH_3(CH_2)_5CH_3 \xrightarrow[\text{Al}_2\text{O}_3\text{ (catalyst)}]{\text{heat, pressure,}} \text{Toluene} \quad 3.8$$

Heptane   Toluene

Gasoline can also be produced by cracking a larger, heavier fraction from the petroleum distillation process into the smaller, lighter gasoline fraction.

Coal and natural gas are also sources of hydrocarbons and substituted hydrocarbons. Coal, in particular, is a source of many aromatic hydrocarbons. Heating coal at very high temperatures in the absence of oxygen gives a mixture of many desired aromatic compounds. For example,

$$\text{Coal} \xrightarrow[\text{1000°C, 15–20 hours, limited air}]{\text{approximately}} \text{aromatic hydrocarbons} + \text{phenols} + \text{cresols} + \text{others} \quad 3.9$$

The gasoline fraction from petroleum is called *straight-run* gasoline. It is a poor grade of gasoline and has poor performance as a fuel. As fuels, branched-chain alkanes are superior to straight-chain compounds because they are more volatile, burn less rapidly in the cylinder, and thus reduce knocking. The alkenes (compounds containing carbon-carbon-double-bonds) and aromatic compounds are also good fuels. Methods have been developed to convert hydrocarbons of higher and lower molecular weights than gasoline to that range or to convert straight-chain hydrocarbons into branched ones. These conversions are important processes because branched-chain hydrocarbons are more volatile than their straight-chain counterparts. For example, the branched-chain alkanes, alkenes, and aromatic compounds are obtained from petroleum hydrocarbons by *catalytic cracking* and *catalytic reforming*.

The antiknock quality of a fuel is measured by its *octane number.* Heptane is a very poor fuel that causes severe knocking in an automobile engine. It is arbitrarily given an octane number of 0. On the other hand, 2,2,4-trimethylpentane (which is *incorrectly* known as *isooctane* in the petroleum industry—a common name that is still used because of the importance of this compound) is an excellent fuel and is given the arbitrary octane number of 100. Fuels are compared in a test engine with varying mixtures of heptane and 2,2,4-trimethylpentane. A gasoline that exhibits a performance equal to that of a mixture of 10 mole % heptane and 90 mole % 2,2,4-trimethylpentane has an octane rating of 90. Some hydrocarbons known today are better fuels than 2,2,4-trimethylpentane, as shown below.

Compound	Octane No.
Heptane	0
2-Methylheptane	24
2-Methylpentane	71
Octane	−20
2-Methylbutane *(isopentane)*	90
2,2,4-Trimethylpentane *(isooctane)*	100
Benzene	101
Toluene	110
2,2,3-Trimethylpentane *(triptane)*	116
Cyclopentane	122
p-Xylene	128

One way to improve the performance of gasoline is to add branched-chain alkanes that have a higher octane rating. Another is to add small amounts of tetraethyl lead (mentioned above), $(CH_3CH_2)_4Pb$, to yield *ethyl* gasoline. Tetraethyl lead was discovered in 1922 and is a highly toxic blue liquid. On combustion, tetraethyl lead forms water, carbon dioxide, and lead oxide, which builds up and is harmful to the cylinder. To avoid this problem, 1,2-dibromoethane, $BrCH_2CH_2Br$, is often added to ethyl gasoline; it combines with the lead oxide to form volatile lead bromide, $PbBr_2$, which passes out with the exhaust fumes. Because of the harmful side effects of using leaded fuels, other, preferred additives, have been developed to improve gasoline performance. These additives, consisting of mixtures of aromatic hydrocarbons, provide today's unleaded gasolines.

In contrast to gasoline, natural gas consists primarily of methane and small amounts of ethane, propane, and butane. The methane is separated and used as a common household and industrial fuel. Propane and butane are sold as bottled gas.

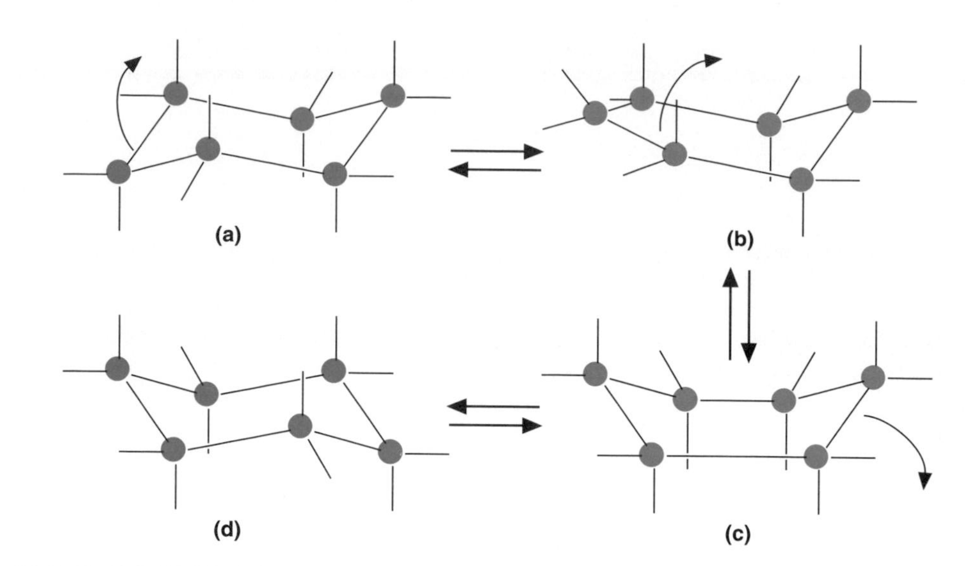

FIGURE 3.4
The principal conformations of cyclohexane are shown: (a) chair form; (b) half-boat form; (c) boat form; and (d) chair form [*identical to conformation (a)*]. Try building these with molecular models.

See Section 12.11 for a specific example related to the action of enzymes.

As we will see, in larger biochemical molecules such as proteins, enzymes, or nucleic acids, the conformation of the molecule (geometry of the molecule in space) may have a critical effect on the activity.

3.6 REACTIONS OF ALKANES AND CYCLOALKANES

Combustion

Alkanes and other hydrocarbons may be oxidized (burned in air), producing carbon dioxide and water; this reaction is called **combustion.** During combustion a large amount of heat energy is released. This fact, coupled with their ready availability and relatively low cost, makes them quite useful as fuels. The combustion reaction for a simple alkane is

$$CH_4 + 2O_2 \longrightarrow CO_2 + 2H_2O \qquad 3.10$$

Combustion is essential to our very existence; it is the process that fuels our bodies through various metabolic pathways, by which we heat our homes, run our automobiles, generate our electricity, and so on. In fact, it is also the mechanism that provides the energy needed for cellular processes through the oxidation of foods. The combustion of coal and petroleum-based fuels (hydrocarbons) plays a vital role in our day-to-day lives and at the same time represents a significant threat to our environment.

see A Human Perspective: The Petroleum and Coal Industries

Combustion is also an example of a degradation reaction, a reaction in which a larger molecule is broken down into smaller molecules.

Halogenation

A second reaction of the alkanes is **halogenation.** In contrast to combustion, halogenation is a *synthetically* useful reaction. That is, it provides a pathway to the preparation of useful molecules from simpler organic starting materials. The reaction does have limitations and provides low yields and a mixture of products that must be separated. Typically, when an alkane reacts with certain halogens (generally bromine, Br_2, and chlorine, Cl_2) in the presence of heat and/or light, a substitution reaction results. Under controlled conditions, one of the C—H bonds of the alkane is broken and replaced with a C—X bond in which a halogen atom (X = Br or Cl) has substituted for a hydrogen atom:

$$\underset{\text{Methane}}{CH_4 + X_2} \xrightarrow{\text{light}} \underset{\substack{\text{Halomethane} \\ (X = Br \text{ or } Cl)}}{CH_3{-}X} + H{-}X \qquad 3.11$$

In more complex alkanes, substitution can occur to some extent at all positions to give a mixture of products. Modification of experimental conditions to favor the formation of a certain product is often possible but can seldom be accomplished with high efficiency. For example, bromination of propane produces a mixture of 1-bromopropane and 2-bromopropane:

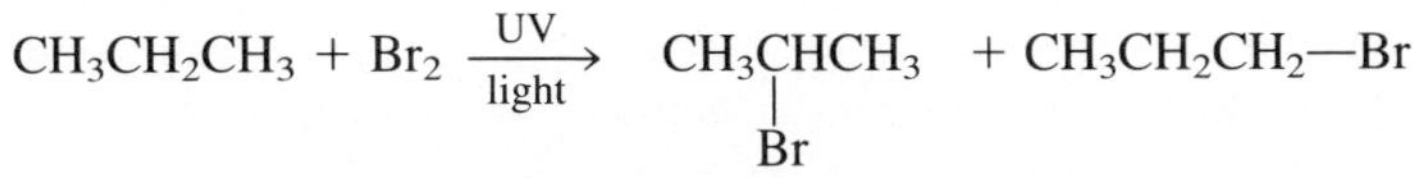

$$CH_3CH_2CH_3 + Br_2 \xrightarrow[\text{light}]{\text{UV}} CH_3\underset{\underset{\displaystyle Br}{|}}{C}HCH_3 + CH_3CH_2CH_2{-}Br$$

Propane (*n*-propane) — 2-Bromopropane (isopropyl bromide) — 1-Bromopropane (*n*-propylbromide)

Cycloalkanes can also be halogenated by this method:

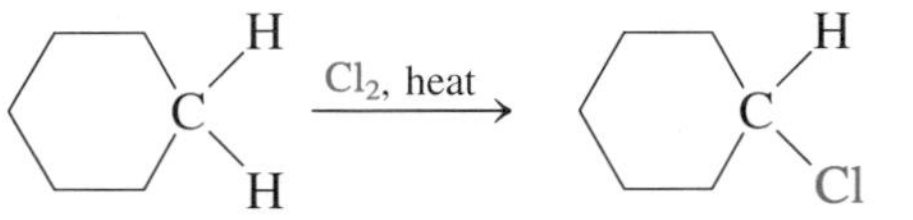

$$C_6H_{12} \xrightarrow{Cl_2,\ \text{heat}} C_6H_{11}Cl$$

Cyclohexane — Chlorocyclohexane

Question 3.5 Assuming that the halogenation of 2-methylbutane using Br_2/light proceeds to give *only monosubstituted product,* how many different monosubstituted products will result? Name them.

3.7 AN INTRODUCTION TO THE CHEMISTRY OF ALKYL HALIDES

Alkyl halides have the general structure R—X, where R represents any alkyl group and X = (a halogen) F—, Cl—, Br—, or I—. For example, R—Cl = CH_3Cl, CH_3CH_2Cl, and so forth. The rules for naming this family have already been discussed, and the substitution of alkanes, used in their preparation, was introduced in the previous section.

Section 3.2

The structure of the members of this family is analogous to the hydrocarbon from which the halide is derived. That is, the structure of an alkyl halide is similar to the tetrahedral structure of the parent alkane:

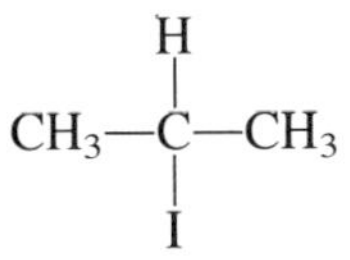

$$CH_3{-}\underset{\underset{\displaystyle I}{|}}{\overset{\overset{\displaystyle H}{|}}{C}}{-}CH_3 \qquad CH_3{-}\underset{\underset{\displaystyle Cl}{|}}{\overset{\overset{\displaystyle CH_3}{|}}{C}}{-}CH_3 \qquad CH_3{-}\underset{\underset{\displaystyle CH_3}{|}}{\overset{\overset{\displaystyle CH_3}{|}}{C}}{-}CH_2{-}Br$$

2-Iodopropane (isopropyl iodide) 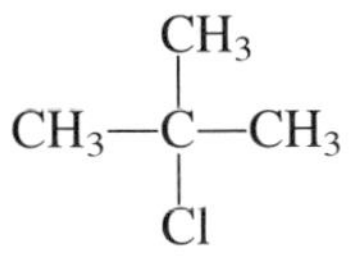 — 2-Chloro-2-methylpropane (*tert*-butyl chloride) 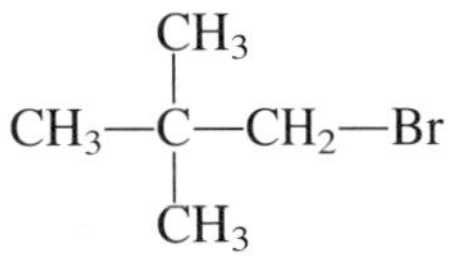 — 1-Bromo-2,2-dimethylpropane (neopentyl bromide)

Question 3.6 Provide the complete structural formula of each of the following molecules:

a. *sec*-butyl chloride
b. *n*-octyl iodide
c. 1,2,3-triiodohexane
d. 1,1-dichlorocyclopentane
e. 3,4,5-tribromooctane
f. 2-chloro-2,4,4-trimethylpentane

The substitution of a halogen atom on a hydrocarbon chain causes a marked change in the physical properties of the compound. See Table 3.4 for a list of some of the physical properties of the more common members of this family.

AN ENVIRONMENTAL PERSPECTIVE

Oil-eating Bacteria

Our highly industrialized society has come to rely more and more on petroleum as a source of energy as well as a raw material source for the manufacture of plastics, drugs, and a host of other consumables. Over 50% of the petroleum consumed in the United States is imported, and the major carrier is the supertanker.

Well-publicized oil spills, such as that from the *Exxon Valdez* (in Alaska in 1989) have fueled research to develop clean-up methods that will help to preserve the fragile aquatic environment.

It has been known for some time that there are strains of bacteria that will accelerate the oxidation of many of the compounds present in unrefined petroleum. These bacteria have been termed "oil-eating bacteria."

Recently, oceanographers at the University of Texas have developed strains of bacteria that will actually "eat" a wide variety of crude oils. At the same time these bacteria have a very short lifetime. It appears that they die shortly after they have operated on an oil slick.

This latter characteristic, a short lifetime in water, is particularly appealing to scientists, who fear that the introduction of nonindigenous (nonnative) bacteria into natural water systems may disrupt the ecology of the water.

Some also fear that the products of these reactions, in which some of the oil is converted to fatty acids, may disperse in water and cause more problems than the original oil spill.

Obviously, a great deal of research involving biodegradation remains to be done. Such technologies offer hope for alleviating many land-based solid waste disposal problems, in addition to petroleum spills.

TABLE 3.4 Physical Properties of Selected Alkyl Halides

Name*	Structure	b.p. (°C) Chloride	Bromide	Iodide
Halomethane (methyl halide)	CH_3—X	−24	5	42
Haloethane (ethyl halide)	CH_3CH_2—X	13	38	72
1-Halopropane (*n*-propyl halide)	$CH_3CH_2CH_2$—X	46	71	102
2-Halopropane (isopropyl halide)	$(CH_3)_2CH$—X	37	60	89
1-Halobutane (*n*-butyl halide)	$CH_3CH_2CH_2CH_2$—X	78	102	130
2-Halobutane (*sec*-butyl halide)	$CH_3CH_2CHCH_3$ (X on C-2)	68	91	119
1-Halo-2-methylpropane (isobutyl halide)	$(CH_3)_2CHCH_2$—X	69	91	120
2-Halo-2-methylpropane (*tert*-butyl halide)	$(CH_3)_3C$—X	51	73	100d
1-Halopentane (*n*-pentyl halide)	$CH_3CH_2CH_2CH_2CH_2$—X	108	130	157
Dihalomethane	CH_2X_2	40	99	180d
Trihalomethane	CHX_3	61	151	solid
Tetrahalomethane	CX_4	77	189	solid

*Common name in parentheses.
d = decomposes on boiling.

Note that the boiling points of alkyl halides are consistently higher than those of the corresponding alkanes. The boiling point increases with increasing carbon content and increasing molecular weight. In considering solubility, alkyl halides have a "hydrocarbon-like" nature: They are not soluble in water and are highly soluble in organic solvents. We also find that monosubstituted alkyl iodides and bromides are more dense than water and that alkyl chlorides are slightly less dense. This observation is due to the much larger atomic mass of iodine and bromine. The density of a substance is mass/volume. The volumes of the individual molecules of CH_3—Br, CH_3—I, and CH_3—Cl do not differ significantly, but their formula weights differ considerably.

Question 3.7 Which member of each of the following pairs has the higher boiling point?

a. butane or octane
b. 1-chloropentane or 1-bromopentane
c. 1,3-dichlorohexane or 1-chlorohexane

3.8 CLASSIFICATION OF ALKYL HALIDES

Carbon atoms may be classified as one of four types depending on the substitution pattern they exhibit. Each type is shown below with its corresponding substitution pattern:

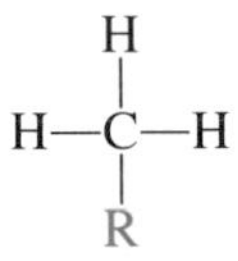

```
    H           H           H           R
    |           |           |           |
 H—C—H       R—C—H       R—C—R       R—C—R
    |           |           |           |
    R           R           R           R
```

Primary carbon (1°) 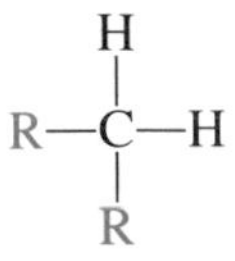 | Secondary carbon (2°) 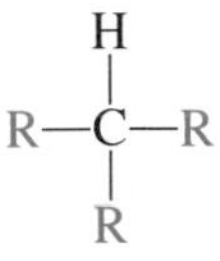 | Tertiary carbon (3°) 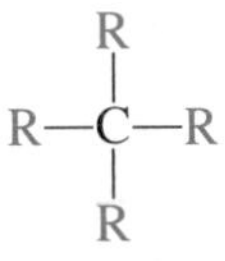 | Quaternary carbon (4°)

1° carbons: Primary carbons have three hydrogen atoms and one other (R) group attached to the central carbon atom.

2° carbons: Secondary carbons have two hydrogens and two (R) groups attached to the central carbon atom.

3° carbons: Tertiary carbons have one hydrogen and three (R) groups attached to the central carbon atom.

4° carbons: Quaternary carbons have four (R) groups and no hydrogens attached to the central carbon atom.

The R groups may be the same or different in each of these examples. For clarity, different R groups are often labeled as R, R', R", and so forth.

The hydrogens attached to each of the carbon atoms above have the same designation as the carbon to which they are attached. That is, a primary hydrogen is attached to a primary carbon, a secondary hydrogen to a secondary carbon, and so on. Similarly, when a hydrogen atom is substituted by some other group, that group takes on the classification of the alkyl group to which it is attached. For example, the attachment of a bromine atom to a primary carbon results in the formation of a primary alkyl bromide. Other examples are provided here:

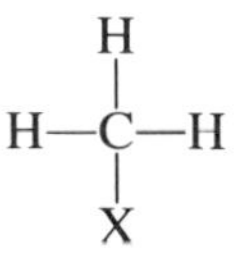

```
    H           H           R           R
    |           |           |           |
 H—C—H       R—C—H       R—C—H       R—C—R
    |           |           |           |
    X           X           X           X
```

Methyl alkyl halide 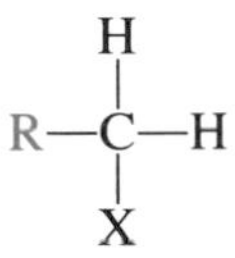 | Primary alkyl halide 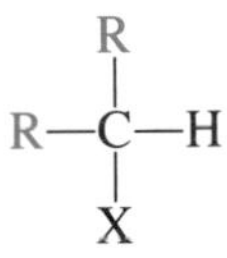 | Secondary alkyl halide 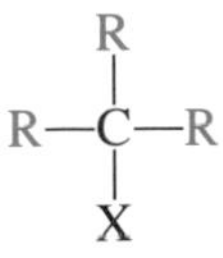 | Tertiary alkyl halide

Question 3.8 Label the carbon shown in bold as methyl, primary, secondary, tertiary, or quaternary:

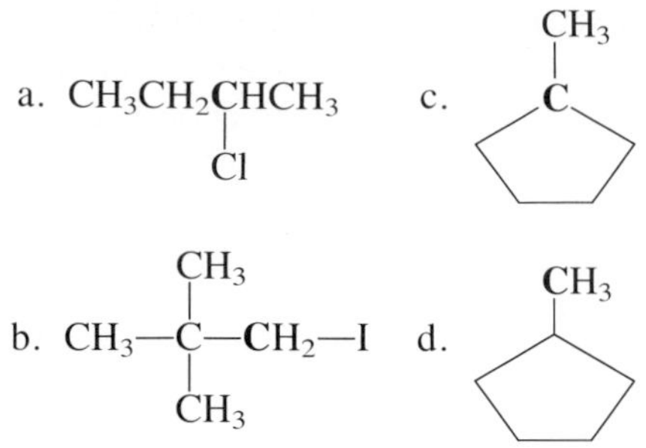

a. $CH_3CH_2\mathbf{C}HCH_3$ (with Cl on the bold C)

b. $CH_3—\mathbf{C}(CH_3)_2—CH_2—I$

c. Cyclopentane ring carbon **C** bearing CH_3

d. Cyclopentane bearing CH_3 (carbon of $\mathbf{CH_3}$)

3.9 REACTIONS OF ALKYL HALIDES

The principal reaction of alkyl halides is **substitution.** To discuss this reaction, we must first discuss the concept of substitution. In a substitution reaction, one or more atoms in a molecule is replaced (substituted) by new atoms. We were introduced to substitution reactions in Section 3.6, in which the halogenation reaction was discussed. In this reaction a halogen atom is substituted for a hydrogen atom, converting a hydrocarbon to an alkyl halide. We also noted that more than one atom in the reactant molecule may be replaced, and as a result, multiple substitutions are also possible.

There are various types of substitution reactions, and in an attempt to classify them, they are often named according to the type of reactant molecules that are involved in the substitution process. Several classes of reactant species may be involved in chemical reactions; two of the more common species encountered are *electrophiles* and *nucleophiles*. Substitution reactions that involve these species are termed **electrophilic substitution** and **nucleophilic substitution.**

The term **electrophile** (Gk: *electro,* electron; *phile,* loving) refers to species that are electron-deficient. **Nucleophiles** (Gk: *nucleo,* nucleus; *phile,* loving) are, by contrast, electron-rich species. An electron-deficient species (for example, H^+) is attracted to electrons because of their negative charge. An electron-rich species, on the other hand (for example, $H\!:\!\ddot{\underset{..}{O}}\!:^-$), is attracted to the nucleus, where there is a positive charge center. Remember—opposite charges attract! Examples of common nucleophiles and electrophiles follow:

In actual practice, the bromonium ion probably does not exist as a discrete, positively charged bromine atom, but rather appears to involve a *polarized* bromine molecule with a partially negative pole and a partially positive pole (i.e., $\overset{\delta^+}{Br}—\overset{\delta^-}{Br}$ polarized bond).

Nucleophiles ($Nu\!:^-$): $:\ddot{\underset{..}{X}}:^-$ (halide), $H\!:\!\ddot{\underset{..}{O}}\!:^-$ (hydroxide), $H\!:^-$ (hydride), $R\!:\!\ddot{\underset{..}{O}}\!:^-$ (alkoxide).

Electrophiles (E^+): H^+ (hydrogen ion), H_3C^+ or R^+ (carbocation or carbonium ion), $:\ddot{\underset{..}{Br}}^+$ (bromonium ion).

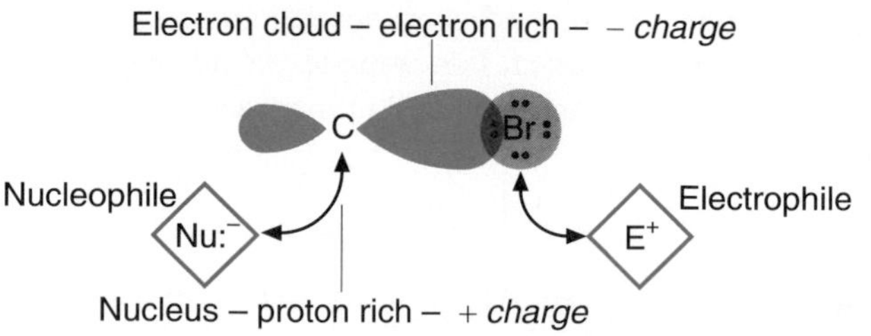

Electron cloud – electron rich – – *charge*

C :Br:

Nucleophile Nu:⁻

Electrophile E⁺

Nucleus – proton rich – + *charge*

For the present we will concern ourselves with only two nucleophilic species: the halide ion ($:\ddot{\underset{..}{X}}:^-$) and the hydroxide ion ($H\!:\!\ddot{\underset{..}{O}}\!:^-$).

Question 3.9 Would each of the following species be considered electrophilic or nucleophilic?

a. H_3O^+
b. H^-
c. H^+
d. $^+CH_3$
e. HO^-
f. $^+NO_2$
g. Br^+

The principal reaction of the alkyl halides is nucleophilic substitution. The generalized reaction is provided below:

$$\underset{\text{Alkyl halide}}{R{-}X} + \underset{\text{Nucleophile}}{Nu{:}^-} \longrightarrow \underset{\text{Product}}{R{-}Nu} + \underset{\text{Halide ion}}{:\ddot{X}:^-} \quad 3.12$$

In this reaction the alkyl halide reacts with a nucleophilic species to give overall substitution in which the nucleophile is substituted for the halide ion. The identity of the resulting product will depend on the nucleophile and alkyl halide used. Such reactions are titled nucleophilic substitution.

Most nucleophiles (including $:\ddot{X}:^-$ and $H{:}\ddot{O}:^-$) react with alkyl halides in substitution reactions. The reaction works particularly well for primary alkyl halides, works for some secondary alkyl halides, but does not work at all for tertiary alkyl halides. For example:

$$\underset{\text{Bromomethane (methyl bromide)}}{CH_3{-}Br} + :\ddot{I}:^- \longrightarrow \underset{\text{Iodomethane (methyl iodide)}}{CH_3{-}\ddot{I}:} + :\ddot{Br}:^- \quad 3.13$$

$$\underset{\text{2-Chloropropane (isopropyl chloride)}}{CH_3\underset{\displaystyle CH_3}{\underset{|}{C}}H{-}Cl} + :\ddot{Br}:^- \longrightarrow \underset{\text{2-Bromopropane (isopropyl bromide)}}{CH_3\underset{\displaystyle CH_3}{\underset{|}{C}}H{-}\ddot{Br}:} + :\ddot{Cl}:^- \quad 3.14$$

$$\underset{\text{Bromoethane (ethyl bromide)}}{CH_3CH_2{-}Br} + H{:}\ddot{O}:^- \longrightarrow \underset{\text{Ethanol (ethyl alcohol)}}{CH_3CH_2{-}\ddot{O}H} + :\ddot{Br}:^- \quad 3.15$$

The first two reactions above illustrate **halide exchange,** in which one halogen is substituted for another. Of course, the reverse reaction, in which the halides exchange again, is also possible. Reaction conditions govern which product predominates. The halide ion in such reactions may come from common salts such as sodium iodide (NaI) or potassium iodide (KI). In reaction 3.15 the substitution results in the conversion of an alkyl halide to an alcohol. The principal functional group has been exchanged. The substitution of alkyl halides is a synthetically useful reaction. It also plays an important role in the biochemistry of living systems, where it is important in chemical carcinogenesis, the ability of certain chemicals to cause cancer.

Section 10.6

Question 3.10 If you were provided with a mixture of all of the continuous-chain alkanes containing between five and ten carbon atoms (pentane through decane), explain how you could separate them into the six pure compounds.

SUMMARY

We begin our study of organic chemistry with a study of the hydrocarbons. In this chapter we focused primarily on the alkanes, the most fundamental of organic compounds. Alkenes, alkynes, and aromatic hydrocarbons will be discussed in subsequent chapters. These families are the four major subdivisions of all organic chemicals. They are principally nonpolar and water-insoluble and have relatively low boiling points. The hydrocarbons can exist in either cyclic or noncyclic (alicyclic or acyclic) forms.

All of the hydrocarbons are named by using either of two principal systems: the common system of nomenclature and the I.U.P.A.C. Nomenclature System. The latter is less ambiguous and is preferred. The I.U.P.A.C. System involves determining the parent compound, the longest continuous carbon chain containing the principal functional group in the molecule; numbering the chain to provide the lowest possible number for all substituents; and placing the substituent names, with their respective substituent numbers, as prefixes in front of the parent compound name to give the complete I.U.P.A.C. name of the compound.

Isomers are molecules that have the same molecular formula but different structures. Their physical and chemical properties reflect these differences. Conformational isomers (conformers) are discrete, distinct isomeric structures that may be interconverted by rotation about the bonds of the molecule.

Hydrocarbons may be oxidized as fuels (combustion), and alkanes can be reacted with bromine or chlorine (halogenation) to give alkyl halides through the process of substitution. Alkyl halides have structures that are similar to the structure of the hydrocarbon (alkane, alkene, alkyne, or aromatic) from which they are derived. They are "hydrocarbon-like," hence water-insoluble, and have reasonably high boiling points and densities.

Alkyl halides are prepared through the halogenation of alkanes and are classified on the basis of the substitution pattern that they exhibit as primary (1°), secondary (2°), tertiary (3°), or quaternary (4°). The principal reaction of alkyl halides is nucleophilic substitution.

GLOSSARY OF KEY TERMS

alkanes (3.1) also called *saturated hydrocarbons,* hydrocarbons that contain only carbon and hydrogen, that are bonded together through carbon-hydrogen and carbon-carbon single bonds, and that have the general molecular formula C_nH_{2n+2}.

alkyl group (3.2) a simple hydrocarbon group that results from the removal of one hydrogen from the original hydrocarbon (e.g. methyl, CH_3-; ethyl, CH_3CH_2-).

alkyl halide (3.7) a substituted hydrocarbon that has the general structure R—X, where R represents any alkyl group and X = (a halogen) F^-, Cl^-, Br^-, or I^-.

catalytic cracking (3.4) a process that results in the decomposition of large hydrocarbons to smaller ones under the influence of heat and/or pressure in the presence of a catalyst.

catalytic reforming (3.4) a process that results in the rearrangement of one organic molecule into another under the influence of heat and/or pressure in the presence of a catalyst.

combustion (3.6) the oxidation of hydrocarbons by burning in the presence of air to produce carbon dioxide and water.

common system of nomenclature (3.2) the nonsystematic and older, though well-established system of nomenclature.

condensed structural formula (3.1) shows all of the atoms in a molecule and places them in a sequential arrangement that details which atoms are bonded to each other; the bonds themselves are not shown.

conformers (3.5) also called *conformational isomers,* discrete, distinct isomeric structures that may be converted, one to the other, by rotation about the bonds in the molecule.

cycloalkanes (3.4) cyclic alkanes (saturated hydrocarbons) that have the general formula C_nH_{2n}.

distillation (3.4) the process of separating materials on the basis of differences in boiling points.

electrophile (3.9) (Gk: *electro,* electron; *phile,* loving) electron-deficient species such as a hydrogen ion (H^+).

electrophilic substitution (3.9) a substitution reaction in which the incoming substituent that reacts with the parent compound is an electrophile.

halide exchange (3.9) a substitution reaction in which one halogen atom is exchanged for another.

halogenation (3.6) a reaction in which one of the C—H bonds of a hydrocarbon is replaced with a C—X bond of a halogen atom (X = Br or Cl, generally).

homolog (3.1) a member of a homologous series.

homologous series (3.1) a group or series of compounds in which each member of the series differs from the immediately preceding or following member by a fixed number of atoms.

isomers (3.3) two molecules that have the same molecular formula but different chemical structures.

I.U.P.A.C. Nomenclature System (3.2) the International Union of Pure and Applied Chemistry (I.U.P.A.C.) standard, universal system for the nomenclature of organic compounds.

molecular formula (3.1) provides the atoms and number of each type of atom in a molecule but gives no information regarding the bonding pattern involved in the molecule's structure.

nucleophile (3.9) (Gk: *nucleo,* nucleus; *phile,* loving), electron-rich species such as the halogen ion (X^-), cyanide ion (CN^-), and hydroxide ion (HO^-).

nucleophilic substitution (3.9) a substitution reaction in which the incoming substituent that reacts with the parent compound is a nucleophile.

parent compound (3.2) in the I.U.P.A.C. Nomenclature System the parent compound is the longest continuous carbon chain containing the principal functional group (e.g., the hydroxyl group) in the molecule that is being named.

structural formula (3.1) shows all of the atoms in a molecule and exhibits all bonds as lines.

substitution reaction (3.9) a reaction that results in the replacement of one group for another.

QUESTIONS AND PROBLEMS

Structure and Properties

3.11 Draw each of the following:

a. 2-bromobutane
b. 2-chloro-2-methylpropane
c. 2,2-dimethylhexane
d. dichlorodiiodomethane
e. 1,4-diethylcyclohexane

3.12 Name each of the following, using the I.U.P.A.C. Nomenclature System:

a. 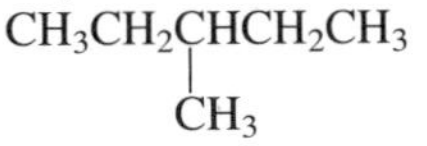 $CH_3CH_2CHCH_2CH_3$ (with CH_3 on C3)
b. $(CH_3)_2CHCH_2CH_2CH(CH_3)_2$
c. 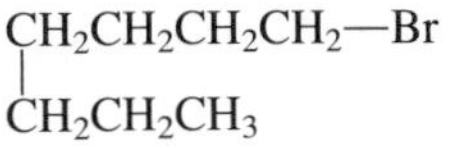 $CH_2CH_2CH_2CH_2—Br$ (with $CH_2CH_2CH_3$ on the first C)
d.  $Cl—CH_2CH_2CHCH_3$ (with CH_3 on the CH)
e.  $CH_3CH_2CHCHCHCH_2CH_3$ (with I, I, CH_2CH_3 on the three CH carbons)
f. 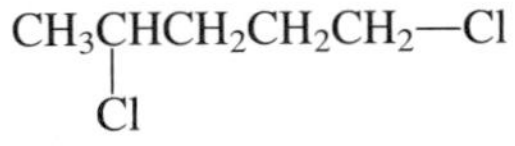 $CH_3CHCH_2CH_2CH_2—Cl$ (with Cl on the CH)
g. $(CH_3)_3C—Br$
h. 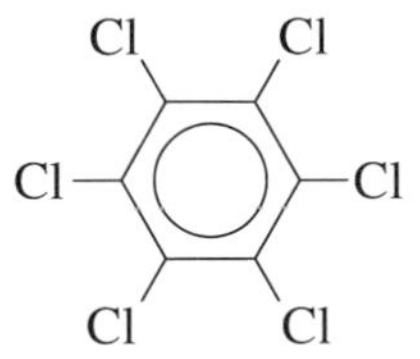 Benzene ring with six Cl substituents (Cl, Cl, Cl, Cl, Cl, Cl)

3.13 Define each of the following terms and, where possible, give an example of each:

a. structural isomers
b. hydrocarbons
c. alkyl groups
d. cracking (catalytic)
e. reforming (catalytic)
f. combustion

3.14 Draw each of the following compounds, using complete structural formulas:

a. 2,2-dibromobutane
b. 2-iododecane
c. 1,2-dichloropentane
d. 1-bromo-2-methylcyclopentane
e. 1,1,1-trichlorodecane
f. 1,2-dibromo-1,1,2-trifluoroethane (a freon)
g. 3,5,5-trimethylheptane
h. 1,3,5-trifluoropentane
i. 2-iodo-2,4,4-trimethylpentane

3.15 Using complete structural formulas, draw all of the structural isomers for each of the following compounds:

a. C_3H_7I
b. C_4H_9F
c. $C_5H_{11}OH$ (alcohols only)

3.16 What is a homologous series? What is a homolog? Give examples of both.

3.17 Draw all the possible linear and cyclic isomers of the alkyl halides having the molecular formula $C_4H_6Cl_2$.

3.18 Draw all the possible isomers of cycloalkanes having the molecular formula C_6H_{12}.

3.19 Identify the class of compounds introduced in this chapter that has each of the following general formulas:

a. C_nH_{2n+2}
b. C_nH_{2n}

3.20 Label each of the following statements as true or false.

a. Branching in a carbon chain lowers the boiling point of a hydrocarbon.
b. Branching in a carbon chain lowers the octane rating.
c. As the length of the carbon chain increases, solubility in water decreases.
d. A straight-chain hydrocarbon is less dense than a branched-chain hydrocarbon.
e. An alkyl halide is less dense than an alkane.

3.21 Fill in the blanks:

a. Alkanes and cycloalkanes are ________ (polar/nonpolar), while alkyl halides tend to be ________ (polar/nonpolar).
b. The aromatic six-carbon unsaturated ring compound is called ________.
c. An aromatic compound is used in unleaded gasolines to ________ (raise/lower) the octane rating.
d. The reference compound with an octane rating of 100 is called ________.

3.22 Which of the following pairs of compounds are identical?

a. 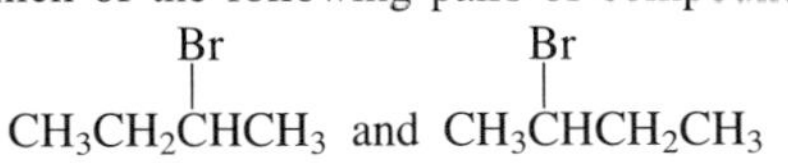 $CH_3CH_2CHCH_3$ (Br on C2) and $CH_3CHCH_2CH_3$ (Br on C2)
b. 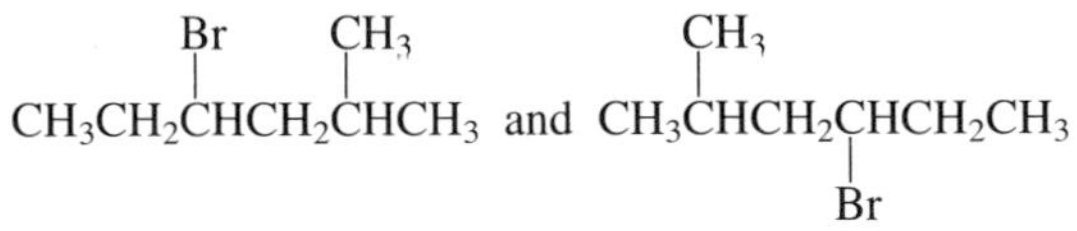 $CH_3CH_2CHCH_2CHCH_3$ (Br and CH_3 on the CH carbons) and $CH_3CHCH_2CHCH_2CH_3$ (CH_3 and Br on the CH carbons)
c. $CH_3CCH_2CH_3$ (Br, Br on C2) and $Br—CCH_2CH_3$ (Br and CH_3 on the C) 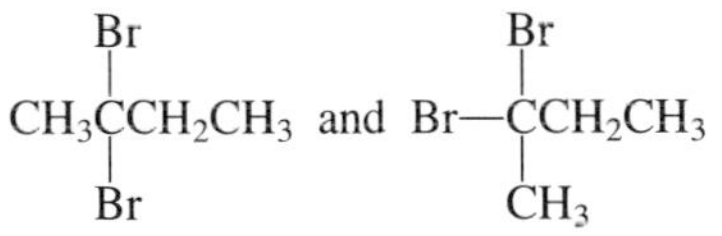
d. 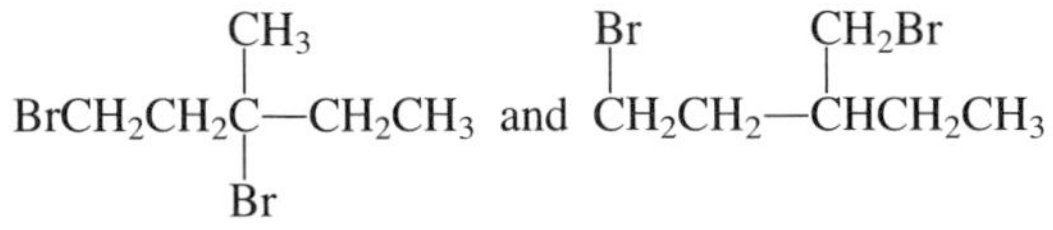 $BrCH_2CH_2C—CH_2CH_3$ (CH_3 and Br on the C) and $CH_2CH_2—CHCH_2CH_3$ (Br on the first CH_2, CH_2Br on the CH)
e. 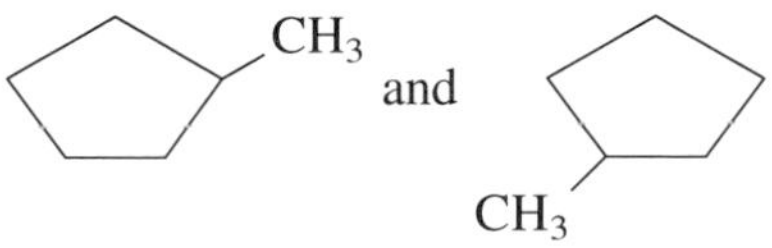 Cyclopentane ring with CH_3 and Cyclopentane ring with CH_3
f. 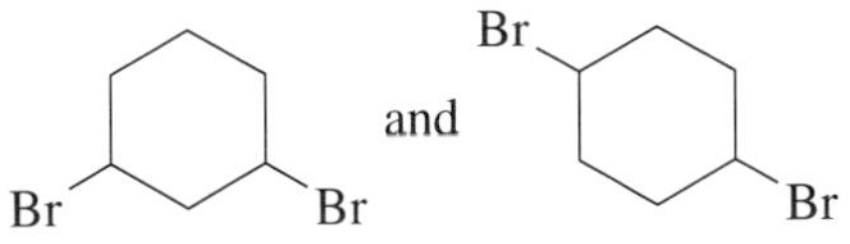 Cyclohexane ring with Br, Br and Cyclohexane ring with Br, Br

3.23 Give the possible arrangements of the $—CH_3$ and $—Br$ groups on a cyclohexane ring.

3.24 Which of the following structures are incorrect?

a. 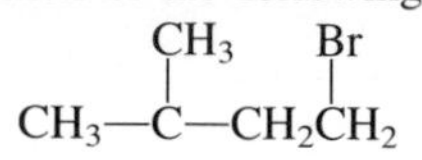
$CH_3 \quad Br$
$CH_3—C—CH_2CH_2$

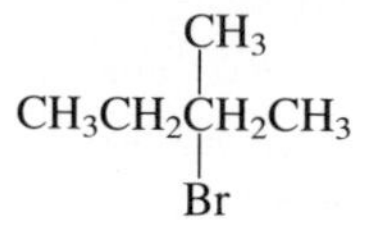

b. 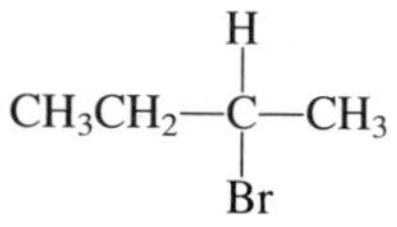
H
$CH_3CH_2—C—CH_3$
Br

c.
CH_3
$CH_3CH_2CH_2CH_3$
Br

d. 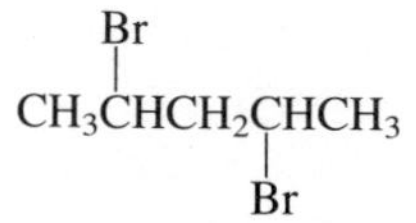
Br
$CH_3CHCH_2CHCH_3$
Br

3.25 Give an example for each of the compounds discussed in Section 3.8.

3.26 For the structures in Problem 3.14, state which compounds have:

a. primary carbons c. tertiary carbons
b. secondary carbons d. quaternary carbons

3.27 Using complete structural formulas, draw all of the possible isomers with the molecular formula $C_4H_8Br_2$.

3.28 Select the correct member in each of the following pairs of compounds.

a. Has the higher boiling point: 1-iodobutane or 1-chlorobutane
b. Is the more dense: 1-chlorobutane or water
c. Is the more dense: 1-iodopentane or 1-iodobutane
d. Is the more dense: 2,3-dibromobutane or water
e. Has the higher boiling point: 1-iodooctane or 1-bromopentane

3.29 What is meant by catalytic cracking? What is meant by catalytic reforming? Give examples of each.

Nomenclature

3.30 Give the common name and the I.U.P.A.C. name for each of the following:

a. 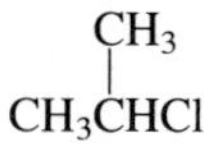
CH_3
CH_3CHCl

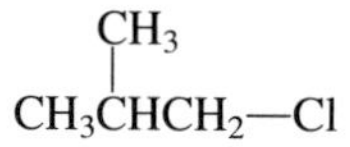

b. 
I
$CH_3CHCH_2CH_3$

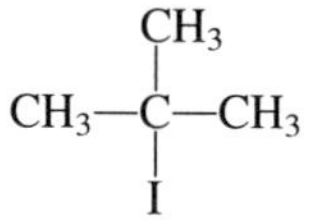

c. 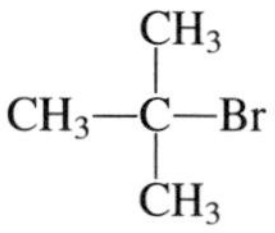
CH_3
$CH_3—C—Br$
CH_3

d.
CH_3
$CH_3CHCH_2—Cl$

e.
CH_3
$CH_3—C—CH_3$
I

3.31 List the steps used to name a compound, using I.U.P.A.C. nomenclature.

3.32 Draw structures of the following compounds. Are the names provided correct or incorrect? If they are incorrect, give the correct name.

a. 2,4-dimethylpentane
b. 1,3-dimethylpentane
c. 1,5-diiodocyclopentane
d. 1,4-diethylheptane
e. 1,6-dibromo-6-methyloctane

3.33 Which of the following are identical compounds? Which are structural isomers? Name each compound using the I.U.P.A.C. Nomenclature System.

a. 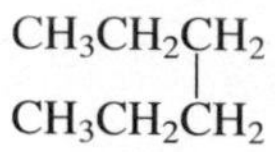
$CH_3CH_2CH_2$
$CH_3CH_2CH_2$

b. 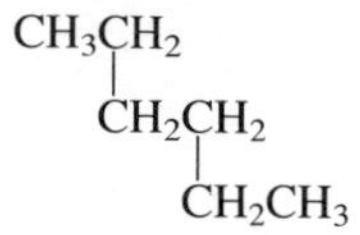
CH_3CH_2
CH_2CH_2
CH_2CH_3

c. 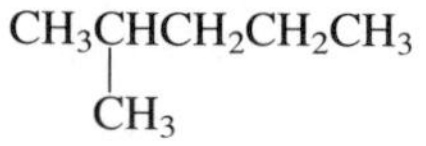
$CH_3CHCH_2CH_2CH_3$
CH_3

3.34 Which of the following are identical compounds? Which are structural isomers? Name each compound using the I.U.P.A.C. Nomenclature System.

a. $CH_3CH_2CH_2CH_2CH_2CH_2CH_3$

b. 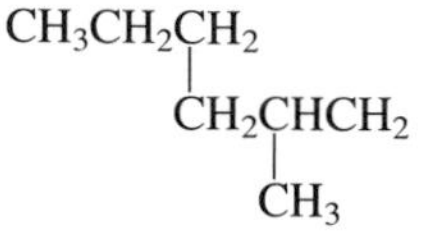
$CH_3CH_2CH_2$
CH_2CHCH_2
CH_3

c. 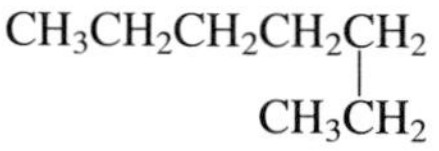
$CH_3CH_2CH_2CH_2CH_2$
CH_3CH_2

3.35 Are the following names correct or incorrect? If they are incorrect, give the correct name.

a. 1,3-dimethylpentane
b. 2-ethylpropane
c. 3-(*n*-butyl)butane
d. 3-ethyl-4-methyloctane
e. 1,5-dibromocyclohexane

Reactions

3.36 What is the difference between a synthesis and a degradation (see Section 4.5 for more information)?

3.37 Label each of the following as a synthesis or a degradation:

a. the combustion of hexane
b. the conversion of bromocyclohexane to chlorocyclohexane
c. the conversion of butane to octane

3.38 Complete each of the following reactions by supplying the missing reactant or product indicated by the question mark:

a. $CH_3CH_2CH_2CH_3 + O_2 \xrightarrow{\text{heat}}$? (Complete combustion)

b. $(CH_3)_3CH + Br_2 \xrightarrow{\text{light}}$? (Give all possible monosubstituted products)

c. 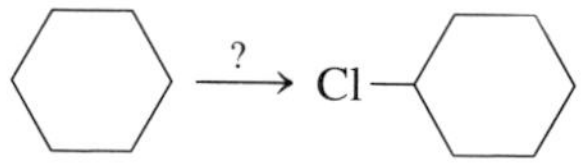 cyclohexane $\xrightarrow{?}$ Cl—cyclohexane

3.39 Give all the possible monochlorinated products for the following reaction:

CH_3
$CH_3CHCH_2CH_3 + Cl_2 \xrightarrow{\text{light}}$?

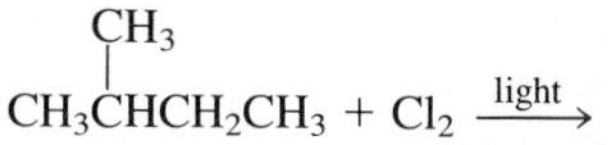

Name the products, using the I.U.P.A.C. Nomenclature System.

3.40 A mole of hydrocarbon formed eight moles of CO_2 and eight moles of H_2O upon combustion. Determine the molecular formula of the hydrocarbon and give the balanced combustion reaction.

3.41 Highly substituted alkyl fluorides, called perfluoroalkanes, are often used as artificial blood substitutes. These perfluoroalkanes have the ability to transport O_2 through the bloodstream as blood does. Some even have twice the O_2 transport capability and are used to treat gangrenous tissue. The structure of perfluorodecalin is shown below. How many moles of fluorine must be reacted with decalin to produce perfluorodecalin?

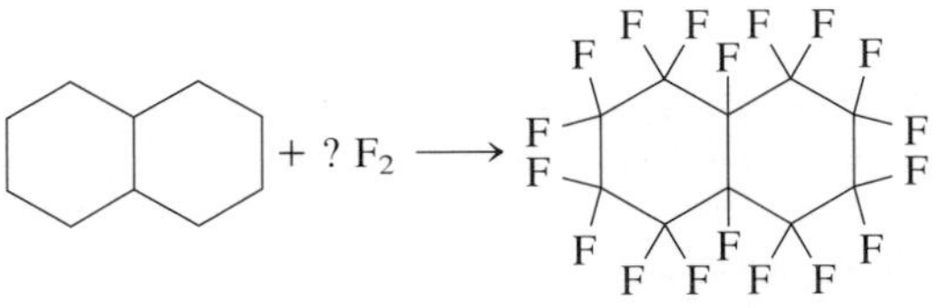

$$\text{Decalin} + ?\ F_2 \longrightarrow \text{Perfluorodecalin}$$

Decalin Perfluorodecalin

3.42 In reality, halogenation of alkanes proceeds to give mixtures of products that are monosubstituted, disubstituted, trisubstituted, and so on. Draw all of the possible products that would result from the bromination of ethane. Draw complete structural formulas and name each product, using the I.U.P.A.C. Nomenclature System.

3.43 Complete each of the following by supplying the missing product indicated by the question mark:

a. 1-bromoheptane $\xrightarrow[\text{acetone (solvent)}]{\text{NaI,}}$?

b. 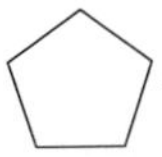 cyclopentane $\xrightarrow{Cl_2,\ \text{light}}$?

c. $CH_3CH_2CH_2CH_2CH_2CH_2CH_2{-}I \xrightarrow[\text{aqueous alcohol}]{\text{NaOH,}}$?

3.44 Why is the halide exchange reaction referred to as a substitution reaction?

Further Problems

3.45 Compare the physical properties (boiling points, melting points, solubilities, and densities) of the halogenated butanes. How do they compare or differ? (*Note:* You will have to use an outside reference to answer this problem.)

3.46 Can conformational isomers be separated from one another?

3.47 There are *two* distinct chair forms of chlorocyclohexane. Draw both of them. Which do you think would be the preferred conformation (more favored)?

3.48 Consider the structural isomers of molecular formula C_6H_{14}.
a. Which one gives two and only two monobromo derivatives when it reacts with Br_2 and light?
b. Which gives three and only three monobromo products?
c. Which gives four and only four monobromo products?
d. Draw the structures of all of the brominated products produced in parts a.–c. Name each of these, using the I.U.P.A.C. Nomenclature System.

3.49 a. Draw and name all of the isomeric products that one would obtain from the bromination of propane with Br_2/light. If halogenation were a completely random reaction and had an equal probability of occurring at any of the C—H bonds in a molecule, what percentage of each of these monobromo products would be expected?
b. Provide the same answers, using 2-methylpropane as the starting material.

3.50 Draw and name all of the isomeric products that one would obtain from the reaction of 2-methylbutane with Cl_2/light. If halogenation were a completely random reaction and had an equal probability of occurring at any of the C—H bonds in a molecule, what percentage of each of these monochloro products would be expected?

3.51 Saturated hydrocarbons are relatively unreactive. They do not, for example, react with strong acids or strong bases. Also, under normal conditions they do not react with common oxidizing or reducing agents. How do you account for these observed properties?

3.52 Considering only rotation about the carbon-carbon bond shown in color below, draw the most stable conformation for 2,3-dimethylbutane.

$$(CH_3)_2CH{-}CH(CH_3)_2$$

2,3-Dimethylbutane

3.53 a. Define oxidation and reduction.
b. Is the combustion of methane (shown below) an oxidation or a reduction? Explain your answer with reference to your definitions in part a.

$$CH_4 + 2O_2 \longrightarrow CO_2 + 2H_2O$$

3.54 Circle and label as many functional groups as you can in the following molecules.
a. Acetylsalicylic acid (aspirin; medicinal used as a pain reliever—analgesic—and to reduce fevers—antipyretic):

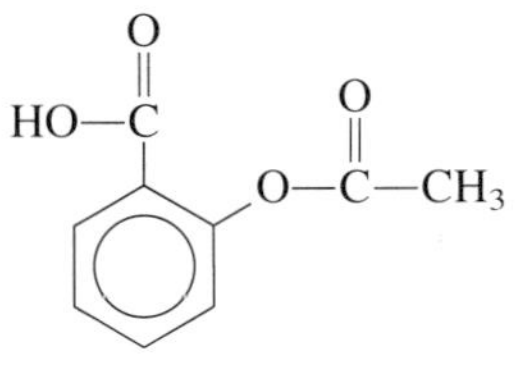

b. Cortisone (a steroid used in the treatment of arthritis):

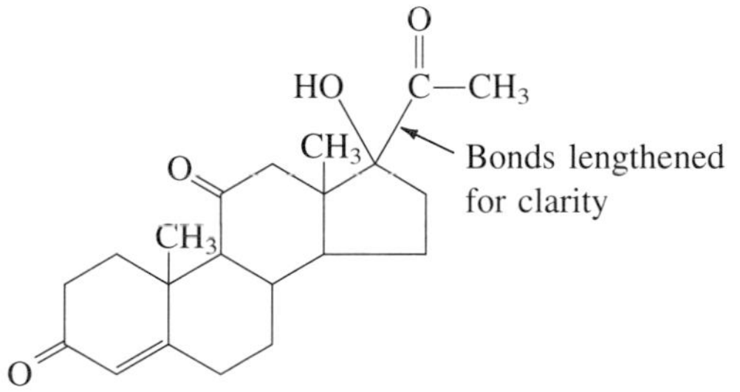

Bonds lengthened for clarity

c. Vitamin E:

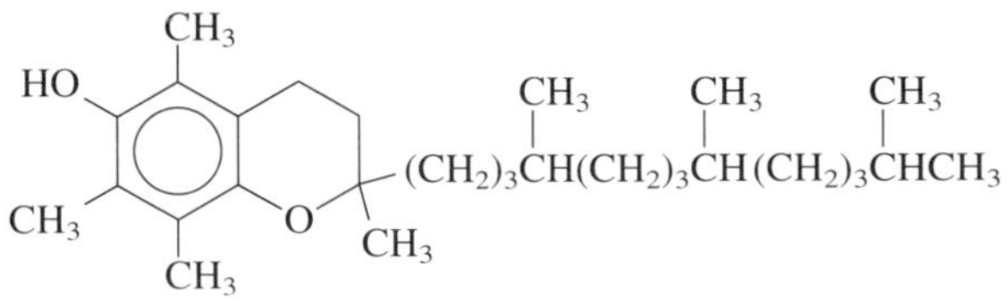

3.55 If 7.45 g of bromocyclopentane are produced through the bromination of cyclopentane with Br_2/heat, how many moles of cyclopentane were reacted? How many grams of cyclopentane were consumed?

3.56 Refer to a full-year organic chemistry textbook, and give the common name for each of the following:

a. 2-chlorobutane
b. 2-chloro-2-methylpropane
c. 2,2-dimethylpropane
d. 1-bromoethane
e. 1-iodo-3-methylbutane

3.57 Condensed structural drawings of several molecules that are relevant to the health professions are provided below. Draw out the complete structural formulas. That is, do not use any shorthand ring notations in your drawings, but rather insert all of the atoms comprising the molecule. Highlight the alkane portion of each molecule.

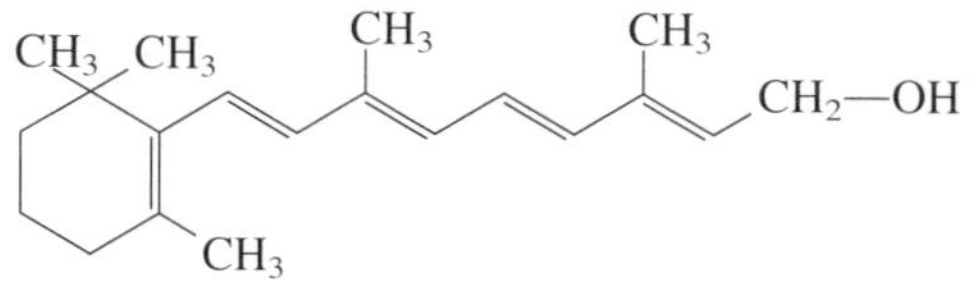

CH3 CH3 CH3 CH3 CH2—OH CH3

Vitamin A

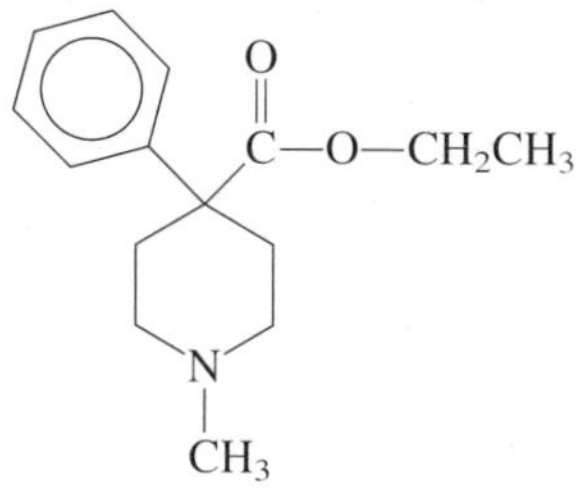

O C—O—CH2CH3 N CH3

Demarol

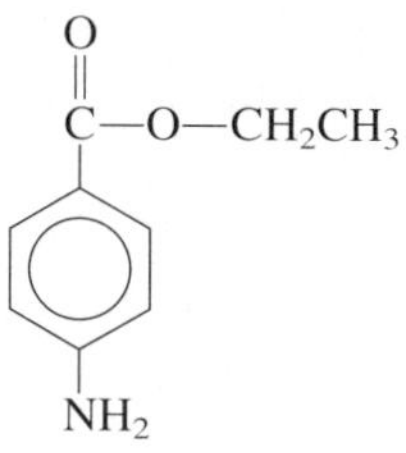

O C—O—CH2CH3 NH2

Benzocaine

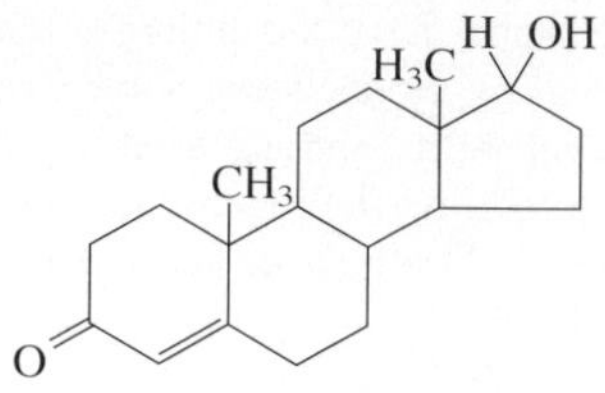

H OH H3C CH3 O

Testosterone

3.58 Dichlorination of propane using Cl_2 and heat produced four isomeric products, each with the formula $C_3H_6Cl_2$. These four products were labeled A, B, C, and D. Further chlorination of A–D yielded one or more trichloropropanes, $C_3H_5Cl_3$. On the basis of the following information, deduce the structures of A, B, C, and D:

A and B each gave three trichloro products, C gave one trichloro product, and D gave two trichloro products. One of the trichloro products from A was identical to the trichloro product from C. (*Hint:* Draw the four possible trichloropropane isomers before beginning the problem.)

C H A P T E R 4

The Unsaturated Hydrocarbons: Alkenes, Alkynes, and Aromatics

OBJECTIVES

- Write structures and name simple alkenes, alkynes, aromatic hydrocarbons, and simple aryl halides.
- Be familiar with the structural features associated with the carbon-carbon double and single bonds and the benzene ring.
- Predict, write structures, and name simple structural and geometric isomers for alkenes.
- Be familiar with the physical properties of the alkenes, alkynes, and aromatic hydrocarbons.
- Write equations predicting the products of the simple addition reactions of alkenes: hydrogenation, hydration, and hydrohalogenation.
- Discuss the addition mechanism for alkenes particularly as it pertains to the hydration reaction.
- Discuss catalysis and the use of catalysts in chemistry and biochemistry.
- Be familiar with important classes of unsaturated hydrocarbons, aromatic hydrocarbons, and polyhalogenated hydrocarbons; provide examples of each of these classes of compounds and discuss their uses.

OUTLINE

Chapter 9

The families of hydrocarbons were introduced in Chapter 3. We shall continue that discussion here and begin the chapter with an introduction to the alkenes and alkynes followed by an introduction to aromatic compounds and the aryl halides (aromatic halides). As will be seen, aromatic compounds and halides are quite important in our day to day lives; several implications regarding food production, health, and disease will be discussed.

4.1 ALKENES AND ALKYNES: STRUCTURE AND PROPERTIES

In Chapter 3 we concentrated on the *saturated hydrocarbons,* compounds that are joined only by carbon-carbon single bonds. In this chapter we introduce the *unsaturated hydrocarbons:* alkenes, alkynes, and aromatic compounds. These families are termed unsaturated because they contain at least one carbon-carbon double bond or carbon-carbon triple bond. As a result of multiple bonding, these compounds contain fewer hydrogens than the corresponding alkanes with the same number of carbon atoms. They are unsaturated in the sense that they do not contain as many hydrogens as their carbon skeleton will allow. We shall also see that the extent of saturation of compounds is important to biological processes as we discuss the family of lipids (fats, oils, and waxes).

Structure of alkenes and alkynes

The following formulas demonstrate the structures of a simple alkane, alkene, and alkyne:

Structural formulas:

```
   H  H          H       H
   |  |           \     /
H—C—C—H            C=C          H—C≡C—H
   |  |           /     \
   H  H          H       H
```

Ethane (ethane)	Ethene (ethylene)	Ethyne (acetylene)

Molecular formulas:

C_2H_6 C_2H_4 C_2H_2

Condensed structural formulas:

CH_3CH_3 $H_2C{=}CH_2$ $HC{\equiv}CH$

Note that each of these compounds contains the same number of carbon atoms but differs in the number of hydrogen atoms. This is true of all alkanes, alkenes, and alkynes that contain the same number of carbon atoms. Alkynes contain two fewer hydrogens than the corresponding alkene, and alkenes contain two fewer hydrogens than the corresponding alkane. In Table 4.1 we see some general formulas for all three families.

Each of these families represents a *homologous series,* a series in which each member differs from the members directly preceding or following it by a constant unit (usually a CH_2 group). The members of such a series are called *homologs*.

The major structural difference among these three families is apparent from the structural formulas presented above. Alkanes contain only single bonds; alkenes have a double bond; and alkynes contain a triple bond. Simply stated, it is the difference in bond order that accounts for differences in the molecular formulas. The difference in bond order is

TABLE 4.1 General Formulas for Alkanes, Alkenes, and Alkynes

	C_nH_{2n+2} Alkanes	C_nH_{2n} Alkenes	C_nH_{2n-2} Alkynes
$n = 1$	C_1H_4	—	—
$n = 2$	C_2H_6	C_2H_4	C_2H_2
$n = 3$	C_3H_8	C_3H_6	C_3H_4
$n = 4$	C_4H_{10}	C_4H_8	C_4H_6

also principally responsible for the variation in chemical reactivity among these three families.

Three-dimensional drawings of several simple members of these families are provided in Figure 4.1. Note that every carbon atom in each of these families is involved in four bonds, while the hydrogens each have one bond. Although there are differences in the types of bonding (single, double, and triple), the total number of bonds (and therefore the number of electrons involved in bonding) remains constant.

Alkanes have carbon atoms that have a tetrahedral electronic geometry. That is, the four bonds to the central carbon in alkanes form a tetrahedron. In alkenes the carbon-to-carbon double bonds have a planar arrangement, while in alkynes the carbon-to-carbon triple bond is linear (see Figure 4.1).

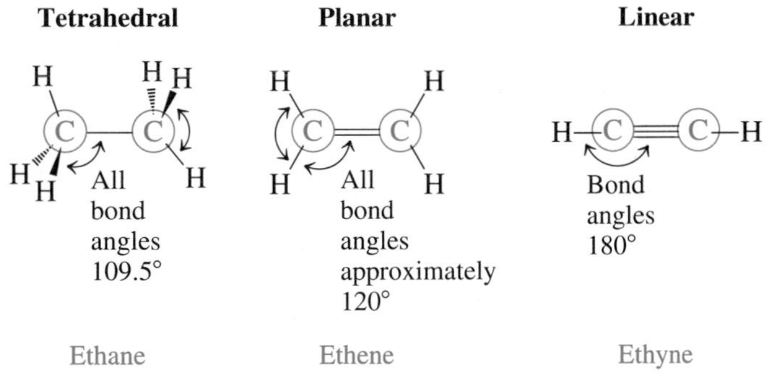

FIGURE 4.1
(a) Three-dimensional molecular model drawings of ethane, ethene, and ethyne. (b) Examples of typical long-chain hydrocarbons, drawn in three dimensions.

	Tetrahedral	**Planar**	**Linear**
	All bond angles 109.5°	All bond angles approximately 120°	Bond angles 180°
	Ethane	Ethene	Ethyne
Bond length:			
C—C	1.54 Å	1.34 Å	1.20 Å
C—H	1.10 Å	1.09 Å	1.06 Å
Bond strength:			
C—C	88 kcal/mole	163 kcal/mole	230 kcal/mole
C—H	98 kcal/mole	103 kcal/mole	103 kcal/mole

(a)

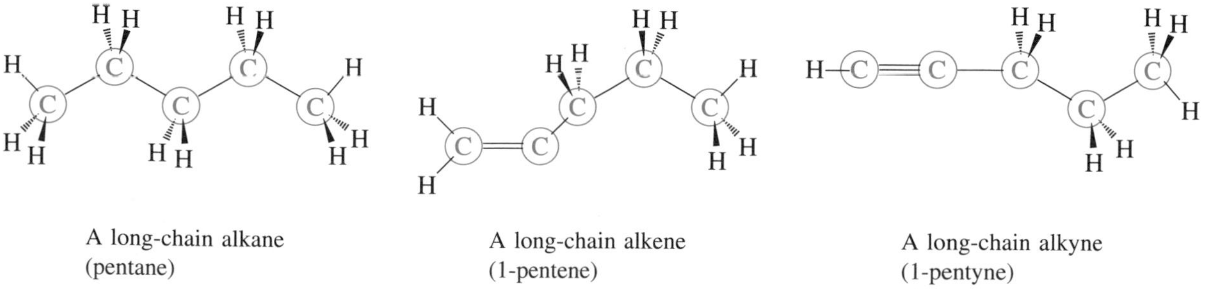

A long-chain alkane (pentane)

A long-chain alkene (1-pentene)

A long-chain alkyne (1-pentyne)

(b)

TABLE 4.2 Physical Properties of Selected Hydrocarbons

Name	Formula	b.p. (°C)	m.p. (°C)
Alkenes			
Ethene	C_2H_4	−102.4	−169.4
Propene	C_3H_6	−47.7	−185.0
1-Butene	C_4H_8	−6.5	−185.4
1-Pentene	C_5H_{10}	30.1	−138.0
1-Octene	C_8H_{16}	123	−101.7
Alkynes			
Ethyne	C_2H_2	−75.0	−82.0
Propyne	C_3H_4	−23.3	−101.5
1-Butyne	C_4H_6	8.0	−122.5
1-Pentyne	C_5H_8	40.0	−98.0
1-Octyne	C_8H_{14}	126.0	−70.0

Physical properties of alkenes and alkynes

Hydrocarbons are nonpolar because they are composed of a nonpolar arrangement of carbon-hydrogen bonds. It follows that they are not water-soluble but are readily soluble in nonpolar solvents such as other hydrocarbons, as well as in many low-polarity organic solvents such as ether or chloroform. This unique solubility is a consequence of the general rule ''like dissolves like.'' Since hydrocarbons are nonpolar in nature, virtually all of the hydrocarbons are less dense than water and have relatively low boiling points (see Table 4.2).

The melting points and boiling points of some alkenes and alkynes are presented in Table 4.2. As is the case with most tables of data, it is not necessary to memorize all of the information provided; the intent is rather to provide a general sense of the properties of the family being discussed. For example, at room temperature (and atmospheric pressure), ethene is a gas and pentene is a liquid. All of these properties vary in some predictable way depending on the carbon chain length and multiple bonds.

4.2 ALKENES AND ALKYNES: NOMENCLATURE

The nomenclature of alkenes and alkynes is analogous to that of the alkanes, with the following exceptions:

Alkenes

1. The parent name is derived from the longest continuous carbon chain containing the double bond. For alkenes, the *-ane* ending of the alkane is replaced with the *-ene* ending of an alkene. For example:

$CH_3—CH_3$ $CH_3—CH_2—CH_3$

Eth*ane*   Prop*ane*

versus

$CH_2{=}CH_2$ $CH_2{=}CH—CH_3$

Eth*ene* (ethylene)   Prop*ene* (propylene)

2. The chain is numbered from end to end to give the lowest numbers for the two carbons containing the double bond. For example:

Refer to Section 3.2 for a complete review of the nomenclature rules that pertain to the alkanes and related families of compounds.

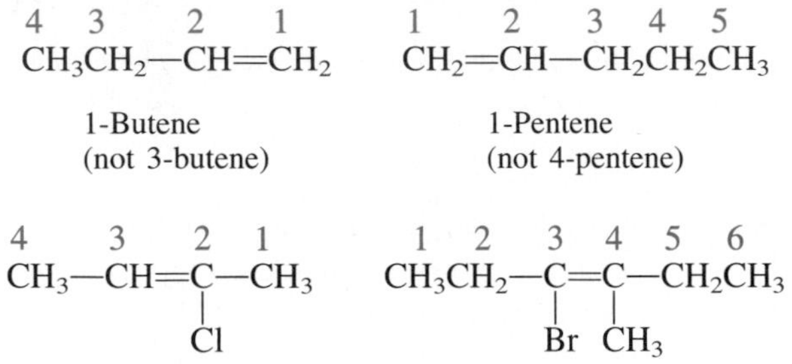

$\overset{4}{C}H_3\overset{3}{C}H_2—\overset{2}{C}H=\overset{1}{C}H_2$ — 1-Butene (not 3-butene)

$\overset{1}{C}H_2=\overset{2}{C}H—\overset{3}{C}H_2\overset{4}{C}H_2\overset{5}{C}H_3$ — 1-Pentene (not 4-pentene)

$\overset{4}{C}H_3—\overset{3}{C}H=\overset{2}{C}(Cl)—\overset{1}{C}H_3$ — 2-Chloro-2-butene

$\overset{1}{C}H_3\overset{2}{C}H_2—\overset{3}{C}(Br)=\overset{4}{C}(CH_3)—\overset{5}{C}H_2\overset{6}{C}H_3$ — 3-Bromo-4-methyl-3-hexene

Alkynes

Alkynes are named in the same way as the alkenes with one exception. The *-ane* ending of the corresponding alkane is replaced with the *-yne* ending of alkynes. The rules used in numbering the alkene chain are also used in alkyne nomenclature. For example:

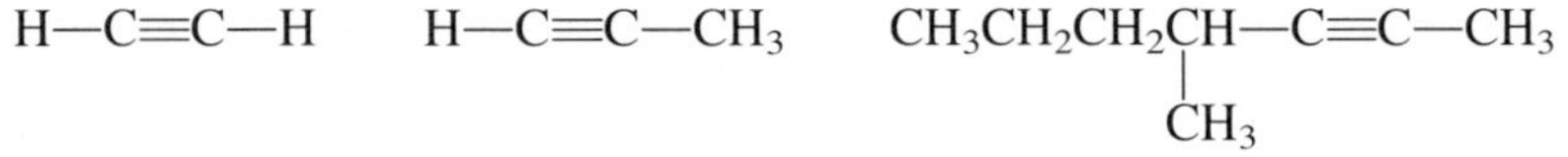

$H—C\equiv C—H$ — Ethyne (acetylene)

$H—C\equiv C—CH_3$ — Propyne (methylacetylene)

$CH_3CH_2CH_2CH(CH_3)—C\equiv C—CH_3$ — 4-Methyl-2-heptyne

Example 4.1

Naming alkenes and alkynes.

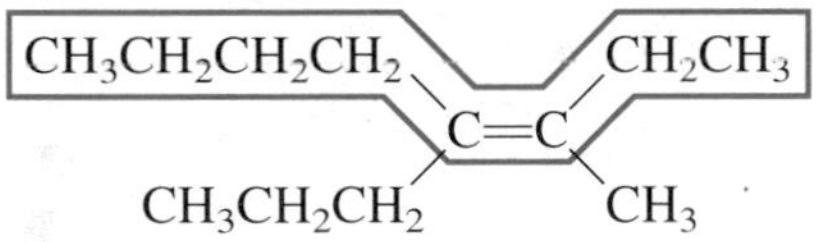

$(CH_3CH_2CH_2CH_2)(CH_3CH_2CH_2)C=C(CH_2CH_3)(CH_3)$

Longest chain containing the double bond: Octene

Position of double bond: 3-Octene (not 5-octene)

Substituents: 3-Methyl and 4-Propyl

Name: 3-Methyl-4-propyl-3-octene

A more complete name is actually *cis*-3-methyl-4-propyl-3-octene; see the following section for a discussion of the prefixes *cis-* and *trans-*.

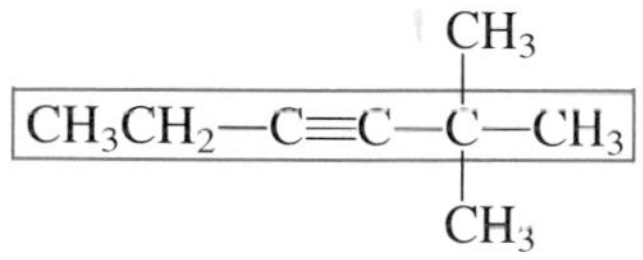

$CH_3CH_2—C\equiv C—C(CH_3)_2—CH_3$

Longest chain containing the triple bond: Hexyne

Position of triple bond: 3-Hexyne (must be!)

Substituents: 2,2-Dimethyl

Name: 2,2-Dimethyl-3-hexyne

Question 4.1 Draw each of the following, using complete structural formulas:

a. 1-bromo-3-hexyne
b. 2-butyne
c. dichloroacetylene
d. ethylene chloride
e. 9-iodo-1-nonyne
f. diethylacetylene

4.3 GEOMETRIC ISOMERS: A CONSEQUENCE OF UNSATURATION

The carbon-carbon double bond is rigid as a result of the shapes of the orbitals involved in its formation. Free rotation of the C—H moiety about the double bond is not possible under normal conditions (in contrast to the carbon-carbon single bond, which does exhibit free rotation). This rigidity of the carbon-carbon double bond in alkenes gives rise to another class of isomers: **geometric isomers.** Geometric isomers are also called *cis-* and *trans-* isomers. The *cis-* and *trans-* prefixes provide an easy method for naming and differentiating between the two possible isomers. Examples are provided here:

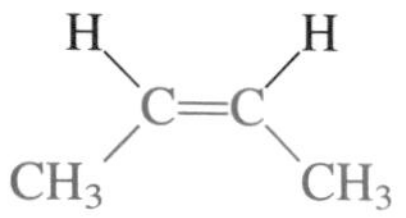

H, CH$_3$ C=C H, CH$_3$	CH$_3$, H C=C H, CH$_3$
I	**II**
cis-2-Butene	*trans*-2-Butene
(b.p. +4°)	(b.p. +1°)

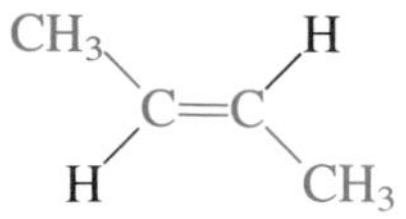

It is important to recognize that *cis-/trans-* isomers are different molecules with different physical and chemical properties. Their properties may be similar but not identical. The degree of similarity depends on the structural similarity of the two isomers.

The prefixes *cis-* and *trans-* refer to the placement of the substituents attached to the carbon-carbon double bond (in this example the H's and CH_3's). When identical groups are on the same side of the double bond (structure **I**), the prefix *cis-* is used; when identical groups are on opposite sides of the double bond (structure **II**) *trans-* is the appropriate prefix. When the groups are not identical, each is assigned a priority, and the *cis-* and *trans-* prefixes are then used to designate the relative orientation of these groups based on this priority. The following examples are illustrative:

H, CH$_3$ C=C CH$_2$CH$_2$CH$_2$CH$_3$, H

trans-2-Heptene

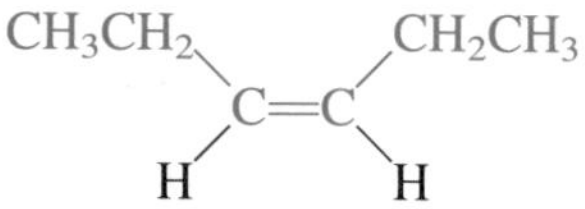

CH$_3$CH$_2$, H C=C CH$_2$CH$_3$, H

cis-3-Hexene

H$_3$C, Br C=C Br, H

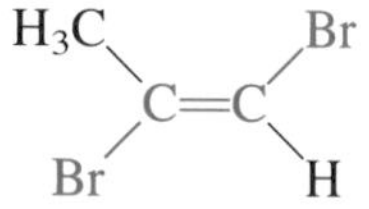

trans-1,2-Dibromopropene

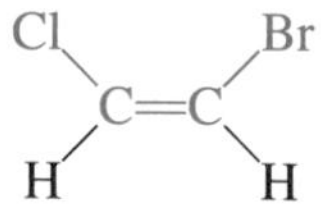

Cl, H C=C Br, H

cis-1-Bromo-2-chloroethene

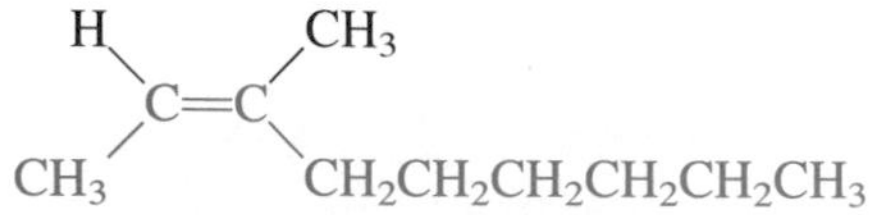

H, CH$_3$ C=C CH$_3$, CH$_2$CH$_2$CH$_2$CH$_2$CH$_2$CH$_3$

cis-3-Methyl-2-nonene

In the last example the two ends of the longest chain of carbon atoms in the molecule dictate which groups are *cis-* or *trans-* to each other. This convention has been defined by the I.U.P.A.C Nomenclature System. In this case the methyl and hexyl groups shown in color dictate the prefix to be used, *cis-* or *trans-* and not the two methyl groups.

Geometric isomers are also possible in certain cycloalkanes. Two substituents on the same side of the plane of the ring give the *cis-* isomer, and two substituents on opposite sides of the plane of the ring give the *trans-* isomer:

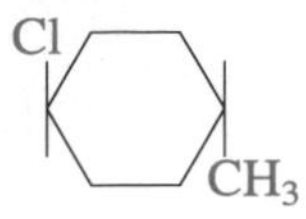

cis-1,2-Dibromocyclopentane
(Both groups on the same side
of the ring—*cis*.)

Cl
CH_3

trans-1-Chloro-4-methylcyclohexane
(Two groups on opposite sides
of the ring—*trans*.)

There is a newer, more systematic nomenclature convention that may also be used to differentiate between two geometric isomers. In this system the prefixes *Z*- and *E*- are used to distinguish between two geometric isomers. The *cis-/trans-* system has been with us for so long and is so widely used that the newer (*Z-/E-*) system is used primarily in examples in which *cis-/trans-* yields ambiguous names. For example, consider the following structure:

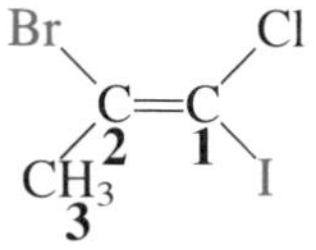

Br, Cl, C=C, 2, 1, CH_3, 3, I

E-2-Bromo-1-chloro-1-iodopropene

It would be impossible to name this compound *unambiguously* using the *cis-/trans-* system. In the *Z-/E-* system, however, this molecule is *E*-2-bromo-1-chloro-1-iodopropene. The prefix *E*- is used because Br— has a higher priority than CH_3— on carbon-2 and I— has a higher priority than Cl— on carbon-1 and because the two higher-priority groups (Br— and I—) are on opposite sides of the double bond. Similarly, *Z*-2-bromo-1-chloro-1-iodopropene is shown below:

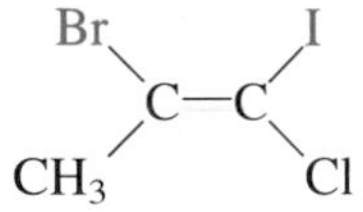

Br, I, C—C, CH_3, Cl

Z-2-Bromo-1-chloro-1-iodopropene

To use the *Z-/E-* system routinely requires one to master a series of rules that are used in setting priorities to the groups attached to the carbon-carbon double bond. Because the system is used only occasionally, we will not devote the time needed to master these rules but will leave further study in this area to the reader's discretion.

Example 4.2

Naming more complex *cis-* and *trans-* compounds

The two ends of the longest chain of carbon atoms in each of the following molecules are set off in color. Note that these two ends *must be part of the chain that also contains the carbon-carbon double bond*. These two ends are used in determining the appropriate prefix, *cis-* or *trans-*, to be used in naming each of the molecules.

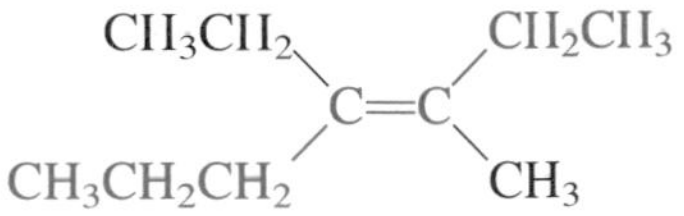

CH_3CH_2, CH_2CH_3, C=C, $CH_3CH_2CH_2$, CH_3

Parent chain: Heptene

Position of double bond: 3-

Configuration: *trans-*

Substituents: 3-Methyl, 4-ethyl

Name: *trans*-4-Ethyl-3-methyl-3-heptene

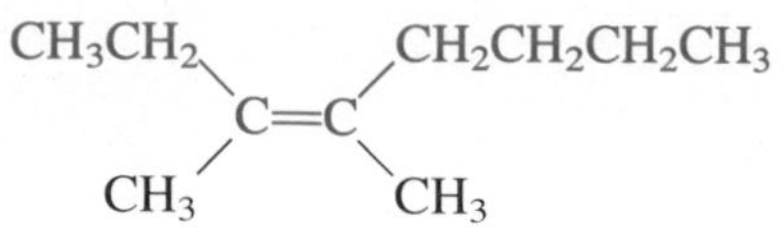

$$\begin{array}{ccc} CH_3CH_2 & & CH_2CH_2CH_2CH_3 \\ & C{=}C & \\ CH_3 & & CH_3 \end{array}$$

Parent chain: Octene

Position of double bond: 3-

Configuration: *cis-*

Substituents: 3,4-Dimethyl

Name: *cis*-3,4-Dimethyl-3-octene

Question 4.2 Draw the following, using complete structural formulas:

a. *cis*-3-octene
b. *trans*-1,2-dihydroxyethene
c. *trans*-2,3-dichloro-2-butene

Restricted rotation around double bonds is partially responsible for the conformation and hence the activity of many biological molecules that we will study later.

4.4 THE PRINCIPAL REACTIONS OF ALKENES

Reactions of alkenes involve the carbon-carbon double bond; the double bond, in effect, is the characteristic functional group of this chemical family. Alkenes, of course, may also contain C—H single bonds and other functional groups. Although reactions of each functional group in the molecule may be influenced by the other groups present, for the sake of simplicity we will assume that they are not. Given that assumption, we will, for the moment, accept that the carbon-carbon double bond will behave and react as a carbon-carbon double bond regardless of the other groups present in the molecule.

Hydrogenation

One important reaction of alkenes is hydrogenation. **Hydrogenation** is the addition of a molecule of hydrogen (H_2) to a carbon-carbon double bond to give an alkane. In this reaction, two new C—H single bonds result as the double bond is broken. Several examples follow:

$$CH_2{=}CH_2 + H_2 \xrightarrow[\text{heat, pressure}]{\text{Pt, Pd, or Ni,}} \underset{\text{H}}{CH_2}{-}\underset{\text{H}}{CH_2} \qquad 4.1$$

Ethene (ethylene) → Ethane

$$CH_3CH_2CH{=}CH_2 + H_2 \xrightarrow[\text{heat}]{\text{Pd and}} CH_3CH_2\underset{\text{H}}{CH}{-}\underset{\text{H}}{CH_2} \qquad 4.2$$

1-Butene (butylene) → Butane

$$CH_3CH{=}CHCH_3 + H_2 \xrightarrow[\text{heat}]{\text{Pd and}} CH_3\underset{\text{H}}{CH}{-}\underset{\text{H}}{CH}CH_3 \qquad 4.3$$

2-Butene (butylene) → Butane

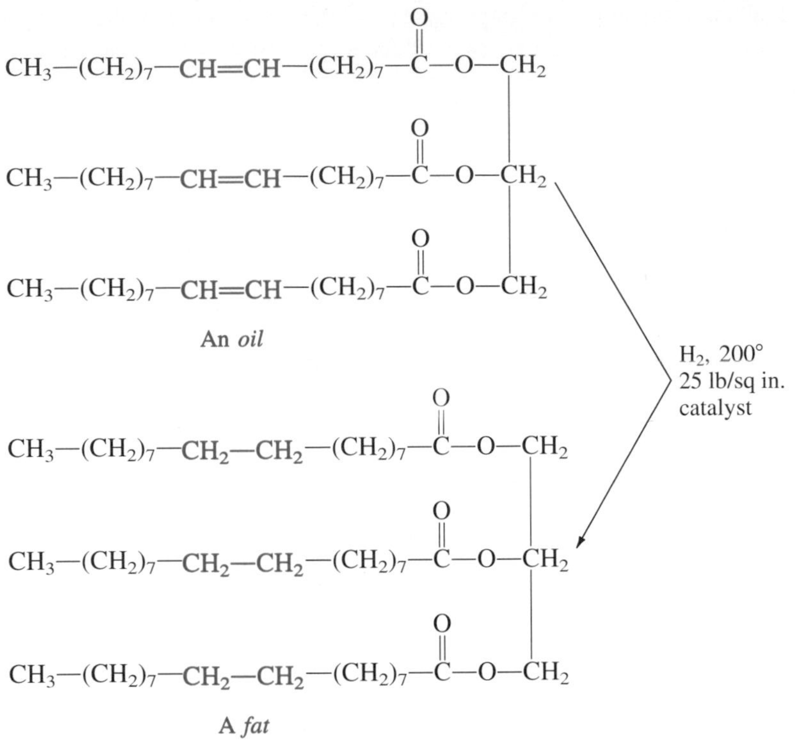

$$CH_3{-}(CH_2)_7{-}CH{=}CH{-}(CH_2)_7{-}\overset{O}{\overset{\|}{C}}{-}O{-}CH_2$$
$$CH_3{-}(CH_2)_7{-}CH{=}CH{-}(CH_2)_7{-}\overset{O}{\overset{\|}{C}}{-}O{-}CH$$
$$CH_3{-}(CH_2)_7{-}CH{=}CH{-}(CH_2)_7{-}\overset{O}{\overset{\|}{C}}{-}O{-}CH_2$$

An *oil*

H_2, 200°
25 lb/sq in.
catalyst

$$CH_3{-}(CH_2)_7{-}CH_2{-}CH_2{-}(CH_2)_7{-}\overset{O}{\overset{\|}{C}}{-}O{-}CH_2$$
$$CH_3{-}(CH_2)_7{-}CH_2{-}CH_2{-}(CH_2)_7{-}\overset{O}{\overset{\|}{C}}{-}O{-}CH$$
$$CH_3{-}(CH_2)_7{-}CH_2{-}CH_2{-}(CH_2)_7{-}\overset{O}{\overset{\|}{C}}{-}O{-}CH_2$$

A *fat*

FIGURE 4.2
The conversion of a typical oil to a fat involves hydrogenation. In this example, triolein (an oil) is converted to tristearin (a fat).

The platinum, palladium, or nickel are present as catalysts to ensure that the reaction occurs in a reasonable time (that is, at a reasonable rate). The conditions employed depend upon the alkene being hydrogenated. Although varying amounts of heat and/or pressure are often required, a metal catalyst is always required. The role of the catalyst in effecting this transformation is not well understood, but it is believed to involve a surface interaction. The metal apparently supplies a surface on which the reaction occurs. The metal itself is not consumed or altered in any way. Recall that a catalyst itself undergoes no change in the course of a chemical reaction.

Hydrogenation is used in the food industry to produce margarine, which is a mixture of hydrogenated vegetable oils. Vegetable oils contain many alkene double bonds. The hydrogenation of these double bonds to single bonds is, in part, responsible for the increase in melting point of the oil and eventually results in a solid fat (for example, Crisco, Spry). Many of these oils are, through further processing, converted to margarines (for example, corn oil and sunflower oil margarines) (Figure 4.2).

Section 9.2

Refer to Chapter 9 for a discussion of certain health issues as they relate to the degree of "saturation and unsaturation" of food products.

The addition of hydrogen to a carbon-carbon double bond to give an alkane is an example of an **addition reaction,** a reaction in which two molecules combine to form a new molecule.

Hydration

The carbon-carbon double bond is an electron-rich region that will attract any electron-deficient species such as a proton (H^+). When an alkene is reacted with water containing a trace of acid, an —OH group becomes attached to one end of the carbon-carbon double bond and an H— atom to the other end. This reaction is termed **hydration;** the result is the addition of water:

$$CH_2{=}CH_2 + H_2O \xrightarrow[\text{trace}]{H^+} \underset{H}{\underset{|}{CH_2}}{-}\underset{OH}{\underset{|}{CH_2}} \qquad 4.4$$

With alkenes in which the groups attached to the two carbons of the double bond are different (unsymmetrical alkenes), two products are possible. For example:

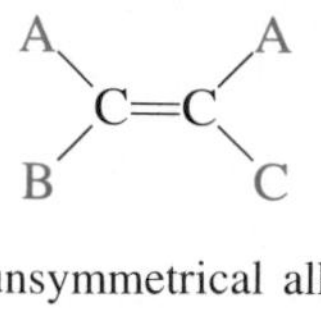

$$\begin{array}{ccc} A & & A \\ & C{=}C & \\ B & & C \end{array}$$

An unsymmetrical alkene

$A = A \qquad B \neq C$

Note that the left side is different from the right side:

$$\begin{array}{c} A \\ \quad C \\ B \end{array} \quad \text{versus} \quad \begin{array}{c} A \\ C \quad \\ C \end{array}$$

For example:

$$\overset{3}{CH_3}{-}\overset{2}{CH}{=}\overset{1}{CH_2} + H_2O\ (H^+) \longrightarrow \overset{3}{CH_3}{-}\underset{\displaystyle OH}{\overset{2}{CH}}{-}\overset{1}{CH_3} \quad \text{(4.5a)}$$

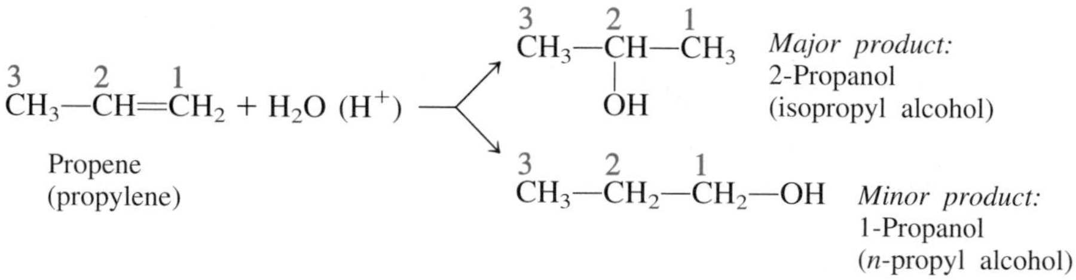

Major product: 2-Propanol (isopropyl alcohol)

$$\longrightarrow \overset{3}{CH_3}{-}\overset{2}{CH_2}{-}\overset{1}{CH_2}{-}OH \quad \text{(4.5b)}$$

Minor product: 1-Propanol (*n*-propyl alcohol)

Propene (propylene)

As is shown above, hydration of an unsymmetrical alkene favors one product over the other. We find that the carbon of the carbon-carbon double bond having the larger number of hydrogen atoms receives the hydrogen atom being added to the double bond. The remaining carbon forms a bond with the —OH. The rule is "the rich get richer"—the carbon with the most hydrogens gets the new one as well. This rule is known as **Markovnikov's Rule.** In the example above, **C1** has two C—H bonds originally and **C2** has only one. The product, 2-propanol, results from the new C—H bond forming on **C1** and the new C—OH bond on **C2.**

Question 4.3 Predict the major product in each of the following reactions:

a. 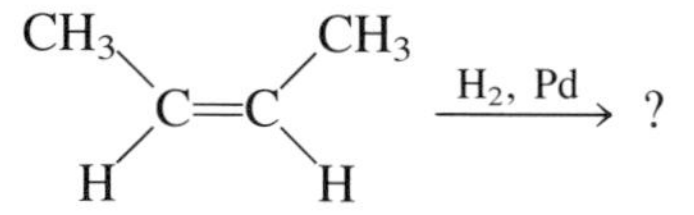 $(CH_3)(H)C{=}C(CH_3)(H)$ (both CH_3 on the same side) $\xrightarrow{H_2,\ Pd}$?

b. 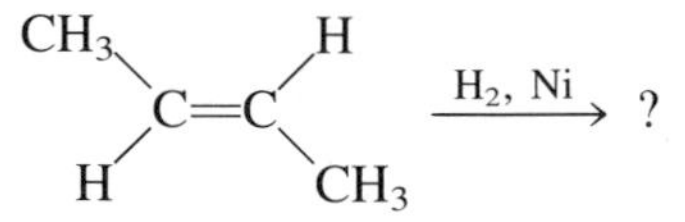 $(CH_3)(H)C{=}C(H)(CH_3)$ (CH_3 groups on opposite sides) $\xrightarrow{H_2,\ Ni}$?

c. $CH_3CH_2CH{=}CH_2 \xrightarrow{H_2O,\ H^+}$?

d. 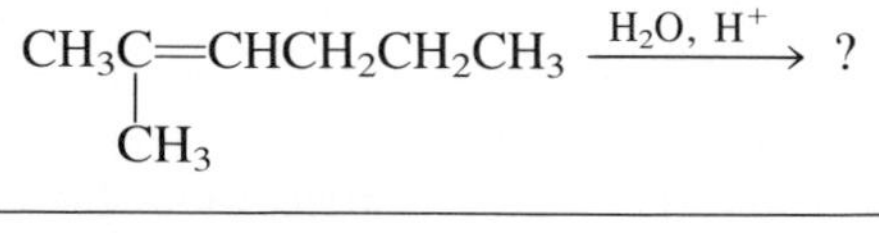 $CH_3\underset{\displaystyle CH_3}{C}{=}CHCH_2CH_2CH_3 \xrightarrow{H_2O,\ H^+}$?

Hydrohalogenation

In the previous section we saw that water could be added to an alkene to produce an alcohol (hydration of an alkene). In a similar fashion a hydrogen halide (H—Br, H—Cl, or H—I) can be added to an alkene to give an alkyl halide, a reaction known as **hydrohalogenation:**

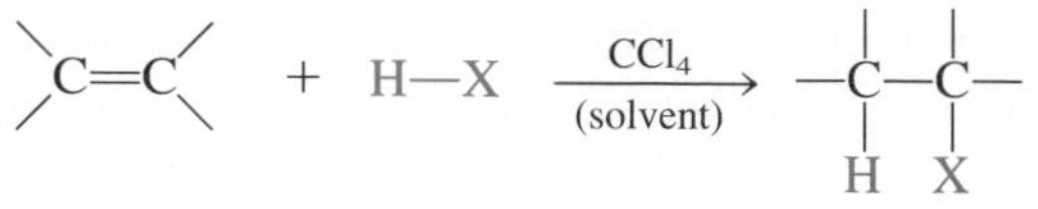

$$\mathrm{>C{=}C<} + \mathrm{H{-}X} \xrightarrow[\text{(solvent)}]{CCl_4} \mathrm{-\overset{|}{\underset{H}{C}}-\overset{|}{\underset{X}{C}}-} \qquad 4.6$$

Alkene — Hydrogen halide — Alkyl halide

The reaction follows the same pattern that was observed for the hydration of alkenes. That is, if H—X is added to an unsymmetrical alkene, "the rich get richer"; Markovnikov's Rule holds for product distribution. The mechanism (discussed in the next section) is the same as that of the hydration reaction. The reaction, though limited because it may be accompanied by a rearrangement of the carbon skeleton of the molecule undergoing reaction, does provide another easy way of preparing alkyl halides. See the examples below.

Section 3.9

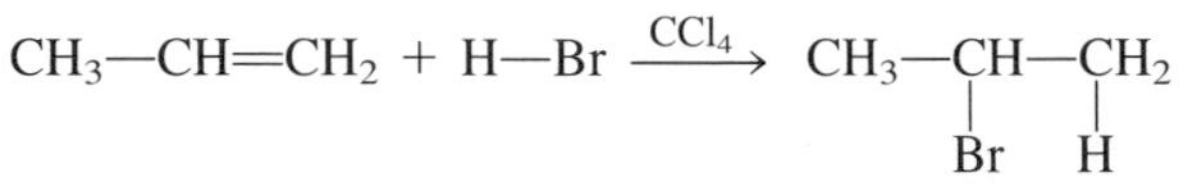

$$CH_3{-}CH{=}CH_2 + H{-}Br \xrightarrow{CCl_4} CH_3{-}\underset{Br}{\underset{|}{CH}}{-}\underset{H}{\underset{|}{CH_2}} \qquad 4.7$$

Propene (propylene) — 2-Bromopropane (isopropyl bromide)

Major products shown

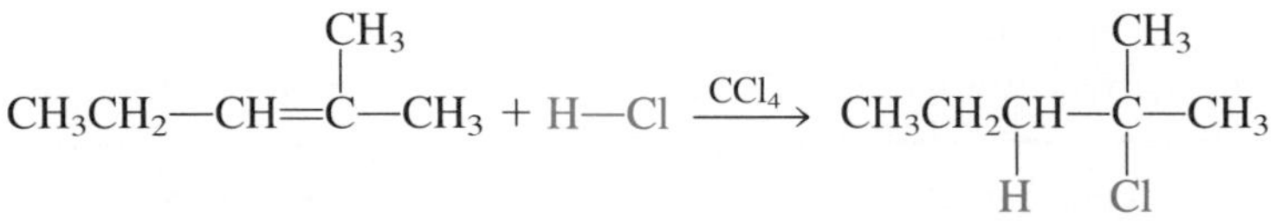

$$CH_3CH_2{-}CH{=}\overset{CH_3}{\overset{|}{C}}{-}CH_3 + H{-}Cl \xrightarrow{CCl_4} CH_3CH_2\underset{H}{\underset{|}{CH}}{-}\overset{CH_3}{\overset{|}{\underset{Cl}{\underset{|}{C}}}}{-}CH_3 \qquad 4.8$$

2-Methyl-2-pentene — 2-Chloro-2-methylpentane

4.5 THE MECHANISM OF HYDRATION

Reaction mechanisms

A **reaction mechanism** is the pictorial, step-by-step explanation of the process by which reactants are converted to products in a chemical reaction. A reaction mechanism is based on all of the experimentally known facts about the reaction.

Let us look at the mechanism of hydration of an alkene in some detail. This process is proposed to occur in three steps.

Step 1. The double bond and proton combine to produce a positively charged, electron-deficient carbon atom intermediate known as a *carbocation*. Electrons flow from the double bond to form a new single bond:

A carbocation is an electron-deficient carbon atom that bears a positive charge and has only six electrons (three bonding pairs) about it.

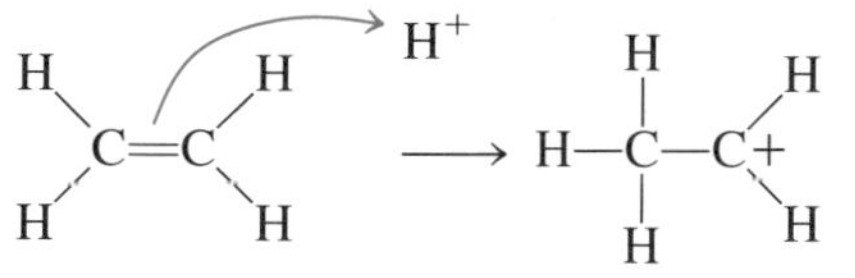

$$\mathrm{H_2C{=}CH_2} \; (\curvearrowright H^+) \longrightarrow \mathrm{H{-}\overset{H}{\overset{|}{\underset{H}{\underset{|}{C}}}}{-}C^{+}H_2}$$

Alkene — A carbocation

The curved arrow shows that two electrons from the double bond flow toward the proton to form the new C—H bond producing the carbocation. These curved arrows are used by organic chemists to keep track of electrons; they are bookkeeping devices. Additional examples of the use of curved arrows will be provided in the chapters ahead.

Step 2. The carbocation reacts with water to form an oxonium ion in which the positive charge is now on the oxygen. Note the conservation of charge throughout these steps; a positively charged molecule and a neutral molecule react to give a new molecule that still bears a positive charge. An oxonium ion is an electron-deficient oxygen atom that bears a positive charge and has three bonding pairs of electrons and one nonbonding pair of electrons about it:

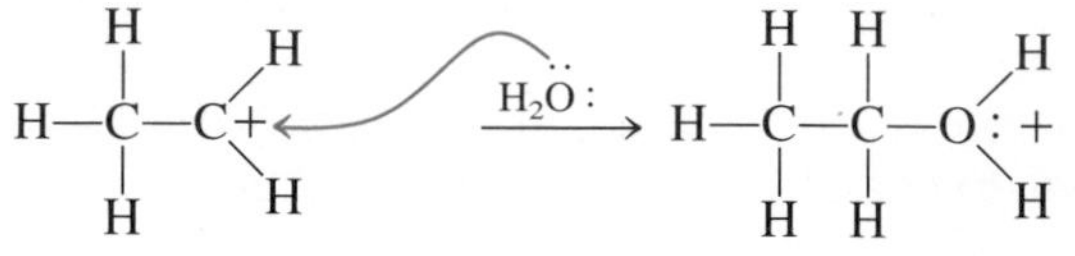

$$CH_3-CH_2^+ \xrightarrow{H_2\ddot{O}:} CH_3-CH_2-\ddot{O}H_2^+$$

A carbocation An oxonium ion

Step 3. All of these charged intermediates are quite reactive. The oxonium ion rapidly reacts to lose a proton to the solvent, water, to yield the neutral alcohol:

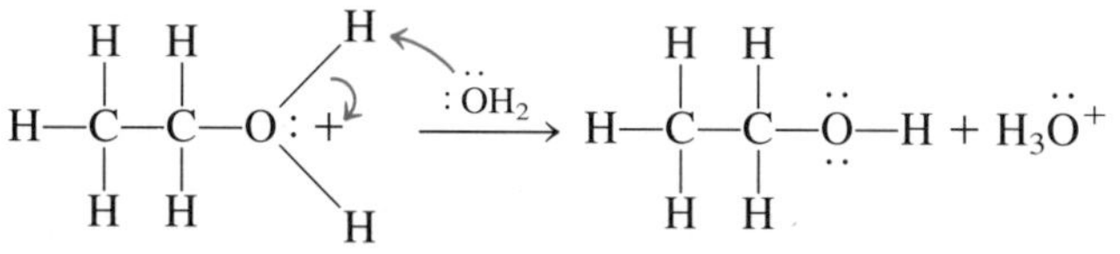

$$CH_3-CH_2-\ddot{O}H_2^+ \xrightarrow{:\ddot{O}H_2} CH_3-CH_2-\ddot{\underset{\cdot\cdot}{O}}-H + H_3\ddot{O}^+$$

Oxonium ion Alcohol

This is an example of an *acid-catalyzed* reaction. The acid is a true catalyst because it is involved in the reaction and speeds up the reaction but is not consumed during the reaction. Hydration, specifically, and acid catalysis in general, are important synthetic and biochemical processes. A variety of biochemical pathways, for example, portions of the mechanism associated with fatty acid degradation, involve these two reactions.

Section 15.2

Differentiating between organic synthesis and degradation

The term synthesis was used throughout the previous section, as it will be throughout this text. It is perhaps wise to spend a few moments discussing the question: What are organic syntheses and, in contrast, what are organic degradations?

A **synthesis** is concerned with the conversion of one molecule into another. This may involve the attachment of additional atoms to the original molecule, the conversion of one functional group into another (for example, $CH_3—H$ to $CH_3—Cl$), or the rearrangement of the molecule into a different form.

At the opposite extreme is a **degradation** reaction. A degradation reaction involves the breakdown of a larger molecule into several smaller molecules. Examples of synthesis and degradation reactions are shown below.

Synthesis:

$$H_2C{=}CH_2 + H_2O \xrightarrow{H^+} \underset{H}{H_2C}-\underset{OH}{CH_2}$$

Degradation:

$$H_2C{=}CH_2 + 3O_2 \xrightarrow[\text{(combustion)}]{\text{heat}} 2CO_2 + 2H_2O$$

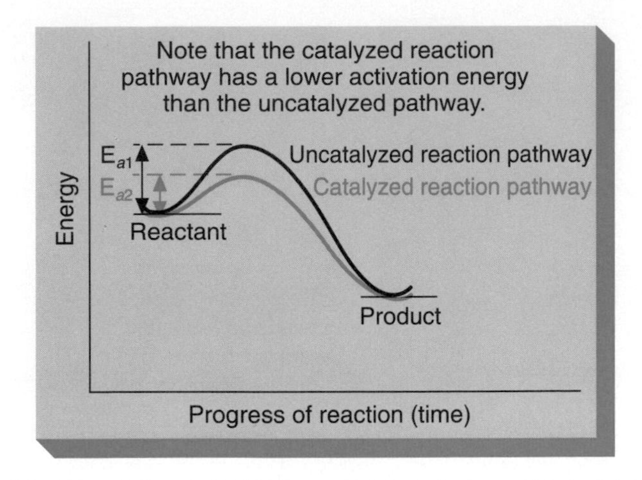

FIGURE 4.3
Catalysis: A generalized one-step reaction (reactant going to product) is shown (a) without a catalyst present and (b) with a catalyst present. The catalyst lowers the energy of activation, facilitating the conversion of reactant to product.

4.6 CATALYSIS

We know that **catalysts** are substances that are used to speed up reactions. Many reactions are amenable to catalysis, but many others are not. Because a catalyst is involved in the intimate transformations that are occurring (bond breaking, bond making) the catalyst is often specific to the particular reaction being attempted. For example, in the hydrogenation reaction discussed in Section 4.4 a metal catalyst is necessary. A metal catalyst, on the other hand, has no effect on the hydration reaction discussed in the same section.

Catalysts operate by lowering the activation energy of the reaction. This is illustrated for a generalized reaction in Figure 4.3. The activation energy is reduced because the catalyst is transiently involved in the reaction, thereby providing an alternative, lower-energy route to the desired product. It is important to note that though catalysts are involved in chemical transformations, they are not consumed in the course of these reactions. Indeed, they emerge from the reactions unchanged and are often recycled and used over and over again to minimize cost. This is particularly important in the industrial sector. Though catalysts speed up reactions, they do not have any effect on the amount of product that results. It is simply the speed with which the product forms that changes.

The method or mechanism through which catalysts accomplish their goal of lowering activation energy is often not completely understood. Many metal catalysts (Ni, Pt, Pd) appear to operate through a surface interaction. That is, the metal provides a surface on which the reaction can occur more easily. This is shown for hydrogenation in Figure 4.4.

In the biochemical world the success of most reactions depends on catalysts called *enzymes*. As we will see, enzymes are proteins, large macromolecules—molecules of

FIGURE 4.4
Hydrogenation of an alkene in the presence of a metal catalyst, a surface interaction.

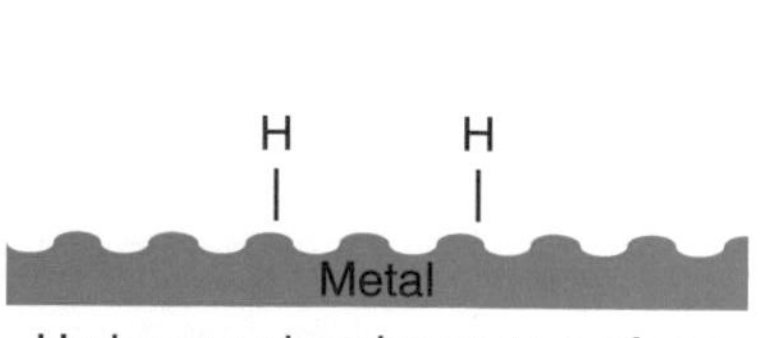

Hydrogen absorbs on to surface of the metal

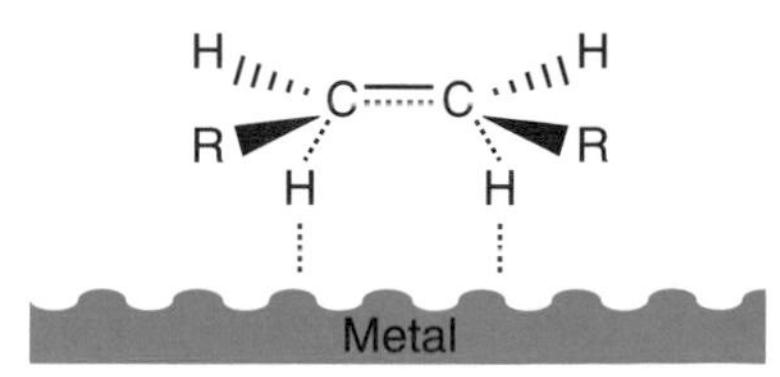

Alkene and metal surface interaction; hydrogen being transferred to alkene; bond making and breaking shown in color

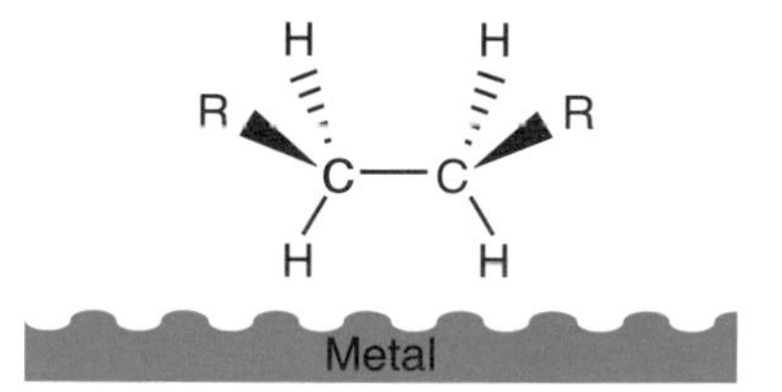

Alkene product forms and metal catalyst regenerated

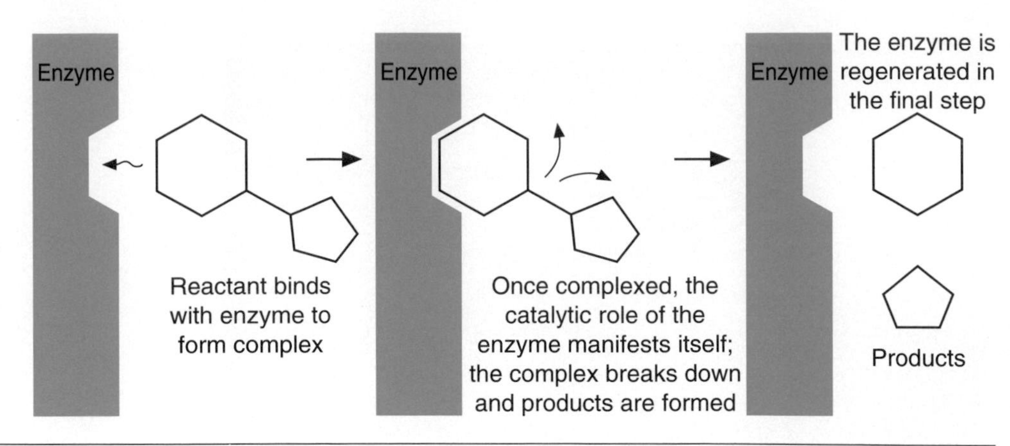

FIGURE 4.5
The "lock and key" analogy for enzyme catalysis: One of the reactants forms a complex with the enzyme; the reactant complex undergoes reaction and products are released; and the enzyme (catalyst) is regenerated in an unchanged form.

high molecular weight. Enzymes catalyze reactions by forming a complex with the one or more reacting molecules, referred to as the substrates. Once the enzyme substrate complex is formed, the reactant(s) is (are) transformed into product(s), and the complex falls apart to give the desired products and the unchanged enzyme, as shown in Figure 4.5. While this ''lock and key'' analogy of enzyme catalysis conveys the idea of specificity of enzyme-substrate binding, we will see in Chapter 12 that enzymes are flexible molecules. They change shape to fit the substrate and, in turn, place stress on the substrate, forcing it into a transition state, a configuration that is somewhere between that of the substrate and that of the product. As a result, the energy of activation is lowered, and product is formed.

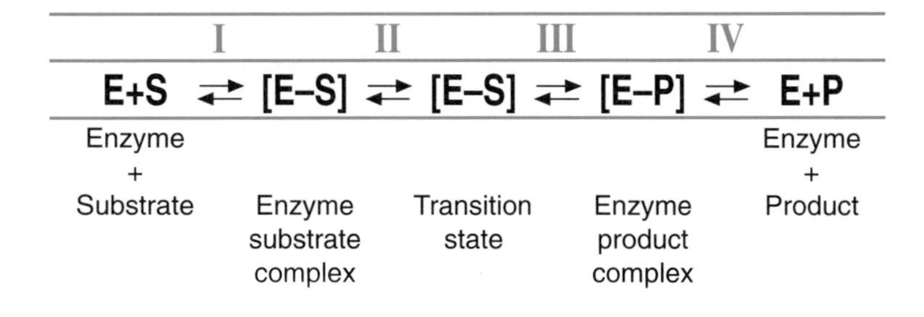

4.7 AN INTRODUCTION TO THE AROMATIC HYDROCARBONS

As we have seen, hydrocarbons are divided into two principal classes: aliphatic and aromatic. We have presented the alkanes, alkenes, and alkynes and will now introduce the **aromatic compounds.**

Aromatic hydrocarbons were originally named for their pleasant aromas. Structurally, we find that each member of this family contains an *aromatic ring*. The fundamental and most common aromatic ring is benzene, whose structure is shown in Figure 4.6.

FIGURE 4.6
The benzene molecule: (a) planar, classical view; (b) orbital view; (c) best, planar representation. The circle represents the six electrons shown as double bonds in part (a) and as overlapped *p*-orbitals in part (b).

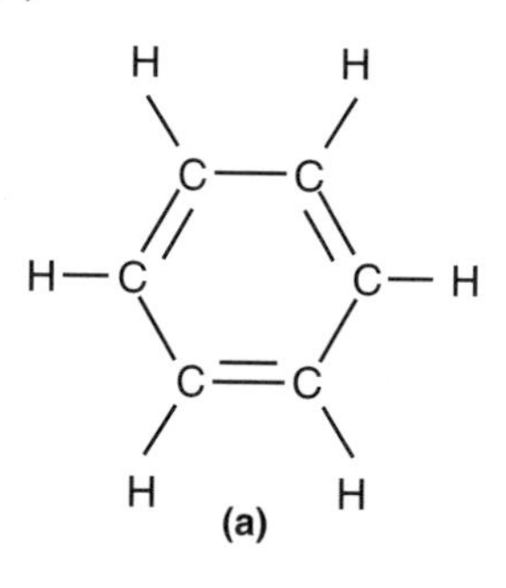

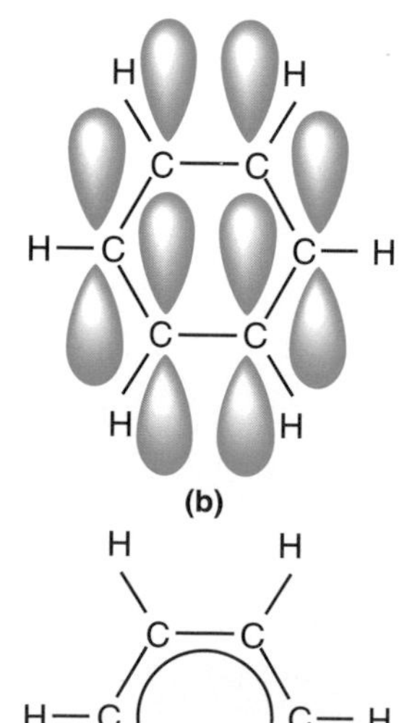

The benzene ring consists of six carbons joined in a hexagonal arrangement. A simplistic, though not completely correct, model (Figure 4.6a) depicts the carbons joined by alternating single and double bonds. In reality the orbitals involved in bonding these carbons (Figure 4.6b) result in six electrons of the three double bonds being shared equally among the six carbons. A more correct way of drawing the structure is provided in Figure 4.6c. The bonding results in a rigid, flat ring structure in contrast to the relatively flexible, nonaromatic cyclohexane ring. Other aromatic rings are possible, but benzene is by far the most common; consequently, we will limit our discussion of aromatic compounds to those containing a benzene ring.

Aromatic hydrocarbons originate from natural sources, as do the other hydrocarbons. Petroleum and coal both contain numerous aromatic hydrocarbons in varying amounts, depending on their source. Additionally, through processes known as **hydroforming,**

platforming, and **reforming,** aliphatic hydrocarbons are converted to aromatic hydrocarbons. These processes require heat, pressure, and a metal catalyst.

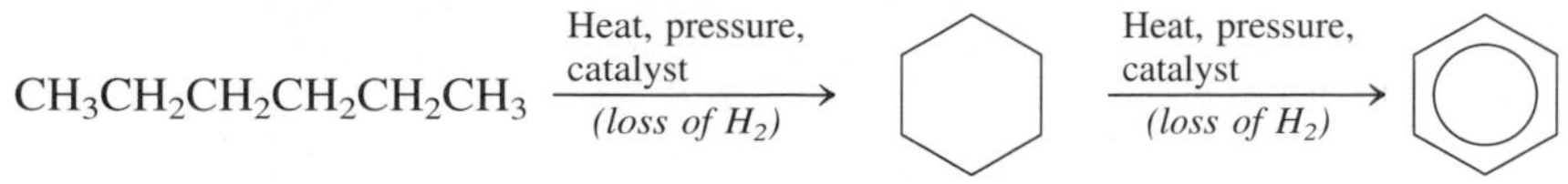

$$CH_3CH_2CH_2CH_2CH_2CH_3 \xrightarrow[\text{(loss of } H_2)]{\text{Heat, pressure, catalyst}} \text{Cyclohexane} \xrightarrow[\text{(loss of } H_2)]{\text{Heat, pressure, catalyst}} \text{Benzene}$$

Hexane Cyclohexane Benzene

Many common aromatic hydrocarbons (for example, benzene, toluene, xylene(s), cresol(s), and phenol(s)) are obtained from one or more of these sources. Others are synthesized commercially from one of these readily available starting materials.

Aromatic compounds tend to be less reactive than aliphatic compounds. For example, no reaction occurs between benzene and Br_2 in the presence of heat or light. Aromatic compounds do undergo unique chemical reactions; examples are presented in Section 4.10.

4.8 AROMATIC COMPOUNDS: NOMENCLATURE

Most simple aromatic compounds are named as derivatives of benzene. Several examples are shown below:

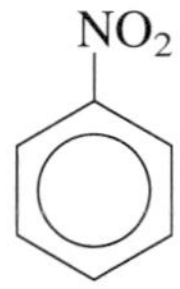

NO_2 CH_2CH_3 Br

Nitrobenzene 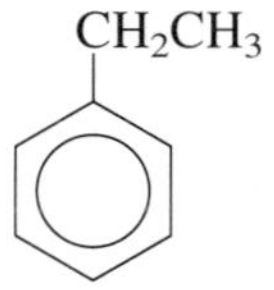 Ethylbenzene 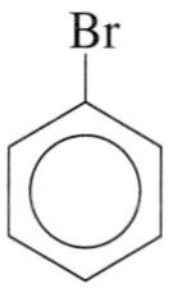 Bromobenzene

Other members of this family have unique names with a historical rather than logical basis:

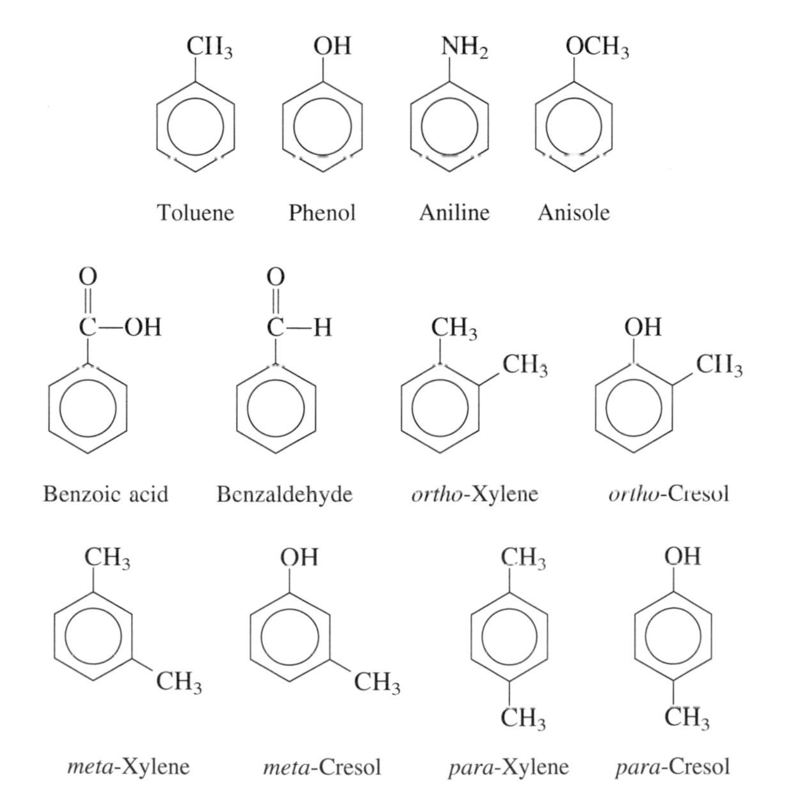

When two groups are present on the ring, three possible orientations exist; on two adjacent carbons (*ortho, o*), on carbons separated by one carbon atom (*meta, m*), or on carbons separated by two carbon atoms (*para, p*), as shown below:

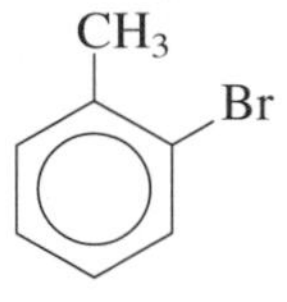

ortho-Bromotoluene 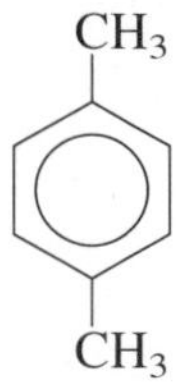 *para*-Xylene 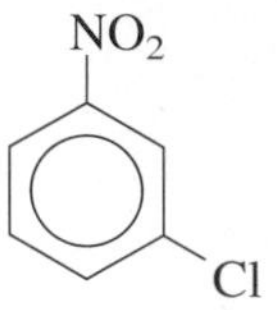 *meta*-Chloronitrobenzene

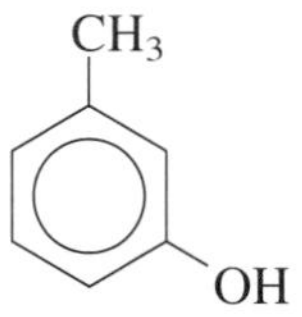

meta-Cresol 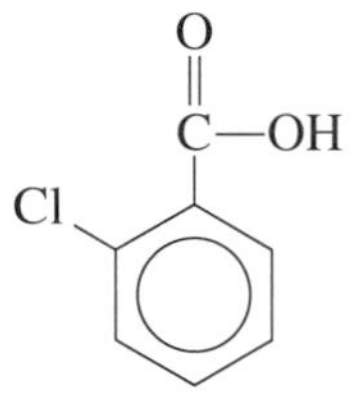 *ortho*-Chlorobenzoic acid

ortho-groups are adjacent, 1,2 to each other on the benzene ring, *meta*-groups are 1,3 to each other, and *para*-groups 1,4 to each other.

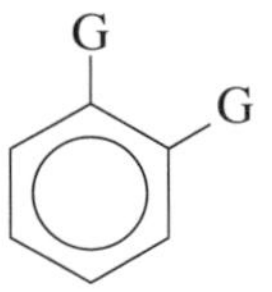

Two groups 1,2 or *ortho* 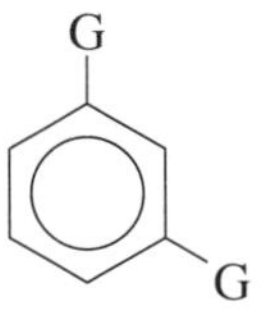 Two groups 1,3 or *meta* 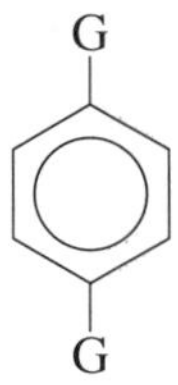 Two groups 1,4 or *para*

G = Any group

Question 4.4 Draw the following compounds:

a. 1,3,5-trichlorobenzene
b. *para*-dinitrobenzene
c. *ortho*-cresol
d. 2-nitroaniline
e. 2,5-dibromophenol
f. *meta*-nitrotoluene

ortho-, *meta*-, and *para*- terminology is restricted to the nomenclature of disubstituted benzenes. In more complex structures a numerical system is used to designate the positions of the various substitutents on the ring.

4.9 AN INTRODUCTION TO ARYL HALIDES

Aryl halides are aromatic compounds in which one of the hydrogens of benzene has been substituted by a halogen atom. The physical properties of aryl halides resemble those of the alkyl halides; they are water-insoluble and are more dense than water. See Table 4.3 for some examples.

Aryl halides, like alkyl halides, are obtained through both industrial and laboratory-scale synthetic reactions; they do not occur in nature to any significant extent. They are

TABLE 4.3 Physical Properties of Selected Aryl Halides

Compound	Boiling Point (°C) (g/100 cc)	Density*
Fluorobenzene	85	1.024
Chlorobenzene	132	1.106
Bromobenzene	155	1.499
Iodobenzene	189	1.832
p-Chlorotoluene	162	1.070
p-Dichlorobenzene	175	1.248

*At 20°C.

normally prepared from the corresponding benzene or substituted benzene precursor. (The **precursor** is the compound that, when reacted with appropriate reagents, gives the desired product.) In the example provided below, benzene is the precursor of bromobenzene, the product.

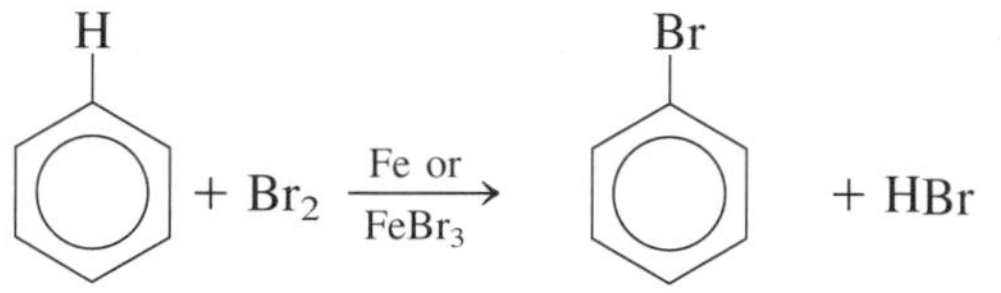

$$C_6H_5\text{–}H + Br_2 \xrightarrow[FeBr_3]{\text{Fe or}} C_6H_5\text{–}Br + HBr \qquad 4.9$$

Benzene Bromobenzene

The bromination of benzene is a substitution reaction. One of the hydrogens attached to the aromatic benzene ring is substituted by a halogen atom; the other product produced is hydrogen halide gas. In order for the reaction to proceed efficiently a catalyst, such as iron, is required.

More specifically, the bromination of benzene is an electrophilic substitution reaction because electrophiles are involved in the substitution process. In this case, bromine and the catalyst first react to form the bromonium ion, an electrophile, which then reacts with the aromatic ring

$$Br_2 + FeBr_3 \longrightarrow :\ddot{\underset{..}{Br}}^+ + FeBr_4^- \qquad 4.10$$

Bromonium ion
An electrophile

When benzene is reacted, only one monosubstituted product is possible. When substituted benzenes are reacted under the same conditions, however, a mixture of products may result. We find that, depending on the substituent present, one (or perhaps several) of the many possible products will predominate.

Example:

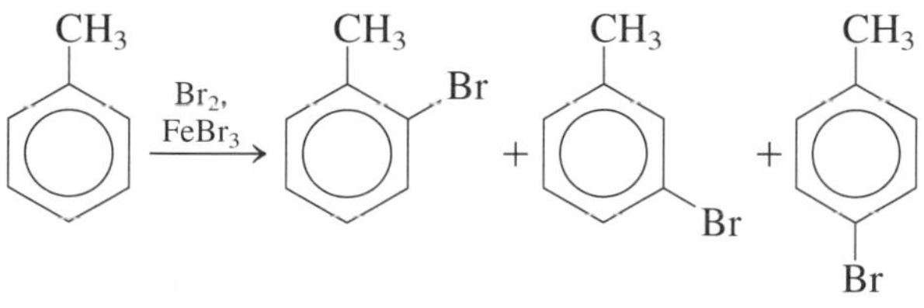

$$C_6H_5CH_3 \xrightarrow{Br_2,\ FeBr_3} o\text{-}BrC_6H_4CH_3 + m\text{-}BrC_6H_4CH_3 + p\text{-}BrC_6H_4CH_3 \qquad 4.11$$

Toluene	40% *ortho-*	0% *meta-*	60% *para-*

Bromotoluenes

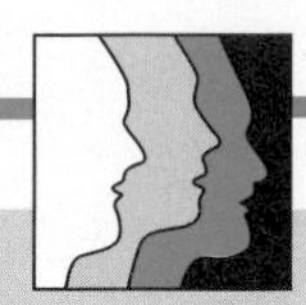

A HUMAN PERSPECTIVE

Aromatic Compounds and Carcinogenesis

Many naturally occurring aromatic compounds play a role in our day-to-day existence. Many food flavorings and fragrances, preservatives, insecticides, pharmaceuticals, medicinals, and toiletries contain aromatic compounds. Selected examples are provided below:

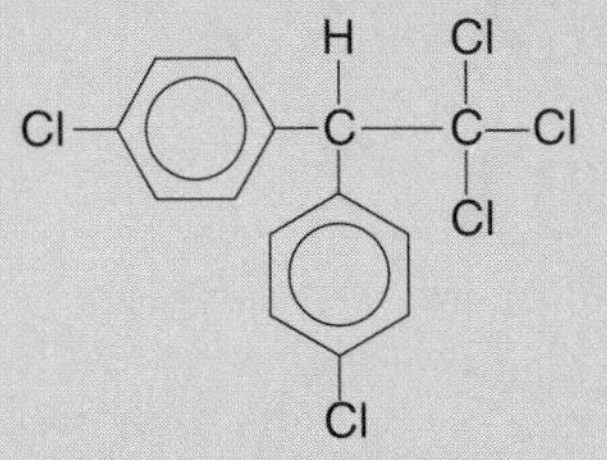

DDT: *Dichlorodiphenyltrichloroethane*

Insecticide; effective against tsetse fly (sleeping sickness), Anopheles mosquito (malaria), and so on; presently banned in the U.S. because it does not biodegrade and is linked to numerous environmental and health problems.

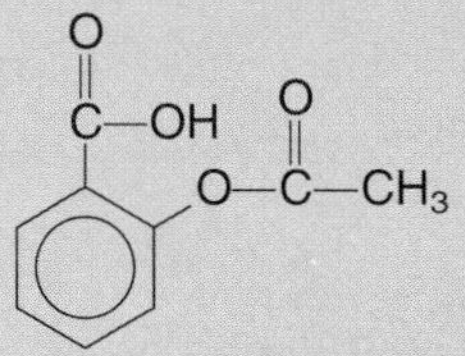

Aspirin: *Acetylsalicylic acid*

Most widely used pain reliever in the world; also useful in reducing fevers; appears to increase the incidences of stomach ulcers and decrease the incidences of heart attacks in people who use the drug regularly—research is ongoing.

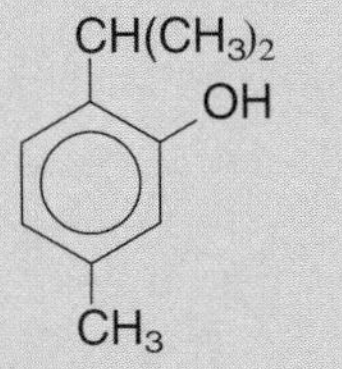

Thymol (fragrance compound)

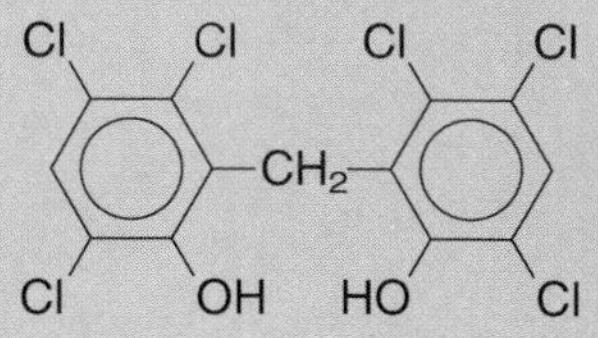

Hexachlorophene:

Antiseptic; used in soaps and face creams; has been linked with cancer (it is a potential carcinogen—cancer causing agent) and is now a controlled substance.

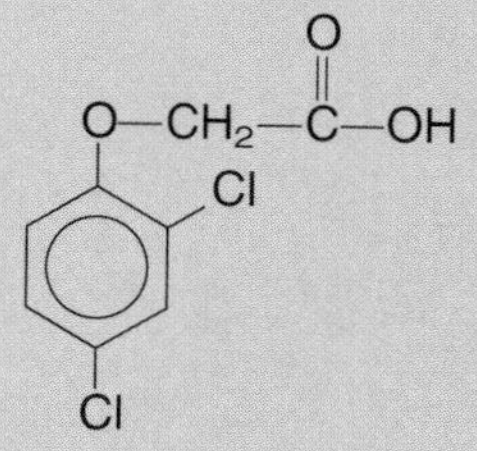

2,4-D

2,4-Dichlorophenoxyacetic acid (herbicide)

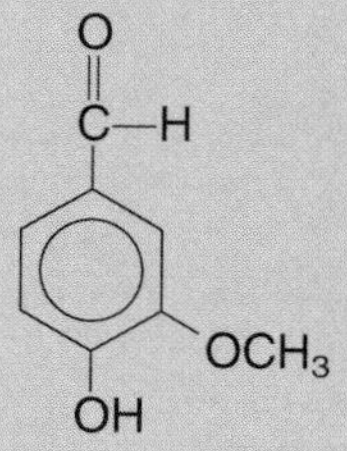

Vanillin (flavoring)

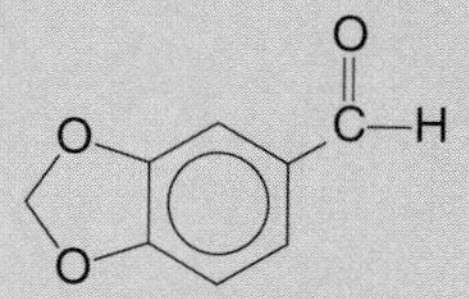

Safrole (flavoring)

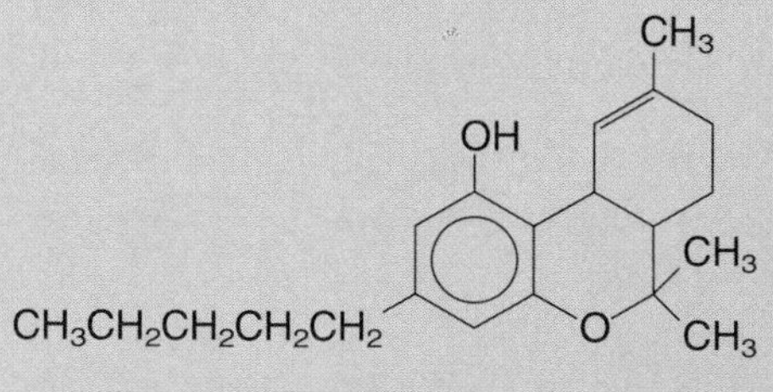

THC (tetrahydrocannabinol)

Question 4.5 Complete each of the following reactions by supplying the resulting product(s). Use complete structural formulas.

a. benzene + Cl_2/Fe
b. 3-hexene + HI/CCl_4
c. 3-methyl-2-hexene + HCl/CCl_4
d. cyclopentene + Br_2/CCl_4
e. cyclohexane + Cl_2/heat
f. 2-methylcyclopentene + HBr/CCl_4

The *polynuclear aromatic hydrocarbons* (PAH) provide an example of yet another important family of aromatic hydrocarbons. PAH are formed from the joining of the rings so that they share a common bond (edge). Several of the more common examples are shown below:

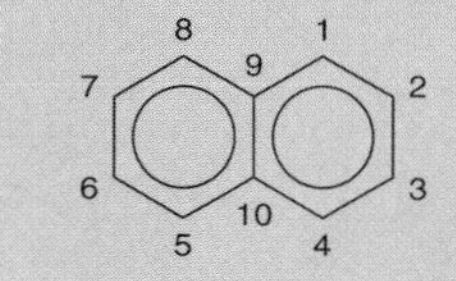

Naphthalene

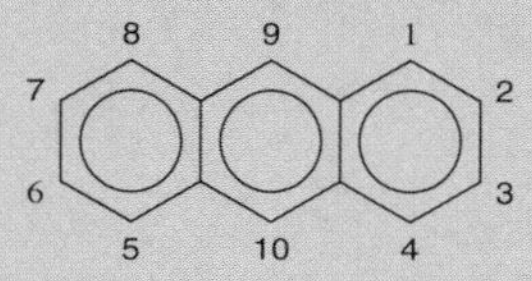

Anthracene

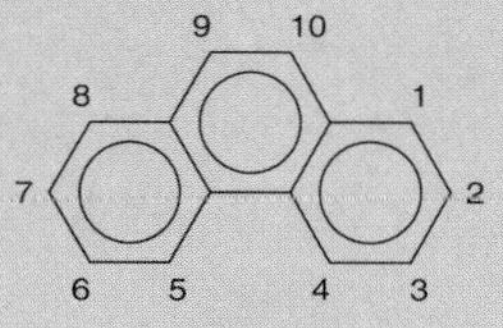

Phenanthrene

The more complex members of this family (typically consisting of five or six rings at a minimum) are among the most potent cancer-causing (carcinogenic) chemicals known. It has been shown that the carcinogenic nature of these chemicals results from their ability to bind to the nucleic acid (DNA) in cells. As we will see in Chapter 16, the ability of the DNA faithfully to guide the cell from generation to generation depends on the appropriate expression of the genetic information, a process called transcription, and the accurate copying (replication) of the DNA so that every generation of cells inherits a completely functional copy of all the genetic information. Covalent binding of these PAH is thought to interfere with correct replication, causing errors to be made that allow the cell to grow out of control.

Polynuclear aromatic hydrocarbons (PAH) are among the strongest cancer-causing chemicals (carcinogens) known. They are thought to cause cancer by reacting with the DNA in cells, "short circuiting" the replication ability of these "chemical templates." A schematic representation of this process is shown above.

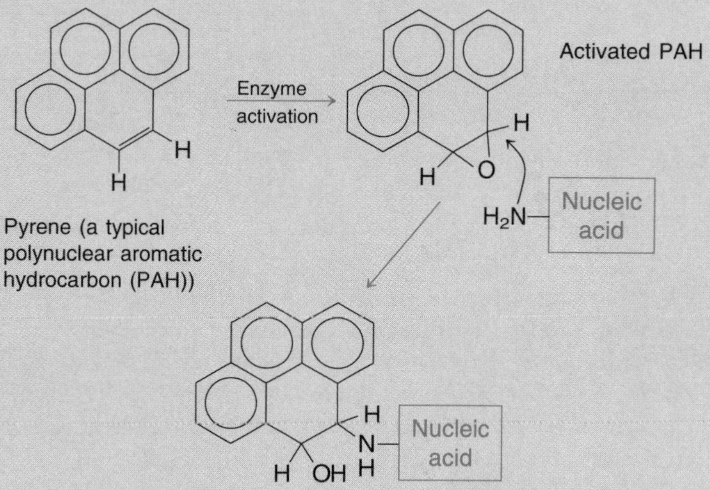

Pyrene (a typical polynuclear aromatic hydrocarbon (PAH))

Enzyme activation

Activated PAH

Nucleic acid

PAH covalently bonded to the nucleic acid

Benzo[a]pyrene, shown below, is found in tobacco smoke, smokestack effluents, charcoal-broiled meat, and automobile exhaust. It is one of the strongest carcinogens known. It is estimated that the vast majority of all cancers are caused by chemical carcinogens, such as PAH, in the environment.

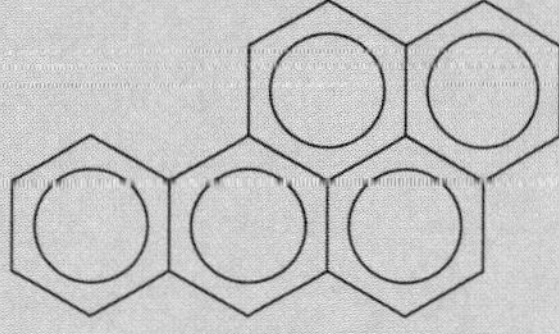

Benzopyrene

4.10 REACTIONS OF ARYL HALIDES

The halide exchange reaction was introduced in Section 3.9. In this reaction, one halogen atom is exchanged for another halogen atom; this is an example of a nucleophilic substitution. One nucleophilic halide ion, $:\ddot{X}:^{-}$, substitutes itself for another halogen atom:

Section 3.9

Br

$\xrightarrow{I^-}$ NO REACTION 4.12

Bromobenzene

Although aryl halides do not normally undergo nucleophilic substitution under normal chemical conditions, electrophilic substitution is possible. In these instances it is not the halide portion of the aryl halide that is reacting, but rather the aromatic ring itself.

Mixtures of products result from substitution, predominantly at the positions *ortho* and *para* to the halogen that is initially present on the ring. For example:

Br $\xrightarrow{Br_2,\ FeBr_3}$ Br, Br + Br, Br 4.13

Bromobenzene *p*-Dibromobenzene *o*-Dibromobenzene

SUMMARY

We have now completed our introduction to the hydrocarbons: the alkanes, alkenes, alkynes, and aromatic hydrocarbons. These families are the four major subdivisions of all organic chemicals. The saturated compounds (alkanes and cycloalkanes) have physical properties similar to those of the unsaturated compounds (alkenes, alkynes, and aromatics; they are nonpolar and water-insoluble and have relatively low boiling points. Their chemical properties, however, are quite different. Alkanes undergo substitution reactions, whereas alkenes and alkynes undergo addition reactions. Aromatic compounds undergo substitution only under their own unique chemical conditions. All of the hydrocarbons are named by using two principal systems: the common system of nomenclature and the I.U.P.A.C. Nomenclature System. The latter is less ambiguous and is preferred over the former.

Hydrocarbons have their origin in what was once living matter: petroleum and coal. Many of the hydrocarbons obtained from these two sources are converted to other compounds through synthesis. Alkanes undergo substitution, while alkenes undergo addition. Alkenes are hydrogenated to alkanes (H_2, heat, catalyst) or hydrated (H_2O, H^+) to give alcohols. Alkenes also react with hydrogen halides (HX) to give alkyl halides.

Aromatic compounds undergo aromatic substitution reactions that often produce a mixture of products. Many of the aromatic hydrocarbons, principally those containing five or more fused benzene rings (polynuclear aromatic hydrocarbons), are cancer-causing chemicals (carcinogens).

Alkyl and aryl halides have structures similar to those of the hydrocarbons (alkane, alkene, alkyne, or aromatic) from which they are derived. They are "hydrocarbon-like" and water-insoluble and have reasonably high boiling points and densities. They are named by using the I.U.P.A.C. system, though common names are still routinely found.

Aryl halides undergo aromatic substitution reactions that are common to all aromatic compounds.

GLOSSARY OF KEY TERMS

addition reaction (4.4) a reaction in which two molecules add together to form a new molecule; often involves the addition of one molecule to a double or triple bond in an unsaturated molecule.

alkene (4.1) a hydrocarbon that contains one or more carbon-carbon double bonds; an unsaturated hydrocarbon with the general formula C_nH_{2n}.

alkyne (4.1) a hydrocarbon that contains one or more carbon-carbon triple bonds; an unsaturated hydrocarbon with the general formula C_nH_{2n-2}.

anesthetic (4.10) any drug that causes lack of sensation in any part of the body (local) or causes unconsciousness (general); an analgesic effect is generally also associated with these compounds.

antiseptic (4.10) any compound that has the effect of preventing or inhibiting bacterial infection of the body.

aromatic compound (4.7) a hydrocarbon that contains a benzene ring or that has properties that are similar to those exhibited by benzene.

aryl halide (4.9) a benzene ring, or other aromatic compound, in which a hydrogen on the ring has been substituted by a halogen atom (F^-, Cl^-, Br^-, or I^-).

biodegradable (4.10) any substance that will breakdown naturally in the atmosphere.

biological magnification (4.10) the process through which chemicals are concentrated in greater and greater quantities as they are processed through the food chain.

AN ENVIRONMENTAL PERSPECTIVE

DDT and Biological Magnification

We have heard the warnings for years: Stop using nonbiodegradable insecticides because they are killing many animals other than their intended victims! Are these chemicals not specifically targeted to poison insects? How then can they be considered a threat to humans and other animals?

DDT, a polyhalogenated hydrocarbon, was discovered in the early 1940s by Paul Muller, a Swiss chemist. Muller showed that DDT is a nerve poison that causes convulsions, paralysis, and eventually death in insects. From the 1940s until 1972, when it was banned in the United States, DDT was sprayed on crops to kill insect pests, sprayed on people as a delousing agent, and sprayed in and on homes to destroy mosquitoes carrying malaria. At first, DDT appeared to be a miraculous chemical, saving literally millions of lives and millions of dollars in crops. However, as time went by, more and more evidence of a dark side of DDT use accumulated. Over time, the chemical had to be sprayed in greater and greater doses as the insect populations evolved to become more and more resistant to it. In 1962, Rachel Carson published her classic work, *The Silent Spring,* which revealed that DDT was accumulating in the environment. In particular, high levels of DDT in birds interfered with their calcium metabolism. As a result, the egg shells produced by the birds were too thin to support development of the chick within. Thus in spring, when the air should have been filled with bird song, there was silence. This is the "silent spring" referred to in the title of Carson's book.

DDT is not biodegradable; furthermore, it is not water-soluble, but it is soluble in nonpolar solvents. Thus if DDT is ingested by an animal, it will dissolve in fat tissue and accumulate there, rather than being excreted in the urine. When DDT is introduced into the food chain, which is inevitable when it is sprayed over vast areas of the country, the result is **biological magnification.** This stepwise process begins when DDT applied to crops is ingested by insects. The insects, in turn, are eaten by birds, and the birds are eaten by a hawk. We can imagine another food chain: Perhaps the insects are eaten by mice, which are in turn eaten by a fox, which is then eaten by an owl. Or to make it more personal, perhaps the grass is eaten by a cow, which then becomes your dinner. With each step up one of these food chains, the concentrationon of DDT in the tissues becomes higher and higher because it is not degraded, it is simply stored. Eventually, the concentration may reach toxic levels in some of the animals in the food chain.

Consider for a moment the series of events that occurred in Borneo in 1955. The World Health Organization elected to spray DDT in Borneo because 90% of the inhabitants were infected with malaria. As a result of massive spraying, the mosquitoes bearing the malaria parasite were killed. If this sounds like the proverbial happy ending, read on. This is just the beginning of the story. In addition to the mosquitoes, millions of other household insects were killed. In tropical areas it is common for small lizards to live in the homes, eating insects found there. The lizards ate the dead and dying DDT-contaminated insects and were killed by the neurotoxic effects of DDT. The house cats ate the lizards, and they, too, died. The number of rats increased dramatically because there were no cats to control the population. The rats and their fleas carried sylvatic plague, a form of bubonic plague. With more rats in contact with the humans came the threat of a bubonic plague epidemic. Happily, cats were parachuted into the affected areas of Borneo, and the epidemic was avoided.

The story has one further twist. Many of the islanders lived in homes with thatched roofs. The vegetation used to make these roofs was the preferred food source for a caterpillar that was not affected by DDT. Normally, the wasp population preyed on these caterpillars and kept the population under control. Unfortunately, the wasps were killed by the DDT. The caterpillars prospered, devouring the thatched roofs, which collapsed on the inhabitants.

Every good story has a moral, and this one is not difficult to decipher. The introduction of large amounts of any chemical into the environment, even to eradicate disease, has the potential for long-term and far-reaching effects that may be very difficult to predict. We must be cautious with our fragile environment. Our well-intentioned intervention all too often upsets the critical balance of nature, and in the end we inadvertently do more harm than good.

catalyst (4.6) any substance that increases the rate of a chemical reaction (by lowering the activation energy of the reaction) and that is not destroyed in the course of the reaction.

degradation (4.5) the breakdown of a larger molecule into smaller molecules.

geometric isomer (4.3) isomers (which by definition have the same molecular formula) that differ from one another owing to the placement of substituents on a double bond or a ring.

hydration (4.4) a reaction in which water is added to a molecule.

hydroforming, platforming, and reforming (4.7) the conversion of aliphatic hydrocarbons to aromatic hydrocarbons under the influence of heat, pressure, and a catalyst.

hydrogenation (4.4) a reaction in which hydrogen (H_2) is added (usually) to a double or a triple bond.

hydrohalogenation (4.4) the addition of a hydrohalogen (HCl, HBr, and so on) to an unsaturated bond.

insecticide (4.10) a compound that is used to kill insects; used in the control of insect populations.

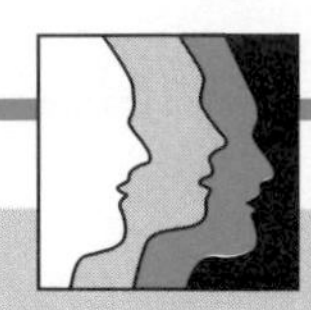

A HUMAN PERSPECTIVE

Polyhalogenated Hydrocarbons: Anesthetics and Insecticides

It is almost impossible to pick up a newspaper or magazine without reading about one of the members of the family of compounds known as **polyhalogenated hydrocarbons,** hydrocarbons containing two or more halogen atoms. Their notoriety is, unfortunately, due to their undesirable properties rather than their many possible benefits for humans. The compounds abbreviated as DDT, PCB, PBB, and EDB, whose complete names and structures are shown below, are among the more common polyhalogenated hydrocarbons encountered. This family of compounds has indeed created problems for humankind, yet without them, our standard of living might be quite different.

Halogenated hydrocarbons were among the first routine **anesthetics** (pain relievers). (A local anesthetic anesthetizes (deadens) a portion of the body, while a general anesthetic anesthetizes the entire body.) These chemicals played a central role as the studies of medicine and dentistry moved from the dark ages into modern times:

$CH_3CH_2—Cl$

Chloroethane
(ethyl chloride)

Local anesthetic: Applied topically (on the skin) to provide analgesic (pain-killing) effect; rapid and of short duration.

$CHCl_3$

Trichloroethane
(chloroform)

General and local anesthetic: Administered through inhalation; rapid, short duration, very powerful. Also was used in the past for treatment for hydrophobia (rabies), tetanus, and colic and in liniments for rheumatism.

$CH_3CH(Cl)—Br$

1-Bromo-1-chloroethane
(halothane)

General anesthetic: Administered by inhalation; considered safe in relation to other general anesthetics; wide applicability.

Many other members of this family are used in a variety of medical applications:

CHI_3

Triiodomethane
(iodoform)

Antiseptic: Applied topically to cuts and/or larger wounds.

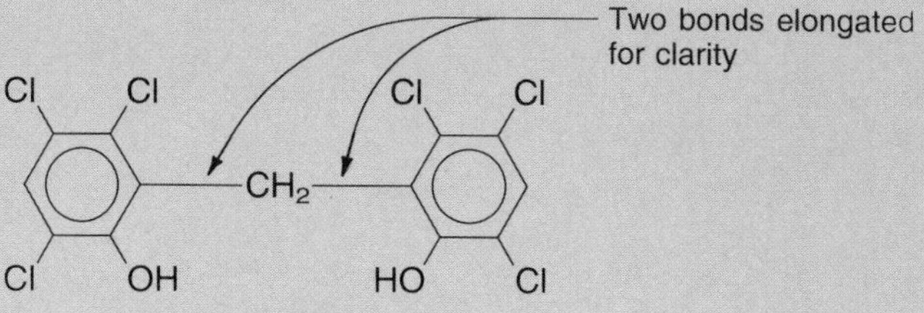

Two bonds elongated for clarity

Hexachlorophene

Antiseptic: Applied topically in soaps and face creams; a suspected carcinogen and can be purchased only by prescription.

The more infamous polyhalogenated hydrocarbons are **insecticides.** DDT, perhaps the best known in this application, provides a good example of a risk/benefit situation relating to the use of synthetic chemicals in our day-to-day lives.

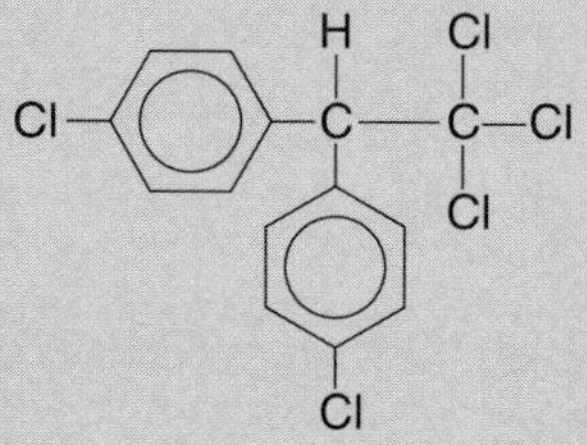

DDT *(dichlorodiphenyltrichloroethane)*

DDT, dichlorodiphenyltrichloroethane (a misnomer; incorrect I.U.P.A.C. name), was discovered in Switzerland in the early part of this century. Its versatility and applicability as an insecticide led to its immediate use throughout Europe. With the outbreak of World War II, Allied soldiers found themselves traveling all over the globe and routinely exposed to diseases that were nonexistent in their own countries: typhus, malaria, and sleeping sickness. All of these diseases are carried by insect pests; typhus is carried by the body louse, malaria by the Anopheles mosquito, and sleeping sickness by the tsetse fly. The wide applicability, economical and simple synthesis, and effectiveness of DDT all contributed to its ready acceptance and wide use.

By the late 1940s, however, problems associated with DDT had begun to appear. DDT-resistent houseflies had been discovered. Higher and higher concentrations of the pesticide were being detected in birds, fish, and shellfish. It was demonstrated that DDT interfered with the normal growth of plankton (the key element in the food chain in the oceans), and hundreds of birds and other wildlife died in areas where heavy spraying of the chemical was practiced. Finally, DDT was implicated in deterioration of the central nervous system, cancer, and even death in humans. Most uses of DDT in the United States were banned in 1972.

Several other insecticides, related to DDT, are shown here.

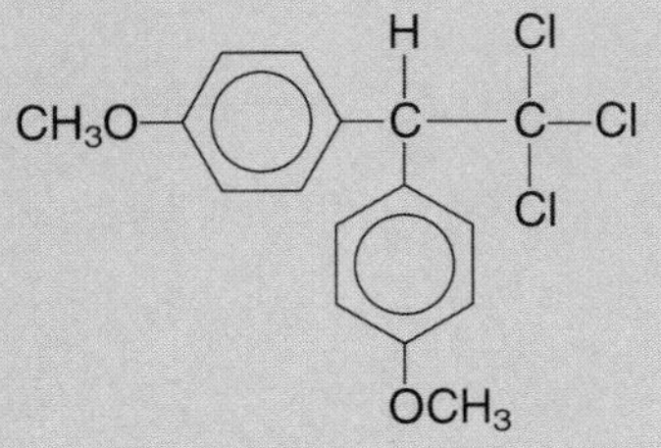

Methoxychlor

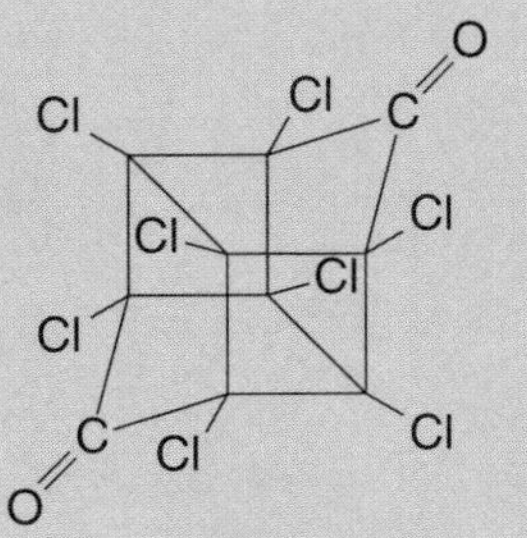

Kepone

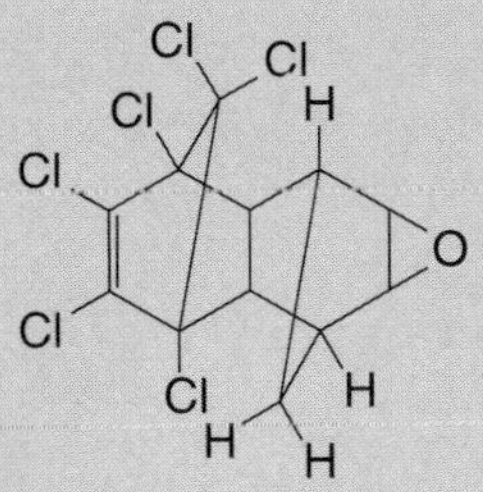

Dieldrin

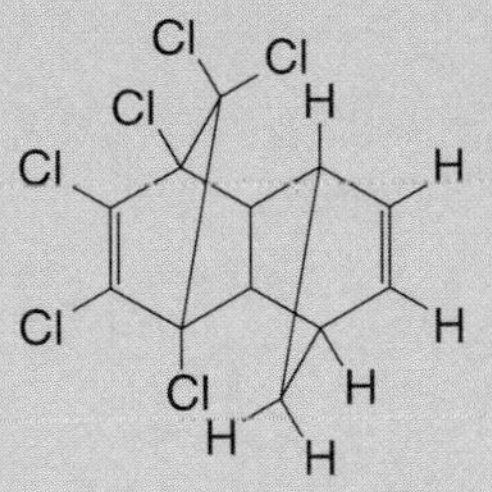

Aldrin

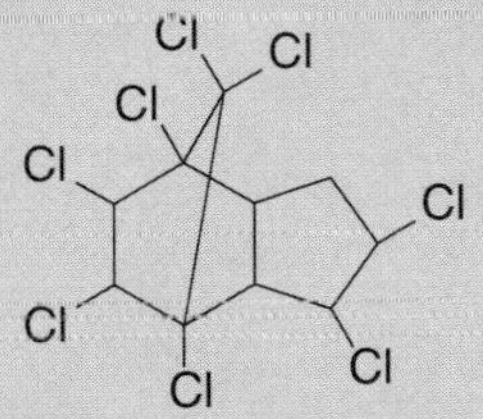

Chlordane

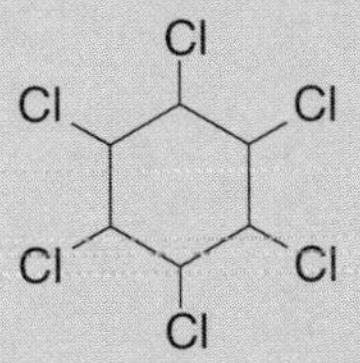

Lindane

None of these insecticides break down rapidly through chemical or biochemical degradation in the environment. They are said to be nonbiodegradable. Thus they persist for a long period of time and build up in soil, water, and plants as additional insecticides are routinely used. These chemicals are also soluble in the fatty tissues in the body of humans and other animals. They accumulate in these tissues as more of the insecticide is ingested. The adverse side effects of the insecticides increase by the process of biological magnification. Once the insecticide enters the food chain, its concentration continually increases: Microscopic plankton ingest it; these are then ingested by fish, the fish by birds, and the birds and fish by humans. The concentration in one fish or shellfish may be thousands of times higher than it was in the water or plankton that inhabit the water simply because one fish eats millions of plankton. As a result, although the concentration at the base of the food chain may be very low, the concentrations ingested by humans may well be beyond safe limits.

Other common polyhalogenated hydrocarbons of interest are provided below:

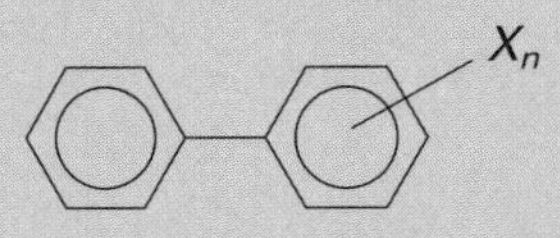

(X_n = 2 or more halogens attached to rings in various locations)

PCC's or PBB's (polychloro(or bromo)biphenyls; used as plasticizers and insulators in electrical equipment)

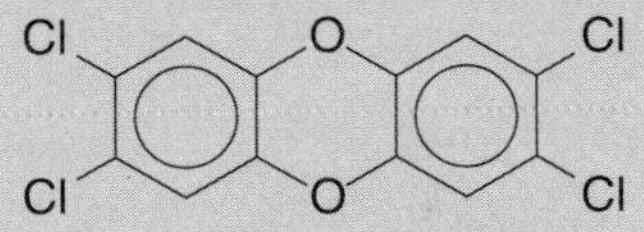

2,3,7,8-Tetrachlorobenz-*p*-dioxin (by-product in 2,3,5,-T synthesis)

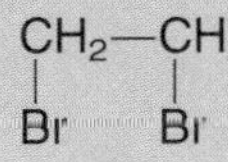

EDB (ethylene dibromide; fumigant for citrus fruit)

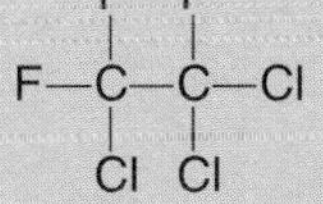

Freons (refrigerants)

(Other structures are also possible; generally halogenated (X = Cl, F) methane or ethane.)

Millions of lives have been saved through the use of these insecticides. It would be virtually impossible to feed the human population without them. But it is important to use these and other chemicals with a clear understanding of the risks and benefits of their use.

Markovnikov's Rule (4.4) The rule that states that a proton, adding to a carbon-carbon double bond, will add to the carbon having the larger number of hydrogens attached to it already.

polyhalogenated hydrocarbon (4.10) a compound that contains several halogen substituents.

precursor (4.9) a molecule that, when reacted with appropriate reagents, gives a desired product; the molecule from which a desired product is immediately derived synthetically.

reaction mechanism (4.5) the pictorial, step-by-step process by which reactants are converted to products in a chemical reaction.

saturated compound (4.1) a hydrocarbon that contains only carbon-hydrogen single bonds, having the general molecular formula C_nH_{2n+2}.

synthesis (4.5) the conversion of one molecule into another molecule.

unsaturated compound (4.1) any hydrocarbon that contains one or more carbon-carbon double or triple bonds.

QUESTIONS AND PROBLEMS

Nomenclature

4.6 Draw a complete structure for each of the following compounds:

a. 2-methyl-2-hexene
b. *trans*-3-heptene
c. *cis*-1-chloro-2-pentene
d. *cis*-2-methyl-2-chloro-3-heptene
e. *trans*-5-bromo-2,6-dimethyl-3-octene

4.7 Draw a complete structure for each of the following compounds:

a. 3-hexyne
b. 4-methyl-1-pentyne
c. 1-chloro-4,4,5-trimethyl-2-heptyne
d. 2-bromo-3-chloro-7,8-dimethyl-4-decyne
e. diphenylacetylene

4.8 Draw a complete structure for each of the following compounds:

a. 2,4-dibromotoluene
b. 1,2,4-triethylbenzene
c. isopropylbenzene (cumene)
d. 2-bromo-5-chlorotoluene
e. 2-bromo-4-chloro-3,5,6-trimethylethylbenzene

4.9 Name each of the following compounds, using the I.U.P.A.C. System. Be sure to indicate *cis* or *trans* where applicable.

a. 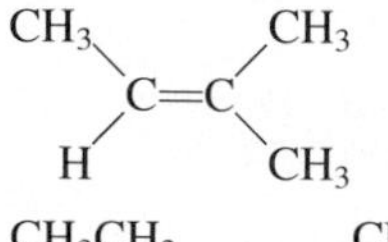 $(CH_3)(H)C{=}C(CH_3)(CH_3)$ — upper: CH_3, CH_3; lower: H, CH_3 (common name: isoprene)

b. 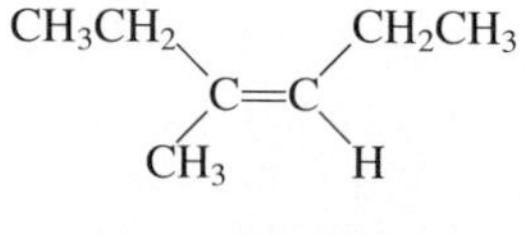 $(CH_3CH_2)(CH_3)C{=}C(CH_2CH_3)(H)$ — upper: CH_3CH_2, CH_2CH_3; lower: CH_3, H

c. $(CH_3)(H)C{=}C(H)(CH_2C(CH_3)_2CH_3)$ — upper: CH_3, H; lower: H, CH_2CCH_3 with CH_3 above and below the quaternary C

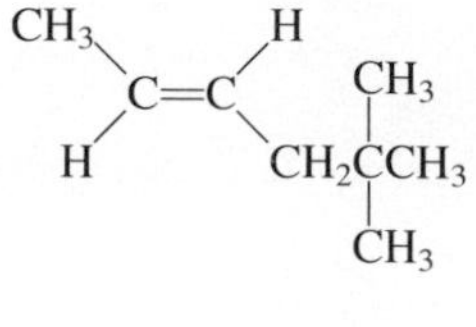

d. 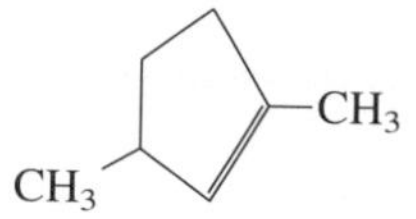 Cyclopentene ring bearing CH_3 on a double-bond carbon and CH_3 on the adjacent ring carbon

e. 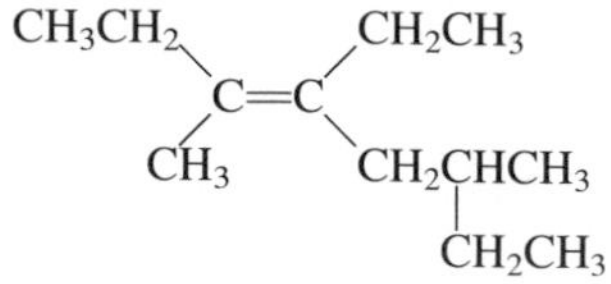 $(CH_3CH_2)(CH_3)C{=}C(CH_2CH_3)(CH_2CHCH_3)$ — upper: CH_3CH_2, CH_2CH_3; lower: CH_3, CH_2CHCH_3 with CH_2CH_3 on the CH

4.10 Name each of the following compounds, using the I.U.P.A.C. System:

a. $CH_3{-}C{\equiv}C{-}CH_2CH_3$

b.  $CH_3{-}CHCHCH_2CH_2C{\equiv}C{-}H$ (Br on each CH: Br Br)

c. $CH_3CH(CH_3){-}C(CH_3){=}C(Br){-}CH(CH_3)CH_3$

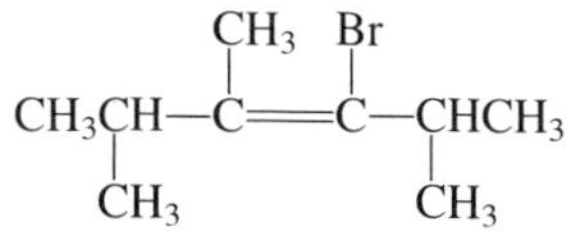

d. 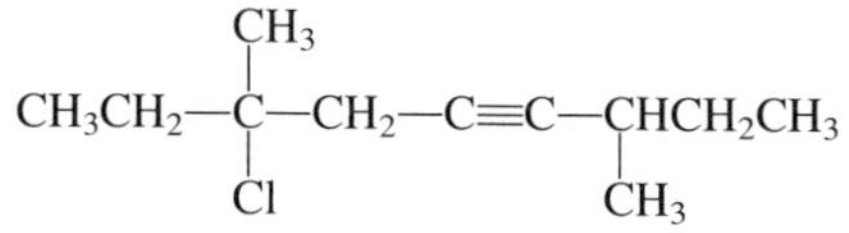 $CH_3CH_2{-}C(CH_3)(Cl){-}CH_2{-}C{\equiv}C{-}CH(CH_3)CH_2CH_3$

e. 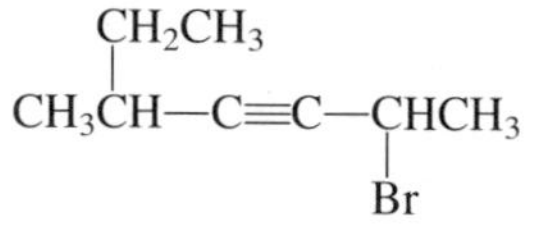 $CH_3CH(CH_2CH_3){-}C{\equiv}C{-}CH(Br)CH_3$

4.11 Name each of the following compounds, using the I.U.P.A.C. System:

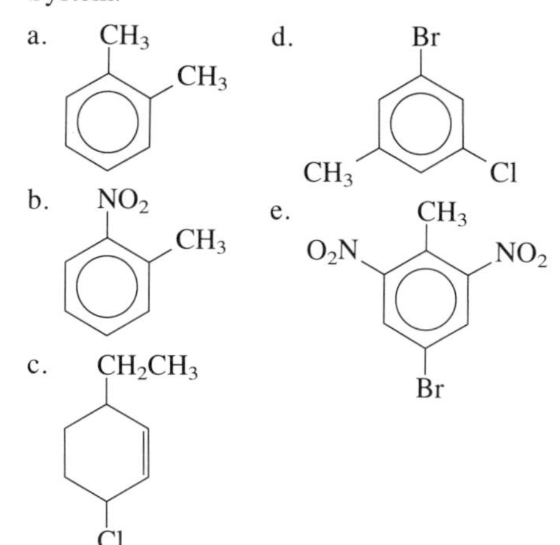

4.12 Which of the following structures have incorrect I.U.P.A.C. names? If incorrect, give the correct I.U.P.A.C. name.

a. 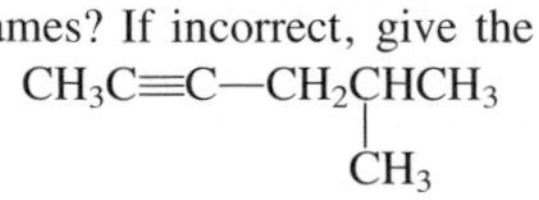 $CH_3C{\equiv}C{-}CH_2CH(CH_3)CH_3$

2-Methyl-4-hexyne

b. 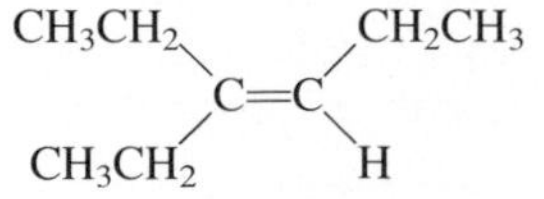 $(CH_3CH_2)(CH_3CH_2)C{=}C(CH_2CH_3)(H)$

3-Ethyl-3-hexene

c. 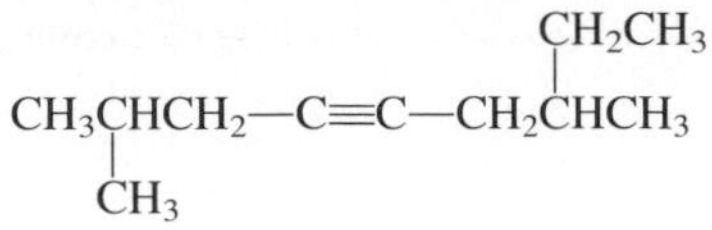 CH_2CH_3
$CH_3CHCH_2—C{\equiv}C—CH_2CHCH_3$
CH_3

2-Ethyl-7-methyl-4-octyne

d. 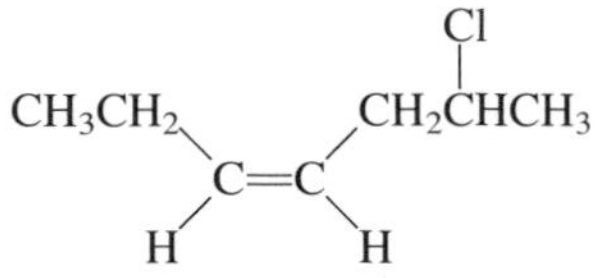 Cl
$CH_3CH_2(H)C{=}C(H)CH_2CHCH_3$

trans-6-Chloro-3-heptene

e. 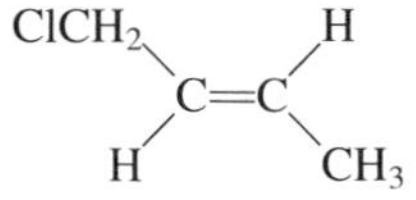 $ClCH_2(H)C{=}C(H)CH_3$

1-Chloro-5-methyl-2-hexene

4.13 Name each of the following, using the I.U.P.A.C. Nomenclature System:

a. 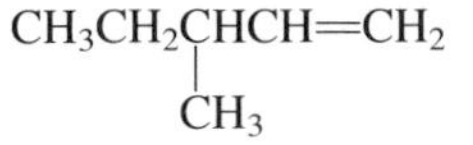 $CH_3CH_2CHCH{=}CH_2$
CH_3

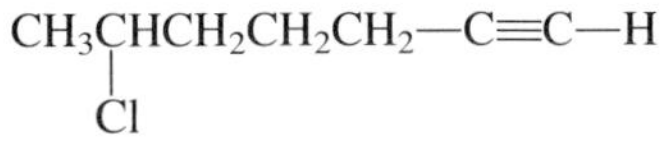

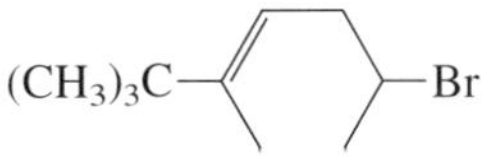

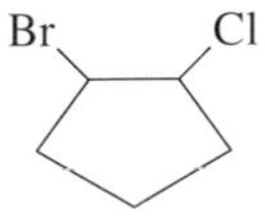

b. $(CH_3)_2CHCH_2CH{=}C(CH_3)_2$

c. 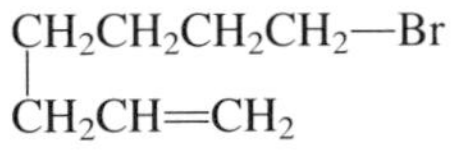 $CH_2CH_2CH_2CH_2—Br$
$CH_2CH{=}CH_2$

d. 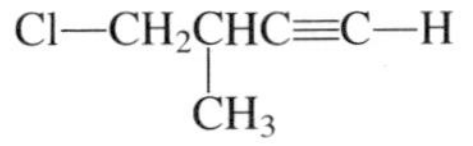 $Cl—CH_2CHC{\equiv}C—H$
CH_3

e. $CH_3CH_2CHCHCHCH_2CH_3$
I I CH_2CH_3

f. $CH_3CHCH_2CH_2CH_2—C{\equiv}C—H$
Cl

g. $(CH_3)_3C$— —Br

h. Br Cl

i. Cl— —CH_3

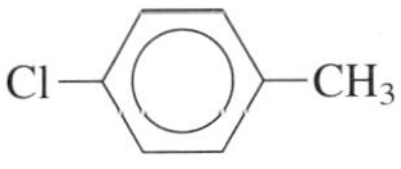

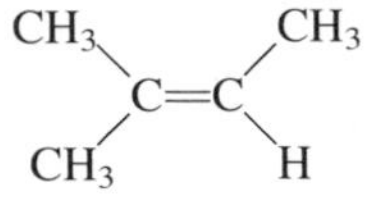

j. 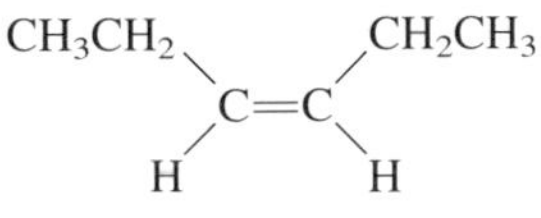 $CH_3CH_2(H)C{=}C(H)CH_2CH_3$

k. 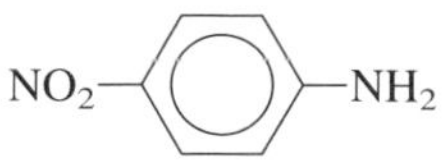 NO_2— —NH_2

l. 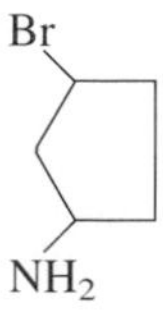 Br
NH_2

Structure, Properties, and Isomers

4.14 Which of the following alkenes would not exhibit *cis-/trans-* geometric isomerism?

a. 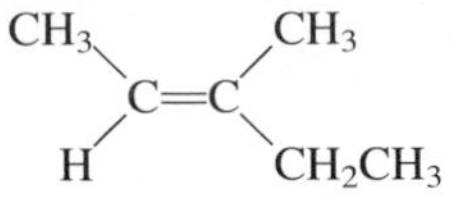 $CH_3(H)C{=}C(CH_3)CH_2CH_3$

b. $CH_3(CH_3)C{=}C(CH_3)H$

c. $CH_3CH_2(CH_3)C{=}C$ with $CHCH_3$ bearing CH_2CH_3 and $CHCH_2CH_3$ bearing CH_3

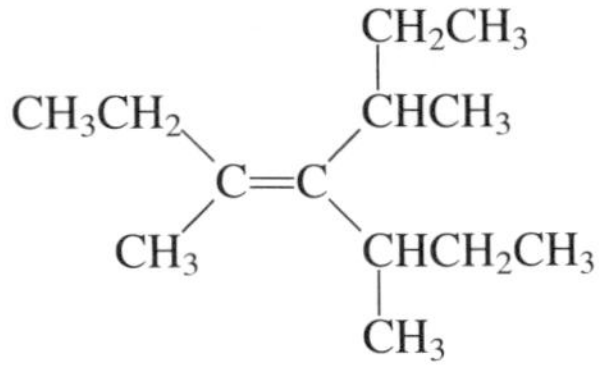

d. 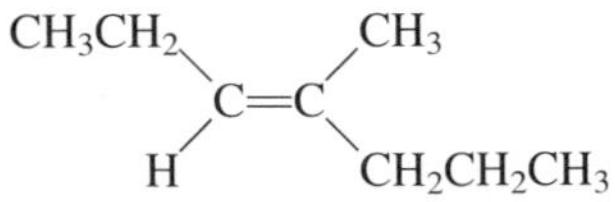 $CH_3CH_2(H)C{=}C(CH_3)CH_2CH_2CH_3$

e. 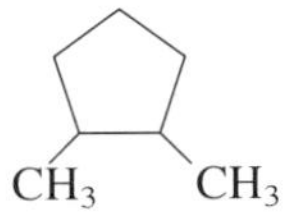 CH_3 CH_3

4.15 Draw each of the following compounds, using complete structural formulas:

a. 1,3,5-trifluoropentane
b. *meta*-cresol
c. *cis*-2-octene
d. 1-octyne
e. di-*n*-propylacetylene
f. isopropylbenzene
g. 3,3,5-trimethyl-1-heptene
h. 1,3,5-trinitrotoluene (TNT)
i. 1-bromo 3 chloro-1-heptyne
j. *cis*-1,2-dichloro-1-fluoroethene

4.16 Give examples of a pair of structural isomers, geometric isomers, and homologs. How do the members of each pair differ trom each other?

4.17 Explain in your own words how polynuclear aromatic hydrocarbons cause cancerous cells to develop.

4.18 Which of the following can exist as *cis-* and *trans-* isomers?

a. $H_2C{=}CH_2$
b. $CH_3CH{=}CHCH_3$
c. $Cl_2C{=}CBr_2$
d. $ClBrC{=}CClBr$
e. $(CH_3)_2C{=}C(CH_3)_2$

4.19 Give the complete structural formula and the name of the product in each of the following reactions:

a. cyclopentene + H_2O (H^+)
b. cyclopentene + HCl
c. cyclopentene + H_2
d. cyclopentene + HI

4.20 Of the following compounds, which can exist in both a *cis–* and a *trans–* isomeric form? Draw the two geometric isomers.

a. 2,3-dibromobutane
b. 2-heptene
c. 2,3-dibromo-2-butene
d. propene
e. 1-bromo-1-chloro-2-methylpropene
f. 1,1-dichloroethene
g. 1,2-dibromoethene
h. 2-methyl-3-*n*-propyl-2-hexene

4.21 For each series below, which compound would be the next homolog?

a. $CH_3CH{=}CH_2$, $CH_3CH_2CH{=}CH_2$, $CH_3CH_2CH_2CH{=}CH_2$, ___?___

b. $BrCH_2CH{=}CH_2$, $BrCH_2CH{=}CHCH_3$, $BrCH_2CH{=}CHCH_2CH_3$, ___?___

c. $CH_3C{\equiv}C{-}CH_3$, $CH_3C{\equiv}CCH_2CH_3$, $CH_3C{\equiv}CCH_2CH_2CH_3$, ___?___

d. $CH_2{=}CH_2$, $CH_2{=}CH{-}CH{=}CH_2$, $CH_2{=}CHCH{=}CHCH{=}CH_2$, ___?___

4.22 Using complete structural formulas, draw all of the structural isomers for each of the following:

a. $C_4H_8Br_2$ c. C_4H_8
b. $C_5H_{10}Cl_2$ d. C_3H_3Br

4.23 List the main families of hydrocarbons and provide the principal features that are characteristic of each of these families.

4.24 What do we mean by the terms ''saturated'' and ''unsaturated'' as they relate to the structure of hydrocarbons?

4.25 Select the correct member in each of the following pairs of compounds:

a. Has the higher boiling point: 1-iodobutane or 1-chlorobutane
b. Is the more dense: 1-chlorobutane or water
c. Is the more dense: 1-iodopentane or 1-iodobutane
d. Is the more dense: chlorobenzene or water
e. Has the higher boiling point: 1-iodooctane or benzene

4.26 What is meant by the term ''biological magnification''?

4.27 Why has DDT been banned from most applications throughout the United States?

4.28 What is an anesthetic? What differentiates a general anesthetic from a local anesthetic? Both chloroform ($CHCl_3$) and diethylether ($CH_3CH_2{-}O{-}CH_2CH_3$) are anesthetics. Why might chloroform be preferred over ether?

Reactions

4.29 What is the difference between a synthesis and a degradation? Give an example of each.

4.30 What is a reaction mechanism? Why is it useful?

4.31 Curved arrows are used for what purpose in discussing reaction mechanisms?

4.32 How does a substitution reaction differ from an addition reaction? Give an example of each.

4.33 Complete each of the following reactions by supplying the missing reactant or product as indicated by the question mark:

a. $CH_3CH_2CH{=}CHCH_2CH_3 \xrightarrow{?} CH_3CH_2CH_2CH_2CH_2CH_3$

b. $CH_3{-}C({=}CH_2){-}CH_3 \xrightarrow{?} CH_3C(CH_3)_2{-}OH$

c. cyclohexene $\xrightarrow{?}$ bromocyclohexane (Br, H)

d. $CH_3CH_2CH_2CH_3 + O_2 \xrightarrow{heat}$? (Complete combustion)

e. cyclohexane $\xrightarrow{?}$ Cl–cyclohexane

f. ? $\xrightarrow{H_2O,\ H^+}$ cyclopentanol (OH)

4.34 Complete each of the following by supplying the missing product indicated by the question mark:

a. 2-butene $\xrightarrow{HBr\ CCl_4}$?

b. 3-methyl-2-hexene $\xrightarrow{HI,\ CCl_4}$?

c. 1-bromoheptane $\xrightarrow[acetone]{NaI,}$?

d. cyclopentane $\xrightarrow{Cl_2,\ light}$?

e. cyclopentene $\xrightarrow{HCl,\ CCl_4}$?

f. 1,4-dichlorobenzene (Cl, Cl) $\xrightarrow{Br_2,\ FeBr_3}$?

g. 1,2-dichlorobenzene (Cl, Cl) $\xrightarrow{Br_2,\ FeBr_3}$?

h. 1,3-dichlorobenzene (Cl, Cl) $\xrightarrow{Br_2,\ FeBr_3}$?

i. $CH_3CH_2CH_2CH_2CH_2CH_2CH_2{-}I \xrightarrow[H_2O,\ alcohol]{NaOH,}$?

Further Problems

4.35 How could you distinguish between two samples of cyclohexane and hexene (both C_6H_{12}), using a simple chemical test?

4.36 Draw out the mechanism for the addition of HI to cyclohexene.

4.37 a. Define geometric isomers and conformational isomers.
b. Can geometric isomers be separated from one another? Can conformational isomers be separated from one another?
c. Explain the answers that you gave in part b.

4.38 Draw all of the structural and geometric isomers possible for dichlorocyclopropane. Name each of the isomers, using the I.U.P.A.C. Nomenclature System.

4.39 In Section 4.4 we read that the hydrohalogenation of an alkene proceeds through a mechanistic pathway analogous to that for the hydration of an alkene. Draw the curved arrow mechanism for the following reaction:

$$CH_3CH_2—CH{=}CH_2 + HBr \longrightarrow CH_3CH_2—\underset{Br}{\underset{|}{CH}}—\underset{H}{\underset{|}{CH_2}}$$

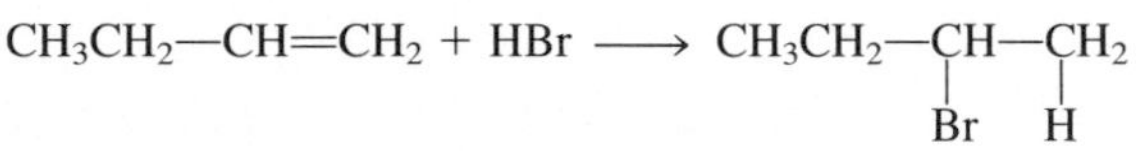

4.40 Using planar rings, draw the possible geometric (*cis* and *trans*) isomers of the following cyclic compounds:

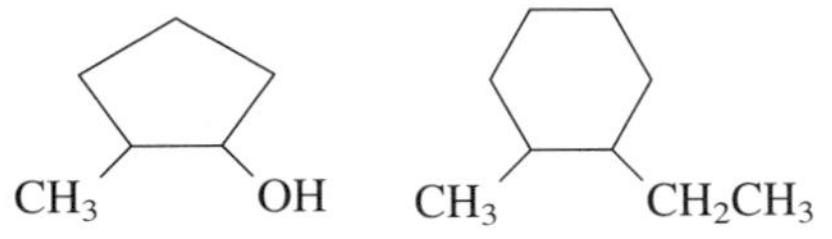

CH_3 OH CH_3 CH_2CH_3

4.41 Quantitatively, one mole of HBr is consumed per mole of alkene, and two moles of HBr is consumed per mole of alkyne. How many moles of HBr would be consumed for each of the following?

a. 2-hexyne

b. cyclohexene

c. 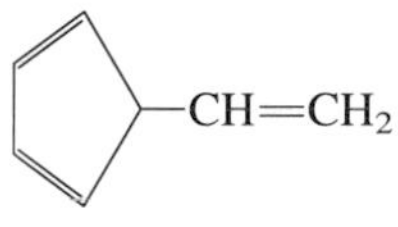 —$CH{=}CH_2$

d. 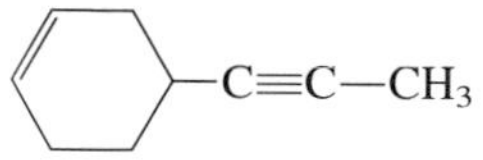 —$C{\equiv}C—CH_3$

4.42 One mole of a compound (C_5H_8) reacts with one mole of H_2 in the presence of a catalyst. Draw a possible structure for this compound.

4.43 A hydrocarbon with a formula C_5H_{10} decolorized Br_2 and consumed one mole of hydrogen upon hydrogenation. Draw all the isomers of C_5H_{10} that are possible on the basis of the above information.

4.44 Draw all of the structural and geometric isomers for the dibromocyclobutanes; name them.

4.45 Draw the most stable geometric isomer for each of the following compounds:

a. 1,2-dimethylcyclohexane

b. 1,3-dimethylcyclohexane

c. 1,4-dimethylcyclohexane

(*Hint:* Build models of each of these isomers and compare them. In general, the less crowding among the atoms, the more favored the structure.)

4.46 Complete the following: Compound **A** (with a molecular formula of C_4H_8) reacts with Br_2 and light to give compound **B** (molecular formula C_4H_7Br). Compound **B** then reacts with HBr to give compound **C** drawn below. Draw the complete structures of and give the names of compounds **A** and **B**.

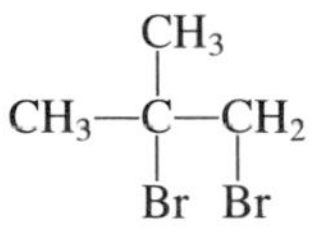

$$CH_3—\overset{CH_3}{\overset{|}{\underset{Br}{\underset{|}{C}}}}—\underset{Br}{\underset{|}{CH_2}}$$

C

This is an example of a road map problem. These problems provide enough information, if used correctly, to derive all of the requested answers by allowing you to travel through the road map in a logical step-by-step sequence. As your foundation in organic chemistry, its reactions, nomenclature, terminology, and so on increases, the complexity of the road map problems will also increase. Still, if approached logically, each individual step can be worked out, and the entire puzzle can be solved.

4.47 There are actually two *trans-* forms and one *cis-* form of 1,2-dibromocyclopropane. Can you draw them? What is the relationship between these forms? (Study the orientation of all of the atoms closely.) Would you predict the physical properties of the forms to be similar or dissimilar? Explain.

4.48 Triple bonds react in a manner analogous to that of double bonds. The extra pair of electrons in the triple bond, however, generally allows two moles of a given reactant to *add* to the triple bond in contrast to one mole with the double bond. The "rich get richer" rule holds. Predict the major product in each of the following reactions:

a. acetylene with 2 moles of HCl

b. propyne with 2 moles of HBr

c. 2-butyne with two moles of HI

4.49 a. Some aromatic compounds have less than desirable physiological effects. Provide an example.

b. Benzene is the simplest aromatic compound, and it has been routinely used as a solvent in many organic chemistry laboratories. Yet it has potentially severe physiological side effects. Go to the library and try to find articles related to the adverse side effects of benzene usage. What alternative solvents are suggested.

4.50 The term LD_{50} is used to indicate the lethal dosage of a given chemical to 50% of the animals being tested. That is, LD_{50} means the dose of a chemical that caused the death of 50% of the test animals. The dosage is normally expressed in units of weight of chemical to weight of test animal (e.g., 0.01 gram of chemical per kilogram of animal body weight). Go to the library and look up the LD_{50} of several of the common pesticides mentioned in this chapter.

4.51 How would you expect that the LD_{50} (see Problem 4.50) of a given chemical is determined for humans?

4.52 a. What are freons?

b. Draw several simple freons.

c. What are the common household uses for this family?

d. What environmental problem do they present?

(*Note:* You will have to do some outside reading to answer this question.)

4.53 Bromine is often used as a laboratory spot test for unsaturation in an aliphatic hydrocarbon. Bromine in CCl_4 is red in color. When bromine reacts with an alkene or alkyne, the alkyl halide formed is colorless; hence a disappearance of the red color is a positive test for unsaturation. A student tested the contents of two vials, **A** and **B**, both containing compounds with a molecular formula, C_6H_{12}. Vial **A** decolorizes bromine, but vial **B** did not. How may the results for vial **B** be explained? What class of compound would account for this?

4.54 Explain in your own words what is meant by the term "aromaticity."

CHAPTER 5

Alcohols, Phenols, Thiols, and Ethers

OBJECTIVES

- Know the names and write structures for the common alcohols, phenols, ethers, and thiols.
- Relate the various members of these four families to their utility—natural, commercial, health, environmental, or industrial.
- Be able to classify alcohols as primary, secondary, or tertiary.
- Be familiar with the physical properties of each of the four families listed above.
- Write equations for the dehydration and oxidation of alcohols.
- Write the mechanism for the dehydration of an alcohol.
- Be able to design simple multistep syntheses involving alcohols and other families.

OUTLINE

Alcohols have the general structure R—OH, where R— represents any alkyl group and —OH is the hydroxyl group. Phenols are similar in structure to alcohols; they contain an aryl group in place of the alkyl group of the alcohols, Ar—OH. The functional group of these two families is the **hydroxyl group:**

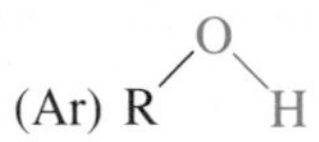

$$\text{(Ar) R} \diagup \overset{\text{O}}{} \diagdown \text{H}$$

The hydroxyl group of alcohols and phenols

Alcohols and phenols may be viewed as substituted water molecules in which one of the H atoms of water has been replaced by an alkyl or aryl group:

$$\underset{\text{Water}}{\text{H—O—H}} \quad \text{versus} \quad \underset{\text{Alcohol}}{\text{R—O—H}}$$

Ethers are similar in structure to the alcohols. They contain two alkyl or aryl groups attached to an oxygen atom as shown here:

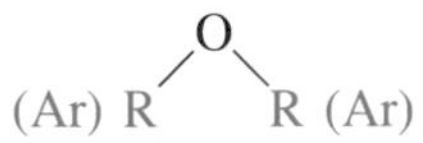

$$\text{(Ar) R} \diagup \overset{\text{O}}{} \diagdown \text{R (Ar)}$$

Ethers

The ether functional group is the C—O—C group. It is common to all ethers. An ether may be viewed as a substituted alcohol in which the H atom of the hydroxyl group has been substituted by an alkyl or aryl group:

$$\underset{\text{Alcohol}}{\text{(Ar) R—O—H}} \quad \text{versus} \quad \underset{\text{Ether}}{\text{(Ar) R—O—R (Ar)}}$$

This seemingly minor modification in structure results in a significant difference in the chemical and physical properties of these two families. The nature and impact of these differences are illustrated in the sections that follow.

Thiols, the family of compounds produced by the substitution of a sulfur atom for the oxygen atom in alcohols and phenols, are also presented:

$$\text{(Ar) R—SH}$$

A thiol

Of these three classes of organic chemicals the functional group of the alcohols, in particular, plays a central role in the structure and chemistry of biological molecules. The hydroxyl group is found in sugars (carbohydrates), proteins (involving amino acids), and lipids. Examples are provided below:

Chapters 7, 9, and 11

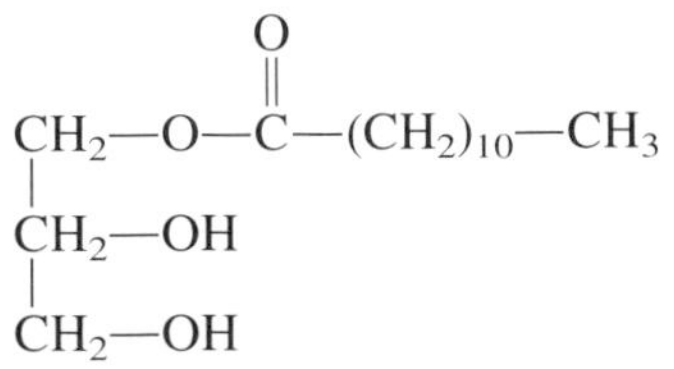

$$\begin{array}{l} \overset{\text{O}}{\|} \\ CH_2—O—C—(CH_2)_{10}—CH_3 \\ | \\ CH_2—OH \\ | \\ CH_2—OH \end{array}$$

Monolaurin, *a lipid*

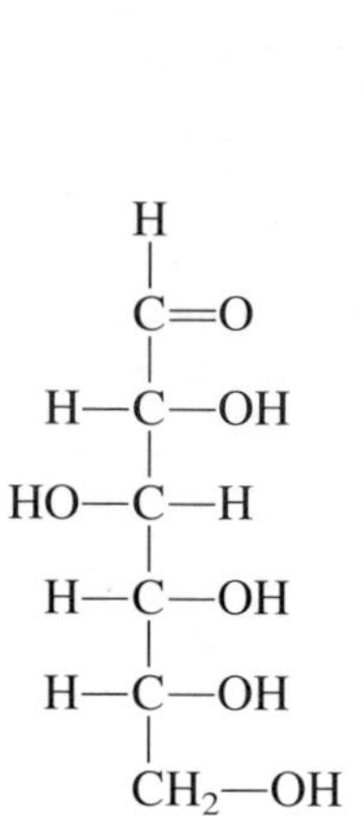

$$\begin{array}{c} H \\ | \\ C{=}O \\ | \\ H{-}C{-}OH \\ | \\ HO{-}C{-}H \\ | \\ H{-}C{-}OH \\ | \\ H{-}C{-}OH \\ | \\ CH_2{-}OH \end{array}$$

D-Glucose, *a sugar*

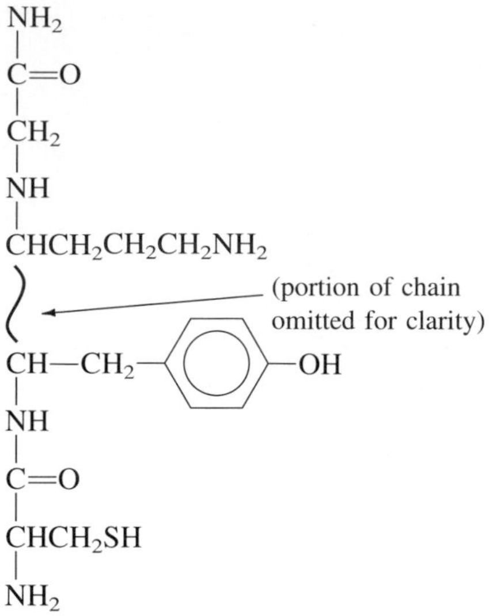

$$\begin{array}{l} NH_2 \\ | \\ C{=}O \\ | \\ CH_2 \\ | \\ NH \\ | \\ CHCH_2CH_2CH_2NH_2 \\ \wr \\ CH{-}CH_2{-}C_6H_4{-}OH \\ | \\ NH \\ | \\ C{=}O \\ | \\ CHCH_2SH \\ | \\ NH_2 \end{array}$$

(portion of chain omitted for clarity)

Lysine vasopressin (partial structure), *a protein*

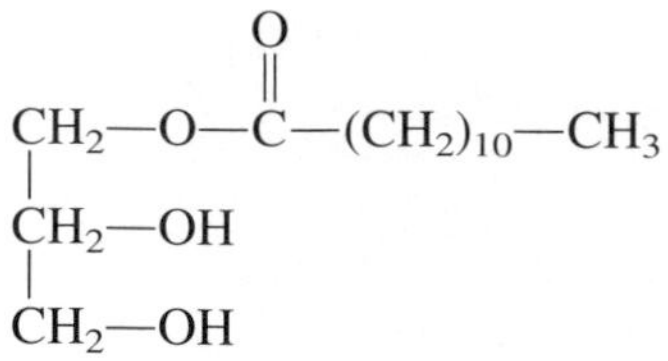

$$\begin{array}{l} CH_2{-}O{-}\overset{\overset{O}{\|}}{C}{-}(CH_2)_{10}{-}CH_3 \\ | \\ CH_2{-}OH \\ | \\ CH_2{-}OH \end{array}$$

Monolaurin, *a lipid*

In these systems the hydroxyl group is often involved in a variety of reactions such as oxidation and reduction and hydration and dehydration. In anaerobic glycolysis (the metabolic pathway via which glucose, a simple sugar, is converted into pyruvate with concomitant production of energy), several steps center on the reactivity of the hydroxyl group.

Section 13.3

One common fermentation process carried out by yeast produces the majority of consumable alcohol (ethanol) in the world. This process and its role in cellular energy metabolism will also be discussed later in the book.

Section 13.4

The majority of this chapter will focus on a discussion of the properties, occurrence, methods of preparation, and reactions of the family of alcohols. Two important and related reactions—hydration and dehydration—will be presented.

5.1 ALCOHOLS: STRUCTURE AND PHYSICAL PROPERTIES

The C—O—H portion of alcohols is planar; this is similar to the structure of water. The bivalent oxygen (an oxygen with two groups attached) as well as the two atoms attached to it (H—, and/or the C— of the R group or the Ar group) are all in the same plane. The bond angles of the C—O—H bond are approximately 104°, the same as the bond angle of water.

Alcohols contain the highly polar hydroxyl group and, as a result, are quite polar. This is reflected in their physical properties and particularly in their ability to participate in intermolecular and intramolecular hydrogen bonding. As a result of the former, they boil at higher temperatures than pure hydrocarbons or ethers having the same carbon content or similar molecular weight (see Table 5.1 and Figure 5.1).

Compare the boiling points of each of the following compounds, all of which have approximately the same molecular weight. Clearly, the functional group (for example, the hydroxyl group in 1-propanol) plays a significant role in determining the properties of each of these families. 1-Propanol (continuing a hydroxyl group) boils at 97.2°, in contrast to butane (containing no polar group), which boils at −0.5°.

$CH_3CH_2CH_2CH_3$	$CH_3{-}O{-}CH_2CH_3$	$CH_3CH_2CH_2{-}O{-}H$
Butane (*n*-butane) mol. wt. 58 b.p. −0.5°	Methoxyethane (ethyl methyl ether) mol. wt. 60 b.p. 7.9°	1-Propanol (*n*-propyl alcohol) mol. wt. 60 b.p. 97.2°

TABLE 5.1 Physical Properties of Alcohols

Name	Structure	m.p. (°C)	b.p. (°C)	Solubility (g/100 g H_2O)
Methanol (methyl alcohol)	CH_3—OH	−97.0	64.7	∞
Ethanol (ethyl alcohol)	CH_3CH_2—OH	−114.0	78.3	∞
1-Propanol (*n*-propyl alcohol)	$CH_3CH_2CH_2$—OH	−126.0	97.2	∞
2-Propanol (isopropyl alcohol)	$(CH_3)_2CH$—OH	−88.5	82.3	∞
1-Butanol (*n*-butyl alcohol)	$CH_3CH_2CH_2CH_2$—OH	−90.0	117.7	7.8
1-Pentanol (*n*-pentyl alcohol)	$CH_3CH_2CH_2CH_2CH_2$—OH	−78.5	138.0	2.3
1-Hexanol (*n*-hexyl alcohol)	$CH_3(CH_2)_4CH_2$—OH	−52.0	155.8	0.6
1-Heptanol (*n*-heptyl alcohol)	$CH_3(CH_2)_5CH_2$—OH	−34.0	176.0	0.2
1-Octanol (*n*-octyl alcohol)	$CH_3(CH_2)_6CH_2$—OH	−16.0	194.0	0.05
1-Nonanol (*n*-nonyl alcohol)	$CH_3(CH_2)_7CH_2$—OH	−5.5	212.0	Insoluble
1-Decanol (*n*-decyl alcohol)	$CH_3(CH_2)_8CH_2$—OH	6.0	233.0	Insoluble
Benzyl alcohol	C_6H_5—CH_2—OH	−15.3	205.0	4.0

$CH_3CH_2CH_2CH_2CH_3$ Pentane (*n*-pentane) mol. wt. 72 b.p. 36°

$CH_3CH_2CH_2CH_2CH_2$—O—H 1-Butanol (*n*-butyl alcohol) mol. wt. 74 b.p. 117.7°

The smaller members of the alcohol family, those that contain fewer than four or five carbon atoms, are highly soluble in water. Many larger alcohols are moderately soluble in water and other polar solvents. The solubility properties and high boiling points that this family exhibits are directly attributable to the hydroxyl group.

The oxygen and hydrogen atoms that form the hydroxyl group have very different electronegativities and the large electronegativity difference results in a polar bond. Therefore alcohols are polar. This charge separation within individual molecules results in strong attractive forces between the molecules (Figure 5.1). Because they are strongly attracted to each other, a reasonable amount of energy is required to separate these molecules; high boiling points result. Polar molecules are not only attracted to each other but are also attracted to other polar molecules (''like dissolves like''). Intermolecular attraction is the reason for the solubility of polar molecules in polar solvents.

The presence of polar hydroxl groups in some of the large biological molecules, for instance, the proteins and nucleic acids, allows hydrogen bonding that keeps these molecules in the shapes needed for biological function.

FIGURE 5.1
Hydrogen bonding in alcohols (interactions between alcohol molecules shown).

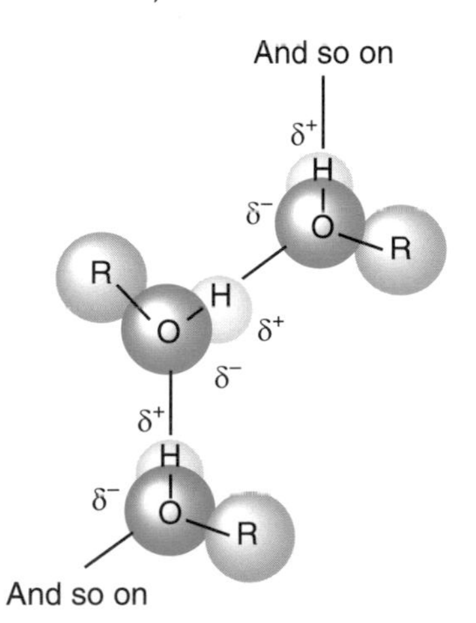

5.2 ALCOHOLS: NOMENCLATURE

I.U.P.A.C. names

Alcohols are named according to the *parent compound,* the longest continuous carbon chain containing the —O—H group. The *-e* ending of the alkane chain is replaced by the *-ol* ending of the alcohols. For example:

Ethan*e* becomes ethan*ol*,

Pentan*e* becomes pentan*ol*,

The longest chain is then numbered to give the hydroxl group the lowest possible number. All substituents are named and numbered appropriately and added as prefixes to the ''alkanol'' name.

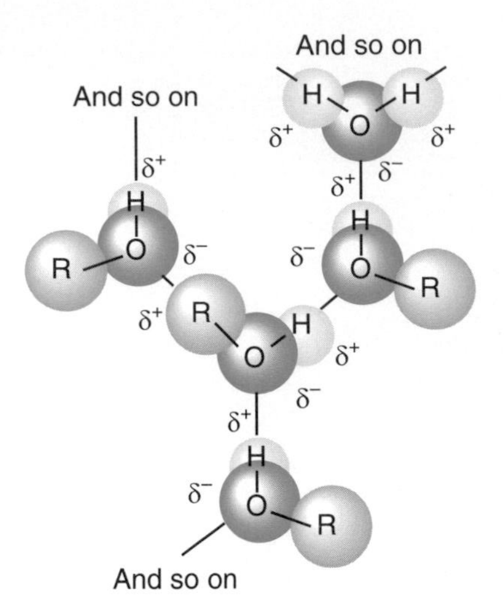

FIGURE 5.2
Hydrogen bonding between alcohols and a polar solvent (interactions between alcohol molecules and other polar molecules shown).

The I.U.P.A.C. nomenclature for several common alcohols is illustrated below:

CH_3CHCH_3 (OH on C-2) — 2-Propanol (isopropyl alcohol)

$CH_3CH_2CH_2CH_2CH_2CHCHCH_3$ (HO and CH_3 on the CH carbons) — 2-Methyl-3-octanol

$HO—CH_2CH_2—OH$ — 1,2-Ethanediol (ethylene glycol)

$HO—CH_2CHCHCHCH_3$ (Cl, Cl, OH on the CH carbons) — 2,3-Dichloro-1,4-pentanediol

$CH_3CHCH_2CHCH_2CHCH_2CH_3$ (OH, Br, CH_2CH_3 on the CH carbons) — 4-Bromo-6-ethyl-2-octanol

Example 5.1

Naming a compound using the I.U.P.A.C. Nomenclature System.

1 2 3 4 5 6 7
$CH_3CHCH_2CH_2CH_2CHCH_3$ (OH on C-2, CH_3 on C-6)

Parent compound: Heptane (becomes heptanol)

Position of —OH: Carbon-2 (*not* carbon-6)

Substituents: 6-Methyl (*not* 2-methyl)

Name: 6-Methyl-2-heptanol

(Cyclohexane ring: OH on carbon 1, Br on carbon 3; ring positions numbered 1, 2, 3)

Parent compound: Cyclohexane (becomes cyclohexanol)

Position of —OH: Carbon-1 (*not* carbon-3)

Substituents: 3-Bromo (*not* 5-bromo)

Name: 3-Bromocyclohexanol (it is assumed that the —OH is on carbon 1 in cyclic structures)

Alcohols containing two hydroxyl groups are named *-diols;* those with three hydroxyl groups are *-triols*. A number giving the position of each of the hydroxyl groups is required.

Common names

Section 3.2

The common names for the alcohols are analogous to those of the corresponding alkyl halides. The halide ending is replaced with the name alcohol. See the examples above and below.

CH_3CHCH_3 (OH on C-2) — Isopropyl alcohol

$CH_3CH_2CH_2CH_2CH_2—OH$ — *n*-Pentyl alcohol

$CH_3—C—CH_3$ (CH_3 above, OH below the central C) — *tert*-Butyl alcohol

Question 5.1 Predict which compound in each of the following pairs has the higher boiling point. Briefly explain the reason for your choice.

a. Methanol or methane
b. Methanol or ethanol
c. Diethyl ether or 1-butanol
d. Diethyl ether or butane
e. 1-Butanol or 2,3-butanediol

Question 5.2 Name each of the following compounds, using the I.U.P.A.C. Nomenclature System:

a. $CH_3CHCH_2CH_2CH_2{-}OH$ (with CH_3 on C-2)

b. $CH_3CHCH_2CHCH_3$ (with OH on C-2 and CH_2CH_3 on C-4)

c. 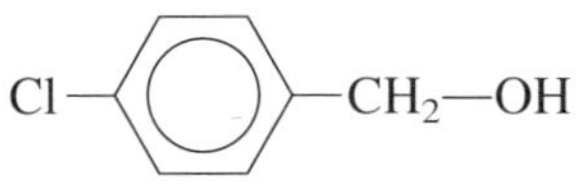 $Cl{-}C_6H_4{-}CH_2{-}OH$ (Cl and CH_2—OH on a benzene ring, para)

d. CH_2CHCH_2 (with OH on each carbon: OH OH OH)

e. $CH_3CH_2CHCHCH_2CH_2{-}OH$ (with Cl and CH_3 on the third and fourth carbons)

5.3 ALCOHOLS: WHERE DO THEY COME FROM?

Industrial sources

Several simpler alcohols are important industrially.

Methanol

Methanol (methyl alcohol), $CH_3{-}OH$, is used as a solvent. It is also the principal starting material for methanal (formaldehyde), a very important industrial chemical. It is a colorless and odorless liquid. Methyl alcohol is often referred to as "wood alcohol" because it is possible to obtain methyl alcohol by heating wood in the absence of air. This process is called **destructive distillation.** Methanol is toxic and causes blindness and perhaps death if ingested. Methanol is prepared from carbon monoxide:

Section 6.3

A Clinical Perspective: Aldehydes in Medicine in Chapter 6

$$CO + 2H_2 \xrightarrow[\text{metal catalysts}]{400°,\ 200\ \text{atm},} CH_3{-}OH \qquad 5.1$$

Methanol (methyl alcohol)

Ethanol

Ethanol (ethyl alcohol), $CH_3CH_2{-}OH$, is the alcohol in alcoholic beverages. It is also widely used as a solvent and as a raw material for the preparation of other organic chemicals, for example, ethanal (acetaldehyde). Ethyl alcohol is a colorless and odorless liquid. It is often called "grain alcohol" because it may be produced by the fermentation of various sugars by yeast. Today it is principally prepared by the hydration of ethene:

Section 6.3

Section 13.4

$$CH_2{=}CH_2 \xrightarrow[325°,\ \text{catalyst}]{H_2O\ \text{(steam)},} CH_3CH_2{-}OH \qquad 5.2$$

Ethene (ethylene) → Ethanol (ethyl alcohol)

The ethanol used in alcoholic beverages comes from the fermentation of the carbohydrates (sugars and starches) that serve as the raw materials in the preparation of the beverages. The beverage produced depends on the raw material used and on the fermentation process: scotch (grain), bourbon (corn), burgundy wine (grapes and grape skins), and chablis wine (grapes without red skins), for example, all result from fermentation processes. The **fermentation** process is illustrated below:

$$(C_6H_{10}O_5)_n + nH_2O \xrightarrow[\text{enzyme action}]{\text{several steps involving}} CH_3CH_2{-}OH + H_2O + CO_2 \qquad 5.3$$

Carbohydrates (starch) (n = a large number, in the thousands) — Ethanol (ethyl alcohol)

The alcoholic beverages listed above have quite different alcohol concentrations. Wines are generally 12–13% alcohol because the yeasts that produce the ethanol are killed by ethanol concentrations of 12–13% (wine producers are always trying to find strains of yeast with greater ethanol tolerance!). To produce bourbon or scotch with an alcohol concentration of 40–45% ethanol (80 or 90 proof), the original fermentation products must be distilled. The details of the ethanol fermentation process are discussed in the section on biochemistry.

Section 13.4

Pure ethanol (100% ethanol) is obtained with great difficulty. Its sale and use are also regulated by the federal government. To prevent illegal use of pure ethanol, it is converted to **denatured alcohol** by the addition of a denaturing agent (methanol is frequently used), which makes it unfit to drink.

2-Propanol

2-Propanol (isopropyl alcohol),

$$CH_3{-}\underset{\displaystyle OH}{\underset{|}{CH}}{-}CH_3$$

is common "rubbing alcohol." It is used as a disinfectant, astringent (skin-drying agent), and industrial solvent or product (for example, dry gas, windshield-cleaning solution) and has wide applicability as a raw material in the synthesis of organic chemicals. It is colorless, has a very slight odor, and is very toxic when ingested.

Patients with high fevers are given alcohol baths to reduce the body temperature. Rapid evaporation of the alcohol from the skin results in skin cooling.

1,2-Ethanediol

1,2-Ethanediol (ethylene glycol),

$$\underset{\displaystyle OH}{\underset{|}{CH_2}}{-}\underset{\displaystyle OH}{\underset{|}{CH_2}}$$

is used as automobile antifreeze. When added to water in the radiator, the ethylene glycol solute lowers the freezing point and raises the boiling point of the resulting solution. Ethylene glycol is one of the principal starting materials in the synthesis of polyesters, the synthetic "double-knit" fibers used by the clothing industry.

1,2,3-Propanetriol

1,2,3-Propanetriol (glycerol),

$$\underset{\displaystyle OH}{\underset{|}{CH_2}}{-}\underset{\displaystyle OH}{\underset{|}{CH}}{-}\underset{\displaystyle OH}{\underset{|}{CH_2}}$$

is a viscous, sweet-tasting, nontoxic liquid. It is very soluble in water and is used in cosmetics, pharmaceuticals, and lubricants. Glycerol is obtained as a by-product of the hydrolysis of fats.

Section 9.2

Laboratory sources

In the laboratory, alcohols are prepared by the **hydration** of alkenes:

Section 4.4

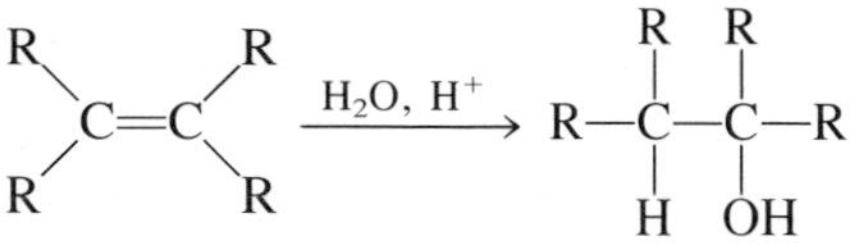

$$R_2C{=}CR_2 \xrightarrow{H_2O,\ H^+} R{-}\underset{H}{\overset{R}{C}}{-}\underset{OH}{\overset{R}{C}}{-}R \qquad 5.4$$

Alkene — Alcohol

The hydration of alkenes was discussed in Section 4.4.

Alcohols may also be prepared via the reduction of aldehydes and ketones, a reaction that is shown below and discussed in Section 6.4:

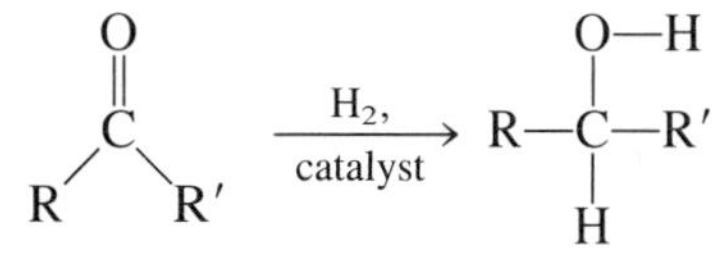

$$R{-}\overset{O}{\overset{\|}{C}}{-}R' \xrightarrow[\text{catalyst}]{H_2,} R{-}\underset{H}{\overset{O-H}{C}}{-}R' \qquad 5.5$$

Aldehyde or ketone — Alcohol

In Section 3.9 we were introduced to the **halide exchange** reaction, an example of a nucleophilic substitution reaction. This reaction can also be used in the preparation of alcohols. The reaction is useful only in the synthesis of primary alcohols or methyl alcohol from the corresponding alkyl halide.

$$HO^- + CH_3CH_2{-}Br \longrightarrow CH_3CH_2{-}OH + Br^- \qquad 5.6$$

Bromoethane (ethyl bromide) — Ethanol (ethyl alcohol)

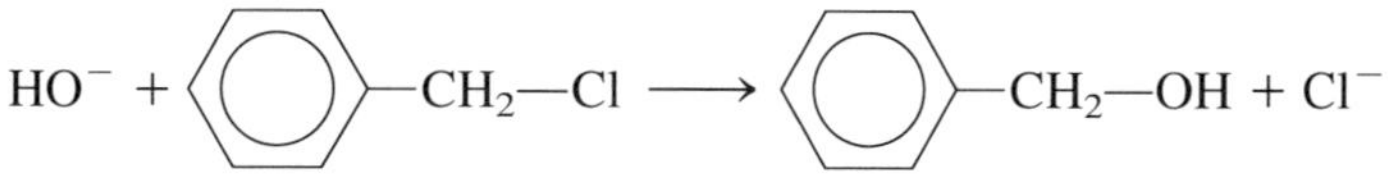

$$HO^- + C_6H_5{-}CH_2{-}Cl \longrightarrow C_6H_5{-}CH_2{-}OH + Cl^- \qquad 5.7$$

Chloromethylbenzene (benzyl chloride) — Hydroxymethylbenzene (benzyl alcohol)

Question 5.3 Predict the major product obtained on reacting each of the following alkenes with water and a trace of acid:

a. Ethene
b. Propene
c. 1-Butene
d. 2-Butene
e. 2-Methylpropene
f. 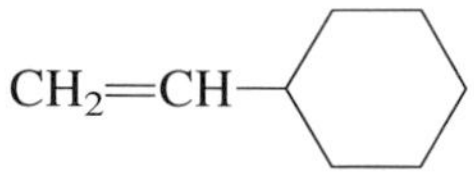 $CH_2{=}CH{-}C_6H_{11}$ (cyclohexyl)

5.4 CLASSIFICATION OF ALCOHOLS

Alcohols are classified as

methyl, primary (1°), secondary (2°), or **tertiary (3°),**

Section 3.8 depending on the structure of the alkyl group to which the hydroxyl —O—H is attached. This classification system is identical to the system utilized for alkyl halides. Examples are provided here:

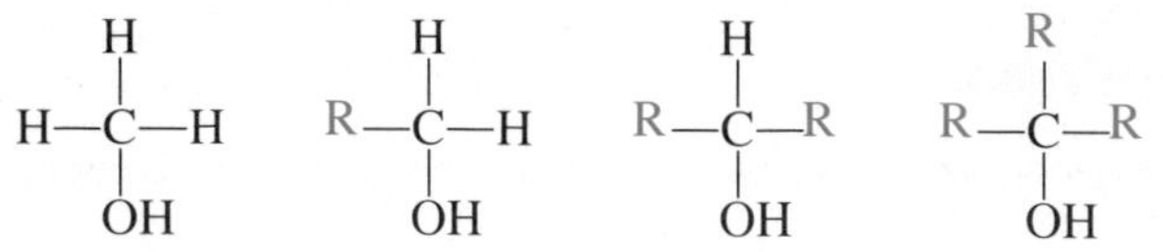

$$\underset{\text{Methyl alcohol}}{\text{H—}\overset{\text{H}}{\underset{\text{OH}}{\text{C}}}\text{—H}} \qquad \underset{1^\circ\text{ alcohol}}{\text{R—}\overset{\text{H}}{\underset{\text{OH}}{\text{C}}}\text{—H}} \qquad \underset{2^\circ\text{ alcohol}}{\text{R—}\overset{\text{H}}{\underset{\text{OH}}{\text{C}}}\text{—R}} \qquad \underset{3^\circ\text{ alcohol}}{\text{R—}\overset{\text{R}}{\underset{\text{OH}}{\text{C}}}\text{—R}}$$

Question 5.4 Classify each of the following alcohols as 1°, 2°, or 3° or aromatic:

a. $CH_3CH_2CH_2CH_2CH_2—OH$

b. $CH_3CH_2\underset{|}{\underset{OH}{C}}HCH_2CH_3$

c. $CH_3\underset{OH}{C}HCH_3$

d. $CH_3CH_2CH_2—\overset{CH_3}{\underset{OH}{C}}—CH_3$

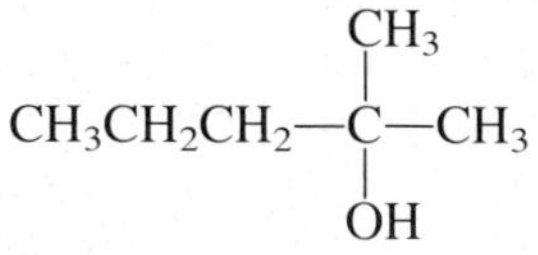

e. 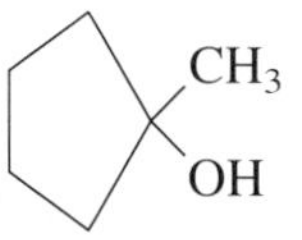 (cyclohexane ring bearing CH_3 and OH on the same carbon)

f. 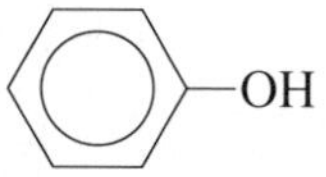 (benzene ring)—OH

g. 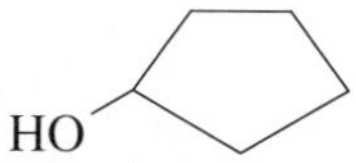 HO—(cyclopentane ring)

5.5 REACTIONS OF ALCOHOLS

Dehydration

Alcohols are **dehydrated** (lose water) when heated with concentrated sulfuric acid (H_2SO_4) or phosphoric acid (H_3PO_4):

$$\underset{\text{Ethanol (ethyl alcohol)}}{\underset{\text{H}\quad\ \text{OH}}{CH_2—CH_2}} \xrightarrow{H^+,\ \text{heat}} \underset{\text{Ethene (ethylene)}}{CH_2{=}CH_2 + H_2O} \qquad 5.8$$

In Section 4.4 we saw that alkenes could be hydrated to give alcohols. Dehydration is the reverse process—the conversion of an alcohol back to an alkene. For example:

$$\underset{\text{1-Propanol }(n\text{-propyl alcohol})}{CH_3CH_2CH_2—OH} \xrightarrow{H^+,\ \text{heat}} \underset{\text{Propene (propylene)}}{CH_3CH{=}CH_2} + H_2O \qquad 5.9$$

$$\underset{\text{2-Butanol}}{CH_3CH_2—\underset{OH}{C}H—CH_3} \xrightarrow[\text{heat}]{H^+,} \underset{\text{1-Butene}}{CH_3CH_2—CH{=}CH_2} + \underset{\textbf{2-Butene (major product)}}{CH_3—CH{=}CH—CH_3} + H_2O \qquad 5.10$$

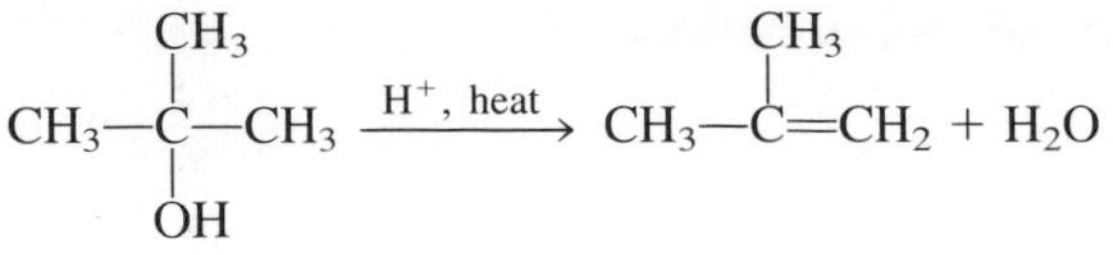

$$CH_3-\overset{\overset{\large CH_3}{|}}{\underset{\underset{\large OH}{|}}{C}}-CH_3 \xrightarrow{H^+,\ heat} CH_3-\overset{\overset{\large CH_3}{|}}{C}=CH_2 + H_2O \qquad 5.11$$

2-Methyl-2-propanol (*tert*-butyl alcohol) → 2-Methylpropene (isobutylene)

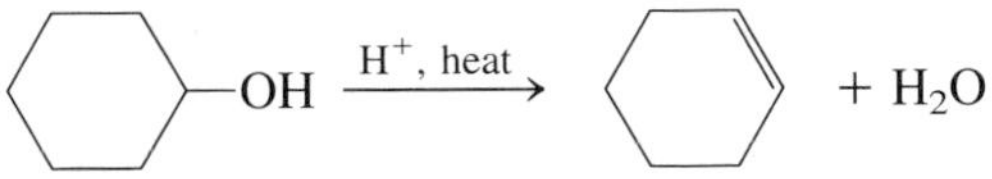

$$\text{Cyclohexanol (C}_6\text{H}_{11}\text{–OH)} \xrightarrow{H^+,\ heat} \text{Cyclohexene} + H_2O \qquad 5.12$$

Cyclohexanol → Cyclohexene

The reaction may produce a mixture of products (for example, reaction 5.10 above). In those instances, when appropriate reaction conditions are used, one product often predominates in the mixture because of the stability of the alkene and the rate at which it is formed.

Example 5.2

Predicting the products of alcohol dehydration.

Assuming that no rearrangement occurs, the product(s) of a dehydration of an alcohol will contain a double bond in which one of the carbons was the original **carbinol carbon**—the carbon to which the hydroxyl group is attached. Consider the following example:

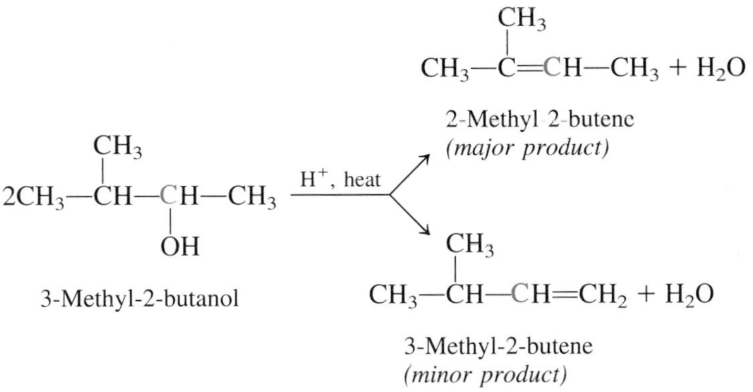

$$2CH_3-\overset{\overset{\large CH_3}{|}}{CH}-\underset{\underset{\large OH}{|}}{CH}-CH_3 \xrightarrow{H^+,\ heat} CH_3-\overset{\overset{\large CH_3}{|}}{C}=CH-CH_3 + H_2O \;\; + \;\; CH_3-\overset{\overset{\large CH_3}{|}}{CH}-CH=CH_2 + H_2O \qquad 12.13$$

3-Methyl-2-butanol → 2-Methyl-2-butene (*major product*) + 3-Methyl-2-butene (*minor product*)

It is clear that both the major and minor products have a double bond to carbon number 2 in the original alcohol (this carbon is set off in color). By realizing these types of structural rules you can more easily predict the products of a given reaction. Alternatively, if given the products, you can determine the needed starting material to effect the conversion. The carbon skeleton of both the reactants and products should always be compared to one another as the first step in designing an organic synthesis or degradation.

Question 5.5 Draw all of the alkene products that would be produced on dehydration of the following alcohols:

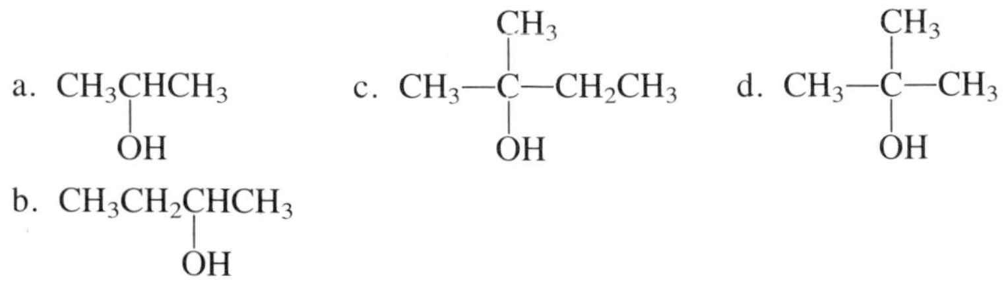

a. $CH_3\underset{\underset{\large OH}{|}}{C}HCH_3$

b. $CH_3CH_2\underset{\underset{\large OH}{|}}{C}HCH_3$

c. $CH_3-\overset{\overset{\large CH_3}{|}}{\underset{\underset{\large OH}{|}}{C}}-CH_2CH_3$

d. $CH_3-\overset{\overset{\large CH_3}{|}}{\underset{\underset{\large OH}{|}}{C}}-CH_3$

The mechanism of dehydration

The mechanism of dehydration is the reverse of that of hydration that was presented in Section 4.5.

Step 1. A proton can be added to the alcohol to produce an oxonium ion:

$$CH_3{-}CH_2{-}\ddot{\underset{..}{O}}H \xrightarrow{H^+} CH_3{-}CH_2{-}\overset{H}{\underset{H}{O}}{:}^{+}$$

Alcohol — An oxonium ion

Electrons flow from the oxygen atom of the alcohol to form a new —O—H bond. The electron flow is shown by the "curved arrow" above.

Step 2. The oxonium ion loses water to give a carbocation:

$$CH_3{-}CH_2{-}\overset{H}{\underset{H}{O}}{:}^{+} \longrightarrow CH_3{-}CH_2^{+} + H_2\ddot{O}{:}$$

An oxonium ion — A carbocation

Carbocations are cations of carbon. They are carbon atoms that carry a positive charge, a result of having only six bonding electrons (three pairs). A neutral carbon atom, in contrast, is surrounded by eight bonding electrons (four pairs). The carbocation has two fewer electrons involved in bonding than the corresponding neutral carbon, hence the positive charge. Similarly, the oxonium ion is a positively charged oxygen, an oxygen that contains six bonding electrons (three pairs) and a nonbonding pair. A neutral oxygen atom normally has four bonding electrons (two pairs) and two nonbonding pairs. We find that the oxygen is also electron-deficient and also bears a positive charge. Though the structural reasons are different, both the carbocation and the oxonium ion are electron-deficient species, and both are positively charged.

Step 3. In the final step the carbocation, a reactive species, loses a proton to the solvent to give the alkene:

$$\underset{H}{CH_2}{-}CH_2^{+} \longrightarrow CH_2{=}CH_2 + H^+$$

A carbocation — An alkene

The dehydration of 2-phosphoglycerate to phosphoenolpyruvate is a critical step in the anaerobic (in the absence of oxygen) metabolism of the sugar glucose. *Section 13.3*

$$O^-H^+{-}C({=}O){-}CH(\text{—P}){-}CH_2{-}OH \longrightarrow O^-H^+{-}C({=}O){-}C(\text{—P}){=}CH_2 + H_2O \qquad 5.14$$

2-Phosphoglycerate — Phosphenolpyruvate; (P) = HO—P(=O)(OH)—O—

Contrast this dehydration mechanism to the hydration mechanism in Section 4.5.

Oxidation reactions

Alcohols are oxidized, using a variety of oxidizing agents, to aldehydes, ketones, and carboxylic acids. The most commonly used oxidizing agents are potassium permanganate ($KMnO_4$) and potassium dichromate ($K_2Cr_2O_7$). The general reaction follows:

Chapter 6

Chapter 8

The oxidizing agent is actually dichromic acid ($H_2Cr_2O_7$). The dichromic acid is produced by dissolving potassium dichromate in an aqueous acid such as phosphoric or sulfuric acid.

Note that the symbol [O] is used throughout this book to designate any oxidizing agent.

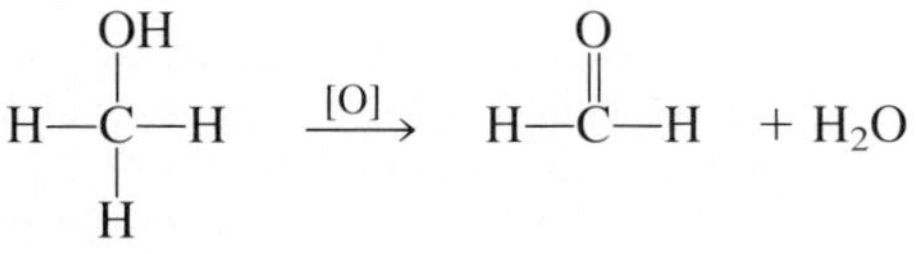

$$\text{H—C(OH)(H)—H} \xrightarrow{[O]} \text{H—C(=O)—H} + H_2O \qquad 5.15$$

Methanol (methyl alcohol) An alcohol — Methanal (formaldehyde) An aldehyde

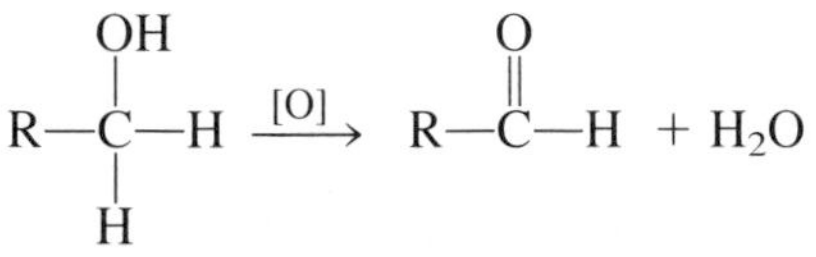

$$\text{R—C(OH)(H)—H} \xrightarrow{[O]} \text{R—C(=O)—H} + H_2O \qquad 5.16$$

1° alcohol — An aldehyde

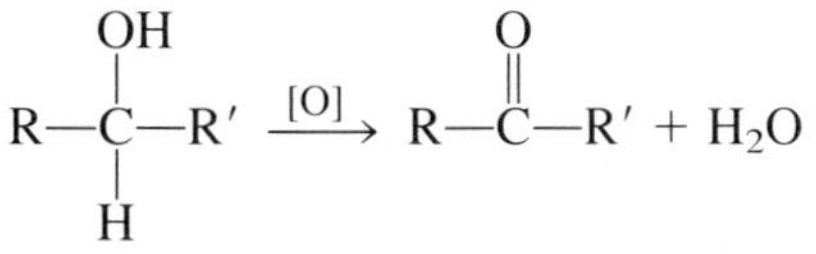

$$\text{R—C(OH)(H)—R}' \xrightarrow{[O]} \text{R—C(=O)—R}' + H_2O \qquad 5.17$$

2° alcohol — A ketone

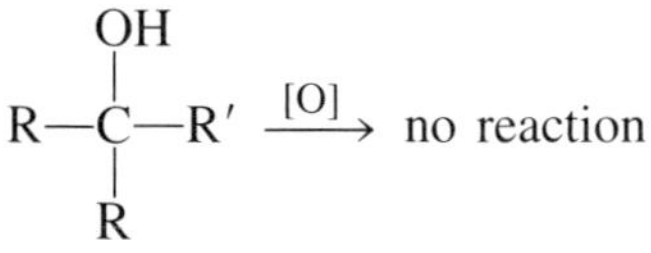

$$\text{R—C(OH)(R)—R}' \xrightarrow{[O]} \text{no reaction} \qquad 5.18$$

3° alcohol

Methanol and primary alcohols produce aldehydes, while secondary alcohols form ketones. Tertiary alcohols cannot be oxidized. The mechanism involved in these oxidations requires that the carbinol carbon contain at least one C—H bond in order for oxidation to occur. Tertiary alcohols contain three C—C bonds to the carbinol carbon. Consequently, they are unable to undergo oxidation. Aldehydes can undergo further reaction to give carboxylic acids, a reaction that we discuss in Sections 6.4 and 8.3. The general reaction is shown below:

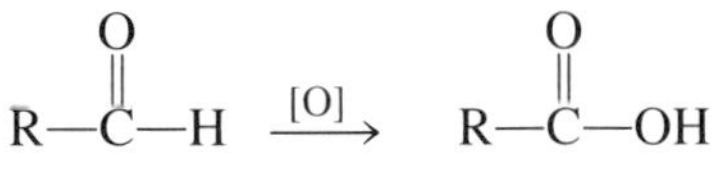

$$\text{R—C(=O)—H} \xrightarrow{[O]} \text{R—C(=O)—OH} \qquad 5.19$$

An aldehyde — A carboxylic acid

Potassium permanganate or potassium dichromate may also be used to effect this transformation. Aldehydes are often difficult to prepare because of the ease with which they are further oxidized to the corresponding carboxylic acid in a secondary reaction. Several examples are provided below:

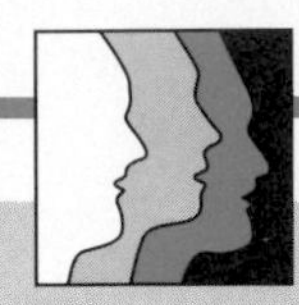

A HUMAN PERSPECTIVE

Alcohol Consumption and the Breathalyzer Test

Ethanol (ethyl alcohol) has been used widely as a beverage, a medicinal, and a solvent in numerous pharmaceutical preparations. Such common usage often overshadows the fact that ethanol, like the other members of its family, is a toxic substance.

Ethanol consumption is associated with numerous long-term effects, chiefly liver and other internal organ damage and alcoholism. For these reasons, medicinal formulations such as cough, cold, and influenza remedies are now manufactured in alcohol-free form.

Short-term effects, linked to the social use of ethanol, center on its effects on behavior, reflexes, and coordination. Blood alcohol levels of 0.05–0.15% seriously inhibit coordination. Blood levels in excess of 0.10% are, in most states, considered evidence of intoxication. Blood alcohol levels in the range of 0.30–0.50% produce unconsciousness and the risk of death.

The loss of some coordination and reflex action is particularly serious when the affected individual attempts to operate a motor vehicle. Law enforcement has come to rely on the "breathalyzer" test to establish the guilt or innocence of individuals suspected of driving while intoxicated.

The suspect is required to exhale into a solution that will react with the unmetabolized alcohol in the breath. The partial pressure of the alcohol in the exhaled air has been demonstrated to be proportional to the blood alcohol level. The solution is an acidic solution of dichromate ion, which is yellow-orange. The alcohol reduces the chromium in the dichromate ion from +6 to +3, the Cr^{3+} ion, which is green. The intensity of the green color is measured and is proportional to the amount of ethanol that was oxidized. The reaction follows:

$$16H^+ + \underset{\text{Yellow-orange}}{2Cr_2O_7^{2-}} + 3CH_3CH_2OH \longrightarrow 3CH_3COOH + \underset{\text{Green}}{4Cr^{3+}} + 11H_2O$$

The body detoxifies ethanol by using an oxidation-reduction reaction as well. Ethanol is oxidized by alcohol dehydrogenase enzymes to acetaldehyde, which is further oxidized to acetic acid, eventually resulting in the formation of carbon dioxide and water:

$$CH_3CH_2OH \xrightarrow[\text{liver enzymes}]{[O]} CH_3{-}\overset{\overset{\displaystyle O}{\|}}{C}{-}H \xrightarrow{[O]} CO_2 + H_2O$$

The breathalyzer test is a technological development based on a scientific understanding of the chemical reactions that ethanol may undergo—a further example of the dependence of technology on science.

The aqueous solution of potassium permanganate is made basic by the addition of a small amount of potassium hydroxide, which facilitates the reaction.

$$\underset{\substack{\text{Ethanol}\\\text{(ethyl alcohol)}}}{CH_3{-}\overset{\overset{\displaystyle OH}{|}}{\underset{\underset{\displaystyle H}{|}}{C}}{-}H} \xrightarrow[\text{basic}]{KMnO_4,\ H_2O} \underset{\substack{\text{Ethanal}\\\text{(acetaldehyde)}}}{CH_3{-}\overset{\overset{\displaystyle O}{\|}}{C}{-}H} + H_2O \qquad 5.20$$

$$\downarrow \text{continued oxidation}$$

$$\underset{\substack{\text{Ethanolic acid}\\\text{(acetic acid)}}}{CH_3{-}\overset{\overset{\displaystyle O}{\|}}{C}{-}OH}$$

A Clinical Perspective: Aldehydes in Medicine, in Chapter 6

When ethanol is metabolized in the liver, it is oxidized to ethanal. The reaction in the liver is brought about by enzyme action. If too much ethanol is present in the body, the alcohol is oxidized, producing an overabundance of ethanal, which causes many of the adverse effects of the "morning-after hangover."

$$CH_3-\underset{\large H}{\overset{\large OH}{C}}-CH_3 \xrightarrow[H_2SO_4]{K_2Cr_2O_7,} CH_3-\overset{\large O}{\overset{\|}{C}}-CH_3 + H_2O \qquad 5.21$$

2-Propanol (isopropyl alcohol) — Propanone (acetone)

$$CH_3-\underset{\large CH_3}{\overset{\large CH_3}{C}}-\underset{\large H}{\overset{\large OH}{C}}-H \xrightarrow[H_2SO_4]{K_2Cr_2O_7,} CH_3-\underset{\large CH_3}{\overset{\large CH_3}{C}}-\overset{\large O}{\overset{\|}{C}}-H + H_2O \qquad 5.22$$

2,2-Dimethyl-1-propanol — 2,2-Dimethylpropanal

$$CH_3-\underset{\large CH_3}{\overset{\large CH_3}{C}}-OH \xrightarrow[basic]{KMnO_4,\ H_2O} \text{no reaction} \qquad 5.23$$

2-Methyl-2-propanol (*t*-butyl alcohol)

Many of the common antiseptics found around the house are mild oxidizing agents. They act by inhibiting the growth of various microorganisms. A dilute solution of potassium permanganate (approximately 0.01%) is effective, as is a dilute solution of hydrogen peroxide (H_2O_2). These solutions are generally applied topically, directly onto the skin. Solutions of iodine (I_2) have also been successfully employed as antiseptics. Common household laundry bleach, sodium hypochloride (NaOCl), has germicidal activity and is one of the best disinfectants known.

Example 5.3

The synthesis of cyclopentanone from cyclopentane.

We are faced with the following conversion of cyclopentane to cyclopentanone. How do we proceed?

Cyclopentane → Cyclopentanone

1. Note first that the carbon skeleton of the reactant and product are identical. Each consists of a cyclopentane ring. The transformation therefore involves only the conversion of functional groups or atoms and does not require an increase or a decrease in chain/ring size.
2. What readily available material might one convert to cyclopentanone? The most obvious is the corresponding alcohol, cyclopentanol. An alcohol can be oxidized to a ketone by using either $KMnO_4$ (basic) or $K_2Cr_2O_7$ (acidic):

$$\text{Cyclopentanol} \xrightarrow[H^+]{K_2Cr_2O_7} \text{Cyclopentanone} + H_2O$$

Cyclopentanol — Cyclopentanone

3. Now, is there a way for us to prepare cyclopentanol from cyclopentane? No, not directly. However, we can convert bromocyclopentane to cyclopentanol, using nucleophilic substitution:

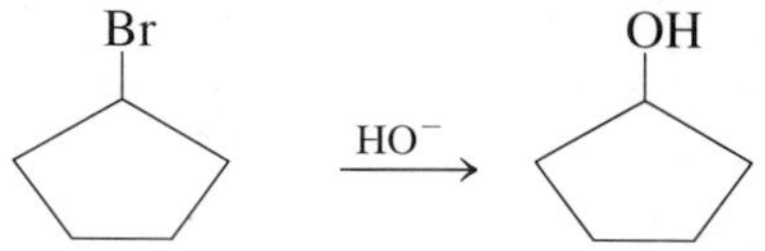

Br — cyclopentane ring $\xrightarrow{HO^-}$ OH — cyclopentane ring

Bromocyclopentane Cyclopentanol

4. Bromocyclopentane can be prepared via the bromination (Br_2, heat) of cyclopentane to complete the synthesis:

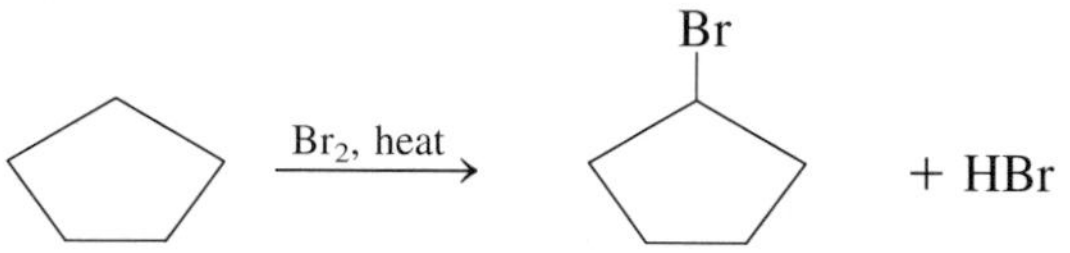

Cyclopentane $\xrightarrow{Br_2,\ heat}$ Bromocyclopentane + HBr

Cyclopentane Bromocyclopentane

Question 5.6 Write the formula of the product obtained through each of the following oxidations:

a. $CH_3CH_2CH_2CH_2{-}OH + K_2Cr_2O_7,\ H^+ \longrightarrow$

b. $CH_3\underset{\displaystyle OH}{\underset{|}{C}}HCH_2CH_2CH_3 + K_2Cr_2O_7,\ H^+ \longrightarrow$

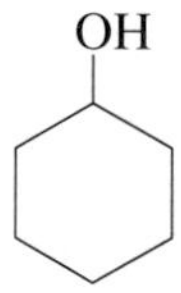

c. Cyclohexanol (OH on cyclohexane ring) $+ KMnO_4,\ HO^- \longrightarrow$

d. $CH_3CH_2\underset{\displaystyle OH}{\underset{|}{C}}(CH_3)_2 + K_2Cr_2O_7,\ H^+ \rightarrow$

5.6 PHENOLS

See Section 4.8 for a review of the discussion of the nomenclature of aromatic compounds.

Phenols are alcohol-like compounds in which the hydroxyl group is attached to benzene. Phenols play important roles in our day-to-day lives. They are found in natural and synthetic flavorings and fragrances [mint (thymol) and savory (carvacrol)], antiseptics (carbolic acid and hexachlorophene), food preservatives (butylated hydroxy toluene, BHT), and cosmetics and health aids (hexylresorcinol), to name a few.

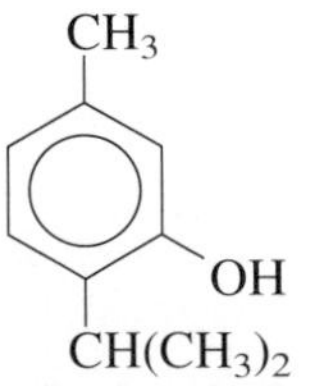

CH_3, OH, $CH(CH_3)_2$

Thymol (mint)

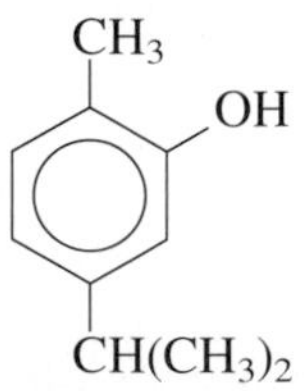

CH_3, OH, $CH(CH_3)_2$

Carvacrol (savory)

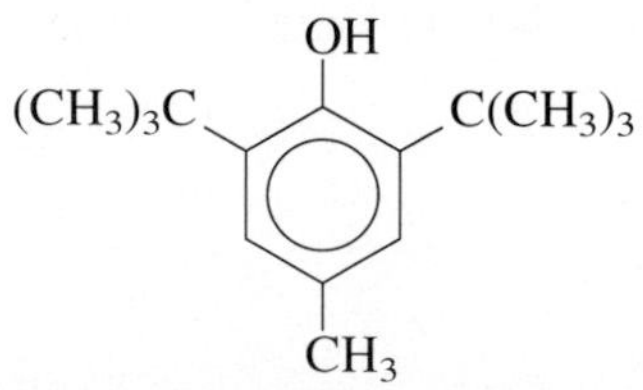

OH, $(CH_3)_3C$, $C(CH_3)_3$, CH_3

Butylated hydroxy toluene, BHT (food preservative)

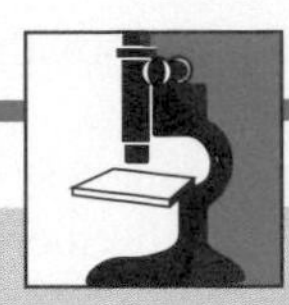

A LABORATORY PERSPECTIVE

Multistep Organic Synthesis

We now have several new reactions in our ever-growing arsenal of chemical knowledge. Let us consider the approaches that one uses in designing a multistep organic synthesis. Consider the following conversion:

$$\underset{\text{2-Butene}}{CH_3CH{=}CHCH_3} \xrightarrow{\text{several steps}} \underset{\text{Butanone}}{CH_3CH_2\overset{\overset{O}{\|}}{C}CH_3}$$

A chemist might approach the problem in the following way. (Keep in mind that the chemist is limited to the number of reactions that he or she has learned in designing syntheses.)

1. Note that the reaction involves a four-carbon compound (2-butene) being converted to another four-carbon compound (butanone). Only the functional group is changing: alkene to ketone.
2. Note that it is possible to convert an alcohol to a ketone via oxidation. In this case, butanone could be prepared from 2-butanol:

$$\underset{\text{2-Butanol}}{CH_3CH_2\overset{\overset{OH}{|}}{C}HCH_3} \xrightarrow{KMnO_4,\ H_2O} \underset{\text{Butanone}}{CH_3CH_2\overset{\overset{O}{\|}}{C}CH_3}$$

Now we continue working backward, one step at a time, always keeping in mind the starting material (2-butene). How can we make 2-butanol? We know, for example, that an alkene can be converted to an alcohol by hydration. Is there an alkene that will work here? Yes—2-butene or 1-butene when hydrated (H_2O, H^+) will give 2-butanol:

$$\underset{\text{2-Butene}}{CH_3CH{=}CHCH_3} \xrightarrow{H_2O,\ H^+} \underset{\text{2-Butanol}}{CH_3CH_2\overset{\overset{OH}{|}}{C}HCH_3}$$

$$\underset{\text{1-Butene}}{CH_3CH_2CH{=}CH_2} \xrightarrow{H_2O,\ H^+} \text{2-Butanol}$$

We have solved the synthesis. We can convert 2-butene to 2-butanol, which can subsequently be oxidized to 2-butanone. We have worked backward from desired product to known starting material, using reasoning and intuition. Sometimes, as the steps become more complex and increase in number, we will encounter roadblocks as we move backward. At that point we simply look for another route. There is usually a workable path to get us the compound we want.

Phenol (carbolic acid; phenol dissolved in water; antiseptic)

Hexachlorophene (antiseptics)

Hexylresorcinol (antiseptic)

o-Phenylphenol (antiseptic)

Phenols, like alcohols, are polar compounds because of the presence of the polar hydroxyl group. The simpler phenols are water soluble.

The chemistry of the phenols (methods of preparation and reactions) will not be discussed here.

5.7 ETHERS

Ethers are structurally related to alcohols. However, they have much lower boiling points than alcohols and display significantly less water solubility because they are much less polar than alcohols or water.

The inertness of ethers is illustrated by the fact that they do not react with oxidizing agents, reducing agents, or bases under normal conditions.

In the common system of nomenclature, ethers are named by placing the names of the two alkyl groups attached to the ether oxygen as prefixes in front of the word ether. For example:

$CH_3—O—CH_3$	$CH_3—O—CH_2CH_3$	$CH_3CH_2—O—CH(CH_3)_2$
Dimethyl ether	Ethyl methyl ether or methyl ethyl ether	Ethyl isopropyl ether

The names of the two groups can be placed either alphabetically or by size (smaller to larger).

In this system the R—O— substituent is named as an alkoxy group; this is analogous to the name hydroxy for the H—O— group. For example, $CH_3—O—$ is methoxy, $CH_3CH_2—O—$ is ethoxy, and so on.

In the I.U.P.A.C. System the ether is named as a substituted hydrocarbon (e.g., alkane). For example:

$$CH_3CH_2\overset{O—CH_3}{\overset{|}{C}}HCH_2CH_2CH_2CH_2CH_2CH_3$$

3-Methoxynonane

Ethers are highly flammable and hence must always be treated with great care. Many ethers that have been left standing for extended periods may also form explosive peroxides. It is wise to date all ethers when they are first opened and to dispose of any ethers that are not used in a timely fashion.

Ethers are important solvents (for example, tetrahydrofuran, THF, a cyclic ether). They are found naturally in a variety of plants [for example: safrole (sassafras); eugenol (cloves); and isoeugenol (nutmeg)] and are found in vitamins (for example, vitamin E) and herbicides (for example, 2,4-dichlorophenoxyethanoic acid, 2,4-D).

Cyclic compounds in which one or more of the carbon atoms in the ring skeleton have been replaced by *hetero* atoms (for example, O, N, P, or S) are called **heterocyclic compounds.**

Tetrahydrofuran, THF (a cyclic ether)

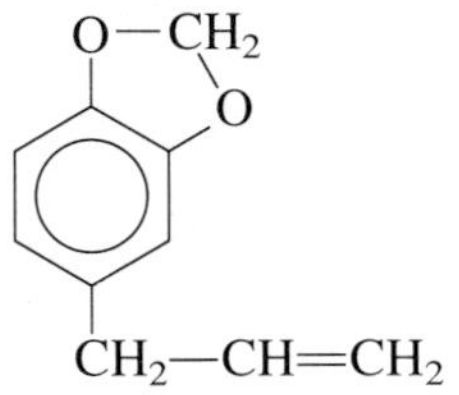

Safrole

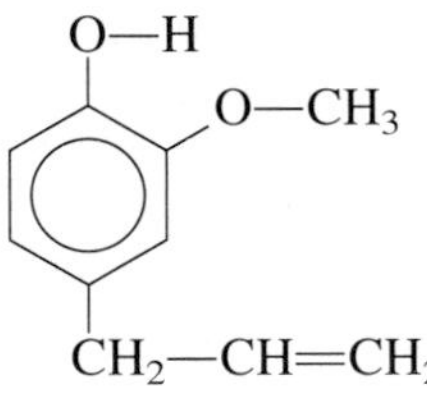

Eugenol

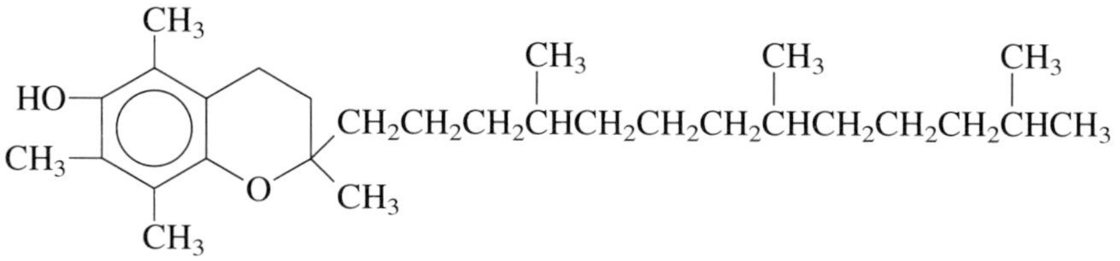

Vitamin E

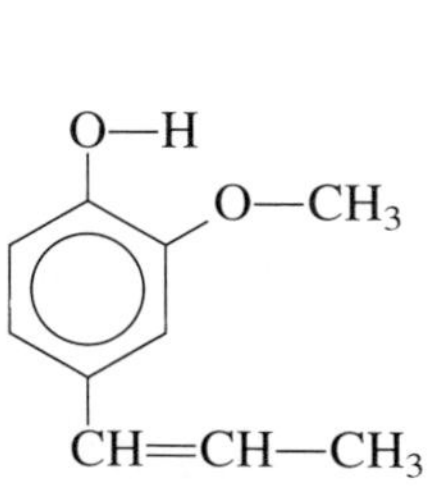

Isoeugenol

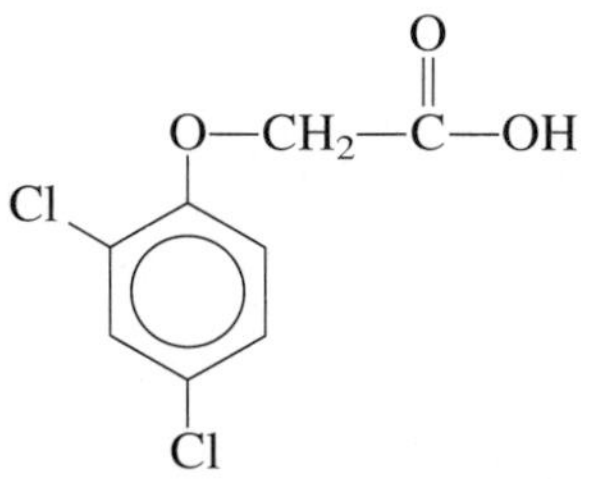

2,4-Dichlorophenoxyethanoic acid, (2,4-D)

Diethyl ether was the first general anesthetic routinely used in the practice of dentistry. A dentist, Dr. William Morton, is credited with its introduction in the late 1800s. Diethyl ether functions as an anesthetic by interacting with the central nervous system

(CNS). It appears that diethyl ether (and many other general anesthetics) function by accumulating in the lipid material of the nerve cells, thereby interfering with nerve impulse transmission. This results in a lessened perception of pain, analgesia.

Halogenated ethers are also routinely used as general anesthetics. They are less flammable than diethyl ether and are therefore safer to store and work with. Penthrane and Enthrane (trade names) are two of the more commonly used members of this family:

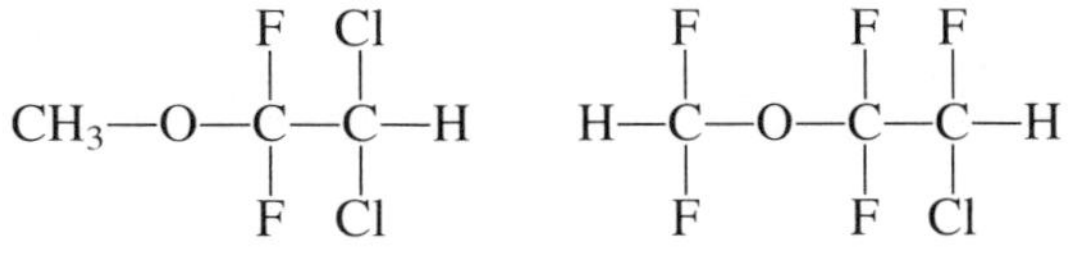

```
           F  Cl             F        F  F
           |  |              |        |  |
CH3—O—C—C—H         H—C—O—C—C—H
           |  |              |        |  |
           F  Cl             F        F  Cl

      Penthrane                  Enthrane
```

Question 5.7 Give the I.U.P.A.C. names for penthrane and enthrane.

See the Human Perspective in Section 4.10 for an expanded discussion of anesthesia.

5.8 OXIDATION AND REDUCTION IN LIVING SYSTEMS

Before beginning a discussion of oxidation-reduction in living systems, we must understand what is meant by the terms "oxidation" and "reduction" as they pertain to organic compounds. With inorganic compounds they are defined as a loss of electrons (**oxidation**) or a gain of electrons (**reduction**). It is ordinarily easy to determine when an oxidation or a reduction occurs because the process is accompanied by a change in charge. For example:

$$Ag^0 \longrightarrow Ag^+ + 1e^-$$

Loss of an electron, neutral atom converted to a positive ion: oxidation

$$:\ddot{\underset{..}{Br}}^+ + 2e^- \longrightarrow :\ddot{\underset{..}{Br}}:^-$$

Gain of two electrons, positive ion converted to a negative ion: reduction

When organic compounds are involved, however, there is no change in charge (see Figure 5.3), and it is often difficult to determine whether oxidation or reduction has occurred. The following simplified view may help.

An organic compound is oxidized as more and more electronegative atoms are attached. In the conversion below, an alkane is oxidized to an alcohol, an alcohol to an aldehyde or ketone, and an aldehyde or ketone to a carboxylic acid. In the first step, a C—O bond replaces a C—H bond. In the second step, the C—O bond is converted to a C═O bond. In the final step, a second C—O bond is substituted for another C—H bond. Slightly polar C—H bonds are being replaced by highly polar C—O and C═O bonds. These are oxidations.

Interconversions of Families of Organic Compounds

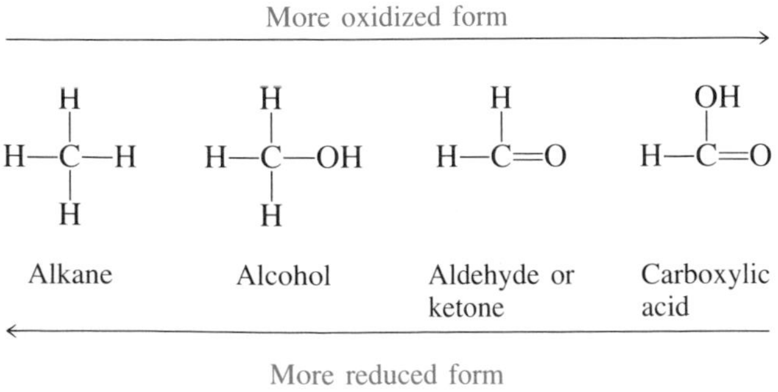

```
                   More oxidized form
  ──────────────────────────────────────────────────→

     H            H              H            OH
     |            |              |            |
  H—C—H       H—C—OH        H—C═O        H—C═O
     |            |
     H            H

  Alkane       Alcohol      Aldehyde or   Carboxylic
                            ketone        acid
  ←──────────────────────────────────────────────────
                   More reduced form
```

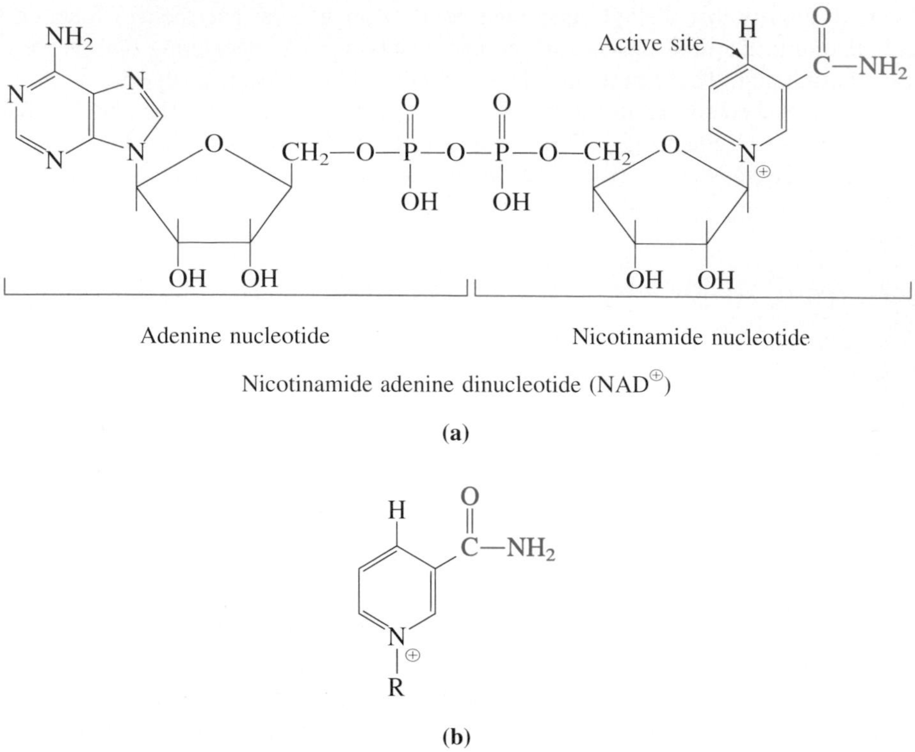

Active site

Adenine nucleotide

Nicotinamide nucleotide

Nicotinamide adenine dinucleotide ($NAD^{\oplus}$)

(a)

(b)

FIGURE 5.3
Nicotinamide adenine dinucleotide: (a) complete structural formula; (b) abbreviated formula.

If we apply similar logic, it is easy to see that the conversions on going from left to right below are also oxidations:

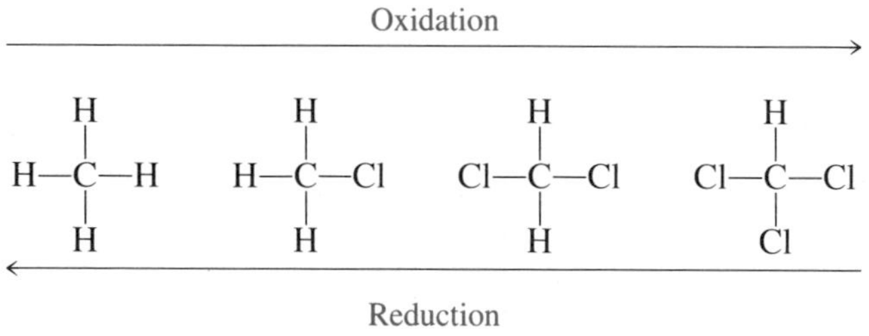

Oxidation →

$$CH_4 \quad CH_3Cl \quad CH_2Cl_2 \quad CCl_4$$

← Reduction

As we go from left to right we are adding more and more chlorine atoms, which are highly electronegative; oxidation results. We will encounter numerous examples of oxidations and reductions in the pages ahead. It is important to have a clear understanding of the oxidation and reduction processes and to be able to recognize when they are playing a role in the reactions being observed.

We have discussed several examples of oxidation and reduction reactions in organic chemistry. The interrelationships of the families of compounds that we have seen thus far are presented below. The conversion of an alkane to an alcohol, an alcohol to a carbonyl compound, and a carbonyl compound to a carboxylic acid are all examples of oxidations. Conversions in the opposite directions are reductions (see above).

Oxidation and reduction reactions also play a preeminent role in the chemistry of living systems. In living systems these reactions are brought about by the action of various enzymes which behave as biochemical catalysts.

See Section 12.7 for a complete discussion of coenzymes.

NAD^+, nicotinamide adenine dinucleotide, is a **coenzyme.** It is a chemical compound, *not* a protein or enzyme itself, but is necessary to the activity of many enzymes. The structure of NAD^+ is given in Figure 5.3.

NAD^+ is involved in several enzymatic oxidation-reduction processes in the body. For example, the conversion of D-ribulose 5-phosphate to D-ribitol 5-phosphate involves the reduction of a carbonyl group to an alcohol (in the sugar) and the oxidation of NADH (the reduced form of NAD^+) to NAD^+.

$$\mathrm{NADH} + \underset{\text{D-Ribulose 5-phosphate}}{\mathrm{CH_2{-}OH;\ C{=}O;\ H{-}C{-}OH;\ H{-}C{-}OH;\ CH_2{-}O{-}P(=O)(O^-Na^+){-}O^-Na^+}} \xrightarrow[\text{action}]{\text{Enzyme}} \underset{\text{D-Ribotol 5-phosphate}}{\mathrm{CH_2{-}OH;\ H{-}C{-}OH;\ H{-}C{-}OH;\ H{-}C{-}OH;\ CH_2{-}O{-}P(=O)(O^-Na^+){-}O^-Na^+}} + \mathrm{NAD^+} \quad 5.24$$

A more detailed view of the oxidation of NADH to NAD^+ is provided in reaction 5.25:

$$\underset{\text{NADH}}{\mathrm{O{=}C\langle}\ +\ \text{H, H; }\mathrm{C(=O){-}NH_2};\ \ddot{N}{-}R} \underset{-\mathrm{H}^{\oplus}}{\overset{\mathrm{H}^{\oplus}}{\underset{\text{enzyme}}{\rightleftharpoons}}} \underset{\mathrm{NAD}^{\oplus}}{\text{H; }\mathrm{C(=O){-}NH_2};\ N^{\oplus}{-}R}\ +\ \mathrm{H{-}\overset{|}{\underset{|}{C}}{-}O{-}H}$$

Another example, the oxidation of ethanol to ethanal during metabolism in the liver, was briefly discussed in Section 5.5. Several additional examples are provided in upcoming chapters. (Glycolysis, the metabolism of carbohydrates, and further actions of NAD^+ are a few of thc topics to be discussed.)

5.9 THIOLS

Compounds that contain the —SH group are known as **thiols.** They are similar to alcohols in structure. In general, they exhibit lower boiling points than corresponding alcohols, though they are higher molecular weight compounds. For example:

$CH_3{-}OH$	$CH_3{-}SH$
Methanol Mol. wt. = 32 b.p. 65°C	Methanethiol Mol. wt. = 48 b.p. 6°C
$CH_3CH_2{-}OH$	$CH_3CH_2{-}SH$
Ethanol Mol. wt. = 46 b.p. 78°C	Ethancthiol Mol. wt. = 62 b.p. 37°C

Thiols and many other sulfur compounds have nauseating aromas. They are commonly found in substances as different as the defensive spray of the North American striped skunk (*trans*-2-butene-1-thiol), onions and garlic (1-propanethiol), cooked meats (3-methyl-1-butanethiol), and oysters (methanethiol).

Question 5.8 Draw the complete structural formulas for each of the thiols named in the preceding paragraph.

Thiols are also involved in protein structure and conformation. It is the ability of two thiol groups to easily undergo oxidation to a —S—S— (disulfide) bond that is responsible for this involvement. Either two different protein chains or two parts of one chain may be linked through this process, as shown in Figure 5.4.

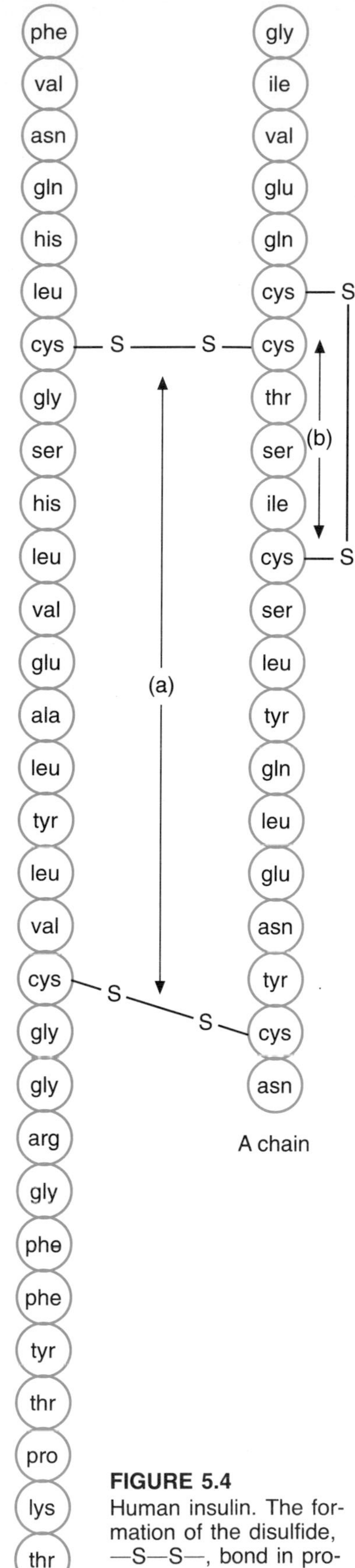

FIGURE 5.4
Human insulin. The formation of the disulfide, —S—S—, bond in proteins: (a) interchain (between two chains) bonding; (b) intrachain (in the same chain) bonding.

Many other thiols play important roles. For example:

BAL (British Anti-Lewisite) is used as an antidote to mercury poisoning. The two thiol groups of BAL complex with mercury and remove it from the system before it can do any damage.

$$\underset{\text{OH}}{\underset{|}{CH_2}}-\underset{\text{SH}}{\underset{|}{CH}}-\underset{\text{SH}}{\underset{|}{CH_2}}$$

BAL

Chapter 11

Cysteine is an amino acid that contains a thiol group. Cysteine is found in many proteins, where it plays a pivotal role in the structure and conformation of the protein.

$$^{+}NH_3-\overset{\text{H}}{\overset{|}{\underset{CH_2-SH}{\underset{|}{C}}}}-\overset{\text{O}}{\overset{\|}{C}}-O^{-}$$

Cysteine

Dihydrolipoic acid is involved in biochemical oxidations and hence energy production in the body.

$$\underset{\text{SH}}{\underset{|}{CH_2}}-CH_2-\underset{\text{SH}}{\underset{|}{CH}}-(CH_2)_4-\overset{\text{O}}{\overset{\|}{C}}-OH$$

Dihydrolipoic acid

Several other members of this family are discussed in forthcoming chapters.

SUMMARY

Alcohols and ethers are structurally related to water. Alcohols, like water, are very polar molecules, while ethers are significantly less polar. These differences in properties are a consequence of the presence of the hydroxyl group in alcohols and water and its absence in ethers. Hydrogen bonding in ethers is not possible; on the other hand, it is the principal intermolecular force of attraction between alcohols and water.

Many alcohols, phenols (aromatic alcohols), and ethers play important roles in our lives. They are present in food products, medicinals, pesticides, cosmetics, deodorants, and health aids. They all are involved in the biochemistry of living organisms—though alcohols are by far the most important of the three classes.

Alcohols are produced industrially principally by the reduction of carbonyl compounds (or carbon monoxide) or by fermentation of carbohydrates and starches. In the laboratory they are prepared by the hydration of alkenes, by the reduction of carbonyl compounds, or through substitution using the halide exchange reaction.

Alcohols can be oxidized to aldehydes, ketones, and carboxylic acids, depending on the structure of the alcohol, reaction conditions, and the oxidizing agent used. The reactions outlined here are simply the reverse of reduction reactions. They can also be dehydrated to give alkenes by the action of concentrated acid and heat.

Several examples of phenols and ethers were presented and briefly discussed in this chapter. Thiols, sulfur analogs of alcohols, were also introduced. The thiols have lower boiling points than alcohols and are found in a wide variety of natural systems. They are involved in protein conformation, food chemistry, metabolism, and medicinal chemistry.

GLOSSARY OF KEY TERMS

alcohol (5.1) an organic compound that contains an hydroxyl group (—OH) attached to an alkyl group

BAL (5.9) British Anti-Lewisite; used as an antidote to mercury poisoning.

carbinol carbon (5.5) in an alcohol the carbon to which the hydroxyl group is attached.

coenzyme (5.8) a cofactor, nonprotein molecule that is necessary to effect many enzymatic transformations.

dehydration (5.5) a reaction that involves the loss of a water molecule, e.g., the loss of water from an alcohol and the concomitant formation of an alkene.

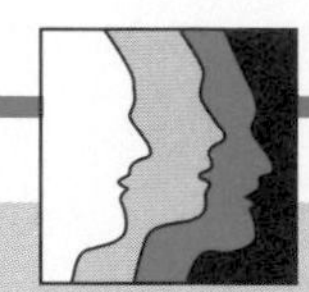

A HUMAN PERSPECTIVE

Fetal Alcohol Syndrome

The first months of pregnancy are a time of great joy and anticipation but are not without moments of anxiety. On the first visit to the obstetrician the mother-to-be is tested for previous exposure to a number of infectious diseases that could harm the fetus. She is provided with information about proper diet, expected weight gain, and avoidance of a variety of drugs that could harm the baby. Among the drugs that should be avoided are alcoholic beverages.

The use of alcoholic beverages by a pregnant woman can cause *fetal alcohol syndrome (FAS).* A syndrome is a set of symptoms that occur together and are characteristic of a particular disease. In this case, physicians noticed that infants born to women with chronic alcoholism showed a reproducible set of abnormalities, including mental retardation, poor growth before and after birth, and facial malformations. The facial abnormalities that are typical of fetal alcohol syndrome are shown in the accompanying figure. The face is somewhat flattened; the head has a smaller circumference; the nasal bridge is lower, and the nose is short; the epicanthic (eye) folds and short eyelid tissues give the eyes a narrow, round appearance; the upper lip is very thin, and the infranal depression, the folds of tissue between the lip and nose, are indistinct.

Mothers who report only social drinking may have children with *fetal alcohol effects,* a less severe form of fetal alcohol syndrome. This milder form is characterized by a reduced birth weight, some learning disabilities, and behavioral problems.

How does alcohol consumption cause these varied symptoms? No one is sure of the precise mechanism, but it is well known that the alcohol consumed by the mother crosses the placenta and enters the bloodstream of the fetus. Within about 15 minutes the concentration of alcohol in the blood of the fetus is *as high as in that of the mother!* The mother has enzymes to detoxify the alcohol in her blood; the fetus does not. Now consider that alcohol can cause cell division to stop or be radically altered. It is thought that even a single night on the town could be enough to cause FAS by blocking cell division during a critical developmental period.

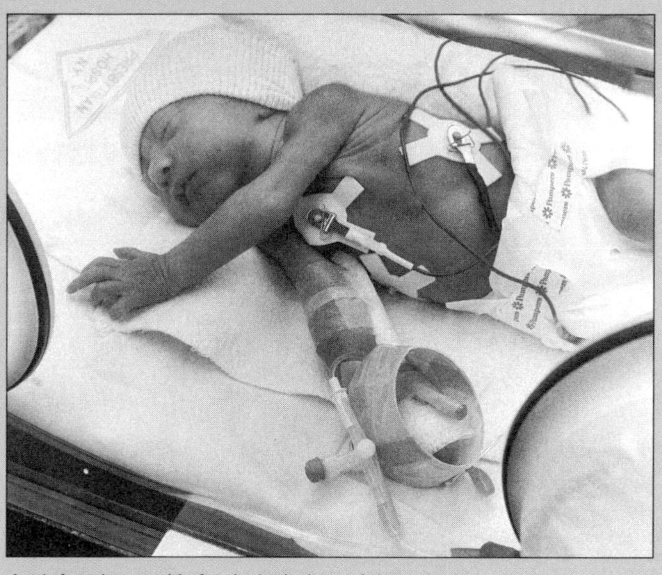

An infant born with fetal alcohol syndrome.

This raises the question: How much alcohol can a pregnant woman safely drink? As we have seen, the severity of the symptoms seems to increase with the amount of alcohol consumed by the mother. However, it is impossible to do the scientific studies that would conclusively determine the risk to the fetus caused by different amounts of alcohol. There is some evidence that suggests that there is a risk associated with drinking even 1 ounce of absolute (100%) alcohol each day. Because of these facts and uncertainties, the American Medical Association and the U.S. Surgeon General recommend that pregnant women completely abstain from alcohol consumption.

Some evidence indicates that the effects of alcohol are much worse when combined with cigarette smoking or a poor diet. It is important for the mother-to-be to eat a well-balanced diet and abstain from alcoholic beverages and cigarettes, at least for the duration of the pregnancy. These sacrifices seem small when you consider the reward: the birth of a beautiful, happy, healthy baby.

denatured alcohol (5.3) ethanol (C_2H_5OH) to which a denaturing agent (often methanol) has been added to make it unfit to drink.

destructive distillation (5.3) a distillation, or separation, of compounds by differences in boiling point, that also involves the decomposition of the compound(s) in the course of the distillation.

disulfide (5.9) an organic compound that contains a disulfide group (—S—S—).

ether (5.7) an organic compound that contains two alkyl and/or aryl groups attached to an oxygen atom; R—O—R, Ar—O—R, and Ar—O—Ar.

fermentation (5.3) the anaerobic (in the absence of oxygen) metabolism or degradation of sugars by microorganisms.

halide exchange (5.3) a substitution reaction in which one halogen atom is exchanged for another.

heterocyclic compounds (5.7) cyclic compounds in which one or more of the carbon atoms in the ring skeleton have been replaced by *hetero* atoms (for example, O, N, P, or S).

hydration (5.3) a reaction in which water is added to a molecule, e.g., the addition of water to an alkene to form an alcohol.

hydroxyl group (Introduction) the —OH functional group.

nicotinamide adenine dinucleotide (NAD^+) (5.8) a coenzyme that is an oxidizing agent used in a variety of metabolic processes.

oxidation (5.8) the loss of electrons by a molecule or atom, e.g., the conversion of an alcohol to an aldehyde or ketone.

phenol (5.6) an organic compound that contains a hydroxyl group (—OH) attached to a benzene ring.

phenyl group (5.1) A benzene ring that has had a hydrogen atom removed, C_6H_5—.

primary (1°) alcohol (5.4) an alcohol with the general formula RCH_2OH.

reaction mechanism (5.5) the pictorial, step-by-step process via which reactants are converted to products in a chemical reaction.

reduction (5.8) a gain of electrons by a molecule or atom, e.g., the conversion of a carbonyl compound to an alcohol via the use of a reducing agent.

secondary (2°) alcohol (5.4) an alcohol with the general formula R_2—CHOH.

tertiary (3°) alcohol (5.4) an alcohol with the general formula R_3C—OH.

thiol (5.9) an organic compound that contains a thiol group (—SH).

QUESTIONS AND PROBLEMS

Structure, Properties, and Nomenclature

5.9 Arrange the following alcohols in order of increasing solubility in water:

a. $CH_3CH_2CH_2OH$

b. *n*-pentyl alcohol

c. $CH_2(OH)CH(OH)CH_2(OH)$ (OH OH OH on CH_2CHCH_2)

d. $CH_2(OH)CH_2CH_2(OH)$ (OH on $CH_2CH_2CH_2$, both ends)

1-hexanol

ethylene glycol

5.10 Which intermolecular interaction is responsible for the water solubility and "higher" boiling point of alcohols?

5.11 Ethyl alcohol, CH_3CH_2OH, boils at 78°C, while ethanethiol, CH_3CH_2SH, boils at 35°C. Although CH_3CH_2SH has a higher molecular weight, its boiling point is significantly lower than that of ethyl alcohol. Explain.

5.12 What are the problems associated with using diethyl ether as an anesthetic?

5.13 Though both ethanol and dimethyl ether have the same molecular weight (they are isomers), they have markedly different boiling points. What are the boiling points? Explain the observed differences in the two.

5.14 Give the common name for each of the following compounds:

a. CH_3CH_2—OH

b. CH_3CHCH_3 with OH on C-2

c. CH_3—C(CH_3)(OH)—CH_3

d. CH_3CH_2—O—CH_2CH_3

e. $CH_3CHCH_2CH_3$ with O—CH_3 on C-2

5.15 Give the I.U.P.A.C. name for each of the following compounds:

a. $CH_3CH_2CH_2CH_2CH_2CH_2CH_2$—OH

b. CH_3CHCH_3 with OH on C-2

c. CH_3—C(CH_3)(CH_2—OH)—CH_3

d. CH_3—O—$CHCH_2CH_3$ with CH_3 on the CH

e. $CH_3CH_2CHCH_2CH_2CH_2$—OH with Br on the CH

f. cyclopentane ring with OH

g. $CH_3CHCCH_2CH_2CH_3$ with OH on C-2 and CH_3 groups (two) on C-3

h. $CH_3CH_2CCH_2CH_3$ with $CH_2CH_2CH_2CH_3$ and OH on C-3

i. cycloheptane ring with OH

j. benzene ring with OH and CH_3 (meta)

5.16 If a bottle of distilled alcoholic spirits—for example, scotch whiskey—is labeled as 80 proof, what is the percentage of alcohol in the scotch?

5.17 Give the I.U.P.A.C. name for each of the following thiols:

a. $CH_3CH_2CH_2$—SH

b. $CH_3CHCH_2CH_3$ with SH on C-2

c. CH_3—C(CH_2CH_3)(SH)—CH_3

d. HS—(cyclohexane ring)—SH

e. CH_2CHCH_3 with SH on C-1 and SH on C-2

f. benzene ring—SH

g. $CH_3CHCH_2CH_2CH_3$ with SH on C-2

h. $CH_3CH_2CH_2CH_2CH_2CH_2CH_2SH$

5.18 What is denatured alcohol? Why is alcohol denatured?

5.19 What is a carbinol carbon?

5.20 Explain in your own words how British Anti-Lewsite (BAL) acts as an antidote to mercury poisoning.

5.21 Arrange the following compounds in order of increasing boiling point beginning with the lowest and work toward the highest:

a. $CH_3CH_2CH_2CH_2CH_3$

b. $CH_3CHCH_2CH_2CH_3$ with OH on C-2

c. CH_3—C(=O)—$CH_2CH_2CH_3$

d. $CH_3CH_2CH_2$—O—CH_2CH_3

5.22 Which member in each of the following pairs is more soluble in water?

a. CH_3CH_2—OH or $CH_3CH_2CH_2CH_2$—OH

b. CH_2CH_2 (with OH on each carbon: OH OH) or CH_3CH_2—OH

c. cyclopentanol (OH on cyclopentane ring) or CH_3CHCH_3 (OH on C-2)

d. $CH_3CH_2CH_2$—O—$CH_2CH_2CH_3$ or $CH_3CH_2CH_2CH_2CH_2CH_2$—OH

e. $CH_3CH_2CH_2CH_2CH_2CH_2CH_2CH_2CH_2CH_3$ or $CH_3CH_2CH_2CH_2CH_2CH_2CH_2CH_2CH_2$—OH

5.23 List three alcohols that are commonly found around the home and list the principal use of each.

5.24 Which member in each of the following pairs has the higher boiling point?

a. CH_3CH_2—OH or $CH_3CH_2CH_2CH_2$—OH

b. CH_3CH_2—SH or CH_3CH_2—OH

c. $CH_3CH_2CH_2$—SH or $CH_3CH_2CH_2CH_2CH_2CH_2$—SH

d. 1,2-ethanedithiol or ethanethiol

5.25 Classify each of the following alcohols as 1°, 2°, or 3°.

a. 3-methyl-1-butanol

b. 2-methylcyclopentanol

c. *t*-butyl alcohol

d. 1-methylcyclopentanol

e. cyclohexane ring—$CH_2CH_2CH_2$—OH

f. 2-methyl-2-pentanol

g. $CH_3CHCHCH_2CH_2CH_2CH_2CH_3$ (CH_3 on C-2, OH on C-3)

5.26 Circle the carbinol carbon in each of the compounds given in Question 5.25.

5.27 Briefly explain why ethers are less soluble in water than the alcohol of corresponding molecular weight.

5.28 Draw each of the following, using complete structural formulas:

a. 3-hexanol	f. cyclohexanol
b. 1,2,3-pentanetriol	g. ethyl isopropyl ether
c. 2-methyl-2-pentanol	h. di-*n*-butyl ether
d. 2,3-dimethyl-2-butanol	i. 3,5-dinitrophenol
e. 1-iodo-2-butanol	j. 3,4-dimethyl-3-heptanol

5.29 2,4,6-Trinitrophenol is known by the common name *picric acid*. Picric acid is a solid but is readily soluble in water. In solution it is used as a biological tissue stain. As a solid, it is also known to be unstable and may explode. In this way it is similar to trinitrotoluene (TNT). Draw the structures of picric acid and TNT. Why is picric acid readily soluble in water while TNT is not?

5.30 Draw all of the alcohols and ethers of molecular formula $C_4H_{10}O$.

5.31 Differentiate between the terms "phenyl" and "phenol."

5.32 Cresols have a hydroxyl group and methyl group as substituents on the benzene ring. Draw the structures of *o*-cresol, *m*-cresol, and *p*-cresol. Name them, using the I.U.P.A.C. System.

5.33 Name the following aromatic compounds, using the I.U.P.A.C. System:

a. benzene ring with OH and NO_2 on adjacent carbons

b. CH_3—CH—CH_3 attached to a benzene ring bearing OH in the para position

c. benzene ring with HO, Br, and Cl (1,3,5 positions)

d. benzene ring with OH, CH_3, and Br

5.34 Give the correct name for each of the following compounds:

a. 2-methyl-3-butanol *incorrect* — $CH_3CHCHCH_3$ (CH_3 on C-2, OH on C-3)

b. 3-chloro-5-pentanol *incorrect* — $CH_3CH_2CHCH_2CH_2OH$ (Cl on C-3)

c. 1-methyl-2-cyclohexanol *incorrect* — cyclohexane ring with OH and CH_3 on adjacent carbons

5.35 Briefly list the I.U.P.A.C. rules for naming alcohols.

Reactions and Mechanisms

5.36 Write an equation, using complete structural formulas, demonstrating each of the following chemical transformations:

a. oxidation of an alcohol to an aldehyde

b. oxidation of an alcohol to a ketone

c. dehydration of a cyclic alcohol to an alkene

d. oxidation of an aldehyde to a carboxylic acid

e. hydrogenation of an alkene to an alkane

5.37 What would be the product(s) resulting from the oxidation of each of the following alcohols with, for example, potassium permanganate? If no reaction occurs, write "N.R."

a. 2-butanol

b. 2-methyl-2-hexanol

c. cyclohexanol

d. 1-methyl-1-cyclopentanol

e. phenol

5.38 We have seen that ethanol is metabolized to acetaldehyde in the liver. What would be the product formed, under the same conditions, of each of the following alcohols:

a. CH_3—OH b. $CH_3CH_2CH_2$—OH

c. $CH_3CH_2CH_2CH_2$—OH

5.39 Ethers may be prepared by the removal of water (dehydration) between two alcohols, as shown below. Give the structure(s) of the ethers formed by the reaction of the following alcohol(s) under acidic conditions with heat:

Example: $CH_3OH + HOCH_3 \xrightarrow[\text{heat}]{H^+,} CH_3OCH_3 + H_2O$

a. $2CH_3CH_2OH \longrightarrow ?$

b. $CH_3OH + CH_3CH_2OH \longrightarrow ?$

c. $(CH_3)_2CHOH + CH_3OH \longrightarrow ?$

d. 2 (cyclopentyl)$-CH_2OH \longrightarrow ?$

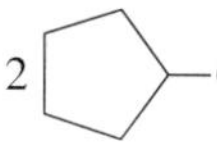

5.40 Give the oxidation products for the following alcohols. If no reaction occurs, write "N.R."

a. $CH_3CH_2CH(OH)CH_2CH_3$

b. $CH_3CH_2CH_2OH$

c. $CH_3CH(OH)CH_2CH(CH_3)CH_3$

d. $CH_3-C(OH)(CH_3)-CH_2CH_3$

e.  (phenyl)$-CH_2CH_2CH_2OH$

f. 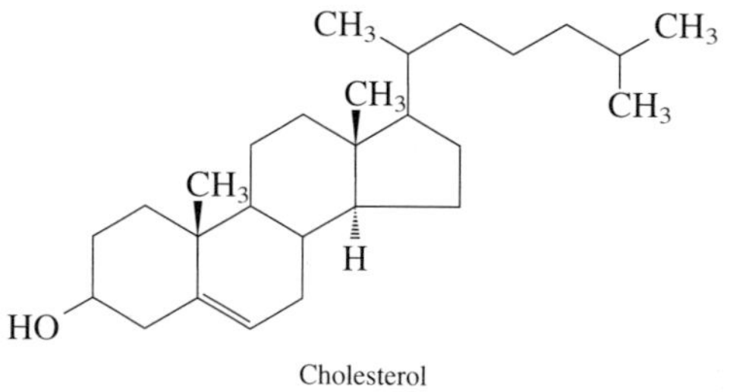 Cholesterol (structure with HO, CH_3, CH_3, CH_3, CH_3, CH_3, H labels)

Cholesterol

5.41 Draw out the complete mechanism for the dehydration of 2-methyl-2-butanol under acidic conditions.

5.42 Complete each of the following syntheses by supplying the missing reactant(s), reagent(s), or product(s) indicated by the question marks.

a. $CH_3CH{=}CHCH_3 \xrightarrow{H_2O,\ H^+} ?(1) \xrightarrow{K_2Cr_2O_7,\ H^+} ?(2)$

b. $CH_3CH_2CH_2CH_2CH_2{-}I \xrightarrow{?} CH_3CH_2CH_2CH_2CH_2{-}OH$

c. $CH_3CH_2CH_2CH_2{-}OH \underset{?(2)}{\overset{?(1)}{\rightleftharpoons}} CH_3CH_2CH_2{-}C(=O){-}H$

d. 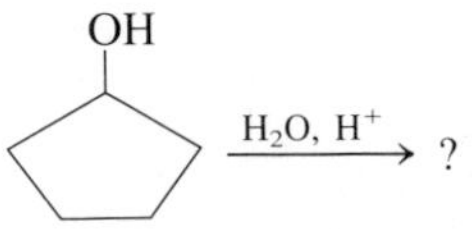 cyclopentanol (OH) $\xrightarrow{H_2O,\ H^+} ?$

e. $CH_3{-}O{-}CH_3 \xrightarrow{K_2Cr_2O_7,\ H^+} ?$

f. $CH_3CH_2{-}C(=O){-}CH_2CH_3 \xrightarrow{?} CH_3CH_2CH(OH)CH_2CH_3$

g. 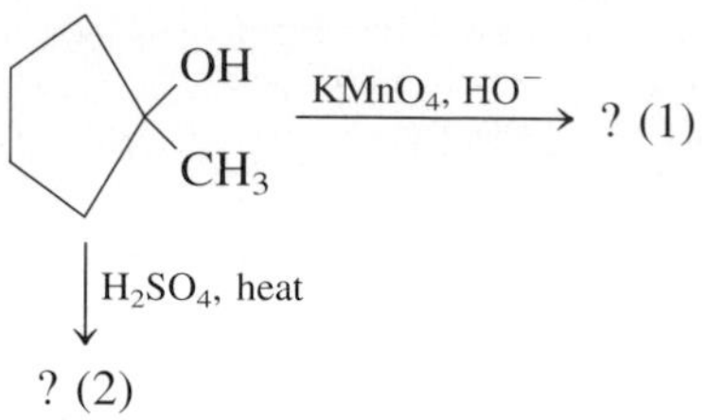 1-methylcyclopentanol (OH, CH_3) $\xrightarrow{KMnO_4,\ HO^-} ?\ (1)$

$\downarrow H_2SO_4$, heat

? (2)

h. cyclopentanone (O) $\xrightarrow{?}$ cyclopentanol (OH)

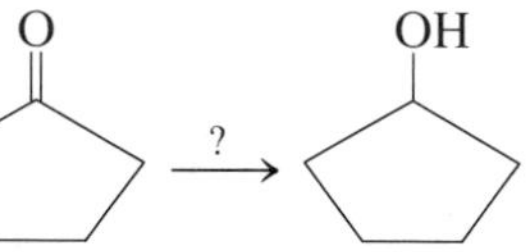

i. $(CH_3)_3C{-}OH \xrightarrow{K_2Cr_2O_7,\ H^+} ?$

j. 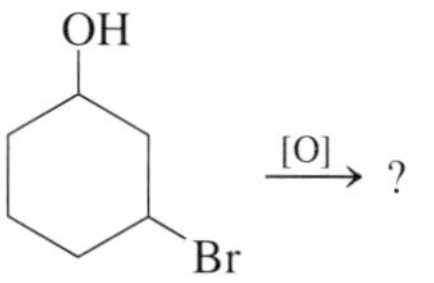 3-bromocyclohexanol (OH, Br) $\xrightarrow{[O]} ?$

Further Problems

5.43 Show how acetone may be prepared from propene.

$CH_3C(=O)CH_3$

Acetone

5.44 Show how bromocyclopentane may be prepared from cyclopentanol.

5.45 Complete the following reactions by supplying those portions missing and indicated with the question marks:

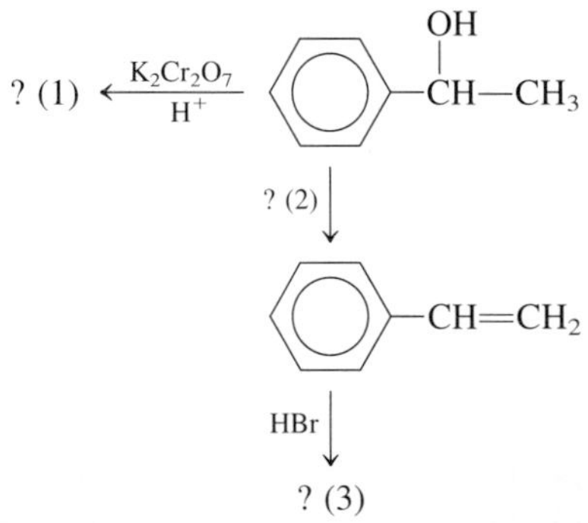

$?\ (1) \xleftarrow[H^+]{K_2Cr_2O_7}$ (phenyl)$-CH(OH)-CH_3$

$\downarrow ?\ (2)$

(phenyl)$-CH{=}CH_2$

$\downarrow HBr$

? (3)

5.46 Complete each of the following syntheses by supplying the reagent(s) necessary to accomplish the indicated transformations. More than one step might be required.

a. $CH_3CH_2CH_2CH_2{-}OH \longrightarrow CH_3CH_2CH_2{-}C(=O){-}H$

b. 1-pentane $\longrightarrow CH_3CH_2CH_2CH_2{-}C(=O){-}OH$

c. 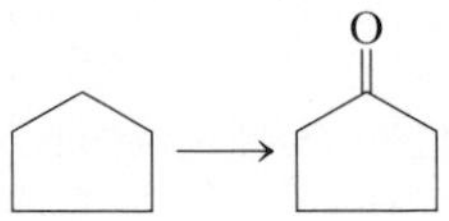 cyclopentane $\longrightarrow$ cyclopentanone (O)

5.47 Sulfides are the sulfur analogs of ethers, that is, ethers in which oxygen has been substituted by a sulfur atom. They are also named in an analogous manner to the ethers with the term "sulfide" replacing "ether." For example, $CH_3—S—CH_3$ is dimethyl sulfide. Draw the sulfides that correspond to the following ethers and name them:

a. diethyl ether
b. methyl propyl ether
c. dibutyl ether
d. ethyl phenyl ether

5.48 Dimethyl sulfoxide (DMSO) has been used by many sports enthusiasts as a liniment for sore joints; it acts as an anti-inflammatory agent and a mild analgesic (pain killer). DMSO is a sulfoxide—it contains the S=O functional group. DMSO is prepared from dimethyl sulfide by mild oxidation, and it has the molecular formula C_2H_6SO. Draw the complete Lewis dot structure of DMSO.

5.49 We learned that various enzymes in the liver are capable of oxidizing alcohols to aldehydes and ketones. For example, ethanol is oxidized to ethanal and methanol to methanal by the action of these enzymes. We find that methanal has an LD_{50} of 0.07 g/kg of body weight, and ethanal has an LD_{50} of 0.9 g/kg of body weight. Which of these compounds is more toxic to humans? Explain.

5.50 We have seen that alcohols are capable of hydrogen bonding to each other. Hydrogen bonding is also possible between alcohol molecules and water molecules or between alcohol molecules and ether molecules. Ether molecules *do not* hydrogen bond to each other, however. Explain.

5.51 Draw the following heterocyclic ethers:

a. 2-chlorotetrahydrofuran
b. furan
c. 2,5-dibromofuran

5.52 Cholic acid, shown below, is a steroid and is a common bile acid in humans. Circle and name as many of the functional groups in cholic acid as you can.

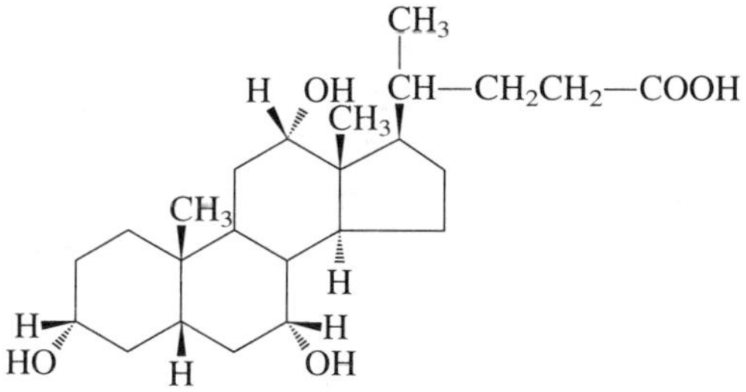

Cholic acid

5.53 Cystine is an amino acid. Cystine is formed from the oxidation in air of cysteine to form a disulfide bond. The molecular formula of cystine is $C_6H_{12}O_4N_2S_2$. Draw the complete structural formula of cystine.

5.54 Cholesterol is an alcohol and a steroid (Section 16.4). Diets that contain large amounts of cholesterol have been linked to heart disease and atherosclerosis, hardening of the arteries. The narrowing of the artery, due to fat buildup, is very apparent. Cholesterol is directly involved in this buildup. The structure of cholesterol is provided below. Describe the various functional groups and principal structural features found in this molecule.

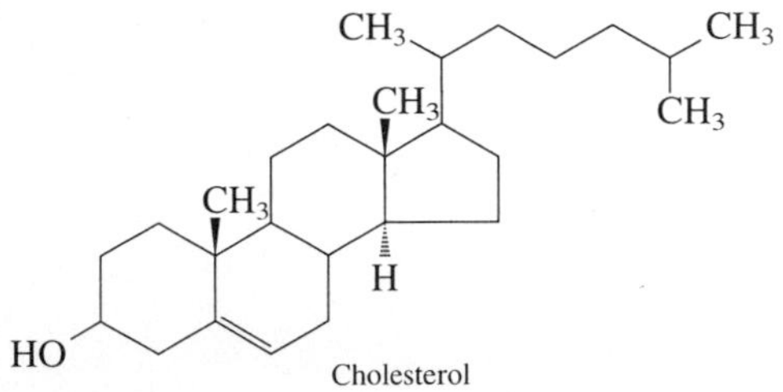

Cholesterol

5.55 An unknown compound **A** is known to be an alcohol with the molecular formula $C_4H_{10}O$. When dehydrated, compound **A** gave only one alkene product, C_4H_8, compound **B**. Compound **A** could not be oxidized. What are the identities of compound **A** and compound **B**?

5.56 Unknown compound **A** had the following percentage composition: 64.80% C, 13.62% H, and the rest due to oxygen. Mild oxidation of compound **A** gives compound **B**, C_4H_8O. Dehydration of compound **A** gave two products, compound **C** and compound **D**, both having the molecular formula C_4H_8. Compound **B** could be further oxidized to compound **E**, $C_4H_8O_2$. Identify compounds **A** through **E**.

CHAPTER 6

Aldehydes and Ketones

OBJECTIVES

- Write the structures and names of the common aldehydes and ketones.
- Recognize the members of these families that are of natural, commercial, health, environmental, or industrial interest.
- Describe the physical properties of each of the two families listed above.
- Write equations for the oxidation and reduction of carbonyl compounds.
- Write equations for the preparation and reactions of hemiacetals, hemiketals, acetals, and ketals.
- Be able to design simple multistep syntheses involving carbonyl compounds.

OUTLINE

Several families of organic compounds contain a common structural feature, the **carbonyl group:**

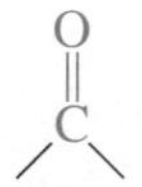

$$\begin{array}{c} O \\ \| \\ C \\ / \quad \backslash \end{array}$$

The carbonyl group

which is recognizable as containing a carbon-oxygen double bond. Depending upon the identity of the particular family, the carbon atom may be attached to two carbons, one carbon and one hydrogen, two hydrogens, or even one carbon or one hydrogen and an oxygen or nitrogen atom. In any event, the carbonyl group and the two atoms attached to the carbonyl group are coplanar (see Figure 6.1). This has a profound effect on the structure, properties, and reactivity of carbonyl compounds.

R and R′ = R, Ar, or H

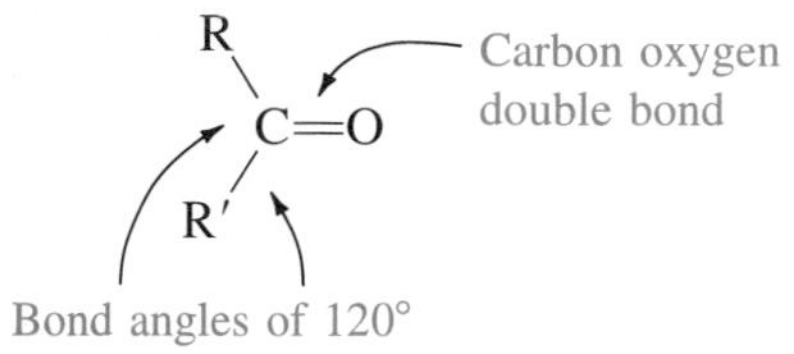

R–C(=O)–R′ : Carbon oxygen double bond; Bond angles of 120°

FIGURE 6.1
The carbonyl group. The coplanar arrangement of the carbonyl group and the two atoms attached to the carbonyl group (shown as R and R′) is illustrated.

Examples of the various families of organic carbonyl-containing compounds are shown in Table 6.1. Carbohydrates, carboxylic acids and esters, and amides, all of which will be discussed in upcoming chapters, are complex carbonyls, containing structural elements other than carbon, hydrogen, and the carbonyl group. Also, note in Table 6.1 that the carbonyl group is only a piece of the functional group associated with these families. That is, the carbonyl group may be combined with other atoms to form the **carboxyl group** of the carboxylic acids or the **acyl group** of the derivatives of the carboxylic acids.

Aldehydes and *ketones,* on the other hand, have a structure and properties that are principally influenced by the carbonyl group. Aldehydes and ketones differ in the type of atom or atoms that are attached to the carbonyl carbon. In ketones the carbonyl carbon is attached to two carbon atoms, while in aldehydes the carbonyl carbon is attached to at

TABLE 6.1 Families of Carbonyl-containing Compounds

General Formula	Name	Family of Organic Compounds
R—C(=O)—H	Carbonyl group	Aldehyde
R—C(=O)—R	Carbonyl group	Ketone
R—C(=O)—OH	Carboxyl group	Carboxylic acid
R—C(=O)—OR′	Acyl group	Ester
R—C(=O)—NH_2	Acyl group	Amide
R—C(=O)—Cl	Acyl group	Acid chloride

R or R′ is the abbreviation for any alkyl or aryl group; aryl is used for aromatic compounds in the same way that alkyl is used for aliphatic compounds. An aryl group (often designated "Ar") is an aromatic compound with one hydrogen removed (for example, phenyl—the phenyl group is benzene with one hydrogen removed, C_6H_5—).

least one hydrogen atom; the second atom attached to the carbonyl carbon in aldehydes may be another hydrogen or a carbon atom. This fundamental structural difference distinguishes these two families of organic compounds:

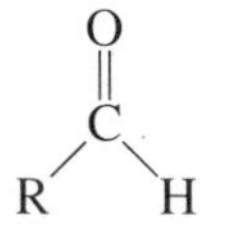

$$\underset{\text{R}\quad\text{H}}{\overset{\text{O}}{\overset{\|}{\text{C}}}} \qquad\qquad \underset{\text{R}\quad\text{R}'}{\overset{\text{O}}{\overset{\|}{\text{C}}}}$$

Aldehyde
R = H, R, or Ar

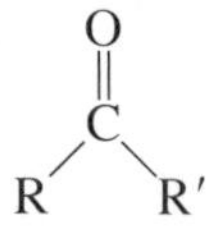

Ketone
R and R′ = R or Ar

6.1 ALDEHYDES AND KETONES: STRUCTURE AND PHYSICAL PROPERTIES

Aldehydes and ketones are moderately polar compounds because of the polar carbonyl group. They boil at higher temperatures than hydrocarbons or ethers that have the same number of carbon atoms or are of equivalent molecular weight. On the other hand, their boiling points are lower than those of alcohols containing the same number of carbon atoms. These trends are clearly demonstrated in the following examples and in Tables 6.2 and 6.3.

$CH_3CH_2CH_2CH_3$
Butane
(*n*-butane)
mol. wt. 58 g/mol
b.p. −0.5°C

$CH_3\text{—}O\text{—}CH_2CH_3$
Methoxyethane
(ethyl methyl ether)
mol. wt. 60 g/mol
b.p. 7.0°C

$CH_3CH_2CH_2\text{—}OH$
1-Propanol
(*n*-propyl alcohol)
mol. wt. 60 g/mol
b.p. 97.2°C

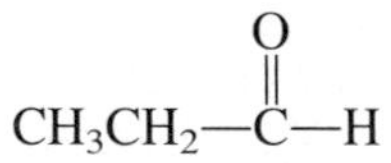

$CH_3CH_2\text{—}\overset{O}{\overset{\|}{C}}\text{—}H$
Propanal
(propionaldehyde)
mol. wt. 58 g/mol
b.p. 49°C

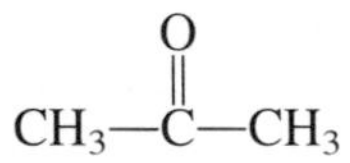

$CH_3\text{—}\overset{O}{\overset{\|}{C}}\text{—}CH_3$
Propanone
(acetone)
mol. wt. 58 g/mol
b.p. 56°C

The smaller members of these two families (containing five or fewer carbon atoms) are reasonably soluble in water. This solubility is a direct result of hydrogen bonding between the carbonyl group and water as shown in Figure 6.2(a).

While simple aldehydes are soluble in aqueous solutions, larger members of these two families are not. As the size of the aldehyde or ketone increases, their water solubility decreases. The larger compounds are less polar and more hydrocarbonlike and thus are soluble in nonpolar organic solvents.

Question 6.1 Simple ketones (for example, acetone (dimethyl ketone) or 3-buten-2-one (methyl vinyl ketone)) are often used as industrial solvents for many organically based products such as adhesives, caulks, and paints. They are often considered ''universal solvents,'' since they dissolve so many diverse materials. Why are these chemicals such good solvents?

It may be helpful to view polar molecules as tiny bar magnets to get a better understanding of the intermolecular attractive forces involving these molecules. Behaving as small magnets, these carbonyl compounds are attracted to one another and to other polar molecules, such as water. This interaction is shown for the carbonyl group in Figure

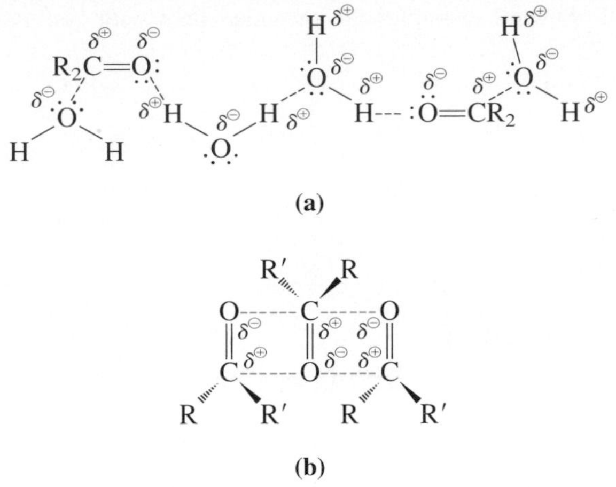

FIGURE 6.2
(a) Hydrogen bonding between the carbonyl group and water;
(b) polar interaction between carbonyl groups.

6.2(b). It is this relatively strong attraction that accounts for the high boiling points and solubility in water (of the simpler) aldehydes and ketones.

6.2 ALDEHYDES AND KETONES: I.U.P.A.C. NOMENCLATURE AND COMMON NAMES

Aldehydes

Aldehydes are named (I.U.P.A.C. System) according to the **parent chain,** the longest continuous carbon chain containing the carbonyl group. The final *-e* of the parent alkane is dropped and replaced by *-al*. The chain is always numbered beginning with the carbonyl

TABLE 6.2 Physical Properties of Selected Aldehydes

Name	Structure	m.p. (°C)	b.p. (°C)	Density*
Methanal (formaldehyde)	CH_2O	−92	−21	0.815
Ethanal (acetaldehyde)	CH_3CHO	−125	21	0.783
Propanal (propionaldehyde)	CH_3CH_2CHO	−81	49	0.807
Butanal (*n*-butyraldehyde)	$CH_3CH_2CH_2CHO$	−99	76	0.817
Pentanal (*n*-valeraldehyde)	$CH_3CH_2CH_2CH_2CHO$	−91	103	0.810
Hexanal (caproaldehyde)	$CH_3CH_2CH_2CH_2CH_2CHO$	−56	128	0.834
Heptanal (*n*-heptaldehyde)	$CH_3CH_2CH_2CH_2CH_2CH_2CHO$	−43	155	0.850
Benzaldehyde		−56	178	1.042

*In g/cm^3 at 20°C.

carbon as carbon-1. Several examples are provided here with common names given in parentheses:

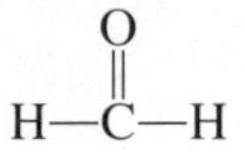

$$\mathrm{H{-}\overset{\displaystyle O}{\overset{\|}{C}}{-}H} \qquad \mathrm{CH_3{-}\overset{\displaystyle O}{\overset{\|}{C}}{-}H} \qquad \mathrm{CH_3CH_2{-}\overset{\displaystyle O}{\overset{\|}{C}}{-}H}$$

Methanal (formaldehyde) — Ethanal (acetaldehyde) — Propanal (propionaldehyde)

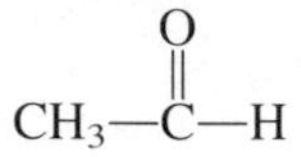

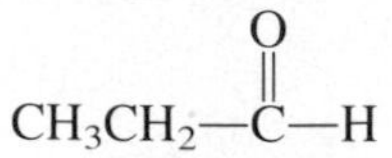

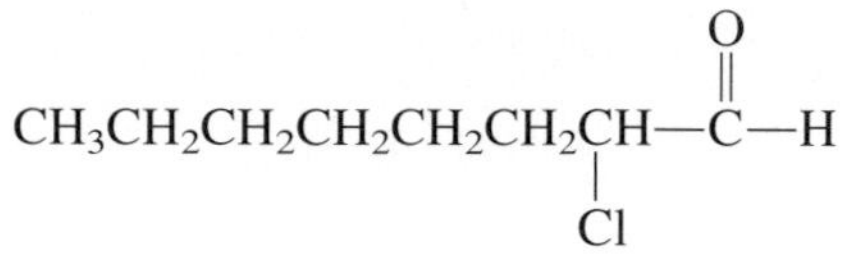

$$\mathrm{CH_3CH_2CH_2CH_2CH_2CH_2\underset{\displaystyle Cl}{\underset{|}{C}}H{-}\overset{\displaystyle O}{\overset{\|}{C}}{-}H}$$

2-Chlorooctanal

$$\mathrm{C_6H_5{-}\overset{\displaystyle O}{\overset{\|}{C}}{-}H}$$

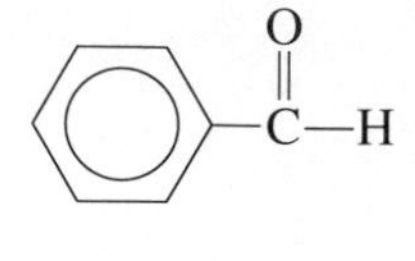

Benzaldehyde, found in almonds

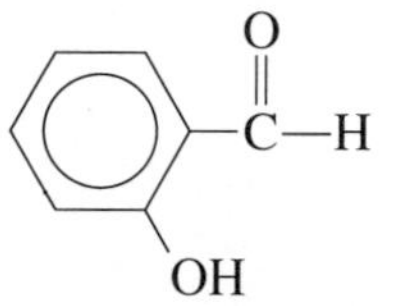

$$\mathrm{C—H}\ (\text{ring with } \mathrm{OH})$$

o-Hydroxybenzaldehyde (salicylaldehyde)

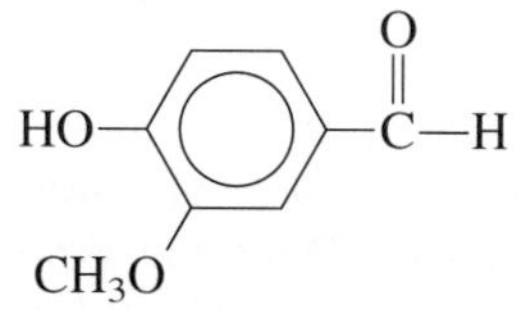

$$\mathrm{HO{-}C_6H_3(OCH_3){-}\overset{\displaystyle O}{\overset{\|}{C}}{-}H}$$

4-Hydroxy-3-methoxybenzaldehyde (vanillin), found in natural vanilla

TABLE 6.3 Physical Properties of Selected Ketones

Name	Structure	m.p. (°C)	b.p. (°C)	Density*
Propanone (acetone)	CH_3COCH_3	−95	56	0.791
Butanone (methyl ethyl ketone)	$CH_3COCH_2CH_3$	−87	80	0.805
2-Pentanone (methyl *n*-propyl ketone)	$CH_3COCH_2CH_2CH_3$	−78	102	0.812
3-Pentanone (diethyl ketone)	$CH_3CH_2COCH_2CH_3$	−42	103	0.816
2-Hexanone (methyl *n*-butyl ketone)	$CH_3COCH_2CH_2CH_2CH_3$	−57	126	0.812
3-Hexanone (ethyl *n*-propyl ketone)	$CH_3CH_2COCH_2CH_2CH_3$	—	123	0.815
4-Heptanone (di-*n*-propyl ketone)	$CH_3CH_2CH_2COCH_2CH_2CH_3$	−33	144	0.817
Cyclohexanone	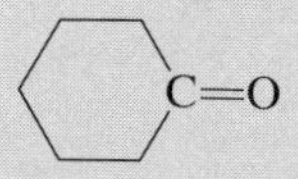 C=O	−16	156	0.998
Acetophenone (methyl phenyl ketone)	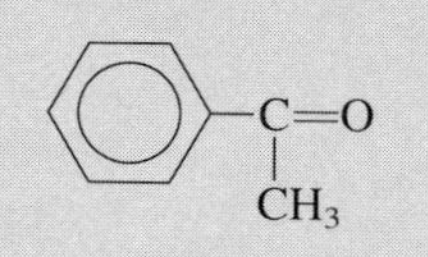 C=O, CH_3	20	202	1.028
Benzophenone (diphenyl ketone)	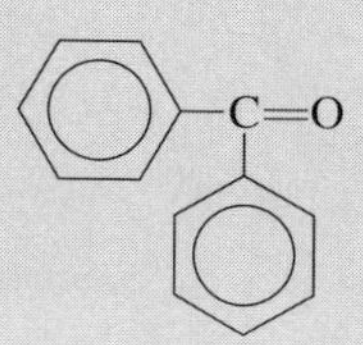 C=O	48	306	1.146

*In g/cm^3 at 20°C.

Ketones

The rules for naming ketones are directly analogous to those for naming aldehydes. In ketones, however, the *-e* ending of the parent alkane is replaced with the *-one* suffix of the ketone family. The longest carbon chain is numbered to give the carbonyl carbon the lowest possible number. For example:

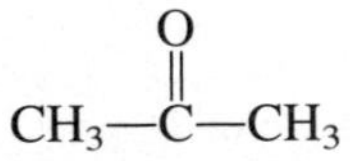

$$CH_3-\overset{\overset{O}{\|}}{C}-CH_3 \qquad CH_3CH_2-\overset{\overset{O}{\|}}{C}-CH_3 \qquad CH_3CH_2CH_2CH_2-\overset{\overset{O}{\|}}{C}-CH_2CH_2CH_3$$

Propanone *no number necessary* (acetone)	Butanone (methyl ethyl ketone)	4-Octanone *not 5-octanone* (*n*-butyl *n*-propyl ketone)

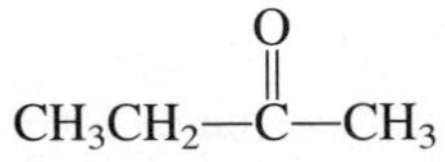

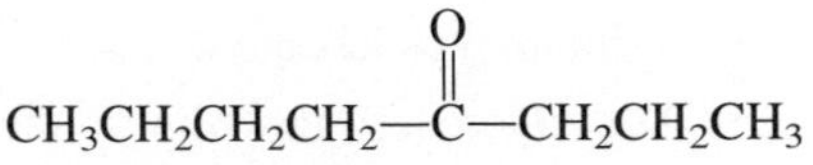

Question 6.2 In the examples provided above, why are propanone and butanone correct names while 2-propanone and 2-butanone are not?

Example 6.1

Naming aldehydes and ketones.

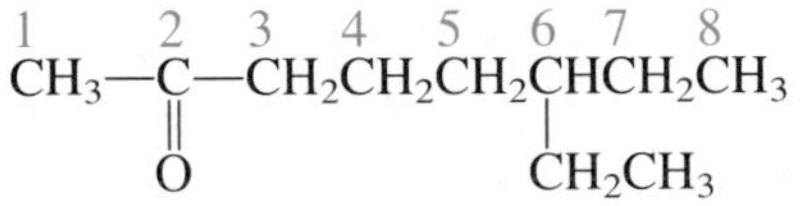

$$\overset{1}{CH_3}-\underset{\underset{O}{\|}}{\overset{2}{C}}-\overset{3}{CH_2}\overset{4}{CH_2}\overset{5}{CH_2}\underset{\underset{CH_2CH_3}{|}}{\overset{6}{C}H}\overset{7}{CH_2}\overset{8}{CH_3}$$

Parent compound: Octane (becomes octanone)

Position of carbonyl group: Carbon-2 (*not* carbon-7)

Substituents: 6-Ethyl (*not* 3-ethyl)

Name: 6-Ethyl-2-octanone

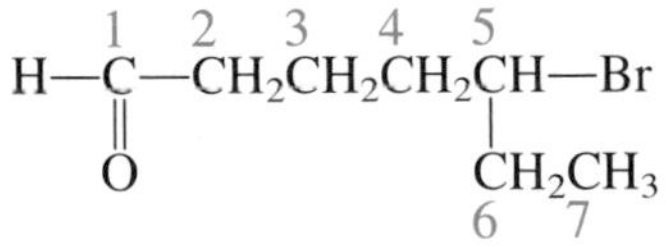

$$H-\underset{\underset{O}{\|}}{\overset{1}{C}}-\overset{2}{CH_2}\overset{3}{CH_2}\overset{4}{CH_2}\underset{\underset{\underset{6\ \ 7}{CH_2CH_3}}{|}}{\overset{5}{C}H}-Br$$

Parent compound: Heptane (to become heptanal) (*not* pentanone)

Position of carbonyl group: Carbon-1 (*not* carbon-7)

Substituents: 5-Bromo (*not* 3-bromo)

Name: 5-Bromoheptanal (*not* 5-bromo-1-heptanal)

Question 6.3 Name the following compounds, using the I.U.P.A.C. nomenclature rules:

a. 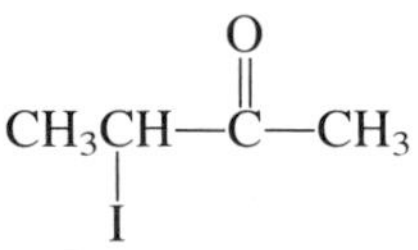 $CH_3\underset{\underset{I}{|}}{C}H-\overset{\overset{O}{\|}}{C}-CH_3$

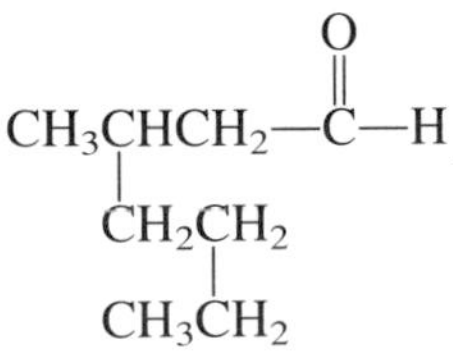

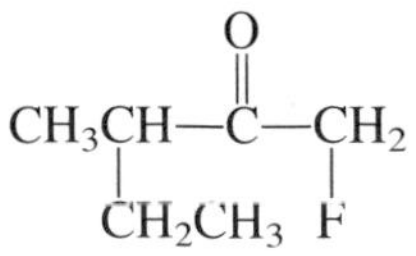

b. 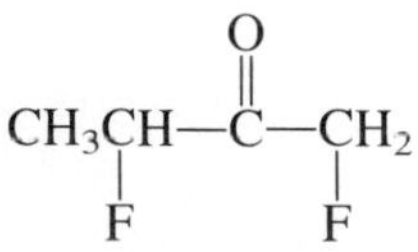 $CH_3\underset{\underset{F}{|}}{C}H-\overset{\overset{O}{\|}}{C}-\underset{\underset{F}{|}}{C}H_2$

c. $CH_3\underset{\underset{CH_2CH_2-CH_3CH_2}{|}}{C}HCH_2-\overset{\overset{O}{\|}}{C}-H$ (branch: CH_2CH_2 bonded to CH_3CH_2)

d. $CH_3\underset{\underset{CH_2CH_2CH_3}{|}}{C}H-\overset{\overset{O}{\|}}{C}-CH_3$

e. $CH_3\underset{\underset{CH_2CH_3}{|}}{C}H-\overset{\overset{O}{\|}}{C}-\underset{\underset{F}{|}}{C}H_2$

f. $Br-\underset{\underset{CH_2\underset{\underset{Br}{|}}{C}H_2}{|}}{C}HCH_2-\overset{\overset{O}{\|}}{C}-H$

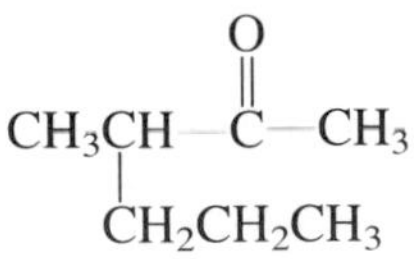

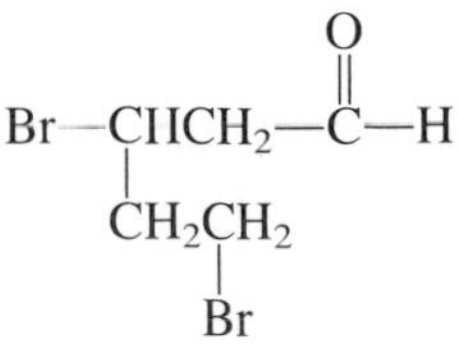

Question 6.4 Write the structural formula for each of the following:

a. methyl isopropyl ketone (What is the I.U.P.A.C. name for this compound?)
b. 4-heptanone
c. nonanal
d. 7-bromoheptanal
e. 2-fluorocyclohexanone
f. hexachloroacetone (what is the I.U.P.A.C. name for this compound?)

6.3 ALDEHYDES AND KETONES: WHERE DO THEY COME FROM?

Industrial sources

Many of the simpler aldehydes and ketones are produced industrially for commercial purposes. The **Fischer-Tropsch Synthesis** produces aldehydes and ketones from the reaction of carbon monoxide (CO) and hydrogen (H_2) at high temperature.

$$CO + H_2 \xrightarrow[\text{coke}]{\text{Heat,}} \text{aldehydes and ketones} \qquad 6.1$$

Coke is simply a term for carbon at high temperature. It reacts with oxygen to produce the carbon monoxide that is necessary for the reaction to occur. The coke behaves as a reducing agent. The three most important industrial carbonyl compounds (methanal, ethanal, and propanone) are all produced by the Fischer-Tropsch process.

Methanal (formaldehyde) is also produced by the **air oxidation** of methanol. The reaction is carried out again at high temperature in the presence of a copper metal catalyst. The copper speeds up the reaction by providing a surface on which the reaction can occur:

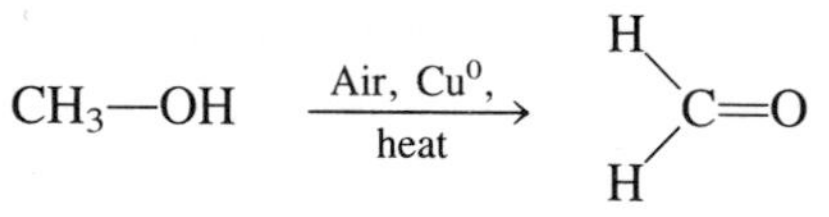

$$CH_3\text{—}OH \xrightarrow[\text{heat}]{\text{Air, Cu}^0\text{,}} H_2C\text{=}O \qquad 6.2$$

Methanol (methyl alcohol) → Methanal (formaldehyde)

Formaldehyde is a gas (b.p. −21°C). It is available commercially as an aqueous solution (trade name: *formalin*) or as solids (trade name: *paraformaldehyde* or *trioxane*). The structures of the solids are provided below. Formalin is used as a preservative for tissues and as an embalming fluid.

For other uses of formaldehyde, see A Clinical Perspective: Aldehydes in Medicine.

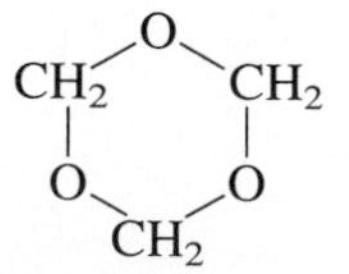

Trioxane (ring of three CH_2 and three O)

Trioxane
(a trimer—composed of three molecules—of formaldehyde)

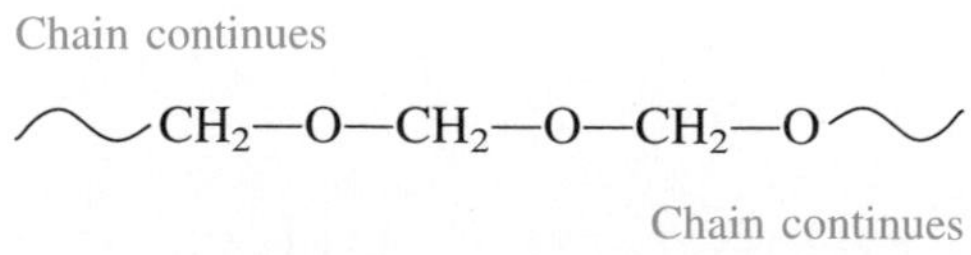

Chain continues

$$\sim CH_2\text{—}O\text{—}CH_2\text{—}O\text{—}CH_2\text{—}O \sim$$

Chain continues

Paraformaldehyde
(a polymer of formaldehyde)

Ethanal is prepared in an analogous manner via the oxidation of ethanol:

$$CH_3CH_2{-}OH \xrightarrow[\text{heat}]{\text{air,}} \underset{H}{\overset{CH_3}{\diagdown}}C{=}O \qquad 6.3$$

Ethanol (ethyl alcohol) → Ethanal (acetaldehyde)

Also produced in the liver, ethanal is responsible for the symptoms of a hangover. It can be condensed to form the polymer paraldehyde similar in structure to paraformaldehyde (shown above).

Propanone (acetone), the simplest ketone, is similarly prepared:

$$CH_3\underset{OH}{CH}CH_3 \xrightarrow[\text{heat}]{\text{air,}} (CH_3)(CH_3)C{=}O \qquad 6.4$$

2-Propanol (isopropyl alcohol) → Propanone (acetone)

Propanone (acetone) is an important and versatile solvent for organic compounds. It has the ability to dissolve organic compounds while being miscible (mixes with) water. As a result, it has a number of industrial applications; it is used as a solvent in adhesives, paints, cleaning solvents, nail polish, and nail polish remover. It has also been used routinely as a cleaning and drying agent for glassware in the organic chemistry laboratory. When a piece of glassware that has just been washed is rinsed with propanone, the propanone mixes with and removes the water. The propanone film then quickly evaporates because of its low boiling point (b.p. 56°), providing a dry piece of glassware. As a result of potential health risks and hazardous waste disposal problems, this practice is no longer recommended. Propanone is flammable and should therefore be treated with appropriate care.

Many complex members of this family are produced industrially as food and fragrance chemicals, medicinals, and agricultural chemicals. Aldehydes and ketones are particularly important to the food industry, in which many of these compounds are used as artificial and/or natural additives to food. Vanillin, a principal component of natural vanilla, is shown in Section 6.2. Artificial vanilla flavoring is simply a dilute solution of synthetic vanillin dissolved in ethanol. See Figure 6.3 for additional examples.

Laboratory sources

As a result of the availability of most alcohols, they provide an easy and accessible route to many aldehydes and ketones that are not routinely available commercially.

In the laboratory, aldehydes and ketones are primarily prepared by the *oxidation* of the corresponding alcohol. Virtually any aldehyde or ketone can be prepared if the alcohol precursor is available. The oxidizing agents most commonly used are potassium permanganate ($KMnO_4$) and potassium dichromate ($K_2Cr_2O_7$). In the examples below and throughout this text, [O] is used to designate any oxidizing agent. A summary follows.

The potassium permanganate is used as a basic, aqueous solution ($pH > 7$).

The oxidizing agent is actually dichromic acid ($H_2Cr_2O_7$). The dichromic acid is produced by dissolving the potassium dichromate in an aqueous acid, for example, sulfuric acid ($pH < 7$).

Oxidation of methyl alcohol gives methanal:

$$H{-}\underset{H}{\overset{H}{C}}{-}OH \xrightarrow{[O]} (H)(H)C{=}O \qquad 6.5$$

Methanol (methyl alcohol) → Methanal (formaldehyde)

See A Clinical Perspective: Aldehydes in Medicine

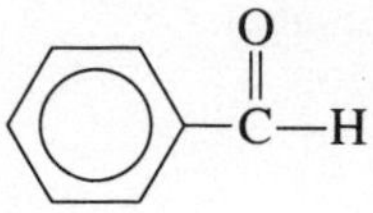

C_6H_5—$\overset{O}{\overset{\|}{C}}$—H Benzaldehyde—oil of almonds

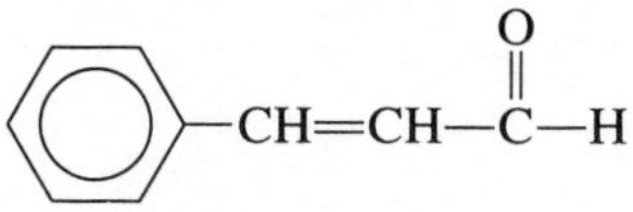

C_6H_5—CH=CH—$\overset{O}{\overset{\|}{C}}$—H Cinnamaldehyde—oil of cinnamon

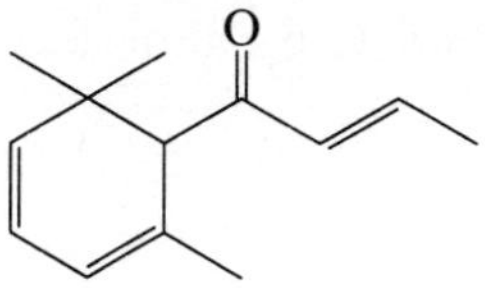

Demascone (alpha)—berry flavoring

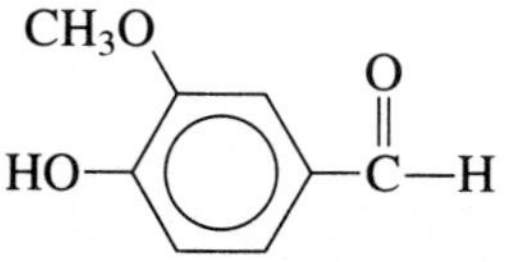

Vanillin—oil of vanilla beans

$CH_3CH_2CH_2CH_2CH_2CH_2—\overset{O}{\overset{\|}{C}}—CH_3$ 2-Octanone—mushroom flavoring

$CH_3—\overset{CH_3}{\overset{|}{C}}=CH—CH_2—CH_2—\overset{CH_3}{\overset{|}{C}}=CH—\overset{O}{\overset{\|}{C}}—H$ Citral—oil of lemongrass

$H—\overset{O}{\overset{\|}{C}}—H$ Methanal—common name formaldehyde—a gas at room temperature; *Formalin* is a 40% solution of methanal in water; it is used to preserve biological specimens and as an embalming fluid.

$CH_3—\overset{O}{\overset{\|}{C}}—H$ Ethanal—common name acetaldehyde—industrial precursor (starting material) for acetic acid (*vinegar*); forms a trimeric solid polymer on heating (paraaldehyde), which has medicinal uses.

$CH_3—\overset{O}{\overset{\|}{C}}—CH_3$ Propanone—common name acetone—an important industrial solvent.

$CH_3CH_2—\overset{O}{\overset{\|}{C}}—CH_3$ Butanone—common name methyl ethyl ketone—another important industrial solvent.

FIGURE 6.3
Industrially important aldehydes and ketones.

1° alcohols give aldehydes:

$$R—\underset{H}{\overset{H}{\overset{|}{\underset{|}{C}}}}—OH \xrightarrow{[O]} \underset{H}{\overset{R}{\ }}\!\!>C=O \qquad (6.6)$$

1° Alcohol — Aldehyde

And 2° alcohols give ketones:

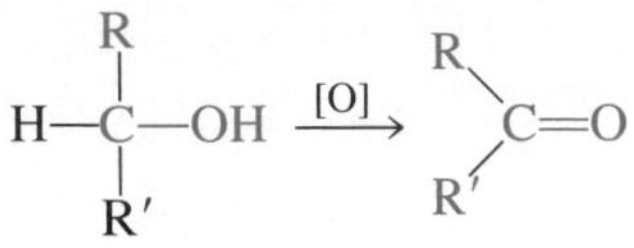

$$\text{H—}\underset{\text{R}'}{\overset{\text{R}}{\text{C}}}\text{—OH} \xrightarrow{[O]} \begin{matrix}\text{R} \\ \text{R}'\end{matrix}\!\!>\!\text{C=O} \qquad 6.7$$

2° Alcohol Ketone

Tertiary alcohols do not react:

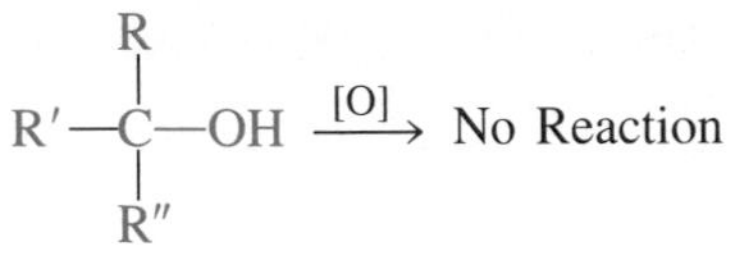

$$\text{R}'\text{—}\underset{\text{R}''}{\overset{\text{R}}{\text{C}}}\text{—OH} \xrightarrow{[O]} \text{No Reaction} \qquad 6.8$$

A careful look at the above general examples provides a partial understanding of the mechanism involved in these oxidation reactions. Since methyl, primary, and secondary alcohol(s) all oxidize to the corresponding aldehydes or ketones while tertiary alcohols do not, it appears that the carbonyl carbon must contain at least one hydrogen substituent for a successful oxidation to result:

$$\text{H—}\underset{|}{\overset{|}{\text{C}}}\text{—OH} \longrightarrow \;>\!\text{C=O}$$

Carbon-hydrogen bond necessary
to a successful reaction

A complete view of the step-by-step mechanism (which we will not cover) substantiates this view. In effect, the elements of hydrogen (two hydrogen atoms) are lost during the oxidation:

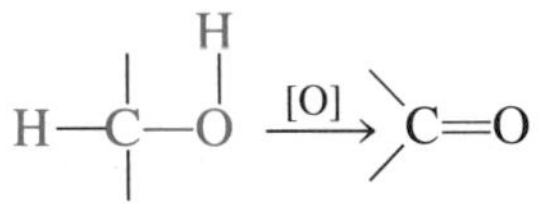

$$\text{H—}\underset{|}{\overset{|}{\text{C}}}\text{—}\overset{\text{H}}{\overset{|}{\text{O}}} \xrightarrow{[O]} \;>\!\text{C=O}$$

The elements of hydrogen
(H^+ and $H:^-$) are lost in
the course of this reaction.

Many other oxidation processes, involving similar functional groups, behave similarly. The biochemical implications of oxidation reactions and their mechanism are discussed in Chapter 13.

Example 6.2

Differences in the oxidation of primary, secondary, and tertiary alcohols.

The oxidation of a primary alcohol to an aldehyde:

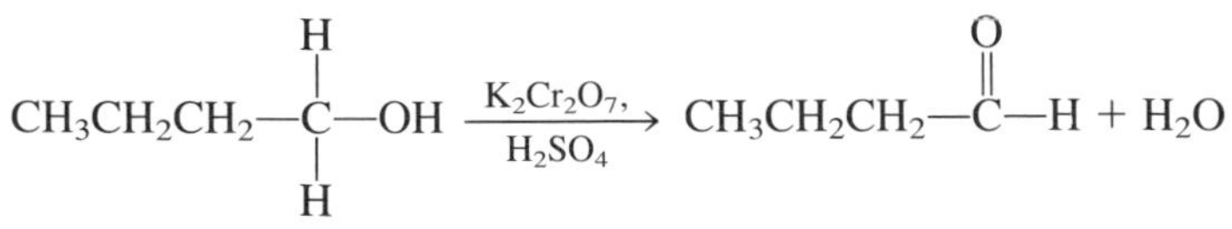

$$CH_3CH_2CH_2\text{—}\underset{\text{H}}{\overset{\text{H}}{\text{C}}}\text{—OH} \xrightarrow[H_2SO_4]{K_2Cr_2O_7,} CH_3CH_2CH_2\text{—}\overset{\text{O}}{\overset{\|}{\text{C}}}\text{—H} + H_2O \qquad 6.9$$

1-Butanol
(n-butyl alcohol)

Butanal
(butyraldehyde)

The oxidation of a secondary alcohol to a ketone:

$$CH_3CH_2CH_2CH_2—\underset{\underset{H}{|}}{\overset{\overset{CH_3}{|}}{C}}—OH \xrightarrow[H_2O]{KMnO_4,\ HO^-,} CH_3CH_2CH_2CH_2—\overset{\overset{O}{||}}{C}—CH_3 + H_2O \qquad 6.10$$

2-Hexanol 2-Hexanone

The oxidation of a cyclic, secondary alcohol to a ketone:

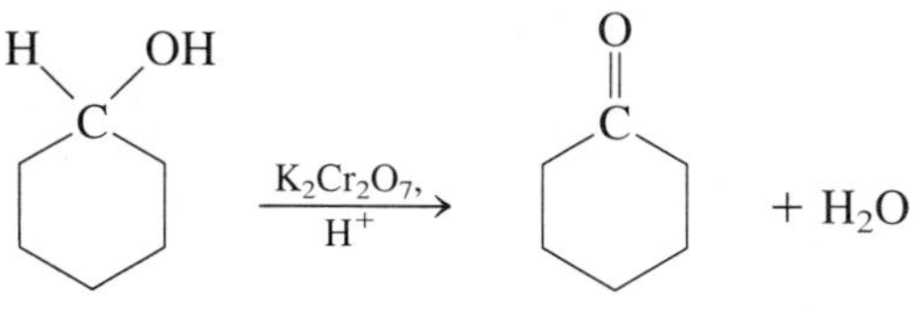

$$\text{Cyclohexanol} \xrightarrow[H^+]{K_2Cr_2O_7,} \text{Cyclohexanone} + H_2O \qquad 6.11$$

Cyclohexanol Cyclohexanone

Tertiary alcohols cannot be oxidized to aldehydes or ketones:

$$CH_3CH_2CH_2—\underset{\underset{CH_3}{|}}{\overset{\overset{CH_3}{|}}{C}}—OH \xrightarrow[H_2SO_4]{K_2Cr_2O_7,} \text{No Reaction} \qquad 6.12$$

2-Methyl-2-pentanol

Any oxidation reaction is a part of a redox couple. Consequently, all oxidations are accompanied by reductions. In the examples provided above (oxidations by $K_2Cr_2O_7$ and $KMnO_4$) the dichromate ($Cr_2O_7^{2-}$, containing Cr^{6+}) is reduced to the Cr^{3+} cation, and permanganate (MnO_4^-, containing Mn^{7+}) is reduced to manganese dioxide (MnO_2, containing Mn^{4+}). The oxidations discussed in the next section also follow this course.

Question 6.5 Draw and name (I.U.P.A.C.) the product that you would expect to produce by oxidizing each of the following alcohols with potassium dichromate in sulfuric acid:

a. 2-butanol
b. 2-methyl-1-propanol
c. cyclopentanol
d. 2-methyl-2-propanol
e. 2-nonanol
f. 1-decanol

6.4 REACTIONS OF ALDEHYDES AND KETONES

Oxidation reactions

Aldehydes are easily **oxidized** further to carboxylic acids, while ketones do not generally undergo further oxidation. In fact, aldehydes are so easily oxidized that it is often very difficult to prepare them—they may continue to react, giving the carboxylic acid rather than the desired aldehyde. Aldehydes are susceptible to air oxidation, even at room temperature, and cannot be stored for long periods of time.

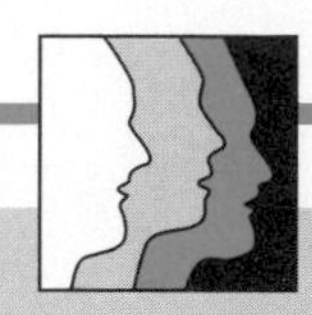

A HUMAN PERSPECTIVE

The Chemistry of Vision

β-Carotene is found in many yellow vegetables. It is also found in tomatoes and spinach. When β-carotene is cleaved, two molecules of vitamin A are produced. Vitamin A is in turn related to 11-*cis*-retinal, an unsaturated aldehyde that is a vital component in the photochemical transformations that make up the vision process.

The retina of the eye contains two types of cells that are responsible for vision: rods and cones. **Rods** are primarily responsible for vision in dim light and cones in bright light. **Cones** are also responsible for the detection of color in vision.

In the retina a protein called **opsin** combines with 11-*cis*-retinal to form a new protein, **rhodopsin.** The 11-*cis*-retinal portion of rhodopsin is a **prosthetic group.** A prosthetic group is a nonprotein portion of a protein that is necessary for the action of the protein.

When light strikes the rods, the light energy is absorbed by the 11-*cis*-retinal, which is photochemically converted to 11-*trans*-retinal. This *cis-/trans-* isomerization begins the visual process. The conversion of 11-*cis*-retinal to 11-*trans*-retinal is accompanied by a change in the shape of the rhodopsin itself. This new isomer of rhodopsin is *metarhodopsin.* This change in shape or conformation results in a nerve impulse that sends a signal to the brain.

The metarhodopsin is then involved in several reaction sequences that culminate with the regeneration of rhodopsin. The entire process is summarized in the accompanying figure.

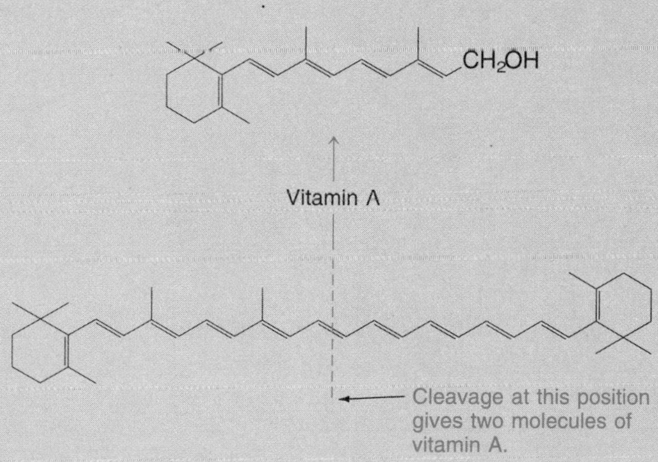

CH_2OH

Vitamin A

Cleavage at this position gives two molecules of vitamin A.

β-Carotene

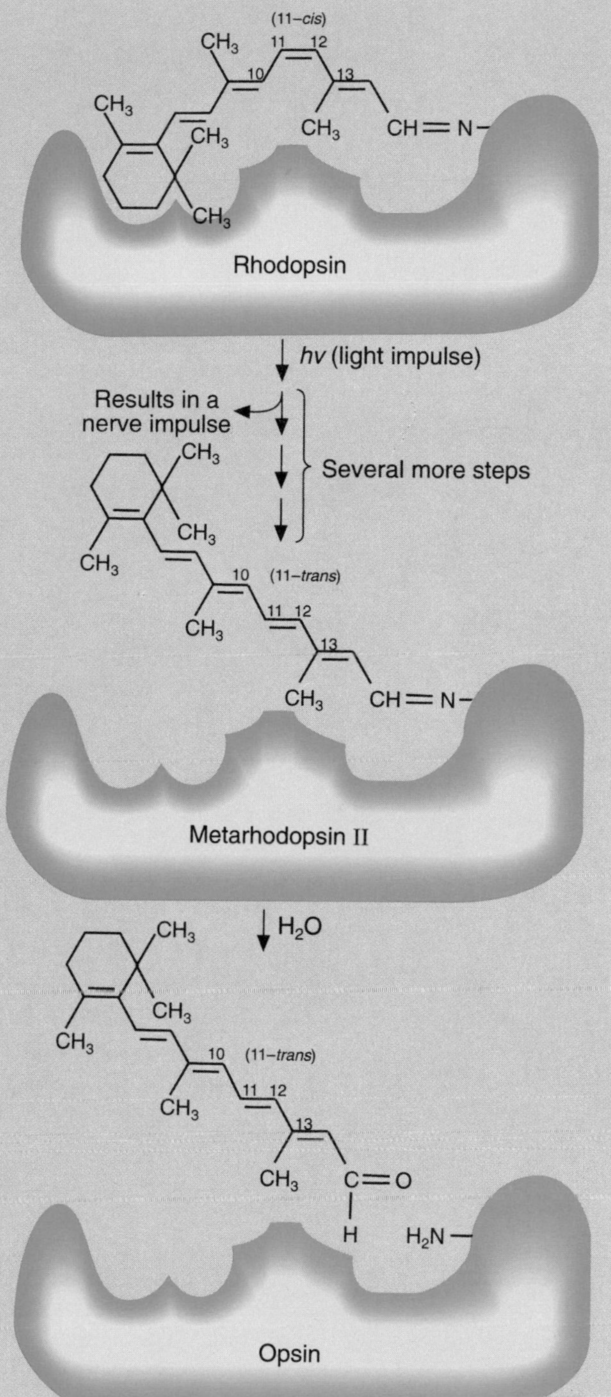

(11-*cis*) Rhodopsin

$h\nu$ (light impulse)

Results in a nerve impulse

Several more steps

(11-*trans*) Metarhodopsin II

H_2O

(11-*trans*) Opsin

The chemistry of vision: (a) Light is absorbed by rhodopsin, *cis-/trans-* isomerization of the 11-*cis*-retinal to 11-*trans*-retinal results, and a nerve impulse is sent to the brain; (b) the metarhodopsin formed in step (a) is converted back, through one or more pathways, to opsin. Opsin is converted to rhodopsin to begin the cycle again.

Oxidation of an aldehyde to a carboxylic acid:

$$\underset{\text{Aldehyde}}{R-\overset{O}{\overset{\|}{C}}-H} \xrightarrow{[O]} \underset{\text{Carboxylic acid}}{R-\overset{O}{\overset{\|}{C}}-OH} \qquad 6.13$$

In synthesis, many oxidizing agents will effect this transformation. Both potassium permanganate and potassium dichromate can be used. For example:

Section 5.5

$$\underset{\substack{\text{Ethanal}\\\text{(acetaldehyde)}}}{CH_3-\overset{O}{\overset{\|}{C}}-H} \xrightarrow[H_2O,\ HO^-]{KMnO_4,} \underset{\substack{\text{Ethanoic acid}\\\text{(acetic acid)}}}{CH_3-\overset{O}{\overset{\|}{C}}-OH} \qquad 6.14$$

The reaction of benzaldehyde to benzoic acid is an example of the conversion of an *aromatic* aldehyde to the corresponding *aromatic* carboxylic acid:

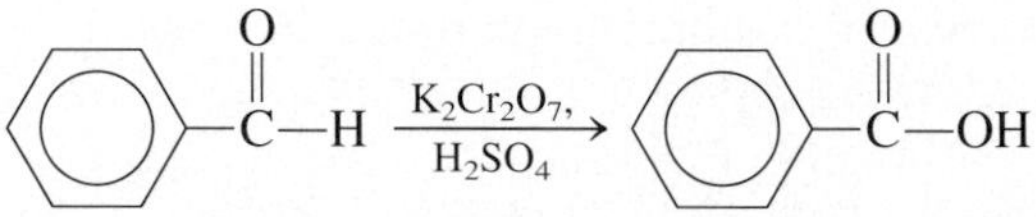

$$\underset{\text{Benzaldehyde}}{C_6H_5-\overset{O}{\overset{\|}{C}}-H} \xrightarrow[H_2SO_4]{K_2Cr_2O_7,} \underset{\text{Benzoic acid}}{C_6H_5-\overset{O}{\overset{\|}{C}}-OH} \qquad 6.15$$

Here too a carbon-hydrogen bond to the carbonyl carbon, present in the aldehyde but not the ketone, is essential to a successful reaction. The mechanism first requires that water add to the carbonyl (carbon-oxygen) double bond in a manner analogous to the addition of water to an alkene.

Section 4.4

Addition of water to a carbon-carbon double bond:

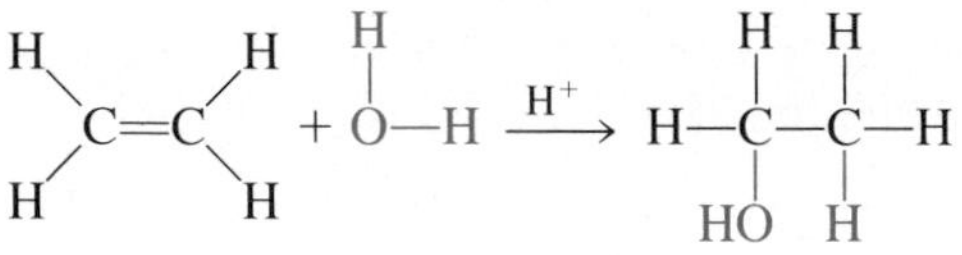

$$\underset{\text{Alkene}}{H_2C{=}CH_2} + H-O-H \xrightarrow{H^+} \underset{\text{Alcohol}}{H-\overset{H}{\underset{HO}{C}}-\overset{H}{\underset{H}{C}}-H} \qquad 6.16$$

Addition of water to a carbon-oxygen double bond:

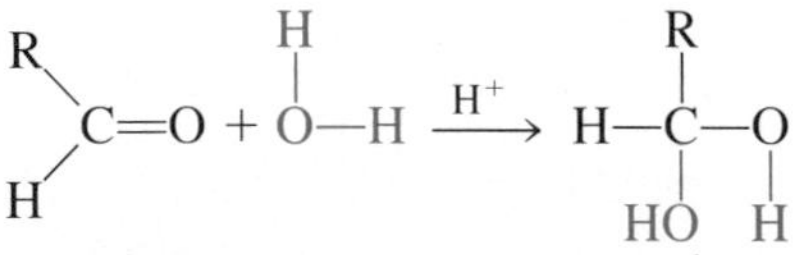

$$\underset{\text{Aldehyde}}{(R)(H)C{=}O} + H-O-H \xrightarrow{H^+} \underset{\text{Diol}}{H-\overset{R}{\underset{HO}{C}}-\underset{H}{O}} \qquad 6.17$$

Oxidation of the resulting unstable diol, which reacts by losing the elements of hydrogen, gives the carboxylic acid.

A carbon-hydrogen bond, involving the carbonyl carbon, is essential to a successful oxidation

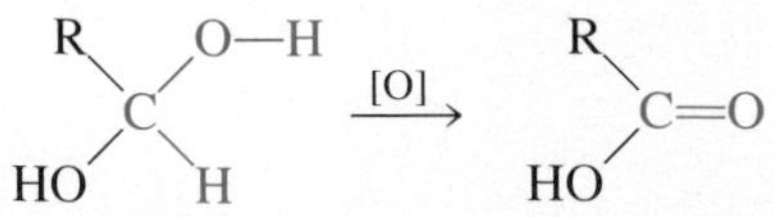

$$\underset{\text{Diol}}{(R)(HO)C(H)(O-H)} \xrightarrow{[O]} \underset{\text{Carboxylic acid}}{(R)(HO)C{=}O} \qquad 6.18$$

Loss of the elements of hydrogen (H^+ and $H:^-$) provides the desired carboxylic acid

As with alcohols, the elements of hydrogen are lost in the course of the oxidation. *Section 5.5*

In reality, when aldehydes are placed into an aqueous solution, very little of the aldehyde is converted to the diol. The two species (aldehyde and diol) are in equilibrium, however. As the small amount of diol reacts (oxidizes) to carboxylic acid, the equilibrium shifts, and more diol is formed. This process continues until all of the aldehyde is converted to carboxylic acid.

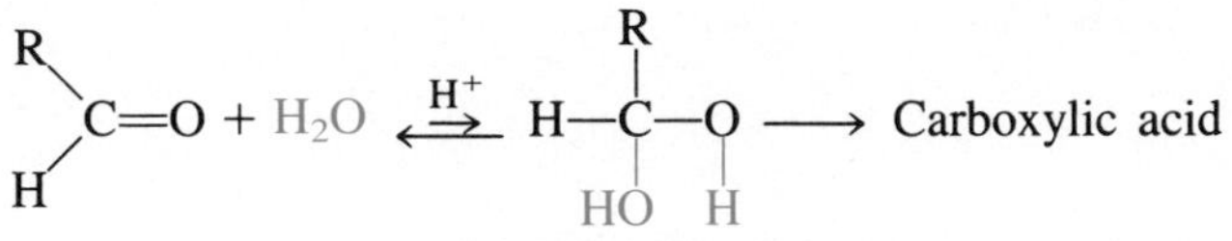

$$\text{RHC=O} + H_2O \underset{}{\overset{H^+}{\rightleftharpoons}} \text{H–C(R)(OH)–O–H} \longrightarrow \text{Carboxylic acid} \qquad 6.19$$

Dynamic equilibrium

It is also possible to differentiate qualitatively between aldehydes and ketones on the basis of differences in their reactivity. The most common laboratory test used to distinguish between an aldehyde and a ketone is the **Tollens' Test.** The Tollens' reagent, $Ag(NH_3)_2OH$, is a basic solution prepared by dissolving silver oxide (Ag_2O) in aqueous ammonium hydroxide (NH_4OH). Treatment of an aldehyde with the Tollens' reagent gives an oxidation-reduction reaction. The silver ion (Ag^+) is reduced to silver metal (Ag^0) as the aldehyde is oxidized to a carboxylic acid:

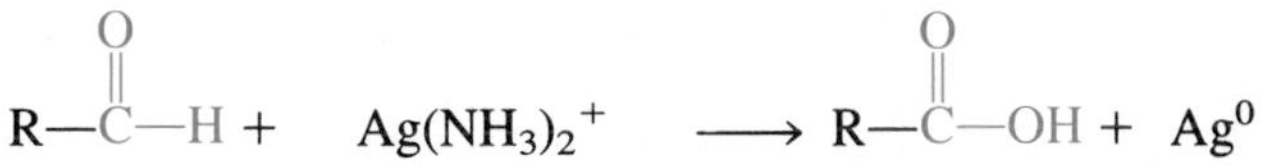

$$\text{R–C(=O)–H} + Ag(NH_3)_2{}^+ \longrightarrow \text{R–C(=O)–OH} + Ag^0 \qquad 6.20$$

Aldehyde; Silver ammonia complex: Tollens' reagent; Carboxylic acid; Silver metal mirror

The extremely fine silver metal precipitates from solution and coats the vessel (usually a test tube), giving a smooth silver mirror. The test is thus often called the Tollens' Silver Mirror Test. Ketones cannot be oxidized to carboxylic acids and do not react with the Tollens' reagent. The commercial manufacture of silver mirrors uses a similar process.

Other examples:

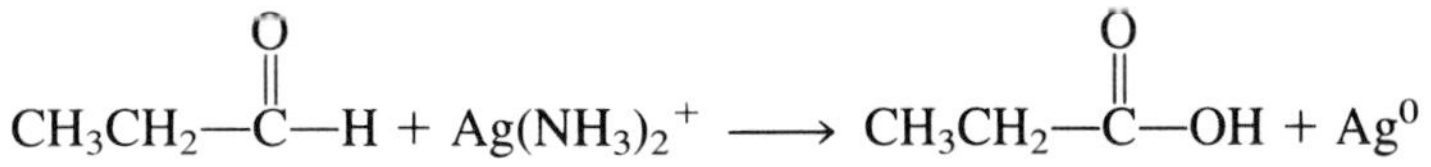

$$CH_3CH_2\text{–C(=O)–H} + Ag(NH_3)_2{}^+ \longrightarrow CH_3CH_2\text{–C(=O)–OH} + Ag^0 \qquad 6.21$$

Propanal; Propanoic acid

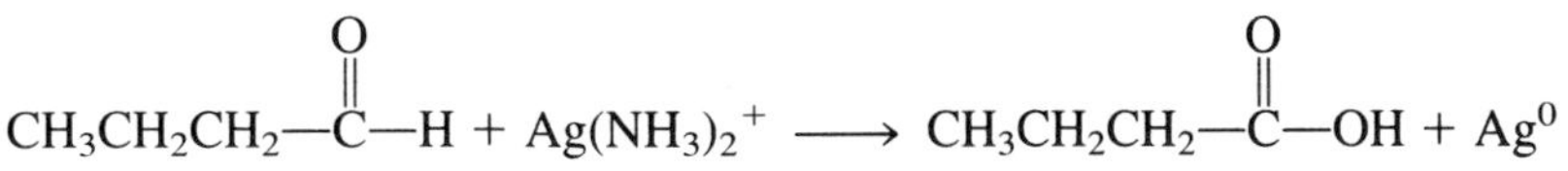

$$CH_3CH_2CH_2\text{–C(=O)–H} + Ag(NH_3)_2{}^+ \longrightarrow CH_3CH_2CH_2\text{–C(=O)–OH} + Ag^0 \qquad 6.22$$

Butanal; Butanoic acid

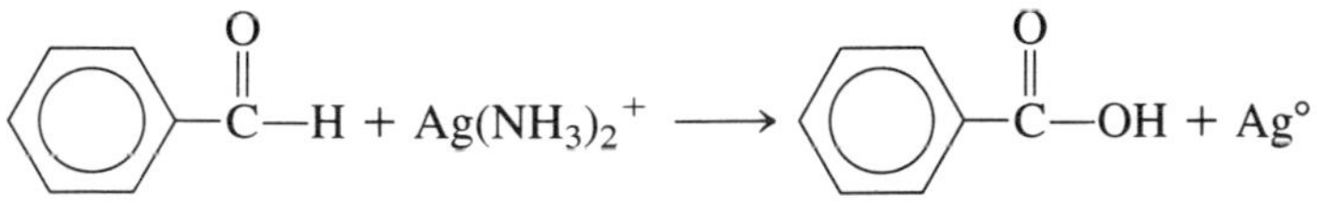

$$C_6H_5\text{–C(=O)–H} + Ag(NH_3)_2{}^+ \longrightarrow C_6H_5\text{–C(=O)–OH} + Ag^\circ \qquad 6.23$$

Benzaldehyde; Benzoic acid

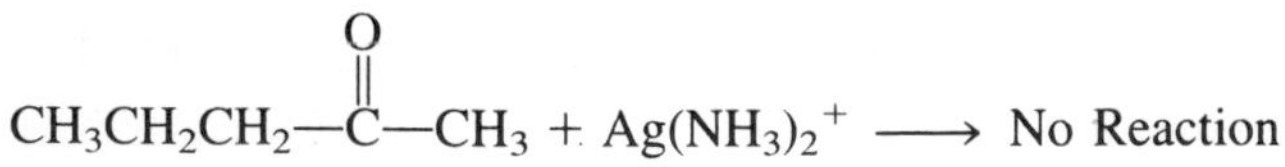

$$CH_3CH_2CH_2\text{–C(=O)–}CH_3 + Ag(NH_3)_2{}^+ \longrightarrow \text{No Reaction} \qquad 6.24$$

2-Pentanone

The sodium citrate helps to dissolve the carbonyl compound and the $Cu(OH)_2$ to afford a homogeneous solution.

Question 6.6 Another test used to distinguish between aldehydes and ketones is the **Benedict's Test.** Here, an aqueous solution of cupric hydroxide and sodium citrate reacts to oxidize aldehydes but does not generally react with ketones. The cupric ion, Cu(II), is reduced to cuprous ion, Cu(I), in the process. Cu(II) is soluble and gives a blue solution, while the Cu(I) precipitates as the red solid Cu_2O.

Which of the following will give a positive Benedict's Test?

a. 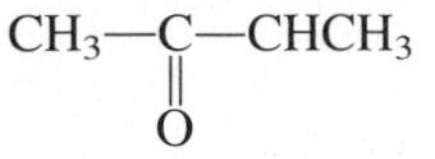 $CH_3-C(=O)-CHCH_3$

b. (cyclopentanone)

c. $H-C(=O)-CH(CH_2CH_2CH_2CH_3)CH_2CH_3$

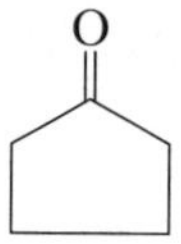

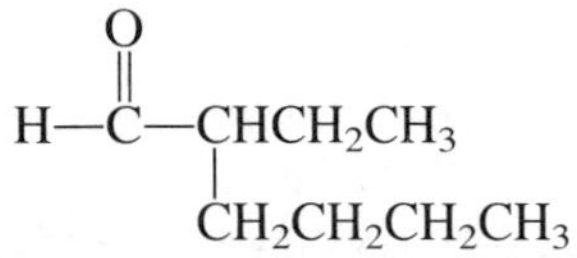

d. 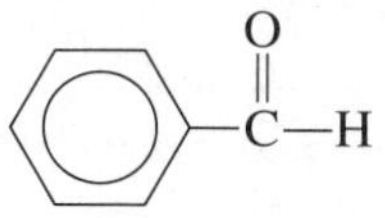 $C_6H_5-C(=O)-H$

e. 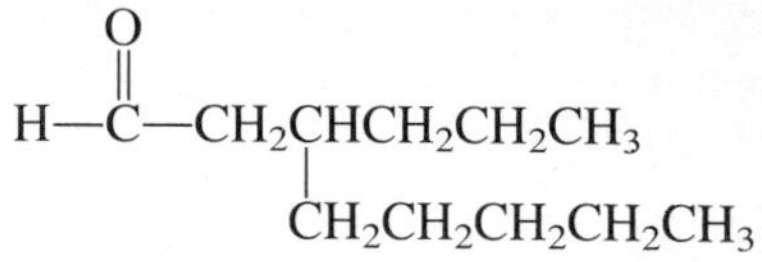 $H-C(=O)-CH_2CH(CH_2CH_2CH_2CH_2CH_3)CH_2CH_2CH_3$

Reduction reactions

Aldehydes and ketones are both readily **reduced** to the corresponding alcohol by a large number of different reducing agents. The general reaction is shown here.

Reduction of carbonyl compounds to alcohols:

[R] = any reducing agent.

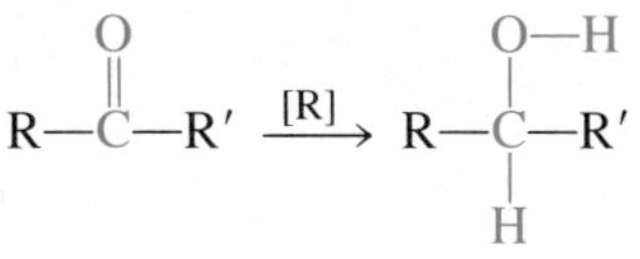

$$\underset{\text{Aldehyde or ketone}}{R-\overset{O}{\overset{\|}{C}}-R'} \xrightarrow{[R]} \underset{\text{Alcohol}}{R-\overset{O-H}{\overset{|}{\underset{H}{\underset{|}{C}}}}-R'} \tag{6.25}$$

The classical method of ketone reduction is **hydrogenation.** The carbonyl compound is reacted with hydrogen gas and a catalyst (Ni, Pt, or Pd metal) in a pressurized reaction vessel. Heating might also be necessary. In this case a carbon-oxygen double bond (the carbonyl group) is reduced to a carbon-oxygen single bond. Contrast this reaction to the reduction of an alkene to an alkane (the reduction of a carbon-carbon double bond to a carbon-carbon single bond). Some examples follow.

Section 4.4

Addition of hydrogen to a carbon-carbon double bond:

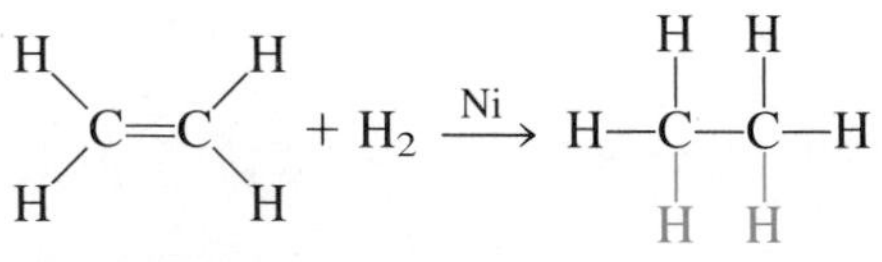

$$\underset{\text{Alkene}}{H_2C=CH_2} + H_2 \xrightarrow{Ni} \underset{\text{Alkane}}{H-CH_2-CH_2-H} \tag{6.26}$$

Addition of hydrogen to a carbon-oxygen double bond. The general reaction is as follows:

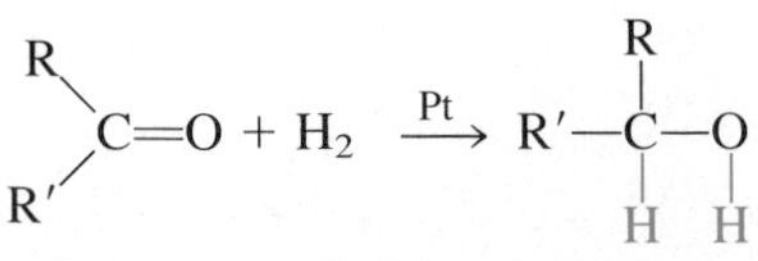

$$\underset{\text{Aldehyde or ketone}}{R(R')C=O} + H_2 \xrightarrow{Pt} \underset{\text{Alcohol}}{R'-\overset{R}{\overset{|}{\underset{H}{\underset{|}{C}}}}-\underset{H}{\underset{|}{O}}} \tag{6.27}$$

Specific examples of the above general reaction are the following:

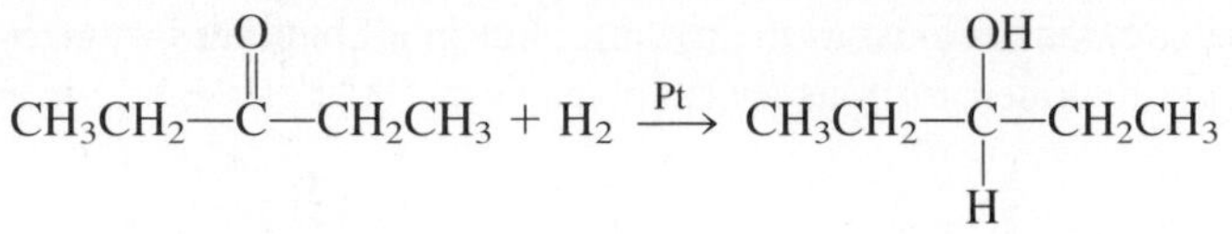

$$CH_3CH_2-\overset{\overset{\displaystyle O}{\|}}{C}-CH_2CH_3 + H_2 \xrightarrow{Pt} CH_3CH_2-\underset{\underset{\displaystyle H}{|}}{\overset{\overset{\displaystyle OH}{|}}{C}}-CH_2CH_3 \qquad 6.28$$

3-Pentanone 3-Pentanol

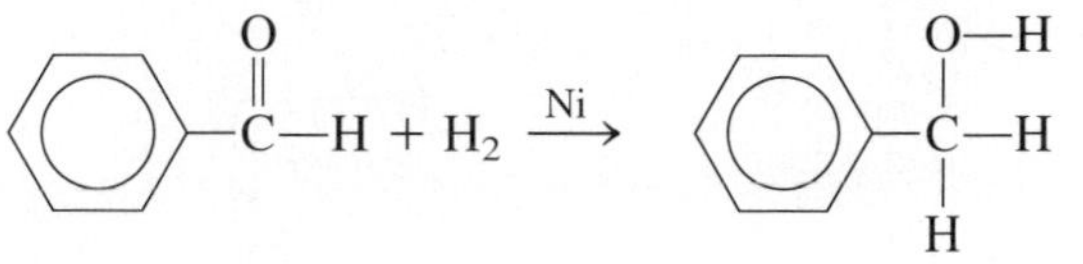

$$C_6H_5-\overset{\overset{\displaystyle O}{\|}}{C}-H + H_2 \xrightarrow{Ni} C_6H_5-\underset{\underset{\displaystyle H}{|}}{\overset{\overset{\displaystyle O-H}{|}}{C}}-H \qquad 6.29$$

Benzaldehyde Hydroxymethylbenzene (benzyl alcohol)

Questions 6.7 Label each of the following as an oxidation or reduction reaction and provide a reagent that would effect each transformation:

a. ethanal to ethanol
b. benzoic acid to benzaldehyde
c. cyclohexanone to cyclohexanol
d. 2-propanol to propanone
e. 2,3-butanedione (found in butter) to 2,3-butanediol

Addition reactions

The addition of H—X and H_2O

The principal reaction of the carbonyl group involves **addition** across the highly polar carbon-oxygen double bond. The reaction is shown here:

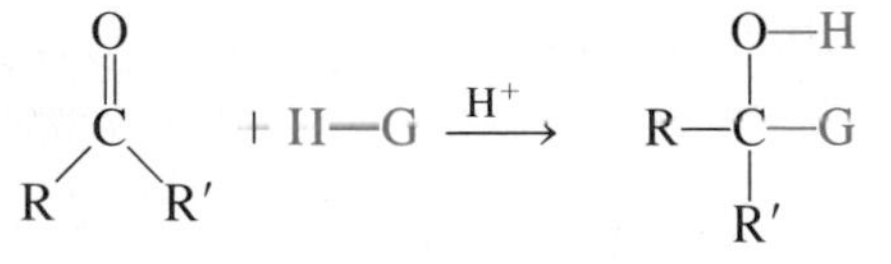

$$\underset{R\quad R'}{\overset{\overset{\displaystyle O}{\|}}{C}} + H-G \xrightarrow{H^+} R-\underset{\underset{\displaystyle R'}{|}}{\overset{\overset{\displaystyle O-H}{|}}{C}}-G \qquad 6.30$$

Carbonyl compound Addition product

In this example, H—G represents a generalized addition reagent; the reagent adds a proton (H^+) to the carbonyl oxygen, the remaining portion of the reagent (G^-) adding to the carbonyl carbon.

The reaction is similar to the hydrohalogenation of alkenes (the addition of H—X to the carbon-carbon double bond). In the latter case, H—X added according to "the rich get richer" (Markovnikov's) rule, while in this case the H^+ adds to the more electronegative oxygen and the X^- to the carbonyl carbon.

Section 4.4

The two reagents of widest applicability are water and alcohols:

$$H-G = H-OH \text{ and } H-OR$$

The reaction actually requires that a catalytic amount of acid be present in the solution. This is shown as H^+ over the arrow for the reaction.

The addition of water to a carbonyl compound produces a **hydrate** of the carbonyl compound. The hydrate of a carbonyl compound contains two HO— groups and is called a diol, as we discussed above. When water is used as the adding reagent, the reaction is reversible and the equilibrium lies far to the side of the carbonyl compound. In solution, many carbonyl compounds exist as the hydrate, but in all but rare instances it is impossible to isolate the hydrated form as a stable species.

Equilibrium lies to the left in most cases. The hydrate is not isolable in most cases.

$$RC(=O)R' + H_2O \overset{H^+}{\rightleftharpoons} R{-}C(OH)(R'){-}OH \qquad 6.31$$

Carbonyl compound — Hydrate (diol) *unstable*

If the carbonyl compound contains a large number of highly electronegative atoms, then it is possible to isolate the hydrated form. Trichloroethanal provides such an example:

Chloral hydrate is a common sedative and the infamous "Mickey Finn" knockout drops of many a Saturday matinee

$$Cl_3C{-}CHO + H_2O \xrightarrow{H^+} Cl{-}CCl_2{-}CH(OH){-}H \qquad 6.32$$

Trichloroethanal — Chloral hydrate (*common name*)

The addition of R—OH

The reaction of an aldehyde with alcohol (with a trace of acid) produces a **hemiacetal**—a half of an **acetal.** The reaction of a ketone with an alcohol produces a **hemiketal**—a half of a **ketal.** Continued reaction of the hemiacetal or hemiketal with excess alcohol results in the addition of a second molecule of alcohol, the loss of water, and the foundation of the acetal or ketal.

Continued reaction of the hemiacetal or hemiketal with excess alcohol results in the addition of a second molecule of alcohol, the loss of water, and the foundation of the acetal or ketal.

$$RC(=O)H \overset{ROH,\ H^+}{\rightleftharpoons} R{-}C(OH)(H){-}OR \overset{ROH,\ H^+}{\rightleftharpoons} R{-}C(OR)(H){-}OR \qquad 6.33$$

Aldehyde — Hemiacetal — Acetal

$$RC(=O)R' \overset{ROH,\ H^+}{\rightleftharpoons} R{-}C(OH)(R'){-}OR \overset{ROH,\ H^+}{\rightleftharpoons} R{-}C(OR)(R'){-}OR \qquad 6.34$$

Ketone — Hemiketal — Ketal

The reaction involves the addition of the elements of H^+ to the carbonyl oxygen and RO^- to the carbonyl carbon. For example:

$$CH_3C(=O)CH_3 + CH_3OH \overset{H^+}{\rightleftharpoons} CH_3{-}C(OH)(CH_3){-}OCH_3 \qquad 6.35$$

Propanone — Propanone methyl hemiketal (common name)

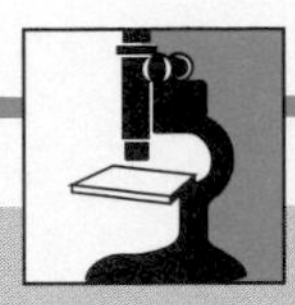

A CLINICAL PERSPECTIVE

Aldehydes in Medicine

As we have seen, most aldehydes have irritating, unpleasant odors. Formalin, a 40% solution of formaldehyde, is used to preserve biological tissues and for embalming. It has also been used to disinfect environmental surfaces, body fluids, and feces. Under no circumstances is it used as an antiseptic on human tissue because of the noxious fumes and skin irritation caused by formaldehyde.

Formaldehyde is useful in the production of killed virus vaccines. A deadly virus, like polio virus, can be treated with heat and formaldehyde. Formaldehyde reacts with the genetic information (RNA) of the virus, damaging it irreparably. It also reacts with the virus proteins but does not change their shape. Therefore, when you are injected with the Salk killed polio vaccine, the dead virus can't replicate and harm you. However, it will be recognized by your immune system, which will produce antibodies that will protect you against polio virus infection.

Formaldehyde can also be produced in the body! As you are aware, drinking wood alcohol (methanol) causes blindness, respiratory failure, convulsions, and death. The liver enzyme, alcohol dehydrogenase, whose function it is to detoxify alcohols, converts methanol to formaldehyde!

$$NAD^+ + CH_3\text{—}OH \xrightarrow[\text{dehydrogenase}]{\text{alcohol}} H\text{—}\overset{\overset{\displaystyle O}{\|}}{C}\text{—}H + NADH$$

The formaldehyde produced reacts with cellular proteins, causing the range of symptoms listed above.

Acetaldehyde is produced from ethanol by the liver enzymes and is largely responsible for the symptoms of hangover that are experienced after a night of too much partying.

$$NAD^+ + CH_3CH_2\text{—}OH \xrightarrow[\text{dehydrogenase}]{\text{alcohol}} CH_3\text{—}\overset{\overset{\displaystyle O}{\|}}{C}\text{—}H + NADH$$

This aldehyde is useful in treating alcoholics because of the unpleasant symptoms that it causes. When the alcoholic takes it orally, in combination with alcohol, acetaldehyde quickly produces symptoms of a violent hangover with none of the *perceived* benefits of drinking alcohol.

The liver enzymes that oxidize ethanol to acetaldehyde are the same as those that oxidize methanol to formaldehyde. Physicians take advantage of this in the treatment of wood alcohol poisoning by trying to keep those enzymes busy with a reactant that produces a *less* toxic (not nontoxic) by-product. In cases of methanol poisoning, the patient receives ethanol intravenously. The intent is that ethanol will be in greater concentration than methanol and will compete successfully for the liver enzymes and be converted to acetaldehyde. This gives the body time to excrete the methanol before it is oxidized to the potentially deadly formaldehyde.

$$C_6H_5\text{—}\overset{\overset{\displaystyle O}{\|}}{C}\text{—}H + CH_3CH_2OH \underset{}{\overset{H^+}{\rightleftharpoons}} C_6H_5\text{—}\underset{\underset{\displaystyle H}{|}}{\overset{\overset{\displaystyle O\text{—}CH_2CH_3}{|}}{C}}\text{—}O\text{—}CH_2CH_3 \qquad 6.36$$

Benzaldehyde | Ethanal (excess) | Benzaldehyde diethyl ketal (common name)

Hemiacetals and hemiketals are frequently found in carbohydrates. Carbohydrates contain many hydroxyl groups and at least one carbonyl group. The carbohydrates undergo an intramolecular reaction in solution to give cyclic hemiacetals or hemiketals. This internal reaction is shown in Figure 6.4 and will be discussed in detail in Section 7.2.

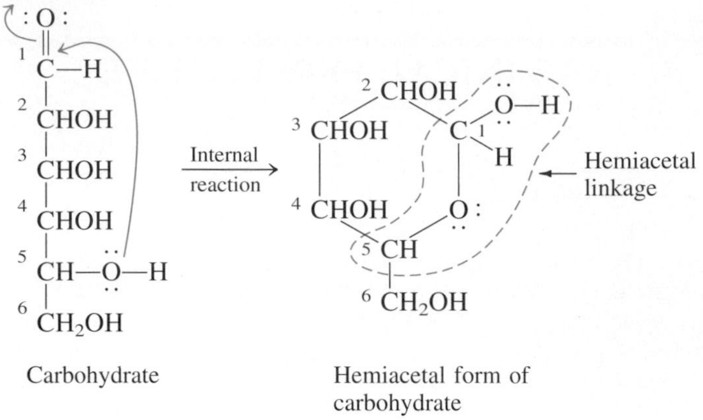

$$\underset{\substack{\\ \text{Carbohydrate}}}{\begin{array}{l} :O: \\ \| \\ {}^{1}C—H \\ | \\ {}^{2}CHOH \\ | \\ {}^{3}CHOH \\ | \\ {}^{4}CHOH \\ | \\ {}^{5}CH—\ddot{O}—H \\ | \\ {}^{6}CH_2OH \end{array}} \xrightarrow[\text{reaction}]{\text{Internal}} \text{Hemiacetal form of carbohydrate}$$

Hemiacetal linkage

FIGURE 6.4
Hemiacetal formation in sugars, shown for the intramolecular reaction of D-glucose.

Question 6.8 An unknown has been determined to be one of the following three compounds:

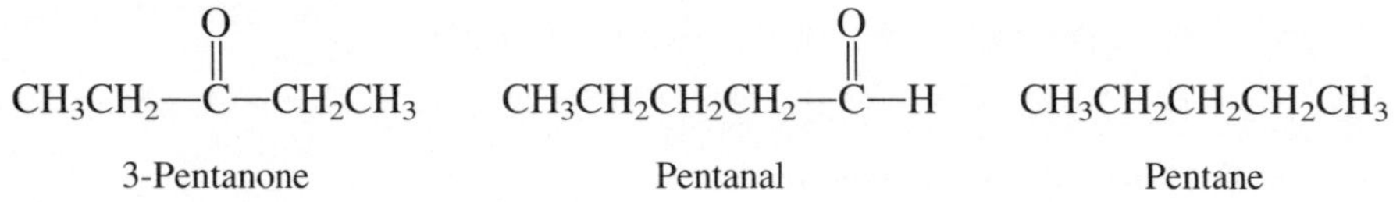

$$CH_3CH_2—\overset{\overset{\large O}{\|}}{C}—CH_2CH_3 \qquad CH_3CH_2CH_2CH_2—\overset{\overset{\large O}{\|}}{C}—H \qquad CH_3CH_2CH_2CH_2CH_3$$

3-Pentanone Pentanal Pentane

The unknown is fairly soluble in water and produces a silver mirror when treated with the silver ammonia complex. A red precipitate appears when it is treated with the Benedict's reagent. Which of the compounds above is the correct structure for the unknown? Explain your reasoning.

SUMMARY

In this chapter we were introduced to the carbonyl group, the functional group of the aldehydes and ketones. The two principal nomenclature systems (common versus I.U.P.A.C.) were applied to naming these two families. Aldehydes and ketones are moderately polar compounds. The simpler ones are soluble in water and have boiling points that are significantly higher than those of hydrocarbons or ethers of comparable molecular weight.

Both industrially and in the research laboratory, aldehydes and ketones are most often prepared via the oxidation of alcohols. This transformation may be effected by using a wide range of oxidizing agents: air (catalytically), potassium permanganate, and potassium dichromate in acid are three principal examples.

Aldehydes can be further oxidized to carboxylic acids, but ketones cannot. Aldehydes and ketones can be differentiated experimentally by using the Tollens' Test or the Benedict's Test. Most common oxidizing agents will effect this transformation. Both families can be reduced (hydrogenation) to alcohols, and both react with water to give the addition product—the hydrate. The hydrates are generally unstable and not isolable. The analogous reaction of the carbonyl compound with alcohol (in the presence of acid) gives the hemiacetal, acetal, hemiketal, or ketal, depending on the reaction conditions.

The carbonyl group is central to biochemistry. It is found in carbohydrates, proteins, nucleic acids, and lipids. It plays a central role in the chemistry of vision.

GLOSSARY OF KEY TERMS

acetal (6.4) the family of organic compounds with the general formula $RCH(OR')_2$; formed via the reaction of two molecules of alcohol with an aldehyde in the presence of an acid catalyst.

acyl group (Introduction) the functional group that contains the carbonyl group attached to one alkyl or aryl group; (Ar)R—C(=O)—; the functional group found in the derivatives of the carboxylic acids.

addition reaction (6.4) a reaction in which two molecules add together to form a new molecule; often involves the addition of one molecule to a double or triple bond in an unsaturated molecule, e.g., the addition of alcohol to an aldehyde or ketone to form a hemiacetal or hemiketal.

air oxidation (6.3) the oxidation of a molecule in the presence of air; air is the oxidizing agent.

Benedict's Test (6.4) use of a test reagent (two solutions: sodium citrate and aqueous cupric hydroxide) to test for the carbonyl functional group, particularly those found in reducing sugars.

carbonyl group (Introduction) the functional group that contains a carbon-oxygen double bond; —C=O; the functional group found in aldehydes and ketones.

carboxyl group (Introduction) the —COOH functional group; the functional group found in carboxylic acids.

cone (6.3) one of the two types of cells in the retina of the eye responsible for vision; primarily responsible for vision in bright light and for the detection of color.

Fischer-Tropsch Synthesis (6.3) an industrial process used in the preparation of various hydrocarbons and their oxygen-containing derivatives that involves the combination of carbon monoxide and hydrogen under controlled conditions of heat pressure and the use of a catalyst.

hemiacetal (6.4) the family of organic compounds with the general formula RCH(OR′)(OH); formed via the reaction of one molecule of alcohol with an aldehyde in the presence of an acid catalyst.

hemiketal (6.4) the family of organic compounds with the general formula RR′C(OR″)(OH); formed via the reaction of one molecule of alcohol with a ketone in the presence of an acid catalyst.

hydrate (6.4) any substance that contains water bound to a molecule; often used to define a product that results from the addition of water to a precursor molecule.

hydrogenation (6.4) a reaction in which hydrogen (H_2) is added to a double or a triple bond.

ketal (6.4) the family of organic compounds with the general formula RR′C(OR″)$_2$; formed via the reaction of two molecules of alcohol with a ketone in the presence of an acid catalyst.

opsin (6.3) a protein found in the eye involved in the vision process; combines with 11-*cis*-retinal to form the protein rhodopsin.

oxidation (6.3) the loss of electrons by a molecule or atom, e.g., the conversion of an alcohol to an aldehyde or ketone via the use of an oxidizing agent.

parent chain (6.2) in the I.U.P.A.C. Nomenclature System the parent chain is the longest carbon-carbon chain containing the principal functional group (e.g., carbon-oxygen double bond) in the molecule that is being named.

prosthetic group (6.3) the nonprotein portion of certain enzymes that is essential to the biological activity of the enzyme; often a complex organic compound.

reduction (6.4) a gain of electrons by a molecule or atom, e.g., the conversion of an aldehyde or ketone to an alcohol via the use of a reducing agent.

rhodopsin (6.3) a protein found in the eye involved in the vision process; formed via a combination of opsin with 11-*cis*-retinal.

rod (6.3) one of the two types of cells in the retina of the eye responsible for vision; primarily responsible for black-and-white vision and vision in dim light.

Tollens' Test (6.4) use of a test reagent (silver oxide in ammonium hydroxide) to test for the carbonyl functional group; also called the Tollens' Silver Mirror Test.

QUESTIONS AND PROBLEMS

Nomenclature and Structure

6.9 Draw each of the following, using complete structural formulas:

a. methanal
b. 7,8-dibromooctanal
c. acetone
d. *o*-bromobenzophenone
e. hydroxyethanal
f. 3-chloro-2-pentanone
g. benzaldehyde
h. triiodoacetone
i. *o*-chloropropionaldehyde
j. butyraldehyde

6.10 Name each of the following, using the I.U.P.A.C. Nomenclature System:

a. 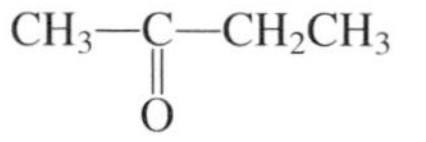 $CH_3—C(=O)—CH_2CH_3$

b. $CH_3CH_2CH(I)CH(I)CH(Br)CH(I)—C(=O)—CH_3$

c. $H—C(=O)—CH(CH_2CH_2CH_2CH_3)CH_2CH_3$

d. (cyclopentanone ring with Cl substituent) O, Cl

e. $Cl—C(Cl)(Cl)—C(=O)—CH_3$

f. $H—C(=O)—$(benzene ring with NO_2)

g. $CH_3—C(=O)—CH_2CH(CH_2CH_2CH_2CH_2CH_3)CH_2CH_2CH_2—Br$

h. (cyclopentanone ring) O, HO, OH

i. $CH_3(H)C=C(H)—C(=O)—H$

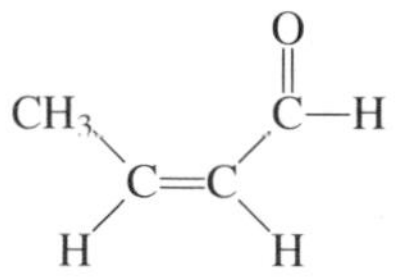

j. 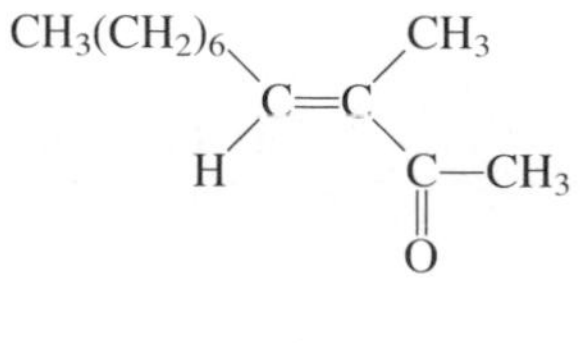 $CH_3(CH_2)_6(H)C=C(CH_3)—C(=O)—CH_3$

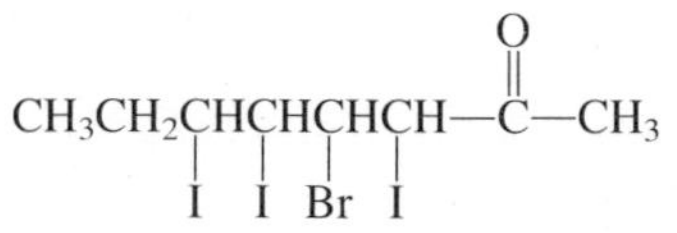

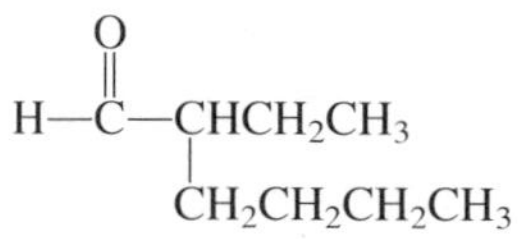

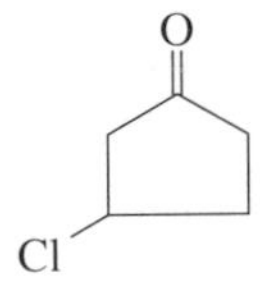

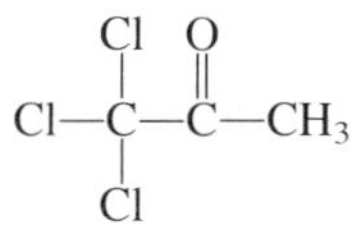

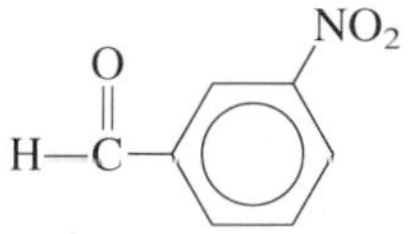

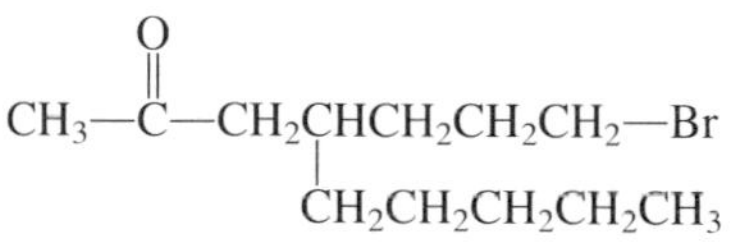

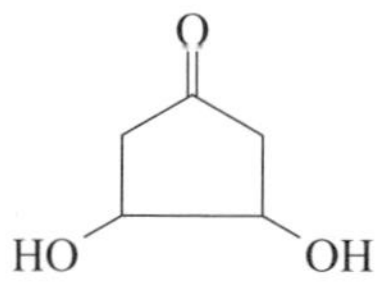

6.11 Give the common name for each of the following compounds:

a. $CH_3-\overset{\overset{O}{\|}}{C}-CH_3$

b. $CH_3CH_2-\overset{\overset{O}{\|}}{C}-CH_3$

c. $CH_3-\overset{\overset{O}{\|}}{C}-H$

d. $CH_3CH_2-\overset{\overset{O}{\|}}{C}-H$

e. $CH_3\underset{\underset{CH_3}{|}}{C}H-\overset{\overset{O}{\|}}{C}-CH_3$

f. $CH_3CH_2-\overset{\overset{O}{\|}}{C}-CH_2CH_3$

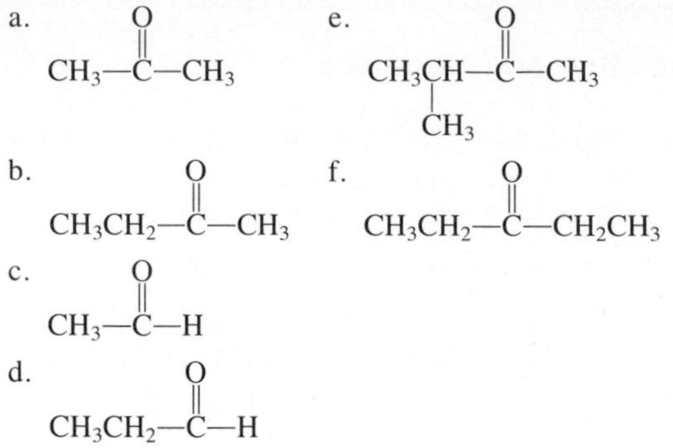

6.12 List the rules for naming ketones using the I.U.P.A.C. Nomenclature System.

6.13 List the rules for naming aldehydes using the I.U.P.A.C. Nomenclature System.

6.14 Give the I.U.P.A.C. name for each of the following compounds:

a. $CH_3\underset{\underset{Br}{|}}{C}HCH_2-\overset{\overset{O}{\|}}{C}-H$

b. $CH_3\underset{\underset{CH_3}{|}}{C}H-\overset{\overset{O}{\|}}{C}-CH_2-\overset{\overset{O}{\|}}{C}-CH_2CH_3$

c. $CH_3\overset{\overset{CH_3}{|}}{\underset{\underset{Cl}{|}}{C}}CH_2-\overset{\overset{O}{\|}}{C}-CH_2CH_2CH_3$

d. $CH_3-\overset{\overset{O}{\|}}{C}-CH_2\overset{\overset{CH_2CH_3}{|}}{\underset{\underset{CH_2CH_3}{|}}{C}}CH_2CH_3$

e. $CH_3\underset{\underset{CH_3}{|}}{C}HCH_2\underset{\underset{CH_3}{|}}{C}H-\overset{\overset{O}{\|}}{C}-CH_2CH_3$

f. (cyclopentanone ring with =O, bearing CH_3 and CH_3 on the same carbon)

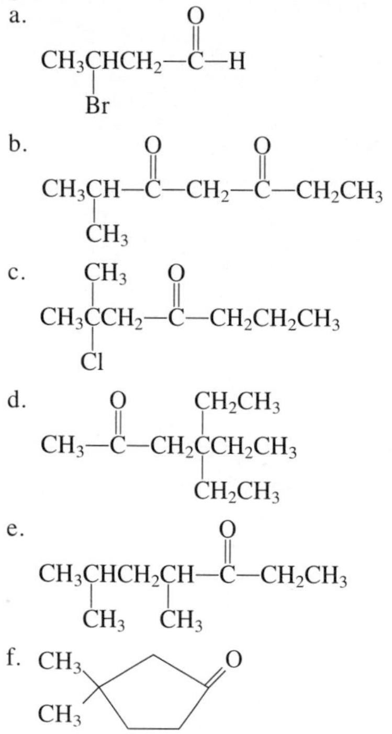

6.15 Draw all of the possible isomeric aldehydes and ketones of molecular formula $C_5H_{10}O$ and name them, using the I.U.P.A.C. Nomenclature System.

6.16 Draw, using dash-wedge drawings, the complete structure of ethanal. Provide all bond angles (approximate).

6.17 What is the bond angle, indicated by **A** in the structure below? **B**? **C**? **D**?

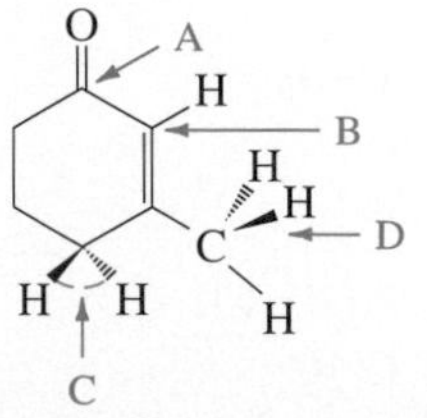

(Structure: cyclohexenone ring with O, H, CH_3 group (C with H, H, H), and H, H; angles labeled A, B, C, D)

6.18 What is the difference between a hemiacetal and a hemiketal? Explain by using an example.

6.19 Draw the hemiacetal or hemiketal that results from the reaction of each of the following aldehydes or ketones with ethanol:

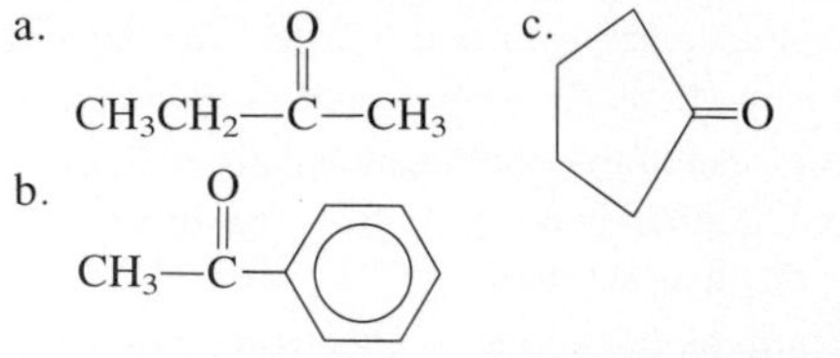

a. $CH_3CH_2-\overset{\overset{O}{\|}}{C}-CH_3$

b. $CH_3-\overset{\overset{O}{\|}}{C}-C_6H_5$ (benzene ring)

c. (cyclopentanone, ring =O)

6.20 Classify the structure of β-D-glucose (shown below) as either a hemiacetal, hemiketal, acetal, or ketal. Explain your choice.

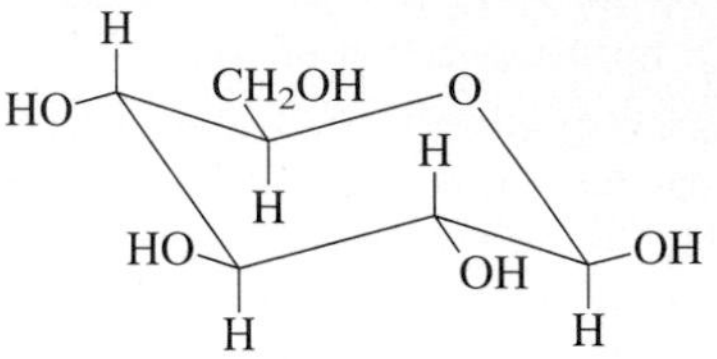

(Ring structure with substituents: H, HO, CH_2OH, O, H, H, HO, OH, OH, H, H)

β-D-Glucose

6.21 Identify each of the following compounds as a hemiacetal, hemiketal, acetal, or ketal:

a. (six-membered ring with O, bearing $-OCH_3$ and CH_3)

b. (cyclopentane bearing OH and OCH_2CH_3 on the same carbon)

c. $CH_3\overset{\overset{OH}{|}}{\underset{\underset{OCH_2CH_3}{|}}{C}}CH_3$

d. (six-membered ring with O, bearing OH and CH_3)

e. $CH_3\overset{\overset{OCH_3}{|}}{\underset{\underset{OCH_2CH_3}{|}}{C}}CH_3$

f. $CH_3CH{=}CH\overset{\overset{OCH_3}{|}}{\underset{\underset{OH}{|}}{C}}CH_3$

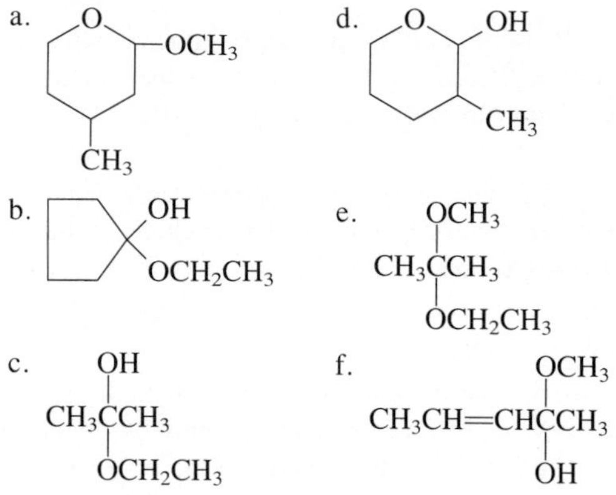

6.22 Draw all of the aldehydes of molecular formula $C_6H_{10}O$.

6.23 2-Heptanone is a common ketone that is used in the food industry as a ''blue cheese'' fragrance chemical. A similar chemical, 1-octen-3-one, is a ''mushroom'' flavoring. Draw both of these chemicals, using complete structural formulas.

Structure and Properties

6.24 Circle the higher boiling member in each of the following pairs of compounds:

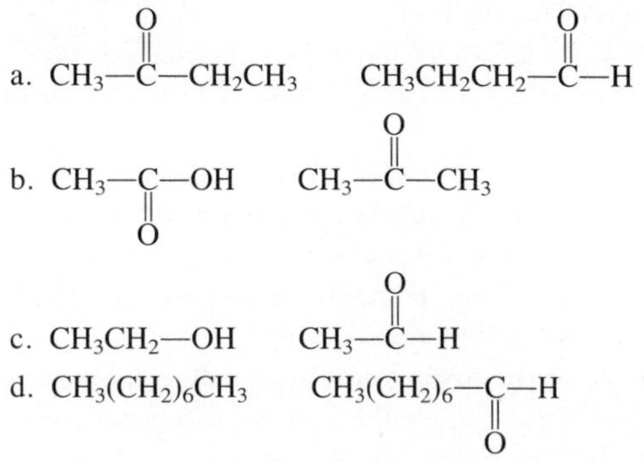

a. $CH_3-\overset{\overset{O}{\|}}{C}-CH_2CH_3$ $\quad CH_3CH_2CH_2-\overset{\overset{O}{\|}}{C}-H$

b. $CH_3-\underset{\underset{O}{\|}}{C}-OH$ $\quad CH_3-\overset{\overset{O}{\|}}{C}-CH_3$

c. CH_3CH_2-OH $\quad CH_3-\overset{\overset{O}{\|}}{C}-H$

d. $CH_3(CH_2)_6CH_3$ $\quad CH_3(CH_2)_6-\underset{\underset{O}{\|}}{C}-H$

e. 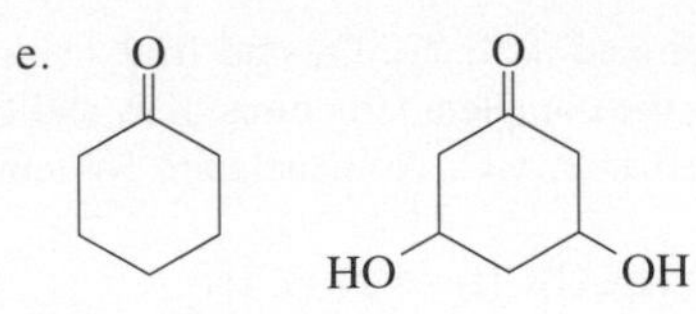

6.25 Explain briefly why simple (containing fewer than five carbon atoms) aldehydes and ketones exhibit appreciable solubility in water.

6.26 Which member in each of the following pairs will be more water soluble?

a. 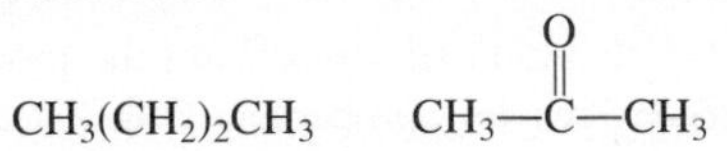 $CH_3(CH_2)_2CH_3$ $\quad$ $CH_3{-}C({=}O){-}CH_3$

b. $CH_3{-}C({=}O){-}CH_2CH_2CH_3$ $\quad$ $CH_3{-}CH(OH){-}CH_2CH_2CH_3$

c. methylcyclopentane (CH_3 on cyclopentane) $\quad$ cyclopentanecarbaldehyde ($H{-}C{=}O$ on cyclopentane)

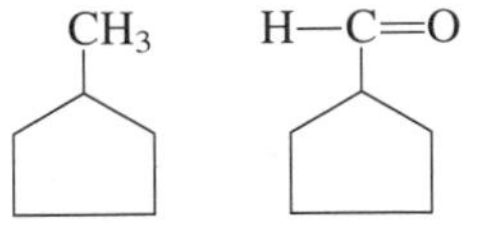

d. 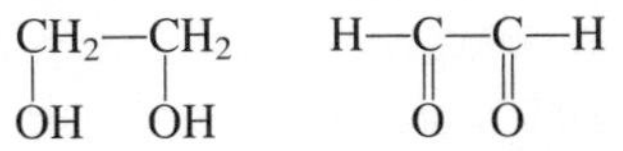 $CH_2(OH){-}CH_2(OH)$ $\quad$ $H{-}C({=}O){-}C({=}O){-}H$

e. 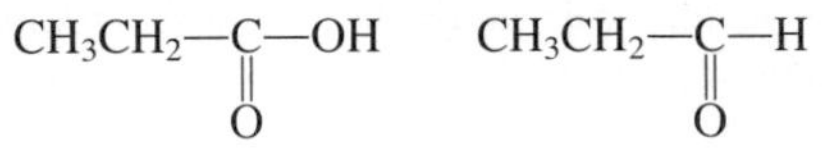 $CH_3CH_2{-}C({=}O){-}OH$ $\quad$ $CH_3CH_2{-}C({=}O){-}H$

Reactions

6.27 Complete the following synthesis by supplying the missing reactant(s), reagent(s), or product(s) indicated by the question marks:

a. 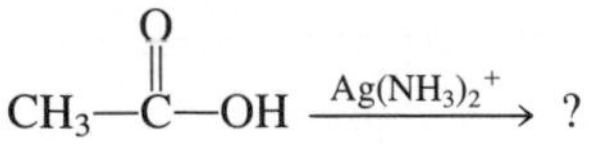 $CH_3{-}C({=}O){-}OH \xrightarrow{Ag(NH_3)_2^+} ?$

b. 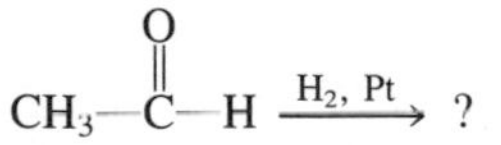 $CH_3{-}C({=}O){-}H \xrightarrow{H_2,\ Pt} ?$

6.28 Complete the following synthesis by supplying the missing reactant(s), reagent(s), or product(s) indicated by the question marks:

$CH_3{-}C({=}O){-}CH_3 \xrightarrow{?(1)} CH_3{-}C(OCH_2CH_3)_2{-}CH_3$

$CH_3{-}CH(OH){-}CH_3 \xrightarrow{?(2)} CH_3{-}C({=}O){-}CH_3$

$CH_3{-}CH(OH){-}CH_3 \xrightarrow{H_2SO_4,\ heat} ?(3)$

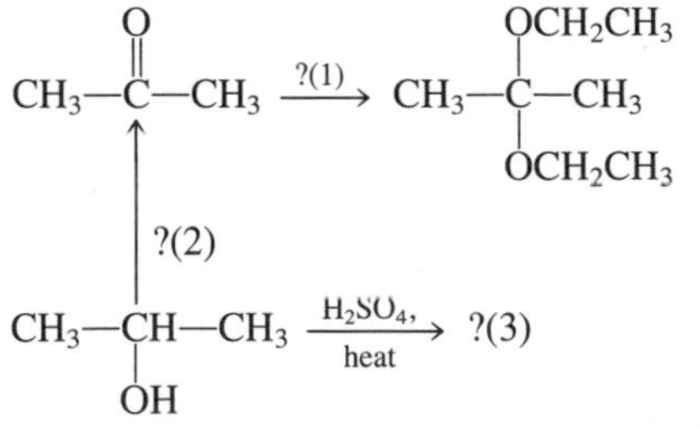

6.29 Complete the following synthesis by supplying the missing reactant(s), reagent(s), or product(s) indicated by the question marks:

$CH_3CH_2CH_2CH_2{-}OH \xrightarrow{?(1)} CH_3CH_2CH_2{-}C({=}O){-}H \underset{?(3)}{\overset{?(2)}{\rightleftarrows}} CH_3CH_2CH_2{-}CH(OH){-}OCH_3$

$CH_3CH_2CH_2{-}C({=}O){-}H \xrightarrow{?(4)} CH_3CH_2CH_2{-}C({=}O){-}OH$

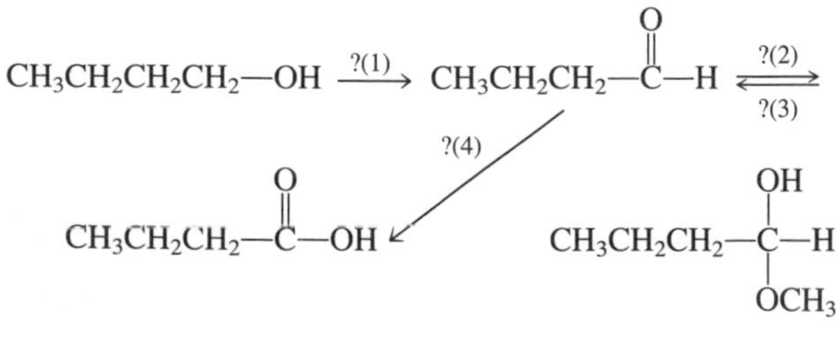

6.30 Show how you could synthesize cyclohexanone from bromocyclohexane.

6.31 Show how you could synthesize 2-pentene from 3-pentanone.

6.32 Cinnamaldehyde (C_9H_8O) is used as a flavoring agent. As the name implies, this compound has the strong odor and taste of cinnamon. Supply the missing components in this synthesis which are indicated by the question marks:

$C_6H_5{-}CH{=}CHCH_2Br \xrightarrow[Aqueous]{HO^-} ?\ (1) \xrightarrow[Acid]{K_2Cr_2O_7} ?\ (2)$

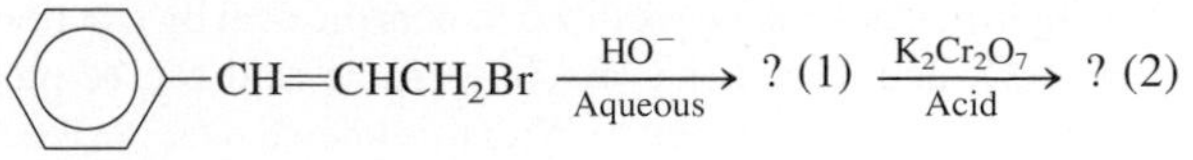

6.33 Complete each of the following reactions by supplying the missing portion indicated by the question marks:

a. 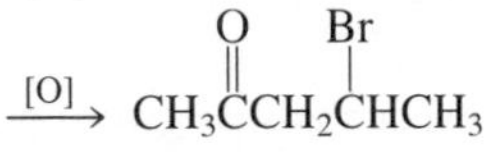 $? \xrightarrow{[O]} CH_3C({=}O)CH_2CH(Br)CH_3$

b. 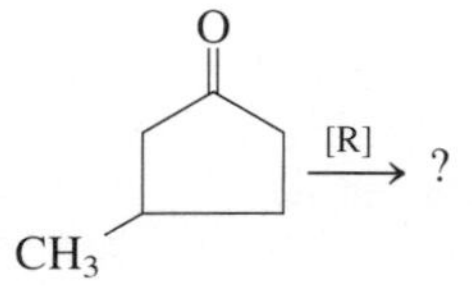 3-methylcyclopentanone $\xrightarrow{[R]} ?$

c. 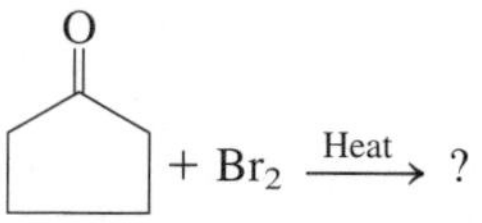 cyclopentanone $+ Br_2 \xrightarrow{Heat} ?$

d. 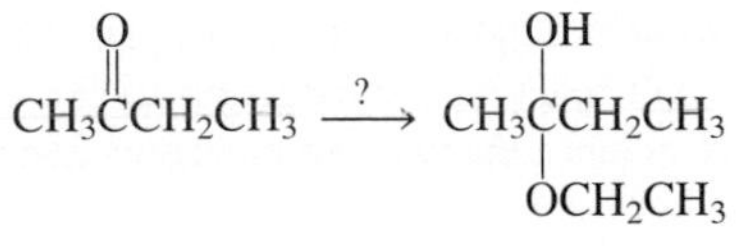 $CH_3C({=}O)CH_2CH_3 \xrightarrow{?} CH_3C(OH)(OCH_2CH_3)CH_2CH_3$

e. 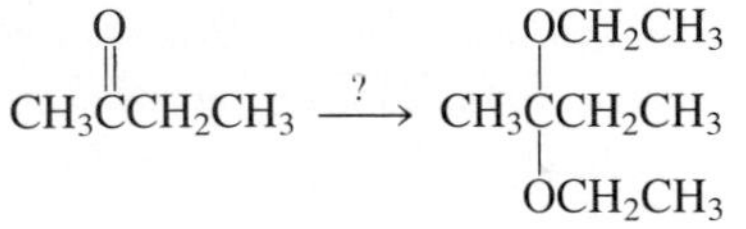 $CH_3C({=}O)CH_2CH_3 \xrightarrow{?} CH_3C(OCH_2CH_3)_2CH_2CH_3$

6.34 Describe a quick laboratory means of distinguishing between:

a. 2-hexanone and hexanal
b. 2-hexanone and 2-hexanol
c. 2-hexanone and 2-hexene
d. hexanal and 2-hexanol

6.35 Which alcohol would one oxidize to produce each of the following compounds?

a. 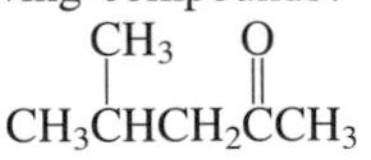 $CH_3CH(CH_3)CH_2C({=}O)CH_3$

b. 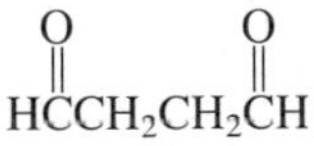 $HC({=}O)CH_2CH_2C({=}O)H$

c. 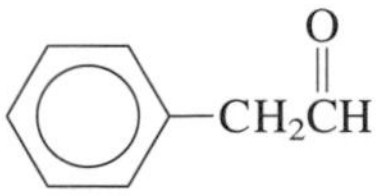 $C_6H_5{-}CH_2C({=}O)H$

d. 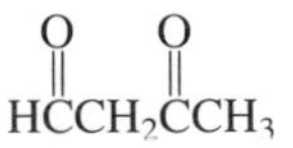 $HC({=}O)CH_2C({=}O)CH_3$

e. 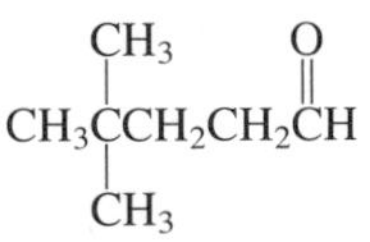 $CH_3C(CH_3)_2CH_2CH_2C({=}O)H$

f. 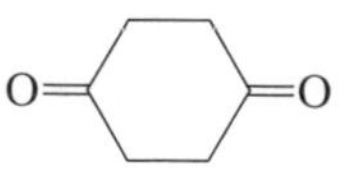 cyclohexane-1,4-dione ($O{=}C_6H_8{=}O$)

6.36 Design a synthesis for each of the following compounds, using any inorganic reagent you wish and any hydrocarbon or alkyl halide you wish:
a. octanal
b. cyclohexanone
c. 2-phenylethanoic acid
d. benzaldehyde
e. 2,2-dimethoxypropane

6.37 Alkylbenzenes can be oxidized to benzoic acid by reaction with $KMnO_4$ and heat. For example, benzoic acid can be prepared from toluene via this route. With this knowledge, design a synthesis for each of the following compounds, using any inorganic reagent you wish and any hydrocarbon you wish:
a. *o*-bromobenzoic acid
b. 2-(bromomethyl)benzoic acid

6.38 Which of the following compounds will give a positive Tollens' Test?
a. 3-pentanone
b. cyclohexanone
c. 3-methoxybutanal
d. cyclopentanol
e. 2,2-dimethyl-1-pentanol
f. acetaldehyde

Further Problems

6.39 When benzaldehyde is treated with the Tollens' reagent or the Benedict's reagent, benzaldehyde oxidizes to benzoic acid. What is reduced in each of these reactions? Explain by drawing both the oxidation half-reaction and reduction half-reaction. Do not bother trying to balance the two. Simply show reactants and products.

6.40 2,4-DNP is a reagent that reacts with carbonyl compounds to produce products that are orange in color. The reagent is therefore useful as a test reagent to determine the presence of carbonyl functional groups.

$$R_2C{=}O + H_2N{-}NH{-}C_6H_3(NO_2)_2$$

Aldehyde or ketone   2,4—Dinitrophenylhydrazine (2,4-DNP)

↓

$$R_2C{=}N{-}NH{-}C_6H_3(NO_2)_2$$

2,4-DNP *Adduct*

With this knowledge, consider the following. Compound **A** ($C_6H_{12}O$) reacts with the 2,4-DNP reagent to give an orange precipitate. Reaction of **A** with the Tollens' reagent, however, gives no reaction. Reduction of **A** with H_2, Ni gives **B** ($C_6H_{14}O$). **B** is dehydrated with H_2SO_4 and heat to give **C** (shown below). Draw the complete structures of **A** and **B** and name them, using the I.U.P.A.C. Nomenclature System.

$$CH_3CH_2CH_2CH_2{-}CH{=}CH_2$$

C

6.41 Draw the complete structural formula for the 2,4-DNP reagent. Discuss the functional groups, bonds, bond order, and bond angles that are present in the structure.

6.42 Draw out the complete structural formulas of the product that would be produced by reacting propanone with 2,4-DNP.

6.43 Repeat Problem 6.42 for each of the following ketones or aldehydes:
a. benzaldehyde
b. 2-pentanone
c. cyclopentanone
d. 2,5-pentanedione

6.44 What is the function of the copper citrate in the Benedict's reagent?

6.45 When alkenes react with ozone, O_3, the double bond is cleaved, and an aldehyde and/or a ketone is produced. The reaction, called *ozonolysis,* is shown in a general sense below:

$$>C{=}C< + O_3 \longrightarrow >C{=}O + O{=}C<$$

Predict the ozonolysis products that are formed when each of the following alkenes is reacted with ozone:
a. 1-butene b. 2-hexene
c. *cis*-3,6-dimethyl-3-heptene

6.46 Repeat Problem 6.45 for each of the following double bond–containing compounds:
a. 2-methyl-2-butene
b. 2-methyl-1-butene
c. cyclohexene
d. 1-methylcyclohexene
e. 1,3-butadiene
f. 1,3-pentadiene

6.47 Look up the mechanism for the ozonolysis reaction and write it out completely for the reaction of ozone with ethene.

6.48 Review the material on the chemistry of vision and, with respect to the isomers of retinal, discuss the changes in structure that occur as the nerve impulses (that result in vision) are produced. Provide complete structural formulas of the retinal isomers that you discuss.

6.49 Compound **A** ($C_5H_{10}O$) contains a C=O but gives a negative Tollens' test. Compound **A** cannot be oxidized with $KMnO_4$ or $K_2Cr_2O_7$. Reduction of **A** gives **B** ($C_5H_{11}OH$). Compound **B** is a secondary alcohol. Dehydration of **B** gives two alkenes, **C** and **D** (both C_5H_{10}). Compound **C** is a terminal alkene, but **D** contains an internal double bond. **C** does not exist as a *cis-/trans-* isomer but **D** does. Give structures and names for **A–D**.

CHAPTER 7
Carbohydrates

OBJECTIVES

- Draw and name the common simple carbohydrates, using structural formulas and Fischer Projection formulas.
- Be familiar with the ways in which carbohydrates are classified.
- Understand the concepts of optical activity, chirality, enantiomers, anomers, stereoisomers, D- and L-configuration and the (+) and (−) rotation of plane polarized light by optically active molecules.
- Using Fischer Projection formulas, be able to differentiate between D- and L-sugars.
- Given the Fischer Projection of a monosaccharide, be able to draw the Haworth Projection of its α- and β-cyclic forms and vice versa.
- Recognize whether a sugar is a reducing or nonreducing sugar.
- Know the general structural features of oligosaccharides and polysaccharides.
- Discuss the structural, chemical, and biochemical properties of the monosaccharides, oligosaccharides, and polysaccharides.
- Understand the difference between galactosemia and lactose intolerance.

OUTLINE

Carbohydrates are a class of organic compounds composed of carbon, hydrogen, and oxygen with the general formula $C_x(H_2O)_y$ (where x and y can be any whole number greater than or equal to 3). This formula is useful for the simple carbohydrates but is often an oversimplification. The world of carbohydrates includes many that have amino groups, such as the carbohydrates of the bacterial cell wall. Others have phosphate groups, especially the intermediates in carbohydrate metabolism. Some carry sulfate groups, particularly the carbohydrates that make up the extracellular matrix around the cells in cartilage.

Chapters 13 and 14

Carbohydrates are the main source of energy for both plants and animals (Figure 7.1). They are found in many natural sources such as grains and cereals, breads, sugar cane and sugar beets, fruits, milk, and honey. The variety of substances that are composed of carbohydrates ranges from the pages of this book (cellulose) to the nucleic acids (DNA and RNA).

Chapter 16

FIGURE 7.1
Carbohydrates are produced by the process of photosynthesis, which uses the energy of sunlight to produce hexoses from CO_2 and H_2O. The plants use these hexoses to generate energy for cellular function and to produce macromolecules, including starch, cellulose, fats, nucleic acids, and proteins. Animals depend on plants as a source of organic carbon. The hexoses are metabolized to generate energy and are used as precursors for the biosynthesis of glycogen, fats, proteins, and nucleic acids.

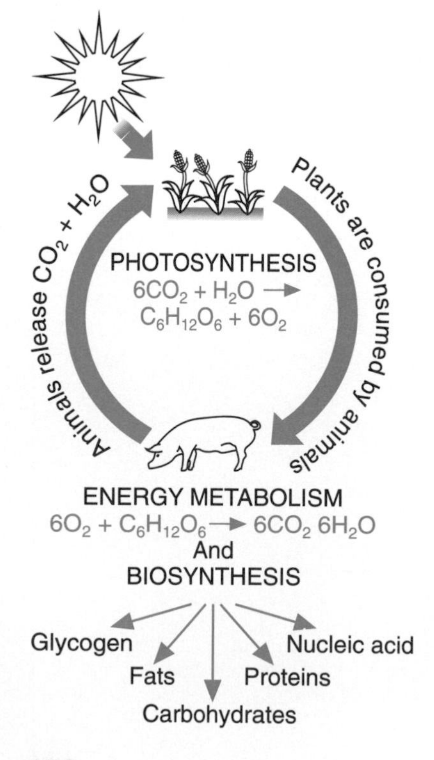

7.1 TYPES OF CARBOHYDRATES

One way of classifying carbohydrates is by size. Carbohydrates are categorized as monosaccharides, oligosaccharides, and polysaccharides. **Monosaccharides** are the simplest carbohydrates; they contain a single (mono-) sugar (**saccharide**) unit. As we will see in Section 7.2, the nature of this unit determines the identity of a particular monosaccharide. The largest and most complex carbohydrates are the **polysaccharides;** they are chains of monosaccharides held together by bonds through ''bridging'' oxygen atoms. Such a bond is termed a **glycosidic bond.** Intermediate in size are the **oligosaccharides,** consisting of two to ten monosaccharide units in the ''chain,'' also held together by bridging oxygen atoms. Among the most important oligosaccharides are disaccharides, consisting of two monosaccharides joined through an oxygen atom bridge.

7.2 MONOSACCHARIDES

Nomenclature

One way of grouping monosaccharides is based on the functional groups that each contains. If the carbohydrate contains a ketone (carbonyl) group, it is called a **ketose.** If an aldehyde (carbonyl) group is present, it is called an **aldose.** Monosaccharides, and carbohydrates in general, are actually *polyhydroxyaldehydes* or *polyhydroxyketones*. That is, they are aldehydes or ketones that also contain a large number of hydroxyl groups.

```
                    H                 CH2OH
                    |                 |
                    C=O               C=O
Aldehyde            |                 |              Ketone
functional      H—C—OH            H—C—OH         functional
group               |                 |              group
                H—C—OH           HO—C—H
                    |                 |
                    CH2OH             CH2OH

              An aldose          A ketose
```

A more detailed system of nomenclature, which tells us the number of carbon atoms in the main carbohydrate skeleton, can also be used. In this system a three-carbon monosaccharide is a ***triose,*** a four-carbon sugar is a ***tetrose,*** a five-carbon sugar is a ***pentose,*** a six-carbon sugar is a ***hexose,*** and so on. A contraction of the two naming systems provides even more information about the structure and composition of a sugar. For example:

H	H	CH_2OH
C=O	C=O	C=O
H—C—OH	H—C—OH	HO—C—H
CH_2OH	H—C—OH	H—C—OH
	HO—C—H	H—C—OH
	HO—C—H	CH_2OH
	CH_2OH	
Aldose	Aldose	Ketose
Triose	Hexose	Hexose
Aldotriose	Aldohexose	Ketohexose
D-Glyceraldehyde	L-Mannose	D-Fructose

These names are used to classify the monosaccharides into broad categories. In addition, each carbohydrate has a specific unique name. These names are shown in color in the preceding examples. Because the carbohydrates often exist in several different isomeric forms, it is important to provide the complete name when discussing a carbohydrate, to avoid ambiguity and misunderstanding. In the structure provided above, the complete name is D-fructose, denoting a specific isomer of fructose, not simply fructose. The various isomeric forms of the carbohydrates, as well as the nomenclature system for this family, are discussed in detail in the next section.

Stereoisomers

Structure

The prefixes D- and L- are used to distinguish specific pairs of isomers from one another. These types of isomers are called **stereoisomers.** By definition, each member of a pair of stereoisomers must have the same molecular formula—all isomers do. How then do D-isomers differ from L-isomers? D- and L-isomers differ in the spatial arrangement of atoms in the molecule.

Stereochemistry is the study of the different spatial arrangements of atoms. In each member of a pair of stereoisomers, all of the atoms are bonded together using exactly the same bonding pattern; they differ only in the arrangements of their atoms in space. An example of a pair of stereoisomers is shown in Figure 7.2. In this example the general molecule Cabcd is formed from the bonding of a central carbon to four different groups—a, b, c, and d. This results in two molecules rather than one. These two isomers are bonded together through the exact *same* bonding pattern, yet they are *not* identical. If they were identical, they would be superimposable one upon the other; *they are not*. They are therefore stereoisomers. These two stereoisomers have a mirror-image relationship that is analogous to the mirror-image relationship of the left and right hands (see Figure 7.2b). Two stereoisomers that are mirror images of one another are called a pair of **enantiomers.** One member of this pair is called the D-enantiomer (or D-isomer), and the other member is called the L-enantiomer (or L-isomer).

Build models of these compounds, using a molecular model set, to prove this to yourself.

Molecules that are capable of existing in two mirror-image, enantiomeric forms are called **chiral.** The term simply means that as a result of certain topological or structural features, the molecule can exist in more than one mirror-image form. For any pair of mirror-image forms (enantiomers), one is always designated the D-isomer and one the L-isomer.

When a molecule contains a carbon atom that has four different groups bonded to it (called an **asymmetric carbon** or *chiral carbon* atom), that molecule can exist as a pair of enantiomers; one D- and one L-. Consider the simplest carbohydrate, glyceraldehyde, which is shown in Figure 7.3. Notice that the second carbon is bonded to four different

There are other molecular arrangements of atoms that result in the formation of enantiomers, but they will not be discussed in this text.

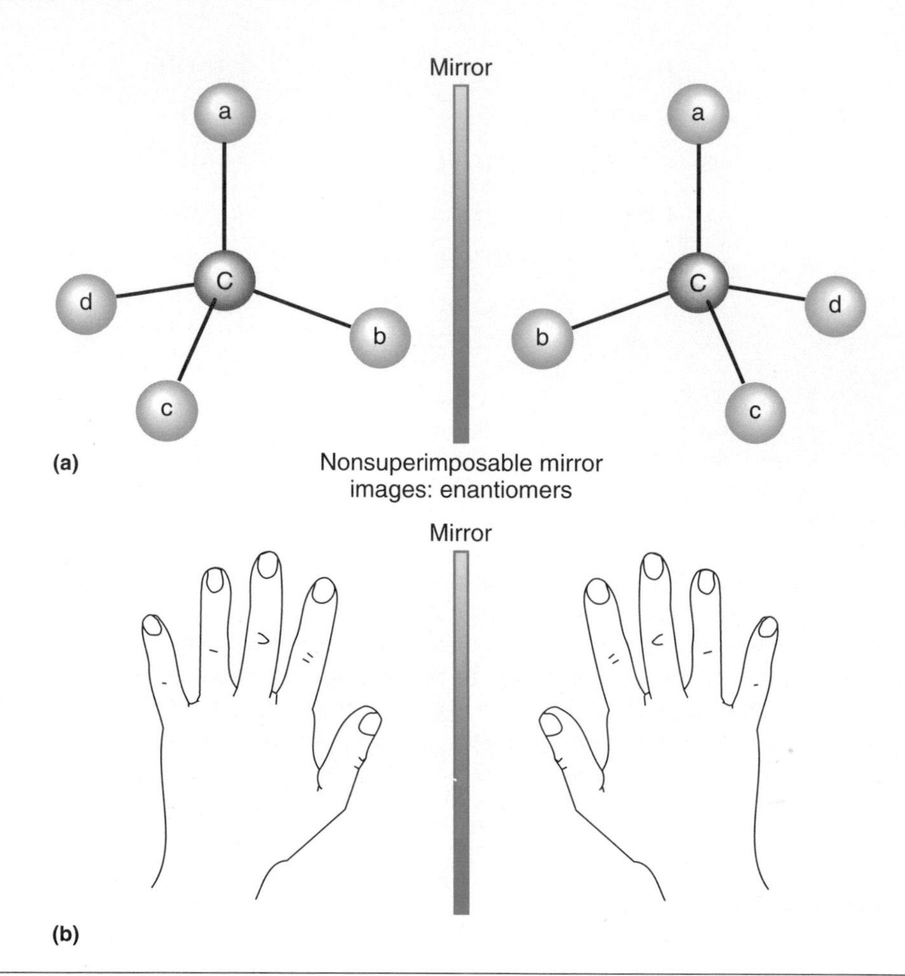

FIGURE 7.2
(a) A pair of stereoisomers for the general molecule Cabcd; (b) analogous mirror-image right and left hands.

groups. It is therefore the chiral carbon. As a result, we can draw two enantiomers of glyceraldehyde, D- and L-, that are mirror images of one another.

In Figure 7.3a the stereoisomers of glyceraldehyde are presented in the two dimensional Fischer Projection. A **Fischer Projection** formula is used to designate a specific three-dimensional arrangement of atoms in space using a two-dimensional drawing (compare Figures 7.3a and 7.3b). The Fischer Projection is drawn as a cross with the asymmetric carbon at the center of the cross, the intersection of the horizontal and vertical lines. The vertical line of the drawing represents two bonds to the asymmetric carbon going away from us, behind the page of the text. The horizontal lines represent two bonds coming toward us, out of the page. In the drawing below, the Fischer Projection is compared with a three-dimensional and a ball and stick drawing for 1-bromo-1-chloroethane.

CH_3 / Br—+—Cl / H = CH_3 / Br—C—Cl / H = (CH_3) / (Br)—(C)—(Cl) / (H)

Fischer projection — 3-D drawing — Ball and stick drawing

When we examine the Fischer Projection for glyceraldehyde, we see that the hydroxyl group of the D-enantiomer is drawn to the right of the chiral carbon, whereas the hydroxyl group is drawn to the left of the chiral carbon in the L-enantiomer. Notice that

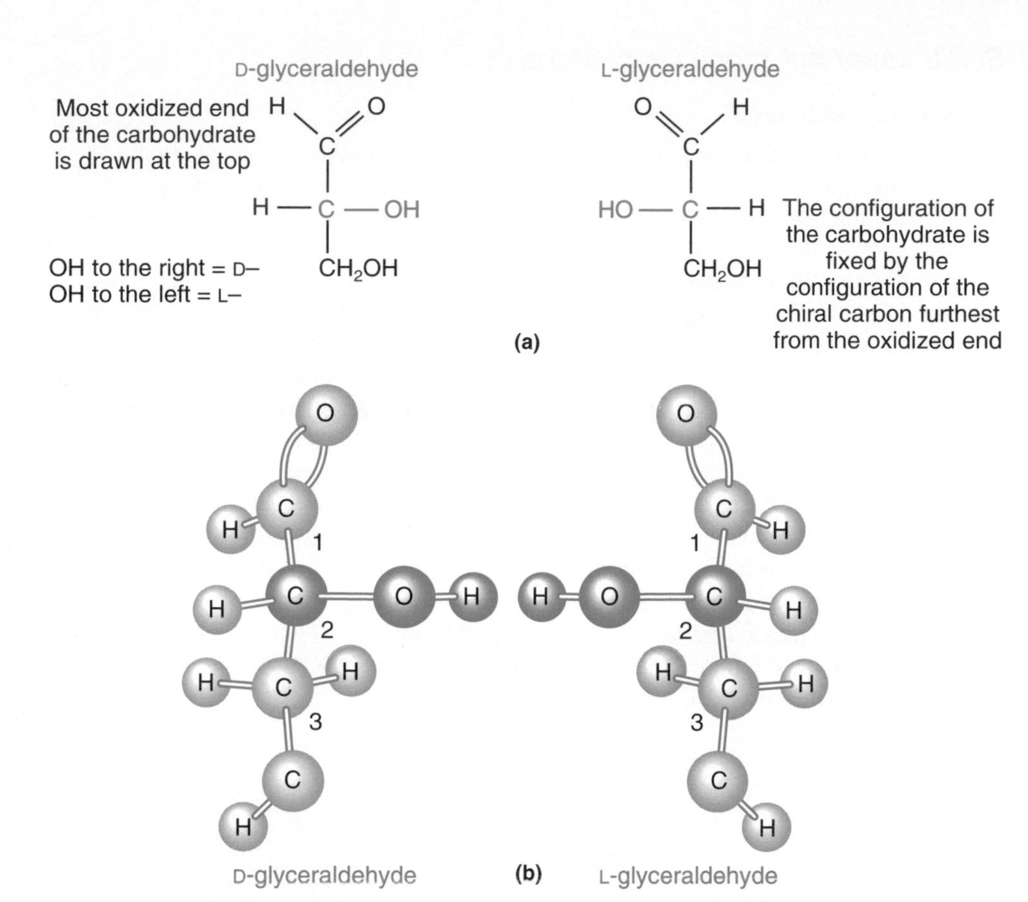

D-glyceraldehyde
L-glyceraldehyde
Most oxidized end of the carbohydrate is drawn at the top
H—C—OH
CH2OH
HO—C—H
The configuration of the carbohydrate is fixed by the configuration of the chiral carbon furthest from the oxidized end
OH to the right = D–
OH to the left = L–
(a)
(b)
D-glyceraldehyde
L-glyceraldehyde

FIGURE 7.3
(a) Fischer Projection of D- and L-glyceraldehyde. (b) A three-dimensional representation of D- and L-glyceraldehyde.

the positions of the groups bonded to carbons 1 and 3 do not matter because they are not chiral carbon atoms. In other words, they do not have four different groups attached to them.

Physical properties

The physical properties of enantiomers are generally identical with one exception. When **plane polarized light** (light that has been filtered through a polaroid lens such as those found in Polaroid sunglasses) is passed through a solution of one enantiomer, the plane of light is rotated in either a clockwise or a counterclockwise direction. If the enantiomer rotates the light in a clockwise direction, it is called the **dextrorotatory** (L: *dexter,* to the right) or (+) enantiomer. The mirror-image enantiomer will then rotate the light in a counterclockwise direction and is called the **levorotatory** (L: *levo,* to the left) or (−) enantiomer. With glyceraldehyde the L-enantiomer rotates light to the left (− isomer) and the D-enantiomer to the right (+ isomer). *This is not necessarily true for other pairs of enantiomers!*

When we begin our study of optical activity, the concepts of D- and L-configuration and (+) and (−) rotation often are confusing. It is important to realize that the two concepts and designations are based on completely different information. The labels (+) and (−) refer *only* to the direction in which plane polarized light is rotated when it interacts with a given chiral molecule. D- and L-, on the other hand, refer *only* to the specific arrangement of the atoms of a chiral molecule in space. Some molecules rotate light to the left (−) and are of the D-configuration, and others rotate light to the right (+) and are of the D-configuration. The same is true of molecules of the L-configuration; some of them rotate light to the left (−), and others rotate light to the right (+). The ability to rotate plane polarized light is called **optical activity.** Molecules that rotate plane polarized light are said to be optically active. Measurement of optical activity is an important analytical technique for differentiating between optical isomers.

The original designations of D- and L- were based on a comparison of the chiral molecule with D-glyceraldehyde. If the principal groups attached to the asymmetric carbon in the two molecules were oriented similarly, the molecule was labeled D-. In contrast, if the groups on the molecule being studied had a mirror-image relationship to those on glyceraldehyde, the molecule was labeled L-. The designations today are related to specific rules relative to the three-dimensional arrangement of molecules in space.

Some important monosaccharides

D-Glyceraldehyde

The simplest carbohydrate is the three-carbon sugar D-glyceraldehyde (Figure 7.3). D-Glyceraldehyde is a triose (3-carbon) and an aldose (aldehyde group at the top); therefore it is an aldotriose. First, note (Figure 7.3) that this structure is drawn with the most oxidized carbon, the aldehyde, at the "top." The most oxidized carbon is always placed at the top when the structures of open-chain carbohydrates are written; this carbon is numbered C-1. C-2 of D-glyceraldehyde is covalently bonded to four different groups. As we have just discussed, such a carbon is chiral. Reflection of D-glyceraldehyde in a mirror gives L-glyceraldehyde. The fact that D-glyceraldehyde and L-glyceraldehyde are nonsuperimposable molecules is very important biologically because the enzymes that act upon D-glyceraldehyde do not interact at all with L-glyceraldehyde. Specificity of this type is often the rule in biochemical systems.

By convention it is the position of the hydroxyl group on the penultimate (next to the last) carbon that determines whether a compound is a D- or L- enantiomer. In other words, the asymmetric carbon farthest from the oxidized end of the chain. If the —OH group is on the right (reader's right), the molecule is in its D-configuration. If the —OH group is on the left, the molecule is in its L-configuration. Though most carbohydrates have more than one asymmetric or chiral carbon, the D- or L- designation refers *only* to the configuration at the penultimate chiral carbon.

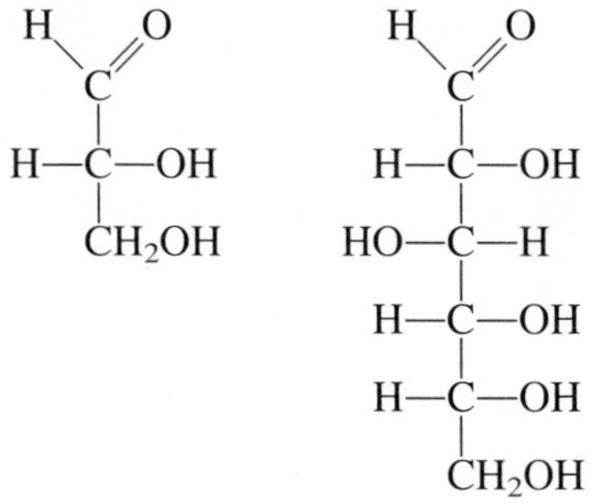

```
  H   O          H   O
   \ //           \ //
    C              C
    |              |
 H—C—OH         H—C—OH
    |              |
    CH2OH      HO—C—H
                   |
                H—C—OH
                   |
                H—C—OH
                   |
                   CH2OH
```

D-Glyceraldehyde D-Glucose

Question 7.1 What is the structural difference between an aldose and a ketose?

Question 7.2 What is the difference between:

a. a ketohexose and an aldohexose?

b. a triose and a pentose?

Question 7.3 Indicate whether each of the following molecules is an aldose or a ketose:

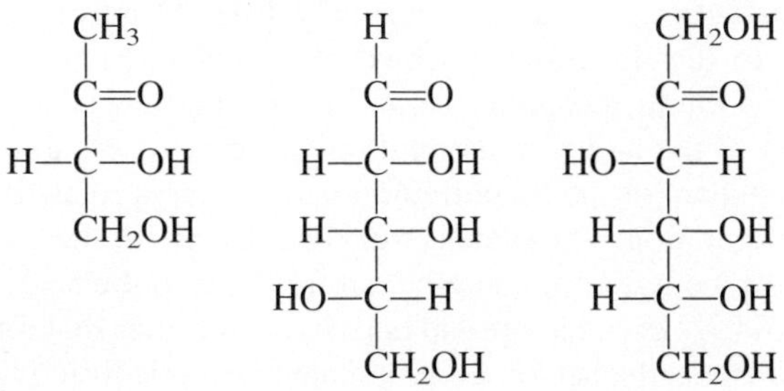

```
    CH3              H                 CH2OH
    |                |                 |
    C=O              C=O               C=O
    |                |                 |
 H—C—OH           H—C—OH          HO—C—H
    |                |                 |
    CH2OH         H—C—OH           H—C—OH
                     |                 |
                 HO—C—H            H—C—OH
                     |                 |
                     CH2OH             CH2OH
```

$$\begin{array}{c} H \\ | \\ C{=}O \\ | \\ H{-}C{-}OH \\ | \\ CH_2OH \end{array} \qquad \begin{array}{c} CH_3 \\ | \\ C{=}O \\ | \\ H{-}C{-}OH \\ | \\ H{-}C{-}OH \\ | \\ H{-}C{-}OH \\ | \\ CH_2OH \end{array} \qquad \begin{array}{c} H \\ | \\ C{=}O \\ | \\ HO{-}C{-}H \\ | \\ H{-}C{-}OH \\ | \\ HO{-}C{-}H \\ | \\ CH_2OH \end{array}$$

Question 7.4 Determine the configuration (D- or L-) for each of the molecules in Question 7.3.

Glucose and fructose, six-carbon sugars

Glucose is the most important sugar in the human diet. It is also the most abundant organic compound found in nature. Glucose has several common names, including dextrose, grape sugar, and blood sugar, and is found in many sweet fruits, in honey, and in nectar. It is also found in blood at concentrations as high as 0.1%.

The following example uses glucose to demonstrate how to draw the linear structure of a monosaccharide.

Example 7.1

Draw the Fischer Projection structure of D-glucose.

Glucose is an aldohexose.

Step 1. Draw six carbons in a straight vertical line; each carbon is separated from the one above and below it by a bond:

$$\begin{array}{rc} 1 & C \\ & | \\ 2 & C \\ & | \\ 3 & C \\ & | \\ 4 & C \\ & | \\ 5 & C \\ & | \\ 6 & C \end{array}$$

Step 2. The most highly oxidized carbon is, by definition, drawn as the uppe carbon (carbon 1). In this case carbon 1 is an aldehyde carbon:

$$\begin{array}{rc} & H \\ & | \\ 1 & C{=}O \\ & | \\ 2 & {-}C{-} \\ & | \\ 3 & {-}C{-} \\ & | \\ 4 & {-}C{-} \\ & | \\ 5 & {-}C{-} \\ & | \\ 6 & {-}C{-} \\ & | \end{array}$$

Most oxidized end of carbon chain; aldehyde

Step 3. The atoms are added to the penultimate carbon atom at the bottom of the chain to give either the D- or L- enantiomer as desired. As was mentioned earlier, this is accomplished by comparing the arrangement of atoms on the penultimate carbon to the asymmetric carbon of D-glyceraldehyde, the reference compound. Effectively, if the —OH group is to the right, you have D-glucose; if it is to the left, the compound is L-glucose.

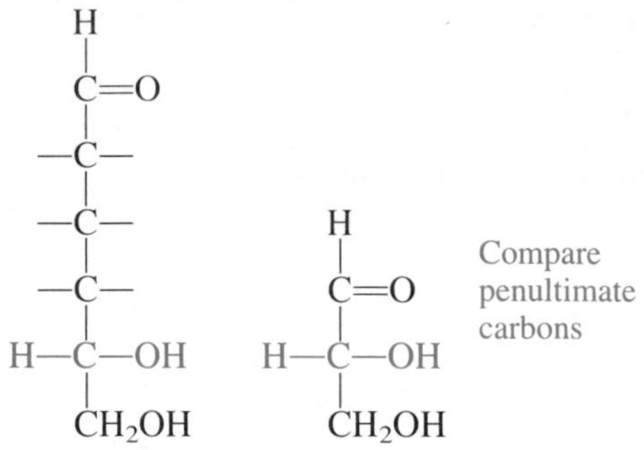

```
      H                    
      |                    
      C=O                  
      |                    
    —C—                    
      |                    
    —C—               H          Compare
      |               |          penultimate
    —C—               C=O        carbons
      |               |
  H—C—OH          H—C—OH
      |               |
      CH2OH           CH2OH

   D-Isomer       D-Glyceraldehyde
```

Step 4. All the remaining atoms are then added to give the desired carbohydrate. For example, for D-glucose, one would draw the following:

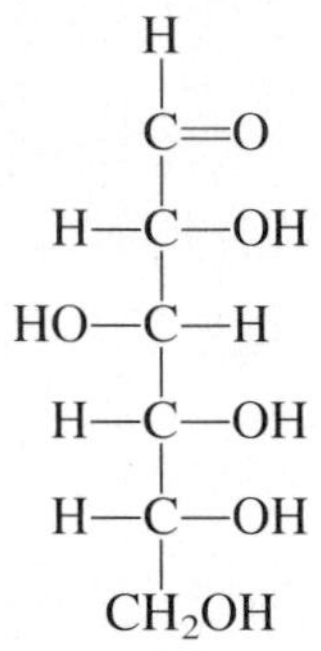

```
        H
        |
        C=O
        |
    H—C—OH
        |
   HO—C—H
        |
    H—C—OH
        |
    H—C—OH
        |
        CH2OH

     D-Glucose
```

The relative positions for the hydrogen atoms and the hydroxyl groups on the remaining carbons would have to be determined from the literature. For instance, the complete structures of D-fructose and D-galactose are seen in Figures 7.6 and 7.7.

Question 7.5 Using complete structural formulas, draw one example of each of the following classes of carbohydrates:

a. Aldose
b. Ketose
c. Pentose
d. Tetrose
e. Ketotetrose
f. Aldopentose
g. Ketohexose

Question 7.6 Provide answers to each of the following:

a. What is the difference between D-glyceraldehyde and L-glyceraldehyde?
b. What is the difference between D-glucose and L-glucose?

If the aldehyde group of glucose at C-1 is reduced to an alcohol, and the hydroxyl group at C-2 is oxidized to a keto group, the product is the ketose called D-fructose.

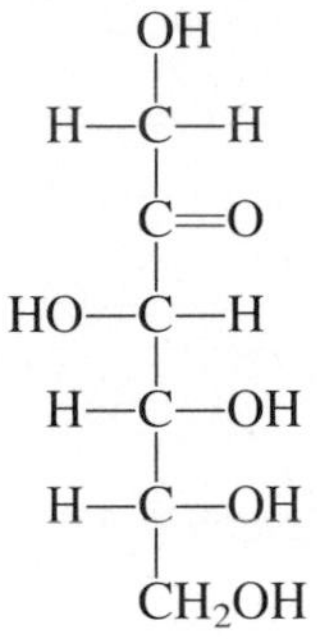

D-Fructose

Fructose is the sweetest of all sugars. It is found in large amounts in honey and in sweet fruits.

In actuality the open chain forms of glucose and fructose are present in very small concentrations in cells. For the most part these sugars exist in cyclic form under physiological conditions. For example, the carbonyl group at C-1 of glucose can react with the hydroxyl group at C-5 to give a six-membered ring. In the discussion of aldehydes we noted that the reaction between an aldehyde and an alcohol yields a **hemiacetal.** When the aldehyde portion of the glucose molecule reacts with the C-5 hydroxyl group, the product is a cyclic *intramolecular hemiacetal*. For D-glucose, two isomers can be formed in this reaction (Figure 7.4). These isomers are called α- and β-D-glucose. Two isomers are formed because the cyclization reaction creates a new asymmetric carbon, in this case C-1. Such isomers, differing in the arrangement of bonds around C-1, are called **anomers.** Like the stereoisomers discussed previously, the α and β forms can be distinguished from one another because they rotate plane polarized light differently. It is of interest to note, however, that in solution the cyclization reaction is readily reversible. Thus the α and β anomers undergo interconversion. For instance, if the pure α-D-glucose anomer is dissolved in water, some of the molecules will linearize and recyclize to yield β-D-glucose. This interconversion is called **mutarotation.**

Section 6.4

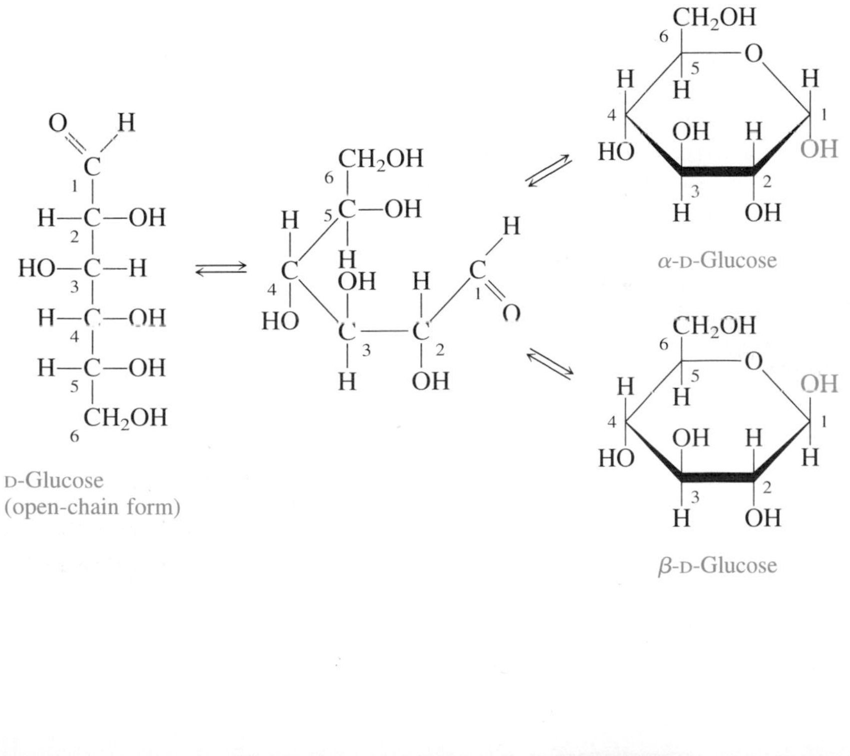

D-Glucose (open-chain form)

α-D-Glucose

β-D-Glucose

FIGURE 7.4
Cyclization of glucose to give α- and β-D-glucose. Note that the carbonyl carbon (C-1) becomes chiral in this process, yielding the two anomeric forms of glucose named above.

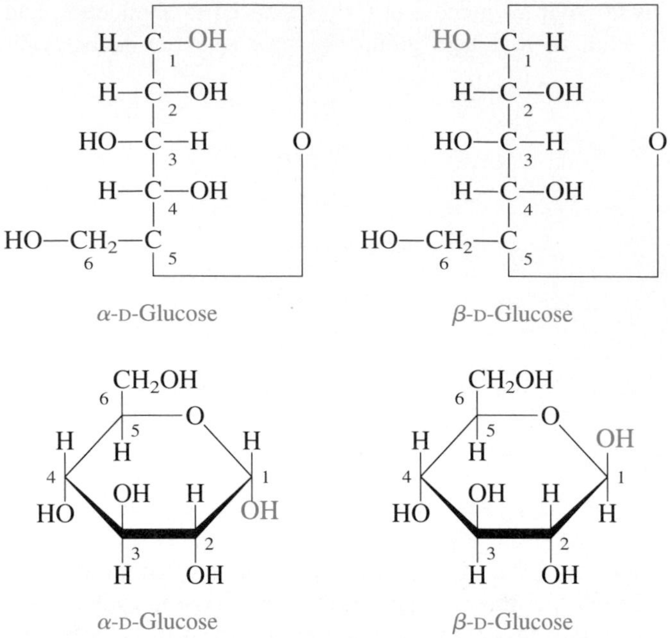

α-D-Glucose β-D-Glucose

α-D-Glucose β-D-Glucose

FIGURE 7.5
The method of converting the Fischer Projection of α- and β-D-glucose into the correct Haworth structures. All groups to the right of the carbon chain in the Fischer Projection are placed below the ring. All functional groups to the left of the carbon backbone are located above the ring structure.

In Figure 7.4 a new type of structural formula is presented. These **Haworth Projections** are useful in interpreting the three-dimensional configuration of molecules. Although on first inspection they appear complicated, it is quite simple to derive a Haworth Projection from a Fischer Projection (Figure 7.5). First, try to imagine that you are seeing the ring in three dimensions. Some of the substituent groups on the molecule are above the ring, and some are beneath it. But how do we determine which groups to place above the ring and which to place beneath the ring? Inspect the two-dimensional Fischer structure. Note the chemical groups to the left of the carbon chain. These are placed above the ring in the Haworth Projection. Now note the groups to the right of the carbon chain. These will be located beneath the carbon ring in the Haworth Projection. Thus in the Haworth Projection of the cyclic form of D-glucose, such as those drawn in Figures 7.4 and 7.5, the CH_2OH group is always "up." When the —OH group at C-1 is also "up," *cis* to the —CH_2OH group, the sugar is β-D-glucose. When the —OH group at C-1 is "down," *trans* to the —CH_2OH group, the sugar is α-D-glucose.

In a similar way, cyclization of fructose produces the anomers α- and β-D-fructose (Figure 7.6). Fructose is a ketose, or ketone sugar. Recall that the reaction between an alcohol and a ketone yields a **hemiketal.** Thus the reaction between the C-2 keto group and the C-5 hydroxyl group in the fructose molecule produces an *intramolecular hemiketal*. Note that although fructose is able to form a six-atom ring structure, the most common ring structure formed by fructose consists of a five-atom skeleton. Thus glucose is represented as a six-membered ring when it cyclizes, and fructose is represented as a five-membered ring.

Section 6.4

Galactose

Another important hexose is **galactose.** The Fisher structure of D-galactose and the Haworth projections of α-D-galactose and β-D-galactose are shown in Figure 7.7. The monosaccharide galactose is not generally found in biological systems, rather it is found as a component of the disaccharide lactose, or milk sugar. This is the principal sugar found in the milk of all mammals. β-D-Galactose and a modified form, N-acetyl-β-D-galactosamine, are also components of the blood group antigens.

Section 7.3

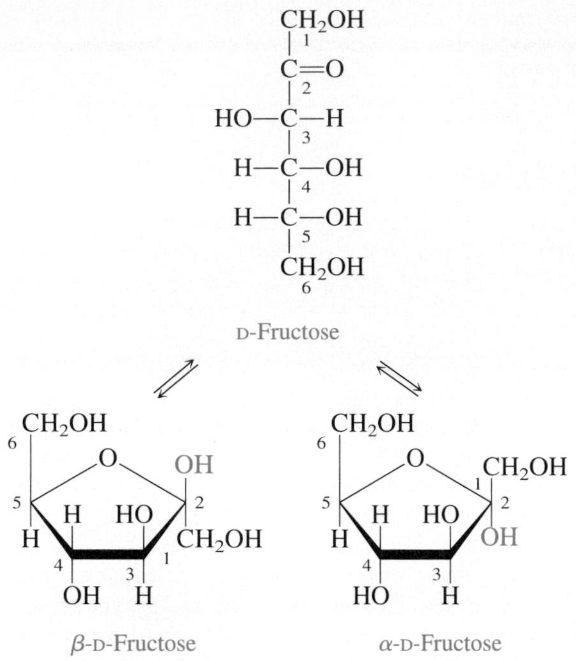

D-Fructose

β-D-Fructose

α-D-Fructose

FIGURE 7.6
Cyclization of D-fructose to give α- and β-D-fructose.

Question 7.7 The structure of the sugar D-galactose differs slightly from that of D-glucose. The —OH group at C-4 is on the right in D-glucose and on the left in D-galactose. Write the structures of the chain form and the cyclic α- and β- forms of D-galactose.

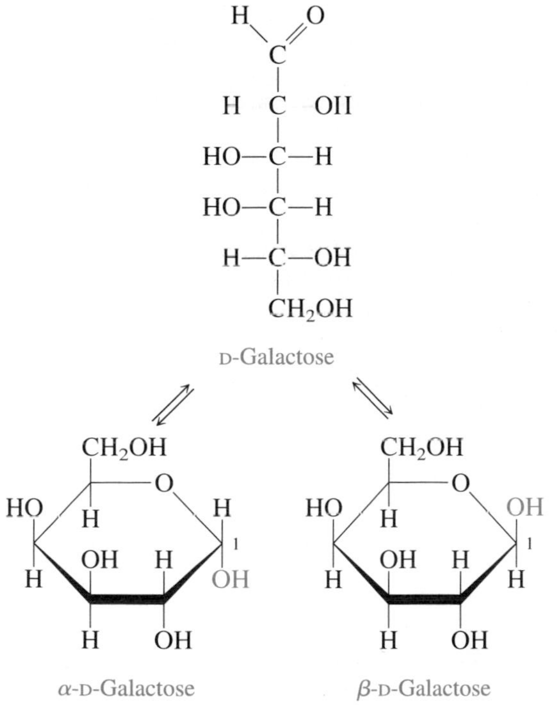

D-Galactose

α-D-Galactose

β-D-Galactose

FIGURE 7.7
Cyclization of D-galactose to α- and β-D-galactose.

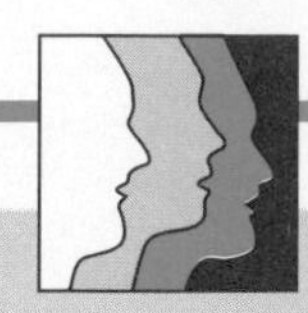

A HUMAN PERSPECTIVE

Blood Transfusions and the Blood Group Antigens

The first attempts to replace blood by transfusion were made in the seventeenth century, when physicians used animal blood to replace human blood lost by hemorrhages. Many people died as a result of this attempted cure, and transfusions were banned in much of Europe. Blood transfusions from human donors to recipients were somewhat less lethal, but violent reactions often led to the death of the recipient, and by the nineteenth century, transfusions had been abandoned as a medical failure.

In 1904, Dr. Karl Landsteiner performed a series of experiments on the blood of workers in his laboratory. His experimental results explained the mysterious fatalities caused by blood transfusions, and blood transfusions were reinstated as a life-saving clinical tool. Landsteiner took blood samples from his co-workers. He separated the blood cells from the serum, the liquid component of the blood, and mixed these samples in test tubes. When he mixed serum from one individual with blood cells of another, Landsteiner observed that, in some instances, the serum samples caused clumping, or agglutination, of red blood cells. The agglutination reaction always indicated that the two bloods were incompatible and that transfusion could lead to life-threatening reactions. As a result of many such experiments, Landsteiner showed that there are four human blood groups, designated A, B, AB, and O. These cross-reactions are summarized in the following table:

Blood Type	Agglutination Reaction with	
	Anti-A Serum	Anti-B Serum
O	−	−
A	+	−
B	−	+
AB	+	+

We now know that differences among blood groups reflect differences among oligosaccharides attached to the membrane lipids, and probably proteins, at the surface of red blood cells. The oligosaccharides on the red blood cell surface have a common core, as shown in the accompanying figure, consisting of N-acetylgalactosamine, galactose, N-acetylneuraminic acid (sialic acid) and L-fucose. It is the terminal monosaccharide of this oligosaccharide that distinguishes among the cells of different blood types.

The end of the blood group oligosaccharides governs the compatibility of the blood types. The A blood group antigen contains β-D-N-acetylgalactosamine at its terminus end, whereas the B blood group antigen contains α-D-galactose. In type O blood, neither of these sugars is found on the cell surface; only the core oligosaccharide is present. As you may already suspect, on type AB blood cells, some of the oligosaccharides have a terminal β-D-N-acetylgalactosamine, while others have a terminal α-D-galactose.

Why does agglutination occur? The clumping reaction that occurs when incompatible bloods are mixed is an antigen-antibody reaction. An antigen is any large molecule that stimulates the immune defenses of the body to produce protective *glycoproteins,* proteins with attached sugar molecules, called antibodies. Frequently, these foreign antigens are bacteria or viruses that are invading the body, and the antibodies are produced as part of our defense against them. Antibodies have the ability to bind to the foreign antigens and facilitate their destruction or removal from the body.

A person with type A blood (A antigens on the red blood cell surface) also has in his or her bloodstream antibodies against type B blood (anti-B antibodies). If the person with type A blood receives a transfusion of type B blood, the anti-B antibodies bind to the type B blood cells. This results in clumping and the destruction of those cells, which can result in death. Individuals who have type B blood also produce anti-A antibodies and are therefore not able to receive a transfusion from a type A individual. Those with type AB blood are considered to be universal recipients because they have neither anti-A nor anti-B antibodies in their blood. (If they did, they would destroy their own red blood cells!) Thus in emergency situations a patient with type AB blood can receive blood from an individual of any blood type without serious transfusion reactions. Type O blood has no A or B antigens on the red blood cell surface but has both anti-A and anti-B antibodies. Because of the presence of both types of antibodies, individuals with type O blood can receive transfusions only from a person who also has type O blood. On the other hand, because of the absence of A and B antigens on the red blood cell surface, type O blood can be safely transfused into patients of any blood type. Hence type O individuals are universal donors. A summary of blood transfusion compatibilities is found in the following table:

Blood Type	Cellular Antigens	Antibodies in Blood	Can Accept Blood from	Can Donate Blood to
O	None	Anti-A, anti-B	O	O, A, B, AB
A	A	Anti-B	A or O	A or AB
B	B	Anti-A	B or O	B or AB
AB	A and B	None	AB, A, B, O	AB

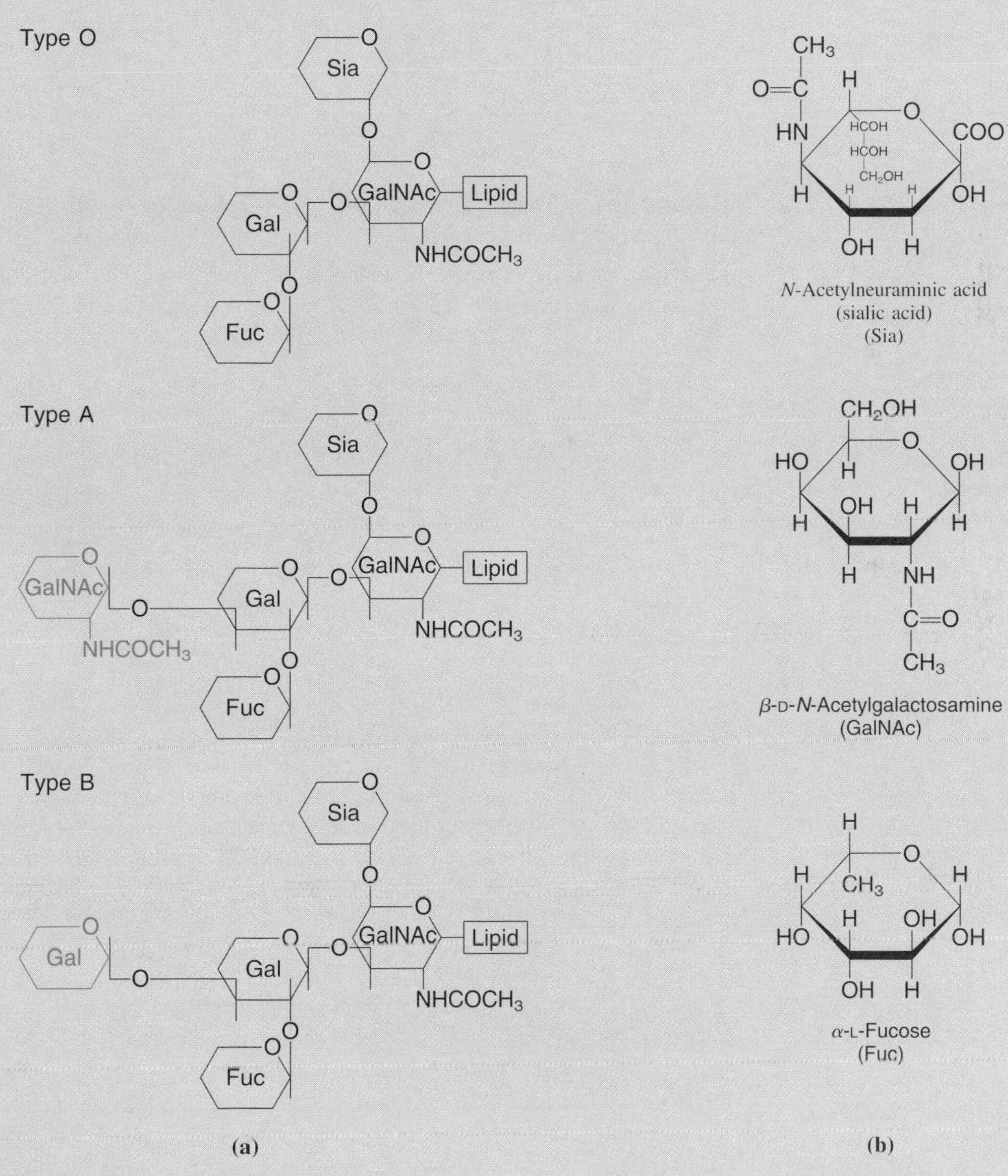

Schematic diagram of the blood group oligosaccharides. (a) Only the core oligosaccharide is found on the surface of type O red blood cells. On type A red blood cells, β-D-N-acetylgalactosamine is linked to the galactose moiety of the core oligosaccharide. On type B red blood cells, a galactose molecule is found attached to the galactose of the core oligosaccharide. (b) The structures of some of the unusual monosaccharides found in the blood group oligosaccharides.

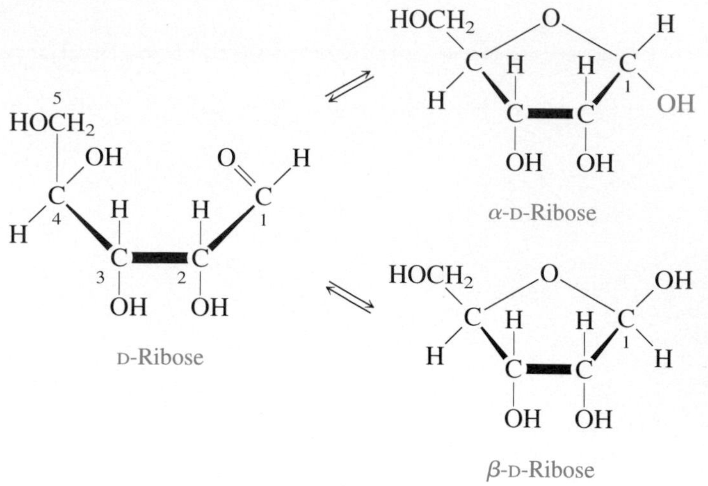

D-Ribose

α-D-Ribose

β-D-Ribose

FIGURE 7.8
Cyclization of D-ribose to α- and β-D-ribose.

Ribose, a five-carbon sugar that can exist as an open-chain or cyclic molecule

Ribose is also a component of many biologically important molecules, including RNA, and various coenzymes, a group of compounds required for many biochemical reactions in the body. The structure of the five-carbon sugar D-ribose is shown in its open-chain form below:

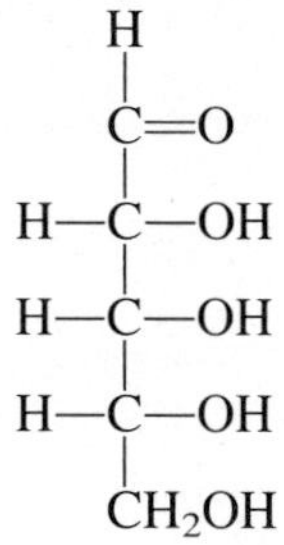

D-Ribose

Note that the —OH groups at the chiral carbons of ribose—C-2, C-3, and C-4—are all on the right, making the structure of ribose particularly easy to remember.

Like glucose, the open chain form of ribose is not its predominant form in cells. The C-4 hydroxyl group can readily react with the aldehyde group at C-1 forming a cyclic intramolecular hemiacetal. In this process a new chiral carbon is created at C-1, resulting in α- and β- forms (Figure 7.8).

2-Deoxyribose is one of the exceptions to the general formula for monosaccharides, $C_x(H_2O)_y$. It is derived from ribose by the replacement of the C-2 hydroxyl group with a hydrogen atom; thus, it is still considered to be a monosaccharide.

In the Haworth projection of the cyclic sugar the $—CH_2OH$ group is "up" and the —OH groups at C-3 and C-2 are "down." The oxygen of the original C-4 hydroxyl group is now the ring oxygen. If the —OH at C-1 is "up," on the same side of the ring as the $—CH_2OH$ group, the sugar is designated β; if the —OH at C-1 is "down," the sugar is designated α. The cyclic forms of ribose are therefore named α-D-ribose and β-D-ribose.

DNA actually contains 2-deoxyribose (Figure 7.9). In this molecule the —OH group at C-2 has been replaced by a hydrogen, hence the designation "2-deoxy."

FIGURE 7.9
Structure of β-D-2-deoxyribose.

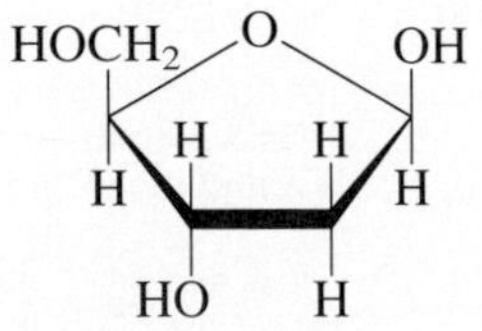

β-D-2-Deoxyribose

Reducing sugars

The aldehyde moiety, or portion, of aldoses is readily oxidized by an aqueous solution of silver ammonia complex, $Ag(NH_3)_2^+$, the **Tollens' reagent.** The reaction of Tollens' reagent with D-glyceraldehyde, typical of the reactions of aldehydes (and aldoses) in general, converts the aldehyde group of the aldose to a carboxylate group. Ag^+ is reduced to Ag^0 in the process. This silver metal is deposited on the sides of the test tube, producing a silver mirror. The formation of the silver mirror indicates that the aldose is a reducing sugar.

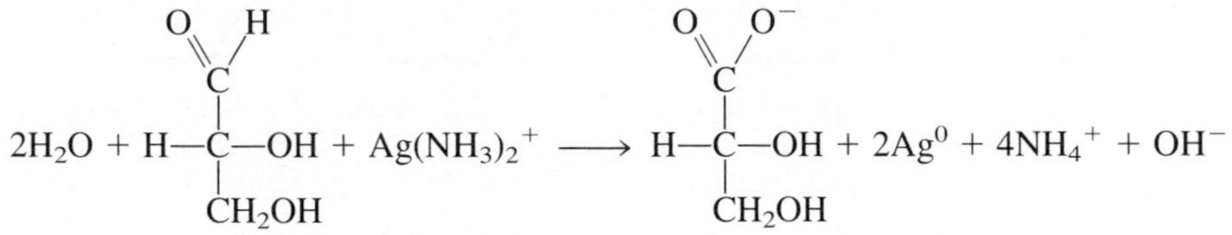

$$2H_2O + \text{H—C(CHO)(OH)—CH}_2\text{OH} + Ag(NH_3)_2^+ \longrightarrow \text{H—C(COO}^-\text{)(OH)—CH}_2\text{OH} + 2Ag^0 + 4NH_4^+ + OH^-$$

Benedict's solution and **Fehling's solution** also react with the aldehyde groups of aldoses. These reagents are prepared in different basic buffer solutions, but each is a solution that contains Cu^{2+} ions. The Cu^{2+} ions are reduced to Cu^{1+}, which precipitates as brick-red Cu_2O. The aldehyde of the aldose is converted to a carboxylate anion.

Section 6.4

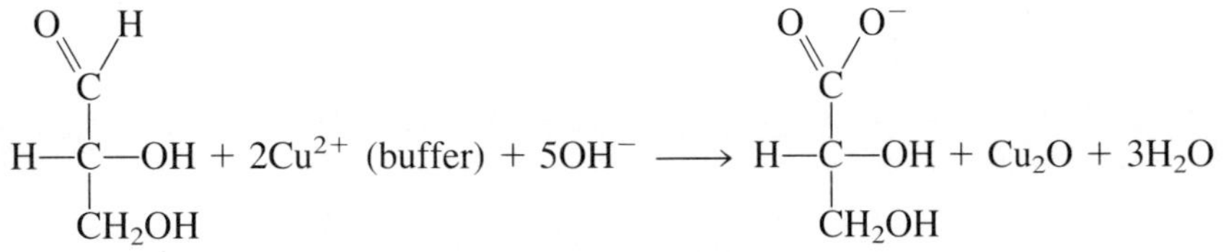

$$\text{H—C(CHO)(OH)—CH}_2\text{OH} + 2Cu^{2+}\ (\text{buffer}) + 5OH^- \longrightarrow \text{H—C(COO}^-\text{)(OH)—CH}_2\text{OH} + Cu_2O + 3H_2O$$

Although ketones generally are not easily oxidized, ketoses are an exception to that rule. Because of the —OH groups on the carbon next to the carbonyl group, ketoses can be converted to aldoses via an **enediol reaction:**

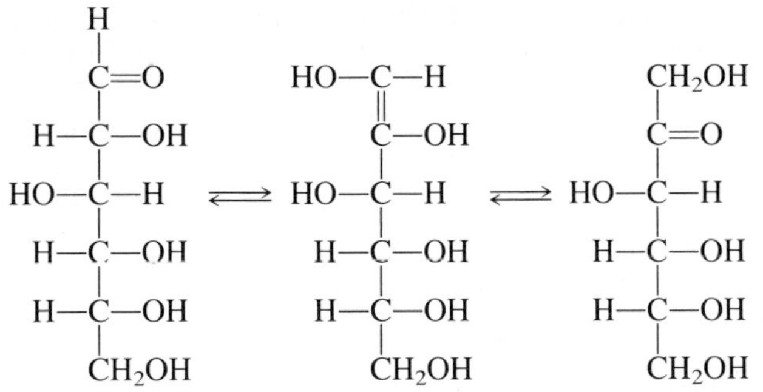

$$\text{D-Glucose: H—C=O, H—C—OH, HO—C—H, H—C—OH, H—C—OH, CH}_2\text{OH} \rightleftharpoons \text{Enediol: HO—C—H = C—OH, HO—C—H, H—C—OH, H—C—OH, CH}_2\text{OH} \rightleftharpoons \text{D-Fructose: CH}_2\text{OH, C=O, HO—C—H, H—C—OH, H—C—OH, CH}_2\text{OH}$$

D-Glucose Enediol D-Fructose

The name of the enediol reaction is derived from the structure of the intermediate through which the ketose is converted to the aldose: It has a double bond (ene), and it has two hydroxyl groups (diol). Because of this enediol reaction, ketoses are also able to react with Tollens', Fehling's, and Benedict's reagents. Since the metal ions in each of these solutions are reduced, the sugars are serving as reducing agents and are called **reducing sugars.** All monosaccharides are reducing sugars.

Benedict's reagent is one of the tests used to detect the presence of excess glucose in the urine, *glucosuria*. Individuals suffering from Type I insulin-dependent diabetes mellitus do not produce the hormone insulin, which controls the uptake of glucose from the blood. When the blood glucose level rises too high, the kidney is unable to reabsorb the excess, and glucose is found in the urine. Although the level of blood glucose can be controlled by the injection of insulin, urine glucose levels must be monitored to ensure that the amount of insulin injected is correct. Benedict's reagent is a useful tool because the amount of Cu_2O formed, and hence the degree of color change in the reaction, is directly proportional to the amount of reducing sugar in the urine. Thus a brick-red color indicates a very high concentration of glucose in the urine. Yellow, green, and blue-green solutions indicate decreasing amounts of glucose in the urine, and a blue solution indicates an insignificant concentration of glucose in the urine.

Chapter 15, A Clinical Perspective: Diabetes Mellitus and Ketone Bodies

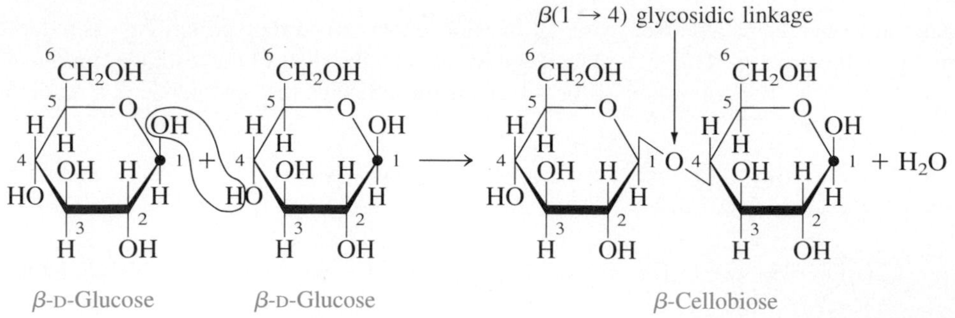

FIGURE 7.10
Reaction of the C-1 hydroxyl group of β-D-glucose with the C-4 hydroxyl group of a second molecule of β-D-glucose to give the disaccharide cellobiose. The prefix β- indicates that the reducing end of the disaccharide has the β-configuration.

β(1 → 4) glycosidic linkage

β-D-Glucose + β-D-Glucose ⟶ β-Cellobiose + H_2O

Question 7.8 Write the balanced chemical equation, including structures of all carbohydrate molecules, for the oxidation of D-galactose with Tollens' reagent. Do the same for D-ribose.

7.3 OLIGOSACCHARIDES

Recall that oligosaccharides are formed by joining two to ten monosaccharides, interconnected by glycosidic bonds. If two sugars are connected, the product is a **disaccharide.** The disaccharides lactose and sucrose, for example, are important sources of energy for many organisms.

The α- or β-hydroxyl group of the cyclic form of D-glucose and other cyclic sugars can react with a hydroxyl group of another sugar to provide an "oxygen bridge" and produce a disaccharide. This process is illustrated in Figure 7.10 for two molecules of β-D-glucose.

The bond formed in the reaction between the C-1 hydroxyl of one β-D-glucose molecule and the C-4 hydroxyl of a second molecule of β-D-glucose is called a **β glycosidic bond** because the original C-1 hydroxyl group is β. If the C-1 hydroxyl group is α, the result is an **α glycosidic bond** (Figure 7.11). Maltose, lactose, and sucrose are the three most common oligosaccharides. The structures and properties of these disaccharides are described below.

Maltose

Section 7.4

If an α-D-glucose and a β-D-glucose are linked, as shown in Figure 7.11, the disaccharide is **maltose,** or malt sugar. This is one of the intermediates in the hydrolysis of starch. The reducing end of maltose has a β hydroxyl group, and the disaccharide is therefore called β-maltose. Since the C-1 hydroxyl group of α-D-glucose is attached to C-4 of another glucose molecule, the disaccharide is linked by an α (1 ⟶ 4) glycosidic bond.

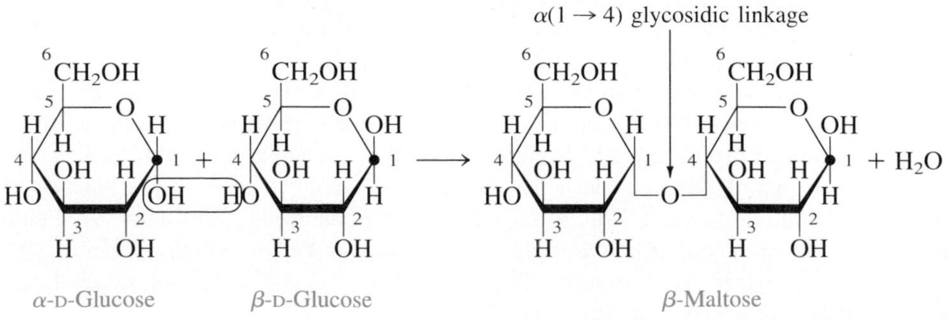

FIGURE 7.11
Glycosidic bond formed between the C-1 hydroxyl group of α-D-glucose and the C-4 hydroxyl group of β-D-glucose. The disaccharide is called β-maltose because the hydroxyl group at the reducing end of the disaccharide has the β-configuration.

α(1 → 4) glycosidic linkage

α-D-Glucose + β-D-Glucose ⟶ β-Maltose + H_2O

If an oligosaccharide possesses an unreacted hemiacetal hydroxyl group, a free —OH group at C-1, it is a reducing sugar. This is because the cyclic structure can break open at this position to form a free aldehyde. The nonreducing end, the end with no free —OH group, is written to the left. Sugars that do not contain an unreacted hemiacetal group on C-1, the original carbonyl carbon of the aldehyde or ketone functional group, do not react with Tollens', Benedict's, or Fehling's solutions and are called **nonreducing sugars.** All disaccharides, except sucrose, are reducing sugars. Sucrose will be discussed below.

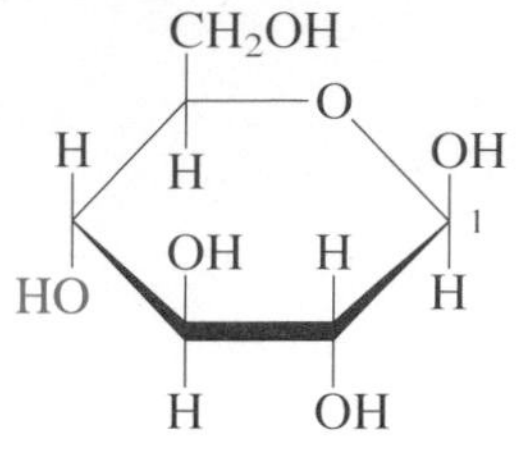

β-D-Glucose

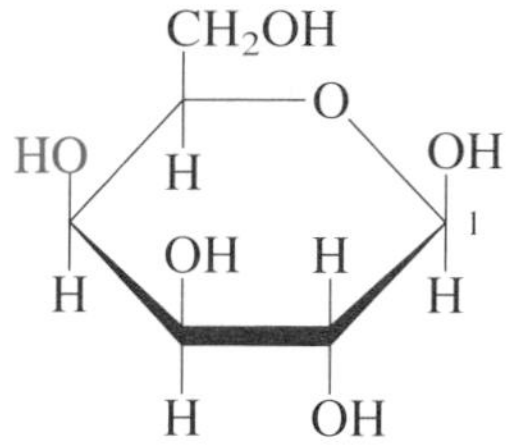

β-D-Galactose

FIGURE 7.12
Comparison of the cyclic forms of glucose and galactose. Note that galactose is identical to glucose except in the position of the C-4 hydroxyl group.

Lactose

Milk sugar, or **lactose,** is a dimer of one molecule of β-D-galactose and one of either α- or β-D-glucose. Galactose differs from glucose only in the configuration of the hydroxyl group at C-4 (Figure 7.12). In the cyclic form of glucose the C-4 hydroxyl group is "down"; in galactose it is "up." In lactose the C-1 hydroxyl group of β-D-galactose reacts with the C-4 hydroxyl group of glucose. The bond between the C-1 oxygen and C-4 is therefore a β (1 ⟶ 4) glycosidic bond (Figure 7.13).

Lactose is the principal sugar in mammalian milk. To be used by the body as an energy source, lactose must be hydrolyzed to produce glucose and galactose. Note that this is simply the reverse of the reaction shown in Figure 7.13. Glucose liberated by the hydrolysis of lactose is used directly in the energy-generating reactions of glycolysis. However, a series of reactions is necessary to convert galactose into a phosphorylated form of glucose that can be used for energy production. In humans, the genetic disease **galactosemia** is caused by the absence of one or more of the enzymes needed for this conversion. Therefore galactose, or one of its metabolic derivatives, cannot be metabolized in the normal way. A reduced form of galactose, called dulcitol, is produced in the body. This toxic metabolite then accumulates in individuals who suffer from galactosemia.

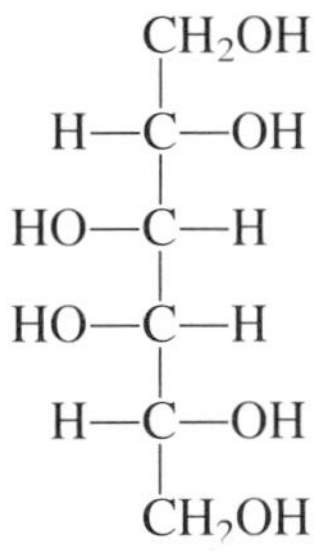

Chapter 13

Dulcitol

If the condition is not treated, galactosemia leads to severe mental retardation, cataracts, and early death. However, the effects of galactosemia can be avoided entirely by providing galactosemic infants with a diet that does not contain galactose. Such a diet, of course, cannot contain lactose and therefore must contain no milk or milk products.

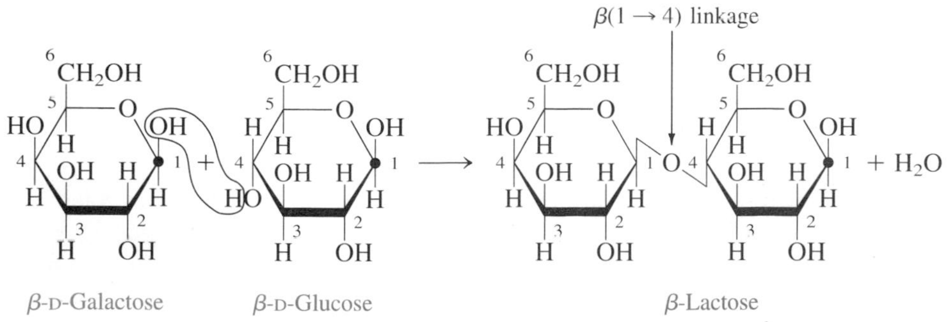

β(1 ⟶ 4) linkage

β-D-Galactose   β-D-Glucose   β-Lactose

FIGURE 7.13
Glycosidic bond formed between the C-1 hydroxyl group of β-D-galactose and the C-4 hydroxyl group of β-D-glucose. The disaccharide is called β-lactose because the hydroxyl group at the reducing end of the disaccharide has the β-configuration.

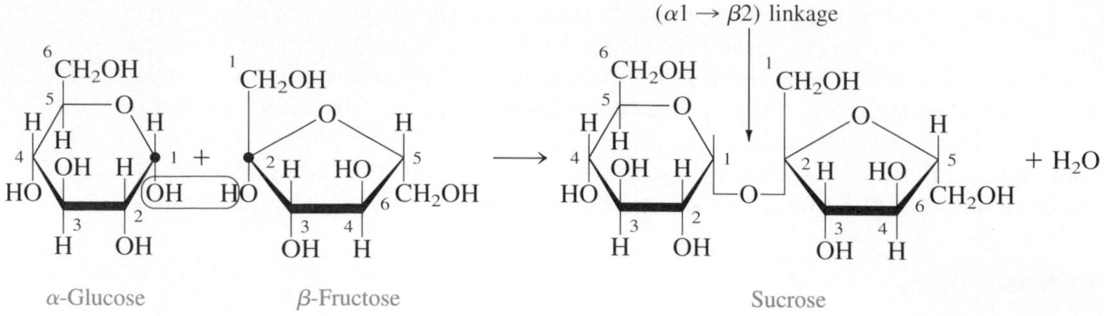

FIGURE 7.14
Glycosidic bond formed between the C-1 hydroxyl of α-D-glucose and the C-2 hydroxyl of β-D-fructose. Because this linkage involves the anomeric carbons of both sugars, the bond formed is called an α,β glycosidic linkage.

Many adults and some children are unable to hydrolyze lactose because they do not make the enzyme *lactase*. This condition is known as **lactose intolerance.** Undigested lactose remains in the intestinal tract, causing cramping and diarrhea that can eventually lead to dehydration. Some of the lactose is metabolized by intestinal bacteria that release organic acids and CO_2 gas into the intestines, causing further discomfort. Lactose intolerance is unpleasant, but its effects can be avoided by a diet that rigorously excludes milk and milk products.

Sucrose

Many sugars are not sweet, but **sucrose,** also called cane sugar or beet sugar, is a conspicuous and important exception. Sucrose is an important carbohydrate in plants. It is water-soluble and can easily be transported through the circulatory system of the plant. It cannot be synthesized by animals. High concentrations of sucrose inhibit the growth of microorganisms, and sucrose is used as a preservative. Of course, it is also widely used as a sweetener. In fact, it is estimated that the average American consumes 86 pounds of sucrose each year. It has been suggested that sucrose in the diet represents a source of empty calories; that is, it contains no vitamins or minerals. However, the only negative association that has been scientifically verified is the link between sucrose in the diet and dental caries, or cavities.

Chapter 17, A Human Perspective: Tooth Decay and Simple Sugars

Sucrose is a disaccharide of α-D-glucose joined to β-D-fructose (Figure 7.14). Take a careful look at the structure of sucrose. The glycosidic linkage between α-D-glucose and β-D-fructose is quite different from those we have examined for lactose and maltose. This linkage involves the anomeric carbons of *both* sugars! Such a bond is called an α,β glycosidic linkage, in this case involving the anomeric C-1 of glucose and the anomeric C-2 of fructose (noted in color in Figure 7.14).

Both of the carbons that were previously part of a hemiacetal or a hemiketal have reacted to form this linkage. Thus the ring structure cannot open up, and no aldehyde or ketone group can be formed. Sucrose therefore will not react with Tollens', Benedict's, or Fehling's reagents and is not a reducing sugar.

7.4 POLYSACCHARIDES

Most carbohydrates that are found in nature are high molecular weight polymers of glucose and its derivatives. A polymer, in a very general sense, is a large unit made up of many small units—the monomers—held together by chemical bonds. By analogy a polysaccharide is a large molecule composed of many monosaccharide units joined in one or more chains. The principal polysaccharides are starch, glycogen, and cellulose. The structures and properties of each are described below.

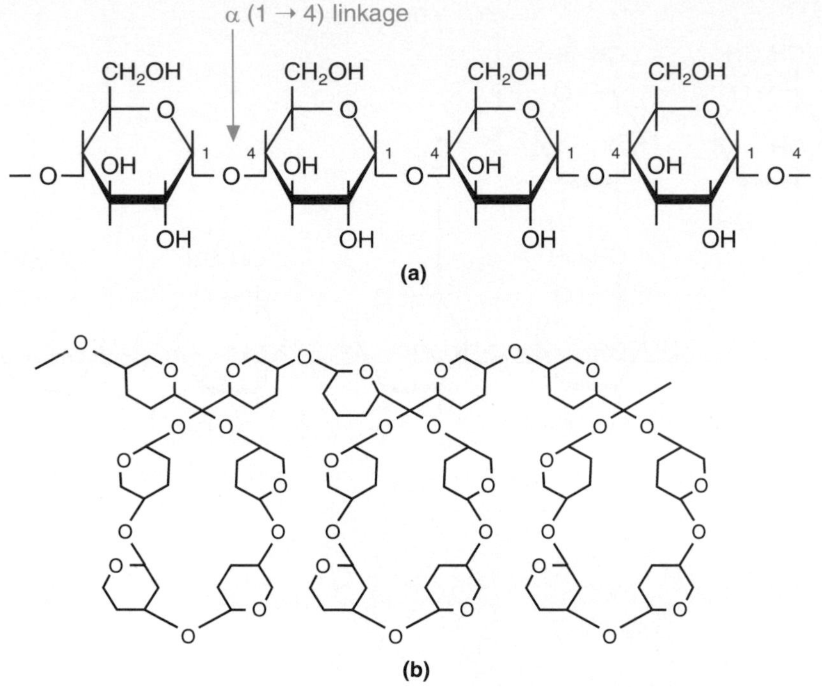

FIGURE 7.15
Structure of amylose. (a) A linear chain of α-D-glucose joined in α $(1 \longrightarrow 4)$ glycosidic linkage makes up the primary structure of amylose. (b) Owing to hydrogen bonding, the amylose chain forms a left-handed helix that contains six glucose units per turn.

Starch

As noted in Figure 7.1, plants have the ability to use the energy of sunlight to produce monosaccharides, principally glucose, from CO_2 and H_2O. Although sucrose is the major transport form of sugar in the plant, the polysaccharide starch is the principal storage form in most plants. These plants store glucose in starch granules. Nearly all plant cells contain some starch granules, but in some seeds, such as corn, as much as 80% of the cell's dry weight is starch. Some plants, the potato plant for instance, have special structures for the storage of starch. The cells of the potato tuber are literally filled with starch.

Starch is a heterogeneous material composed of the glucose polymers **amylose** and **amylopectin.** Amylose, which accounts for about 80% of the starch of a plant cell, is a linear polymer of α-D-glucose residues connected by glycosidic bonds between C-1 and C-4. A single chain can contain up to 4000 glucose residues. Amylose coils up into a helix that repeats every six glucose residues.

Amylose is degraded by two types of enzymes. These are produced in the pancreas, from which they are secreted into the small intestines, and the salivary glands, from which they are secreted into the saliva. *α-Amylase* cleaves the glycosidic bonds of amylose chains at random along the chain, producing shorter polysaccharide chains. The enzyme *β-amylase* sequentially cleaves dimers of glucose, maltose, from the reducing end of the amylose chain. The structure of amylose is provided in Figure 7.15.

Section 7.3

Amylopectin is a highly branched amylose in which the branches are attached to the C-6 hydroxyl groups by α $(1 \longrightarrow 6)$ glycosidic bonds (see Figure 7.16). The main chain consists of α $(1 \longrightarrow 4)$ glycosidic bonds. Each branch contains 20–25 glucose residues, and there are so many branches that the main chain can scarcely be distinguished.

Glycogen

Glycogen, the glucose storage molecule of animals, is stored in granules in liver and muscle cells. The structure of glycogen is similar to that of amylopectin. Glycogen differs only by having more and shorter branches. Otherwise, the two molecules and their metabolism are virtually identical (Figure 7.17).

Section 13.6

The liver supply of glycogen is important in the regulation of blood glucose levels. After a meal the liver cells take up the excess glucose from the blood and convert it to glycogen for storage, thereby lowering the blood glucose concentration. Between meals,

We will discuss the synthesis and degradation of glycogen in Chapter 13.

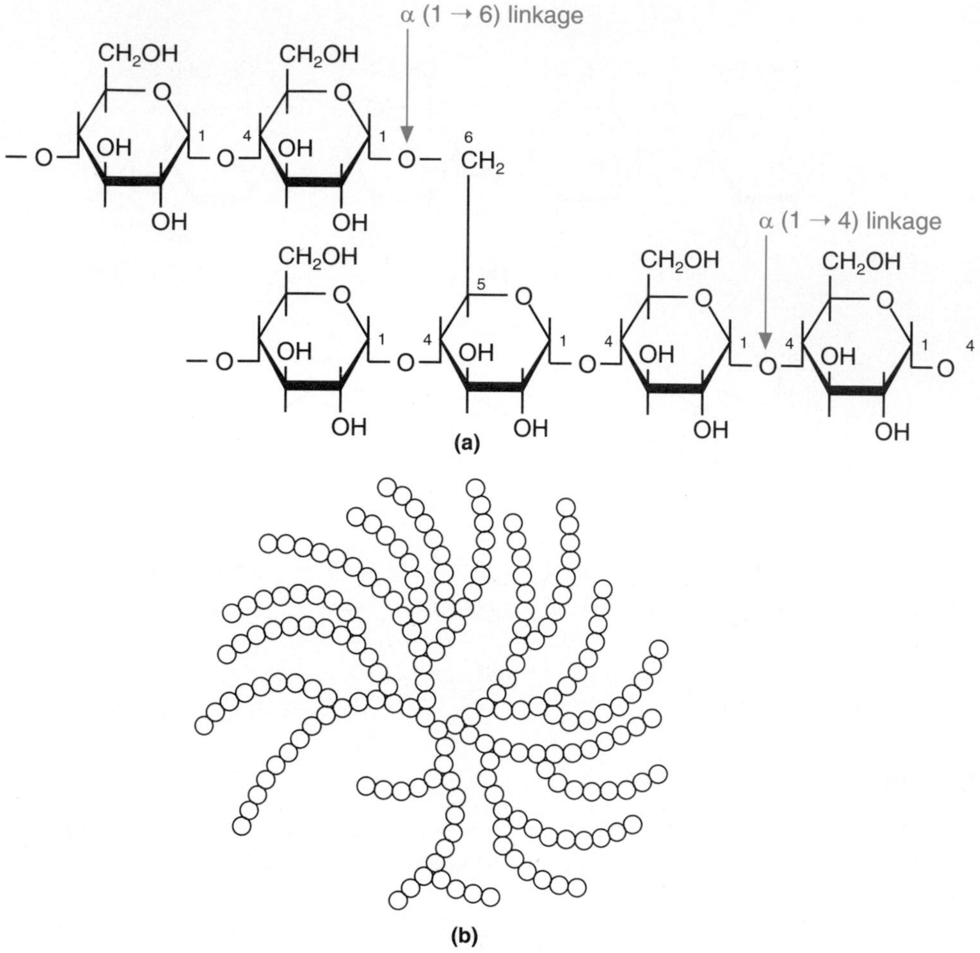

FIGURE 7.16
Structure of amylopectin. (a) Amylopectin consists of branched chains of amylose. Branching in amylopectin occurs by α (1 $\longrightarrow$ 6) glycosidic bonds between glucose units. The main chain is bonded α (1 $\longrightarrow$ 4). (b) A representation of the branched-chain structure of amylopectin.

when the blood glucose levels fall, the liver cells break glycogen down to glucose. The glucose is released into the bloodstream, thereby increasing blood glucose levels.

Cellulose

The most abundant polysaccharide, indeed the most abundant organic molecule in the world, is **cellulose,** a polymer of β-D-glucose units linked by β (1 $\longrightarrow$ 4) glycosidic bonds (Figure 7.18). A molecule of cellulose typically contains about 3000 glucose residues, but

FIGURE 7.17
The structure of glycogen.

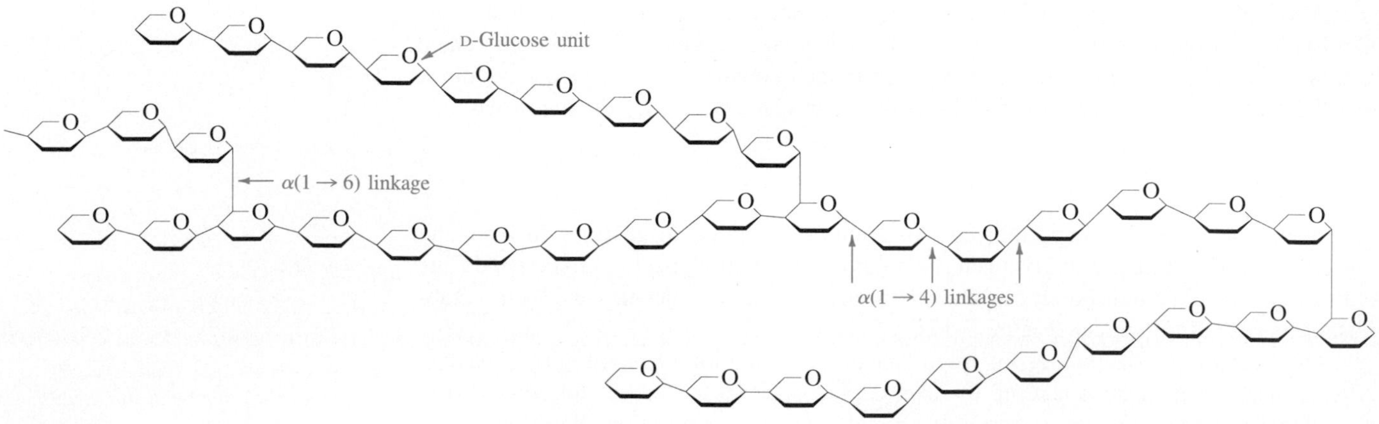

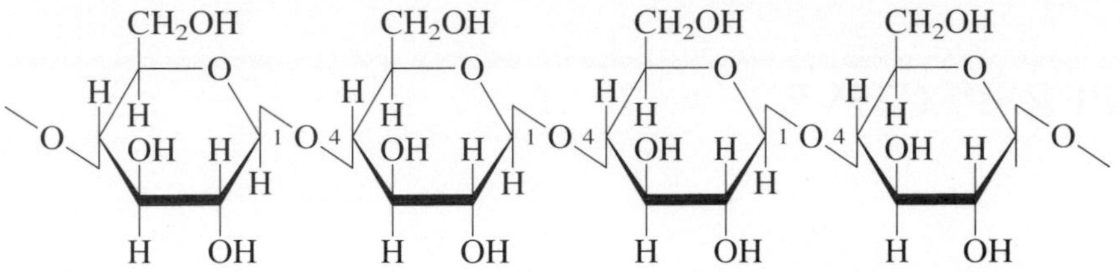

FIGURE 7.18
The structure of cellulose.

the largest known cellulose, produced by species of the alga *Valonia,* contains 26,000 glucose residues.

Cellulose is a structural component of the plant cell wall. The unbranched structure of the cellulose polymer and the $\beta\,(1 \longrightarrow 4)$ glycosidic linkages allow cellulose molecules to form long fibrils. These consist of long, straight chains of parallel cellulose molecules. Such fibrils are quite rigid; therefore it is not surprising that cellulose is a cell wall structural element.

In contrast to glycogen, amylose, and amylopectin, cellulose *cannot* be digested by humans. The reason for this is that we are unable to synthesize the enzyme cellulase, which can hydrolyze the $\beta\,(1 \longrightarrow 4)$ glycosidic linkages of the cellulose polymer. Indeed, only a few animals, such as termites, cows, and goats, are able to digest cellulose. These animals have, within their digestive tracts, microorganisms that produce the enzyme cellulase. The sugars released by this microbial digestion can then be absorbed and used by these animals.

SUMMARY

Carbohydrates are found in a wide variety of naturally occurring substances and serve as principal energy sources for the body. Carbohydrates are an important class of polyhydroxyaldehydes and polyhydroxyketones and have the empirical formula $(CH_2O)_n$. Carbohydrates that have an aldehyde as their most oxidized functional group are aldoses, and those having a ketone group as their most oxidized functional group are ketoses. They may be classified as trioses, tetroses, pentoses, and so forth, depending upon the number of carbon atoms in the main skeletal backbone of the carbohydrate.

Stereoisomers of carbohydrates, due to the presence of chiral carbon atoms, are classified as D- or L- depending on the arrangement of the atoms on the penultimate chiral carbon of the skeletal backbone. In a Fischer Projection, if the —OH on this terminal carbon is to the right, the stereoisomer is of the D-configuration. If the —OH group is to the left, the stereoisomer is of the L-configuration. Depending on the direction in which they rotate plane polarized light, steroisomers may also be classified as (+) or dextrorotatory (rotates light to the right) and (−) or levorotatory (rotates light to the left).

Carbohydrates are classified as monosaccharides (one sugar unit), oligosaccharides (two to ten sugar units), or polysaccharides (many sugar units).

Important monosaccharides include glyceraldehyde, glucose, fructose, and ribose. Monosaccharides containing five or six carbon atoms can exist as five-membered or six-membered rings. Formation of a ring produces a new chiral carbon, termed an anomeric carbon, at the original carbonyl carbon, which is designated either α or β depending on the orientation of the groups. The cyclization of an aldose produces an intramolecular hemiacetal, and the cyclization of a ketose yields an intramolecular hemiketal.

Reducing sugars are oxidized by Tollens' reagent and by Fehling's or Benedict's solution.

Important oligosaccharides include lactose and sucrose. Lactose is a disaccharide of β-D-galactose bonded $(1 \longrightarrow 4)$ with D-glucose. In galactosemia, defective metabolism of galactose leads to accumulation of the toxic by-product dulcitol. The ill effects of galactosemia are avoided by exclusion of galactose from the diet of galactosemic infants. Sucrose is a dimer of α-D-glucose bonded $(1 \longrightarrow 2)$ with β-D-fructose.

Starch, the storage polysaccharide of plant cells, is composed of approximately 80% amylose and 20% amylopectin. Amylose is a polymer of α-D-glucose residues bonded $\alpha\,(1 \longrightarrow 4)$. Amylose forms a helix. Amylopectin has many branches. Its main chain consists of α-D-glucose units bonded $(1 \longrightarrow 4)$. The branches are connected by $\alpha\,(1 \longrightarrow 6)$ glycosidic bonds. Glycogen, the major storage polysaccharide of animal cells, resembles amylopectin, but it has more and shorter branches.

Cellulose, an important polysaccharide, is the major structural molecule of plants. It is a $\beta\,(1 \longrightarrow 4)$ polymer of D-glucose that can contain thousands of glucose monomers. Cellulose cannot be digested by most animals because they do not produce an enzyme that is capable of cleaving the $\beta\,(1 \longrightarrow 4)$ glycosidic linkage.

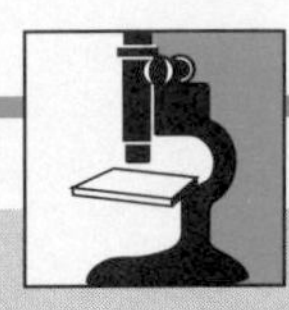

A CLINICAL PERSPECTIVE

The Bacterial Cell Wall

The major component of bacterial cell walls is a complex polysaccharide known as a *peptidoglycan.* The carbohydrate portion of the peptidoglycan is a polymer of alternating residues of N-acetylglucosamine and N-acetylmuramic acid.

In N-acetylglucosamine (NAG) an amino (NH_2) group has replaced the C-2 hydroxyl group of glucose, and the amino group has in turn been converted to an amide with acetic acid. The sugar is therefore N-acetylglucosamine.

N-acetylmuramic acid (NAM) is produced by modifying NAG. Thus it too has an N-acetyl group at C-2, but in addition, it has an ether linkage between lactic acid and the C-3 hydroxyl group. In the alternating polymer, C-1 and C-4 are connected by a β (1 $\longrightarrow$ 4)-glycosidic bond. The lactic acid side chain of N-acetylmuramic acid is connected to four amino acids. Amino acids, as we will see in Chapter 11, are the building blocks of proteins.

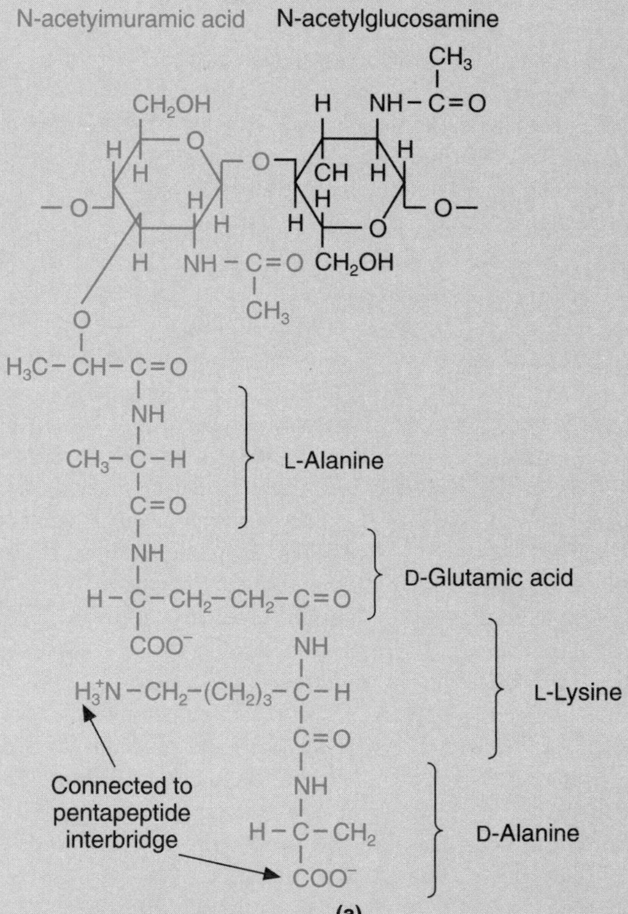

N-acetyimuramic acid N-acetylglucosamine

L-Alanine

D-Glutamic acid

L-Lysine

D-Alanine

Connected to pentapeptide interbridge

(a)

Structures of N-acetylglucosamine and N-acetylmuramic acid in β (1 $\longrightarrow$ 4) glycosidic linkage. Note the tetrapeptide bridge linked to the NAM.

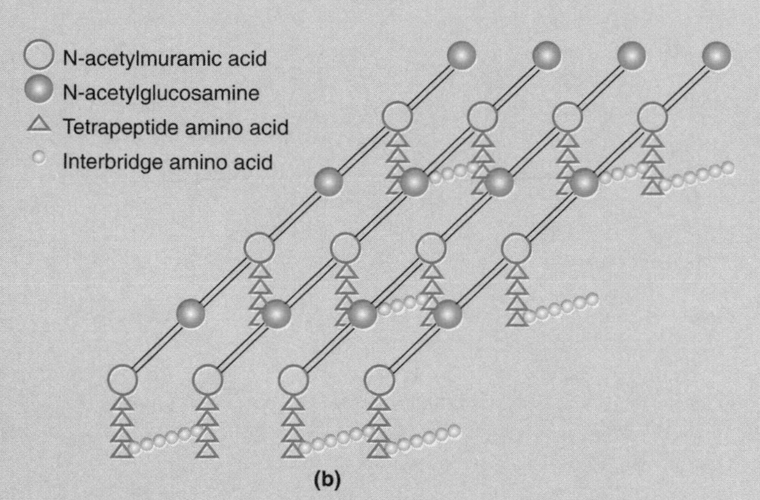

The three-dimensional structure of one layer of peptidoglycan.

The three-dimensional structure of the cell wall is a result of cross-linkages between these basic repeat units by the peptide interbridges, as shown in the above figure. Millions of such cross-linking events produce an enormous peptidoglycan molecule, many layers thick, around the bacterium. A single layer of the cell wall is represented in the figure. This thick cell wall is quite rigid and allows the bacterium to maintain its shape and to resist changes in the osmotic pressure of the environment. In other words, the bacteria are protected from bursting if the salt concentration of the environment is too low.

Our bodies are constantly being assaulted by a variety of bacteria, and as you would expect, we have evolved protective mechanisms to minimize the damage. One such protective mechanism is the presence of the enzyme *lysozyme* in our tears and saliva. Enzymes are proteins that act as biological catalysts in biochemical reactions. The reaction catalyzed by lysozyme is the hydrolysis of the β $(1 \longrightarrow 4)$-glycosidic linkages of peptidoglycan. As can be seen in the following figure, the enzyme has an active site, a deep groove on the surface, which holds a six-sugar unit of the cell wall.

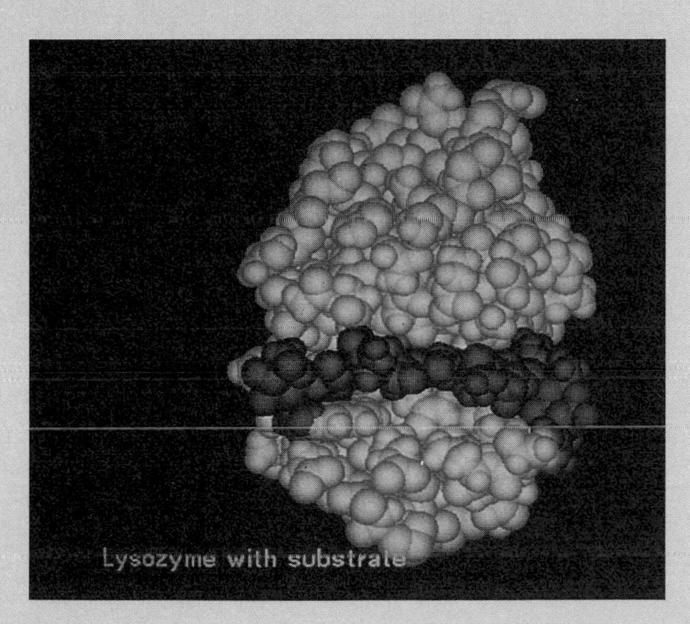

Conformation of lysozyme bound to its substrate. The enzyme binds with a six-sugar portion of the bacterial cell wall and cleaves it. The substrate fits into a deep crevice on the surface of the enzyme.

When the sugar is bound within the active site, the enzyme catalyzes the hydrolysis reaction. Once the bacterial cell wall has been cleaved, unfavorable osmotic conditions cause the cell membrane to rupture, and the cell dies.

Many antibiotics that interfere with bacterial cell wall synthesis have been discovered. Among these are penicillin and the penicillin derivatives:

$$C_6H_5{-}CH_2{-}\overset{O}{\overset{\|}{C}}{-}NH{-}CH{-}CH{-}S{-}C(CH_3)_2{-}CH{-}COOH \quad (\text{β-lactam ring: } CH{-}C(=O){-}N)$$

Penicillin G

Because the human body has no structural components similar to the bacterial cell wall, treatment with the penicillins selectively destroys the bacteria, causing no harm to the patient. In practice, however, it must always be remembered that some individuals develop an allergy to the penicillins.

The principal mode of action of the penicillins is the inhibition of the enzyme that catalyzes the cross-linkage reaction. Penicillins bind irreversibly to the active site of that enzyme. Because the active site is occupied, the enzyme cannot bind to the amino acid tail of the NAG-NAM repeat unit, and no cross-linkage can be made. Without the rigid, highly cross-linked peptidoglycan, the bacterial cells rupture and die.

GLOSSARY OF KEY TERMS

aldose (7.2) a sugar that contains an aldehyde (carbonyl) group.

amylopectin (7.4) a highly branched form of amylose; the branches are attached to the C-6 hydroxyl by α (1 $\longrightarrow$ 6) glycosidic linkage; a component of starch.

amylose (7.4) a linear polymer of α-D-glucose in α (1 $\longrightarrow$ 4) glycosidic linkage that is a major component of starch; a polysaccharide storage form.

anomers (7.2) isomers of cyclic sugars that differ in the arrangement of groups around an asymmetric carbon.

asymmetric carbon (7.2) a chiral carbon; a carbon bonded to four different groups.

Benedict's solution (7.2) a buffered solution of Cu^{2+} ions that can be used to test for reducing sugars.

cellulose (7.4) the most abundant organic compound in the world; a polymer of β-D-glucose linked by β (1 $\longrightarrow$ 4) glycosidic bonds.

chiral (7.2) molecules that are capable of existing in mirror image forms.

dextrorotatory (7.2) the enantiomer that rotates plane polarized light in a clockwise direction.

disaccharide (7.3) a sugar composed of two monosaccharides joined through an oxygen atom bridge.

enantiomers (7.2) stereoisomers that are mirror images of one another.

enediol reaction (7.2) the reaction by which ketoses are converted to aldoses.

Fehling's solution (7.2) a buffered solution of Cu^{2+} ions that is used to test for reducing sugars.

Fischer Projection (7.2) a two-dimensional formula used to designate the three-dimensional structure of a molecule. It is drawn as a cross with the chiral carbon in the center.

galactose (7.2) an aldohexose that is a component of lactose, milk sugar.

galactosemia (7.3) a human genetic disease caused by the inability to convert galactose to glucose-1-phosphate.

glucose (7.2) an aldohexose, the most abundant monosaccharide; it is a component of many disaccharides, such as lactose and sucrose, and polysaccharides, such as cellulose, starch, and glycogen.

glycogen (7.4) the glucose storage form of animals; a linear backbone of α-D-glucose in α (1 $\longrightarrow$ 4) linkage, with numerous short branches attached to the C-6 hydroxyl group by α (1 $\longrightarrow$ 6) linkage.

glycosidic bond (7.3) the bond between the hydroxyl group of the anomeric carbon of one sugar and a hydroxyl group of another sugar.

Haworth Projections (7.2) a means of representing the orientation of substituent groups around a cyclic sugar molecule.

hemiacetal (7.2) the product of the reaction between an aldehyde and an alcohol.

hemiketal (7.2) the product of the reaction between a ketone and an alcohol.

hexose (7.2) a six-carbon monosaccharide.

ketose (7.2) a sugar that contains a ketone (carbonyl) group.

lactose (7.3) a disaccharide composed of β-D-galactose and either α- or β-D-glucose in β (1 $\longrightarrow$ 4) glycosidic linkage; milk sugar.

lactose intolerance (7.3) the inability to produce the enzyme lactase, which degrades lactose to galactose and glucose.

levorotatory (7.2) the enantiomer that rotates plane polarized light in a counterclockwise direction.

maltose (7.3) a disaccharide composed of two glucose molecules in α (1 $\longrightarrow$ 4) glycosidic linkage.

monosaccharide (7.1) the simplest type of carbohydrate, consisting of a single saccharide unit.

mutarotation (7.2) the interconversion of α and β anomers.

nonreducing sugar (7.3) a sugar that cannot be oxidized by Fehling's, Benedict's, or Tollens' reagents.

oligosaccharide (7.1) an intermediate-sized carbohydrate composed of from two to ten monosaccharides.

optical activity (7.2) the ability to rotate plane polarized light.

pentose (7.2) a five-carbon monosaccharide.

plane polarized light (7.2) light filtered through a polaroid lens.

polysaccharide (7.1) a large, complex carbohydrate composed of long chains of monosaccharides.

reducing sugar (7.2) a sugar that can be oxidized by Fehling's, Benedict's, or Tollens' reagents. This includes all monosaccharides and most disaccharides.

ribose (7.2) a five-carbon monosaccharide that is a component of RNA and many coenzymes.

saccharide (7.1) a sugar molecule.

stereochemistry (7.2) the study of the spatial arrangement of atoms in a molecule.

stereoisomers (7.2) a pair of molecules having the same structural formulas and bonding patterns but differing in the arrangement of the atoms in space.

sucrose (7.3) a disaccharide composed of α-D-glucose and β-D-fructose in α,β glycosidic linkage; table sugar.

tetrose (7.2) a four-carbon monosaccharide.

Tollens' reagent (7.2) an aqueous solution of $Ag(NH_3)_2^+$ that can be used to test for reducing sugars.

triose (7.2) a three-carbon monosaccharide.

QUESTIONS AND PROBLEMS

Structure, Nomenclature, and Properties

7.9 Define what is meant by the term "saccharide."

7.10 What is the major structural difference between an aldose and a ketose?

7.11 Identify the following sugars:

a. CH_2OH; O; H; OH; H; OH; H; HO; H; H; OH

b. $HOCH_2$; O; OH; H; HO; H; CH_2OH; OH; H

c. CH_2OH; O; HO; H; H; OH; H; H; OH; H; OH

7.12 Draw the open chain form of the sugars in Problem 7.11.

7.13 Identify the sugars in Problem 7.11 as either hemiacetals or hemiketals.

7.14 β-D-Ribose units are components of ribonucleic acids, and 2-deoxy-β-D-ribose units are components of DNA molecules. Draw the structures of these sugars.

7.15 What is the name of the type of bond by which the monosaccharides are linked in a disaccharide?

7.16 Maltose is a disaccharide isolated from amylose that consists of two glucose units linked α (1 ⟶ 4). Draw the structure of this molecule.

7.17 What is the major biological source of lactose?

7.18 Sucrose is a disaccharide formed by linking α-D-glucose and α-D-fructose by a 1 ⟶ 2 glycosidic bond. Draw the structure of this disaccharide.

7.19 What metabolic defect causes galactosemia?

7.20 What simple treatment prevents most of the ill effects of galactosemia?

7.21 What toxic metabolite accumulates in galactosemia?

7.22 What are the major physiological effects of galactosemia?

7.23 Why are individuals who have type O blood universal donors but not universal recipients?

7.24 Which of the following blood groups are compatible?
a. A and O
b. A and B
c. B and O

7.25 The Incas successfully performed blood transfusions hundreds of years before they were possible in Europe. Provide a plausible explanation for the Incas' success.

7.26 How does penicillin kill bacteria?

7.27 What is the function of cellulose in the plant cell?

7.28 What is the difference between the structure of cellulose and the structure of amylose?

7.29 How does the structure of amylose differ from that of amylopectin and glycogen?

7.30 What is the major physiological purpose of glycogen?

7.31 How does lysozyme kill bacteria?

7.32 Some oligosaccharides are often referred to by their common names. Give the chemical names of the following:
a. milk sugar
b. beet sugar
c. cane sugar

7.33 Draw all of the different possible aldotrioses of molecular formula $C_3H_6O_3$.

7.34 Draw all of the different possible aldopentoses of molecular formula $C_5H_{10}O_5$.

7.35 Draw all of the different possible aldohexoses of molecular formula $C_6H_{12}O_6$.

7.36 Is there any difference between dextrose and D-glucose?

7.37 What is the difference between a starch and a sugar?

Reactions

7.38 Which of the following would give a positive Benedict's Test?
a. sucrose c. β-maltose
b. glycogen d. α-lactose

7.39 Write a balanced chemical equation for the reaction of Tollens' reagent with D-glyceraldehyde. Why does a "silver mirror" form in this reaction?

7.40 Benedict's reagent and Fehling's solution both react with reducing sugars, such as D-glucose. Write the balanced chemical equation for the oxidation of D-glucose with Benedict's reagent.

7.41 Explain why sucrose is not a reducing sugar.

7.42 Draw the hemiacetal that would result from the reaction of D-glucose with methanol.

Stereochemistry

7.43 Define the term D- as it relates to the stereochemistry of carbohydrates.

7.44 How are D- and L-glyceraldehyde related?

7.45 How is the stereochemistry of L-alanine related to that of L-glyceraldehyde?

7.46 Why does cyclization of D-glucose give two isomers, α- and β-D-glucose?

7.47 How is the structure of β-D-galactose related to that of β-D-glucose?

7.48 Draw the structure of the open chain form of D-fructose and show how it cyclizes to form α- and β-D-fructose.

7.49 Explain in your own words the difference between a D-isomer and an L-isomer.

Further Problems

7.50 The structure of D-glucose is provided below. Draw its mirror image.

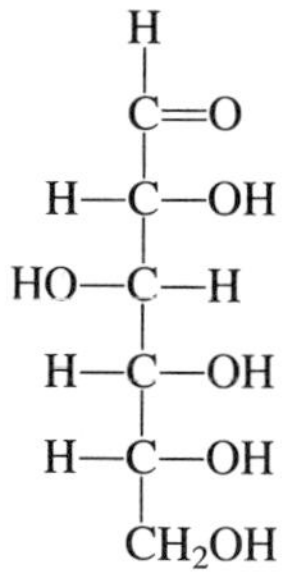

```
      H
      |
      C=O
      |
   H—C—OH
      |
  HO—C—H
      |
   H—C—OH
      |
   H—C—OH
      |
      CH2OH
```

7.51 Find the chiral centers in the following compounds:

a. 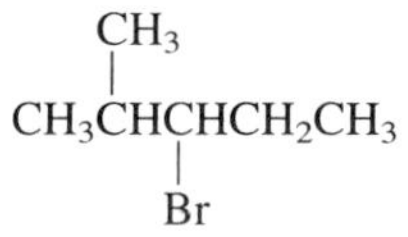
```
     CH3
     |
  CH3CHCHCH2CH3
        |
        Br
```

b. 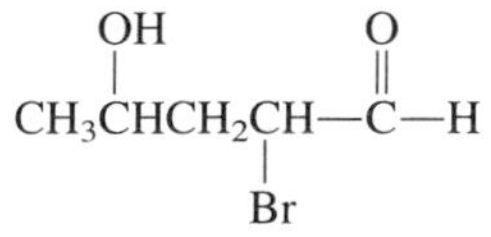
```
     OH          O
     |           ||
  CH3CHCH2CH—C—H
            |
            Br
```

c.
```
        OH
        |
  CH2=CHCHCH3
```

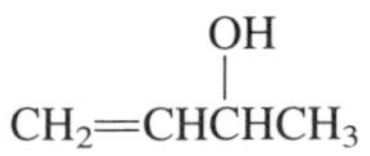

d. 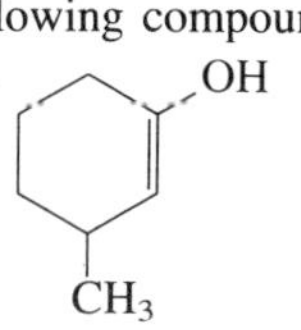 (cyclohexene ring bearing OH and CH_3 substituents)
```
   ___
  /   \\ OH
  |   ||
  \\___/
    |
    CH3
```

e. 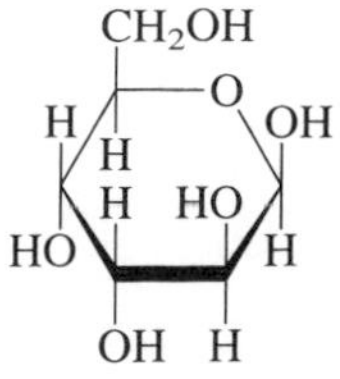 (pyranose ring)
```
        CH2OH
     H /----O
      / H    \\ OH
      \\ H  HO /
   HO  \\____/ H
       OH   H
```

7.52 How many chiral centers are present in each of the following compounds?

a. $CH_3CH(CH_3)CH_2CH_3$

(drawn as: CH_3 over the second carbon; $CH_3CHCH_2CH_3$)

b. $BrCH_2CH(OH)CH_2CH(CH_3)CH_2OH$

(drawn as: OH and CH_3 over the chain; $BrCH_2CHCH_2CHCH_2OH$)

c. $CH_3CH_2CH(CH_3)CH_2$—C(=O)—H

(drawn as: $CH_3CH_2CHCH_2$—C—H, with O above the C and CH_3 below the CH)

d. $CH_3CH(Br)CH(CH_2CH(OH)CH_2CH(OH)CH_3)CH_2$—C(=O)—H

(drawn as: Br above; $CH_3CHCHCH_2$—C—H with O above; below: $CH_2CHCH_2CHCH_3$ with OH and OH below)

e. Cyclohexane ring bearing CH_3 and OH on one carbon, and Br on the opposite carbon.

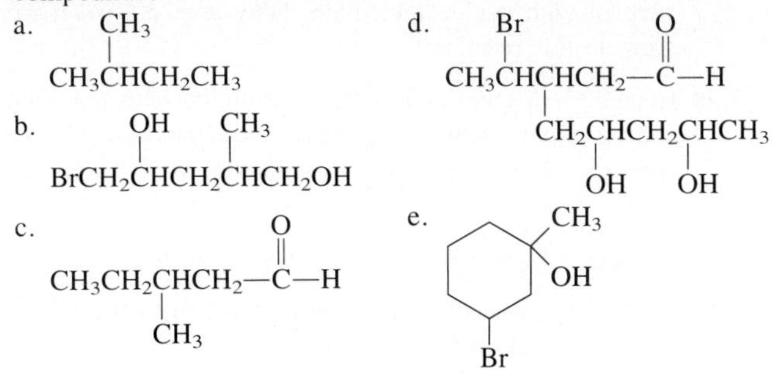

7.53 Aspartame, the sweetener marketed under the trade name NutraSweet, has the following structure. How many chiral centers does the molecule contain?

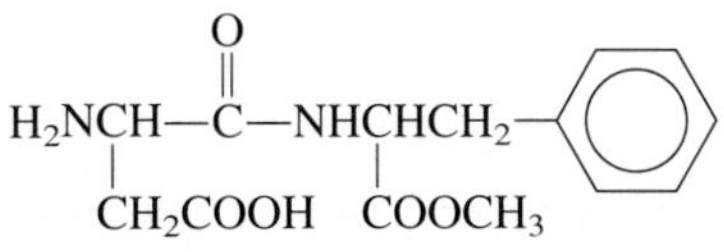

H_2NCH—C(=O)—$NHCHCH_2$—(benzene ring)

with CH_2COOH below the first CH and $COOCH_3$ below the second CH.

7.54 Determine whether each of the following compounds is a D- or an L-sugar:

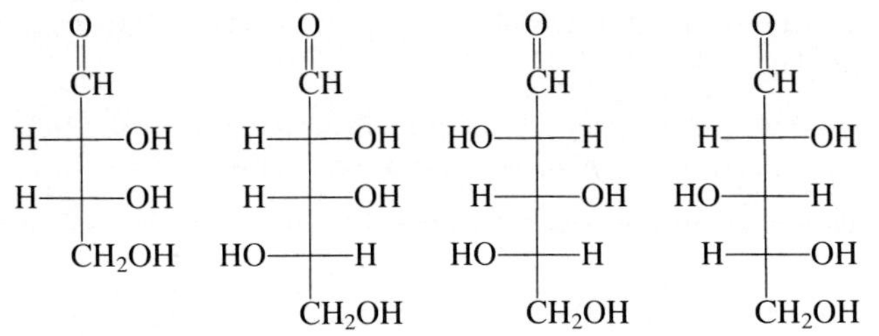

O=CH	O=CH	O=CH	O=CH
H—C—OH	H—C—OH	HO—C—H	H—C—OH
H—C—OH	H—C—OH	H—C—OH	HO—C—H
CH_2OH	HO—C—H	HO—C—H	H—C—OH
	CH_2OH	CH_2OH	CH_2OH

7.55 Draw the mirror image of each of the following compounds:

a.

CH_3
H—C—OH
Br

b.

CH_3
H—C—OH
H—C—OH
CH_2Br

c.

$CH{=}CH_2$
CH_3CH_2—C—H
OH

d.

HC=O
H—C—OH
HO—C—H
CH_2OH

e.

CH_3
H—C—Br
HO—C—H
CH_3

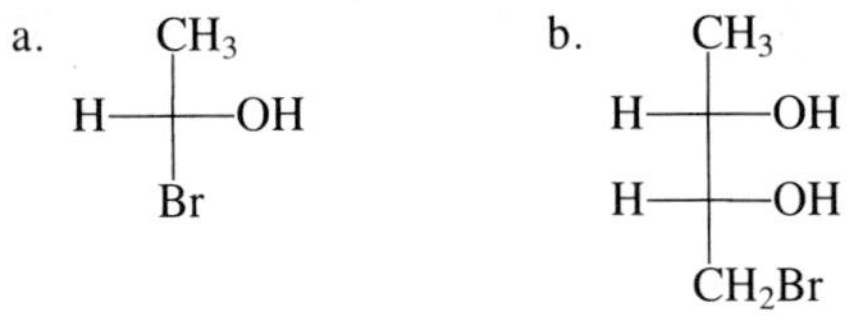

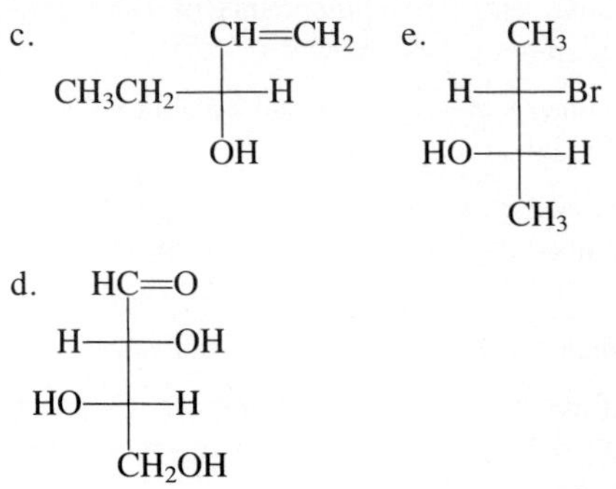

7.56 For the following pairs of compounds, which pairs are identical and which pairs are mirror images?

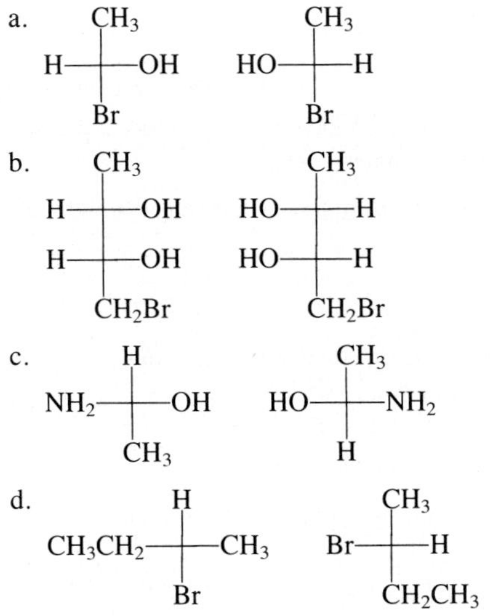

a.

CH_3	CH_3
H—C—OH	HO—C—H
Br	Br

b.

CH_3	CH_3
H—C—OH	HO—C—H
H—C—OH	HO—C—H
CH_2Br	CH_2Br

c.

H	CH_3
NH_2—C—OH	HO—C—NH_2
CH_3	H

d.

H	CH_3
CH_3CH_2—C—CH_3	Br—C—H
Br	CH_2CH_3

7.57 Read the labels on some of the foods in your kitchen and see how many you can find that list one or more carbohydrates among the ingredients in the package. Make a list of these compounds and attempt to classify them according to parent structures (for example, glucoses, fructoses).

CHAPTER 8

Carboxylic Acids and Carboxylic Acid Derivatives

OBJECTIVES

- Draw and name the common carboxylic acids and carboxylic acid derivatives.
- Know the common members of these two families that are of natural, commercial, health, environmental, and industrial interest.
- Be familiar with the physical properties of each of the two families listed above.
- Understand the methods of preparation of carboxylic acids and methods for the preparation of the carboxylic acid derivatives.
- Write equations for the hydrolysis of carboxylic acid derivatives.
- Design simple multistep syntheses involving acyl compounds.
- Be familiar with polymers, their general structure(s), and the way in which their structure contributes to their potential use(s).

OUTLINE

Carboxylic acids have the following general structure:

$$(\text{Ar—})\ \text{R—}\overset{\displaystyle \text{O}}{\overset{\|}{\text{C}}}\text{—OH}$$

Carboxylic acid

They are characterized by the **carboxyl group,** shown in bold in the structure above, which is the functional group characteristic to this family. It may also be written in the condensed form, —COOH. The name ''carboxylic acid'' describes this family of compounds quite well. ''Carboxylic'' is taken from carbonyl and hydroxyl, the two structural units that make up the carboxyl group, while ''acid'' points out one of their more important properties.

In this chapter we will introduce the nomenclature, chemical and physical properties, preparations, and some of the more important reactions of this family. The *esters, acid chlorides,* and *anhydrides* of the *carboxylic acids*—the carboxylic acid derivatives—will also be discussed. **Carboxylic acid derivatives** have the following general structure:

$$(\text{Ar—})\ \mathbf{R—}\overset{\displaystyle \mathbf{O}}{\overset{\|}{\mathbf{C}}}\text{—Z}$$

General structure of carboxylic acid derivatives

The Z-group may be any one of the following:

—Z = —Cl	Acid chloride
= —OR or —OAr	Ester
= $—NH_2$, —NHR or —NHAr, $—NR_2$ or $—NAr_2$	Amide
$= —O—\overset{O}{\overset{\|}{C}}—R$ or $—O—\overset{O}{\overset{\|}{C}}—Ar$	Acid anhydride

In effect, the —OH group of the carboxylic acid is replaced by some other substituent: —X, —OR, or NR_2, for example. Other substituents (not listed above) are also possible, but we will limit our discussion to the structures listed here.

The group shown in bold $[(\text{Ar—})\ \text{R—}\overset{\text{O}}{\overset{\|}{\text{C}}}\text{—}]$ is called the **acyl group.** The acyl group is the functional group of the carboxylic acid derivatives.

The anhydride functional group is central to an understanding of the structure of ATP, adenosine triphosphate, the principal molecule involved in energy storage and generation in the body. Esters are found in many naturally occurring materials including flavors, waxes, and *pheromones,* chemicals involved in communication among various species of insects.

Section 13.1

Section 9.2

The *prostaglandins,* an important class of carboxylic acids involved in the regulation of many biological processes, are described in this chapter and discussed in greater detail in Chapter 9. *Fatty acids,* another important subclass of carboxylic acids, are also discussed in Chapter 9.

8.1 CARBOXYLIC ACIDS AND DERIVATIVES: NOMENCLATURE

Carboxylic acids: I.U.P.A.C. names

The I.U.P.A.C. System for naming acids uses the longest continuous carbon chain, containing the carboxyl group, as the parent compound. The *-e* ending of the parent alkane is dropped and replaced with *-oic acid*. The chain is always numbered beginning with the carboxyl carbon as carbon-1. Substituents are named in the usual manner. For example:

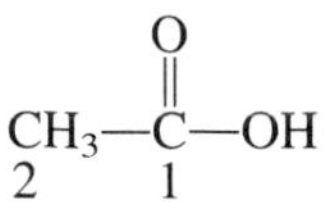

$$\underset{2}{CH_3}-\underset{1}{\overset{O}{\overset{\|}{C}}}-OH \qquad \underset{3\ \ 2}{CH_3CH_2}-\underset{1}{\overset{O}{\overset{\|}{C}}}-OH \qquad \overset{4\ \ 3\ \ 2}{CH_3\underset{CH_3}{\underset{|}{C}}HCH_2}-\overset{O}{\overset{1\|}{C}}-OH$$

Ethanoic acid 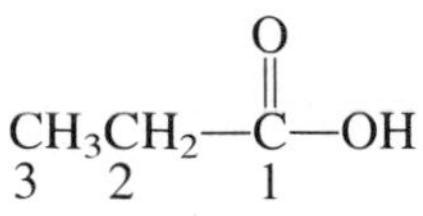 — Propanoic acid 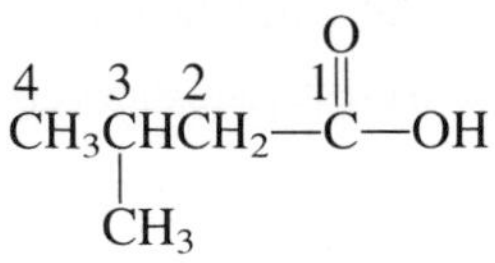 — 3-Methylbutanoic acid

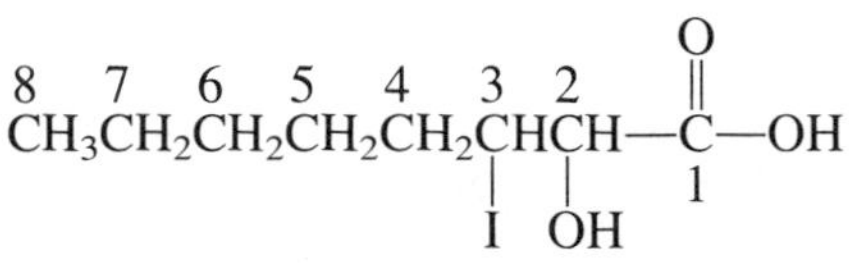

$$\overset{8\ \ 7\ \ 6\ \ 5\ \ 4\ \ 3\ \ 2}{CH_3CH_2CH_2CH_2CH_2\underset{I}{\underset{|}{C}}H\underset{OH}{\underset{|}{C}}H}-\underset{1}{\overset{O}{\overset{\|}{C}}}-OH \qquad HO-\overset{3\ \ 2}{CH_2\underset{CH_3}{\underset{|}{C}}H}-\underset{1}{\overset{O}{\overset{\|}{C}}}-OH$$

2-Hydroxy-3-iodooctanoic acid 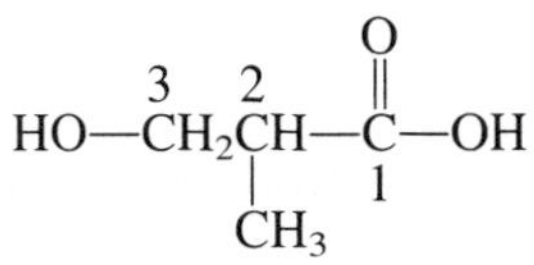 — 3-Hydroxy-2-methylpropanoic acid

Example 8.1

Naming carboxylic acids using the I.U.P.A.C. Nomenclature System.

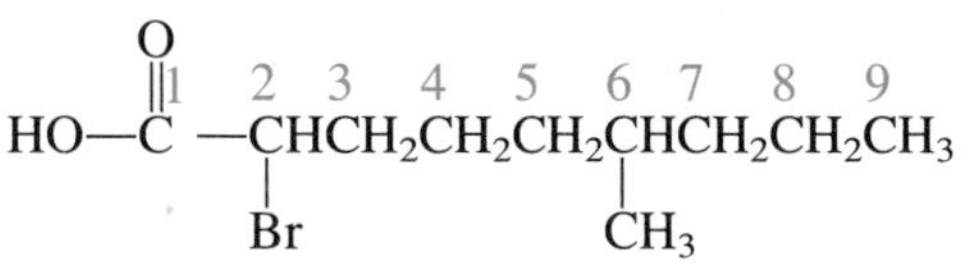

$$HO-\overset{O}{\overset{\|1}{C}}-\overset{2\ \ 3\ \ 4\ \ 5\ \ 6\ \ 7\ \ 8\ \ 9}{\underset{Br}{\underset{|}{C}}HCH_2CH_2CH_2\underset{CH_3}{\underset{|}{C}}HCH_2CH_2CH_3}$$

Parent compound: Nonane (becomes nonanoic acid)

Position of —COOH: Carbon-1 (must be!)

Substituents: 2-Bromo and 6-methyl

Name: 2-Bromo-6-methylnonanoic acid

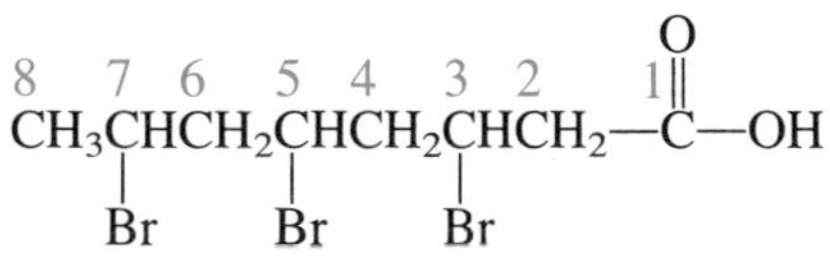

$$\overset{8\ \ 7\ \ 6\ \ 5\ \ 4\ \ 3\ \ 2}{CH_3\underset{Br}{\underset{|}{C}}HCH_2\underset{Br}{\underset{|}{C}}HCH_2\underset{Br}{\underset{|}{C}}HCH_2}-\overset{O}{\overset{1\|}{C}}-OH$$

Parent compound: Octane (becomes octanoic acid)

Position of —COOH: Carbon-1 (must be!)

Substituents: 3,5,7-Tribromo

Name: 3,5,7-Tribromooctanoic acid

Additional examples are provided in the following section and in Table 8.1.

Question 8.1 The carboxylic derivatives of cyclohexane are named by adding the suffix carboxylic acid to the name of the cyclohexane or substituted cyclohexane. That is, compound **I** is named cyclohexanecarboxylic acid. The carboxyl group is defined to be on carbon-1, so no number is necessary. Write the structure for each of the following substituted cyclohexanecarboxylic acids:

$$\text{C}_6\text{H}_{11}\text{—}\overset{\text{O}}{\overset{\|}{\text{C}}}\text{—OH}$$

I

a. 2-bromocyclohexanecarboxylic acid
b. 2,6-dichlorocyclohexanecarboxylic acid
c. 1,4-cyclohexanedicarboxylic acid
d. 4-hydroxycyclohexanecarboxylic acid

Carboxylic acids: common names

As we have seen, the use of common names, rather than systematic, still persists and the common names of several of the carboxylic acids are provided in Table 8.1. Often, these names have evolved from the source of a given compound. This is certainly true of the carboxylic acids. Formic acid, HCOOH, is used by ants as a defensive chemical agent (L: *formica,* ant). Acetic acid, CH_3COOH, is found in vinegar (L: *acetum,* vinegar), and

TABLE 8.1 Physical Properties of Selected Carboxylic Acids

Name*	Structure	m.p. (°C)	b.p. (°C)
Methanoic (formic)	HCOOH	8.4	100.7
Ethanoic (acetic)	CH_3COOH	17.0	118.0
Propanoic (propionic)	CH_3CH_2COOH	−21.0	141.0
Butanoic (butyric)	$CH_3CH_2CH_2COOH$	−6.5	163.5
Pentanoic (valeric)	$CH_3CH_2CH_2CH_2COOH$	−34.5	187.0
Hexanoic (caproic)	$CH_3CH_2CH_2CH_2CH_2COOH$	−1.5	205.0
Octanoic (caprylic)	$CH_3(CH_2)_6COOH$	16.0	237.0
Decanoic (capric)	$CH_3(CH_2)_8COOH$	31.0	270.0
Dodecanoic (lauric)	$CH_3(CH_2)_{10}COOH$	44.0	—
Tetradecanoic (myristic)	$CH_3(CH_2)_{12}COOH$	54.0	—
Hexadecanoic (palmitic)	$CH_3(CH_2)_{14}COOH$	63.0	350.0
Octadecanoic (stearic)	$CH_3(CH_2)_{16}COOH$	70.0	—

*The appendage ''acid'' is omitted for simplification.

butyric acid, $CH_3CH_2CH_2COOH$, is responsible for the stench of rancid butter (L: *butyrum,* butter). The six-, eight-, and ten-carbon acids (caproic, caprylic, and capric acids) are all found in goat fat (L: *caper,* goat). Palmitic acid is found in palm oil, and stearic acid is found in tallow (Gk: *stear,* tallow).

Question 8.2 Write the structure for each of the following compounds, using complete structural formulas:

a. 2-bromopentanoic acid
b. 2-bromo-3-methylbutanoic acid
c. 4,6,8-timethylstearic acid
d. propenoic acid

Question 8.3 Write complete structural formulas for all of the carboxylic acids of molecular formula $C_4H_8O_2$; $C_5H_{10}O_2$.

The aromatic carboxylic acids are usually named, in either system, as derivatives of benzoic acid, though other ''common names'' of substituted benzoic acids (for example, toluic acid, phthalic acid, and so on) may also be used.

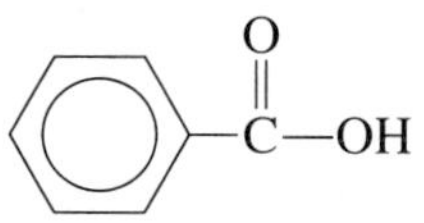

Benzoic acid

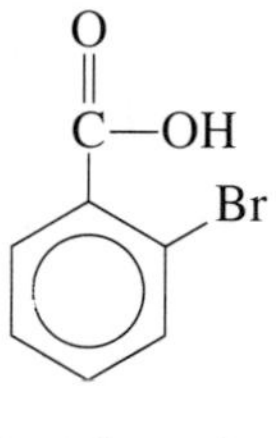

o-Bromobenzoic acid

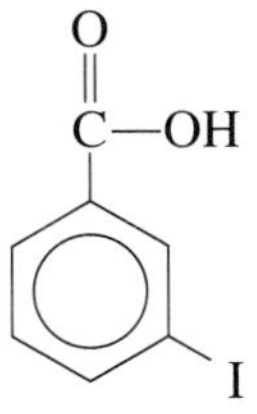

m-Iodobenzoic acid

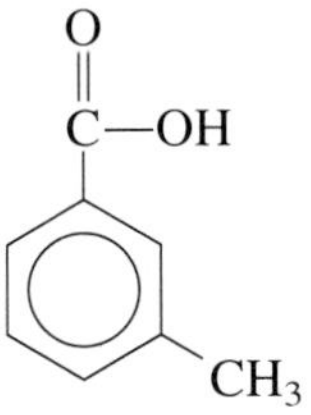

m-Toluic acid (*not m*-methylbenzoic acid)

Often, the phenyl group is simply treated as a substituent, and the name is derived from the appropriate alkanoic acid parent chain. For example:

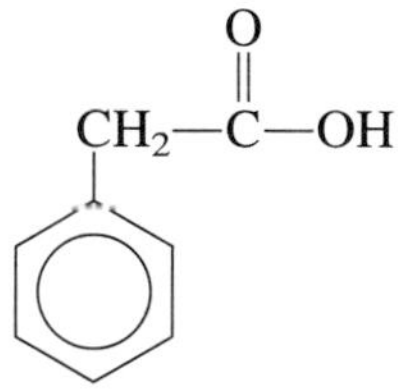

2-Phenylethanoic acid (α-phenylacetic acid)

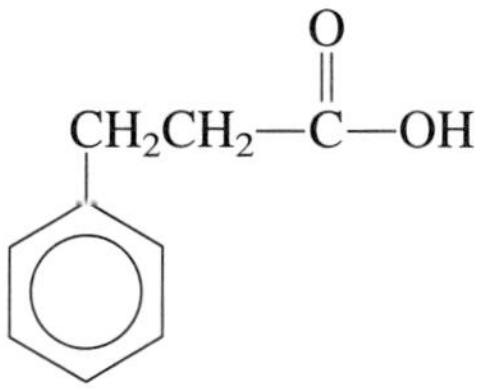

3-Phenylpropanoic acid (β-phenylpropionic acid)

Question 8.4 Draw each of the following:

a. *o*-toluic acid
b. 2-chloro-*p*-toluic acid
c. 2,4,6-tribromobenzoic acid
d. 3-phenylhexanoic acid
e. 2,2,2-triphenylethanoic acid
f. 3-phenylcyclohexanecarboxylic acid

Carboxylic acid derivatives: I.U.P.A.C. and common names

Acid chlorides

Acid chlorides are named by dropping the *-ic acid* ending of the common or the *-oic acid* ending of the I.U.P.A.C. name and replacing it with *-oyl chloride*. For example:

$$CH_3CH_2CH_2{-}\overset{\overset{\Large O}{\|}}{C}{-}OH \xrightarrow{PCl_3} CH_3CH_2CH_2{-}\overset{\overset{\Large O}{\|}}{C}{-}Cl \qquad 8.1$$

Butyr**ic acid** → Butan**oyl chloride**

Also:

2-bromoethan**oic acid** ⟶ 2-bromoethan**oyl chloride**

p-chlorobenz**oic acid** ⟶ *p*-chlorobenz**oyl chloride**

Esters

Esters are formed from the reaction of a carboxylic acid with an alcohol, and both of these families are reflected in the naming of the ester. The *alkyl* or *aryl* portion of the alcohol name is used as a prefix and is followed by the name of the carboxylic acid in which the *-ic acid* (common or I.U.P.A.C.) is replaced with *-ate*. For example:

$$CH_3{-}\overset{\overset{\Large O}{\|}}{C}{-}OH + CH_3{-}OH \xrightarrow{H^+,\ heat} CH_3{-}\overset{\overset{\Large O}{\|}}{C}{-}OCH_3 \qquad 8.2$$

Ethano**ic acid** + **Methanol** → **Methyl** ethano**ate**

Also:

Acet**ic acid** + **ethanol** ⟶ **ethyl** acet**ate**

benzo**ic acid** + **isopropyl** alcohol ⟶ **isopropyl** benzo**ate**

Acid anhydrides

Anhydrides are formed by the combination of two carboxylic acids with the concomitant loss of water. They are named by replacing the *acid* ending in the name with *anhydride*. For example:

$$CH_3{-}\overset{\overset{\Large O}{\|}}{C}{-}O{-}\overset{\overset{\Large O}{\|}}{C}{-}CH_3$$

Ethanoic anhydride (acetic anhydride)

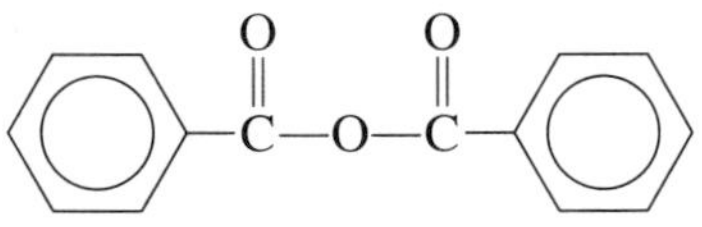

$$C_6H_5{-}\overset{\overset{\Large O}{\|}}{C}{-}O{-}\overset{\overset{\Large O}{\|}}{C}{-}C_6H_5$$

Benzoic anhydride

Also:

butanoic **acid** ⟶ butanoic **anhydride**

hexanoic **acid** ⟶ hexanoic **anhydride**

2-bromopentanoic **acid** ⟶ 2-bromopentanoic **anhydride**

Question 8.5 Write structures for all of the carboxylic acid derivatives named in the preceding section.

8.2 CARBOXYLIC ACIDS AND DERIVATIVES: PHYSICAL PROPERTIES

Carboxylic acids

Because of the presence of both a polar carbonyl group and a polar hydroxyl group, the carboxylic acids are very polar compounds. They boil at higher temperatures than aldehydes or ketones; even alcohols of comparable molecular weight have lower boiling points. For example:

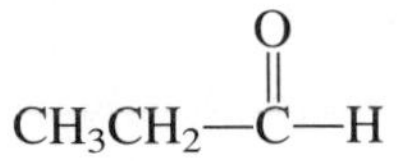

$CH_3CH_2—\overset{O}{\overset{\parallel}{C}}—H$	$CH_3CH_2CH_2—OH$
Propanal (propionaldehyde) mol. wt. 58 g/mol b.p. 49°C	1-Propanol (*n*-propyl alcohol) mol. wt. 60 g/mol b.p. 97.2°C
$CH_3—\overset{O}{\overset{\parallel}{C}}—CH_3$	$CH_3—\overset{O}{\overset{\parallel}{C}}—OH$
Propanone (acetone) mol. wt. 58 g/mol b.p. 56°C	Ethanoic acid (acetic acid) mol. wt. 60 g/mol b.p. 118°C

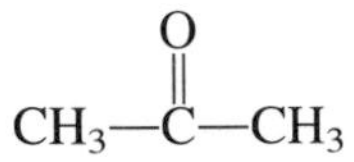

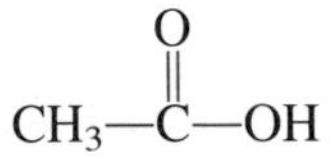

All of the simpler carboxylic acids are reasonably soluble in water (see Figure 8.1). They are similar to alcohols of comparable molecular weight in this regard. Solubility falls off dramatically as the carbon content of the acid increases. For example, butanoic acid is infinitely soluble in water, while pentanoic acid can be solubilized only to the extent of 4 g/100 mL of water.

The lower molecular weight carboxylic acids have sharp, sour tastes and unpleasant aromas. The longer fatty acids, important in biochemistry, are discussed in Chapter 9. As a general rule, carboxylic acids become more "organic," more hydrocarbonlike, as the length of the carbon chain increases.

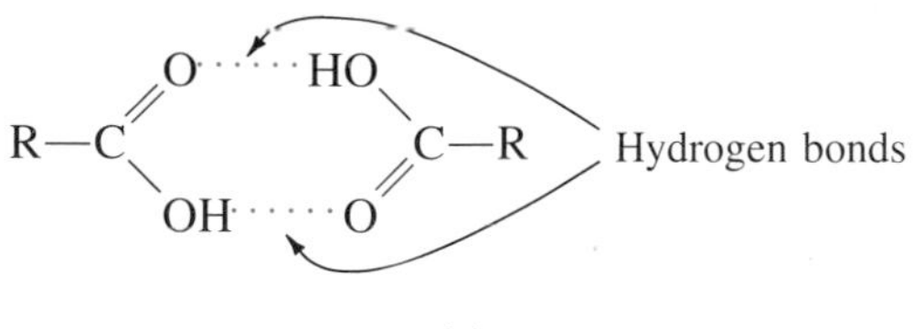

Hydrogen bonds

(a)

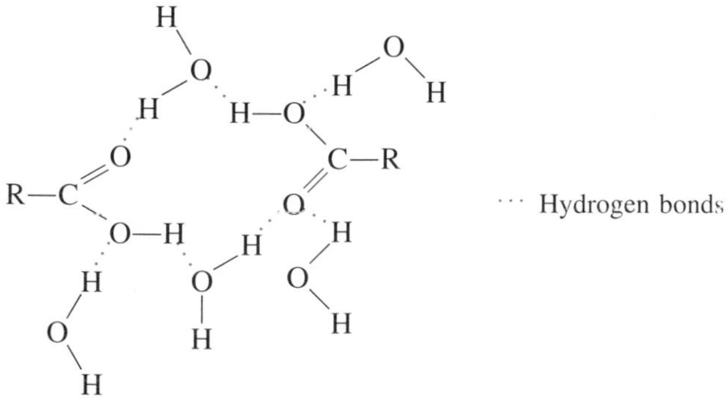

··· Hydrogen bonds

(b)

FIGURE 8.1
Hydrogen bonding (a) in carboxylic acids and (b) between carboxylic acids and water.

Question 8.6 Which member in each of the following pairs has the lower boiling point?

a. hexanoic acid or 3-hexanone
b. 3-hexanone or 3-hexanol
c. 3-hexanol or hexane
d. di-*n*-propyl ether or hexanal
e. hexanal or hexanoic acid

Question 8.7 The functional group is largely responsible for the physical and chemical properties that we observe in the various chemical families. Why would one predict that a carboxylic acid would be more polar and have a higher boiling point than an alcohol of comparable molecular weight?

Carboxylic acid derivatives

Acid chlorides

Acid chlorides are noxious, irritating chemicals. They are slightly polar and boil at approximately the same temperature as the corresponding aldehyde or ketone of comparable molecular weight. They react violently with water and therefore cannot be dissolved in that solvent.

Esters

A Human Perspective: Carboxylic Acid Derivatives of Special Interest

Esters are also mildly polar. Unlike acid chlorides, however, they have pleasant aromas. Many esters are found in natural foodstuffs; banana oil (pentyl ethanoate), pineapples (ethyl butanoate), and raspberries (isobutyl methanoate) are but a few examples. They boil at approximately the same temperature as the carbonyl compound (aldehyde or ketone) of comparable molecular weight. The simpler ones are reasonably soluble and nonreactive in water.

Acid anhydrides

Anhydrides boil at much lower temperatures than the carboxylic acids of corresponding molecular weight. They are also less soluble in water than the corresponding acids and most often react with water, a reaction that is discussed in the next section.

8.3 CARBOXYLIC ACIDS: WHERE DO THEY COME FROM?

Industrial and natural sources

Many carboxylic acids occur in nature. Several natural sources were presented in Section 8.1. The fatty acids can be isolated from a variety of sources, including palm and coconut oil, butter, milk, lard, and tallow (beef fat). Many more complex carboxylic acids are also found in a variety of foodstuffs. Several examples follow:

$$\begin{array}{ccc}
\begin{array}{r}
CH_2\text{—}COOH \\
| \\
HO\text{—}C\text{—}COOH \\
| \\
CH_2\text{—}COOH
\end{array}
&
\begin{array}{c}
COOH \\
| \\
H\text{—}C\text{—}OH \\
| \\
CH_3
\end{array}
&
\begin{array}{c}
COOH \\
| \\
HO\text{—}C\text{—}H \\
| \\
HO\text{—}C\text{—}H \\
| \\
COOH
\end{array}
\end{array}$$

Citric acid	Lactic acid	Tartaric acid
Citrus fruit	*Spoiled milk*	*Grape juice*

Two other important classes of naturally occurring carboxylic acids are the prostaglandins and the lipids. They are discussed in Chapter 9.

Several of the simpler carboxylic acids are prepared on a commercial scale. For example, ethanoic (acetic) acid, CH_3COOH, is found in vinegar and a variety of alcoholic beverages (produced by the enzymatic **oxidation** of ethanol). It is produced commercially by the oxidation of either ethanol or ethanal as shown here:

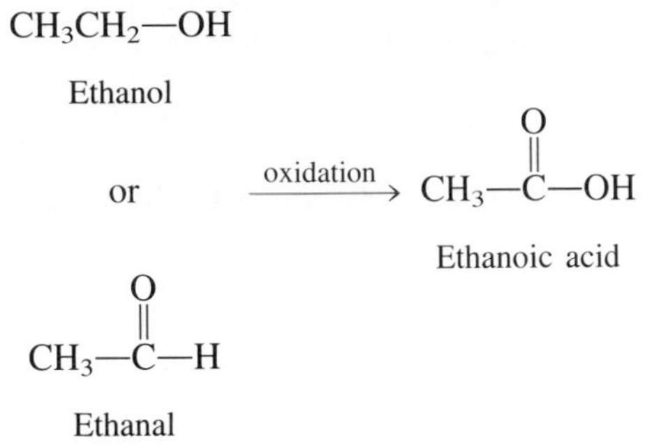

$$\begin{array}{c} CH_3CH_2\text{—}OH \\ \text{Ethanol} \\ \text{or} \\ CH_3\text{—}\overset{\overset{\displaystyle O}{\|}}{C}\text{—}H \\ \text{Ethanal} \end{array} \xrightarrow{\text{oxidation}} \underset{\text{Ethanoic acid}}{CH_3\text{—}\overset{\overset{\displaystyle O}{\|}}{C}\text{—}OH} \qquad 8.3$$

A variety of oxidizing agents, including oxygen, can be used in these reactions. Catalysts are often required to provide acceptable yields. Other simple carboxylic acids may be prepared by using a similar reaction via oxidation of the appropriate alcohol or aldehyde.

Laboratory sources

Carboxylic acids are prepared in the laboratory by the oxidation of aldehydes or (primary) alcohols. Most common oxidizing agents, such as potassium permanganate, can be used. The general reaction is as follows:

These reactions were discussed in Sections 5.5 and 6.4, respectively.

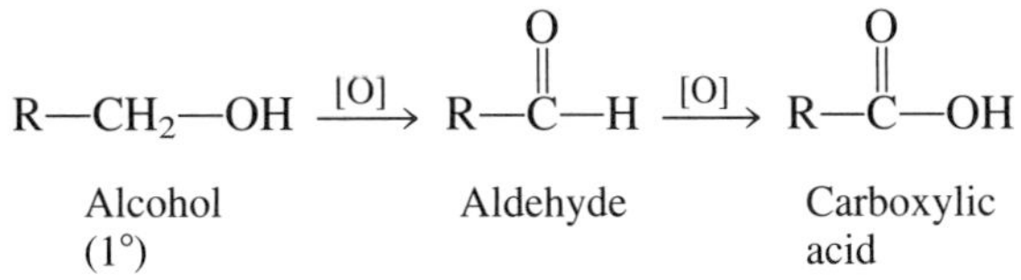

$$\underset{\substack{\text{Alcohol}\\(1°)}}{R\text{—}CH_2\text{—}OH} \xrightarrow{[O]} \underset{\text{Aldehyde}}{R\text{—}\overset{\overset{\displaystyle O}{\|}}{C}\text{—}H} \xrightarrow{[O]} \underset{\substack{\text{Carboxylic}\\\text{acid}}}{R\text{—}\overset{\overset{\displaystyle O}{\|}}{C}\text{—}OH} \qquad 8.4$$

For example:

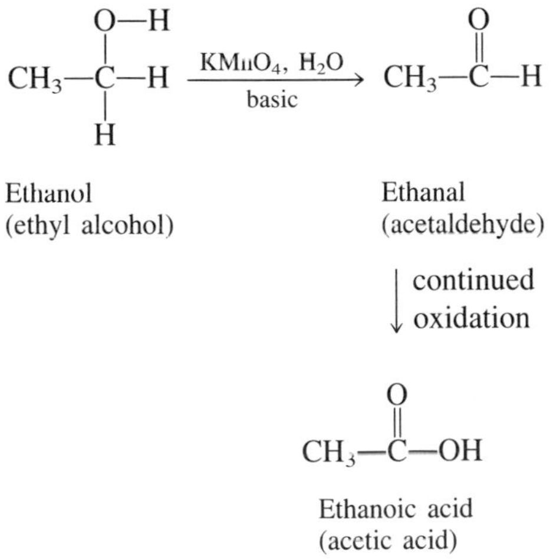

$$\underset{\substack{\text{Ethanol}\\\text{(ethyl alcohol)}}}{CH_3\text{—}\overset{\overset{\displaystyle O\text{—}H}{|}}{\underset{\underset{\displaystyle H}{|}}{C}}\text{—}H} \xrightarrow[\text{basic}]{KMnO_4,\ H_2O} \underset{\substack{\text{Ethanal}\\\text{(acetaldehyde)}}}{CH_3\text{—}\overset{\overset{\displaystyle O}{\|}}{C}\text{—}H} \qquad 8.5$$

$$\Big\downarrow \text{continued oxidation}$$

$$\underset{\substack{\text{Ethanoic acid}\\\text{(acetic acid)}}}{CH_3\text{—}\overset{\overset{\displaystyle O}{\|}}{C}\text{—}OH}$$

$$\underset{\substack{\text{2-Propanol}\\\text{(isopropyl alcohol)}}}{CH_3\text{—}\overset{\overset{\displaystyle O\text{—}H}{|}}{\underset{\underset{\displaystyle H}{|}}{C}}\text{—}CH_3} \xrightarrow[H_2SO_4]{K_2Cr_2O_7,} \underset{\substack{\text{Propanone}\\\text{(acetone)}}}{CH_3\text{—}\overset{\overset{\displaystyle O}{\|}}{C}\text{—}CH_3} \qquad 8.6$$

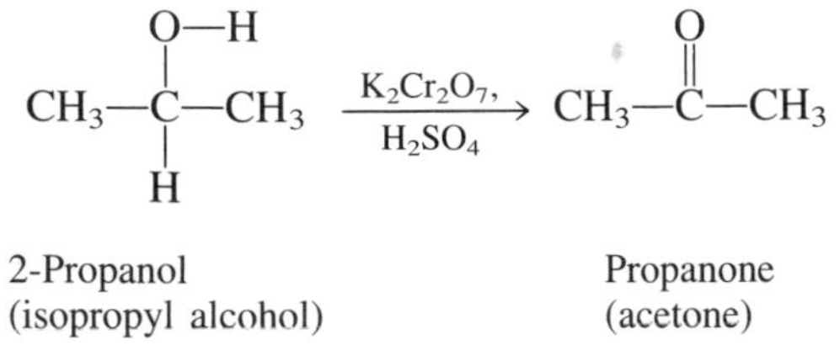

Example 8.2

Designing the synthesis of a carboxylic acid derivative—an ester.

The ease with which alcohols are oxidized to aldehydes, ketones, or carboxylic acids (depending on the alcohol that you start with and the conditions that you employ), coupled with the ready availability of alcohols, provides the pathway necessary to many successful synthetic transformations. For example, let us develop a method for synthesizing ethyl propanoate, using any inorganic reagent you wish but limiting yourself to organic alcohols that contain three or fewer carbon atoms:

$$CH_3CH_2{-}\overset{\overset{\displaystyle O}{\|}}{C}{-}O{-}CH_2CH_3$$

Ethyl propanoate
(ethyl propionate)

Section 8.4 Ethyl propanoate can be made from propanoic acid and ethanol:

$$CH_3CH_2{-}\overset{\overset{\displaystyle O}{\|}}{C}{-}OH + HO{-}CH_2CH_3 \xrightarrow{H^+} CH_3CH_2{-}\overset{\overset{\displaystyle O}{\|}}{C}{-}O{-}CH_2CH_3 \qquad 8.7$$

Propanoic acid (propionic acid) | Ethanol (ethyl alcohol) | Ethyl propanoate (ethyl propionate)

Ethanol is a two-carbon alcohol that is an allowed starting material, but propanoic acid is not. Can we now make propanoic acid from an alcohol of three or fewer carbons? Yes!

$$CH_3CH_2CH_2{-}OH \xrightarrow[HO^-]{KMnO_4,} CH_3CH_2{-}\overset{\overset{\displaystyle O}{\|}}{C}{-}OH \qquad 8.8$$

Propanol (*n*-propyl alcohol) | Propanoic acid (propionic acid)

Propanol is a three-carbon alcohol, an allowed starting material. The synthesis is now complete. By beginning with ethanol and propanol, ethyl propanoate can be synthesized easily.

Example 8.3

Designing a synthesis of more complex carboxylic acid derivatives.

Using the same limitations with regard to starting materials as we applied in the previous example, show how you could synthesize the diethyl acetal of propanal, propanal diethyl acetal:

$$CH_3CH_2{-}\underset{\underset{\displaystyle H}{|}}{\overset{\overset{\displaystyle O{-}CH_2CH_3}{|}}{C}}{-}O{-}CH_2CH_3$$

Propanal diethyl acetal
(common name)

Section 6.4 The diethyl acetal of propanal can be easily synthesized from propanal by treatment with an excess of ethanol in the presence of a trace amount of acid. The question therefore becomes: How can we make propanal from an alcohol of three or fewer carbons? The answer is to oxidize 1-propanol to propanal. The entire reaction sequence is shown below:

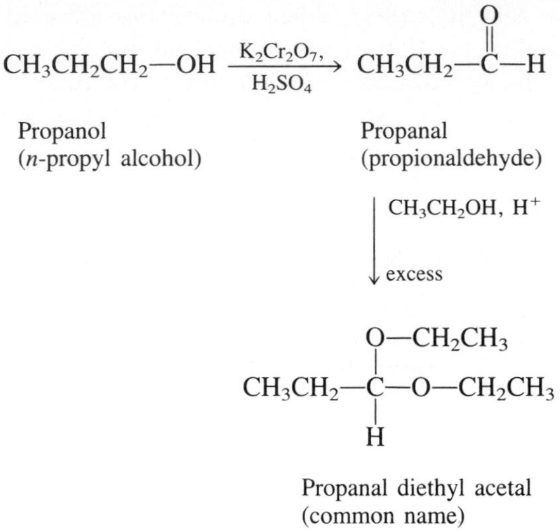

$$CH_3CH_2CH_2—OH \xrightarrow[H_2SO_4]{K_2Cr_2O_7,} CH_3CH_2—\overset{\overset{\displaystyle O}{\|}}{C}—H \quad 8.9$$

Propanol (*n*-propyl alcohol) — Propanal (propionaldehyde)

$\downarrow$ CH_3CH_2OH, H^+ excess

$$CH_3CH_2—\underset{\underset{\displaystyle H}{|}}{\overset{\overset{\displaystyle O—CH_2CH_3}{|}}{C}}—O—CH_2CH_3$$

Propanal diethyl acetal (common name)

Question 8.8 Write the formula of the organic product obtained through each of the following oxidations:

a. $CH_3CH_2CH_2—OH \xrightarrow{K_2Cr_2O_7,\ H^+}$

b. $CH_3\underset{\underset{\displaystyle OH}{|}}{C}HCH_2CH_2CH_3 \xrightarrow{K_2Cr_2O_7,\ H^+}$

c. 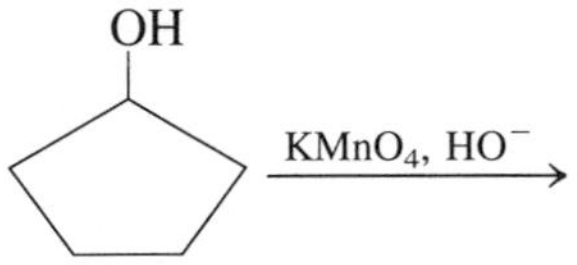 Cyclopentanol (OH on cyclopentane ring) $\xrightarrow{KMnO_4,\ HO^-}$

d. $(CH_3)_2CH_2\underset{\underset{\displaystyle OH}{|}}{C}(CH_3)_2 \xrightarrow{K_2Cr_2O_7,\ H^+}$

8.4 REACTIONS OF CARBOXYLIC ACIDS

We will discuss two of the more important reactions of the carboxylic acids: their reactions as acids and the formation of carboxylic acid derivatives.

Acidity

The carboxylic acids behave like weak acids. They are proton donors. They react with bases to form the corresponding carboxylate salts. In the example shown here, water is a weak base—the proton acceptor.

$$R—\overset{\overset{\displaystyle O}{\|}}{C}—OH + H—O—H \rightleftharpoons R—\overset{\overset{\displaystyle O}{\|}}{C}—O^- + H—\overset{\overset{\displaystyle H}{|}}{\underset{+}{O}}—H \quad 8.10$$

Carboxylic acid — Water (weak base) — Carboxylate ion — Hydronium ion

Carboxylic acids are weakly dissociated in solution. Only a small percentage of hydronium ions results. The majority of the acid remains in solution in the unionized form. Typically, less than 5% of the acid is ionized (approximately five carboxylate ions to every 95 carboxylic acid molecules). When strong bases are used, however, the acid protons are abstracted by the HO^- to form water and the carboxylate ion. The acid is essentially 100% in the ionized, carboxylate ion form. The equilibrium is shifted to the right, owing to removal of OH^-, which is an illustration of LeChatelier's Principle. For example:

Note that the salt of a carboxylic acid is named by replacing the *-ic acid* ending with *-ate* (acetic acid becomes acetate). This name is preceded by the name of the appropriate cation (for example, sodium).

$$CH_3—\overset{\overset{\displaystyle O}{\|}}{C}—OH + NaOH \rightleftharpoons CH_3—\overset{\overset{\displaystyle O}{\|}}{C}—O^-Na^+ + H_2O \qquad 8.11$$

Acetic acid; Sodium hydroxide *Strong base*; Sodium acetate; Water

$$C_6H_5—\overset{\overset{\displaystyle O}{\|}}{C}—OH + NaHCO_3 \rightleftharpoons \qquad 8.12$$

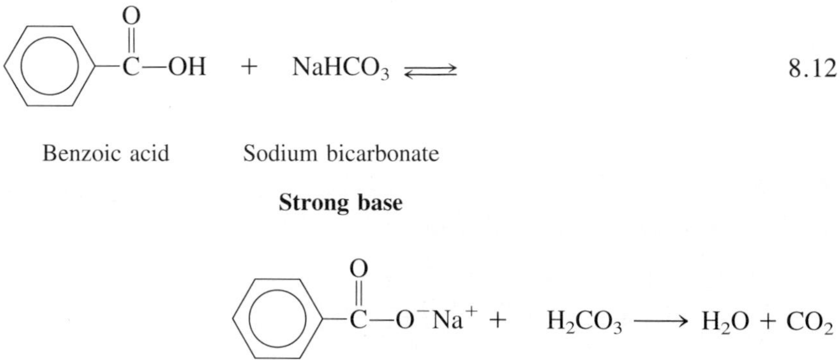

Benzoic acid; Sodium bicarbonate **Strong base**

$$C_6H_5—\overset{\overset{\displaystyle O}{\|}}{C}—O^-Na^+ + H_2CO_3 \longrightarrow H_2O + CO_2$$

Sodium benzoate; Carbonic acid

Question 8.9 Complete each of the following reactions by supplying the missing product(s):

a. $CH_3CH_2COOH + KOH \longrightarrow ?$

b. $CH_3CH_2CH_2COOH + Ba(OH)_2 \longrightarrow ?$

c. 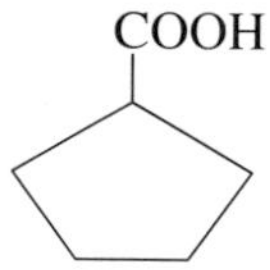 (cyclopentane–COOH) $+ Na_2CO_3 \longrightarrow ?$

d. benzoic acid + sodium hydroxide $\longrightarrow ?$

Salts are ionic substances and hence are quite soluble in water. The long-chain acid salts (fatty acid salts) are good **soaps.** The lore of soap making dates back to the times of the ancient Romans. Originally, it involved the hydrolysis of a fat or oil with an aqueous alkali (base), which was usually a mixture of potassium carbonate and potassium hydroxide obtained from the leaching of wood ashes with water.

Today, soap is still obtained from animal or vegetable fats or oils. The carbon content of the fatty acid salts governs the solubility of a soap. The lower molecular weight acid salts (up to C_{12}) have greater solubility in water and give a lather containing large bubbles. The higher molecular weight acid salts (C_{14}–C_{20}) are much less soluble in water and give a lather with fine bubbles. The potassium salts of the acids are more soluble in water than the corresponding sodium salts.

$$\begin{array}{l} CH_2-O-\overset{O}{\overset{\|}{C}}-R \\ | \\ CH-O-\overset{O}{\overset{\|}{C}}-R' \\ | \\ CH_2-O-\overset{O}{\overset{\|}{C}}-R'' \end{array} \xrightarrow[H_2O,\ heat]{M^{\oplus}:\ddot{O}H^{\ominus}} \begin{array}{l} CH_2-OH \\ | \\ CH-OH \\ | \\ CH_2-OH \end{array} + R-\overset{O}{\overset{\|}{C}}-\ddot{O}:^{\ominus}M^{\oplus} + R'-\overset{O}{\overset{\|}{C}}-\ddot{O}:^{\ominus}M^{\oplus} + R''-\overset{O}{\overset{\|}{C}}-\ddot{O}:^{\ominus}M^{\oplus}$$

Fat or oil (triglyceride) — Glycerol — Soap (Mixture of carboxylic acid salts)

where $M^{\oplus} = Na^{\oplus}$ or $K^{\oplus}$

The question of how soap works can be answered by considering the functional groups in soap and how they interact with oil and water. The long, continuous side chains of carbon atoms in a soap molecule resemble an alkane, and they dissolve other nonpolar compounds such as oils and greases. The adage "like dissolves like" applies here. The large nonpolar hydrocarbon part of the molecule is described as *hydrophobic,* which means "water-fearing." The highly polar carboxylate end of the molecule is called *hydrophilic,* which means "water-loving." When soap is dissolved in water, the carboxylate end actually dissolves. The hydrocarbon part is repelled by the water molecules, so a thin film of soap is formed on the surface of the aqueous layer, with the hydrocarbon chains protruding outward. This greatly lowers the surface tension of the water. When soap solution comes in contact with oil or grease, the hydrocarbon part dissolves in the oil or grease, but the polar carboxylate group remains dissolved in water. When particles of oil or grease are surrounded by soap molecules, the resulting "units" formed are called *micelles.* A simplified view of this phenomenon is shown in Figure 8.2. Keep in mind that this view is highly simplified because it does not show hydrogen bonding and solvation. These micelles repel one another because they are surrounded on the surface by the negatively charged carboxylate ions. Mechanical action (for example, scrubbing or tumbling in a washing machine) causes oil or grease to be broken into small droplets so that relatively small micelles are formed with soap solution. Careful examination of this solution shows that it is an *emulsion* containing suspended micelles.

FIGURE 8.2
Simplified view of the action of a soap on an oil in aqueous solution. The wiggly line represents the long, continuous carbon chain of each soap molecule.

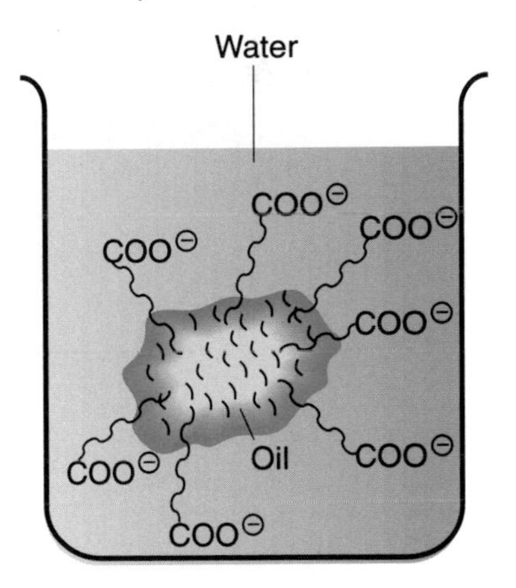

Carboxylic acid derivatives

Acid chlorides

Acid chlorides are prepared from the corresponding carboxylic acid by reaction with a variety of reagents: PCl_3, PCl_5, or $SOCl_2$. Examples are provided below:

An emulsion is a suspension of very fine droplets of an oil in water.

$$CH_3-\overset{O}{\overset{\|}{C}}-OH \xrightarrow[\text{Phosphorus trichloride}]{PCl_3} CH_3-\overset{O}{\overset{\|}{C}}-Cl + \text{inorganic products} \qquad 8.13$$

Ethanoic acid (acetic acid) — Ethanoyl chloride (acetyl chloride)

$$C_6H_5-\overset{O}{\overset{\|}{C}}-OH + SOCl_2 \longrightarrow C_6H_5-\overset{O}{\overset{\|}{C}}-Cl + \text{inorganic products} \qquad 8.14$$

Benzoic acid — Sulfonyl chloride — Benzoyl chloride

Acid chlorides are noxious, irritating chemicals and must be handled with great care. They have little commercial value other than their utility in the synthesis of esters and amides, two of the other carboxylic acid derivatives.

Esters

Direct conversion of a carboxylic acid to an ester is effected by heating the carboxylic acid and the appropriate alcohol with a trace of acid (H^+) (for example, reaction 8.15). Alternatively, the acid may first be converted to an acid chloride. Subsequent treatment of the acid chloride with the alcohol of choice gives the ester directly. Both methods work well.

If esters are prepared directly from the corresponding carboxylic acid, the reaction is acid catalyzed and accompanied by the loss of water. For example:

$$CH_3CH_2-\overset{\overset{\displaystyle O}{\|}}{C}-OH \xrightarrow[H^+,\ \text{heat}]{CH_3-OH} CH_3CH_2-\overset{\overset{\displaystyle O}{\|}}{C}-OCH_3 + H_2O \qquad 8.15$$

Propanoic acid (propionic acid) → Methyl propanoate (methyl propionate)

When the acid chloride is reacted with a nucleophilic species, such as an alcohol or alkoxide ion, the nucleophile displaces the chloride of the acid chloride in a substitution reaction; the ester results.

$$CH_3CH_2CH_2-\overset{\overset{\displaystyle O}{\|}}{C}-OH \xrightarrow{SOCl_2} CH_3CH_2CH_2-\overset{\overset{\displaystyle O}{\|}}{C}-Cl \xrightarrow{CH_3CH_2OH} CH_3CH_2CH_2-\overset{\overset{\displaystyle O}{\|}}{C}-OCH_2CH_3 + HCl \qquad 8.16$$

Butanoic acid (butyric acid) → Butanoyl chloride (butyryl chloride) → Ethyl butanoate (ethyl butyrate)

Acid anhydrides

Acid anhydrides are also prepared in a manner similar to esters. An acid chloride is reacted with a nucleophilic species, in this case the carboxylate ion:

$$R-\overset{\overset{\displaystyle O}{\|}}{C}-Cl \xrightarrow{R'-\overset{\overset{\displaystyle O}{\|}}{C}-O^-} R-\overset{\overset{\displaystyle O}{\|}}{C}-O-\overset{\overset{\displaystyle O}{\|}}{C}-R' + H_2O \qquad 8.17$$

Acid halide; Carboxylate ion; Acid anhydride; Water

The name for this family is quite fitting. The anhydride of the acid is literally two moles of the acid combined with the concomitant loss of water; the word "anhydride" means "without water."

$$R-\overset{\overset{\displaystyle O}{\|}}{C}-O-H \quad H-O-\overset{\overset{\displaystyle O}{\|}}{C}-R \longrightarrow H-OH + R-\overset{\overset{\displaystyle O}{\|}}{C}-O-\overset{\overset{\displaystyle O}{\|}}{C}-R$$

Specific examples follow:

$$CH_3-\overset{\overset{O}{\|}}{C}-OH \xrightarrow{SOCl_2} CH_3-\overset{\overset{O}{\|}}{C}-Cl \xrightarrow{CH_3-\overset{\overset{O}{\|}}{C}-O^-} CH_3-\overset{\overset{O}{\|}}{C}-O-\overset{\overset{O}{\|}}{C}-CH_3 \quad (8.18)$$

Ethanoic acid (acetic acid) → Ethanoyl chloride (acetyl chloride) → Ethanoic anhydride (acetic anhydride)

$$CH_3CH_2CH_2-\overset{\overset{O}{\|}}{C}-OH \xrightarrow{SOCl_2} CH_3CH_2CH_2-\overset{\overset{O}{\|}}{C}-Cl \xrightarrow{CH_3CH_2CH_2-\overset{\overset{O}{\|}}{C}-O^-} CH_3CH_2CH_2-\overset{\overset{O}{\|}}{C}-O-\overset{\overset{O}{\|}}{C}-CH_2CH_2CH_3 + H_2O \quad (8.19)$$

Butanoic acid (butyric acid) → Butanoyl chloride (butyryl chloride) → Butanoic anhydride (butyric anhydride)

Question 8.10 Supply the missing reagents (indicated by the question mark) necessary to complete each of the following transformations:

a. $CH_3-\overset{\overset{O}{\|}}{C}-OH \xrightarrow{?} CH_3-\overset{\overset{O}{\|}}{C}-Cl$

b. cyclopentyl$-\overset{\overset{O}{\|}}{C}-OH \xrightarrow{?}$ cyclopentyl$-\overset{\overset{O}{\|}}{C}-OCH(CH_3)_2$

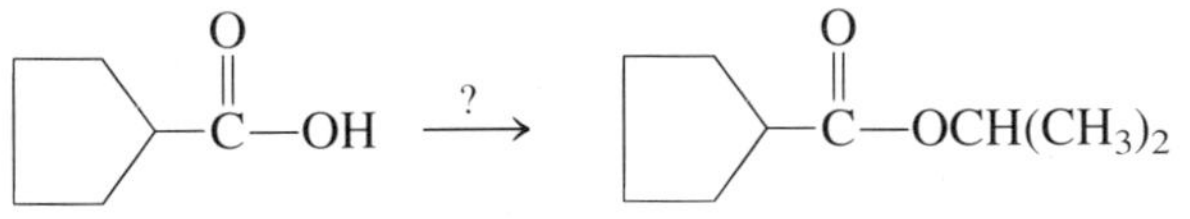

c. phenyl$-\overset{\overset{O}{\|}}{C}-OH \xrightarrow{?}$ cyclohexyl$-\overset{\overset{O}{\|}}{C}-O-\overset{\overset{O}{\|}}{C}-$phenyl

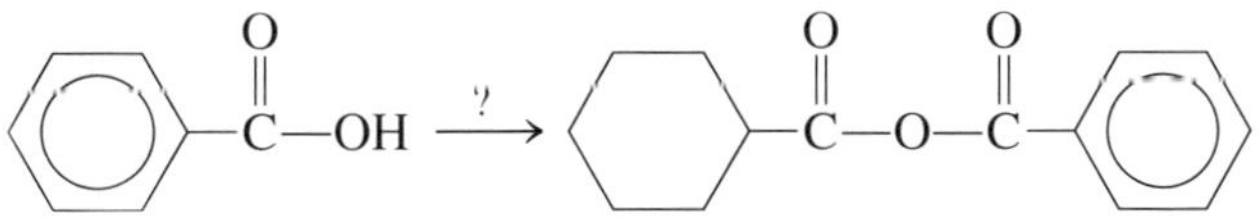

Question 8.11 Supply the missing reagents (indicated by the question mark) necessary to complete each of the following transformations. Some of the reactions may require more than one step to complete.

a. $CH_3CH_2CH_2CH_2-OH \xrightarrow{?} CH_3CH_2CH_2-\overset{\overset{O}{\|}}{C}-Cl$

b. $CH_3CH_2-OH \xrightarrow{?} CH_3-\overset{\overset{O}{\|}}{C}-OCH_2CH_2CH_2CH_3$

c. ethanol $\xrightarrow{?}$ ethanoic anhydride

A HUMAN PERSPECTIVE

Phosphoric Acid Esters and Anhydrides

The reaction between an alcohol and a phosphoric acid produces a phosphate ester. The steps are summarized here:

$$R{-}OH + HO{-}\overset{\overset{\displaystyle O}{\|}}{\underset{\underset{\displaystyle OH}{|}}{P}}{-}OH \longrightarrow R{-}O{-}\overset{\overset{\displaystyle O}{\|}}{\underset{\underset{\displaystyle OH}{|}}{P}}{-}OH + H_2O \qquad 8.27$$

Alcohol Phosphoric acid Phosphate ester (Phosphate ester bond) Water

Some of the most important phosphate esters come from natural products that have various phosphate bonds. One example is **adenosine triphosphate (ATP).** The structure of ATP is shown below. The phosphate ester functional group is shown in bold. Note also that ATP contains two phosphoric anhydride bonds; these are shown in color. ATP is the body's main energy storehouse and, in fact, is the universal energy storage form for all living organisms. The energy is made available through hydrolysis of the phosphoric anhydride bonds. This is an exothermic process; energy is given off. It is estimated that we synthesize and break down an amount of ATP equivalent to our body weight each day. The hydrolysis of the phosphoric anhydride bond is also shown below. The function and properties of ATP will be discussed in greater detail in Chapter 13.

ATP (Bond cleaved on hydrolysis) $\xrightarrow{H_2O}$ ADP $+ H_3PO_4 +$ *energy*

Nicotinamide adenine dinucleotide (NAD$^+$) is a coenzyme that is similar in structure to ATP. NAD$^+$ is one of the more important coenzymes in the body and plays a central role in the metabolism of carbohydrates. The function of NAD$^+$ in biochemical systems will be discussed in detail in Chapters 13 and 14, and its structure is provided below. Note the two phosphate ester bonds as well as the phosphoric anhydride bonds that are present.

We will be studying these and other molecules of similar structure in future chapters.

Ester bonds to phosphorous

Adenine nucleotide — Nicotinamide nucleotide

Anhydride bonds to phosphorous

Nicotinamide adenine dinucleotide (NAD$^\oplus$)

8.5 CARBOXYLIC ACID DERIVATIVES: HYDROLYSIS

All of the carboxylic acid derivatives undergo a common reaction with water: **hydrolysis.** Acid chlorides react violently with water; acid anhydrides react readily with water. On the other hand, esters and amides must be heated (in water) to effect a similar change. The rate of hydrolysis of anhydrides is increased by the addition of a trace of acid or hydroxide base to the solution. The hydrolysis of esters behaves similarly. In the latter case the presence of a small amount of acid (H^+) or base (HO^-) catalyzes the reaction. Amides are normally hydrolyzed only under acid conditions. Several examples are provided here:

$$CH_3{-}\overset{\displaystyle O}{\overset{\|}{C}}{-}Cl + H_2O \longrightarrow CH_3{-}\overset{\displaystyle O}{\overset{\|}{C}}{-}OH + HCl \qquad 8.20$$

Ethanoyl chloride (acetyl chloride) → Ethanoic acid (acetic acid)

$$C_6H_5{-}\overset{\displaystyle O}{\overset{\|}{C}}{-}OCH_3 + H_2O \xrightarrow[\text{heat}]{H^+,} C_6H_5{-}\overset{\displaystyle O}{\overset{\|}{C}}{-}OH + CH_3{-}OH \qquad 8.21$$

Methyl benzoate → Benzoic acid

$$CH_3CH_2{-}\overset{\displaystyle O}{\overset{\|}{C}}{-}NH_2 + H_2O \xrightarrow{H^+,\ \text{heat}} CH_3CH_2{-}\overset{\displaystyle O}{\overset{\|}{C}}{-}OH + NH_3 \qquad 8.22$$

Propanamide (propionamide) → Propanoic acid (propionic acid)

$$CH_3{-}\overset{\displaystyle O}{\overset{\|}{C}}{-}OCH_2CH_2CH_2CH_3 \xrightarrow{HO^-,\ \text{heat}} CH_3{-}\overset{\displaystyle O}{\overset{\|}{C}}{-}OH + CH_3CH_2CH_2CH_2{-}OH \qquad 8.23$$

Butyl ethanoate (butyl acetate) → Ethanoic acid (acetic acid) + 1-Butanol (*n*-butyl alcohol)

$$CH_3(CH_2)_5{-}\overset{\displaystyle O}{\overset{\|}{C}}{-}NHCH_3 + H_2O \xrightarrow{H^+,\ \text{heat}} CH_3(CH_2)_5{-}\overset{\displaystyle O}{\overset{\|}{C}}{-}OH + NH_2CH_3 \qquad 8.24$$

n-Methylheptanamide → Heptanoic acid + *n*-Methylamine

$$CH_3CH_2{-}\overset{\displaystyle O}{\overset{\|}{C}}{-}O{-}\overset{\displaystyle O}{\overset{\|}{C}}{-}CH_2CH_3 + H_2O \xrightarrow{\text{heat}} 2CH_3CH_2{-}\overset{\displaystyle O}{\overset{\|}{C}}{-}OH \qquad 8.25$$

Propanoic anhydride (propionic anhydride) → Propanoic acid (propionic acid)

The base-catalyzed hydrolysis of an ester (reaction 8.23) is called **saponification.** Under basic conditions it is actually the sodium salt of the carboxylic acid that is produced. The carboxylic acid itself is obtained once the reaction mixture is neutralized with mineral acid (e.g., HCl).

(continues on page 198)

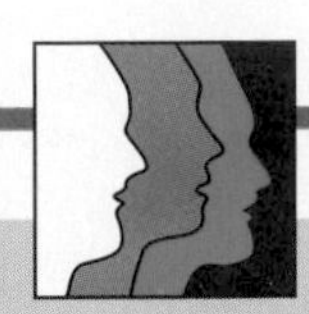

A HUMAN PERSPECTIVE

Carboxylic Acid Derivatives of Special Interest

Polymers

Polymers are *macromolecules* that result from the combination of many smaller molecules, usually in a repeating pattern, a process known as **polymerization,** to give molecules whose molecular weight may be 10,000 g/mol or greater. The small molecules that make up the polymer are called **monomers.** See the accompanying figures.

If **A** represents a molecule (*a monomer*) that you want to polymerize, the polymer would have the following general structure:

Chain continues ∽**A—A—A—A—A—A—A—A**∽ Chain continues

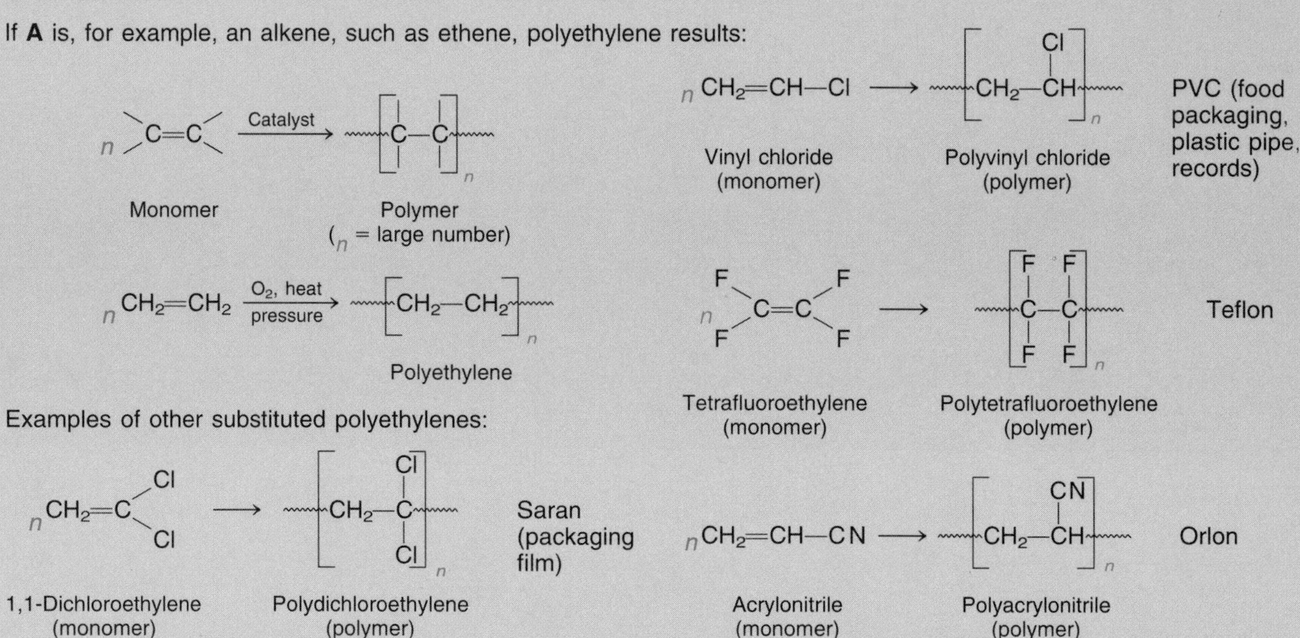

Polymers and the process of polymerization are central to the chemical industry.

Two important classes of synthetic polymers are the *polyesters* and *nylons (polyamides).* The most familiar polyesters are the "double-knit" fabrics.

If **A** and **B** represent two molecules (*monomers*) that you want to copolymerize, the polymer may have the following general structure (other arrangements and mixtures are also possible):

Chain continues ∽**A—B—A—B—A—B—A—B**∽ Chain continues

In the following examples, **A** and **B** are joining to form polyesters and polyamides respectively:

$$\sim CH_2CH_2-O-\left[\overset{O}{\overset{\|}{C}}-C_6H_4-\overset{O}{\overset{\|}{C}}-OCH_2CH_2O\right]_n-\overset{O}{\overset{\|}{C}}-C_6H_4-\overset{O}{\overset{\|}{C}}-OCH_2CH_2\sim$$

Dacron (polymeric ester)

(a) Dacron—a polyester

$$\sim\overset{O}{\overset{\|}{C}}-(CH_2)_4-\overset{O}{\overset{\|}{C}}-\left[NH-(CH_2)_6-NH-\overset{O}{\overset{\|}{C}}-(CH_2)_4-\overset{O}{\overset{\|}{C}}\right]_n-NH-(CH_2)_6\sim$$

Nylon 66 (n = 50–65)

(b) Nylon 66—a polyamide

A common (a) polyester (Dacron) and (b) polyamide Nylon 66.

Polyesters are synthesized by reacting a dicarboxylic acid and a dialcohol (diol). Each of the combining molecules has two reactive functional groups. The polymer chain grows when the "head" of one chain combines with the "tail" of a second chain. Each time a pair of molecules react, using one functional group from each, a new molecule is formed that still contains two reactive functional groups. In theory, the process could continue indefinitely. However, depending on conditions, side reactions, called **termination reactions,** occur, halting the process. See the figure below:

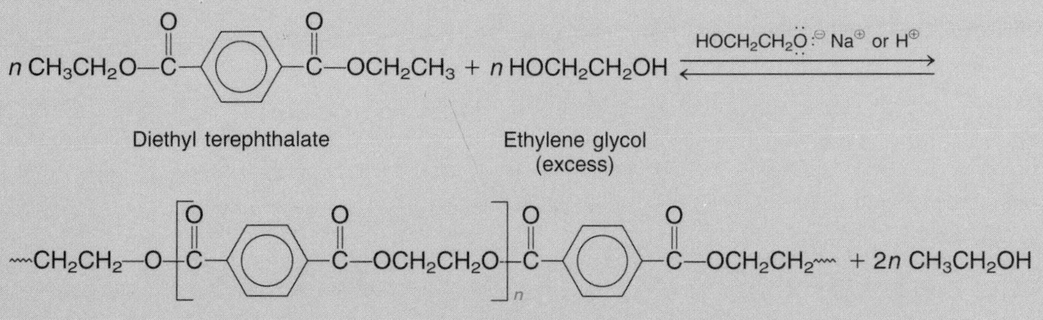

$$n\,CH_3CH_2O{-}\overset{O}{\overset{\|}{C}}{-}C_6H_4{-}\overset{O}{\overset{\|}{C}}{-}OCH_2CH_3 + n\,HOCH_2CH_2OH \xrightleftharpoons{HOCH_2CH_2\ddot{O}:^{\ominus}\,Na^{\oplus}\text{ or }H^{\oplus}}$$

Diethyl terephthalate — Ethylene glycol (excess)

$$\sim CH_2CH_2{-}O{-}\left[\overset{O}{\overset{\|}{C}}{-}C_6H_4{-}\overset{O}{\overset{\|}{C}}{-}OCH_2CH_2O\right]_n{-}\overset{O}{\overset{\|}{C}}{-}C_6H_4{-}\overset{O}{\overset{\|}{C}}{-}OCH_2CH_2\sim + 2n\,CH_3CH_2OH$$

Dacron
(polymeric ester)

The polymerization process for difunctional compounds (shown for Dacron).

Amides, another class of derivatives of the carboxylic acids are discussed in Chapter 10. As with esters, amides (Chapter 10) may be polymerized to form polyamides (nylons). The **polyamides** are formed in a manner analogous to the polyesters. Typically, a diacid reacts with a diamine. The reaction here is analogous to the esterification reaction used in the preparation of esters and polyesters. In this case an amine reacts with an acid; in the former case an alcohol reacts with an acid. (See the figure below.)

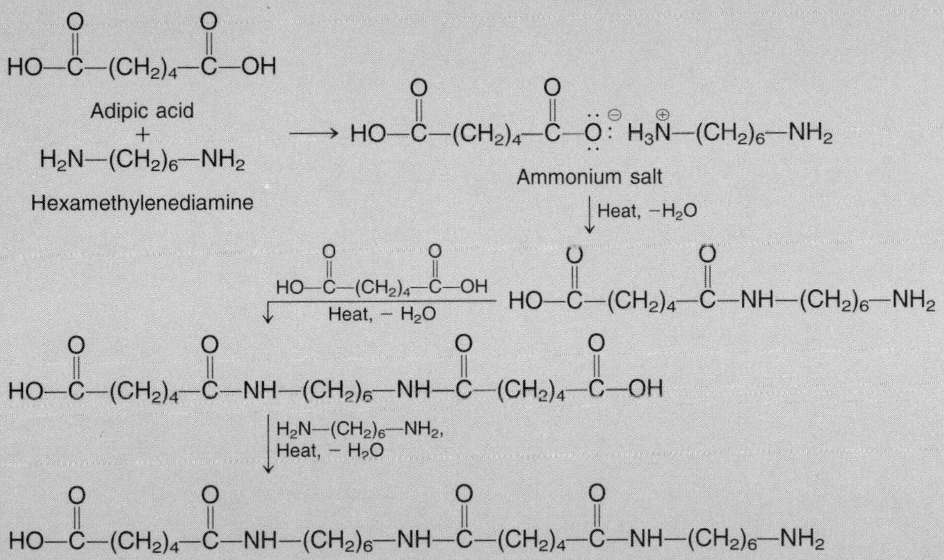

$$HO{-}\overset{O}{\overset{\|}{C}}{-}(CH_2)_4{-}\overset{O}{\overset{\|}{C}}{-}OH \text{ (Adipic acid)} + H_2N{-}(CH_2)_6{-}NH_2 \text{ (Hexamethylenediamine)}$$

$$\longrightarrow HO{-}\overset{O}{\overset{\|}{C}}{-}(CH_2)_4{-}\overset{O}{\overset{\|}{C}}{-}\ddot{\underset{..}{O}}:^{\ominus}\ H_3\overset{\oplus}{N}{-}(CH_2)_6{-}NH_2$$

Ammonium salt

$$\downarrow \text{Heat, } -H_2O$$

$$HO{-}\overset{O}{\overset{\|}{C}}{-}(CH_2)_4{-}\overset{O}{\overset{\|}{C}}{-}NH{-}(CH_2)_6{-}NH_2$$

$$\downarrow HO{-}\overset{O}{\overset{\|}{C}}{-}(CH_2)_4{-}\overset{O}{\overset{\|}{C}}{-}OH,\ \text{Heat, } -H_2O$$

$$HO{-}\overset{O}{\overset{\|}{C}}{-}(CH_2)_4{-}\overset{O}{\overset{\|}{C}}{-}NH{-}(CH_2)_6{-}NH{-}\overset{O}{\overset{\|}{C}}{-}(CH_2)_4{-}\overset{O}{\overset{\|}{C}}{-}OH$$

$$\downarrow H_2N{-}(CH_2)_6{-}NH_2,\ \text{Heat, } -H_2O$$

$$HO{-}\overset{O}{\overset{\|}{C}}{-}(CH_2)_4{-}\overset{O}{\overset{\|}{C}}{-}NH{-}(CH_2)_6{-}NH{-}\overset{O}{\overset{\|}{C}}{-}(CH_2)_4{-}\overset{O}{\overset{\|}{C}}{-}NH{-}(CH_2)_6{-}NH_2$$

These reactions continue because each time a molecule of diacid or diamine is added, the ends of the new molecule contain a —COOH group or an —NH_2 group that reacts further with another molecule of diamine or diacid, respectively. The general structure of the nylon polymer is the following:

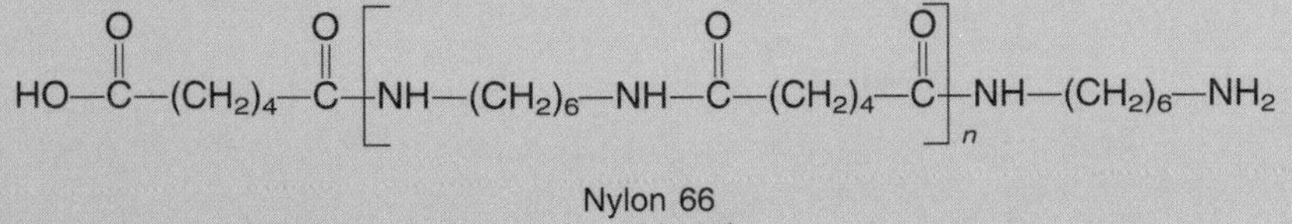

$$HO{-}\overset{O}{\overset{\|}{C}}{-}(CH_2)_4{-}\overset{O}{\overset{\|}{C}}\left[NH{-}(CH_2)_6{-}NH{-}\overset{O}{\overset{\|}{C}}{-}(CH_2)_4{-}\overset{O}{\overset{\|}{C}}\right]_n NH{-}(CH_2)_6{-}NH_2$$

Nylon 66
($n = 50–65$)

The formation of polyamides (shown for Nylon 66).

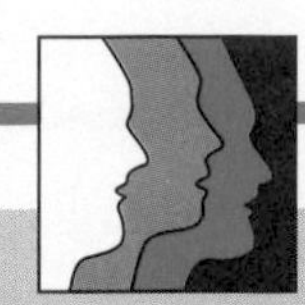

A HUMAN PERSPECTIVE

Carboxylic Acid Derivatives of Special Interest (Continued)

In addition to the synthetic polymers introduced above, there are also several classes of naturally occurring polymers of vital biochemical importance: (1) proteins (polyamino acids), (2) polysaccharides (polycarbohydrates), and (3) nucleic acids (polynucleotides).

Analgesics (pain killers) and antipyretics (fever reducers)

Aspirin (acetylsalicylic acid) is the most widely used drug in the world. Hundreds of millions of dollars are spent annually on this compound. It is used primarily as a pain reliever (**analgesic**) and in the reduction of fever (antipyretic). Aspirin's side effects are a problem for some individuals. Because aspirin inhibits clotting, it is not recommended during pregnancy, nor should it be used by individuals with ulcers. In those instances, two related analgesics, *acetaminophen* (e.g., Tylenol) and *phenacetin* (once used in Excedrin), are often prescribed.

Acetylsalicylic acid (*aspirin*) | Phenacetin (*APC tablets*) | Acetaminophen (*tylenol*)

Some common analgesics.

Prostaglandins

Prostaglandins are naturally occurring carboxylic acids that are widely distributed in mammalian tissue. Like the steroids, they play a vital role in many biological processes, including blood platelet aggregation (clotting), the regulation of blood pressure, the pain response, and fertility control. The prostaglandins are all C_{20} compounds related to eicosanic acid and prostanoic acid in structure:

Eicosanoic acid

(a)

Prostanoic acid

(b)

PGE_1 PGF_1 PGE_2 PGF_2 PGE_3 PGF_3

(c)

(a) Eicosonic acid; (b) prostanoic acid; (c) six common prostaglandins.

Aspirin functions by inhibiting prostaglandin synthesis. Inhibition of prostaglandin synthesis is responsible for benefits of aspirin in alleviating pain and inflammation, but it is also responsible for the side effects: impaired clotting and tendency to produce stomach ulcers.

Pheromones

Pheromones, chemicals secreted by animals, influence the behavior of other members of the same species. They often represent the major means of communication among simple animals. The term "pheromone" literally means "to carry" and "to excite" (Gk: *pherein,* to carry; Gk: *horman,* to excite). They are chemicals carried by one member of the species and used to excite other members of the species.

Pheromones may be involved in sexual attraction, trail marking, aggregation or recruitment, territorial marking, or signaling alarm. Other pheromones may be involved in defense or in species socialization—for example, designating various classes within the species as a whole. Among all of the pheromones, insect pheromones have been the most intensely studied. Many of the insect pheromones are carboxylic acids or acid derivatives. Examples of members of this class of chemicals are provided in the figure below, along with the principal function of that compound.

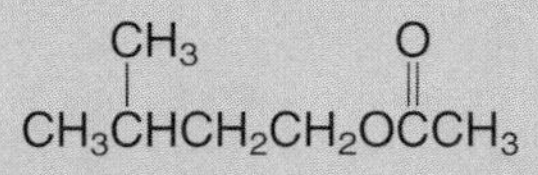

$CH_3CH(CH_3)CH_2CH_2OC(=O)CH_3$

Isoamyl acetate
(honey bee alarm pheromone)

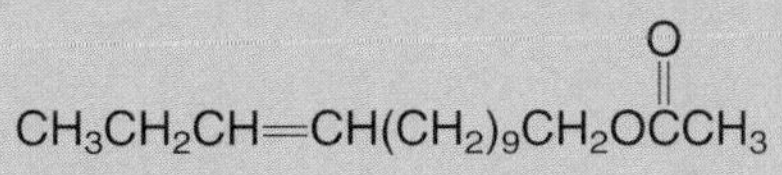

$CH_3CH_2CH{=}CH(CH_2)_9CH_2OC(=O)CH_3$

Tetracecenyl acetate
(European corn borer sex pheromone)

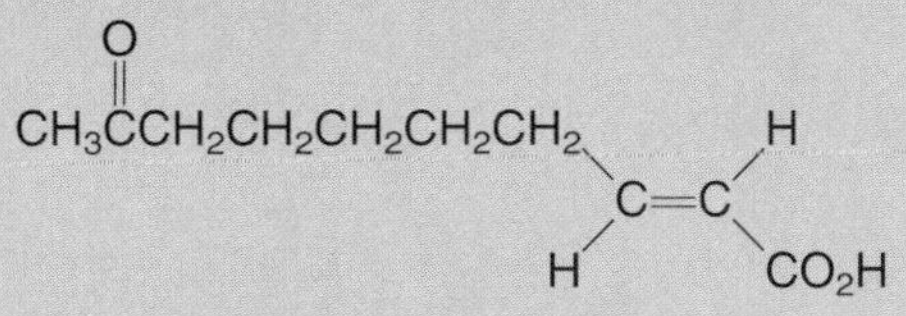

$CH_3C(=O)CH_2CH_2CH_2CH_2CH_2CH{=}CHCO_2H$ (trans)

9-Keto-*trans*-2-decenoic acid
(queen bee socializing/royalty pheromone)

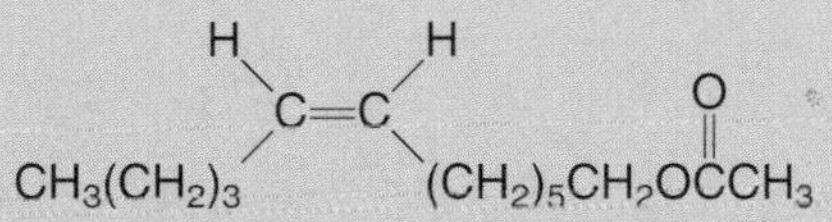

$CH_3(CH_2)_3CH{=}CH(CH_2)_5CH_2OC(=O)CH_3$ (cis)

cis-7-Dodecenyl acetate
(cabbage looper sex pheromone)

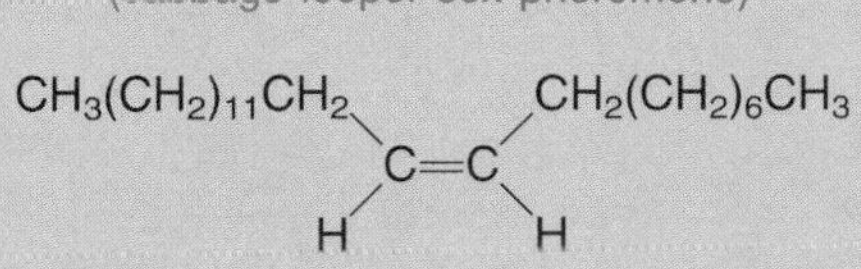

$CH_3(CH_2)_{11}CH_2CH{=}CHCH_2(CH_2)_6CH_3$ (cis)

cis-Tricosene
(common housefly sex attractant)

Some common insect pheromones.

Flavor and fragrance chemicals

Esters are often pleasant, in both aroma and flavor. They are found in many natural products (foodstuffs, for example). They are also widely used in the food and perfume industries as flavor and/or fragrance compounds.

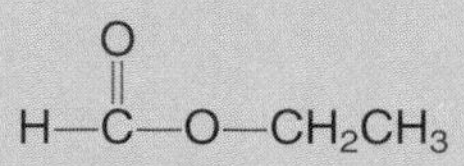

$H{-}C(=O){-}O{-}CH_2CH_3$ Rum

Ethyl methanoate
(ethyl formate)

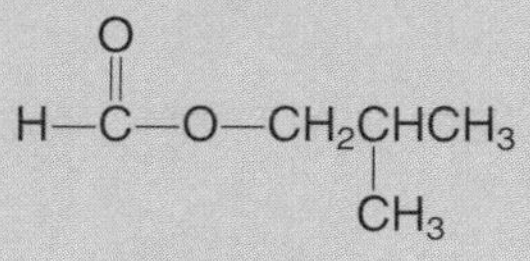

$H{-}C(=O){-}O{-}CH_2CH(CH_3)CH_3$ Raspberries

Isobutyl methanoate
(isobutyl formate)

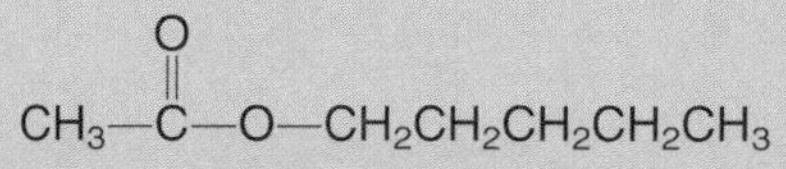

$CH_3{-}C(=O){-}O{-}CH_2CH_2CH_2CH_2CH_3$ Bananas

Pentyl ethanoate
(pentyl acetate)

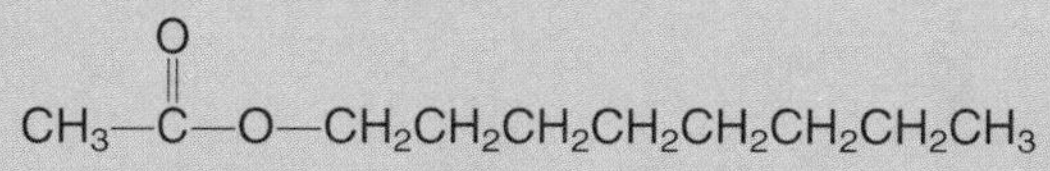

$CH_3{-}C(=O){-}O{-}CH_2CH_2CH_2CH_2CH_2CH_2CH_2CH_3$ Oranges

Octyl ethanoate
(octyl acetate)

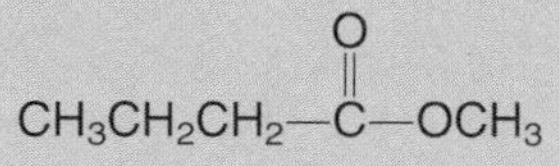

$CH_3CH_2CH_2{-}C(=O){-}OCH_3$ Apples

Methyl butanoate
(methyl butyrate)

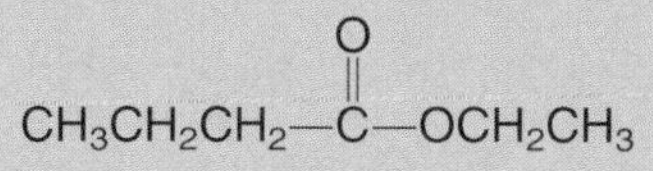

$CH_3CH_2CH_2{-}C(=O){-}OCH_2CH_3$ Pineapples

Ethyl butanoate
(ethyl butyrate)

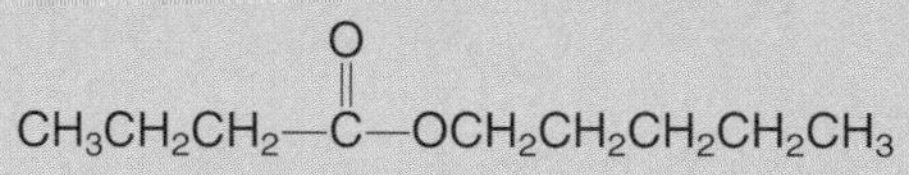

$CH_3CH_2CH_2{-}C(=O){-}OCH_2CH_2CH_2CH_2CH_3$ Apricots

Pentyl butanoate
(pentyl butyrate)

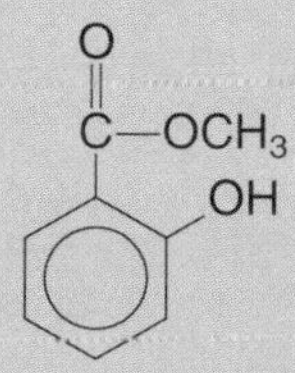

$o{-}HOC_6H_4{-}C(=O){-}OCH_3$ Oil of wintergreen

Methyl salicylate

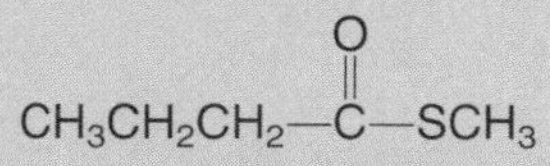

$CH_3CH_2CH_2{-}C(=O){-}SCH_3$ Strawberries

Methyl thiobutanoate
(methyl thiobutyrate)

(a thioester in which sulfur replaces oxygen)

Some interesting flavor and fragrance compounds.

$$CH_3-\overset{\overset{\displaystyle O}{\|}}{C}-OCH_2CH_2CH_2CH_3 \xrightarrow{\text{NaOH, heat}} \quad 8.26$$

Butyl ethanoate
(butyl acetate)

$$CH_3-\overset{\overset{\displaystyle O}{\|}}{C}-O^-Na^+ + CH_3CH_2CH_2CH_2OH \xrightarrow{\text{HCl (aq)}}$$

Sodium ethanoate (sodium acetate) — 1-Butanol (n-butyl alcohol)

$$CH_3-\overset{\overset{\displaystyle O}{\|}}{C}-OH + NaCl$$

Ethanoic acid (acetic acid) — Sodium chloride

Saponification is used to hydrolyze fats and oils, which are esters, to the salts of long chain fatty acids—*soaps*.

Section 8.4

Question 8.12 Complete each of the following reactions by supplying the missing product. Name the product, using the I.U.P.A.C. Nomenclature System.

a. $C_6H_5-\overset{\overset{\displaystyle O}{\|}}{C}-Cl + H_2O \longrightarrow ?$

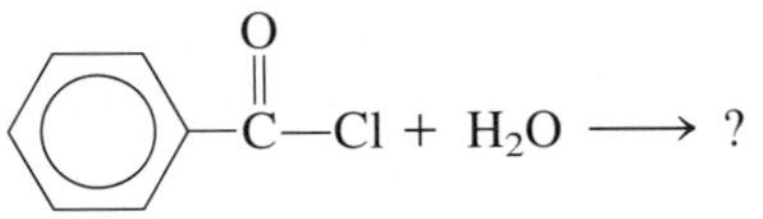

b. $CH_3CH_2CH_2CH_2-\overset{\overset{\displaystyle O}{\|}}{C}-OCH(CH_3)_2 + H_2O \xrightarrow{HO^-} ?$

c. $CH_3\underset{\displaystyle Br}{\underset{|}{C}}H\underset{\displaystyle Br}{\underset{|}{C}}HCH_2CH_2-\overset{\overset{\displaystyle O}{\|}}{C}-NH_2 + H_2O \xrightarrow{H^+} ?$

d. $C_5H_9-\overset{\overset{\displaystyle O}{\|}}{C}-OCH_2CH_3 + H_2O \xrightarrow{H^+} ?$ (cyclopentyl)

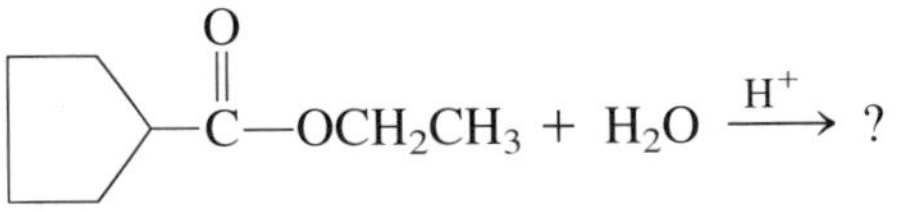

e. $CH_3-\overset{\overset{\displaystyle O}{\|}}{C}-O-\overset{\overset{\displaystyle O}{\|}}{C}-CH_3 + H_2O \xrightarrow{\text{heat}} ?$

Question 8.13 Using complete structural formulas, show all of the products that result from the saponification of methyl benzoate with sodium hydroxide followed by neutralization with hydrochloric acid.

SUMMARY

The carboxylic acids and their derivatives are central to industrial and medicinal chemistry, biochemistry, and the food, flavor, and fragrance industries. Their chemistry is intimately related to that of the alcohols and carbonyl compounds that we studied in earlier chapters.

The functional group of the carboxylic acids is the carboxyl group

$$-\overset{\overset{\displaystyle O}{\|}}{C}-OH$$

and that of the carboxylic acid derivatives is the acyl group

$$R{-}\overset{\overset{\large O}{\|}}{C}{-}$$

Carboxylic acids are prepared by the oxidation of either aldehydes or primary alcohols.

Three of the derivatives of the carboxylic acids are the esters, acid chlorides, and anhydrides. Esters are prepared from carboxylic acids by reaction with alcohols or via the acid chloride, again by reaction with alcohol. Anhydrides result from the reaction of two molecules of carboxylic acid with concurrent loss of one molecule of water. All of these families can be converted back to the acid by hydrolysis. The hydrolysis of esters under basic conditions is called saponification.

GLOSSARY OF KEY TERMS

acyl group (Introduction) the functional group that contains the carbonyl group attached to one alkyl or aryl group; $(Ar)R{-}\underset{|}{C}{=}O$; the functional group found in the derivatives of the carboxylic acids.

adenosine triphosphate (ATP) (8.5) the triphosphate ester of the nucleoside adenosine; the primary energy transport molecule used by the cells in cellular metabolism.

analgesic (8.5) any drug that acts as a pain killer; e.g., aspirin, phenacetin, acetaminophen.

carboxyl group (Introduction) the —COOH functional group; the functional group found in carboxylic acids.

carboxylic acid (Introduction) the family of organic compounds that contains the —COOH functional group.

carboxylic acid derivative (Introduction) any of several families of organic compounds that are derived from carboxylic acids and which have the general formula

$$(Ar)R{-}\underset{\underset{\large Z}{|}}{C}{=}O$$

($Z = X$, OR, NH_2, and so on); includes the acid chlorides, esters, amides, and acid anhydrides.

fatty acid (Introduction) any member of the family of continuous-chain carboxylic acids that generally contain from four to 20 carbon atoms; the naturally occurring members of this class contain an even number of carbon atoms.

hydrolysis (8.5) a chemical reaction that involves the reaction of a molecule with water; results in the splitting off of the water molecule as new bonds are formed.

monomers (8.5) the individual molecules from which a polymer is formed.

nicotinamide adenine dinucleotide (NAD^+) (8.5) a coenzyme that is an oxidizing agent used in a variety of metabolic processes.

oxidation (8.3) the loss of electrons by a molecule or atom; e.g., the conversion of an alcohol to a carboxylic acid via the use of an oxidizing agent.

pheromone (8.5) any compound involved in chemical communication.

polyamide (8.5) a polymer in which the monomeric units result in a series of amide linkages (bonds); nylons.

polyester (8.5) a polymer in which the monomeric units result in a series of ester linkages (bonds); double-knits.

polymer (8.5) a very large molecule formed by the combination of many very small molecules (called monomers) (e.g., polyamides, nylons).

polymerization (8.5) a reaction that produces a polymer.

prostaglandin (8.5) a class of lipids that are derived from prostanoic acid and that have hormone-like activity.

saponification (8.5) a reaction in which a soap is produced; more generally, the hydrolysis of an ester by an aqueous base.

soap (8.4) any of a variety of the alkali metal salts of fatty acids.

termination reaction (8.5) a reaction that is used to ''terminate'' a polymerization.

QUESTIONS AND PROBLEMS

Structure and Nomenclature

8.14 Name each of the following compounds, using the I.U.P.A.C. Nomenclature System:

a. $H{-}\overset{\overset{\large O}{\|}}{C}{-}OH$

b. $H{-}\overset{\overset{\large O}{\|}}{C}{-}OCH_3$

c. $CH_3CH_2{-}\overset{\overset{\large O}{\|}}{C}{-}OCH_2CH_3$

d. $CH_3\underset{\underset{\large I}{|}}{C}HCH_2{-}\overset{\overset{\large O}{\|}}{C}{-}Cl$

e. 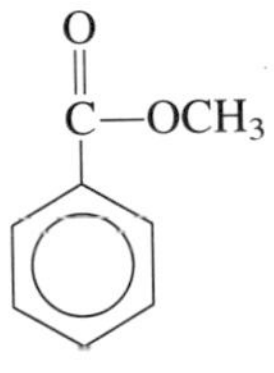 $C_6H_5{-}\overset{\overset{\large O}{\|}}{C}{-}OCH_3$ (benzene ring attached to the carbonyl carbon)

f. 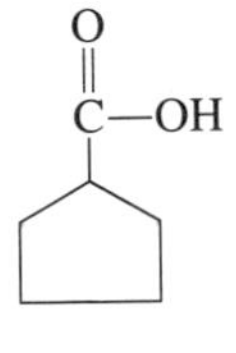 $C_5H_9{-}\overset{\overset{\large O}{\|}}{C}{-}OH$ (cyclopentane ring attached to the carbonyl carbon)

g. $CH_3CH_2CH_2{-}\overset{\overset{\large O}{\|}}{C}{-}O\underset{\underset{\large CH_3}{|}}{C}HCH_3$

h. $\underset{\underset{\large CH_2CH_2CH_2CH_3}{|}}{C}H_2\overset{\overset{\large Cl}{|}}{C}HCH_2{-}\overset{\overset{\large O}{\|}}{C}{-}OCH_2CH_3$

8.15 Write the three-dimensional structures of an aldehyde, a ketone, a carboxylic acid, and an ester, showing only the carbon atoms. What similarities exist among the various structures?

8.16 Write each of the following, using complete structural formulas:

a. methyl benzoate
b. propanoic acid
c. 2-chlorooctanoic acid
d. *n*-butyl decanoate
e. methyl propionate
f. isopropyl acetate
g. ethyl propionate
h. methyl butyrate
i. valeric acid
j. ethyl *m*-nitrobenzoate
k. decanoic anhydride
l. acetic anhydride
m. valeric anhydride
n. benzoyl chloride

8.17 Name each of the following carboxylic acids, using the I.U.P.A.C. Nomenclature System:

a. $CH_3CH(CH_3)CH_2COOH$

b. 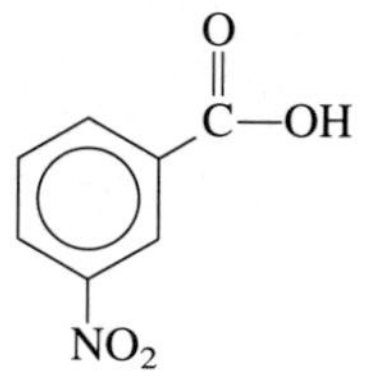 (benzene ring with –COOH and, meta to it, NO_2)

c. $CH_3CH_2CH(Br)CH(CH_3)CH_2COOH$

d. 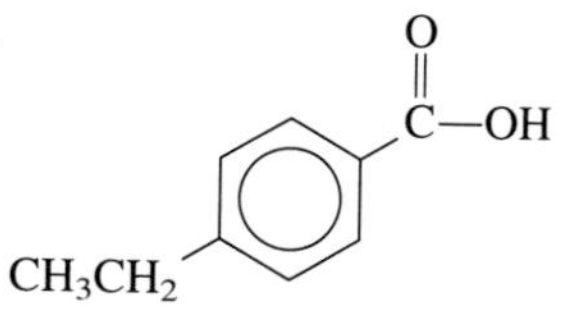 (benzene ring with –COOH and, para to it, CH_3CH_2)

e. $CH_3CH_2CH(CH_2CH_3)CH_2CH_2COOH$

f. 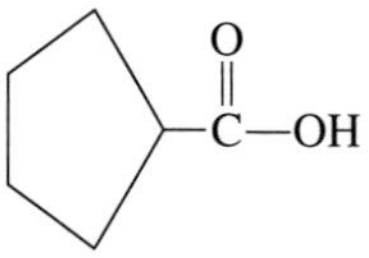 (cyclopentane ring with –COOH)

8.18 Name the following esters, using the I.U.P.A.C. Nomenclature System:

a. $CH_3COOCH_2CH_3$

b. $CH_3CH_2COOCH_3$

c. $CH_3CH(CH_3)CH_2COOCH_3$

d. 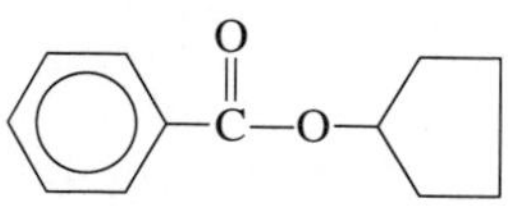 (benzene ring)–C(=O)–O–(cyclopentyl)

e. 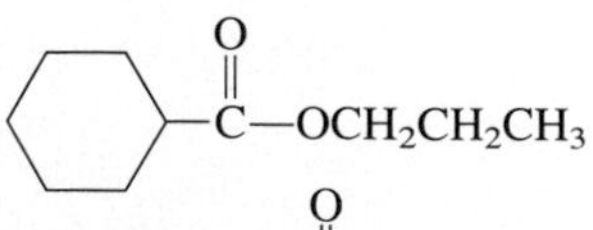 (cyclohexyl)–C(=O)–$OCH_2CH_2CH_3$

f. $CH_2(Br)CH(Br)CH_2CH_2COOCH_2CH_3$

Structure and Properties

8.19 Write the condensed structure of each of the following carboxylic acids:
a. 4,4-dimethylhexanoic acid
b. 3-bromo-4-methylpentanoic acid
c. 2,3-dinitrobenzoic acid
d. 3-methylcyclohexanecarboxylic acid

8.20 How are prostaglandins and steroids related?

8.21 Give the structure of
a. a typical polyester
b. a nylon

8.22 Which member in each of the following pairs has the higher boiling point?
a. heptanoic acid or 1-heptanol
b. propanal or 1-propanol
c. methyl pentanoate or pentanoic acid
d. 1-butanol or butanoic acid
e. pentanoic anhydride or pentanoic acid

8.23 Which member in each of the following pairs is more soluble in water?

a. $CH_3CH_2CH_2CH_2CH_2{-}C(=O){-}OH$ or $CH_3CH_2CH_2CH_2CH_2{-}C(=O){-}O^-Na^+$

b. $CH_3CH_2CH_2CH_2CH_2CH_2CH_2CH_2CH_3$ or $CH_3CH_2CH_2CH_2CH_2CH_2CH_2CH_2{-}OH$

c. $CH_3CH_2{-}O{-}CH_2CH_3$ or $CH_3CH_2{-}C(=O){-}OCH_3$

d. $CH_3CH_2{-}O{-}CH_2CH_3$ or $CH_3CH_2CH_2CH_2CH_2CH_3$

e. decanoic acid or ethanoic acid

f. $CH_3CH_2CH_2{-}C(=O){-}OH$ or $CH_3CH_2{-}C(=O){-}OCH_3$

Reactions

8.24 Complete each of the following reactions by supplying the missing part(s) indicated by the question marks:

a. $CH_3{-}C(=O){-}H \xrightarrow{KMnO_4,\ aq} ?$

b. $CH_3CH_2CH_2{-}C(=O){-}OH + CH_3{-}OH \xrightarrow[H^+,\ heat]{CH_3OH,} ?$

c. 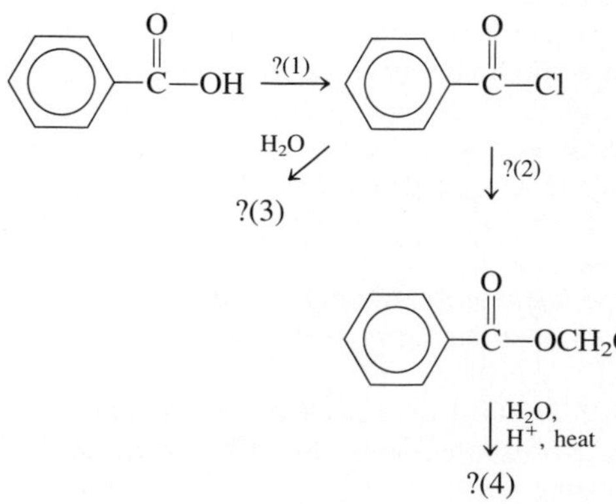 (benzene ring)–C(=O)–OH $\xrightarrow{?(1)}$ (benzene ring)–C(=O)–Cl

(benzene ring)–C(=O)–Cl $\xrightarrow{H_2O}$?(3)

(benzene ring)–C(=O)–Cl $\xrightarrow{?(2)}$ (benzene ring)–C(=O)–OCH_2CH_3 $\xrightarrow[H^+,\ heat]{H_2O,}$?(4)

d. 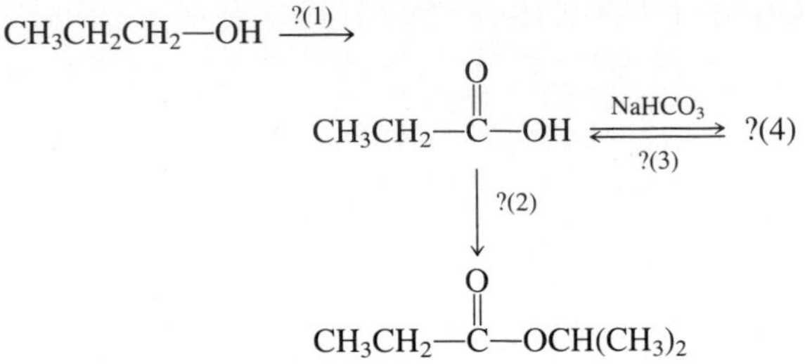 $CH_3CH_2CH_2{-}OH \xrightarrow{?(1)}$

$$CH_3CH_2{-}\overset{O}{\overset{\|}{C}}{-}OH \underset{?(3)}{\overset{NaHCO_3}{\rightleftharpoons}} ?(4)$$

$$\downarrow ?(2)$$

$$CH_3CH_2{-}\overset{O}{\overset{\|}{C}}{-}OCH(CH_3)_2$$

e. $CH_3O{-}\overset{O}{\overset{\|}{C}}{-}CH_2CH_2{-}\overset{O}{\overset{\|}{C}}{-}OCH_3 \xrightarrow[heat]{H_2O,\ HO^-} ?$

f. $CH_3{-}COOH + NaOH \longrightarrow ?$

g. $CH_3CH_2CH_2CH_2CH_2{-}COOH + Na_2CO_3 \longrightarrow ?$

h. $CH_3{-}\overset{O}{\overset{\|}{C}}{-}CH_3 + K_2Cr_2O_7 \xrightarrow{H^+} ?$

i. $CH_3CH_2NH{-}\underset{\underset{O}{\|}}{C}{-}CH_2CH_2CH_2CH_2CH_3 \xrightarrow{?}$

$$CH_3CH_2CH_2CH_2CH_2{-}\underset{\underset{O}{\|}}{C}{-}OH$$

8.25 Complete each of the following reactions by supplying the missing portion indicated with the question marks:

a. $CH_3CH_2CH_2{-}\overset{O}{\overset{\|}{C}}{-}Cl + CH_3CH_2OH \xrightarrow{H^+} ?$

b. 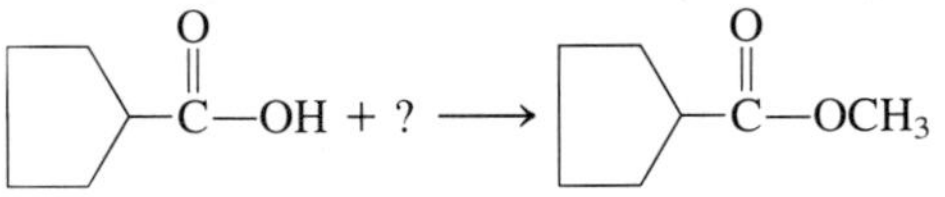 (cyclopentyl)$-\overset{O}{\overset{\|}{C}}{-}OH + ? \longrightarrow$ (cyclopentyl)$-\overset{O}{\overset{\|}{C}}{-}OCH_3$

c. $CH_3CH_2{-}\overset{O}{\overset{\|}{C}}{-}OCH_2CH_3 + H_2O \xrightarrow{H^+} ?$

d. $? + CH_3CH_2\overset{CH_3}{\overset{|}{C}H}{-}OH \xrightarrow{H^+} CH_3{-}\overset{O}{\overset{\|}{C}}{-}O\underset{CH_3}{\underset{|}{C}H}CH_2CH_3$

8.26 What is saponification? Give an example using actual molecules.

8.27 When the methyl ester of hexanoic acid is hydrolyzed in aqueous sodium hydroxide and heat, a homogeneous solution results. When the solution is acidified with dilute aqueous hydrochloric acid, a precipitate forms. What is the precipitate? Draw its structure.

Further Problems

8.28 The structure of salicylic acid is shown below. If this acid reacts with methanol, the product is an ester, methyl salicylate. Methyl salicylate is known as oil of wintergreen and is often used as a flavoring agent. Write the equation for this reaction.

(benzene ring with OH)$-\overset{O}{\overset{\|}{C}}{-}OH + CH_3OH \longrightarrow ?$

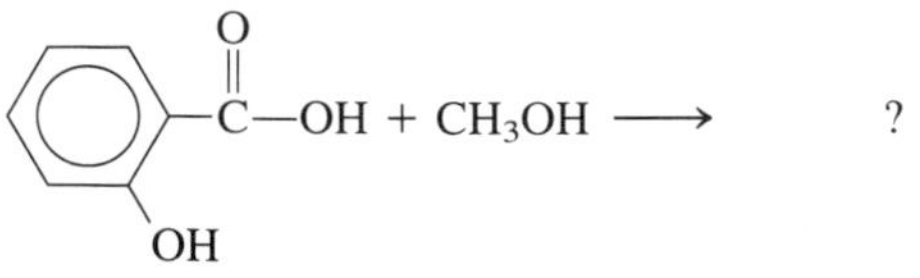

Salicylic acid — Methyl salicylate, oil of wintergreen

8.29 When salicylic acid reacts with acetic anhydride, one of the products is an ester, acetylsalicylic acid. Acetylsalicylic acid is the active ingredient in aspirin. Write the equation for this reaction.

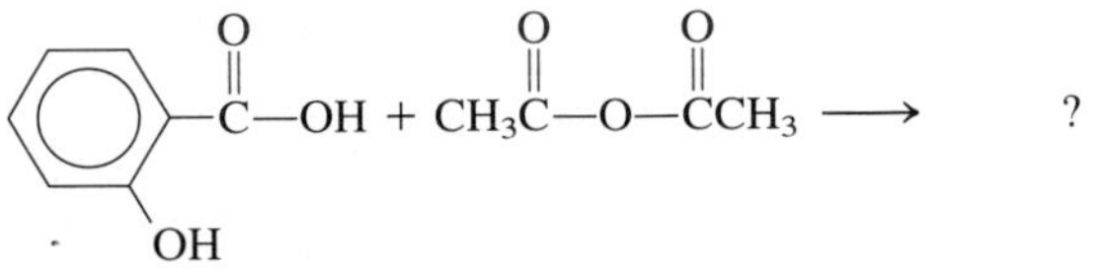

(benzene ring with OH)$-\overset{O}{\overset{\|}{C}}{-}OH + CH_3\overset{O}{\overset{\|}{C}}{-}O{-}\overset{O}{\overset{\|}{C}}CH_3 \longrightarrow ?$

Salicylic acid — Acetylsalicylic acid

8.30 Discuss the difference in the manner in which salicylic acid is reacting in the previous two questions and explain the observed products.

8.31 Describe how insect pheromones can be used as part of a series of "natural" insecticides. (*Note:* You should be able to find many interesting articles on this subject in the popular literature.)

8.32 Compound **A** ($C_6H_{12}O_2$) reacts with water, acid, and heat to yield compound **B** ($C_5H_{10}O_2$) and compound **C** (CH_4O). Compound **B** is acidic. Both compounds **A** and **B** contain a tetrahedral carbon. Deduce possible structures of compounds **A**, **B**, and **C**.

8.33 Copolymers are polymers formed from two or more different monomers. An example of a copolymer is Dacron, a copolymer of terephthalic acid and ethylene glycol. Dacron is often used for woven and knitted fabrics. In medicine, Dacron is used for sutures and as substitutes for blood vessels or the esophagus. The structures of terephthalic acid and ethylene glycol are provided below. Draw the general structure of Dacron.

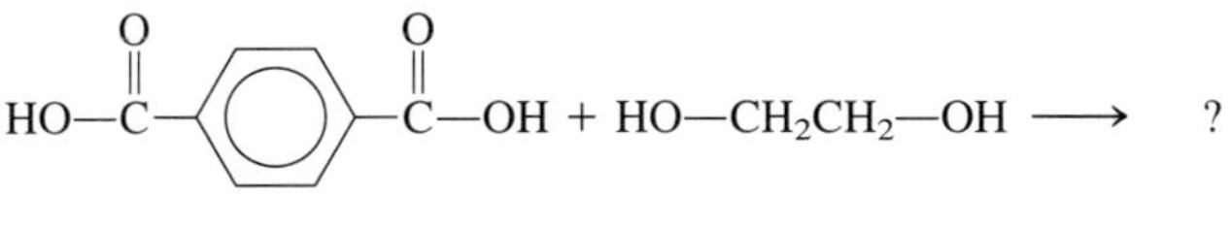

$HO{-}\overset{O}{\overset{\|}{C}}{-}$(benzene ring)$-\overset{O}{\overset{\|}{C}}{-}OH + HO{-}CH_2CH_2{-}OH \longrightarrow ?$

Terephthalic acid — Ethylene glycol — Dacron

8.34 Design a step-by-step synthesis that will allow you to convert 1-propanol to propanoic anhydride. Be specific as to the reagents that are required to accomplish each step of the synthesis.

8.35 Design a step-by-step synthesis that will allow you to convert butanoic acid to butyl butanoate. Be specific as to the reagents that are required to accomplish each step of the synthesis.

8.36 Design a step-by-step synthesis that will allow you to convert 2-methylcyclopentanecarboxylic acid to its N,N-diethylamide. Be specific as to the reagents that are required to accomplish each step of the synthesis.

8.37 We have described the molecule ATP as the body's energy storehouse. What do we mean by this designation? How does ATP actually store energy and provide it to the body as needed?

8.38 When butanedioic acid, shown below, is heated, water and an organic product of molecular formula $C_4H_4O_3$ are produced. When this product is reacted with excess water with a trace of acid, the butanedioic acid is reformed. Explain these observations.

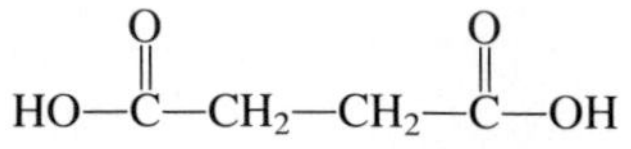

$$\mathrm{HO{-}\overset{\overset{\Large O}{\|}}{C}{-}CH_2{-}CH_2{-}\overset{\overset{\Large O}{\|}}{C}{-}OH}$$

Butanedioic acid

8.39 What products are formed when hexanoyl chloride reacts with each of the following?

a. water and heat
b. excess methylamine
c. *p*-bromophenol
d. 2-phenylethanol
e. excess dilute base (NaOH)
f. cold, concentrated ammonia

8.40 What products are formed when methyl *o*-bromobenzoate reacts with each of the following?

a. aqueous acid and heat
b. aqueous base and heat
c. aqueous acid and heat followed by $SOCl_2$
d. aqueous acid and heat followed by $SOCl_2$ and then $CH_3CH_2CH_2CH_2OH$

8.41 How might $CH_3CH_2CH_2CH_2CH_2OH$ be converted to each of the following products in good yield?

a. $CH_3CH_2CH_2CH_2CHO$
b. $CH_3CH_2CH_2CH_2COOH$
c. $CH_3CH_2CH_2CH_2COCl$
d. $CH_3CH_2CH_2CH_2CONH(CH_3)_2$
e. $CH_3CH_2CH_2CH_2CH_3$

8.42 By reacting phosphoric acid with an excess of ethanol, it is possible to obtain the mono-, di-, and triesters of phosphoric acid. Draw all three of these products.

8.43 It is also possible to form esters of other inorganic acids such as sulfuric acid and nitric acid. One particularly noteworthy product is nitroglycerine, which is both highly unstable (explosive) and widely used in the treatment of the heart condition known as angina, a constricting pain in the chest that usually results from coronary heart disease. In the latter case its function is to alleviate the pain associated with angina. Nitroglycerine may be administered as a tablet (usually placed just beneath the tongue when needed) or as a salve or paste that can be applied to and absorbed through the skin. Nitroglycerine is the trinitroester of glycerol. Draw this structure of nitroglycerine.

CHAPTER 9

Lipids and Their Functions in Biochemical Systems

OUTLINE

OBJECTIVES

- List some examples of the common classes of lipids; describe the principal structural features that differentiate the various classes from each other.
- Be familiar with the general physical and chemical properties of each of the families of lipids.
- Write the structures of simple examples of each of the classes of lipids discussed.
- Understand the biological function and origin of the common members of each of the lipid families.
- Be able to name the common lipids.
- Know the method of synthesizing glycerides and the reactions of glycerides: esterification, hydrolysis, saponification, and hydrogenation.
- Be familiar with the structure and function of cell membranes; be able to discuss these characteristics as a function of the structure of the lipid molecules that make up the membrane.
- Understand the processes of diffusion, osmosis, and active transport.

Lipids are organic molecules of varying chemical composition that are grouped together on the basis of their solubility. They dissolve in nonpolar solvents such as benzene, carbon tetrachloride, and diethyl ether. For the purpose of discussion, lipids are commonly subdivided into four main groups:

1. *Fatty acids* (both saturated and unsaturated)
2. *Glycerides* (glycerol-containing lipids)
3. *Nonglyceride lipids* (sphingolipids, steroids, waxes)
4. *Complex lipids* (lipoproteins)

9.1 SUMMARY OF LIPID FUNCTIONS

The function of a lipid is determined by its properties, which are a direct consequence of the lipid structure. The following brief list will give you an idea of the functions and importance of lipids in biological processes.

1. **Cell membrane structural components.** The cell membrane creates a barrier between the cell and its environment. However, this boundary is not passive; rather, the membrane controls the flow of metabolites into and out of the cell, participates in the generation of metabolic energy, and contains molecules that provide for cellular recognition and communication. Cell membranes are intimately involved in most aspects of the life of the cell.
2. **Energy storage.** Most of the energy stored in the body is in the form of lipids (fats). Stored in cells called adipocytes, these fats are a particularly rich source of energy for the body. These lipids also play an important role in the physical structure of the bodies of animals and humans.
3. **Hormones and vitamins.** Some lipids function as hormones or vitamins. The lipid hormones, such as the steroids, are critical chemical messengers that allow tissues of the body to communicate with one another. The vitamins are essential for the regulation of biological processes, including blood clotting and vision.

In the following sections we shall examine the structure, properties, chemical reactions, and functions of each of the lipid groups shown in Figure 9.1.

9.2 FATTY ACIDS

Structure and properties

Fatty acids are long-chain (12–24 carbons) monocarboxylic acids. As a consequence of their biosynthesis, fatty acids generally contain an *even number* of carbon atoms. The general formula for a **saturated fatty acid** is

$$CH_3(CH_2)_nCOOH$$

(where n is an even integer between 10 and 22). If $n = 16$, an 18-carbon saturated fatty acid, stearic acid, is produced that has the following structural formula:

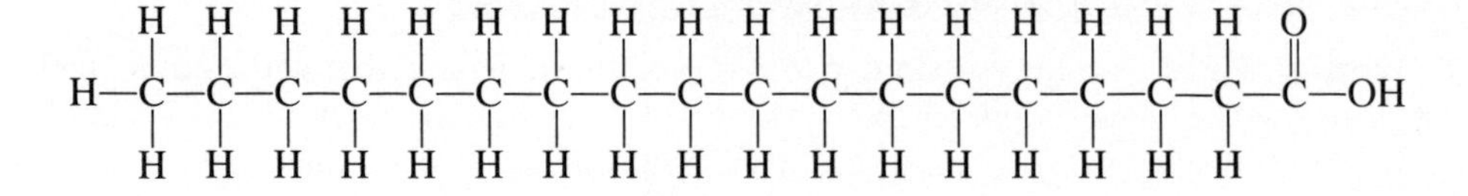

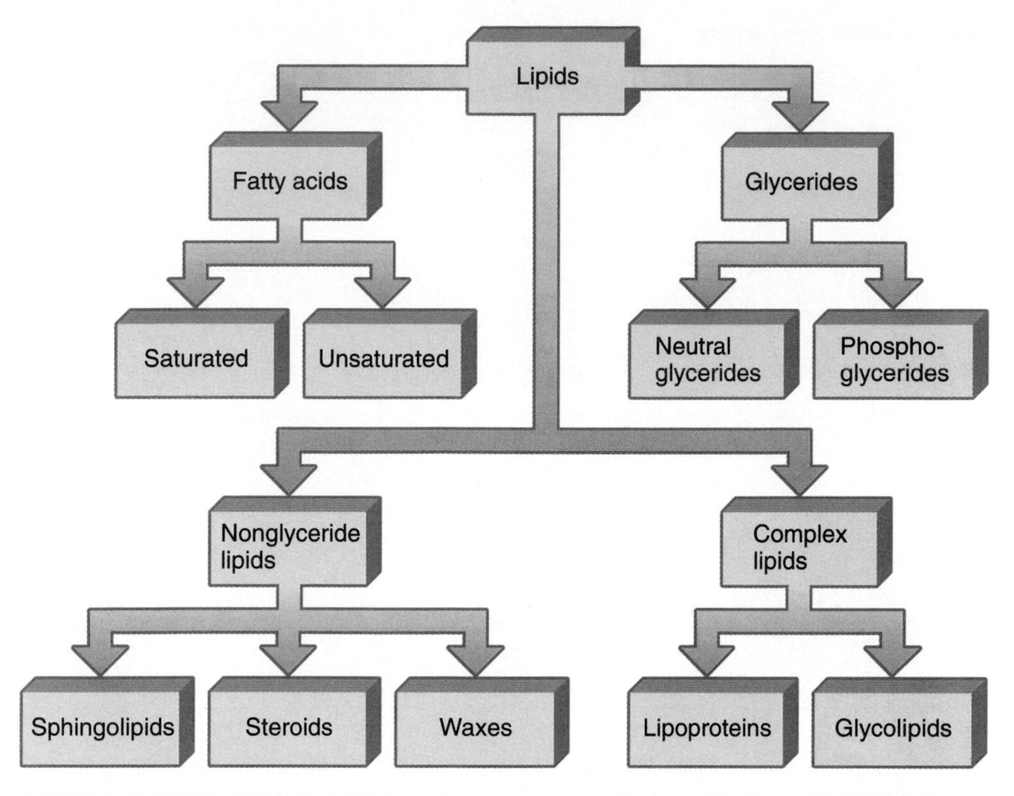

FIGURE 9.1
The classification of lipids.

Notice that each of the carbons in the chain is bonded to the maximum number of hydrogen atoms. To help remember the structure of a saturated fatty acid, you might think of each carbon in the chain being "saturated" with hydrogen atoms. Examples of common saturated fatty acids are given in Table 9.1.

The saturated fatty acids may be thought of as derivatives of alkanes, the saturated hydrocarbons described in Chapter 3.

An **unsaturated fatty acid** is represented by the general formula

$$CH_3(CH_2)_nCH{=}CH(CH_2)_{n'}COOH$$

(where n and n' are integers between 7 and 20). If $n = 7$ and $n' = 7$, the result is an 18-carbon unsaturated fatty acid, oleic acid, having the following structural formula:

```
     H  H  H  H  H  H  H  H  H  H  H  H  H  H  H  H  H  O
     |  |  |  |  |  |  |  |  |  |  |  |  |  |  |  |  |  ||
H—C—C—C—C—C—C—C—C—C=C—C—C—C—C—C—C—C—C—OH
     |  |  |  |  |  |  |  |        |  |  |  |  |  |  |
     H  H  H  H  H  H  H  H        H  H  H  H  H  H  H
```

In the case of unsaturated fatty acids there is at least one carbon-to-carbon double bond. By virtue of the double bonds, these carbons are not "saturated" with hydrogen atoms. Examples of common unsaturated fatty acids are also given in Table 9.1.

The unsaturated fatty acids may be thought of as derivatives of the alkenes, the unsaturated hydrocarbons discussed in Chapter 4.

Inspection of Table 9.1 will reveal a number of significant points. First, the melting points of saturated fatty acids increase with increasing carbon number, as is the case with alkanes. Saturated fatty acids containing ten or more carbons are solids at room temperature. The same effect is seen with unsaturated fatty acids; however, the melting point also decreases markedly as the number of C=C units increases. The trends noted above are graphically illustrated in Figure 9.2.

Section 3.1

Chemical reactions of fatty acids

The reactions of fatty acids are similar to those of short-chain carboxylic acids, which were discussed in the previous chapter. The major reactions that they undergo include esterification, hydrolysis, acid-base reactions, and addition at the double bond.

TABLE 9.1 Common Saturated and Unsaturated Fatty Acids

Common Saturated Fatty Acids

Common Name	I.U.P.A.C. Name	Melting Point (°C)	RCOOH Formula	Condensed Formula
Capric	*n*-Decanoic	32	$C_9H_{19}COOH$	$CH_3(CH_2)_8COOH$
Lauric	*n*-Dodecanoic	44	$C_{11}H_{23}COOH$	$CH_3(CH_2)_{10}COOH$
Myristic	*n*-Tetradecanoic	54	$C_{13}H_{27}COOH$	$CH_3(CH_2)_{12}COOH$
Palmitic	*n*-Hexadecanoic	63	$C_{15}H_{31}COOH$	$CH_3(CH_2)_{14}COOH$
Stearic	*n*-Octadecanoic	70	$C_{17}H_{35}COOH$	$CH_3(CH_2)_{16}COOH$
Arachidic	*n*-Eicosanoic	77	$C_{19}H_{39}COOH$	$CH_3(CH_2)_{18}COOH$

Common Unsaturated Fatty Acids

Common Name	I.U.P.A.C. Name	Melting Point (°C)	RCOOH Formula	Number of Double Bonds	Position of Double Bonds
Palmitoleic	*cis*-9-Hexadecenoic	0	$C_{15}H_{29}COOH$	1	9
Oleic	*cis*-9-Octadecenoic	16	$C_{17}H_{33}COOH$	1	9
Linoleic	*cis,cis*,9,12-Octadecadienoic	5	$C_{17}H_{31}COOH$	2	9, 12
Linolenic	All *cis*-9,12,15-Octadecatrienoic	−11	$C_{17}H_{29}COOH$	3	9, 12, 15
Arachidonic	All *cis*-5,8,11,14-Eicosatetraenoic	−50	$C_{19}H_{31}COOH$	4	5, 8, 11, 14

Condensed Formula

	Condensed Formula
Palmitoleic	$CH_3(CH_2)_5CH{=}CH(CH_2)_7COOH$
Oleic	$CH_3(CH_2)_7CH{=}CH(CH_2)_7COOH$
Linoleic	$CH_3(CH_2)_4CH{=}CH{-}CH_2{-}CH{=}CH(CH_2)_7COOH$
Linolenic	$CH_3CH_2CH{=}CH{-}CH_2{-}CH{=}CH{-}CH_2{-}CH{=}CH(CH_2)_7COOH$
Arachidonic	$CH_3(CH_2)_4CH{=}CH{-}CH_2{-}CH{=}CH{-}CH_2{-}CH{=}CH{-}CH_2{-}CH{=}CH{-}(CH_2)_3COOH$

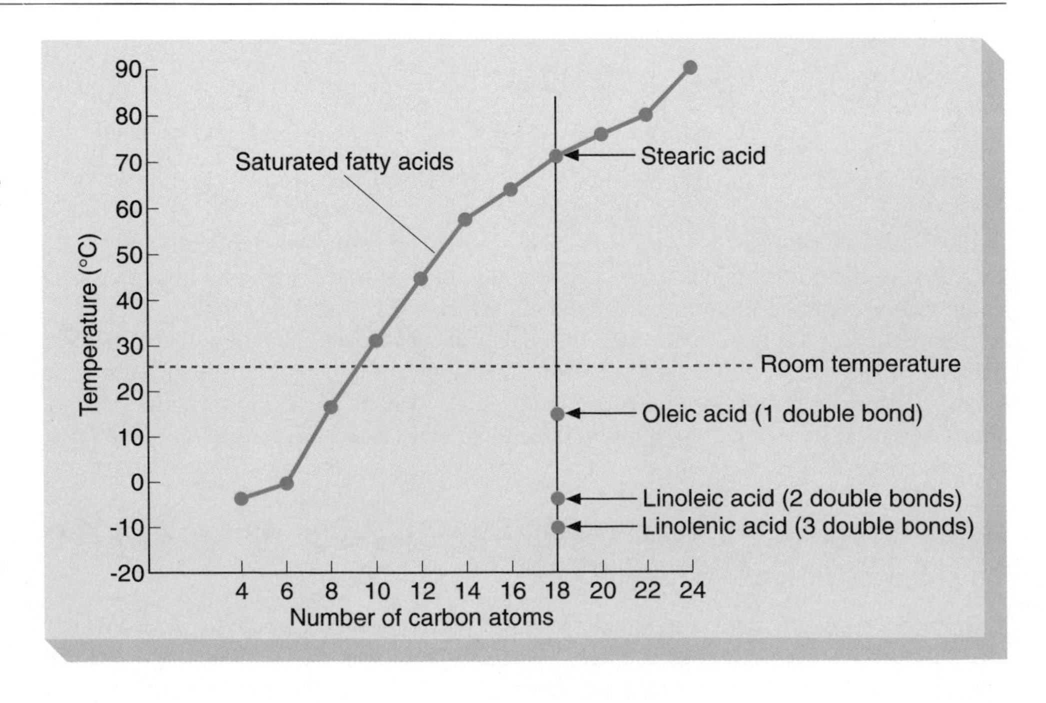

FIGURE 9.2
The melting points of fatty acids. Melting points of both saturated and unsaturated fatty acids increase as the number of carbon atoms in the chain increases. The melting points of unsaturated fatty acids are lower than those of the corresponding fatty acid with the same number of carbon atoms. Also, as the number of double bonds in the chain increases, the melting points decrease.

Esterification

Fatty acids react with alcohols to form esters and water according to the following general reaction:

Section 8.4

$$R{-}\overset{\overset{\displaystyle O}{\|}}{C}{-}OH + HO{-}R' \longrightarrow R{-}\overset{\overset{\displaystyle O}{\|}}{C}{-}OR' + H_2O$$

Acid Alcohol Ester Water

Hydrolysis

Recall that this reaction is the reverse of esterification, producing fatty acids from esters:

Section 8.5

$$R{-}\overset{\overset{\displaystyle O}{\|}}{C}{-}OR' + HO{-}H \longrightarrow R{-}\overset{\overset{\displaystyle O}{\|}}{C}{-}OH + R'OH$$

Ester Water Acid Alcohol

Acid-base reactions (saponification)

As one would expect, these reactions yield a salt and water:

Section 8.4

$$R{-}\overset{\overset{\displaystyle O}{\|}}{C}{-}OH + NaOH \longrightarrow R{-}\overset{\overset{\displaystyle O}{\|}}{C}{-}O^-Na^+ + H_2O$$

Acid Base Salt Water

The product of this reaction, an ionized salt, is a soap. Soap molecules have a long, uncharged carbon tail and a negatively charged terminus. This allows soap molecules to form layers (micelles) around oil and dirt particles. A micelle forms because the nonpolar tails of the soap molecules aggregate around the oil particle, exposing the charged termini to the surrounding water. As a result, the dirt is emulsified, broken into small particles, and can be rinsed away.

An example of a micelle formed by phospholipids is seen in Figure 15.1.

Problems can arise when ''hard'' water is used for cleaning because the high concentrations of Ca^{2+} and Mg^{2+} in such water cause fatty acids to precipitate. Not only does this interfere with the emulsifying action of the soap, it also leaves a hard scum on the surface of sinks and tubs.

Reaction at the double bond (unsaturated fatty acids)

Hydrogenation is an example of an addition reaction. The following is a typical example of the addition of hydrogen to a double bond of a fatty acid:

Section 4.4

$$CH_3(CH_2)_4CH{=}CHCH_2CH{=}CH(CH_2)_7COOH$$

Linoleic acid

$$\Big\downarrow \; 2H_2 \text{ (hydrogen gas), Ni (catalyst)}$$

$$CH_3(CH_2)_{16}COOH$$

Stearic acid

Hydrogenation is used in the food industry to convert polyunsaturated vegetable oils into saturated solid fats. Margarine would be liquid at room temperature, but hydrogenation converts it to a spreadable solid with properties similar to those of butter. The hydrogenated fats are as detrimental in the diet as natural saturated fats. But margarine is recommended over butter because it contains both polyunsaturated and saturated fats. Careful inspection of the label will reveal the amounts of both types of fat and allow the consumer to evaluate the health benefits of different brands.

Question 9.1 Draw complete structural formulas of:

a. Oleic acid c. Linoleic acid
b. Lauric acid d. Stearic acid

Question 9.2 What is the I.U.P.A.C. name of each of the fatty acids in Question 9.1?

Question 9.3 What physical property most clearly distinguishes a saturated fatty acid from an unsaturated fatty acid with the same number of carbon atoms?

Question 9.4 Is the fatty acid $C_{13}H_{25}COOH$ saturated or unsaturated?

Question 9.5 Write out the complete reaction described below for each acid in Question 9.1.

a. Esterification b. Acid/base reaction with NaOH

Question 9.6 Why would neither stearic nor lauric acid undergo an addition reaction?

Eicosanoids: prostaglandins, leukotrienes, and thromboxanes

Some of the unsaturated fatty acids containing more than one double bond cannot be synthesized by the body. For many years it has been known that linolenic acid and linoleic acid, called the **essential fatty acids,** are necessary for specific biochemical functions and must be supplied in the diet (see Table 9.1). The function of linoleic acid became clear in the 1960s when it was discovered that linoleic acid is required for the biosynthesis of arachidonic acid, the precursor of a class of hormonelike molecules known as **eicosanoids.** The name is derived from the Greek work *eikos,* meaning "twenty," because they are all derivatives of 20 carbon fatty acids. The eicosanoids include three groups of structurally related compounds: the prostaglandins, the leukotrienes, and the thromboxanes.

See Chapter 8, "A Human Perspective: Carboxylic Acid Derviatives of Special Interest."

The **prostaglandins** are extremely potent biological molecules with hormonelike activity. They received this name because they were originally isolated from seminal fluid produced in the prostate gland, but they have since been isolated from most animal tissues. All the prostaglandins are derived from the C_{20} polyunsaturated, essential fatty acid **arachidonic acid** and have certain structural similarities. They are carboxylic acids, but in addition they have a five-carbon ring and are composed of the basic C_{20} carbon skeleton of prostanoic acid. Several general classes of prostaglandins are grouped under the designations A, B, E, and F, among others, depending on the basic carbon skeletal arrangement and the number and orientation of double bonds, hydroxyl groups, and ketone groups. A compound is given the more complete symbolism PGE_1 or PGF_2, for example, where PG stands for prostaglandin, E or F indicates the prostaglandin group, and 1 and 2 indicate the number of carbon-carbon double bonds in the compound. The examples in Figure 9.3 illustrate the general structure of prostaglandins and the current nomenclature system.

Prostaglandins are made in all tissues and exert their biological effects on the cells that produce them and on other cells in the immediate vicinity. The extraordinary range of prostaglandin functions include the following.

- Stimulation of smooth muscle
- Regulation of steroid biosynthesis
- Inhibition of gastric secretion
- Inhibition of hormone-sensitive lipases
- Inhibition of platelet aggregation
- Stimulation of platelet aggregation
- Regulation of nerve transmission
- Sensitization to pain
- Mediation of inflammatory response

Because the prostaglandins and the closely related leukotrienes and thromboxanes affect so many body processes and because they often cause opposing effects in different tissues, it can be difficult to keep track of their many regulatory functions. The following is a brief summary of some of the biological processes that are thought to be regulated by the prostaglandins, leukotrienes, and thromboxanes.

1. **Blood clotting.** Blood clots form when a blood vessel is damaged, yet such clotting along the walls of undamaged vessels could result in heart attack or stroke. **Thromboxane A_2** is produced by platelets in the blood and stimulates constriction of the blood vessels and aggregation of the platelets. Conversely, PGI_2 (prostacyclin) is produced by the cells lining the blood vessels and has precisely the opposite effect of thromboxane A_2. Prostacyclin inhibits platelet aggregation and causes dilation of blood vessels and thus prevents the untimely production of blood clots.

2. **The inflammatory response.** The inflammatory response is another of the body's protective mechanisms. When tissue is damaged by mechanical injury, burns, or invasion by microorganisms, a variety of white blood cells descend on the damaged site to try to minimize the tissue destruction. The result of this response is swelling, redness, fever, and pain. Prostaglandins are thought to promote certain aspects of the inflammatory response, especially pain and fever. Drugs such as aspirin block prostaglandin synthesis and help to relieve the symptoms. We will examine the mechanism of action of these drugs later in this section.

3. **Reproductive system.** PGE_2 stimulates smooth muscle contraction, particularly uterine contractions. An increase in the level of prostaglandins has been noted immediately before the onset of labor. PGE_2 has also been used to induce second trimester abortions. There is strong evidence that dysmenorrhea (painful menstruation) suffered by many women may be the result of an excess of two prostaglandins. Indeed, drugs that inhibit prostaglandin synthesis have been approved by the FDA and are found to provide virtually complete relief from these symptoms.

4. **Gastrointestinal tract.** Prostaglandins have been shown to both inhibit the secretion of acid and increase the secretion of a protective mucous layer into the stomach. In this way, prostaglandins help to protect the stomach lining. Consider for a moment the possible side effect that prolonged use of a drug such as aspirin might have on the stomach—ulceration of the stomach lining. Because aspirin inhibits prostaglandin synthesis, it may actually encourage stomach ulcers by inhibiting the formation of the normal protective mucous layer, while simultaneously allowing increased secretion of stomach acid.

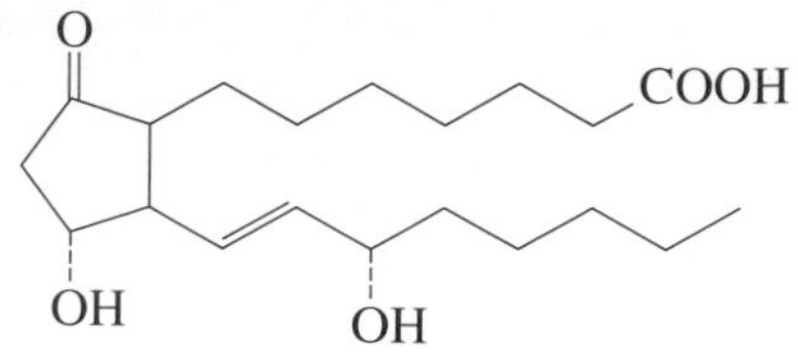

O COOH OH OH

Prostaglandin E_1

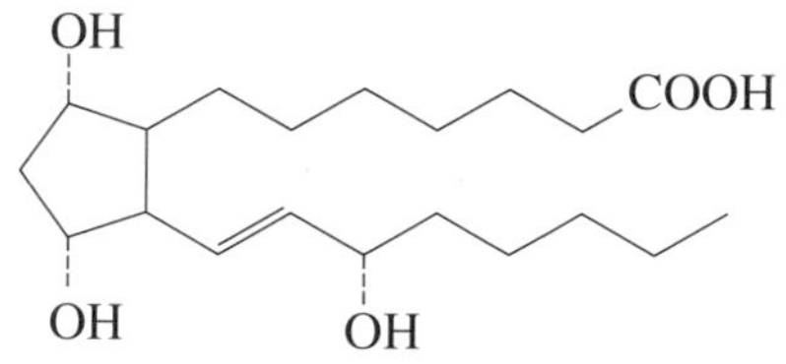

OH COOH OH OH

Prostaglandin F_1

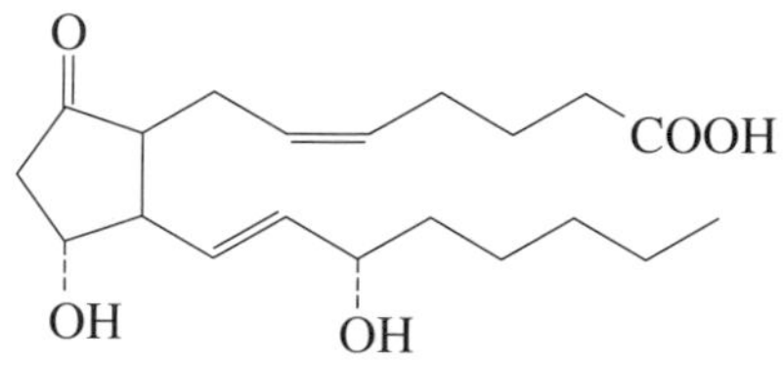

O COOH OH OH

Prostaglandin E_2

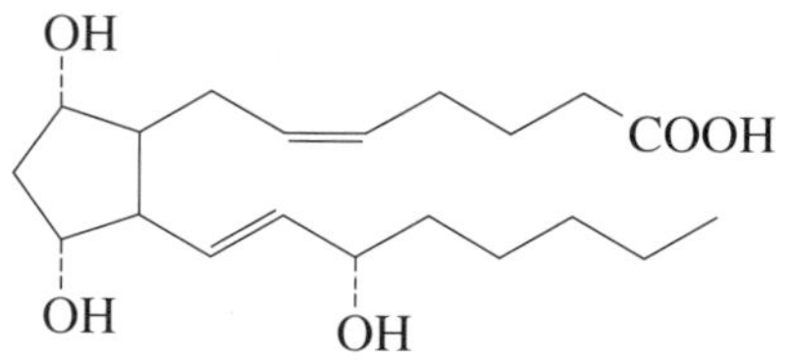

OH COOH OH OH

Prostaglandin F_2

FIGURE 9.3
The structures of four prostaglandins.

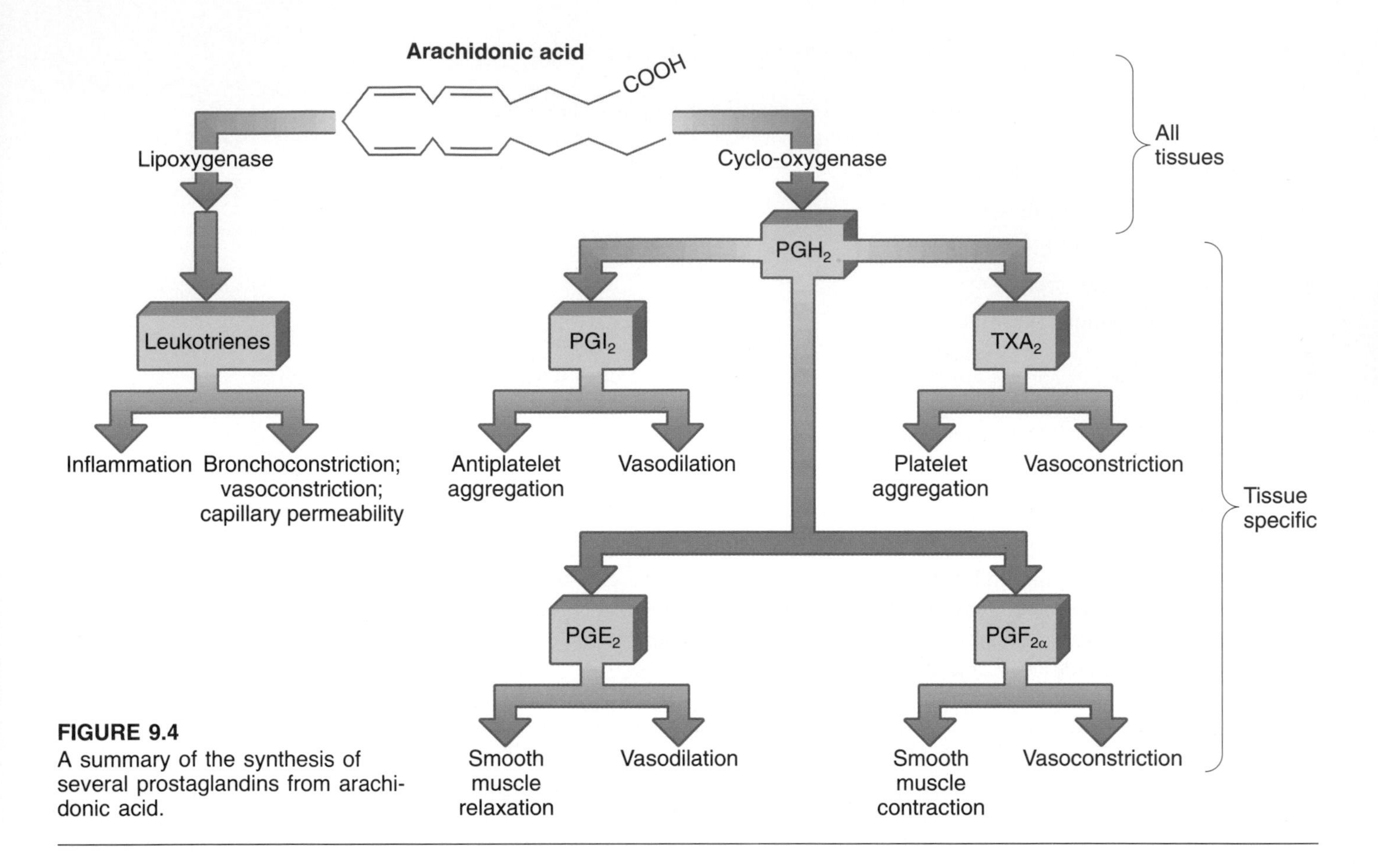

FIGURE 9.4
A summary of the synthesis of several prostaglandins from arachidonic acid.

5. **Kidneys.** Prostaglandins produced in the kidneys cause the renal blood vessels to dilate. The greater flow of blood through the kidney results in increased water and electrolyte excretion.
6. **Respiratory tract.** Eicosanoids produced by certain white blood cells, the **leukotrienes,** promote the constriction of the bronchi associated with asthma. Other prostaglandins promote bronchodilation.

As this brief survey suggests, the prostaglandins have numerous, often antagonistic effects. Although they do not fit the formal definition of a hormone (a substance produced in a specialized tissue and transported by the circulatory system to target tissues elsewhere in the body), the prostaglandins are clearly strong biological regulators with far-reaching effects.

As was mentioned above, prostaglandins stimulate the inflammatory response and, as a result, are partially responsible for the cascade of events that cause pain. Aspirin has long been known to alleviate such pain, and we now know that it does so by inhibiting the synthesis of prostaglandins.

The first two steps of prostaglandin synthesis (Figure 9.4), the release of arachidonic acid from the membrane and its conversion to PGH_2 by the enzyme cyclooxygenase, occur in all tissues that are able to produce prostaglandins. The conversion of PGH_2 into the other biologically active forms is tissue-specific and requires the appropriate enzymes, which are found only in certain tissues.

Aspirin works by inhibiting the cyclooxygenase, which catalyzes the first step in the pathway leading from arachidonate to PGH_2. The enzyme becomes covalently bound to the acetyl group of aspirin and is inactivated (Figure 9.5). Because cyclooxygenase is one of the steps that occurs in all cells, aspirin effectively inhibits synthesis of all of the prostaglandins.

9.3 GLYCERIDES

Glycerides, lipids which contain glycerol, may be subdivided into two classes: neutral glycerides and phosphoglycerides. Neutral glycerides are so named because they are nonionic and nonpolar. As we will see, phosphoglyceride molecules have a polar region, due to the phosphate group, in addition to the nonpolar fatty acid tails. The structures of each of these types of glycerides are critical to their function.

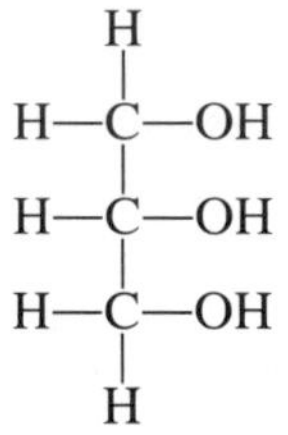

Glycerol

Neutral glycerides

The esterification of glycerol with a fatty acid produces a **neutral glyceride.** Esterification may occur at one, two, or all three positions, producing a **monoglyceride, diglyceride,** or **triglyceride.** You will also see these referred to as mono-, di-, or triacylglycerols. The following example shows the synthesis of a monoglyceride:

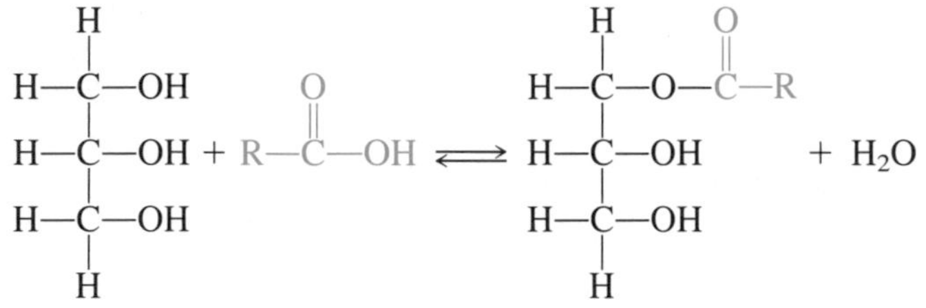

$$\text{Glycerol} + R\text{—}C(=O)\text{—}OH \rightleftharpoons \text{Monoglyceride} + H_2O$$

Glycerol Fatty acid Monoglyceride Water

Although monoglycerides and diglycerides are present in nature, the most important species is the triglyceride, the major component of fat cells. The triglyceride consists of a glycerol backbone (shown in black below) joined to three fatty acid units through ester bonds (shown in color below). The formation of a triglyceride is shown in the following reaction:

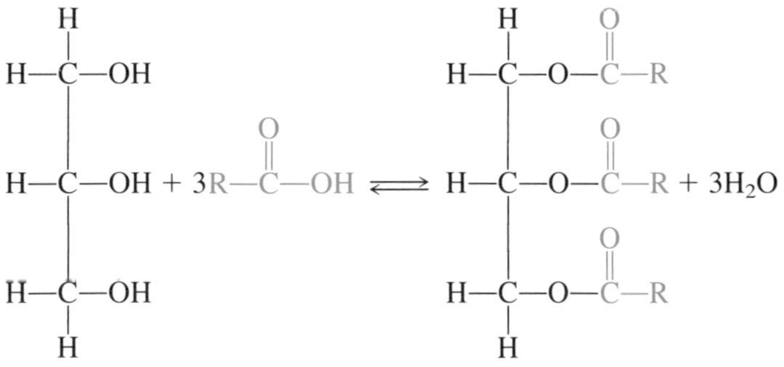

$$\text{Glycerol} + 3R\text{—}C(=O)\text{—}OH \rightleftharpoons \text{Triglyceride} + 3H_2O$$

Glycerol Fatty acid Triglyceride Water

A common example of a triglyceride is glyceryl tristearate:

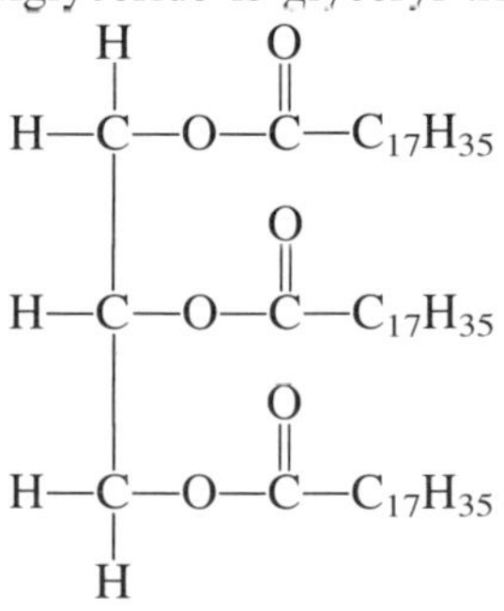

$$H\text{—}C(H)\text{—}O\text{—}C(=O)\text{—}C_{17}H_{35}$$
$$H\text{—}C\text{—}O\text{—}C(=O)\text{—}C_{17}H_{35}$$
$$H\text{—}C(H)\text{—}O\text{—}C(=O)\text{—}C_{17}H_{35}$$

Glyceryl tristearate
(tristearin)

FIGURE 9.5
Aspirin inhibits the synthesis of prostaglandins by acetylating the enzyme cyclooxygenase. The acetylated enzyme is no longer functional.

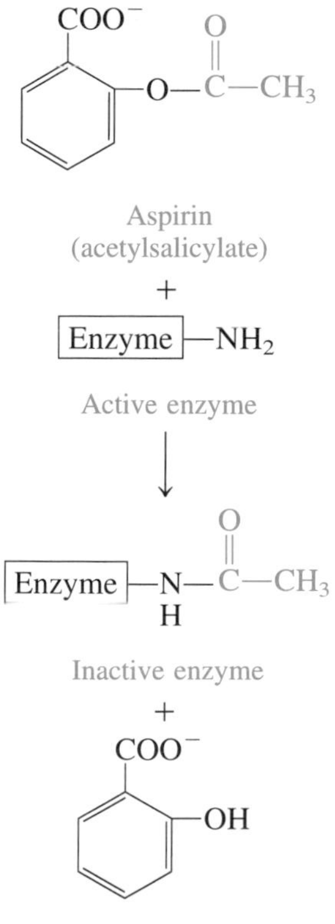

Aspirin
(acetylsalicylate)
+
Enzyme—NH_2
Active enzyme
↓
Enzyme—N(H)—C(=O)—CH_3
Inactive enzyme
+
Salicylate

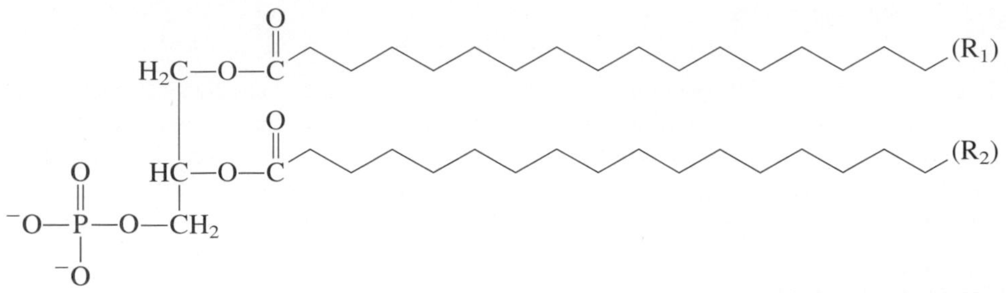

$$
\begin{array}{l}
\mathrm{H_2C{-}O{-}C(=O){-}(R_1)} \\
\mathrm{HC{-}O{-}C(=O){-}(R_2)} \\
\mathrm{^{-}O{-}P(=O)(O^{-}){-}O{-}CH_2}
\end{array}
$$

Phosphatidate

FIGURE 9.6
The structure of a phosphoglyceride or phosphatidate.

A fatty acid is named by combining the ''backbone'' name, glyceryl (from glycerol), with the fatty acid name, stearate. The prefix tri- is used because three stearic acid units are attached to the glycerol backbone. It should be noted that glyceryl tristearate is a *simple* triglyceride, one in which all three groups attached to the glycerol backbone are identical. Mixed triglycerides, in which two or more different fatty acids are attached to the glycerol backbone, are also known but will not be discussed.

Neutral glycerides are, in fact, neutral because they do not dissociate into charged species. Bonding throughout the molecule is covalent and nonpolar. As a result, these long nonpolar molecules readily stack with one another and constitute the majority of the material stored in the body's fat cells. In contrast to the phosphoglycerides the neutral glycerides are not found in the cell membrane.

Sections 15.1 and 15.2

Their principal function in biochemical systems is the storage of energy. If more energy-rich nutrients are consumed than are required for metabolic processes, much of the excess is converted to neutral glycerides and is *stored* in triglyceride molecules in fat cells (adipose tissue). When energy is needed, the triglycerides are metabolized, releasing their energy. It is for precisely this reason that exercise, along with moderate reduction in caloric intake, is recommended for overweight individuals. Exercise, an energy-demanding process, causes the metabolism of fats, resulting in weight loss.

See Chapter 15, A Human Perspective: Losing those Unwanted Pounds of Adipose Tissue.

Phosphoglycerides

Phospholipids are a group of lipids containing a phosphate group (PO_4^{3-}). Introduction of such a group results in a molecule with a polar head (the phosphate group) and a nonpolar tail (the alkyl chain of the fatty acid). Because the phosphate ionizes to a degree in solution, a charged lipid results.

The most abundant membrane lipids are derived from glycerol-3-phosphate and are known as **phosphoglycerides.** Phosphoglycerides contain acyl groups derived from long-chain fatty acids esterified at C-1 and C-2 of glycerol-3-phosphate. The simplest phosphoglyceride contains a free phosphoryl group and is known as a **phosphatidate** (Figure 9.6). When the phosphoryl group is attached to another hydrophilic molecule, a more complex phosphoglyceride is formed. For example, **phosphatidylcholine (lecithin)** and **phosphatidylethanolamine (cephalin)** are found in many cell membranes (Figure 9.7).

Phosphatidylcholine (lecithin) possesses a polar ''head'' and nonpolar ''tail,'' a structure that is similar to the structure of soap and detergent molecules, discussed earlier. The ionic ''head'' is hydrophilic and interacts with water molecules, while the nonpolar ''tail'' is hydrophobic and interacts with nonpolar molecules. This bipolar nature is central to the structure and function of cell membranes.

Section 9.6

In addition to being a component of cell membranes, lecithin is found in egg yolks and soybeans. It is also used as an emulsifying agent in ice cream. An **emulsifying agent** aids in the suspension of fats in water. The bipolar lecithin serves as a bridge, holding together two noncompatible substances. This is directly analogous to the action of soaps.

Phosphatidylethanolamine, cephalin, is similar in general structure to lecithin; the amino-alcohol group bonded to the phosphate group is the only difference. Cephalins are

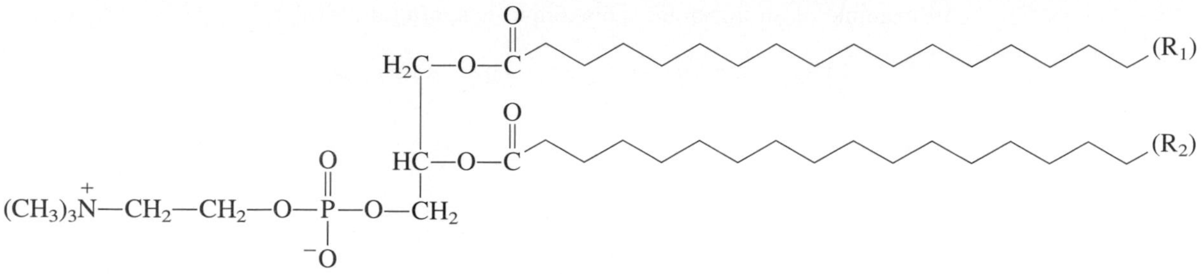

$$
\begin{array}{l}
\quad\quad\quad\quad\quad\quad\quad\quad\quad\quad\quad\quad\quad\quad\quad O \\
\quad\quad\quad\quad\quad\quad\quad\quad\quad\quad\quad\quad\quad\quad\quad \| \\
\quad\quad\quad\quad\quad\quad\quad\quad\quad\quad\quad H_2C{-}O{-}C{\sim\sim\sim}(R_1) \\
\quad\quad\quad\quad\quad\quad\quad\quad\quad\quad\quad\quad\quad\quad\quad O \\
\quad\quad\quad\quad\quad\quad\quad\quad\quad\quad\quad\quad\quad\quad\quad \| \\
\quad\quad\quad\quad\quad\quad\quad\quad\quad O\quad\quad HC{-}O{-}C{\sim\sim\sim}(R_2) \\
\quad\quad\quad\quad\quad\quad\quad\quad\quad \| \\
(CH_3)_3\overset{+}{N}{-}CH_2{-}CH_2{-}O{-}P{-}O{-}CH_2 \\
\quad\quad\quad\quad\quad\quad\quad\quad\quad | \\
\quad\quad\quad\quad\quad\quad\quad\quad\quad {}^{-}O
\end{array}
$$

Phosphatidylcholine (lecithin)

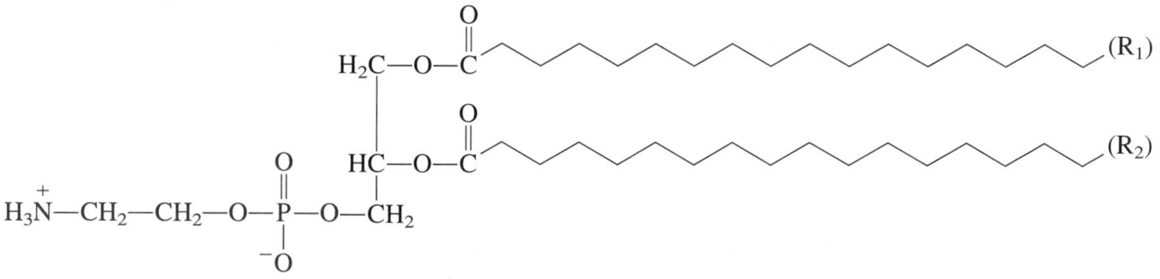

$$
\begin{array}{l}
\quad\quad\quad\quad\quad\quad\quad\quad\quad\quad\quad\quad\quad\quad\quad O \\
\quad\quad\quad\quad\quad\quad\quad\quad\quad\quad\quad\quad\quad\quad\quad \| \\
\quad\quad\quad\quad\quad\quad\quad\quad\quad\quad\quad H_2C{-}O{-}C{\sim\sim\sim}(R_1) \\
\quad\quad\quad\quad\quad\quad\quad\quad\quad\quad\quad\quad\quad\quad\quad O \\
\quad\quad\quad\quad\quad\quad\quad\quad\quad\quad\quad\quad\quad\quad\quad \| \\
\quad\quad\quad\quad\quad\quad\quad\quad\quad O\quad\quad HC{-}O{-}C{\sim\sim\sim}(R_2) \\
\quad\quad\quad\quad\quad\quad\quad\quad\quad \| \\
H_3\overset{+}{N}{-}CH_2{-}CH_2{-}O{-}P{-}O{-}CH_2 \\
\quad\quad\quad\quad\quad\quad\quad\quad\quad | \\
\quad\quad\quad\quad\quad\quad\quad\quad\quad {}^{-}O
\end{array}
$$

Phosphatidylethanolamine (cephalin)

FIGURE 9.7
The structures of two common phospholipids found in membranes: phosphatidylcholine (lecithin) and phosphatidylethanolamine (cephalin).

also components of cell membranes, comprising 18–20% of the phospholipids in the membranes of human cells.

Question 9.7 Using structural formulas, draw the mono-, di-, and triglycerides that would result from the reaction of glycerol with each of the following acids:

a. Oleic acid
b. Capric acid
c. Palmitic acid
d. Lauric acid

Question 9.8 Name the triglycerides that are produced in the reactions discussed in Question 9.7.

9.4 NONGLYCEROL LIPIDS

Sphingolipids

Sphingolipids are lipids that are not derived from glycerol. They are, however, phospholipids because they contain the phosphate group. They are derived from sphingosine, a nitrogen-containing (amino) alcohol:

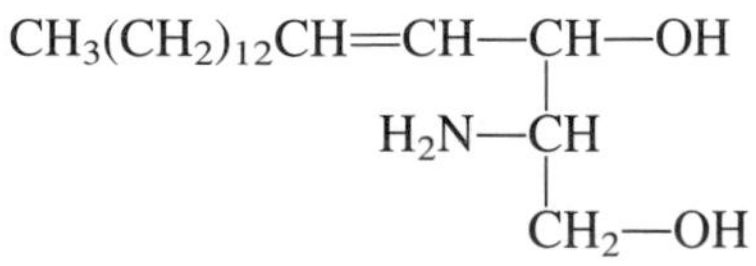

$$
\begin{array}{l}
CH_3(CH_2)_{12}CH{=}CH{-}CH{-}OH \\
\quad\quad\quad\quad\quad\quad\quad\quad\quad\quad | \\
\quad\quad\quad\quad\quad\quad\quad H_2N{-}CH \\
\quad\quad\quad\quad\quad\quad\quad\quad\quad\quad | \\
\quad\quad\quad\quad\quad\quad\quad\quad\quad CH_2{-}OH
\end{array}
$$

Sphingosine

An example of an important sphingolipid is **sphingomyelin:**

$$
\begin{array}{l}
CH_3(CH_2)_{12}CH{=}CH{-}CH{-}OH \\
\quad R{-}\underset{\|}{\overset{}{C}}{-}NH{-}CH \qquad\qquad \overset{O}{\|} \\
\qquad\ \ O \qquad\qquad CH_2{-}O{-}P{-}O{-}CH_2{-}CH_2{-}N^+(CH_3)_3 \\
\qquad\qquad\qquad\qquad\qquad\quad\ \ |\\
\qquad\qquad\qquad\qquad\qquad\quad\ O^-
\end{array}
$$

Sphingosine; Fatty acid; Phosphate; Choline

Sphingomyelin

Sphingomyelins are located throughout the body but function principally in brain and nerve tissue. They are found in abundance in the myelin sheath that surrounds and insulates cells of the central nervous system. In humans, about 25% of the lipids of the myelin sheath are sphingomyelins. Their role is essential to proper cerebral function and nerve transmission.

Steroids

Steroids are a naturally occurring family of organic molecules of biochemical and/or medicinal interest. Members of this family derived from cholesterol include some sex hormones and bile salts that function in digestion. For many years a great deal of controversy has surrounded various steroids. We worry about the amount of cholesterol in the diet and the possible health effects. We are concerned about the use of anabolic steroids by athletes who wish to build muscle mass and improve their performance. However, members of this family of molecules derived from cholesterol have many important functions in the body. The bile salts that aid in the emulsification and thus digestion of lipids are steroid molecules, as are the sex hormones testosterone and estrone.

Sections 9.4 and 15.1

All steroids are structured around the steroid nucleus (steroid carbon skeleton) as shown here:

HO; carbons numbered 1–19

Steroid nucleus

The steroid carbon skeleton consists of four fused rings. Two fused rings share one or more common bonds as part of their ring backbones. For example, rings A and B, B and C, and C and D are all fused in the preceding structure. Note also that many steroids commonly contain two methyl groups, bonded to carbons 18 and 19, as a part of their carbon skeleton.

Cholesterol, a common steroid, is found in the membranes of most animal cells (Figure 9.8). It is readily soluble in the hydrophobic region of membranes and is the principal membrane lipid involved in regulation of the fluidity of the membrane. There is a strong correlation between the concentration of cholesterol found in the blood plasma and heart disease, particularly hardening of the arteries (atherosclerosis). Cholesterol, in combination with other substances, appears to coat the arteries, resulting in a narrowing. As this narrowing increases, more and more pressure is necessary to ensure adequate blood flow. The pressure in the blood vessels increases, and high blood pressure (hypertension) develops. Hypertension is also linked to heart disease.

FIGURE 9.8
Cholesterol crystals removed from biological tissue.

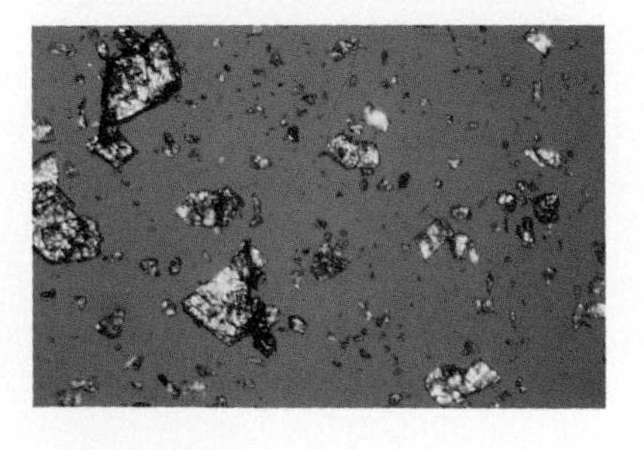

Because egg yolks contain a high concentration of cholesterol, as do many other dairy products and animal fats, it has been recommended that the amounts of these products in the diet be regulated to moderate the dietary intake of cholesterol.

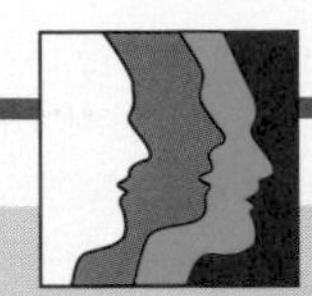

A HUMAN PERSPECTIVE

Anabolic Steroids and Athletics

In the 1988 Summer Olympics, Ben Johnson of Canada ran the fastest 100-meter race in history, 9.79 seconds. He was awarded the Gold Medal for this achievement. Little more than two days later, Michele Verdier of the International Olympic Committee stood before a press conference and read the following statement: "The urine sample of Ben Johnson, Canada, Athletics, 100 meters, collected Saturday, 24th September 1988, was found to contain the metabolites of a banned substance, namely stanozolol, an anabolic steroid." Johnson was disqualified, and Carl Lewis of the United States became the Olympic Gold Medalist in the 100-meter race.

Why do athletes competing in power sports take anabolic steroids? Use of anabolic steroids has a number of desirable effects for the athlete. First, they help to build the muscle mass needed to succeed in sprints or weight lifting. They hasten the healing of muscle damage caused by the intense training of the competitive athlete. Finally, anabolic steroids may help the athlete maintain an aggressive attitude, not just during the competition but also throughout training.

If these hormones have such beneficial effects, why not allow all athletes to use them? Unfortunately, the beneficial effects are far outweighed by the negative side effects. These include kidney and liver damage, stroke, impotence and infertility, an increase in cardiovascular disease, and extreme aggressive behavior.

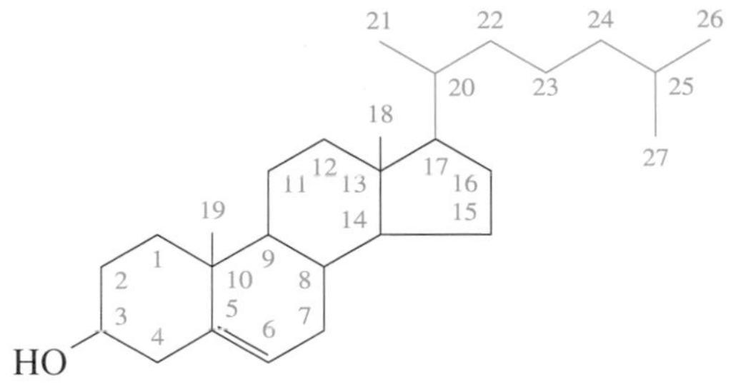

C_{27}
Cholesterol

Many steroids play roles in the reproductive cycle. Cholesterol, in a series of chemical reactions, is converted to the steroid **progesterone,** the most important hormone associated with pregnancy. Produced in the ovaries and in the placenta, progesterone is responsible for both the successful initiation and completion of pregnancy. Progesterone prepares the lining of the uterus to accept the fertilized egg. Once the egg is attached, progesterone is involved in the development of the fetus and also plays a role in the suppression of further ovulation during pregnancy.

Testosterone, a male sex hormone found in the testes, and **estrone,** a female sex hormone, are both produced by the chemical modification of progesterone. Both hormones are involved in the development of the traditional male and female sex characteristics.

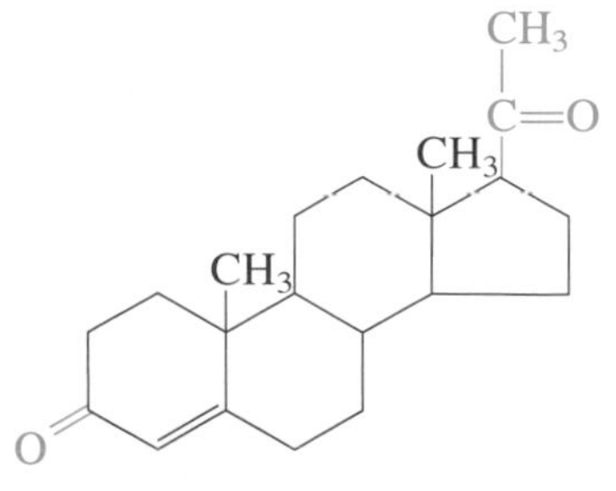

Progesterone

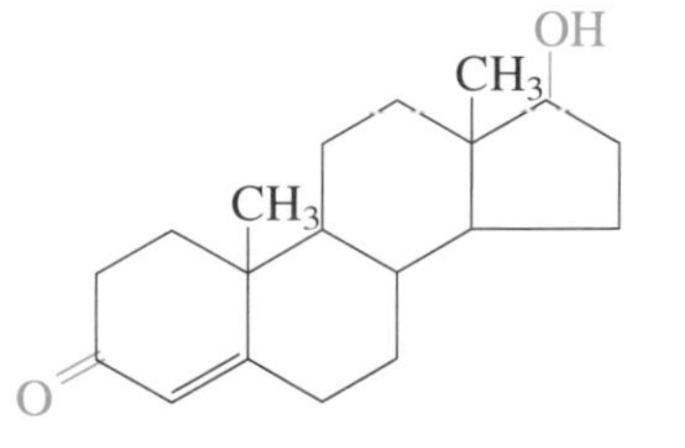

Testosterone

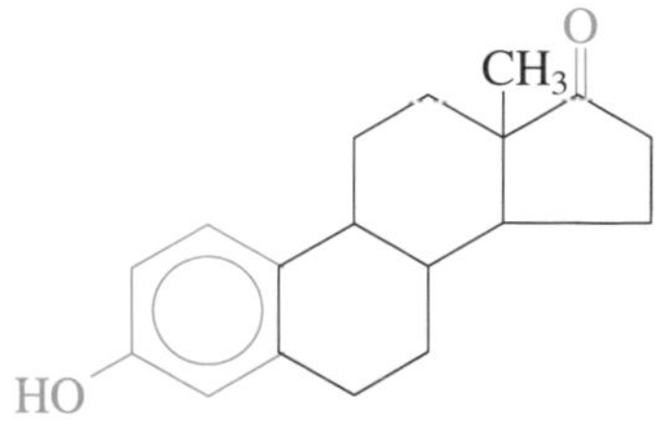

Estrone

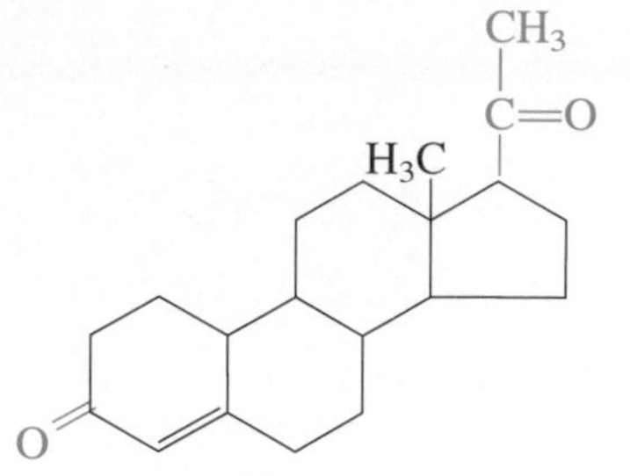

19-Norprogesterone

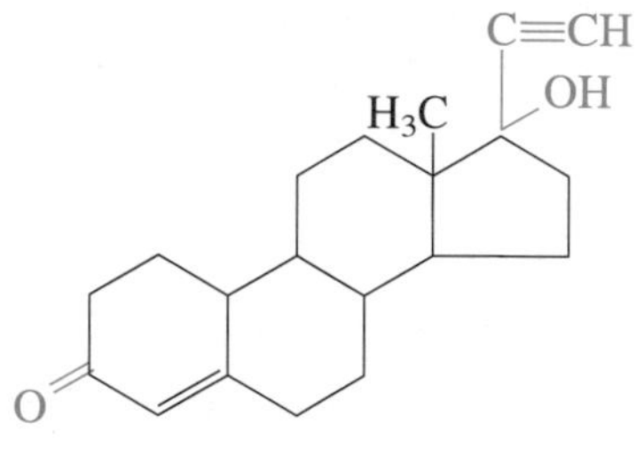

Norlutin

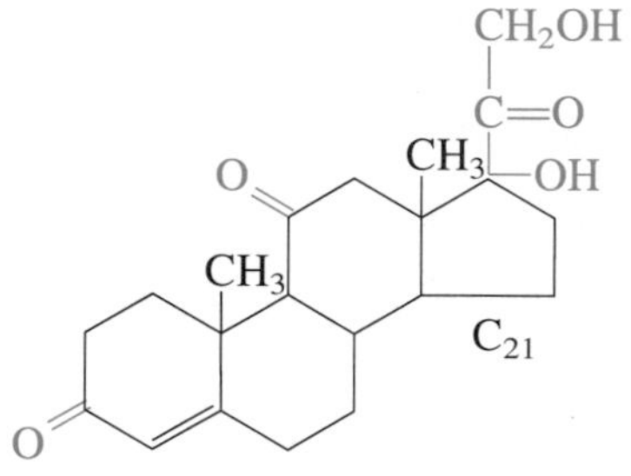

Cortisone

Many steroids, for example, progesterone itself, have played important roles in the development of birth control agents. 19-**Norprogesterone,** for example, was one of the first synthetic birth control agents. It is approximately ten times as effective as progesterone in providing birth control. However, because this compound could not be administered orally and had to be taken by injection, its utility was severely limited. A related compound, **norlutin** (chemical name: 17-α-ethynyl-19-nortestosterone), was found to provide both the strength and effectiveness of 19-norprogesterone and could be taken orally. A number of oral birth control agents have been marketed that are similar to norlutin in structure. All of these compounds act by inducing a false pregnancy in the female. The false pregnancy prevents the ovulation process. Though there have been problems associated with ''the pill,'' for the majority of the population it appears to be an effective and safe method of family planning.

Cortisone is also important to the proper regulation of a number of biochemical processes. For example, it is involved in the metabolism of carbohydrates. Cortisone is also a very important medicinal agent that is used in the treatment of rheumatoid arthritis, asthma, gastrointestinal disorders, many skin conditions, and a variety of other diseases. However, treatment with cortisone is not without risk. Some of the possible side effects of cortisone therapy include fluid retention, sodium retention, and potassium loss, which can lead to congestive heart failure, muscle weakness, osteoporosis, gastrointestinal upsets including peptic ulcers, and neurological symptoms including headaches, vertigo, and convulsions.

Question 9.9 Draw the structure of the steroid nucleus. Note the locations of the A, B, C, and D steroid rings. What is meant by the term ''fused ring''?

Waxes

Waxes are derived from many different sources and have a variety of chemical compositions, depending on the source. *Paraffin wax,* for example, is composed of a mixture of solid hydrocarbons (usually straight-chain compounds). *Carbowax,* in contrast, is a polyether that is prepared synthetically. More generally, the natural **waxes** are composed of a long-chain fatty acid esterified to a long-chain alcohol. Because the head group (COO^-) is only very weakly hydrophilic, while the long hydrocarbon tails are extremely hydrophobic, waxes are completely insoluble in water. Waxes are also solid at room temperature, owing to their high molecular weight. Two examples of waxes are myricyl palmitate, a major component of *beeswax,* and *spermaceti wax* from the head of the sperm whale, which is composed primarily of cetyl palmitate.

$$CH_3(CH_2)_{14}-\overset{\overset{\large O}{\|}}{C}-O-(CH_2)_{29}CH_3$$

Myricyl palmitate
(beeswax)

$$CH_3(CH_2)_{14}-\overset{\overset{\large O}{\|}}{C}-O-(CH_2)_{15}CH_3$$

Cetyl palmitate
(spermaceti wax)

Other naturally occurring waxes include lanolin, which serves as a protective coating for hair and skin; carnauba wax, used in automobile polish; and whale oil, once used as a fuel, in ointments, and in candles. Synthetic waxes have replaced whale oil to a large extent, resulting from efforts to ban the hunting of whales.

9.5 COMPLEX LIPIDS

Complex lipids are lipids that are bonded to other types of molecules. The most common and important complex lipids are plasma lipoproteins, which are responsible for the transport of other lipids in the body.

Lipids are only sparingly soluble in water, and the movement of lipids from one organ to another through the blood stream requires a transport system that operates via **plasma lipoproteins.** Lipoprotein particles consist of a core of hydrophobic lipids. The shell around the core consists of polar lipids and proteins (Figure 9.9).

There are four major classes of human plasma lipoproteins. **Chylomicrons** carry triglycerides from the intestines to other tissues, except kidney. The remaining lipopro-

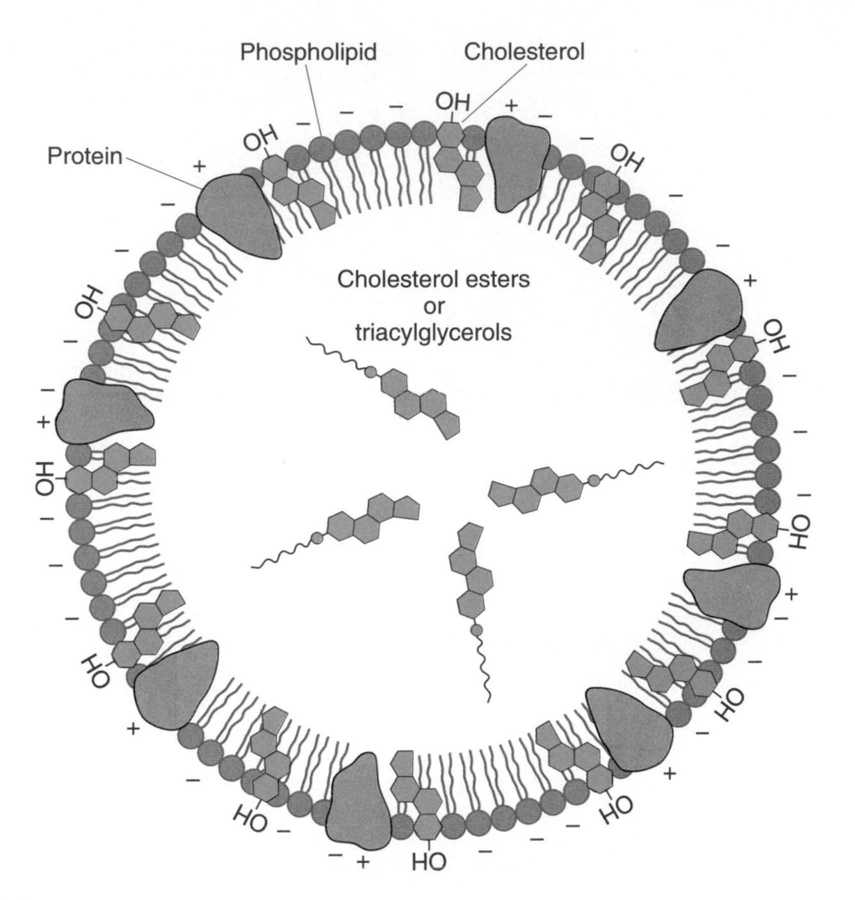

FIGURE 9.9
A model for the structure of a plasma lipoprotein. The various lipoproteins are composed of a shell of protein, cholesterol, and phospholipids surrounding more hydrophobic molecules such as triglycerides or cholesterol esters (cholesterol esterified to fatty acids).

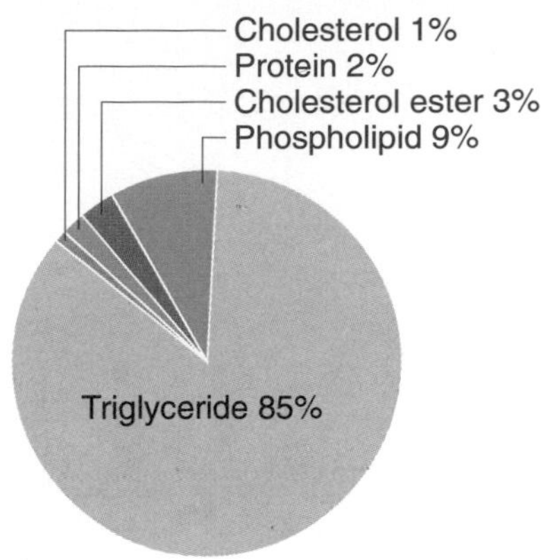

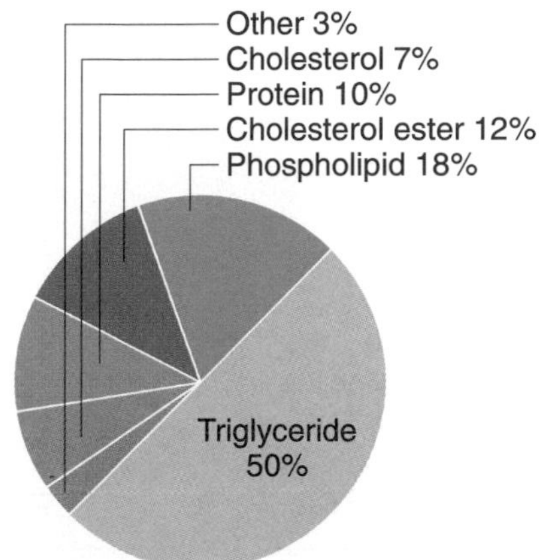

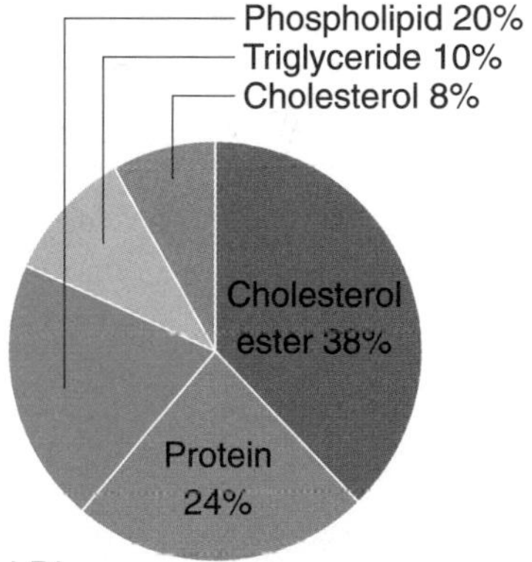

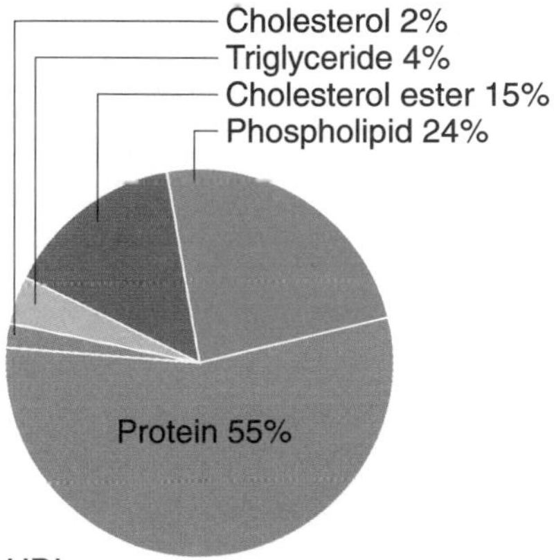

teins are classified by their densities. **Very low-density lipoproteins (VLDL)** bind triglycerides that are synthesized in the liver and carry them to adipose and other tissues for storage. **Low-density lipoproteins (LDL)** carry cholesterol to peripheral tissues and help to regulate cholesterol levels in those tissues. These are richest in cholesterol, frequently carrying 40% of the plasma cholesterol. **High-density lipoproteins (HDL)** are bound to plasma cholesterol; however, they transport cholesterol from peripheral tissues to the liver. A summary of the composition of each of the plasma lipoproteins is presented in Figure 9.10.

A general scheme for transport of lipids in lipoprotein complexes between tissues is summarized in Figure 9.11.

The details of the digestion of dietary fats are described in Chapter 15. Briefly, dietary fat is emulsified in the small intestines and triglycerides are hydrolyzed by pancreatic lipase to liberate free fatty acids and monoglycerides. These are absorbed by the cells lining the intestines and reassembled into triglycerides. Aggregates of triglycerides and protein (chylomicrons) are produced that eventually enter the bloodstream for transport to cells throughout the body. Not all lipids in the blood are derived directly from the diet. Triglycerides and cholesterol are also synthesized in the liver, and these lipids, too, are

FIGURE 9.10
A summary of the relative amounts of cholesterol, phospholipid, protein, triglycerides, and cholesterol esters in the four classes of lipoproteins.

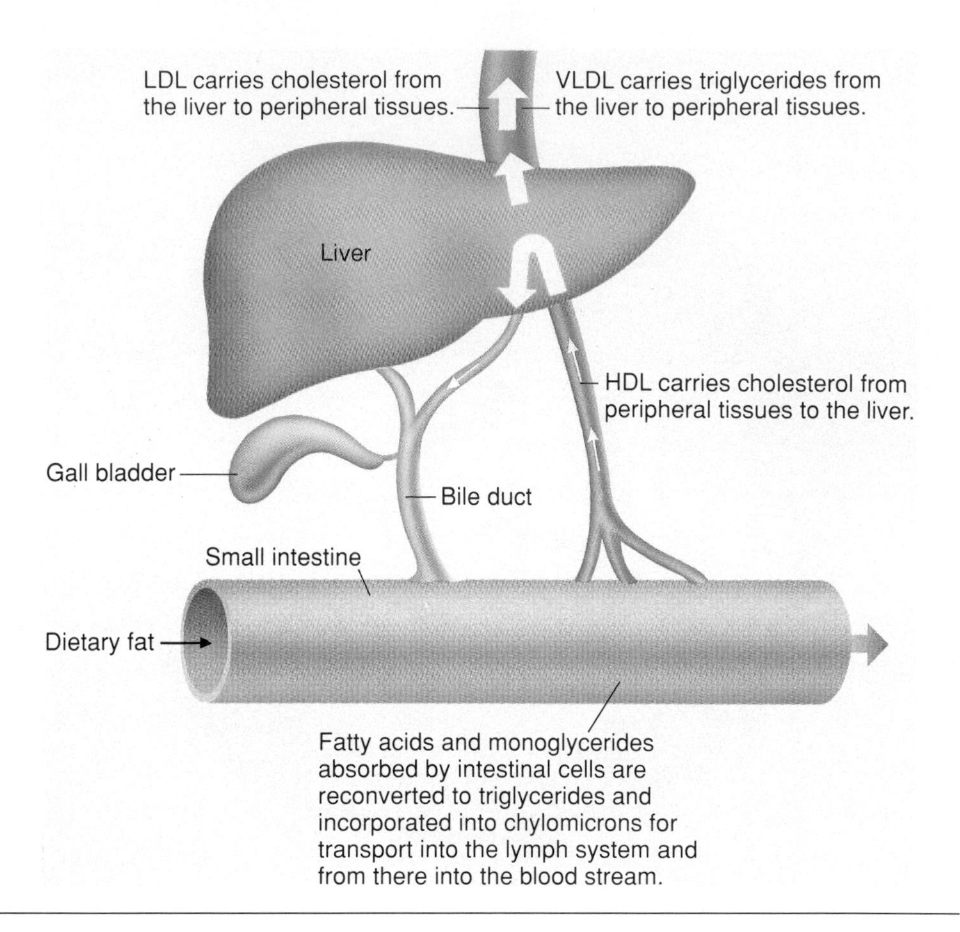
LDL carries cholesterol from the liver to peripheral tissues.
VLDL carries triglycerides from the liver to peripheral tissues.
Liver
HDL carries cholesterol from peripheral tissues to the liver.
Gall bladder
Bile duct
Small intestine
Dietary fat
Fatty acids and monoglycerides absorbed by intestinal cells are reconverted to triglycerides and incorporated into chylomicrons for transport into the lymph system and from there into the blood stream.

FIGURE 9.11
General scheme for the transport of lipoproteins between tissues in the human body.

transported through the blood in lipoprotein packages. Triglycerides are assembled into VLDL particles, which carry the energy-rich lipid molecules either to tissues requiring an energy source or to adipose tissue for storage. Similarly, cholesterol is assembled into LDL particles for transport from the liver to peripheral tissues.

Entry of LDL particles into the cell depends on a specific recognition event and binding between the LDL particle and a protein receptor embedded within the cell membrane. Low-density lipoprotein receptors (LDL receptors) are responsible for the uptake of cholesterol by the cells of various tissues. LDL (lipoprotein bound to cholesterol) binds specifically to the LDL receptor and the complex is taken into the cell by a process called **receptor-mediated endocytosis** (Figure 9.12). The membrane begins to invaginate; in other words, it is pulled into the cell at the site of the LDL receptor complex. This draws the entire LDL molecule and its receptor into the cell. Eventually, the invagination becomes very deep, and the portion of the membrane surrounding the LDL particle pinches away from the cell membrane and forms a membrane around the LDL particle. As we will see in Section 9.6, membranes are fluid and readily flow to form such vesicles or endosomes. Digestive cell organelles known as **lysosomes** fuse with the endosomes. This fusion is accomplished when the membranes of the endosome and the lysosome flow together to create one larger membrane-bound body or vesicle. Hydrolytic enzymes from the lysosome then digest the entire complex to release cholesterol into the cytoplasm of the cell. There, cholesterol inhibits its own biosynthesis and activates an enzyme that stores cholesterol in cholesterol ester droplets. Cholesterol also inhibits the synthesis of LDL receptors to ensure that the cell will not take up too much cholesterol. People who have a genetic defect in the gene coding for the LDL receptor accumulate LDL-cholesterol in the plasma. This excess plasma cholesterol is then deposited in the artery wall, inducing **atherosclerosis.**

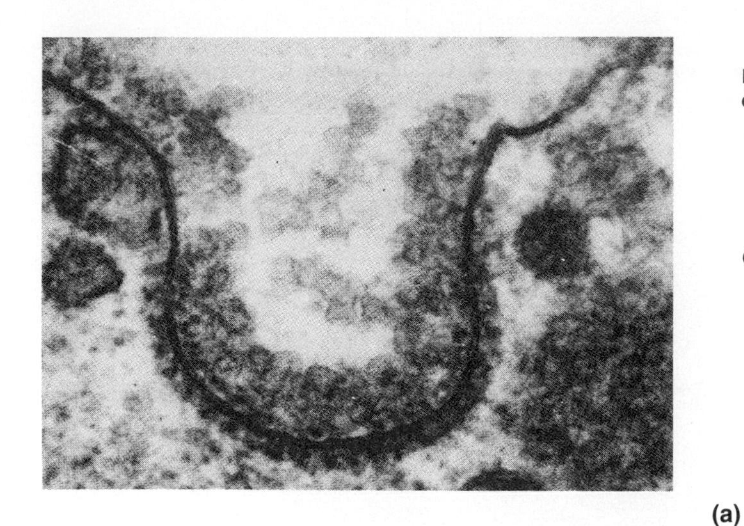

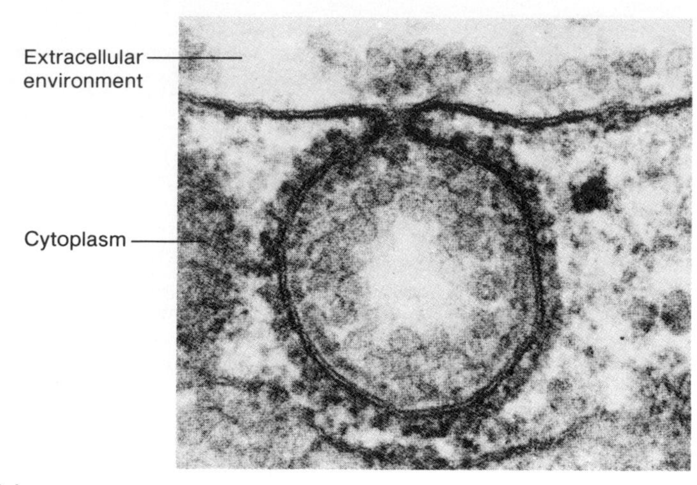

(a)

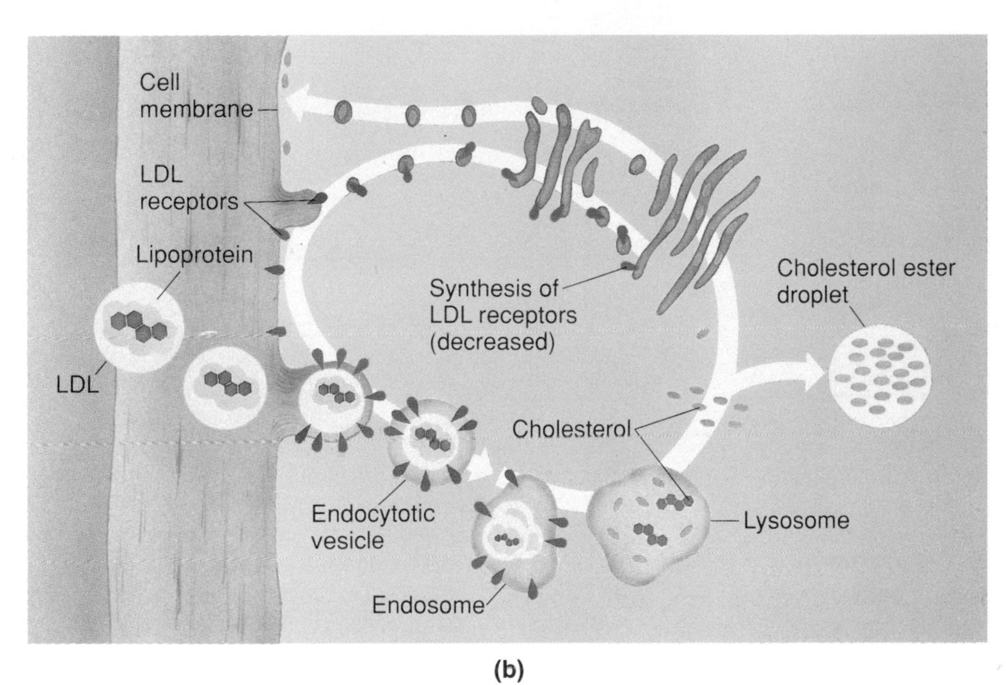

(b)

FIGURE 9.12
Receptor-mediated endocytosis. (a) Electron micrographs of the process of receptor-mediated endocytosis. (b) Summary of the events of receptor-mediated endocytosis of LDL and the fate of the cholesterol brought into the cell.

Liver LDL receptors enable large amounts of cholesterol to be removed from the blood, thus ensuring low concentrations of cholesterol in plasma. Other factors being equal, the person with the most lipoprotein receptors will be the least vulnerable to a high-cholesterol diet and will have the least likelihood of developing atherosclerosis.

There is also evidence that high levels of HDL in the blood help to reduce the incidence of atherosclerosis. This may be due to the fact that HDL carries cholesterol from the peripheral tissues back to the liver. In the liver, some of the cholesterol is used for bile synthesis and secreted into the intestines, from which it is excreted.

Sections 9.5 and 15.1

A final correlation has been made between diet and atherosclerosis. Those whose diet is high in saturated fats also tend to have high levels of cholesterol in the blood. Although

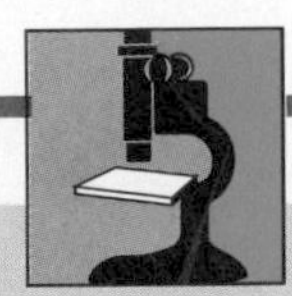

A CLINICAL PERSPECTIVE

Steroids and the Treatment of Heart Disease

The foxglove plant *(Digitalis purpurea)* is an herb that produces one of the most powerful stimulants of heart muscle known. It is used in the preparation of "May wine" and largely accounts for the stimulating effects of that particular wine. The active ingredients of the foxglove plant (digitalis) are the so-called cardiac glycosides or cardiotonic steroids, which include digitoxin, digoxin, and gitalin.

Digitoxin

The structure of digitoxin, one of the cardiotonic steroids produced by the foxglove plant.

Digitalis purpurea, the foxglove plant.

These drugs are used clinically in the treatment of congestive heart failure. Congestive heart failure results when the heart is not beating with strong, efficient strokes. When the blood is not propelled through the cardiovascular system efficiently, fluid builds up in the lungs and lower extremities (edema). The major symptoms of congestive heart failure are an enlarged heart, weakness, edema, shortness of breath, and fluid accumulation in the lungs.

This condition was originally described in 1785 by a physician, William Withering, who found a peasant woman whose folk medicine was famous far and wide as a treatment for chronic heart problems. Her potion contained a mixture of more than 20 herbs, but Dr. Withering, a botanist as well as a physician, quickly discovered that foxglove was the active ingredient of this mixture. Withering used *Digitalis purpurea* successfully to treat congestive heart failure and even described some cautions in its use.

The cardiac glycosides are extremely strong heart stimulants. A dose as low as 1 mg increases the stroke volume of the heart (volume of blood per contraction), increases the strength of the contraction, and reduces the heart rate. When the heart is pumping more efficiently as a result of the stimulation of digitalis, the edema disappears. A long train of biochemical events is set in motion by digitalis. Its first effect is inhibition of the Na^+-K^+ ATPase, a membrane protein described in Section 9.6 that pumps Na^+ out of the cell and K^+ into the cell. When the Na^+-K^+ ATPase is inhibited, the Na^+ concentration in the cells increases. This event triggers the release of Ca^{2+}, which, in turn, stimulates heart muscle contraction.

Digitalis can be used to control congestive heart failure, but the dose must be carefully determined and monitored because the therapeutic dose is close to the dose that causes toxicity. The symptoms that result from high body levels of cardiac glycosides include vomiting, blurred vision and lightheadedness, increased water loss, convulsions, and death. Only a physician can determine the initial dose and maintenance schedule for an individual to control congestive heart failure and avoid the toxic side effects.

the relationship between saturated fatty acids and cholesterol metabolism is unclear, it is known that a diet rich in unsaturated fats results in decreased cholesterol levels. In fact, the use of unsaturated fat in the diet results in a decrease in the level of LDL and an increase in the level of HDL. With the positive correlation between heart disease and high cholesterol levels, current dietary recommendations include a diet that is low in fat and the substitution of unsaturated fats, such as vegetable oils, for saturated fats, such as animal fats.

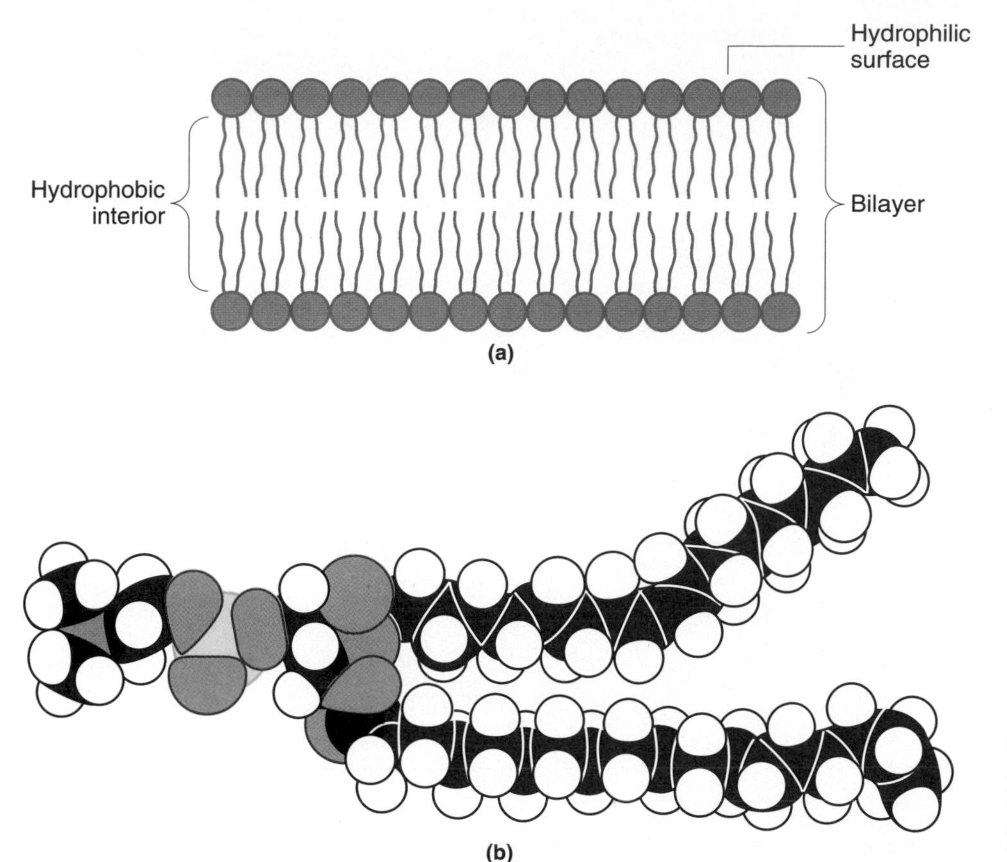

FIGURE 9.13
Schematic diagram of a lipid bilayer. (a) Schematic drawing; (b) view using drawings of "space-filling" models.

9.6 THE STRUCTURE OF BIOLOGICAL MEMBRANES

Biological membranes are *lipid bilayers* in which the hydrophobic hydrocarbon tails are packed in the center of the bilayer and the ionic head groups are exposed on the surface (Figure 9.13). The formation of lipid bilayers is driven by the tendency of the hydrocarbon tails of the lipids to escape from water. The hydrocarbon tails of membrane phospholipids provide a thin shell of nonpolar material that prevents mixing of molecules on either side. The nonpolar tails of membrane phospholipids thus provide a barrier between the cell and its surroundings. The polar heads of lipids are exposed to water, and they are highly solvated. Little exchange, known colloquially as "flip-flop," occurs between lipids on the outer and inner halves of the bilayers (Figure 9.14). The movement of a lipid molecule within one sheath of the bilayer, by contrast, is rapid. A bacterial cell is about 2 μm long, and a lipid molecule diffuses from one end of the cell to the other in a second.

FIGURE 9.14
Lateral diffusion in a biological membrane is rapid, but "flip-flop" across the membrane is very slow and almost never occurs.

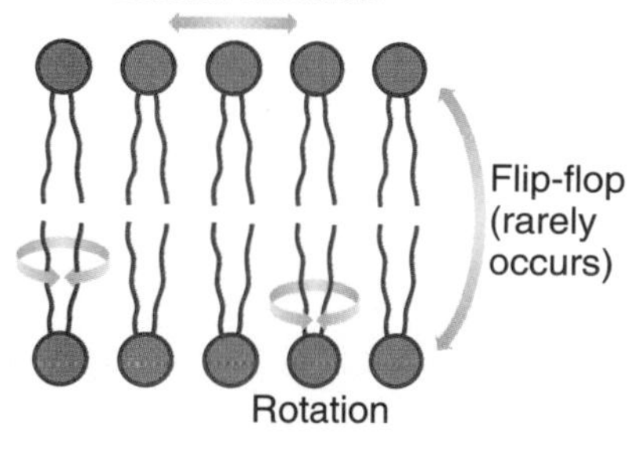

Fluid mosaic structure of biological membranes

As we have just noted, membranes are not static, they are composed of molecules in motion. The fluidity of biological membranes is determined by the proportions of saturated and unsaturated fatty acid groups in the membrane phospholipids. About half of the fatty acids that are isolated from membrane lipids from all sources are unsaturated. We also find that the percentage of unsaturated fatty acid groups in membrane lipids is inversely proportional to the temperature of the environment. Bacteria, for example, have different ratios of saturated and unsaturated fatty acid moieties in their membrane lipids, depending on the temperatures of their surroundings. For instance, the membranes of

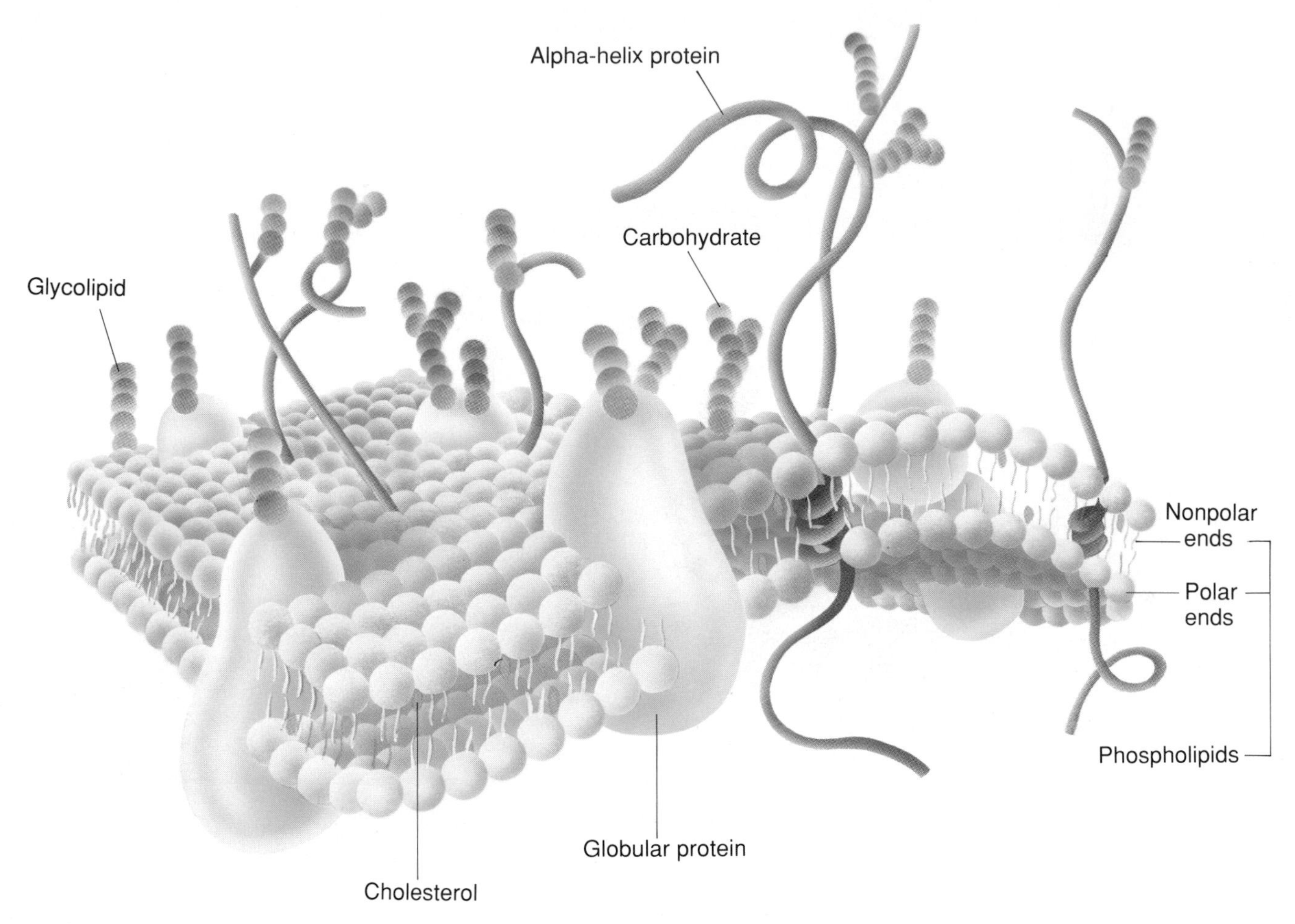

FIGURE 9.15
The fluid mosaic model of membrane structure.

bacteria that grow in the Arctic Ocean have high levels of unsaturated fatty acids so that their membranes remain fluid even at these frigid temperatures. Conversely, the organisms that live in the hot springs of Yellowstone National Park, with temperatures near the boiling point of water, have membranes with high levels of saturated fatty acids. This flexibility in fatty acid content enables the bacteria to maintain the same membrane fluidity over a temperature range of almost 100°C.

Generally, the body temperatures of mammals are quite constant, and the fatty acid composition of their membrane lipids is therefore usually very uniform. One interesting exception is the reindeer. Much of the year the reindeer must travel through ice and snow. Thus the hooves and lower legs must continue to function at much colder temperatures than the rest of the body. Because of this, the percentage of unsaturation in the membranes varies along the length of the reindeer leg. We find that the proportion of unsaturated fatty acid groups increases closer to the hoof. The lower freezing points and greater fluidity of lipids that contain a high proportion of unsaturated fatty acid groups permit the membranes to function in the low temperatures of ice and snow to which the hoof is exposed.

Thus membranes are fluid, regardless of the environmental temperature conditions. In fact, it has been estimated that membranes have the consistency of light machine oil.

Although the hydrophobic barrier created by the fluid lipid bilayer is an important feature of membranes, the proteins embedded within the lipid bilayer are equally important and are responsible for critical cellular functions. The presence of these membrane proteins was revealed by an electron microscopic technique called freeze-fracture. Cells are frozen to very cold temperatures and then fractured with a very fine diamond knife. Some of the cells are fractured between the two layers of the lipid bilayer. When viewed with the electron microscope, the membrane appeared to be a mosaic, studded with proteins. Because of the consistency of membranes and the physical appearance by electron microscopy, our concept of membrane structure is called the **fluid mosaic theory** (Figure 9.15).

Some of the observed proteins, called **peripheral** membrane proteins, are bound only to one of the surfaces of the membrane by interactions between ionic head groups of the membrane lipids and ionic residues on the surface of the peripheral protein. Other membrane proteins, termed **integral** membrane proteins, are found to be embedded within the membrane and to extend completely through it, being exposed both inside and outside the cell. Since the lipid bilayer is fluid, there can be rapid lateral diffusion of membrane proteins through the lipid bilayer; but membrane proteins, like membrane lipids, do not "flip-flop" across the membrane or turn in the membrane like a revolving door of a department store.

Membranes are dynamic structures, as we may infer from our knowledge about the mobility of membrane proteins and lipids. The mobility of proteins embedded in biological membranes was studied by labeling certain proteins in human and mouse cell membranes with red and green fluorescent dyes. The human and mouse cells were fused; in other words, special techniques were used to cause the membranes of the mouse and human cell to flow together to create a single cell. The new cell was observed through a special ultraviolet or fluorescence microscope. The red and green patches were localized within regions of their original cell membranes when the experiment began. An hour later the color patches were uniformly distributed in the fused cellular membrane (Figure 9.16). This experiment suggests that we can think of the fluid mosaic membrane as an ocean filled with mobile, floating icebergs.

Membrane transport

The cell membrane mediates the interaction of the cell with its environment and is responsible for the controlled passage of material into and out of the cell. The external cell membrane controls the entrance of fuel and the exit of waste products and internal cellular membranes partition metabolites among cell organelles. Most of these transport processes are controlled by integral membrane transport proteins. These transport proteins are the

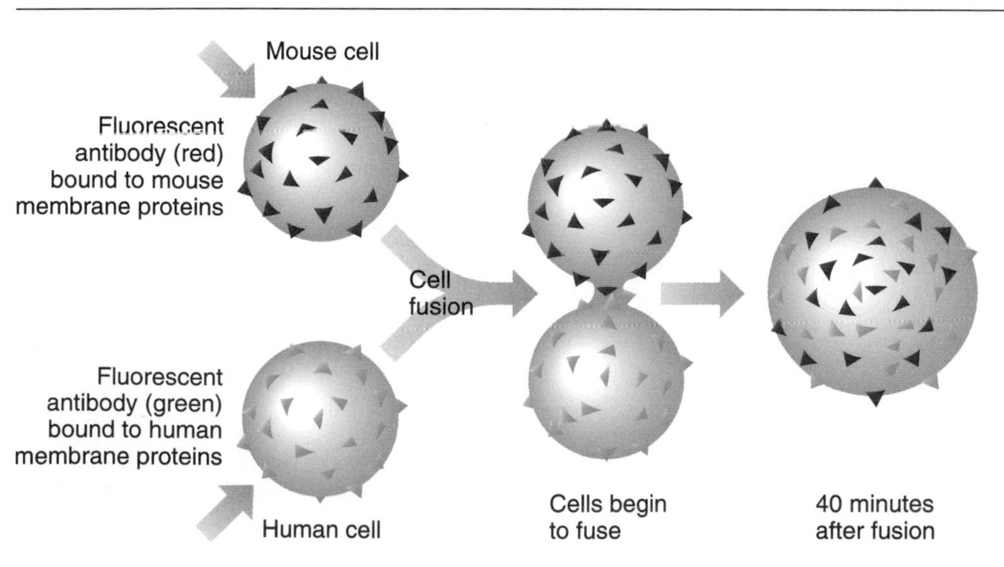

FIGURE 9.16
Demonstration that membranes are fluid and that proteins move freely in the plane of the lipid bilayer.

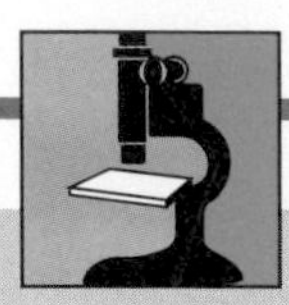

A CLINICAL PERSPECTIVE

Antibiotics That Destroy Membrane Integrity

The Age of Antibiotics began in 1927, when Alexander Fleming discovered, quite by accident, that a product of the mold *Penicillium* can kill susceptible bacteria. We now know that penicillin inhibits bacterial growth by interfering with cell wall synthesis (see Chapter 7, "A Clinical Perspective: The Bacterial Cell Wall"). Since Fleming's time, hundreds of antibiotics, microbial products that either kill or inhibit the growth of susceptible bacteria or fungi, have been developed. The key to antibiotic therapy is to find a "target" in the microbe, some metabolic process or structure that the human does not have. In this way the antibiotic will selectively inhibit the disease-causing organism without harming the patient.

Many antibiotics disrupt cell membranes. Because human cells also have membranes, these antibiotics are often used to combat infections on body surfaces, so that damage to the host is minimized but the inhibitory effect on the microbe is maximized. When these antibiotics are used in cases of life-threatening illness, they exhibit a wide range of toxic side effects.

Polymyxins (A, B, C, D, and E) are antibiotics produced by the bacterium *Bacillus polymyxa*. They are protein derivatives having one end that is highly hydrophobic because of an attached fatty acid. The opposite end is highly positively charged and therefore hydrophilic. Because of these properties, the polymyxins bind to membranes with the hydrophobic end embedded within the membrane, while the hydrophilic end remains outside the cell.

As a result, the integrity of the membrane is disrupted, and leakage of cellular constituents occurs, causing cell death. Although the polymyxins have been found to be useful in treating some urinary tract infections, pneumonias, and infections of burn patients, other antibiotics are now favored because of the toxic effects of the polymyxins on the kidney and central nervous system. Polymyxin B is still used topically (on body surfaces) and is available as an over-the-counter ointment in combination with neomycin and bacitracin.

Amphotericin B

Nystatin

The structures of amphotericin B and nystatin, two antifungal antibiotics.

Both amphotericin B and nystatin are large ring structures that are used in treating serious systemic fungal infections.

These antibiotics form complexes with cholesterol in the fungal cell membrane, disrupting the membrane permeability and causing leakage of cellular constituents. Neither is useful in treating bacterial infections because most bacteria have no cholesterol in their membranes. Both amphotericin B and nystatin are extremely toxic, causing symptoms that include nausea and vomiting, fever and chills, anemia, and renal failure. It is easy to understand why the use of these drugs is restricted to life-threatening fungal disease.

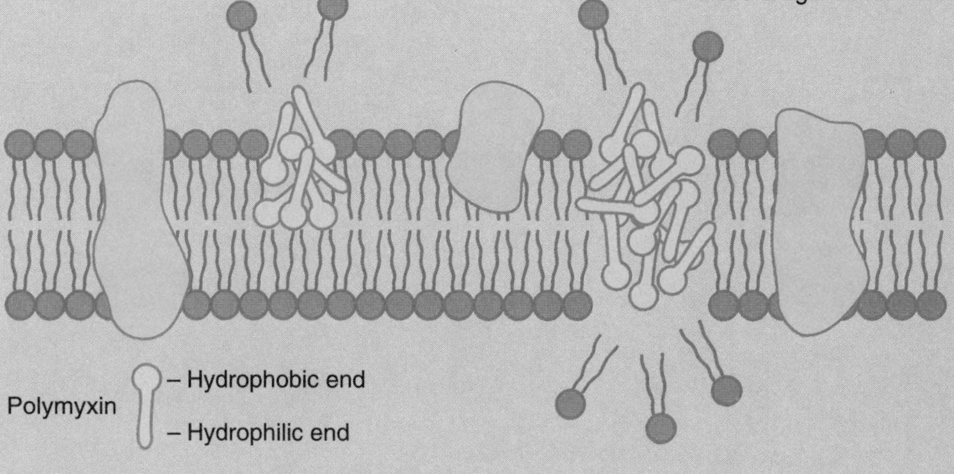

The mechanism of action of polymyxins. The hydrophilic end of these peptide derivatives remains outside the cell, while the hydrophobic end is embedded within the membrane. In this way it acts like a detergent, disrupting membrane integrity and killing the cell.

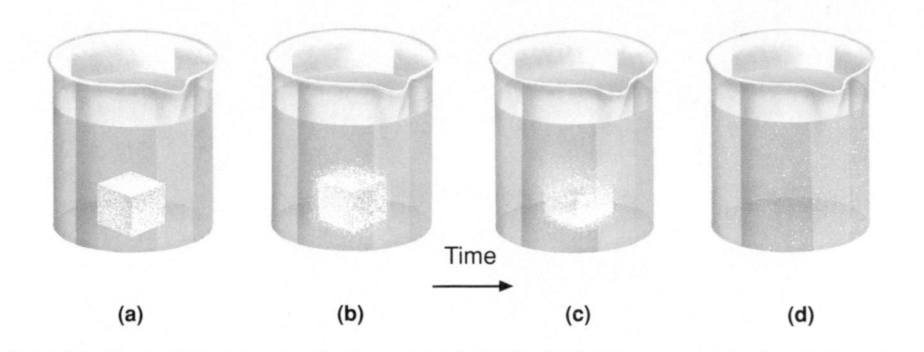

FIGURE 9.17
Diffusion results in the *net* movement of sugar and water molecules from the area of high concentration to the area of low concentration. Eventually, the concentration of sugar throughout the beaker will be equal.

cellular gate-keepers, whose function in membrane transport is analogous to the function of enzymes in carrying out cellular chemical reactions. However, some molecules pass through the membranes unassisted, by the **passive transport** processes of diffusion and osmosis. These are referred to as passive processes because they do not require any energy expenditure by the cell.

Passive diffusion: the simplest form of membrane transport

If we put a teaspoon of instant iced tea on the surface of a glass of water, the tea molecules soon spread throughout the solution. The molecules of both the solute (tea) and the solvent (water) are propelled by the random molecular motion. The initially concentrated tea becomes more and more dilute. This process of the *net* movement of a solute with the gradient (from an area of high concentration to an area of low concentration) is called *diffusion* (Figure 9.17).

Diffusion is one means of passive transport across membranes. Let us now suppose that a biological membrane is present and that a substance is found at one concentration outside the membrane and at half that concentration inside the membrane. Assuming that the solute can pass through the membrane, diffusion will occur with net transport of material from the region of initial high concentration to the region of initial low concentration, and the substance will equilibrate across the cell membrane (Figure 9.18). After a while, the concentration of substance will be the same on both sides of the membrane, the system will be at equilibrium, and no more spontaneous change will occur.

Of what practical value is the process of diffusion to the cell? Certainly, diffusion is able to effectively distribute metabolites throughout the interior of the cell. But what about the movement of molecules through the membrane? Because of the lipid bilayer structure of the membrane, only a few molecules are able to diffuse freely across a membrane. These include small molecules such as O_2, CO_2, and H_2O. Any large or highly charged molecules will not be able to pass through the lipid bilayer directly. Such molecules require an assist from some cell membrane proteins in order to pass through the membrane. Any membrane that allows the diffusion of some molecules but not others is said to be *selectively permeable*.

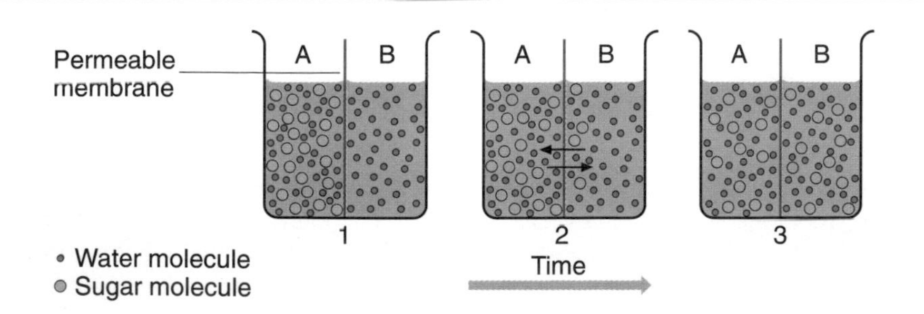

FIGURE 9.18
Diffusion of a solute through a membrane.

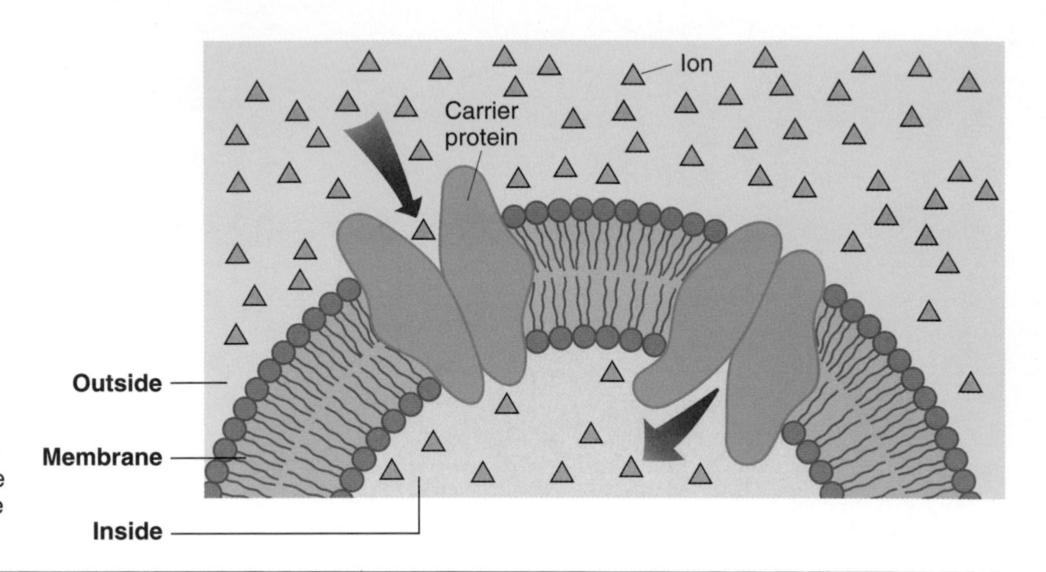

FIGURE 9.19
Transport by facilitated diffusion occurs through pores in the transport protein whose size and shape are complementary to those of the transported molecule.

Facilitated diffusion: specificity of molecular transport

Most molecules are transported across biological membranes by specific protein carriers known as **permeases.** When a solute diffuses through a membrane from an area of high concentration to an area of low concentration by passing through a channel within a permease, the process is known as **facilitated diffusion.** No energy is consumed by facilitated diffusion; thus it is another means of passive transport, and the direction of transport will depend upon the concentrations of metabolite on each side of the membrane.

Transport across cell membranes by facilitated diffusion occurs through pores within the permease that have conformations, or shapes, that are complementary to those of the transported molecules. The charge and conformation of the pore define the stereospecificity of the carrier (Figure 9.19). Only molecules that have the correct shape can enter the pore. As a result, the rate of diffusion for any molecule is limited by the number of carrier permease molecules in the membrane that are responsible for the passage of that molecule.

The transport of glucose illustrates the specificity of carrier permease proteins. D-Glucose is transported by the glucose carrier, but its enantiomer, L-glucose, is not. Thus the glucose permease exhibits stereospecificity. In other words, the solute to be brought into the cell must "fit" precisely into a recognition site within the structure of the permease. In Chapter 12 we will see that enzymes that catalyze the biochemical reactions within cells show this same type of specificity.

Section 12.5

The rate of transport of metabolites into the cell has profound effects on the net metabolic rate of many cells. Insulin, a polypeptide hormone synthesized by the islet β-cells of the pancreas, increases the maximum rate of glucose transport by a factor of three to four. The result is that the metabolic activity of the cell is greatly increased.

Sections 13.6 and 15.6

Red blood cells use the same anion channel to transport Cl^- into the cell in exchange for HCO_3^- ions (Figure 9.20). This two-way transport is known as *antiport*. This transport occurs by facilitated diffusion, so that Cl^- flows from a high exterior concentration to a low interior one, and bicarbonate flows from a high interior concentration to a low exterior one.

Osmosis: passive movement of a solvent across a membrane

Because a cell membrane is selectively permeable, it is not always possible for solutes to pass through it in response to a concentration gradient. In such cases the solvent diffuses through the membrane. Such membranes, permeable to solvent but not to solute, are specifically called **semipermeable membranes. Osmosis** is the diffusion of a solvent

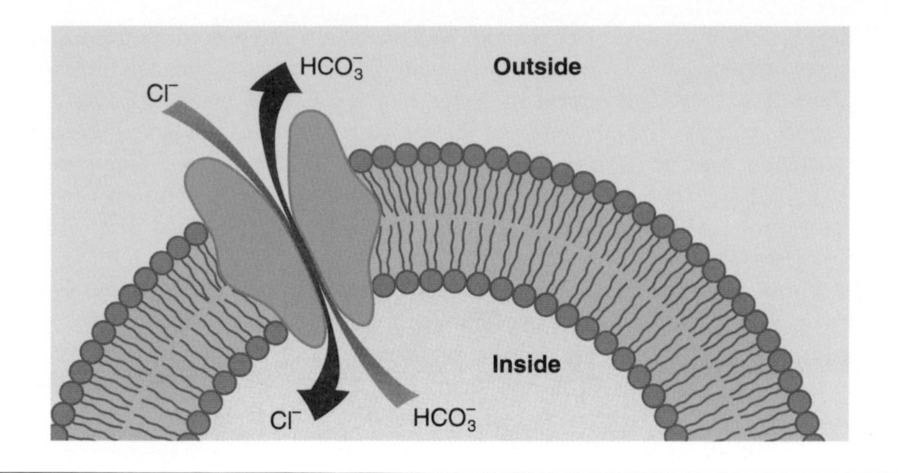

FIGURE 9.20
Transport of Cl^- and HCO_3^- ions in opposite directions across the red blood cell membrane.

(water in biological systems) through a semipermeable membrane in response to a water concentration gradient.

Suppose that we place a 0.5 M glucose solution in a dialysis bag that is composed of a membrane with pores that allow the passage of water molecules but not glucose molecules. Consider what will happen when we place this bag into a beaker of pure water. We have created a gradient in which there is a higher concentration of glucose inside the bag than outside but the glucose cannot diffuse through the bag to achieve equal concentration on both sides of the membrane.

Now let's think about this situation in another way. We have a higher concentration of water molecules outside the bag (where there is only pure water) than inside the bag (where some of the water molecules are occupied in the hydration of solute particles and are consequently unable to move freely in the system). Since water can diffuse through the membrane, a net diffusion of water will occur through the membrane into the bag. This is the process of osmosis (Figure 9.21).

Recall that there is an inverse relationship between the osmotic (solute) concentration of a solution and the water concentration of that solution.

As you have probably already guessed, this system can never reach equilibrium (equal concentrations inside and outside the bag) because regardless of how much water diffuses into the bag, diluting the glucose solution, the concentration of glucose will always be higher inside the bag (and the accompanying free water concentration will always be lower).

What happens when the bag has taken in as much water as it can, when it has expanded as much as possible? Now the walls of the bag exert pressure that will stop the *net* flow of water into the bag. **Osmotic pressure** is the force or pressure that must be exerted to stop the flow of water across a selectively permeable membrane by osmosis. Stated more precisely, the osmotic pressure of a solution is the net force with which water enters it by osmosis from a pure water compartment when the two compartments are separated by a semipermeable membrane.

Osmotic concentration or *osmolarity* is the term used to describe the osmotic strength of a solution. It depends only on the ratio of the number of solute particles to the number of solvent particles. Thus the chemical nature and size of the solute are not important, only the concentration, expressed in molarity. For instance, a 2 M solution of glucose (a sugar of molecular weight 180) has the same osmolarity as a 2 M solution of albumin (a protein of molecular weight 60,000).

Blood plasma has an osmolarity equivalent to a 0.30 M glucose solution or a 0.15 M NaCl solution. This latter is true because in solution NaCl dissociates into Na^+ and Cl^- and thus contributes twice the number of solute particles as a molecule that does not ionize. If red blood cells, which have an osmolarity equal to blood plasma, are placed in a 0.30 M glucose solution, no net osmosis will occur because the osmolarity and water concentration inside the red blood cell are equal to that of the 0.30 M glucose solution. The solutions inside and outside the red blood cell are said to be **isotonic** (*iso* means "same," and *tonic* means "strength"). Since the osmolarity is the same inside and outside, the red blood cell will remain the same size (Figure 9.22).

FIGURE 9.21
Osmosis across a membrane. The solvent, water, diffuses from an area of lower solute concentration to an area of higher solute concentration.

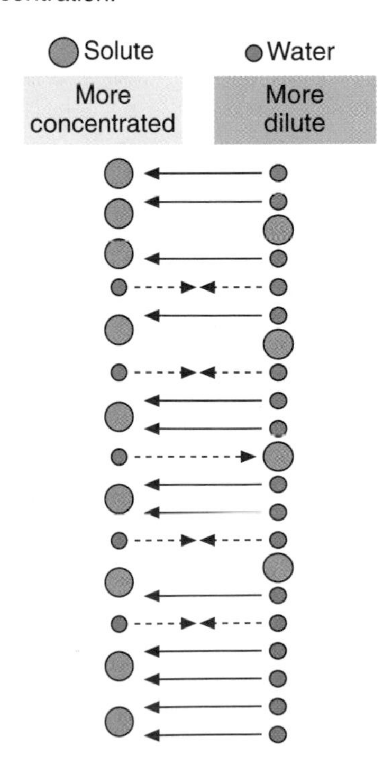

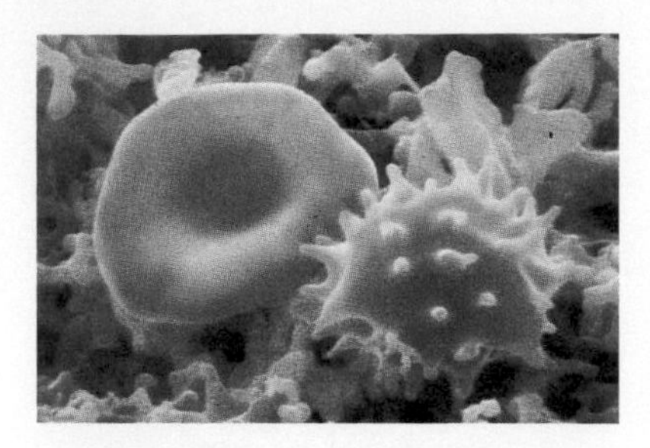

FIGURE 9.22
A colorized scanning electron micrograph of red blood cells exposed to an isotonic solution (left) and a hypertonic solution (right).

What happens if we now place the red blood cells into a **hypotonic solution,** in other words, a solution having a lower osmolarity than the cytoplasm of the cell? In this situation there will be a net movement of water into the cell as water diffuses down its concentration gradient. The membrane of the red blood cell does not have the strength to exert a sufficient pressure to stop this flow of water, and the cell will swell and burst. Alternatively, if we place the red blood cells into a **hypertonic solution** (one with a greater osmolarity than the cell), water will pass out of the cells, and they will shrink dramatically in size.

These principles have important applications in the delivery of intravenous (IV) solutions into an individual. Normally, any fluids infused intravenously must have the correct osmolarity; they must be isotonic with the blood cells and the blood plasma. Such infusions are frequently either 5% dextrose (glucose) or ''normal saline.'' The first solution is composed of 5.0 g of glucose per 100 mL of solution (0.30 M), and the latter of 9.0 g of NaCl per 100 mL of solution (0.15 M). In either case they have the same osmotic pressure and osmolarity as the plasma and blood cells and can therefore be safely administered without upsetting the osmotic balance between the blood and the blood cells.

Body fluids must be maintained within a fairly narrow range of osmotic concentration. For this reason the body has evolved a very complex system to regulate the water balance. This system involves **osmoreceptors** in the hypothalamus of the brain, which

FIGURE 9.23
The response of the body to dehydration. Osmoreceptors in the hypothalamus detect the rise in osmotic pressure and trigger the thirst response. The hypothalamus also stimulates the anterior pituitary to secrete antidiuretic hormone, which decreases the amount of water excreted by the kidneys.

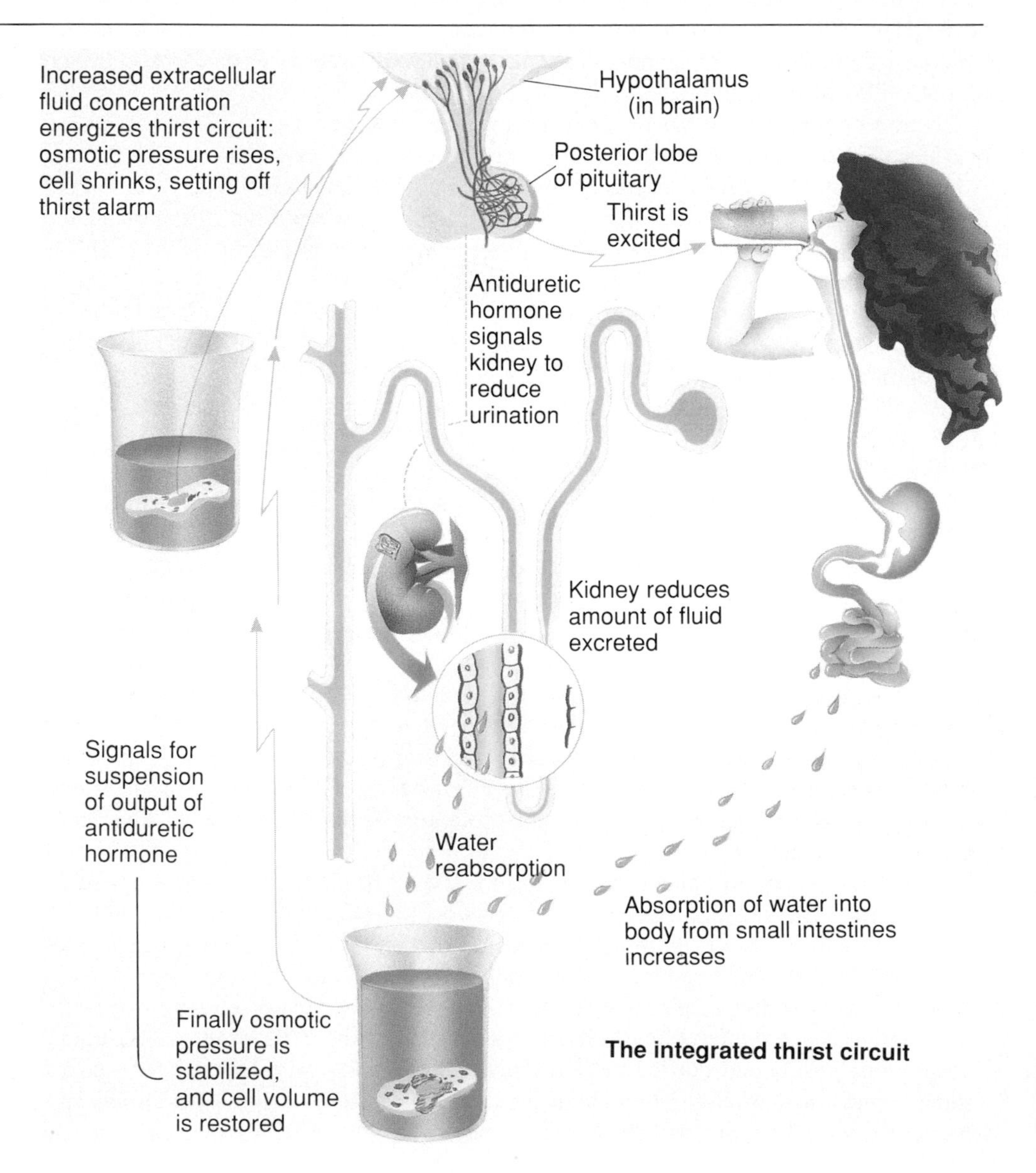

sense changes in blood osmolarity. For instance, when a person becomes dehydrated because of illness or physical exertion, the hypothalamus triggers the thirst response, which stimulates increased water consumption (Figure 9.23). In addition, the hypothalamus stimulates the anterior pituitary to secrete a hormone called the **antidiuretic hormone,** which acts on the kidney, causing it to excrete less water. By these two means the correct osmolarity of the blood will be restored.

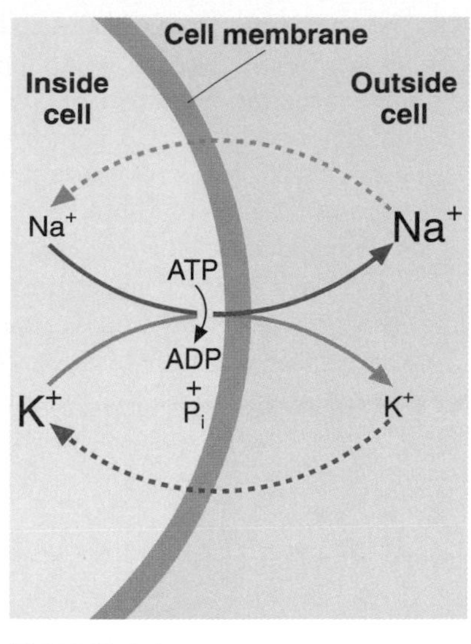

FIGURE 9.24
Schematic diagram of the operation of the Na^+-K^+ ATPase.

Energy requirements for transport

The processes of membrane transport that we have previously considered—simple diffusion, facilitated diffusion, and osmosis—involve the spontaneous flow of material between regions of differing concentration (a concentration gradient). However, to survive, cells must frequently maintain unfavorable concentration gradients across their plasma membranes. In other words, if nutrients are in low concentration outside the cell, in order to provide itself with a sufficient energy supply the cell must move nutrients against the gradient. The transport of a metabolite ''uphill,'' against a concentration gradient, requires energy. This phenomenon is called **active transport.** Many ions and food molecules are imported by active transport through the cell membrane. The energy expended for these transport processes often consumes more than half the total energy produced by cellular metabolic processes. We will consider the important example of the transport of Na^+ and K^+ ions against their concentration gradients.

Most cells generate Na^+-K^+ gradients across their cell membranes; that is, they maintain a high concentration of Na^+ outside the cell and a high concentration of K^+ inside the cell. The maintenance of this gradient requires a continuous expense of cellular energy. The cellular energy currency is adenosine triphosphate, ATP. Hydrolysis of ATP provides the energy that drives the pumping of Na^+ and K^+ ions across cell membranes (Figure 9.24).

Section 13.1

Over a third of the total energy supply of the cell is consumed in maintaining the Na^+-K^+ gradient. The transport system that generates this ion imbalance is known as the *Na^+-K^+ ATPase*. The term ATPase refers to an enzymatic activity of the transport system that hydrolyzes ATP during the transport process.

See A Clinical Perspective: Steroids and the Treatment of Heart Disease.

The Na^+-K^+ ATPase catalyzes exchange of Na^+ and K^+ ions. Two K^+ ions enter the cell exchange for three Na^+ ions. Each exchange of three Na^+ ions for two K^+ ions is accompanied by hydrolysis of one molecule of ATP.

SUMMARY

Lipids are organic molecules that are characterized by their solubility in nonpolar solvents. Lipids are subdivided into classes based upon structural characteristics: fatty acids, glycerides, nonglycerides, and complex lipids.

Lipids serve many functions ranging from candle wax and furniture polish to various biochemical processes. Cells store chemical energy in lipids, and the cell membrane is, in fact, a lipid bilayer.

Fatty acids are carboxylic acids, either saturated or unsaturated, containing between 12 and 24 carbon atoms. Fatty acids with even numbers of carbon atoms occur most frequently in nature.

Glycerides are the most abundant lipids; the triesters of glycerol, or triglycerides, are of greatest importance. Neutral, nonionic triglycerides are important because of their ability to store energy. The phospholipids, ionic in nature, are the major components of all membranes.

Nonglycerides consist of sphingolipids, steroids, and waxes. Sphingomyelin is a component of the myelin sheath around cells of the central nervous system. The steroids are important for many biochemical functions: cholesterol is a membrane component, testosterone and estrone are sex hormones, and cortisone is an anti-inflammatory steroid that is important in the regulation of many biochemical pathways.

Complex lipids, chiefly plasma lipoproteins, are massive molecules that are responsible for the transport of other lipids through the bloodstream.

The fluid mosaic theory of membrane structure states that biological membranes are composed of lipid bilayers in which proteins are imbedded. Membrane lipids contain polar head groups and nonpolar hydrocarbon tails. The hydrocarbon tails of phospholipids are derived from saturated and unsaturated long-chain fatty acids that contain an even number of carbon atoms. The lipids and proteins diffuse rapidly in the lipid bilayer but seldom cross from one side to the other.

The simplest type of membrane transport is passive diffusion of a metabolite across the lipid bilayer from the region of highest concentration to that of lowest concentration. Many metabolites are transported across biological membranes by permeases that form pores through the membrane. The conformation of the pore is complementary to that of the substrate to be transported. Cells use energy to transport molecules across the plasma membrane against their concentration gradients, a process known as active transport.

The Na^+-K^+ ATPase hydrolyzes one molecule of ATP to provide the driving force for pumping three Na^+ ions out of the cell in exchange for two K^+ ions.

GLOSSARY OF KEY TERMS

active transport (9.6) the movement of molecules across a membrane against a concentration gradient.

antidiuretic hormone (9.6) a hormone secreted from the anterior pituitary that decreases water excretion by the kidney.

arachidonic acid (9.2) a fatty acid that is derived from linolenic acid and is the precursor of the prostaglandins.

atherosclerosis (9.5) deposition of excess plasma cholesterol and other lipids and proteins on the walls of arteries. The decrease in artery diameter results in increased blood pressure.

cephalin (9.3) one of the major phospholipids. The molecule is a complex phosphoglyceride with the amino alcohol ethanolamine attached to the phosphoryl group. (See also phosphatidylethanolamine.)

cholesterol (9.4) a 27-carbon steroid ring structure that serves as the precursor of the steroid hormones.

chylomicron (9.5) a plasma lipoprotein that carries triglycerides from the intestines to most other tissues.

complex lipid (9.5) a lipid that is bonded to other types of molecules.

cortisone (9.4) a steroid used to suppress the inflammatory response in the treatment of rheumatoid arthritis, asthma, and many other diseases.

diglyceride (9.3) the product of esterification of glycerol at two positions.

eicosanoid (9.2) any of the derivatives of 20-carbon fatty acids, including the prostaglandins, leukotrienes, and thromboxanes.

emulsifying agent (9.3) a bipolar molecule that aids in the suspension of fats in water.

essential fatty acid (9.2) the fatty acids linolenic and linoleic acids, which must be supplied in the diet because they cannot be synthesized by the body.

estrone (9.4) a female sex hormone produced by modification of progesterone.

facilitated diffusion (9.6) movement of a solute across a membrane from an area of high concentration to an area of low concentration through an integral membrane protein, or permease.

fatty acid (9.2) a long-chain (12–24 carbons) monocarboxylic acid.

fluid mosaic theory (9.6) the model of membrane structure that describes the fluid nature of the lipid bilayer and the presence of numerous proteins embedded within the membrane.

glyceride (9.3) a lipid that contains glycerol.

high-density lipoprotein (HDL) (9.5) a plasma lipoprotein that transports cholesterol from peripheral tissue to the liver.

hypertonic solution (9.6) a solution with a greater solute concentration and hence a greater osmotic pressure.

hypotonic solution (9.6) a solution with a lesser solute concentration and hence a lesser osmotic pressure.

integral proteins (9.6) a protein that is embedded within a membrane, traverses the lipid bilayer, and protrudes from the membrane both inside and outside the cell.

isotonic solution (9.6) a solution that has the same solute concentration (water activity and osmotic pressure) as another solution with which it is being compared.

lecithin (9.3) a complex phosphoglyceride with the phosphoryl group attached to the hydrophilic amino alcohol choline. (See also phosphatidylcholine.)

leukotriene (9.2) an eicosanoid produced by white blood cells that causes bronchial constriction, such as is associated with asthma.

lipid (Introduction) a member of the group of organic molecules of varying composition that are classified together on the basis of their solubility in nonpolar solvents.

low-density lipoprotein (LDL) (9.5) a plasma lipoprotein that carries cholesterol to peripheral tissues and helps to regulate cholesterol levels in those tissues.

lysosome (9.5) a membrane-bound vesicle in the cell cytoplasm that contains hydrolytic enzymes.

monoglyceride (9.3) the product of the esterification of glycerol at one position.

neutral glyceride (9.3) the product of the esterification of glycerol at one, two, or three positions.

norlutin (9.4) 17-α-ethynyl-19-nortestosterone, a synthetic steroid hormone administered orally for birth control.

norprogesterone (9.4) one of the first synthetic steroid hormones used as a birth control agent; it had to be administered by injection.

osmoreceptor (9.6) a receptor in the hypothalamus that detects changes in the osmotic concentration of the blood.

osmosis (9.6) movement of a solvent across a semipermeable membrane in response to a water chemical activity gradient.

osmotic pressure (9.6) the net force with which water enters a solution through a semipermeable membrane from a region of pure water.

passive diffusion (9.6) the net movement of a solute from an area of high concentration to an area of low concentration.

peripheral protein (9.6) a protein that is bound to either the inner or outer surface of a membrane.

permease (9.6) an integral membrane protein that transports molecules across biological membranes.

phosphatidate (9.3) a molecule with fatty acids esterified to C-1 and C-2 of glycerol and a free phosphoryl group esterified at C-3.

phosphatidylcholine (9.3) a complex phosphoglyceride, found in cell membranes, in which the phosphoryl group is attached to the hydrophilic amino alcohol choline. The result is a molecule with a polar "head" and a nonpolar "tail" that can be used as an emulsifying agent to suspend fats in water. (See also lecithin.)

phosphatidylethanolamine (9.3) a complex phosphoglyceride with the phosphoryl group attached to the hydrophilic amino alcohol ethanolamine. This is one of the major membrane phospholipids. (See also cephalin.)

phosphoglyceride (9.3) a molecule with fatty acids esterified at the C-1 and C-2 positions of glycerol and a phosphate group esterified at the C-3 position.

phospholipid (9.3) a lipid containing a phosphoryl group.

plasma lipoprotein (9.5) a complex composed of lipid and protein that is responsible for the transport of lipids throughout the body.

progesterone (9.4) one of the most important hormones associated with pregnancy. It is produced by chemical modification of cholesterol.

prostaglandins (9.2) a family of hormonelike substances derived from the 20-carbon fatty acid arachidonic acid. They are produced by many cells of the body and regulate a great many body functions.

receptor-mediated endocytosis (9.6) the process by which molecules attach to receptors in the cell surface and are brought into the cell by invagination of the cell membrane.

saturated fatty acid (9.2) a long-chain monocarboxylic acid in which each carbon of the chain is bonded to the maximum number of hydrogen atoms.

semipermeable membrane (9.6) a membrane that is permeable to the solvent but not the solute.

sphingolipid (9.4) a phospholipid that is derived from the amino alcohol sphingosine, rather than from glycerol.

sphingomyelin (9.4) a sphingolipid that is found in abundance in the myelin sheath that surrounds and insulates cells of the central nervous system.

steroid (9.4) a lipid that is derived from cholesterol and composed of one five-sided ring and three six-sided rings. The steroids include sex hormones and anti-inflammatory compounds.

testosterone (9.4) a male sex hormone produced by chemical modification of progesterone.

thromboxane A_2 (9.2) an eicosanoid produced by blood platelets that stimulates vasoconstriction and platelet aggregation.

triglyceride (9.3) the product of the esterification of glycerol at all three carbons.

unsaturated fatty acid (9.2) a long-chain monocarboxylic acid having at least one carbon-to-carbon double bond.

very low-density lipoprotein (VLDL) (9.5) a plasma lipoprotein that binds triglycerides synthesized by the liver and carries them to adipose tissue for storage.

wax (9.4) a collection of lipids that are generally considered to be esters of long-chain alcohols.

QUESTIONS AND PROBLEMS

Fatty Acids: Structure, Properties, and Chemical Reactions

9.10 Draw the structures of each of the following molecules:

a. decanoic acid
b. stearic acid
c. *trans*-5-decenoic acid
d. *cis*-5-decenoic acid

9.11 A wax found in beeswax is myricyl palmitate. What fatty acid(s) and fatty alcohol(s) are used to form myricyl palmitate? The structure of myricyl palmitate is shown below:

$$CH_3(CH_2)_{14}{-}\overset{\overset{\displaystyle O}{\|}}{C}{-}O{-}(CH_2)_{29}CH_3$$

Myricyl palmitate

9.12 The *trans*-isomer of oleic acid is also called elaidic acid. It is synthesized via the hydrogenation of vegetable oil in the production of margarine. Draw the structures of these two fatty acids.

9.13 What is the most important difference in the physical properties of decanoic acid, *trans*-5-decenoic acid, and *cis*-5-decenoic acid?

9.14 Draw the structure of the triacylglycerol (triglyceride) molecule formed by esterification at C-1, C-2, and C-3 with hexadecanoic acid, *trans*-9-hexadecenoic acid, and *cis*-9-hexadecenoic acid, respectively.

9.15 Draw the structure of the phosphatidate molecule formed when glycerol-3-phosphate is esterified at C-1 and C-2 with *trans*-9-hexadecenoic acid and *cis*-9-hexadecenoic acid, respectively.

9.16 Write an equation for the reaction of lauric acid with potassium hydroxide.

Fatty Acids: Eicosanoids

9.17 What is the function of essential fatty acids?

9.18 What molecules are formed from arachidonic acid?

9.19 What is the biochemical basis for the effectiveness of aspirin in decreasing inflammation?

9.20 What property of some prostaglandins makes them useful for therapeutic abortions?

9.21 List four major effects of prostaglandins.

9.22 What are the functions of thromboxane A_2 and the leukotrienes?

Phospholipids

9.23 Draw the structure of a lecithin molecule in which the fatty acyl groups are derived from the C-18 carboxylic acid stearic acid.

9.24 Draw the structure of a cephalin molecule in which the fatty acyl tails are derived from stearic acid.

Neutral Glycerides

9.25 Draw the structure of a triglyceride that contains the three fatty acids stearic acid, palmitic acid, and oleic acid.

9.26 What is the major lipid component of adipose cells?

Membrane Structure

9.27 What is the major effect of cholesterol on the properties of biological membranes?

9.28 How will properties of a biological membrane change if the fatty acyl groups of the membrane phospholipids are converted from saturated to unsaturated chains?

9.29 What is the basic structure of a biological membrane?

9.30 What is the difference between a peripheral membrane protein and an integral membrane protein?

9.31 Explain why lipids diffuse rapidly in one plane of a lipid bilayer but cross to the other side of the bilayer very slowly.

9.32 What is the function of unsaturation in the hydrocarbon tails of membrane lipids?

9.33 What are the structural and functional differences between triglycerides (triacylglycerols) and phospholipids?

9.34 What experimental observation shows that proteins diffuse within the lipid bilayers of biological membranes?

9.35 Why don't proteins turn around in biological membranes like revolving doors?

9.36 Why do the hydrocarbon tails of membrane phospholipids provide a barrier between the inside and outside of the cell?

Complex Lipids

9.37 What are the four major types of plasma lipoproteins?

9.38 What are the functions of the plasma lipoproteins?

9.39 What is the role of lysosomes in the metabolism of plasma lipoproteins?

9.40 What is the mechanism of uptake of cholesterol from plasma?

9.41 What is the result of a person's inability to synthesize receptors for plasma lipoproteins?

9.42 How does cholesterol regulate its own metabolism?

Membrane Transport

9.43 Explain the difference between simple diffusion across a membrane and facilitated diffusion.

9.44 Explain what would happen to a red blood cell placed in each of the following solutions:
a. hypotonic c. hypertonic
b. isotonic

9.45 How does active transport differ from facilitated diffusion?

9.46 Why is the function of the Na^+-K^+ ATPase an example of active transport?

9.47 What is the stoichiometry of the Na^+-K^+ ATPase?

9.48 What is the meaning of the term "antiport"?

9.49 What properties of a transport protein (permease) determine its specificity?

9.50 Explain why ions cross plasma membranes very slowly by simple diffusion.

9.51 What is the physiological effect of digitalis?

9.52 How will the Na^+ and K^+ concentrations of a cell change if the Na^+-K^+ ATPase is inhibited?

9.53 Why is digitalis called a cardiotonic steroid?

9.54 By what mechanism are Cl^- and HCO_3^- ions transported across the red blood cell membrane?

9.55 How does insulin affect the transport of glucose?

9.56 How does membrane transport resemble enzyme catalysis?

9.57 Why is D-glucose transported by the glucose transport protein whereas L-glucose is not?

9.58 Explain why drinking May wine increases the rate at which the heart beats.

Further Problems

9.59 Draw the structure of prostaglandin E_1. To what do the designations "E" and "1" refer?

9.60 Write a balanced equation for the reaction of palmitoleic acid with NaOH.

9.61 Write a balanced equation for the complete hydrogenation of arachidonic acid. What is the common name of the product?

9.62 What is the principal function of sphingomyelin?

9.63 What is the medical value of cortisone? What are some of the possible side effects of its use?

9.64 What are the biological functions of progesterone?

CHAPTER 10

Amines and Amides

OBJECTIVES

- Draw and name the common amines and amides.
- Draw and discuss the structure of the amide bond.
- Be familiar with the common members of these two families that are of natural, commercial, health, environmental, and industrial interest.
- Be familiar with the physical properties of each of the two families.
- Be familiar with the methods of preparation of both the amines and amides.
- Be prepared to discuss the basicity of amines relative to other bases that we have encountered.
- Design a simple multistep synthesis involving amines and amides.
- Be familiar with polymers, their general structure(s), and their potential use(s).

OUTLINE

Aliphatic and aromatic **amines** contain the amino or substituted amino group $—NH_2$:

$$(\text{Ar—})\ \text{R—NH}_2$$

Amines

They may be viewed as substituted ammonia molecules in which one or more of the ammonia hydrogens has been substituted by a more complex organic moiety:

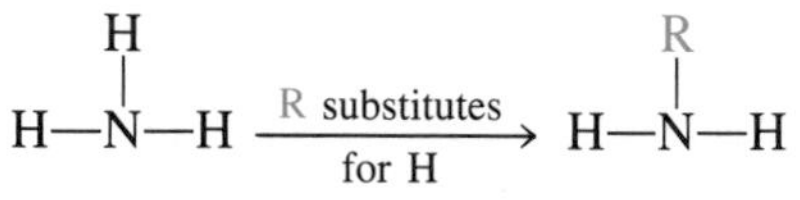

$$\underset{\text{Ammonia}}{\text{H—}\overset{\text{H}}{\overset{|}{\text{N}}}\text{—H}} \xrightarrow[\text{for H}]{\text{R substitutes}} \underset{\text{An amine}}{\text{H—}\overset{\text{R}}{\overset{|}{\text{N}}}\text{—H}}$$

Alkaloids are a family of naturally occurring nitrogen heterocyclic compounds derived from plants. Their name, alkaloid, literally means "alkali-like" (basic). The alkaloids include caffeine, nicotine, morphine, and coniine (the hemlock plant poison that Socrates drank).

The amino group is an important functional group that is found in amino acids, proteins and enzymes, nucleic acids (DNA and RNA), and pharmaceuticals and medicinals such as amphetamines, antibiotics, anesthetics, and analgesics. They are also found in many naturally occurring compounds known as **alkaloids.** Amines also have many industrially important applications including synthetic fibers (polymers such as the nylons), dyes, and food additives (artificial sweeteners).

In this chapter we shall discuss the structure and physical properties of amines and amine salts, the nomenclature system for these families, and methods for their preparation and the reactions they undergo. **Heterocyclic amines,** cyclic amines in which one of the atoms making up the backbone of the ring is nitrogen, will also be introduced. The heterocyclic compounds play a key role in the structure of the nucleic acids and proteins.

10.1 AMINES: STRUCTURE AND PHYSICAL PROPERTIES

Amines are pyramidal. The nitrogen atom has three groups attached and a nonbonding pair of electrons. The pyramidal structure that results, shown in Figure 10.1, is directly analogous to the structure of ammonia.

Amines are classified according to the number of alkyl or aryl groups that are attached to the nitrogen. The general structures of each of the classes are provided here:

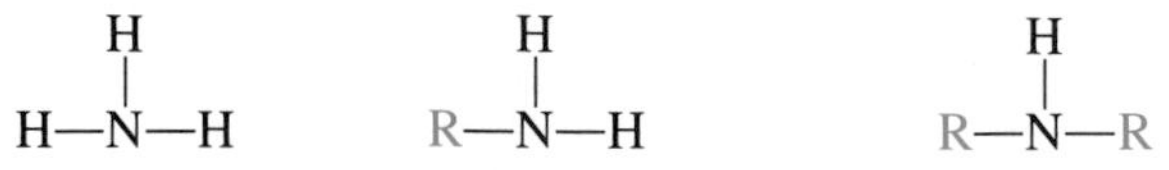

$$\underset{\text{Ammonia}}{\text{H—}\overset{\text{H}}{\overset{|}{\text{N}}}\text{—H}} \qquad \underset{\text{1° amine (a primary amine)}}{\text{R—}\overset{\text{H}}{\overset{|}{\text{N}}}\text{—H}} \qquad \underset{\text{2° amine (a secondary amine)}}{\text{R—}\overset{\text{H}}{\overset{|}{\text{N}}}\text{—R}}$$

$$\underset{\text{3° amine (a tertiary amine)}}{\text{R—}\overset{\text{R}}{\overset{|}{\text{N}}}\text{—R}} \qquad \underset{\text{Quaternary ammonium salt}}{\text{R—}\overset{\text{R}}{\overset{|+}{\underset{\text{H}}{\underset{|}{\text{N}}}}}\text{—R X}^-}$$

1° amine (a **primary amine**); 2° amine (a **secondary amine**); 3° amine (a **tertiary amine**); **Quaternary ammonium salt**

FIGURE 10.1
The pyramidal structure of amines. Note the similarities in structure between an amine and the ammonia molecule.

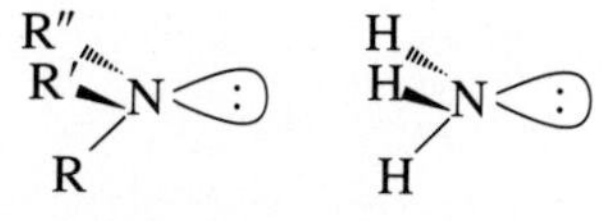

$$\underset{\text{An amine}}{\text{R}''\text{, R}'\text{, R}\ \text{N}\langle:\rangle} \qquad \underset{\text{Ammonia}}{\text{H, H, H}\ \text{N}\langle:\rangle}$$

The nitrogen atom in amines is electronegative; as a result, the N—H bond is polar. Hydrogen bonding is therefore possible, and this polarity is reflected in the boiling points and solubility properties of this family. Consider the following compounds:

$CH_3CH_2CH_3$	$CH_3—O—CH_3$	$CH_3CH_2—NH—H$	$CH_3CH_2—OH$	H—O—H
Propane (*n*-propane)	Dimethyl ether	Aminoethane (ethylamine)	Ethanol (ethyl alcohol)	Water
mol. wt. 44	mol. wt. 46	mol. wt. 45	mol. wt. 46	mol. wt. 18
b.p. −42.2°	b.p. −24°	b.p. 17°	b.p. 78°	b.p. 100°

The greater polarity of the —O—H bond is reflected in the boiling points of both water and ethanol. Both of these compounds have a much higher boiling point than the corresponding amine, ether, or alkane. The amine, however, has a much higher boiling point than the corresponding ether or alkane of comparable molecular weight. This difference is a result of the polarity of the —N—H bond. This polarity is illustrated in Figure 10.2.

Additional examples of amines and their physical properties are given in Table 10.1. The lower molecular weight amines are soluble in water, although, as was observed with alcohols and carbonyl compounds, their solubility decreases as the size of the hydrocarbon portion of the amine increases and the overall polarity of the molecule decreases progressively. The hydrogen bonding of methylamine with water is shown in Figure 10.2b.

Question 10.1 Which compound in each of the following pairs has the higher boiling point? Can you explain why?

a. methanol or methylamine
b. methylamine or ethylamine
c. dimethylamine or ethanol
d. *n*-propylamine or butane
e. water or ethylene glycol
f. ethylene glycol or 1,2-diaminoethane

Question 10.2 Refer to Figure 10.2 and draw a similar figure showing the hydrogen bonding that results between:

a. water and a 2° or 3° amine
b. two secondary amines

Question 10.3 Compare the boiling points of methylamine, dimethylamine, and trimethylamine. Explain why they are different.

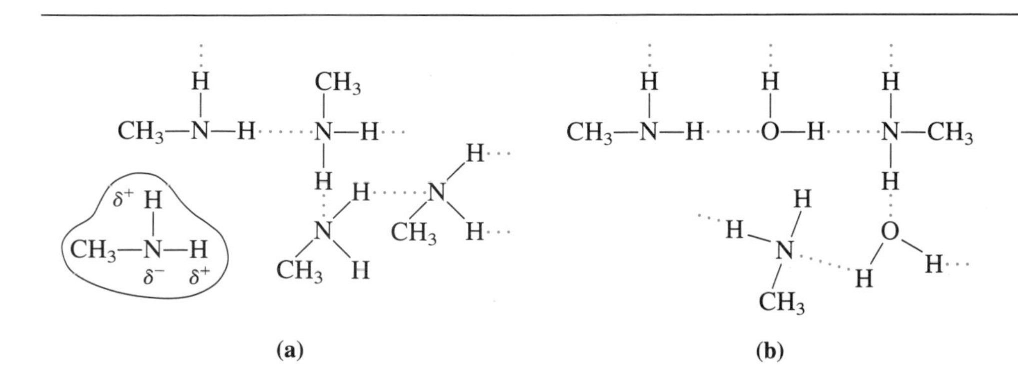

FIGURE 10.2
Hydrogen bonding (a) in methylamine and (b) between methylamine and water.

TABLE 10.1 Physical Properties of Amines

Name	Structure	m.p. (°C)	b.p. (°C)	Density at 20°C (g/cc)
Aminomethane (methylamine)	CH_3NH_2	−93	−6.3	0.699
Aminoethane (ethylamine)	$CH_3CH_2NH_2$	−84	17.0	0.689
1-Aminopropane (*n*-propylamine)	$CH_3CH_2CH_2NH_2$	−83	49.0	0.719
1-Aminobutane (*n*-butylamine)	$CH_3(CH_2)_2CH_2NH_2$	−50	78.0	0.740
N-Methylaminomethane (dimethylamine)	$(CH_3)_2NH$	−96	7.0	0.680
N-Ethylaminoethane (diethylamine)	$(CH_3CH_2)_2NH$	−48	56.0	0.711
N-(*n*-propyl)-1-aminopropane (di-*n*-propylamine)	$(CH_3CH_2CH_2)_2NH$	−40	110.0	0.738
N-(*n*-butyl)-1-aminobutane (di-*n*-butylamine)	$(CH_3CH_2CH_2CH_2)_2NH$	−59	159.0	0.767
N,*N*-dimethylaminomethane (trimethylamine)	$(CH_3)_3N$	−117	3.5	0.662
N,*N*-diethylaminoethane (triethylamine)	$(CH_3CH_2)_3N$	−115	90.0	0.728

10.2 AMINES: NOMENCLATURE

I.U.P.A.C. names

The guidelines for applying the I.U.P.A.C. rules to the amines are analogous to those for other substituted hydrocarbons such as haloalkanes. The specific rules follow:

1. The parent compound is the longest continuous carbon chain to which the amino group is bonded. The alkane name of the parent compound is used as a suffix for the amine name. The alkane name is also preceded by amino and a number that designates the position of the amino group on the chain:

$CH_3{-}NH_2$ — Aminomethane (no numbers necessary)

$CH_3CH_2CH_2{-}NH_2$ — 1-Aminopropane

$CH_3CH_2CH_2CH(NH_2)CH_3$ — 2-Aminopentane

$CH_3{-}CH(NH_2){-}CH_3$ — 2-Aminopropane

$CH_3(CH_2)_5CH_2{-}NH_2$ — 1-Aminoheptane

$CH_3CH_2CH(NH_2)CH(CH_3)CH_3$ — 2-Methyl-3-aminopentane

Note: If a substituent is present on the nitrogen, it is designated by the prefix *N*-:

$CH_3CH_2CH_2—NH—CH_3$ *N-Methyl*-1-aminopropane

$CH_3CH_2CH_2CH_2CH_2CH_2—NH—CH_3$ *N-Methyl*-1-aminohexane

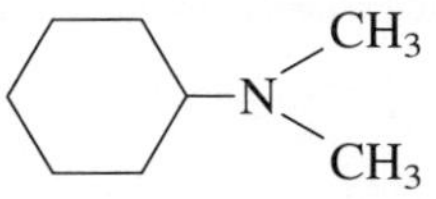

N,N-dimethylaminocyclohexane

2. As the amine becomes more complex, all substituents are named and numbered in the usual way. The amino group (or substituted amino group) may remain the principal functional group used in the naming or simply become one more substituent when present in conjunction with other functional groups. For example:

Section 3.2

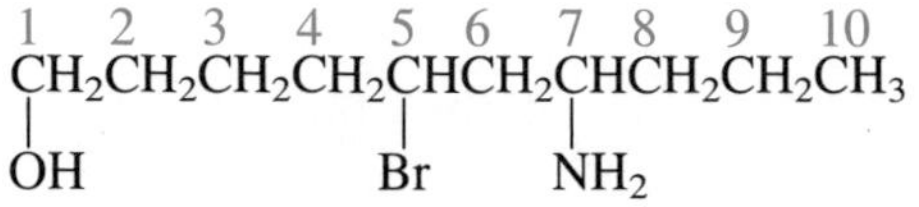

7-Amino-5-bromo-1-decanol
(*named as an alcohol*)

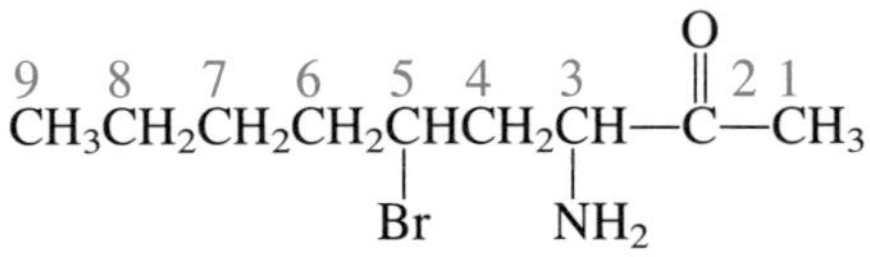

3-Amino-5-bromo-2-nonanone
(*named as a ketone*)

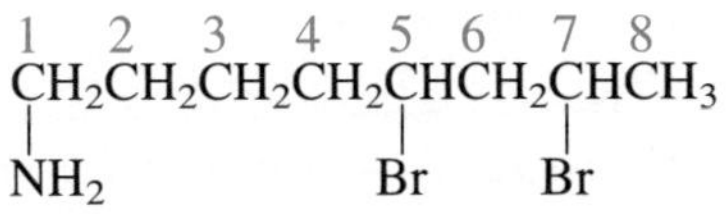

5,7-Dibromo-1-aminooctane
(*named as an amine*)

3. Several amines, particularly among the aromatic amines, have special names that are also approved for use in the I.U.P.A.C. system. Examples follow:

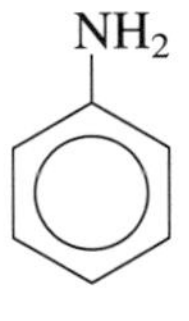

Aniline 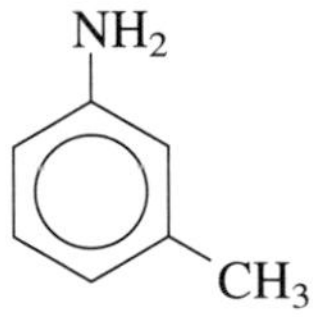 *m*-Toluidine or (*meta*-toluidine)

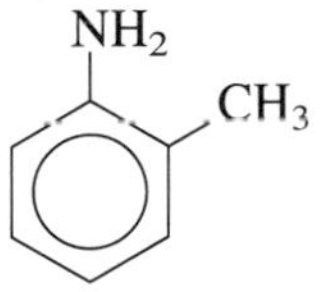

o-Toluidine or (*ortho*-toluidine) 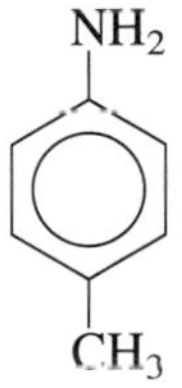 *p*-Toluidine or (*para*-toluidine)

Question 10.4 The structure of aniline is provided above. Draw the complete structural formulas for each of the following amines:

a. *N*-methylaniline
b. *N,N*-dimethylaniline
c. *N*-ethylaniline
d. *N*-isopropylaniline

Question 10.5 The name aniline is accepted in the I.U.P.A.C. System, and phenylamine is not. Yet diphenylamine and triphenylamine are accepted. Draw these two compounds.

Example 10.1

Name amines using the I.U.P.A.C. Nomenclature System.

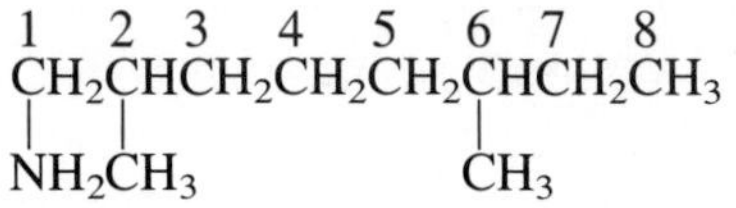

```
1  2  3   4   5   6  7  8
CH2CHCH2CH2CH2CHCH2CH3
 |  |              |
NH2CH3            CH3
```

Parent compound: Octane (becomes aminooctane)

Position of —NH_2: Carbon-1 (*not* carbon-8)

Substituents: 2,6-Dimethyl (*not* 3,7-dimethyl)

Name: 2,6-Dimethyl-1-aminooctane

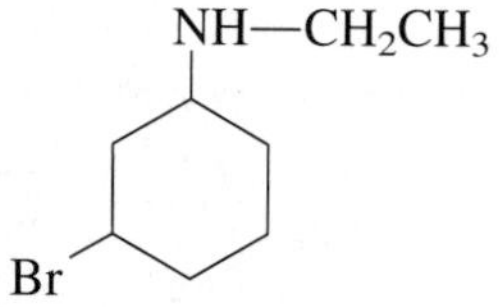

NH—CH_2CH_3 on cyclohexane ring, Br at position 3

Parent compound: Cyclohexane

Position of —N: Carbon-1 (*not* carbon-3)

Substituents: 3-Bromo (*not* 5-bromo) and N-ethyl

Name: 3-Bromo-N-ethylaminocyclohexane

Common names

The common names of the amines are derived from the various alkyl groups that are bonded to the amine nitrogen. Each group is listed alphabetically or in order of size (smallest to largest) in one continuous word ending with ''amine.''

$CH_3—NH_2$	$CH_3—NH—CH_3$	$CH_3—N(CH_3)—CH_3$
Methylamine	Dimethylamine	Trimethylamine

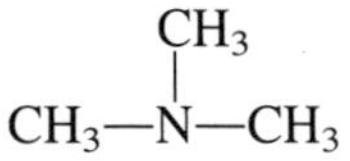

$CH_3CH_2—NH_2$	$CH_3CH_2CH_2CH_2CH_2—NH_2$	$CH_3CH_2CHCH_3$ with H—N—CH_3 on C-2
Ethylamine	*n*-Pentylamine	Methyl-*sec*-butylamine

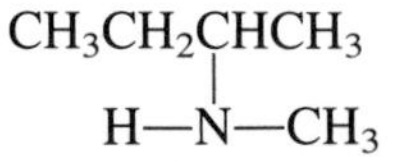

The common system of nomenclature is still used widely, and both systems are used in the examples provided in this text.

Question 10.6 Draw complete structural formulas for each of the following compounds:

a. 2-aminobutane
b. 2,3-diaminopentane
c. 2-aminooctane
d. 4-chloro-5-iodo-1-aminononane
e. *N*-ethyl-2-aminoheptane
f. *N*,*N*-diethyl-1-aminopentane

Question 10.7 Draw complete structural formulas for each of the following compounds:

a. methyldiethylamine
b. tri-*sec*-butylamine
c. 4-aminopentanal
d. isopentylamine
e. *N*-methylaniline
f. *N*-ethyl-1-aminohexane

10.3 HETEROCYCLIC AMINES

Heterocyclic compounds were introduced in Section 5.7. We will now discuss nitrogen heterocycles, the most common and important of all the heterocyclic compounds. They are found throughout nature: in DNA, RNA, many alkaloids (cocaine, nicotine, quinine, morphine, heroin, and LSD for example), and in many enzymes and coenzymes (NAD^+ and NADPH, for example). They also play important roles in the manufacture of polymers, medicinals, pharmaceuticals, and food additives; some examples of products are Vitamin B_6, nylon, and saccharin. The common names for several of the more common nitrogen heterocycles are presented here:

Sections 5.7 and 5.8

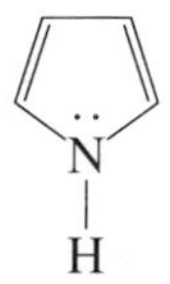

Pyrrole 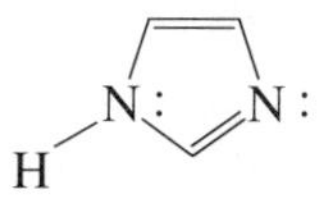 Imidazole 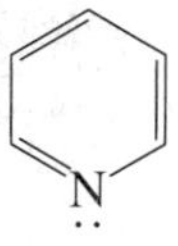 Pyridine 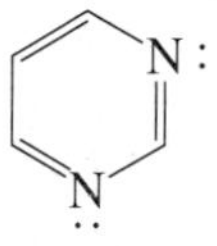 Pyrimidine

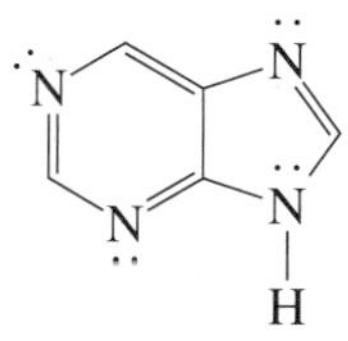

Metal atom

Purine 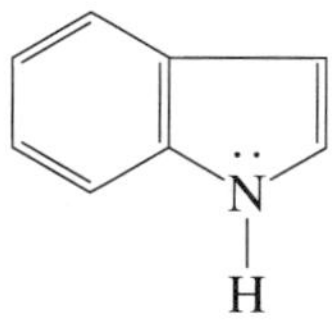 Indole 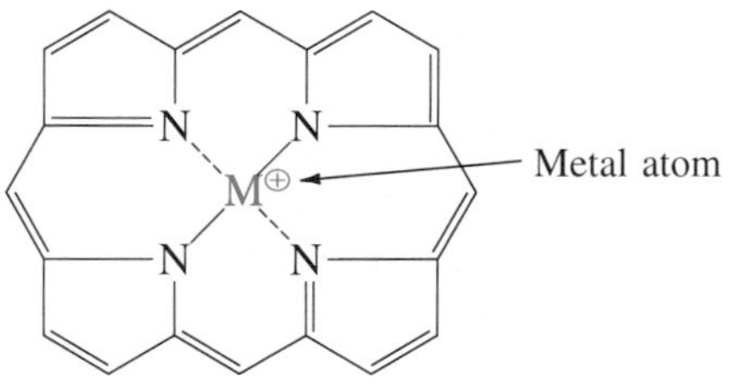 Porphyrin

The pyrimidine and purine rings are found in the nucleic acids (DNA and RNA). The porphyrin ring is found in hemoglobin (a blood protein), myoglobin (a protein found in muscle tissue), and chlorophyll (a plant pigment), while imidazole is found in amino acids, peptides, and proteins. The indole and pyridine rings are found in many alkaloids such as strychnine and nicotine. Several examples are provided in Figure 10.3.

Section 16.1

Chapter 11

Question 10.8 The general structure of a porphyrin ring system is provided in the previous section. Review this structure and refer to a reference text to find the complete structure of chlorophyll. Explain in your own words the bonding that is present between the porphyrin ring and the magnesium metal ion in chlorophyll.

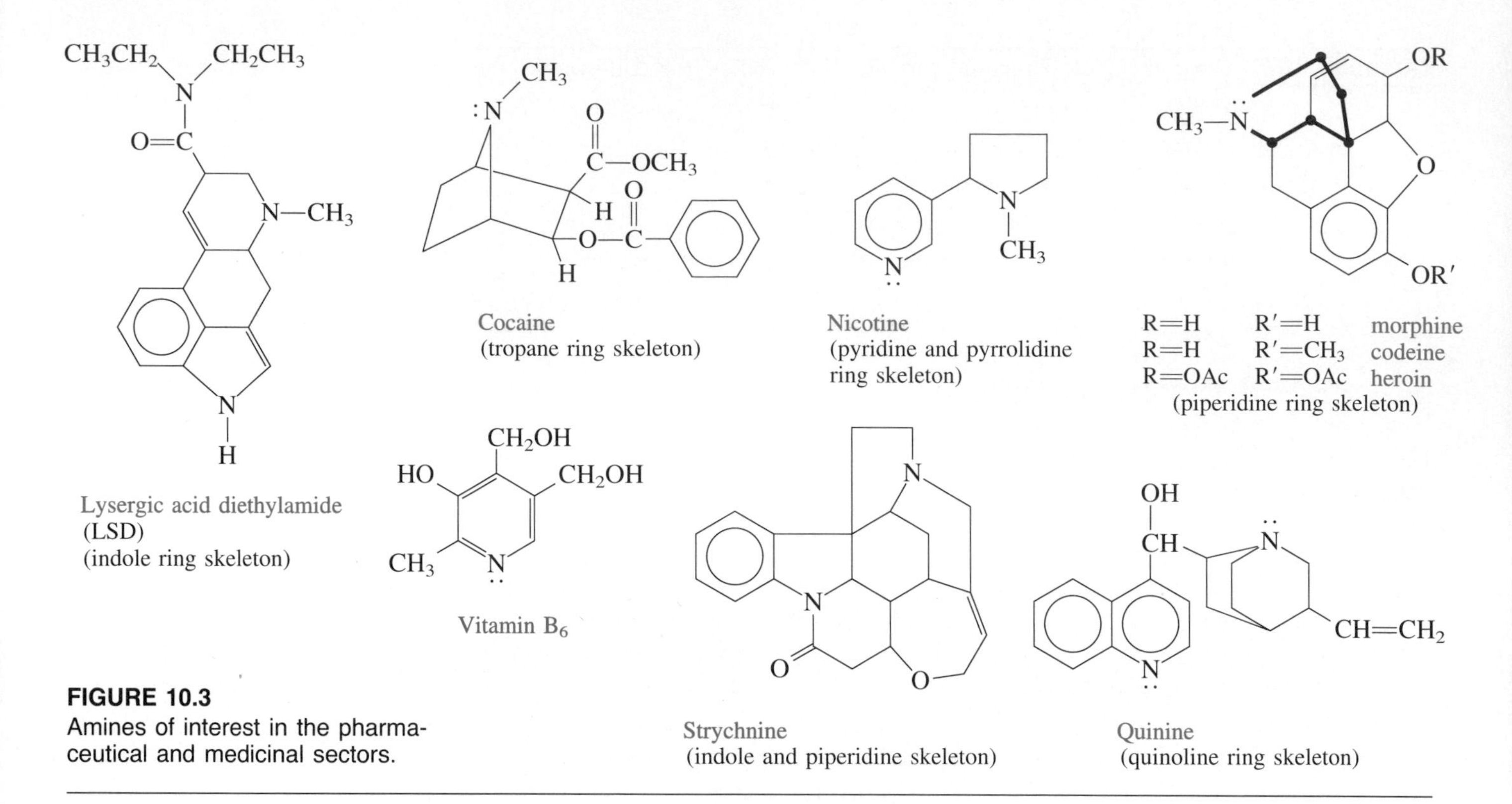

FIGURE 10.3
Amines of interest in the pharmaceutical and medicinal sectors.

10.4 AMINES: WHERE DO THEY COME FROM?

Amines come to us from a variety of natural and synthetic sources. An overview of several of the more important sources is provided below.

Industrial sources

Industrially, the simple amines are synthesized from readily available starting materials: methanol and ammonia. The reaction to produce trimethylamine is shown here:

$$NH_3 \xrightarrow[\text{Al}_2\text{O}_3,\ \text{heat}]{CH_3OH} CH_3\text{—}NH_2 \xrightarrow[\text{Al}_2\text{O}_3,\ \text{heat}]{CH_3OH,} CH_3\text{—}NH\text{—}CH_3 \xrightarrow[\text{Al}_2\text{O}_3,\ \text{heat}]{CH_3OH} CH_3\text{—}\overset{\overset{\displaystyle CH_3}{|}}{N}\text{—}CH_3 \quad 10.1$$

Other, more complex amines are prepared on an industrial scale by using reactions developed in the laboratory.

Laboratory sources

In the laboratory, amines are prepared via the reduction of various functional groups; *amides* and *nitro compounds* are commonly used.

$$\text{(Ar) } R\text{—}\overset{\overset{\displaystyle O}{\|}}{C}\text{—}NH_2 \qquad \text{(Ar) } R\text{—}NO_2$$

An amide $\qquad$ A nitro compound

Amides are discussed in detail in Section 10.6. The reduction of these two classes proceeds as shown in the following examples:

$$CH_3{-}\overset{O}{\overset{\|}{C}}{-}NH_2 \xrightarrow{[R]} CH_3{-}CH_2{-}NH_2 \qquad 10.2$$

Ethanamide → Aminoethane (ethylamine)

$$CH_3CH_2CH_2{-}\overset{O}{\overset{\|}{C}}{-}NH_2 \xrightarrow{[R]} CH_3CH_2CH_2CH_2{-}NH_2 \qquad 10.3$$

Butanamide → Aminobutane (*n*-butylamine)

$$CH_3CH_2{-}\overset{O}{\overset{\|}{C}}{-}NH{-}CH_3 \xrightarrow{[R]} CH_3CH_2CH_2NH{-}CH_3 \qquad 10.4$$

N-Methylpropanamide → *N*-Methylaminopropane (methyl-*n*-propylamine)

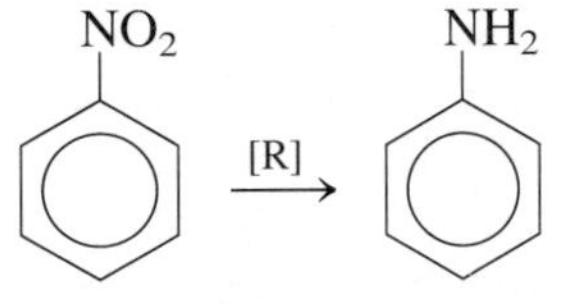

$$C_6H_5{-}NO_2 \xrightarrow{[R]} C_6H_5{-}NH_2 \qquad 10.5$$

Nitrobenzene → Aniline

As we have seen, [R] is used by organic chemists as the general symbol for any reducing agent. Several different reducing agents may be used to effect the changes shown here; for example, metallic iron and acid may be used to reduce aromatic nitro compounds, and $LiAlH_4$ in ether reduces amides.

Example 10.2

Designing the synthesis of an amine.

How do we plan the synthesis of an amine? In the same way that we have approached all synthesis: by working in reverse. Consider the following amine:

$$CH_3CH_2CH_2CH_2CH_2CH_2{-}NH_2$$

1-Aminohexane

In planning the synthesis of this compound we would try to determine the identity of the immediate precursor of the desired product. We know that two principal pathways are used in the synthesis of amines: the reduction of amides and the reduction of nitro compounds. We can plan accordingly. The immediate precursor of the desired 1-aminohexane is 1-nitrohexane or hexanamide:

$$CH_3CH_2CH_2CH_2CH_2CH_2{-}NO_2$$

1-Nitrohexane

or $\xrightarrow{[R]}$ $CH_3CH_2CH_2CH_2CH_2CH_2{-}NH_2$

1-Aminohexane

$$CH_3CH_2CH_2CH_2CH_2{-}\overset{O}{\overset{\|}{C}}{-}NH_2$$

Hexanamide

We now have two alternative reaction pathways that we can follow. As we study more of the chemistry of amides and nitro compounds, we will be able to continue this process backward until a simple, inexpensive, readily available starting material is obtained.

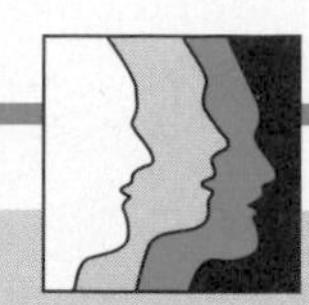

A HUMAN PERSPECTIVE

Some Amines of Special Interest

Although amines play many different roles in our day-to-day lives, two of the foremost areas of interest involve medically related amines. A host of medicinal chemicals derived from amines are responsible for improving the quality of life, while others are among the most powerful cancer-causing substances known. One industrially important class of amides, derived from amines, has already been mentioned—the nylons. These polyamides are an important and versatile class of synthetic polymers (see Section 8.5, A Human Perspective: Carboxylic Acid Derivatives of Special Interest").

Medicinal chemicals

Amphetamines, barbiturates, analgesics, anesthetics, decongestants, and antibiotics number among the medicinal compounds that are amines or amine derivatives.

1. *Amphetamines.* **Amphetamines** ("uppers") stimulate the central nervous system (CNS) and are often used in diet pills to help decrease appetite. They are also used to treat psychological disorders such as severe depression. Two of the more common amphetamines are shown here:

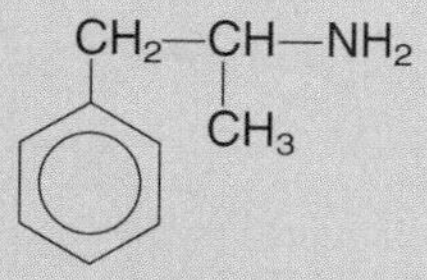

$C_6H_5-CH_2-CH(CH_3)-NH_2$

2-Amino-1-phenylpropane
(Amphetamine)

Benzedrine

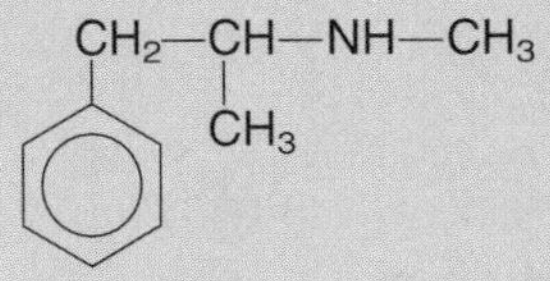

$C_6H_5-CH_2-CH(CH_3)-NH-CH_3$

2-Methylamino-1-phenylpropane
(Methamphetamine)

Methedrine

2. *Barbiturates.* In contrast to amphetamines, **barbiturates** ("downers"), which are derived from amides, are used as sedatives; they are also used as anticonvulsants for epileptics and for people suffering from a variety of brain disorders that manifest themselves in neurosis, anxiety, and tension.

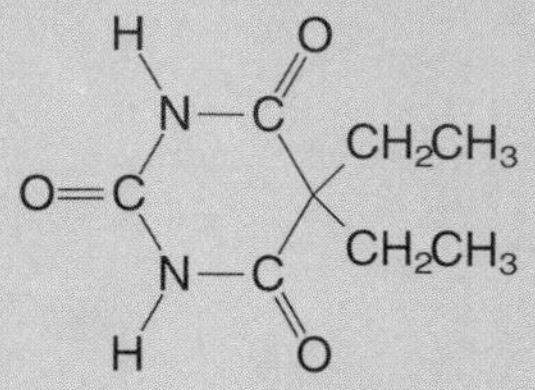

Barbital—a barbiturate

3. *Analgesics and anesthetics.* Many of the medicinal amines are also used as **analgesics** (pain relievers) and **anesthetics** (pain blockers). Still others are hallucinogenic and are abused "street drugs." Novocaine and related compounds, for example, are used as local anesthetics, and Demerol is a very strong pain reliever. Phenacetin and acetaminophen, common substitutes for aspirin, are also amine derivatives. In addition, many of the alkaloids have medicinal uses, and others are hallucinogenic and addictive; their use is allowed only in certain prescribed cases. Several of these are shown in Figure 10.3 and the figure below. Quinine, for example, is an antimalarial drug, and morphine and codeine are extremely strong pain killers. LSD, on the other hand, is a hallucinogenic chemical that can cause severe mental disorders; cocaine is one of the most abused of the "social drugs."
4. *Decongestants.* Ephedrine and neosynephrine are used as **decongestants** in cough syrups and nose drops. They act by causing a shrinking of the membranes that line the nasal passages. Both of these compounds are related to several chemicals that are important to the functioning of the central nervous system: dopa and dopamine. These chemicals are described in A Human Perspective: Amines and the Central Nervous System.

Analgesics

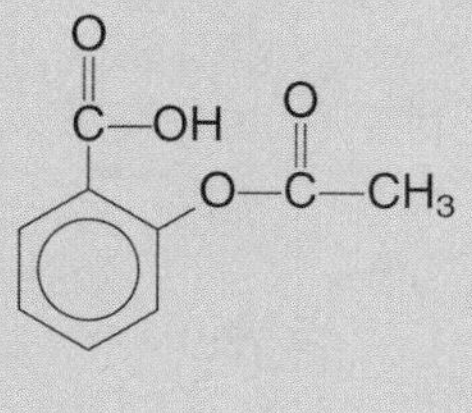

Aspirin

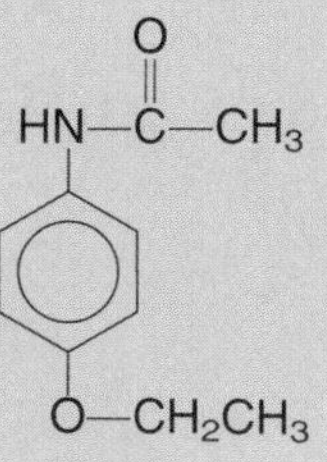

Phenacetin

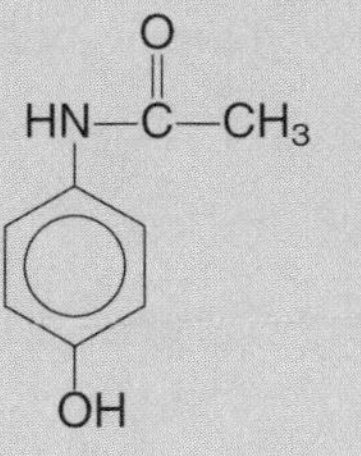

Acetaminophen

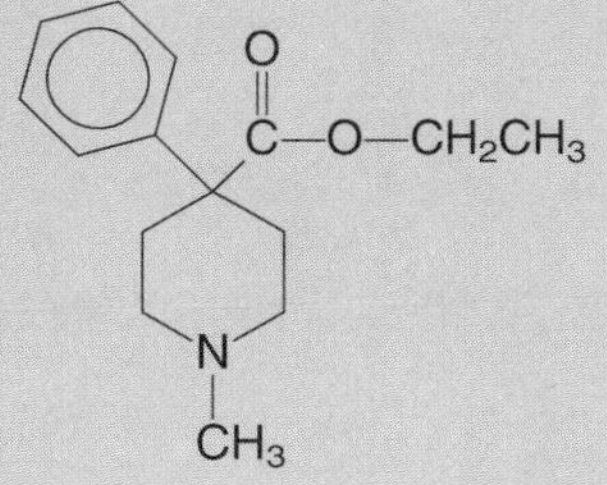

Demerol

Decongestants

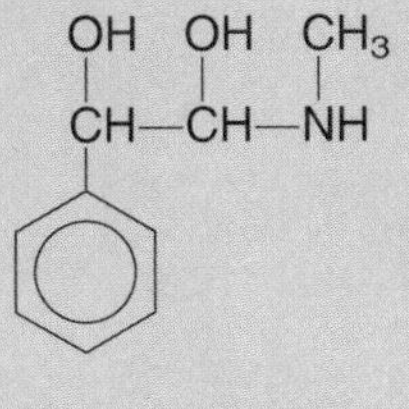

Ephedrine

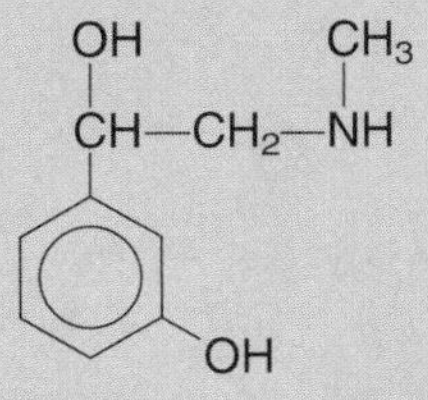

Neosynephrine

Anesthetic

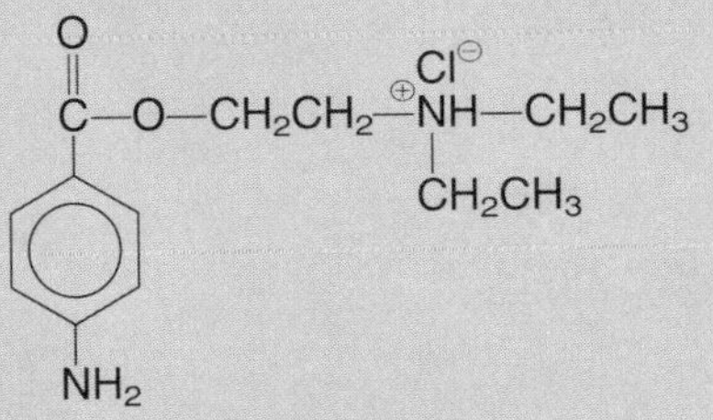

Novocaine

5. *Antibiotics.* The **sulfa drugs,** an important class of antibiotics (Section 17.3), are synthesized from amines.

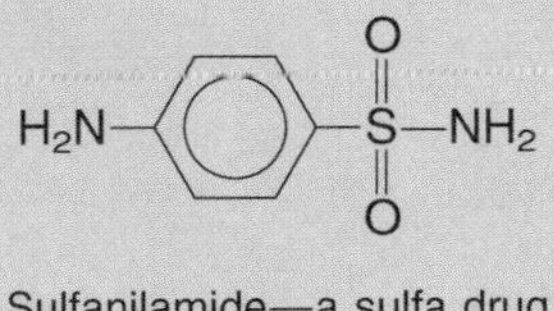

Sulfanilamide—a sulfa drug

Carcinogens

Many of the amines and amine derivatives have been linked to cancer in animals and, in several cases, humans. Little is known about the mode of action of many of these chemicals, and a great deal of research is underway to determine how they are involved in the development of cancer. It appears that these carcinogens (cancer-causing chemicals) damage the DNA, causing mutations in genes that control cell division. This damage results in the formation of "outlaw" cancer cells. It appears that many of these compounds act by alkylating (adding an alkyl group to) the DNA chain.

It is known that secondary amines can react with substances such as nitrous acid (HNO_2) to form nitrosoamines. Nitrosoamines, in turn, can react further to give diazocompounds (see the structure below). Diazocompounds act as the alkylating agents to alkylate the DNA. The reaction is shown, in part, here:

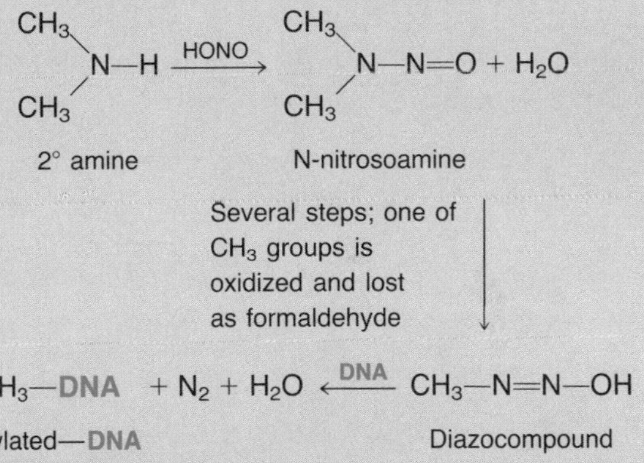

$$(CH_3)_2N{-}H \xrightarrow{HONO} (CH_3)_2N{-}N{=}O + H_2O$$

2° amine — N-nitrosoamine

Several steps; one of CH_3 groups is oxidized and lost as formaldehyde

$$CH_3{-}DNA + N_2 + H_2O \xleftarrow{DNA} CH_3{-}N{=}N{-}OH$$

Alkylated—DNA — Diazocompound

Many of the medicinals and foodstuffs that we consume contain secondary amines. In addition, nitrite ion is widely used as a preservative for bacon, ham, sausage, and other meat products. It is quite conceivable that the nitrite ions that we ingest, reacting with the acid found in the saliva and stomach juices, forms nitrous acid, which in turn can undergo the reactions shown above. Research is still in progress in search of the answers to many of these questions.

Remember, in designing a synthesis, one works in reverse, through the various pathways that develop, always keeping in mind the end product that you desire and the types of starting materials that you prefer. As your knowledge of reactions and your experience increase, your ability to create synthetic routes will increase proportionately.

Question 10.9 Draw the complete structural formula of a compound that would yield the following amines when reduced:

a. $CH_3CH_2NH_2$

b. $CH_3CH_2CHNH_2$ (with CH_3 on the CH)

c. 2-bromoaniline (benzene ring with NH_2 and Br on adjacent carbons)

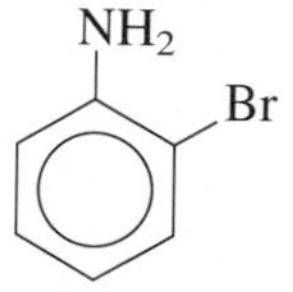

d. CH_3—N—CH_2CH_3 (with CH_3 on N)

e. 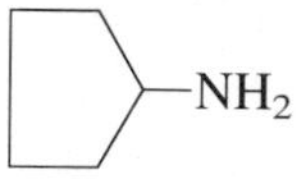 cyclopentyl—NH_2

f. $CH_3CH_2CH_2CH_2CH_2NH_2$

10.5 BASICITY OF AMINES: SALT FORMATION

The amines contain a nonbonding pair of electrons, an unshared pair, that can be shared with electron-deficient groups to form a new bond. For example, if an amine reacts with a proton (H^+), a new N—H bond results. The nonbonding pair of electrons on nitrogen have been shared with the electron-deficient proton to form the new bond:

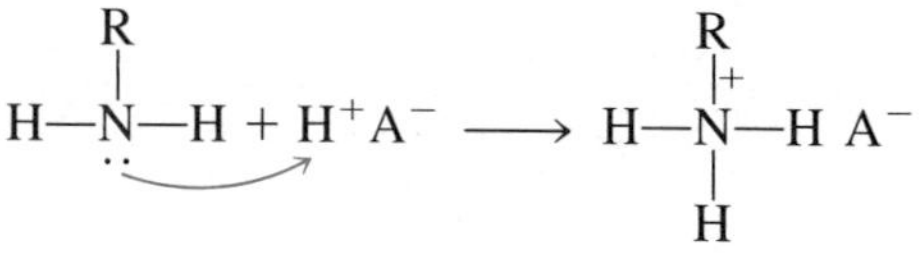

$$R\ddot{N}H_2 + H^+A^- \longrightarrow RNH_3^+\ A^- \qquad (10.6)$$

Amine — Proton (acid) — Ammonium salt

electrons flow to form new bond

In this type of reaction the amine is behaving as a **Lewis base.** In acid-base reactions, Lewis bases are electron pair donors, donating a pair of electrons to form a bond with another atom, often a proton. Many other species that we encounter regularly may also behave as Lewis bases under appropriate reaction conditions. For example:

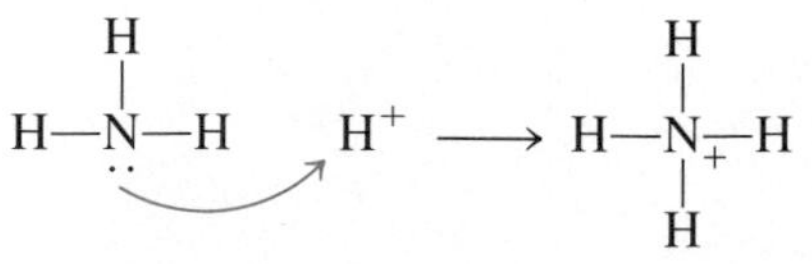

$$H\text{—}\ddot{N}H_2 \quad H^+ \longrightarrow NH_4^+ \qquad (10.7)$$

$$H_2\ddot{O}: \quad H^+ \longrightarrow H_3O^+ \qquad (10.8)$$

$$:\ddot{Cl}:^- \quad H^+ \longrightarrow H\text{—}\ddot{Cl}: \qquad (10.9)$$

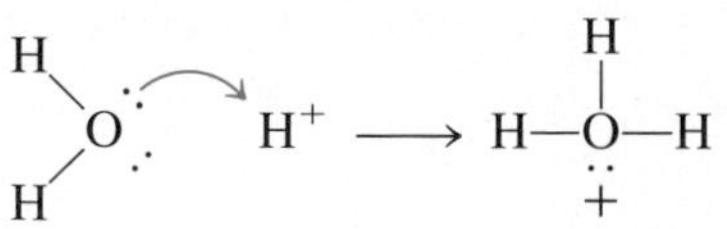

$$H\ddot{O}:^- \quad H^+ \longrightarrow H\text{—}\ddot{O}\text{—}H \qquad (10.10)$$

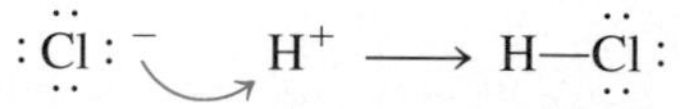

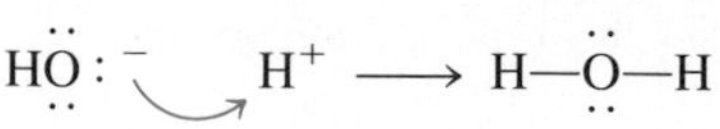

The latter two examples (chloride ion and hydroxide ion) are also examples of Brönsted-Lowry bases—they are proton acceptors. But they are also behaving as Lewis bases—they are electron pair donors. An electron pair from the Lewis base is shared with the proton to form the new bond to the hydrogen atom. The definition used is the one that is most applicable to the situation. In the discussion of amines it is advantageous to consider their reactions from the view of electron transfer. We have encountered numerous examples of acids and bases in previous chapters, and there are many more to come. An understanding of acid-base chemistry is *essential* to an understanding of chemistry.

Amines are moderately strong bases and will react with most acids to form salts (recall that an acid + a base ⟶ a salt). The reaction of methylamine with hydrochloric acid, shown below, is typical of these reactions.

Section 8.5

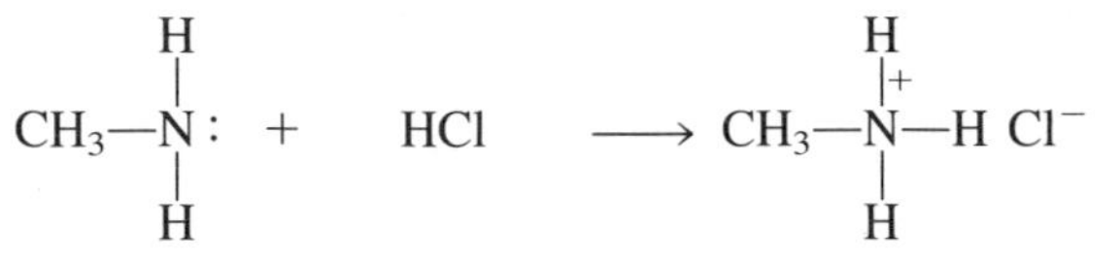

$$CH_3\text{—}\overset{\overset{\displaystyle H}{|}}{\underset{\underset{\displaystyle H}{|}}{N}}\!: \; + \; HCl \longrightarrow CH_3\text{—}\overset{\overset{\displaystyle H}{|+}}{\underset{\underset{\displaystyle H}{|}}{N}}\text{—}H \; Cl^- \qquad 10.11$$

Methylamine | Hydrochloric acid | Methylammonium chloride

The salt is named by replacing amine with ammonium followed by the name of the anion. The salts are ionic species and hence are quite soluble in water.

Recall that water is amphiprotic, that is, it can behave as an acid or a base. Amines, being basic substances, can react with water to produce a salt. Recall also that the reaction of a basic (or acidic) substance with water is called a hydrolysis reaction. The reaction of methylamine with water is shown as an example:

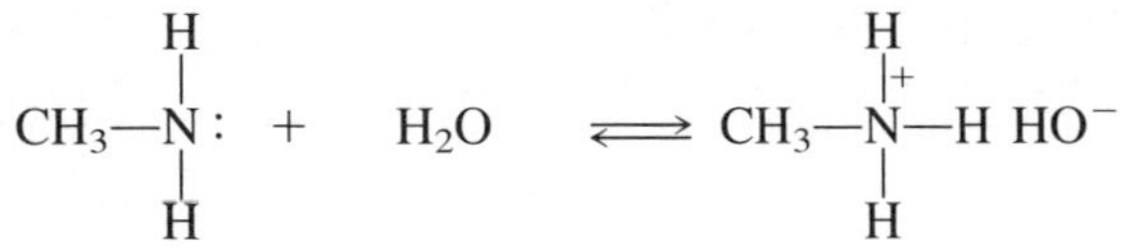

$$CH_3\text{—}\overset{\overset{\displaystyle H}{|}}{\underset{\underset{\displaystyle H}{|}}{N}}\!: \; + \; H_2O \rightleftharpoons CH_3\text{—}\overset{\overset{\displaystyle H}{|+}}{\underset{\underset{\displaystyle H}{|}}{N}}\text{—}H \; HO^- \qquad 10.12$$

Methylamine *behaving as a base* | Water *behaving as an acid* | Methylammonium hydroxide

The product of the reaction is a hydroxide salt; therefore the solution is basic. The reaction, as shown, is also reversible. Under suitable reaction conditions the salt may be, at least partially, converted back to the free amine. Bases other than HO^- may convert the salt back to the free amine. As shown below, even water may react in this way:

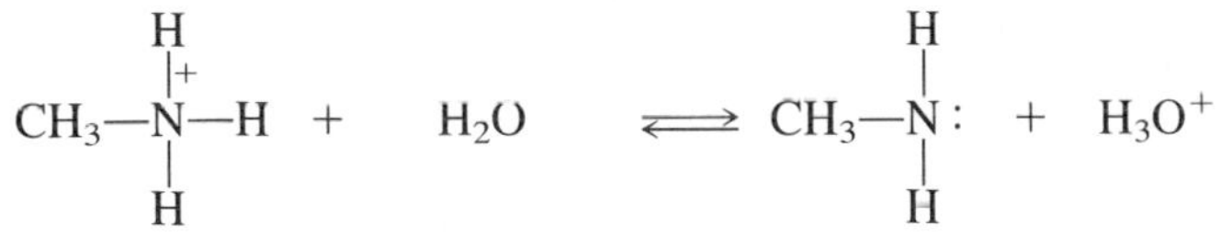

$$CH_3\text{—}\overset{\overset{\displaystyle H}{|+}}{\underset{\underset{\displaystyle H}{|}}{N}}\text{—}H \; + \; H_2O \rightleftharpoons CH_3\text{—}\overset{\overset{\displaystyle H}{|}}{\underset{\underset{\displaystyle H}{|}}{N}}\!: \; + \; H_3O^+ \qquad 10.13$$

Ammonium salt *behaving as an acid* | Water *behaving as a base* | Amine | Hydronium ion

The general reactions, involving amines and their salts, may be summarized as follows:

amine + hydronium ion ⇌ ammonium salt + water

Base | Acid | Salt | Water

The ability of the amines to act as bases (proton acceptors) and of amine salts to act as acids (proton donors) makes the amines useful as *buffers*. In fact, many of the naturally occurring amines that are found in living systems do indeed serve as natural buffers to

Section 11.10

maintain the acid/base (pH) balance of the blood. The molecule hemoglobin is perhaps the best example of such a buffer.

The formation of water-soluble amine salts has important consequences in medicinal chemistry, in which this solubility provides a means of delivering drugs to parts of the body that would be inaccessible otherwise.

Question 10.10 Complete the following reactions by supplying the missing product(s):

a. $CH_3NH_2 + HI \longrightarrow$

b. $CH_3CH_2NH_2 + H_2SO_4 \longrightarrow$

c. $(CH_3CH_2)_2NH + HCl \longrightarrow$

d. 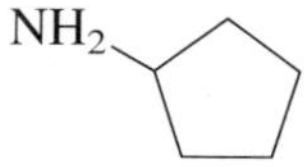 NH_2–(cyclopentane) $+ HBr \longrightarrow$

e. $NH_4Cl + HO^- \longrightarrow$

f. $(CH_3)_3NH^+Cl^- + NaOH \longrightarrow$

10.6 REACTIONS OF AMINES: AMIDE FORMATION

The most important reaction of the amines was discussed in Section 10.4: salt formation due to the basicity of amines. A second reaction of the amines has been alluded to several times in the preceding sections: the formation of amides. **Amides,** an important class of nitrogen-containing organic compounds, contain the functional group shown here:

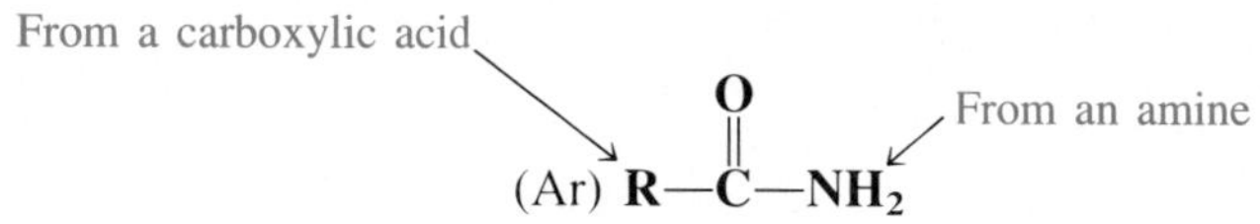

$$\text{(Ar) } \mathbf{R{-}\overset{\overset{\Large O}{\|}}{C}{-}NH_2}$$

From a carboxylic acid → (Ar) R; From an amine → NH_2

An amide—the amide functional group

The amide group is composed of two portions: one portion from a carboxylic acid and a second portion from an amine. Amides are members of the general family of acyl compounds: Amides are derivatives of carboxylic acids.

Section 8.4

Structure and properties of amides

Most amides are solids at room temperature. They have very high boiling points (even higher than the corresponding acid), and the simpler ones are quite soluble in water. Both of these properties are a result of strong intermolecular hydrogen bonding between the N—H bond of one amide and the C═O group of a second amide. This is shown in Figure 10.4. As we have seen with the other families of organic compounds, their solubility in water decreases with increasing molecular weight, as the ''organic'' or nonpolar character of the molecule increases.

FIGURE 10.4
Hydrogen bonding in amides.

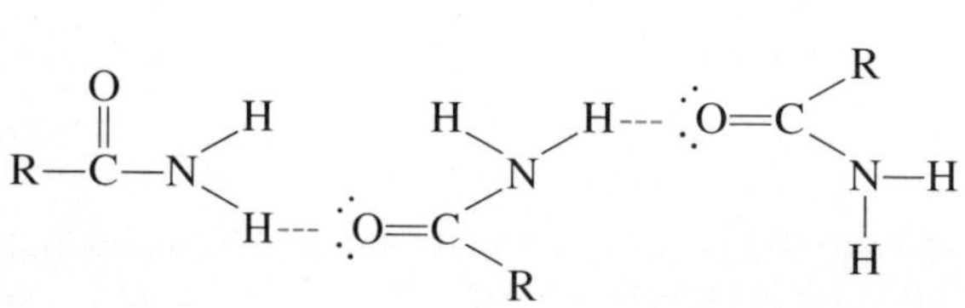

$$R{-}\overset{\overset{O}{\|}}{C}{-}N(H){-}H\cdots :O{=}C(R){-}N(H){-}H\cdots :O{=}C(R){-}NH_2$$

Nomenclature of amides

The common and I.U.P.A.C. names of the amides are derived from the common and I.U.P.A.C. names of the corresponding carboxylic acids and diacids. *Section 8.1*

Amides are named by removing the *-ic acid* ending of the common name or the *-oic acid* ending of the I.U.P.A.C. name of the carboxylic acid or diacid and replacing it with *-amide*. For example:

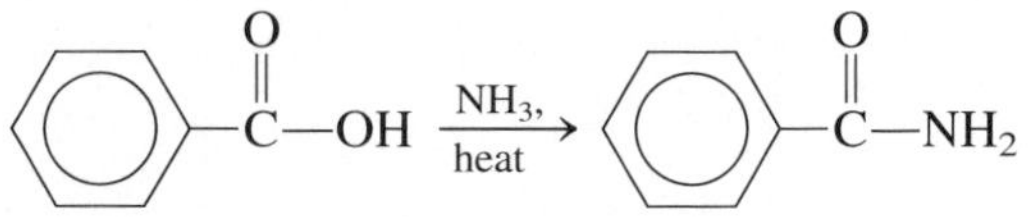

$$C_6H_5-\overset{O}{\overset{\|}{C}}-OH \xrightarrow[\text{heat}]{NH_3,} C_6H_5-\overset{O}{\overset{\|}{C}}-NH_2 \qquad 10.14$$

Benzoic acid — Benzamide

Also

Acet**ic acid** ⟶ Acet**amide**

Propan**oic acid** ⟶ Propan**amide**

Substituents on the nitrogen are placed as prefixes and are indicated by *N*- followed by the name of the substituent. There are no spaces between the prefix and the amide name. For example:

$$CH_3CH_2CH_2CH_2CH_2-\overset{O}{\overset{\|}{C}}-NH-CH_2CH_2CH_3$$

N-Propylhexanamide

Other examples of amide nomenclature include:

$$CH_3-\overset{O}{\overset{\|}{C}}-NH_2 \qquad CH_3CH_2CH_2-\overset{O}{\overset{\|}{C}}-NH_2 \qquad Cl-C_6H_4-\overset{O}{\overset{\|}{C}}-NH_2$$

Ethanamide (acetamide) — Butanamide (butyramide) — *p*-Chlorobenzamide

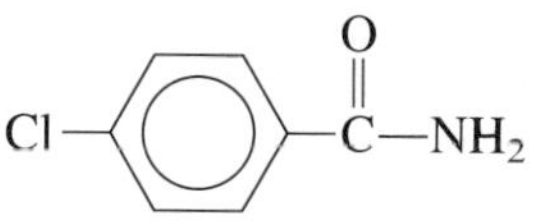

$$CH_3CH_2CH{=}CHCH_2CH_2-\overset{O}{\overset{\|}{C}}-NH_2 \qquad CH_3\underset{C_6H_5}{\underset{|}{C}}HCH_2CH_2-\overset{O}{\overset{\|}{C}}-NHCH_3$$

4-Heptenamide — N-Methyl-4-phenylpentanamide

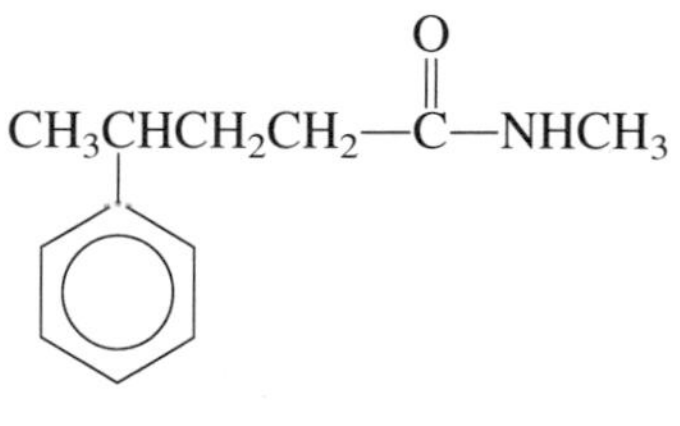

Preparation of amides

Amides, like esters, can be prepared directly from the carboxylic acid or via the **acid chloride.** The necessary conditions are provided in the following examples:

$$R-\overset{O}{\overset{\|}{C}}-OH + NH_3 \xrightarrow{\text{heat}} R-\overset{O}{\overset{\|}{C}}-NH_2 + H_2O \qquad 10.15$$

An acid — Ammonia — An amide — Water

$$R-\overset{O}{\overset{\|}{C}}-Cl + NH_3 \xrightarrow{\text{heat}} R-\overset{O}{\overset{\|}{C}}-NH_2 + HCl \qquad 10.16$$

An acid chloride — Ammonia — An amide — Hydrochloric acid

Four specific examples follow:

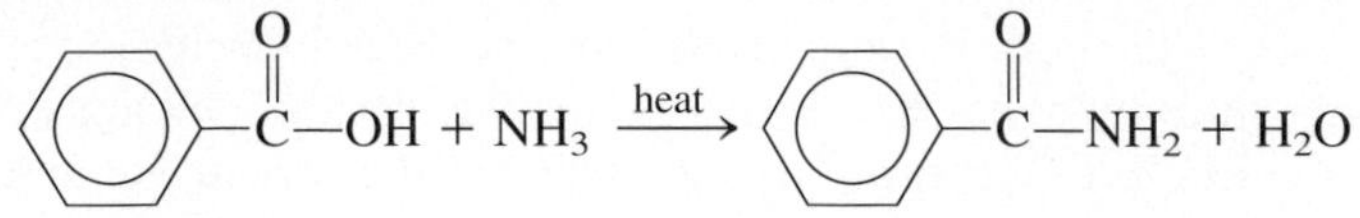

$$C_6H_5{-}\overset{O}{\overset{\|}{C}}{-}OH + NH_3 \xrightarrow{\text{heat}} C_6H_5{-}\overset{O}{\overset{\|}{C}}{-}NH_2 + H_2O \qquad 10.17$$

Benzoic acid — Benzamide

$$CH_3CH_2CH_2\underset{\underset{CH_3}{|}}{CH}{-}\overset{O}{\overset{\|}{C}}{-}Cl + NH_3 \xrightarrow{\text{heat}} CH_3CH_2CH_2\underset{\underset{CH_3}{|}}{CH}{-}\overset{O}{\overset{\|}{C}}{-}NH_2 + HCl \qquad 10.18$$

2-Methylpentanoyl chloride — 2-Methylpentanamide

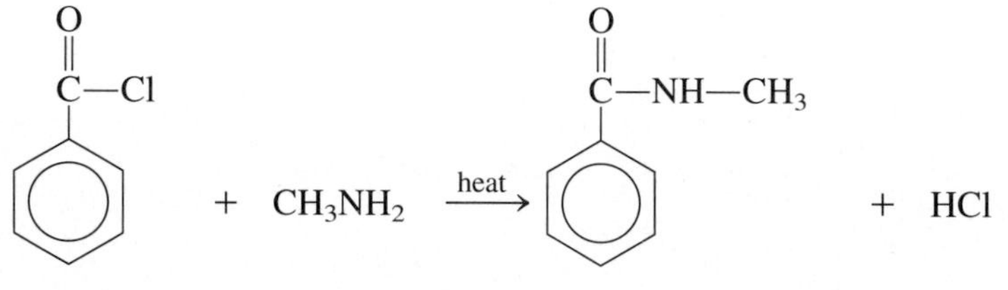

$$C_6H_5{-}\overset{O}{\overset{\|}{C}}{-}Cl + CH_3NH_2 \xrightarrow{\text{heat}} C_6H_5{-}\overset{O}{\overset{\|}{C}}{-}NH{-}CH_3 + HCl \qquad 10.19$$

Benzoyl chloride — Methylamine — N-methylbenzamide

$$CH_3CH_2CH_2CH_2{-}\overset{O}{\overset{\|}{C}}{-}OH \xrightarrow{PCl_5} CH_3CH_2CH_2CH_2{-}\overset{O}{\overset{\|}{C}}{-}Cl$$

Pentanoic acid (valeric acid) — Pentanoyl chloride (valeryl chloride)

$$\xrightarrow{CH_3NH_2} CH_3CH_2CH_2CH_2{-}\overset{O}{\overset{\|}{C}}{-}NHCH_3 + HCl \qquad 10.20$$

N-methylpentanamide (N-methyl valeramide)

Question 10.11 What is the structure of the amine that, on reaction with the acid chloride shown, would give the indicated products?

a. $$C_6H_5{-}\overset{O}{\overset{\|}{C}}{-}Cl \longrightarrow C_6H_5{-}\overset{O}{\overset{\|}{C}}{-}NH_2$$

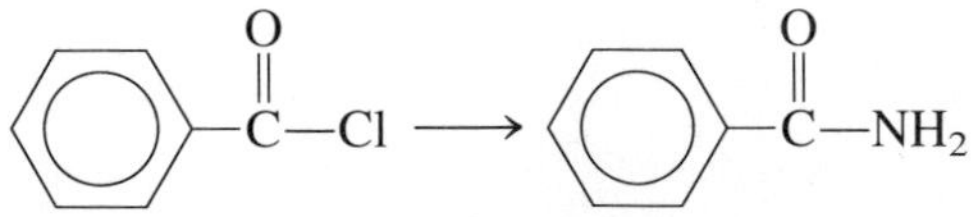

b. $$CH_3{-}\overset{O}{\overset{\|}{C}}{-}Cl \longrightarrow CH_3{-}\overset{O}{\overset{\|}{C}}{-}NHCH_3$$

c. $$CH_3CH_2CH_2CH_2\underset{\underset{CH_3CH_2}{|}}{CH}{-}\overset{O}{\overset{\|}{C}}{-}Cl \longrightarrow (CH_3)_2N{-}\overset{O}{\overset{\|}{C}}{-}\underset{\underset{CH_2CH_3}{|}}{C}HCH_2CH_2CH_2CH_3$$

We discuss the peptides, proteins, and enzymes in Chapter 11.

The amide bond is the central feature in the structure of peptides and proteins. By joining the amino acids, one to another, we create larger and larger molecules. Initially, we produce *peptides,* and at molecular weights above 50,000, *proteins* result. Both

Amide bond

$$H_2N-\underset{CH_2-COOH}{\overset{H}{C}}-\overset{O}{\overset{\|}{C}}-\underset{H}{N}-\underset{CH_2C_6H_5}{\overset{H}{C}}-\overset{O}{\overset{\|}{C}}-OCH_3$$

As methyl ester

Aspartic acid Phenylalanine

A dipeptide

FIGURE 10.5
The amide bond. NutraSweet®, the dipeptide aspartame is a molecule composed of two amino acids joined by an amide (peptide) bond.

peptides and proteins are involved in many of the intricate biochemical processes that are necessary to life functions. A simple dipeptide is shown in Figure 10.5. It is the compound marketed as aspartame or NutraSweet®, the artificial sweetener that is rapidly replacing saccharin in many foods and beverages. It is composed of two amino acids, aspartic acid and the methyl ester of phenylalanine, which are joined via an amide bond.

See Figure 12.15 for examples of complex proteins, the enzymes chymotrypsin and elastase.

SUMMARY

Amines are polar molecules. This polarity is due to the presence of polar N—H bond(s) as well as the way in which these groups are incorporated into various molecules. Hydrogen bonding is possible and is responsible for the relatively high boiling points and water solubilities of this family.

Amines are synthesized principally by the reduction of nitro compounds and amides. A variety of reducing agents may be used to effect these transformations. Two principal

A HUMAN PERSPECTIVE

Amines and the Central Nervous System

The **central nervous system (CNS)** refers to the brain, the spinal cord, and all of the nerves that radiate from the spinal cord. The molecules that are principally responsible for transmitting messages—signals—between the various nerve cells are called **neurotransmitters.** Serotonin, a heterocyclic amine, and acetylcholine, a quaternary ammonium salt, are two of the most important neurotransmitters in the body. Serotonin appears to mediate certain types of behavior. A deficiency in serotonin, for example, has been linked to mental depression, a disease that some believe is becoming a national epidemic. Acetylcholine, in contrast, is involved in the transmission of nerve impulses that control the contraction of muscles (Section 12.15). As a result of these properties, it is often used in the treatment of Alzheimer's Disease.

HO, $CH_2CH_2NH_2$, N, H

Serotonin

$$CH_3-\underset{CH_3}{\overset{CH_3}{N}}-CH_2CH_2-O-\overset{O}{\overset{\|}{C}}-CH_3$$

Acetylcholine

Some important amines in our day-to-day lives.

Dopa, dopamine, and the related compounds epinephrine and norepinephrine are also critical to good mental health. A deficiency in dopamine, for example, results in Parkinson's Disease. Dopa is used as a medicinal in the treatment of this disorder. Both epinephrine (adrenaline) and norepinephrine are hormones that are involved in the "fight-or-flight" response. Epinephrine is involved in the breakdown of glycogen to glucose, while norepinephrine is involved with the CNS in the stimulation of other glands and the constriction of blood vessels.

reactions of amines are salt formation (due to their basicity) and the formation of amides. The heterocyclic amines are of particular interest because of the wide spectrum of important compounds and families that contain nitrogen heterocyclic rings as a principal part of their skeletal framework. The alkaloids are perhaps the most important of these families.

Amines and amides play important roles in our day-to-day lives. They are found in living systems, medicinals, industrial chemicals, and foodstuffs.

GLOSSARY OF KEY TERMS

acid chlorides (10.6) the family of organic compounds with the general formula

$$\text{(Ar) } R-\overset{\overset{\displaystyle O}{\|}}{C}-Cl$$

alkaloids (10.3) a class of naturally occurring compounds that contain one or more nitrogen heterocyclic rings; many of the alkaloids have medicinal and other physiological effects.

amides (10.6) the family of organic compounds with the general formula

$$\text{(Ar) } R-\overset{\overset{\displaystyle O}{\|}}{C}-NH_2$$

amines (Introduction) the family of organic molecules with the general formula $R-NH_2$, R_2NH, or R_3N (R— can equal R— or Ar—); they may be viewed as substituted ammonia molecules in which one or more of the ammonia hydrogens has been substituted by a more complex organic moiety.

amphetamines (10.6) a family of compounds, many of which are physiologically active and stimulate the central nervous system; they are often called ''uppers'' and are often used in diet pills to help decrease appetite; they are also used to treat psychological disorders such as severe depression.

analgesic (10.6) any drug that acts as a pain killer.

anesthetic (10.6) any drug that causes lack of sensation in any part of the body (local) or causes unconsciousness (general); an analgesic effect is generally also associated with these compounds.

barbiturates (10.6) a class of physiologically active compounds that are derived from amides (which in turn are prepared from amines); the barbiturates are ''downers'' and are used as sedatives; they are also used as anticonvulsants for epileptics and for people suffering from a variety of brain disorders.

carcinogen (10.1) any cancer-causing substance.

central nervous system (CNS) (10.1) the brain, the spinal cord, and all of the nerves that radiate from the spinal cord column.

decongestants (10.6) a group of compounds that are useful in the treatment of the symptoms associated with colds and many allergies; they act by causing a shrinking of the membranes that line the nasal passages.

heterocyclic amine (Introduction) a heterocyclic compound that contains nitrogen in at least one position in the ring skeleton.

Lewis base (10.5) Lewis bases are electron pair donors, donating a pair of electrons to form a bond with another atom, often a proton.

neurotransmitter (10.1) a chemical substance that acts as a chemical ''bridge'' in nerve impulse transmission; these chemicals are released at the nerve ending of one nerve, travel across the synaptic gap, and react with an adjacent nerve or muscle cell.

primary (1°) amine (10.1) an amine with the general formula RNH_2.

quaternary ammonium salt (10.1) an amine salt with the general formula $R_4N^+A^-$ (in which R— can be an alkyl or aryl group or a hydrogen atom and A^- can be any anion).

secondary (2°) amine (10.1) an amine with the general formula R_2-NH.

sulfa drugs (10.6) a family of drugs that are derived from the compound sulfanilamide and that have antibacterial properties.

tertiary (3°) amine (10.1) an amine with the general formula R_3-N.

QUESTIONS AND PROBLEMS

Nomenclature and Structure

10.12 Name each of the following amines, using the I.U.P.A.C. Nomenclature System. Where appropriate, give the common name also.

a. $CH_3CH_2\underset{\underset{\displaystyle Br}{|}}{C}H-NH_2$

b. $CH_3CH_2CH_2\underset{\underset{\displaystyle NH_2}{|}}{C}HCH_2CH_3$

c. (cyclopentane ring)$-NH_2$

d. $(CH_3)_3C-NH_2$

10.13 Name each of the following amines, using the I.U.P.A.C. Nomenclature System. Where appropriate, give the common name also.

a. $CH_3CH_2CH_2CH_2CH_2CH_2CH_2CH_2NH_2$

b. Cl—(benzene ring)—NH_2

c. $CH_3\underset{\underset{\displaystyle NH_2}{|}}{C}HCH_2CH_3$

d. $CH_3CH_2\underset{\underset{\displaystyle NH_2}{|}}{\overset{\overset{\displaystyle CH_3}{|}}{C}}CH_2CH_2CH_3$

10.14 Draw each of the following, using complete structural formulas:

a. diethylamine
b. *n*-butylamine
c. 3-aminodecane
d. 3-bromo-2-aminopentane
e. triphenylamine
f. *N,N*-di-*n*-propylaniline
g. aminocyclohexane
h. 4,5-diaminononane
i. tetraethylammonium iodide
j. 3,5-diaminobromobenzene

10.15 Name the following amides, using the I.U.P.A.C. Nomenclature System:

a. $CH_3CH_2-\overset{\overset{\displaystyle O}{\|}}{C}-NH_2$

b. $CH_3-\overset{\overset{\displaystyle O}{\|}}{C}-N(CH_3)_2$

c. 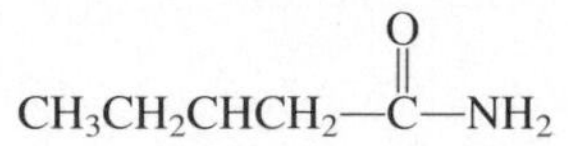 $CH_3CH_2CHCH_2—\overset{O}{\overset{\|}{C}}—NH_2$

d. 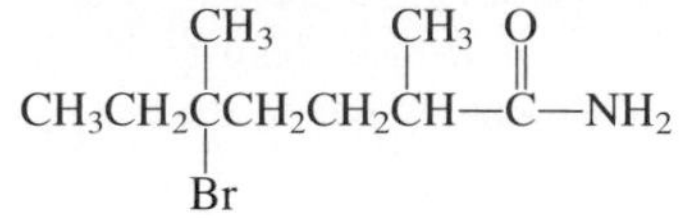 $CH_3CH_2\overset{CH_3}{\overset{|}{C}}(Br)CH_2CH_2\overset{CH_3}{\overset{|}{C}H}—\overset{O}{\overset{\|}{C}}—NH_2$

e. 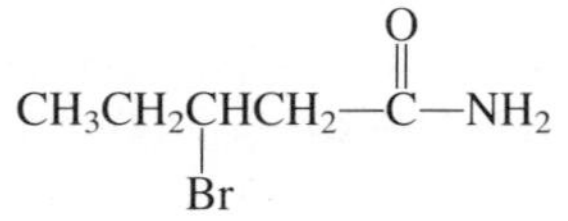 $CH_3CH_2\underset{Br}{\underset{|}{C}H}CH_2—\overset{O}{\overset{\|}{C}}—NH_2$

f. 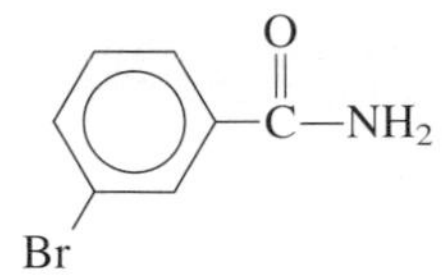 (benzene ring, Br substituent)$—\overset{O}{\overset{\|}{C}}—NH_2$

Structure and Properties

10.16 Draw the condensed structure of each of the following amides:
a. 4-methylpentanamide
b. *N*-methylpropanamide
c. *N,N*-diethylbenzamide
d. 3-bromo-4-methylhexanamide
e. *N,N*-dimethylacetamide

10.17 Draw structural formulas for the eight isomeric amines that have the molecular formula $C_4H_{11}N$. Name each of the isomers, using the I.U.P.A.C. System, and determine whether each isomer is a 1°, 2°, or 3° amine.

10.18 Lidocaine is often used as a local anesthetic. For medicinal purposes it is often used in the form of its hydrochloride salt, since the salt is water-soluble. The structure of lidocaine hydrochloride is provided below. Locate the amide functional group. What other functional groups are present?

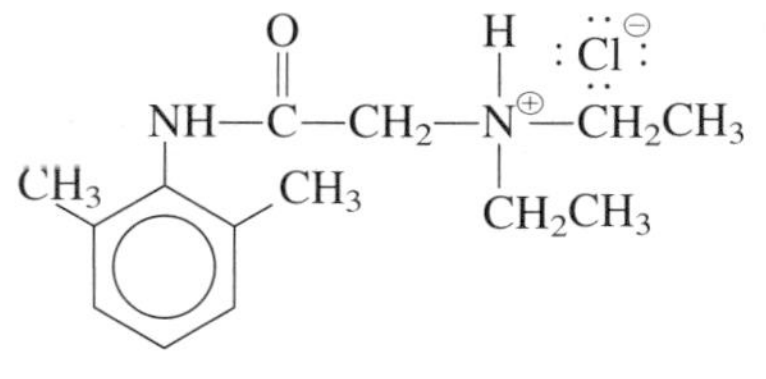

$NH—\overset{O}{\overset{\|}{C}}—CH_2—\overset{H}{\overset{|}{N^{\oplus}}}(CH_2CH_3)—CH_2CH_3 \quad :\ddot{Cl}:^{\ominus}$ (NH attached to a benzene ring bearing two CH_3 groups)

Lidocaine hydrochloride

10.19 Putrescine and cadaverine are two odoriferous amines that are produced by decaying flesh. Putrescine is 1,4-diaminobutane, and cadaverine is 1,5-diaminopentane. Draw the complete structural formulas for each of these two compounds. Which would have the higher boiling point? Which would be more soluble in water?

10.20 Draw each of the following, using complete structural formulas:
a. 2-aminopentane   c. ethylisopropylamine
b. 1-bromo-2-aminobutane   d. aminocyclopentane

10.21 Draw each of the following using complete structural formulas:
a. di-*n*-pentylamine
b. 3,4-dinitroaniline
c. 3,4-dimethyl-4-aminoheptane
d. *t*-butyl-*n*-pentylamine
e. 2,3-diaminohexane
f. triethylammonium iodide

10.22 Name each of the following salts:
a. $(CH_3)_4N^+Br^-$ c. $(CH_3)_2NH(CH_2CH_3)^+Cl^-$
b. $NH_4^+I^-$ d. $CH_3CH_2CH_2NH_3^+HSO_4^-$

10.23 Draw all of the amines of molecular formula $C_5H_{13}N$.

10.24 a. What are heterocyclic compounds?
b. Draw an example of a heterocyclic oxygen molecule, a heterocyclic sulfur molecule, and a heterocyclic phosphorous molecule.

10.25 Determine the structure of the monomer that may be condensed to form the following polymer:

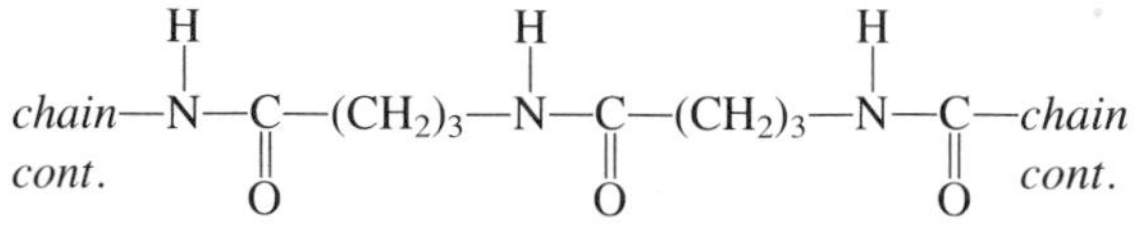

$$\text{chain cont.}—\overset{H}{\overset{|}{N}}—\underset{O}{\underset{\|}{C}}—(CH_2)_3—\overset{H}{\overset{|}{N}}—\underset{O}{\underset{\|}{C}}—(CH_2)_3—\overset{H}{\overset{|}{N}}—\underset{O}{\underset{\|}{C}}—\text{chain cont.}$$

10.26 The antibiotic penicillin BT contains functional groups that were discussed in this chapter. The structure of penicillin BT is provided below. Locate and name as many functional groups as you can.

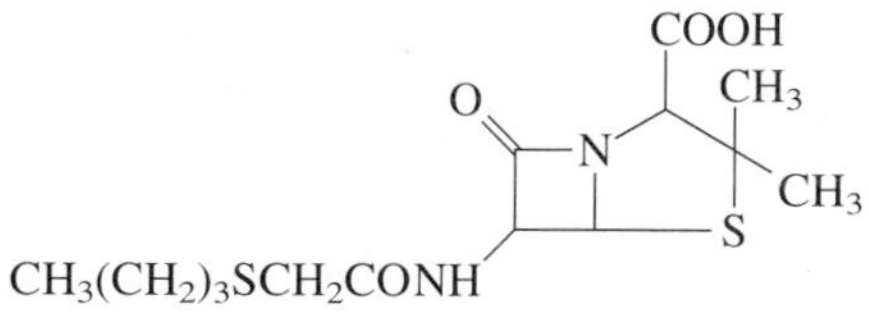

$CH_3(CH_2)_3SCH_2CONH$ — (β-lactam ring with O, N; thiazolidine ring with S, COOH, CH_3, CH_3)

Penicillin BT

10.27 The structure of saccharin, an artificial sweetener, is shown below. Circle the amide group.

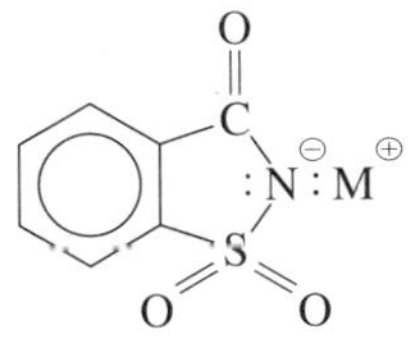

(benzene ring fused to ring containing C=O, $:\ddot{N}:^{\ominus}M^{\oplus}$, and $S(=O)_2$)

Saccharin

10.28 Explain chemically how ammonia can act as an acid or a base. Can the ammonium ion also exhibit both of these types of behavior?

10.29 Using the Lewis definition of a base, discuss the basicity of trimethylamine in its reaction with, for example, HCl. Does trimethylamine also meet the criteria associated with the Brönsted-Lowry definition of a base? Explain your reasoning.

10.30 Arrange the following in order of increasing boiling point:
a. 1-aminohexane   c. hexane
b. 1-hexanol   d. di-*n*-propyl ether

10.31 Select the stronger base in each of the following pairs:
a. $CH_3CH_2CH_2NH_2$ or $(CH_3CH_2CH_2)_3N$
b. $CH_3CH_2CH_2NH_2$ or $NH_2—C_6H_5$ (aniline)
c. NH_3 or NH_4^+
d. $CH_3CH_2CH_2CH_2NH_2$ or $CH_3CH_2—O—CH_2CH_3$

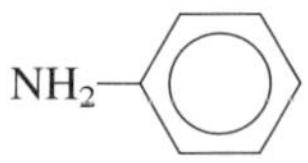

10.32 Would you expect an amine such as 1-aminohexane to be more soluble in water or in dilute acid (for example 0.1 M HCl)?

10.33 Which of the following would be more soluble in water?
a. 1-aminobutane c. butane
b. 1-butanol
Which would have the higher boiling point? Explain your reasoning.

10.34 What is the difference between a barbiturate and an amphetamine both structurally and functionally?

10.35 What is a buffer? Why is a buffer used?

10.36 Why are amines well suited to function as buffers?

10.37 Classify each of the following amines as 1°, 2°, or 3°:
a. aminocyclohexane
b. *t*-butylamine
c. 2-methyl-2-aminoheptane
d. neopentylamine
e. *p*-bromoaniline
f. 3-ethyl-2-aminopentane
g. methylethylamine
h. tri-*n*-propylamine
i. *m*-chloroaniline
j. *N*-ethylaniline

10.38 Briefly explain why the smaller (lower molecular weight) amines exhibit appreciable solubility in water.

10.39 The salt of an amine is appreciably more soluble in water than the amine precursor (for example, dimethylammonium chloride versus dimethylamine). Briefly explain this observation.

10.40 *N,N*-dimethylacetamide (m.p. −20°C, b.p. 165°C) is a liquid at room temperature, whereas acetamide (m.p. 82°C, b.p. 222°C) is a solid. Explain these facts, keeping in mind that the molecular weight of acetamide is considerably less than that of the *N,N*-dimethylacetamide.

Reactions

10.41 Complete each of the following syntheses by supplying the missing reactant(s), reagent(s), or product(s) indicated by the question marks:

a. $CH_3CH_2CH_2CH_2CH_2NH_2 \xrightarrow{?} CH_3CH_2CH_2CH_2CH_2NH_3{}^+Cl^-$

b. 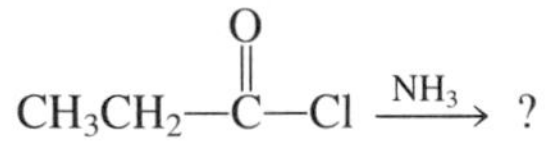 $CH_3CH_2-\overset{O}{\overset{\|}{C}}-Cl \xrightarrow{NH_3} ?$

c. $? \xrightarrow{HBr} C_6H_5-NH_3{}^+\ Br^-$

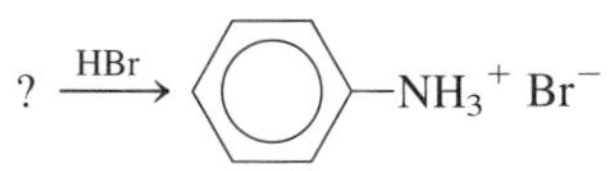

d. $? \xrightarrow{[R]} CH_3CH_2CH_2CH_2CH_2CH_2CH_2NH_2$

e. $(CH_3)_3CNH_2 \xrightarrow{H_2SO_4} ?$

f. 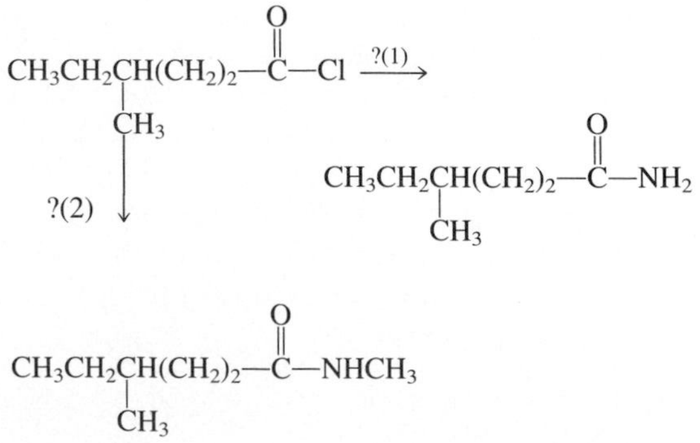 $CH_3CH_2\underset{CH_3}{\underset{|}{C}}H(CH_2)_2-\overset{O}{\overset{\|}{C}}-Cl \xrightarrow{?(1)} CH_3CH_2\underset{CH_3}{\underset{|}{C}}H(CH_2)_2-\overset{O}{\overset{\|}{C}}-NH_2$

$\downarrow ?(2)$

$CH_3CH_2\underset{CH_3}{\underset{|}{C}}H(CH_2)_2-\overset{O}{\overset{\|}{C}}-NHCH_3$

Further Problems

10.42 Spermine is a polyamine found in human seminal fluid. It has the molecular formula $C_{10}H_{26}N_4$. Spermine can be prepared by the total reduction of compound I below with, for example, $LiAlH_4$. Can you draw the complete structural formula of spermine?

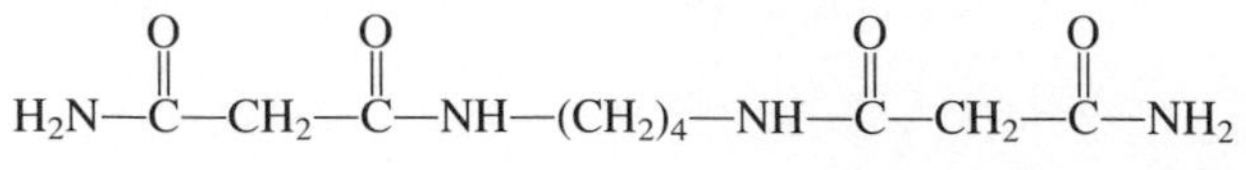

$$H_2N-\overset{O}{\overset{\|}{C}}-CH_2-\overset{O}{\overset{\|}{C}}-NH-(CH_2)_4-NH-\overset{O}{\overset{\|}{C}}-CH_2-\overset{O}{\overset{\|}{C}}-NH_2$$

I

10.43 Complete the following by supplying the missing reagents:
a. ethanoic acid to 1-aminoethane
b. nitrobenzene to aniline
c. nitrobenzene to *m*-bromonitrobenzene
d. 3-hexanoic acid to 3,4-dibromo-1-aminohexane
e. trimethylamine to trimethylammonium hydrogen sulfate
f. pyridine to pyridinium chloride

10.44 Refer to the Human Perspective feature in Section 10.6 in which we discuss the mechanism through which amines may cause cancer. Using whatever reference texts you wish, draw the *complete* step-by-step mechanism for each of the transformations that were summarized.

10.45 When comparing aliphatic amines to aromatic amines, one finds that the latter are significantly weaker bases than the former. Aniline, for example, has a pK_a of 9.4, while methylamine has a pK_a of 3.36. Can you account for the large difference in observed basicity? (*Note:* Refer to an organic chemistry textbook for background related to this question.)

10.46 The most important industrial application involving amines is the synthesis of the polymer nylon—a polyamide. This application was discussed in Chapter 8 (see Section 8.5, "Human Perspective: Carboxylic Acid Derivatives of Special Interest"). The polymer nylon-66 is shown here:

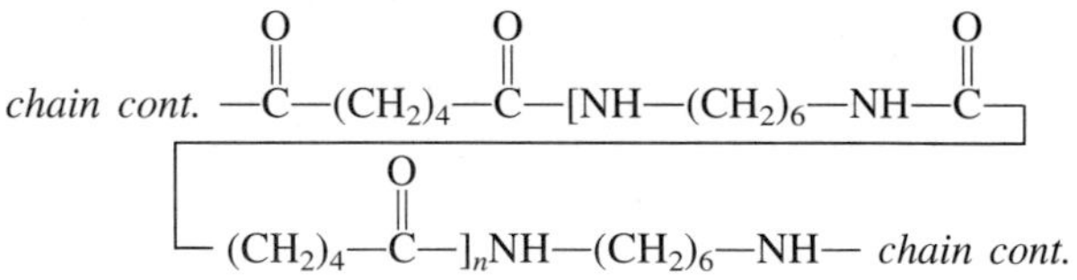

$$\textit{chain cont.}\ -\overset{O}{\overset{\|}{C}}-(CH_2)_4-\overset{O}{\overset{\|}{C}}-[NH-(CH_2)_6-NH-\overset{O}{\overset{\|}{C}}-(CH_2)_4-\overset{O}{\overset{\|}{C}}-]_nNH-(CH_2)_6-NH-\ \textit{chain cont.}$$

$(n = 50–65)$

Nylon-66

Using structural formulas, draw the monomeric units that combine to form nylon-66.

10.47 A polymer called nylon-6 is prepared by heating the cyclic amide caprolactam (compound I) in the presence of a trace of water. Suggest a structure for nylon-6.

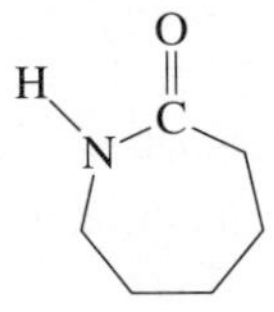

(seven-membered ring: H–N–C(=O) within ring)

I

10.48 Nylon-6–10 is produced from hexamethylenediamine and a diacid. Hexamethylenediamine is shown below. What might the structure of the diacid be? The structure of nylon-6–10?

$$H_2NCH_2CH_2CH_2CH_2CH_2CH_2NH_2$$

10.49 Barbiturates have the general structure shown below for barbituric acid. Barbituric acid is prepared from urea and diethylmalonate as follows:

$$H_2N\text{—}\overset{O}{\overset{\|}{C}}\text{—}NH_2 + CH_3CH_2O\text{—}\overset{O}{\overset{\|}{C}}\text{—}CH_2\text{—}\overset{O}{\overset{\|}{C}}\text{—}OCH_2CH_3 \longrightarrow$$

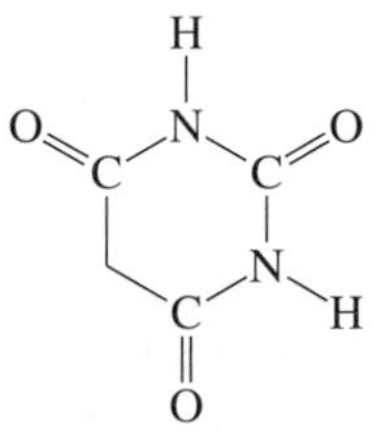

Urea — Diethyl malonate

Barbituric acid
$R^1 = R^2 = H$

What kind of reaction is involved in this synthesis? What functional groups are involved? Be as explicit as possible.

10.50 Most drugs containing amine groups are not administered as the amine but rather as the ammonium salt. Can you suggest a reason why?

10.51 Why does aspirin upset the stomach while acetaminophen (Tylenol) does not?

10.52 The active ingredient in many insect repellents is *N,N*-diethyltoluamide. Draw out the complete structural formula of this compound. Which carboxylic acid and amine could be reacted to give this compound?

10.53 When carboxylic acids and amines are combined, they neutralize each other, forming salts. If the salt is heated, an amide is formed. This procedure may be used to synthesize acetaminophen, the active ingredient in Tylenol. Complete the following reaction:

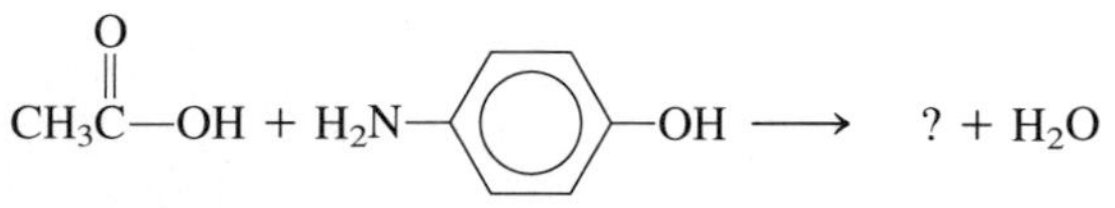

$$CH_3\overset{O}{\overset{\|}{C}}\text{—}OH + H_2N\text{—}C_6H_4\text{—}OH \longrightarrow \; ? + H_2O$$

Acetaminophen

C H A P T E R 1 1

Protein Conformation and Function

O U T L I N E

O B J E C T I V E S

- Be familiar with the cellular functions of proteins.
- Draw the general structure of an amino acid and know the classification of side chains or R groups of the amino acids.
- Describe the primary structure of proteins and the structure of the peptide bond.
- Be familiar with the rules for the structure and nomenclature of small peptides.
- Describe the types of secondary structure of a protein.
- Understand the forces that maintain secondary structure.
- Be familiar with the structures and functions of the fibrous proteins, like collagen.
- Understand the role of vitamin C in collagen metabolism.
- Be familiar with tertiary and quaternary structure.
- Understand the R group interactions that maintain the three-dimensional conformation of a protein.
- Define conjugated proteins and prosthetic groups.
- Describe the roles of hemoglobin and myoglobin in oxygen transport and storage.
- Know why the correct three-dimensional structure of a protein is essential to its function.
- Be familiar with the consequences of altering the amino acid sequence of a protein.
- Understand how extremes of pH and temperature cause denaturation of proteins.

About 200 years ago, chemists began the systematic study of matter isolated from living organisms. One of the first substances that they isolated had an unusual property. When most liquids are heated, they vaporize. But the liquid substance isolated by these early biochemists turned solid when heated! You observe this state change every time you cook an egg. These chemists had isolated a solution of protein.

Another century elapsed before Johannes Mulder gave the name ''protein'' to this group of biological molecules. The term protein is derived from a Greek word meaning ''of first importance.'' Indeed, proteins are a particularly important class of food molecules because they provide an organism not only with carbon and hydrogen, but also with nitrogen and sulfur. These latter two elements are unavailable from fats and carbohydrates, the other major classes of food molecules.

In this chapter we will examine the molecular nature and the variety of important functions of protein molecules. The basic subunits from which proteins are made are the α-amino acids. Although more than 100 amino acids have been isolated, most proteins in an organism are made from a set of 20 of these structurally similar α-amino acids (Figure 11.1). The proteins are composed of a large number of amino acids and therefore have very large molecular weights. A small protein like the hormone insulin has a molecular weight of 5000, and the molecular weights of some proteins are well over 100,000. Proteins were the first cellular macromolecules to be isolated, purified, and studied in detail. They are the most abundant macromolecules in cells, and they typically account for the majority of the dry weight of cells.

Recall that macromolecules are molecules of very large molecular weight.

11.1 CELLULAR FUNCTIONS OF PROTEINS

Proteins have many biological functions, as the following short list suggests.

1. All of the biochemical catalysts known as **enzymes** are proteins. The enzymes catalyze almost all of the chemical reactions that occur in living cells. Reactions that would take days or weeks without enzymes are completed in an instant. Without this remarkable catalytic power, by which reaction rates are accelerated a millionfold or more, life would not be possible.

Enzymes are described in detail in Chapter 12.

2. The proteins known collectively as **immunoglobulins,** or antibodies, are specific protein molecules produced by specialized cells of the immune system in response to foreign antigens. In the broadest sense, an **antigen** is any substance that stimulates an immune response. These foreign invaders include bacteria and viruses that enter the body and initiate infection. Each antibody is custom-made to bind to a single foreign antigen, although it may also bind to closely related antigens. Antibodies help to terminate infection by binding specifically to an antigen and facilitating its destruction or removal from the body.

The properties of antibodies and the immune system are described in A Human Perspective: Immunoglobulins: Proteins That Defend the Body.

3. **Transport proteins** carry materials from one place to another in the body. Often this involves transport across cell membranes. For instance, most of the food molecules, or energy sources, for the cell are either too large or too highly charged to simply diffuse into the cell. Entry into the cell is facilitated by binding of the molecule to a specific **receptor** at the cell surface. The receptor then helps the food molecule to cross the cell membrane and enter the cytoplasm, where it will be utilized to provide energy for cellular functions. Without such transport, cells would die of starvation. Transport is not limited to movement across membranes. The protein transferrin transports iron from the liver to the bone marrow, where it is used to synthesize the heme group for hemoglobin. As we will see later in this chapter, the proteins hemoglobin and myoglobin are responsible for transport and storage of oxygen in higher organisms.

The importance and function of transport proteins and cell surface receptors are discussed in Sections 9.5 and 9.6.

4. **Regulatory proteins** control many aspects of cell function, including metabolism and reproduction. Organisms such as ourselves can function only within a limited set

The opposing roles of insulin and glucagon in human metabolism are described in Sections 13.6, 15.6, and 15.7.

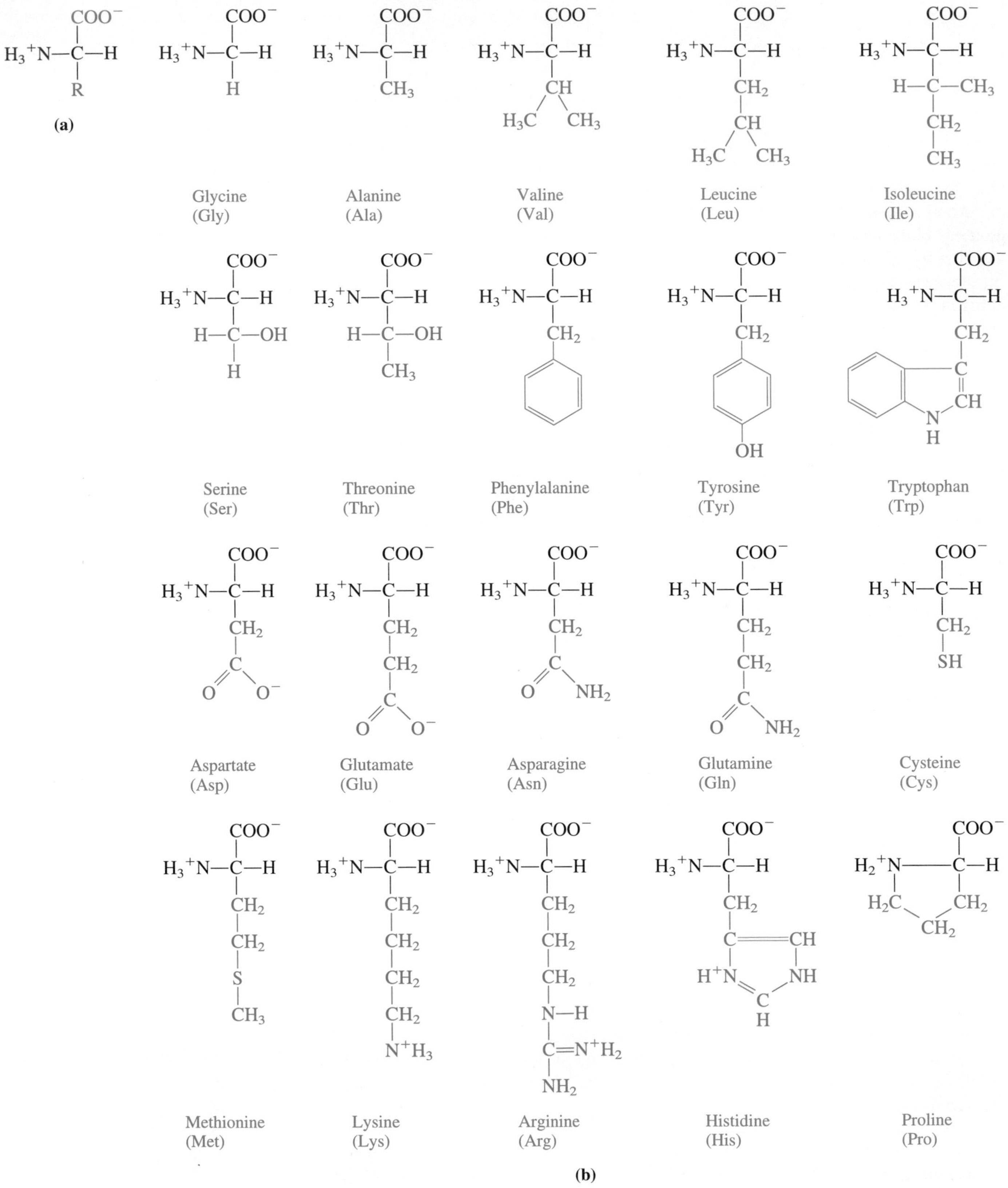

(a)

Glycine (Gly) | Alanine (Ala) | Valine (Val) | Leucine (Leu) | Isoleucine (Ile)

Serine (Ser) | Threonine (Thr) | Phenylalanine (Phe) | Tyrosine (Tyr) | Tryptophan (Trp)

Aspartate (Asp) | Glutamate (Glu) | Asparagine (Asn) | Glutamine (Gln) | Cysteine (Cys)

Methionine (Met) | Lysine (Lys) | Arginine (Arg) | Histidine (His) | Proline (Pro)

(b)

FIGURE 11.1
Structure of the amino acids. (a) The general structure of an amino acid. (b) Structures of the 20 α-amino acids isolated from proteins. The side chains are classified by their polarity.

of conditions. The body temperature, the pH of the blood, and blood glucose levels are just a few of the conditions that must be carefully regulated. If any of these parameters fluctuates much beyond the normal range, the organism will die. Many of the hormones that regulate body function, such as insulin, human growth hormone, and glucagon, are proteins.

5. **Structural proteins** provide mechanical support to large animals and provide them with their outer coverings. Our hair and fingernails are largely composed of the protein keratin. As we will see later in this chapter, another protein, collagen, provides mechanical strength for our bones, tendons, and skin. Without such support, large multicellular organisms like ourselves could not exist.

The nature of these structural proteins is discussed in Section 11.6.

6. Proteins are necessary for all forms of movement. Our muscles—including that most important muscle, the heart—contract and expand through the interaction of actin and myosin proteins. Sperm are motile because they have long flagella that are an intricate aggregation of the protein tubulin. Bacterial motility also occurs by means of flagella, which are an assembly of the protein flagellin.

The proteins involved in motility are covered in Section 11.5.

11.2 THE COMPLEX DESIGN OF PROTEINS

α-Amino acids: structure

The French chemist Henri Braconnot discovered in 1820 that the polysaccharide cellulose can be broken down into its constituents by heating in acid. When he tried the same approach with the protein gelatin, a white crystalline solid with a sweet taste precipitated from solution. This molecule, which he called glycine (Gk.: *glycos,* sweet), proved to have the structure shown to the right; it is an ***α*-amino acid.**

$H_3{}^+N—CH_2—CO_2{}^-$

Glycine

It turns out that gelatin is largely composed of the structural protein collagen and that collagen contains more glycine than any other protein.

Section 11.6

The process by which the bonds in proteins are broken into their constituents by water is called **hydrolysis.** Controlled hydrolysis of most proteins gives a mixture of its amino acid constituents (Figure 11.2).

The mixture of amino acids obtained by hydrolysis defines the amino acid composition of the protein. Every protein has a unique amino acid composition. We find that 19 of the 20 amino acids that are commonly isolated from proteins have the same general structure (Figure 11.3). The α-carbon in the general structure is attached to a carboxylate group ($CO_2{}^-$), a protonated amino group ($NH_3{}^+$), a hydrogen, and an R group. These structural features of α-amino acids are particularly important because the carboxylate group and protonated amino groups are necessary for the covalent binding of amino acids to one another to form a protein. Furthermore, the R groups cause the proteins to fold into precise, three-dimensional configurations that will determine their ultimate function.

The pH within the cell must be about 7 for life functions to occur. At this pH the carboxylic acid group exists as a carboxylate anion, $CO_2{}^-$, and the amino group is protonated, $-H_3N^+$. The molecular species in which the carboxylate group is protonated, CO_2H, and the amino group unprotonated, NH_2, does not exist in aqueous solution because the acidic carboxyl group ionizes and the basic amino group picks up the proton that is released. As a result, amino acids in water at pH 7 exist as dipolar ions called *zwitterions*.

FIGURE 11.2
Hydrolysis of a protein breaks the bonds holding the protein together and releases its constituent amino acids. Quantitative analysis of the resulting mixture enables one to determine the amino acid composition of the protein. Twenty different amino acids are commonly isolated from proteins.

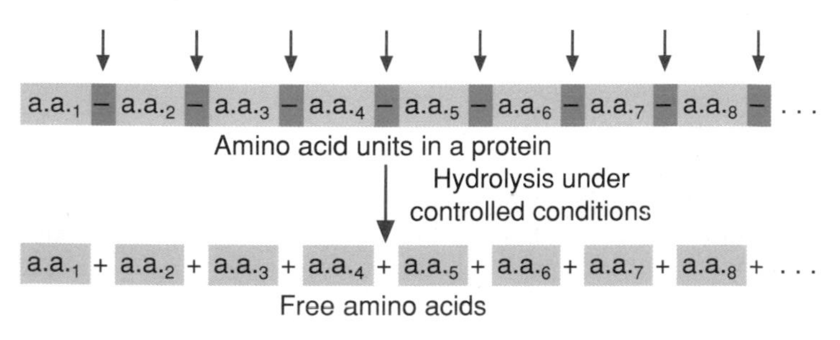

FIGURE 11.3
General structure of an α-amino acid. All amino acids isolated from proteins, except proline, have this general structure.

α-Carboxylate group
α-Amino group → $H_3\overset{+}{N}—C_\alpha(CO_2{}^-)(R)—H$
α-Carbon
Side chain "R"-group

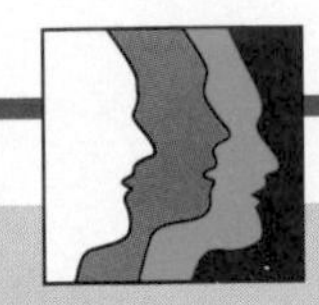

A HUMAN PERSPECTIVE

The Opium Poppy and Peptide Synthesis in the Brain

The seeds of the Oriental poppy contain morphine. **Morphine** is a narcotic that has a variety of effects on the body and the brain, including drowsiness, euphoria, mental confusion, and chronic constipation. Although morphine was first isolated in 1805, not until the 1850s and the advent of the hypodermic was it effectively used as a pain killer. During the American Civil War, morphine was used extensively to relieve the pain of wounds and amputations. It was at this time that the addictive properties were noticed. By the end of the Civil War, over 100,000 soldiers were addicted to morphine.

As a result of the Harrison Act (1914), morphine came under government control and was made available only by prescription. Although morphine is addictive, **heroin,** a derivative of morphine, is much more addictive and induces a greater sense of euphoria that lasts for a longer time.

Why do heroin and morphine have such powerful effects on the brain? Both drugs have been found to bind to **receptors** on the surface of the cells of the brain. The function of these receptors is to bind specific chemical signals and to direct the brain cells to respond. Yet it seemed odd that the cells of our brain should have receptors for a plant chemical. This mystery was solved in 1975, when John Hughes discovered that the brain itself synthesizes small peptide hormones with a morphinelike structure. Two of these opiate peptides are called *methionine enkephalin,* or met-enkephalin, and *leucine enkephalin,* or leu-enkephalin.

These neuropeptide hormones have a variety of effects. They inhibit intestinal motility and blood flow to the gastrointestinal tract. This explains the chronic constipation of morphine users. In addition, it is thought that these **enkephalins** play a role in pain perception, perhaps serving as a pain blockade. This is supported by the observation that they are found in higher concentrations in the bloodstream following painful stimulation. It is further suspected that they may play a role in mood and mental health. The so-called "runner's high" is thought to be a euphoria brought about by an excessively long or strenuous run!

Unlike morphine, the action of enkephalins is short-lived. They bind to the cellular receptor and thereby induce the cells to respond. Then they are quickly destroyed by enzymes in the brain that hydrolyze the peptide bonds of the enkephalin. Once destroyed, they are no longer able to elicit a cellular response. Morphine and heroin bind to these same receptors and induce the cells to respond. However, these drugs are not destroyed and therefore persist in the brain for long periods of time at concentrations high enough to continue to cause biological effects.

Many researchers are working to understand why drugs like morphine and heroin are addictive. Studies with cells in culture have suggested one mechanism for morphine tolerance and addiction. Normally, when the cell receptors bind to enkephalins, this signals the cell to decrease the production of a chemical messenger called **cyclic AMP,** or simply cAMP. (This compound is very closely related to the nucleotide adenosine-5′-monophosphate; see Chapter 16). The decrease in cAMP level helps to block

Heroin

Morphine

The structures of heroin and morphine.

α-Amino acids: configuration

Section 7.2

The α-carbon is attached to four different groups for all amino acids except glycine. The α-carbon of α-amino acids is therefore chiral. That is, an α-amino acid isolated from a protein cannot be superimposed on its mirror image.

The configuration of α-amino acids isolated from proteins is L-. This designation is based upon comparison of amino acids with the D-glyceraldehyde (Figure 11.4). The configuration of α-amino acids isolated from proteins is opposite to that of

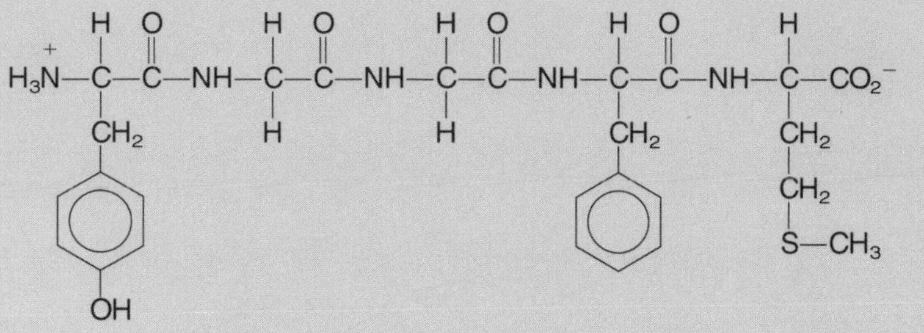

Tyr-Gly-Gly-Phe-Met
Methionine enkephalin

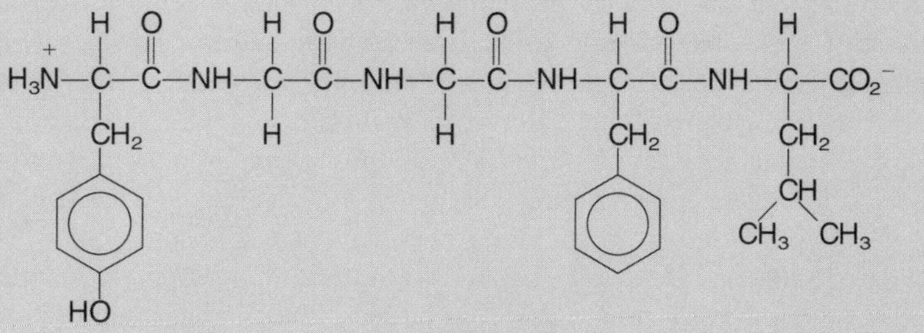

Tyr-Gly-Gly-Phe-Leu
Leucine enkephalin

Structures of the peptide opiates leucine enkephalin and methionine enkephalin. These are the body's own opiates.

pain and elevate one's mood. When morphine is applied to these cells, they become desensitized; that is, they do not decrease cAMP production and thus behave as though there were no morphine present. However, a greater amount of morphine will once again cause the decrease in cAMP levels. Thus addiction and the progressive need for more of the drug seem to result from biochemical reactions in the cells.

This logic can be extended to understand withdrawal symptoms. When an addict stops using the drug, he or she exhibits withdrawal symptoms that include excessive sweating, anxiety, and tremors. The cause of this may be that the high levels of morphine were keeping the cAMP levels low, thus reducing pain and causing euphoria. When morphine is removed completely, the cells overreact and produce huge quantities of cAMP. The result of this is all of the unpleasant symptoms known collectively as the withdrawal syndrome.

Clearly, morphine and heroin have demonstrated the potential for misuse and are a problem for society in several respects. Often, the money needed to support a drug habit is acquired by illegal means such as robbery, theft, and prostitution. More recently, it has become apparent that the use of shared needles for the injection of drugs is resulting in the alarming spread of the virus responsible for Acquired Immune Deficiency Syndrome (AIDS). Nonetheless, morphine remains one of the most effective pain killers known. Certainly, for people suffering from cancer or painful burns or injuries the risk of addiction is far outweighed by the benefits of relief from excruciating pain.

D-glyceraldehyde; that is, the orientation of the four groups around L-alanine resembles the orientation of the four substituents around the chiral carbon of L-glyceraldehyde.

α-Amino acids: substituents on the α-carbon

Because all of the amino acids have an α-carboxylate group and an α-amino group, all differences between amino acids depend upon their side-chain R groups. The amino acids are grouped in Figure 11.1 according to the polarity of their side chains.

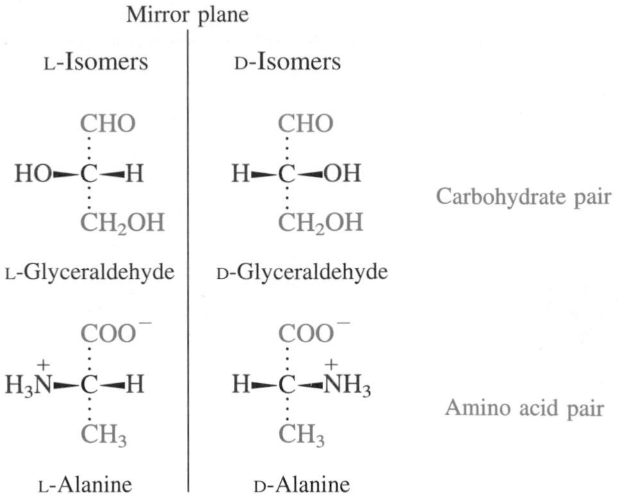

Mirror plane

L-Isomers	D-Isomers	
CHO $HO-C-H$ CH_2OH	CHO $H-C-OH$ CH_2OH	Carbohydrate pair
L-Glyceraldehyde	D-Glyceraldehyde	
COO^- $H_3\overset{+}{N}-C-H$ CH_3	COO^- $H-C-\overset{+}{N}H_3$ CH_3	Amino acid pair
L-Alanine	D-Alanine	

FIGURE 11.4
Structure of D- and L-glyceraldehyde and their relationship to D- and L-alanine. (The student should build models of these compounds, from which it will be immediately apparent that the members of each pair are nonsuperimposable mirror images.)

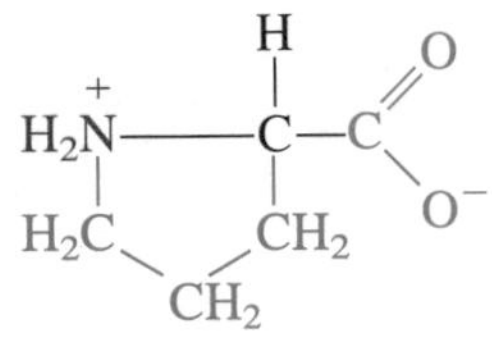

Proline (Pro)

The side chains of some amino acids are nonpolar. They prefer contact with one another to contact with water and are said to be **hydrophobic** *(water-fearing)*. These amino acids are generally found buried in the interior of proteins, where they can associate with one another and remain isolated from water. This hydrophobic interaction is one of the forces that helps to maintain the proper three-dimensional folding of a protein into its characteristic shape. Ten amino acids fall into this category: alanine, valine, leucine, isoleucine, proline, glycine, cysteine, methionine, phenylalanine, and tryptophan. The side chain of proline is unique; it is actually bonded to the α-amino group (in margin).

The side chains of the remaining amino acids are polar, and they have a high affinity for water. These side chains are **hydrophilic** *(water-loving)*. The hydrophilic side chains are often found on the surfaces of proteins. These polar amino acids can be subdivided into three classes:

1. *Polar, neutral amino acids* have side chains that have a high affinity for water but are not ionic. Serine, threonine, tyrosine, asparagine, and glutamine fall into this category. These amino acids can associate with one another by hydrogen bonding. This is another interaction that helps to maintain the proper three-dimensional structure of a protein.
2. *Negatively charged amino acids* have ionized carboxylate groups in their side chains. At pH 7 these amino acids have a net charge of -1. Aspartate and glutamate are the two amino acids in this category. They are acidic amino acids because ionization of the carboxylic acid releases a proton.
3. *Positively charged amino acids*. At pH 7, lysine, arginine, and histidine have a net positive charge because their side chains contain positive groups. These amino acids are basic because the side chain reacts with water, picking up a proton and releasing an hydroxide anion. The positively and negatively charged amino acids within a protein can interact with one another to form ionic bridges. This is yet another force that helps to keep the protein chain folded in a precise way.

The names of the amino acids are abbreviated by their first three letters. These abbreviations are shown in Table 11.1.

TABLE 11.1 Names and Three-Letter Abbreviations of the α-Amino Acids

Amino Acid	Three-Letter Abbreviation
Alanine	ala
Arginine	arg
Asparagine	asn
Aspartic acid	asp
Cysteine	cys
Glutamine	gln
Glutamic acid	glu
Glycine	gly
Histidine	his
Isoleucine	ile
Leucine	leu
Lysine	lys
Methionine	met
Phenylalanine	phe
Proline	pro
Serine	ser
Threonine	thr
Tryptophan	trp
Tyrosine	tyr
Valine	val

Question 11.1 Write the three-letter abbreviations and draw the structures of each of the following amino acids:

a. Glycine
b. Proline
c. Phenylalanine
d. Alanine
e. Cysteine
f. Cystine

11.3 THE PEPTIDE BOND

Proteins are polymers of L-α-amino acids. The α-carboxylate group of one amino acid is linked to the α-amino group of another amino acid. This bond is formed by a condensation reaction. The condensation of two amino acids is illustrated in Figure 11.5. The link between two amino acids is called a **peptide bond.** The molecule formed by condensing two amino acids is called a **dipeptide.** The amino acid units that remain after a peptide bond has formed are known as **amino acid residues.** The residue with a free α-$N^{+}H_3$ group is known as the amino terminal, or simply the **N-terminal residue** (glycine in Figure 11.5), and the residue with a free α-CO_2^{-} group is known as the carboxyl, or **C-terminal residue** (alanine in Figure 11.5). Structures of proteins are conventionally written with the N-terminal residue on the left.

Why was this convention chosen? When we look at the actual process of protein synthesis, we see that the N-terminal amino acid is the first amino acid of the protein. The first amino acid forms a peptide bond involving its carboxylate group and the protonated amino group of the second amino acid in the protein. Thus there is a free protonated amino group literally projecting from the left end of the protein. Similarly, the C-terminal amino acid is the last amino acid added to the protein during protein synthesis. Since the peptide bond is formed between the protonated amino group of this amino acid and the carboxylate group of the previous amino acid, there is a free carboxylate group projecting from the right end of the protein chain.

Section 16.11

The number of amino acids in small peptides is indicated by the prefixes di- (two units), tri- (three units), tetra- (four units), and so forth. Peptides are named as derivatives of the C-terminal residue, which receives its entire name. For all other residues the ending *-ine*, is changed to *-yl*. Thus the dipeptide alanyl-glycine has glycine as its C-terminal amino acid residue, as is indicated by its full name ''glycine.''

FIGURE 11.5
(a) Condensation of two α-amino acids to give a dipeptide. The two amino acids shown are glycine and alanine. (b) Structure of a pentapeptide. Amino acid residues are enclosed in boxes. Glycine is the amino-terminal amino acid, and alanine is the carboxy-terminal amino acid.

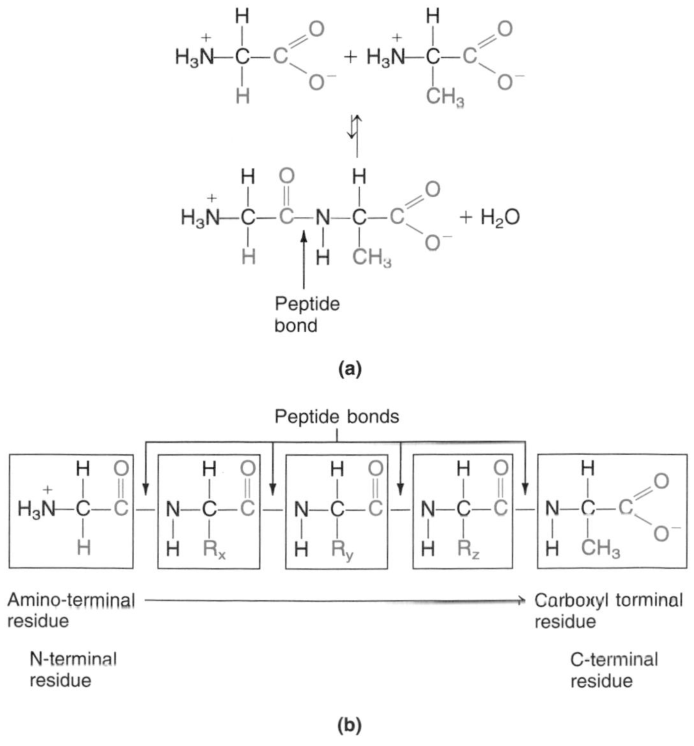

Peptide bond

(a)

Peptide bonds

Amino-terminal residue → Carboxyl terminal residue

N-terminal residue

C-terminal residue

(b)

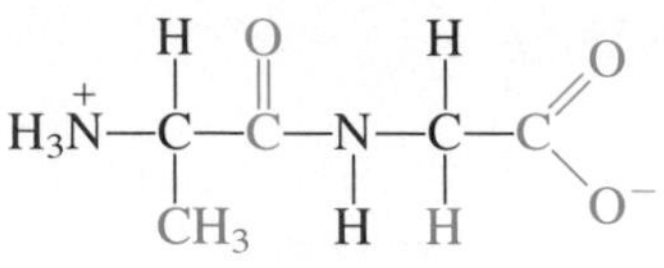

```
      H  O     H
  +   |  ‖     |    //O
H3N—C—C—N—C—C
      |     |  |    \
     CH3    H  H     O⁻
```

Alanyl-glycine
ala-gly

The dipeptide formed from alanine and glycine that has alanine as its C-terminal amino acid is glycyl-alanine:

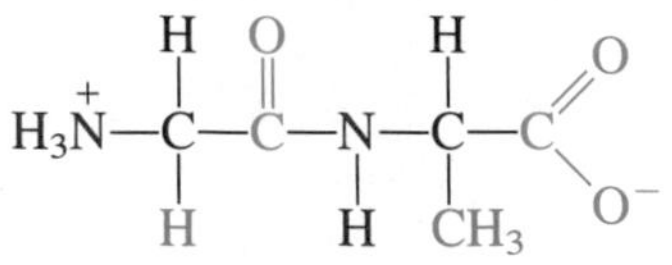

```
      H  O     H
  +   |  ‖     |    //O
H3N—C—C—N—C—C
      |     |  |    \
      H     H  CH3   O⁻
```

Glycyl-alanine
gly-ala

The structures of small peptides can easily be drawn with practice if certain rules are followed. First, note that the backbone of the peptide contains the repeating sequence

```
N—C—C—N—C—C—N—C—C
  1 2     1 2     1 2
```

where *N* is the α-amino group, *carbon-1* is the α-carbon, and *carbon-2* is the carboxyl group. *Carbon-1* is always bonded to an H atom and to the R group side chain that is unique to each amino acid. Continue drawing as outlined in the following example.

Example 11.1

Write the structure of the tripeptide alanyl-glycyl-valine.

Step 1. Write out the backbone for a tripeptide. It will contain three sets of three atoms, nine atoms in all. Remember that the N-terminal amino acid residue is written to the left.

```
N—C—C     N—C—C     N—C—C
Set 1     Set 2     Set 3
```

Step 2. Add oxygens to the carboxyl carbons and hydrogens to the amino nitrogens:

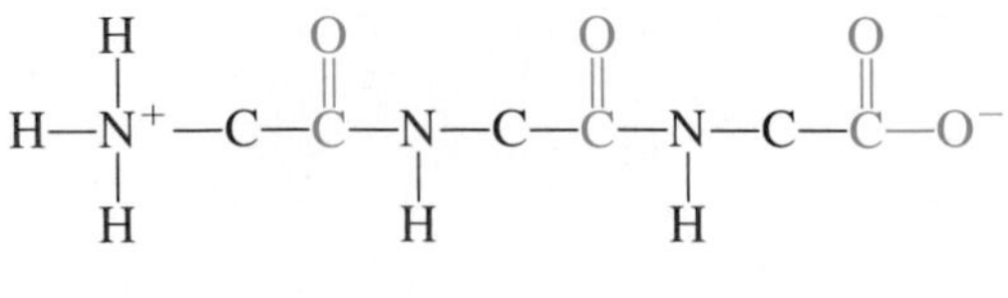

```
  H       O        O        O
  |       ‖        ‖        ‖
H—N⁺—C—C—N—C—C—N—C—C—O⁻
  |         |        |
  H         H        H
```

Step 3. Add hydrogens to the α-carbons:

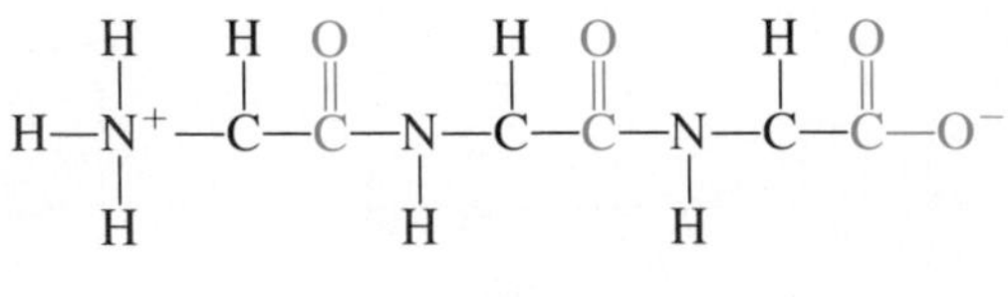

```
  H    H  O     H  O     H  O
  |    |  ‖     |  ‖     |  ‖
H—N⁺—C—C—N—C—C—N—C—C—O⁻
  |         |        |
  H         H        H
```

Step 4. Add the side chains. In this example (ala-gly-val) they are, from left to right, "-CH_3," "H," and "-$CH(CH_3)_2$."

```
      H    H  O       H  O       H  O
      |    |  ||      |  ||      |  ||
   H—N  —C—C—N—C—C—N—C—C—O⁻
      |    |      |   |      |   |
      H   CH₃     H   H      H   CH
                                / \
                             CH₃   CH₃
```

Question 11.2 Write the structures of:

a. Alanyl-phenylalanine c. Phenylalanyl-tyrosyl-leucine

b. Lysyl-alanine

The peptide bond is planar, and the two adjacent α-carbons lie *trans* to it (Figure 11.6). The hydrogen of the amide nitrogen is also *trans* to the oxygen of the carbonyl group. Almost all of the peptide bonds in proteins are planar and have a *trans* configuration. This is quite important physiologically because it makes protein structures relatively rigid. If they could not "hold their shapes," they could not function.

Question 11.3 What properties of the peptide bond are responsible for its geometry? (*Hint:* Recall the discussion of amides in Section 10.6.)

11.4 THE PRIMARY STRUCTURE OF PROTEINS

The sequence of amino acid residues in a protein is known as the **primary structure** of the protein. The primary structures of proteins are translations of information contained in genes. Each protein has a different primary structure with different amino acids in different places within the chain.

Genes can change by the process of mutation during the course of evolution. A mutation in a gene can result in a change in the primary amino acid sequence of a protein. Over longer periods of time, more of these changes will occur. If two species of organisms diverged (became new species) very recently, the differences in the amino acid sequences of their proteins will be few. On the other hand, if they diverged millions of years ago, there will be many more differences in the amino acid sequences of their

Section 16.5

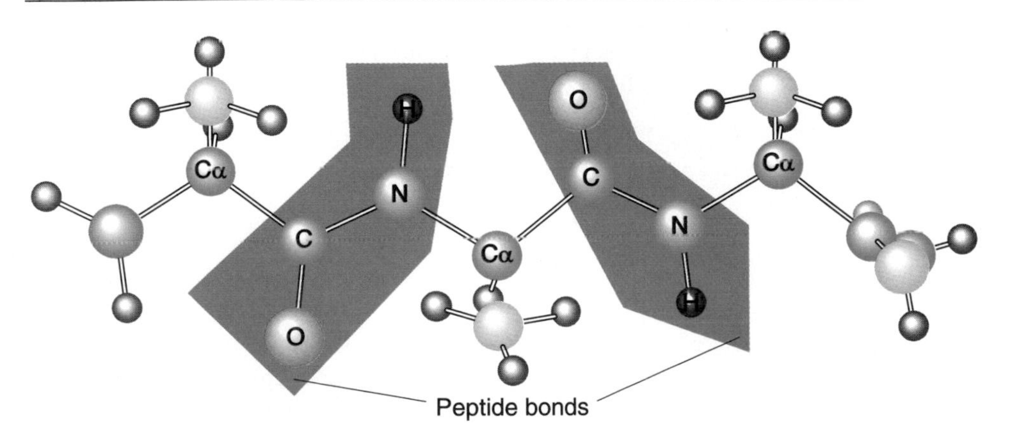

FIGURE 11.6
Conformation of peptide bond is planar. The C═O and N—H groups of the peptide bond are *trans* to one another. These groups are in boxes in the structure.

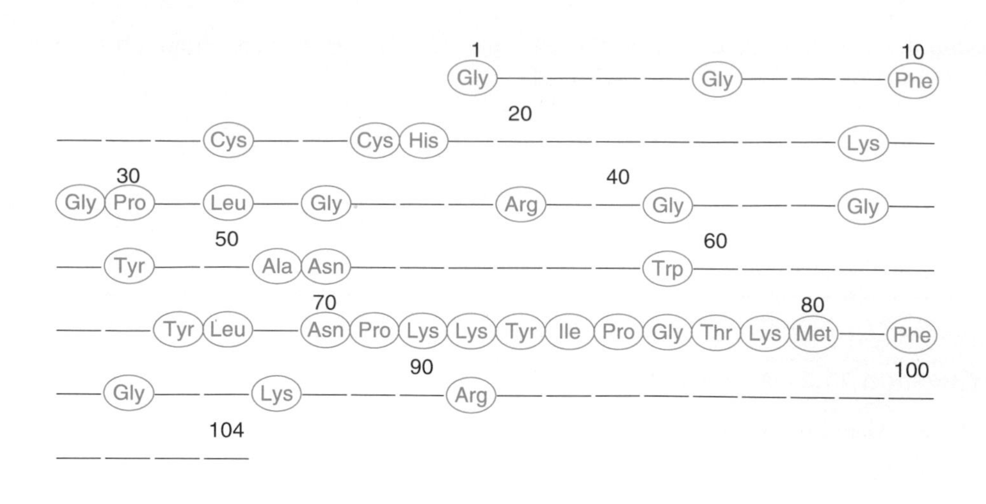

FIGURE 11.7
Cytochrome *c*, a protein of 104 amino acids, is required for respiration in all aerobic organisms. The amino acids shown in blue are conserved in all species that have been studied. The remaining amino acids, designated by a dashed line are highly variable. The conserved amino acid residues are essential for the functioning of the protein.

proteins. As a result, we can compare evolutionary relationships between species by comparing the primary structures of proteins present in both species.

Chapter 14

One protein that has been used extensively for evolutionary studies of the relatedness of organisms is **cytochrome *c*.** Cytochrome *c* is necessary for respiration and is found in all aerobic organisms. It is therefore ideally suited for studies of evolution. In most species, cytochrome *c* contains 104 amino acid residues (Figure 11.7). The amino acid residues shown in blue in Figure 11.7 are the *same* in all species for which sequences are known. These residues are essential for the function of the protein and have been conserved for tens of millions of years of evolution. Why have these amino acid residues remained unchanged over evolutionary time while others have changed a great deal? Changes in the amino acids that are essential to the correct functioning of this critical protein would result in a lethal mutation. That is, the organisms bearing such changes would die, and those changes would thus be lost from the population. Table 11.2 lists the number of residues by which the primary structure of a given cytochrome *c* differs from human cytochrome *c*.

TABLE 11.2 Number of Residues by Which the Primary Structure of a Given Cytochrome c Differs from Human Cytochrome c

Species	No. of Residues
Chimpanzee	0
Rhesus monkey	1
Rabbit	9
Kangaroo	10
Whale	10
Cow	10
Pig	10
Sheep	10
Dog	11
Donkey	11
Horse	12
Chicken	13
Turkey	13
Rattlesnake	14
Turtle	15
Tuna	21
Dogfish	23
Screwworm fly	25
Moth	31
Wheat	43
Yeast	43

As species diverge, sequence differences become more pronounced. We see that human cytochrome *c* is identical to chimpanzee cytochrome *c*, which implies a close evolutionary relationship between *Homo sapiens* and chimpanzees.

An evolutionary ''tree'' for cytochrome *c* is shown in Figure 11.8. On the evolutionary time scale, the greater the time interval since ancestral lines diverged, the greater the number of sequence variations. This generalization seems to apply to all proteins, although the rate of sequence divergence varies from one protein to another. As a rule, proteins whose primary structures have changed very little for millions of years of evolutionary development have functions that are essential for survival, whereas those whose primary structures diverge rapidly have less important functions.

11.5 THE SECONDARY STRUCTURE OF PROTEINS

The primary sequence of a protein, the chain of covalently linked amino acids, folds into regularly repeating structures that resemble designs in a tapestry. These repeating conformations define the **secondary structure** of the protein, which is a very important part of protein structure. Proteins are rather fragile molecules that are held in the correct three-dimensional shape by weak attractions between amino acid residues. The weak attraction responsible for the secondary structure of a protein is known as hydrogen bonding. Many

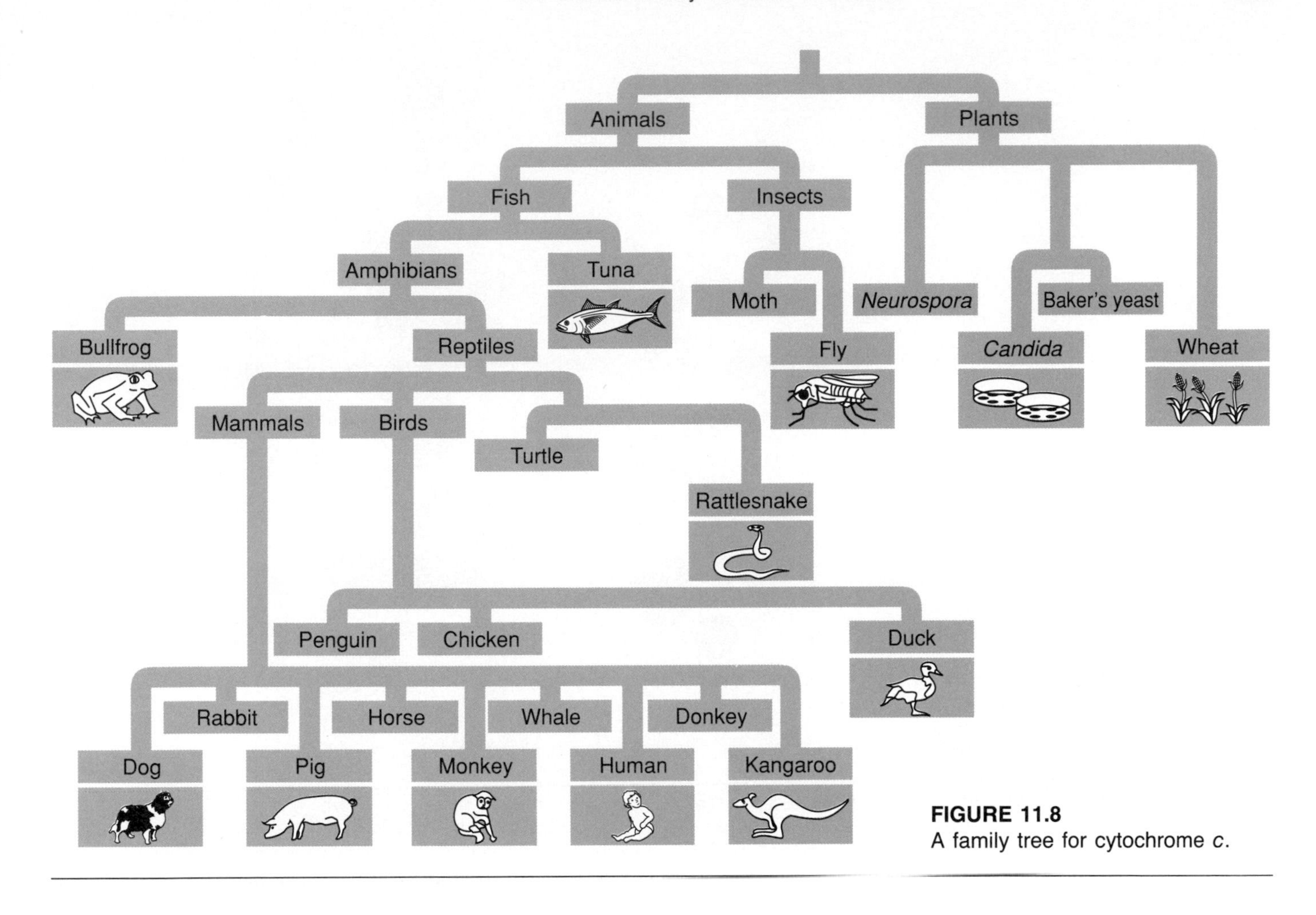

FIGURE 11.8
A family tree for cytochrome *c*.

hydrogen bonds are required to maintain the secondary structures and thereby the overall structure of the protein.

α-Helical conformations

The most common type of secondary structure is a coiled, helical conformation known as the **α-helix** (Figures 11.9 and 11.10). The α-helix has several important characteristics:

- Every amide hydrogen and carbonyl oxygen is involved in a hydrogen bond when the chain coils into an α-helix. These hydrogen bonds lock the α-helix into place.
- Every carbonyl oxygen is hydrogen-bonded to an amide hydrogen four residues away in the chain.
- The hydrogen bonds of the α-helix are parallel to the long axis of the helix (Figure 11.9a).
- The polypeptide chain in an α-helix is right-handed. It is oriented like a normal screw. If you turn a screw clockwise it goes into the wall; turned counterclockwise, it comes out of the wall.
- The repeat distance of the helix, or its pitch, is 5.4 Å, and there are 3.6 amino acid residues per turn of the helix.

The **α-keratins** are a group of fibrous proteins that form the covering (hair, wool, and fur) of most land animals. Human hair provides a typical example of the structure of the α-keratins. The proteins of hair consist almost exclusively of polypeptide chains coiled up into α-helices. A single α-helix is coiled in a bundle with two other helices to give a three-stranded **protofibril** that is part of an array known as a **microfibril**

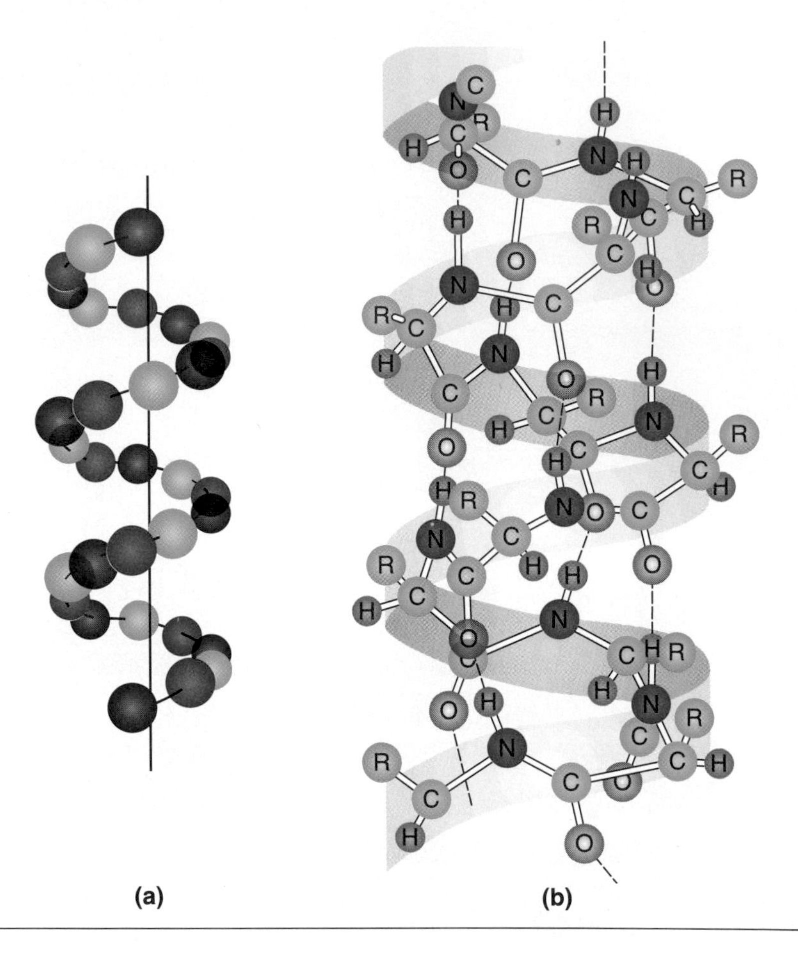

FIGURE 11.9
The α-helix. (a) Schematic diagram showing only the helical backbone; (b) molecular model representation. Note that all of the hydrogen bonds between C═O and N—H groups are parallel to the long axis of the helix. The pitch of the helix is 5.4 Å (0.54 nm), and there are 3.6 amino acid residues per turn.

(Figure 11.11). These structures, which resemble "molecular pigtails," possess great mechanical strength and are virtually insoluble in water.

The fibrous proteins of muscle are also composed of proteins that contain considerable numbers of α-helices. For example, **myosin,** one of the major proteins of muscle, is a rodlike structure in which two α-helices are coiled around one another (Figure 11.12).

The major structural property of a coil of α-helices is its great mechanical strength. This property is applied very efficiently in both the fibrous proteins of skin and those of muscle. As you can imagine, these proteins must be very strong to carry out their functions of mechanical support and muscle contraction.

FIGURE 11.10
Top view of an α-helix. The side chains of the helix point away from the long axis of the helix. The view is into the barrel of the helix.

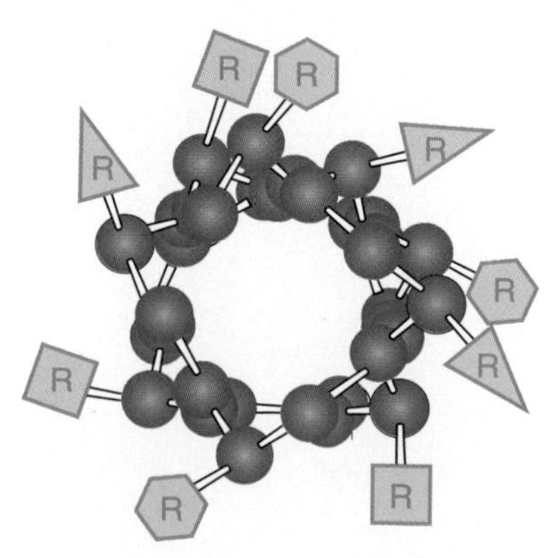

β-Pleated sheet

The second common secondary structure in proteins resembles the pleated folds of drapery and is known as ***β*-pleated sheet** (Figure 11.13). All of the carbonyl oxygens and amide hydrogens in a β-pleated sheet are involved in hydrogen bonds, and the polypeptide chain is nearly completely extended. The polypeptide chains in a β-pleated sheet can have two orientations. If the N-termini are aligned (head to head), the structure is known as a *parallel* β-pleated sheet. And if the N-terminus of one chain is aligned with the C-terminus of a second chain (head to tail), the structure is known as an *antiparallel* pleated sheet.

Some fibrous proteins are composed of β-pleated sheets. For example, the silkworm produces **silk fibroin,** a protein whose structure is an antiparallel β-pleated sheet (Figure 11.14). The polypeptide chains of a β-pleated sheet are almost completely extended, and silk does not stretch easily. Glycine accounts for nearly half of the amino acid residues of silk fibroin. Alanine and serine account for most of the others. The methyl groups of alanine residues and the hydroxymethyl groups of serine residues lie on opposite sides of the sheet. Thus the stacked sheets nestle comfortably, like sheets of corrugated cardboard.

11.6 COLLAGEN, AN IMPORTANT PROTEIN IN HUMANS

Collagen accounts for about a third of the total protein content in humans. It is a structural protein that provides mechanical strength to bone, tendon, and skin. Collagen fibers in bone provide a scaffolding around which **hydroxyapatite** (a calcium phosphate polymer) crystals are arranged. Skin contains loosely woven collagen fibers that can expand in all directions. Blood vessels also contain collagen fibers arranged in helical networks. Collagen is actually a family of related proteins. The flexible and expandable collagen of cowhide is structurally similar to the much less elastic collagen of the same cow's tendons.

FIGURE 11.11
Structure of the α-keratins. These proteins are assemblies of triple-helical protofibrils that are assembled in an array known as a microfibril. These in turn are assembled into macrofibrils. Hair is a collection of macrofibrils and hair cells.

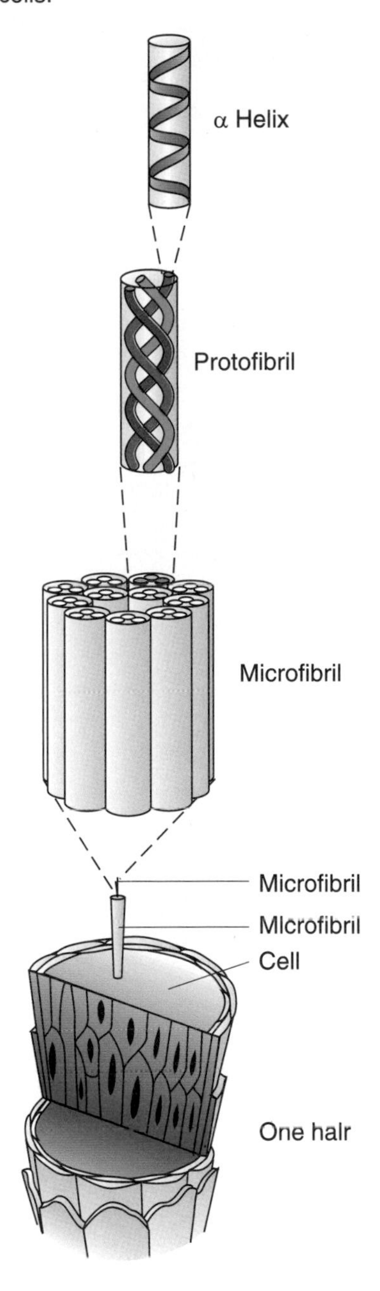

FIGURE 11.12
Schematic diagram of the structure of myosin. This muscle protein consists of a rodlike coil of α-helices with two globular heads, also composed of protein, attached to myosin at its C-terminus. In muscle, myosin molecules are assembled into thick filaments that alternate with thin filaments composed of the proteins actin, troponin, and tropomyosin. Working together, these filaments allow muscles to contract and relax.

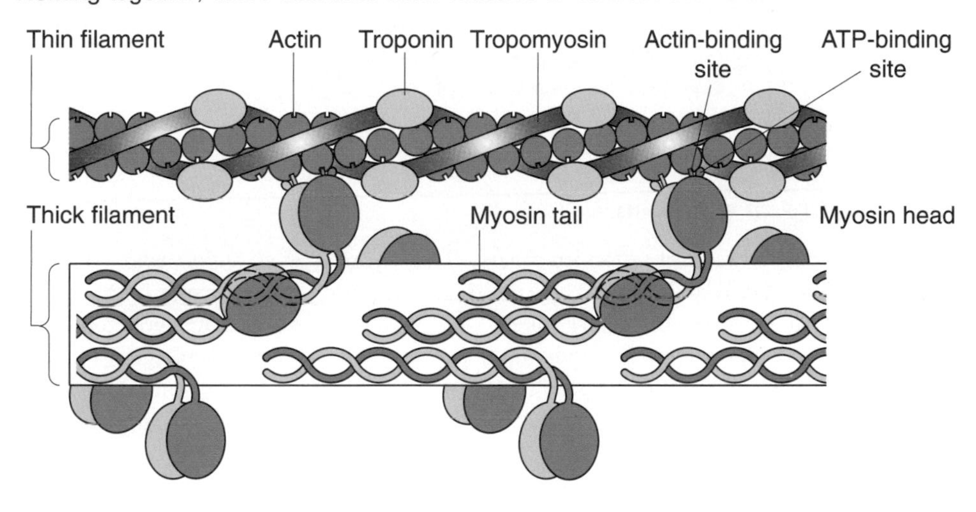

FIGURE 11.13
Structure of the β-pleated sheet. The polypeptide chains are nearly completely extended, and hydrogen bonds between C=O and N—H groups are at right angles to the long axis of the polypeptide chains.

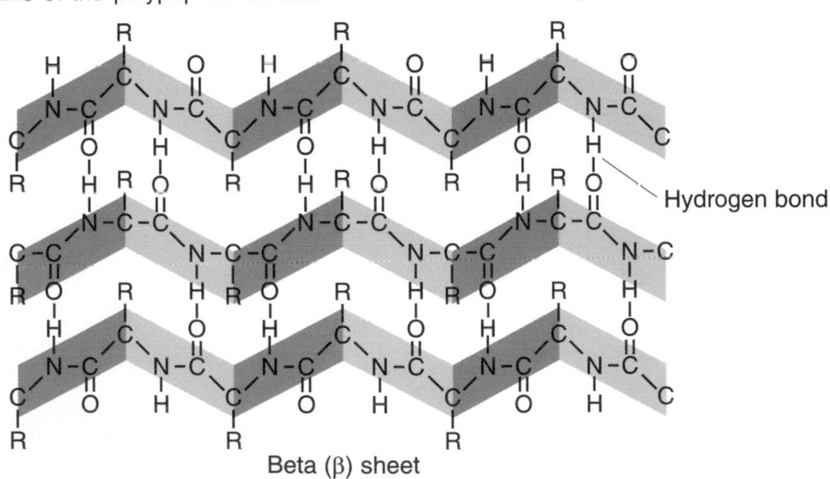

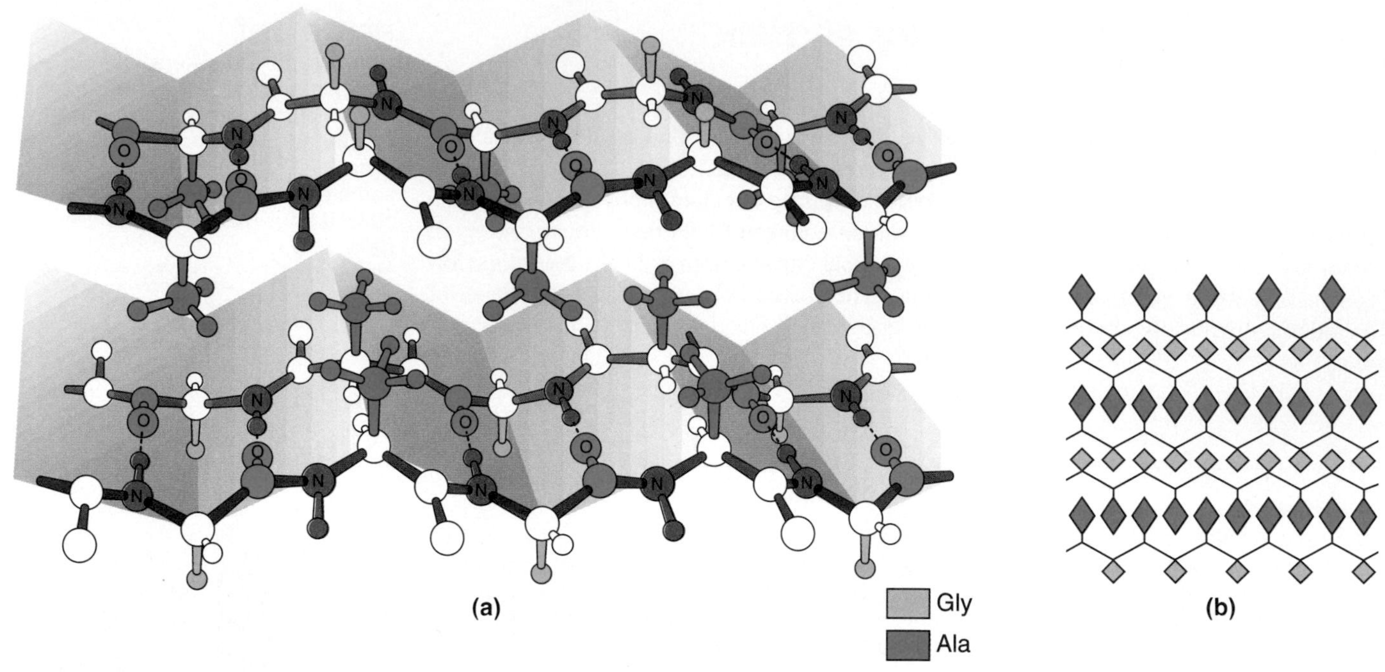

FIGURE 11.14
The structure of silk fibroin is almost entirely antiparallel β-pleated sheet. (a) The molecular structure of a portion of the silk fibroin protein. (b) A schematic representation of the antiparallel β-pleated sheet with the nestled R groups.

The structure of collagen

Collagen is made from triple-stranded fibers. Two hydroxylated amino acid residues, *4-hydroxyproline* (Hyp) and *5-hydroxylysine* (Hyl), account for nearly one fourth of the amino acid composition of collagen (Figure 11.15), and glycine makes up about a third of the amino acid composition. The individual peptide chains of tropocollagen are left-handed helices. These left-handed strands are wrapped around one another in a right-handed sense. The triple-stranded helical fiber of collagen is known as **tropocollagen** (Figure 11.16).

FIGURE 11.15
Structures of 4-hydroxyproline and 5-hydroxylysine, two amino acids found only in collagen.

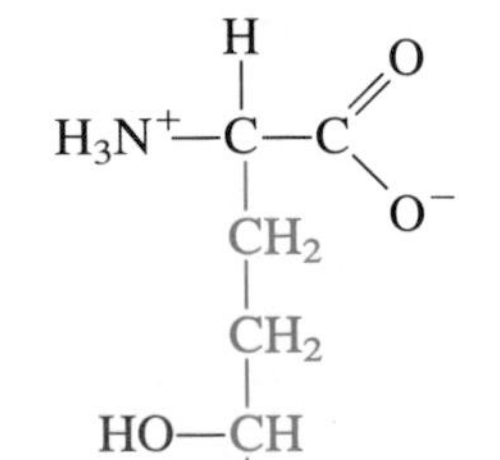

4-Hydroxyproline

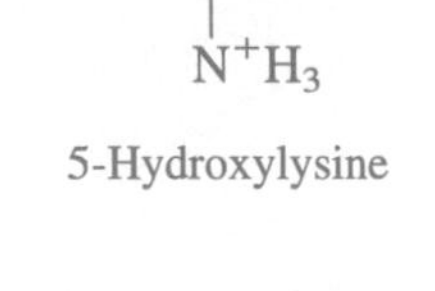

5-Hydroxylysine

The role of vitamin c in collagen metabolism

The major known physiological function of **vitamin C** is in the hydroxylation of collagen. When the collagen protein is made, the amino acids proline and lysine are incorporated into the chain of amino acids. These are later modified by two enzymes to form 4-hydroxyproline and 5-hydroxylysine. These enzymes, proline hydroxylase and lysine hydroxylase, require vitamin C to carry out these important hydroxylation reactions. People who are deprived of vitamin C, as were sailors on long voyages prior to the eighteenth century, develop **scurvy,** a disease of collagen metabolism. The symptoms of scurvy include skin lesions, fragile blood vessels, and bleeding gums.

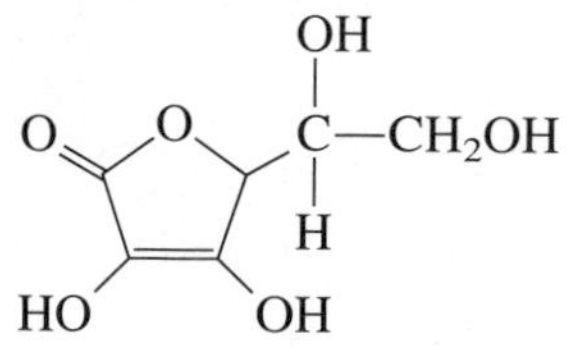

Ascorbic acid

The British Navy provided the antidote to scurvy by including limes, a fruit that is rich in vitamin C (as are all citrus fruits) in the diets of its sailors. The epithet ''limey,'' a slang term for ''British,'' entered the English language as a result.

11.7 THE TERTIARY STRUCTURE OF PROTEINS

Most **fibrous proteins,** such as silk, collagen, and the α-keratins, are almost completely insoluble in water. (Our skin would do us very little good if it dissolved in the rain.) The majority of cellular proteins, however, are soluble in the cell cytoplasm. Soluble proteins are usually **globular.** This globular, three-dimensional structure is called the **tertiary structure** of the protein. The peptide chain with its regions of secondary structure, α-helix and β-pleated sheet, further folds on itself to achieve the tertiary structure.

We have seen that the forces that maintain the secondary structure of a protein are hydrogen bonds between the amide hydrogen and the carbonyl oxygen of the peptide bond. What are the forces that maintain the tertiary structure of a protein? The globular tertiary structure forms spontaneously and is maintained as a result of interactions among the side chains, the R groups, of the amino acids. The structure is maintained by the following molecular interactions:

- Hydrophobic attractions between the R groups of nonpolar amino acids
- Hydrogen bonds between the polar R groups of the polar amino acids
- Ionic bonds between the R groups of oppositely charged amino acids
- Covalent bonds between the sulfhydryl containing amino acids. Two of the polar cysteine residues can be oxidized to a dimeric amino acid called cystine (Figure 11.17). The disulfide bond of cystine can be a cross-link between different proteins, or it can tie two segments within a protein together.

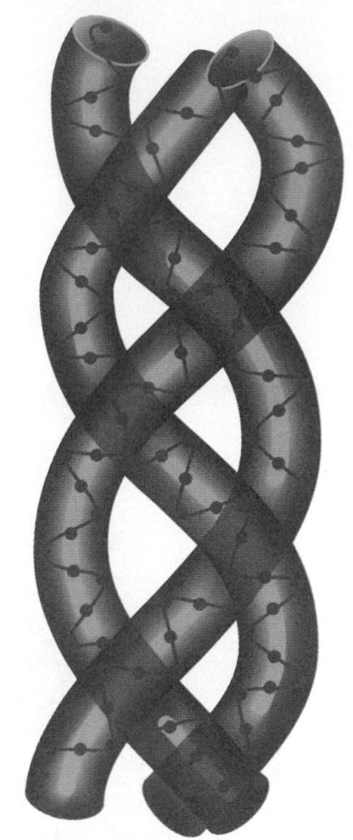

FIGURE 11.16
Structure of the collagen triple helix.

The bonds that maintain the tertiary structure of proteins are shown in Figure 11.18. The importance of these bonds becomes clear when we realize that it is the tertiary structure of the protein that determines its biological function. Most of the time, nonpolar amino acid residues are buried, closely packed, in the interior of a globular protein, out of contact with water. Polar and charged amino acid residues lie on the surfaces of globular proteins. Globular proteins are extremely compact, and there are no holes anywhere inside them. The structure of the globular protein myoglobin is seen in Figure 11.19. The tertiary structure can contain regions of α-helix and regions of β-pleated sheet. Other secondary structures are also possible. The exact amount of each type of secondary structure varies from one protein to the next.

11.8 THE QUATERNARY STRUCTURE OF PROTEINS

For many proteins the functional form is not composed of a single peptide, but is rather an aggregate of smaller globular proteins. For instance, the protein hemoglobin is composed of four individual protein chains: two identical α-chains and two identical β-chains. Only when the four peptides are bonded to one another is the protein molecule fully functional

```
     H  H  O                        H  H  O
     |  |  ||                       |  |  ||
  ---N--C--C---                  ---N--C--C---
        |                              |
        CH2                            CH2
        |                              |
        S--H       Oxidation           S
                  ---------->          |          + 2H+ + 2e-
        S--H      <----------          S
        |          Reduction           |
        CH2                            CH2
        |                              |
  ---N--C--C---                  ---N--C--C---
     |  |  ||                       |  |  ||
     H  H  O                        H  H  O

     Cysteine                       Cystine
```

FIGURE 11.17
Oxidation of two cysteines to give the dimer cystine. This reaction occurs in cells and is readily reversible.

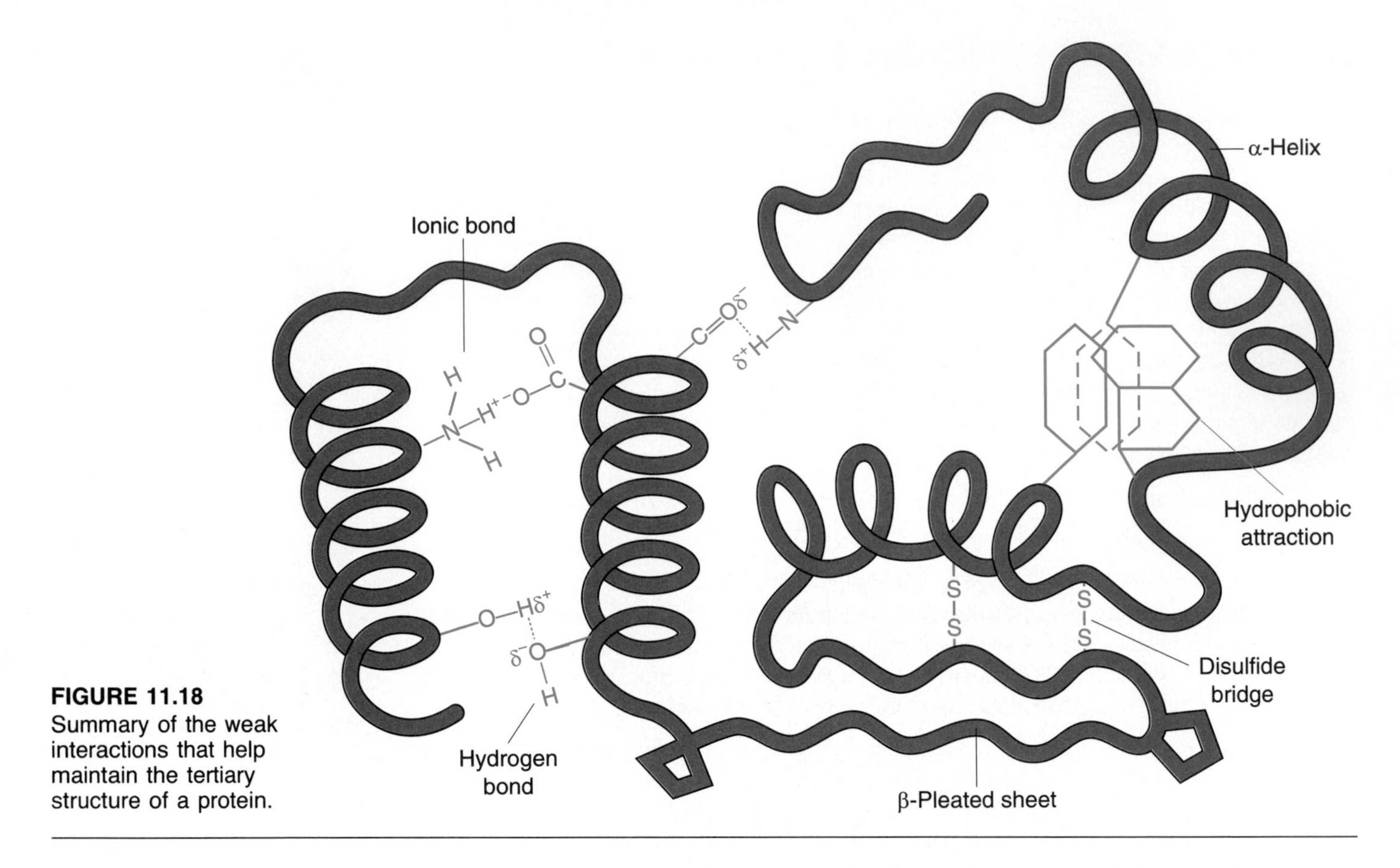

FIGURE 11.18
Summary of the weak interactions that help maintain the tertiary structure of a protein.

in oxygen transport. The binding of several peptides to produce a functional protein defines the **quaternary structure** of a protein.

The forces that maintain the quaternary structure of a protein are the same as those that hold the tertiary structure. Those include hydrogen bonds between polar amino acids, ionic bridges between oppositely charged amino acids, hydrophobic attraction between nonpolar amino acids, disulfide bridges, and van der Waals forces.

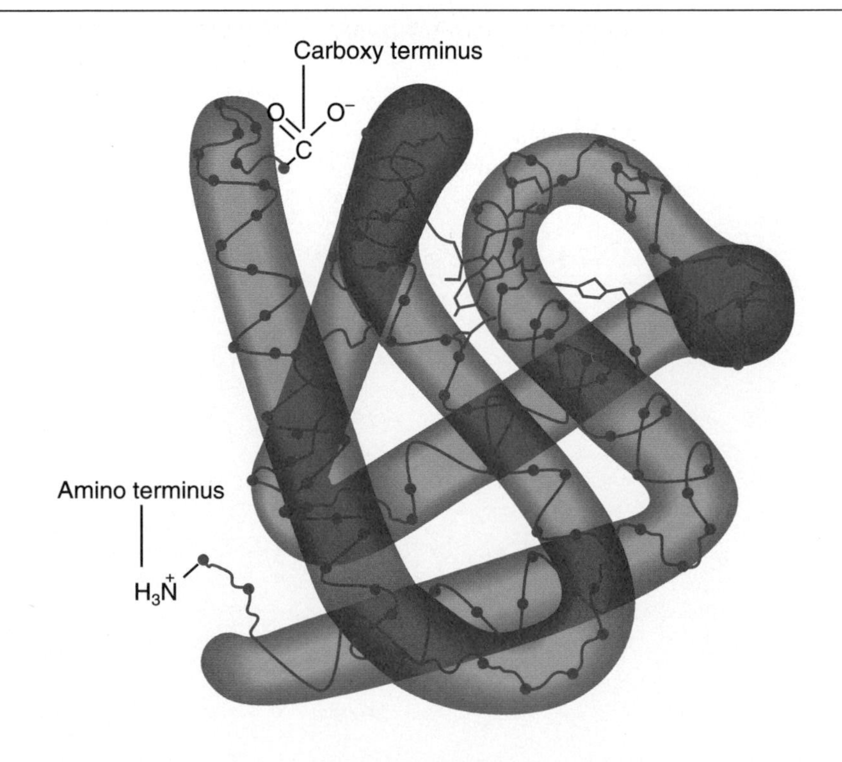

FIGURE 11.19
Myoglobin. The heme group, shown in color, has an iron atom to which oxygen binds.

```
            CH3    CH=CH2
            |      |
            C======C
            |      |
     HC—C         C=CH
     ||    \\   /      |
CH3—C—C       N        C=C—CH3
    ||  \      ¦      /      |
    ||   N--- Fe(2+)—N        |
    ||  //     |      \      |
CH2—C—C          N       C=C—CH=CH2
|      |     /   \       |
CH2    HC=C       C=CH
|          |      |
COOH       C======C
           |      |
           CH2    CH3
           |
           CH2
           |
           COOH
```

FIGURE 11.20
Structure of heme prosthetic group bound to myoglobin.

In some cases a protein must be bound to something extra in order for it to be functional. This additional group is called a **prosthetic group,** and the functional protein carrying a prosthetic group is called a **conjugated protein.** Many of the receptor proteins on cell surfaces are **glycoproteins.** These are conjugated proteins with sugar residues covalently attached. As we will see in Sections 11.9 and 11.10, hemoglobin is also a conjugated protein. Each of the subunits is bound to an iron-containing heme group. As in the case of hemoglobin, the prosthetic group often determines the function of a protein. For instance, in hemoglobin it is the iron-containing heme groups that have the ability to bind reversibly to oxygen.

11.9 MYOGLOBIN AND OXYGEN STORAGE

A single-celled organism can obtain its fuel directly from the environment. However, many of the cells in a multicellular organism are buried in the interior and unable to obtain fuel or eliminate waste directly. The circulatory system provides part of the solution to this problem by delivering nutrients and oxygen to body cells and carrying away wastes. One essential requirement of complex aerobic organisms is, of course, a steady supply of oxygen. Oxygen is only very slightly soluble in aqueous solutions, and the circulatory system is not by itself adequate to supply oxygen to bodily tissues. **Hemoglobin,** the major protein component of red blood cells, is the oxygen transport protein of higher animals, and **myoglobin** is the oxygen storage protein of skeletal muscle.

The structure of myoglobin (Mb) is shown in Figure 11.19. The **heme** group (Figure 11.20) is an essential component of this protein. Heme is the binding site for oxygen in both myoglobin and hemoglobin. Heme contains an iron atom in the 2+ oxidation state, to which oxygen binds.

11.10 HEMOGLOBIN AND OXYGEN TRANSPORT

Hemoglobin (Hb) is the oxygen transport protein of the blood. Hemoglobin is composed of four peptide subunits, designated α and β (Figure 11.21). The hemoglobin tetramer has the subunit composition α_2-β_2. In addition, each subunit of hemoglobin contains a heme group. A hemoglobin molecule therefore has the ability to bind four molecules of oxygen:

$$\underset{\text{Deoxyhemoglobin}}{\text{Hb}} + 4O_2 \longrightarrow \underset{\text{Oxyhemoglobin}}{\text{Hb}(O_2)_4}$$

FIGURE 11.21
Structure of hemoglobin. The protein contains four subunits, designated α and β. The α and β subunits face each other across a central cavity. Each subunit in the tetramer contains a heme group that binds oxygen.

Hemoglobin
Alpha chains
Beta chains
Heme groups

A HUMAN PERSPECTIVE

Immunoglobulins: Proteins That Defend the Body

A living organism is subjected to a constant barrage of bacterial, viral, parasitic, and fungal diseases. Without a defense against such perils we would soon perish. All vertebrates possess an **immune system.** In humans the immune system is composed of about 10^{12} cells, about as many as the brain or liver, that protect us from foreign invaders. This immune system has three important characteristics.

1. **It is highly specific.** The immune response to each infection is specific to, or directed against, only one disease organism or similar, related organisms.
2. **It has a memory.** Once the immune system has responded to an infection, the body is protected against reinfection by the same organism. This is the reason that we seldom suffer from the same disease more than once. Most of the diseases that we suffer recurrently, such as the common cold and flu, are actually caused by many different strains of the same virus. Each of these strains is "new" to the immune system.
3. **It can recognize "self" from "nonself."** When we are born, our immune system is already aware of all the antigens of our bodies. These it recognizes as "self" and will not attack. Every antigen that is not classified as "self" will be attacked by the immune system when it is encountered. Some individuals suffer from a defect of the immune response that allows it to attack the cells of one's own body. The result is an **autoimmune reaction** that can be fatal.

One facet of the immune response is the synthesis of *immunoglobulins,* or **antibodies,** that specifically bind a single macromolecule called an *antigen.* These antibodies are produced by specialized white blood cells called **B lymphocytes.** We are born with a variety of B lymphocytes that are capable of producing antibodies against perhaps a million different antigens. When a foreign antigen enters the body, it binds to the B lymphocyte that was preprogrammed to produce antibodies to destroy it. This stimulates the B cell to grow and divide. Then all of these new B cells produce antibodies that will bind to the disease agent and facilitate its destruction. Each B cell produces only one type of antibody with an absolute specificity for its target antigen. Many different B cells respond to each infection because the disease-causing agent is made up of many different antigens. Antibodies are made that bind to all of the antigens of the invader. This primary immune response is rather slow. It can take a week or two before there are enough B cells to produce a high enough level of antibodies in the blood to combat an infection.

Because the immune response has a memory, the second time we encounter a disease-causing agent the antibody response is immediate. This is why it is extremely rare to suffer from mumps or measles or chickenpox a second time. We take advantage of this property of the immune system to protect ourselves against many diseases. In the process of **vaccination** an animal can be immunized against an infectious disease by injection of a small amount of the antigens of the virus or microorganism (the vaccine). The B lymphocytes of the host organism then manufacture antibodies against the antigens of the infectious agent. If the animal comes into contact with the disease-causing microorganism at some later time, the sensitized B lymphocytes "remember" the antigen and very quickly produce a large amount of specific antibody to overwhelm the microorganism or virus before it can cause overt disease.

Immunoglobulin molecules contain four peptide chains that are connected by disulfide bonds and arranged in a Y-shaped quaternary structure.

Each immunoglobulin has two identical antigen-binding sites located at the tips of the Y and is therefore bivalent. Since most antigens have three or more antigen-binding sites, immunoglobulins can form large cross-linked antigen-antibody complexes that precipitate from solution.

Immunoglobulin G (IgG) is the major serum immunoglobulin. Some immunoglobulin G molecules can cross cell membranes and thus can pass between mother and child through the placenta, before birth. This is important because the immune system of a fetus is immature and cannot provide adequate protection from disease.

Oxygen is efficiently transferred from hemoglobin in the red blood cells to myoglobin in the muscle tissue because myoglobin has a greater affinity for oxygen than hemoglobin does.

Oxygen transport from mother to fetus

A fetus receives its oxygen from its mother by simple diffusion across the placenta. If both the fetus and the mother had the same type of hemoglobin, this transfer process would not

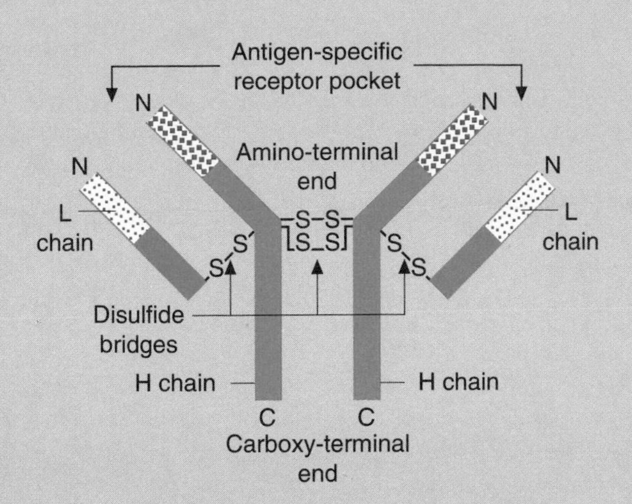

Schematic diagram of a Y-shaped immunoglobulin molecule. The binding sites for antigens are at the tips of the Y.

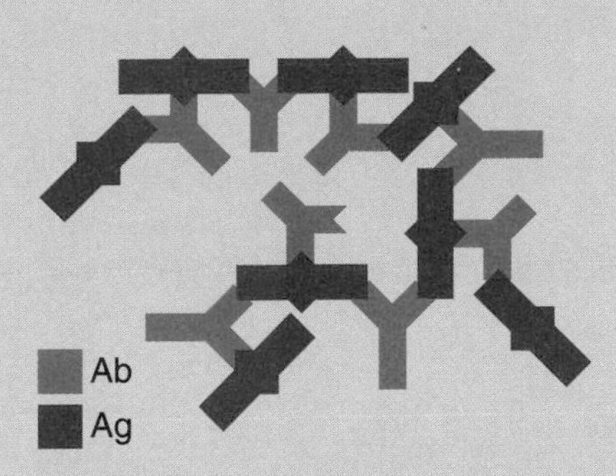

Schematic diagram of cross-linked immunoglobulin-antigen lattice.

Fortunately, the IgG acquired from the mother protects the infant against most bacterial and viral infections that the infant might encounter before birth.

There are four additional types of antibody molecules that vary in their protein composition, but all have the same general Y shape. One of these is **IgM,** which is the first antibody produced in response to an infection. Secondarily, the B cell produces **IgG** molecules with the same antigen-binding region but a different protein composition in the rest of the molecule. **IgA** is the immunoglobulin responsible for protecting the body surfaces, such as the mucous membranes of the gut, the oral cavity, and the genitourinary tract. IgA is also found in mother's milk, protecting the newborn against diseases during the first few weeks of life. **IgD** is found in very small amounts and is thought to be involved in the regulation of antibody synthesis. The last type of immunoglobulin is **IgE.** For many years the function of IgE was unknown. It is found in large quantities in the blood of individuals who suffer from allergies and is therefore thought to be responsible for this "overblown" immunological reaction to dust particles and pollen grains.

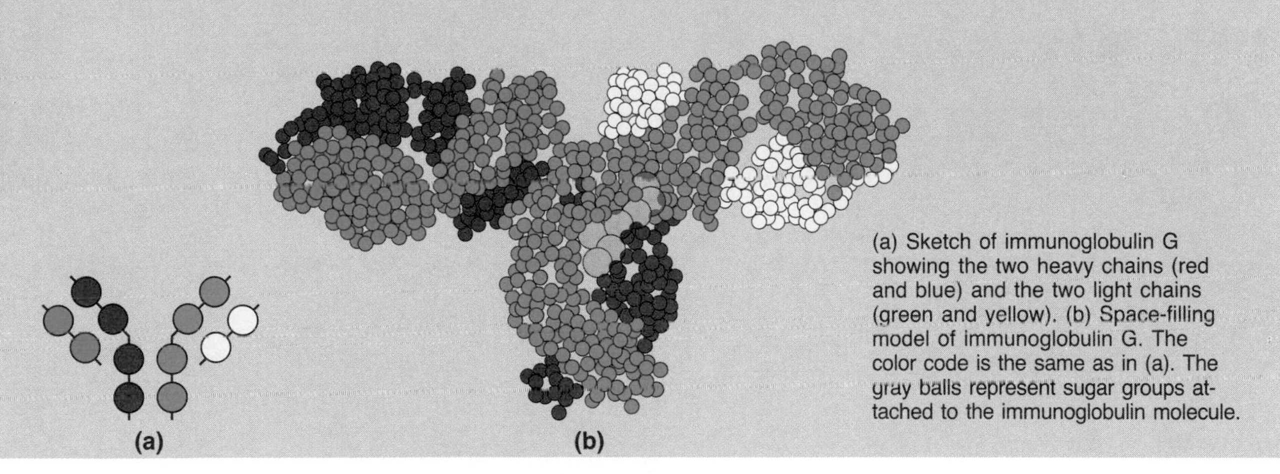

(a) Sketch of immunoglobulin G showing the two heavy chains (red and blue) and the two light chains (green and yellow). (b) Space-filling model of immunoglobulin G. The color code is the same as in (a). The gray balls represent sugar groups attached to the immunoglobulin molecule.

be efficient, since the hemoglobin of the fetus and that of the mother would have the same affinity for oxygen. However, the fetus possesses a unique type of hemoglobin, called **fetal hemoglobin** (Hb F). This unique hemoglobin molecule has a greater affinity for oxygen than does the hemoglobin of the mother, which is designated hemoglobin, Hb A. Oxygen is therefore efficiently transported, via the circulatory system, from the lungs of the mother to the fetus. The biosynthesis of fetal hemoglobin stops shortly after birth, when the genes encoding fetal hemoglobin are switched "off" and the genes coding for adult hemoglobin are switched "on."

Sickle cell anemia

Sickle cell anemia is a human genetic disease that first appeared in tropical west and central Africa. It afflicts about 0.4% of Black Americans. These individuals produce a mutant hemoglobin known as sickle cell hemoglobin (Hb S). Sickle cell anemia receives its name from the sickled appearance of the red blood cells that form in this condition (Figure 11.22). The sickled cells are unable to pass through the small capillaries of the circulatory system, and circulation is hindered. This results in damage to many organs, especially bone and kidney, and leads to death at an early age.

The genetic basis of this alteration is discussed in Chapter 16.

Sickle cell hemoglobin differs from normal hemoglobin by a single amino acid residue. In the β-chain of sickle cell hemoglobin a valine residue (a hydrophobic amino acid) has replaced a glutamic acid residue (a negatively charged amino acid). This substitution provides a basis for the binding of hemoglobin S molecules to one another. When oxyhemoglobin S unloads its oxygen, individual deoxyhemoglobin S molecules bind to one another as long polymeric fibers. This occurs because the valine residue fits into a pocket on the surface of a second deoxyhemoglobin S molecule. The fibers generated in this way radically alter the shape of the red blood cell, resulting in the sickling effect.

When hemoglobin is carrying O_2, it is called oxyhemoglobin. When it is not bound to O_2, it is called deoxyhemoglobin.

Sickle cell anemia occurs in individuals who have inherited the gene for sickle cell hemoglobin from both parents. Afflicted individuals produce 90–100% defective β-chains. Individuals who inherit one normal gene and one defective gene produce both normal and altered β-chains. They are therefore less severely affected. Their condition is known as **sickle cell trait.**

An interesting relationship exists between sickle cell trait and resistance to malaria. In some parts of Africa, up to 20% of the population has sickle cell trait. In those same parts of Africa, one of the leading causes of death is malaria. The presence of sickle cell trait is linked to an increased resistance to malaria because the malarial parasite cannot feed efficiently on sickled red blood cells. People who have sickle cell disease die young; those without sickle cell trait have a high probability of succumbing to malaria. Occupying the middle ground, people who have sickle cell trait do not suffer much from sickle cell anemia and simultaneously resist deadly malaria. Because those with sickle cell trait have a greater chance of survival, the sickle hemoglobin gene is maintained in the population.

FIGURE 11.22
(a) Sickled and normal red blood cells photographed with a light microscope. Scanning electron micrographs of (b) normal and (c) sickled red blood cells.

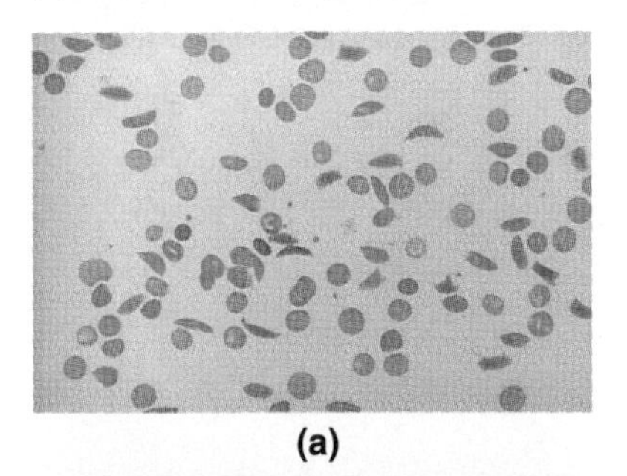

(a)

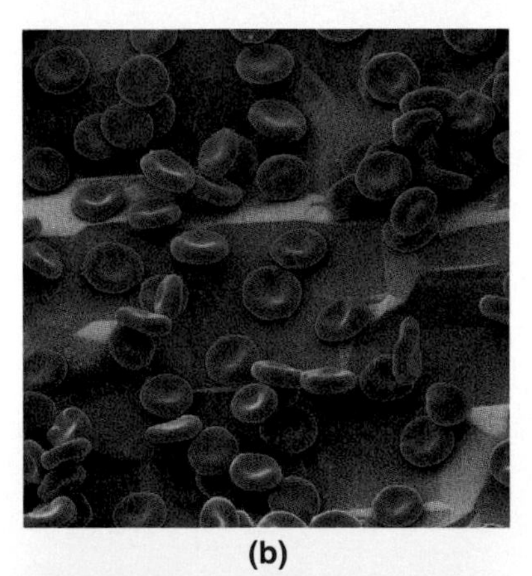

(b)

(c)

11.11 AN OVERVIEW OF PROTEIN STRUCTURE AND FUNCTION

Let's try to summarize the various types of protein structure and their relationship to one another (Figure 11.23).

1. *Primary Structure:* The primary structure of the protein is the amino acid sequence of the protein. The primary structure results from the covalent bonding between the various amino acid residues in the chain.
2. *Secondary Structure:* As the protein chain grows, numerous opportunities for noncovalent interactions become available. These cause the chain to fold and orient itself in a variety of gross conformational arrangements. The secondary level of structure includes the α-helix and the β-pleated sheet, which are the result of hydrogen bonding between the amide hydrogens and carbonyl oxygens of the peptide bonds. Different portions of the chain may be involved in different types of secondary structure arrangements; some regions might be α-helix, while others might be a β-pleated sheet.
3. *Tertiary Structure:* When we discuss tertiary structure, we are interested in the overall folding of the entire chain. In other words, we are concerned with the further folding of the secondary structure. Are the two ends of the chain close together or far apart? What general shape is involved? Both noncovalent interactions between the R groups of the amino acids and covalent —S—S— bridges play a role in determining the

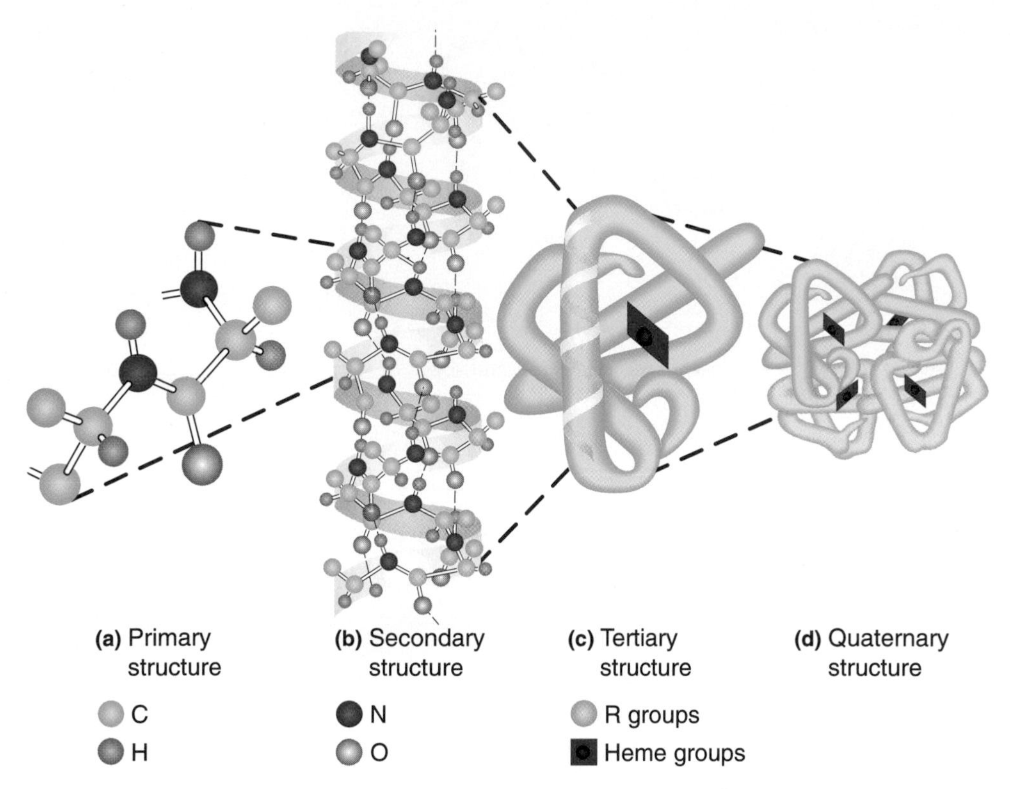

FIGURE 11.23
Summary of the four levels of protein structure, using hemoglobin as an example.

tertiary structure. The noncovalent interactions include hydrogen bonding, ionic bonding, hydrophobic attractions, and van der Waals forces.

4. *Quaternary Structure:* Like tertiary structure, quaternary structure is concerned with the topological, spatial arrangements of two or more peptide chains with respect to each other. How is one chain oriented with respect to another? What is the overall shape of the final functional protein?

The quaternary structure is maintained by the same forces that are responsible for tertiary structure. It is the tertiary and quaternary structure of the protein that ultimately determine the function of the protein. Some have a fibrous structure with great mechanical strength. These make up the major structural components of the cell and the organism. Often they are also responsible for the movement of the organism. Others fold into globular shapes. Most of the transport proteins, regulatory proteins, and enzymes are globular proteins. The very precise three-dimensional structure of the transport proteins allows them to recognize a particular food molecule and facilitate its entry into the cell. Similarly, it is the specific three-dimensional shape of regulatory proteins that allows them to bind to their receptors on the surfaces of the target cell. In this way they can communicate with the cell, instructing it to take some course of action. In the next chapter we will see that the three-dimensional structure of enzyme active sites allows them to bind to their specific reactants and speed up biochemical reactions.

As we have seen with the example of sickle cell hemoglobin, an alteration of just a single amino acid within the primary structure of a protein can have far-reaching implications. When an amino acid replaces another in a peptide, there is a change in the R groups of the protein. This leads to different tertiary and perhaps quaternary folding because the nature of the noncovalent interactions is altered by changing the R groups that are available for that bonding. If the protein does not fold in just the right three-dimensional structure, it will be nonfunctional. In the case of sickle cell hemoglobin this can lead to death.

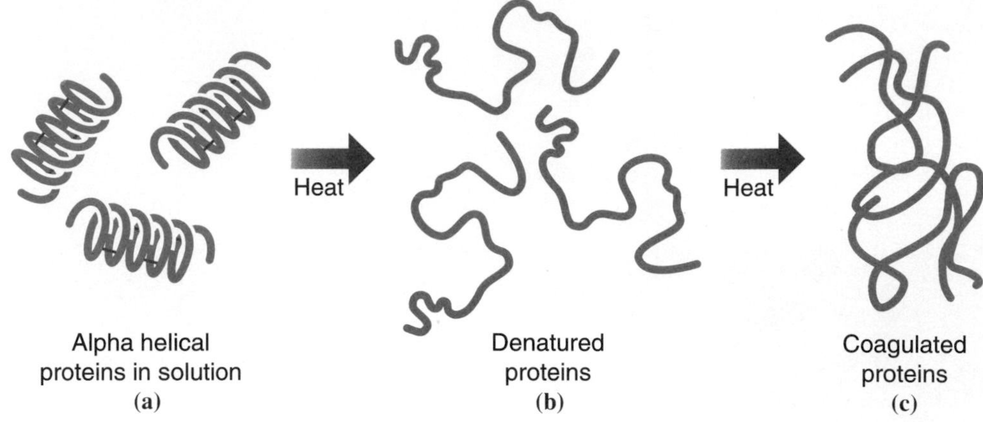

FIGURE 11.24
The denaturation of proteins by heat. (a) The α-helical proteins are in solution. (b) As heat is applied, the hydrogen bonds maintaining the secondary structure are disrupted, and the protein structure becomes disorganized. The protein is denatured. (c) The denatured proteins clump together, or coagulate, and are now in an insoluble form.

11.12 THE EFFECT OF TEMPERATURE ON PROTEINS

We have seen that the shape of a protein is absolutely essential to its function. We have also mentioned that life can exist only within a rather narrow range of temperature and pH. How are these two concepts related? In the next two sections we will see that extremes of pH or temperature have a drastic effect on protein conformation, causing the molecules to lose their characteristic three-dimensional shape. When the organized structures of a protein, the α-helix, the β-pleated sheet, and tertiary folds become completely disorganized, the protein is said to be **denatured.** Denaturation of an α-helical protein is shown in Figure 11.24.

Consider the effect of increasing temperature on a solution of proteins, for instance, egg white. At first, increasing the temperature simply increases the rate of molecular movement, the movement of the individual molecules within the solution. Then, as the temperature continues to increase, the bonds within the proteins begin to vibrate more violently. Eventually, the weak interactions, like the hydrogen bonds, that maintain the protein structure are disrupted. The protein molecules are denatured as they lose their characteristic three-dimensional conformation and become completely disorganized. The protein molecules then **coagulate** as they clump together. At this point they are no longer in solution; they have aggregated to become a solid (Figure 11.24). Our egg white began as a viscous solution of egg albumins; but when we finished cooking it, the proteins had been denatured and had coagulated to become solid.

Many of the proteins of our cells, for instance, the enzymes, are in the same kind of viscous solution within the cytoplasm. To continue to function properly, they must remain in solution and maintain the correct three-dimensional configuration. If the body temperature becomes too high, or if local regions of the body are subjected to very high temperatures, as when you touch a hot cookie sheet, cellular proteins become denatured. They lose their function, and the cell or the organism dies.

11.13 THE EFFECT OF pH ON PROTEINS

Because of the R groups of the amino acids, all proteins have a characteristic electric charge. Since every protein has a different amino acid composition, each will have a different net electric charge on its surface. The positively and negatively charged R groups on the surface of the molecule interact with ions and water molecules, and these interactions keep the protein in solution within the cytoplasm.

The protein shown in Figure 11.25a has a net charge of 2+ because it has two extra $-N^+H_3$ groups. If we add a base, such as NaOH, the protonated amino groups lose their protons and thus become electrically neutral. Now the net charge of the protein is zero.

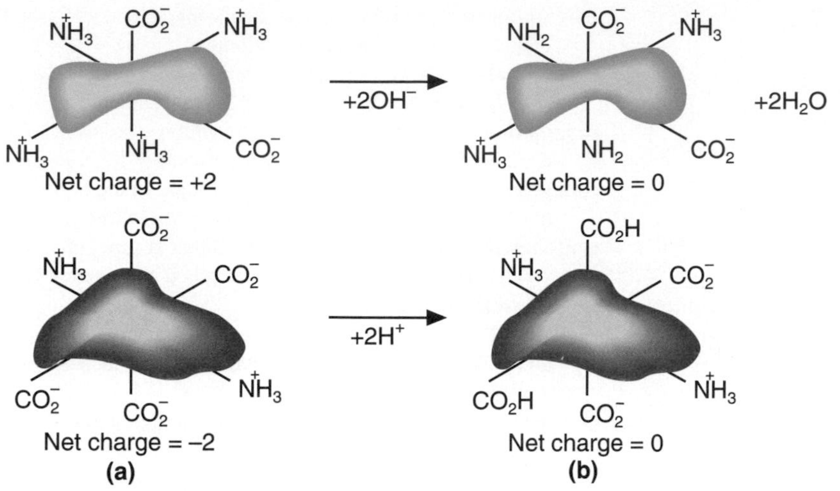

FIGURE 11.25
The effect of pH on proteins. (a) This protein has an overall charge of 2+. When a base is added, some of the protonated amino groups lose their protons. Now the protein is isoelectric; it has an equal number of positive and negative charges. (b) This protein has an overall charge of 2−. As acid is added, some of the carboxylate groups are protonated. The result is that the protein becomes isoelectric.

When the polypeptide has an equal number of positive and negative charges, it is said to be **isoelectric.** The protein shown in Figure 11.25b has a net charge of 2− because of two additional carboxylate groups. When an acid is added, the carboxylate groups become protonated. They are now electrically neutral, and the net charge on the protein is zero. As in the above example, the protein is now isoelectric.

Once a protein has no charge on its surface, it no longer has a means of interacting with the surrounding water molecules, and it cannot remain in solution. Under these conditions the protein molecules aggregate with one another, and coagulation occurs.

This is a reaction that you have probably observed in your own kitchen. When milk sits in the refrigerator for a prolonged period, the bacteria in the milk begin to grow. They use the milk sugar, lactose, as an energy source in the process of fermentation and produce lactic acid as a by-product. As the bacteria continue to grow, the concentration of lactic acid increases. The additional acid results in the protonation of exposed carboxylate groups on the surface of the dissolved milk proteins. They become isoelectric and coagulate into a solid curd.

Section 13.4

Imagine for a moment what would happen if the pH of the blood were to become too acidic or too basic. Blood is a fluid that contains water and dissolved electrolytes, a variety of cells, including the red blood cells responsible for oxygen transport, and many different proteins. These proteins include fibrinogen, which is involved in the clotting reaction; immunoglobulins, which protect us from disease; and albumins, which carry hydrophobic molecules in the blood. If the pH of the blood were to become too acidic or too basic, the proteins dissolved in the blood and the proteins of the cells in the blood would become isoelectric. In this extreme condition the blood proteins would denature and would no longer be able to carry out their required functions. The blood cells would also die as their critical enzymes were denatured. The hemoglobin in the red blood cells would become denatured and would no longer be able to transport oxygen. Fortunately, the body has a number of mechanisms to avoid the radical changes in the blood pH that can occur as a result of metabolic or respiratory difficulties.

Temperature and pH changes are only two of the means by which proteins can be denatured. Ultraviolet light and organic solvents, such as alcohols, can denature proteins and are used in the health care field as sterilizing agents. In addition, whipping or shaking can also denature proteins. This is the reason that whipping egg whites produces a stiff meringue.

SUMMARY

Proteins are made from 20 different amino acids, each having an α-CO_2^- group and an α-N^+H_3 group. They differ only in their side chain R groups. All α-amino acids are chiral except glycine. Naturally occurring amino acids have the same chirality, designated L.

Amino acids are joined by peptide bonds to produce peptides and proteins. Proteins are linear polymers of amino acid residues whose sequence of amino acids defines the primary structure of the protein. The secondary structure of a protein consists of regions of α-helix and β-pleated sheet. Most of the structural proteins, such as α-keratins, wool, silk fibroin, and collagen, have three-dimensional structures composed almost entirely of α-helix or β-pleated sheet.

Globular proteins contain varying amounts of α-helix and β-pleated sheets folded into higher levels of structure called the tertiary structure. Some proteins are composed of more than one peptide. The bound subunit of these proteins is the quaternary structure. The forces that maintain tertiary and quaternary structure include a variety of weak interactions between the R groups of the constituent amino acids.

Conjugated proteins require an attached, nonprotein prosthetic group. Myoglobin, the oxygen storage protein of skeletal muscle, is a conjugated protein with a prosthetic group called the heme group. The heme group is the site of oxygen binding. Hemoglobin, another conjugated protein, is a tetramer with four heme groups that transports oxygen from the lungs to the tissues. Myoglobin has a greater affinity for oxygen than does hemoglobin, and so oxygen is efficiently transferred from hemoglobin to myoglobin in tissues in which the oxygen pressure is low. Fetal hemoglobin has a greater affinity for oxygen than does maternal hemoglobin, and oxygen transfer occurs efficiently across the placenta from the mother to the fetus. A mutant hemoglobin is responsible for the molecular disease sickle cell anemia.

When a protein is subjected to very high temperatures or extremes of pH, the well-ordered secondary, tertiary, and quaternary structure is destroyed. When a protein loses its structural organization, it is denatured. Such denatured proteins can then aggregate into an insoluble form; they become coagulated.

GLOSSARY OF KEY TERMS

α-amino acid (11.2) the basic subunits of proteins. Each is composed of an α-carbon bonded to a carboxylate group, a protonated amino group, a hydrogen atom, and a variable R group.

amino acid residues (11.3) the amino acid units that remain after a peptide bond has been formed.

antibodies (11.8) specific proteins produced by cells of the immune system in response to invasion by infectious agents.

antigen (11.1) any substance able to stimulate the immune system. Antigens are usually proteins or large carbohydrates.

autoimmune reaction (11.8) a reaction of the immune system against one's own tissues.

B lymphocytes (11.8) specialized white blood cells of the immune system that produce antibodies.

C-terminal residue (11.3) the amino acid residue with a free α-CO_2^- group. This is the last amino acid in a peptide.

coagulation (11.12) the process by which proteins in solution are denatured and aggregate with one another to produce a solid.

collagen (11.6) the most abundant protein in the body. It confers mechanical stability to skin, bone, and tendons.

conjugated protein (11.8) a protein that is functional only when it carries other chemical groups attached by covalent linkages or by weak interactions.

cyclic AMP (11.1) a nucleotide that is a chemical messenger that allows cells to respond to certain stimuli, such as binding of enkephalins, to cell surfaces.

cytochrome *c* (11.4) a protein of 104 amino acid residues that is required for aerobic respiration.

denaturation (11.12) the process by which the organized structure of a protein is disrupted, resulting in a completely disorganized, nonfunctional form of the protein.

dipeptide (11.3) a molecule formed by condensing two amino acids.

enkephalins (11.1) peptide opiates that are involved in the perception of pain.

enzymes (11.1) proteins that serve as biological catalysts.

fetal hemoglobin (11.10) the form of hemoglobin produced by the fetus; it has a greater affinity for oxygen than does the mother's adult hemoglobin.

fibrous protein (11.7) proteins composed of peptides arranged in long sheets or fibers.

globular protein (11.7) proteins composed of polypeptide chains that are tightly folded into a compact spherical shape.

glycoprotein (11.8) proteins conjugated with sugar groups. Often these are receptors on the cell surface.

α-helix (11.5) a right-handed coiled secondary structure maintained by hydrogen bonds between the amide hydrogen of one amino acid and the carbonyl oxygen of an amino acid four residues away.

heme (11.9) the chemical group found in hemoglobin and myoglobin that is responsible for the ability to carry oxygen.

hemoglobin (11.9) the major protein component of red blood cells. The function of this iron-containing protein is transport of oxygen.

heroin (11.1) a derivative of morphine that is much more addictive than morphine and induces a longer-lived euphoria.

hydrolysis (11.2) the process by which molecules are broken into their constituents by addition of water.

hydrophilic (11.2) "water loving," polar and ionic amino acids that have a high affinity for water.

hydrophobic (11.2) "water fearing," a nonpolar amino acid that prefers contact with other nonpolar amino acids over contact with water.

hydroxyapatite (11.6) a calcium phosphate polymer that crystallizes on collagen and makes up the mineral portion of bone.

IgA (11.8) the type of immunoglobulin that protects body surfaces, such as the gut, oral cavity, and genitourinary tract.

IgD (11.8) an immunoglobulin that is thought to regulate antibody synthesis.

IgE (11.8) an immunoglobulin that is thought to be responsible for allergies.

IgG (11.8) the major immunoglobulin found in blood serum.

IgM (11.8) the first immunoglobulin produced by the B cells in response to an infection.

immune system (11.8) an organized group of cells that defend the body against bacterial, viral, parasitic, and fungal infections.

immunoglobulins (11.1) antibodies; very specific proteins that are formed in response to invasion of the body by infectious agents such as bacteria and viruses.

isoelectric (11.13) the protein has an equal number of positive and negative charges and therefore has an overall net charge of zero.

α-keratin (11.5) fibrous proteins that form the covering of most land animals. They are major components of fur, skin, beaks, and nails.

microfibril (11.5) aggregates of α-keratin protofibrils that possess great mechanical strength.

morphine (11.1) a narcotic that causes drowsiness, euphoria, mental confusion, and chronic constipation; it is used as a pain killer.

myoglobin (11.9) the oxygen storage protein found in muscle.

myosin (11.5) one of the major proteins of muscle tissue. It has a rodlike structure of two α-helices coiled around one another.

N-terminal residue (11.3) the amino acid residue with a free α-N^+H_3 group. This is the first amino acid of a peptide.

peptide bond (11.3) the covalent linkage between two amino acids in a peptide chain. It is formed by a condensation reaction.

β-pleated sheet (11.5) a common secondary structure that resembles the pleats of an Oriental fan.

primary protein structure (11.4) the sequence of amino acids in a protein. This is determined by the genetic information of the gene for each protein.

prosthetic group (11.8) the chemical group found attached to a conjugated protein.

protein (11.3) macromolecules whose primary structure is a linear sequence of α-amino acids and whose final structure results from folding of the chain into a specific three-dimensional structure. They have many functions in the organism, including catalysis, structural components, and nutritional elements.

protofibril (11.5) three single α-keratin helices coiled around one another in a bundle.

quaternary protein structure (11.8) functional proteins that are composed of several globular proteins aggregated together.

receptors (11.1) proteins on the cell surface that bind to specific food molecules and facilitate their entry into the cell. Other receptors bind specific chemical signals and direct the cell to respond appropriately.

regulatory proteins (11.1) proteins that control cell functions such as metabolism and reproduction.

scurvy (11.6) a disease of collagen metabolism resulting from a deficiency of vitamin C. The symptoms include skin lesions, fragile blood vessels, and bleeding gums.

secondary protein structure (11.5) folding of the primary structure of a protein into an α-helix or a β-pleated sheet. The forces that maintain the folding are hydrogen bonds between the amide hydrogen and the carbonyl oxygen of the peptide bond.

sickle cell anemia (11.10) a human genetic disease resulting from inheriting mutant hemoglobin genes from both parents. The disease is fatal because of poor oxygen transport to the tissues.

sickle cell trait (11.10) the condition of having one normal and one mutant hemoglobin gene. These individuals are less severely affected than those with two mutant genes.

silk fibroin (11.5) fibrous protein that is produced by silkworms and whose structure is an antiparallel β-pleated sheet.

structural proteins (11.1) proteins that provide mechanical support for large plants and animals.

tertiary protein structure (11.7) the globular, three-dimensional structure of a protein that results from folding the regions of secondary structure. This folding occurs spontaneously as a result of interactions of the side chains or R groups of the amino acids.

transport proteins (11.1) proteins that transport materials across the cell membrane or from tissue to tissue within the body.

tropocollagen (11.6) a triple-stranded helical fiber of collagen.

vaccination (11.8) the process of immunizing an animal against a particular infectious agent by injecting small amounts of the antigens from the virus or organism causing the disease.

vitamin C (11.6) a water-soluble vitamin whose physiological function is the hydroxylation of collagen.

QUESTIONS AND PROBLEMS

Amino Acids

11.4 Write the basic general structure of an L-α-amino acid.

11.5 What is the meaning of the prefix L in the name of an amino acid?

11.6 Write the structures of D- and L-alanine and show how they are related to the structures of D- and L-glyceraldehyde.

11.7 Why are the α-amino acids chiral?

11.8 Why is the structure of an α-amino acid that does not have an ionizing side chain always written as a dipolar ion?

11.9 Write the structures of the ten amino acids that have hydrophobic side chains.

11.10 Write the structures of the aromatic amino acids. Indicate whether you would expect to find each on the surface or buried in a globular protein.

11.11 Write the structure of the amino acid produced by the oxidation of cysteine.

11.12 What is the role of cystine in maintaining protein structure?

Peptides

11.13 Write the structures of the following peptides:
a. his-trp-cys c. arg-ile-val
b. gly-leu-ser

11.14 What is the net charge at pH 7 of each of the following peptides?
a. ile-leu-phe c. asp-glu-ser-lys
b. his-arg-lys

11.15 Write the peptide that would result from the condensation of alanine and serine. Identify the peptide bond, and write the structure with alanine as the N-terminal residue.

Protein Structure

11.16 A certain protein contains a buried aspartate residue. This residue is found to be protonated in the biologically active conformation of the protein. Explain this observation.

11.17 State the most conspicuous structural feature of:
a. wool fibers b. silk fibroin

11.18 Define each of the following terms and give an example of each:
a. primary structure of proteins
b. secondary structure of proteins
c. tertiary structure of proteins
d. quaternary structure of proteins

11.19 Describe the forces that maintain the two types of secondary structure, α-helix and β-pleated sheet.

11.20 Describe the forces that maintain tertiary and quaternary structure.

11.21 What is a conjugated protein?

11.22 What is a prosthetic group?

Collagen

11.23 What is the structure of tropocollagen?

11.24 What is the role of vitamin C in the biosynthesis of collagen?

11.25 Write the structures of hydroxylysine and hydroxyproline.

Hemoglobin and Myoglobin

11.26 What is the function of heme in hemoglobin and myoglobin?

11.27 How does oxidation of myoglobin differ from oxygenation of myoglobin?

11.28 How do sickled red blood cells hinder circulation?

11.29 What is the difference between sickle cell disease and sickle cell trait?

11.30 How is it possible for sickle cell trait to confer a survival benefit on the person who possesses it?

Immunoglobulins

11.31 What is the basic structure of immunoglobulin G?

11.32 Define the following terms:
a. antibody b. antigen

11.33 What is the major function of IgA?

11.34 Describe the three characteristics of the immune response?

11.35 What are the biological roles of IgM, IgD, and IgE?

11.36 Why do we seldom suffer from the same disease more than once?

11.37 What is the function of the B cell in the immune response?

11.38 What is the major function of immunoglobulin G?

11.39 Explain why a single antigen can bind to many different immunoglobulin molecules.

Further Problems

11.40 The primary structure of a protein known as histone H1, which tightly binds DNA, is identical in all mammals and differs by only one amino acid residue between the calf and pea seedlings. What does this extraordinary conservation of primary structure imply about the importance of that one amino acid?

11.41 What does it mean to say that the structure of proteins is genetically determined?

11.42 Carbon monoxide and cyanide bind tightly to the heme groups of hemoglobin and myoglobin. How does this affinity reflect the toxicity of carbon monoxide and cyanide?

11.43 What is the purpose of the coils of fibrous proteins such as α-keratins and collagen?

11.44 List five biological functions of proteins and provide an example for each.

11.45 What is the major structural feature of each of the following fibrous proteins?
a. α-keratins c. collagen
b. silk fibroin d. myosin

11.46 Explain why the primary structures of cytochrome *c* isolated from a mouse and a rat are related to the evolution of the two species from a common ancestor.

11.47 How does the structure of the peptide bond make the structure of proteins relatively rigid?

11.48 Why are several immunoglobulins synthesized to combat the same invading microorganism?

11.49 Why does the replacement of a glutamic acid residue with a valine residue so radically alter the structure of hemoglobin?

11.50 Explain why α-keratins that have many disulfide bonds between adjacent polypeptide chains are much less elastic and much harder than those without disulfide bonds.

11.51 Why is hydrogen bonding so important in discussing the structure of proteins?

11.52 As you increase the temperature of an enzyme-mediated reaction, the rate of the reaction initially increases. It then reaches a maximum rate and finally dramatically declines. Keeping in mind that enzymes are proteins, how do you explain these changes in reaction rate?

11.53 Every enzyme has a pH at which it functions optimally. The proteolytic enzyme pepsin, which hydrolyzes proteins in the stomach, functions optimally at a pH of 2. The enzyme trypsin, which degrades proteins in the intestines, functions optimally at a pH of 8.5.
a. How can you explain these results in terms of protein structure?
b. What is the physiological relevance of the pH optima of these two enzymes?

CHAPTER 12

Enzyme Catalysis

OBJECTIVES

- Recognize the correlation between an enzyme's common name and its function.
- Know how to classify enzymes according to the type of reaction catalyzed and the type of specificity.
- Describe the effect that enzymes have on the activation energy of a reaction.
- Understand the effect of substrate concentration on enzyme-catalyzed reactions.
- Discuss the role of the active site and the importance of enzyme specificity.
- Describe the difference between the lock-and-key model and the induced fit model of enzyme-substrate complex formation.
- Discuss the roles of cofactors and coenzymes in enzyme activity.
- Understand the mechanisms used by cells to regulate enzyme activity.
- Discuss the mechanisms by which certain chemicals inhibit enzyme activity.
- Recognize how pH and temperature affect the rate of an enzyme-catalyzed reaction.
- Discuss the role of the enzyme chymotrypsin and other serine proteases.
- Describe the process of blood coagulation and the role of vitamin K in the formation of blood clots.
- Explain the role of acetylcholinesterase in nerve transmission.
- Provide examples of medical and industrial uses of enzymes.

OUTLINE

We have seen that proteins can exist in many different shapes and perform many different functions. In this chapter we shall consider some aspects of the conformation and function of **enzymes,** which are proteins that act as catalysts in cell processes. The life of the cell depends upon the simultaneous occurrence of hundreds of chemical reactions that must take place rapidly under mild conditions. It is possible, for example, to add water to an alkene. However, this reaction is usually carried out at a temperature of 100°C in aqueous sulfuric acid. Such conditions would kill a cell. The fragile cell must carry out its chemical reactions at pH 7, at body temperature (37°C), and in the absence of any strong acids or bases. How can this be accomplished? Catalysts lower the energy of activation of a chemical reaction and thereby increase the rate of the reaction. This allows reactions to occur under milder conditions. The cell uses biological catalysts, called enzymes, to solve the problem of allowing chemical reactions to occur rapidly under the mild conditions found within the cell. The enzyme *facilitates* a biological chemical reaction, lowering the energy of activation and increasing the rate of the reaction. The efficient functioning of enzymes is essential for the life of the cell and of the organism.

A typical cell contains thousands of molecules, each of which is important to the chemistry of life processes. Any given enzyme ''recognizes'' only one, or occasionally a few, of these molecules. One of the most remarkable features of enzymes is this **specificity.** Each can recognize and bind to a single type of **substrate** or reactant. The molecular size, shape, and charge distribution of both the enzyme and substrate must be compatible for this selective bonding process to occur. The enzyme then transforms the substrate into the **product** with lightning speed. In fact, enzyme-catalyzed reactions often occur from one million to 100 million times faster than the corresponding uncatalyzed reaction. The enzyme **catalase** provides one of the most spectacular examples of the increase in reaction rates brought about by enzymes. This enzyme is required for life in an oxygen-containing environment. In this environment the process of the aerobic (oxygen-requiring) breakdown of food molecules produces hydrogen peroxide. Since H_2O_2 is toxic to the cell, it must be destroyed. One molecule of catalase converts *40 million* molecules of hydrogen peroxide to harmless water and oxygen every second:

$$2H_2O_2 \xrightarrow[\text{(an enzyme)}]{\text{catalase}} 2H_2O + O_2$$

Reaction occurs millions of times every second!

This is the same reaction that you witness when you pour hydrogen peroxide on a wound. The catalase released from injured cells rapidly breaks down the hydrogen peroxide. The bubbles that you see are oxygen gas released as a product of the reaction. These twin phenomena of high specificity and rapid reaction rates are the cornerstones of enzyme activity and the topic of this chapter.

12.1 NOMENCLATURE AND CLASSIFICATION

Nomenclature of enzymes

One of the most straightforward concepts in chemistry is enzyme nomenclature. The common name of an enzyme is derived from the name of the substrate with which the enzyme interacts and/or the type of reaction that it mediates. Because of this, the function of the enzyme is generally conveyed directly by its common name.

Let's look at a few examples of this simple concept. *Urea* is the substrate acted on by the enzyme *urease*:

urea − *a* + ase = urease

Substrate Enzyme

Lactose is the substrate of *lactase*:

lactose − *ose* + ase = lactase

Substrate Enzyme

Other enzymes may be named for the reactions they catalyze. For example:

Dehydrogenases remove hydrogen.

Decarboxylases remove carboxyl groups.

The prefix *de-* indicates that a functional group is being removed. Hydrogenases and carboxylases, on the other hand, add hydrogen or carboxyl groups. Some enzyme names include *both* the substrate and reaction type. For example:

Lactate dehydrogenase removes hydrogen atoms from lactate ions.

Acetoacetate decarboxylase removes carboxyl groups from acetoacetate.

As in other areas of chemistry, historical names, having no relationship to either substrate or reaction, continue to be used. Often these names do not reveal the nature of the substrate or of the reaction, and their substrates and reactions must simply be memorized. Examples of some historical common names include catalase, pepsin, chymotrypsin, and trypsin. We will learn more about these enzymes shortly.

Question 12.1 What is the substrate for each of the following enzymes?

a. Sucrase

b. Pyruvate decarboxylase

c. Succinate dehydrogenase

Question 12.2 What chemical reaction is mediated by each of the above three enzymes?

Classification of enzymes

Enzymes may be classified according to the reaction type in which they are involved. These six classes are as follows:

Oxireductases

Oxireductases are enzymes that catalyze oxidation-reduction (redox) reactions. Redox reactions, as you recall, involve electron transfer from one molecule to another. A common example is lactate dehydrogenase. The function of this enzyme is to facilitate the removal of hydrogen from a molecule of lactate. Other subclasses of the oxireductases include oxidases and reductases.

$$\underset{\text{Lactate}}{HO-\overset{\overset{\displaystyle COO^-}{|}}{\underset{\underset{\displaystyle CH_3}{|}}{C}}-H} + NAD^+ \rightleftharpoons \underset{\text{Pyruvate}}{\overset{\overset{\displaystyle COO^-}{|}}{\underset{\underset{\displaystyle CH_3}{|}}{C}}=O} + NADH + H^+$$

Transferases

Transferases are enzymes that catalyze the transfer of functional groups from one molecule to another. For example, a transaminase catalyzes the transfer of an amine functional group, and a kinase catalyzes the transfer of a phosphate group. Kinases play a major role in energy production processes involving ATP. In the adrenal glands, norepinephrine is converted to epinephrine by the enzyme phenylethanolamine-N-methyltransferase (PNMT), a transmethylase.

Section 14.5

See Chapter 10, "A Human Perspective: Amines and the Central Nervous System"

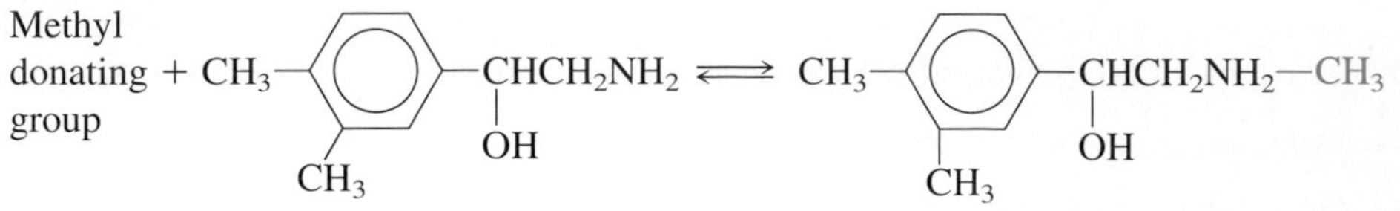

Methyl donating group + CH_3–(benzene ring, CH_3)–$CHCH_2NH_2$ (OH) ⇌ CH_3–(benzene ring, CH_3)–$CHCH_2NH_2$–CH_3 (OH)

Norepinephrine Epinephrine

Hydrolases

Hydrolases catalyze hydrolysis reactions, that is, the addition of a water molecule to a bond resulting in bond breakage. These reactions are of importance in the digestive process. For example, lipases catalyze the hydrolysis of triglycerides:

Section 15.1

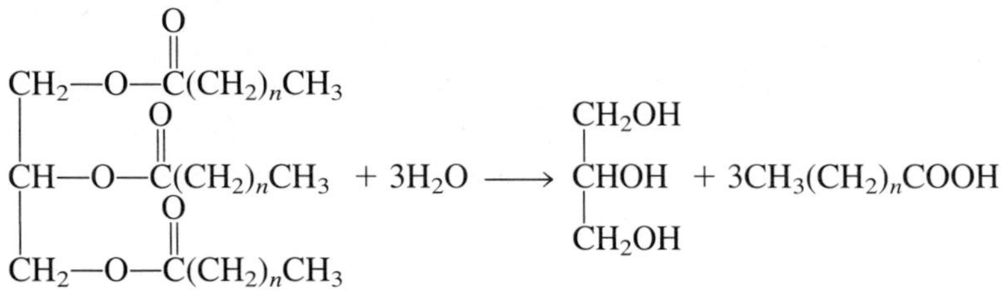

$$\begin{array}{l} CH_2{-}O{-}C(=O)(CH_2)_nCH_3 \\ CH{-}O{-}C(=O)(CH_2)_nCH_3 \\ CH_2{-}O{-}C(=O)(CH_2)_nCH_3 \end{array} + 3H_2O \longrightarrow \begin{array}{l} CH_2OH \\ CHOH \\ CH_2OH \end{array} + 3CH_3(CH_2)_nCOOH$$

Triglyceride Glycerol Fatty acids

Lyases

Lyases are enzymes that catalyze the cleavage of C—O, C—C, or C—N bonds. In the process, a double bond is formed. Citrate lyase catalyzes the removal of an acetate group from a molecule of citrate. The products of the reaction include oxaloacetate, acetyl CoA, ADP, and an inorganic phosphate group (P_i).

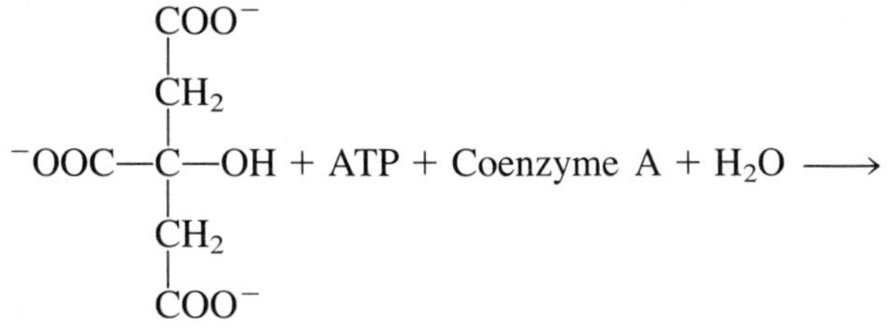

$$\begin{array}{c} COO^- \\ | \\ CH_2 \\ | \\ {}^-OOC{-}C{-}OH \\ | \\ CH_2 \\ | \\ COO^- \end{array} + ATP + \text{Coenzyme A} + H_2O \longrightarrow$$

Citrate

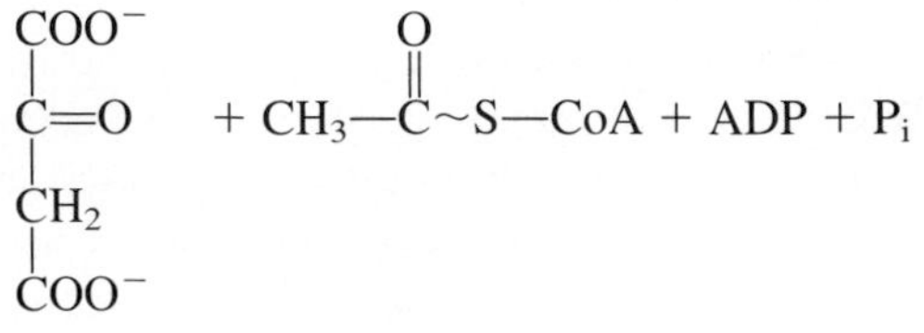

$$\begin{array}{c} COO^- \\ | \\ C{=}O \\ | \\ CH_2 \\ | \\ COO^- \end{array} + CH_3{-}\overset{O}{\overset{\|}{C}}{\sim}S{-}CoA + ADP + P_i$$

Oxaloacetate Acetyl CoA

Isomerases

Isomerases rearrange the functional groups within a molecule and catalyze the conversion of one isomer into another. For example, phosphoglyceromutase converts one structural isomer, 3-phosphoglycerate, into another, 2-phosphoglycerate:

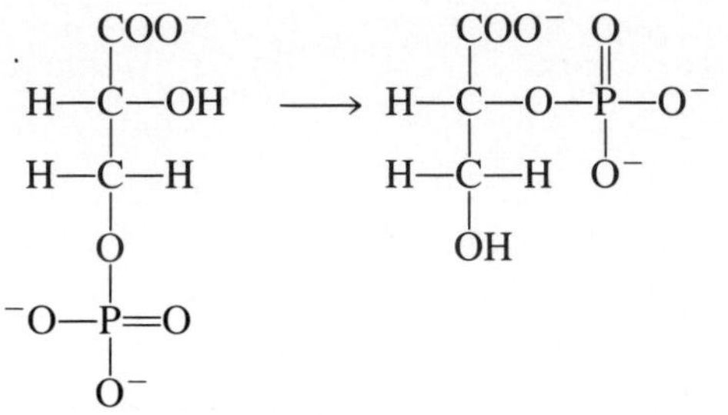

$$\begin{array}{c} COO^- \\ | \\ H-C-OH \\ | \\ H-C-H \\ | \\ O \\ | \\ ^-O-P=O \\ | \\ O^- \end{array} \longrightarrow \begin{array}{c} COO^- \quad O \\ | \qquad \| \\ H-C-O-P-O^- \\ | \qquad | \\ H-C-H \quad O^- \\ | \\ OH \end{array}$$

3-Phosphoglycerate 2-Phosphoglycerate

Ligases

Ligases are enzymes that catalyze the condensation or joining of two molecules. This new bond formation requires energy, usually provided in the form of hydrolysis of ATP. For example, DNA ligase catalyzes the joining of the hydroxyl group of a nucleotide in a DNA strand with the phosphate group of the adjacent nucleotide to form a phosphoester linkage.

Section 16.4

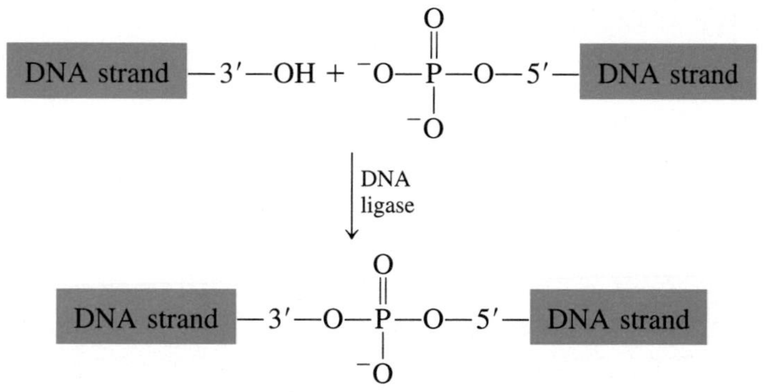

$$\boxed{\text{DNA strand}}-3'-OH + {}^-O-\overset{\overset{O}{\|}}{\underset{\underset{^-O}{|}}{P}}-O-5'-\boxed{\text{DNA strand}}$$

$$\Big\downarrow \text{DNA ligase}$$

$$\boxed{\text{DNA strand}}-3'-O-\overset{\overset{O}{\|}}{\underset{\underset{^-O}{|}}{P}}-O-5'-\boxed{\text{DNA strand}}$$

Question 12.3 To which class of enzymes does each of the following belong?

a. Pyruvate kinase
b. RNA ligase
c. Triose isomerase
d. Pyruvate dehydrogenase
e. Pyruvate carboxylase
f. Maltase

12.2 THE EFFECT OF ENZYMES ON THE ACTIVATION ENERGY OF A REACTION

We recall that every chemical reaction is characterized by an equilibrium constant. Consider, for example, the simple equilibrium

$$A \overset{K_{eq}}{\rightleftharpoons} B$$

The equilibrium constant for this reaction, K_{eq}, is defined as

$$K_{eq} = \frac{[B]}{[A]} = \frac{[\text{product}]}{[\text{reactant}]}$$

This equilibrium constant is actually a reflection of the difference in energy between reactants and products. It is a measure of the relative stabilities of the reactants and products. No matter how the chemical reaction occurs (which path it follows), the differ-

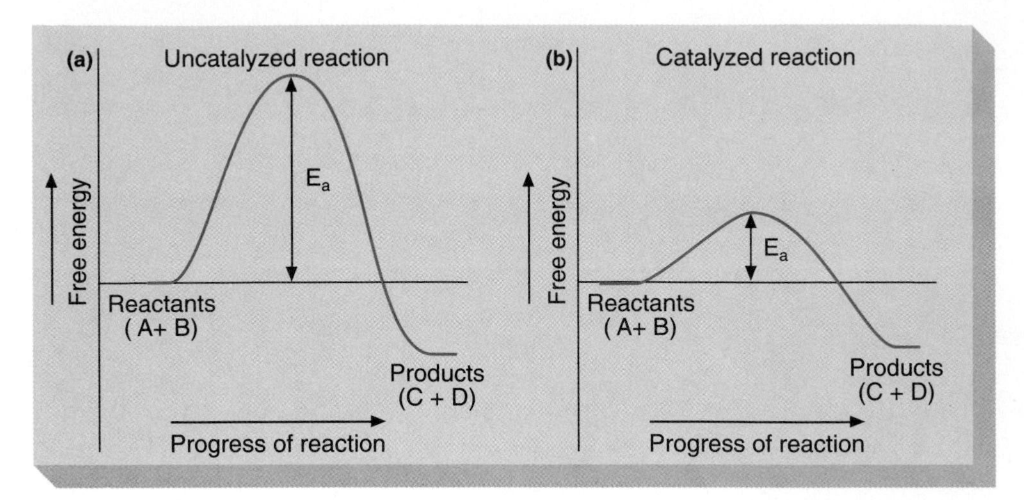

FIGURE 12.1
Diagram of the difference in energy between the reactants (*A* and *B*) and products (*C* and *D*) for a reaction. Enzymes cannot change this energy difference but act by lowering the activation energy (E_a) for the reaction, thereby speeding up the reaction.

ence in energy between the reactants and the products is always the same. An enzyme cannot therefore alter the equilibrium constant for the reaction that it catalyzes. An enzyme does, however, change the path by which the process occurs, providing a lower-energy route for the conversion of the substrate into the product. An enzyme increases the rate of a chemical reaction by lowering the activation energy for the reaction (Figure 12.1). An enzyme thus increases the rate at which the reaction it catalyzes reaches equilibrium.

12.3 THE EFFECT OF SUBSTRATE CONCENTRATION ON ENZYME-CATALYZED REACTIONS

The rates of chemical reactions (reactions that are not enzymatically catalyzed) often double every time the substrate concentration is doubled. Therefore as long as the substrate concentration increases, there is a direct increase in the rate of the reaction. For enzyme-catalyzed reactions, however, this is not the case. Although the rate of the reaction is initially responsive to the substrate concentration, at a certain concentration of substrate the rate of the reaction reaches a maximum. A graph of the rate of reaction, V, versus the substrate concentration, $[S]$, is shown in Figure 12.2. We see that the rate of the reaction initially increases rapidly as the substrate concentration is increased but that the

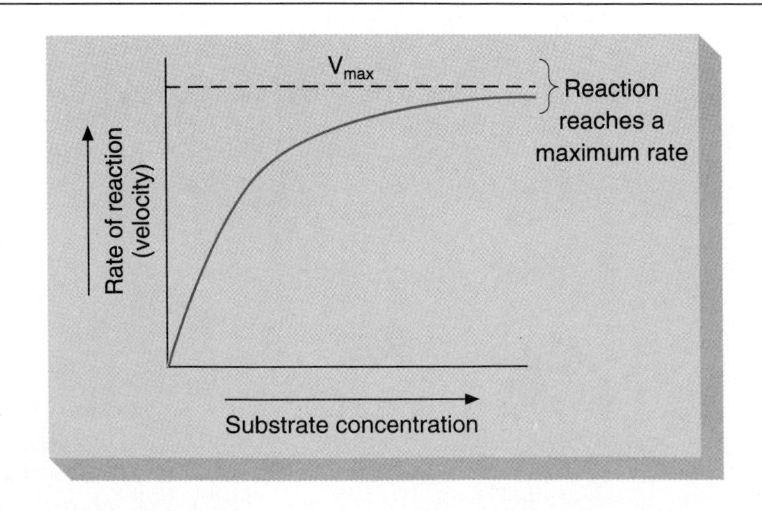

FIGURE 12.2
Plot of the rate or velocity, V, of an enzyme-catalyzed reaction versus the concentration of substrate, $[S]$. The rate is at a maximum when all of the enzyme molecules are bound to the substrate. Beyond this concentration of substrate, further increases in substrate concentration have no effect on the rate of the reaction.

increase in rate levels off at a maximum value. At its maximum rate the active sites of all the enzyme molecules in solution are occupied by a substrate molecule. A new molecule of substrate cannot bind to a given enzyme molecule until another substrate molecule already bound to the enzyme is converted to product. Thus it appears that the enzyme-catalyzed reaction occurs in two stages. The first, rapid step is the formation of the *enzyme-substrate complex*. The second step is slower and thus controls the rate at which the reaction can occur. It is said to be rate-limiting and involves conversion of the substrate to product and the release of the product and enzyme from the resulting enzyme-product complex. In effect, the rate of the reaction is limited by the speed with which the substrate is converted into product and thus on the availability of the enzyme.

12.4 THE ENZYME-SUBSTRATE COMPLEX

The first step in an enzyme-catalyzed reaction involves the encounter of the enzyme with its substrate and the formation of an **enzyme-substrate complex:**

$$\underset{\text{Enzyme}}{E} + \underset{\text{Substrate}}{S} \longrightarrow \underset{\text{Enzyme-substrate complex}}{ES}$$

The portion of the enzyme that is in contact with the substrate is called the **active site.** We find that the properties of the active site are crucial to the function of the enzyme and have the following general characteristics:

1. *The active site is small in relation to the size of the enzyme*. For example, an enzyme that contains 200 amino acid residues may devote as few as a dozen residues directly to catalysis. The R groups of these amino acid residues are the functional groups involved in catalysis and are called the **catalytic groups.** The rest of the enzyme is by no means superfluous. The events of the catalytic process require the precise positioning of several amino acid residues around the substrate, and a large structure is required to provide the ''scaffolding'' for positioning of the residues at the active site. The correct structure, including the shape of the active site and the nature of the surrounding scaffolding, is a function of the secondary, tertiary, and perhaps even quaternary folding. Thus the structure is maintained by the variety of weak interactions described in Section 11.11. When the enzyme binds to its substrate, the amino acid residues must be positioned in the correct orientation for the reaction to occur. All subsequent catalytic steps depend upon this correct orientation. The lowering of the activation energy and the correct orientation between enzyme and substrate are inseparable. Rate enhancement and specificity go hand in hand, and both depend on the principles of protein structure and folding.
2. *An enzyme attracts and holds its substrate by weak, noncovalent interactions*. The functional groups of some of the amino acids are involved only in substrate binding, not necessarily in catalysis. These amino acids make up the **binding site.** The substrate is initially attached to the binding groups of the enzyme active site by a combination of hydrogen bonds, the attraction of ionic positive and negative charges, and van der Waals forces.

Sections 11.5 and 11.7

3. *The shape or conformation of the active site is complementary to the conformation of the substrate*. That is, the substrate fits the enzyme as a key fits a lock. This model of enzyme-substrate binding is called the **lock-and-key model** (Figure 12.3). In reality, enzymes are flexible molecules, however, and we must imagine that the enzyme and the substrate both change their shapes to accommodate one another as the enzyme-substrate complex forms.
4. *Enzyme active sites are pockets or clefts in the surface of the enzyme*. This aspect of enzyme structure is revealed particularly clearly by the interaction of *lysozyme*, an

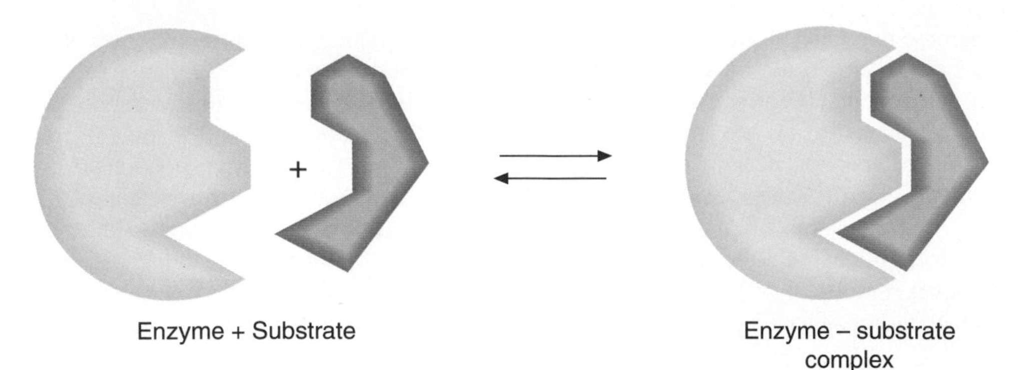

FIGURE 12.3
The lock-and-key model of enzyme-substrate binding assumes that the enzyme active site has a rigid structure that is precisely complementary in shape and charge distribution to the substrate.

enzyme that hydrolyzes certain bonds in the cell walls of bacteria (Figure 12.4). The portion of the cell wall that is cleaved by lysozyme fits into a cleft on the surface of the enzyme as a letter slides into a slot in a mailbox.

5. *The conformation of the active site determines the specificity of the enzyme.* Only those substrates whose shapes are complementary to the active site can bind the enzyme. Often only one or a very few molecules have just the right shape to match the shape of a given enzyme, a feature that accounts for the high specificity.

FIGURE 12.4
Space-filling model of lysozyme bound to its substrate. The substrate, a portion of the cell wall of a bacterium, fits precisely into a groove or cleft on the enzyme's surface.

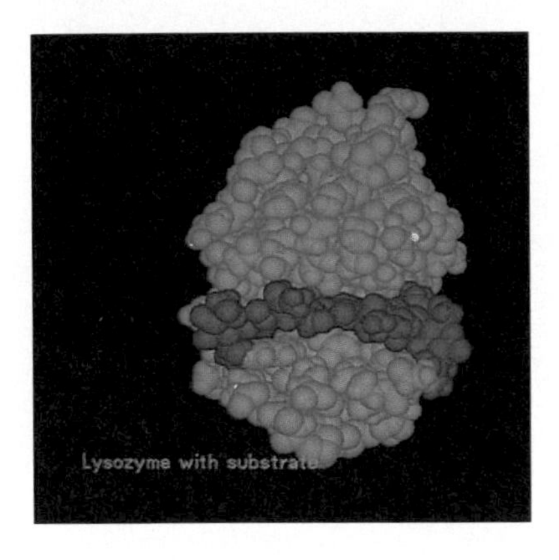

Question 12.4 Distinguish between the catalytic and binding groups of an enzyme active site.

12.5 ENZYME-CATALYZED REACTIONS: SPECIFICITY OF THE ENZYME-SUBSTRATE COMPLEX

The lock-and-key model of enzyme activity shown in Figure 12.3 was devised by Emil Fischer in 1894. At that time, proteins were considered to be rigid molecules, and it seemed reasonable to assume that the enzyme-substrate complex was simple; for example, it was thought that the substrate snaps into place like a piece of a jigsaw puzzle or a key into a lock.

Today we know a great deal more about protein structure and thus about enzyme structure. For instance, we know that proteins are flexible molecules. The overall conformation of a protein is maintained by a series of weak intramolecular interactions. At any time a few of these weak interactions may be disrupted by thermal energy (heat) or local chemical disturbances (a transient change in pH or electrical charge). If the number of disrupted weak interactions are few, they will reform very quickly. The overall result is that a transient alteration in the protein structure is seen, but the normal three-dimensional configuration is quickly reestablished. Thus the protein or enzyme can be viewed as a flexible molecule, changing shape slightly in response to minor environmental perturbations.

Our understanding that proteins are flexible molecules has led to a more sophisticated concept of the way in which enzymes and substrates interact. This model, called the **induced fit model,** is shown in Figure 12.5. In this model, the active site of the enzyme is not a rigid pocket into which the substrate fits precisely; rather, it is a flexible pocket that approximates the shape of the substrate. The substrate, bearing some combination of charged, polar, and nonpolar R groups on its surface, enters the active site. This interaction between the substrate and the active site of the enzyme can be viewed as a local chemical perturbation in the environment of the enzyme. The attraction between oppositely charged groups and hydrogen bonds between polar groups on the enzyme and

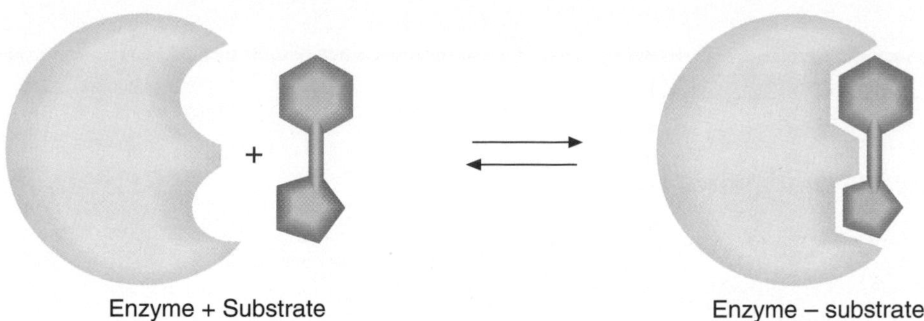

FIGURE 12.5
The induced fit model of enzyme substrate binding. As it binds to the substrate, the shape of the active site conforms precisely to the shape of the substrate. The shape of the substrate may also change.

substrate cause a change in the three-dimensional structure of the active site, resulting in an active site that conforms to the surface of the substrate (Figure 12.5).

For this enzyme-substrate interaction to occur, the surface characteristics of the enzyme and substrate must still be complementary to one another. It is this critical requirement for specific complementarity that will determine whether an enzyme will bind to a particular substrate and mediate a chemical reaction.

Question 12.5 Compare the lock-and-key and induced fit models of enzyme-substrate binding.

To illustrate the specificity of enzymes, consider the following reactions:

The enzyme urease catalyzes the decomposition of urea to carbon dioxide and ammonia as follows:

$$\underset{\text{Urea}}{H_2N{-}\overset{\overset{\displaystyle O}{\|}}{C}{-}NH_2} + H_2O \xrightarrow{\text{urease}} \underset{\text{Carbon dioxide}}{CO_2} + \underset{\text{Ammonia}}{2NH_3}$$

Methylurea, in contrast, though structurally similar to urea, is catalytically unaffected by urease:

$$\underset{\text{Methylurea}}{H_2N{-}\overset{\overset{\displaystyle O}{\|}}{C}{-}NHCH_3} + H_2O \xrightarrow{\text{urease}} \text{NO REACTION}$$

Not all enzymes exhibit the same degree of specificity. For convenience we generally classify enzymes into one of four groups:

1. *Absolute specificity*. An enzyme that catalyzes the reaction of only one substrate is **absolutely specific.** Aminoacyl-tRNA synthetases exhibit absolute specificity. Each must attach the correct amino acid to the appropriate transfer RNA molecule. If the wrong amino acid were attached to the transfer RNA, it would be mistakenly added to a peptide chain. Our study of sickle cell anemia reveals how serious the consequences of the alteration of a single amino acid can be.

Section 16.11

2. *Group specificity*. An enzyme that catalyzes processes involving similar molecules containing the same functional group is **group-specific.** One such enzyme is hexokinase, which catalyzes the phosphorylation of the hexose sugar glucose in the first step of glycolysis. Hexokinase also catalyzes the phosphorylation of several other six-carbon sugars.

Section 16.3

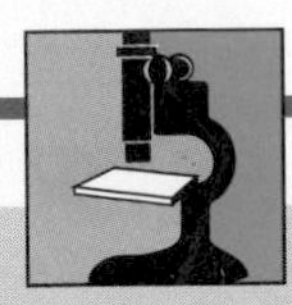

A CLINICAL PERSPECTIVE

Enzymes, Isoenzymes, and Myocardial Infarction

A patient is brought into the emergency room with acute, squeezing chest pains, shallow and irregular breathing, and pale, clammy skin. The immediate diagnosis is myocardial infarction, a heart attack. The first thoughts of the attending nurses and physicians concern the series of treatments and procedures that will save the patient's life. It is a short time later, when the patient's condition has stabilized, that the doctor begins to consider the battery of enzyme assays that will confirm the diagnosis and perhaps even help to predict the prognosis.

Myocardial infarction occurs when the blood supply to the heart muscle is blocked for an extended period of time. If this lack of blood supply, called **ischemia,** is prolonged, the myocardium suffers irreversible cell damage and muscle death, or infarction. When this happens, the concentration of cardiac enzymes in the blood rises dramatically as the dead cells release their contents into the bloodstream. Although many enzymes are liberated, three are of prime importance. These three enzymes, creatine phosphokinase (CPK), lactate dehydrogenase (LDH), and aspartate aminotransferase/serum glutamate-oxaloacetate transaminase (AST/SGOT), show a very characteristic sequential rise in blood serum level following myocardial infarction and then return to normal. This enzyme profile, seen in the accompanying figure, is characteristic of, and is the basis for the diagnosis of, a heart attack.

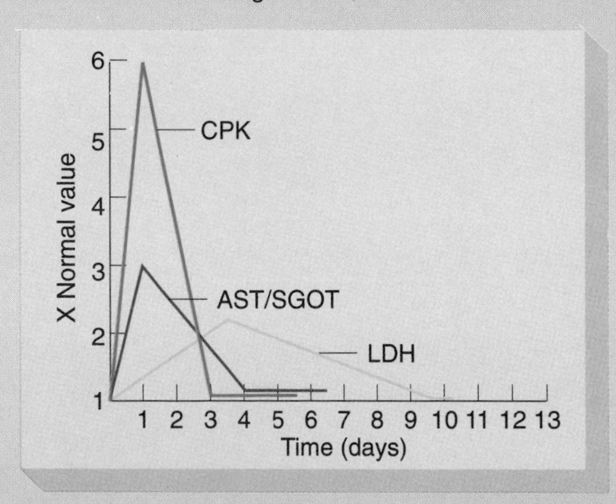

Characteristic pattern of serum cardiac enzyme concentrations following a myocardial infarction.

To ensure against misdiagnosis due to tissue damage in other organs, the levels of other serum enzymes, including alanine aminotransferase/serum glutamate-pyruvate transaminase (ALT/SGPT), and isocitrate dehydrogenase (ICD) are also measured. ALT and AST are usually determined simultaneously to differentiate between cardiac and hepatic disease. The concentration of ALT is higher in liver disease, while the serum concentration of AST is higher following acute myocardial infarction. ICD is found primarily in the liver, and serum levels would not be elevated after a heart attack.

The use of LDH and CPK levels alone can also lead to a misdiagnosis because these enzymes are produced by many tissues. How can a clinician diagnose heart disease with confidence when the elevated serum enzyme levels could indicate coexisting disease in another tissue? The physician is able to make such a decision because of the presence of **isoenzymes,** which provide diagnostic accuracy because they reveal the tissue of origin.

Isoenzymes are forms of the same enzyme with slightly different amino acid sequences. The binding and catalytic sites are the same, but there are differences in the scaffolding sequences of the enzyme that maintain the three-dimensional structure of the protein. Each of the cells of the body contains the genes that could direct the production of all of the different forms of these enzymes, yet the expression is *tissue-specific.* This means that the genes for certain isoenzymes are expressed preferentially in different types of tissue.

It is not clear why a certain isoenzyme is "turned on" in the liver while another predominates in the heart, but it is known that we can distinguish among the different forms in the laboratory on the basis of their migration through a gel placed in an electric field. This process is called *gel electrophoresis.* This test is based upon the fact that each protein has a characteristic surface charge resulting from the R groups of the amino acids. If these proteins are placed in a gel matrix and an electrical current is applied, the proteins will migrate as a function of that charge. In the figure on the next page, we see the position of the five isoenzymes of LDH following electrophoresis.

Imagine a mixture of serum proteins, each with a different overall charge, subjected to an electric field. The proteins with the greatest negative charge will be most strongly attracted to the positive pole and will migrate rapidly toward it, while those with a lesser negative charge will migrate much more slowly. Once electrophoresis is terminated, the enzyme assay is carried out within the gel. The result is a stained band that can be seen visually and measured spectrophotometrically. By inspecting the positions of the bands of enzymatic activity, one can determine which tissue isoenzymes are present. This gives the clinician a very accurate picture of the nature of the diseased tissues. The accompanying table shows the serum enzyme and isoenzyme changes that follow an acute myocardial infarction.

The physician also has enzymes available to treat a heart attack patient. Most myocardial infarctions are the result of a *thrombus,* or clot, within a coronary blood vessel. The clot restricts blood flow to the heart muscle. One technique that shows promise for treatment following a coronary thrombosis, a heart attack caused by the forma-

Serum Isoenzymes Characteristic of Myocardial Infarction

Enzyme	Normal Values*	% Normal Values	Period of Elevation	Peak	Primary Site of Production
Creatine phosphokinase (CPK)	80–780		4–72 hr	12–36 hr	
Isoenzymes:					
CPK I		0–3%	0	0	Brain
CPK II		0–5%	4–72 hr	12–36 hr	Heart
CPK III		90–100%	0	0	Skeletal muscle
Aspartate transaminase (AST)	117–450		6 hr–6 d	36–48 hr	Heart
Lactate dehydrogenase (LDH)	750–1500		12 hr–14 d	24–96 hr	
Isoenzymes:					
LDH1		20–27%	12 hr–14 d	24–96 hr	Heart, RBC
LDH2		25–37%	0	0	Immune system
LDH3		16–25%	0	0	Lungs
LDH4		3–8%	0	0	Kidney Pancreas Brain
LDH5		0–5%	0	0	Liver Skeletal muscle

*Normal values are expressed in nmol × s^{-1}/L and are given for an adult male. CPK and AST levels are somewhat lower for normal females, and the values for children and newborns are generally much higher.

tion of a clot, is destruction of the clot by intravenous or intracoronary injection of an enzyme called **streptokinase.** This enzyme, formerly purified from the pathogenic bacterium *Streptococcus pyogenes* but now available through recombinant DNA techniques catalyzes the production of the proteolytic enzyme plasmin. Plasmin has the ability to degrade a fibrin clot into subunits. Of course, this has the effect of dissolving the clot that is responsible for restricted blood flow to the heart, but there is an additional protective function as well. The subunits produced by plasmin degradation of fibrin clots are able to inhibit further clot formation by inhibiting thrombin.

Recombinant DNA technology has provided medical science with yet another, perhaps more promising, clot-dissolving enzyme. **Tissue-type plasminogen activator (TPA)** is an enzyme that occurs naturally in the body as a part of the anticlotting mechanisms. Injection of TPA within two hours of the initial chest pain can significantly improve the circulation to the heart and greatly improve the patient's chances of survival.

Fraction	%	Reference range
LD1	19.8	13.5 – 28.8
LD2	37.3	26.9 – 43.5
LD3	19.4	15.9 – 25.2
LD4	11.1	6.0 – 14.0
LD5	12.5	4.9 – 20.8

Enzyme Activity

Migration

(a)

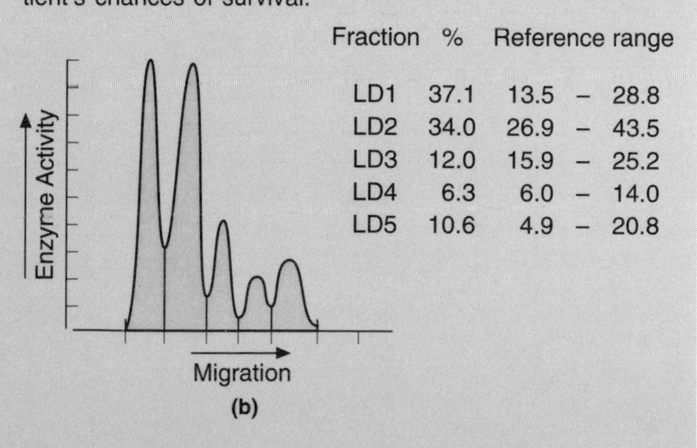

Fraction % Reference range
LD1 37.1 13.5 - 28.8
LD2 34.0 26.9 - 43.5
LD3 12.0 15.9 - 25.2
LD4 6.3 6.0 - 14.0
LD5 10.6 4.9 - 20.8
Enzyme Activity
Migration
(b)

A profile of the isoenzymes of lactate dehydrogenase. (a) The pattern of LDH isoenzymes from a normal individual. (b) The pattern of LDH isoenzymes from an individual suffering from a myocardial infarction.

Sections 12.11 and 12.12

3. *Linkage specificity*. An enzyme that preferentially catalyzes the formation or breakage of only certain bonds in a molecule is **linkage-specific.** Proteases, such as trypsin, chymotrypsin, and elastin, are enzymes that selectively hydrolyze peptide bonds. Thus these enzymes are linkage-specific.

Sections 7.2 and 11.2

4. *Stereochemical specificity*. An enzyme that is capable of distinguishing one stereoisomer from another is **stereochemically specific.** This last group, stereochemical specificity, is particularly interesting. Recall that stereoisomers are molecules with identical molecular formulas but that the atoms of the two are arranged in such a way that they are mirror images of one another. Two molecules differing only in shape interact with an enzyme in a totally different fashion. L-Aspartate, for example, is catalytically converted to the ammonium ion and fumarate by the enzyme aspartase, while the D-isomer is unaffected:

$$\begin{array}{c} COO^- \\ | \\ H_3{}^+N—C—H \\ | \\ H—C—H \\ | \\ COO^- \end{array} \xrightarrow{\text{aspartase}} \begin{array}{c} H \quad COO^- \\ \diagdown \diagup \\ C \\ \| \\ C \\ \diagup \diagdown \\ ^-OOC \quad H \end{array} + NH_4{}^+$$

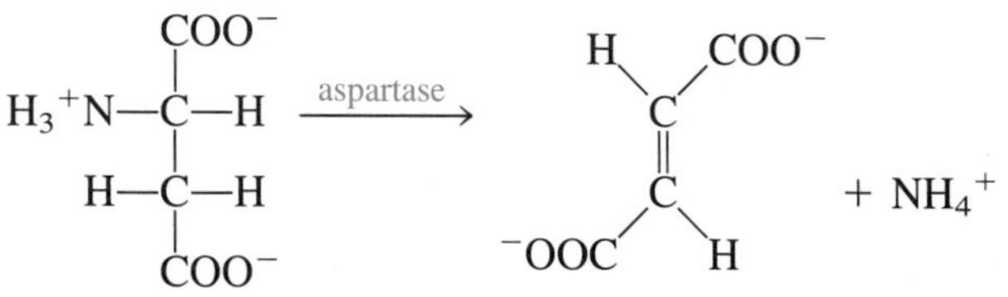

L-Aspartate Fumarate

But

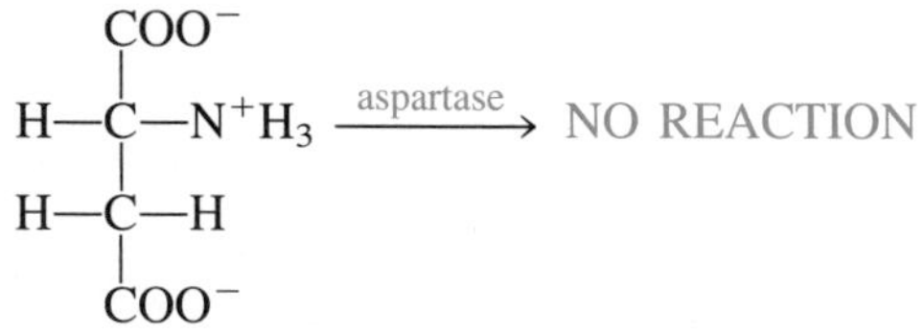

$$\begin{array}{c} COO^- \\ | \\ H—C—N^+H_3 \\ | \\ H—C—H \\ | \\ COO^- \end{array} \xrightarrow{\text{aspartase}} \text{NO REACTION}$$

D-Aspartate

12.6 ENZYME-CATALYZED REACTIONS: THE TRANSITION STATE AND PRODUCT FORMATION

We have seen the way in which an enzyme and substrate interact to form an enzyme-substrate complex, but how does this binding result in a faster chemical reaction? The precise answer to this question is probably different for each enzyme-substrate pair, and indeed, we understand the exact mechanism of catalysis for very few enzymes. Nonetheless, we can look at the general features of enzyme-substrate interactions that result in enhanced reaction rate and product formation. This overall process is described by the following series of reversible reactions:

$$E + S \underset{}{\overset{\text{step I}}{\rightleftharpoons}} ES \underset{}{\overset{\text{step II}}{\rightleftharpoons}} ES^* \underset{}{\overset{\text{step III}}{\rightleftharpoons}} EP \underset{}{\overset{\text{step IV}}{\rightleftharpoons}} E + P$$

Enzyme + substrate	Enzyme-substrate complex	Transition state	Enzyme product complex	Enzyme + product

In Section 12.5 we examined the events of step I by which the enzyme and substrate interact to form the enzyme-substrate complex. In this section we will look at the events that lead to product formation. We have already described the enzyme as a flexible

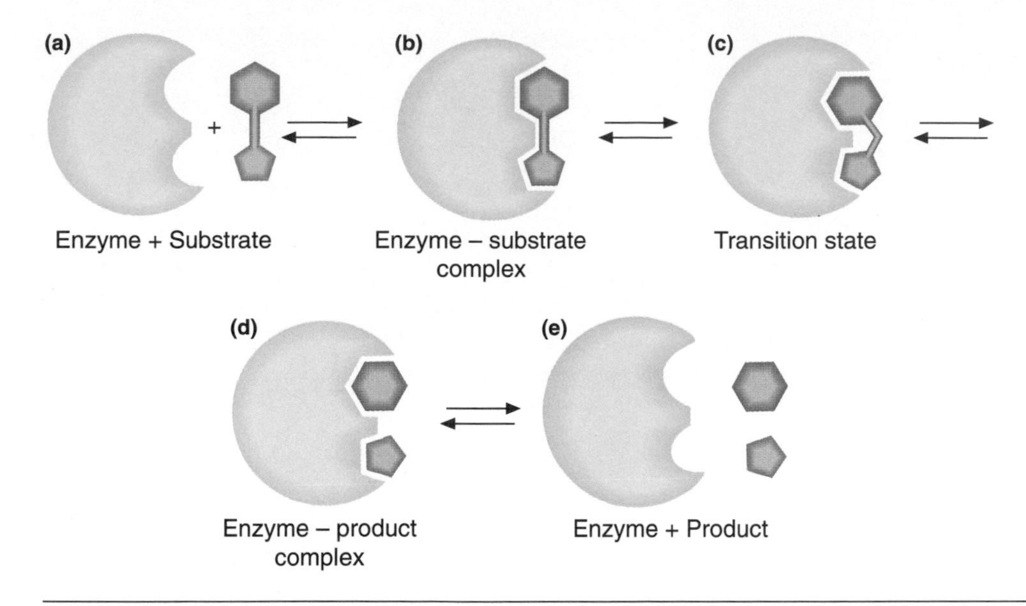

FIGURE 12.6
Bond breakage is facilitated by the enzyme as a result of stress on a bond. (a, b) The enzyme-substrate complex is formed. (c) In the transition state, the enzyme changes shape and thereby puts stress on the O-glycosidic linkage holding the two monosaccharides together. This lowers the energy of activation of this reaction. (d, e) The bond is broken, and the products are released.

molecule; the substrate also has a degree of flexibility. The continued interaction between the enzyme and substrate changes the shape or position of the substrate in such a way that the molecular configuration is no longer energetically stable (step II). In this state, the **transition state,** the shape of the substrate is altered, because of its interaction with the enzyme, into an intermediate form having features of both the substrate and the final product. This transition state, in turn, favors the conversion of the substrate into product (step III). The product remains bound to the enzyme for a very brief time, then in step IV the product and enzyme dissociate from one another, leaving the enzyme completely unchanged.

What kinds of transition state changes might occur in the substrate that would make a reaction proceed more rapidly?

1. The enzyme might put ''pressure'' on a bond and thereby facilitate bond breakage. Consider the hydrolysis of the sugar sucrose by the enzyme sucrase. The enzyme facilitates the hydrolysis of the disaccharide sucrose into the monosaccharides glucose and fructose. The formation of the enzyme-substrate complex (Figures 12.5 and 12.6a) results in a change in the shape of the enzyme. This, in turn, may stretch or put pressure on one of the bonds of the substrate. Such a stress would weaken the bond, allowing it to be broken much more easily than in the absence of the enzyme. This is represented in Figure 12.6b as the bending of the O-glycosidic bond between the fructose and the glucose. In the transition state the substrate has a molecular form resembling both the disaccharide, the original substrate, and the two monosaccharides, the eventual products. Clearly, the stress placed on the bond will weaken it, and much less energy will be required to break the bond to form products (Figures 12.6c and 12.6d). This would also have the effect of speeding up the reaction.
2. An enzyme may facilitate a reaction by bringing two reactants into close proximity and in the proper orientation for reaction to occur. Consider now the dehydration reaction between glucose and fructose to produce sucrose (Figure 12.7a). Each of the sugars has five hydroxyl groups that could undergo condensation to produce a disaccharide. But the purpose is to produce sucrose, not some other disaccharide. By random molecular collision there is a 1 in 25 chance that the two molecules will collide in the proper orientation to produce sucrose. The probability that the two will react is actually much less than that because of a variety of conditions in addition to orientation that must be satisfied for the reaction to occur. For example, at body temperature, most molecular collisions will not have a sufficient amount of energy to overcome the energy of activation, even if the molecules are in the proper orientation.

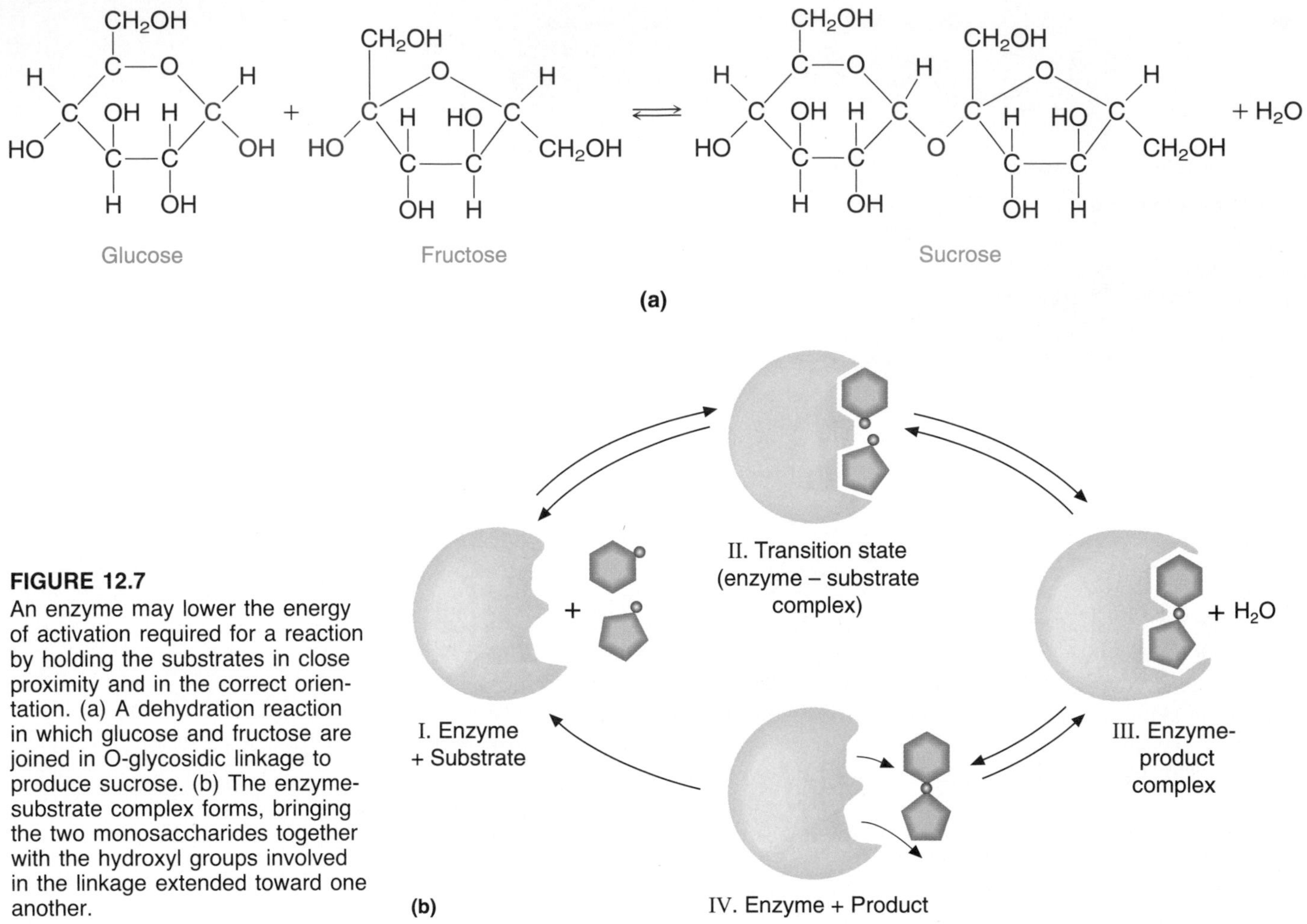

FIGURE 12.7
An enzyme may lower the energy of activation required for a reaction by holding the substrates in close proximity and in the correct orientation. (a) A dehydration reaction in which glucose and fructose are joined in O-glycosidic linkage to produce sucrose. (b) The enzyme-substrate complex forms, bringing the two monosaccharides together with the hydroxyl groups involved in the linkage extended toward one another.

The enzyme can facilitate the reaction by bringing the two molecules close together in the correct alignment (Figure 12.7b), thereby forcing the desired reactive groups of the two molecules together in the transition state and greatly speeding up the reaction.

3. The active site of an enzyme may modify the pH of the microenvironment surrounding the substrate. To accomplish this, the enzyme may, for example, serve as a donor or an acceptor of H^+. As a result, there would be a change in the pH in the vicinity of the substrate without disturbing the normal pH of the cellular milieu.

Question 12.6 Summarize three ways in which an enzyme might lower the energy of activation of a reaction.

12.7 ENZYME-CATALYZED REACTIONS: COFACTORS AND COENZYMES

In Section 11.8 we saw that some proteins, the conjugated proteins, require an additional nonprotein prosthetic group in order to function. The same is true of some enzymes. The protein portion of such an enzyme is called the **apoenzyme,** and the nonprotein prosthetic group is called the **cofactor.** Cofactors are generally metal ions that must be bound to the enzyme to maintain the correct configuration of the enzyme active site (Figure 12.8).

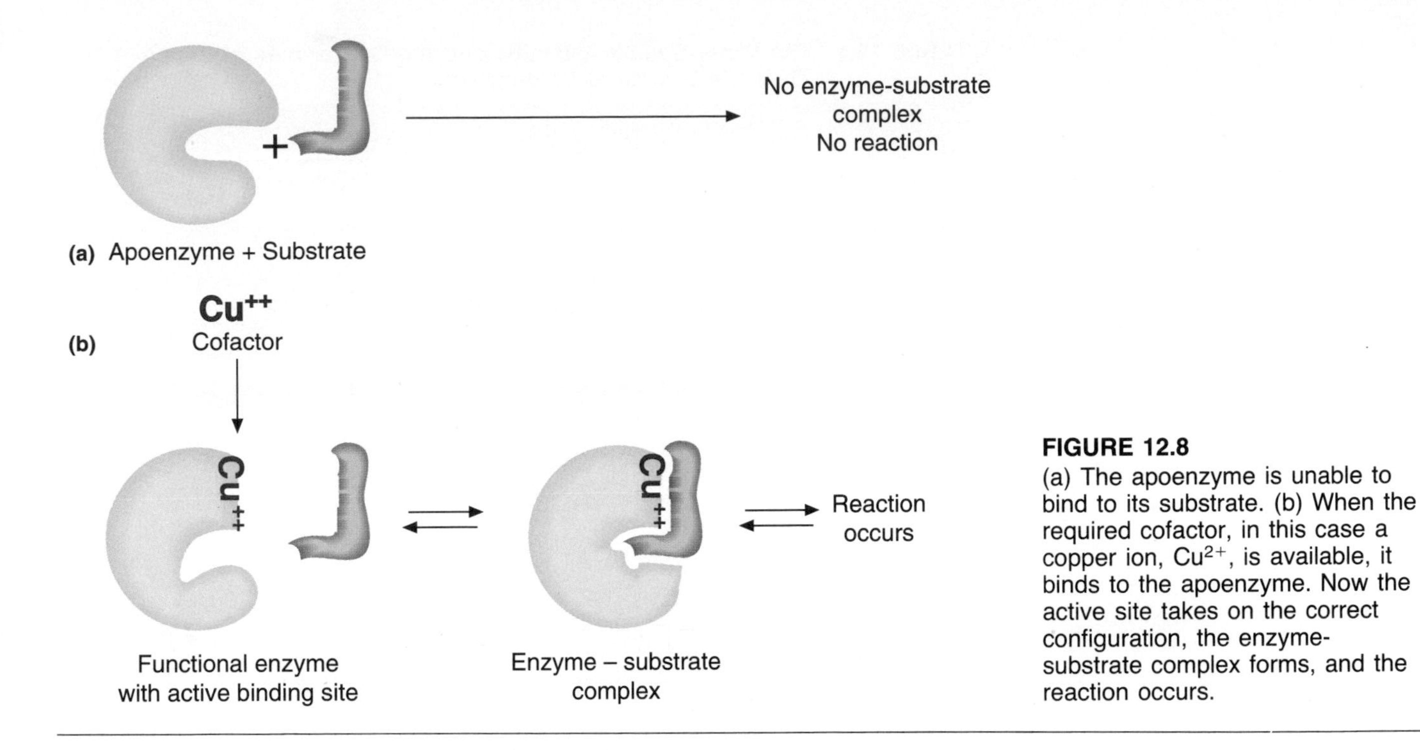

FIGURE 12.8
(a) The apoenzyme is unable to bind to its substrate. (b) When the required cofactor, in this case a copper ion, Cu^{2+}, is available, it binds to the apoenzyme. Now the active site takes on the correct configuration, the enzyme-substrate complex forms, and the reaction occurs.

When the cofactor is bound and the active site is in the proper conformation, the enzyme can bind the substrate and mediate the reaction.

Other enzymes require the transient binding of a **coenzyme.** Such binding is generally mediated by weak interactions like hydrogen bonds. In these instances the coenzymes are organic groups that generally serve as carriers of electrons or chemical groups. In chemical reactions they may either donate chemical groups or serve as recipients of chemical groups that are removed from the substrate. In this capacity they serve as catalytic groups for the enzyme (Figure 12.9).

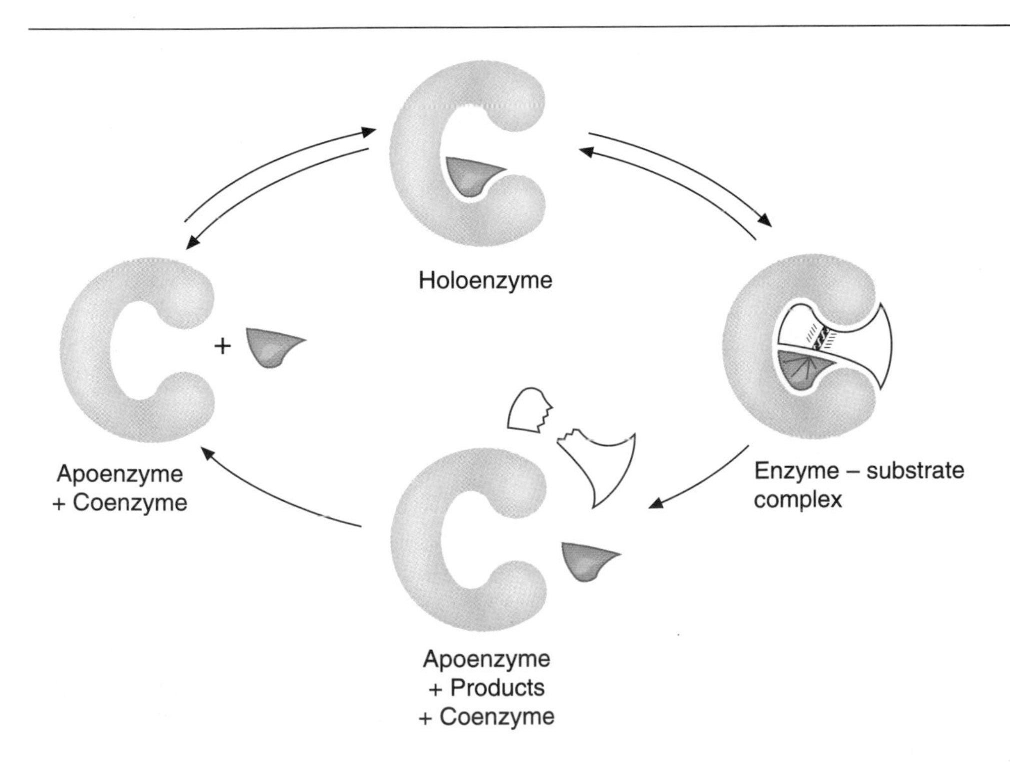

FIGURE 12.9
Some enzymes require a coenzyme to facilitate the reaction. The apoenzyme binds the coenzyme and then the substrate. The coenzyme is a part of the catalytic domain and will either donate or accept functional groups, allowing the reaction to occur. Once the product is formed, both the product and the coenzyme are released.

TABLE 12.1 The Water-Soluble Vitamins and the Coenzymes of Which They Are Structural Components

Vitamin	Coenzyme	Function
Thiamine (B_1)	Thiamine pyrophosphate	Decarboxylation reactions
Riboflavin (B_2)	Flavin mononucleotide (FMN)	Carrier of H atoms
	Flavin adenine dinucleotide (FAD)	
Niacin (B_3)	Nicotinamide adenine dinucleotide (NAD^+)	Carrier of hydride ions
	Nicotinamide adenine dinucleotide phosphate ($NADP^+$)	
Pyridoxine (B_6)	Pyridoxal phosphate	Carriers of amino and carboxyl groups
	Pyridoxamine phosphate	
Cyanocobalamin (B_{12})	Deoxyadenosyl cobalamin	Coenzyme in amino acid metabolism
Folic acid	Tetrahydrofolic acid	Coenzyme for 1-C transfer
Pantothenic acid	Coenzyme A	Acyl group carrier
Biotin	Biocytin	Coenzyme in CO_2 fixation
Ascorbic acid	Unknown	Hydroxylation of proline and lysine in collagen

Section 11.6

Often coenzymes contain modified vitamins as part of their structure. Indeed, of the water-soluble vitamins, only vitamin C, which has a critical role in collagen metabolism, has not been associated with a coenzyme. Table 12.1 is a summary of the water-soluble vitamins and the coenzymes of which they are a part.

Section 14.3

Nicotinamide adenine dinucleotide (NAD^+), seen in Figure 12.10, is an example of a coenzyme that is of critical importance in the oxidation reactions of the cellular energy-generating processes. The NAD^+ molecule has the ability to accept a hydride ion, a hydrogen atom with two electrons, from the substrate of the energy-generating reactions. The substrate is oxidized, and the portion of NAD^+ that is derived from the vitamin niacin is reduced to produce NADH. The NADH subsequently yields the hydride ion to the first acceptor in an electron transport chain. This regenerates the NAD^+ and provides electrons for the generation of ATP, the chemical energy required by the cell. Also shown in Figure 12.10 are the hydride carriers $NADP^+$ and FAD, which are also used in the oxidation-reduction reactions that provide energy for the cell.

Question 12.7 Why does the body require the water-soluble vitamins?

Question 12.8 What are the coenzymes formed from each of the following vitamins?

a. Pantothenic acid
b. Niacin
c. Riboflavin
d. Ascorbic acid

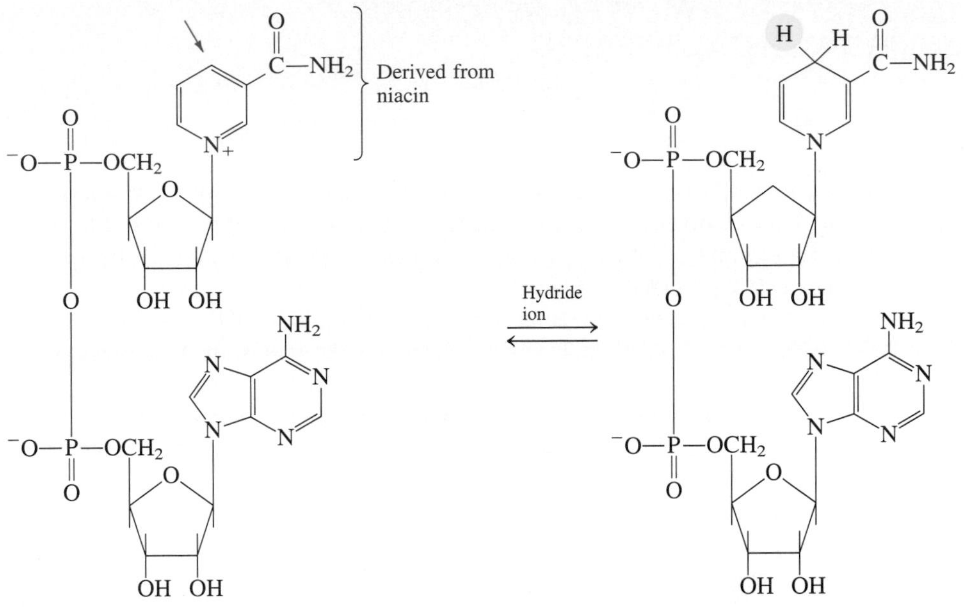

Derived from niacin

Hydride ion

NAD^+
Nicotinamide adenine dinucleotide
(oxidized form)

NADH
(reduced form)

(a)

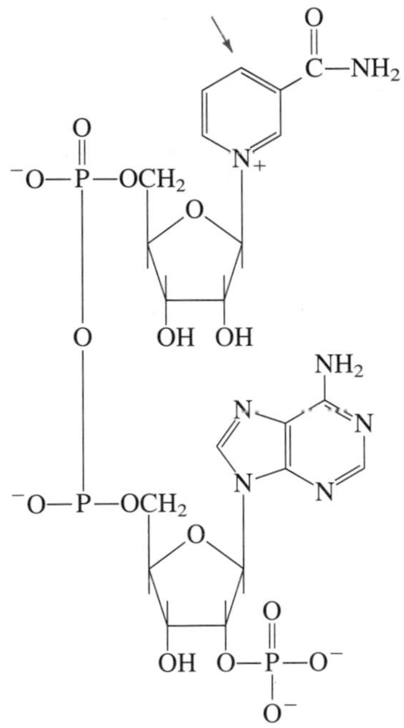

$NADP^+$
Nicotinamide adenine
dinucleotide phosphate
(oxidized form)

(b)

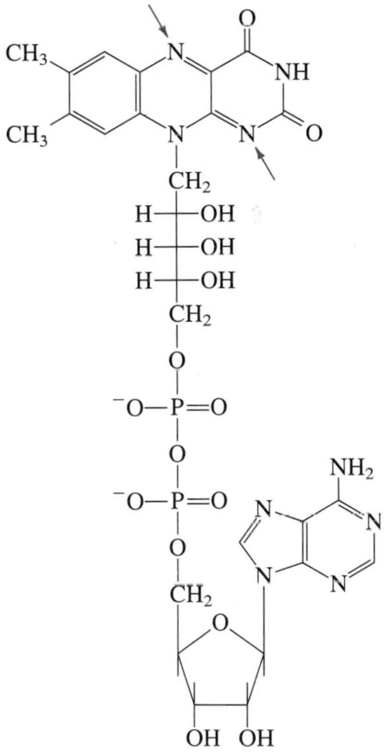

FAD (flavin adenine dinucleotide)
(oxidized form)

(c)

FIGURE 12.10
The structure of three coenzymes. (a) The oxidized and reduced forms of nicotinamide adenine dinucleotide. (b) The oxidized form of closely related hydride ion carrier, nicotinamide adenine dinucleotide phosphate ($NADP^+$) accepts hydride ions at the same position as NAD^+ (colored arrow). (c) The oxidized form of flavin adenine dinucleotide (FAD) accepts hydrogen atoms at the positions indicated by the colored arrows.

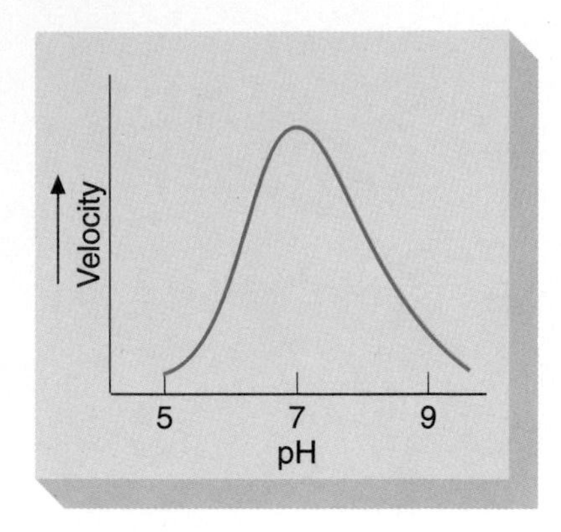

FIGURE 12.11
Effect of pH on the rate of an enzyme-catalyzed reaction. The enzyme functions most efficiently at pH 7. The rate of the reaction falls rapidly as the solution is made either more acidic or more basic.

12.8 ENZYME-CATALYZED REACTIONS: ENVIRONMENTAL EFFECTS

Effect of pH

Most enzymes are active only within a very narrow pH range. The cellular cytoplasm has a pH of 7, and most cytoplasmic enzymes function at a maximum efficiency at this pH. A plot of the relative rate at which a typical cytoplasmic enzyme catalyzes its specific reaction versus pH is provided in Figure 12.11.

The pH at which an enzyme functions optimally is called the **pH optimum.** Making the solution more basic or more acidic sharply decreases the rate of the reaction. As was discussed in Section 11.13, at extremes of pH the enzyme actually loses its biologically active conformation and is *denatured.* This is because pH changes will alter the degree of ionization of the R groups of the amino acids within the protein chain, as well as the extent to which they can hydrogen bond. Just as these interactions can drastically alter the overall configuration of a protein, changes in the R groups of the binding or catalytic domains of an enzyme active site will destroy the ability to form the enzyme-substrate complex.

Although the cytoplasm of the cell and the fluids that bathe the cells have a pH that is carefully controlled so that it remains at about pH 7, there are environments within the body in which enzymes must function at a pH far from 7. Protein sequences have evolved that can maintain the proper three-dimensional structure under extreme conditions of pH. For instance, the pH of the stomach is approximately 2 as a result of the secretion of hydrochloric acid by specialized cells of the stomach lining. The proteolytic digestive enzyme **pepsin** must effectively degrade proteins at this extreme pH. In the case of pepsin the enzyme has evolved in such a way that it can maintain a stable tertiary structure at a pH of 2 and is catalytically most active in the hydrolysis of peptides that have been denatured by very low pH. Thus pepsin has a pH optimum of 2.

In a similar fashion, another proteolytic enzyme, *trypsin,* functions under the conditions of higher pH found in the intestines. Both pepsin and trypsin cleave peptide bonds by virtually the identical mechanisms, yet their amino acid sequences have evolved so that they are stable and active in very different environments.

The body has used adaptation of enzymes to different environments to protect itself against one of its own destructive defense mechanisms. Within the cytoplasm of a cell are organelles called **lysosomes.** Christian de Duve, who discovered lysosomes in 1956, called them ''suicide bags'' because they are membrane-bound vesicles containing about 50 different kinds of hydrolases or hydrolytic enzymes. The purpose of these enzymes is to degrade large molecules into small molecules that are useful for energy-generation processes. For instance, some of the enzymes in the lysosomes can degrade proteins to amino acids, while others hydrolyze polysaccharides into monosaccharides. Certain cells of the immune defense system engulf foreign invaders, such as bacteria and viruses. They then use the hydrolytic enzymes in the lysosomes to degrade and destroy the invaders and use the simple sugars, amino acids, and lipids that are produced as energy sources.

What would happen if the hydrolytic enzymes of the lysosome were accidentally released into the cytoplasm of the cell? Certainly, the result would be the destruction of cellular macromolecules and death of the cell. Because of this danger, the cell invests a great deal of energy in maintaining the integrity of the lysosomal membranes. An additional protective mechanism relies on the fact that lysosomal enzymes function optimally at an acid pH (pH 4.8). Should some of these enzymes leak out of the lysosome or should a lysosome accidentally rupture, the cytoplasmic pH of 7.0–7.3 renders them inactive, and there is no degradation of the cellular constituents.

Question 12.9 How does a decrease in pH alter the activity of an enzyme?

Effect of temperature

Enzymes are rapidly destroyed if the temperature of the solution rises much above 37°C, but they remain stable at much lower temperatures. It is for this reason that solutions of enzymes used for clinical assays are stored in refrigerators or freezers prior to use. Figure 12.12 shows the effects of temperature on enzyme-catalyzed and uncatalyzed reactions. The rate of the uncatalyzed reaction steadily increases with increasing temperature because more collisions occur with sufficient energy to surmount the energy barrier for the reaction. The rate of an enzyme-catalyzed reaction also increases with modest increases in temperature because there are increasing numbers of collisions between the enzyme and the substrate. But before long, increasing temperature begins to increase the vibrational energy of the bonds within the enzyme. Eventually, so many bonds and weak interactions are disrupted that the enzyme becomes denatured, and the reaction stops.

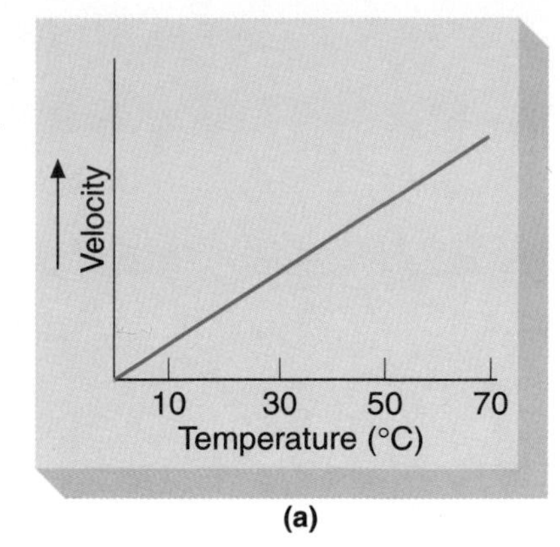

(a)

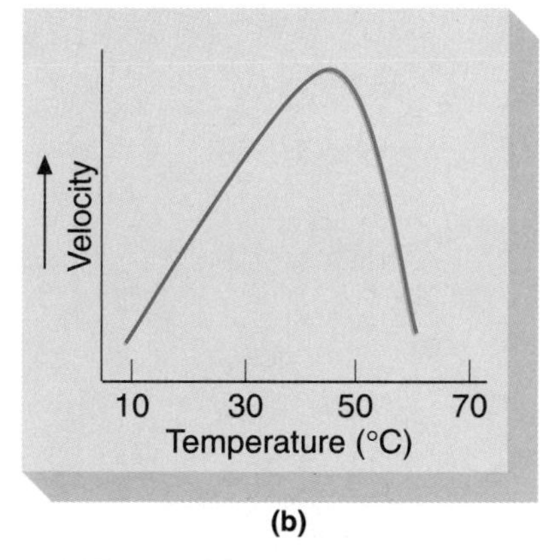

(b)

FIGURE 12.12
Effect of temperature on (a) uncatalyzed reactions and (b) enzyme-catalyzed reactions.

Since heating enzymes and other proteins destroys their characteristic three-dimensional structure, and hence their activity, a cell cannot survive very high temperatures. Since cells cannot function without proper enzyme function, heat is an effective means of sterilizing medical instruments and solutions for transfusion or clinical tests. Although instruments can be sterilized by dry heat (160°C) applied for at least 2 hours in a dry air oven, autoclaving is a quicker, more reliable procedure. The autoclave works on the principle of the pressure cooker. Air is pumped out of the chamber, and steam under pressure is pumped into the chamber until a pressure of 15 lb/in^2 above atmospheric pressure is achieved. The pressure causes the temperature of the steam, which would be 100°C at atmospheric pressure, to rise to 121°C. Within 20 minutes, all the bacteria and viruses are killed. This is the most effective means of destroying the very heat-resistant endospores that are formed by many bacteria of clinical interest. These bacteria include the genera *Bacillus* and *Clostridium,* which are responsible for such unpleasant and deadly diseases as anthrax, gas gangrene, tetanus, and botulism food poisoning.

However, not all enzymes are inactivated by heating, even to rather high temperatures. Certain bacteria live in such out-of-the-way places as coal slag heaps, which are actually burning. Others live in deep vents on the ocean floor where temperatures and pressures are extremely high. Still others grow in the hot springs of Yellowstone National Park, where some bacteria thrive at temperatures near the boiling point of water. These organisms, along with their enzymes, survive under such incredible conditions because the amino acid sequences of their proteins dictate structures that are stable at such seemingly impossible temperature extremes.

Question 12.10 Heating is an effective mechanism for killing bacteria on surgical instruments. How does elevated temperature result in cellular death?

12.9 REGULATION OF ENZYME ACTIVITY

One of the major ways in which enzymes differ from nonbiological catalysts is that the activity of the enzyme is often regulated by the cell. There are many reasons for this control of enzyme function. Some involve energy considerations. If the cell runs out of chemical energy, it will die; therefore many mechanisms exist to conserve cellular energy. For instance, it is a great waste of energy to produce an enzyme if the substrate is not available. Similarly, if the product of an enzyme-mediated reaction is present in excess, it is a waste of energy for the enzyme to continue to facilitate the reaction, thereby producing more of the unwanted product.

Just as there are many reasons for regulation of enzyme activity, there are many mechanisms for such regulation. The simplest mechanism is to produce the enzyme only when the substrate is present. This mechanism is used by the bacteria to regulate the enzymes needed to break down various sugars to yield energy for cellular work. The bacteria have no control over their environment or over what food sources, if any, might be available. It would be an enormous waste of energy to produce all of the enzymes that are needed to break down all the possible sugars. Thus the bacteria save energy by producing the enzymes only when a specific sugar substrate is available. Other mechanisms for regulating enzyme activity, discussed below, include end product inhibition, use of allosteric enzymes, feedback inhibition, and production of zymogens.

End product inhibition

Another simple mode of regulation is **end product inhibition.** Most enzymatic reactions, like any other chemical reaction, can proceed in the forward or the reverse direction. As the enzyme-substrate complex forms and the reaction occurs, there is a buildup of the product. If the product is not removed from the site of the reaction or used as the substrate for another reaction, the product will continue to accumulate. As a consequence of LeChatelier's Principle, the concentration of product can become great enough to cause the reaction to occur in reverse, from product to substrate. Thus with reversible reactions the product can inhibit its own synthesis by direct interaction with the enzyme that produces it.

Allosteric enzymes

A more complex level of enzyme regulation involves enzymes that have more than a single binding site. These enzymes, called **allosteric enzymes,** meaning ''other forms,'' are composed of more than a single peptide. In other word, they have quaternary structure. One peptide subunit has the ability to bind a substrate in the active site and catalyze a biochemical reaction. The second subunit has a pocket for the binding of an *effector molecule*. As shown in Figure 12.13, the effector binding to the second subunit alters the shape of the active site of the enzyme. The result can be to convert the active site to an inactive configuration, **negative allosterism,** or to convert the active site to the active

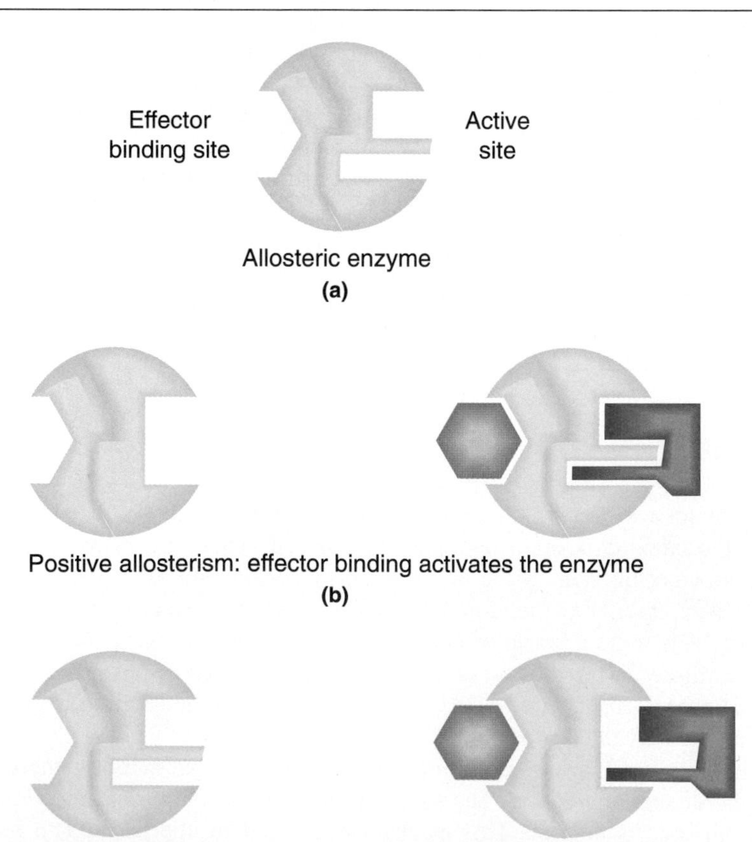

FIGURE 12.13
(a) Allosteric enzymes regulate a great many biochemical pathways. The allosteric enzyme has an active site and an effector binding site. (b) Positive allosterism. (c) Negative allosterism.

configuration, **positive allosterism.** In either case, binding of the effector molecule regulates enzyme activity by determining whether it will be active or inactive.

Feedback inhibition

Allosteric enzymes are the basis of yet another type of enzyme regulation, **feedback inhibition.** This system functions on the same principle as the thermostat on your furnace. You set the thermostat at 70°F; the furnace turns on and produces heat until the sensor in the thermostat registers a room temperature of 70°F. It then signals the furnace to shut off.

Feedback inhibition usually regulates pathways of enzymes involved in the synthesis of a biological molecule. Such a pathway can be shown schematically as follows:

$$A \xrightarrow{E_1} B \xrightarrow{E_2} C \xrightarrow{E_3} D \xrightarrow{E_4} E \xrightarrow{E_5} F$$

In this pathway the starting material, A, is converted to B by the enzyme E_1. Enzyme E_2 immediately converts B to C, and so on until the final product, F, has been synthesized. If F is no longer needed, it is a waste of cellular energy to continue to produce it.

To avoid this waste of energy, the cell uses feedback inhibition, in which the product can shut off the entire pathway for its own synthesis. This is the result of the fact that the product, F, acts as a negative allosteric effector on one of the early enzymes of the pathway. For instance, enzyme E_1 may have an effector-binding site for F in addition to the active site that binds to A. When F is present in excess, it binds to the effector-binding site. This binding causes the active site to close so that it cannot bind to substrate A. Thus A is not converted to B. If no B is produced, there is no substrate for enzyme E_2, and the entire pathway ceases to operate. The product, F, has turned off all the steps involved in its own synthesis, just as the heat produced by the furnace is ultimately responsible for turning off the furnace itself.

Zymogens

The final means of regulating enzyme activity involves the production of the enzyme in an inactive form called a **zymogen.** It is then converted, usually by proteolysis (hydrolysis of the protein), to the active form when it has reached the site of its activity. What is the purpose of this type of mechanism? On first examination it seems wasteful to add a step to the synthesis of an enzyme. But consider for a moment the very destructive nature of some of the enzymes that are necessary for life. The enzymes pepsin, trypsin, and chymotrypsin are all proteolytic enzymes of the digestive tract. They are necessary to life because they degrade dietary proteins into amino acids that are used by the cell. But what would happen to the cells that produce these enzymes if they were synthesized in active form? Those cells would be destroyed. Thus the cells of the stomach that produce pepsin actually produce an inactive zymogen, called **pepsinogen.** Pepsinogen has an additional 42 amino acids. In the presence of stomach acid and previously activated pepsin, the extra 42 amino acids are cleaved off, and the zymogen is transformed into the active enzyme. Table 12.2 lists several other zymogens and the enzymes that convert them to active form.

TABLE 12.2 Zymogens of the Digestive Tract

Zymogen	Activator	Enzyme
Proelastase	Trypsin	Elastase
Trypsinogen	Trypsin	Trypsin
Chymotrypsinogen A	Trypsin + chymotrypsin	Chymotrypsin
Pepsinogen	Acid pH + pepsin	Pepsin
Procarboxypeptidases	Trypsin	Carboxypeptidase A, carboxypeptidase B

12.10 INHIBITION OF ENZYME ACTIVITY

Many chemicals can bind to enzymes and either eliminate or drastically reduce their catalytic ability. These chemicals, called *enzyme inhibitors,* have been used for hundreds of years. When she poisoned her victims with arsenic, Lucretia Borgia was unaware that it was binding to the sulfhydryl groups of cysteine residues of the proteins of her victims and thus interfering with the formation of disulfide bonds needed to stabilize the tertiary structure of enzymes. However, she was well aware of the deadly toxicity of heavy metal salts like arsenic and mercury. When you take penicillin for a bacterial infection, you are taking another enzyme inhibitor. Penicillin inhibits several enzymes that are involved in the synthesis of bacterial cell walls.

The effect of penicillin on bacterial cell wall biosynthesis is discussed in Chapter 7: "A Clinical Perspective: The Bacterial Cell Wall."

Enzyme inhibitors are classified on the basis of whether the inhibition is reversible or irreversible, competitive or noncompetitive. Reversibility deals with whether or not the inhibitor will eventually dissociate from the enzyme, releasing it in the active form. Competition refers to whether the inhibitor is a structural analogue, or look-alike, of the natural substrate. If the latter is the case, the inhibitor and substrate will compete for the enzyme-active site.

Irreversible inhibitors

Irreversible inhibitors, such as arsenic, usually bind very tightly, sometimes even covalently, to the enzyme. This generally involves binding of the inhibitor to one of the R groups of an amino acid in the active site. Inhibitor binding may block the active site binding groups so that the enzyme-substrate complex cannot form. Alternatively, an inhibitor may interfere with the catalytic groups of the active site, thereby effectively eliminating catalysis. Irreversible inhibitors generally inhibit many different enzymes.

Question 12.11 Why are irreversible inhibitors considered to be poisons?

Reversible, noncompetitive inhibitors

Reversible, noncompetitive inhibitors bind to R groups of amino acids or perhaps to the metal ion cofactors. Unlike the situation of irreversible inhibition, however, the binding is weak, and the enzyme activity is restored when the inhibitor dissociates from the enzyme-inhibitor complex. Although these inhibitors generally do not bind to the active site, they do modify the shape of the active site by binding elsewhere in the protein structure. Keep in mind that the entire three-dimensional structure of an enzyme is needed to maintain the correct shape of the active site. Ionic bonding or weak bonding of the inhibitor to one or more sites on the enzyme surface can alter the shape of the active site in a fashion analogous to that of an allosteric effector. These inhibitors also inactivate a broad range of enzymes.

Reversible, competitive inhibitors

Reversible, competitive inhibitors are often referred to as **structural analogues,** that is, they are molecules that resemble the structure and charge distribution of the natural substrate for a particular enzyme. Because of this resemblance, the inhibitor can occupy the enzyme-active site. However, no reaction can occur, and enzyme activity is inhibited. This inhibition is said to be competitive because the enzyme-inhibitor complex is maintained by weak interactions and readily dissociates. The empty active site is now available for substrate binding. Thus the active site can alternatively bind the inhibitor or the normal substrate. The two will compete for this binding, and so the degree of inhibition depends on their relative concentrations. If the inhibitor is in excess, it will occupy the active site

See Chapter 16, "A Human Perspective: Fooling the AIDS Virus with Look-Alike Nucleotides"

more frequently, and enzyme activity will be greatly decreased. On the other hand, if the natural substrate is present in excess, it will more frequently occupy the active site, and there will be little inhibition. The sulfa drugs, the first antibiotics to be discovered, are competitive inhibitors of a bacterial enzyme required for synthesis of a required vitamin.

Section 17.3 describes the mechanism of action of the sulfa drugs in greater detail.

Question 12.12 What is a structural analogue?

Question 12.13 How can structural analogues serve as enzyme inhibitors?

12.11 PROTEOLYTIC ENZYMES: CHYMOTRYPSIN

Proteolytic enzymes are protein-cleaving enzymes, that is they break the peptide bonds that maintain the primary protein structure. **Chymotrypsin,** for example, is an enzyme that hydrolyzes dietary proteins in the small intestine. It acts specifically at peptide bonds on the carbonyl side of the peptide bond. The C-terminal residues (amino acids and/or peptides) of the fragments produced by bond cleavage are tyrosine, tryptophan, and phenylalanine. The specificity of chymotrypsin depends upon the presence of a *hydrophobic pocket,* a cluster of hydrophobic amino acids brought together by the three-dimensional folding of the protein chain. The flat aromatic side chains of certain amino acid residues (tyrosine, tryptophan, phenylalanine) slide into this pocket, providing the binding specificity required for catalysis (Figure 12.14).

How can we determine which bond is cleaved by a protease such as chymotrypsin? To know which bond is cleaved, we must write out the sequence of amino acids in the region of the peptide that is being cleaved. This can be determined by amino acid sequencing techniques. Remember that the N-terminal residue is written to the left and the C-terminal residue to the right. Consider a protein having within it the sequence —Ala-Phe-Gly—. A reaction is set up in which the enzyme, chymotrypsin, is mixed with the protein substrate. After the reaction has occurred, the proteolytically cleaved products are purified, and their amino acid sequences are determined. Experiments of this sort show that chymotrypsin cleaves the bond between phenylalanine and glycine, which is the peptide bond on the side having an aromatic side chain.

$$H_3{}^{+}N\text{—}\underset{CH_3}{\overset{H}{C}}\text{—}\overset{O}{\overset{\|}{C}}\text{—NH—}\underset{CH_2C_6H_5}{\overset{H}{C}}\text{—}\overset{O}{\overset{\|}{C}}\text{—NH—}\underset{H}{\overset{H}{C}}\text{—}\overset{O}{\overset{\|}{C}}\text{—}O^{-}$$

Peptide bond cleaved (between the Phe carbonyl C and the Gly NH)

Ala ——— Phe ——— Gly

Question 12.14 Draw out the complete structural formulas of the following peptides.

a. ala-phe-ala
b. tyr-ala-tyr
c. trp-val-gly
d. phe-ala-pro

Question 12.15 For the peptides drawn in Question 12.14, show which bond would be cleaved on reaction with chymotrypsin.

FIGURE 12.14
The specificity of chymotrypsin is determined by a hydrophobic pocket that holds the aromatic side chain of the substrate. This brings the peptide bond to be cleaved into the catalytic domain of the active site.

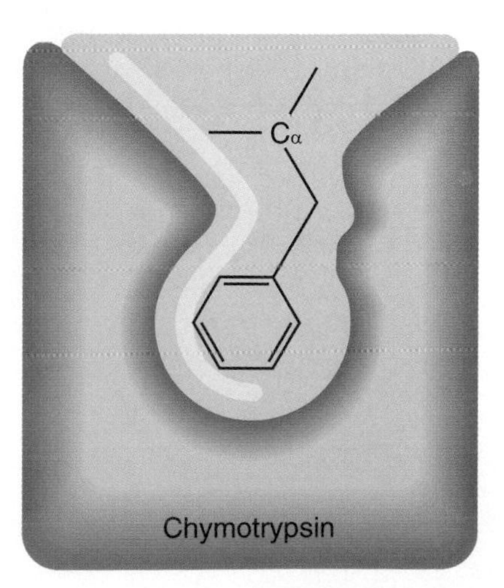

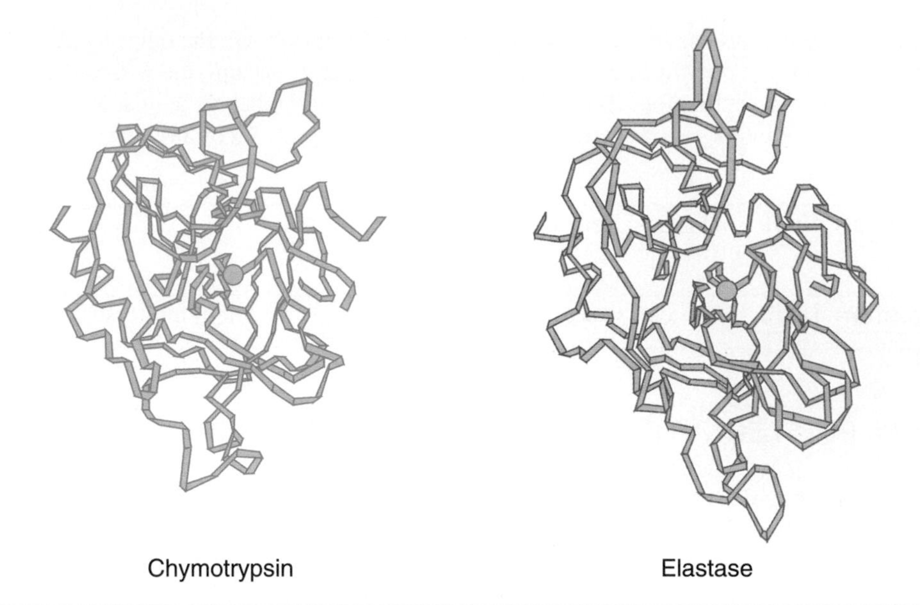

FIGURE 12.15
Structures of chymotrypsin and elastase are virtually identical, indicating that these enzymes have evolved from a common ancestral protease.

12.12 PANCREATIC SERINE PROTEASES

These enzymes are called serine proteases because they have a serine residue in the catalytic region of the active site that is essential for hydrolysis of the peptide bond.

The **pancreatic serine proteases** trypsin, chymotrypsin, and elastase all hydrolyze peptide bonds. These enzymes are the result of **divergent evolution** in which a single ancestral gene first duplicated and then evolved along separate lines. They have similar primary structures, similar tertiary structures (Figure 12.15), and virtually identical mechanisms of action. However, as a result of evolution, these enzymes all have different specificities:

- Chymotrypsin cleaves peptide bonds on the carboxyl side of aromatic residues.
- Trypsin cleaves peptide bonds on the carboxyl side of basic amino acid residues.
- Elastase cleaves peptide bonds on the carboxyl side of glycine and alanine residues.

These enzymes have different pockets for the side chains of their substrates; *different keys fit different locks*. This difference manifests itself in the substrate specificity alluded to above. Yet while the binding pocket has undergone divergent evolution, the catalytic site has remained unchanged, and the mechanism of proteolytic action is the same for all the serine proteases.

Question 12.16 Draw the structural formula of the peptide val-phe-ala-gly-leu. Which bond would be cleaved if this peptide were reacted with chymotrypsin? With elastase?

12.13 PROTEOLYTIC ENZYMES AND BLOOD COAGULATION

We have seen that many enzymes have very powerful physiological effects and thus are often synthesized in an inactive form to prevent them from acting where they might harm the cell. The enzymes that are responsible for blood coagulation are an important example of this phenomenon. The formation of a *(fibrin) blood clot* requires a sequential series or cascade of proteolytic reactions. The process begins with the formation of a prothrombin activator by either the *intrinsic pathway* or the *extrinsic pathway*. The **intrinsic pathway** is activated by damage within (intrinsic) the blood vessel—for instance, to the endothe-

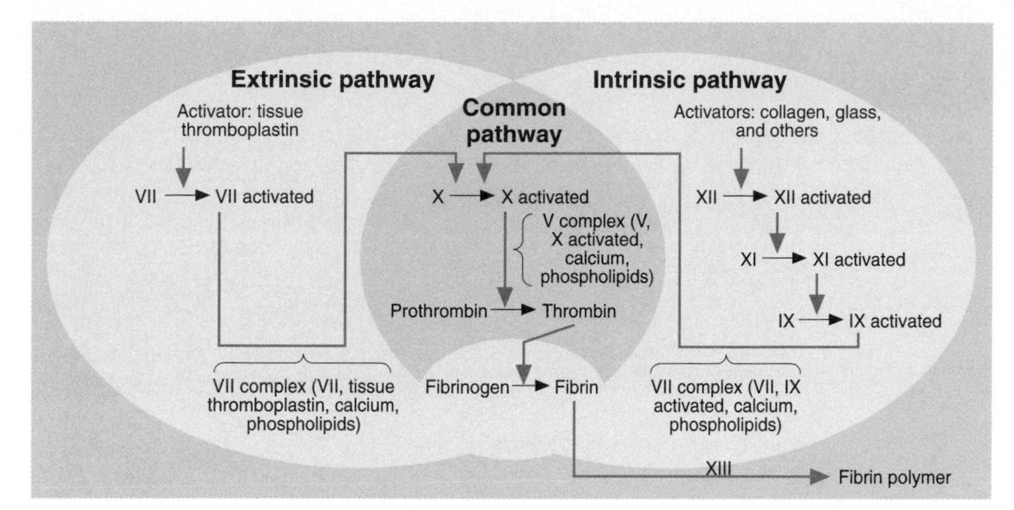

FIGURE 12.16
Formation of a blood clot may be initiated by either the intrinsic pathway or the extrinsic pathway. Both induce the common pathway through the activation of the prothrombin activator complex. Clotting results from the action of thrombin (an enzyme) on fibrinogen (a soluble blood protein) to produce fibrin (an insoluble blood protein—the clot).

lium—while the **extrinsic pathway** is activated by tissue damage outside (extrinsic) the blood vessel. The details of the two pathways are quite different, but each involves a series of proteolytic reactions that results in activation of factor X and formation of the prothrombin activator complex. This complex initiates the final **common pathway** (Figure 12.16).

The final step leading to clot formation is the enzymatic conversion of the soluble fibrous protein **fibrinogen** to the insoluble protein **fibrin.** This is accomplished by the serine protease **thrombin.** Thrombin, produced from prothrombin by the enzymatic action of the prothrombin activator complex, cleaves four peptide bonds between arginine-glycine in fibrinogen to produce fibrin. The structure and function of thrombin closely resemble those of trypsin. Thrombin, then, is a protease. It is, however, much more specific than trypsin, since it ignores all peptide bonds except the four in fibrinogen that it hydrolyzes. This reflects the precise fit of thrombin to its substrate, fibrinogen. No other molecule in the cell is able to bind thrombin in a way that permits catalysis to occur.

Many of the proteins in the blood clotting cascade are referred to as "factors." They are distinguished from one another by a roman numeral designation.

When fibrinogen has been converted to fibrin, the clot that forms is a meshwork of polymerized fibrin threads that become attached to blood cells, blood vessel walls, and plasma proteins. This meshwork is unstable. The blood factor, factor XIII, is a transglutaminase, an enzyme that stabilizes the fibrin clot by forming covalent cross-linkages between fibrin threads in the clot.

Hemophilia is a genetic deficiency of one of the protein factors involved in blood clotting. About 85% of all hemophilia is hemophilia A, which results from the production of an abnormal factor VIII that cannot promote clotting. This is a sex-linked, recessive trait that affects males primarily. The condition is characterized by spontaneous intramuscular or subcutaneous pain; bleeding following even minor wounds; hematuria (blood in the urine); bleeding from the gums, lips, and mouth; and joint hemorrhages that are extremely painful and disfiguring. Bleeding can be stopped temporarily by transfusion of purified factor VIII. A deficiency of factor IX is responsible for hemophilia B. This is also a sex-linked disorder and accounts for about 10% of all hemophilia. Hemophilia C is a mild form of the disease and is caused by a deficiency of factor XI. This disorder accounts for about 2% of all hemophiliacs.

A sex-linked trait is located on the X-chromosome. Males have only one X-chromosome. Thus, if they have the defective gene, it will be expressed. Females having the defect on only one X-chromosome will have normal blood clotting because the defect will be "masked" by the normal copy of the gene on the other X-chromosome.

Question 12.17 Describe several types of hemophilia.

Question 12.18 Differentiate between the intrinsic and extrinsic pathways for blood clotting.

FIGURE 12.17
Ten glutamate residues near the amino terminus of prothrombin are converted to γ-carboxyglutamates. This reaction requires vitamin K as a coenzyme.

12.14 VITAMIN K AND ITS ROLE IN BLOOD CLOTTING

Prothrombin, the inactive precursor of thrombin, and three other clotting factors are synthesized in the liver. The synthesis of these four proteins requires **vitamin K.**

$$\text{(naphthoquinone ring with } CH_3 \text{ and } (CH_2{-}CH{=}\overset{CH_3}{C}{-}CH_2)_6{-}H)$$

Vitamin K

Prothrombin contains a sequence near its amino terminus in which ten glutamate residues have been converted to the unusual amino acid γ-carboxyglutamate (Figure 12.17).

$$H_3\overset{+}{N}{-}\underset{\substack{|\\CH_2\\|\\CH\\ \swarrow\ \searrow \\ {}^{-}OOC \quad COO^{-}}}{\overset{\substack{COO^{-}\\|}}{C}}{-}H$$

γ-Carboxyglutamate

Vitamin K is required as a coenzyme by the enzyme that catalyzes the γ-carboxylation of these glutamate residues of prothrombin and of glutamate residues on factors VII, IX, and X. The γ-carboxyglutamate residues bind Ca^{2+} ions as shown in Figure 12.18.

The complex of Ca^{2+} and prothrombin is needed for the cleavage of prothrombin by the prothrombin activator complex. The N-terminal region of prothrombin is released in this step, producing thrombin, which then cleaves fibrinogen and causes clotting (Figure 12.16).

Vitamin K is a fat-soluble vitamin that is produced by our intestinal bacteria. A significant dietary source of vitamin K is provided by leafy green vegetables and liver. A deficiency of vitamin K leads to clotting disorders and hemorrhage. Deficiency can occur as a result of obstructive liver disease in which the obstruction blocks the flow of bile salts that are required for the uptake of all fat-soluble vitamins. Newborns are generally deficient in vitamin K because of the lack of intestinal bacteria and the immature state of the liver. For this reason, newborns are frequently given injections of vitamin K to prevent any serious bleeding that might occur, for instance, during circumcision.

Question 12.19 What are the major sources of vitamin K?

The **coumarin** anticoagulants, sometimes called ''blood thinners,'' are often prescribed to patients following a heart attack or stroke that results from a clot. These anticoagulants reduce the amount of clotting by acting as competitive inhibitors of vitamin K. In this way they reduce the amount of prothrombin and factors VII, IX, and X synthesized by the liver.

Question 12.20 Coumarin acts as a competitive inhibitor of vitamin K. What effect does this have physiologically?

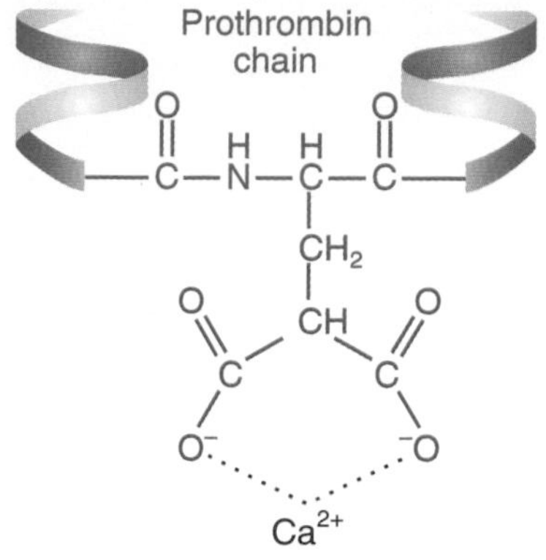

Prothrombin chain

Carboxyglutamate complexed with Ca^{2+}

FIGURE 12.18
Structure of the complex between γ-carboxyglutamate and Ca^{2+} ions. The ability to bind Ca^{2+} is required for the proteolytic cleavages involved in the clotting mechanism.

12.15 ACETYLCHOLINESTERASE AND NERVE TRANSMISSION

The transmission of nerve impulses at the *neuromuscular junction* involves many steps, one of which is hydrolysis of **acetylcholine** by the enzyme **acetylcholinesterase** (Figure 12.19). The need for this enzyme activity becomes clear when we consider the events that begin with a message from the nerve cell and end in the appropriate response by the muscle cell. Acetylcholine is a chemical messenger that transmits a message from the nerve cell to the muscle cell. Such a molecule is known as a *neurotransmitter*. Acetylcholine is stored in membrane-bound bags, called synaptic vesicles, in the nerve cell ending (see Figure 12.20).

Acetylcholinesterase comes into play in the following way. The arrival of a nerve impulse at the end plate of the *nerve axon* results in an influx of Ca^{2+}. This causes the acetylcholine-containing vesicles to migrate to the nerve cell membrane that is in contact with the muscle cell. This is called the presynaptic membrane. The vesicles fuse with the presynaptic membrane and release the neurotransmitter. The acetylcholine then diffuses across the *nerve synapse* (the space between the nerve and muscle cells) and binds to the acetylcholine receptor protein in the *postsynaptic membrane* of the muscle cell. This receptor then opens pores in the membrane through which Na^+ and K^+ ions flow into and out of the cell, respectively. This generates the nerve impulse and causes the muscle to contract (Figure 12.20). If acetylcholine remains at the neuromuscular junction, it will continue to stimulate the muscle contraction. To stop this continued stimulation, acetylcholine is hydrolyzed, and hence destroyed, by acetylcholinesterase.

Inhibitors of acetylcholinesterase are used both as poisons and as drugs. Among the most important poisons of acetylcholinesterase are a class of compounds known as *organic fluorophosphates*. One of these is diisopropyl fluorophosphate (DIFP). This molecule forms a covalently bonded intermediate with the active site of acetylcholinesterase. Thus it acts as an irreversible, noncompetitive inhibitor.

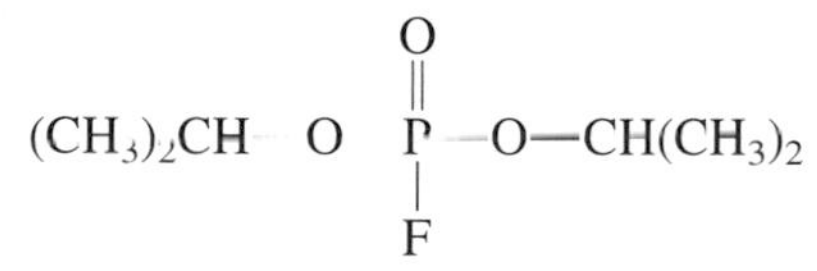

$$(CH_3)_2CH—O—\overset{\overset{O}{\|}}{\underset{\underset{F}{|}}{P}}—O—CH(CH_3)_2$$

Diisopropyl fluorophosphate (DIFP)

$$H_3C—\overset{\overset{O}{\|}}{C}—O—CH_2—CH_2—\overset{+}{N}—(CH_3)_3 + H_2O$$

Acetylcholine

$$\Big\Updownarrow \text{ Acetylcholinesterase}$$

$$H_3C—C\langle^{O}_{O^-} + HO—CH_2—CH_2—\overset{+}{N}—(CH_3)_3 + H^+$$

Acetate Choline

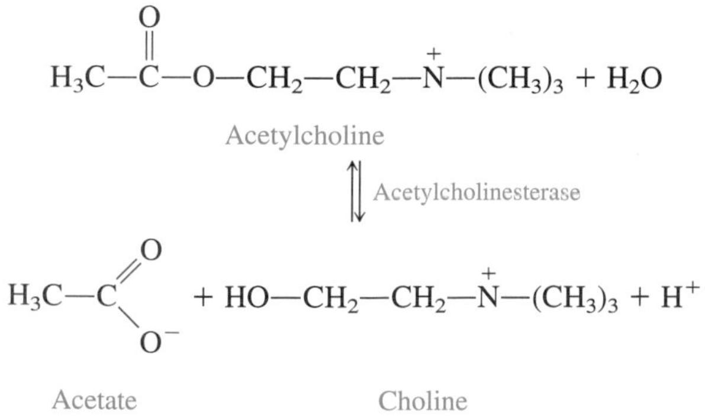

FIGURE 12.19
Acetylcholinesterase hydrolyzes acetylcholine to produce choline. The choline is not able to bind the acetylcholine receptor and continue the stimulation.

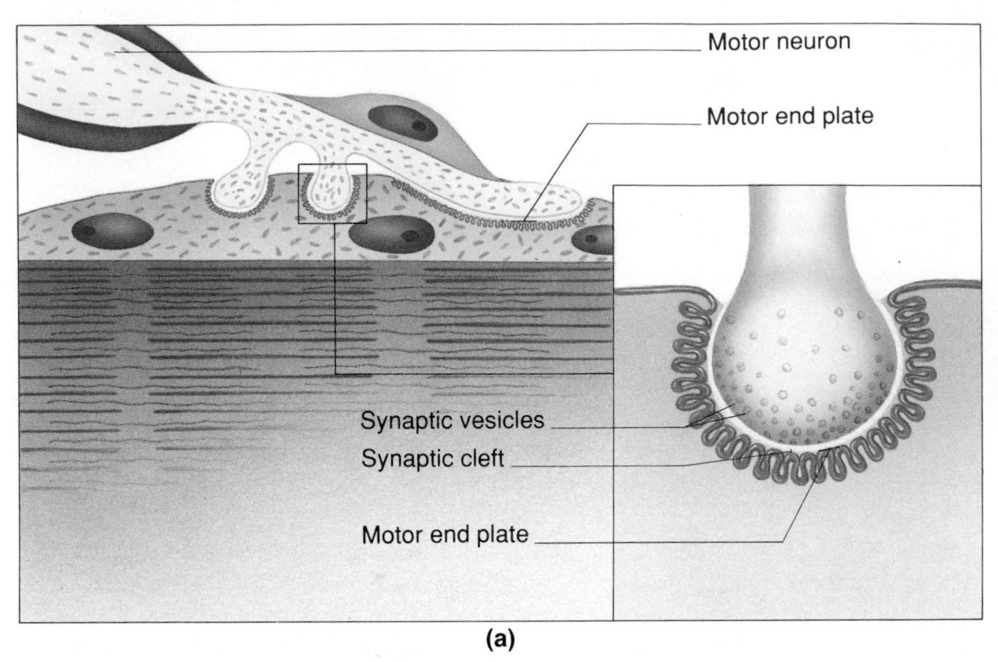

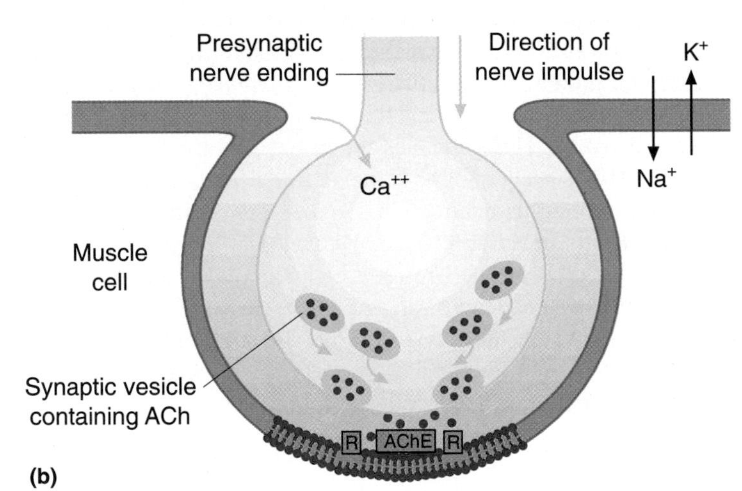

FIGURE 12.20
(a) Schematic diagram of the synapse at the neuromuscular junction. (b) The nerve impulse causes acetylcholine (ACh) to be released from synaptic vesicles. Acetylcholine diffuses across the synaptic cleft and binds to a specific receptor protein (R) on the postsynaptic membrane. A channel opens that allows Na^+ ions to flow into the cell and K^+ ions to flow out of the cell. This results in the muscle contraction. Any ACh remaining in the synaptic cleft is destroyed by acetylcholinesterase (AChE) to terminate the stimulation of the muscle cell.

The covalent intermediate is stable, and acetylcholinesterase is therefore inactive. It is unable to react with other substrates. Nerve transmission continues, resulting in muscle spasm. Death may occur as a result of laryngeal spasm. Antidotes for poisoning by organophosphates, which include many insecticides and nerve gases, have been developed. The antidotes work by reversing the effects of the inhibitor. One of these antidotes is known as PAM, an acronym for pyridine aldoxime methiodide. This molecule displaces the organophosphate group from the active site of the enzyme, alleviating the effects of the poison.

$$^{-}I \quad \text{(pyridinium ring, } N^{+}\text{–}CH_3\text{)}\text{–}\underset{}{\overset{H}{\overset{|}{C}}}{=}N{-}OH$$

Pyridine aldoxime methiodide
(PAM)

A structural analogue of acetylcholine called succinylcholine exhibits competitive inhibition of acetylcholine binding to the receptor and can be used as a muscle relaxant in surgical procedures. This compound has a structure that resembles acetylcholine closely enough that it can bind to the acetylcholine receptor protein. However, it does not have the ability to stimulate muscle contraction. The result is that muscles relax. Normal muscle contraction resumes after infusion of the drug ceases.

$$CH_3-\overset{+}{N}(CH_3)_2-CH_2-CH_2-O-\overset{O}{\overset{\|}{C}}-CH_2-CH_2-\overset{O}{\overset{\|}{C}}-CH_2-CH_2-\overset{+}{N}(CH_3)_2-CH_3$$

Succinylcholine

12.16 USE OF ENZYMES

Today, research in enzyme chemistry, **enzymology,** is an area of great interest in biochemistry. Because of the specificity of enzymes and their ability to affect the rates of chemical reactions, an understanding of the mechanisms of enzyme function is a fundamental part of understanding chemistry itself. We defined chemistry as the study of matter and the changes it undergoes; enzymes are at the very heart of biochemical change.

From a more practical point of view, enzymes are tools for diagnosis and treatment in medicine and have been commonplace in the food industry for decades. A few examples of each are considered in the following sections.

Enzymes in medicine

Analysis of blood serum for levels (concentrations) of certain enzymes can provide a wealth of information about the medical condition of the patient. Often, such tests are used to confirm a preliminary diagnosis based on the disease symptoms or clinical picture.

For example, when a heart attack occurs, a lack of blood supplied to the heart muscle causes some of the heart muscle cells to die. These cells release their contents, including their enzymes, into the bloodstream. Simple tests can be done to measure the amounts of certain enzymes in the blood. Such tests, called **enzyme assays,** are very precise and specific because they are based on the specificity of the enzyme-substrate complex. If you wish to test for the enzyme lactate dehydrogenase (LDH), you need only add the appropriate substrate, in this case pyruvate and NADH + H^+. The reaction that occurs is the oxidation of NADH + H^+ to NAD^+ and the reduction of pyruvate to lactate. To measure the rate of the chemical reaction, one can measure the disappearance of the substrate or the accumulation of one of the products. In the case of LDH spectrophotometric methods are available to measure the rate of production of NAD^+. The role of LDH and other enzymes in disease diagnosis was discussed in ''A Clinical Perspective: Enzymes, Isoenzymes, and Myocardial Infarction.''

Of course, the pharmaceutical companies have devised many clever and simple enzyme assays, but each of these tests is based on measuring the rate at which the enzyme converts the substrate into product. All of these assays must be standardized in some way so that the enzyme levels determined in one clinical laboratory can be understood and interpreted by health care professionals anywhere in the world. Thus from the raw data obtained from the enzyme assay, the concentration or activity of the enzyme is calculated and reported in *international units*. The international unit, which is the reference standard, is equal to the amount of enzyme that will catalyze conversion of 1 μmol (10^{-6} mole) of the appropriate substrate to product in one minute at standard conditions of temperature and pH. These latter conditions must be determined for each enzyme.

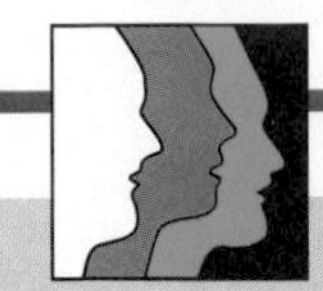

A HUMAN PERSPECTIVE

The AIDS Test

In 1981 the Center for Disease Control in Atlanta, Georgia, recognized a new disease syndrome, Acquired Immune Deficiency Syndrome (AIDS). The syndrome is characterized by an impaired immune system, a variety of opportunistic infections and cancers, and brain damage that results in dementia. It soon became apparent that the disease was being transmitted by blood and blood products, as well as by sexual contact. The threat of contamination of blood supplies worldwide resulted in a multinational effort to elucidate the cause of AIDS and to develop a suitable test for the presence of the virus. As a part of this effort, Francoise Barre-Sinoussi and her colleagues at the Pasteur Institute first isolated the virus, now called human immunodeficiency virus, in 1983.

By April 1985 a test for virus infection was available for testing blood products. The test, called an **enzyme-linked immunosorbent assay (ELISA),** is based on the specificity of antigen-antibody binding and uses a specific enzyme reaction to detect the presence of the virus in the blood. Because it is quite difficult and expensive to test for the virus itself, scientists actually test for the presence of the antibodies *produced* by the body in response to the virus infection.

The ELISA test is performed by coating the wells of a plastic microtiter plate with viral antigens (see figure at right). These protein antigens are produced by growing the virus in tissue culture and purifying the viral proteins. A series of dilutions of the patient serum is prepared and placed in the wells of the microtiter plate. All samples are tested in duplicate, consistent with good analytical technique. If there are antibodies against HIV in the blood, they will bind to the viral antigens on the surface of the plastic. However, this binding is invisible. How can we visualize whether or not the binding reaction has occurred? This involves the use of an additional antibody to which an enzyme has been covalently linked. The second antibody reacts with human IgG antibody molecules. Thus antibodies from the blood that have bound to the viral antigen on the plate will now bind to the second antibody-enzyme. Enzymes that are commonly used for this are horseradish peroxidase and alkaline phosphatase. A substrate for the enzyme is chosen that will produce a colored product. For horseradish peroxidase the substrate is ortho-phenylenediamine, and the product is blue. For

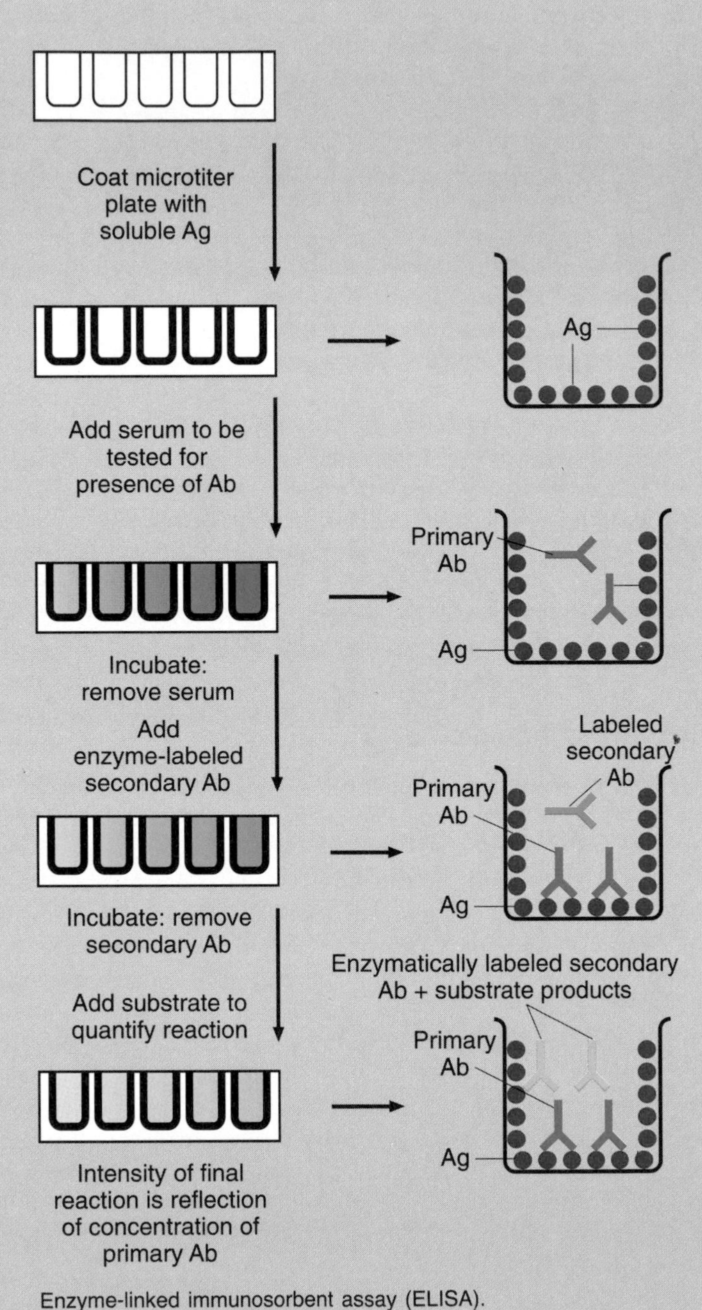

Enzyme-linked immunosorbent assay (ELISA).

Isoenzymes are forms of the same enzyme with slightly different amino acid sequences.

Liver diseases such as cirrhosis and hepatitis result in elevated levels of one of the isoenzymes of lactate dehydrogenase (LDH_5), as well as elevated levels of alanine aminotransferase/serum glutamate-pyruvate transaminase (ALT/SGPT) and aspartate aminotransferase/serum glutamate-oxaloacetate transaminase (AST/SGOT) in blood serum. In fact, these latter two enzymes also increase in concentration following a heart attack, but the physician can differentiate between these two conditions by considering the relative the substrate is para-nitrophenylphosphate, and the product is yellow. If the second antibody-enzyme has been bound by the sample, a colored product will appear when the substrate is added, and it can be concluded that the test is positive for the presence of HIV infection. If no color change is observed, the patient has not been exposed to HIV, and the test is negative. The more intense the color observed, the greater the amount of HIV antibodies in the test serum:

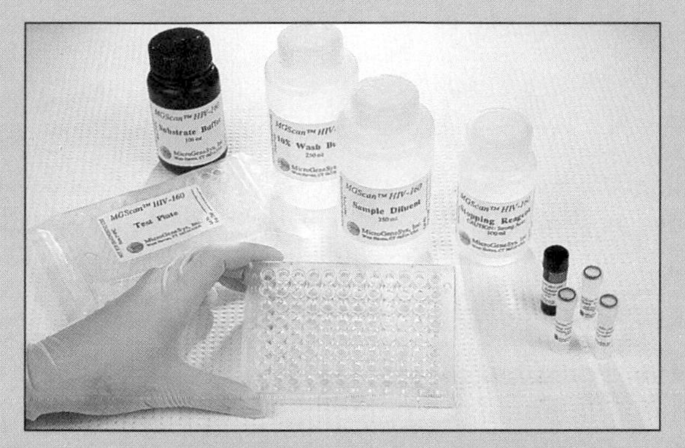

ELISA test for the presence of HIV antibodies in the blood serum. The more intense the color reaction, the greater the concentration of antibodies in the serum.

It would seem that the advent of this test would remove all threat of a contaminated blood supply, but this is unfortunately not the case. Sometimes, individuals who are not infected with the virus show positive results with the test: false positives. This is caused by other antibodies in the blood of the individual that react with other antigens that contaminate the HIV protein preparation. Other individuals who are infected demonstrate a negative result: false negatives. One reason for a false negative is the fact that it can take up to six months before the body produces antibodies against the virus. (The longest lag reported was 42 months.) During this period the individual tests negative but is infectious. Such an individual could donate contaminated blood that would then be available for transfusion.

Because of the incidence of false positive results, any blood that tests positive is tested, in duplicate, a second time. If the result is positive in the second test, then a more accurate test is done to determine with certainty whether the subject has been infected with HIV. This test is called a **Western blot** and relies on the use of gel electrophoresis to separate HIV proteins according to their size. The proteins are then detected by using antigen-antibody binding and enzyme assays, as with the ELISA test. Proteins of the HIV virus are electrophoresed and transferred to a membrane. The membrane is then covered with the serum of the test subject. If antibodies are present that can bind to the individual HIV proteins on the membrane, antigen-antibody complexes will form. The membrane is then treated with an antibody-enzyme complex similar to that used in the ELISA test. Finally the enzyme substrate is added. At any position on the membrane where the necessary antigen-antibody-antibody-enzyme complexes have formed, a colored band will appear, as seen in the figure below:

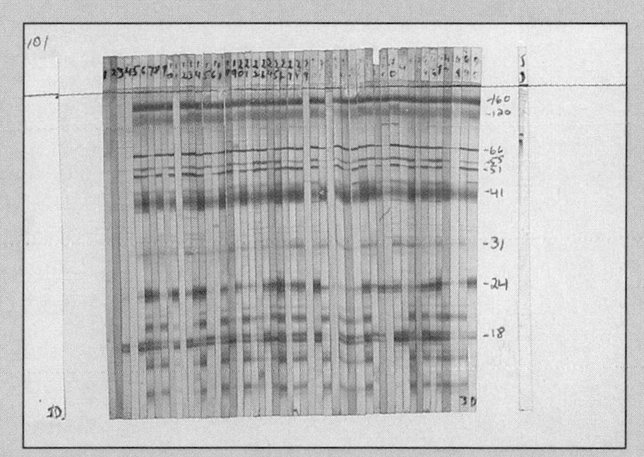

Western blot analysis of serum for the presence of HIV antibodies.

Seven different tests are on the market for HIV testing, and at least ten new tests are undergoing clinical trials. Many of these new generation tests will use HIV proteins that have been produced by recombinant DNA technology. These antigens will be much purer than those prepared from whole virus particles. It is hoped that, with fewer non-viral antigens present in the test, the incidence of false positives will dramatically decrease.

increase in the two enzymes. If ALT/SGPT is elevated to a greater extent than AST/SGOT, it can be concluded that the problem is liver dysfunction.

Elevated blood serum concentrations of the enzymes amylase and lipase are indications of pancreatitis, or an inflammation of the pancreas. Pancreatitis is characterized by extensive hemorrhage and tissue death in the pancreas as a result of the release of active hydrolytic enzymes. The pancreas produces over 20 hydrolytic enzymes and secretes them as zymogens. In pancreatitis, trypsin is activated early and, in turn, proteolytically cleaves the precursors of many of the other zymogens (see Table 12.2). As a result of the release of these degradative enzymes into the pancreas, there is extensive cell death and necrosis. Obviously, this cell death will release many enzymes into the bloodstream, including amylase and lipase, which are used as the major diagnostic indices in pancreatitis.

Enzymes are also used as analytical reagents in the clinical laboratory because of their specificity. They will often selectively react with one substance of interest, producing a product that is easily measured.

A common example involves the clinical analysis of urea in blood. The measurement of urea levels in blood is difficult, owing to the complexity of blood. However, if urea is converted to ammonia, the ammonia becomes an *indicator* of urea, since it is produced from urea, and it is easily measured. The ammonia bears a *stoichiometric relationship* to the urea. Two molecules of ammonia are produced for each molecule of urea that reacts. The enzyme, urease, catalyzes this reaction:

$$\underset{\text{Urea}}{H_2N{-}\overset{\overset{\displaystyle O}{\|}}{C}{-}NH_2} + H_2O \xrightarrow{\text{urease}} \underset{\text{Ammonia}}{2NH_3} + \underset{\text{Carbon dioxide}}{CO_2}$$

This test, called the BUN test (blood urea nitrogen test), is useful in the diagnosis of kidney malfunction and serves as one example of the utility of enzymes in clinical chemistry.

Enzymes in food chemistry

Enzymes are used in foods and food preparation; their properties of specificity and reaction rate enhancement are useful in many ways. For example:

1. *Amylases* catalyze the hydrolysis of long-chain polysaccharides, such as starch, into free glucose and short oligosaccharides of glucose, primarily maltose. Starch is not sweet-tasting, but the glucose that is produced is. Thus one can enhance the sweetness of bakery goods, cereals, and syrups without adding large amounts of sugar simply by treating with amylase.
2. *Glucose oxidase* removes glucose from egg-containing products (e.g., cake mixes, instant scrambled eggs). This is useful because glucose interacts with egg albumin, resulting in a bad flavor.
3. *Rennin* is a proteolytic enzyme purified from the stomach of calves. It has the ability to break down the milk protein casein. This results in the production of curds (denatured milk protein) and whey. The curd is then used in the production of cheese.
4. *Papain* from the papaya fruit and a similar enzyme from pineapples hydrolyze protein. As such, they make good meat tenderizers. This is the reason that in some cultures a meal with several meat courses is followed by a serving of fresh pineapple to aid in the digestion of the meat proteins. In fact, you have witnessed this reaction if you have ever tried to prepare Jell-O with fresh pineapple. The enzyme in the fruit hydrolyzes the gelatin protein so that it cannot gel, and the fruit salad remains liquid!

The above examples are but a few of the many uses of enzymes in the food industry. The development of new ways to process and package foods is an area of great commercial interest, and enzyme research is an important part of this effort.

SUMMARY

We have seen that enzymes are the biological catalysts of cells. They lower the activation energies but do not alter the equilibrium constants of the reactions they catalyze. Formation of an enzyme-substrate complex is the first step of an enzyme-catalyzed reaction. This involves the binding of the substrate to the active site of the enzyme. The active site has catalytic groups that are involved in catalysis and binding groups that are responsible for binding of a particular substrate. The chemical reaction is then mediated through a transition state.

Enzymes are most frequently named by using the common system of nomenclature. The names are useful because they are often derived from the name of the substance and/or a reaction of the substrate catalyzed by the enzyme.

Enzymes are classified according to function. The six general classes include oxireductases, transferases, hydrolases, lyases, isomerases, and ligases. They are also classified on the basis of their specificity. The four classifications of specificity are absolute, group, linkage, and stereochemical specificity.

Enzymes are sensitive to pH and temperature. For example, high temperatures or extremes of pH rapidly inactivate most enzymes by denaturing the enzyme.

Enzymes differ from inorganic catalysts in that they are regulated by the cell. Some of the means of enzyme regulation include end product inhibition, allosteric regulation, feedback inhibition, and production of inactive forms, or zymogens.

Enzyme activity can be destroyed by a variety of inhibitors. These chemicals can cause reversible or irreversible inhibition, depending on the strength of their binding to the enzyme. In addition, the inhibition may be competitive or noncompetitive, depending on whether the inhibitor is a structural analogue of the natural substrate.

The pancreatic proteases chymotrypsin, trypsin, and elastase have similar structures and mechanisms of action and have apparently evolved from a common ancestral protease.

Blood clotting requires the activation of inactive proteases that are synthesized as zymogens and are activated only when they are needed. Vitamin K has an important role in blood clotting.

Acetylcholinesterase is required for nerve transmission at the neuromuscular junction. Many drugs act by specific enzyme inhibition. Inhibition of acetylcholinesterase can be both helpful and harmful. Medically, succinylcholine (an inhibitor) causes muscle relaxation during surgery. In contrast, diisopropyl phosphofluoridate (an inhibitor) can lead to death. Enzymes have proven useful in medicine as diagnostic indicators of disease and as useful reagents in the clinical laboratory. Other uses of enzymes involve the preparation of a wide range of products in the food industry.

GLOSSARY OF KEY TERMS

absolute specificity (12.5) the property of certain enzymes that allows them to bind to, and catalyze the reaction of, only one substrate.

acetylcholine (12.15) a chemical messenger that transmits a message from the nerve cell to the muscle cell.

acetylcholinesterase (12.15) an enzyme that destroys acetylcholine in the neuromuscular junction and thereby stops the nerve impulse.

active site (12.4) the cleft in the surface of an enzyme that is the site of substrate binding and catalysis.

allosteric enzymes (12.9) enzymes that have an effector binding site as well as an active site. Effector binding changes the shape of the active site, rendering it either active or inactive.

apoenzyme (12.7) the protein portion of an enzyme that requires a cofactor in order to function in catalysis.

binding site (12.4) the chemical groups of the active site that are involved in the specific substrate binding.

catalase (Introduction) an enzyme that mediates the conversion of two molecules of hydrogen peroxide into water and oxygen.

catalytic groups (12.4) the chemical groups of the active site that are involved in catalysis.

chymotrypsin (12.11) a proteolytic enzyme that is produced in the pancreas and secreted into the small intestines, where it hydrolyzes dietary protein.

coenzyme (12.7) an organic group that is required by some enzymes; generally a donor or acceptor of electrons or functional groups in a reaction.

cofactor (12.7) an inorganic group, usually a metal ion, that must be bound to an apoenzyme to maintain the correct configuration of the active site.

common clotting pathway (12.13) the final stages of the formation of a fibrin clot that are the same regardless of whether the intrinsic or extrinsic pathway is used to initiate clot formation.

coumarin (12.14) an anticoagulant that reduces blood clotting by reducing the amount of prothrombin and factors VII, IX, and X. It is a competitive inhibitor of vitamin K.

divergent evolution (12.12) the process whereby copies of an identical gene evolve to become dissimilar.

end product inhibition (12.9) a means of enzyme regulation in which the product of a reaction inhibits its own synthesis by interacting with the enzyme and forcing the reaction to proceed in the reverse, from product to substrate.

enzyme (Introduction) a protein that acts as a biological catalyst.

enzyme assays (12.16) tests to measure the amount of enzyme in a sample, for instance, in the bloodstream.

enzyme-linked immunosorbent assay (ELISA) (12.16) an assay that is used to detect the presence of human immunodeficiency virus (HIV) antibodies in blood serum. It is based on antigen-antibody binding and the detection system is an enzyme reaction.

enzyme specificity (Introduction) the ability of an enzyme to bind to only one, or a very few, substrates and thus catalyze only a single reaction.

enzyme-substrate complex (12.4) a molecular aggregate formed when the substrate binds to the active site of the enzyme.

enzymology (12.16) the study of the function and structure of enzymes.

extrinsic pathway of blood clotting (12.13) the initial stages of the clotting mechanism that are induced by damage to the tissues surrounding the blood vessel.

feedback inhibition (12.9) when produced in excess, the product of a biosynthetic pathway can turn off the entire pathway for its own synthesis. This is an example of negative allosterism. The product of the pathway is a negative allosteric effector of one of the enzymes early in the pathway.

fibrin (12.13) the insoluble blood protein that makes up a blood clot.

fibrinogen (12.13) the soluble blood protein that is converted to insoluble fibrin by the enzyme thrombin in the last stage of blood clotting.

group specificity (12.5) the property of certain enzymes that allows them to catalyze reactions involving similar substrate molecules having the same functional groups.
hemophilia (12.13) a genetic disorder in the blood clotting mechanism in which one of the clotting factors is missing or inactive. It is characterized by excessive bleeding that may occur spontaneously or following even minor wounds.
hydrolase (12.1) an enzyme that catalyzes a hydrolysis reaction.
induced fit model (12.5) the theory of enzyme-substrate binding that assumes that the enzyme is a flexible molecule and that both the substrate and enzyme change their shapes to accommodate one another as the enzyme-substrate complex forms.
international unit (12.16) the amount of enzyme needed to catalyze conversion of 1 μmol of substrate to product in one minute at standard conditions of temperature and pH.
intrinsic pathway of blood clotting (12.13) the initial stages of the clotting mechanisms that are induced by damage within the blood vessel.
irreversible enzyme inhibitor (12.10) a chemical that binds strongly to the R group of an amino acid in the active site and eliminates enzyme activity.
ischemia (12.5) interrupted blood flow to an organ.
isoenzymes (12.5) forms of the same enzyme with slightly different amino acid sequences.
isomerase (12.1) an enzyme that catalyzes the conversion of one isomer to another.
ligase (12.1) an enzyme that catalyzes the joining of two molecules.
linkage specificity (12.5) the property of certain enzymes that allows them to catalyze reactions involving only one kind of bond in the substrate molecule.
lock-and-key model (12.4) the theory of enzyme-substrate binding that supposes that enzymes are inflexible molecules and that the substrate fits into the rigid active site in the same way that a key fits into a lock.
lyase (12.1) an enzyme that catalyzes the cleavage of C—O, C—C, or C—N bonds, thereby producing a product containing a double bond.
lysosomes (12.8) membrane-bound vesicles that contain numerous hydrolytic enzymes. The cell uses these enzymes to break down macromolecules into their simple subunits, which can then be used as an energy source or as substrates in biosynthetic reactions.
myocardial infarction (12.5) heart attack, damage to the heart muscle caused by an interrupted blood supply.
negative allosterism (12.9) effector binding inactivates the active site of an allosteric enzyme.
neurotransmitter (12.15) a chemical signal that transmits a message from a nerve cell to a muscle cell.
oxireductases (12.1) enzymes that catalyze oxidation-reduction reactions.
pancreatic serine proteases (12.12) a family of proteolytic enzymes, including trypsin, chymotrypsin, and elastase that arose by divergent evolution.
pepsin (12.8) a proteolytic enzyme found in the stomach. It catalyzes the breakdown of dietary proteins.
pepsinogen (12.9) the inactive form of pepsin produced in cells lining the stomach. It is converted to pepsin by proteolytic cleavage in the stomach.
pH optimum (12.8) the pH at which an enzyme catalyzes the reaction at maximum efficiency.
positive allosterism (12.9) effector binding activates the active site of an allosteric enzyme.
product (Introduction) the chemical species that results from an enzyme-catalyzed reaction.
proteolytic enzymes (12.11) enzymes that hydrolyze the peptide bonds between amino acids in a protein chain.
prothrombin (12.14) the inactive precursor of thrombin.
reversible, competitive inhibitor (12.10) a chemical that resembles the structure and charge distribution of the natural substrate and competes with it for the active site.
reversible, noncompetitive inhibitor (12.10) a chemical that binds weakly to an amino acid R group or cofactor and inhibits activity. When the inhibitor dissociates, the enzyme is restored to its active form.
stereochemical specificity (12.5) the property of certain enzymes that allows them to catalyze reactions involving only one stereoisomer of the substrate.
streptokinase (12.5) an enzyme produced by the bacterium *Streptococcus pyogenes* that destroys fibrin clots. It is used to treat heart attack victims.
structural analogue (12.10) a chemical having a structure and charge distribution that are very similar to those of a natural enzyme substrate.
substrate (Introduction) the reactant in a chemical reaction that binds to the enzyme active site and is converted to the product.
thrombin (12.13) the proteolytic enzyme that converts fibrinogen to fibrin in the last stage of clot formation.
tissue-type plasminogen activator (TPA) (12.5) an enzyme that is a part of the natural anticlotting mechanism of the human body and that is used to treat heart attack victims.
transferase (12.1) an enzyme that catalyzes the transfer of a functional group from one molecule to another.
transition state (12.6) the unstable intermediate in catalysis in which the enzyme has altered the form of the substrate so that it now shares properties of both the substrate and the product.
vitamin K (12.14) a fat-soluble vitamin that is required for the synthesis of prothrombin and factors VII, IX, and X in the liver.
Western blot (12.16) a test that involves the separation of viral proteins by electrophoresis and the detection of those proteins by antigen-antibody binding and enzyme assay.
zymogen (12.9) the inactive form of a proteolytic enzyme.

QUESTIONS AND PROBLEMS

Nomenclature and Classification

12.21 Match each of the following substrates with its corresponding enzyme:

1. urea	a. lipase
2. hydrogen peroxide	b. glucose-6-phosphatase
3. lipid	c. peroxidase
4. aspartic acid	d. sucrase
5. glucose-6-phosphate	e. urease
6. sucrose	f. aspartase

12.22 Give a systematic name for the enzyme that would act upon each of the following substrates:

a. alanine
b. citrate
c. ampicillin
d. ribose
e. methyl amine

12.23 Describe the function implied by the name of each of the following enzymes:
a. citrate decarboxylase
b. tyrosine kinase
c. oxalate reductase
d. nitrite oxidase
e. *cis-trans* isomerase

12.24 List the classes of enzymes and describe the function of each.

The Effect of Enzymes on the Activation Energy of a Reaction

12.25 Write and explain the equation for the equilibrium constant of an enzyme mediated reaction. Does the enzyme alter the K_{eq}?

12.26 If an enzyme does not alter the equilibrium constant of a reaction, how does it speed up the reaction?

The Effect of Substrate Concentration on Enzyme-Catalyzed Reactions

12.27 What is the effect of doubling the substrate on the rate of a chemical reaction?

12.28 Why is this not the case for enzyme-catalyzed reactions? In other words, why doesn't the rate of an enzyme-catalyzed reaction increase indefinitely when the substrate concentration is made very large?

12.29 a. Draw a graph that describes the effect of increasing the concentration of the substrate on the rate of an enzyme-catalyzed reaction.
b. What does this graph tell us about the nature of enzyme-catalyzed reactions?

The Enzyme-Substrate Complex

12.30 Name three major properties of enzyme-active sites.

12.31 Since enzyme-active sites are small, why are enzymes so large?

12.32 What is the lock-and-key model of enzyme-substrate binding?

12.33 Why is the induced fit model of enzyme-substrate binding a much more accurate model than the lock-and-key model?

12.34 List and define four classes of enzyme specificities.

Enzyme-Catalyzed Reactions

12.35 Outline the four general stages in an enzyme-mediated reaction.

12.36 What types of transition states might be envisioned that would decrease the energy of activation of an enzyme?

12.37 What is the role of a cofactor in enzyme activity?

12.38 How does a coenzyme function in an enzyme-mediated reaction?

12.39 What is the structure of NAD^+? What is the function of NAD^+? What class of enzymes would require a coenzyme of this sort?

12.40 How will each of the following changes in conditions alter the rate of an enzyme-catalyzed reaction?
a. decreasing the temperature to 10°C
b. increasing the pH of the solution from 7 to 11
c. heating the enzyme to 100°C

12.41 Label each of the following statements as true or false:
a. Enzymes are specific for the reactions that they catalyze.
b. Enzymes are not affected by temperature.
c. Enzymes have an optimum pH for operation at maximum activity.
d. Ligases are enzymes that join two molecules.
e. The active site of an enzyme must have the correct shape and charge in order to interact with its corresponding substrate.

12.42 Why does an enzyme lose activity when the pH is drastically changed from optimum pH?

12.43 Define the optimum pH for enzyme activity? How does this pH compare to physiological pH?

12.44 List the factors that affect enzyme activity.

12.45 An increase in temperature will increase the rate of a reaction if a nonenzymatic catalyst is used; however, an increase in temperature will eventually *decrease* the rate of a reaction when an enzyme catalyst is used. Explain the apparent contradiction of these two statements.

12.46 What is a lysosome?

12.47 Of what significance is it that lysosomal enzymes have a pH optimum of 4.8?

Regulation of Enzyme Activity

12.48 Why is it important for cells to regulate the level of enzyme activity?

12.49 Why does an excess of the product of a reaction cause the reaction to proceed in reverse?

12.50 What is an allosteric enzyme?

12.51 What is the difference between positive and negative allosterism?

12.52 Define feedback inhibition. Describe the role of allosteric enzymes in feedback inhibition. Is this positive or negative allosterism?

12.53 What is a zymogen?

12.54 Three zymogens that are involved in digestion of proteins in the stomach and intestines are pepsinogen, chymotrypsinogen, and trypsinogen. What is the advantage of producing these enzymes as inactive peptides?

Inhibition of Enzyme Activity

12.55 Suppose that a certain drug company manufactured a compound that had nearly the same structure as a substrate for a certain enzyme but that could not be acted upon chemically by the enzyme. What type of interaction would the compound have with the enzyme?

12.56 The addition of phenylthiourea to a preparation of the enzyme polyphenoloxidase completely inhibits the activity of the enzyme. Knowing that phenylthiourea binds all copper ions, what conclusion can you draw about whether polyphenoloxidase requires a cofactor? What kind of inhibitor is phenylthiourea?

12.57 Define competitive, reversible enzyme inhibition.

Proteases

12.58 What do the similar structures of chymotrypsin, trypsin, and elastase suggest about their evolutionary relationship?

12.59 What properties are shared by chymotrypsin, trypsin, and elastase?

12.60 Draw the complete structural formula for the peptide tyr-lys-ala-phe. Show which bond would be broken when this peptide is reacted with chymotrypsin.

12.61 Repeat Question 12.15 for the peptide trp-pro-gly-tyr.

12.62 The sequence of a peptide that contains ten amino acid residues is shown below. Indicate with arrows and labels the peptide bond(s) that are cleaved by

a. elastase
b. trypsin
c. chymotrypsin

ala-gly-val-leu-trp-lys-ser-phe-arg-pro

12.63 What structural features of trypsin, chymotrypsin, and elastase account for their different specificities?

Blood Clotting

12.64 Explain the relationships between the following pairs of proteins:

a. thrombin and prothrombin
b. fibrin and fibrinogen

12.65 Draw the structure of γ-carboxyglutamate in a complex with a Ca^{2+} ion.

12.66 A structural analogue of vitamin K is used as an anticoagulant to treat persons prone to form blood clots. What is the biochemical function of the drug?

12.67 Cattle that are inadvertently fed spoiled sweet clover often die of hemorrhages. This clover contains a compound that is a structural analogue of vitamin K. Suggest a reason for the death of the cattle.

Acetylcholinesterase

12.68 How does pyridine aldoxime methiodide serve as an antidote for diisopropyl fluorophosphate (DIFP) poisoning?

Further Problems

12.69 List the enzymes whose levels are elevated in blood serum following a myocardial infarction.

12.70 Trypsin and thrombin are serine proteases that have very different functions in the body. Why are trypsin and thrombin thought to be the products of divergent evolution?

12.71 Explain why enzymes do not alter the equilibrium constants of the reactions that they catalyze.

12.72 Ethylene glycol is a poison that causes about 50 deaths a year in the United States. Treatment of individuals who have drunk ethylene glycol with massive doses of ethanol can save their lives. The structures of ethylene glycol and ethanol are shown below. Suggest a reason for the effect of ethanol.

$HO—CH_2—CH_2—OH$	$CH_3—CH_2—OH$
Ethylene glycol	Ethanol

12.73 Why are enzymes that are used for clinical assays in hospitals stored in refrigerators?

12.74 Why do extremes of pH inactivate enzymes?

12.75 Why is succinylcholine a muscle relaxant?

12.76 How is an enzyme-active site able to distinguish between enantiomers?

12.77 Organofluorophosphates are often used as insecticides. How do they work?

12.78 What is the function of factor X in blood clotting?

12.79 Describe the use of enzymes in the ELISA test for AIDS.

C H A P T E R 1 3

Carbohydrate Metabolism

O U T L I N E

O B J E C T I V E S

- Understand the importance of ATP in cellular energy transfer processes.
- Describe the three steps in the degradation of glucose.
- Understand the way in which proteins, fats, and carbohydrates are digested.
- Know what is meant by a catabolic process and summarize the three major steps in catabolism.
- Describe glycolysis in terms of its two segments.
- List the conditions under which pyruvate is converted to lactate, ethanol, or acetyl coenzyme A.
- Understand the practical and the metabolic roles of fermentation reactions.
- Explain the process of gluconeogenesis.
- Note the major difference between glycolysis and gluconeogenesis.
- Compare the processes of glycogenesis and glycogenolysis.
- Understand the role of insulin and glucagon in glycogen metabolism.
- Summarize how glycogen synthesis and degradation are compatible.
- Name the enzymes responsible for glycogenesis and glycogenolysis.
- Describe several glycogen storage diseases and their relationship to glycogen metabolism.

Just as we need energy to run, jump, and think, the cell needs a ready supply of cellular energy for the many functions that support these activities. Cells need energy for *active transport,* to move molecules between the environment and the cell, across cells, or within cells. Energy is also needed for *biosynthesis* of small metabolic molecules and production of macromolecules from these intermediates. Finally, energy is required for *mechanical work,* including muscle contraction, motility of sperm cells, and movement of organelles within the cytoplasm. Table 13.1 lists some examples of each of these energy-requiring processes.

We need a supply of energy-rich food molecules that can be degraded, or oxidized, to provide the needed cellular energy. Our diet includes three major sources of energy: carbohydrates, fats, and proteins. Each of these groups of macromolecules must be broken down into their substituents—simple sugars, fatty acids and glycerol, and amino

TABLE 13.1 The Types of Cellular Work That Require Energy

Biosynthesis: Synthesis of Metabolic Intermediates and Macromolecules
Synthesis of glucose from CO_2 and H_2O in the process of photosynthesis in plants
Synthesis of amino acids
Synthesis of nucleotides
Synthesis of lipids
Protein synthesis from amino acids
Synthesis of nucleic acids
Synthesis of organelles and membranes
Active Transport: Movement of Ions and Molecules
Transport of H^+ to maintain constant pH
Transport of food molecules into the cell
Transport of Na^+ and K^+ into and out of nerve cells for transmission of nerve impulses
Secretion of HCl from parietal cells into the stomach
Transport of waste from the blood into the urine in the kidneys
Transport of amino acids and most hexose sugars into the blood from the intestines
Accumulation of calcium ions in the mitochondria
Motility
Contraction and flexion of muscle cells
Separation of chromosomes during cell division
Ability of sperm to swim via flagella
Movement of foreign substances out of the respiratory tract by cilia on the epithelial lining of the trachea
Translocation of eggs into the Fallopian tubes by cilia in the female reproductive tract

acids—before they can be taken into the cell and used to produce cellular energy. Of these classes of food molecules, carbohydrates are the most readily used. In evolutionary terms the pathway for the first stages of carbohydrate degradation, called *glycolysis*, seems to be the first successful energy-generating pathway on earth. So successful is it that it has been faithfully conserved in organisms as widely separated in evolutionary time as the simple bacterium and the human. Comparison of the amino acid sequences of the glycolytic enzymes from these diverse species shows that the amino acids within the critical active sites have changed little in 3.5 billion years, although the scaffolding sequences are considerably more divergent. So important is glycolysis that there is no known genetic defect of this pathway. Presumably, such a defect would be fatal before the embryo could undergo the first cell division, which requires enormous amounts of energy.

Sections 11.4 and 12.5

In this chapter we are going to examine the steps of this ancient energy-generating pathway. We will see that it is responsible for the capture of some of the bond energy of carbohydrates and the storage of that energy in the molecular form of *adenosine triphosphate (ATP)*. Glycolysis actually releases and stores very little (2.2%) of the potential energy of glucose, but the pathway also serves as a source of biosynthetic intermediates, and it modifies the carbohydrates in such a way that other pathways are able to release as much as 38–40% of the potential energy.

Chapter 14

13.1 ATP: THE CELLULAR ENERGY CURRENCY

The degradation of fuel molecules, called **catabolism,** provides the energy that enables the cell to function. Actually, the energy of a food source can be released in one of two ways: as heat or, more important to the cell, as chemical bond energy. We can envision two alternative modes of aerobic degradation of the simple sugar glucose. We can, metaphorically, simply set the glucose afire. This would result in its complete oxidation to CO_2 and H_2O and would release 686 kilocalories per mole of glucose. Yet in terms of a cell, what would be accomplished? Nothing; all of the potential energy of the bonds of glucose is lost as heat and light. Such a violent release of energy would denature enzymes and other proteins and would destroy the cell.

Consider the alternative, the step-by-step oxidation of glucose accompanied by the release of small amounts of energy at several points in the pathway. If this is to be a successful strategy for the cell, it will require a series of enzymes to catalyze the stepwise oxidation and a means of saving that chemical energy for use by the energy-requiring reactions.

Theoretically, this could be achieved by a direct coupling of **exothermic reactions** (energy-producing reactions) with **endothermic reactions** (energy-requiring reactions). When the cell has an energy demand, it could simply burn an energy source and generate energy for the job at hand. This strategy requires that an energy source is always available, and that is far from true. It would also require that the energy-generating and energy-requiring reactions to be coupled are compatible with respect to the amount and form of the energy. Each set of reactions would have to be custom designed, which would probably necessitate a very large number of diverse reactions. The synthesis of the enzymes required to sustain such a system would be an overwhelming energy expense for the cell.

It would be far more efficient to have a single ''go-between'' to couple all of the exothermic and endothermic reactions in the cell; a single molecule for the storage of chemical energy that could serve as a universal energy currency. This is exactly what happens in the cell! The molecule that serves this function is **adenosine triphosphate (ATP).** If ATP is to be a useful storage form of chemical energy, it must be able to accept the energy harvested in the energy-releasing pathways and yield that energy to the energy-requiring pathways. The secret to the function of ATP as a go-between lies in its chemical structure, seen in Figure 13.1.

An N-glycosidic bond results when the anomeric carbon of a cyclic sugar is linked to an amine nitrogen. In the case of ATP the sugar is ribose and the amine nitrogen is a component of the heterocyclic compound adenine. Further details of nucleotide structure are described in Section 16.1.

ATP is a **nucleotide** composed of the nitrogenous base adenine bonded in N-glycosidic linkage to the sugar ribose. Ribose, in turn, is bonded to one (AMP), two (ADP), or three (ATP) phosphoryl groups by phosphoester bonds. The molecule is a high-energy

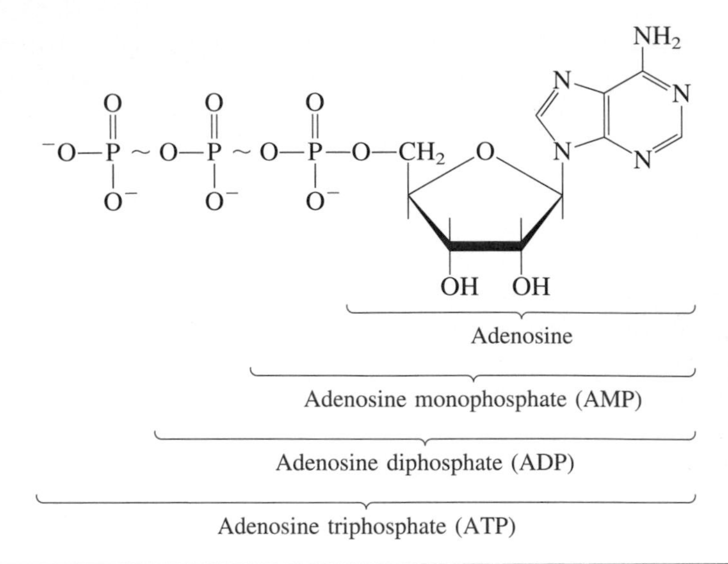

FIGURE 13.1
The structure of the universal energy currency, ATP.

compound because of the bonds holding the terminal phosphoryl groups. When these bonds are broken, or hydrolyzed, they release a large amount of energy that can be used for cellular work. For this reason they are called *high-energy bonds*. The high-energy bonds of the ATP molecule are indicated as squiggles (~) in Figure 13.1.

The structure of ATP is only a part of the reason that the molecule is a good go-between in energy transformations in the cell. ATP must have a higher energy content than the compounds to which it will donate energy, but it must also contain less energy than the compounds that are involved in forming it. In this way, all the reactions are favored because both the reactions that produce ATP and the hydrolysis of ATP to provide energy for cellular work are exothermic. Figure 13.2 shows the relative energies of some phosphorylated compounds, including ATP, that are involved in energy metabolism.

Hydrolysis of ATP yields adenosine-5′-diphosphate (ADP), an inorganic phosphate group (P_i), and energy (Figure 13.3). The energy released by this hydrolysis of ATP is then used to drive biological processes, for instance, the phosphorylation of glucose or fructose.

An example of the way in which the energy of ATP is used can be seen in the first step of glycolysis, the anaerobic degradation of glucose to produce chemical energy. The first step involves the transfer of a phosphoryl group, $-PO_3^{2-}$, from ATP to the C-6

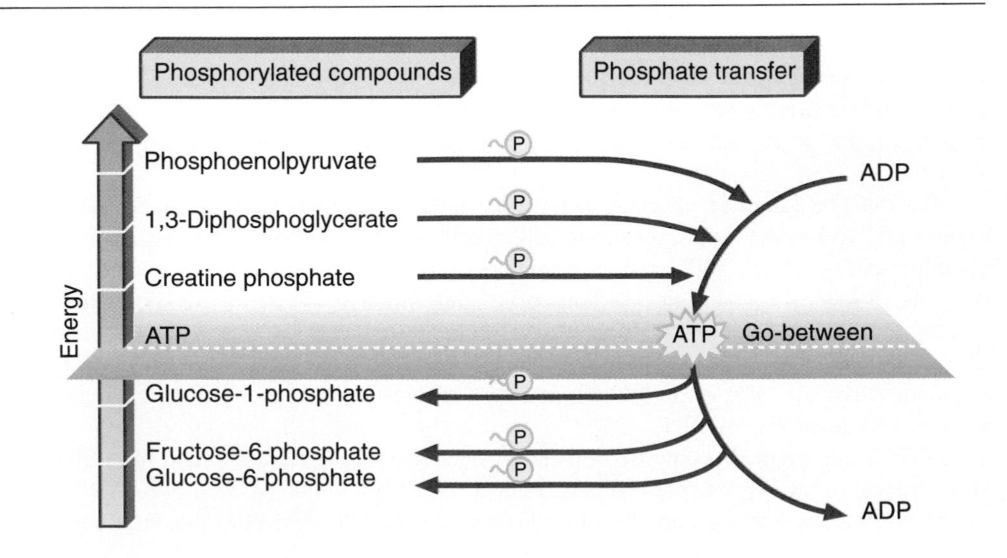

FIGURE 13.2
An energy comparison of several phosphorylated compounds that are important in cellular metabolic processes.

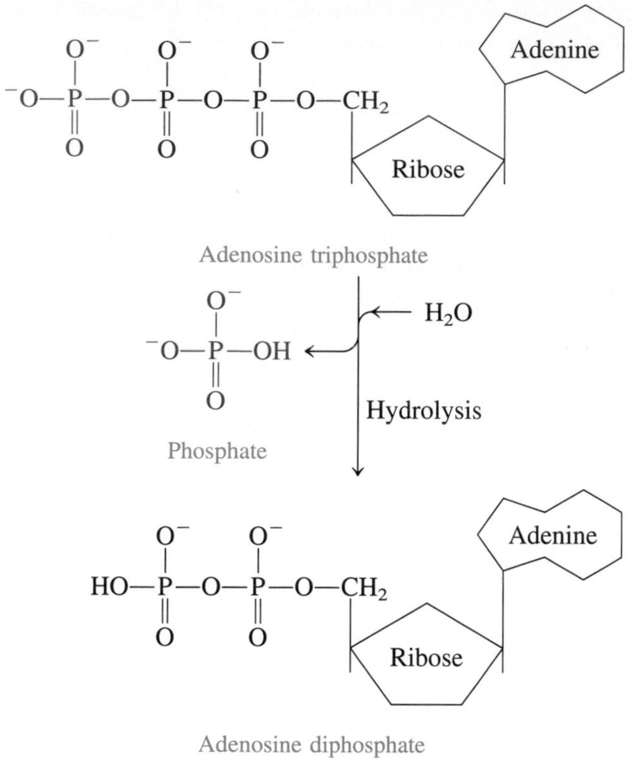

$$^{-}O-\overset{O^-}{\underset{O}{P}}-O-\overset{O^-}{\underset{O}{P}}-O-\overset{O^-}{\underset{O}{P}}-O-CH_2-\text{Ribose}-\text{Adenine}$$

Adenosine triphosphate

$\downarrow$ H_2O, Hydrolysis $\rightarrow$ $^{-}O-\overset{O^-}{\underset{O}{P}}-OH$ Phosphate

$$HO-\overset{O^-}{\underset{O}{P}}-O-\overset{O^-}{\underset{O}{P}}-O-CH_2-\text{Ribose}-\text{Adenine}$$

Adenosine diphosphate

FIGURE 13.3
Hydrolysis of ATP to ADP breaks the phosphoester linkage and releases an inorganic phosphate group.

hydroxyl group of glucose (Figure 13.4). This reaction is catalyzed by the enzyme hexokinase.

This reaction can be dissected to reveal the role of ATP as a source of energy. The first reaction is the hydrolysis of ATP to ADP and phosphate, abbreviated P_i. This reaction *releases* about 7 kcal/mol of energy:

$$\text{ATP} + H_2O \longrightarrow \text{ADP} + P_i + 7\,\frac{\text{kcal}}{\text{mol}}\ \text{energy released}$$

Second, the synthesis of glucose-6-phosphate from glucose and phosphate *requires* 3.0 kcal/mol:

$$3.0\,\frac{\text{kcal}}{\text{mol}}\ \text{energy required} + \text{glucose} + P_i \longrightarrow \text{glucose-6-phosphate} + H_2O$$

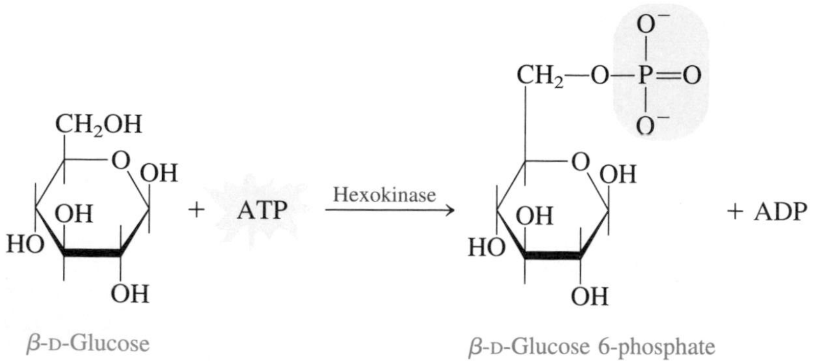

β-D-Glucose + ATP $\xrightarrow{\text{Hexokinase}}$ β-D-Glucose 6-phosphate + ADP

FIGURE 13.4
Phosphoryl group transfer from ATP to the C-6 hydroxyl group of glucose.

These two chemical reactions can then be added to give the equation showing the way in which ATP hydrolysis is *coupled* to the phosphorylation of glucose:

$$\begin{array}{rl} \text{ATP} + H_2O & \longrightarrow \text{ADP} + P_i \\ \text{glucose} + P_i & \longrightarrow \text{glucose-6-phosphate} + H_2O \\ \hline \text{Net: ATP} + \text{glucose} & \longrightarrow \text{glucose-6-phosphate} + \text{ADP} \end{array}$$

Since the hydrolysis of ATP releases more energy than is required to synthesize glucose-6-phosphate from glucose and phosphate, there is an overall energy gain in this process: 3 kcal/mol consumed and 7 kcal/mol released for a gain of 4 kcal/mol.

The primary function of all catabolic pathways is the generation of ATP energy. This continuous generation of ATP provides the energy that powers most cellular functions.

13.2 OVERVIEW OF CATABOLIC PROCESSES

Although carbohydrates, fats, and proteins can all be degraded to release energy, the carbohydrates are the most readily used energy source. We will begin by examining the

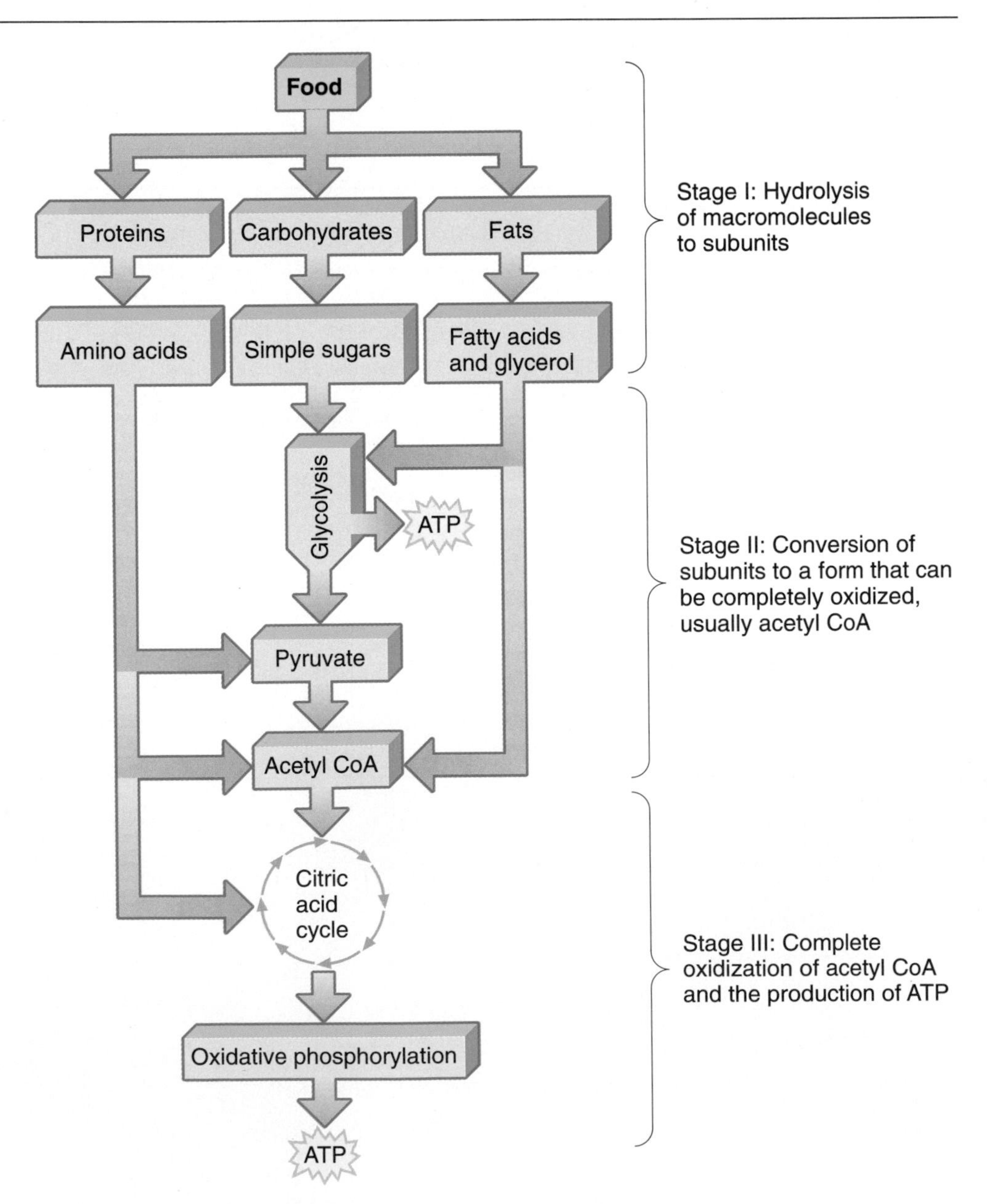

FIGURE 13.5
The three stages of the conversion of food into cellular energy in the form of ATP.

oxidation of the hexose glucose. In Chapters 14 and 15 we will see the way in which the pathways for glucose oxidation are also used for the degradation of fats and proteins.

Any catabolic process must begin with a supply of nutrients. When you eat a meal, you are consuming quantities of carbohydrates, fats, and proteins. From this point the catabolic processes can be broken down into a series of stages. The three stages of catabolism are summarized in Figure 13.5.

Stage I: hydrolysis of dietary macromolecules into small subunits

The purpose of this stage of catabolism is to degrade the large polymeric molecules into their constituent monomeric subunits so that the latter can be taken into the cells of the body for use as an energy source. The process of digestion is summarized in Figure 13.6.

Polysaccharides are hydrolyzed to monosaccharides. This process begins in the mouth, where the enzyme amylase begins the hydrolysis of starch, and continues in the small intestines, where pancreatic amylase further hydrolyzes the starch into maltose (a dissacharide of glucose). Maltase hydrolyzes the maltose, producing two glucose molecules. Similarly, sucrose is hydrolyzed to glucose and fructose by the enzyme sucrase, and lactose (milk sugar) is degraded into the monosaccharides glucose and galactose by the enzyme lactase. The monosaccharides are taken up by the epithelial cells of the intestines in an energy-requiring process, active transport. The digestion of proteins be-

FIGURE 13.6
An overview of the digestive processes that hydrolyze carbohydrates, proteins, and fats.

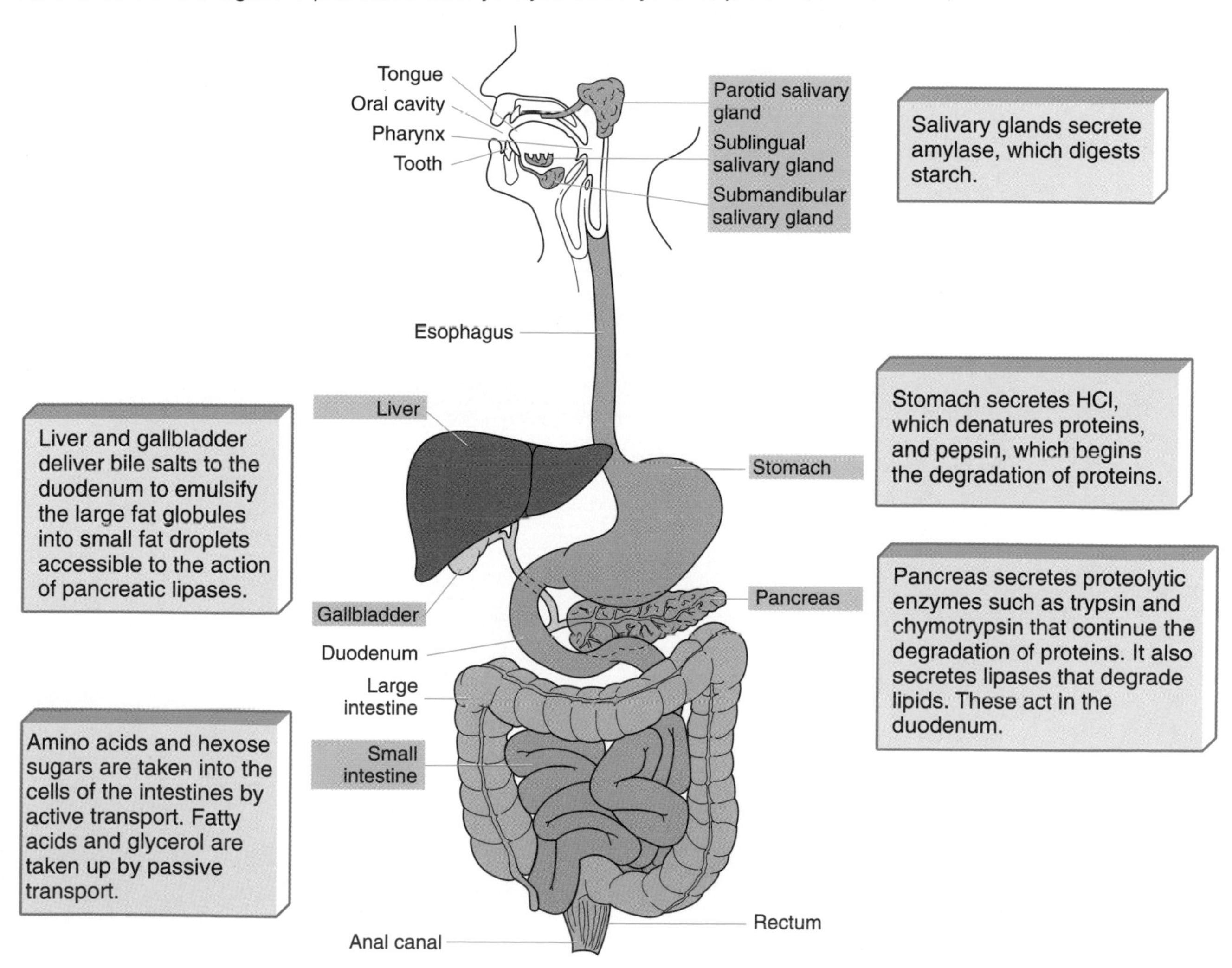

gins in the stomach, where the low pH (2) denatures the proteins, rendering them much more susceptible to hydrolysis by the enzyme pepsin. They are further degraded in the intestines by carboxypeptidase and the serine proteases trypsin, chymotrypsin, and elastase into amino acids and short oligopeptides that are taken up by the cells lining the intestines. Again this involves an active transport mechanism. Digestion of fats does not begin until the food reaches the small intestines, although there are lipases in both saliva and stomach fluid. Fats arrive in the duodenum, the first portion of the small intestines, in the form of large fat globules. Bile salts produced by the liver break these up into smaller fat droplets, which, because of the greater surface area, are now more accessible to the action of pancreatic lipase and phospholipase A. These enzymes hydrolyze the fats into fatty acids, monoglycerides, and glycerol, which are taken up by intestinal cells by passive (non-energy-requiring) transport. A summary of these hydrolysis reactions is shown in Figure 13.7.

Sections 12.11 and 12.12

The digestion and transport of fats are considered in greater detail in Chapter 15.

FIGURE 13.7
A summary of the hydrolysis reactions of carbohydrates, proteins, and fats.

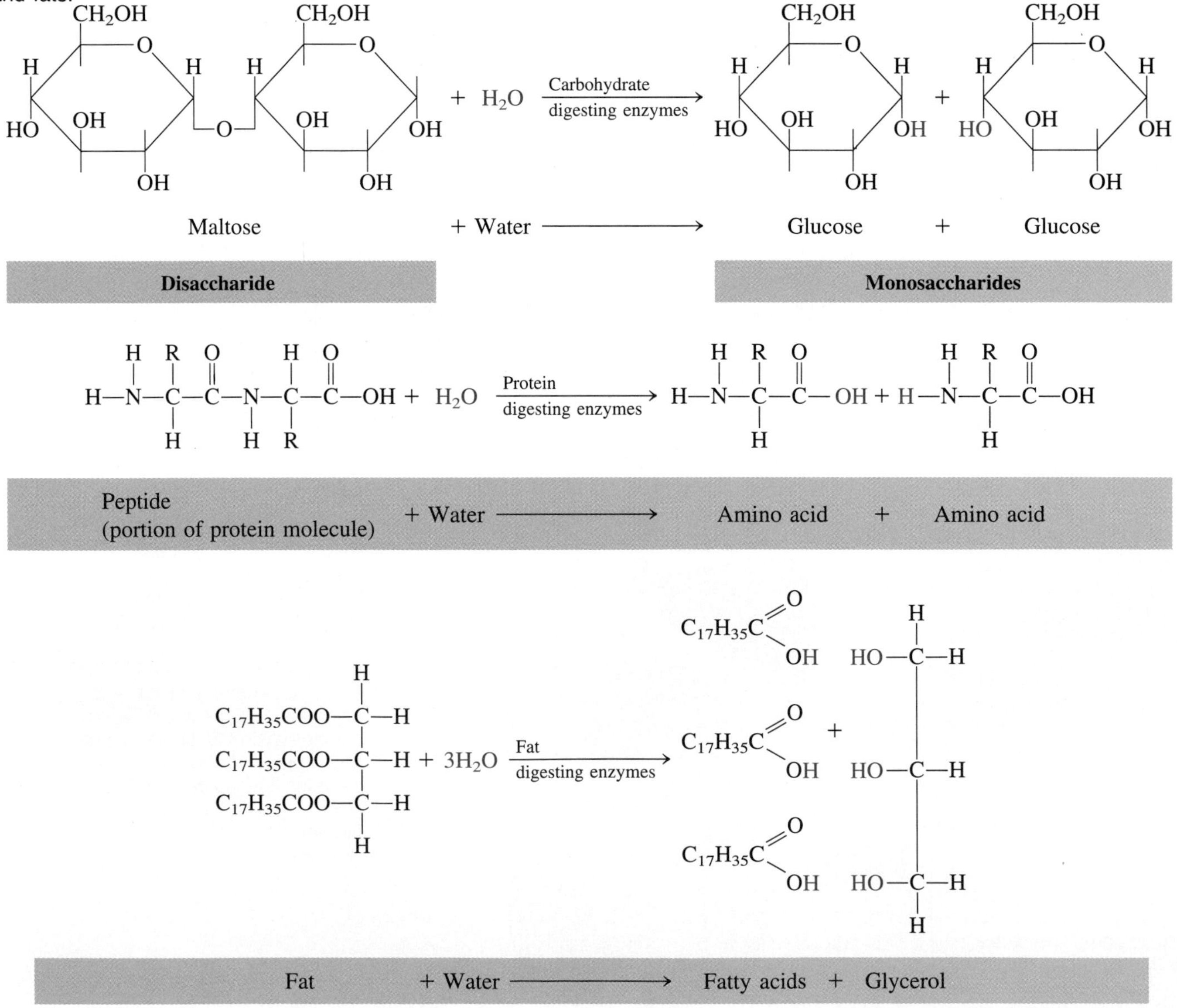

Maltose + H_2O $\xrightarrow{\text{Carbohydrate digesting enzymes}}$ Glucose + Glucose

Maltose + Water ⟶ Glucose + Glucose

Disaccharide **Monosaccharides**

$$H{-}N(H){-}C(R)(H){-}C(=O){-}N(H){-}C(H)(R){-}C(=O){-}OH + H_2O \xrightarrow{\text{Protein digesting enzymes}} H{-}N(H){-}C(R)(H){-}C(=O){-}OH + H{-}N(H){-}C(R)(H){-}C(=O){-}OH$$

Peptide (portion of protein molecule) + Water ⟶ Amino acid + Amino acid

$$\begin{array}{l} C_{17}H_{35}COO{-}CH_2 \\ C_{17}H_{35}COO{-}CH \\ C_{17}H_{35}COO{-}CH_2 \end{array} + 3H_2O \xrightarrow{\text{Fat digesting enzymes}} 3\,C_{17}H_{35}COOH + \begin{array}{l} HO{-}CH_2 \\ HO{-}CH \\ HO{-}CH_2 \end{array}$$

Fat + Water ⟶ Fatty acids + Glycerol

Stage II: conversion of monomers into a form that can be completely oxidized

The monosaccharides, amino acids, and fatty acids and glycerol must now be assimilated into the pathways of energy metabolism (Figure 13.5). Sugars usually enter the glycolysis pathway in the form of glucose or fructose. They are eventually converted to acetyl CoA, which is a form that can be completely oxidized in the citric acid cycle. Amino acids are deaminated and can enter the catabolic processes at many stages. Some enter at the level of pyruvate, others enter at the level of acetyl CoA, and still others are converted directly into intermediates of the citric acid cycle. Fatty acids are converted to acetyl CoA and enter the citric acid cycle in that form. Glycerol derived from the hydrolysis of fats is converted to glyceraldehyde-3-phosphate, one of the intermediates of glycolysis, and enters energy metabolism at that level.

Stage III: the complete oxidation of nutrients and the production of ATP

Acetyl CoA carries two-carbon remnants of the dietary nutrients, acetyl groups, to the citric acid cycle. Here electrons and hydrogen atoms are harvested during the complete oxidation of the nutrients to CO_2. These electrons and hydrogen atoms are used in the process of oxidative phosphorylation to produce ATP.

13.3 GLYCOLYSIS

General considerations

Glycolysis is the first stage of carbohydrate catabolism. The first substrate in the pathway is D-glucose, the most abundant organic molecule in the biosphere. As we noted earlier, the very fact that all organisms are able to use glucose as an energy source for the glycolytic pathway suggests that glycolysis was the first successful pathway for energy generation that evolved on earth. The process evolved at a time when the earth's atmosphere was a reducing or *anaerobic* atmosphere; no free oxygen was available. As a result, the glycolytic pathway requires no oxygen; it is an anaerobic process. In fact, it is the only anaerobic pathway for the production of energy in the form of ATP. Further, it must have evolved in very simple, single-celled organisms, much like the bacterium. These organisms did not have complex organelles in the cytoplasm to carry out specific cellular functions. Thus glycolysis was a process carried out by enzymes free in the cytoplasm. To this day, glycolysis remains an anaerobic process that is carried out by cytoplasmic enzymes, even in cells as complex as our own.

Inspection of the outline of glycolysis seen in Figure 13.8 reveals that glycolysis requires the following materials: glucose (a hexose sugar), two inorganic phosphate groups, four ADP molecules, two **NAD^+** (the oxidized form of the coenzyme **nicotinamide adenine dinucleotide**), two ATP molecules, and ten enzymes to catalyze the individual reactions of the pathway.

Further inspection of Figure 13.8 reveals that there are three major products of glycolysis. These are chemical energy in the form of ATP, chemical energy in the form of NADH, and two pyruvate molecules.

Chemical energy as ATP

Four ATP molecules are formed by the process of **substrate-level phosphorylation.** This means that high-energy phosphoryl groups from one of the substrates in glycolysis are transferred to ADP to form ATP. The two substrates that effect this transfer are 1,3-diphosphoglycerate and phosphoenolpyruvate (steps 6 and 9 in Figure 13.8). Both of these substrates are of higher energy than ATP (Figure 13.2). Thus substrate-level phosphorylation is an exothermic reaction and, as such, is energetically favored. Although four ATP are produced during glycolysis, the *net* gain is only two ATP. This is due to the fact that two ATP are used early in glycolysis. The two ATP produced represent only

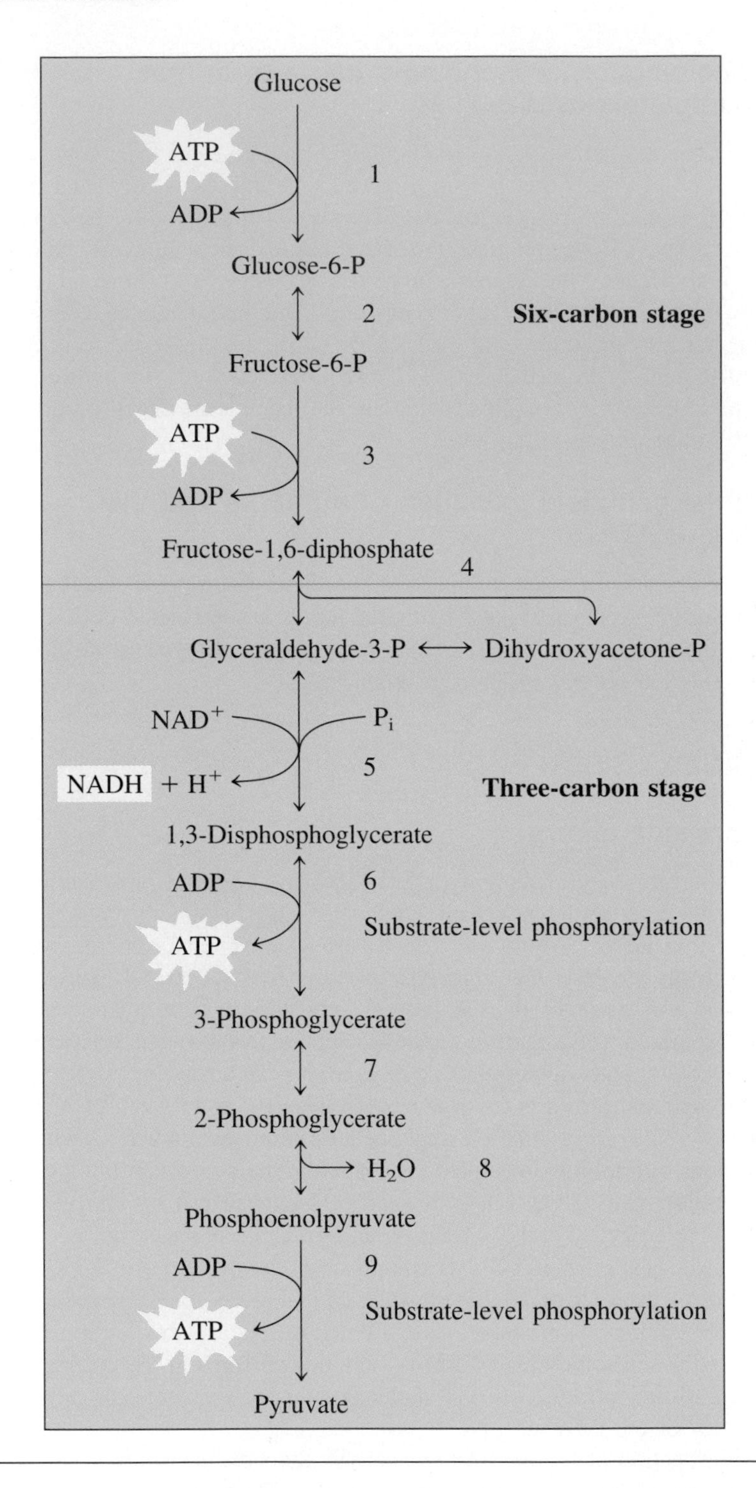

FIGURE 13.8
A summary of the nine reactions of glycolysis.

about 2% of the potential energy of the glucose molecule; therefore glycolysis is a rather inefficient energy-generating process.

The structure of NAD^+ and the way in which it functions as a hydride anion carrier was shown in Figure 12.10 and described in Section 12.7.

Chemical energy in the form of reduced NAD^+, NADH

NADH carries a hydride anion, hydrogen with two electrons ($H:^-$), removed during the oxidation of one of the substrates, glyceraldehyde-3-phosphate (step 5 in Figure 13.8). Under aerobic conditions the electrons and hydrogen atom are donated to an electron transport system for the generation of ATP by **oxidative phosphorylation.** Under anaerobic conditions, NADH is used as reducing power, a source of electrons, in fermentation reactions.

Section 14.3

Two pyruvate molecules

At the end of glycolysis the six-carbon glucose molecule has been converted into two three-carbon pyruvate molecules. The fate of the pyruvate also depends on whether the reactions are occurring in the presence or absence of oxygen. Under aerobic conditions it is used to produce acetyl CoA destined for the citric acid cycle and complete oxidation. Under anaerobic conditions it is used as the substrate in fermentation reactions.

In any event, these last two products, NADH and pyruvate, must be utilized in some way so that glycolysis can continue to function and produce ATP. Inspection of Figure 13.8 reveals the reason for this. In the first place, if pyruvate were allowed to build up, it would cause glycolysis to cease, thereby stopping the generation of energy. Thus pyruvate must be utilized in some kind of follow-up reaction, aerobic or anaerobic. In step 5 we see that glyceraldehyde-3-phosphate is oxidized and NAD^+ serves as the acceptor of the hydride anion in this reaction. The cell has a limited supply of NAD^+. If all the NAD^+ is reduced, none will be available for this reaction, and glycolysis will cease. Therefore NADH must be reoxidized so that glycolysis can continue to produce energy for the cell.

Reactions of glycolysis

The structures of the intermediates of glycolysis are seen in Figure 13.9, along with a concise description of the reactions that occur at each step, and the names of the enzymes that catalyze each reaction. For convenience the structural modifications that occur in each reaction are highlighted.

Glycolysis can be divided into two major segments. The first is the investment of ATP energy. Without this investment, glucose would not have enough energy for glycolysis to continue, and there would be no net energy gain. This segment includes the first four reactions of the pathway. The second major segment involves the remaining reactions of the pathway (5–9), those that result in a net energy gain.

Reaction 1

The substrate, glucose, is phosphorylated by the enzyme **hexokinase.** This reaction is seen in Figure 13.4. The name ''hexokinase'' tells us that this is an enzyme that adds phosphate groups to a six-carbon sugar, glucose. The source of the phosphate group is ATP. On first inspection this reaction seems contrary to the overall purpose of catabolism, the *production* of ATP. The expenditure of ATP in these early reactions must be thought of as an ''investment.'' The cell actually goes into energy ''debt'' in these early reactions, but this is absolutely necessary for two reasons. The first is that glucose is freely permeable to cells, but phosphorylated sugars are not. Thus the phosphorylation of glucose provides a way of trapping it within the cell. The second reason is that the enzyme that splits fructose-1,6-diphosphate into glyceraldehyde-3-phosphate in reaction 4 will not accept any other substrate; the energy must be provided to effect this conversion if we are to produce the desired intermediate, glyceraldehyde-3-phosphate:

$$\text{Glucose} + \text{ATP} \xrightarrow{\text{Hexokinase}} \text{Glucose-6-phosphate} + \text{ADP} + H^+$$

Reaction 2

The glucose-6-phosphate formed in the first reaction undergoes a rearrangement. The enzyme **phosphoglucoisomerase** rearranges the hydrogens and oxygens to produce fructose-6-phosphate. The result is that the C-1 carbon of the six-carbon sugar is exposed; it

Section 7.1

Glucose

ATP → ADP, Hexokinase (1)

1. Glucose is phosphorylated at the expense of ATP to produce glucose-6-phosphate.

Glucose-6-phosphate

Phosphogluco-isomerase (2)

2. Glucose-6-phosphate is rearranged to produce fructose-6-phospate.

Fructose-6-phosphate

ATP → ADP, Phospho-fructokinase (3)

3. Fructose-6-phosphate is phosphorylated to produce fructose-1,6-diphosphate at the expense of another ATP. The expenditure of 2 ATP represents an energy investment to "activate" the glucose for its eventual oxidation.

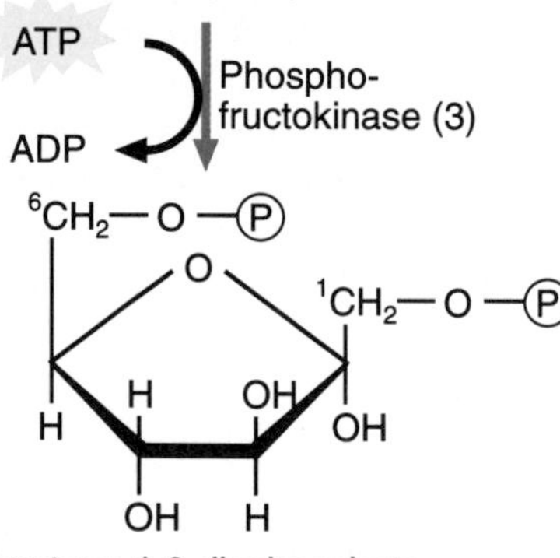

Fructose-1,6-disphosphate

Aldolase (4)

4. Aldolase cleaves the six carbon fructose-1,6-diphosphate into two nonidentical three-carbon molecules, dihydroxyacetone phosphate and glyceraldehyde-3-phosphate. The dihydroxyacetone phosphate is converted to glyceraldehyde-3-phosphate by the enzyme triose isomerase.

Dihydroxyacetone phosphate ⇌ Glyceraldehyde-3-phosphate

FIGURE 13.9
The intermediates and enzymes of glycolysis.

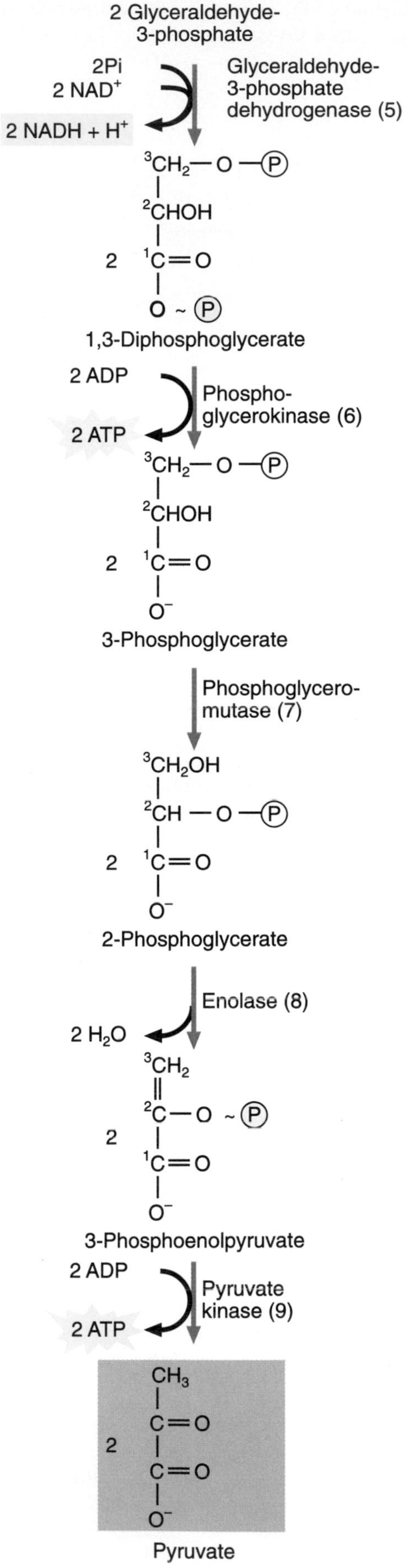

2 Glyceraldehyde-3-phosphate

2Pi
2 NAD^+
2 NADH + H^+
Glyceraldehyde-3-phosphate dehydrogenase (5)

$^3CH_2$—O—Ⓟ
^{2}CHOH
2 ^{1}C=O
O ~ Ⓟ

1,3-Diphosphoglycerate

5. Glyceraldehyde-3-phosphate is oxidized and NADH is produced. An inorganic phosphate group is transferred to the carboxylate group to produce 1,3-diphosphoglycerate.

2 ADP
2 ATP
Phosphoglycerokinase (6)

$^3CH_2$—O—Ⓟ
^{2}CHOH
2 ^{1}C=O
O^-

3-Phosphoglycerate

6. ATP energy is produced in the first substrate level phosphorylation in the pathway. The phosphoryl group is transferred from the substrate to ADP to produce ATP.

Phosphoglyceromutase (7)

$^3CH_2OH$
^{2}CH—O—Ⓟ
2 ^{1}C=O
O^-

2-Phosphoglycerate

7. The C-3 phosphoryl group of 3-phosphoglycerate is transferred to the second carbon.

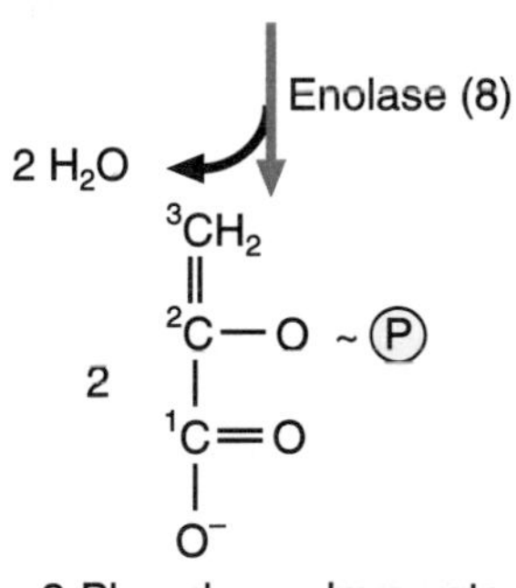

3-Phosphoenolpyruvate

8. Dehydration of 2-phosphoglycerate generates the energy-rich molecule phosphoenolpyruvate.

2 ADP
2 ATP
Pyruvate kinase (9)

CH_3
C=O
2
C=O
O^-

Pyruvate

9. The final substrate level phosphorylation produces ATP and pyruvate.

is no longer part of the ring structure. Examination of the open-chain structures reveals that this isomerization converts an aldose sugar into a ketose sugar:

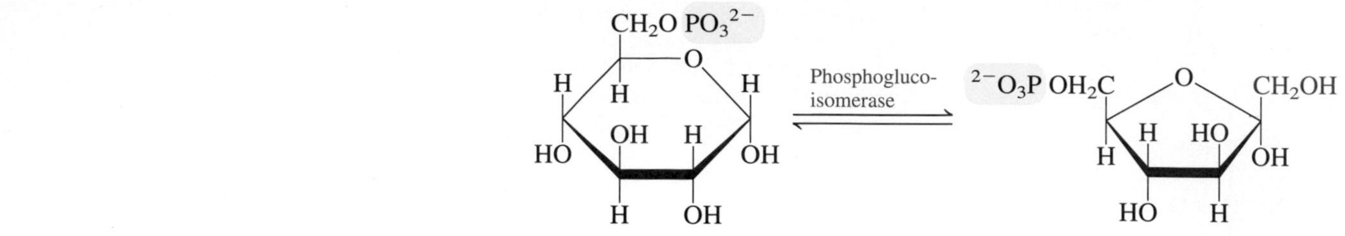

Glucose-6-phosphate $\xrightleftharpoons{\text{Phosphogluco-isomerase}}$ Fructose-6-phosphate

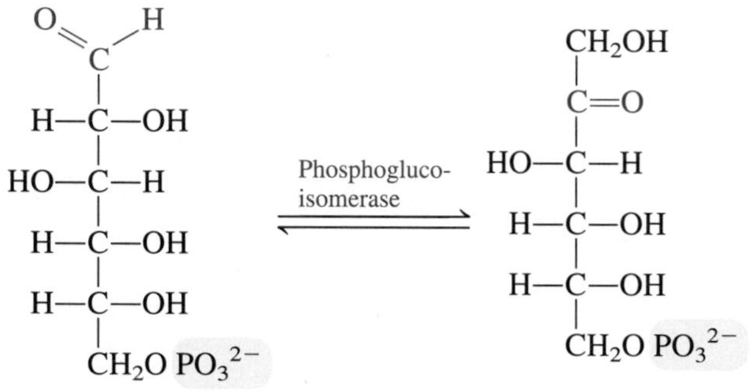

Glucose-6-phosphate (an aldose) $\xrightleftharpoons{\text{Phosphogluco-isomerase}}$ Fructose-6-phosphate (a ketose)

Reaction 3

A second energy ''investment'' is made by the enzyme **phosphofructokinase.** ATP is hydrolyzed, and a phosphoester linkage between the phosphoryl group and the now exposed C-1 hydroxyl group of fructose-6-phosphate is formed. The product is fructose-1,6-diphosphate. At this point, our energy expenditures have produced the appropriate substrate for the final reaction of the energy-investment segment of glycolysis:

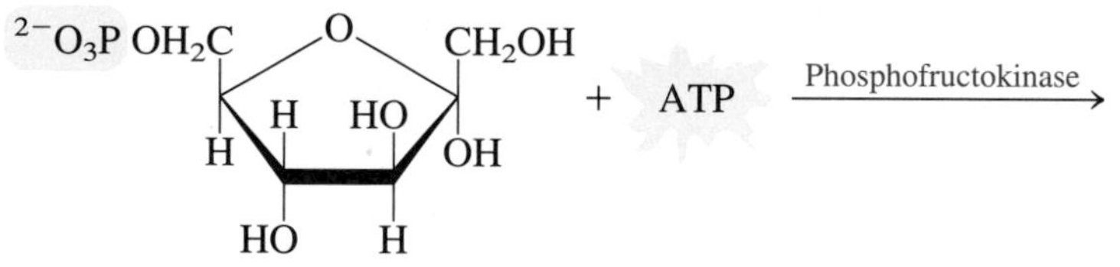

Fructose-6-phosphate + ATP $\xrightarrow{\text{Phosphofructokinase}}$ Fructose-1,6-disphosphate + ADP + H^+

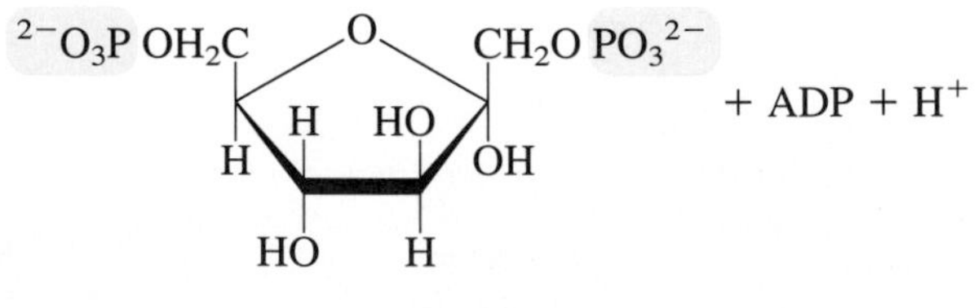

Reaction 4

Fructose-1,6-diphosphate is split by the enzyme **aldolase** into two three-carbon intermediates: glyceraldehyde-3-phosphate (3-PGAL) and dihydroxyacetone phosphate. Since 3-PGAL is the only substrate that can be used by the next enzyme in the pathway, the dihydroxyacetone phosphate is rearranged to become a second molecule of 3-PGAL. The enzyme that mediates this isomerization is **triose isomerase:**

$$\begin{array}{c} CH_2OPO_3^{2-} \\ | \\ C{=}O \\ | \\ HO{-}C{-}H \\ | \\ H{-}C{-}OH \\ | \\ H{-}C{-}OH \\ | \\ CH_2OPO_3^{2-} \end{array} \quad \underset{}{\overset{\text{Aldolase}}{\rightleftharpoons}} \quad \begin{array}{c} CH_2OPO_3^{2-} \\ | \\ C{=}O \\ | \\ HO{-}C{-}H \\ | \\ H \end{array} \quad + \quad \begin{array}{c} H \diagdown \; \diagup\!\!\!=O \\ C \\ | \\ H{-}C{-}OH \\ | \\ CH_2OPO_3^{2-} \end{array}$$

Fructose-1,6-disphosphate — Dihydroxyacetone phosphate — Glyceraldehyde-3-phosphate

$$\begin{array}{c} CH_2OH \\ | \\ C{=}O \\ | \\ CH_2OPO_3^{2-} \end{array} \quad \overset{\text{Triose isomerase}}{\rightleftharpoons} \quad \begin{array}{c} O{=}C{-}H \\ | \\ H{-}C{-}OH \\ | \\ CH_2OPO_3^{2-} \end{array}$$

Dihydroxyacetone phosphate — Glyceraldehyde 3-phosphate

Reaction 5

This reaction involves the oxidation of the aldehyde of glyceraldehyde-3-phosphate to a carboxylic acid by **glyceraldehyde-3-phosphate dehydrogenase.** This is the first step in glycolysis that harvests energy, and it involves the coenzyme nicotinamide adenine dinucleotide (NAD^+). NAD^+ is reduced to NADH in step 5 of glycolysis. This reaction occurs in two steps. First, NAD^+ is reduced to NADH as the aldehyde group of glyceraldehyde-3-phosphate is oxidized to a carboxylic acid. Second, an inorganic phosphate group is transferred to the carboxylate group to give 1,3-diphosphoglycerate. Notice that the phosphoester bond is denoted with a squiggle (~), indicating that this is a high-energy bond:

$$\begin{array}{c} O{=}C{-}H \\ | \\ H{-}C{-}OH \\ | \\ CH_2OPO_3^{2-} \end{array} \quad + NAD^+ + P_i \quad \overset{\text{Glyceraldehyde 3-phosphate dehydrogenase}}{\rightleftharpoons} \quad \begin{array}{c} O{=}C{-}OPO_3^{2-} \\ | \\ H{-}C{-}OH \\ | \\ CH_2OPO_3^{2-} \end{array} \quad + NADH + H^+$$

Glyceraldehyde 3-phosphate — 1,3-Diphosphoglycerate

Reaction 6

This reaction is the first step of the pathway in which energy is harvested in the form of *ATP*. The phosphoryl group of 1,3-diphosphoglycerate is transferred to ADP in the first substrate level phosphorylation of glycolysis. This reaction is mediated by the enzyme **phosphoglycerokinase:**

$$\begin{array}{c} O{=}C{-}OPO_3^{2-} \\ | \\ H{-}C{-}OH \\ | \\ CH_2OPO_3^{2-} \end{array} \quad + ADP \quad \overset{\text{Phosphoglycerokinase}}{\rightleftharpoons} \quad \begin{array}{c} O{=}C{-}O^- \\ | \\ H{-}C{-}OH \\ | \\ CH_2OPO_3^{2-} \end{array} \quad + ATP$$

1,3-Diphosphoglycerate — 3-Phosphoglycerate

Reaction 7

This is the last of the isomerization reactions in the pathway and is mediated by the enzyme **phosphoglyceromutase.** In this reaction the high-energy phosphoryl group attached to the third carbon of 3-phosphoglycerate is transferred to the second carbon and the product is 2-phosphoglycerate:

$$\begin{array}{c} O{=}C{-}O^- \\ | \\ H{-}C{-}OH \\ | \\ H{-}C{-}O\,PO_3^{2-} \\ | \\ H \end{array} \underset{\text{Phosphoglyceromutase}}{\rightleftharpoons} \begin{array}{c} O{=}C{-}O^- \\ | \\ H{-}C{-}O\,PO_3^{2-} \\ | \\ H{-}C{-}OH \\ | \\ H \end{array}$$

3-Phosphoglycerate 2-Phosphoglycerate

Reaction 8

In this step the enzyme **enolase** mediates an unusual reaction in which a dehydration reaction, the removal of a water molecule, generates an energy-rich product, phosphoenolpyruvate:

$$\begin{array}{c} O{=}C{-}O^- \\ | \\ H{-}C{-}O\,PO_3^{2-} \\ | \\ H{-}C{-}OH \\ | \\ H \end{array} \overset{\text{Enolase}}{\rightleftharpoons} \begin{array}{c} O{=}C{-}O^- \\ | \\ C{-}O\,PO_3^{2-} \\ \| \\ H{-}C \\ | \\ H \end{array} + H_2O$$

2-Phosphoglycerate Phosphoenolpyruvate

Reaction 9

Here we see the final substrate-level phosphorylation in the pathway. **Pyruvate kinase** mediates the reaction in which phosphoenolpyruvate serves as a donor of the phosphoryl group that is transferred to ADP to produce ATP. The final product of glycolysis is pyruvate:

$$\begin{array}{c} O{=}C{-}O^- \\ | \\ C{-}O\,PO_3^{2-} \\ \| \\ H{-}C \\ | \\ H \end{array} + ADP + H^+ \xrightarrow{\text{Pyruvate kinase}} \begin{array}{c} O{=}C{-}O^- \\ | \\ C{=}O \\ | \\ CH_3 \end{array} + ATP$$

Phosphoenolpyruvate Pyruvate

It should be noted that each of the reactions 5–9 occurs twice per glucose molecule, since the starting six-carbon sugar is split into two three-carbon molecules. Thus in reaction 5, two NADH molecules are generated, and in reactions 6 and 9, two ATP are generated, for a total of four ATP. However, the net ATP gain from this pathway is only two ATP. In steps 1–4 there was an energy investment of two ATP. This was paid back in step 6, in which two ATP were produced by substrate-level phosphorylation. The actual energy yield is produced in reaction 9 when two phosphoenolpyruvate molecules donate high-energy phosphoryl groups to two ADP molecules and produce two ATP.

Question 13.1 In what respect is generation of ATP by glycolysis unique?

Question 13.2 What is the importance of glycolysis? Why is ATP production necessary?

Question 13.3 What is the evidence that glycolysis is the most ancient energy-generating pathway in existence?

Question 13.4 How do the names of the enzymes of the glycolytic pathway relate to the reactions that they mediate?

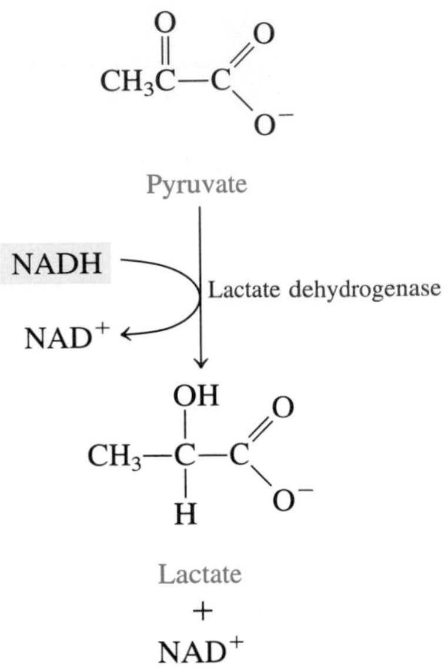

$CH_3C(=O)-COO^-$

Pyruvate

NADH → NAD^+ (Lactate dehydrogenase)

$CH_3-CH(OH)-COO^-$

Lactate + NAD^+

FIGURE 13.10
Lactate fermentation.

13.4 FERMENTATIONS

In the general considerations of glycolysis we noted that the pyruvate that is generated must be utilized so that the pathway will continue to produce energy. Similarly, the NADH produced by glycolysis in step 5 (Figure 13.9) must be regenerated at a later time, or glycolysis will grind to a halt as the available NAD^+ is used up. If the cell is functioning under aerobic conditions, NADH will be reoxidized, and pyruvate will be completely oxidized by respiration. Under anaerobic conditions, however, cells of different types employ fermentation reactions to accomplish these purposes. We will examine two types of fermentation pathways in detail: lactate fermentation and alcoholic fermentation.

Sections 14.2 and 14.3

Lactate fermentation

This is a fermentation familiar to anyone who has performed strenuous exercise. Under these conditions you may exercise beyond the capacity of your lungs and circulatory system to deliver oxygen to the working muscles. The aerobic energy-generating pathways will no longer be able to supply enough ATP, but the muscles still demand energy. Under these anaerobic conditions, pyruvate can be reduced to lactate by the enzyme **lactate dehydrogenase.** NADH is the actual reducing agent for this process, which regenerates the NAD^+ required for the continued anaerobic functioning of glycolysis (Figure 13.10). As pyruvate is reduced, NADH is oxidized, and NAD^+ is again available, permitting glycolysis to continue producing two ATP per glucose.

The lactate produced in the working muscle passes into the blood. Eventually, if strenuous exercise is continued, the concentration of lactate becomes so high that this fermentation can no longer proceed. Glycolysis, and thus energy production, ceases. The muscle, deprived of energy, can no longer function. This point of exhaustion is called the **anaerobic threshold.**

Of course, most of us do not exercise to this point. When exercise is finished, the body begins the process of reclaiming all of the potential energy that was lost in the form of lactate. This requires oxygen and is part of the reason that we continue to breathe heavily for several minutes after strenuous exercise. We are paying back the **oxygen debt.** The liver takes up the lactate from the blood and converts it back to pyruvate. Now that a sufficient supply of oxygen is available, the pyruvate can be completely oxidized in the much more efficient aerobic energy-generating reactions to replenish the store of ATP. Alternatively, the pyruvate may be used to restore the supply of liver and muscle glycogen through the Cori Cycle.

Section 13.5

A variety of bacteria are able to carry out lactate fermentation under anaerobic conditions. This is of great importance in the dairy industry, since these organisms are used to produce yogurt and some cheeses. The tangy flavor of yogurt is contributed by the lactate

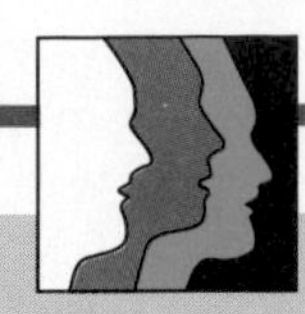

A HUMAN PERSPECTIVE

Fermentations: The Good, the Bad, and the Ugly

In this chapter we have seen that fermentation is an anaerobic, cytoplasmic process that allows continued energy generation by glycolysis. Energy production can continue because the pyruvate produced by the pathway is utilized in the fermentation and because NAD^+ is regenerated.

The stable end products of alcohol fermentation are CO_2 and ethyl alcohol. These have been used by humankind in a variety of ways, including the production of alcoholic beverages, bread making, and as alternative fuel sources.

The production of bread, wine, and cheese depends on the fermentation process.

If alcohol fermentation is carried out by using fruit juices in a vented vat, the CO_2 will escape and the result will be a still wine (not bubbly). But conditions must remain anaerobic; otherwise, fermentation will stop, and aerobic energy-generating reactions will ruin the wine. Fortunately for vintners (wine makers), when a vat is fermenting actively, enough CO_2 is produced to create a layer that keeps the oxygen-containing air away from the fermenting juice, thus maintaining an anaerobic atmosphere.

Now suppose we want to make a sparkling wine, such as champagne. To do this, we simply have to trap the CO_2 produced. In this case the fermentation proceeds in a sealed bottle, a very strong bottle. Both the fermentation products, CO_2 and ethyl alcohol, accumulate. Under pressure within the sealed bottle the CO_2 remains in solution. When the top is "popped," the pressure is released, and the CO_2 comes out of solution in the form of bubbles.

In either case the fermentation continues until the alcohol concentration reaches 12–13%. At that point the yeast "stews in its own juices"! That is, 12–13% ethanol kills the yeast that produces it. This points out a last generalization about fermentations. The stable fermentation end product, whether it is lactate or ethyl alcohol, eventually accumulates to a concentration that is toxic to the organism. Muscle fatigue is the early effect of lactate buildup in the working muscle. In the same way, continued accumulation of the fermentation product can lead to concentrations that are fatal if there is no means of getting rid of the toxic product or of getting away from it. For single-celled organisms the result is generally death. Our bodies have evolved in such a way that lactate buildup contributes to muscle fatigue that causes the exerciser to stop

produced by these bacteria. Unfortunately, similar organisms are also responsible for spoilage of milk. The lactate produced by oral bacteria is responsible for the gradual removal of calcium from tooth enamel and the resulting dental cavities.

See Chapter 17, "A Human Perspective: Tooth Decay and Simple Sugars"

Alcohol fermentation

This type of fermentation has been appreciated, if not understood, since the dawn of civilization. The fermentation process itself was discovered by Louis Pasteur during his studies of the chemistry of wine making and "diseases of wines." Under anaerobic conditions, yeasts are able to ferment the sugars produced by fruit and grains. The sugars are dismantled from glucose to pyruvate in glycolysis. The subsequent fermentation is catalyzed by two enzymes. First, **pyruvate decarboxylase** removes CO_2 from the pyruvate, producing CO_2 and acetaldehyde (Figure 13.11). Second, **alcohol dehydrogenase** catalyzes the reduction of acetaldehyde to ethyl alcohol but, more important, reoxidizes NADH in the process. The regeneration of NAD^+ allows glycolysis to continue, just as in the case of lactate fermentation. The two products of alcoholic fermentation, then, are ethyl alcohol and CO_2. We have taken advantage of this fermentation in the production of the exercise. Then the lactate is removed from the blood by the process of gluconeogenesis.

Another application of the alcohol fermentation is the use of yeast in bread making. When we mix the water, sugar, and dried yeast, the yeast cells begin to grow and carry out the process of fermentation. This mixture is then added to the flour, milk, shortening, and salt, and the dough is placed in a warm place to rise. The yeast continues to grow and ferment the sugar, producing CO_2 that causes the bread to rise. Of course, when we bake the bread, the yeast is killed, and the ethanol evaporates, but we are left with a light and airy loaf of bread.

Today, alcohol produced by fermentation is being considered as an alternative fuel to replace the use of some fossil fuels. Geneticists and bioengineers are trying to develop strains of yeast that can survive higher alcohol concentrations and thus convert more of the sugar of corn and other grains into alcohol.

Bacteria perform a variety of other fermentations. The propionibacteria produce propionic acid and CO_2. The acid gives Swiss cheese its characteristic flavor, and the CO_2 gas produces the characteristic holes in the cheese. Other bacteria, the clostridia, perform a fermentation that is responsible in part for the horrible symptoms of gas gangrene. When these bacteria are inadvertently introduced into deep tissues by a puncture wound, they find a nice anaerobic environment in which to grow. In fact, these organisms are **obligate anaerobes,** that is, they are killed by even a small amount of oxygen. As they grow, they perform a fermentation called the butyric acid, butanol, acetone fermentation. This results in the formation of CO_2, the gas associated with gas gangrene. The CO_2 infiltrates the local tissues and helps to maintain an anaerobic environment because oxygen from the local blood supply cannot enter the area of the wound. Now able to grow well, these bacteria produce a variety of toxins and enzymes that cause extensive tissue death and necrosis. In addition, the fermentation produces acetic acid, ethyl alcohol, acetone, isopropyl alcohol, butanol, and butyric acid (which is responsible, along with the necrosis, for the characteristic foul smell of gas gangrene). Certainly, the presence of these organic chemicals in the wound causes enhanced tissue death.

Gas gangrene is very difficult to treat. Because the bacteria establish an anaerobic region of cell death and cut off the local circulation, systemic antibiotics do not infiltrate the wound and kill the bacteria. Even our immune response is stymied. Treatment usually involves surgical removal of the necrotic tissue accompanied by antibiotic therapy. In some cases a hyperbaric oxygen chamber is employed. The infected extremity is placed in an environment with a very high partial pressure of oxygen. The oxygen forced into the tissues is poisonous to the bacteria, and they die.

These are but a few examples of the fermentations that have an impact on humans. Regardless of the specific chemical reactions, all fermentations share the following traits:

- They use pyruvate produced in glycolysis.
- They reoxidize the NADH.
- They are self-limiting because the accumulated stable fermentation end product eventually kills the cell that produces it.

wines and other alcoholic beverages and in the process of bread making. These applications and other fermentations are described in "A Human Perspective: Fermentations: The Good, the Bad, and the Ugly."

FIGURE 13.11
Alcohol fermentation.

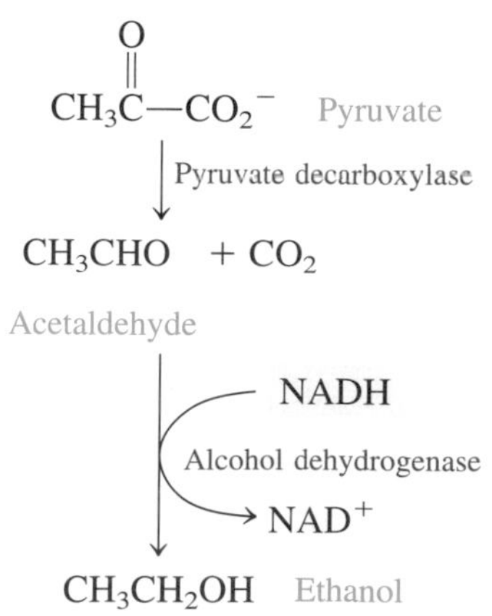

$$CH_3C(=O)—CO_2^-$$ Pyruvate

↓ Pyruvate decarboxylase

$$CH_3CHO + CO_2$$ Acetaldehyde

↓ NADH → Alcohol dehydrogenase → NAD^+

$$CH_3CH_2OH$$ Ethanol

Question 13.5 How is the alcohol fermentation in yeast similar to lactate production in skeletal muscle?

13.5 GLUCONEOGENESIS: THE SYNTHESIS OF GLUCOSE

Under some conditions the body must make glucose. This is necessary after strenuous exercise to replenish the liver and muscle stores of glycogen. It also occurs under starvation conditions to maintain adequate blood glucose levels because brain cells and red blood cells rely exclusively on glucose as their energy source.

Under these conditions, glucose is produced by the process of **gluconeogenesis,** the production of glucose from noncarbohydrate precursors. Gluconeogenesis occurs primarily in the liver. One precursor is lactate, as when the body cleanses the blood of this fermentation product and uses it to replace glycogen stores. Other precursors include all the amino acids, except leucine and lysine, and glycerol from fats. The amino acids and glycerol are generally used only under starvation conditions.

On first examination it appears that gluconeogenesis is simply the reverse of glycolysis (Figure 13.12) because the intermediates of the two pathways are identical. However, this is not the case because steps 1, 3, and 9 of glycolysis are irreversible, and therefore the reverse reactions must be mediated by other enzymes. Step 1 of glycolysis is mediated by hexokinase, which phosphorylates glucose. In gluconeogenesis the dephosphorylation of glucose-6-phosphate is carried out by the enzyme **glucose-6-phosphatase,** which is found in the liver but not in muscle. Similarly, reaction 3, the phosphorylation of fructose-6-phosphate by phosphofructokinase, is irreversible. That step is bypassed in gluconeogenesis by using the enzyme **fructose diphosphatase.** Finally, the phosphorylation of ATP by pyruvate kinase, step 9 of glycolysis, cannot be reversed. The reaction that converts pyruvate to phosphoenolpyruvate actually involves two enzymes and some unu-

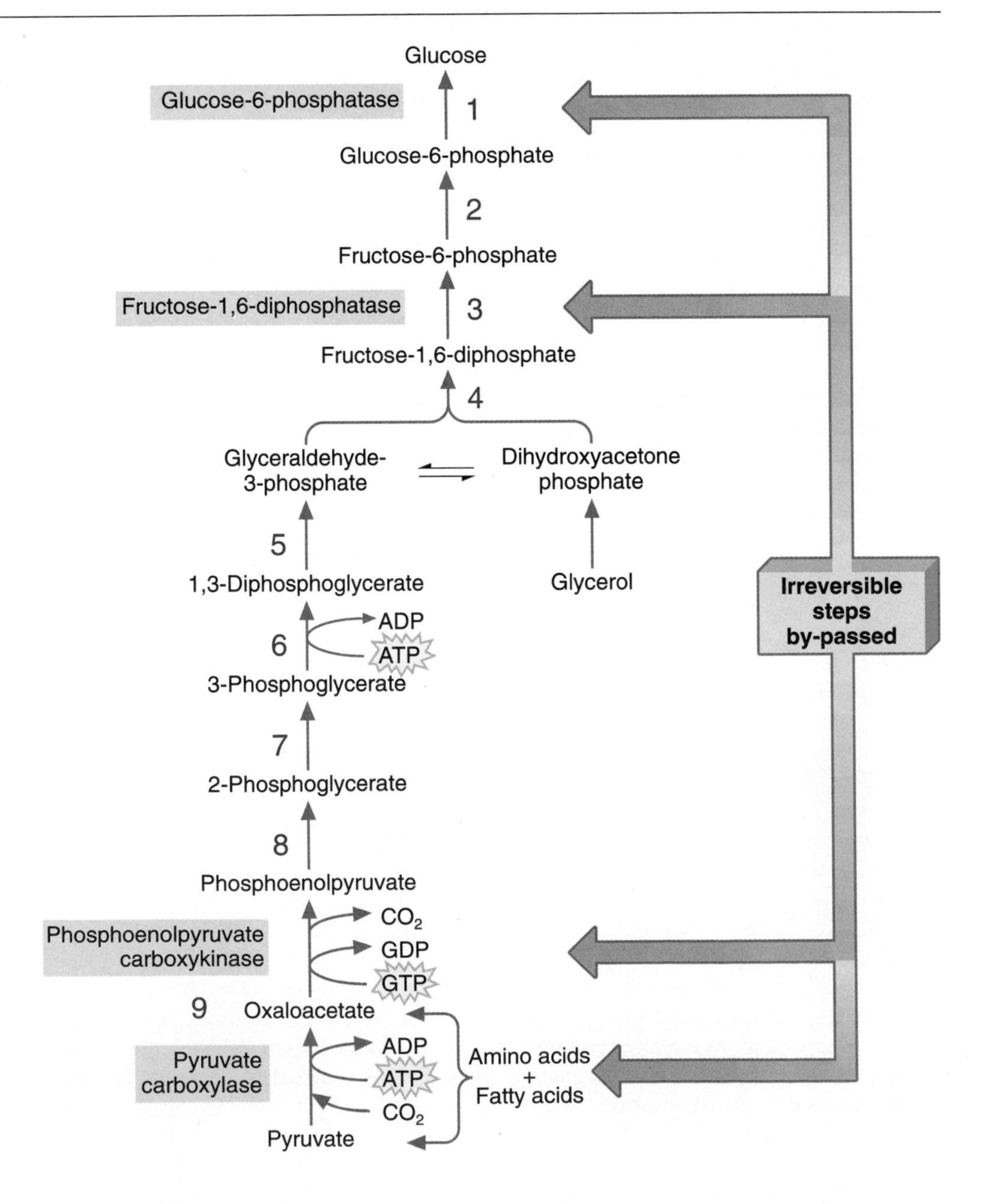

FIGURE 13.12
Comparison of glycolysis and gluconeogenesis.

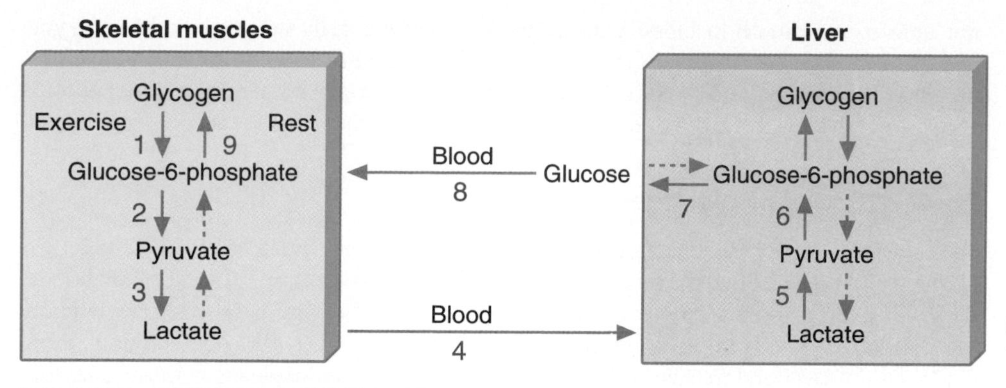

FIGURE 13.13
The Cori Cycle.

sual mechanisms. First, the pyruvate is carboxylated by the addition of atmospheric CO_2, and the product is the four-carbon compound oxaloacetate. This is mediated by the enzyme **pyruvate carboxylase.** Then **phosphoenolpyruvate carboxykinase** removes the CO_2 and adds a phosphoryl group to the oxaloacetate. The donor of the phosphoryl group in this unusual reaction is **guanosine triphosphate** (GTP). This is a nucleotide like ATP, except that the nitrogenous base is guanine.

Section 16.1

If glycolysis and gluconeogenesis were not regulated in some fashion, the two pathways would occur simultaneously, with the disastrous effect that nothing would get done. Three convenient sites for this regulation are the three bypass reactions. Step 3 of glycolysis is mediated by the enzyme phosphofructokinase. This enzyme is stimulated by high concentrations of AMP, ADP, and inorganic phosphate, signals that the cell needs energy. When the enzyme is active, glycolysis proceeds. On the other hand, when ATP is plentiful, phosphofructokinase is inhibited, and fructose diphosphatase is stimulated. The net result is that in times of energy excess (high concentrations of ATP), gluconeogenesis will occur.

As wc have seen, the conversion of lactate into glucose is important in mammals. As the muscles work, they produce lactate, which is converted back to glucose in the liver. The glucose is transported into the blood and from there back to the muscle. In the muscle it can be used to generate energy, or it can be used to replenish the muscle stores of glycogen. This cyclic process between the liver and skeletal muscles is called the **Cori Cycle,** shown in Figure 13.13. Through this cycle, gluconeogenesis produces enough glucose to restore the depleted muscle glycogen reservoir within 48 hours.

Question 13.6 What are the major differences between gluconeogenesis and glycolysis?

13.6 GLYCOGEN SYNTHESIS AND DEGRADATION

Glucose is the sole source of energy of mammalian red blood cells and the major source of energy for the brain. Neither red blood cells nor the brain can store glucose; thus a constant supply must be available as blood glucose. This is provided by dietary glucose and by the production of glucose either by gluconeogenesis or by **glycogenolysis,** the degradation of glycogen. Glycogen is a long-branched-chain polymer of glucose. Stored in the liver and skeletal muscles, it is the principal storage form of glucose.

The total amount of glucose in the blood of a 70-kg (approximately 150-lb) adult is about 20 g, but the brain alone consumes 5–6 g of glucose per hour. Breakdown of glycogen in the liver mobilizes the glucose when hormonal signals register a need for increased levels of blood glucose. Skeletal muscle also contains substantial stores of glycogen, which provide energy for rapid energy contraction. However, this glycogen is not able to contribute to blood glucose because muscle cells do not have the enzyme glucose-6-phosphatase. Because glucose cannot be formed from the glucose-6-phosphate, it cannot be released into the bloodstream. The total glycogen content of a 70-kg adult is about 200 g, enough to provide for metabolic needs for about one day.

The structure of glycogen

Glycogen is a highly branched glucose polymer in which the ''main chain'' is linked by α $(1 \longrightarrow 4)$ glycosidic bonds. The polymer also has numerous α $(1 \longrightarrow 6)$ glycosidic bonds, which provide many branch points along the chain. This structure is shown schematically in Figure 13.14. **Glycogen granules** with a diameter of 100–400 Å are found in the cytoplasm of liver and muscle cells. These granules exist in complexes with the enzymes that are responsible for glycogen synthesis and degradation. The structure of such a granule is seen also in Figure 13.14.

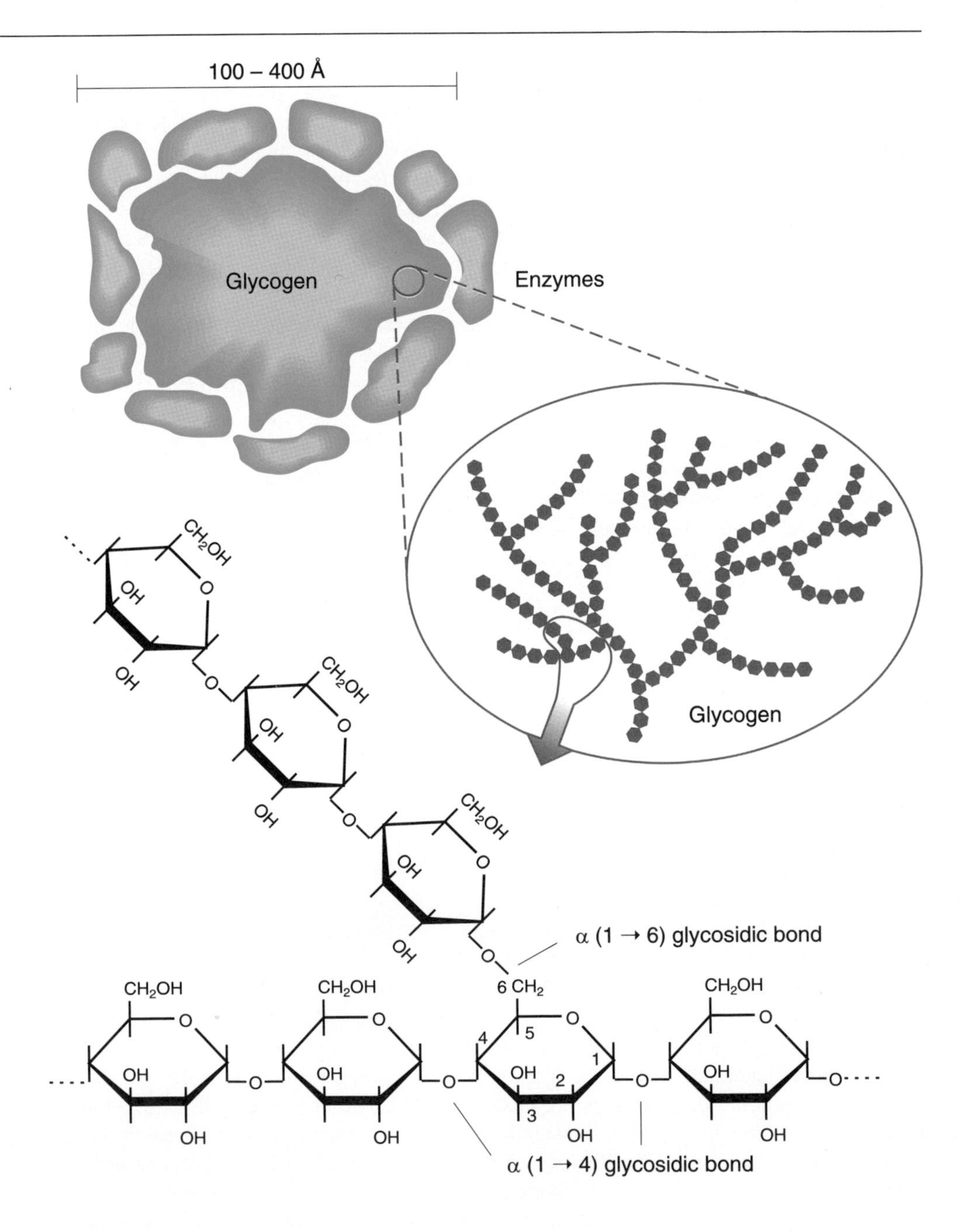

FIGURE 13.14
The structure of glycogen and a glycogen granule.

The enzymatic degradation of glycogen

Two hormones control glycogenolysis, the degradation of glycogen. These are **glucagon,** a peptide hormone synthesized in the pancreas, and **epinephrine,** produced in the adrenal glands. Glucagon is released from the pancreas in response to low blood glucose, and epinephrine is released from the adrenal glands in response to a threat or a stress. Both situations require an increase in blood glucose, and both hormones function by altering the activity of two enzymes, glycogen phosphorylase and glycogen synthetase. **Glycogen phosphorylase** is involved in glycogen degradation and is activated; **glycogen synthetase** is involved in glycogen synthesis and is inactivated. The steps in glycogen degradation are summarized below:

Step 1. The enzyme glycogen phosphorylase catalyzes phosphorolysis of a glucose residue at one end of glycogen (Figure 13.15). The reaction involves the displacement of a glucose unit of glycogen by a phosphate group. As a result of phosphorolysis, glucose-1-phosphate is produced without using ATP as the phosphoryl group donor.

FIGURE 13.15
The action of glycogen phosphorylase in glycogenolysis.

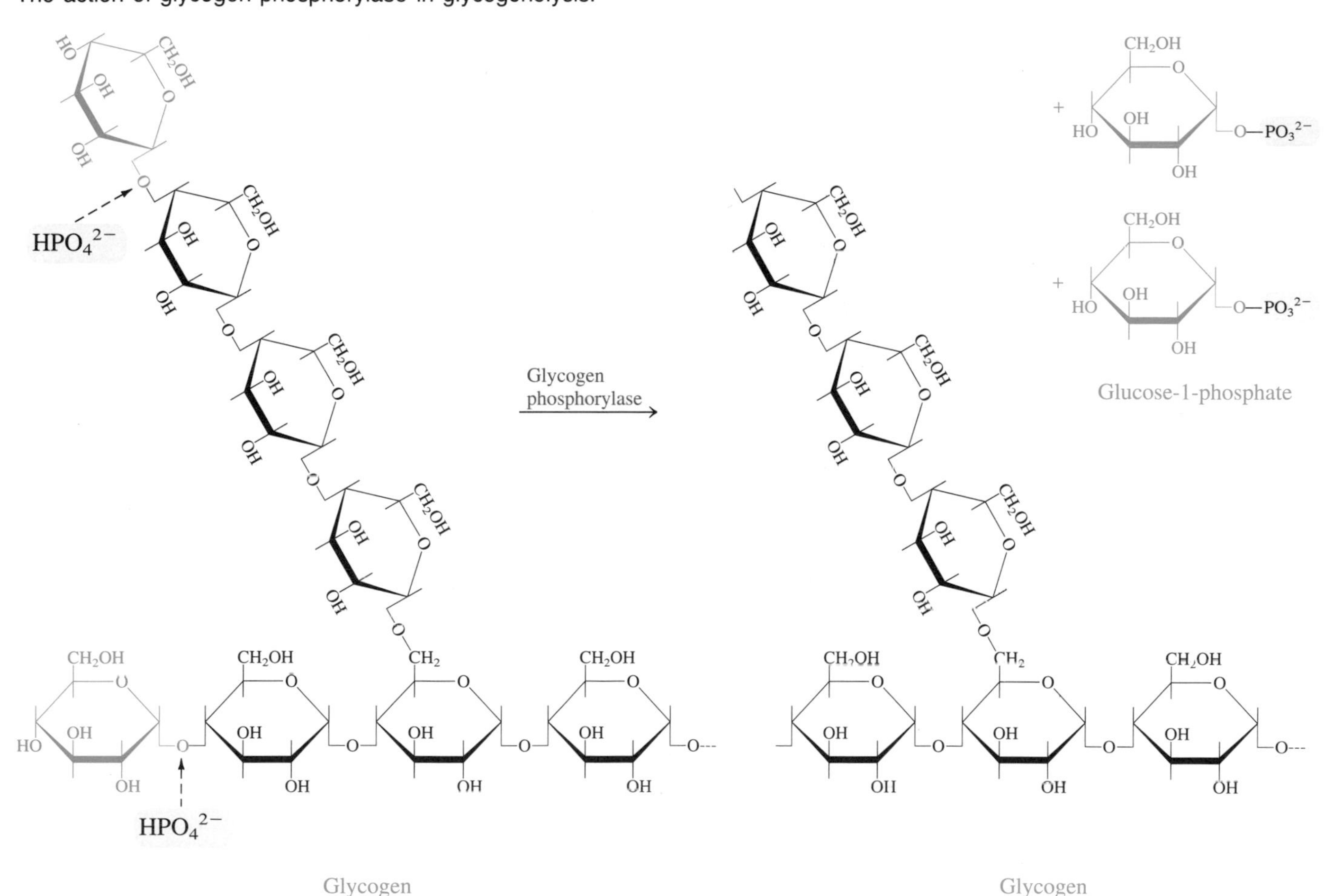

$$\text{Glycogen (glucose)}_x + n\ HPO_4^{2-} \xrightarrow[\text{phosphorylase}]{\text{Glycogen}} (\text{glucose})_{x-n} + n \text{ glucose-1-phosphate}$$

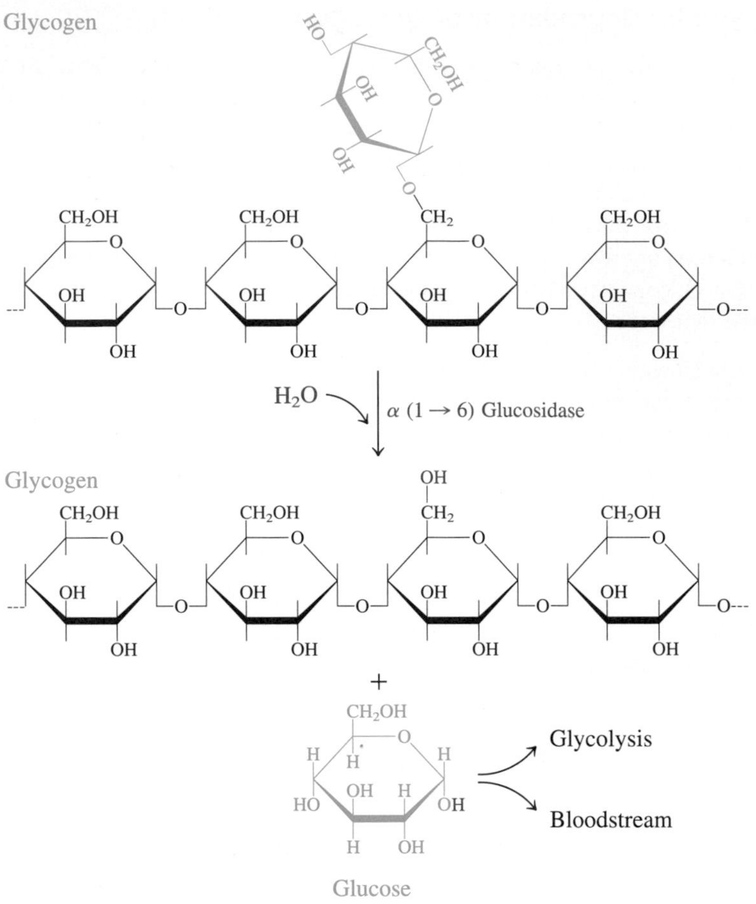

FIGURE 13.16
The action of the debranching enzyme in glycogen degradation.

Step 2. Glycogen contains many branches bound to the α (1 ⟶ 4) backbone by α (1 ⟶ 6) glycosidic bonds. These branches must be removed to allow the complete degradation of glycogen. The extensive action of glycogen phosphorylase produces a smaller polysaccharide with a single glucose residue bound by an α (1 ⟶ 6) glycosidic bond to the main chain. The **debranching enzyme, α (1 ⟶ 6) glycosidase,** hydrolyzes the α (1 ⟶ 6) glycosidic bond at a branch point and frees one molecule of glucose (Figure 13.16). This molecule of glucose can be phosphorylated and utilized in glycolysis, or it may be released into the bloodstream for use elsewhere. Hydrolysis of the branch bond liberates another stretch of α-linked glucose for the action of glycogen phosphorylase.

Step 3. Glucose-1-phosphate is converted to glucose-6-phosphate by **phosphoglucomutase** (Figure 13.17). Glucose originally stored in glycogen enters glycolysis through the action of phosphoglucomutase. Alternatively, in the liver and kidneys it may be dephosphorylated for transport into the bloodstream.

Question 13.7 Explain how glucagon affects the synthesis and degradation of glycogen.

Insulin and glycogenesis

The hormone **insulin,** produced by the pancreas in response to high blood glucose levels, stimulates the synthesis of glycogen, **glycogenesis.** Insulin is perhaps one of the most

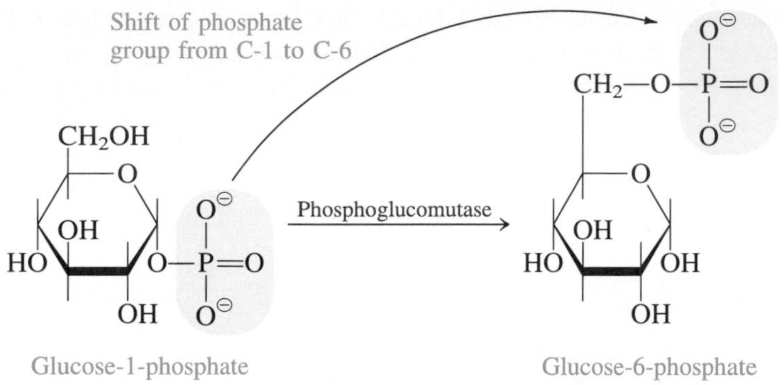

Shift of phosphate group from C-1 to C-6

Phosphoglucomutase

Glucose-1-phosphate

Glucose-6-phosphate

FIGURE 13.17
The action of phosphoglucomutase in glycogen degradation.

influential hormones in the body because it directly alters the metabolism and uptake of glucose in all but a few cells of the body.

When blood glucose rises, as after a meal, the beta cells of the pancreas secrete insulin. It immediately accelerates the uptake of glucose by all the cells of the body except the brain and certain blood cells. In these cells the uptake of glucose is insulin-independent. The increased uptake of glucose is especially marked in the liver, heart, skeletal muscle, and adipose tissue.

In the liver, insulin promotes glycogen synthesis and storage. It does so by influencing the enzymes involved in glycogen metabolism in a fashion that is opposed to the action of glucagon. Insulin inhibits glycogen phosphorylase, thus inhibiting glycogenolysis. It also stimulates glycogen synthetase and glucokinase, two enzymes that are involved in glycogen synthesis.

Although glycogenesis and glycogenolysis share some reactions in common, the two pathways are not simply the reverse of one another. Glycogenesis involves some very unusual reactions, which we will now examine in detail.

The first reaction of glycogen synthesis traps glucose within the cell by phosphorylating it. In this reaction, mediated by the enzyme **glucokinase,** ATP serves as a phosphoryl donor, and glucose-6-phosphate is formed:

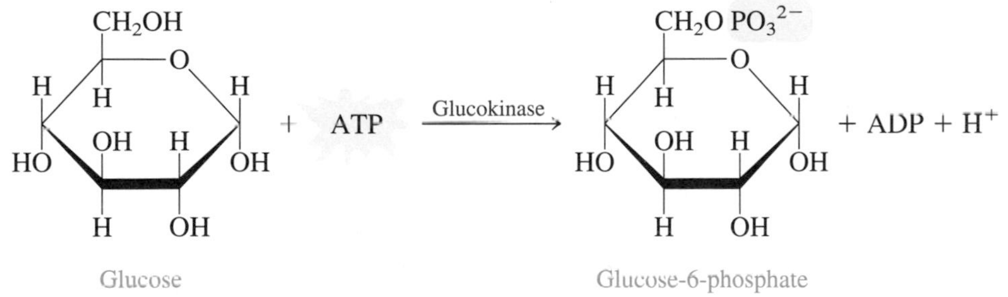

Glucose + ATP —Glucokinase→ Glucose-6-phosphate + ADP + H^+

The second reaction of glycogenesis is the reverse of one of the glycogenolytic reactions. The glucose-6-phosphate formed in the first step is isomerized to glucose-1-phosphate. The enzyme that mediates this step is phosphoglucomutase:

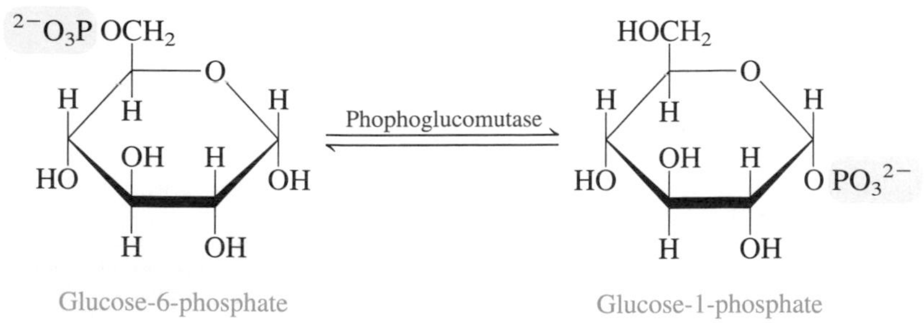

Glucose-6-phosphate ⇌ (Phophoglucomutase) Glucose-1-phosphate

The glucose-1-phosphate must now be activated before it can be added to the growing glycogen chain. The high-energy compound that accomplishes this is the nucleotide **uridine triphosphate** (UTP). In this reaction, mediated by the enzyme **pyrophosphorylase,** the C-1 phosphoryl group of glucose is linked to the α phosphoryl group of UTP to produce UDP-glucose:

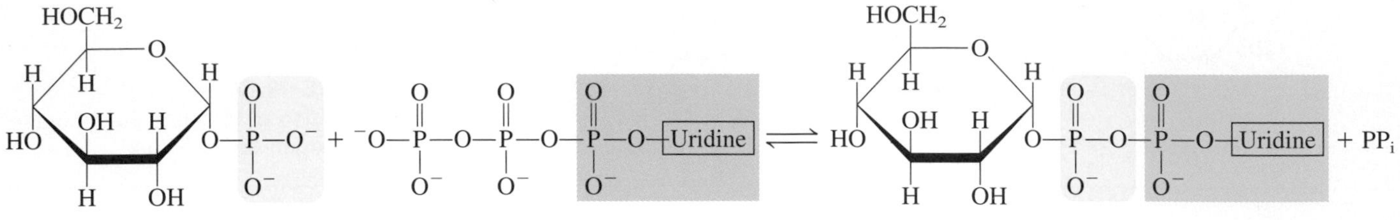

Glucose-1-phosphate

UDP-glucose

$$\text{Glucose-1-phosphate} + \text{UTP} \xrightarrow{\text{Pyrophosphorylase}} \text{UDP-glucose} + \text{pyrophosphate}$$

This is accompanied by the release of a pyrophosphate group (PP_i). The structure of UDP-glucose is seen in Figure 13.18.

The UDP-glucose can now be used to extend glycogen chains. The enzyme glycogen synthetase breaks the phosphoester linkage of UDP-glucose and forms an α $(1 \longrightarrow 4)$ glycosidic bond between the glucose moiety and the growing glycogen chain. UDP is released in the process.

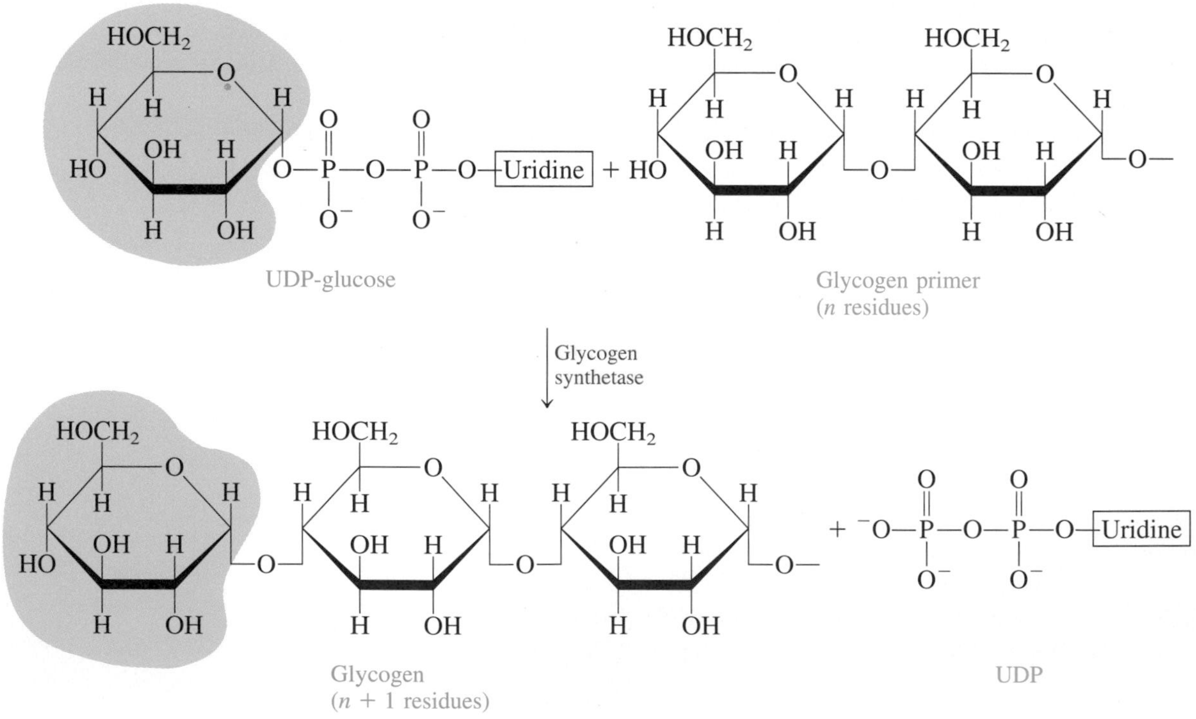

UDP-glucose

Glycogen primer
(n residues)

Glycogen synthetase

Glycogen
(n + 1 residues)

UDP

Finally, we must introduce the α $(1 \longrightarrow 6)$ glycosidic linkages to form the branches. The branches are quite important to proper glycogen utilization. As seen in Figure 13.19, the **branching enzyme** removes a section of the linear α $(1 \longrightarrow 4)$ linked glycogen and reattaches it in α $(1 \longrightarrow 6)$ glycosidic linkage elsewhere in the chain.

Question 13.8 Explain how glycogenesis and glycogenolysis differ.

Glucose Uridine diphosphate

FIGURE 13.18
The structure of UDP-glucose.

Question 13.9 How does insulin affect the storage and degradation of glycogen?

FIGURE 13.19
The action of the branching enzyme in glycogen synthesis.

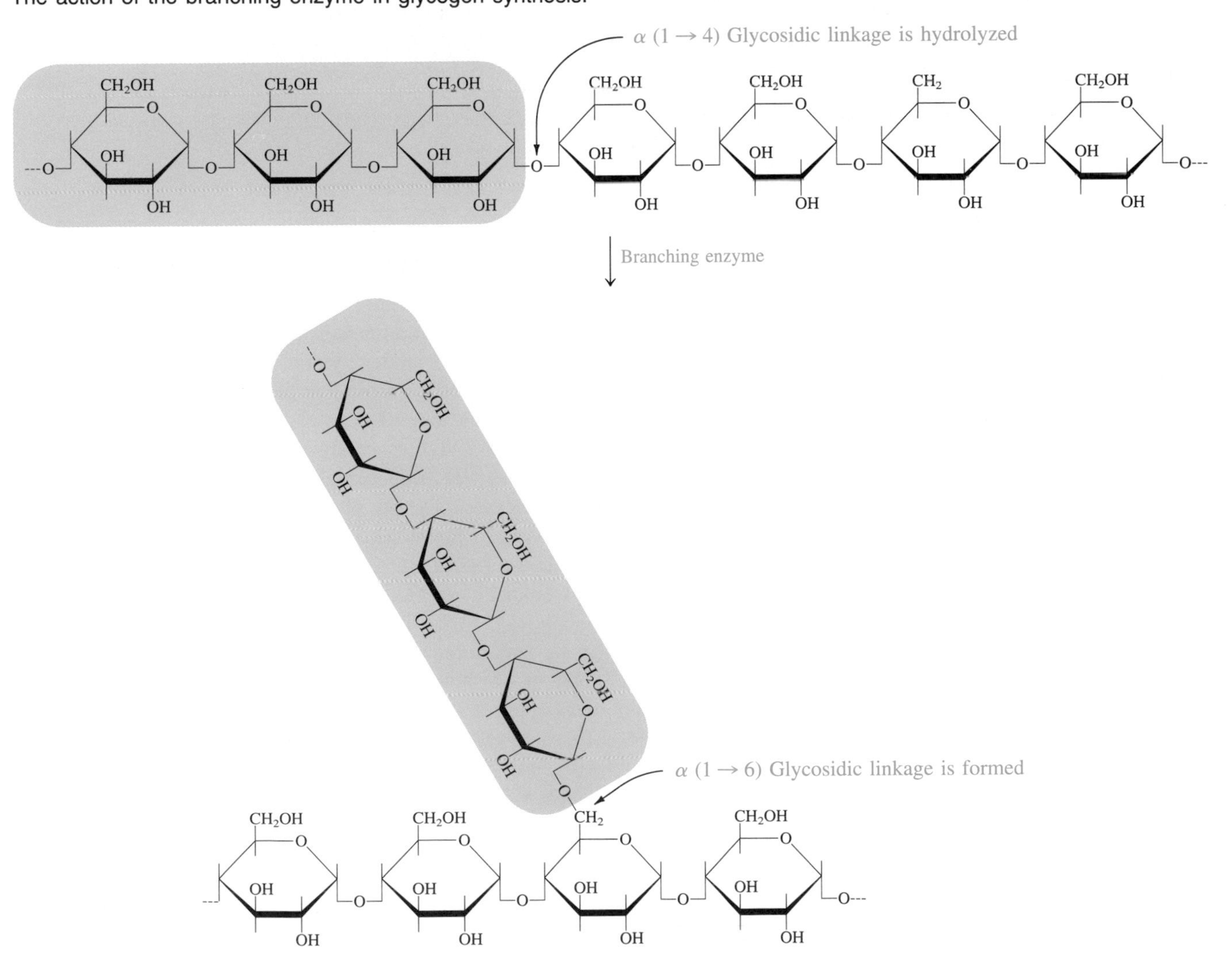

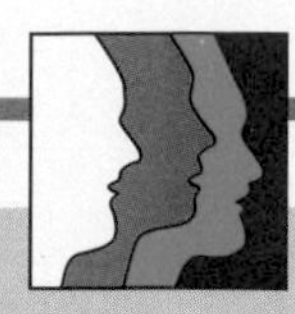

A HUMAN PERSPECTIVE

Glycogen Storage Diseases

Glycogen metabolism is important for the proper function of many aspects of cellular metabolism. Many diseases of glycogen metabolism have been discovered. Generally, these are diseases that result in the excessive accumulation of glycogen in the liver, muscle, and tubules of the kidneys. Often they are caused by defects in one of the enzymes involved in the degradation of glycogen.

One example is an inherited defect of glycogen metabolism known as **von Gierke's disease.** This disease results from a defective gene for glucose-6-phosphatase, which catalyzes the final step of gluconeogenesis. People who lack glucose-6-phosphatase cannot convert glucose-6-phosphate to glucose. As we have seen, the liver is the primary source of blood glucose, and much of this glucose is produced by gluconeogenesis. Glucose-6-phosphate, unlike glucose, cannot cross the cell membrane, and the liver of a person suffering from von Gierke's disease cannot provide the organism with glucose. The blood sugar level falls precipitously low between meals. In addition, the lack of glucose-6-phosphatase also affects glycogen metabolism. Since glucose-6-phosphatase is absent, the supply of glucose-6-phosphate in the liver is large. This glucose-6-phosphate can also be converted to glycogen. A person suffering from von Gierke's disease has a massively enlarged liver as a result of enormously increased stores of glycogen.

Defects in other enzymes of glycogen metabolism also exist. **Cori's Disease** is caused by a genetic defect in the debranching enzyme. As a result, individuals who have this disease cannot completely degrade glycogen and thus use their glycogen stores very inefficiently.

On the other side of the coin, **Andersen's Disease** results from a genetic defect in the branching enzyme. Individuals who have this disease produce very long, unbranched glycogen chains. These unbranched chains cannot be properly degraded by the glycogenolytic enzymes, again resulting in inefficient use of glycogen stores.

One final example of a glycogen storage disease is **McArdle's Disease.** In this syndrome the muscle cells lack the enzyme glycogen phosphorylase and cannot degrade glycogen to glucose. Individuals who have this disease have little tolerance for physical exercise because their muscles cannot provide enough glucose for the necessary energy-generating processes. It is interesting to note that the liver enzyme glycogen phosphorylase is perfectly normal, and these people respond appropriately with a rise of blood glucose levels under the influence of glucagon or epinephrine.

Compatibility of glycogenesis and glycogenolysis

As was the case with glycolysis and gluconeogenesis, it would be futile for the cell to carry out glycogen synthesis and degradation simultaneously. The results achieved by the action of one pathway, glycogenesis, would be undone by the other, glycogenolysis. This problem is avoided by a series of hormonal controls that activate the enzymes of one pathway while simultaneously inactivating the enzymes of the other pathway.

When the blood glucose level is too high, a condition known as **hyperglycemia,** insulin stimulates the uptake of glucose via a transport mechanism. It further stimulates the trapping of the glucose by the elevated activity of glucokinase. Finally, it activates glycogen synthetase, the last enzyme in the synthesis of glycogen chains. To further accelerate storage, insulin *inhibits* the first enzyme in glycogen degradation, glycogen phosphorylase. The net effect, seen in Figure 13.20, is that glucose is removed from the bloodstream and converted into glycogen in the liver. When the glycogen stores are filled, excess glucose is converted to fat and stored in adipose tissue!

Glucagon is produced in response to low blood glucose levels, a condition known as **hypoglycemia,** and has an effect opposite to that of insulin. It stimulates glycogen phosphorylase, which catalyzes the first stage of glycogen degradation. This accelerates glycogenolysis and release of glucose into the bloodstream. The effect is further enhanced because glucagon inhibits glycogen synthetase. The opposing effects of insulin and glucagon are summarized in Figure 13.20.

This elegant system of hormonal control ensures that the reactions involved in glycogen degradation and synthesis do not compete with one another. In this way they provide glucose when the blood level is too low, and they cause the storage of glucose in times of excess.

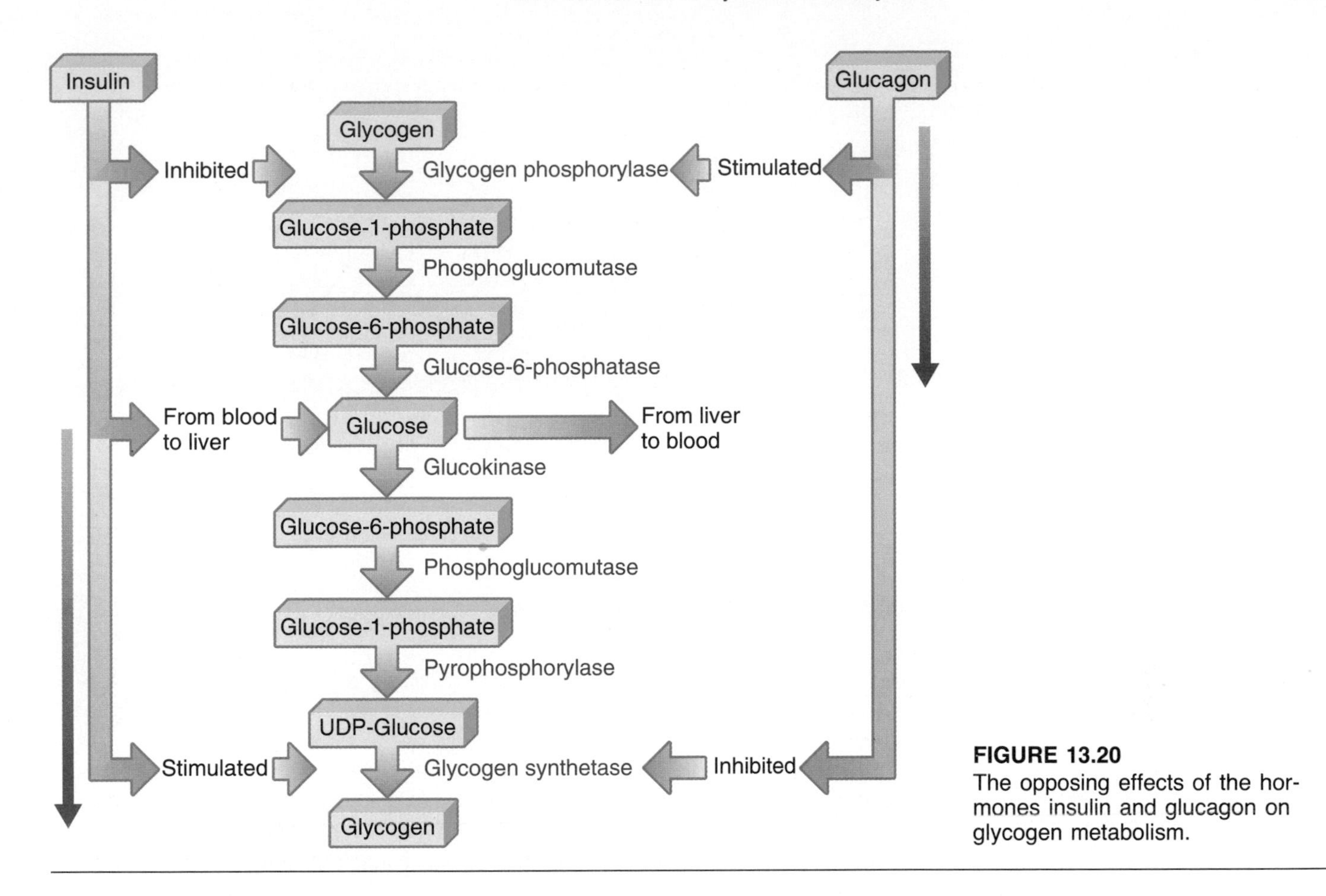

FIGURE 13.20
The opposing effects of the hormones insulin and glucagon on glycogen metabolism.

Question 13.10 How do the opposing actions of glucagon and insulin help to modulate blood glucose levels?

13.7 CONVERSION OF PYRUVATE TO ACETYL CoA

In this chapter we have seen that, under anaerobic conditions, glucose is metabolized to two pyruvate molecules, which are subsequently converted to a stable fermentation product. This limited degradation of glucose releases very little of the potential energy of glucose. Under aerobic conditions the cells can use oxygen and completely oxidize glucose to CO_2 in a metabolic pathway called the citric acid cycle. The last topic in this chapter is the production of the intermediate that carries two carbon fragments, acetyl groups, into the aerobic energy generating pathway. That intermediate is **acetyl CoA.**

The citric acid cycle and the electron transport chain responsible for the production of large quantities of ATP are discussed in detail in Sections 14.2 and 14.3.

The structure of acetyl CoA is seen in Figure 13.21. The **coenzyme A** portion of this molecule is derived from ATP, the vitamin pantothenic acid, and the amino acid cysteine. It serves as an acceptor of acetyl groups, shown in blue in Figure 13.21, which are linked to the cysteine moiety by a thioester bond. We can consider acetyl CoA to be an ''activated'' form of the acetyl group.

The reaction that converts pyruvate to acetyl CoA is shown in Figure 13.22. The pyruvate is decarboxylated, which liberates a molecule of CO_2. It is also oxidized, and the hydride anion that is removed is accepted by NAD^+, which is thus reduced. Finally, the remaining acetyl group, $CH_3CO_2^-$, is linked to coenzyme A by a thioester bond. This very complex reaction is mediated by three enzymes and five coenzymes. To economize the process, these enzymes and coenzymes are all localized in a single bundle called the

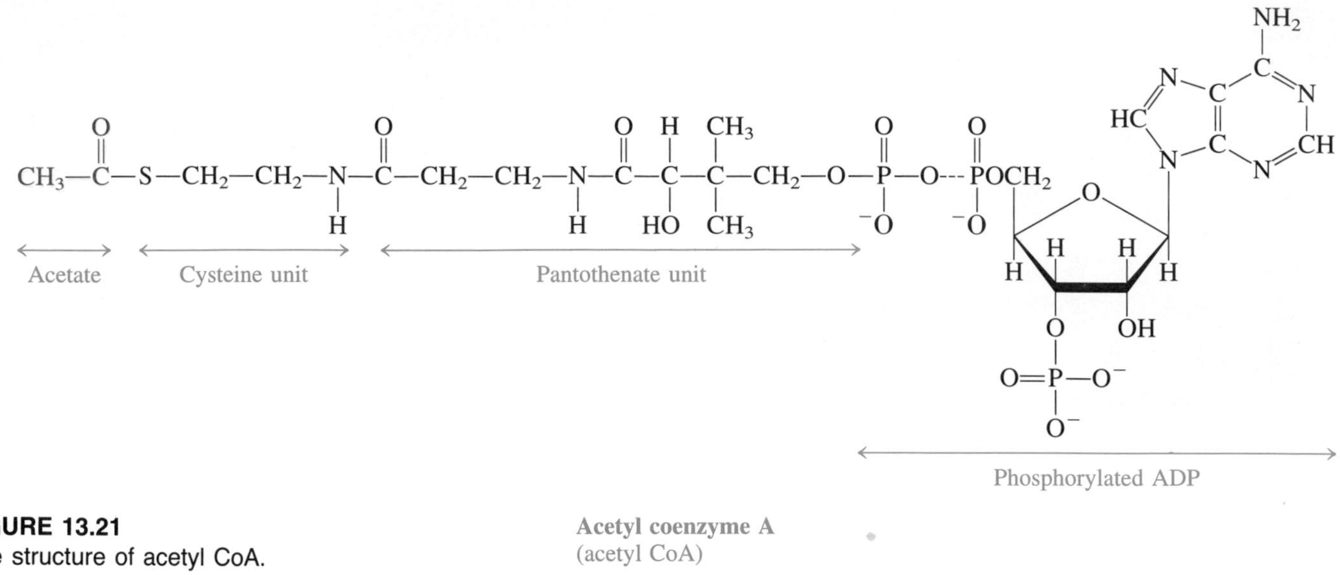

FIGURE 13.21
The structure of acetyl CoA.

pyruvate dehydrogenase complex (Figure 13.22b). In this way the substrate can be passed from one enzyme to the next as each modification occurs. This ''disassembly line'' is shown in Figure 13.22.

This single reaction requires four vitamin derivatives. These are **thiamine pyrophosphate,** derived from vitamin B_1 or thiamine; FAD, derived from vitamin B_2 or riboflavin; NAD^+, derived from niacin; and coenzyme A, derived from pantothenic acid. Obviously, a deficiency in any of these vitamins would seriously hamper the ability to produce acetyl

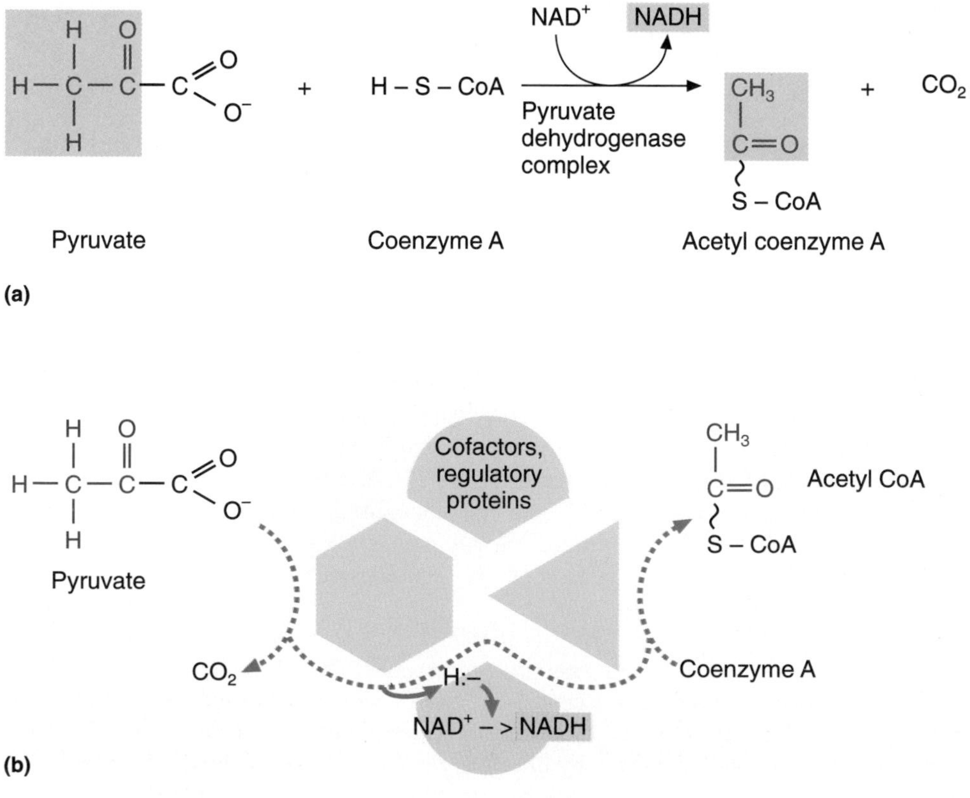

FIGURE 13.22
The decarboxylation and oxidation of pyruvate to produce acetyl CoA. (a) The overall reaction in which CO_2 and an H:$^-$ are removed from pyruvate and the remaining acetyl group is attached to coenzyme A. This requires the concerted action of three enzymes and five coenzymes. (b) The pyruvate dehydrogenase complex that carries out this reaction is actually a cluster of enzymes and coenzymes. The substrate is passed from one enzyme to the next as the reaction occurs.

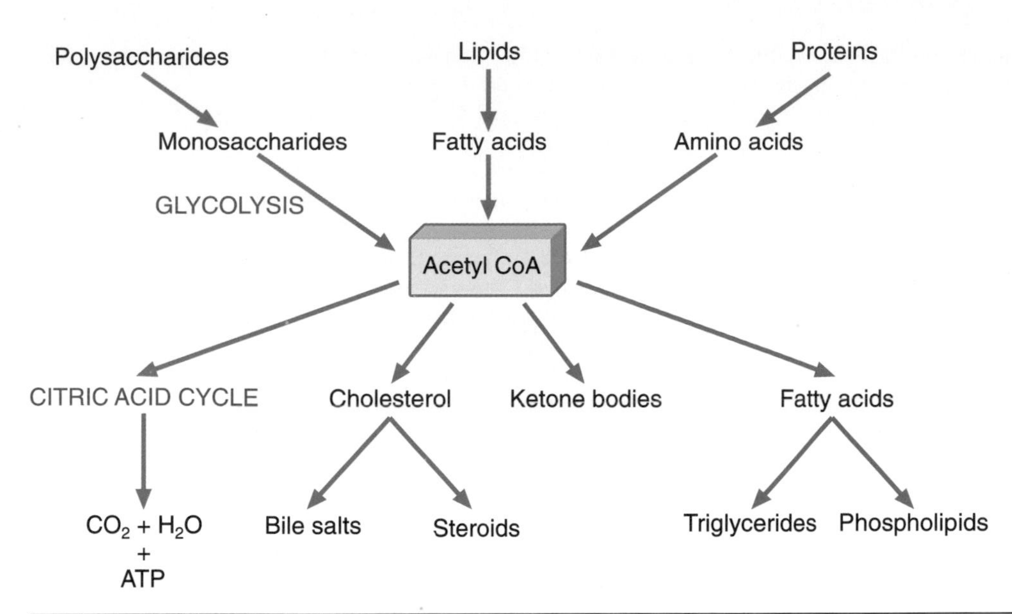

FIGURE 13.23
The central role of acetyl CoA in cellular metabolism.

CoA and hence to generate the large amount of ATP that is required by organisms such as ourselves. Fortunately, as we will see in Chapter 17, a well-balanced diet provides an adequate supply of these and other vitamins.

Inspection of Figure 13.23 reveals that acetyl CoA is a central character in cellular metabolism. Not only is it produced from glucose, but degradation of fatty acids and some amino acids produces acetyl CoA. The principal function of acetyl CoA in energy-generating pathways is to carry the acetyl group to the citric acid cycle, where it will be used to produce large amounts of ATP energy. In addition to these catabolic duties the acetyl group of acetyl CoA can also be used for *anabolic* or biosynthetic reactions to produce cholesterol and fatty acids. It is through this intermediate, acetyl CoA, that all the energy sources (fats, proteins, and carbohydrates) are interconvertible, for both energy-generating and biosynthetic reactions.

Question 13.11 What vitamins are required for acetyl CoA production from pyruvate?

SUMMARY

The body needs a supply of energy to carry out life processes. To provide this energy, we consume a variety of energy-rich food molecules, particularly carbohydrates, lipids, and proteins. In the digestive tract these large molecules are degraded into smaller molecules (monosaccharides, glycerol, fatty acids, and amino acids) that can be absorbed by our cells. These molecules are further broken down to produce ATP, the universal energy currency.

Glycolysis is the pathway for the catabolism of glucose that leads to pyruvate. Two molecules of ATP are produced for each molecule of glucose that is converted to pyruvate. Two molecules of ATP are consumed in the conversion of glucose to fructose-1,6-diphosphate. The first ATP-producing step of glycolysis is phosphoryl group transfer from 1,3-diphosphoglycerate to ADP. The second ATP-yielding step of glycolysis is phosphoryl group transfer from phosphoenolpyruvate to ADP.

The NADH produced by phosphoglycerate dehydrogenase is used to reduce pyruvate to lactate in skeletal muscle and to convert acetaldehyde to ethanol in yeast under anaerobic conditions. Pyruvate is oxidized by pyruvate dehydrogenase under aerobic conditions with formation of acetyl CoA.

Gluconeogenesis is the pathway for glucose synthesis from noncarbohydrate precursors in mammalian liver. Gluconeogenesis is not simply the reversal of glycolysis. Three steps in glycolysis in which ATP is produced or consumed are bypassed by separate enzymes in gluconeogenesis. All other enzymes in gluconeogenesis are shared with glycolysis.

Glucagon stimulates gluconeogenesis and glycogen degradation in liver. Insulin stimulates glycogen synthesis. Together these two hormones regulate blood glucose levels.

Glycogen degradation and glycogen synthesis are reciprocally regulated pathways. When glycogen synthesis is "on," glycogen degradation is "off," and vice versa. The concentration of blood glucose is controlled by the liver. A high glucose concentration inhibits glycogen degradation and stimulates glycogen synthesis.

GLOSSARY OF KEY TERMS

acetyl CoA (13.7) a molecule composed of coenzyme A and an acetyl group. This intermediate provides acetyl groups for complete oxidation by aerobic respiration.

adenosine triphosphate (ATP) (13.1) a nucleotide composed of the purine adenine, the sugar ribose, and three phosphoryl groups. It serves as the major energy storage form of the cell.

alcohol dehydrogenase (13.4) the enzyme that reduces acetaldehyde to produce ethanol in the alcohol fermentation.

aldolase (13.3) the enzyme that cleaves fructose-1,6-diphosphate into dihydroxyacetone phosphate and glyceraldehyde-3-phosphate.

anaerobic threshold (13.4) the point at which the level of lactate in the exercising muscle inhibits glycolysis. When deprived of energy the muscle ceases to function.

Andersen's Disease (13.6) a genetic defect in the glycogen branching enzyme resulting in difficulty using glycogen stores efficiently.

branching enzyme (13.6) the enzyme in glycogenesis that makes the α (1 $\longrightarrow$ 6) glycosidic linkages at the branch points.

catabolism (13.1) the degradation of fuel molecules and production of energy for cellular functions.

coenzyme A (13.7) a molecule derived from ATP, the vitamin pantothenic acid, and the amino acid cysteine that functions in the transfer of acetyl groups.

Cori Cycle (13.5) a metabolic pathway in which the lactate produced by working muscle is converted back to glucose by gluconeogenesis in the liver.

Cori's Disease (13.6) a genetic defect in the glycogen debranching enzyme that results in inefficient utilization of stored glycogen.

debranching enzyme (13.6) an α (1 $\longrightarrow$ 6) glycosidase that hydrolyzes the α (1 $\longrightarrow$ 6) glycosidic bonds at the branch points of glycogen. The enzyme is used during the process of glycogenolysis.

endothermic reaction (13.1) a reaction that requires an input of energy in order to occur.

enolase (13.3) the enzyme that catalyzes the dehydration of 3-phosphoglycerate, which produces phosphoenolpyruvate.

epinephrine (13.6) a hormone produced by the adrenals in times of stress, also called adrenaline. One of its functions is to stimulate glycogenolysis and thereby increase the concentration of blood glucose.

exothermic reaction (13.1) a reaction that liberates energy.

fructose diphosphatase (13.5) the enzyme that catalyzes the removal of the C-1 phosphoryl group from fructose-1,6-diphosphate in gluconeogenesis.

glucagon (13.6) a peptide hormone released by the pancreas in response to low blood sugar. It promotes glycogenolysis and thereby increases the concentration of blood glucose.

glucokinase (13.6) an enzyme that catalyzes the phosphorylation of glucose in the first step of glycogenesis.

gluconeogenesis (13.5) the synthesis of glucose from noncarbohydrate precursors.

glucose-6-phosphatase (13.5) the enzyme that catalyzes the dephosphorylation of glucose-6-phosphate in the last step of gluconeogenesis.

glyceraldehyde-3-phosphate dehydrogenase (13.3) the enzyme that catalyzes the oxidation of glyceraldehyde-3-phosphate to 1,3-diphosphoglycerate and the concomitant reduction of NAD^+ in glycolysis.

glycogen (13.6) a highly branched polymer of glucose stored in the liver and in muscles.

glycogen granule (13.6) a core of glycogen surrounded by enzymes that are responsible for glycogen synthesis and degradation.

glycogen phosphorylase (13.6) an enzyme that mediates the displacement of a molecule of glucose from glycogen by the addition of a phosphate group. This is the first step in glycogenolysis.

glycogen synthetase (13.6) an enzyme that catalyzes the formation of an α (1 $\longrightarrow$ 4) glycosidic bond between the glucose moiety of UDP-glucose and the growing glycogen polymer.

glycogenesis (13.6) the metabolic pathway that results in the addition of glucose to growing glycogen polymers when blood glucose levels are high.

glycogenolysis (13.6) the biochemical pathway that results in the removal of glucose molecules from glycogen polymers when blood glucose levels are low.

glycolysis (13.3) the enzymatic pathway that converts a glucose molecule into two molecules of pyruvate. This anaerobic process generates energy in the form of two ATP and two NADH.

α (1 $\longrightarrow$ 6) glycosidase (13.6) the enzyme that hydrolyzes the α (1 $\longrightarrow$ 6) glycosidic bond at the branch points of glycogen.

guanosine triphosphate (13.5) a nucleotide composed of the purine guanosine, the sugar ribose, and three phosphoryl groups. It serves as a phosphoryl group donor in the conversion of oxaloacetate to phosphoenolpyruvate during gluconeogenesis.

hexokinase (13.3) the enzyme that catalyzes the phosphorylation of glucose in the first reaction of glycolysis.

hyperglycemia (13.6) blood glucose levels that are higher than normal.

hypoglycemia (13.6) blood glucose levels that are lower than normal.

insulin (13.6) a hormone released from the pancreas in response to high blood glucose levels. Insulin promotes glycogenesis and fat storage.

lactate dehydrogenase (13.4) the enzyme that catalyzes the reduction of pyruvate to produce lactate in the lactate fermentation.

McArdle's Disease (13.6) a genetic defect in the muscle enzyme glycogen phosphorylase that results in an inability to degrade glycogen in muscle tissue.

nicotinamide adenine dinucleotide (NAD^+) (13.3) a molecule synthesized from the vitamin niacin and the nucleotide ATP and that serves as a carrier of hydride anions.

nucleotide (13.1) a molecule composed of a nitrogenous base, a five-carbon sugar, and one, two, or three phosphoryl groups.

obligate anaerobe (13.4) an organism that cannot live in an environment containing molecular oxygen (O_2).

oxidative phosphorylation (13.3) production of ATP using the energy of electrons harvested during biological oxidation-reduction reactions.

oxygen debt (13.4) the O_2 required after strenuous exercise to restore ATP levels and to convert lactate back to glucose.

phosphoenolpyruvate carboxykinase (13.5) the enzyme that catalyzes the removal of CO_2 from and the addition of a phosphoryl group to oxaloacetate in gluconeogenesis.

phosphofructokinase (13.3) the enzyme that catalyzes the phosphorylation of fructose-6-phosphate to produce fructose-1,6-diphosphate in glycolysis.

phosphoglucoisomerase (13.3) the enzyme that catalyzes the rearrangement of glucose-6-phosphate to fructose-6-phosphate in glycolysis.

phosphoglucomutase (13.6) the enzyme that catalyzes the isomerization of glucose-1-phosphate to glucose-6-phosphate in glycogen degradation.

phosphoglycerokinase (13.3) the enzyme that catalyzes the first substrate-level phosphorylation in glycolysis, the transfer of a phosphoryl group from 1,3-diphosphoglycerate to ADP to produce ATP.

phosphoglyceromutase (13.3) the enzyme that catalyzes the isomerization of 3-phosphoglycerate to 2-phosphoglycerate in glycolysis.

pyrophosphorylase (13.6) the enzyme that mediates bond formation between the C-1 phosphoryl group of glucose-1-phosphate and the α phosphoryl group of uridine triphosphate in glycogenesis.

pyruvate carboxylase (13.5) the enzyme that catalyzes the addition of atmospheric CO_2 to pyruvate to produce oxaloacetate in gluconeogenesis.

pyruvate decarboxylase (13.4) the enzyme that catalyzes the removal of CO_2 from pyruvate to produce acetaldehyde in the alcohol fermentation.

pyruvate dehydrogenase complex (13.7) a complex of all the enzymes and coenzymes required for the synthesis of CO_2 and acetyl CoA from pyruvate.

pyruvate kinase (13.3) the enzyme that catalyzes the transfer of a phosphoryl group from phosphoenolpyruvate to ADP and the production of ATP and pyruvate in the last reaction of glycolysis.

substrate-level phosphorylation (13.3) the production of ATP by the transfer of a phosphoryl group from the substrate of a reaction to ADP.

thiamine pyrophosphate (13.7) a coenzyme derived from vitamin B_1 (thiamine) that is part of the pyruvate dehydrogenase complex.

triose isomerase (13.3) the enzyme that catalyzes the isomerization of dihydroxyacetone phosphate to glyceraldehyde-3-phosphate in glycolysis.

uridine triphosphate (13.6) a nucleotide composed of the pyrimidine uracil, the sugar ribose, and three phosphate groups and that serves as a carrier of glucose-1-phosphate in glycogenesis.

von Gierke's Disease (13.6) a genetic defect in the enzyme glucose-6-phosphatase. The result is that the liver cannot provide a supply of blood glucose. The liver stores the excess glucose-6-phosphate as glycogen with the result that the liver becomes greatly enlarged.

QUESTIONS AND PROBLEMS

Role of ATP in Metabolism

13.12 What molecule is primarily responsible for conserving the energy released in catabolism?

13.13 List five cellular functions that require ATP energy.

Glycolysis

13.14 An enzyme that hydrolyzes ATP (an ATPase) bound to the plasma membrane of certain tumor cells has an abnormally high activity. How will this activity affect the rate of glycolysis?

13.15 What effect would inhibition of lactate dehydrogenase have on glycolysis under anaerobic conditions?

13.16 Why does glycolysis require a supply of NAD^+ to function?

13.17 What is the net energy yield of ATP in glycolysis?

13.18 Write a balanced chemical equation for the conversion of acetaldehyde to ethanol by alcohol dehydrogenase.

13.19 Under what metabolic conditions is pyruvate converted to acetyl CoA?

13.20 Explain how muscle is able to carry out rapid contraction for prolonged periods even though its supply of ATP is sufficient only for a fraction of a second of rapid contraction.

13.21 Write the balanced chemical equation for glycolysis.

13.22 After running a 100-yard dash, a sprinter had a high concentration of muscle lactate. What process is responsible for production of lactate?

13.23 Diving seals produce considerable lactate. What is the origin of this metabolic product?

13.24 If a mutation in an animal were to provide the animal with muscle that did not contain lactate dehydrogenase, would the muscles be capable of anaerobic glycolysis? Why or why not?

13.25 Explain why no net oxidation occurs during anaerobic glycolysis followed by lactate fermentation.

13.26 The eye is an anaerobic organ. How does it obtain ATP?

13.27 Write a chemical equation for the transfer of a phosphoryl group from ATP to fructose-6-phosphate.

Gluconeogenesis

13.28 What organ is primarily responsible for gluconeogenesis?

13.29 What is the physiological function of gluconeogenesis?

13.30 Explain why gluconeogenesis is not simply the reversal of glycolysis.

13.31 Lactate can be converted to glucose by gluconeogenesis. To what metabolic intermediate must lactate be converted so that it can be a substrate for the enzymes of gluconeogenesis?

13.32 L-Alanine can be converted to pyruvate. Can L-alanine also be converted to glucose?

13.33 Phosphoenolpyruvate is converted to pyruvate by pyruvate kinase with concomitant production of ATP in glycolysis. How is this reaction bypassed in gluconeogenesis?

Glycogen Metabolism

13.34 Why is phosphorolysis of glycogen by glycogen phosphorylase an energy-efficient process?

13.35 A certain person was found to have a defect in glycogen metabolism. The liver of this person could (a) make glucose-6-phosphate from lactate and (b) synthesize glucose-6-phosphate from glycogen but (c) could not synthesize glycogen from glucose-6-phosphate. What enzyme is defective?

13.36 What condition exists in individuals who have hypoglycemia?

13.37 Explain how a defect in glycogen metabolism can cause hypoglycemia.

13.38 What defects of glycogen metabolism lead to a large increase in the concentration of liver glycogen?

13.39 What enzymes are stimulated by insulin? What effect does this have on glycogen metabolism? What effect does this have on blood glucose levels?

13.40 What enzymes are stimulated by glucagon? What effect does this have on glycogen metabolism? What effect does this have on blood glucose levels?

Further Problems

13.41 Define each of the following terms:

a. gluconeogenesis
b. catabolism
c. glycolysis
d. aerobic energy production
e. anaerobic energy production
f. fermentation

13.42 Label each of the following statements as true or false:

a. Glycolysis is an aerobic process.
b. The metabolism of most organisms revolves around the synthesis and hydrolysis of ATP.
c. In the first step of glycolysis a phosphoryl group, $-PO_3^{2-}$, is transferred from ATP to the C-6 hydroxyl group of glucose.
d. The enzyme that catalyzes the first step of glycolysis is called hexokinase.
e. Hydrolysis of ATP absorbs energy.

13.43 Which organ of the body is the primary source of blood glucose?

13.44 Fill in the blanks:

a. ________ molecules of ATP are produced per molecule of glucose that is converted to pyruvate.
b. Two molecules of ATP are consumed in the conversion of ________ to fructose-1,6-diphosphate.
c. NAD^+ is ________ to NADH in the first energy-producing step of glycolysis.
d. The second substrate level phosphorylation in glycolysis is phosphoryl group transfer from phosphoenolpyruvate to ________.

13.45 Which steps in the glycolysis pathway are irreversible?

13.46 What enzyme catalyzes the reduction of pyruvate to lactate?

13.47 Which glycolysis reactions are catalyzed by each of the following enzymes?

a. hexokinase
b. pyruvate kinase
c. phosphoglyceromutase
d. glyceraldehyde-3-phosphate dehydrogenase

13.48 What four vitamins are required by the pyruvate dehydrogenase complex?

13.49 Explain von Gierke's Disease. What are the symptoms and which organ of the body is affected?

13.50 A child was brought to the doctor's office suffering from a strange set of symptoms. When the child exercised hard, she became giddy and behaved as though drunk. What do you think is the metabolic basis of these symptoms?

13.51 A family started a batch of wine by adding yeast to grape juice and placing the mixture in a sealed bottle. Two weeks later, the bottle exploded. Why?

13.52 Write a balanced equation for the production of acetyl CoA from pyruvate. Under what conditions does this reaction occur?

CHAPTER 14

Metabolism and Energy Production

OUTLINE

OBJECTIVES

- Recognize that energy production in the cell is principally due to oxidative metabolic reactions.
- Know that these aerobic energy-generating reactions occur in the mitochondria.
- Be able to name the major regions of the mitochondria and their function.
- Be aware of the importance of the citric acid cycle in the digestive process.
- Summarize the major reactions of the citric acid cycle.
- Recognize the advantages of compartmentalization of metabolic pathways.
- Describe oxidative phosphorylation and its relationship to the respiratory electron transport system.
- Describe the chemiosmotic theory.
- Understand the ways in which the citric acid cycle is regulated.
- Describe the fate of amino acids in the citric acid cycle.
- Know the importance of the urea cycle and describe its essential steps.
- Summarize the role of the citric acid cycle in catabolism and anabolism.

The oxidative reactions of metabolism provide for most cellular energy production. These reactions occur in metabolic pathways located in the *mitochondria,* the cellular "power plant." Mitochondria are a type of **organelle,** a membrane-enclosed compartment within the cytoplasm that has a specialized function. In contrast to the rest of the cell, which is anaerobic, mitochondria are aerobic. They are responsible for the oxidation of the fuel molecules produced by preliminary degradation via glycolysis, fatty acid degradation, and amino acid catabolism. If they did not have mitochondria, eukaryotic cells would be anaerobic and would have to depend on the very small amount of ATP that they could make by glycolysis.

The conversion of glucose to lactate by glycolysis is an anaerobic pathway that occurs in the cytoplasm of eukaryotic cells. Only two molecules of ATP are generated by anaerobic glycolysis per molecule of glucose. This energy yield represents only a small fraction of the energy that is available in glucose. In contrast, the oxidative degradation of one molecule of glucose by successive action of the citric acid cycle and oxidative phosphorylation, processes that occur in the mitochondria, can yield 36 molecules of ATP.

Section 13.3

14.1 MITOCHONDRIA

Structure and function

Mitochondria are football-shaped organelles that range in size from 0.2 to 0.8 μm in diameter and from about 0.5 to 1.0 μm in length, roughly the size of a bacterial cell. They are bounded by an **outer membrane** and an **inner membrane** (Figure 14.1). The region between the two membranes is known as the **intermembrane space,** and the region enclosed by the inner membrane is known as the **matrix space.**

FIGURE 14.1
Structure of the mitochondrion. (a) Electron micrograph of mitochondria. (b) Schematic drawing of the mitochondrion.

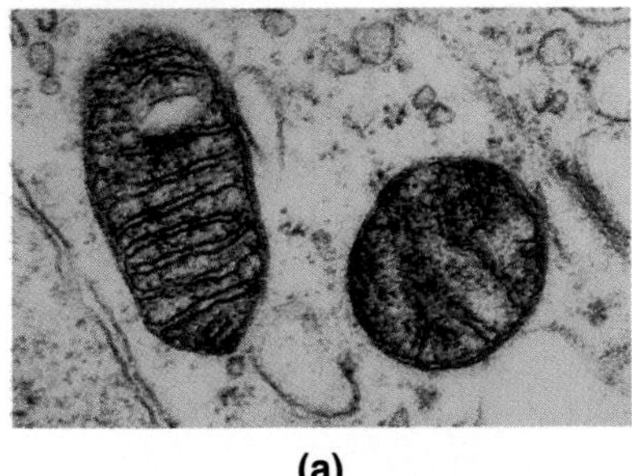

(a)

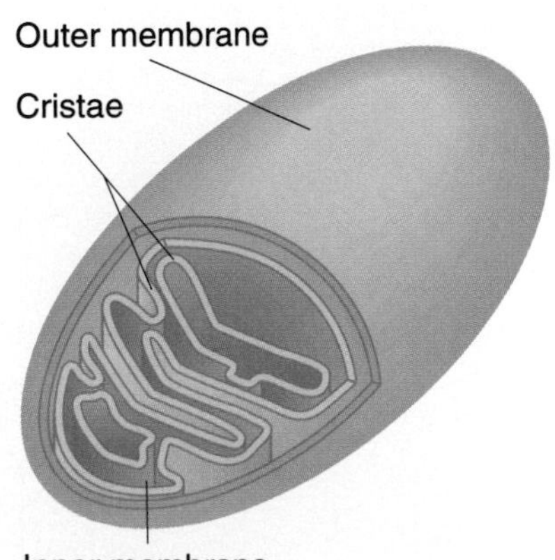

(b)

The outer mitochondrial membrane consists mostly of phospholipids, such as phosphatidylcholine and phosphatidylethanolamine, and contains considerable cholesterol. The outer membrane also contains many copies of a transport protein that forms large pores in the membrane. These pores cause the membrane to be freely permeable to substances whose molecular weights are less than 10,000. Metabolites produced by glycolysis, for example, can freely enter the intermembrane space through these pores.

The inner membrane is a continuous structure that is highly folded. The folded membranes are known as **cristae.** The lipid composition of the inner membrane is somewhat different from that of the outer membrane. A major phospholipid found there is cardiolipin, an unusual phospholipid that makes up about 20% of the inner membrane. In part as a result of the cardiolipin, the inner mitochondrial membrane is virtually impermeable to most substances. Three types of proteins are found in the inner membrane.

1. *Transport proteins*. Since the inner mitochondrial membrane is impermeable to most substances, a method to allow the transport of metabolites across the inner mitochondrial membrane is an essential feature of the mitochondria.
2. *Respiratory chain proteins*. The protein complexes that are responsible for respiration are embedded in the inner mitochondrial membrane. Respiration is the ATP-producing process involving electron transfers for which O_2 serves as the terminal electron acceptor.
3. A very large multiprotein complex known as *ATP synthase,* which is responsible for phosphorylation of ADP.

The enzymes that are responsible for the citric acid cycle (the primary energy-producing pathway in the body) and for fatty acid oxidation are located in the mitochondrial matrix space.

The number of mitochondria in a eukaryotic cell varies widely and usually reflects the energy needs of the cell. Certain algae contain but one mitochondrion, whereas amphibian eggs, which need a great deal of energy for the period of rapid cell division that follows fertilization, possess as many as 100,000 mitochondria. Vertebrate cells typically contain 500–2000 mitochondria. A liver cell, for example, contains between 1000 and 2000 mitochondria. The liver is the powerhouse of the human organism, much as the mitochondrion is the power plant of the cell. As a rule, aerobic cells, such as heart and slow twitch skeletal muscles, contain many mitochondria, whereas anaerobic cells, like the fast twitch muscle of the sprinter, contain few mitochondria.

Question 14.1 What is the function of the mitochondrion?

Question 14.2 How does the mitochondrion differ from the other organelles of eukaryotic cells?

Origin of mitochondria

Not only are mitochondria roughly the size of a bacterium, they have several other features that have led researchers to suspect that they may once have been free-living bacteria that were captured by eukaryotic cells. These organelles have their own genetic information in the form of a circular chromosome. They also make their own ribosomes, small platforms for protein synthesis, which are much more like bacterial ribosomes than eukaryotic ribosomes. These ribosomes allow the mitochondria to synthesize some of their own proteins. Rather than being produced by the cell, the mitochondria are actually self-replicating. They grow in size and divide to produce new mitochondria. All of these characteristics are strong evidence that the mitochondria that produce the majority of the ATP for our cells evolved from captured bacteria, perhaps as long as 1.5×10^9 years ago.

Section 16.11

Question 14.3 Draw a schematic diagram of the mitochondrion and label the parts of this organelle.

Question 14.4 Describe the evidence that suggests that mitochondria evolved from free-living bacteria.

14.2 THE CITRIC ACID CYCLE (THE KREBS CYCLE)

Reactions of the citric acid cycle

The **citric acid cycle** is sometimes called the Krebs Cycle, in honor of its discoverer, Sir Hans Krebs. It is the final stage of the degradation of carbohydrates, fats, and amino acids released from dietary proteins (Figure 14.2).

To gain a better knowledge of this important cycle, let us follow the fate of the acetyl group of acetyl CoA as it passes through the citric acid cycle. The numbered steps listed below correspond to the steps in the citric acid cycle, which are summarized in Figure 14.2.

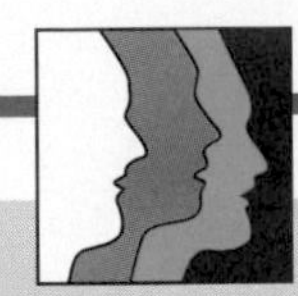

A HUMAN PERSPECTIVE

Exercise and Energy Metabolism

The Olympic sprinters get set in the blocks. The gun goes off, and roughly ten seconds later the 100-m dash is over. Elsewhere the marathoners line up. They will run 26 miles and 385 yards in a little over 2 hours. Both these sports involve running, but they utilize very different sources of energy.

Let's look at the sprinter first. The immediate source of energy for the sprinter is stored ATP. But the quantity of stored ATP is very small, only about three ounces. This allows the sprinter to run as fast as he or she can for about 3 seconds. Obviously, another source of stored energy must be tapped, and that energy store is **creatine phosphate.**

$$^{-}O-\underset{\underset{O^-}{|}}{\overset{\overset{O}{\|}}{P}}-\overset{\overset{H}{|}}{N}-\overset{\overset{NH}{\|}}{C}-\underset{\underset{CH_3}{|}}{N}-CH_2-C\overset{O^-}{\underset{O}{\lessgtr}}$$

The structure of creatine phosphate.

Creatine phosphate, stored in the muscle, donates its high-energy phosphate to ADP to produce new supplies of ATP.

This will keep our runner in motion for another 5 or 6 seconds before the store of creatine phosphate is also depleted. This is almost enough energy to finish the 100-m dash, but in reality, all the runners are slowing down owing to energy depletion, and the winner is the sprinter who is slowing down the least!

Consider a longer race, the 400-meter or the 800-meter. These runners run at maximum capacity for much longer times. When they have depleted their ATP and creatine phosphate stores, they must synthesize more ATP. Of course, the cells have been making ATP all the time, but now the demand for energy is much greater. To supply this increased demand, the anaerobic energy-generating reactions (glycolysis and lactate fermentation, Chapter 20) and aerobic processes (citric acid cycle and oxidative phosphorylation) begin to function much more rapidly. Often, however, these athletes are running so strenuously that they cannot provide enough oxygen to the exercising muscle to allow oxidative phosphorylation to function efficiently. When this happens, the muscles must rely on glycolysis and lactate fermentation to provide *most* of the energy requirement. The chemical by-product of these anaerobic processes, lactate, builds up in the muscle and diffuses into the bloodstream. However, the concentration of lactate inevitably builds up in the working muscle and

$$^{-}O-\underset{\underset{O^-}{|}}{\overset{\overset{O}{\|}}{P}}-\overset{\overset{H}{|}}{N}-\overset{\overset{NH}{\|}}{C}-\underset{\underset{CH_3}{|}}{N}-CH_2-C\overset{O^-}{\underset{O}{\lessgtr}} + ADP \xrightarrow{\text{Creatine kinase}} H_2N-\overset{\overset{NH}{\|}}{C}-\underset{\underset{CH_3}{|}}{N}-CH_2-C\overset{O^-}{\underset{O}{\lessgtr}} + ATP$$

Phosphoryl group transfer from creatine phosphate to ADP is catalyzed by the enzyme creatine kinase.

Step 1: The acetyl group of acetyl CoA is transferred to oxaloacetate by the enzyme **citrate synthase** in the first step of the citric acid cycle. The product that is formed is citrate:

$$\underset{\text{Oxaloacetate}}{\begin{matrix} COO^- \\ | \\ C{=}O \\ | \\ CH_2 \\ | \\ COO^- \end{matrix}} + \underset{\text{Acetyl CoA}}{H_3C-\overset{\overset{O}{\|}}{C}\sim S-CoA} + H_2O \xrightarrow{\text{Citrate synthase}} \underset{\text{Citrate}}{\begin{matrix} COO^- \\ | \\ CH_2 \\ | \\ HO-C-COO^- \\ | \\ H-C-H \\ | \\ COO^- \end{matrix}} + \underset{\text{Coenzyme A}}{HS-CoA} + H^+$$

causes muscle fatigue and, eventually, muscle failure. Thus exercise that depends primarily on anaerobic energy generation cannot continue for very long periods.

The marathoner presents us with a different scenario. This runner will deplete his or her stores of ATP and creatine phosphate as quickly as a short-distance runner. The anaerobic glycolytic pathway will begin to degrade glucose provided by the blood at a more rapid rate, as will the citric acid cycle and oxidative phosphorylation. The major difference in energy generation between the long-distance runner and the short- or middle-distance runner is that the muscles of the long-distance runner derive almost all the energy through aerobic pathways. These individuals continue to run long distances at a pace that allows them to supply virtually all the oxygen needed by the exercising muscle. In fact, only aerobic pathways can provide a constant supply of ATP for exercise that goes on for hours. Theoretically, under such conditions our runner could run indefinitely, utilizing first his or her stored glycogen and eventually stored lipids. Of course, in reality, other factors such as dehydration and fatigue place limits on the athlete's ability to continue.

We have seen, then, that long-distance runners must have a great capacity to produce ATP aerobically, in the mitochondria, while short- and middle-distance runners need a great capacity to produce energy anaerobically, in the cytoplasm of the muscle cell. It is interesting to note that the muscles of these runners reflect these diverse needs.

When one examines muscle tissue that has been surgically removed, one finds two predominant types of muscle fibers. **Fast twitch muscle fibers** are large, relatively plump cells that are pale in color. These cells have only a few mitochondria but contain a large reserve of glycogen and high concentrations of the enzymes that are needed for glycolysis and lactate fermentation. These muscle fibers fatigue rather quickly because anaerobic energy generation is inefficient, quickly depleting the cell's glycogen store and causing the accumulation of lactate. **Slow twitch muscle fiber** cells are about half the diameter of fast twitch muscle cells and are red in color. The red color is a result of the high concentrations of myoglobin in these cells. Recall that myoglobin stores oxygen for the cell (Section 11.9) and facilitates rapid diffusion of oxygen throughout the cell. In addition, slow twitch muscle fiber cells are packed with mitochondria. With this abundance of oxygen and mitochondria these cells have the capacity for extended energy generation via aerobic pathways—ideal for endurance sports like marathon racing.

It is not surprising, then, that researchers have found that the muscles of sprinters have many more fast twitch muscle fibers and those of endurance athletes have many more slow twitch muscle fibers. One question that many researchers are trying to answer is whether the type of muscle fibers an individual has is a function of genetic makeup or training. Is a marathon runner born to be a long-distance runner, or are his or her abilities due to the type of training the runner undergoes? There is no doubt that the training regimen for an endurance runner does indeed increase the number of slow twitch muscle fibers and that of a sprinter increases the number of fast twitch muscle fibers. But there is intriguing new evidence to suggest that the muscles of endurance athletes have a greater proportion of slow twitch muscle fibers before they ever begin training. Thus it appears that some of us truly were born to run.

Step 2: The enzyme **aconitase** catalyzes the dehydration of citrate, producing *cis*-aconitate:

$$\begin{array}{c} COO^- \\ | \\ CH_2 \\ | \\ HO-C-COO^- \\ | \\ H-C-H \\ | \\ COO^- \\ \text{Citrate} \end{array} \xrightarrow{\text{Aconitase}} \begin{array}{c} COO^- \\ | \\ CH_2 \\ | \\ C-COO^- \\ \| \\ C-H \\ | \\ COO^- \\ \textit{cis}\text{-Aconitate} \end{array} + H_2O$$

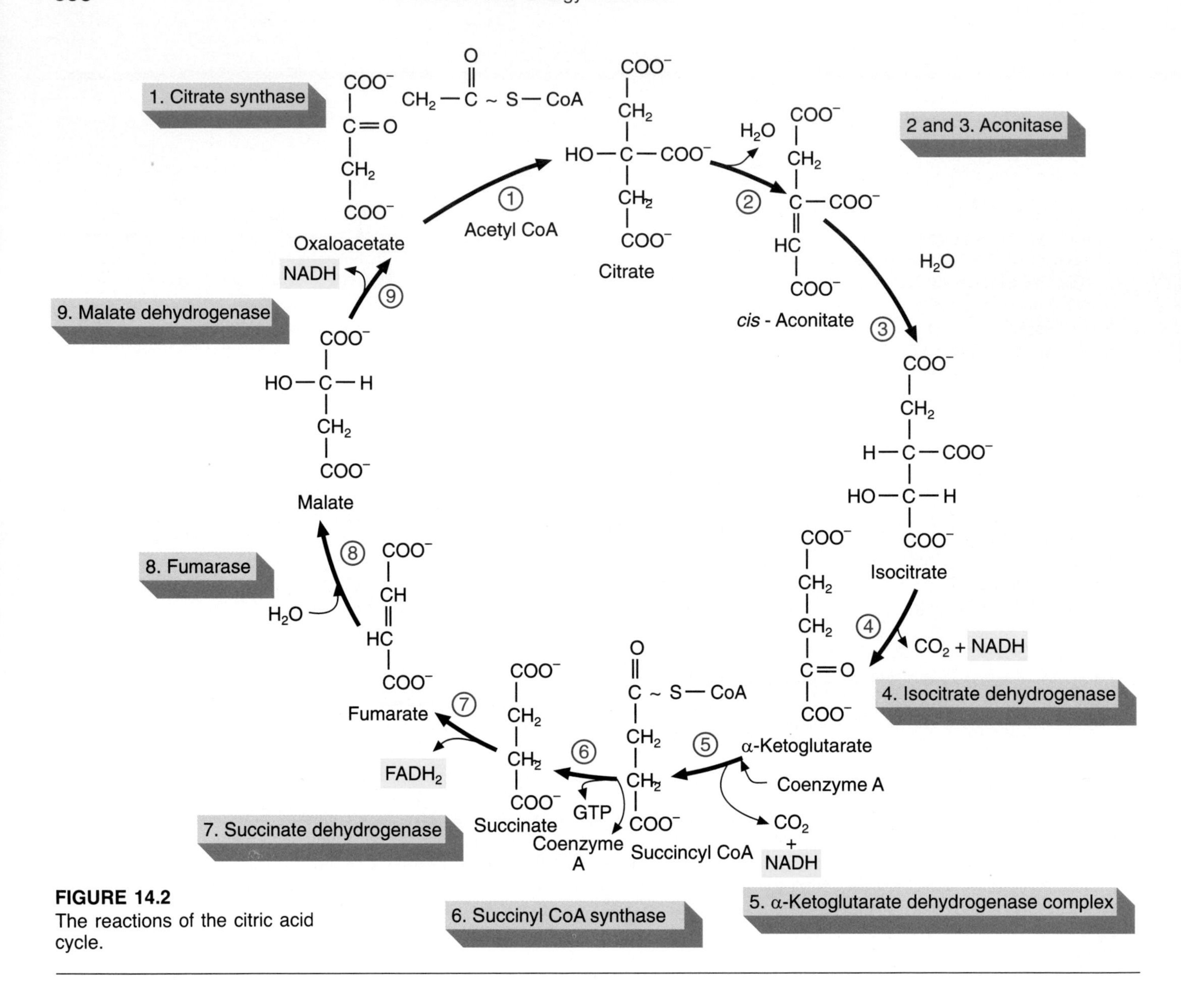

FIGURE 14.2
The reactions of the citric acid cycle.

Step 3: The same enzyme, aconitase, then adds a water molecule to the *cis*-aconitate, converting it to isocitrate. The net effect of the last two steps is the isomerization of citrate to isocitrate:

$$\begin{array}{l} COO^- \\ | \\ CH_2 \\ | \\ C{-}COO^- \\ \| \\ C{-}H \\ | \\ COO^- \end{array} + H_2O \xrightarrow{\text{Aconitase}} \begin{array}{c} COO^- \\ | \\ CH_2 \\ | \\ H{-}C{-}COO^- \\ | \\ HO{-}C{-}H \\ | \\ COO^- \end{array}$$

cis-Aconitate Isocitrate

Step 4: The first oxidative step of the citric acid cycle is catalyzed by an **isocitrate dehydrogenase.** It is a complex reaction in which three things happen:

- the hydroxyl group of isocitrate is oxidized to a ketone;
- carbon dioxide is released; and
- NAD^+ is reduced to NADH.

The reduction of NAD^+ is shown in Figure 12.10.

The product of this oxidative decarboxylation reaction is α-ketoglutarate:

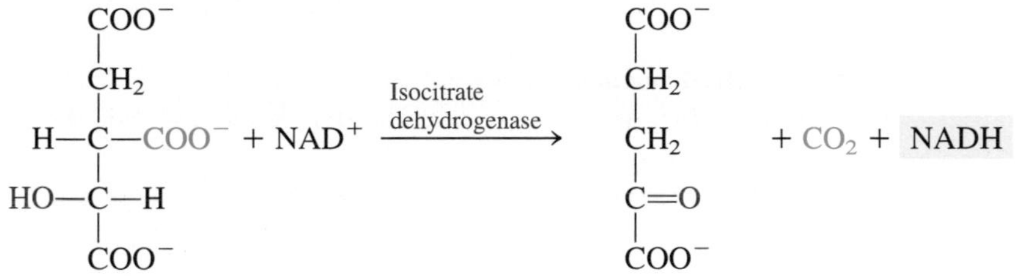

$$\text{Isocitrate } (^-OOC-CH_2-CH(COO^-)-CH(OH)-COO^-) + NAD^+ \xrightarrow{\text{Isocitrate dehydrogenase}} \alpha\text{-Ketoglutarate } (^-OOC-CH_2-CH_2-C(=O)-COO^-) + CO_2 + NADH$$

Isocitrate α-Ketoglutarate

Step 5: Coenzyme A enters the picture again in step 4 of the citric acid cycle. The **α-ketoglutarate dehydrogenase** enzyme complex mediates this series of reactions. This complex is very similar to the pyruvate dehydrogenase complex and requires the same coenzymes. Once again, three chemical events occur:

The pyruvate dehydrogenase complex was discussed in Section 13.7 and shown in Figure 13.22.

- α-ketoglutarate loses a carboxylate group as CO_2;
- NAD^+ is reduced to NADH; and
- coenzyme A combines with the product to form succinyl CoA. The bond thus formed between succinate and coenzyme A is a high-energy thioester linkage:

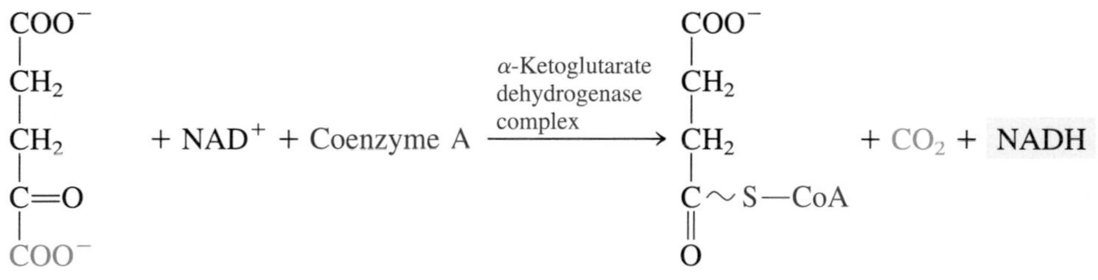

$$^-OOC-CH_2-CH_2-C(=O)-COO^- + NAD^+ + \text{Coenzyme A} \xrightarrow{\alpha\text{-Ketoglutarate dehydrogenase complex}} {}^-OOC-CH_2-CH_2-C(=O)\sim S-CoA + CO_2 + NADH$$

α-Ketoglutarate Succinyl CoA

Step 6: Succinyl CoA is converted to succinate in this step, which once more is chemically very involved. The enzyme **succinyl CoA synthase** removes the CoA group, but this is not a simple hydrolysis reaction. The enzyme removes the CoA and uses the energy of the thioester bond to add an inorganic phosphate group to GDP to give GTP:

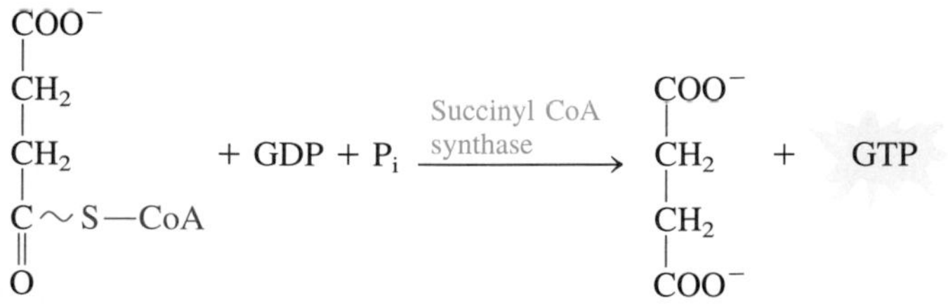

$$^-OOC-CH_2-CH_2-C(=O)\sim S-CoA + GDP + P_i \xrightarrow{\text{Succinyl CoA synthase}} {}^-OOC-CH_2-CH_2-COO^- + GTP$$

Succinyl CoA Succinate

Another enzyme, **nucleotide diphosphokinase,** then shifts a phosphoryl group from GTP to ADP:

$$\text{GTP} + \text{ADP} \xrightarrow[\text{}]{\text{Nucleotide diphosphokinase}} \text{GDP} + \text{ATP}$$

The reduction of FAD was shown in Figure 12.10.

Step 7: **Succinate dehydrogenase** then catalyzes the oxidation of succinate to fumarate. The oxidizing agent, *flavin adenine dinucleotide,* is reduced in this step:

$$\underset{\text{Succinate}}{^{-}OOC{-}CH_2{-}CH_2{-}COO^{-}} + \text{FAD} \xrightarrow{\text{Succinate dehydrogenase}} \underset{\text{Fumarate}}{^{-}OOC{-}CH{=}CH{-}COO^{-}} + FADH_2$$

Step 8: Addition of H_2O to the double bond of fumarate produces malate. The enzyme **fumarase** catalyzes this reaction:

$$\underset{\text{Fumarate}}{^{-}OOC{-}CH{=}CH{-}COO^{-}} + H_2O \xrightarrow{\text{Fumarase}} \underset{\text{Malate}}{^{-}OOC{-}CH(OH){-}CH_2{-}COO^{-}}$$

Step 9: In the final step of the citric acid cycle, **malate dehydrogenase** reduces NAD^+ to NADH and oxidizes malate to oxaloacetate:

$$\underset{\text{Malate}}{^{-}OOC{-}CH(OH){-}CH_2{-}COO^{-}} + NAD^+ \xrightarrow{\text{Malate dehydrogenase}} \underset{\text{Oxaloacetate}}{^{-}OOC{-}C({=}O){-}CH_2{-}COO^{-}} + \text{NADH}$$

Since the citric acid cycle ''began'' with the addition of an acetyl group to oxaloacetate, we have come full circle.

Summary of the energy yield

The complete oxidation of an acetyl group by the citric acid cycle results in the evolution of two molecules of CO_2 and in the production of three molecules of NADH and one of $FADH_2$. A molecule of GTP is produced, and its phosphoryl group is transferred to ADP. Three molecules of ATP are produced by the oxidation of each NADH, and two molecules of ATP are produced by the oxidation of each $FADH_2$. The only exception to this rule is the NADH generated in the cytoplasm during glycolysis. Energy must be expended to

Section 14.3

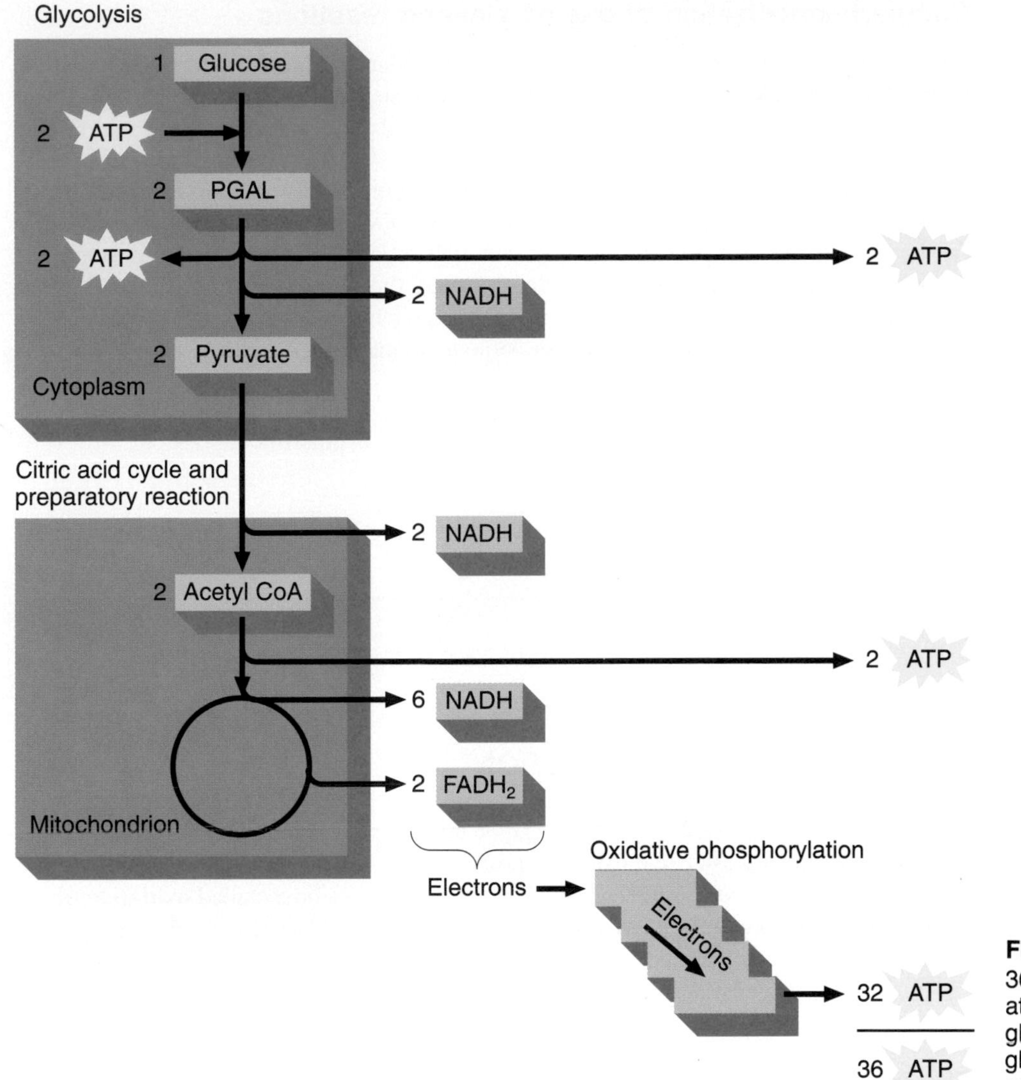

FIGURE 14.3
36 molecules of ATP are generated by the complete oxidation of glucose via the combined action of glycolysis, the citric acid cycle, and oxidative phosphorylation.

shuttle electrons from NADH in the cytoplasm to $FADH_2$ in the mitochondrion. As a result, the energy yield from the oxidation of one cytoplasmic NADH is only two ATP.

Knowing this, we can sum up the total energy yield that results from the complete oxidation of one glucose molecule. In glycolysis, two ATP are produced by substrate-level phosphorylation, and two NADH are generated. This is equivalent to a total of six ATP. The conversion of pyruvate to acetyl CoA yields another two NADH or six more ATP. Finally, in the citric acid cycle, the production of six NADH and two $FADH_2$ generates 22 ATP, and two ATP are produced by the conversion of succinyl CoA to succinate. Thus the net result of the complete oxidation of glucose is production of 36 molecules of ATP (Figure 14.3). This represents an energy harvest of about 40% of the potential energy of glucose.

Obviously, aerobic metabolism is very much more efficient than anaerobic energy production. The abundant energy released by aerobic metabolism has had enormous consequences for the biological world. Much of the energy released by the oxidation of fuels is not dissipated as heat, but conserved in the form of ATP. Organisms that possess abundant energy have evolved into multicellular organisms and have developed specialized functions. As a consequence of their energy requirement, all multicellular organisms are aerobic.

Compartmentalization of energy-yielding reactions

The reactions of the citric acid cycle, fatty acid oxidation, and respiration occur in the mitochondria of eukaryotic cells. In contrast, glycolysis and fatty acid biosynthesis occur in the cytoplasm. Similarly, oxidation of amino acids also occurs in the mitochondria, and amino acid biosynthesis occurs in the cytoplasm.

Because different metabolic pathways are located in different regions (compartments) of the cell, they can be easily and independently regulated. Diverse metabolic processes can be carried out simultaneously in different cell organelles. Such reactions would severely interfere with one another were they to share the same space in the cell. Thus the cell can simultaneously synthesize necessary components and oxidize fuel molecules by carrying out these processes in different cellular organelles.

Question 14.5 What is the metabolic advantage of compartmentalization?

Question 14.6 Suggest a reason for the observation that multicellular organisms function aerobically.

14.3 RESPIRATION: OXIDATIVE PHOSPHORYLATION

In the last section we noted that NADH can be used to produce three ATP molecules and $FADH_2$ to produce two ATP. What is the process by which the electrons carried by these coenzymes is converted to ATP energy? It is a series of reactions called **oxidative phosphorylation,** which couples the oxidation of NADH and $FADH_2$ to the phosphorylation of ADP to generate ATP.

Electron carriers and ATP synthase

Before we try to understand the mechanism of oxidative phosphorylation, let us first look at the molecules that carry out this complex process. Embedded within the mitochondrial inner membrane are a series of electron carriers. Prominent among these electron carriers are the cytochromes, which carry a heme group. All these molecules are arranged within the membrane in an array that allows them to pass electrons from one to the next. This array of electron carriers is called the **respiratory electron transport system** (Figure 14.4). As you would expect in such sequential oxidation-reduction reactions, the electrons lose some energy with each transfer. Some of this energy will be used to make ATP.

Another important feature of these electron carriers is that some are able to carry hydrogen atoms, while others can carry only electrons. They are arranged in the membrane in such a way that those that carry electrons alone alternate with those that carry hydrogen atoms. When a hydrogen atom carrier picks up a pair of hydrogen atoms from the mitochondrial matrix, it can pass the electrons, but not the protons, to the next carrier. However, to allow the respiratory electron transport system to continue to function, the hydrogen atom carrier must get rid of the protons. It does so by depositing them on the other side of the inner membrane, within the intermembrane space.

The last component needed for oxidative phosphorylation is a multiprotein complex called **ATP synthase** or **F_0F_1 complex** (Figure 14.4). Spanning the inner mitochondrial membrane is a protein complex (F_0) that provides a channel through which protons may pass. Protruding into the mitochondrial matrix is a spherical protein complex (F_1) with the enzymatic ability to phosphorylate ADP to produce ATP.

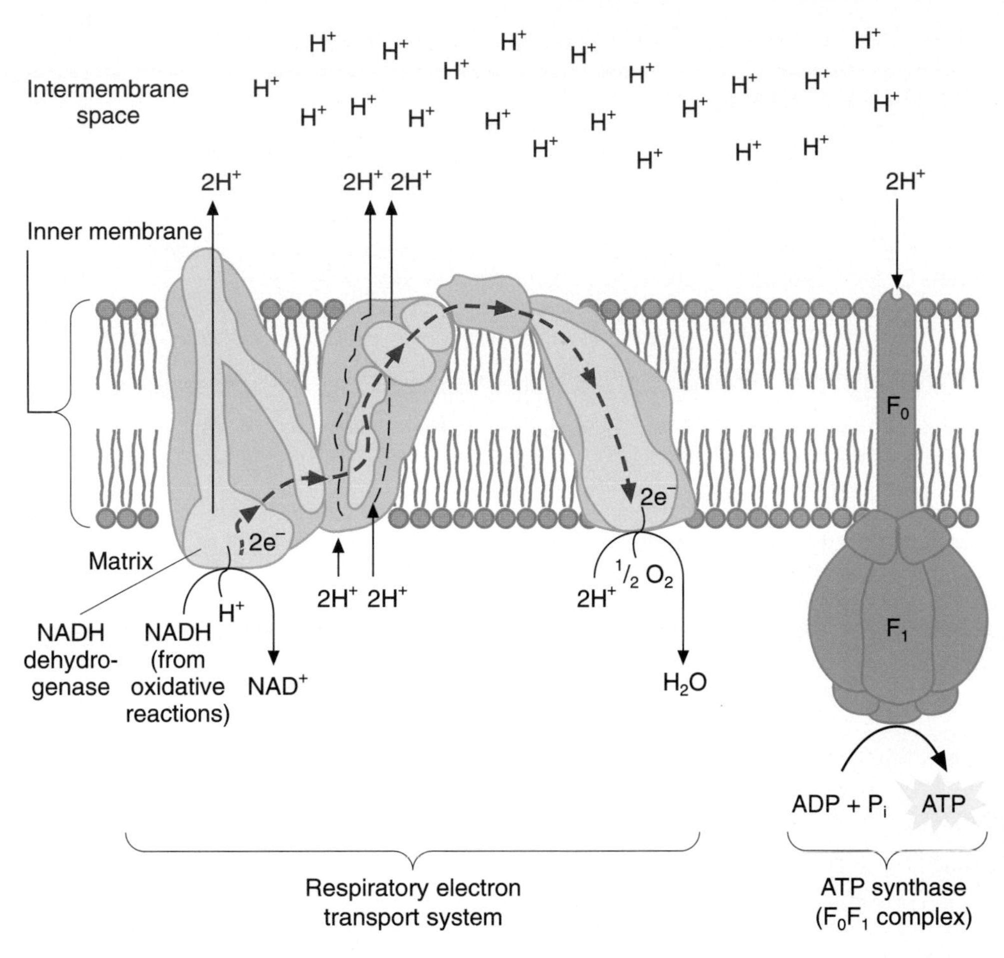

FIGURE 14.4
Electrons flow from NADH to molecular oxygen through a series of electron carriers embedded in the inner mitochondrial membrane. Protons are pumped out of the mitochondrial matrix space, generating a proton gradient between the matrix and intermembrane space. As protons pass through the channel in ATP synthase, their energy is used to phosphorylate ADP and produce ATP.

The chemiosmotic theory

How does all this complicated machinery actually function? NADH carries electrons, originally from glucose, to the first carrier of the electron transport system (NADH dehydrogenase, Figure 14.4). There it is oxidized to NAD^+, donating a pair of hydrogen atoms, and returns to the site of the citric acid cycle to be reduced again. As Figure 14.4 shows (dashed red line), the pair of electrons is passed to the next electron carrier, but the protons are transported to the intermembrane compartment. The electrons are passed sequentially through the electron transport system. Notice that at two additional points, protons from the matrix are transported into the intermembrane compartment. Finally, the electrons arrive at the last carrier. They now have too little energy to accomplish any more work, but they *must* be donated to some final electron carrier so that the electron transport system can continue to function. (Consider that cyanide poison binds to the heme group iron of some of the cytochromes, instantly stopping electron transfers and causing death within minutes!) In aerobic organisms the **terminal electron acceptor** is molecular oxygen, O_2, and the product is water.

Thus the respiratory electron transport chain carries out the oxidation of NADH and $FADH_2$ with molecular oxygen as the final electron acceptor. The reaction is summarized here:

$$\boxed{\text{NADH}} + H^+ + \tfrac{1}{2}O_2 \longrightarrow NAD^+ + H_2O$$

As was noted above, $FADH_2$ donates its electrons to a carrier of lower energy. The result of this is that fewer protons are pumped into the mitochondrial intermembrane space, and ultimately, less ATP is produced.

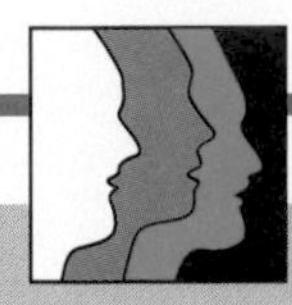

A HUMAN PERSPECTIVE

Brown Fat: The Fat That Makes You Thin?

Humans have two types of fat, or adipose, tissue. *White fat* is distributed throughout the body and is composed of aggregations of cells having membranous vacuoles containing stored triglycerides. The size and number of these storage vacuoles determines whether a person is overweight or not. The other type of fat is **brown fat.** Brown fat is a specialized tissue for heat production, called **nonshivering thermogenesis.** As the name suggests, this is a means of generating heat in the absence of the shivering response. The cells of brown fat look nothing like those of white fat. They do contain small fat vacuoles; however, the distinguishing feature of brown fat is the huge number of mitochondria within the cytoplasm. In addition, brown fat tissue contains a great many blood vessels. These provide oxygen for the thermogenic metabolic reactions.

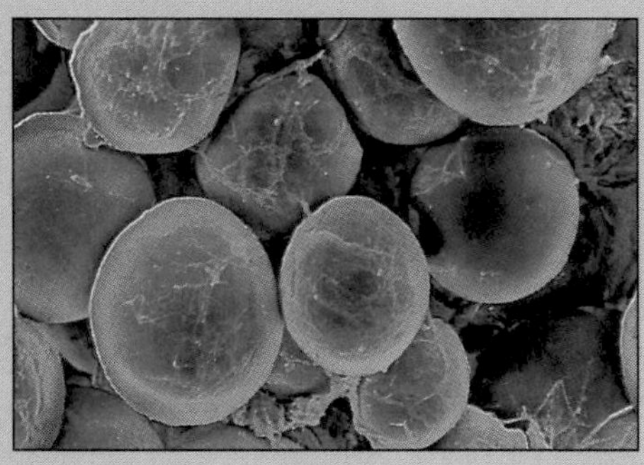

(a)

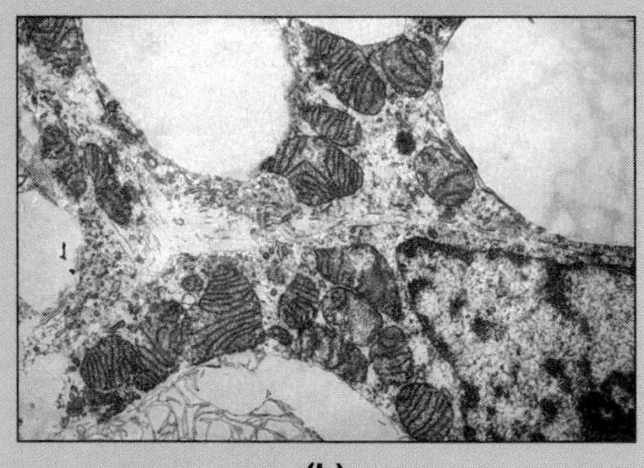

(b)

(a) A light micrograph of white fat cells. (b) An electron micrograph of brown fat cells. The cytoplasm contains few lipid storage vacuoles and a large number of mitochondria.

Brown fat is most pronounced in newborns, cold-adapted mammals, and hibernators. One major difficulty faced by a newborn is temperature regulation. The baby leaves an environment in which he or she was bathed in fluid of a constant 37°C, body temperature. Suddenly the child is thrust into a world that is much colder and in which he or she must generate his or her own warmth internally. By having a good reserve of active brown fat to generate that heat, the newborn is protected against cold shock at the time of birth. However, this thermogenesis literally burns up most of the brown fat tissue, and adults typically have so little brown fat that it can be found only by using a special technique called thermography, which detects temperature differences throughout a body. However, in some individuals, brown fat is very highly developed. For instance, the Korean diving women who spend 6–7 hours every day diving for pearls in cold water have a massive amount of brown fat to warm them by nonshivering thermogenesis. Thus, development of brown fat is a mechanism of cold adaptation.

When it was noticed that such cold-adapted individuals were seldom overweight, a correlation was made between the amount of brown fat in the body and the tendency to become overweight. Studies done with rats suggest that, to some degree, fatness is genetically determined. In other words, you are as lean as your genes allow you to be. In these studies, cold-adapted and non-cold-adapted rats were fed cafeteria food—as much as they wanted—and their weight gain was monitored. In every case the cold-adapted rats, with their greater quantity of brown fat, gained significantly less weight than their non-cold-adapted counterparts, despite the fact that they ate as much as the non-cold-adapted rats. This and other studies led researchers to conclude that brown fat burns excess fat in a highly caloric diet, a phenomenon called **thermogenic hyperphagia.**

How does brown fat generate heat and burn excess calories? For the answer we must turn to the mitochondrion. In addition to the ATP synthase and the electron transport system proteins that are found in all mitochondria, there is a protein in the inner mitochondrial membrane of brown fat tissue called **thermogenin.** This protein has a channel in the center through which the protons (H^+) of the intermembrane space could pass back into the mitochondrial matrix. Under normal conditions this channel is plugged by a GDP molecule so that it remains closed and the proton gradient can continue to drive ATP synthesis by oxidative phosphorylation.

When brown fat is turned on, by cold exposure or in response to certain hormones, there is an immediate increase in the rate of glycolysis and β-oxidation of the stored fat (Chapter 15). These reactions produce acetyl CoA, which then fuels the citric acid cycle. The citric acid cycle, of course, produces NADH and $FADH_2$, which carry electrons to the electron transport system. Finally, the electron transport system pumps protons into the intermembrane space. Under usual conditions the energy of the proton gradient would be used to synthesize ATP. However, when brown fat is stimulated, the GDP that had plugged the pore in thermogenin is lost. Now protons pass freely back into the matrix space, and the proton gradient is dissipated. The energy of the gradient, no longer useful for generating ATP, is released as *heat,* the heat that warms and protects newborns and cold-adapted individuals.

Brown fat is just one of the body's many systems for maintaining a constant internal environment regardless of the conditions in the external environment. Such mechanisms, called **homeostatic mechanisms,** are absolutely essential to allow the body to adapt to and survive in an ever-changing environment.

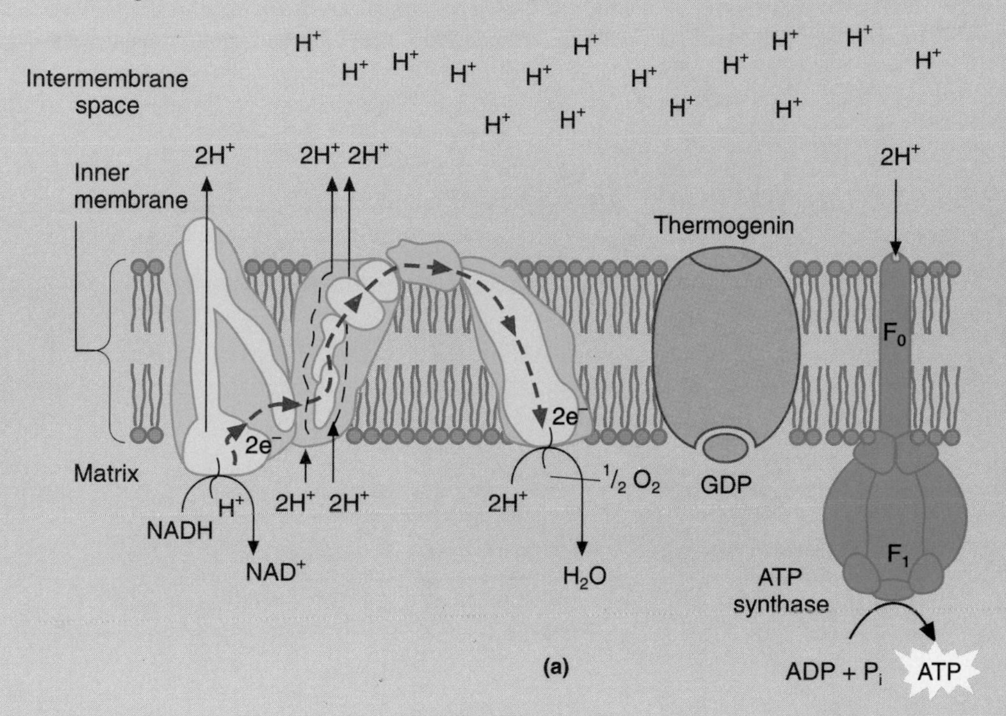

(a)

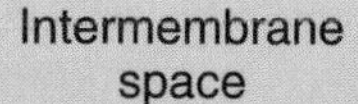

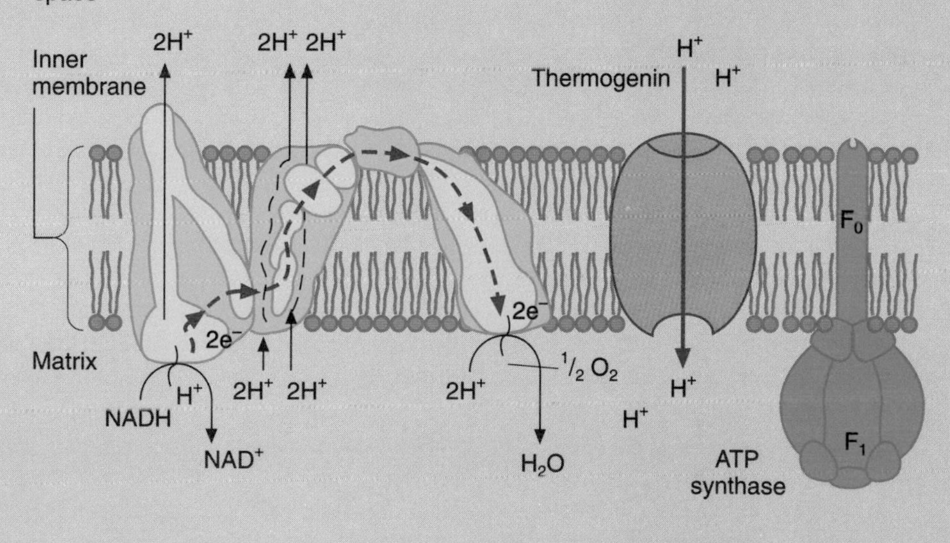

(b)

(a) The inner membrane of brown fat mitochondria contains thermogenin. In the normal state the pore in the center of thermogenin is plugged by a GDP molecule. (b) When brown fat is activated for thermogenesis, the GDP molecule is removed from the pore, and the protons from the H^+ reservoir are free to flow back into the matrix of the mitochondrion. As the gradient dissipates, the energy is used to generate heat.

As the electron transport system continues to function, it is apparent that a large excess of protons will accumulate in the intermembrane space. This creates a concentration gradient (a progressively larger concentration of protons in one region) across the inner membrane because the concentration of protons in the intermembrane space is much greater than the concentration of protons in the matrix. This is a dual gradient. It is chemical because the pH in the intermembrane space is lower than that in the matrix, and it is an electrical gradient because the concentration of positive charges is greater in the intermembrane space than in the matrix. Such a gradient is an enormous energy source, like water stored behind a dam. The mitochondria make use of the potential energy of the gradient to synthesize ATP energy.

It is the ATP synthase that has the ability to harvest the energy of this gradient and use it to produce ATP. Although the inner mitochondrial membrane is quite impermeable, the ATP synthase provides a channel (F_0) through which the protons can return to the matrix. As the protons pass into the matrix, some of their energy is used by the enzymatic portion of ATP synthase (F_1) to catalyze the phosphorylation of ADP to ATP.

This is called the **chemiosmotic theory.** It explains the way in which the oxidation of glucose is coupled to the phosphorylation of ADP. In reality, then, the chemiosmotic production of ATP is an indirect process consisting of two related events (Figure 14.4):

1. **H^+ transport:** Movement of the protons across the inner mitochondrial membrane and into the intermembrane space.
2. **F_0F_1 event:** Phosphorylation of ADP to ATP by the ATP synthase, which is able to harvest the energy of the gradient as the protons pass through its channel back into the matrix.

Question 14.7 Write a balanced chemical equation for the reduction of NAD^+.

Question 14.8 Write a balanced chemical equation for the reduction of FAD.

14.4 CONTROL OF THE CITRIC ACID CYCLE

The control of the citric acid cycle depends upon many factors, including the availability of oxygen and the energy demands of the cell. It is logical that this pathway should be able to respond to greater energy demands by producing more ATP. Similarly, if energy demands decrease and energy is in excess, the pathway should respond by slowing energy generation. What kinds of chemical signals could indicate that the cell requires more ATP? One reliable indicator is a high concentration of ADP. On the other hand, high concentrations of ATP and NADH signal that energy is abundant.

Section 12.9

How are these chemical signals recognized and translated into increased or decreased energy production? Some of the enzymes of the citric acid cycle are allosteric enzymes. Binding of the chemical signal, or effector, alters the shape of the active site. Effector binding may turn the enzyme on (positive allosterism), or it may inhibit the enzyme (negative allosterism). Because the control must be precise, there are several enzymatic steps that are regulated (Figure 14.5).

1. *Conversion of pyruvate to acetyl CoA*. The pyruvate dehydrogenase complex is inhibited by high concentrations of ATP, acetyl CoA, and NADH. Of course, the presence of these compounds in abundance signals that the cell has an adequate supply of energy, and thus energy metabolism is slowed.
2. *Synthesis of citrate from oxaloacetate and acetyl CoA*. The enzyme citrate synthase is an allosteric enzyme. In this case the negative effector is ATP. Again, this is logical because an excess of ATP indicates that the cell has an abundance of energy.

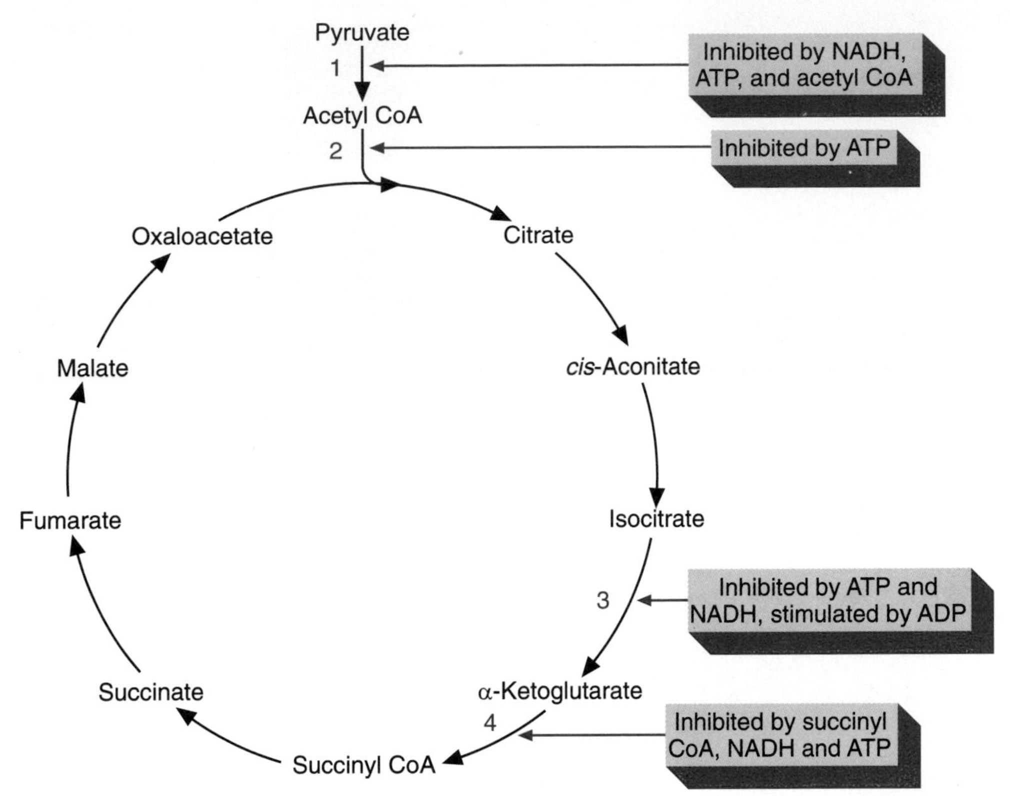

FIGURE 14.5
Regulation of the citric acid cycle.

3. *Oxidation and decarboxylation of isocitrate to α-ketoglutarate.* Isocitrate dehydrogenase is also an allosteric enzyme; however, the enzyme is controlled by the positive allosteric effector, ADP. ADP is a signal that the levels of ATP must be low, and therefore the rate of energy production by the citric acid cycle should be increased. Interestingly, isocitrate dehydrogenase is also *inhibited* by high levels of NADH and ATP.
4. *Conversion of α-ketoglutarate to succinyl CoA.* The α-ketoglutarate dehydrogenase complex is inhibited by high levels of the products of the reactions that it catalyzes, namely, NADH and succinyl CoA. It is further inhibited by high concentrations of ATP.

Question 14.9 What is the importance of the regulation of the citric acid cycle?

14.5 THE DEGRADATION OF AMINO ACIDS

Carbohydrates are not our only source of energy. Dietary protein is digested to amino acids, which may also be used as an energy source, although this is not their major metabolic function. Most of the amino acids that are used for energy come from the diet. In fact, only under starvation conditions, when stored glycogen and lipid have been depleted, does the body begin to burn its own protein—for instance, from muscle—as a fuel.

The fate of the mixture of amino acids provided by digestion of protein depends upon a balance between the requirement for amino acids for biosynthesis and the need for cellular energy. Only the amino acids that are not needed for protein synthesis are eventually converted into citric acid cycle intermediates and used as fuel.

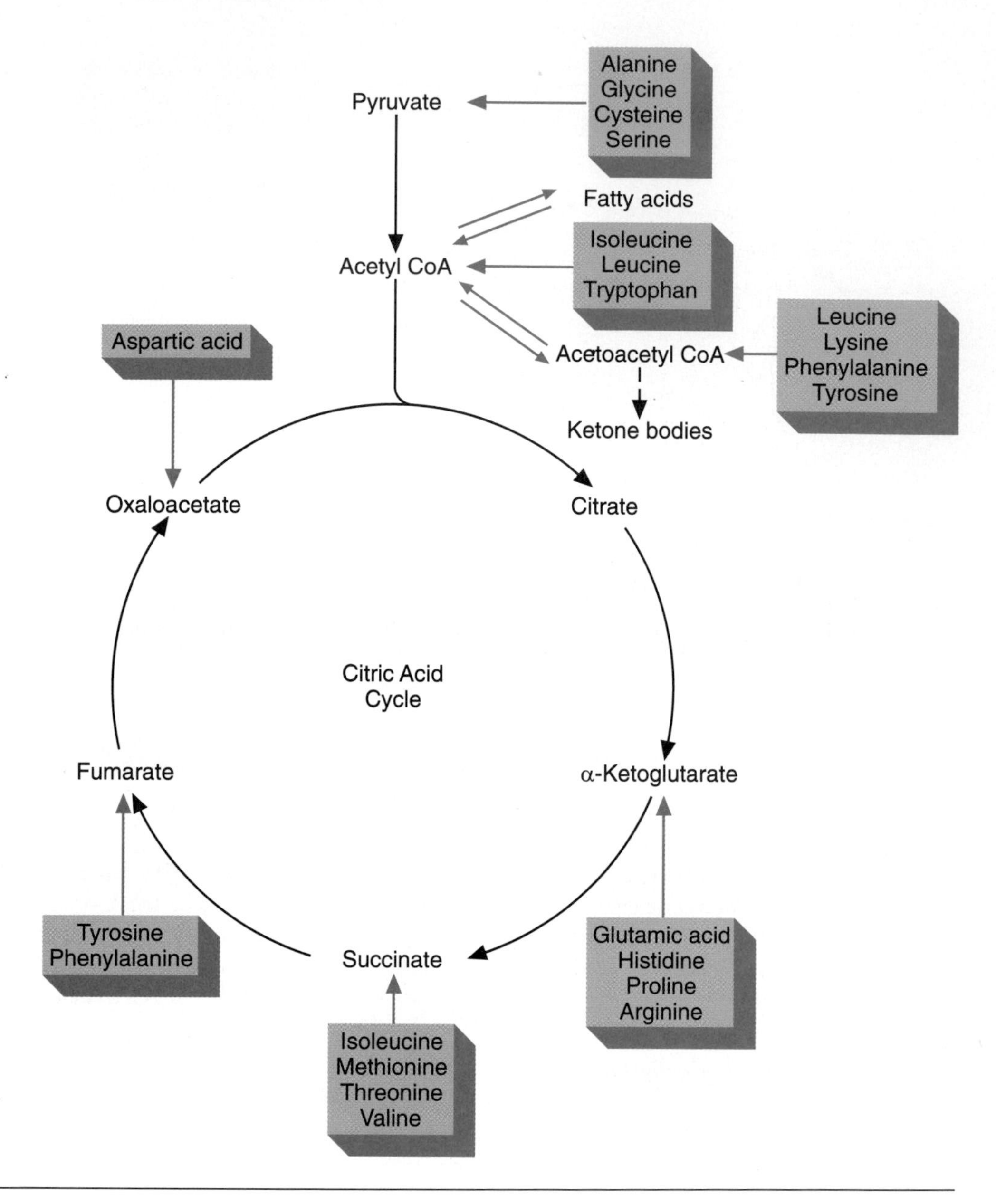

FIGURE 14.6
The carbon skeletons of amino acids can be converted to citric acid cycle intermediates and completely oxidized to produce ATP energy.

The degradation of amino acids occurs primarily in the liver and takes place in two stages. The first stage is the removal of the α-amino group, and the second is the degradation of the carbon skeleton. In land mammals the amino group generally ends up in urea, which is excreted in the urine. The carbon skeletons can be converted into a variety of compounds, including citric acid cycle intermediates, pyruvate, acetyl CoA, and acetoacetyl CoA. The degradation of the carbon skeletons is summarized in Figure 14.6. Deamination reactions and the fate of the carbon skeletons of amino acids are described in detail below.

Removal of α-amino groups: transamination

The first stage of amino acid degradation, the removal of the α-amino group, is usually accomplished by a **transamination** reaction. **Aminotransferases,** also called **transaminases,** catalyze the transfer of the α-amino group from an α-amino acid to an α-keto acid. This general reaction is summarized below:

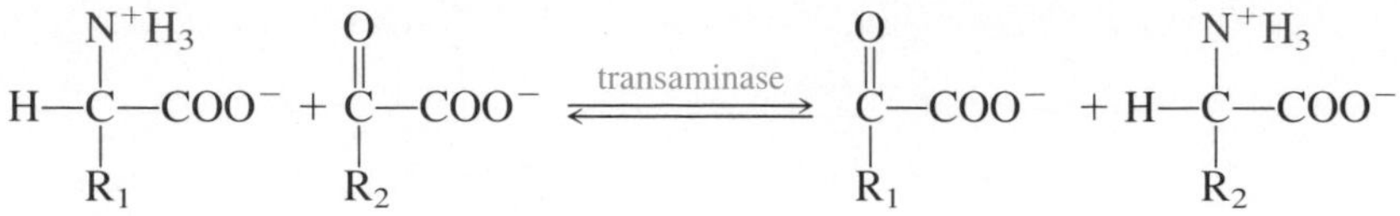

$$H-\underset{R_1}{\overset{N^+H_3}{C}}-COO^- + \underset{R_2}{\overset{O}{\overset{\|}{C}}}-COO^- \xrightleftharpoons{\text{transaminase}} \underset{R_1}{\overset{O}{\overset{\|}{C}}}-COO^- + H-\underset{R_2}{\overset{N^+H_3}{C}}-COO^-$$

Donor amino acid — Acceptor keto acid — Carbon skeleton of amino acid — New amino acid

The α-amino group of a great many amino acids are transferred to α-ketoglutarate to produce the amino acid glutamate. This glutamate family of aminotransferases is especially important because the α-keto acid corresponding to glutamate is the citric acid cycle intermediate, α-ketoglutarate. The glutamate aminotransferases thus provide a direct link between amino acid degradation and the citric acid cycle.

One of the most important aminotransferases is **aspartate aminotransferase,** which catalyzes the transfer of the α-amino group of aspartate to α-ketoglutarate, producing oxaloacetate and glutamate:

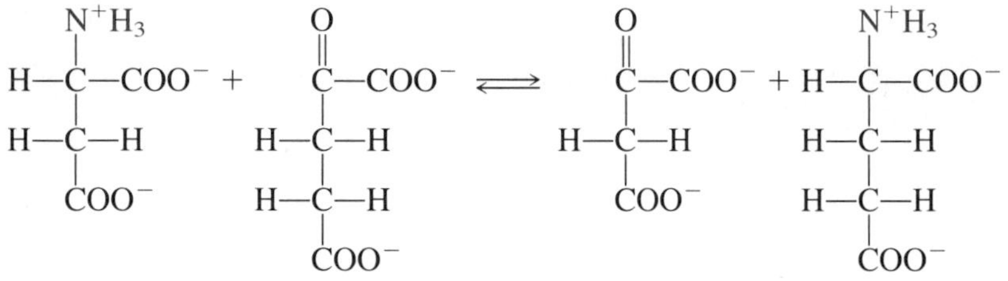

$$H-\overset{N^+H_3}{C}(CH_2COO^-)-COO^- + O{=}C(CH_2CH_2COO^-)-COO^- \rightleftharpoons O{=}C(CH_2COO^-)-COO^- + H-\overset{N^+H_3}{C}(CH_2CH_2COO^-)-COO^-$$

Aspartate — α-Ketoglutarate — Oxaloacetate — Glutamate

Another important transaminase in mammalian tissues is **alanine aminotransferase,** which catalyzes the transfer of the α-amino group of alanine to α-ketoglutarate and produces pyruvate and glutamate:

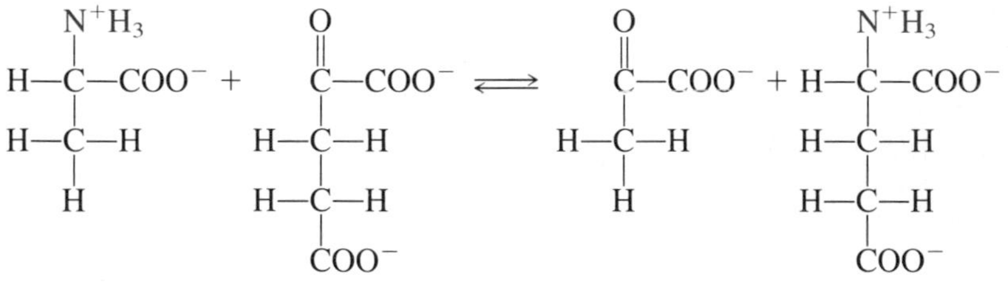

$$H-\overset{N^+H_3}{C}(CH_3)-COO^- + O{=}C(CH_2CH_2COO^-)-COO^- \rightleftharpoons O{=}C(CH_3)-COO^- + H-\overset{N^+H_3}{C}(CH_2CH_2COO^-)-COO^-$$

Alanine — α-Ketoglutarate — Pyruvate — Glutamate

All of the more than 50 aminotransferases that have been discovered require the prosthetic group pyridoxal phosphate. **Pyridoxal phosphate** is a coenzyme derived from vitamin B_6 or pyridoxine (Figure 14.7).

The transamination reactions shown above appear to be a simple transfer, but in reality, the reaction is much more complex. The α-amino group is first transferred to pyridoxal phosphate and then from pyridoxal phosphate to the α-keto acid. This series of reactions is summarized in Figure 14.8. The α-amino group of aspartate is transferred to pyridoxal phosphate, producing pyridoxamine phosphate and oxaloacetate (Figure 14.8b). The amino group is then transferred to an α-keto acid, in this case, α-ketoglutarate (Figure 14.8c), to produce the amino acid glutamate (Figure 14.8d). Next we will examine the fate of the amino group that has been transferred to glutamate.

Question 14.10 What is the role of pyridoxal phosphate in transamination reactions?

FIGURE 14.7
The structure of pyridoxal phosphate, the coenzyme required for all transamination reactions, and pyridoxine, vitamin B_6, the vitamin from which it is derived.

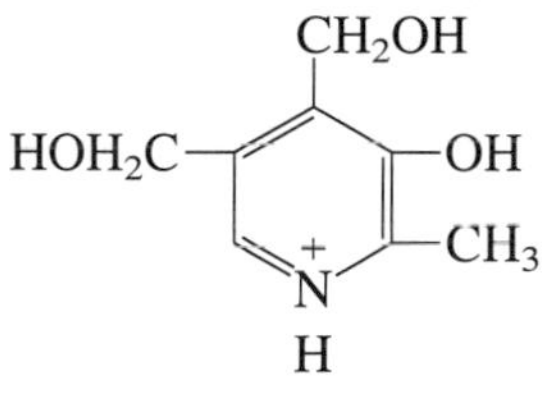

CH_2OH; HOH_2C; OH; CH_3; $\overset{+}{N}$–H

Pyridoxine (vitamin B_6)

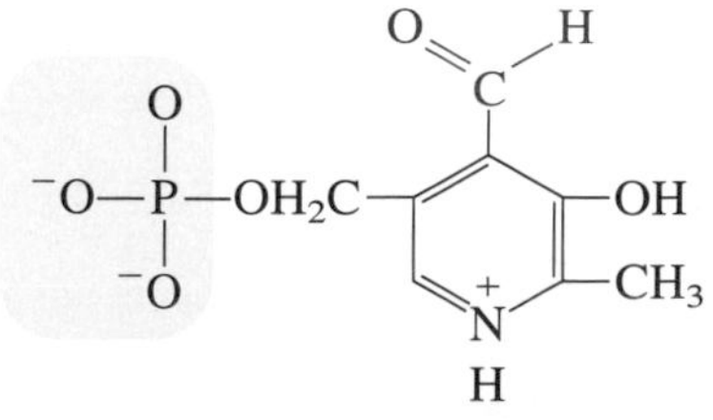

$O{=}C-H$; $^-O-\overset{O}{P}(O^-)-OH_2C$; OH; CH_3; $\overset{+}{N}$–H

Pyridoxal phosphate

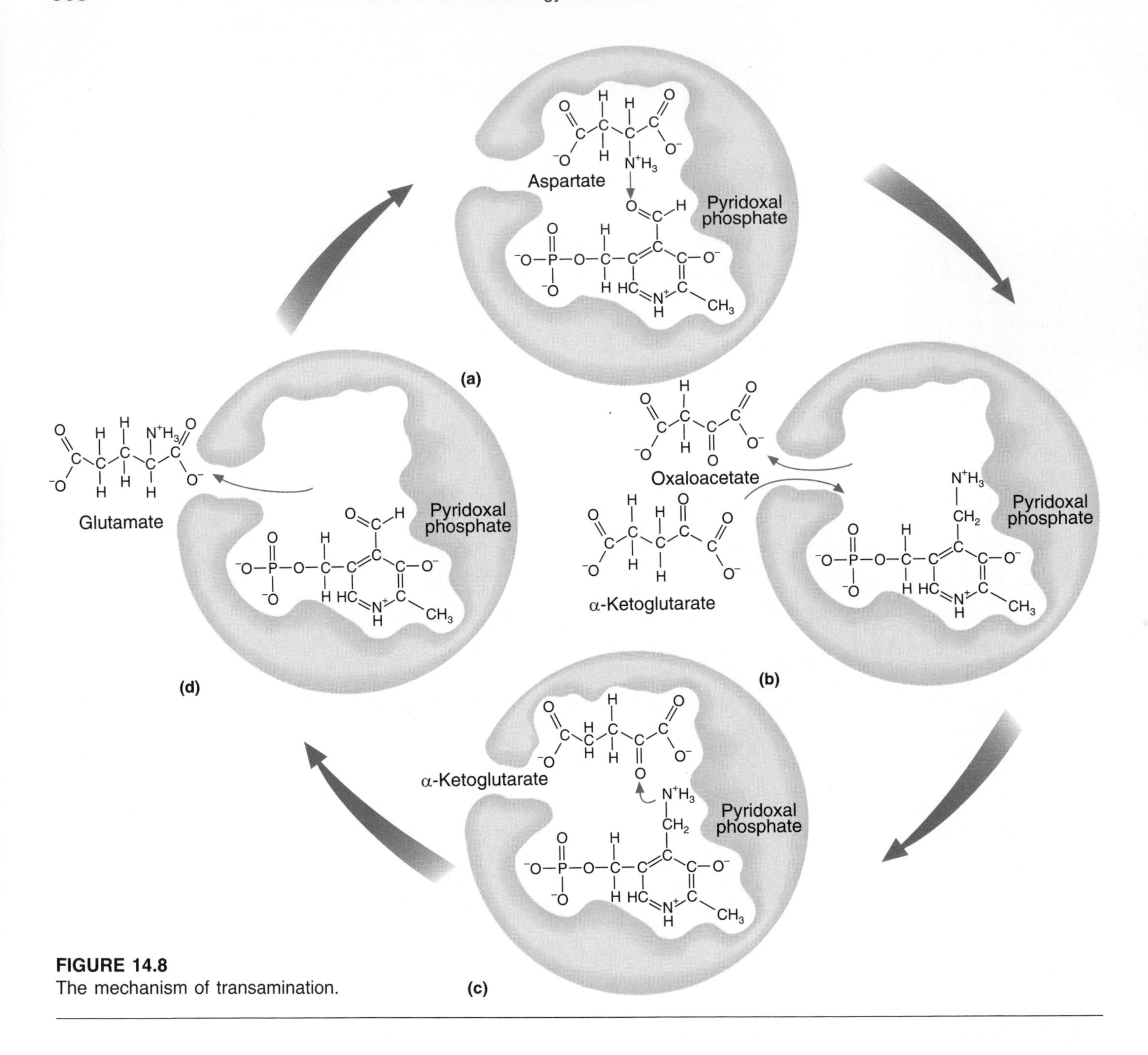

FIGURE 14.8
The mechanism of transamination.

Removal of α-amino groups: oxidative deamination

In the next stage of amino acid degradation, ammonium ion is liberated from the glutamate formed by the aminotransferase. This oxidative deamination of glutamate, catalyzed by the enzyme **glutamate dehydrogenase,** is shown below:

$$\underset{\text{Glutamate}}{\begin{array}{c} N^+H_3 \\ | \\ H-C-COO^- \\ | \\ H-C-H \\ | \\ H-C-H \\ | \\ COO^- \end{array}} + NAD^+ + H_2O \rightleftharpoons NH_4^+ + \underset{\alpha\text{-Ketoglutarate}}{\begin{array}{c} O \\ \| \\ C-COO^- \\ | \\ H-C-H \\ | \\ H-C-H \\ | \\ COO^- \end{array}} + \boxed{NADH} + H^+$$

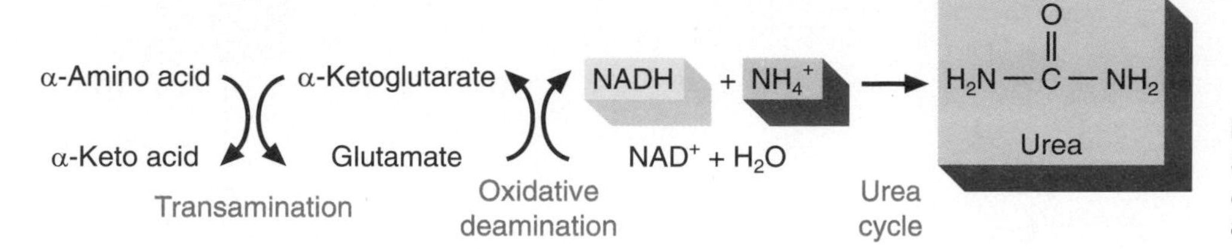

FIGURE 14.9
Summary of the deamination of an α-amino acid and the fate of the ammonium ion (NH_4^+).

This is an example of an oxidative deamination, an oxidation-reduction process. NAD^+ is reduced to NADH, and the amino acid is deaminated. The coenzyme shown in the reaction above is NAD^+, although glutamate dehydrogenase is unusual in that it can also utilize $NADP^+$. In addition, the enzyme is allosterically regulated; it is inhibited by ATP and GTP and activated by ADP and GDP. Thus the reaction shown above is favored when the cell needs energy. A summary of the deamination reactions described above is shown in Figure 14.9.

The fate of amino acid carbon skeletons

The carbon skeletons produced by these and other deamination reactions enter the energy-generating pathways at many steps. For instance, we have seen that transamination converts aspartate to oxaloacetate and alanine to pyruvate. Serine and threonine can be directly deaminated to yield ammonium ion and pyruvate or α-ketobutyrate, respectively. The positions at which the carbon skeletons of various amino acids enter the energy-generating pathways is summarized in Figure 14.6.

Question 14.11 Summarize the degradation of the amino acid aspartate and show where the carbon skeleton enters the energy generating pathways.

Both oxidative deamination of glutamate and deamination of serine and threonine produce considerable quantities of ammonium ion. If this were not incorporated into a biological molecule and removed from the body, it would quickly reach toxic levels. Thus it is of critical importance to the survival of the organism to have a mechanism for the excretion of ammonium ions, regardless of the energy required. In humans and most terrestrial vertebrates the means of ammonium ion removal is the urea cycle.

14.6 THE UREA CYCLE

As was noted above, ammonium ions are extremely toxic. The reason for this toxicity is that they shift the equilibrium position of glutamate dehydrogenase toward glutamate, drastically lowering the concentration of α-ketoglutarate. Without an adequate supply of α-ketoglutarate the rates of the citric acid cycle and respiration decrease, and thus the ability to generate ATP is greatly diminished. In addition, some of the glutamate that is produced as a result of the equilibrium shift is converted to glutamine. This latter amino acid seems to accumulate in the brain under the above conditions and may directly cause some toxicity. The conversion of ammonium ions to urea by the **urea cycle** occurs in the liver, and ammonium ions are thereby kept out of the blood. The excess ammonium ions incorporated in urea are excreted by terrestrial vertebrates (Figure 14.10).

Question 14.12 Explain why the accumulation of ammonium ion is toxic to the body.

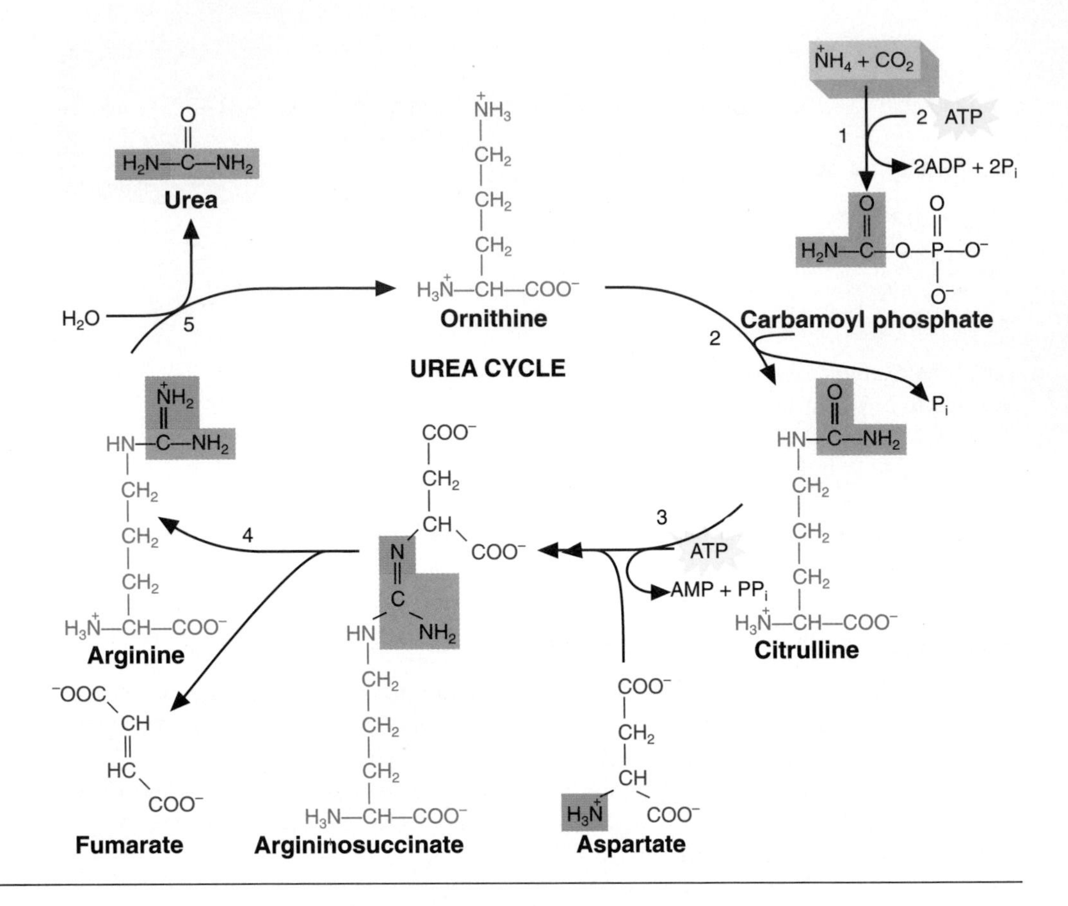

FIGURE 14.10
The reactions of the urea cycle.

Reactions of the urea cycle

The five reactions of the urea cycle are shown in Figure 14.10, and details of the reactions are summarized below.

Step 1: The first step of the cycle is a reaction in which CO_2 and NH_4^+ produce carbamoyl phosphate. This reaction also requires ATP and H_2O. In the reactions shown below, the abbreviation P_i is used to designate an inorganic phosphate group.

$$CO_2 + NH_4^+ + 2ATP + H_2O \longrightarrow H_2N\text{—}\overset{\overset{\displaystyle O}{\|}}{C}\text{—}O\text{—}\underset{\underset{\displaystyle O^-}{|}}{\overset{\overset{\displaystyle O}{\|}}{P}}\text{—}O^- + 2ADP + P_i + 3H^+$$

Carbamoyl phosphate

Step 2: The carbamoyl phosphate thus produced condenses with the amino acid ornithine to produce the amino acid citrulline:

$$\begin{array}{c} N^+H_3 \\ | \\ H\text{—}C\text{—}H \\ | \\ H\text{—}C\text{—}H \\ | \\ H\text{—}C\text{—}H \\ | \\ H\text{—}C\text{—}N^+H_3 \\ | \\ COO^- \end{array} + H_2N\text{—}\overset{\overset{\displaystyle O}{\|}}{C}\text{—}O\text{—}\underset{\underset{\displaystyle O^-}{|}}{\overset{\overset{\displaystyle O}{\|}}{P}}\text{—}O^- \longrightarrow \begin{array}{c} H\text{—}N\text{—}\overset{\overset{\displaystyle O}{\|}}{C}\text{—}NH_2 \\ | \\ H\text{—}C\text{—}H \\ | \\ H\text{—}C\text{—}H \\ | \\ H\text{—}C\text{—}H \\ | \\ H\text{—}C\text{—}N^+H_3 \\ | \\ COO^- \end{array} + P_i$$

Ornithine Carbamoyl phosphate Citrulline

Step 3: Citrulline now condenses with aspartate to produce arginosuccinate. This reaction requires energy released by the hydrolysis of ATP:

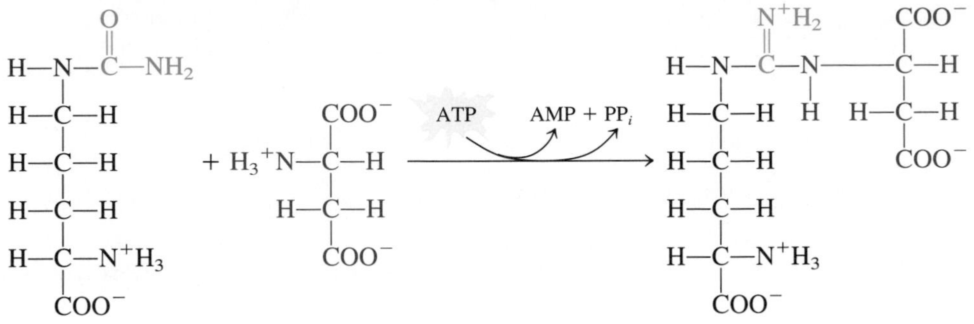

$$\text{H—N—C(=O)—NH}_2\text{; H—C—H; H—C—H; H—C—H; H—C—N}^+\text{H}_3\text{; COO}^- \;+\; \text{H}_3{}^+\text{N—C(COO}^-\text{)—H; H—C—H; COO}^- \xrightarrow{\text{ATP} \;\rightarrow\; \text{AMP} + \text{PP}_i} \text{H—N—C(=N}^+\text{H}_2\text{)—N(H)—C(COO}^-\text{)—H; H—C—H (H—C—H, COO}^-\text{); H—C—H; H—C—H; H—C—N}^+\text{H}_3\text{; COO}^-$$

Citrulline Aspartate Arginosuccinate

Step 4: Now the arginosuccinate is cleaved to produce the amino acid arginine and the citric acid cycle intermediate fumarate:

$$\text{Arginosuccinate} \longrightarrow \text{H—N—C(=N}^+\text{H}_2\text{)—NH}_2\text{; H—C—H; H—C—H; H—C—H; H—C—N}^+\text{H}_3\text{; COO}^- \;+\; \text{H(COO}^-\text{)C=C(H)OOC}^-$$

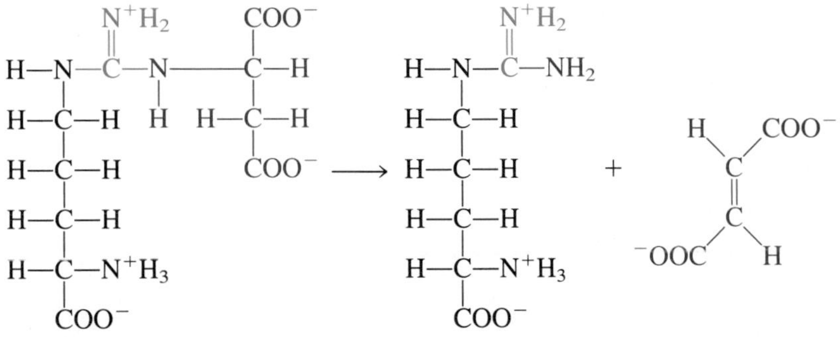

Arginosuccinate Arginine Fumarate

Step 5: Finally, arginine is hydrolyzed to generate urea, the product of the reaction to be excreted, and ornithine, the original reactant in the cycle. Note that one of the amino groups is the original ammonium ion and the second is derived from the amino acid aspartate.

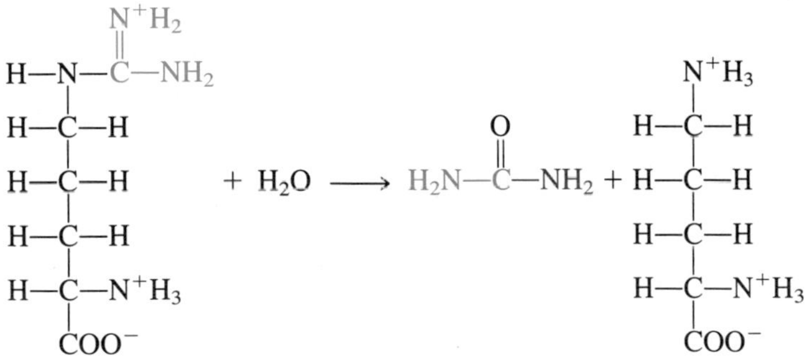

$$\text{H—N—C(=N}^+\text{H}_2\text{)—NH}_2\text{; H—C—H; H—C—H; H—C—H; H—C—N}^+\text{H}_3\text{; COO}^- \;+\; \text{H}_2\text{O} \longrightarrow \text{H}_2\text{N—C(=O)—NH}_2 \;+\; \text{N}^+\text{H}_3\text{; H—C—H; H—C—H; H—C—H; H—C—N}^+\text{H}_3\text{; COO}^-$$

Arginine Water Urea Ornithine

Interaction between the urea cycle and the citric acid cycle

One connection between the urea cycle and the citric acid cycle has to do with the energy requirement of the urea cycle. The urea cycle is an energy-expensive series of reactions, using three ATP produced by the citric acid cycle and electron transport system. In addition to the energy connection the urea cycle generates a citric acid cycle intermediate, fumarate, and utilizes an indirect product of the citric acid cycle (Figure 14.11).

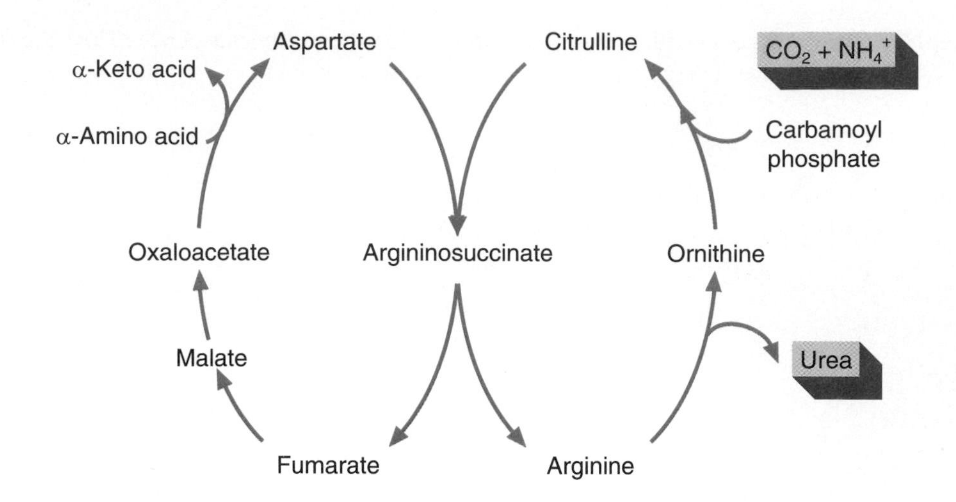

FIGURE 14.11
Interaction between the urea cycle and the citric acid cycle.

In step 4 we saw that arginosuccinate was cleaved to produce arginine and fumarate. Enzymes of the citric acid cycle convert the fumarate to malate and then oxidize the malate to oxaloacetate. Oxaloacetate may, of course, condense with acetyl CoA to produce citric acid that will continue through the citric acid cycle. Alternatively, it may be transaminated to produce aspartate, which, in turn, is involved in step 3 of the urea cycle, the condensation of aspartate and citrulline to generate arginosuccinate. The constant release of fumarate by the urea cycle and its subsequent use to generate the aspartate needed by the urea cycle keep the two cycles in balance. The drain of oxaloacetate from the citric acid cycle is always counteracted by the production of fumarate by the urea cycle.

Question 14.13 Why doesn't the removal of oxaloacetate from the citric acid cycle for use in the urea cycle diminish the ability of the cell to generate energy?

The enzymatic reactions of the urea cycle are located in mammalian liver, where they are distributed between the cytoplasm and the mitochondrial matrix space. The urea cycle consists of the sequential action of five enzymes. Enzymes that catalyze the first four steps of the urea cycle are found in all mammalian tissues, but arginase, which catalyzes the final step of the urea cycle, is found only in the liver.

There are genetically transmitted diseases that result from a deficiency of one of the enzymes of the urea cycle. The importance of the urea cycle is apparent when we consider the dire effects suffered by afflicted persons. A deficiency of urea cycle enzymes causes an elevation of the concentration of NH_4^+, a condition known as **hyperammonemia.** If there is a complete deficiency of one of the enzymes of the urea cycle, the result is death in early infancy. However, if there is a partial deficiency of one of the enzymes of the urea cycle, the result may be retardation, convulsions, and vomiting. In these milder forms of hyperammonemia a low-protein diet leads to a lower concentration of NH_4^+ in blood and less severe clinical symptoms.

14.7 THE CITRIC ACID CYCLE AS A SOURCE OF BIOSYNTHETIC INTERMEDIATES

So far we have treated the citric acid cycle solely as an energy-generating mechanism. We have seen that dietary carbohydrates and amino acids enter the pathway at various stages and are oxidized to generate NADH and $FADH_2$, which, via oxidative phosphorylation, are used to make ATP.

However, the role of the citric acid cycle in cellular metabolism involves more than just catabolism, or energy-generating reactions. It plays a key role in biosynthesis as well. Figure 14.12 shows the central role of glycolysis and the citric acid cycle as energy-generating reactions, as well as sources of major biosynthetic precursors.

As you might already suspect from the fact that amino acids can be converted into citric acid cycle intermediates, these same citric acid cycle intermediates can also be used as precursors for the synthesis of amino acids. Oxaloacetate provides the carbon skeleton for the one-step synthesis of the amino acid aspartate by the transamination reaction shown below:

$$\text{oxaloacetate} + \text{glutamate} \rightleftharpoons \text{aspartate} + \alpha\text{-ketoglutarate}$$

Aside from providing aspartate for protein synthesis, this reaction provides aspartate for the urea cycle.

Asparagine is synthesized from aspartate by the amidation reaction shown below:

$$\text{aspartate} + NH_4^+ + \text{ATP} \longrightarrow \text{asparagine} + \text{AMP} + PP_i + H^+$$

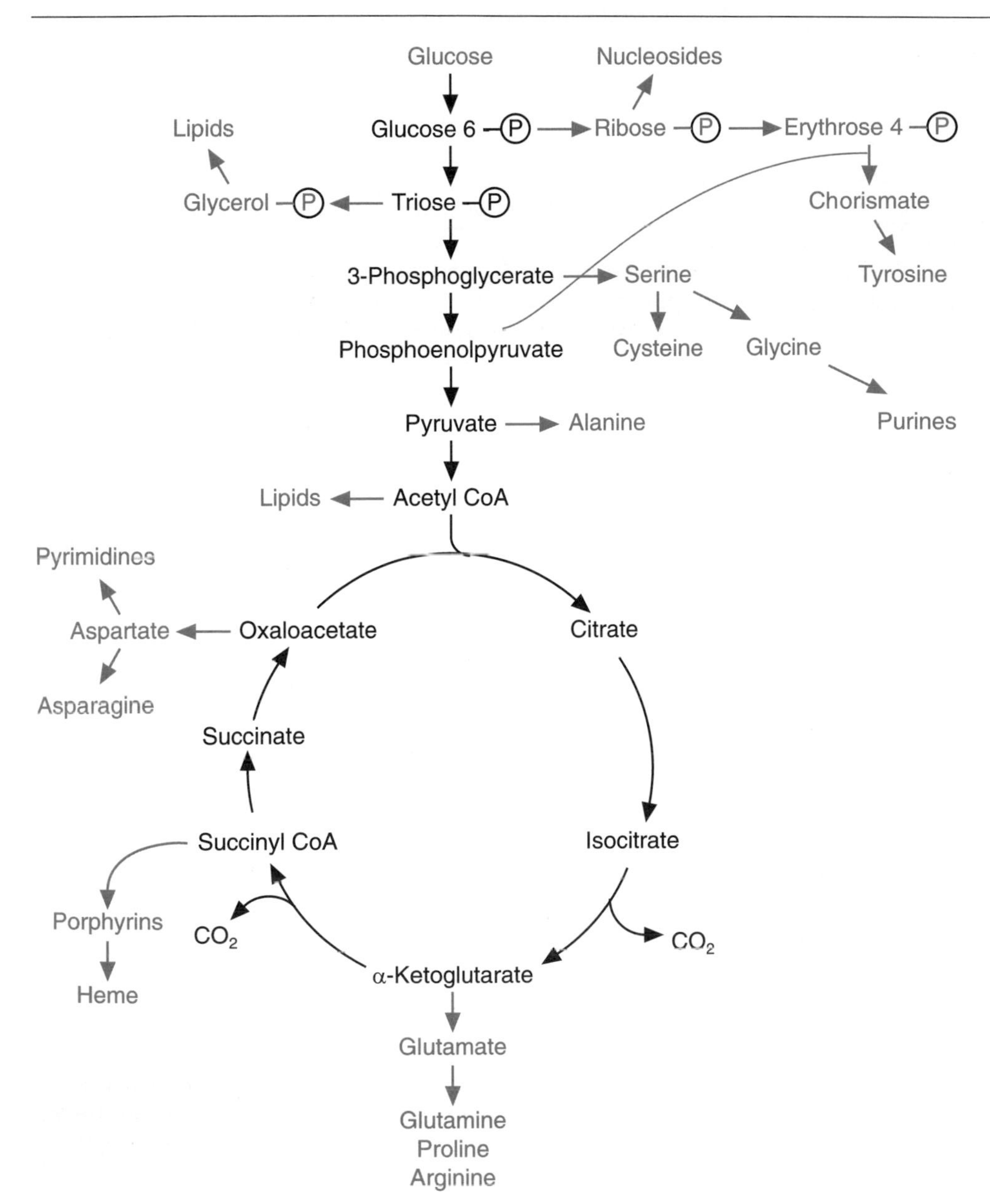

FIGURE 14.12
Glycolysis and the citric acid cycle also provide a variety of precursors for the biosynthesis of amino acids, nitrogenous bases, and porphyrins.

α-Ketoglutarate serves as the precursor carbon chain for the family of amino acids including glutamate, glutamine, proline, and arginine. Glutamate is of particular interest because it serves as the donor of the α-amino group for almost all other amino acids. It is synthesized from NH_4^+ and α-ketoglutarate in a reaction that is mediated by glutamate dehydrogenase. This is the reverse of the reaction shown in Figure 14.9 and described above. In this case the coenzyme that serves as the reducing agent is NADPH in the reaction that is summarized below:

$$NH_4^+ + \alpha\text{-ketoglutarate} + NADPH \rightleftharpoons \text{L-glutamate} + NADP^+ + H_2O$$

Glutamine, proline, and arginine are synthesized from glutamate.

Examination of Figure 14.12 reveals that serine, glycine, and cysteine are synthesized from 3-phosphoglycerate; alanine is synthesized from pyruvate; and tyrosine is produced from phosphoenolpyruvate and the four-carbon sugar erythrose-4-phosphate, which, in turn, is synthesized from glucose-6-phosphate. The nine amino acids not shown in Figure 14.12 (histidine, isoleucine, leucine, lysine, methionine, phenylalanine, threonine, tryptophan, and valine) are called the essential amino acids because they cannot be synthesized by humans and other mammals. In addition to the amino acid precursors, glycolysis and the citric acid cycle also provide precursors for lipids and the nitrogenous bases that are required to make DNA, the genetic information. They also generate precursors for heme, the prosthetic group required for hemoglobin, myoglobin, cytochrome *c*, and catalase.

Section 17.1

Clearly, the reactions of glycolysis and the citric acid cycle are central to both **anabolic** (biosynthetic) and **catabolic** (energy-generating) cellular activities. Metabolic pathways that function in both anabolism and catabolism are termed **amphibolic pathways.** Consider for a moment the difficulties that the dual nature of these pathways could present to the cell. When the cell is actively growing, there is a great demand for biosynthetic precursors to build new cell structures. The demand for amino acids may draw a great deal of oxaloacetate and α-ketoglutarate away from the citric acid cycle. The dilemma that arises is that periods of active cell growth also demand enormous amounts of energy, but depletion of mitochondrial oxaloacetate would reduce the cell's capacity to produce that energy.

Question 14.14 Explain how the citric acid cycle serves as an amphibolic pathway.

The solution to this problem is to have an alternative pathway for oxaloacetate synthesis that can produce enough oxaloacetate to supply the anabolic and catabolic requirements of the cell. Although bacteria and plants have several mechanisms, the only means by which mammalian cells can produce more oxaloacetate is by the carboxylation of pyruvate, a reaction that is also important in gluconeogenesis. This reaction is summarized below:

Section 13.5

$$\text{pyruvate} + CO_2 + ATP + H_2O \longrightarrow \text{oxaloacetate} + ADP + P_i + H^+$$

The enzyme that catalyzes this reaction is *pyruvate carboxylase*. It is a conjugated protein having as its covalently linked prosthetic group the vitamin *biotin*. The enzyme is also an allosteric enzyme; its positive effector is acetyl CoA. The presence of high concentrations of acetyl CoA is a signal that the cell requires high levels of the citric acid cycle intermediates, particularly oxaloacetate, the beginning substrate.

As was mentioned above, conversion of pyruvate to oxaloacetate is also an important reaction in gluconeogenesis. How does the cell know whether the oxaloacetate produced by pyruvate carboxylase should be used for gluconeogenesis (to store energy) or the citric acid cycle (to generate ATP energy)? The answer, in this case, is straightforward. If concentrations of ATP are high, the oxaloacetate is used in gluconeogenesis. However, if the concentration of ATP is low, the oxaloacetate will enter the citric acid cycle.

The reaction catalyzed by pyruvate carboxylase is called an **anaplerotic reaction.** The term ''anaplerotic'' means ''to fill up.'' Indeed, this critical enzyme must constantly replenish the oxaloacetate and thus indirectly all the citric acid cycle intermediates that are withdrawn as biosynthetic precursors for the reactions summarized in Figure 14.12.

Question 14.15 What is the function of an anaplerotic reaction?

SUMMARY

The complete oxidation of glucose, which occurs under aerobic conditions, provides most of the energy for the cell. These reactions occur in the mitochondria.

The mitochondrion is an aerobic cell organelle that is responsible for most energy production in eukaryotic cells. It is enclosed by a double membrane. The outer membrane permits low molecular weight molecules to pass through. The inner mitochondrial membrane, by contrast, is almost completely impermeable to most molecules. The inner mitochondrial membrane is the site where oxidative phosphorylation occurs. The enzymes of the citric acid cycle, of amino acid catabolism, and of fatty acid oxidation are located in the matrix space of the mitochondrion.

The citric acid cycle is the final common pathway for the degradation of carbohydrates, amino acids, and fatty acids. The citric acid cycle occurs in the matrix of the mitochondrion. It is a cyclic series of biochemical reactions that accomplishes the complete oxidation of the carbon skeletons of food molecules.

Oxidative phosphorylation, or respiration, is the process by which NADH and $FADH_2$ are oxidized, with concomitant production of ATP. Two molecules of ATP are produced when $FADH_2$ is oxidized, and three molecules of ATP are produced when NADH is oxidized. The complete oxidation of one glucose molecule by the citric acid cycle and oxidative phosphorylation yields 36 molecules of ATP, versus two molecules of ATP for anaerobic degradation of glucose by glycolysis.

Because the rate of energy production by the cell must vary with the amount of available oxygen and the energy requirements of the body at any particular time, the citric acid cycle is regulated at several steps. This allows the cell to generate more energy when needed, as for exercise, and less energy when the body is at rest.

Amino acids are oxidized in the mitochondria. The first step of amino acid catabolism is deamination. The carbon skeletons of amino acids are converted to citric acid cycle intermediates. In the urea cycle the toxic ammonium ions that are released by deamination of amino acids are incorporated in urea and excreted.

In addition to its role in catabolism, the citric acid cycle also plays an important role in cellular anabolism, or biosynthetic reactions. Many of the citric acid cycle intermediates are precursors for the synthesis of amino acids and other macromolecules that are required by the cell. A pathway that functions in both catabolic and anabolic reactions is called an amphibolic pathway.

GLOSSARY OF KEY TERMS

aconitase (14.2) the enzyme that catalyzes the isomerization of citrate to isocitrate. This is a two-step reaction in which citrate is dehydrated to produce *cis*-aconitate, which is then rehydrated to produce isocitrate.

alanine aminotransferase (14.5) the enzyme that catalyzes the transfer of the α-amino group of alanine to α-ketoglutarate, producing pyruvate and glutamate.

aminotransferase (14.5) an enzyme that catalyzes the transfer of an amino group from one molecule to another. (Also called transaminase.)

amphibolic pathways (14.7) metabolic pathways that function in both anabolism and catabolism.

anabolism (14.7) energy-requiring biosynthetic pathways.

anaplerotic reactions (14.7) a reaction that replenishes a substrate needed for a biochemical pathway.

aspartate aminotransferase (14.5) an enzyme that mediates the transfer of the α-amino group of aspartate to α-ketoglutarate, producing oxaloacetate and glutamate.

ATP synthase (14.3) a multiprotein complex within the inner mitochondrial membrane that uses the energy of the proton gradient to produce ATP. (Also called F_0F_1 complex.)

brown fat (14.3) specialized tissue that burns lipids and produces heat.

catabolism (14.7) energy-generating degradative pathways.

chemiosmotic theory (14.3) the theory that explains the way in which the oxidation of glucose is coupled to the phosphorylation of ADP to generate ATP.

citrate synthase (14.2) the enzyme that catalyzes the condensation of the acetyl group of acetyl CoA with oxaloacetate in the first reaction of the citric acid cycle.

citric acid cycle (14.2) a cyclic biochemical pathway that is the final stage of degradation of carbohydrates, fats, and amino acids. It results in the complete oxidation of acetyl groups derived from these dietary fuels. Also called the Krebs Cycle.

creatine phosphate (14.1) a molecule in muscle tissue that serves as a storage reservoir for high-energy phosphate groups.

cristae (14.1) the highly folded inner membrane of the mitochondria.

fast twitch muscle fibers (14.1) muscle fibers that are specialized to provide ATP energy principally by anaerobic glycolysis.

F_0F_1 complex (14.3) the multiprotein complex in the inner mitochondrial membrane that uses the energy of the proton gradient to produce ATP. (Also called ATP synthase.)

F_0F_1 event (14.3) the stage of oxidative phosphorylation in which ATP synthase (F_0F_1 complex) phosphorylates ADP to produce ATP, using the energy of the proton gradient.

fumarase (14.2) an enzyme that catalyzes the hydration of fumarate to produce malate.

glutamate dehydrogenase (14.5) an enzyme that catalyzes the oxidative deamination of the amino acid glutamate.

H^+ transport (14.3) the stage of oxidative phosphorylation in which electrons pass from NADH or $FADH_2$ through a series of electron carriers, causing H^+ to be transported from the mitochondrial matrix to the intermembrane space.

homeostatic mechanisms (14.3) mechanisms that work to maintain a constant internal environment within the body.

hyperammonemia (14.6) a genetic defect in one of the enzymes of the urea cycle that results in toxic or even fatal elevations of the concentration of ammonium ions in the body.

inner mitochondrial membrane (14.1) the highly folded, impermeable membrane within the mitochondrion that is the location of the electron transport system and ATP synthase.

intermembrane space (14.1) the region between the outer and inner mitochondrial membranes that is the location of the proton reservoir that drives ATP synthesis.

isocitrate dehydrogenase (14.2) an enzyme that catalyzes the oxidation and decarboxylation of isocitrate to yield α-ketoglutarate.

α-ketoglutarate dehydrogenase complex (14.2) a multienzyme complex that accomplishes the oxidation and decarboxylation of α-ketoglutarate to produce succinyl CoA.

malate dehydrogenase (14.2) an enzyme that catalyzes the oxidation of malate to produce oxaloacetate.

matrix space (14.1) the region of the mitochondrion within the inner membrane. It is the location of the enzymes that carry out the reactions of the citric acid cycle.

mitochondria (14.1) the cellular ''power plants'' in which the reactions of the citric acid cycle, the electron transport system, and ATP synthase function to produce ATP.

nonshivering thermogenesis (14.3) production of heat by the mitochondria of brown fat tissue.

nucleotide diphosphokinase (14.2) an enzyme that catalyzes the transfer of a phosphoryl group from GTP to ADP.

organelles (Introduction) a membrane-bound structure within the cytoplasm that carries out a specialized, compartmentalized function for the cell.

outer mitochondrial membrane (14.1) the membrane that surrounds the mitochondrion and separates it from the contents of the cytoplasm. It is highly permeable to small ''food'' molecules.

oxidative deamination (14.5) an oxidation-reduction reaction in which NAD^+ is reduced and the amino acid is deaminated.

oxidative phosphorylation (14.3) a series of reactions that couples the oxidation of NADH and $FADH_2$ to the phosphorylation of ADP to produce ATP.

pyridoxal phosphate (14.5) a coenzyme derived from vitamin B_6 that is required for all transamination reactions.

respiratory electron transport system (14.3) the series of electron transport proteins embedded in the inner mitochondrial membrane that accepts high-energy electrons from NADH and $FADH_2$.

slow twitch muscle fibers (14.1) muscle fibers that are specialized to produce ATP by oxidative phosphorylation.

succinate dehydrogenase (14.2) an enzyme that catalyzes the oxidation of succinate to fumarate.

succinyl CoA synthase (14.2) an enzyme that catalyzes a substrate-level phosphorylation in which the thioester bond of succinyl CoA is cleaved to yield succinate and coenzyme A and GDP is phosphorylated to produce GTP.

terminal electron acceptor (14.3) the final electron acceptor in an electron transport system that removes the low-energy electrons from the system. In aerobic organisms the terminal electron acceptor is molecular oxygen.

thermogenic hyperphagia (14.3) the ability of brown fat tissue to burn excess calories in the diet.

thermogenin (14.3) a protein in the inner mitochondrial membrane of brown fat that allows the dissipation of the proton gradient and thereby produces heat.

transaminase (14.5) an enzyme that catalyzes the transfer of an amino group from one molecule to another. (Also called aminotransferase.)

transamination (14.5) a reaction in which an amino group is transferred from one molecule to another.

urea cycle (14.6) a cyclic series of reactions that detoxifies ammonium ions by incorporating them into urea, which is excreted from the body.

QUESTIONS AND PROBLEMS

The Mitochondria

14.16 What biochemical processes occur in the matrix space of the mitochondrion?

14.17 How is the intermembrane space defined?

14.18 What is the name of the infoldings of the inner mitochondrial membrane?

14.19 In what major way do the inner and outer mitochondrial membranes differ?

14.20 In what major way do metabolic processes in the mitochondrion differ from those in other cell organelles?

14.21 What three types of proteins are found in the inner membrane of the mitochondria?

14.22 What is the distinguishing feature of all subcellular organelles, including the mitochondrion?

Citric Acid Cycle

14.23 Label each of the following statements as true or false:

a. Both glycolysis and the citric acid cycle are aerobic processes.

b. Both glycolysis and the citric acid cycle are anaerobic processes.

c. Glycolysis occurs in the cytoplasm, and the citric acid cycle occurs in the mitochondrion.

d. The inner membrane of the mitochondrion is virtually impermeable to most substances.

14.24 To what metabolic intermediate is the acetyl group of acetyl CoA transferred in the citric acid cycle? What is the product of this reaction?

14.25 How many ions of NAD^+ are reduced to molecules of NADH during one turn of the citric acid cycle?

14.26 How many molecules of FAD are converted to $FADH_2$ by one turn of the citric acid cycle?

14.27 To what final products is the acetyl group of acetyl CoA converted during oxidation in the citric acid cycle?

14.28 What is the chemical meaning of the term "decarboxylation"?

14.29 GTP is formed in one step of the citric acid cycle. How does this GTP result in production of a molecule of ATP?

14.30 What is the net yield of ATP for anaerobic glycolysis?

14.31 How many molecules of ATP are produced by the complete oxidation of glucose via glycolysis, the citric acid cycle, and oxidative phosphorylation?

14.32 What is the function of acetyl CoA in the citric acid cycle?

14.33 What is the function of oxaloacetate in the citric acid cycle?

14.34 Fill in the blanks:

a. The three types of proteins found in the inner membrane of mitochondria are called ______ proteins, ______ proteins, and a large multiprotein complex called ______.

b. The outer mitochondrial membrane consists mostly of ______ and contains considerable ______.

c. Energy released by oxidation in the citric acid cycle is conserved in the form of phosphate bonds in ______.

d. The urea cycle converts excess ______ ions into urea in order to excrete nitrogen.

Amino Acid Metabolism

14.35 What chemical transformation is carried out by aminotransferases?

14.36 Write a balanced chemical equation for an aminotransferase that shifts an amino group from alanine to α-ketoglutarate.

14.37 Why is the glutamate family of aminotransferases so important?

14.38 What biochemical reaction is catalyzed by glutamate dehydrogenase?

14.39 Into which citric acid cycle intermediate is each of the following amino acids converted?

a. alanine
b. glutamate
c. aspartate

Respiration

14.40 How many molecules of ATP are produced when one molecule of NADH is oxidized by oxidative phosphorylation?

14.41 How many ATP are produced for each $FADH_2$ generated in the citric acid cycle?

14.42 What is the source of energy for the synthesis of ATP in respiring mitochondria?

14.43 What is the name of the enzyme that catalyzes ATP synthesis in mitochondria?

14.44 What is the function of respiratory electron transport chain complexes?

14.45 What is the cellular location of respiratory electron transport chain complexes?

14.46 Explain why chemicals that abolish the proton gradient across the inner mitochondrial membrane can uncouple oxidation of NADH from phosphorylation of ADP.

14.47 What is the net yield of ATP for degradation of alanine by the citric acid cycle and oxidative phosphorylation?

14.48 Why is the coupling of oxidation of NADH to phosphorylation of ADP said to occur indirectly?

Urea Cycle

14.49 Why is the urea cycle found only in the liver?

14.50 What metabolic condition is produced if the urea cycle does not function properly?

14.51 What metabolic intermediate couples the urea cycle to the citric acid cycle?

14.52 In what subcellular compartment does the urea cycle occur?

14.53 The structure of urea is given below:

$$NH_2-\overset{\overset{\displaystyle O}{\|}}{C}-NH_2$$

a. What substances are the source of both amino groups?
b. What substance is the source of the carbonyl group?

14.54 What is hyperammonemia? How are mild forms of this disease treated?

Further Problems

14.55 Compare the number of molecules of ATP produced by glycolysis to the number of ATP molecules produced by oxidation of glucose via the citric acid cycle. Which pathway produces more energy? Explain.

14.56 Write a balanced equation for the synthesis of glutamate that is mediated by the enzyme glutamate dehydrogenase.

14.57 ______ muscle fibers are pale in color and have high concentrations of enzymes for glycolysis and lactate fermentation; ______ muscle fibers are red and have large numbers of mitochondria.

14.58 Write a balanced equation for the reaction catalyzed by pyruvate carboxylase and describe how this reaction allows the citric acid cycle to fulfill its role in catabolism and anabolism.

14.59 List the positive and negative effectors that regulate enzymes of the citric acid cycle and list the enzymes involved in regulation.

CHAPTER 15

Fatty Acid Metabolism

OUTLINE

OBJECTIVES

- Recognize the pivotal role of acetyl CoA in fatty acid and lipid metabolism.
- Summarize the role of bile salts and lipases in the digestion of lipids.
- Understand the importance of fatty acid metabolism in energy production in the cell.
- Describe the degradation of fatty acids, β-oxidation.
- Understand the role of ''ketone body'' production in β-oxidation.
- Describe the major differences between β-oxidation and fatty acid biosynthesis.
- Describe the regulation of lipid and carbohydrate metabolism in relation to the liver, adipose tissue, muscle tissue, and the brain.
- Summarize the antagonistic effects of glucagon and insulin.
- Relate insulin and insulin production to the disease diabetes mellitus.

The metabolism of fatty acids and lipids revolves around the fate of acetyl CoA. In Chapters 13 and 14 we saw that pyruvate is converted to acetyl CoA and that the acetyl CoA feeds into the citric acid cycle. Fatty acids are also degraded to acetyl CoA and oxidized by the citric acid cycle, as are certain amino acids. Moreover, acetyl CoA is itself the starting material for the biosynthesis of fatty acids, fully half of the amino acids, cholesterol, and steroid hormones. Acetyl CoA is thus one of the major metabolites of intermediary metabolism.

15.1 LIPID METABOLISM IN ANIMALS

Digestion and absorption of dietary fats

Since fats are highly hydrophobic (water-dreading), they must be extensively processed before they can be absorbed, digested, and metabolized. For this reason the **lipases** that are found in the stomach and in the saliva are not very effective. In fact, most dietary fat arrives in the duodenum, the first part of the small intestines, in the form of fat globules. These fat globules stimulate the secretion of bile from the gall bladder. **Bile** is composed of micelles of lecithin, cholesterol, protein, bile salts, inorganic ions, and bile pigments. **Micelles** (Figure 15.1) are aggregations of molecules having a polar region and a nonpolar region. The nonpolar ends of bile salts or phospholipids like lecithin tend to bunch together when placed in water. The hydrophilic regions of these molecules will interact with water.

Lipases are enzymes that hydrolyze triglycerides to glycerol and fatty acids.

Bile salts are produced in the liver and stored in the gall bladder, awaiting the stimulus to be secreted into the duodenum. The major bile salts in humans are cholic acid and chenodeoxycholic acid (Figure 15.2).

FIGURE 15.1
The structure of a micelle formed from the phospholipid lecithin. The straight lines represent the long hydrophobic fatty acid tails, and the spheres represent the hydrophilic heads of the phospholipid.

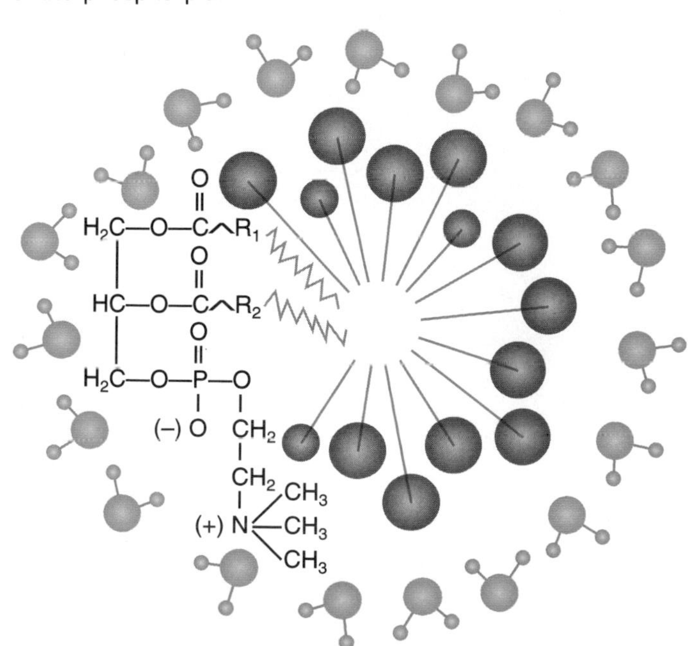

FIGURE 15.2
Structures of the most common bile acids in human bile, cholic acid, and chenodeoxycholic acid.

OH CH₃ CH—CH₂—CH₂—COOH HO OH

Cholic acid

CH₃ CH—CH₂—CH₂—COOH HO OH

Chenodeoxycholic acid

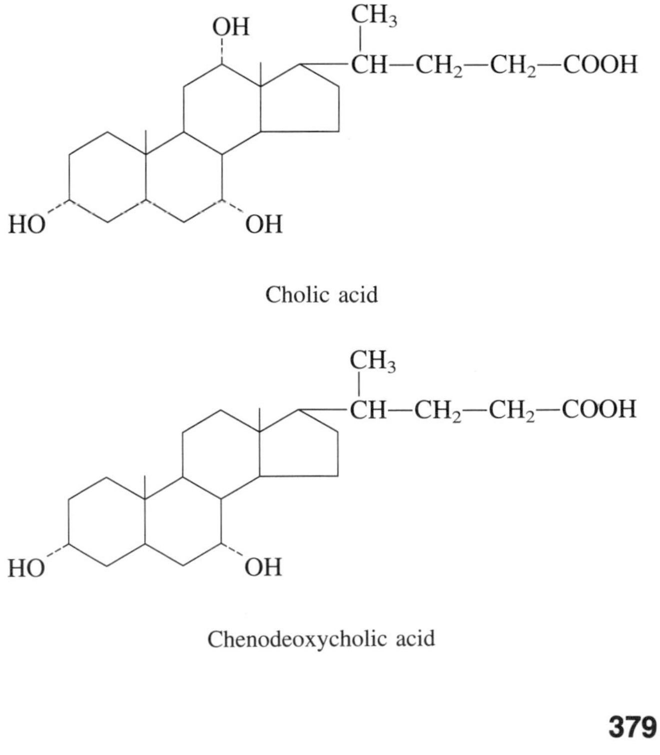

A HUMAN PERSPECTIVE

Losing Those Unwanted Pounds of Adipose Tissue

Weight, or overweight, is a topic of great concern to the American populace. A glance through almost any popular magazine quickly informs us that by today's standards, "beautiful" is synonymous with "thin." The models in all these magazines are extremely thin, and there are literally dozens of ads for weight-loss programs. Americans spend millions of dollars each year trying to attain this slim ideal of the fashion models.

Studies have revealed that this slim ideal is often below a desirable, healthy body weight. In fact, the suggested weight for a 6 foot tall male between 18 and 39 years of age is 179 pounds. For a 5′6″ female in the same age range, the desired weight is 142 pounds. For a 5′1″ female, 126 pounds is recommended. Just as being too thin can cause health problems, so too can obesity.

What is obesity and does it have disadvantages beyond aesthetics? An individual is considered to be obese if his or her body weight is more than 20% above the ideal weight for his or her height. The accompanying table lists desirable body weights, according to sex, age, height, and body frame.

Overweight carries with it a wide range of physical problems, including elevated blood cholesterol levels; high blood pressure; increased incidence of diabetes, cancer, and heart disease; and increased probability of early death. It often causes psychological problems as well, such as guilt and low self-esteem.

Many factors may contribute to obesity. These include genetic factors, a sedentary lifestyle, and a preference for high-calorie, high-fat foods. However, the real concern is how to lose weight. How can we lose weight wisely and safely and keep the weight off for the rest of our lives? Unfortunately, the answer is *not* the answer that most people want to hear. The prevalence and financial success of the quick-weight-loss programs suggest that the majority of people want a program that is rapid and effortless. Unfortunately, most programs that promise dramatic weight reduction with little effort are usually ineffective or, worse, unsafe. The truth is that weight loss and management are best obtained by a program involving three elements.

Men*					Women**				
Height					Height				
Feet	Inches	Small Frame	Medium Frame	Large Frame	Feet	Inches	Small Frame	Medium Frame	Large Frame
5	2	128–134	131–141	138–150	4	10	102–111	109–121	118–131
5	3	130–136	133–143	140–153	4	11	103–113	111–123	120–134
5	4	132–138	135–145	142–156	5	0	104–115	113–126	122–137
5	5	134–140	137–148	144–160	5	1	106–118	115–129	125–140
5	6	136–142	139–151	146–164	5	2	108–121	118–132	128–143
5	7	138–145	142–154	149–168	5	3	111–124	121–135	131–147
5	8	140–148	145–157	152–172	5	4	114–127	124–138	134–151
5	9	142–151	148–160	155–176	5	5	117–130	127–141	137–155
5	10	144–154	151–163	158–180	5	6	120–133	130–144	140–159
5	11	146–157	154–166	161–184	5	7	123–136	133–147	143–163
6	0	149–160	157–170	164–188	5	8	126–139	136–150	146–167
6	1	152–164	160–174	168–192	5	9	129–142	139–153	149–170
6	2	155–168	164–178	172–197	5	10	132–145	142–156	152–173
6	3	158–172	167–182	176–202	5	11	135–148	145–159	155–176
6	4	162–176	171–187	181–207	6	0	138–151	148–162	158–179

*Weights at ages 25–59 based on lowest mortality. Weight in pounds according to frame (in indoor clothing weighing 5 lbs., shoes with 1″ heels).
**Weights at ages 25–59 based on lowest mortality. Weight in pounds according to frame (in indoor clothing weighing 3 lbs., shoes with 1″ heels).
Reprinted with permission of the Metropolitan Insurance Companies.

Cholesterol is virtually insoluble in water, but the metabolic conversion of cholesterol to bile salts creates **detergents** whose polar ''heads'' make them soluble in the aqueous phase of the cytoplasm and whose hydrophobic ''tails'' bind lipids. After a meal is consumed, bile flows through the common bile duct into the duodenum where bile salts

1. *Reduced caloric intake.* A pound of body fat is equivalent to 3500 Calories (kilocalories). So if you want to lose 2 pounds each week, a reasonable goal, you must reduce your caloric intake by 1000 Calories per day. Remember that diets recommending fewer than 1200 Calories per day are difficult to maintain because they are not very satisfying and may be unsafe because they don't provide all the required vitamins and minerals (Chapter 17). The best way to decrease Calories is to reduce fat and increase complex carbohydrates in the diet.
2. *Exercise.* Increase energy expenditures by 200–400 Calories each day. You may choose walking, running, or mowing the lawn; the type of activity doesn't matter, as long as you get moving. Exercise has additional benefits. It increases cardiovascular fitness, provides a psychological lift, and may increase the base rate at which you burn calories after exercise is finished. The following table summarizes the caloric expenditure of several activities.
3. *Behavior modification.* Overweight is as much a psychological problem as it is a physical problem, and half the battle is learning to recognize the triggers that cause eating behavior. Several principles of behavior modification have been found to be very helpful.
   a. Keep a diary. Record the amount of foods eaten and the circumstances—for instance, a meal at the kitchen table or a bag of chips in the car on the way home.
   b. Identify your eating triggers. Do you eat when you feel stress, boredom, fatigue, joy?
   c. Develop a plan for avoiding or coping with your trigger situations or emotions. You might exercise when you feel that stress-at-the-end-of-the-day trigger or carry a bag of carrot sticks for the midmorning-boredom trigger.
   d. Set realistic goals and reward yourself when you reach them. The reward should be a new necklace or a movie, not a hot fudge sundae.

As you can see, there is no "quick fix" for safe, effective weight control. A commitment must be made to modify existing diet and exercise habits. Most important, those habits must be avoided forever and replaced by new, healthier behaviors and attitudes.

Activity	Kcal per Hour*
Badminton, competitive singles	480
Basketball	360–660
Bicycling	
10 mph	420
11 mph	480
12 mph	600
13 mph	660
Calisthenics, heavy	600
Handball, competitive	660
Rope skipping, vigorous	800
Rowing machine	840
Running	
5 mph	600
6 mph	750
7 mph	870
8 mph	1,020
9 mph	1,130
10 mph	1,285
Skating, ice or roller, rapid	700
Skiing, downhill, vigorous	600
Skiing, cross-country	
2.5 mph	560
4 mph	600
5 mph	700
8 mph	1,020
Swimming, 25–50 yards per min.	360–750
Walking	
Level road, 4 mph (fast)	420
Upstairs	600–1,080
Uphill, 3.5 mph	480–900
Gardening, much lifting, stooping, digging	500
Mowing, pushing hand mower	450
Sawing hardwood	600
Shoveling, heavy	660
Wood chopping	560

*Caloric expenditure is based on a 150-lb person. There is a 10% increase in caloric expenditure for each 15 lbs over this weight and a 10% decrease for each 15 lbs under.

From E. L. Wynder, *The Book of Health: The American Health Foundation.* © 1981 Franklin Watts, Inc., New York. Used with permission.

emulsify the fat globules into tiny droplets. This increases the surface area of the lipid molecules to facilitate chemical digestion (Figure 15.3).

Much of the lipid in these droplets is in the form of **triglycerides**, or triacylglycerols, which are fatty acyl esters of glycerol. A protein called **colipase** binds to the surface of the

Sections 9.1 and 9.3

FIGURE 15.3
Stages of lipid digestion in the intestinal tract. Step 1 is the emulsification of fat droplets by bile salts. Step 2 is the hydrolysis of triglycerides in emulsified fat droplets into fatty acid and monoglycerides. Step 3 involves dissolving fatty acids and monoglycerides into micelles to produce "mixed micelles."

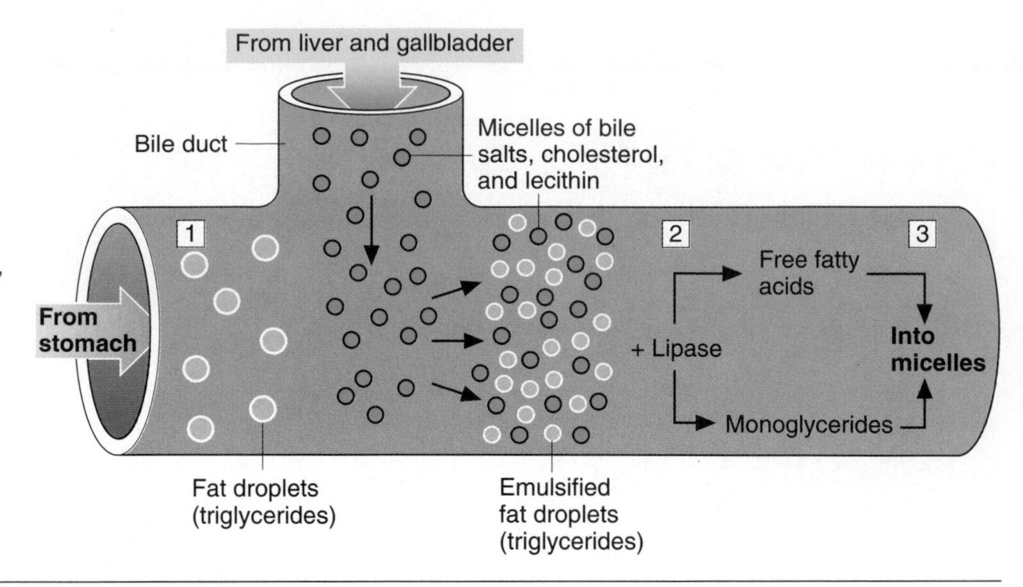

FIGURE 15.4
The action of pancreatic lipase in the hydrolysis of dietary lipids.

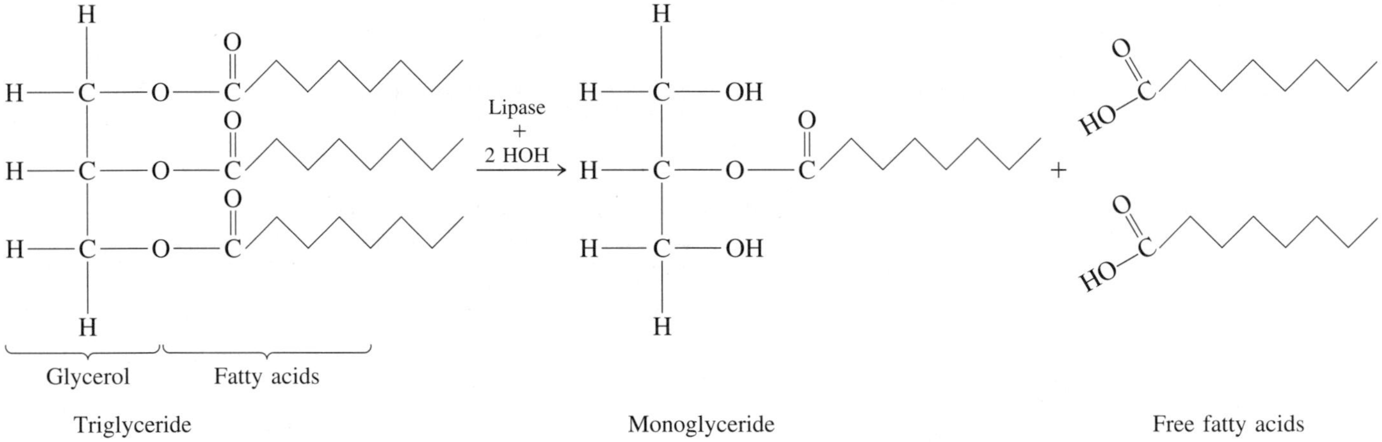

$$\text{Triglyceride (Glycerol + Fatty acids)} \xrightarrow{\text{Lipase} + 2\,\text{HOH}} \text{Monoglyceride} + \text{Free fatty acids}$$

Glycerol Fatty acids

Triglyceride Monoglyceride Free fatty acids

lipid droplets and helps pancreatic lipases to adhere to the surface and hydrolyze the ester linkages between the glycerol and fatty acid moieties of the triglycerides (Figure 15.4). In this process, two of the three fatty acids are liberated, and the monoglycerides and free fatty acids produced are able to mix freely with the micelles of bile. These micelles are readily absorbed through the membranes of the intestinal epithelial cells (Figure 15.5).

Surprisingly, the monoglycerides and fatty acids are then reassembled into triglycerides, which are combined with protein to produce the class of plasma lipoproteins called **chylomicrons** (Figure 15.5). These aggregations of lipid and protein are secreted into small lymphatic vessels and eventually arrive in the bloodstream. In the bloodstream the triglycerides are once again hydrolyzed to produce glycerol and free fatty acids, which can then be absorbed by the cells of the body either for energy or for lipid storage.

Section 9.5

Approximately 90% of the ingested lipids are transported via chylomicrons. However, very short-chain fatty acids, such as those found in buttermilk, are not readily reassembled into triglycerides. These pass directly into the bloodstream.

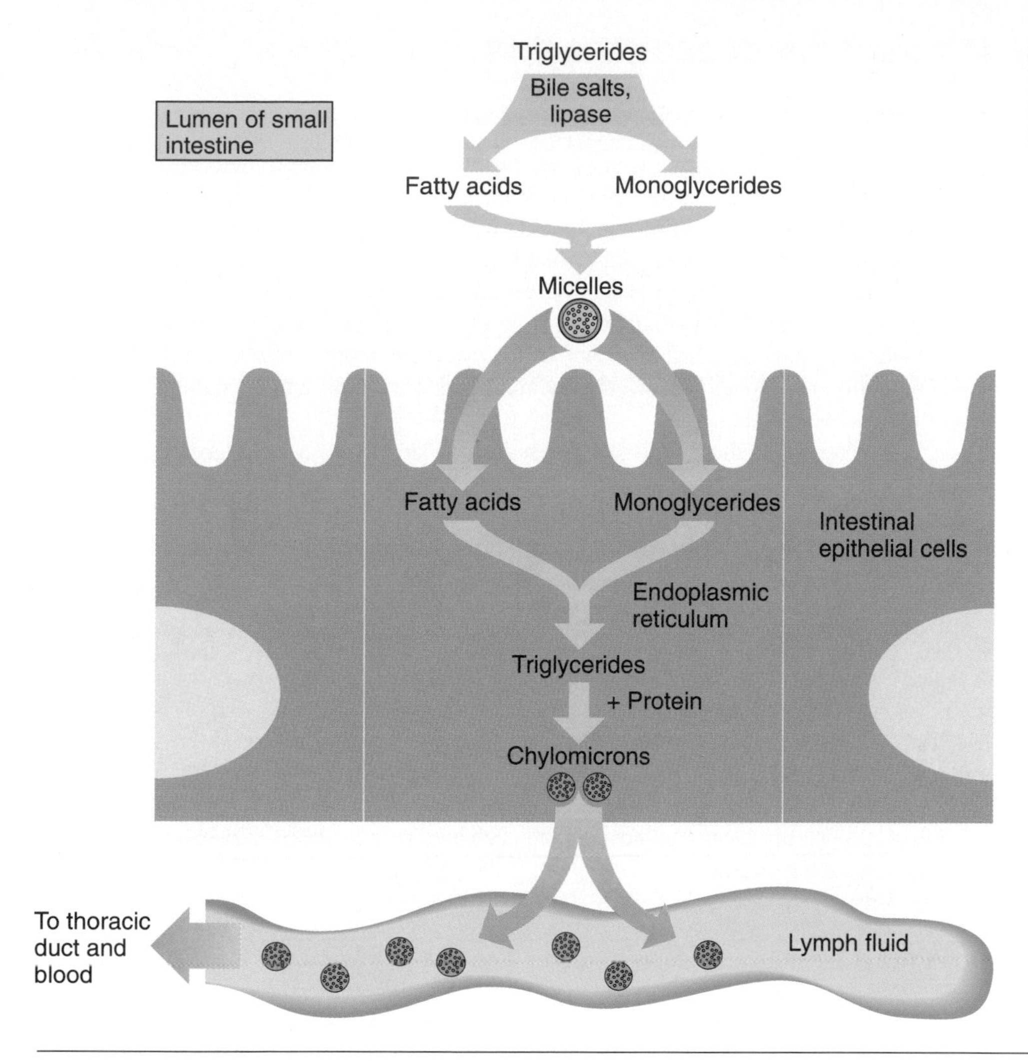

FIGURE 15.5
Passage of triglycerides in micelles into the cells of the intestinal epithelium.

Lipid storage

Fatty acids are stored in the form of triglycerides. Most of the body's triglyceride molecules are stored as fat droplets in the cytoplasm of **adipose tissue**. A fat cell, or **adipocyte**, contains a large fat droplet that accounts for nearly the entire volume of the cell. The fat droplets of other cells, such as those of cardiac muscle, are surrounded by mitochondria. The enzymatic hydrolysis of triglycerides releases fatty acids that are transported into the matrix space of the mitochondria, where they are completely oxidized.

The fatty acids provided by the hydrolysis of triglycerides are highly reduced sources of fuel for the oxidative reactions of mitochondria. The complete oxidation of fatty acids releases much more energy than the oxidation of a comparable mass of glycogen because glycogen is about 70% water by weight, whereas fat is almost totally unsolvated. An adult has approximately a one-day supply of glycogen but typically has enough fat to last a month. Fat is released into the blood and distributed to other tissues after fasting. The acetyl CoA that is metabolized in the morning before breakfast is mostly derived from fatty acids, not from glycogen.

See Chapter 14, "A Human Perspective: Brown Fat: The Fat That Makes You Thin?"

Question 15.1 What is the major storage form of fatty acids and where are they found?

15.2 FATTY ACID DEGRADATION

An overview of fatty acid degradation

Early in the twentieth century a very clever experiment was done to determine how fatty acids are degraded. Radioactive elements can be attached to biological molecules and followed through the body. A German biochemist, Franz Knoop, devised a similar kind of labeling experiment long before radioactive tracers were available. Knoop fed dogs fatty acids in which the usual terminal methyl group was replaced by a phenyl group. Such molecules are called ω-labeled fatty acids (Figure 15.6). He found that phenyl acetate was formed when the fatty acid had an even number of carbon atoms in the chain. On the other hand, oxidation of ω-phenyl-labeled fatty acids that contained an odd number of carbon atoms yielded benzoate.

Knoop interpreted these data to suggest that the degradation of ω-labeled fatty acids occurs via the removal of acetate from the carboxyl end of the acid and the subsequent removal of two carbon fragments. The sequential removal of these two-carbon fragments is called ***β*-oxidation**. Fifty years elapsed before Lynen and Reichart showed that the two-carbon fragment produced by β-oxidation is not acetate, but acetyl CoA. These researchers and others described the series of reactions that occur during the β-oxidation of fatty acids.

They found that the β-oxidation cycle consists of a set of five repeated reactions whose overall form is similar to the reactions of the citric acid cycle (Figure 15.7). Each trip through the set of five reactions releases acetyl CoA and returns a fatty acyl CoA molecule that contains two fewer carbons. One molecule of $FADH_2$, equivalent to two ATP, and one of NADH, equivalent to three ATP, are produced for each cycle of β-oxidation. As we noted in Chapter 14, the enzymes responsible for β-oxidation are located in the mitochondria.

Question 15.2 What products would be formed by β-oxidation of 9-phenylnonanoic acid? 8-phenyloctanoic acid? 7-phenylheptanoic acid? 12-phenyldocecanoic acid?

FIGURE 15.6
The last carbon of the chain is called the ω carbon, so the attached phenyl group is an ω-phenyl group. (a) Oxidation of ω-phenyl-labeled fatty acids occurs two carbons at a time. Fatty acids having an even number of carbon atoms are degraded to phenylacetate and "acetate." (b) Oxidation of ω-phenyl-labeled fatty acids that contain an odd number of carbon atoms yields benzoic acid and "acetate."

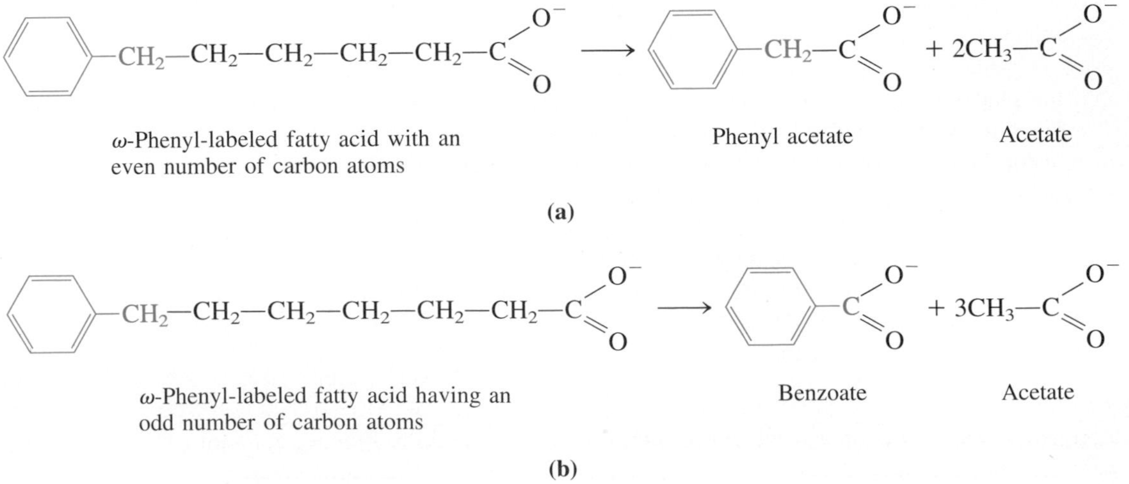

$$C_6H_5{-}CH_2{-}CH_2{-}CH_2{-}CH_2{-}CH_2{-}COO^- \longrightarrow C_6H_5{-}CH_2{-}COO^- + 2CH_3{-}COO^-$$

ω-Phenyl-labeled fatty acid with an even number of carbon atoms — Phenyl acetate — Acetate

(a)

$$C_6H_5{-}CH_2{-}CH_2{-}CH_2{-}CH_2{-}CH_2{-}CH_2{-}COO^- \longrightarrow C_6H_5{-}COO^- + 3CH_3{-}COO^-$$

ω-Phenyl-labeled fatty acid having an odd number of carbon atoms — Benzoate — Acetate

(b)

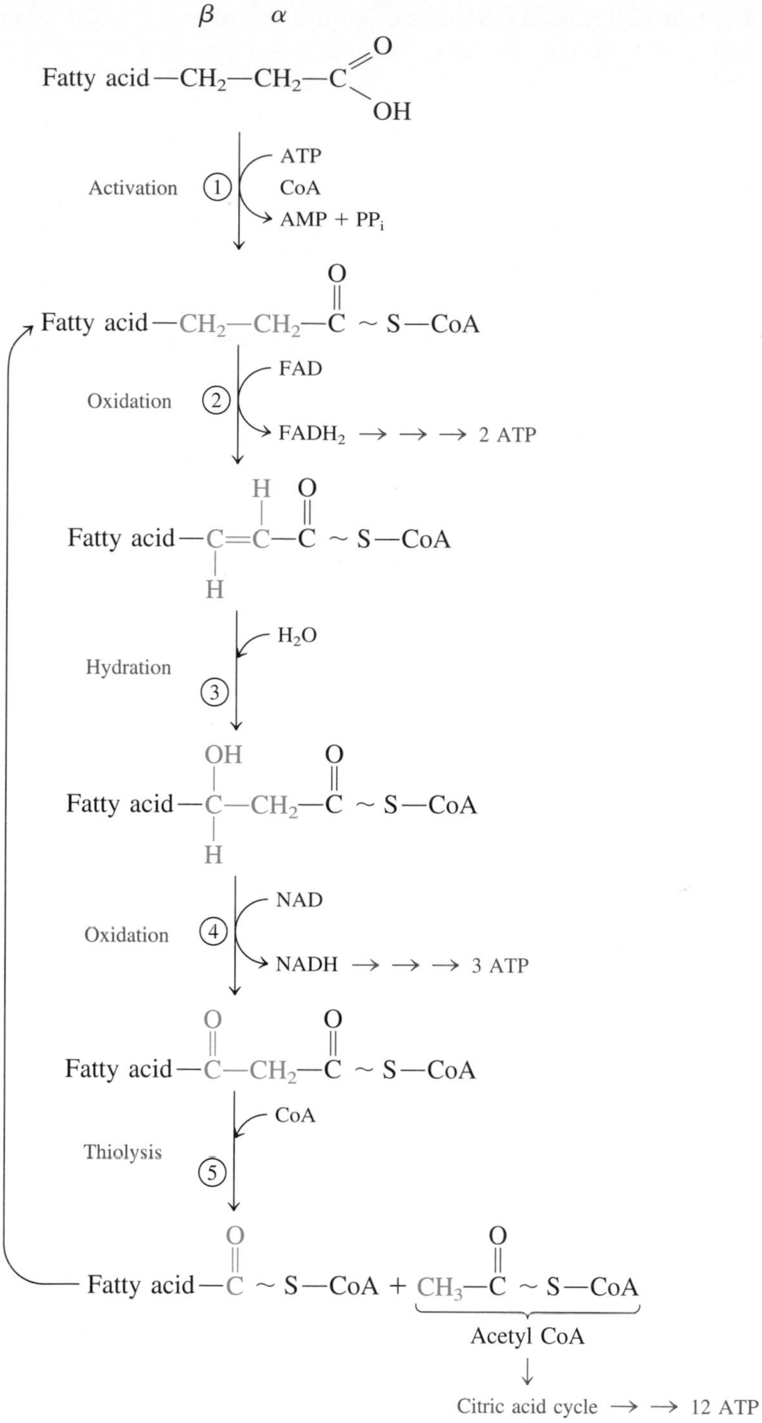

β α

Fatty acid—CH_2—CH_2—C(=O)OH

Activation ① ATP, CoA → AMP + PP_i

Fatty acid—CH_2—CH_2—C(=O) ~ S—CoA

Oxidation ② FAD → $FADH_2$ → → → 2 ATP

Fatty acid—C(H)=C(H)—C(=O) ~ S—CoA

Hydration ③ H_2O

Fatty acid—C(OH)(H)—CH_2—C(=O) ~ S—CoA

Oxidation ④ NAD → NADH → → → 3 ATP

Fatty acid—C(=O)—CH_2—C(=O) ~ S—CoA

Thiolysis ⑤ CoA

Fatty acid—C(=O) ~ S—CoA + CH_3—C(=O) ~ S—CoA

Acetyl CoA

↓

Citric acid cycle → → 12 ATP

FIGURE 15.7
The cycle of reactions in β-oxidation of fatty acids.

Question 15.3 What does ω refer to in the naming of ω-phenyl-labeled fatty acids?

The reactions of β-oxidation

Special transport mechanisms are required to bring the fatty acid molecules into the mitochondrial matrix. This is due to the impermeability of the inner mitochondrial membrane. We will now take a look at the individual reactions of β-oxidation that occur within the mitochondrial matrix, and we will see the way in which these reactions interact with oxidative phosphorylation and the citric acid cycle to produce ATP. Figure 15.7 summarizes the β-oxidation pathway and the products that are involved in ATP synthesis.

Step 1. The first step is an *activation* reaction that results in the production of a fatty acyl CoA molecule by a thioester linkage between coenzyme A and the fatty acid:

$$\underset{\text{Fatty acid}}{CH_3-(CH_2)_n-CH_2-CH_2-\overset{\overset{\displaystyle O}{\|}}{\underset{\underset{\displaystyle OH}{|}}{C}}} \xrightarrow[\text{Coenzyme A}]{\text{ATP} \quad \text{AMP} + \text{PP}_i} \underset{\text{Fatty acyl CoA}}{CH_3-(CH_2)_n-CH_2-CH_2-\overset{\overset{\displaystyle O}{\|}}{C} \sim S-CoA}$$

This activation reaction requires energy in the form of ATP, which is cleaved to AMP and pyrophosphate. Here again we see the need to invest some energy in order to achieve a much greater energy gain. Coenzyme A is also required for this step. The product, a fatty acyl CoA, has a high-energy thioester bond between the fatty acid and the CoA moieties.

Step 2. The second reaction is an *oxidation* reaction that removes a pair of hydrogen atoms from the fatty acid. These are used to reduce FAD to produce $FADH_2$:

$$CH_3-(CH_2)_n-CH_2-CH_2-\overset{\overset{\displaystyle O}{\|}}{C} \sim S-CoA \xrightarrow{\text{FAD} \quad \text{FADH}_2} CH_3-(CH_2)_n-\underset{\underset{\displaystyle H}{|}}{C}=\overset{\overset{\displaystyle H}{|}}{C}-\overset{\overset{\displaystyle O}{\|}}{C} \sim S-CoA$$

The electrons carried by $FADH_2$ are used to transport H^+ from the mitochondrial matrix into the intermembrane space. The energy of the protons is harvested in the F_0F_1 event and used to produce two ATP molecules.

Step 3. This reaction involves the *hydration* of the double bond produced in step 2. As a result the β-carbon is hydroxlated:

$$CH_3-(CH_2)_n-\underset{\underset{\displaystyle H}{|}}{C}=\overset{\overset{\displaystyle H}{|}}{C}-\overset{\overset{\displaystyle O}{\|}}{C} \sim S-CoA \xrightarrow{H_2O} CH_3-(CH_2)_n-\overset{\overset{\displaystyle OH}{|}}{\underset{\underset{\displaystyle H}{|}}{C}}-CH_2-\overset{\overset{\displaystyle O}{\|}}{C} \sim S-CoA$$

Step 4. In this *oxidation* reaction, the hydroxyl group of the β-carbon is now dehydrogenated. NAD^+ is reduced to form NADH, which is subsequently used to produce three ATP molecules:

$$CH_3-(CH_2)_n-\overset{\overset{\displaystyle OH}{|}}{\underset{\underset{\displaystyle H}{|}}{C}}-CH_2-\overset{\overset{\displaystyle O}{\|}}{C} \sim S-CoA \xrightarrow{\text{NAD}^+ \quad \text{NADH}} CH_3-(CH_2)_n-\overset{\overset{\displaystyle O}{\|}}{C}-CH_2-\overset{\overset{\displaystyle O}{\|}}{C} \sim S-CoA$$

Step 5. The final step is the cleavage that releases acetyl CoA. This is accomplished by *thiolysis,* the attack of a molecule of coenzyme A on the β-carbon. The result is the release of acetyl CoA and a fatty acyl CoA that is two carbons shorter than the beginning fatty acid:

$$CH_3-(CH_2)_n-\overset{O}{\overset{\|}{C}}-CH_2-\overset{O}{\overset{\|}{C}}\sim S-CoA \xrightarrow{CoA} CH_3-(CH_2)_{n-2}-CH_2-CH_2-\overset{O}{\overset{\|}{C}}\sim S-CoA + CH_3-\overset{O}{\overset{\|}{C}}\sim S-CoA$$

The shortened fatty acyl CoA is further oxidized by cycling through steps 2–5 until the fatty acid carbon chain is completely degraded to acetyl CoA. The acetyl CoA produced is completely oxidized by the reactions of the citric acid cycle. Of course, this eventually results in the production of 12 ATP per acetyl CoA released during β-oxidation.

The balance sheet for ATP production of the C_{16}-fatty acid palmitic acid when it is degraded by β-oxidation is summarized in Figure 15.8. Complete oxidation of palmitate

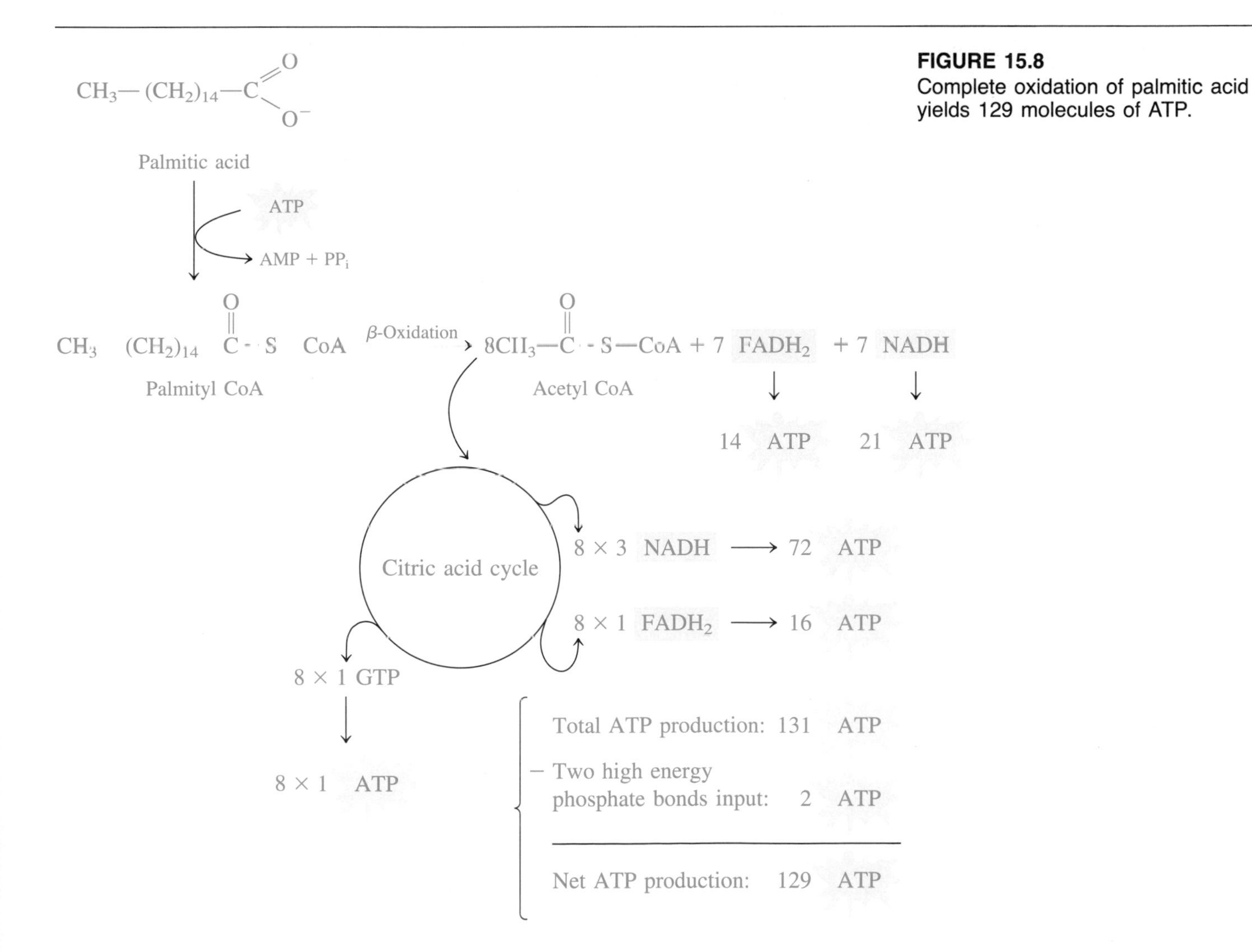

FIGURE 15.8
Complete oxidation of palmitic acid yields 129 molecules of ATP.

results in production of 129 molecules of ATP, three and a half times *more energy than results from the complete oxidation of glucose.*

Question 15.4 Write out the sequence of steps for β-oxidation of butyryl-CoA.

15.3 KETONE BODIES

The oxidation of the acetyl CoA produced by the β-oxidation of fatty acids occurs via the citric acid cycle, which in turn depends upon an adequate supply of oxaloacetate. If glycolysis and β-oxidation are functioning at the same rate, there will be a steady supply of pyruvate (from glycolysis), which can be converted to oxaloacetate. But what happens if the supply of oxaloacetate is too low to allow all of the acetyl CoA to enter the citric ketone acid cycle? Under these conditions, acetyl CoA is converted to the so-called **ketone bodies**: β-hydroxybutyrate, acetoacetate, and acetone (Figure 15.9).

Ketosis

Diabetes mellitus is a disease characterized by the appearance of glucose in the urine as a result of high blood glucose levels. The disease is usually caused by the inability to produce the hormone insulin.

Ketosis, the abnormal rise in the concentration of blood ketone bodies, is a situation that arises under some pathological conditions, such as starvation, a diet that is extremely low in carbohydrates (as with the high-protein liquid diets), or **diabetes mellitus**. The carbohydrate intake of a diabetic is normal, but the carbohydrates cannot get into the cell to be used as fuel. Thus diabetes amounts to starvation in the midst of plenty. In diabetes the very high concentration of ketone acids in the blood leads to **ketoacidosis**. The ketone acids are relatively strong acids and therefore dissociate to release H^+. Under these conditions the blood pH becomes more acidic.

Ketogenesis

The pathway for the production of ketone bodies (Figure 15.10) begins with a ''reversal'' of the last step of β-oxidation. When oxaloacetate levels are low, the enzyme that mediates the last reaction of β-oxidation, thiolase, now mediates the fusion of two acetyl CoA molecules to produce acetoacetyl CoA:

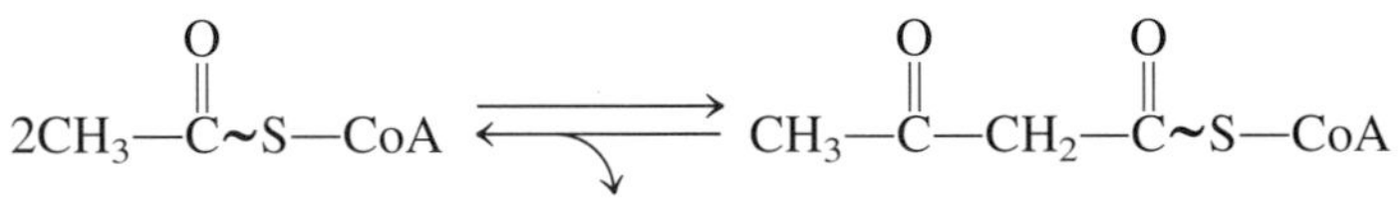

$$2CH_3{-}\overset{\overset{\Large O}{\|}}{C}{\sim}S{-}CoA \rightleftharpoons CH_3{-}\overset{\overset{\Large O}{\|}}{C}{-}CH_2{-}\overset{\overset{\Large O}{\|}}{C}{\sim}S{-}CoA + CoA$$

Acetyl CoA → Acetoacetyl CoA (releasing CoA)

$$CH_3{-}\underset{H}{\overset{OH}{C}}{-}CH_2{-}COO^- \qquad CH_3{-}\overset{\overset{\Large O}{\|}}{C}{-}CH_3 \qquad CH_3{-}\overset{\overset{\Large O}{\|}}{C}{-}CH_2{-}COO^-$$

β-Hydroxybutyrate — Acetone — Acetoacetate

FIGURE 15.9
Structures of ketone bodies.

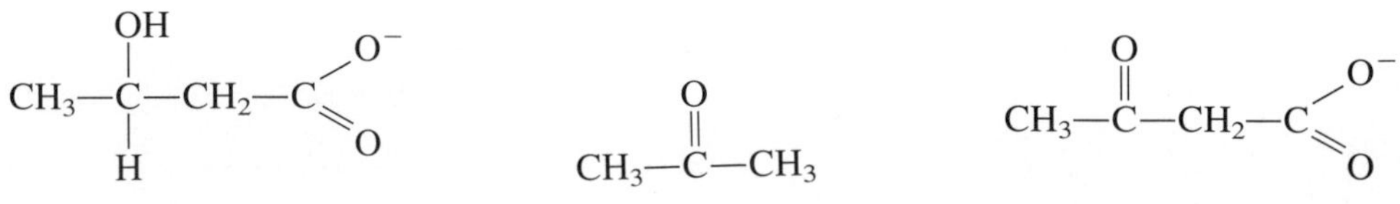

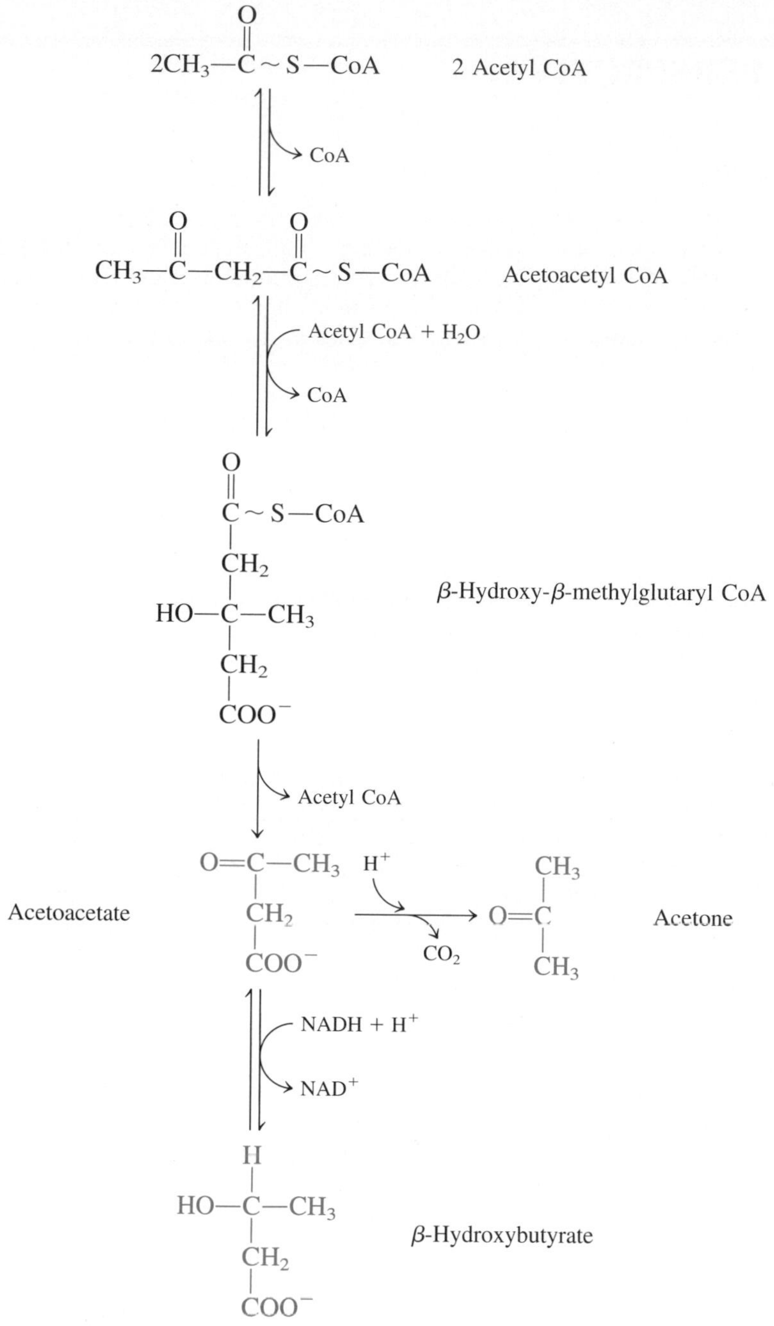

$$2CH_3-\overset{\overset{O}{\|}}{C}\sim S-CoA \quad \text{2 Acetyl CoA}$$

$$\rightleftharpoons \; (\to CoA)$$

$$CH_3-\overset{\overset{O}{\|}}{C}-CH_2-\overset{\overset{O}{\|}}{C}\sim S-CoA \quad \text{Acetoacetyl CoA}$$

$$\rightleftharpoons \; (\text{Acetyl CoA} + H_2O \to CoA)$$

$$\begin{array}{c} O \\ \| \\ C\sim S-CoA \\ | \\ CH_2 \\ | \\ HO-C-CH_3 \\ | \\ CH_2 \\ | \\ COO^- \end{array} \quad \beta\text{-Hydroxy-}\beta\text{-methylglutaryl CoA}$$

$$\downarrow \; (\to \text{Acetyl CoA})$$

$$\text{Acetoacetate} \quad \begin{array}{c} O{=}C-CH_3 \\ | \\ CH_2 \\ | \\ COO^- \end{array} \xrightarrow[CO_2]{H^+} \begin{array}{c} CH_3 \\ | \\ O{=}C \\ | \\ CH_3 \end{array} \quad \text{Acetone}$$

$$\rightleftharpoons \; (NADH + H^+ \to NAD^+)$$

$$\begin{array}{c} H \\ | \\ HO-C-CH_3 \\ | \\ CH_2 \\ | \\ COO^- \end{array} \quad \beta\text{-Hydroxybutyrate}$$

FIGURE 15.10
Summary of the reactions involved in ketogenesis.

Acetoacetyl CoA can react with a third acetyl CoA to yield β-hydroxy-β-methylglutaryl CoA (HMG-CoA):

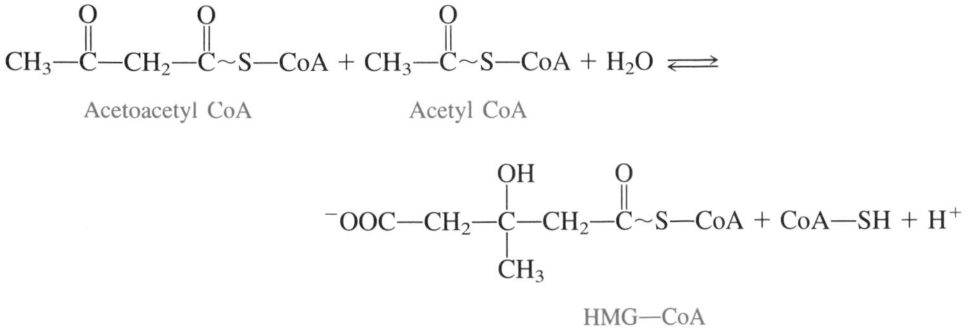

$$\underset{\text{Acetoacetyl CoA}}{CH_3-\overset{\overset{O}{\|}}{C}-CH_2-\overset{\overset{O}{\|}}{C}\sim S-CoA} + \underset{\text{Acetyl CoA}}{CH_3-\overset{\overset{O}{\|}}{C}\sim S-CoA} + H_2O \rightleftharpoons$$

$$\underset{\text{HMG—CoA}}{{}^-OOC-CH_2-\overset{\overset{OH}{|}}{\underset{\underset{CH_3}{|}}{C}}-CH_2-\overset{\overset{O}{\|}}{C}\sim S-CoA} + CoA-SH + H^+$$

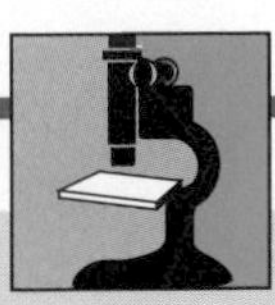

A CLINICAL PERSPECTIVE

Diabetes Mellitus and Ketone Bodies

More than one person, found unconscious on the streets of some metropolis, has been carted to jail only to die of complications arising from uncontrolled diabetes mellitus. Others are fortunate enough to arrive in hospital emergency rooms. A quick test for diabetes mellitus–induced coma is the odor of acetone on the breath of the afflicted person. Acetone is one of several metabolites produced by diabetics that are known collectively as *ketone bodies.*

The term "diabetes" was used by the ancient Greeks to designate diseases in which excess urine is produced. Two thousand years later, in the eighteenth century, the urine of certain individuals was found to contain sugar, and the name "diabetes mellitus" (L.: *mellitus*, sweetened with honey) was given to this disease. People suffering from diabetes mellitus waste away as they excrete large amounts of sugar-containing urine.

The cause of insulin-dependent-diabetes mellitus is an inadequate production of insulin by the body. Insulin is secreted in response to high blood glucose levels. It binds to the membrane receptor protein on its target cells. Binding increases the rate of transport of glucose across the membrane and stimulates glycogen synthesis, lipid biosynthesis, and protein synthesis. As a result, the blood glucose level is reduced. Clearly, the inability to produce sufficient insulin seriously impairs the body's ability to regulate metabolism.

Individuals suffering from diabetes mellitus do not produce enough insulin to properly regulate blood glucose levels. This generally results from the destruction of the β cells of the islets of Langerhans. One theory to explain the mysterious disappearance of these cells is that a virus infection stimulates the immune system to produce antibodies that cause the destruction of the β cells.

In the absence of insulin the uptake of glucose into the tissues is not stimulated, and a great deal of glucose is eliminated in the urine. Without insulin, then, adipose cells are unable to take up the glucose required to synthesize triglycerides. As a result, the rate of fat hydrolysis is much greater than the rate of fat resynthesis, and large quantities of free fatty acids are liberated into the bloodstream. Because glucose is not being efficiently taken into cells, carbohydrate metabolism slows, and there is an increase in the rate of lipid catabolism. In the liver this lipid catabolism results in the production of ketone bodies: acetone, acetoacetate, and β-hydroxybutyrate.

A similar situation can develop from improper eating, fasting, or dieting—any situation in which the body is not provided with sufficient energy in the form of carbohydrates. These ketone bodies cannot all be oxidized by the citric acid cycle, which is limited by the supply of oxaloacetate. The acetone concentration in blood rises to levels so high that acetone can be detected in the breath of untreated diabetics. The elevated concentration of ketones in the blood can overwhelm the buffering capacity of the blood, resulting in ketoacidosis. Ketones, too, will be excreted through the kidney. In fact, the presence of excess ketones in the urine can raise the osmotic concentration of the urine so that it behaves as an "osmotic diuretic," causing the excretion of enormous amounts of water. As a result, the patient may become severely dehydrated. In extreme cases the combination of dehydration and ketoacidosis may lead to coma and death.

It has been observed that diabetics also have a higher than normal level of glucagon in the blood. As we have seen, glucagon stimulates lipid catabolism and ketogenesis. It may be that the symptoms described above may result from both the deficiency of insulin and the elevated glucagon levels. The absence of insulin may cause the elevated blood glucose and fatty acid levels, while the glucagon, by stimulating ketogenesis, may be responsible for the ketoacidosis and dehydration.

There is no cure for diabetes. However, when the problem is the result of the inability to produce active insulin, blood glucose levels can be controlled moderately well by the injection of either animal insulin or human insulin produced from the cloned insulin gene (See Chapter 16, "A Human Perspective: The New Genetics and Human Genetic Disease"). Unfortunately, one or even a few injections of insulin each day cannot mimic the precise control of blood glucose accomplished by the pancreas.

As a result, diabetics suffer progressive tissue degeneration that leads to early death. One primary cause of this degeneration is atherosclerosis, the deposition of plaque on the walls of blood vessels. This causes a high frequency of strokes, heart attack, and gangrene of the feet and lower extremities, often necessitating amputation. Kidney failure causes the death of about 20% of diabetics under 40 years of age, and diabetic retinopathy (various kinds of damage to the retina of the eye) ranks fourth among the leading causes of blindness in the United

States. Nerves are also damaged, resulting in neuropathies that can cause pain or numbness, particularly of the feet.

There is no doubt that insulin injections prolong the life of diabetics, but only the presence of a fully functioning pancreas can allow a diabetic to live a life free of the complications noted here. At present, pancreas transplants do not have a good track record. Only about 50% of the transplants are functioning after one year. It is hoped that improved transplantation techniques will be developed so that diabetics can live a normal life span, free of debilitating disease.

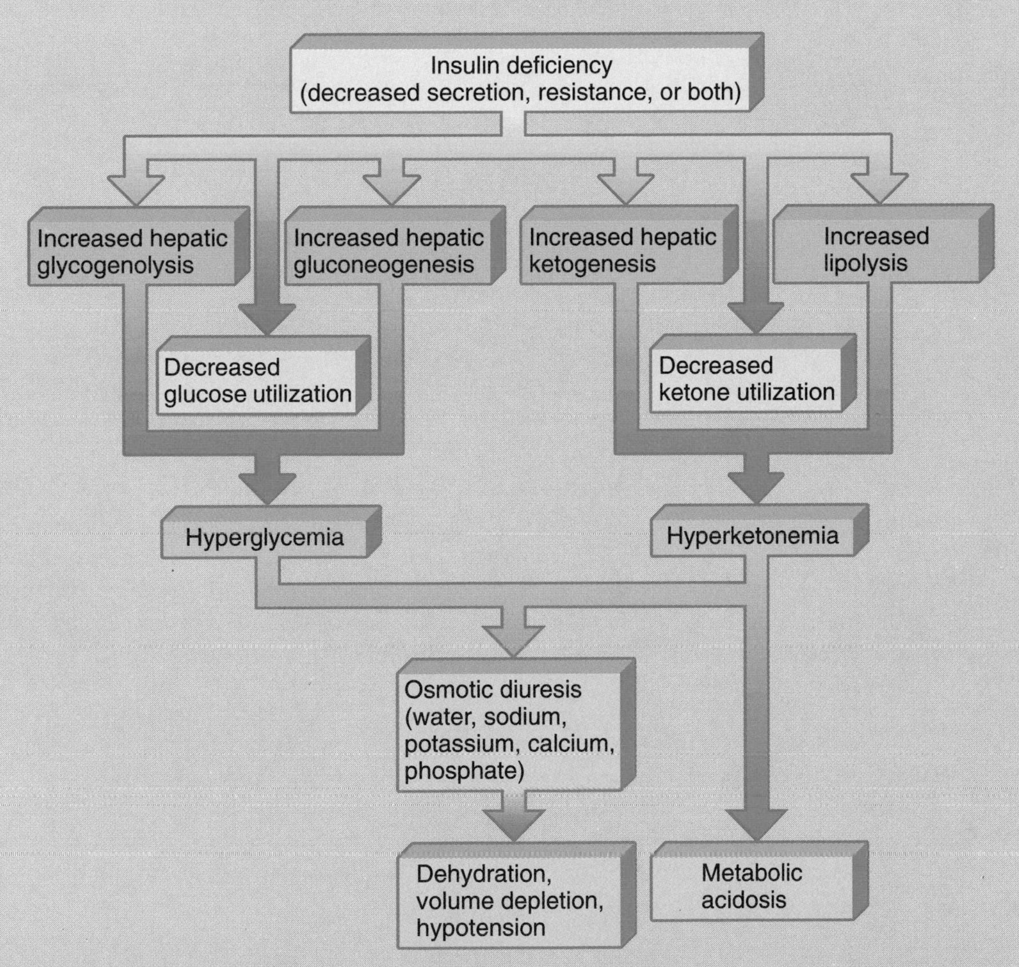

The array of metabolic events that occur in diabetes that can lead to coma and death.

If HMG-CoA were formed in the cytoplasm, it would serve as a precursor for cholesterol biosynthesis. But ketogenesis, like β-oxidation, occurs in the mitochondrial matrix, and here HMG-CoA is cleaved to yield acetoacetate and acetyl CoA:

$$^{-}OOC{-}CH_2{-}\overset{\overset{\displaystyle OH}{|}}{\underset{\underset{\displaystyle CH_3}{|}}{C}}{-}CH_2{-}\overset{\overset{\displaystyle O}{\|}}{C}{\sim}S{-}CoA \longrightarrow {}^{-}OOC{-}CH_2{-}\overset{\overset{\displaystyle O}{\|}}{\underset{\underset{\displaystyle CH_3}{|}}{C}} + CH_3{-}\overset{\overset{\displaystyle O}{\|}}{C}{\sim}S{-}CoA$$

HMG-CoA   Acetoacetate + Acetyl CoA

In very small amounts, acetoacetate spontaneously loses carbon dioxide to give acetone. This is the reaction that causes the "acetone breath" often associated with diabetes mellitus. More frequently it undergoes NADH-dependent reduction to produce β-hydroxybutyrate:

$$^{-}OOC{-}CH_2{-}\overset{\overset{\displaystyle O}{\|}}{\underset{\underset{\displaystyle CH_3}{|}}{C}} + H^+ \xrightarrow[\searrow CO_2]{} CH_3{-}\overset{\overset{\displaystyle O}{\|}}{C}{-}CH_3$$

Acetoacetate   Acetone

or

$$^{-}OOC{-}CH_2{-}\overset{\overset{\displaystyle O}{\|}}{\underset{\underset{\displaystyle CH_3}{|}}{C}} \xrightarrow{NADH \quad NAD^+} {}^{-}OOC{-}CH_2{-}\overset{\overset{\displaystyle OH}{|}}{\underset{\underset{\displaystyle H}{|}}{C}}{-}CH_3$$

Acetoacetate β-Hydroxybutyrate

Acetoacetate and β-hydroxybutyrate are produced primarily in the liver. These metabolites diffuse into the blood and are circulated to other tissues, where they may be reconverted to acetyl CoA and used to produce energy. In fact, the heart muscle derives most of its metabolic energy from the oxidation of ketone bodies, not from the oxidation of glucose. Other tissues that are best adapted to the use of glucose will increasingly rely on ketone bodies for energy when glucose becomes unavailable or limited. This is particularly true of the brain.

Question 15.5 What condition leads to excess production of ketone bodies?

15.4 FATTY ACID SYNTHESIS

All organisms possess the ability to synthesize fatty acids. In humans the excess acetyl CoA produced by carbohydrate degradation is stored in the fatty "acyl" moieties of triglycerides as a long-term energy reserve. Some organisms also require fatty acid synthesis to provide membrane lipids.

On first examination, fatty acid synthesis appears to be simply the reverse of β-oxidation. Specifically, the fatty acid chain is constructed by the sequential addition of two-carbon acetyl groups (Figure 15.11). Although the chemistry of fatty acid synthesis and that of breakdown are similar, there are several major differences between β-oxidation and fatty acid biosynthesis. These are summarized below:

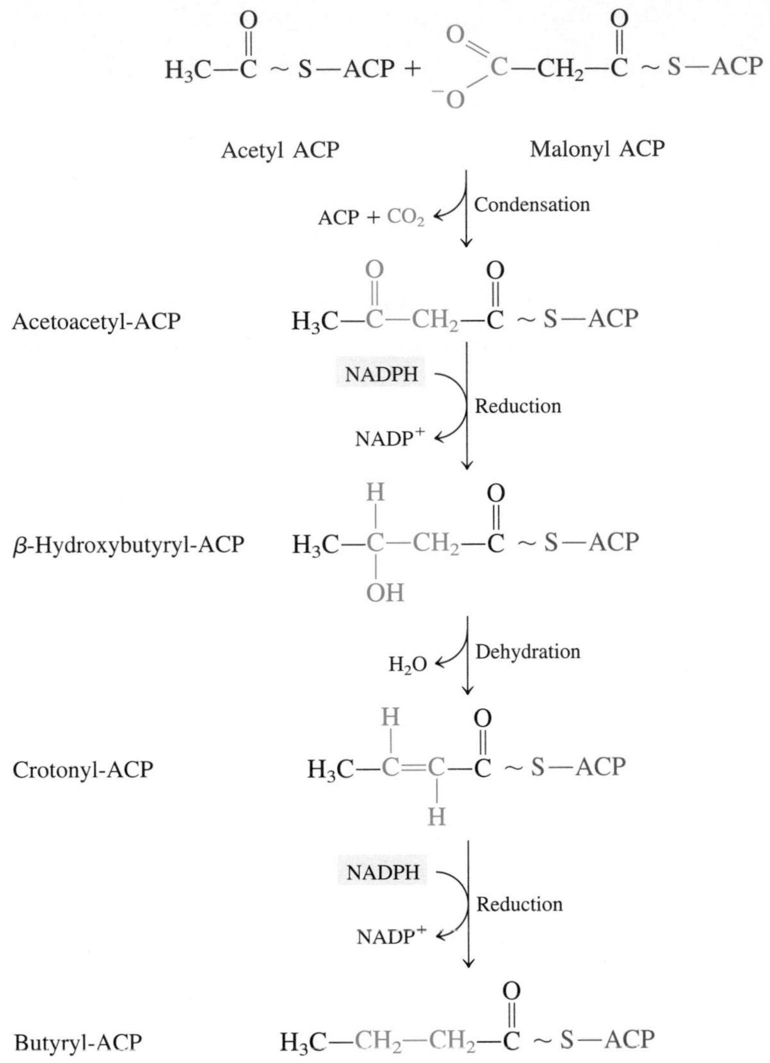

Acetyl ACP: $H_3C{-}C({=}O) \sim S{-}ACP$ + Malonyl ACP: $^-O{-}C({=}O){-}CH_2{-}C({=}O) \sim S{-}ACP$

Condensation ($\rightarrow$ ACP + CO_2)

Acetoacetyl-ACP: $H_3C{-}C({=}O){-}CH_2{-}C({=}O) \sim S{-}ACP$

Reduction (NADPH $\rightarrow$ $NADP^+$)

β-Hydroxybutyryl-ACP: $H_3C{-}CH(OH){-}CH_2{-}C({=}O) \sim S{-}ACP$

Dehydration ($\rightarrow$ H_2O)

Crotonyl-ACP: $H_3C{-}CH{=}CH{-}C({=}O) \sim S{-}ACP$

Reduction (NADPH $\rightarrow$ $NADP^+$)

Butyryl-ACP: $H_3C{-}CH_2{-}CH_2{-}C({=}O) \sim S{-}ACP$

FIGURE 15.11
Summary of the reactions of fatty acid synthesis.

- *Intracellular location*. The enzymes that are responsible for fatty acid biosynthesis are located in the cytoplasm of the cell, whereas those responsible for the degradation of fatty acids are in the mitochondria.

- *Acyl group carriers*. The activated intermediates of fatty acid biosynthesis are bound to a sulfhydryl group derived from a **phosphopantetheine** moiety bound covalently to a protein known as the **acyl carrier protein (ACP)** (Figure 15.12). Thus the thioester intermediates of fatty acid biosynthesis are not derivatives of coenzyme A. These carriers have a common reactive group. Note that the reactive sulfhydryl group in both coenzyme A and ACP is the phosphopantetheine group derived from the vitamin pantothenic acid.

- *Enzymes involved*. All the reactions of fatty acid biosynthesis are carried out by a multienzyme complex known as **fatty acid synthase**. The enzymes that are responsible for fatty acid degradation are not physically associated in such complexes.

- *Electron carriers*. NADH and $FADH_2$ are produced by fatty acid oxidation, whereas NADPH is the reducing agent for fatty acid biosynthesis. As a general rule, *NADH is*

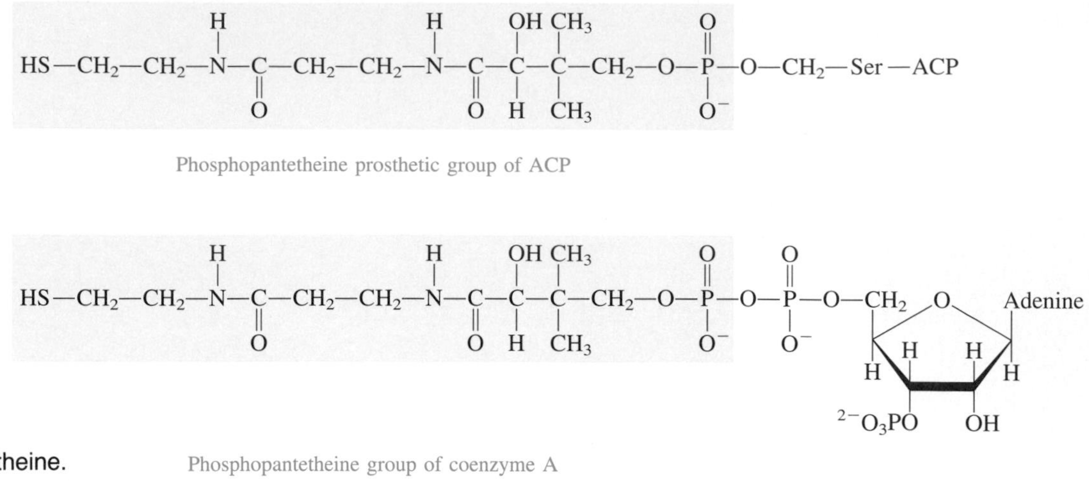

$$HS{-}CH_2{-}CH_2{-}N(H){-}C(=O){-}CH_2{-}CH_2{-}N(H){-}C(=O){-}C(OH)(H){-}C(CH_3)_2{-}CH_2{-}O{-}P(=O)(O^-){-}O{-}CH_2{-}Ser{-}ACP$$

Phosphopantetheine prosthetic group of ACP

$$HS{-}CH_2{-}CH_2{-}N(H){-}C(=O){-}CH_2{-}CH_2{-}N(H){-}C(=O){-}C(OH)(H){-}C(CH_3)_2{-}CH_2{-}O{-}P(=O)(O^-){-}O{-}P(=O)(O^-){-}O{-}CH_2\text{-ribose(3'-}OPO_3^{2-}\text{, 2'-OH)-Adenine}$$

Phosphopantetheine group of coenzyme A

FIGURE 15.12
Structure of phosphopantetheine.

FIGURE 15.13
Structure of NADPH. The phosphate group shown in blue is the structural feature that distinguishes NADH from NADPH.

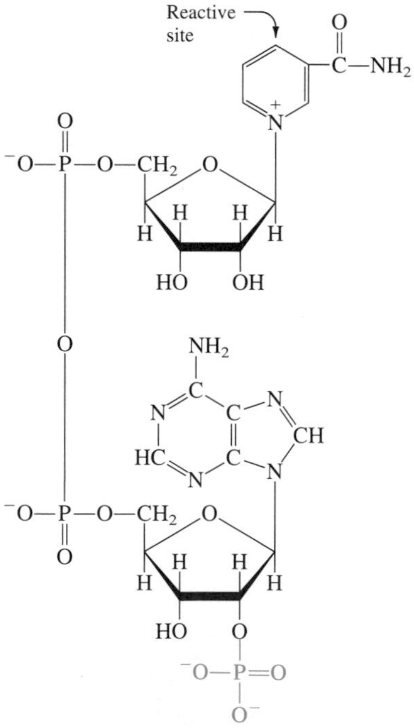

produced by catabolic reactions, and NADPH is the reducing agent of biosynthetic reactions. These two coenzymes differ only by the presence of a phosphate group bound to the ribose ring of NADPH (Figure 15.13). However, the enzymes that use these coenzymes are easily able to distinguish them on this basis.

15.5 THE REGULATION OF LIPID AND CARBOHYDRATE METABOLISM

The metabolism of fatty acids and carbohydrates occurs to a different extent in different organs. The regulation of these two related aspects of metabolism is of great physiological importance.

The liver: regulation of the flux of metabolites to brain, muscle, and adipose tissue

The liver provides a steady supply of glucose for muscle and brain and plays a major role in the regulation of blood glucose concentration. This regulation is under hormonal control. Blood glucose can be imported into the liver and stored as glycogen (glycogenesis). We also recall that muscles produce lactate under anaerobic conditions and that this ''muscle poison'' can be converted to glucose in the liver by gluconeogenesis. Both glycogen degradation (glycogenolysis) and gluconeogenesis provide pathways for producing glucose for export to other organs (Figure 15.14).

The liver also plays a central role in lipid metabolism. When excess fuel is available, the liver synthesizes fatty acids. These are esterified to produce triglycerides that are transported from the liver to adipose tissues by very low-density lipoprotein (VLDL) complexes. In fact, VLDL complexes provide adipose tissue with its major source of fatty acids. This transport is particularly active when more Calories are eaten than are burned! During fasting or starvation conditions, however, the liver converts fatty acids to acetoacetate and other ''ketone bodies.'' The liver cannot use these ketone bodies because it lacks an enzyme for the conversion of acetoacetate to acetyl CoA. Therefore the ketone bodies produced by the liver are exported to other organs. Surprisingly, the preferred fuel for the liver is not glucose. Rather, the liver obtains most of its metabolic energy from the carbon skeletons of such amino acids as alanine.

Section 9.5

Section 14.5

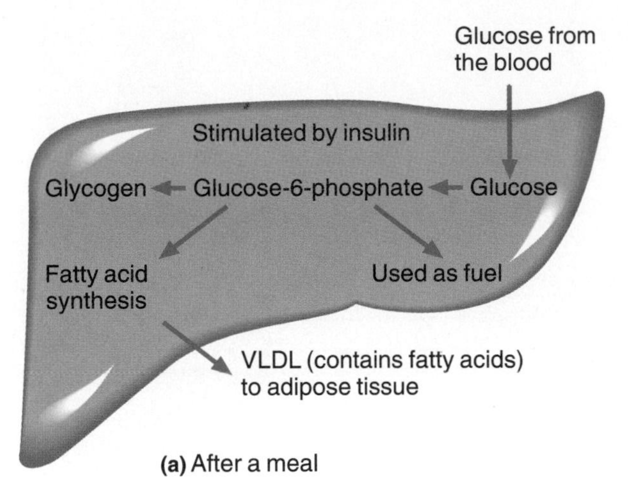

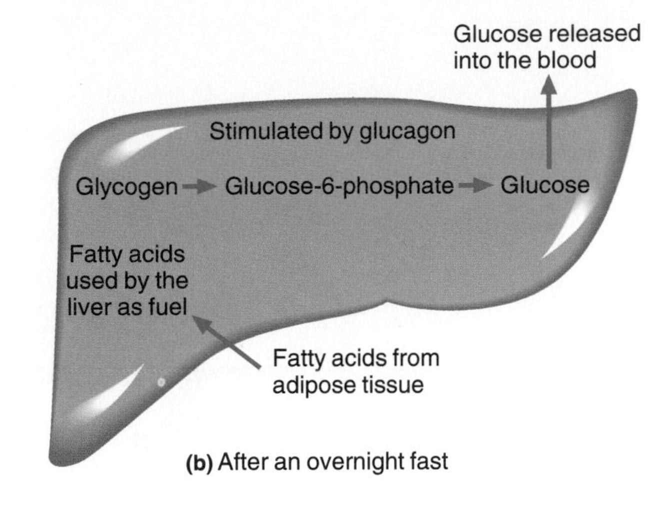

FIGURE 15.14
The liver controls the concentration of blood glucose.

Adipose tissue: coordination of glycolysis and fatty acid metabolism

Adipose tissue is the major storage depot of fatty acids. Esterified fatty acids (triglycerides) produced by the liver are transported through the bloodstream as components of VLDL complexes. The triglycerides are hydrolyzed by the same lipases that act on chylomicrons, and the fatty acids are absorbed by adipose tissue. The synthesis of triglycerides in adipose tissue requires glycerol-3-phosphate.

$$
\begin{array}{l}
\quad\ \ \mathrm{H} \\
\quad\ \ | \\
\mathrm{H{-}C{-}OH} \\
\quad\ \ | \\
\mathrm{H{-}C{-}OH} \qquad \mathrm{O} \\
\quad\ \ | \qquad\qquad\ \ \| \\
\mathrm{H{-}C{-\!-\!-}O{-}P{-}O^-} \\
\qquad\qquad\qquad\ \ | \\
\qquad\qquad\qquad\ \mathrm{O^-}
\end{array}
$$

Glycerol-3-phosphate

However, adipose tissue is unable to make glycerol-3-phosphate and depends upon glycolysis for its supply of this precursor. Thus adipose cells must have a ready source of glucose in order to synthesize and store triglycerides.

Triglycerides are constantly being hydrolyzed and resynthesized in the cells of adipose tissue. Lipases that are under hormonal control determine the rate of hydrolysis, that is, the release of fatty acids from triglycerides. Complete hydrolysis of triglycerides produces glycerol and fatty acids. If glucose is in limited supply, there will not be sufficient glycerol-3-phosphate for the resynthesis of triglycerides, and the fatty acids and glycerol are exported to the liver for further processing (Figure 15.15).

FIGURE 15.15
Synthesis and degradation of triacylglycerols in adipose tissue.

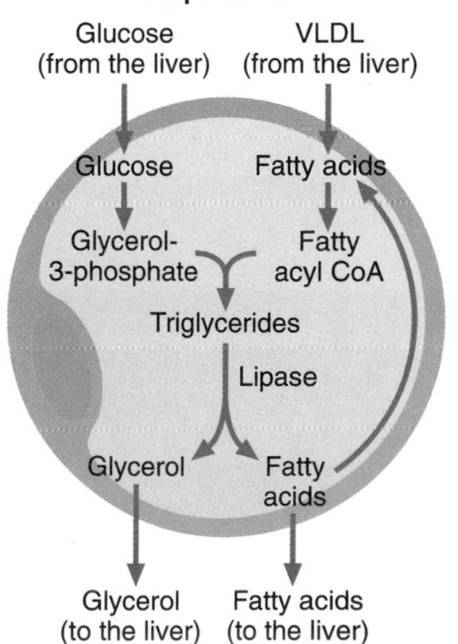

Muscle tissue: oxidization of glucose, fatty acids, and ketone bodies

The energy demand of resting muscle is generally supplied by the β-oxidation of fatty acids. The heart muscle actually prefers ketone bodies over glucose.

Working muscle, however, obtains energy by degradation of its own supply of glycogen. Glycogen degradation produces glucose-6-phosphate, which is directly funneled into

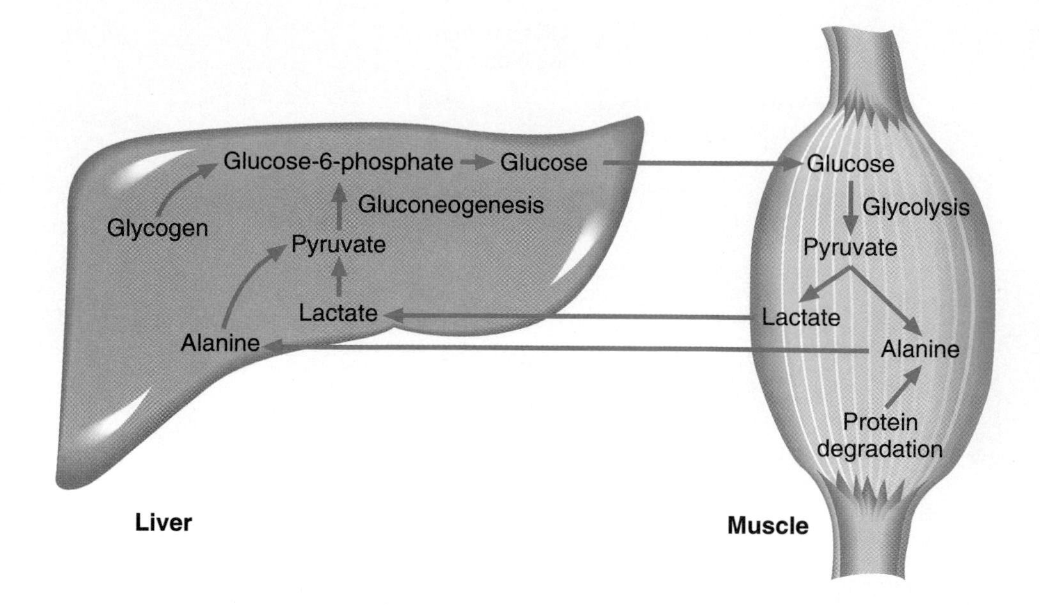

FIGURE 15.16
Metabolic relationships between liver and muscle.

glycolysis. Since muscle does not contain glucose-6-phosphatase, the cells cannot convert glucose-6-phosphate to glucose, and muscle therefore cannot export glucose. We recall, however, that muscle soon outstrips its oxygen supply in periods of rapid contraction and produces large amounts of lactate. This glycolytic end product, as well as the alanine derived from catabolism of proteins and from the transamination of pyruvate, is exported to the liver. These metabolic by-products are then converted to glucose by gluconeogenesis (Figure 15.16).

Sections 13.5 and 13.6

The brain: utilization of glucose as a fuel

Under normal conditions the brain uses glucose as its sole source of metabolic energy. When the body is in the resting state, about 60% of the free glucose of the body is used by the brain. Starvation depletes glycogen stores, and the amount of glucose available to the brain drops sharply. The ketone bodies acetoacetate and β-hydroxybutyrate, derived from fatty acid degradation, then provide an alternative source of metabolic energy. Fatty acids themselves are transported in the blood in complexes with serum albumin and cannot cross the **blood-brain barrier**. But ketone bodies, which have a free carboxylate group, are soluble in blood and can enter the brain. This mechanism enables the body to degrade lipids before proteins during starvation.

15.6 INSULIN: BIOSYNTHESIS AND CATABOLISM IN LIVER, MUSCLE, AND ADIPOSE TISSUE

Section 13.6

The polypeptide hormone *insulin* received its name late in the nineteenth century, more than 20 years before the actual substance was discovered in 1920. Insulin is secreted by the β-cells of the islets of Langerhans in the pancreas in response to an increase in the blood sugar level.

The function of insulin is to reduce the blood glucose levels back to normal, and the simplest way to do that is to stimulate storage of glucose, both as glycogen and as triglycerides. *Insulin therefore activates biosynthetic processes and inhibits catabolic processes.*

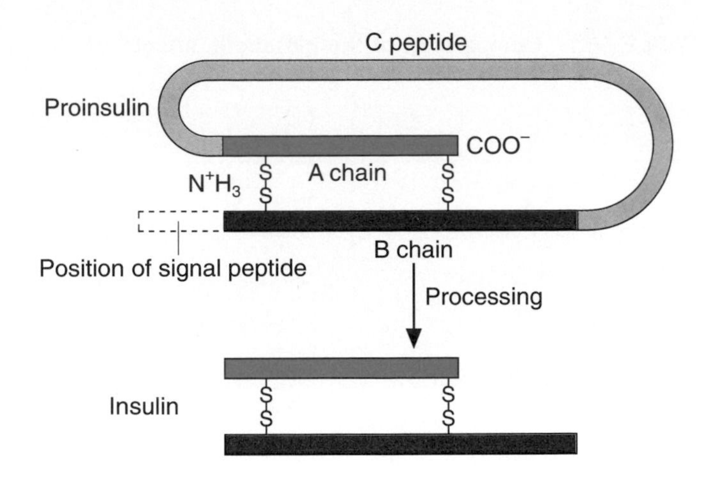

FIGURE 15.17
The structures of proinsulin and insulin.

Insulin activity

Insulin is synthesized in an inactive form. The inactive precursor of insulin is activated in three stages. These stages are summarized below:

The stages are also described in detail in Chapter 16, "A Human Perspective: The New Genetics and Human Genetic Disease."

1. The first precursor of insulin is known as **preproinsulin**. It has an amino-terminal peptide containing 23 amino acid residues, called a **signal peptide**, which is needed to direct the movement of the protein through cell membranes within the cell in the early stages of synthesis. The signal peptide is removed from preproinsulin by a specific proteolytic enzyme.
2. Removal of the signal peptide produces **proinsulin** (Figure 15.17), the immediate precursor of insulin. Proinsulin is an 84-residue polypeptide that consists of an A chain, a B chain, and a connecting, or C, peptide.
3. Proinsulin is converted to insulin by proteolytic cleavages that remove the entire C peptide. The product is the active hormone insulin, which is shown in Figure 15.17.

The insulin receptor protein

Insulin acts only on those cells, known as *target cells*, that possess a specific insulin receptor protein in their plasma membranes. The major target cells for insulin are liver, adipose, and muscle cells. Insulin binds very tightly to its receptor protein because the concentration of insulin in blood is very low, in the range of 10^{-10} M, and a fat cell contains only about 10^4 molecules of receptor. Extremely tight binding of insulin to the receptor is required for insulin to exert its metabolic effects.

Insulin and cellular metabolism

The blood sugar level is normally about 10 mM. At this level the glucose dissolved in the blood plasma is largely reabsorbed by the kidneys, and sugar does not appear in the urine. However, a substantial meal increases the concentration of blood sugar considerably and stimulates insulin synthesis. Subsequent binding of insulin to the plasma membrane insulin receptor protein increases the rate of transport of glucose across the membrane.

Insulin exerts a variety of effects on all aspects of cellular metabolism:

- *Carbohydrate metabolism*. Insulin stimulates glycogen formation while simultaneously inhibiting glycogenolysis and gluconeogenesis.

Section 13.6

- *Protein metabolism*. Insulin stimulates transport and uptake of amino acids, as well as incorporation of amino acids into proteins.

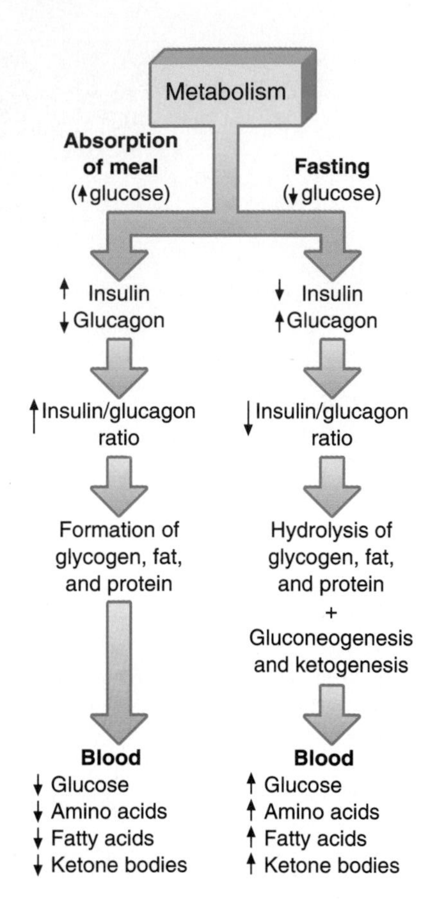

FIGURE 15.18
A summary of the antagonistic effects of insulin and glucagon.

TABLE 15.1 Comparison of the metabolic effects of insulin and glucagon

Actions	Insulin	Glucagon
Cellular glucose transport	Increased	No effect
Glycogen synthesis	Increased	Decreased
Glycogenolysis in liver	Decreased	Increased
Gluconeogenesis	Decreased	Increased
Amino acid uptake and protein synthesis	Increased	No effect
Inhibition of amino acid release and protein degradation	Decreased	No effect
Lipogenesis	Increased	No effect
Lipolysis	Decreased	Increased
Ketogenesis	Decreased	Increased

- *Lipid metabolism.* Insulin-stimulated uptake of glucose by adipose cells stimulates lipogenesis, the formation and storage of triglycerides. As we have seen, storage of lipids requires a ready source of glucose, and insulin helps the process by increasing the available glucose. At the same time, insulin inhibits lipolysis.

15.7 THE EFFECTS OF GLUCAGON ON CELLULAR METABOLISM

As you might have already guessed, insulin is only part of the overall regulation of cellular metabolism in the body. A second hormone, **glucagon**, is secreted by the α-cells of the islets of Langerhans in response to decreased blood glucose levels. The effects of glucagon, generally the opposite of the effects of insulin, are summarized in Table 15.1.

Although it has no direct effect on glucose uptake, glucagon inhibits glycogen synthesis and stimulates glycogenolysis and gluconeogenesis. It also stimulates the breakdown of fats (lipolysis) and ketogenesis. The antagonistic effects of these two hormones, seen in Figure 15.18, are critical for the maintenance of adequate blood glucose levels.

During fasting, low blood glucose levels stimulate production of glucagon, which increases blood glucose by inducing the breakdown of glycogen and the production of glucose by gluconeogenesis. This ensures a ready supply of glucose for the tissues, especially the brain. On the other hand, when blood glucose levels are too high, insulin is secreted. It facilitates the removal of the excess glucose by enhancing uptake and inducing pathways for storage.

SUMMARY

Lipids, like carbohydrates and proteins, can be oxidized in energy yielding reactions in the mitochondria. Fatty acids are stored as triglycerides (triacylglycerols) in fat droplets in the cytoplasm of adipose cells. Lipases hydrolyze triglycerides to produce fatty acids and glycerol. Fatty acids are degraded to acetyl CoA in the mitochondria of eukaryotic cells by the β-oxidation pathway, which involves the repetition of four steps: (1) oxidation of the fatty acid by an FAD-dependent dehydrogenase, (2) hydration, (3) oxidation by an NAD^+-dependent dehydrogenase, and (4) cleavage of the chain with release of acetyl CoA and a saturated fatty acyl CoA that is two carbons shorter.

Under some conditions fatty acid degradation occurs

more rapidly than glycolysis. As a result, a large amount of acetyl CoA is produced from the fatty acids, but little oxaloacetate is generated from pyruvate. When oxaloacetate levels are too low, the excess acetyl CoA is converted to ketone bodies—acetone, acetoacetate, and β-hydroxybutyrate.

Fatty acid biosynthesis occurs by sequential addition of acetyl groups, and on first inspection appears to be a simple reversal of the β-oxidation pathway. Although the chemical reactions are similar, fatty acid synthesis differs from β-oxidation in the following ways: It occurs in the cytoplasm, utilizes acyl carrier protein and NADPH, and is carried out by a multienzyme complex, fatty acid synthase.

Lipid and carbohydrate metabolism occur to different extents in different organs. The liver regulates the flux of metabolites to brain, muscle, and adipose tissue and ultimately controls the concentration of blood glucose. Adipose tissue is the major storage depot for fatty acids. Triglycerides are constantly hydrolyzed and resynthesized in adipose tissue. Muscle oxidizes glucose, fatty acids, and ketone bodies. The brain uses glucose as a fuel except in prolonged fasting or starvation, when it will use ketone bodies as an energy source.

Insulin stimulates biosynthetic processes and inhibits catabolism in liver, muscle, and adipose tissue. Insulin is synthesized as preproinsulin that is converted to the active hormone by proteolysis. It is secreted when the blood glucose levels become too high. The insulin receptor protein binds to the insulin and mediates a variety of responses in target tissues. The effects of insulin encourage the storage of glucose and lipids. Glucagon is secreted when blood glucose levels are too low. It has the opposite effects on metabolism, including the breakdown of lipids and glycogen.

GLOSSARY OF KEY TERMS

acyl carrier protein (ACP) (15.4) the protein that forms a thioester linkage with fatty acids during fatty acid synthesis.

adipocyte (15.1) a fat cell.

adipose tissue (15.1) fatty tissue that stores most of the body lipids.

bile (15.1) micelles of lecithin, cholesterol, bile salts, protein, inorganic ions, and bile pigments that aid in lipid digestion by emulsifying fat droplets.

blood-brain barrier (15.5) structures and cells in the brain that select which molecules may enter the brain.

chylomicron (15.1) an aggregate of protein and triglycerides that carries triglycerides from the intestines to all body tissues via the bloodstream.

colipase (15.1) a protein that aids in lipid digestion by binding to the surface of lipid droplets and facilitating binding of pancreatic lipase.

detergent (15.1) a molecule having a highly charged (polar) group and a long hydrocarbon chain. In water the charged end dissolves, but the hydrophobic chain does not. When fats are present, the hydrocarbon chain binds to them, forming micelles.

diabetes mellitus (15.3) the appearance of glucose in the urine caused by high blood glucose levels. The disease generally results from the inability to produce insulin.

fatty acid synthase (15.4) the multienzyme complex that carries out fatty acid biosynthesis.

glucagon (15.7) a peptide hormone synthesized by the α-cells of the islets of Langerhans in the pancreas in response to low blood glucose levels.

ketoacidosis (15.3) a drop in the pH of the blood caused by elevated ketone levels.

ketone bodies (15.3) acetone, acetoacetone, and β-hydroxybutyrate produced from fatty acids in the liver via acetyl CoA.

ketosis (15.3) an abnormal rise in the level of ketone bodies in the blood.

lipase (15.1) an enzyme that hydrolyzes the ester linkage between glycerol and the fatty acids of triglycerides.

micelle (15.1) an aggregation of molecules having a nonpolar and a polar region. The nonpolar regions of the molecules aggregate, leaving the polar regions facing the surrounding water.

β-oxidation (15.2) the biochemical pathway that results in the oxidation of fatty acids and the production of acetyl CoA.

phosphopantetheine (15.4) the portion of coenzyme A and the acyl carrier protein that is derived from the vitamin pantothenic acid. It forms a thioester linkage with fatty acids.

preproinsulin (15.6) the first precursor in insulin biosynthesis.

proinsulin (15.6) an intermediate in insulin biosynthesis that has lost the signal peptide.

signal peptide (15.6) a sequence of hydrophobic amino acids at the amino terminus of a protein that guides the protein through cellular membranes.

triglyceride (15.1) a molecule composed of glycerol esterified to three fatty acids. (Also called triacylglycerol.)

QUESTIONS AND PROBLEMS

Lipid Metabolism

15.6 What is the major metabolic function of adipose tissue?

15.7 What is the outstanding structural feature of an adipocyte?

15.8 What is the source of the glycerol unit of triglycerides?

15.9 What is the general reaction catalyzed by lipases?

15.10 List three major metabolic products for which acetyl CoA is a precursor.

15.11 What are chylomicrons and what is their function?

15.12 What is the function of the bile salts?

15.13 What is the function of colipase?

Fatty Acid Degradation

15.14 What products are formed when the ω-phenyl-labeled carboxylic acid 10-phenyldecanoic acid is degraded by β-oxidation?

15.15 What products are formed when 5-phenylpentanoic acid is degraded by β-oxidation?

15.16 Calculate the number of ATP molecules produced for complete β-oxidation of a 16-carbon saturated fatty acid (palmitic acid).

15.17 Write the sequence of steps that would be followed for one round of β-oxidation of hexanoic acid.

15.18 How many molecules of ATP are produced for each molecule of $FADH_2$ that is generated by β-oxidation?

15.19 How many molecules of ATP are produced for oxidation of each molecule of NADH generated by β-oxidation?

15.20 What are the parallels of β-oxidation of fatty acids and the citric acid cycle?

Ketone Bodies

15.21 What are the structures of the ketone bodies?

15.22 How will the concentration of ketone bodies change if the concentration of oxaloacetate increases?

15.23 How does inhibition of the enzyme that converts acetyl CoA and oxaloacetate to citrate affect fatty acid metabolism?

15.24 Describe the formation of ketone bodies through β-oxidation.

15.25 When does the body *use* ketone bodies?

Fatty Acid Synthesis

15.26 What is the role of phosphopantetheine in fatty acid biosynthesis?

15.27 How does the biochemical function of NADPH differ from that of NADH?

15.28 How does the structure of fatty acid synthase differ from that of the enzymes that carry out β-oxidation?

15.29 In what cellular compartments do fatty acid biosynthesis and β-oxidation occur?

15.30 In what respects does fatty acid biosynthesis differ from β-oxidation?

Regulation of Lipid and Carbohydrate Metabolism

15.31 What is the major metabolic function of the liver?

15.32 What is the fate of lactate produced in skeletal muscle during rapid contraction?

15.33 What are the major fuels of the heart, brain, and liver?

15.34 Why can't the brain use fatty acids as fuel?

15.35 Explain how acetyl CoA derived from glucose can be converted to lipid.

Insulin: Biosynthesis and Catabolism in Liver, Muscle, and Adipose Tissue

15.36 What is the name of the first precursor of insulin?

15.37 How does preproinsulin differ from proinsulin?

15.38 What steps lead to conversion of proinsulin to insulin?

15.39 What is the function of the signal sequence, the hydrophobic peptide of preproinsulin that is removed to form proinsulin?

15.40 Where is insulin produced?

15.41 How does insulin affect carbohydrate metabolism?

15.42 How does insulin affect lipid metabolism?

15.43 What is the primary cause of diabetes mellitus?

15.44 What is the origin of the ''acetone breath'' associated with diabetes mellitus?

15.45 What causes ketoacidosis?

Effects of Glucagon on Cellular Metabolism

15.46 What is the effect of glucagon on carbohydrate metabolism?

15.47 What is the effect of glucagon on lipid metabolism?

15.48 What role is glucagon thought to play in diabetes mellitus?

Further Questions

15.49 Fill in the blanks:

a. Fatty acids are stored as ______ in adipose tissue.
b. An enzyme that catalyzes the release of fatty acids from triglycerides is called a ______.
c. For each acetyl CoA produced by β-oxidation ______ ATP will be synthesized.
d. Glucose, fatty acids, and ketone bodies are oxidized by ______ tissue.

15.50 Label each of the following statements as true or false.

a. Diabetes mellitus is caused by insulin deficiency.
b. Acetyl CoA is one of the minor metabolites of fatty acids.
c. The complete oxidation of fatty acids releases more energy than the oxidation of a comparable mass of glycogen.
d. β-Oxidation of fatty acids involves the sequential removal of single-carbon fragments.

15.51 How do bile salts alter cholesterol to make it water-soluble?

15.52 What is the most effective means of weight control?

C H A P T E R 1 6

Introduction to Molecular Genetics

O U T L I N E

O B J E C T I V E S

- Understand the role of DNA in the transmission of genetic information.
- Describe the general composition of DNA.
- Know the role of nucleosides and nucleotides in DNA structure.
- Recognize the major properties of the double helix, the basic structure of DNA.
- Describe the template model of DNA replication.
- Know what is meant by mutation and how mutations can cause cancer.
- Describe how ultraviolet light functions as a germicide, a mutagen, and a carcinogen.
- Recognize the consequences of thymine dimers and other mutations.
- Explain how visible radiation can be used by bacteria to repair ultraviolet radiation-induced damage.
- Summarize the central dogma, the path of information flow in organisms.
- Describe the structure of RNA and how it differs from that of DNA.
- List three classes of RNA molecules.
- Describe the biosynthesis of messenger RNA.
- Explain the difference between continuous and discontinuous genes.
- Describe the essential elements of the genetic code and try to develop a "feel" for its elegance.
- Understand, in general terms, the fundamental processes involved in protein synthesis.

DNA is the carrier of the genetic information that is ultimately responsible for every characteristic of the cell. All of this information is encoded within the structure of DNA. In this chapter we will explore the structure of DNA and the processes by which the genetic information is interpreted to produce proteins. The relationship between structure and properties, first mentioned in our discussion of atoms early in the book, is taken to its pinnacle in this discussion of molecular genetics.

In the preceding chapters we explored many important biochemical pathways that are absolutely essential to life. In addition to the enzymes that mediate these pathways, we studied many proteins that serve the body in other necessary capacities. A question arises. Where is the information that determines the amino acid sequence of all these proteins stored? This is obviously vital information, since the primary sequence of a protein dictates the secondary and ultimately the precise tertiary folding of the protein. Only if this precise shape is attained will the enzyme or protein function properly. When we consider the thousands of chemical reactions that occur within the body and the thousands of proteins that accomplish this work, we become aware that the information storage molecule is extremely important. It must be a very large molecule with almost infinite variety to be able to store so much information. It must also be a very stable molecule because an error in just one gene can result in a fatal or life-threatening genetic disease. This vital molecule, the carrier of the genetic information, is *deoxyribonucleic acid (DNA)*.

DNA is the genetic material of all organisms, with the exception of a few viruses. Although DNA was discovered by Friedrich Meischer in 1869, not until 1950 was it recognized as the genetic information of the cell. Indeed, the vital function of DNA lay hidden until its structure was elucidated by James Watson and Francis Crick in 1953. These discoveries opened the door to an understanding of genetics at the level of molecular structure and function and launched the field of molecular biology that has revolutionized biology.

Recent laboratory advances in genetic engineering, or gene splicing, have already provided us with safer and less expensive vaccines and products for the treatment of genetic diseases such as diabetes and hemophilia. Furthermore, genetic engineering offers the possibility of the eradication of other genetic diseases. In this chapter we will explore the basic structure and expression of the genetic information so that we can appreciate the manipulation of DNA by recombinant DNA technology and the promise this new technology holds.

See "A Human Perspective: The New Genetics and Human Genetic Diseases"

16.1 NUCLEIC ACIDS

Chemical composition of DNA and RNA

Two types of nucleic acids are important to the cell. The first is **deoxyribonucleic acid (DNA),** which carries all of the genetic information for an organism. The second type is **ribonucleic acid (RNA),** which is responsible for interpreting the genetic information into proteins that will carry out essential cellular functions.

The components of these nucleic acids are released by treatment with an acid and can then be identified by chromatographic techniques. Treatment of DNA with a strongly acidic solution releases the sugar 2′-deoxyribose, phosphoric acid, and four *heterocyclic bases,* called *nitrogenous bases*. These bases are divided into two families known as **pyrimidines** and **purines** (Figure 16.1). Hydrolysis of RNA also releases purine and pyrimidine bases, phosphate, and a sugar, in this case ribose. The pyrimidine bases of DNA are **cytosine** (C) and **thymine** (T). The pyrimidines found in RNA are cytosine and **uracil** (U) (Figure 16.1). Notice that the three pyrimidines differ from one another only in the positioning of certain functional groups around the pyrimidine ring.

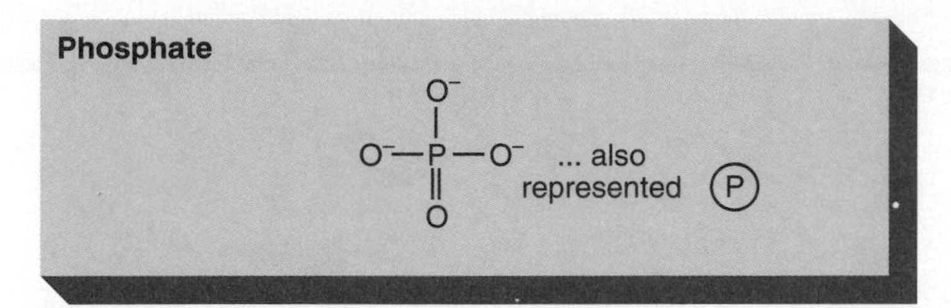

FIGURE 16.1
The components of nucleic acids include phosphate groups, the five-carbon sugars ribose and deoxyribose, and purine and pyrimidine nitrogenous bases. The ring positions of the sugars are designated with primes to distinguish them from the ring positions of the bases.

FIGURE 16.2
General structures of a purine and a pyrimidine nucleoside. Notice that the N-glycosidic linkage involves the 1′ carbon of the sugar and either the N-1 of the pyrimidine or N-9 of the purine.

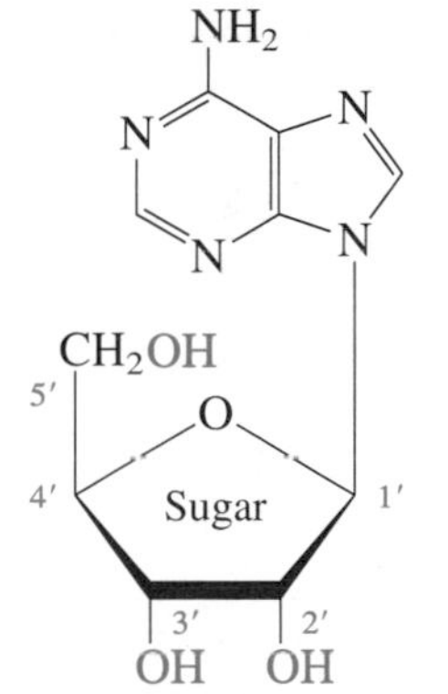

The major purine bases of both DNA and RNA are **adenine** (A) and **guanine** (G) (Figure 16.1). As with the pyrimidines, the purines differ from one another only in the location of functional groups around the purine ring.

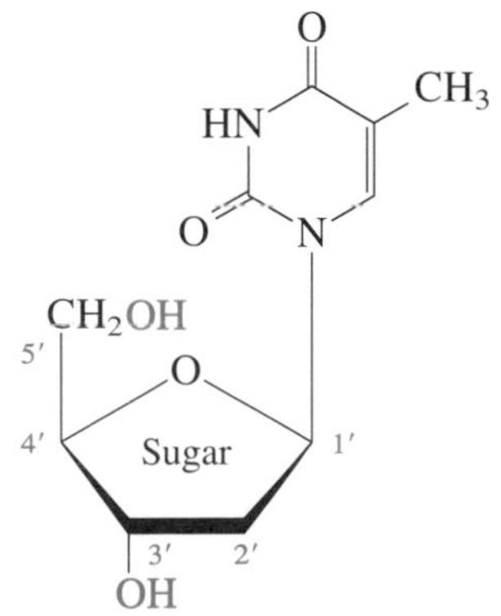

Nucleosides

Nucleosides are produced by the combination of a sugar, either ribose (in RNA) or 2′-deoxyribose (in DNA), with a purine or a pyrimidine base. The ring atoms of the sugar are designated with a prime to distinguish them from atoms in the base (Figure 16.1). The covalent bond between the sugar and the base is called a *β-N-glycosidic linkage*. The general structures of a purine nucleoside and a pyrimidine nucleoside are shown in Figure 16.2.

N-1 of pyrimidines and N-9 of purines participate in the glycosidic bonds of nucleosides. The nucleosides formed with ribose and adenine or guanine are called **adenosine** or **guanosine,** respectively. If 2′-deoxyribose is the sugar in these nucleosides, they are

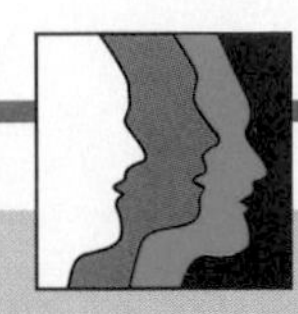

A HUMAN PERSPECTIVE

Fooling the AIDS Virus with "Look-Alike" Nucleotides

The virus responsible for the **Acquired Immune Deficiency Syndrome (AIDS)** is called the **human immunodeficiency virus,** or HIV. Members of this class of viruses, **retroviruses,** have single-stranded RNA as their genetic material. The genetic information in this RNA is converted into a DNA copy in a process that is the opposite of the central dogma. The central dogma, discussed in Section 16.6, states that the flow of genetic information is unidirectional, from DNA to RNA, but these viruses reverse that flow, RNA to DNA. For this reason these viruses are called retroviruses, which literally means "backward viruses." The process of producing a DNA copy of the RNA is called reverse transcription. The enzyme required for this must be able to read a single strand of RNA to produce a single strand of DNA. Then it must destroy the RNA strand and read the newly synthesized DNA strand to produce the complementary DNA strand. The ultimate product is a double-stranded DNA copy of the original single-stranded RNA. This multitalented enzyme is called **reverse transcriptase.**

Because our genetic information is DNA and it is expressed by the classical DNA to RNA to protein pathway, our cells have no need for a reverse transcriptase enzyme. Thus the HIV reverse transcriptase is a good target for antiviral chemotherapy because inhibition of the reverse transcriptase should kill the virus but have no effect on the human host. Many drugs have been tested for the ability to selectively inhibit the HIV reverse transcriptase. Among these is 3′-azido-2′,3′-dideoxythymidine, commonly called **AZT** or **zidovudine.** AZT tests on people with AIDS began in February 1986. Two hundred eighty AIDS patients were split into two groups. Approximately half were treated with AZT, and half were given a placebo. Within six months it was apparent that AZT had a dramatic effect. In the untreated group, 19 patients had died, while in the treated group, only one patient had died and many others reported fewer complications of AIDS, weight gain, enhanced immune function, and improved quality of life. For humanitarian reasons the tests were halted, and AZT was made available as a prescription drug for severe HIV infection.

How does AZT work? It is one of many drugs that looks like one of the normal nucleosides. These are called *nucleoside analogues.* The analogue is phosphorylated by the cell and then tricks a polymerase, in this case reverse transcriptase, into incorporating it into the growing DNA chain in place of the normal phosphorylated nucleoside. AZT is a nucleoside analogue that looks like the nucleoside thymidine except that in the 3′ position of the ribose sugar there is an azido group (—N_3) rather than the 3′-OH group. Compare the structures of thymidine and AZT shown in the accompanying figure. The 3′-OH group is necessary for further DNA polymerization because it is there that the phosphodiester linkage must be made between the growing DNA strand and the next nucleotide. If an azido group, or some other group, is present at the 3′

Comparison of the structures of the normal nucleoside, 2′-deoxythymidine, and the nucleoside analogue, 3′-azido-2′,3′-dideoxythymidine.

called **2′-deoxyadenosine** and **2′-deoxyguanosine.** The ribonucleosides formed from cytosine and uracil are called **cytidine** and **uridine,** respectively. The deoxyribonucleosides of cytosine and thymine are called **2′-deoxycytidine** and **2′-deoxythymidine,** respectively.

Nucleotides

A nucleotide is produced if a hydroxyl group of the sugar portion of a nucleoside is converted into a phosphate ester. The nucleosides that have the sugar ribose are **ribonucleotides,** and those that have the sugar 2′-deoxyribose are **deoxyribonucleotides.** Phosphate groups can be esterified at any hydroxyl group of the sugar, but 5′-phosphate esters are the most common. The structures and names of the most common deoxynucleoside monophosphates are shown in Figure 16.3.

position, the nucleotide analogue can be incorporated into the growing DNA strand, but further chain elongation is blocked. If the viral RNA cannot be reverse transcribed into the DNA form, the virus will not be able to replicate and can be considered to be dead.

AZT is particularly effective because the HIV reverse transcriptase actually prefers it over the normal nucleoside, thymidine. Nonetheless, AZT is not a cure. At best it prolongs the life of an AIDS patient for a year or two. Eventually, however, AZT has a negative effect on the body. The cells of our bone marrow are constantly dividing to produce new blood cells: red blood cells to carry oxygen to the tissues, white blood cells of the immune system, and platelets for blood clotting. For cells to divide, they must replicate their DNA. The DNA polymerases of these dividing cells also accidentally incorporate AZT into the growing DNA chains with the result that cells of the bone marrow begin to die. This can result in anemia and even further depression of the immune response. It is hoped that research with other nucleoside analogues will provide a drug that is as effective as AZT against the virus but without the toxic side effect.

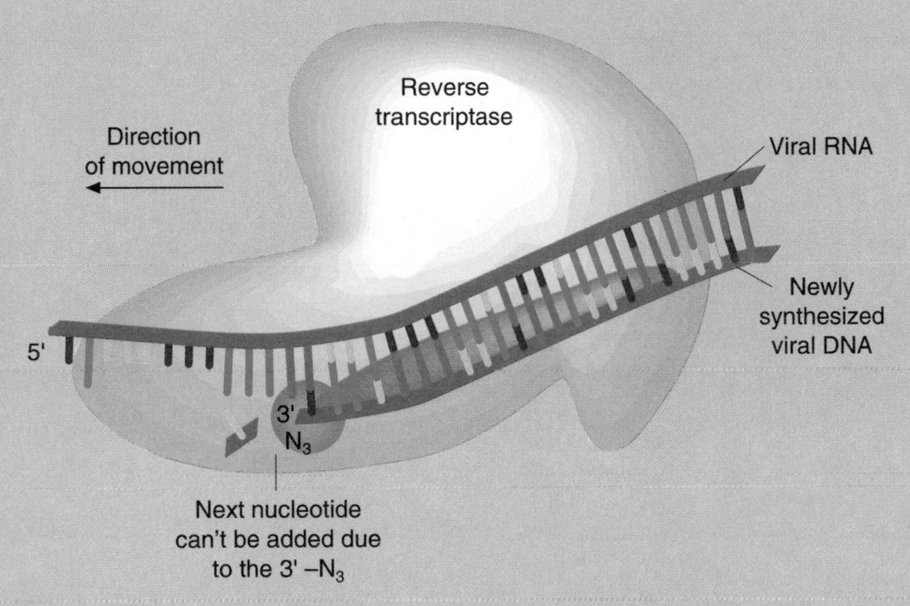

The mechanism by which AZT inhibits HIV reverse transcriptase. Incorporation of AZT into the growing HIV DNA strand in place of deoxythymidine results in DNA chain termination; the azido group on the 3′ carbon of the sugar cannot react to produce the phosphodiester linkage required to add the next nucleotide.

Question 16.1 Draw the general structure of a purine and a pyrimidine.

Question 16.2 Now complete the structures of the following by adding the functional groups at the proper locations:

a. Adenine
b. Guanine
c. Cytosine
d. Thymine
e. Uracil

On first exposure the nomenclature of the nucleosides and nucleotides is quite confusing. The following examples reexamine that nomenclature and show the common abbreviations that are used.

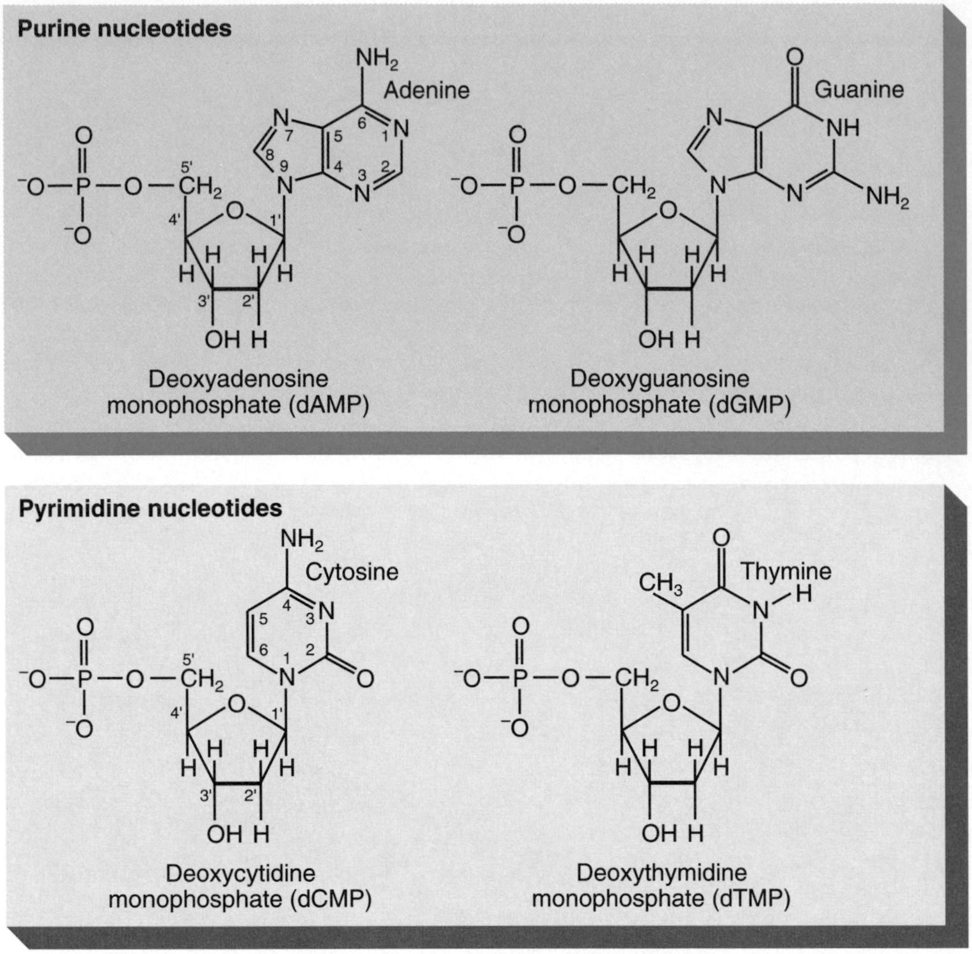

Purine nucleotides

Adenine — Deoxyadenosine monophosphate (dAMP)

Guanine — Deoxyguanosine monophosphate (dGMP)

Pyrimidine nucleotides

Cytosine — Deoxycytidine monophosphate (dCMP)

Thymine — Deoxythymidine monophosphate (dTMP)

FIGURE 16.3
Structures and names and common abbreviations of four deoxynucleotides or deoxyribonucleoside 5′-phosphates.

Example 16.1

Write the names of the nitrogenous bases and the DNA and RNA nucleosides.

Base	Nucleoside (DNA)	Nucleoside (RNA)	Abbreviation
Adenine	Deoxyadenosine	Adenosine	A
Guanine	Deoxyguanosine	Guanosine	G
Cytosine	Deoxycytidine	Cytosine	C
Thymine	Deoxythymidine		T
Uracil		Uridine	U

Example 16.2

Write the names and abbreviations of the DNA nucleotides of adenine, including the nucleoside monophosphate, nucleoside diphosphate, and nucleoside triphosphate.

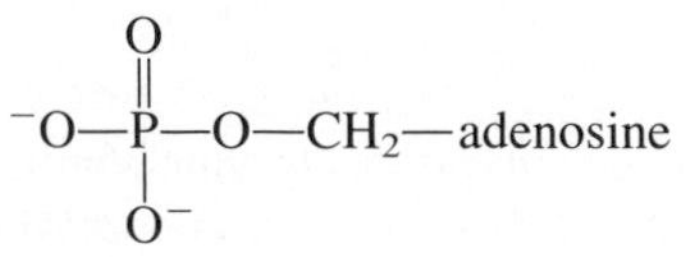

$$^{-}O-\overset{O}{\overset{\|}{P}}(O^{-})-O-CH_2-\text{adenosine}$$

AMP = adenosine-5′-monophosphate

$$^{-}O-\overset{\overset{O}{\|}}{\underset{\underset{O^-}{|}}{P}}-O-\overset{\overset{O}{\|}}{\underset{\underset{O^-}{|}}{P}}-O-CH_2-\text{adenosine}$$

ADP = adenosine-5′-diphosphate

$$^{-}O-\overset{\overset{O}{\|}}{\underset{\underset{O^-}{|}}{P}}-O-\overset{\overset{O}{\|}}{\underset{\underset{O^-}{|}}{P}}-O-\overset{\overset{O}{\|}}{\underset{\underset{O^-}{|}}{P}}-O-CH_2-\text{adenosine}$$

ATP = adenosine-5′-triphosphate

Question 16.3 What is the difference between a nucleoside and a nucleotide?

16.2 BONDING IN DNA

The *primary structure* of DNA is the linear sequence of its 2′-deoxyribonucleotide residues. The nucleotide residues of DNA are linked by 3′-5′ phosphodiester bonds. In Figure 16.4a we see the esterification reaction between two nucleotides that results in the phosphodiester linkage. In Figure 16.4b, several nucleotides are shown, each linked to the next by a phosphodiester bond. An even longer array of nucleotides is schematically represented in Figure 16.4c. As you can see, each single strand of DNA is characterized by a backbone of alternating deoxyribose and phosphate groups in phosphodiester linkage. Each sugar is also linked to one of the nitrogenous bases by an N-glycosidic linkage.

DNA molecules are enormous. The smallest DNA molecules of viruses contain several thousand deoxyribonucleotides, those of bacteria contain several million residues, and mammalian DNA molecules contain a billion nucleotide residues. A single molecule of bacterial DNA is about 1 mm long, and the length of a DNA molecule in a human chromosome would be over 6 cm in length if fully extended.

Question 16.4 Draw the structure of a 3′-5′ phosphodiester bond between a deoxyadenosine-5′-phosphate molecule and a thymidine-5′-phosphate molecule in which adenine is the 5′-nucleotide unit.

Question 16.5 Draw a similar structure using the ribonucleotides adenosine-5′-phosphate and uridine-5′-phosphate.

16.3 DNA AND THE DOUBLE HELIX

DNA in cells consists of two strands wound around each other. Each strand has a helical conformation, and the resultant structure is a **double helix** (Figure 16.5). In three dimensions the structure of the double helix resembles that of a spiral staircase.

The major properties of the DNA double helix are summarized below:

- *Sugar-phosphate backbone*. The sugar phosphate backbone winds around the *outside* of the bases like the handrail of a spiral staircase.
- *Base pairing*. Each purine base is hydrogen-bonded to a pyrimidine base in the interior of the double helix. Adenine is always paired with thymine, and guanine is always paired with cytosine (Figure 16.6). Each **base pair** lies at nearly right angles

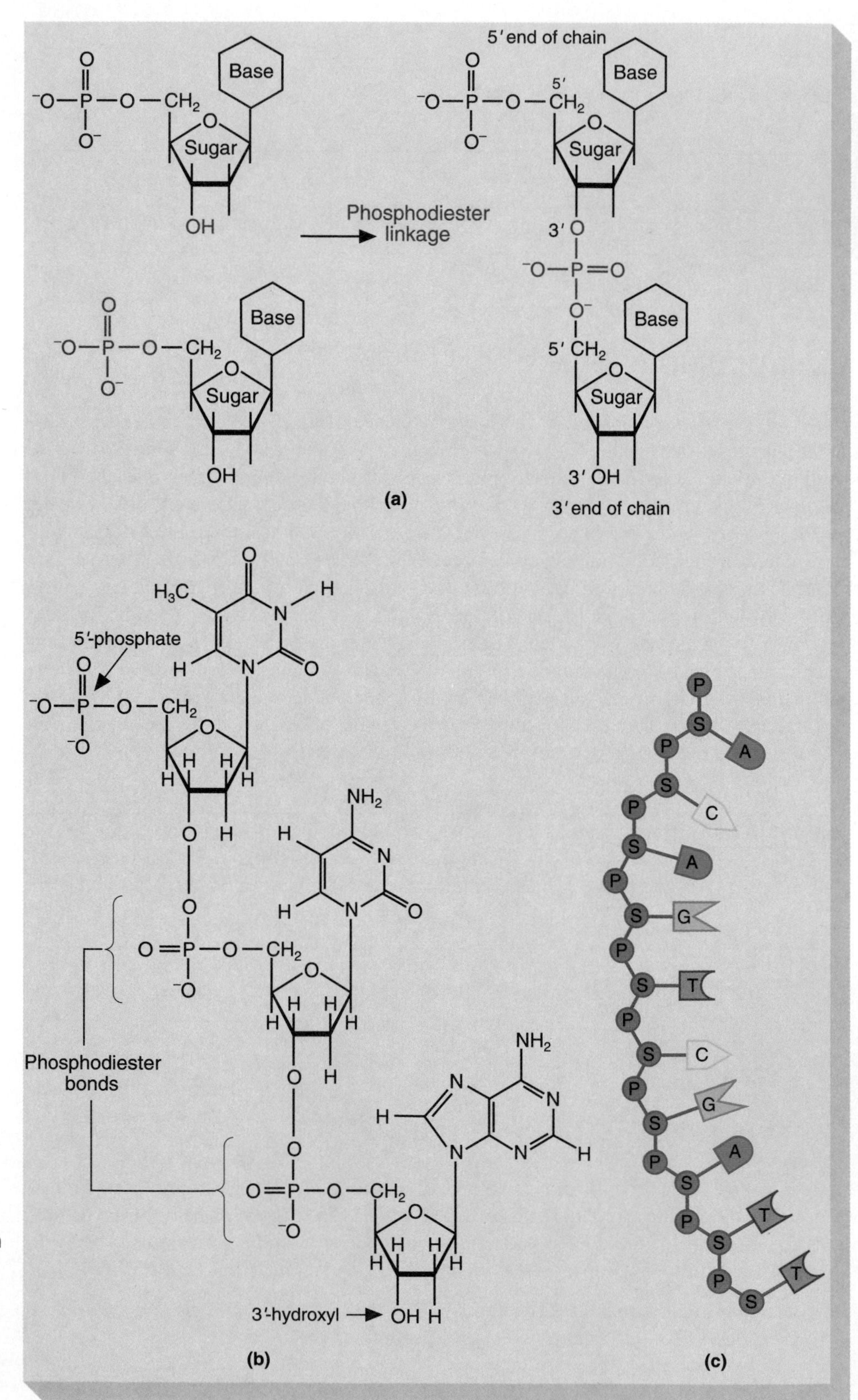

FIGURE 16.4
The covalent, primary structure of DNA. (a) The esterification reaction by which two nucleotides become linked by a phosphodiester bond. (b) A series of three covalently linked deoxyribonucleotides. (c) A stylized representation of the covalent primary structure of a longer piece of DNA.

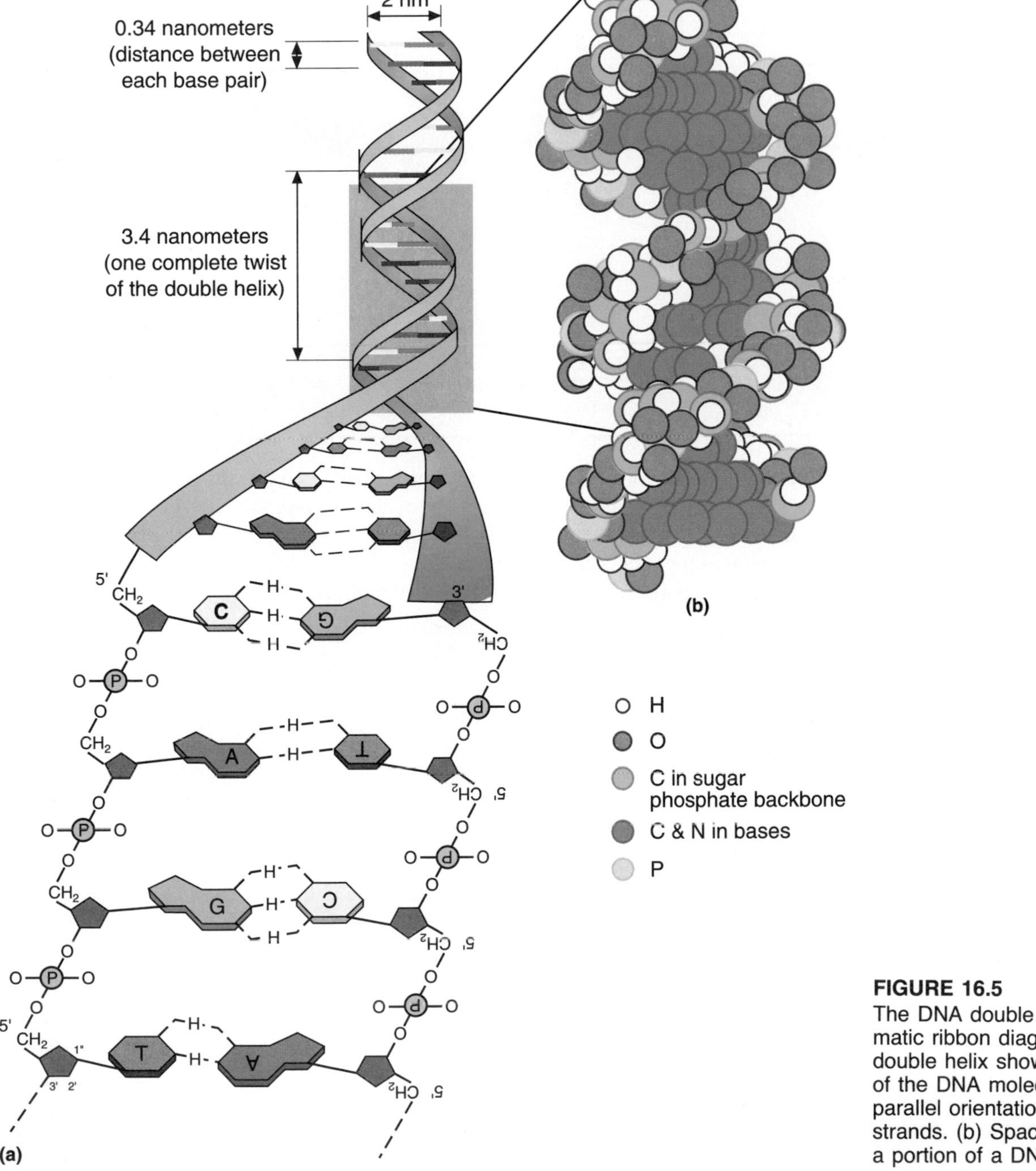

FIGURE 16.5
The DNA double helix. (a) Schematic ribbon diagram of the DNA double helix shows the dimensions of the DNA molecule and the antiparallel orientation of the two strands. (b) Space-filling model of a portion of a DNA molecule.

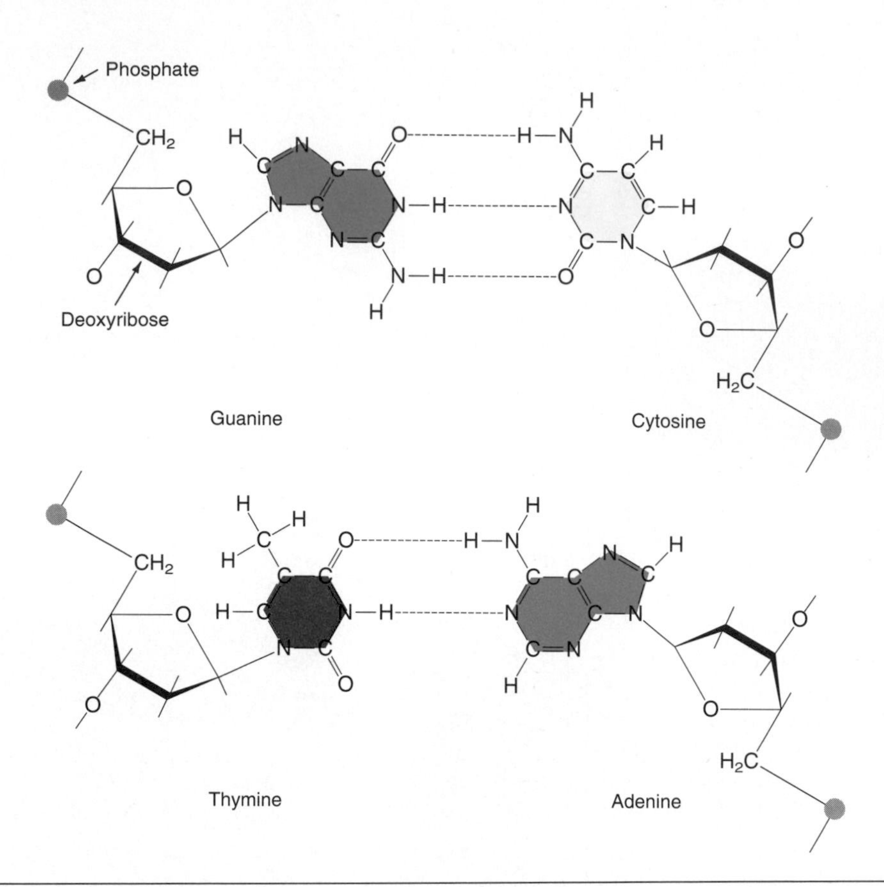

FIGURE 16.6
Base pairing in DNA. Adenine is always paired with thymine (A—T), and guanine is paired with cytosine (G—C).

to the long axis of the helix, like the stairs of the spiral staircase. Since adenine is always paired with thymine and guanine with cytosine, the two strands are **complementary** to one another. Therefore the sequence of bases on one strand *automatically* determines the sequence of the other. We shall see in Section 16.4 that this structural property is critical in the process of DNA replication.

- *Orientation of strands*. The two strands of the DNA double helix are **antiparallel** (head to tail). One strand advances in the $5' \longrightarrow 3'$ direction, and the second strand advances in the $3' \longrightarrow 5'$ direction. This orientation is necessary to allow formation of hydrogen-bonded base pairs (Figure 16.5).
- *Dimensions of DNA*. The double helix of DNA completes one turn every ten nucleotide residues. One complete turn of the helix advances the helix 3.4 nm, and each base pair therefore advances the helix by 0.34 nm. The diameter of the helix is 2.0 nm. This is dictated by the dimensions of the purine-pyrimidine base pairs.

Question 16.6 Write the complete structures of the base pairs that form between:

a. adenine and thymine
b. guanine and cytosine

16.4 DNA REPLICATION

DNA replication occurs each time a cell divides. In this way, all of the genetic information is passed from one generation to the next. As a result of base pairing, the sequence of bases along the sugar-phosphate backbone of each strand of DNA *automatically* specifies the sequence of bases in the complementary strand. The simplest mechanism of DNA replication involves separation of the two strands and synthesis of two new strands, each of which is complementary to an original, parental strand. Each strand, then, acts as a template, or form, upon which the second strand is synthesized.

Experimental evidence for this mechanism of DNA replication was provided by an experiment designed by Matthew Meselson and Franklin Stahl in 1958. *Escherichia coli* cells were grown in a medium in which $^{15}NH_4^+$ was the sole nitrogen source. Recall that the chemical behavior of the isotopes of a single element is identical. Therefore, $^{15}NH_4^+$ will function in the same way as the "normal" ammonium ion, $^{14}NH_4^+$. All cellular DNA thus contained this nonradioactive, heavy isotope. The cells containing only $^{15}NH_4^+$ were then added to a medium containing only the abundant isotope of nitrogen, $^{14}NH_4^+$, and were allowed to grow for one cycle of cell division. When the daughter DNA molecules were isolated and analyzed, it was found that each was made up of one strand of "heavy" DNA, the parental strand, and one strand of "light" DNA, the new daughter strand. After a second round of cell division, one half of the isolated DNA contained no ^{15}N, and one half of the isolated DNA contained a 50/50 mixture of ^{14}N and ^{15}N-labeled DNA (Figure 16.7). This demonstrated conclusively that each parental strand of the DNA molecule serves as the template for the synthesis of a daughter strand. Thus each newly synthesized DNA molecule is composed of one parental strand and one newly synthesized daughter strand.

All the genetic information of bacteria like *E. coli* is contained on a single circular piece of DNA made up of about three million nucleotides and called the chromosome. DNA replication in *E. coli* begins at a unique sequence on the circular chromosome known as the **replication origin.** Replication occurs bidirectionally at the rate of about 500 new nucleotide residues every second! The point at which the new deoxyribonucleotides are added to the growing daughter strand is known as the **replication fork** (Figure 16.8). It is here that the DNA has been opened to allow binding of the various proteins and enzymes responsible for DNA replication. Since DNA synthesis occurs bidirectionally, there are two replication forks moving in opposite directions. The circular DNA molecule thus resembles the Greek letter theta (Θ).

DNA synthesis in *E. coli* is catalyzed by a large number of enzymes. Some of these help to separate the two parental strands for the enzyme that is responsible for "reading" the sequence of nucleotides and synthesizing the complementary strand. The major enzyme involved in this function is DNA polymerase III. Not only does DNA polymerase III read the parental strand and produce the daughter strand, it also has the ability to proofread the newly synthesized strand to ensure that no errors are made. If an incorrect nucleotide has been inserted into the growing DNA strand, it is removed and replaced by the correct one.

FIGURE 16.7
Semiconservative replication of DNA was demonstrated by the experiment of Meselson and Stahl. Parent DNA molecules were produced by growing the cells in medium containing a heavy isotope of nitrogen (blue DNA molecules). The cultures were then switched to a medium containing the more abundant light isotope of nitrogen, ^{14}N. The daughter molecules produced in the first generation of growth in light medium were hybrid, consisting of one heavy (blue) strand and one light (green) strand. The conclusion of the experiment was that each parental DNA strand directs the synthesis of a new daughter strand.

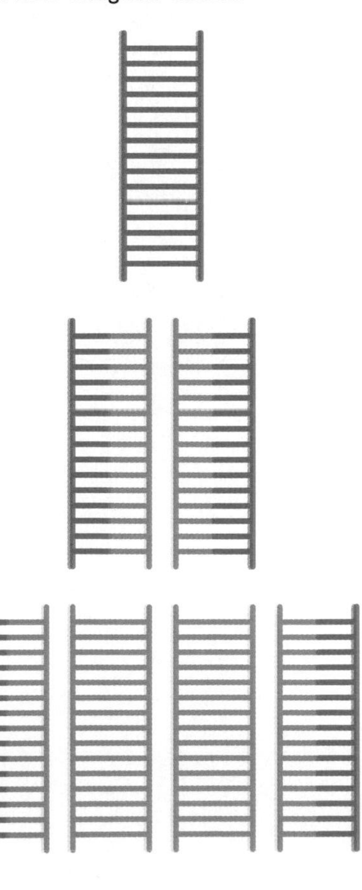

Question 16.7 Relate each of the following terms to the discussion of DNA structure:

a. double helix
b. template
c. the experiment of Meselson and Stahl

16.5 MUTATIONS IN DNA

Although DNA replication is very accurate, mistakes are sometimes made. Such a genetic change is called a **mutation** (L.: *mutare,* to change). An individual is constantly subject to a barrage of chemicals and other **mutagens** in the environment that introduce mutations in DNA. Mutations can be as simple as the change of a single nucleotide, or they may involve the deletion or rearrangement of entire sequences of DNA. Although mutations do provide the raw material for evolution, the process by which species change over time, many mutations are harmful to the organism. Consider for a moment that cancer results from a mutation in the DNA of a single cell that causes it to grow uncontrollably, threatening the life of the individual. The integrity and proper development of an organism depend upon maintaining a stable set of genetic information each time the cell divides.

Cells go to great lengths to provide mechanisms for DNA repair. We have already seen that DNA polymerase III corrects errors made during the process of replication. In addition, cells have a variety of other mechanisms to repair damage caused by mutagens. For instance, the simple yeast cell produces at least 50 DNA repair enzymes to maintain the integrity of its DNA. Despite this effort, mutations do occur, and they can have serious implications for the individual.

Consider the following examples of mutation in human populations. There are about 4000 human genetic diseases, many of which result from a mutation in a single gene. While some of these can be treated, many result in retardation, developmental abnormalities, physical deformities, and death. Currently, only 250 of these genetic disorders can be detected by genetic screening procedures. Sickle cell anemia is one such genetic disease. Recall that the hemoglobin protein is composed of two α-chains and two β-chains. There is a gene that carries the genetic code for each of these proteins. An individual with sickle cell anemia carries a mutation in the gene for the β-peptide. The mutation is so small that it seems almost insignificant; it is the change of just one of the 438 bases of the gene, from an adenine to a thymine. When the abnormal gene is translated into protein, the single alteration results in the substitution of a valine in place of glutamic acid in the protein. As a result, the protein folds improperly and does not carry oxygen as effectively. Individuals with sickle cell anemia suffer recurrent pain, general weakness, and ulcers in tissues that are not receiving enough oxygen for normal respiration. Inevitably, these individuals have a much decreased life expectancy.

As was stated above, most cancers result from one or more mutations in a single normal cell. These mutations result in the loss of normal growth control, causing the abnormal cell to proliferate. If that growth is not controlled or destroyed, it will result in the death of the individual. It has been estimated that one third of the children born in the late 1980s will contract cancer during their lifetimes and that perhaps one quarter will eventually die of cancer. We are exposed to many chemicals or mutagens that are known as **carcinogens** because they induce cancer by altering DNA structure. Sometimes we are exposed to a carcinogen by accident, but in some cases it is by choice. There are about 3000 chemical components in cigarette smoke, and several are potent mutagens. As a result, people who smoke have a much greater chance of lung cancer than those who don't.

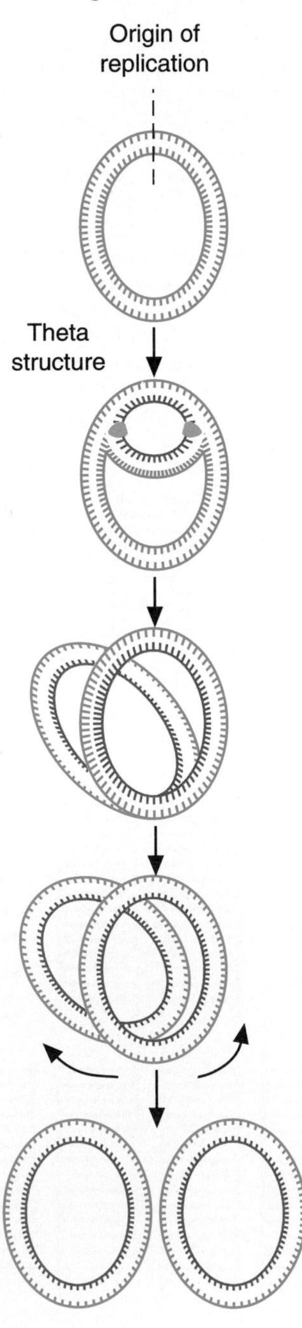

FIGURE 16.8
Summary of the events of DNA replication. The DNA replication enzyme complex recognizes a specific DNA sequence called the origin of replication. An enzyme in the replication complex opens the DNA strands so that the DNA polymerase will be able to read the sequence of nucleotides and produce a complementary daughter strand. Replication proceeds bidirectionally, with two growing forks proceeding in opposite directions. This generates a DNA molecule that resembles the Greek letter theta (Θ). Eventually, the two replication forks collide, releasing two circular daughter DNA molecules.

Ultraviolet light and thymine dimerization

One of the mutagens that causes damage to the DNA is ultraviolet (UV) light. The interaction of the ultraviolet light with the pyrimidine bases, especially adjacent thymines, results in one of the most important biochemical reactions of DNA, **photodimerization** (Figure 16.9). The most common product of dimerization is the **thymine dimer,** two adjacent thymine bases that are covalently linked to one another. As a result of dimerization there is no hydrogen bonding between opposite, complementary bases. Mutations occur when the UV damage repair mechanisms make an error and cause a change in the nucleotide sequence of the DNA.

In medicine the photodimerization reaction is used to advantage in hospitals where germicidal (UV) light is used to kill bacteria in the air and on environmental surfaces such as in a vacant operating room. This cell death is caused by photodimerization on a massive scale. Even though bacteria such as *E. coli* have four different mechanisms for the repair of UV light damage, the photodimerization is so extensive that it overwhelms the repair systems, and the cells die. Of course, the same type of photodimerization can occur in human cells as well. Lying out in the sun all day to acquire a fashionable tan exposes the skin to large amounts of UV light. This damages the skin by formation of many thymine dimers. Exposure to high levels of UV from sunlight or tanning booths has been linked to a rising incidence of skin cancer in human populations.

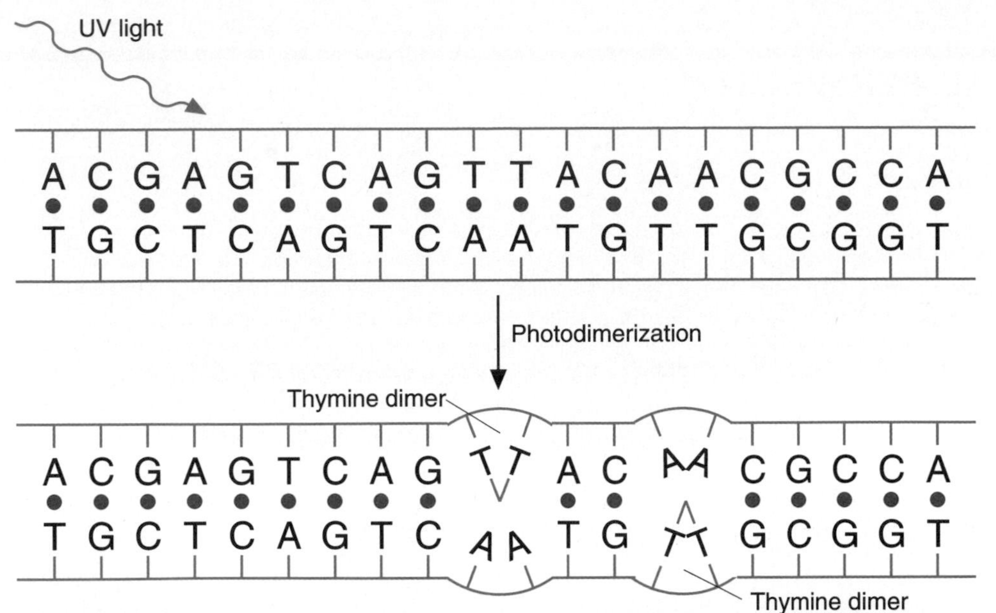
UV light
A C G A G T C A G T T A C A A C G C C A
T G C T C A G T C A A T G T T G C G G T
Photodimerization
Thymine dimer
A C G A G T C A G T T A C A A C G C C A
T G C T C A G T C A A T G T T G C G G T
Thymine dimer

FIGURE 16.9
Photodimerization of adjacent thymines results from the absorption of ultraviolet light.

Thymine dimer repair

One of the bacterial mechanisms for the repair of thymine dimers is called **photoreactivation.** This process uses visible light to repair thymine dimers formed by ultraviolet light. A *photoreactivation enzyme* binds to the distorted double helix at the site of the thymine dimer. The DNA-enzyme complex absorbs visible light and reverses the dimerization reaction. The photoreactivation enzyme dissociates from the repaired DNA, and the adenine and thymine base pair again normally (Figure 16.10).

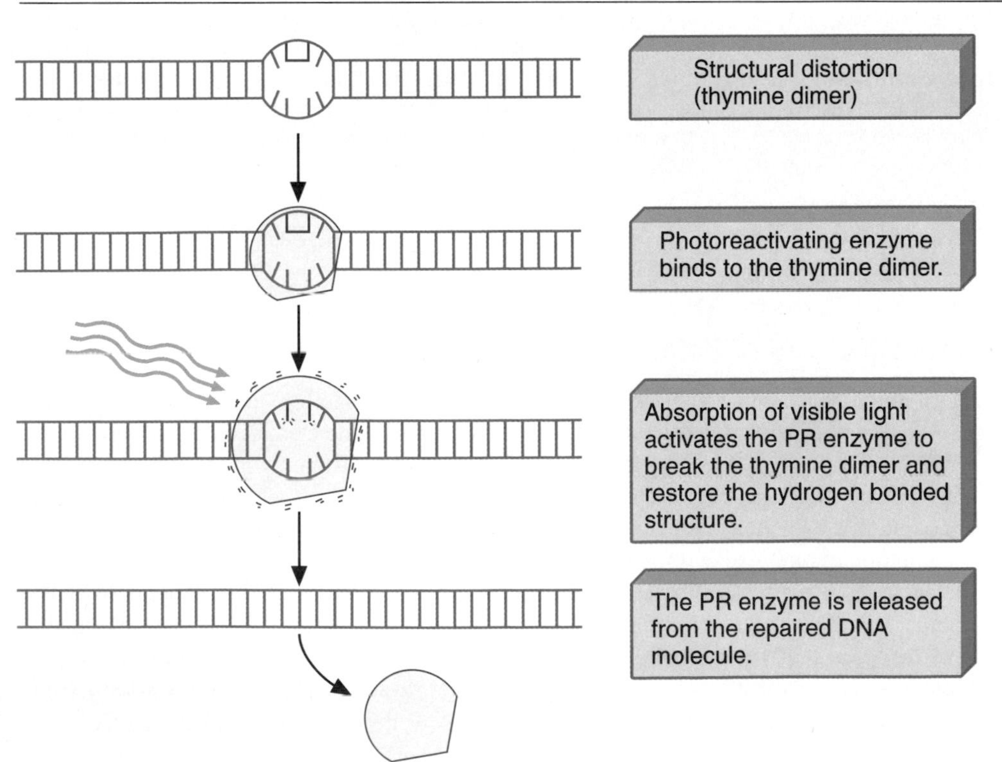
Structural distortion (thymine dimer)
Photoreactivating enzyme binds to the thymine dimer.
Absorption of visible light activates the PR enzyme to break the thymine dimer and restore the hydrogen bonded structure.
The PR enzyme is released from the repaired DNA molecule.

FIGURE 16.10
Mechanism of action of the photoreactivating enzyme repairing thymine dimers.

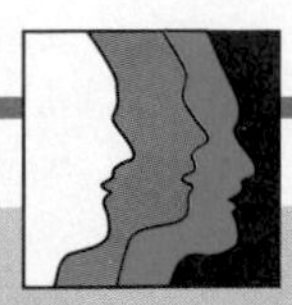

A CLINICAL PERSPECTIVE

The Ames Test for Carcinogens

Each day, we come into contact with a variety of chemicals, including insecticides, food additives, hair dyes, automobile emissions, and cigarette smoke. Some of these chemicals have the potential to cause cancer. How do we determine whether these agents are harmful, and, more particularly, how do we determine whether they cause cancer?

If we consider the example of cigarette smoke, we see that it can be years, even centuries, before a relationship is seen between a chemical and cancer. Europeans and Americans have been smoking since Sir Walter Raleigh introduced tobacco into England in the seventeenth century. However, not until three centuries later did physicians and scientists demonstrate the link between smoking and lung cancer. Obviously, this epidemiological approach takes too long, and too many people die. Alternatively, we can test chemicals by treating laboratory animals, such as mice, and observing them for various kinds of cancer. However, this, too, can take years, is expensive, and requires the sacrifice of many laboratory animals. How, then, can chemicals be tested for carcinogenicity (the ability to cause cancer) quickly and inexpensively? In the 1970s it was recognized that most carcinogens are also mutagens. That is, they cause cancer by causing mutations in the DNA, and the mutations cause the cells of the body to lose growth control. Bruce Ames, a bacterial geneticist, developed a test using mutants of the bacterium *Salmonella typhimurium* that can demonstrate in 48–72 hours whether a chemical is a mutagen and thus a suspected carcinogen.

Ames chose several mutants of *S. typhimurium* that cannot grow unless the amino acid histidine is added to the growth medium. The Ames Test involves subjecting these bacteria to a chemical and determining whether the chemical causes reversion of the mutation. In other words, the researcher is looking for a mutation that reverses the original mutation. When a reversion occurs, the bacteria will be able to grow in the absence of histidine.

The details of the test are shown in the accompanying figure. The mutant bacteria and the chemical to be tested are placed in a tube. As a negative control, a second tube containing the bacteria but lacking the chemical is prepared. The contents of the two tubes are placed on a solid agar growth medium that contains a very small amount of histidine. This added histidine is just enough to allow a mutation to occur but is not enough to produce colonies on the solid medium in the plates. The agar plates are incubated for 48–72 hours at 37°C and then examined for the presence of colonies on the surface. Any bacterial colony that forms must have arisen from a single cell that has regained the ability to make histidine; in other words, a reversion has occurred. If a large number of colonies are found on the surface of the experimental plate and a very few colonies are found on the negative control plate, it can be concluded that the chemical tested is a mutagen. It is therefore very likely that the chemical is also a carcinogen.

Some chemicals are not carcinogens until they are chemically altered by enzymes in the liver. Bacteria are unable to chemically modify chemicals in this way. To test for these potential carcinogens, Ames devised an additional test. In this case the *S. typhimurium* mutant and the chemical compound are mixed together with an extract prepared from mammalian livers. The enzymes in this extract can convert the potential carcinogen into a carcinogen, and bacterial revertants will then be formed. By using this additional test, the researcher can distinguish between chemicals that are direct mutagens and those that must

Consequences of defects in DNA repair

The human repair system for thymine dimers is more complex, requiring at least five enzymes. The first step in repair of the thymine dimer is the cleavage of the sugar-phosphate backbone of the DNA near the site of the damage. The enzyme that performs this cleavage of the sugar-phosphate backbone is called a repair endonuclease. If the gene encoding this enzyme is defective, thymine dimers cannot be repaired. The accumulation of mutations combined with a simultaneous decrease in the efficiency of DNA repair mechanisms leads to an increased incidence of cancer. For example, a mutation in the repair endonuclease gene and in other genes in the repair pathway results in the genetic skin disorder called **xeroderma pigmentosum.** People suffering from xeroderma pigmentosum are extremely sensitive to the ultraviolet rays of sunlight and develop multiple skin cancers, usually before the age of 20.

be activated after they have entered the body.

The Ames Test has greatly accelerated our ability to test new compounds for mutagenic and possibly carcinogenic effects. However, once the Ames Test identifies a mutagenic compound, testing in animals must be done to show conclusively that the compound also causes cancer.

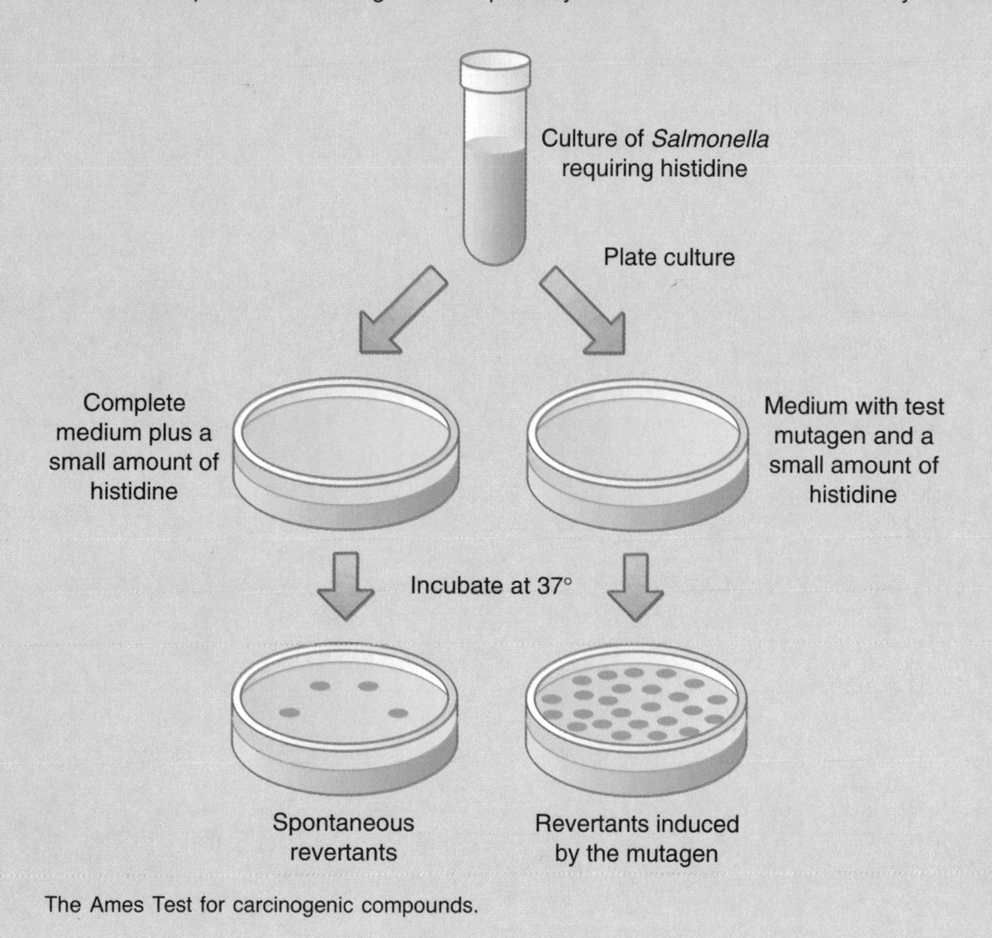

The Ames Test for carcinogenic compounds.

16.6 THE FLOW OF BIOLOGICAL INFORMATION

The **central dogma** of molecular biology states that the flow of genetic information contained in DNA is a one-way street that leads from DNA to RNA to protein (Figure 16.11). The process by which a single strand of DNA serves as a template for the synthesis of an RNA molecule is called **transcription.** The word ''transcription'' is derived from the Latin word *transcribere* and simply means ''to make a copy.'' Thus in this process, part of the information in the DNA is copied into a strand of RNA. The process by which the message is converted into protein is called **translation.** Unlike transcription the process of translation involves converting the information from one language to another. In this case the genetic information in the linear sequence of nucleotides is being translated into a protein, a linear sequence of amino acids. The expression of the informa-

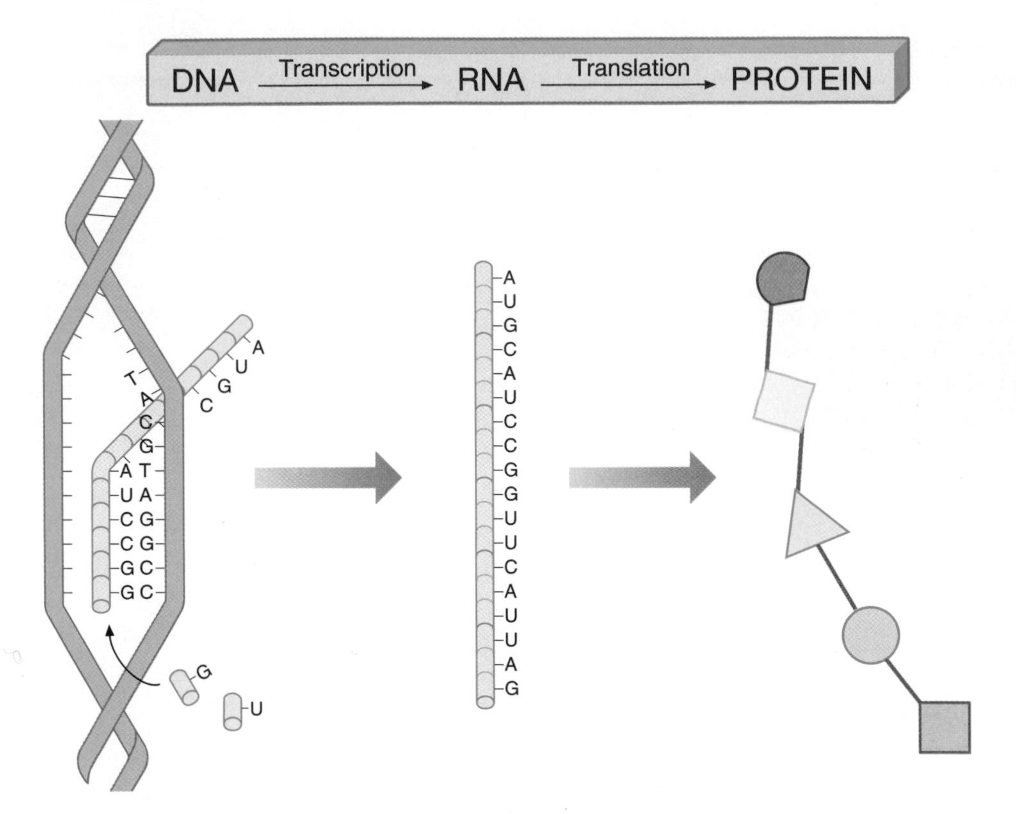

FIGURE 16.11
The central dogma of molecular biology. Information flows from DNA to RNA to protein. All the genetic information for all the structures and functions of the cell is carried by the DNA. The process of transcription copies a portion of the genetic information into a messenger RNA molecule. In the process of translation the genetic code carried by the mRNA is decoded into a linear sequence of amino acids. Each group of three nucleotides is called a codon, and each codes for a particular amino acid.

tion contained in DNA is fundamental to the growth, development, and maintenance of all organisms.

Question 16.8 Define each of the following terms:

a. Central dogma
b. Transcription
c. Translation

16.7 RNA MOLECULES

RNA molecules are produced by transcription of genes along the DNA. Some RNA species, called messenger RNA, carry the genetic code for a protein. Others, termed transfer RNA, serve to decode the sequence of ribonucleotides along the messenger RNA into the primary sequence of a protein. Still others play a structural role. Regardless of the function, all RNA species have many structural features in common.

Covalent structure of RNA and DNA

The covalent backbone of RNA consists of ribonucleotide residues linked by 3′-5′ phosphodiester bonds. The phosphodiester bonds are identical to those found in DNA (Figure 16.4). However, the structure of RNA differs in three major ways from that of DNA. First, the sugar unit of RNA molecules is ribose rather than 2′-deoxyribose. Second, uracil (U) replaces thymine as one of the four common bases. Finally, RNA molecules are single-stranded. Base pairing between adenine and uracil and between guanine and cytosine is possible for single-stranded RNA molecules and as a result these molecules can coil into three-dimensional structures.

Classes of RNA molecules

Ribonucleic acids are found in multiple copies and in multiple forms within a given cell. An RNA molecule is classified by its cellular location and by its function.

Messenger RNA (mRNA) carries the genetic information for a protein from DNA to the ribosomes. The **ribosomes** are particles composed of RNA and protein and are the sites of protein synthesis. A **messenger RNA** molecule is a faithful copy of the information encoded in a gene and, along with the other classes of RNA molecules, is synthesized by the process of transcription. mRNA molecules are synthesized in the nucleus and transported to the cytoplasm, where they are translated.

Ribosomal RNA (rRNA) is a structural and functional component of the ribosomes. The ribosomes serve as ''platforms'' on which protein synthesis occurs. **Ribosomal RNA** is a structural component of the ribosomes but also participates in the synthesis of the protein from the code carried by the mRNA. The cellular RNA is about 75% rRNA. There are three rRNA molecules in bacterial ribosomes and four in eukaryotic ribosomes. rRNA is also synthesized, and ribosomal subunits are assembled, in the nucleus. The subunits are then shuttled to the cytoplasm, where they participate in translation.

Transfer RNA (tRNA) is responsible for translating the genetic code of the mRNA into the primary sequence of amino acids in the protein. **Transfer RNA** molecules are very small, having only 70–90 nucleotide residues. They carry the needed amino acids to the ribosome for use in protein synthesis. Like the other classes of RNA, tRNA is produced by transcription in the nucleus and ferried to the cytoplasm.

The primary, secondary, and tertiary structures of one tRNA, the yeast phenylalanyl tRNA ($tRNA^{phe}$), are shown in Figure 16.12. This tRNA can carry only the amino acid

FIGURE 16.12
Structure of tRNA. (a) The primary structure of a tRNA is the linear sequence of ribonucleotides. Here we see the hydrogen bonded secondary structure of a tRNA showing the three loops and the amino acid accepting end. (b) The three-dimensional structure of a tRNA. (c) A schematic diagram that will be used to represent a tRNA throughout the chapter.

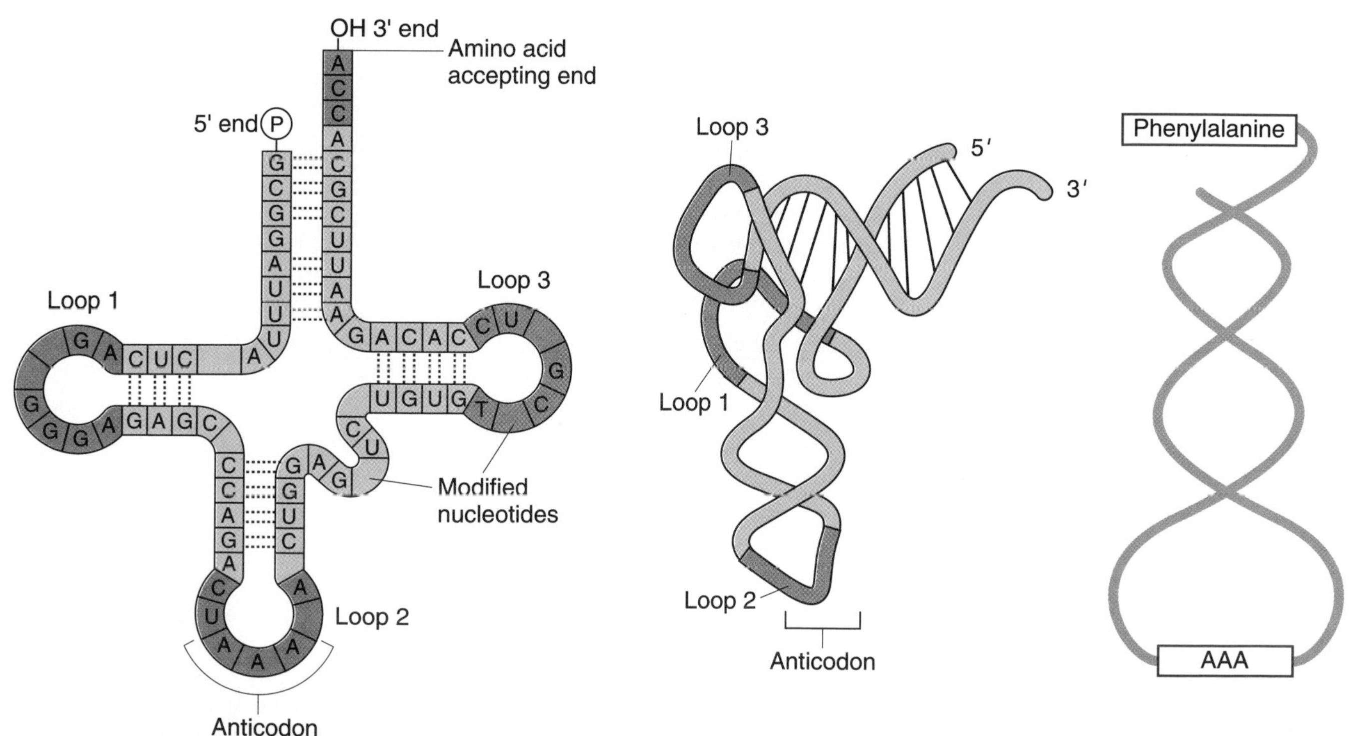

phenylalanine and transfers it to the growing peptide chain on the ribosome. In Figure 16.12a this tRNA is drawn as a "cloverleaf structure" in which base-paired regions are shown. The 3′ end of the tRNA molecule, which has the base sequence CCA in all tRNA molecules, is the site at which the amino acid is covalently attached. The three nucleotides (triplet) labeled as the **anticodon** at the base of the cloverleaf structure form hydrogen bonds to a **codon** (complementary sequence of bases) in a messenger RNA (mRNA) molecule on the surface of a ribosome during protein synthesis. In this way the correct amino acid is brought to the site of protein synthesis at the appropriate location in the growing peptide chain. Once the amino acid is transferred to the growing protein chain, the tRNA departs so that a new tRNA, carrying another amino acid, can attach and continue the process.

Section 16.11

The conformation of a tRNA molecule is shown in a ribbon diagram in Figure 16.12b. All tRNA molecules have this same basic shape, although the nucleotide sequence, in particular the anticodon sequence, varies from one tRNA to the next. We will use the sketch form shown in Figure 16.12c to depict the events of protein synthesis schematically.

Question 16.9 What are the major classes of RNA and their functions?

16.8 BIOSYNTHESIS OF MESSENGER RNA

As was noted above, the process by which RNA is copied from a DNA template is called *transcription,* and the product of this transcription process is one of the three classes of RNA molecules. The enzyme that catalyzes the transcription of DNA is RNA polymerase (Figure 16.13).

RNA polymerase binds to a specific nucleotide sequence known as the **promoter.** Promoter sequences indicate where the RNA polymerase should begin transcription and point it in "the right direction" to transcribe the correct strand of DNA. Unlike DNA replication, transcription produces a complementary copy of only one of the two DNA strands.

DNA is double-stranded at the moment when RNA polymerase binds tightly to a promoter sequence. In the initial interaction of DNA with RNA polymerase, the enzyme separates the two strands of DNA so that it can "read" the sequence of deoxyribonucleotides and synthesize a complementary sequence of ribonucleotides, the RNA molecule.

Chain elongation begins after the RNA polymerase has bound to the DNA and opened the double helix. Once again the DNA will serve as a template, and with each catalytic step the RNA polymerase "reads" a nucleotide of the DNA. It then transfers a complementary ribonucleotide to the end of the growing RNA chain by catalyzing the formation of a 3′–5′-phosphodiester bond between the 5′-phosphate of the incoming ribonucleotide and the 3′-hydroxyl of the last ribonucleotide of the growing RNA chain. The process of transcription is finished when the RNA polymerase reaches a specific termination sequence at the end of the gene. At this point, RNA polymerase dissociates from the DNA, releasing the newly synthesized RNA molecule (Figure 16.14).

Question 16.10 What is the function of the promoter in transcription?

16.9 EUKARYOTIC GENES

In prokaryotic organisms like the bacteria, which have no true membrane-bound nucleus, all of the nucleotide sequences of the gene on the DNA are copied, from beginning to end, and are found in the mRNA. These genes are said to be continuous. Until recently, there

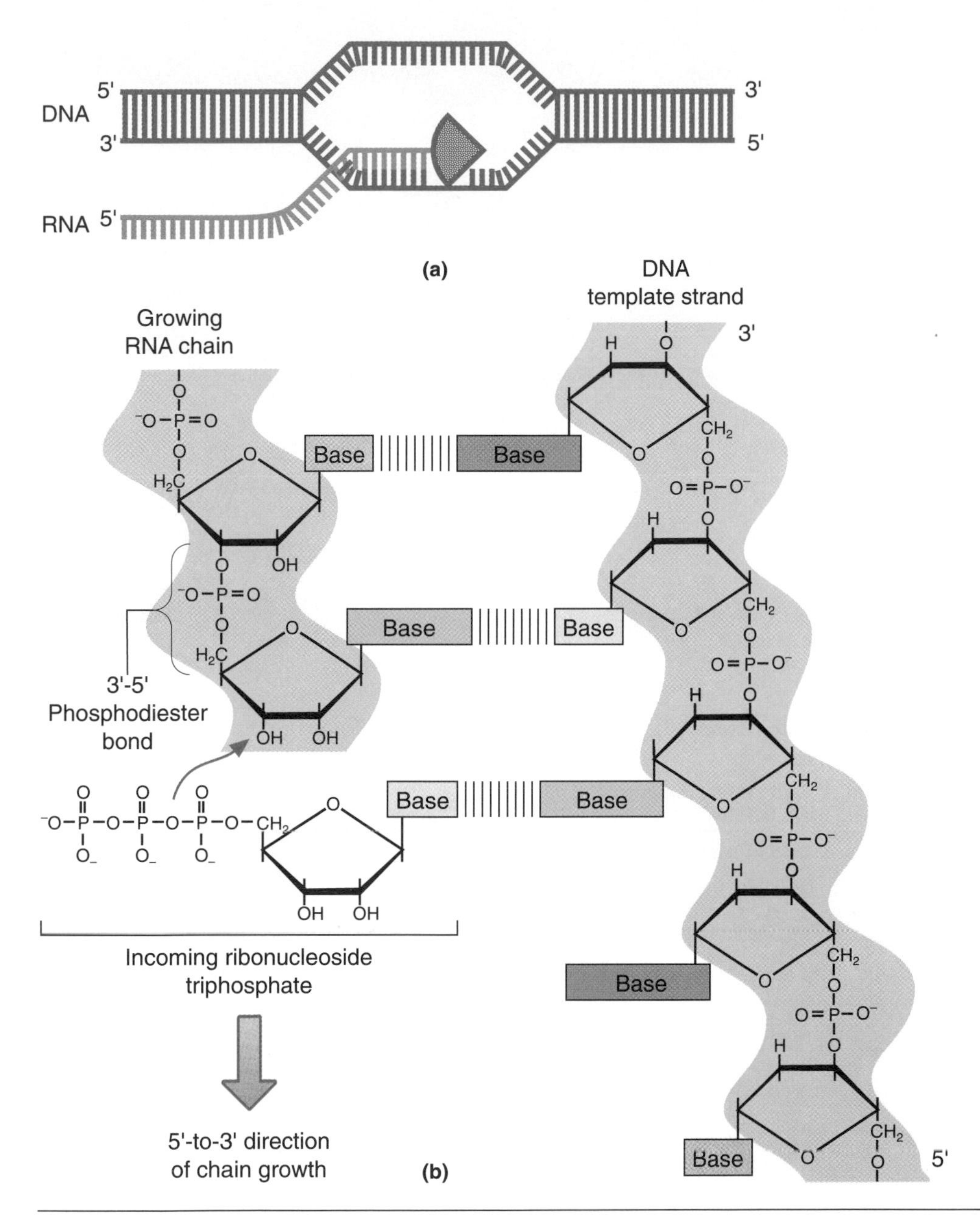

FIGURE 16.13
The reaction catalyzed by RNA polymerase. (a) RNA polymerase separates the two strands of DNA and produces an RNA copy of one of the two DNA strands. Transcription obeys the same laws of base pairing that are observed during DNA replication, except that the nitrogenous base uracil replaces thymine. (b) Phosphodiester bond formation occurs as a nucleotide is added to the growing RNA chain.

was no reason to suspect that the genes of eukaryotes, organisms with cells that have true membrane-bound nuclei, were any different. However, in 1977 it was found that eukaryotic genes are discontinuous. Within the sequence of the gene are regions of DNA that do not appear in the final mature mRNA! The initial mRNA, called the primary transcript, carries the sequences that encode the protein, but those sequences are interrupted periodically by additional, noncoding sequences, called *intervening sequences* or **introns.** Obviously, these must be enzymatically removed at a later stage of mRNA synthesis because their presence would interfere with synthesis of the correct protein. After removal the coding sequences of the mRNA, called *exons,* must be spliced together to give the final mature mRNA that will ultimately be translated.

The basic mechanism of mRNA splicing is summarized in Figure 16.15. The mRNA being spliced here is that which will carry the genetic code for β-globin.

In the mRNA splicing process the primary transcript, carrying the introns, folds into a loop that brings together the coding sequences to be spliced together. The process requires

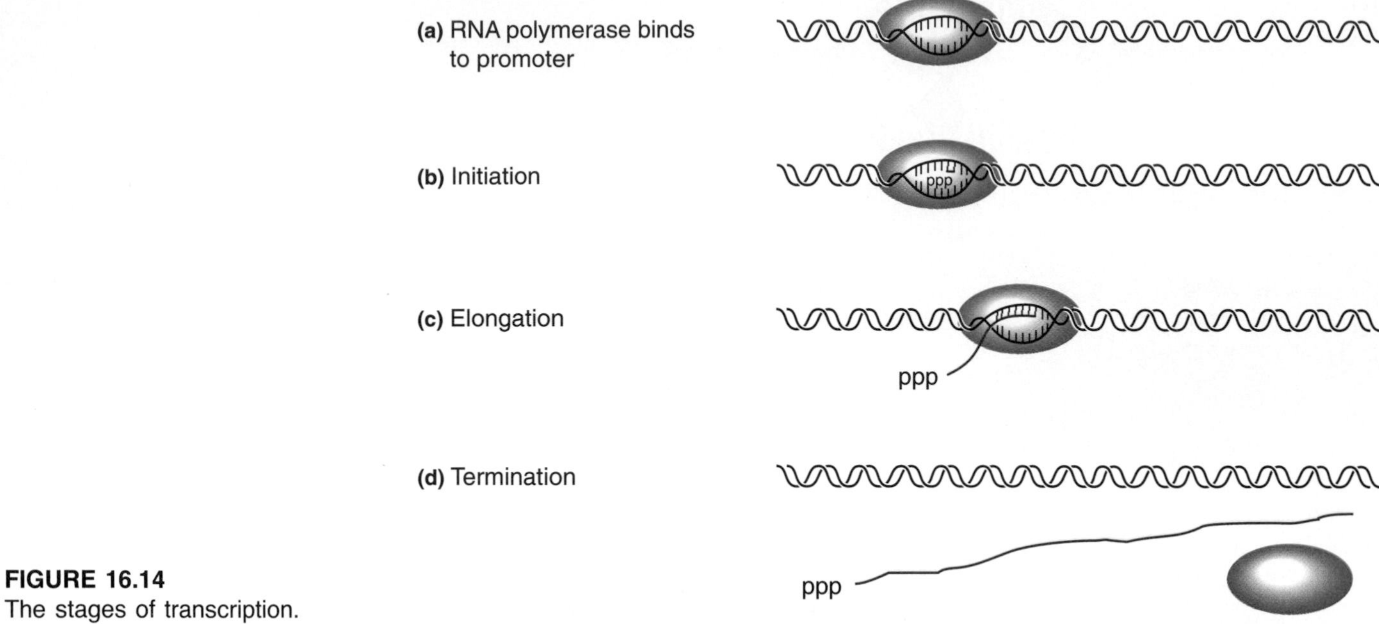

FIGURE 16.14
The stages of transcription.

two breaks in the phosphodiester backbone and perfect alignment during resealing. One can imagine how precise the process must be, since the shift of a single base will alter the ''message'' of the mRNA, resulting in synthesis of a nonfunctional protein.

One of the first eukaryotic genes shown to contain introns was the gene for the beta subunit of adult hemoglobin. We have devoted considerable attention to hemoglobin in our discussion of protein structure and mutation because of its physiological importance and because it reveals many principles of the relationship between protein conformation and function. We encounter hemoglobin again within the context of gene splicing. On the DNA the gene for β-hemoglobin is 1200 nucleotides long, but only 438 nucleotides carry the genetic information for protein. The remaining sequences are found in two introns of 116 and 646 nucleotides that are removed by splicing before translation. It is interesting that the larger intron is longer than the final β-globin mRNA. Introns range in size from

Section 11.10

FIGURE 16.15
Schematic diagram of mRNA splicing. (a) The β-globin gene contains protein coding exons, as well as noncoding sequences, called introns. (b) The primary transcript of the DNA carries both the introns and the exons. (c) The introns are looped out, the phosphodiester backbone of the mRNA is cut twice, and the pieces are tied together. (d) The final mature mRNA now carries only the coding sequences (exons) of the gene.

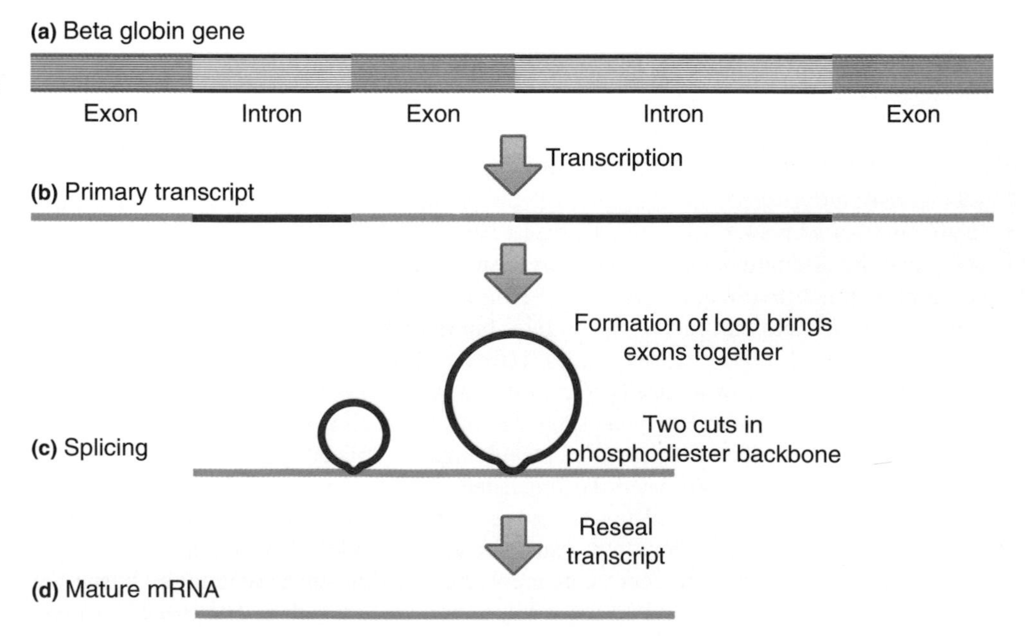

50 to 20,000 nucleotides in length, and there may be many throughout a gene. Thus a typical human gene might be 10–30 times longer than the final mature mRNA.

16.10 THE GENETIC CODE

The final mature mRNA carries the genetic code, which will direct the synthesis of a protein. Before we examine the process by which this occurs, we must be able to decipher the genetic code.

Imagine that you wish to send a message to a friend in a secret code. A cryptographer, suspecting mischief, examines the coded message, detects certain clues, and is able to deduce the message from the cryptogram. This is the ordinary procedure; the decoding proceeds from the cryptogram to the message. Now imagine a situation in which the cryptographer possesses the message but must guess the cryptogram. This is a much more difficult problem, since the possible combinations used to construct a code from the Roman alphabet with its 26 letters is limitless. Cracking the genetic code is an example of this latter type of problem. The message is known; it consists of the primary structure of proteins. But what is the code by which the message is conveyed? In 1954, George Gamow proposed that since there are only four ''letters'' in the DNA alphabet (A, T, G, and C) and since there are 20 amino acids, the genetic code must contain words made of at least three letters taken from the four letters in the DNA alphabet. How did he come to this conclusion? He reasoned that a code of two-letter words constructed from any combination of the four letters has a ''vocabulary'' of only 16 words (4^2). That is not enough to encode all 20 amino acids. A code of four-letter words gives 256 words (4^4), far more than are needed. A code of three-letter words, however, has a possible vocabulary of 64 words (4^3), sufficient to encode the 20 amino acids but not too excessive.

A series of elegant experiments proved that Gamow was correct by demonstrating that the genetic code is indeed a triplet code. Mutations that inserted (or deleted) one, two, or three nucleotides into a gene were introduced into the DNA of a bacterial virus. The researchers then looked for the protein encoded by that gene. When one or two nucleotides were inserted, no protein was produced. However, when a third base was inserted, the sense of the mRNA was restored, and the protein was made. You can imagine this experiment by using a sentence composed of only three-letter words. For instance,

Section 16.5

THE CAT RAN OUT

What happens to the sense of the sentence if we insert one letter?

THE FCA TRA NOU T

The reading frame of the sentence has been altered, and the sentence is now nonsense. Can we now restore the sense of the sentence by inserting a second letter?

THE FAC ATR ANO UT

No, we have not restored the sense of the sentence. Once again we have altered the reading frame, but since our code has only three-letter words, the sentence is still nonsense. If we now insert a third letter, it should restore the correct reading frame.

THE FAT CAT RAN OUT

Indeed, by inserting three new letters we have restored the sense of the message by restoring the reading frame. This is exactly the way in which the message of the mRNA is interpreted. Each group of three nucleotides in the sequence of the mRNA is called a codon, and each encodes a single amino acid. If the sequence is interrupted or changed, it can change the amino acid composition of the protein that is produced or even result in the production of no protein at all.

We noted above that a three-letter genetic code contains 64 words, called codons, but there are only 20 amino acids. Thus there are 44 more codons than are required to specify all of the amino acids incorporated in proteins. Three of the codons (UAG, UAA, and

UGA) specify termination signals for the process of translation. But this still leaves us with 41 additional codons. What is the function of the "extra" code words? Crick (recall Watson and Crick and the double helix) proposed that the genetic code is **degenerate.** The term degenerate is used to indicate that different triplet codons may serve as code words for the same amino acid.

The complete genetic code is shown in Figure 16.16. Several patterns emerge from an analysis of the genetic code.

- *The genetic code is highly degenerate*. Methionine and tryptophan are the only amino acids that have a single codon. All others have at least two codons, and serine and leucine each have six codons.
- *The genetic code is somewhat mutation-resistant*. For amino acids that have multiple codons the first two bases define the amino acid, and the third position is variable. Mutations in the third position therefore often have no effect upon the amino acid that is incorporated in a protein.

The substitution of a single base in a codon, known as a **point mutation,** is not always harmless, however. We have seen that a point mutation in the beta chain of

FIGURE 16.16
The genetic code. The table shows the possible codons found in mRNA. To read the universal biological language from this chart, find the first base in the column on the left, the second base from the row across the top, and the third base from the column to the right. This will direct you to one of the 64 squares in the matrix. Within that square you will find the codon and the amino acid that it specifies. In the cell this message is decoded by tRNA molecules like those shown to the right of the table.

FIRST BASE	SECOND BASE				THIRD BASE
	U	C	A	G	
U	UUU Phenylalanine	UCU Serine	UAU Tyrosine	UGU Cysteine	U
U	UUC Phenylalanine	UCC Serine	UAC Tyrosine	UGC Cysteine	C
U	UUA Leucine	UCA Serine	UAA STOP	UGA STOP	A
U	UUG Leucine	UCG Serine	UAG STOP	UGG Tryptophan	G
C	CUU Leucine	CCU Proline	CAU Histidine	CGU Arginine	U
C	CUC Leucine	CCC Proline	CAC Histidine	CGC Arginine	C
C	CUA Leucine	CCA Proline	CAA Glutamine	CGA Arginine	A
C	CUG Leucine	CCG Proline	CAG Glutamine	CGG Arginine	G
A	AUU Isoleucine	ACU Threonine	AAU Asparagine	AGU Serine	U
A	AUC Isoleucine	ACC Threonine	AAC Asparagine	AGC Serine	C
A	AUA Isoleucine	ACA Threonine	AAA Lysine	AGA Arginine	A
A	AUG *(START)* Methionine	ACG Threonine	AAG Lysine	AGG Arginine	G
G	GUU Valine	GCU Alanine	GAU Aspartic acid	GGU Glycine	U
G	GUC Valine	GCC Alanine	GAC Aspartic acid	GGC Glycine	C
G	GUA Valine	GCA Alanine	GAA Glutamic acid	GGA Glycine	A
G	GUG Valine	GCG Alanine	GAG Glutamic acid	GGG Glycine	G

Cysteine
ACG
Asparagine
UUG

hemoglobin results in sickle cell hemoglobin. Let's take a closer look at the related codons for glutamic acid and valine:

Glutamic Acid: GAA or GAG   Valine: GUG, GUC, GUA, GUG

A point mutation A $\longrightarrow$ U in the second nucleotide changes some codons for glutamic acid to a codon for valine:

$$\begin{array}{rcl} \text{GAA} & \longrightarrow & \text{G}\underline{\text{U}}\text{A} \\ \text{GAG} & \longrightarrow & \text{G}\underline{\text{U}}\text{G} \\ \text{glutamic acid codon} & \longrightarrow & \text{valine codon} \end{array}$$

This mutation in a single codon leads to the substitution of valine for glutamic acid in position 6 of the β-chain human hemoglobin, resulting in sickle cell hemoglobin. Taken as a whole, however, the genetic code is remarkably resistant to the synthesis of nonfunctional proteins as a result of point mutations.

Section 11.10

Question 16.11 Why is the genetic code said to be degenerate?

16.11 PROTEIN SYNTHESIS

Protein synthesis, also called translation, is carried out on ribosomes. Ribosomes are complexes of ribosomal RNA and proteins. Each ribosome is made up of two subunits, a small and a large ribosomal subunit (Figure 16.17a). About half of the mass of eukaryotic ribosomes is ribosomal RNA (rRNA). The small ribosomal subunit contains one rRNA molecule and 33 different ribosomal proteins, and the large subunit contains three rRNA molecules and about 49 different proteins.

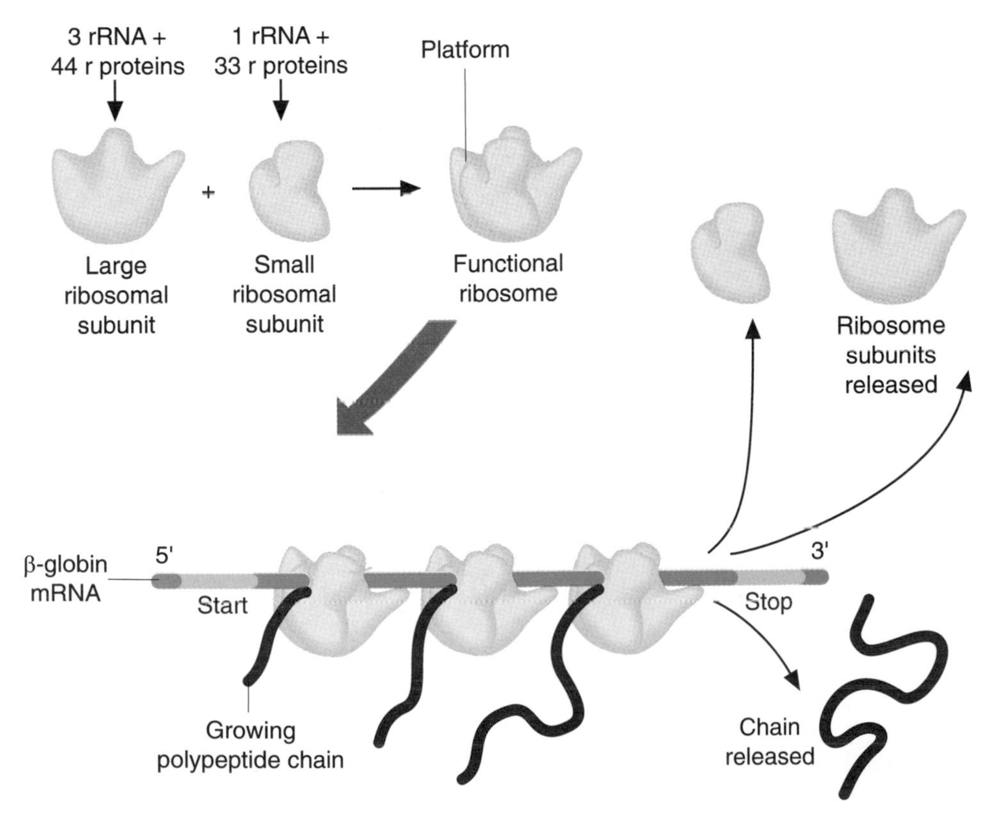

FIGURE 16.17
Structure of the ribosome. The large and small subunits form the functional complex in association with an mRNA molecule. A polyribosome translating the mRNA for a β-globin chain of hemoglobin.

Protein synthesis involves the simultaneous action of many ribosomes on a single mRNA molecule. These complexes of many ribosomes along a single mRNA are known as polyribosomes or **polysomes** (Figure 16.17b). Each ribosome is synthesizing one copy of the protein molecule encoded by the mRNA. Thus many copies of a protein are simultaneously produced as each ribosome moves along the mRNA molecule until the entire sequence has been translated. Individual ribosomes are separated by about 80 nucleotide residues on the mRNA. For example, the mRNA encoding hemoglobin contains about 500 nucleotide residues and is usually bound to five ribosomes while it is being translated (Figure 16.17b).

Question 16.12 What is a polysome?

Question 16.13 How many ribosomes would you expect to find in association with a 7200-nucleotide poliovirus mRNA?

The role of transfer RNA

The codons of mRNA must be sequentially read if the genetic message is to be translated into protein. The molecule that decodes the information in the mRNA molecule into the primary structure of a protein is transfer RNA (tRNA). To decode the genetic message into the primary sequence of a protein, the tRNA must faithfully perform two functions.

First, the tRNA must covalently bind one, and only one, specific amino acid. There is at least one transfer RNA for each amino acid. The common structural features of tRNA molecules, which contain 70–90 nucleotide residues, were summarized in Figure 16.12. All tRNA molecules have the sequence CCA at their 3′ ends. This is the site where the amino acid will be covalently attached to the tRNA molecule. Each tRNA is specifically recognized by the active site of an enzyme called an **aminoacyl synthetase.** This enzyme also recognizes the correct amino acid and covalently links the amino acid to the 3′ end of the tRNA molecule. Figure 16.18a shows the recognition of the amino acid methionine and its cognate tRNA by the methionyl tRNA synthetase. The resulting structure is called an **aminoacyl tRNA,** or charged tRNA, in this case, methionyl tRNA. In Figure 16.18b the reaction that results in the attachment of the aminoacyl group to the tRNA is shown.

FIGURE 16.18
Methionyl tRNA synthetase. (a) The enzyme specifically recognizes the amino acid methionine in one region of the active site and the methionyl tRNA in another. (b) The acylation reaction that results in a covalent linkage of the amino acid to the tRNA.

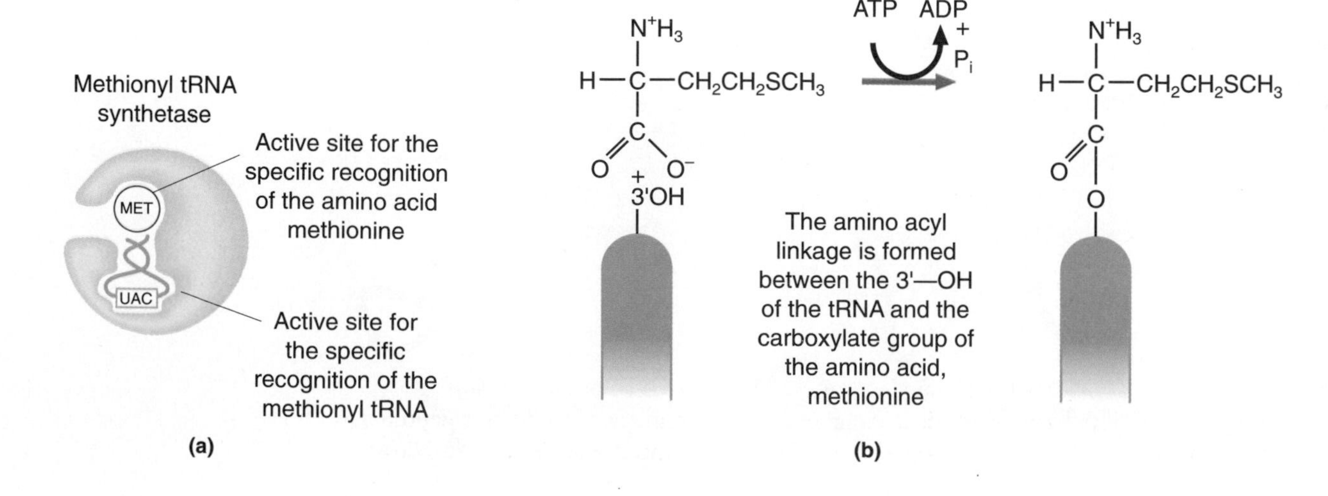

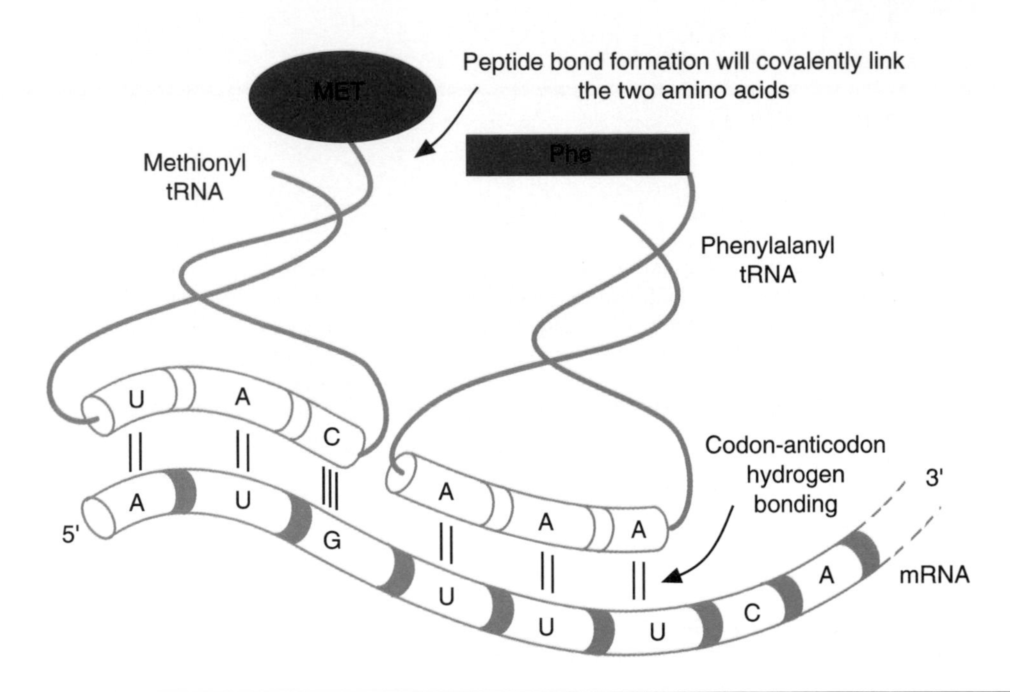

FIGURE 16.19
Codon-anticodon binding.

The amino acid is transferred from the tRNA to a growing polypeptide chain during protein synthesis.

Second, the tRNA must be able to recognize the appropriate codon on the mRNA that calls for that amino acid. This is mediated through a sequence of three bases called the anticodon, which is located at the bottom of the tRNA cloverleaf (refer to Figure 16.12). The anticodon sequence for each tRNA is complementary to the codon on the mRNA that specifies a particular amino acid. As you can see in Figure 16.19, the codon-anticodon complementary hydrogen bonding will bring the correct amino acid to the site of protein synthesis.

Question 16.14 How are codons related to anticodons?

The process of protein synthesis

The first stage of protein synthesis is *initiation*. Proteins called **initiation factors** are required to mediate the formation of a translation complex composed of an mRNA molecule, the small and large ribosomal subunits, and the initiator tRNA. This initiator tRNA recognizes the codon AUG, the **initiation codon,** and is charged with the modified amino acid, N-formylmethionine. In bacteria, N-formylmethionine is always the first amino acid in a protein chain, while in eukaryotes the initiator tRNA is charged with methionine.

The ribosome has two sites for binding tRNA molecules. The first site, called the **peptidyl tRNA binding site** (P-site), holds the growing peptide bound to a tRNA molecule. The second site, called the **aminoacyl tRNA binding site** (A-site), holds the aminoacyl, or charged, tRNA carrying the next amino acid to be added to the peptide chain. Each of the tRNA molecules is hydrogen-bonded to the mRNA molecule by codon-anticodon complementarity. The entire complex is further stabilized by the fact that the mRNA is also bound to the ribosome. Figure 16.20a shows the series of events that result in the formation of the initiation complex. The initiator methionyl tRNA occupies the P-site in this complex.

The second stage of translation is **chain elongation.** This occurs in three steps that are repeated until protein synthesis is complete. We enter the action after a tripeptide has already been assembled, and a peptidyl-tRNA occupies the P-site (Figure 16.20b).

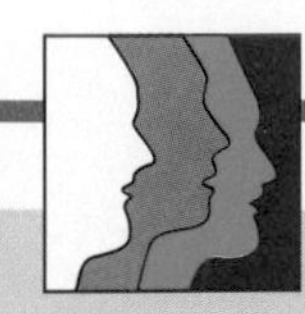

A HUMAN PERSPECTIVE

The New Genetics and Human Genetic Disease

It is estimated that about half a million babies are born each year suffering from one of almost 4000 known genetic defects. The care and treatment of these children is a tremendous drain on both the parents and the health care systems.

The revolution in molecular genetics that began in the early 1960s has led to important advances in our understanding of human genetics and human genetic diseases. As a result, genetic counseling has become a major component of the health care system and through genetic screening about 250 of these genetic defects can be detected in fetuses, children, and adults.

Recombinant DNA technology has been a critical component of these advances. For instance, the genes for human insulin, human growth hormone, and the blood clotting protein factor VIII have been cloned, and the protein products have been purified and are available for the treatment of diabetes, pituitary dwarfism, and hemophilia. The significance of the availability of these human gene products is made more evident when we consider some of the problems faced by sufferers of these genetic defects.

Many people who suffer from **diabetes** are insulin-dependent; in other words, they require injections of insulin to control their blood glucose levels. Previously, only insulin from the pancreas of pigs and cows was available. These hormones differ from the human hormone by only a few amino acids, but this slight difference caused many diabetics to develop immunity to their insulin. As a result, the injected hormone could not control blood glucose. This problem has been alleviated by the fact that the human insulin gene was cloned and human insulin (Humalin) was approved for human use in 1982.

At least 3500 children in the United States suffer from **pituitary dwarfism.** Without treatment these children can expect to grow to no more than four feet in height. However, treatment with **human growth hormone** (HGH) purified from the pituitary glands of cadavers allowed these children to attain nearly normal height. In 1985 a medical crisis arose. Some of the cadavers were infected with the virus causing Creutzfeldt-Jacob syndrome, a degenerative brain infection, and children who had received HGH from these cadavers were contracting this fatal disease. Recombinant DNA–produced HGH was immediately approved by the FDA for safe treatment of these individuals.

It is estimated that one in 10,000 males is born with classical **hemophilia.** The disease primarily affects males because the gene for the defect is on the X chromosome. Males have only one X chromosome, thus a mutation in the Factor VIII gene on that chromosome results in hemophilia. Females must have the mutation on both X chromosomes. Individuals with hemophilia were faced with a dilemma similar to the children with pituitary dwarfism. **Factor VIII,** which is needed to arrest severe bleeding from even minor wounds, is purified from human blood plasma. In the early 1980s this plasma supply became contaminated with the virus causing Acquired Immune Deficiency Syndrome (AIDS), with the result that a great many hemophiliacs contracted this fatal disease. Recombinant DNA–produced factor VIII is currently being tested for use by hemophiliacs.

These examples clearly demonstrate the value of recombinant DNA technology to humans. The power of this amazing technology is apparent when one looks at the production of human **insulin.** In the cells of the pancreas the synthesis of insulin involves a long, complex series of steps that are outlined in the accompanying figure. Transcription produces a messenger RNA that encodes a protein called **preproinsulin.** The primary transcript is longer than the mature mRNA because the preproinsulin gene has an intron. The preproinsulin protein is much longer than the final hormone, insulin. Since the protein must be secreted into the bloodstream to exert its effect, there is a signal sequence at the amino terminus of the protein, the "presequence," that directs transport of the polypeptide chain across cellular membranes. The signal sequence is immediately removed, and the resulting peptide, **proinsulin,** is encased and stored within a membrane vesicle. Proinsulin is then cleaved in two places by a proteolytic enzyme. This releases a 33-amino-acid fragment, the C chain, and produces the A and B chains, which are held together by disulfide bonds.

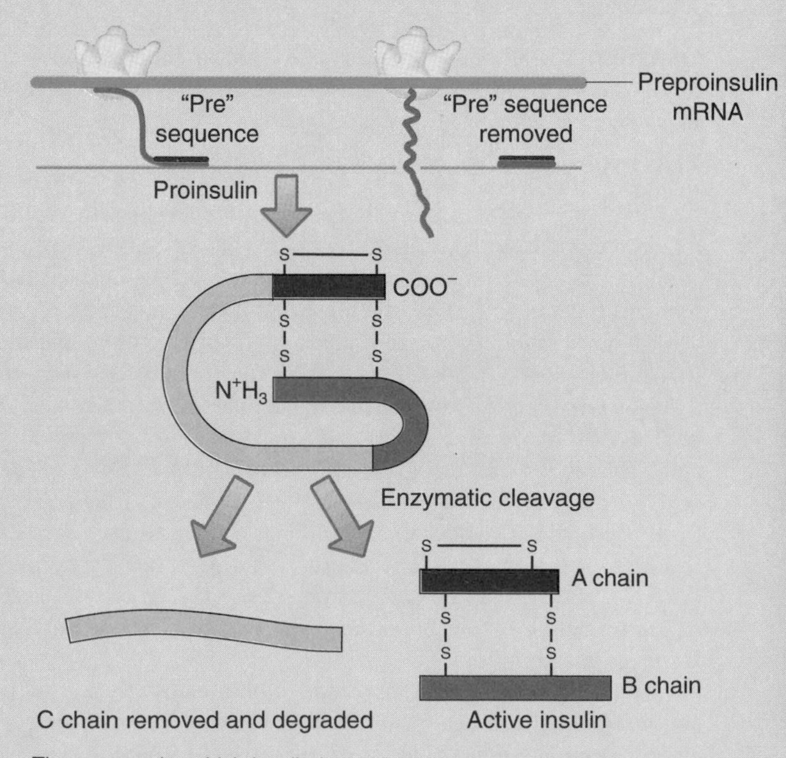

The process by which insulin is produced in the pancreas.

Consider for a moment the obstacles that must be overcome to get expression of this gene in a bacterial cell. If we start with the preproinsulin gene from the human chromosome, we have the problem that bacteria can't remove the intron. Neither do bacteria have the proper proteolytic enzyme to produce insulin from proinsulin. In addition, if we use the natural insulin gene promoter, the bacterial RNA polymerase will never recognize it, so we won't get any transcription at all. It seems remarkable that genetic engineers were able to solve all of these problems to produce insulin from bacterial cells.

The cloning of the insulin A and B chains is shown in the figure below. Since the complete amino acid sequences of the A and B chains were known, scientists simply predicted what the genetic code must be that would encode insulin. They synthesized genes for the A and B chains in a test tube. This was fairly easy to do because the genes are very small. They placed an AUG initiator codon at the beginning of each chain and a UGA terminator codon at the end, and then they inserted each gene into a bacterial plasmid. A plasmid is an extra piece of circular DNA that can replicate and be expressed in a bacterial cell. The site of insertion was near a bacterial promoter that can be turned on or off at will by the simple addition or removal of an inducer compound. Each plasmid was then inserted into the bacterium *Escherichia coli,* and when the chemical inducer was added, the bacterial RNA polymerase recognized the plasmid promoter and began producing enormous quantities of the A or B chains of insulin. The A and B chains were purified and treated to remove any bacterial protein, and finally they were mixed and allowed to self-assemble into functional insulin protein.

Genetic engineers are trying many different strategies to clone genes to treat and detect human genetic diseases. The eventual goal of this type of research would be to actually replace the mutated gene that is responsible for the genetic defect with a normal functional copy of the gene. This ultimate goal is many years in the future. We must first have a better understanding of the way in which normal genes are controlled by the cell before we can begin to tamper with genes in the intact organism.

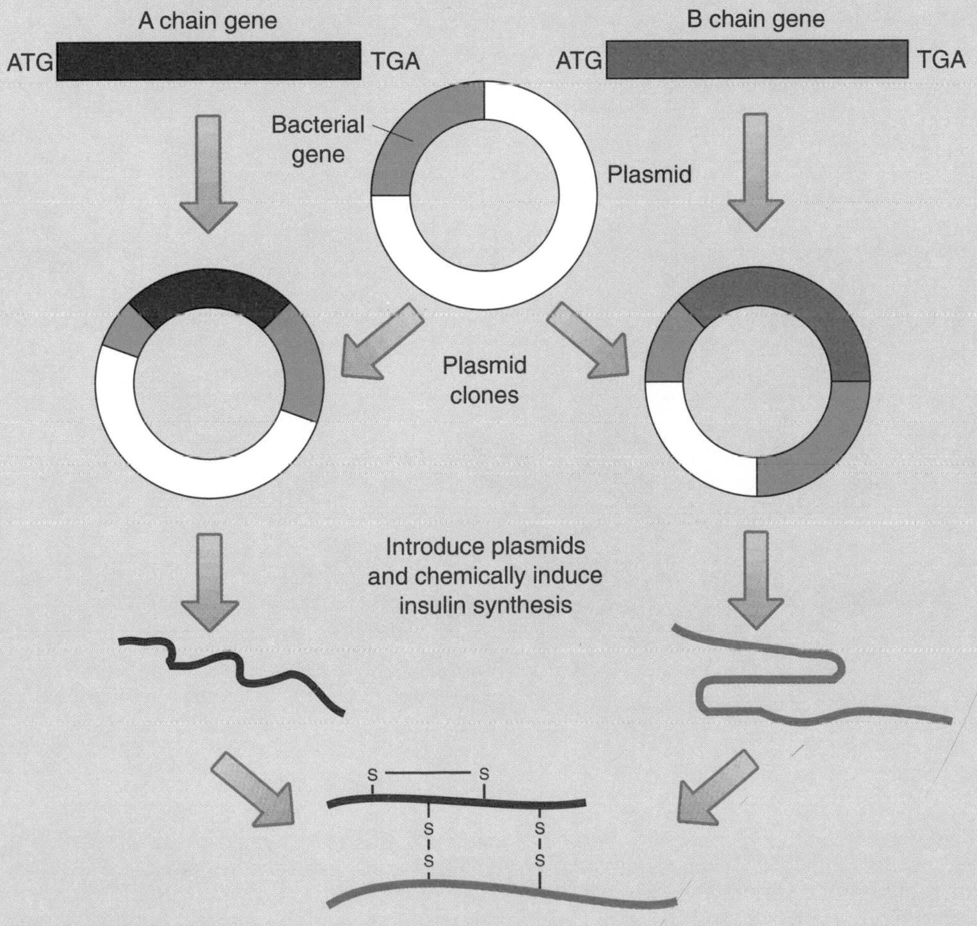

The cloning of synthetic insulin genes.

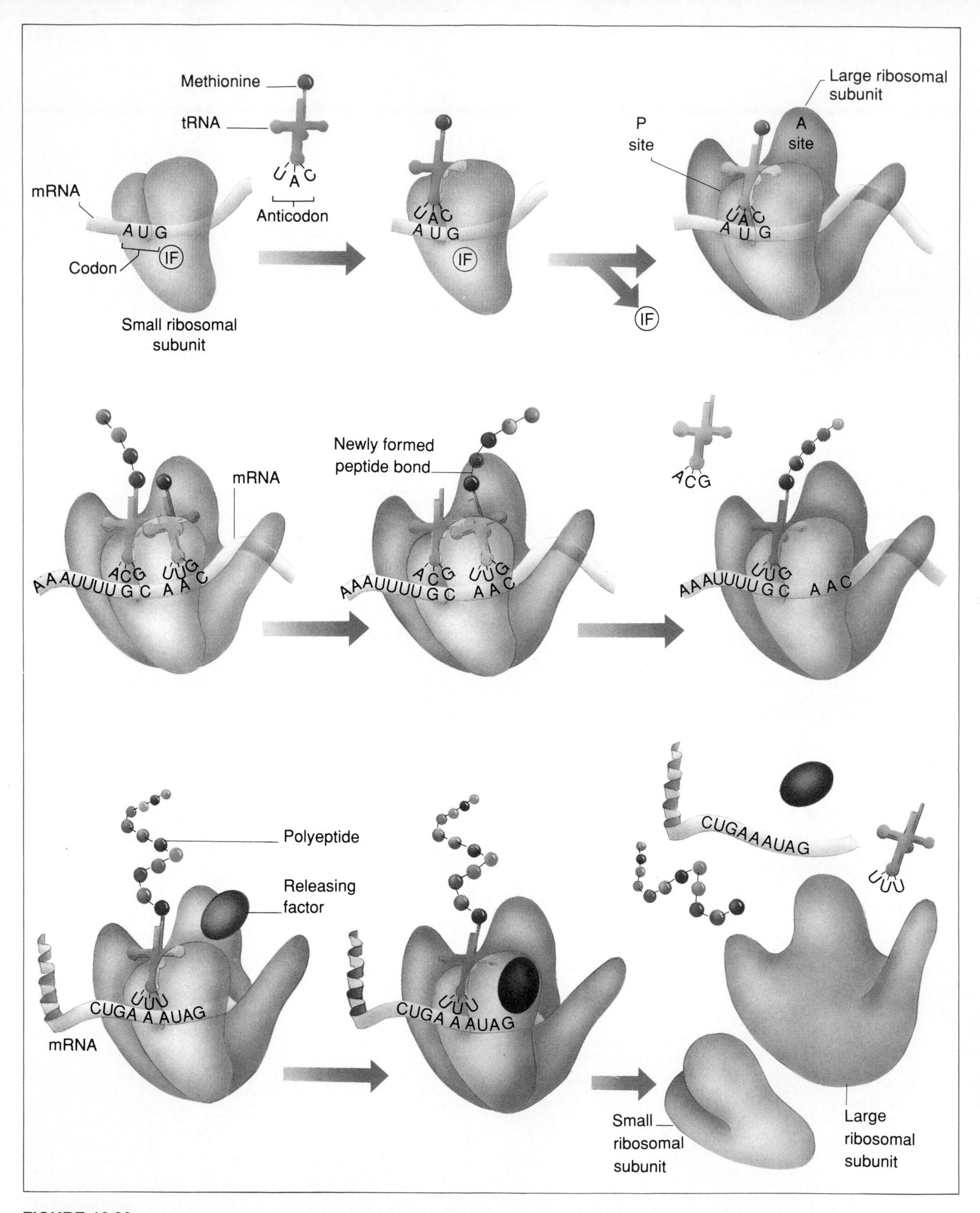

FIGURE 16.20
Formation of an initiation complex sets protein synthesis in motion. The mRNA and proteins called initiation factors bind to the small ribosomal subunit. Next, a charged methionyl tRNA molecule binds, and finally, the initiation factors are released and the large subunit binds. The elongation phase of protein synthesis involves addition of new amino acid residues to the C-terminus of the growing peptide. An amino acyl-tRNA molecule binds at the empty A-site, and the peptide bond is formed. The uncharged tRNA molecule is released, and the peptidyl-tRNA is shifted to the P-site as the ribosome moves along the mRNA. Termination of protein synthesis occurs when a release factor binds the stop codon on mRNA. This leads to the hydrolysis of the ester bond linking the peptide to the peptidyl-tRNA molecule in the P-site. The ribosome then dissociates into its two subunits, releasing the mRNA and the newly synthesized peptide.

The first event is binding of an amino acyl-tRNA molecule to the empty A-site. Next, peptide bond formation occurs. This is catalyzed by an enzyme called *peptidyl transferase* that is part of the ribosome. Now the peptide chain is shifted to the tRNA that occupies the A-site. Finally, the uncharged tRNA molecule is discharged, and the ribosome changes positions so that the next codon on the mRNA occupies the A-site. This movement of the ribosome is called **translocation.** The process shifts the new peptidyl-tRNA from the A-site to the P-site.

The last stage of translation is *termination*. There are three codons—UAA, UAG, and UGA—for which there are no corresponding tRNA molecules. When one of these **termination codons** is encountered, translation is terminated. A **release factor** binds the empty A-site. The ribosomal protein that had previously catalyzed peptide bond formation hydrolyzes the ester link between the peptidyl-tRNA and the last residue of the newly synthesized protein (Figure 16.20c). At this point the tRNA, the newly synthesized peptide, and the two ribosomal subunits are released.

Question 16.15 What are the functions of the ribosomal P-site and the A-site in protein synthesis?

The peptide that is released following translation is not necessarily in its final functional form. In some cases the peptide is cleaved and processed, as with insulin. (See "A Human Perspective: The New Genetics and Human Genetic Disease.") Sometimes it must associate with other peptides to form a functional protein, as in the case of hemoglobin. Cellular enzymes add carbohydrate or lipid groups to some proteins. These final modifications are specific for particular proteins and, like the sequence of the protein itself, are directed by the cellular genetic information.

Section 11.10

SUMMARY

The genetic information is ultimately responsible for every trait of an organism. The structural elements that create the familiar form of an organism and the enzymes that drive the energy-generating and biosynthetic reactions are all products of the genes. DNA, the molecule that carries the genetic information, is a linear, double-stranded polymer of nucleotides.

The basic components of all nucleic acids include pyrimidine and purine bases, five-carbon sugars, and phosphate groups. The major purines are adenine and guanine; the major pyrimidines are thymine, uracil, and cytosine. Nucleosides are β-N-glycosides formed between a purine or pyrimidine and a sugar. The sugar is 2′-deoxyribose in DNA and ribose in RNA. Nucleotides, the monomers from which nucleic acids are made, are sugar-phosphate esters of nucleosides.

DNA is a double-stranded polymer of 2′-deoxyribonucleotide residues linked by 3′–5′ phosphodiester bonds. The two antiparallel strands are wound around each other to form a double helix. Adenine is always base-paired with thymine, and guanine is always base-paired with cytosine in DNA. Base-pairing is stabilized by hydrogen bonds between adenine and thymine or guanine and cytosine.

In DNA replication, each strand of the parent molecule serves as a template for the synthesis of a complementary daughter strand. DNA polymerase catalyzes DNA replication on a DNA template and proofreads the sequences that it has synthesized to detect and repair any errors.

Environmental hazards continually damage DNA and produce mutations in DNA that are repaired by repair enzymes. Thymine dimers are produced as a result of ultraviolet irradiation. In bacteria thymine dimers are repaired by photoreactivation. In human beings the genetic disease xeroderma pigmentosum results from the failure of the repair system for thymine dimers.

Transcription is the process in which the genetic message in DNA is copied by RNA polymerase to produce three types of RNA. These are messenger RNA, which carries the code for a protein; transfer RNA, which decodes the genetic code within the mRNA; and ribosomal RNA, which forms a structural and functional part of the ribosomes.

The genetic code consists of 64 nucleotide triplet codons. Three of these—UAA, UAG, and UGA—signal the termination of protein synthesis, and one, AUG, is a "start" signal. The remainder encode particular amino acids. The sequence of codons in mRNA is colinear with the sequence of amino acid residues in the protein synthesized from the mRNA template.

Translation, the process of protein synthesis, occurs on ribosomes, which are complexes of RNA and protein that bind the mRNA and provide a platform on which translation occurs. During translation the sequence of nucleotides in the mRNA is translated into the sequence of amino acids in the protein by the transfer RNA molecules. This is specified by the interaction between the codon on the mRNA and the anticodon on the tRNA. The major steps of protein synthesis are initiation, chain elongation, and chain termination.

GLOSSARY OF KEY TERMS

Acquired Immune Deficiency Syndrome (AIDS) (16.1) a set of symptoms that results from infection with the human immunodeficiency virus. These symptoms include a variety of opportunistic infections and cancers, an impaired immune response, and dementia.

adenine (16.1) one of the purine nitrogenous bases in DNA and RNA.

adenosine (16.1) a nucleoside formed from adenine and ribose.

aminoacyl synthetase (16.11) an enzyme that recognizes one tRNA and covalently links the appropriate amino acid to it.

aminoacyl tRNA (16.11) a charged tRNA. The transfer RNA is covalently linked to the correct amino acid.

aminoacyl tRNA binding site of ribosome (16.11) a pocket on the surface of a ribosome that holds the amino acyl tRNA during translation.

anticodon (16.7) a sequence of three ribonucleotides on a tRNA that are complementary to a codon on the mRNA. Codon-anticodon binding results in the tRNA bringing the correct amino acid to the site of protein synthesis.

antiparallel (16.3) a term describing the polarities of the two strands of the DNA double helix; on one strand the sugar-phosphate backbone advances in the $5' \longrightarrow 3'$ direction, while on the opposite, complementary strand the sugar phosphate backbone advances in the $3' \longrightarrow 5'$ direction.

AZT (16.1) 3′-azido-2′,3′-dideoxythymidine, a nucleoside analogue of deoxythymidine that is used to inhibit the growth of the human immunodeficiency virus.

base pair (16.3) a hydrogen-bonded pair of bases within the DNA double helix. The standard base pairs always involve a purine and a pyrimidine; in particular, adenine always base pairs with thymine and cytosine with guanine.

carcinogen (16.5) any chemical or physical agent that causes mutations in the DNA that lead to uncontrolled cell growth or cancer.

central dogma (16.6) a statement of the directional transfer ot the genetic information in cells: DNA $\longrightarrow$ RNA $\longrightarrow$ protein.

chain elongation (16.11) the sequential addition of amino acids to the growing peptide chain during translation.

codon (16.7) a group of three ribonucleotides on the mRNA that specifies the addition of a specific amino acid onto the growing peptide chain.

complementary strands (16.3) the opposite strands of the double helix are hydrogen-bonded to one another such that adenine and thymine or guanine and cytosine are always paired.

cytidine (16.1) a nucleoside composed of the pyrimidine cytosine and the sugar ribose.

cytosine (16.1) one of the pyrimidine nitrogenous bases found in DNA and RNA.

degenerate code (16.10) a term used to describe the fact that different triplet codons may be used to specify a single amino acid.

deoxyadenosine (16.1) a nucleoside made up of the purine adenine and the sugar 2′-deoxyribose.

deoxycytidine (16.1) a nucleoside composed of the pyrimidine cytosine and the sugar 2′-deoxyribose.

deoxyguanosine (16.1) a nucleoside made up of the purine guanine and the sugar 2′-deoxyribose.

deoxyribonucleic acid (DNA) (16.1) the nucleic acid molecule that carries all of the genetic information of an organism. The DNA molecule is a double helix composed of two strands, each of which is composed of phosphate groups, deoxyribose, and the nitrogenous bases thymine, cytosine, adenine, and guanine.

deoxyribonucleotide (16.1) a nucleoside phosphate or nucleotide composed of a nitrogenous base in β-N-glycosidic linkage to the 1′-carbon of the sugar 2′-deoxyribose and with one, two, or three phosphoryl groups esterified at the hydroxyl of the 5′-carbon.

deoxythymidine (16.1) a nucleoside composed of the pyrimidine thymine and the sugar 2′-deoxyribose.

diabetes (16.11) a disease caused by the production of insufficient levels of insulin and characterized by the appearance of very high levels of glucose in the blood and urine.

double helix (16.3) the spiral staircaselike structure of the DNA molecule characterized by two sugar-phosphate backbones wound around the outside and nitrogenous bases extending into the center.

factor VIII (16.11) one of the proteins that is necessary for proper blood clotting.

guanine (16.1) one of the purine nitrogenous bases found in DNA and RNA.

guanosine (16.1) a nucleoside composed of the purine guanine and the sugar ribose.

hemophilia (16.11) one of several genetic disorders that results in abnormally slow blood clotting.

human growth hormone (16.11) a hormone secreted by the anterior pituitary that stimulates the growth of an organism.

human immunodeficiency virus (16.1) the virus that produces the set of symptoms called the Acquired Immune Deficiency Syndrome (AIDS).

initiation codon (16.11) the triplet of ribonucleotides, AUG, on the mRNA that specifies binding of the initiator tRNA, N-formylmethionyl tRNA. This is always the first codon of a gene.

initiation factors (16.11) proteins that are required for formation of the translation initiation complex, which is composed of the large and small ribosomal subunits, the mRNA, and the initiator tRNA, N-formylmethionyl tRNA.

insulin (16.11) a hormone released by the pancreas in response to high blood glucose levels. It stimulates cellular uptake and storage of glucose from the blood.

intron (16.9) a noncoding sequence within a eukaryotic gene. It must be removed from the primary transcript to produce a functional mRNA.

messenger RNA (16.7) an RNA species that is produced by transcription and that specifies the amino acid sequence for a protein.

mutagen (16.5) any chemical or physical agent that causes changes in the nucleotide sequence of a gene.

mutation (16.5) any change in the nucleotide sequence of a gene.

nucleoside (16.1) a five-carbon sugar, ribose or 2′-deoxyribose, linked by a β-N-glycosidic linkage to either a purine or a pyrimidine base.

peptidyl tRNA binding site of ribosome (16.11) a pocket on the surface of the ribosome that holds the tRNA bound to the growing peptide chain.

photodimerization (16.5) the photochemical reaction in which the absorption of ultraviolet light causes the production of thymine dimers.

photoreactivation (16.5) a mechanism for repair of thymine dimers. The enzyme that cleaves the dimers, photoreactivating enzyme, uses visible light as an energy source.

pituitary dwarfism (16.11) a genetic disease caused by the inability to produce human growth hormone and characterized by very short stature.

point mutation (16.10) the substitution of a single base in a codon. This may or may not alter the genetic code of the mRNA resulting in the substitution of one amino acid in the protein.

polysome (16.11) complexes of many ribosomes all simultaneously translating a single mRNA.

preproinsulin (16.11) a precursor of the hormone insulin that is composed of a signal sequence and the unprocessed form of insulin, proinsulin.

proinsulin (16.11) a precursor of the hormone insulin that must be proteolytically cleaved to become the active hormone.

promoter (16.8) the sequence of nucleotides immediately before a gene that is recognized by the RNA polymerase and signals the start point and direction of transcription.

purines (16.1) a family of nitrogenous bases that are components of DNA and RNA and consist of a six-sided ring fused to a five-sided ring. The common purines in nucleic acids are adenine and guanine.

pyrimidines (16.1) a family of nitrogenous bases that are components of nucleic acids and consist of a single six-sided ring. The common pyrimidines of DNA are cytosine and thymine; the common pyrimidines of RNA are cytosine and uracil.

release factors (16.11) proteins that bind to the termination codons in the empty A-site and cause the peptidyl transferase to hydrolyze the bond between the peptide and the peptidyl tRNA.

replication fork (16.4) the point at which new nucleotides are added to the growing daughter DNA strand.

replication origin (16.4) the region of a DNA molecule where DNA replication always begins.

retrovirus (16.1) a class of animal viruses with genetic information that is RNA. The RNA is converted to DNA before it can be expressed to produce new virus particles.

reverse transcriptase (16.1) the enzyme produced by retroviruses that synthesizes double-stranded DNA from a single strand of RNA.

ribonucleic acid (RNA) (16.1) single-stranded nucleic acid molecules that are composed of phosphate groups, ribose, and the nitrogenous bases uracil, cytosine, adenine, and guanine.

ribonucleotide (16.1) a ribonucleoside phosphate or nucleotide composed of a nitrogenous base in β-N-glycosidic linkage to the 1′-carbon of the sugar ribose and with one, two, or three phosphoryl groups esterified at the hydroxyl of the 5′-carbon of the ribose.

ribosomal RNA (16.7) the RNA species that are structural and functional components of the small and large ribosomal subunits.

ribosome (16.7) an organelle composed of a large and a small subunit, each of which is made up of ribosomal RNA and proteins. It functions as a platform on which translation can occur and has the enzymatic activity that forms peptide bonds.

termination codon (16.11) a triplet of ribonucleotides with no corresponding anticodon on a tRNA. As a result, translation will end, since there is no amino acid to transfer to the peptide chain.

thymine (16.1) one of the pyrimidine nitrogenous bases found in DNA but not in RNA.

thymine dimers (16.5) two adjacent thymine bases in a DNA strand become covalently linked to one another. As a result of ultraviolet light–induced damage, there can be no hydrogen bonding to the opposite, complementary strand.

transcription (16.6) the synthesis of RNA from a DNA template.

transfer RNA (16.7) small RNAs that bind to a specific amino acid at the 3′-end and mediate its addition at the appropriate site in a growing peptide chain. This is accomplished by recognition of the correct codon on the mRNA by the complementary anticodon on the tRNA.

translation (16.6) the synthesis of a protein from the genetic code carried on the mRNA. The process occurs on ribosomes, and the code of the mRNA is decoded by the anticodon of the tRNA.

translocation (16.11) movement of the ribosome along the mRNA during translation. Each time an amino acid is added to the chain, the ribosome moves to the next codon on the mRNA.

uracil (16.1) one of the pyrimidine nitrogenous bases found in RNA but not in DNA.

uridine (16.1) a nucleoside composed of the pyrimidine uracil and the sugar ribose.

xeroderma pigmentosum (16.5) a human genetic disease caused by a mutation in a gene for one of the enzymes involved in a DNA repair pathway. Sufferers usually die of skin cancer at an early age.

zidovudine (16.1) trade name for the drug that is used to block the replication of human immunodeficiency virus (see AZT).

QUESTIONS AND PROBLEMS

DNA Structure

16.16 Write the structure of the purine ring and indicate the nitrogen that is bound to sugars in nucleosides and nucleotides.

16.17 What is the ring structure of the pyrimidines?

16.18 Which nitrogen atom of pyrimidine rings is bound to the sugar of nucleosides and nucleotides?

16.19 How many hydrogen bonds link the adenine-thymine base pair?

16.20 How many hydrogen bonds link the guanine-cytosine base pair?

16.21 Explain why an adenine-thymine base pair is weaker and therefore easier to disrupt than a guanine-cytosine base pair.

16.22 Define the term β-N-glycoside and illustrate with an example between 2′-deoxyribose and guanine.

16.23 Write the structure of the dinucleotide that results when cytosine 5′-phosphate is linked by a 3′–5′ phosphodiester bond to thymidine-5′-phosphate.

16.24 Why does the sequence of one strand of DNA automatically determine the sequence of the other strand?

DNA Replication

16.25 Illustrate by a diagram that DNA undergoes semiconservative replication?

16.26 Why is DNA polymerase said to be template-directed?

16.27 What is a replication fork?

16.28 What catalytic activity does DNA polymerase have in addition to its ability to synthesize DNA?

16.29 If a DNA strand had the following nucleotide sequence, ATGCGGCTAGAATATTCCATATGGCCGGTATTCG, what would the sequence of the complementary daughter strand be?

16.30 Why is DNA replication a self-correcting process?

Mutations

16.31 Why is UV light a mutagen?

16.32 What is a carcinogen?

16.33 What reaction does UV light cause in DNA?

16.34 Describe the process by which photoreactivation repair of DNA occurs.

16.35 What is responsible for the genetically transmitted disease xeroderma pigmentosum?

16.36 What genetic defect is responsible for sickle cell anemia?

16.37 Why do you think a single amino acid change would make such a difference in the function of the hemoglobin molecule. (*Hint:* Review the discussion of the levels of protein structure in Chapter 11.)

16.38 Define the term "point mutation."

Transcription

16.39 What is the central dogma of molecular biology?

16.40 Describe three differences between RNA and DNA molecules.

16.41 What are the functions of the following:
a. messenger RNA
b. ribosomal RNA
c. transfer RNA

16.42 By what type of bond is an amino acid linked to a tRNA molecule in an aminoacyl tRNA molecule?

16.43 How many codons constitute the genetic code?

16.44 On what molecule is the anticodon found?

16.45 Draw the structure of an alanine residue bound to the 3′ position of adenine at the 3′ end of alanyl tRNA.

16.46 What sequence of DNA is recognized by RNA polymerase?

16.47 If a gene had the following nucleotide sequence, ATGCCTAGAGGCTGATATTGCTAGCTCTGGTCAT, what would the sequence of the mRNA be?

RNA Processing

16.48 What is meant by the term "gene splicing"?

16.49 What is an intervening sequence?

16.50 The following is the primary transcript of a gene:

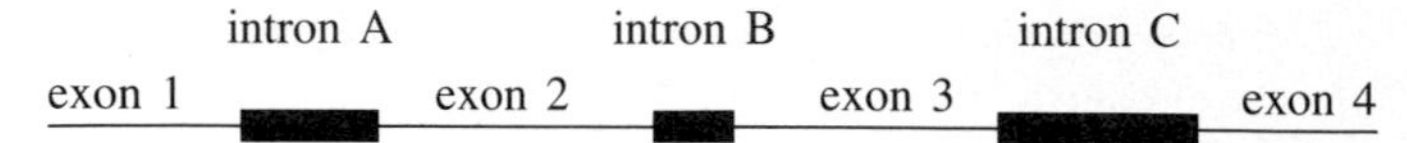

What would the structure of the final mature mRNA look like and which of the above sequences would be found in the mature mRNA?

Protein Synthesis

16.51 What are two properties of the genetic code?

16.52 What is the function of ribosomes?

16.53 What function does an aminoacyl tRNA synthetase have in protein synthesis?

16.54 How does an aminoacyl tRNA recognize its binding site on mRNA?

16.55 What is the role of N-formylmethionyl tRNA in bacterial protein synthesis?

16.56 What steps are required to form the initiation complex of protein synthesis?

16.57 How does elongation of polypeptide chains occur?

16.58 How is protein synthesis terminated?

16.59 What are the two tRNA binding sites on the ribosome?

16.60 What molecule couples transcription of mRNA to its translation?

16.61 What peptide sequence would be formed from the following mRNA?
AUGUGUAGUGACCAACCGAUUUCACUGUGA

Further Problems

16.62 What is a nucleoside analogue?

16.63 What is the function of reverse transcriptase in HIV replication?

16.64 Draw the structure of 3′-azido-2′,3′-dideoxythymidine. What is the common name of this nucleoside?

16.65 Why are many mutagens also carcinogens?

16.66 What is a reversion?

16.67 Describe the means by which human insulin is produced by recombinant DNA technology.

CHAPTER 17

Nutrition

OBJECTIVES

- Describe the four major food groups.
- List and describe the three major dietary energy sources.
- Discuss the advantages of complex carbohydrates over simple carbohydrates.
- Describe the functions of fats in the body.
- Understand the difference between essential and nonessential amino acids.
- Understand the difference between complete and incomplete proteins.
- Distinguish between major and trace minerals and give examples of each.
- List the five major minerals and describe their biological and biochemical functions.
- List several important trace minerals and describe the functions of each.
- Distinguish between water-soluble and lipid-soluble vitamins and give some examples of each.
- Describe how the Recommended Dietary Allowance is determined for a vitamin.
- Describe the sources and functions of each of the water-soluble vitamins discussed in the text.
- Describe the functions and list the sources of each of the lipid-soluble vitamins.
- Be aware of the possible side effects of ingestion of excessive amounts of fat-soluble vitamins.
- Become familiar with the structures of the various nutrients discussed in this chapter.

OUTLINE

Among the major health problems of the world, none is as pressing as the clash between rising populations and the available food supply. In fact, it is hardly possible to pick up a newspaper without reading about catastrophic famines in various parts of the world. Thousands of people starve in these famines, and the survivors, especially the children, suffer irreparable damage. It is estimated that one in ten people in the world are malnourished and that 100,000 people starve to death each day. Tragically, 40,000 are babies. Additional millions of children are blinded or mentally impaired owing to malnourishment. Even the wealthy countries of Europe and North America contain areas of severe malnutrition.

A healthful human diet must consist of a variety of substances, including carbohydrates, vitamins, essential fatty acids, essential amino acids, and minerals, to provide for all of the body's metabolic needs. The chief requirement is that of energy. The major sources of metabolic energy in the human diet are fats, carbohydrates, and proteins.

What is a healthy diet? Many people echo the words of Mark Twain's Pudd'nhead Wilson: ''The only way to keep your health is to eat what you don't want, drink what you don't like, and do what you'd rather not.'' But a diet consisting primarily of ''junk'' food is not healthy, and it is easy to design a well-balanced diet of appetizing, interesting foods by simply including selections from the four major food groups. The nutrients that each of these food groups provides are discussed in this chapter. The following is a list of the four major food groups and the recommended number of servings of each for an adult:

- *Milk* and dairy products provide protein, calcium, phosphorus, and riboflavin. In addition, fortified milk provides vitamins A and D. Two to three servings per day are recommended.
- *Meats* and the high-protein vegetables (dried beans, peas, and nuts) provide protein, thiamine, niacin, vitamin B_{12}, iron, and minerals. Two to three servings per day are recommended.
- *Fruits* and *vegetables* provide carotene (a precursor of vitamin A), vitamin C, other water-soluble vitamins, and ''fiber.'' These foods also contain protein. Although no single fruit or vegetable provides an adequate supply of dietary protein, it is possible to mix fruits and vegetables to supply most dietary needs. Four to five servings per day are recommended.
- *Cereals* and *breads* provide carbohydrates, fiber, and protein. If the wheat and cereals are not refined, they also provide vitamins and iron. Four to five servings per day are recommended.

The words of Pudd'nhead Wilson notwithstanding, a combination of foods from each of these groups will provide all of the components known to be required in the human diet and will prevent excessive intake of potentially harmful substances such as saturated fats and cholesterol.

17.1 MAJOR DIETARY ENERGY SOURCES

Carbohydrates

Section 13.3

Section 14.2, Section 14.3

A kilocalorie is the same as the calorie referred to in the count-your-calories books.

Carbohydrates are the primary source of energy for the brain and nervous system and can also be used by many other tissues, including muscle. Their degradation by glycolysis and their oxidation by the citric acid cycle and respiration release 4 kilocalories of energy for every gram of carbohydrate utilized. Some of the intermediates in the glycolytic and citric acid pathways can be used to synthesize amino acids, fats, and nucleic acids. Thus carbohydrates also provide some of the molecules that are required for biosynthesis of various cell components.

A healthy diet should contain both complex carbohydrates, such as starches and cellulose, and simple sugars, such as fructose and sucrose. However, the quantity of simple sugars, especially sucrose, should be minimized. There has been a great deal of controversy recently concerning the use of raw sugar instead of refined sugars. Although raw sugar may have a few additional minerals, the principal component is still sucrose, and large quantities of sucrose in the diet promote obesity and tooth decay. It is clear that complex carbohydrates are better for us than the simple sugars like sucrose. Starch, such as we find in rice, potatoes, breads, and cereals, is an excellent energy source. In addition, the complex carbohydrates, such as cellulose, provide us with an important supply of dietary fiber.

Section 7.4, 7.2, and 7.3

See A Human Perspective: Tooth Decay and Simple Sugars

Question 17.1 What is the advantage of eating complex carbohydrates?

Question 17.2 What are some sources of complex carbohydrates?

It is difficult to say exactly what percentage of the daily diet *should* consist of carbohydrates. The actual percentage varies widely throughout the world, from 80% in the Far East, where rice is the principal component of the diet, to 40–50% in the United States. The current U.S. Dietary Guidelines recommend that the well-balanced diet should contain 58% carbohydrates and that no more than 10% of the daily caloric intake should be sucrose (see Figure 17.1).

Fats

Fatty acids derived from triglycerides are also a major source of energy. When oxidized, each gram of fat releases 9 kilocalories of energy, more than twice the energy released by oxidation of a gram of carbohydrate.

Section 15.2

In addition to serving as an energy source, fats have a number of other functions in the body. The degradation of fats provides the acetyl CoA that is used for the synthesis of such complex molecules as cholesterol. Fatty acids are the precursors of many hormones, like the prostaglandins. Although humans can synthesize many fatty acids, there are two **essential fatty acids,** *linoleic acid* and *linolenic acid,* that must be obtained from triglycerides found in plants. Linoleic acid is required as a precursor of the prostaglandins. Linolenic acid is one of the so-called ω-3 fatty acids that are abundant in fish oils.

Section 9.2

FIGURE 17.1
Components of a well-balanced diet. Nutritionists recommend that we eat a diet composed of 55–60% carbohydrates, preferably complex carbohydrates, 10–15% protein, and not more than 30% fats.

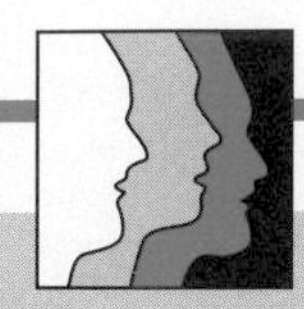

A HUMAN PERSPECTIVE

Tooth Decay and Simple Sugars

How many times have you heard the lecture from parents or your dentist about brushing your teeth after a sugary snack? Annoying as this lecture might be, it is based on sound scientific data that demonstrate that the cause of tooth decay is plaque and acid formed by the bacterium *Streptococcus mutans* using sucrose as its substrate.

Saliva is teeming with bacteria. There are about 20 species of oral bacteria in concentrations up to one hundred million (10^8) per milliliter of saliva! Within minutes of brushing your teeth, glycoproteins in the saliva adsorb, or stick to, tooth surfaces. This is called an **acquired pellicle.** The glycoproteins of the acquired pellicle are very sticky, and millions of oral bacteria immediately adhere to this surface. Although all oral bacteria adhere to the tooth surface, only *S. mutans* causes dental caries, or cavities.

Why can this bacterium cause cavities when all the others do not? First, tooth enamel cannot be dissolved unless the pH is less than 5.0. *S. mutans* grows well in the pH range 4.5–5.0. Second, *S. mutans* produces large amounts of lactic acid from carbohydrates (Section 13.4). Third, of all the oral bacteria, only *S. mutans* has a special enzyme, **glucosyl transferase,** on its cell surface.

Glucosyl transferase is an enzyme with absolute specificity (Section 12.5), and its sole substrate is sucrose, table sugar. The enzyme cleaves the disaccharide into its two component monosaccharides, glucose and fructose, and adds the glucose to a growing polysaccharide chain called **dextran.** Dextran adheres tightly to both the tooth enamel and the bacteria. **Dental plaque** is made up of huge masses of bacteria, embedded in dextran, adhering to the tooth surface.

This is just the first stage of cavity formation. Note in the figure that the second sugar released by the cleavage of sucrose is fructose. The bacteria utilize the fructose in the energy-generating pathways of glycolysis and lactic acid fermentation. Production of lactic acid decreases the pH on the tooth surface and begins to dissolve caicium from the tooth enamel.

Why is the acid not washed away from the tooth surface? After all, we produce about one liter of saliva each day, which should dilute the acid and remove it from the tooth surface. The problem is the dextran plaque, which is not permeable to saliva. Thus plaque keeps the bacteria and their fermentation product, lactic acid, localized on the enamel.

What measures can we take to prevent tooth decay? Practice good oral hygiene; brushing teeth after each meal and flossing regularly reduces plaque buildup. Eat a diet rich in calcium; this helps to build strong tooth enamel. Include many complex carbohydrates in the diet; these are not substrates for glucosyl transferase and will not lead to the formation of acid. Further, the complex carbohydrates from fruits and vegetables help to prevent decay by mechanically removing plaque from tooth surfaces. Avoid sucrose-containing snacks between meals. Studies have shown that the consumption of a sucrose-rich dessert with a meal followed by brushing does not produce many cavities. However, even small amounts of sugar ingested between meals are very cariogenic.

Treatment of teeth with fluoride also helps to prevent dental caries. In many communities, fluoride is added to the community water supply. This is most beneficial to

Increasing the amount of fish in the diet, and hence the amount of ω-3 fatty acids in the diet, appears to decrease the level of serum cholesterol and triglycerides. Linolenic acid and the other ω-3 fatty acids may play a role in preventing the overproduction of prostaglandins. Dietary fat serves as a carrier of the fat-soluble vitamins A, D, E, and K. All of these are transported into cells in association with fat molecules. Thus a diet that is too low in fat can result in a deficiency of these four vitamins. Fats also serve as a shock absorber, or protective layer, for the vital organs. About 4% of the total body fat is reserved for this critical function. Finally, fat stored beneath the skin (subcutaneous fat) serves to insulate the body from extremes of cold temperatures.

There has been considerable controversy over what types of dietary fat should be consumed. Although it has not been demonstrated conclusively, a strong correlation has been found between heart disease and a high level of saturated fatty acids and cholesterol in the diet. High levels of dietary saturated fatty acids may also predispose an individual to certain kinds of cancers, including colon, esophageal, stomach, and breast cancers. Such results suggest that a diet that is low in cholesterol and saturated fats may be beneficial in the prevention of heart disease and these cancers.

Standards of fat intake have not been experimentally determined. However, the most recent U.S. Dietary Guidelines recommend that dietary fat not exceed 30% of the daily caloric intake and that no more than 10% should be saturated fats.

small children whose teeth are developing. The incorporation of fluoride into tooth enamel converts hydroxyapatite into fluorapatite, which is much more resistant to being dissolved by acids. Fluoride has also been added to many toothpastes. There is suggestive evidence that this can even help to prevent tooth decay in adults. The proposed mechanism of action for this effect is that fluoride may build up in plaque to levels that are high enough to inhibit the growth of oral microorganisms.

It would be most beneficial to have a vaccine against tooth decay. Researchers have developed a vaccine that prevents tooth decay in rats. The vaccination involves injection of *S. mutans* into the salivary glands. The immune system of the body then produces antibodies against the bacteria. In this case the class of antibodies is IgA, which is the class that protects the body surfaces and mucous membranes from bacterial and viral invasion. These antibodies produced a marked reduction in the number of cavities in the rat. It is hoped that such a vaccine will one day be available for human beings.

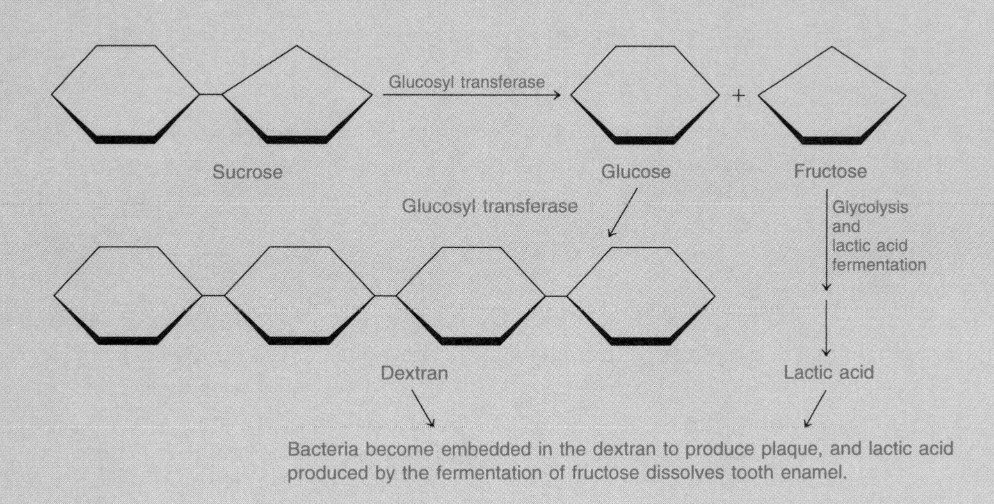
Glucosyl transferase
Sucrose
Glucose
Fructose
Glucosyl transferase
Glycolysis and lactic acid fermentation
Dextran
Lactic acid
Bacteria become embedded in the dextran to produce plaque, and lactic acid produced by the fermentation of fructose dissolves tooth enamel.

Action of the glucosyl transferase of *Streptococcus mutans*, which is responsible for tooth decay.

Question 17.3 Why do physicians recommend a diet that is low in saturated fats?

Proteins

Proteins are the third major type of energy source in the diet. Like carbohydrates and fats, proteins serve several dietary purposes. They can be oxidized to provide energy. In addition, the amino acids liberated by the hydrolysis of proteins are used directly in biosynthesis. The protein synthetic machinery of the cell can incorporate amino acids, released by the digestion of dietary protein, directly into the cellular proteins. In addition to being used directly in protein synthesis, amino acids are also the precursors of a large, diverse group of compounds collectively referred to as the nitrogen compounds. This group includes some hormones and the heme groups of hemoglobin, myoglobin, and the cytochromes. A more detailed listing of the nitrogen compounds is shown in Table 17.1.

Section 14.5

Section 16.11

Amino acids can be divided into two major nutritional classes. **Essential amino acids** are amino acids that cannot be synthesized by the body and are required in the diet. **Nonessential amino acids** are amino acids that can be synthesized by the body and need not be included in the diet. The term "nonessential" is actually a misnomer because a

TABLE 17.1 Nitrogen-Containing Compounds Derived from Amino Acids

Amino Acid Precursor	Nitrogen Compound	Function
Glycine, glutamine, and aspartic acid	Purines	Nucleic acid molecules (DNA and RNA)
Glycine	Pyrimidines	Nucleic acid molecules (DNA and RNA)
Methionine, serine	Choline, sphingosine	Complex lipids of nerve tissue
Tyrosine, tryptophan	Thyroxine, epinephrine	Hormones
Tyrosine	Melanin	Skin pigmentation
Tryptophan	Niacin (NAD^+), porphyrins	Enzyme cofactors and coenzymes
Histidine	Histamine	Vasodilator

sufficient supply of protein is required to supply all the components needed for their synthesis.

Table 17.2 lists the essential and nonessential amino acids.

Proteins are also classified as **complete** or **incomplete.** Protein derived from animal sources is generally complete. That is, it provides all of the essential and nonessential amino acids in approximately the correct amounts for biosynthesis. In contrast, proteins derived from vegetable sources are generally incomplete because they lack a sufficient amount of one or more amino acids. People who try to maintain a strictly vegetarian diet or for whom animal protein is often not available have the problem that no single high-protein vegetable has all of the essential amino acids to ensure a sufficient daily intake. For example, the major protein of beans contains abundant lysine and tryptophan but very little methionine, whereas corn contains considerable methionine but very little tryptophan or lysine. A mixture of corn and beans, however, satisfies both requirements. This combination, called succotash, was a staple of the diet of American Indians for centuries.

TABLE 17.2 The Essential and Nonessential Amino Acids

Essential Amino Acids	Nonessential Amino Acids
Isoleucine	Alanine
Leucine	Arginine
Lysine	Asparagine
Methionine	Aspartate
Phenylalanine	Cysteine[2]
Threonine	Glutamate
Tryptophan	Glutamine
Valine	Glycine
	Histidine[1]
	Proline
	Serine
	Tyrosine[2]

[1]Histidine is an essential amino acid for infants but not for adults.
[2]Cysteine and tyrosine are considered to be semiessential amino acids. They are required by premature infants and adults who are ill.

A few vegetarian meals each week can provide all the required amino acids and simultaneously help to reduce the amount of saturated fats in the diet. Many enjoyable ethnic foods apply the principle of mixing protein sources. Mexican foods such as tortillas and refried beans, Cajun dishes of spicy beans and rice, Indian cuisine of rice and lentils, and even the traditional peanut butter sandwich are all examples of ways to mix foods to provide complete protein.

Question 17.4 Why must foods be mixed to provide an adequate diet?

Question 17.5 How can vegetable proteins be mixed to provide all of the essential amino acids?

Question 17.6 Think of ethnic foods other than those noted in Section 17.1 that apply the principle of mixing vegetable proteins to provide all of the essential amino acids.

Question 17.7 What are some common dietary sources of protein?

Question 17.8 What are some sources of dietary fat?

17.2 MINERALS AND CELLULAR FUNCTION

Many minerals are required in the human diet. These may be divided into two nutritional classes:

- **Major minerals** must be consumed in amounts greater than 100 mg/day.
- **Trace minerals,** in contrast, are required in much smaller amounts (less than 100 mg/day). In some cases the required levels are so small that they cannot be very accurately measured.

The major minerals are identified in the periodic table in Figure 17.2 in dark blue and the trace minerals in light blue.

Calcium and phosphorus

The development of bones and teeth depends upon the adequate dietary intake of calcium and phosphorus. Calcium is a major component of bones and teeth. It is part of a crystalline calcium phosphate mineral known as **hydroxyapatite,** whose formula is $[Ca_{10}(PO_4)_6(OH)_2]$. In addition, calcium is required for normal blood clotting and muscle function. The Recommended Dietary Allowance (RDA) for calcium is 1200 mg/day for adults between 19 and 24 years of age and 800 mg/day for adults over age 25. Milk, cheese, canned salmon, and dark green leafy vegetables are all rich sources of dietary calcium.

Section 11.6, Section 12.13
Section 12.14

Phosphorus is required not only as a component of hydroxyapatite in bone, but also as a component of nucleic acids and many other biologically important molecules. Without phosphorus we would have no energy storage molecules, such as ATP and creatine phosphate, for the energy derived from glycolysis and the citric acid cycle. The RDA for phosphorus is the same as that for calcium. Because it is abundant in most foods, a deficiency of phosphorus in the presence of an otherwise adequate diet is virtually impossible.

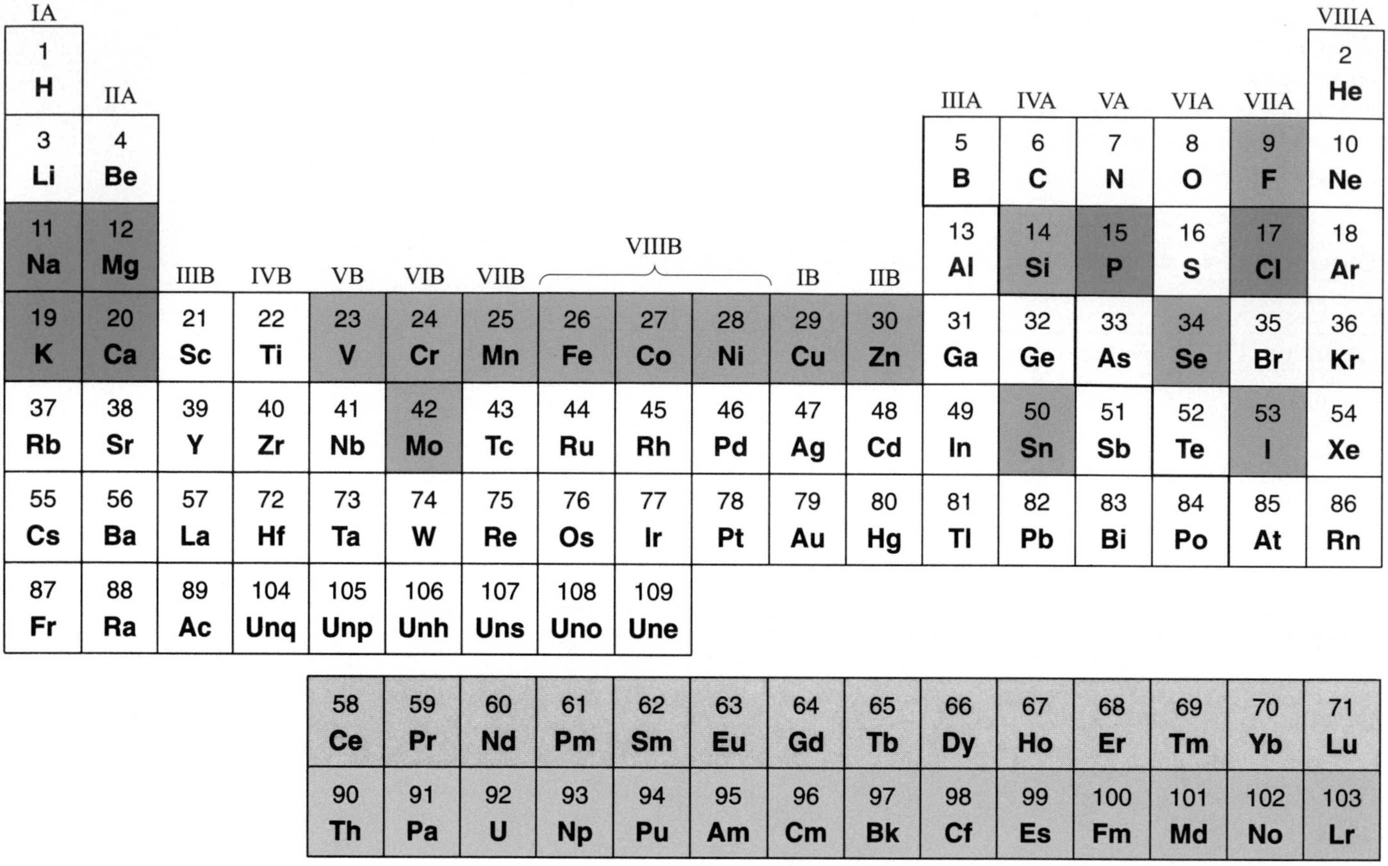

IA	IIA	IIIB	IVB	VB	VIB	VIIB	VIIIB			IB	IIB	IIIA	IVA	VA	VIA	VIIA	VIIIA
1 H																	2 He
3 Li	4 Be											5 B	6 C	7 N	8 O	9 F	10 Ne
11 Na	12 Mg											13 Al	14 Si	15 P	16 S	17 Cl	18 Ar
19 K	20 Ca	21 Sc	22 Ti	23 V	24 Cr	25 Mn	26 Fe	27 Co	28 Ni	29 Cu	30 Zn	31 Ga	32 Ge	33 As	34 Se	35 Br	36 Kr
37 Rb	38 Sr	39 Y	40 Zr	41 Nb	42 Mo	43 Tc	44 Ru	45 Rh	46 Pd	47 Ag	48 Cd	49 In	50 Sn	51 Sb	52 Te	53 I	54 Xe
55 Cs	56 Ba	57 La	72 Hf	73 Ta	74 W	75 Re	76 Os	77 Ir	78 Pt	79 Au	80 Hg	81 Tl	82 Pb	83 Bi	84 Po	85 At	86 Rn
87 Fr	88 Ra	89 Ac	104 Unq	105 Unp	106 Unh	107 Uns	108 Uno	109 Une									

58 Ce	59 Pr	60 Nd	61 Pm	62 Sm	63 Eu	64 Gd	65 Tb	66 Dy	67 Ho	68 Er	69 Tm	70 Yb	71 Lu
90 Th	91 Pa	92 U	93 Np	94 Pu	95 Am	96 Cm	97 Bk	98 Cf	99 Es	100 Fm	101 Md	102 No	103 Lr

FIGURE 17.2
The periodic table showing elements required in the human diet. The elements in dark blue are major minerals, and those in light blue are trace elements.

Sodium, potassium, and chloride

Sodium, potassium, and chloride ions are all required in the human diet. When dissolved in water, sodium and potassium are positively charged ions, while chloride is a negatively charged ion. These three minerals are called blood electrolytes because the ions can conduct electrical currents. Sodium is found primarily in the extracellular fluids, and potassium is found predominantly within the cell. Both these elements are needed to maintain a proper fluid balance inside and outside of the cell. Because these three minerals are found in most foods, deficiency is rare.

The Food and Nutrition Board of the National Research Council has removed the three electrolytes from its table of *estimated safe and adequate daily dietary intake (ESADDI)* because there is not sufficient information available to establish a recommended amount. The major dietary source of sodium and chloride is table salt (40% sodium and 60% chloride). Physicians still recommend that the intake of sodium be restricted to 1–2 g daily. The recommended intake of chloride is approximately 1.7–5.1 g daily. However, getting enough sodium and chloride is not a problem. In fact, sodium intake in the United States is about 5–7 g/day, far in excess of the 1–2 g/day that a normal adult requires.

Potassium is the major intracellular cation. It is found in citrus fruits, bananas, and tomatoes. Dietary intake of potassium is about 1.9–5.6 g/day in the United States. Potassium deficiency is rare, but loss of potassium in severe diarrhea, such as can occur in cholera, and the excretion of potassium by a person suffering from diabetes mellitus can lead to a debilitating deficiency. However, potassium deficiency is seen most commonly in individuals who are taking diuretics.

A high intake of table salt, sodium chloride, the major source of sodium in the diet, is one factor that may cause high blood pressure, *hypertension,* in susceptible persons. There has been considerable emphasis on ''low-salt'' diets as a means of avoiding hypertension. However, it appears that sodium is not the only culprit. It is the sodium-to-potassium ratio that appears to be important in controlling blood pressure. Ideally, the Na/K ratio should be about 0.6, but the Na/K ratio consumed by the average American is greater than 1.0. To avoid hypertension in later life, it is important to both reduce the amount of sodium in the diet *and* increase the amount of potassium.

Magnesium

Magnesium ions are vital to cellular metabolism. They are required for the reactions in the liver that convert glycogen to glucose. They are also important in normal muscle function, nerve conductance, and bone development. Mg^{2+} ions bind to AMP, ADP, ATP, and nucleic acids. Many enzymes that are involved in the catabolic breakdown of glucose require magnesium ions as cofactors. A typical adult contains about 25 g of magnesium, and the recommended daily intake is about 300 mg/day. Magnesium is plentiful in leafy green vegetables, legumes, cereal grains, and lean meats.

Trace minerals

Iron is a required mineral for heme-containing proteins and is an element that is absolutely essential for normal physiological functioning. It is found in the iron transport and storage proteins, hemoglobin and myoglobin, and is also a component of the cytochromes that participate in the respiratory electron transport chain. The requirement for iron is so well known that it might be thought that no one would suffer the effects of iron deficiency. In fact, iron deficiency is rather common in the United States, especially among women. Iron can be absorbed by the body only in its ferrous, Fe^{2+}, oxidation state. The iron in meat is absorbed more efficiently than that in most other foods. Vegetarians whose protein intake is mostly in the form of cereal grains run a risk of iron deficiency because iron in grains is absorbed poorly by the body.

Sections 11.9 and 11.10

Hemoglobin, myoglobin, and the cytochromes of the respiratory electron transport chain all contain heme. Heme, of course, contains iron, and it is this need that must be satisfied by the diet. Deficiency of iron leads to *iron-deficiency anemia,* a condition in which the amount of hemoglobin in red blood cells is abnormally low.

Copper is a mineral that is required for many essential enzymes. The respiratory electron transport chain contains an enzyme, *cytochrome oxidase,* that contains both heme groups and copper ions. Copper is therefore required in the diet for the function of this essential enzyme. Copper is also required by some of the enzymes that are responsible for the synthesis of connective tissue proteins. Seafood, vegetables, nuts, and meats such as liver are excellent sources of copper ions. The ESADDI for adults is 1.5–3.0 mg. Copper ions in high concentrations are also toxic. In fact, mental retardation and death in early adolescence result from an inability to remove excess copper ions from the body. Here, as everywhere in life, it is the balance of the system that is critical to its function.

Iodine is a mineral that is required for the proper function of the thyroid gland. The thyroid gland extracts iodine from nutrients and incorporates it into various hormones. The once-common condition of goiter, an enlargement of the thyroid gland, is an abnormality that results from an effort to compensate for low iodine intake. Goiter can be prevented if iodine is included in the diet. Seafood is one of the best sources of iodine. But in areas where seafood is not available, dietary iodine is easily obtained in the form of iodized salt, which is found in most grocery stores.

Fluoride aids in the prevention of dental caries. The presence of fluoride ions in the water supplies of many cities has dramatically reduced the incidence of dental caries (cavities), the most widespread ''disease'' in the United States. Popular resistance to water fluoridation, spurred by the debate over its safety, has caused many municipalities to abandon this practice. An excess of fluoride is toxic, but at the level found in fluoridated water supplies, approximately 1 part per million, no toxic effects are observed. Fluoride works by displacing hydroxide in calcium hydroxyapatite to give a crystalline mineral in teeth known as *fluorapatite,* $[Ca_3(PO_4)_2.CaF_2]$, which is far more resistant to the bacteria responsible for caries than hydroxyapatite itself.

Many other trace minerals are required in the diet. Among these are zinc, nickel, vanadium, tin, silicon, chromium, selenium, and cobalt. Zinc and molybdenum are required by various enzymes, while cobalt is a component of vitamin B_{12}. Deficiencies of trace minerals are virtually nonexistent, since they are needed in such small quantities that requirements for them are likely to be met in nearly every diet. As in the case of copper, however, most trace minerals are extremely toxic if ingested in large quantities, and ''heavy metal poisoning'' has been a scourge of industrial cities throughout the world.

Question 17.9 What are the physiological functions of each of the following major minerals?

a. Calcium
b. Phosphorus
c. Sodium
d. Potassium

17.3 WATER-SOLUBLE VITAMINS

Vitamins (L.: *vita,* life + amine) are organic substances required in the diet that promote a variety of essential enzymatic reactions in cells. Since they are not an energy source, they are required only in small amounts. If a vitamin is absent from the diet, the results are often catastrophic. Tables 17.3 and 17.4 list the vitamins that are required in the human diet, their major nutritional sources, and the clinical conditions that result from their deficiency. In the affluent countries of the world, major malnutrition is a problem primarily in urban slums. But in the poor, undeveloped countries of the world, deficiencies of folic acid, vitamin C, niacin, riboflavin, and thiamine are a more general problem and often lead to either slow death or, at best, serious dysfunction. Vitamin deficiencies are particularly common in regions where protein is in scarce supply. This is due in part to the fact that foods that are high in protein are often rich in vitamins as well. It also reflects the fact that some vitamins can be synthesized from amino acid precursors. For instance, human beings have all the enzymes that are needed to synthesize nicotinamide from the amino acid tryptophan. However, we cannot synthesize the tryptophan (Table 17.2); it must be provided by the diet. Thiamine is also found in foods that are rich in protein, and a diet that is low in protein is also likely to be low in thiamine as well. Thus in a region where the diet contains little protein, vitamin deficiency is generally a serious problem.

Section 12.7

Vitamins are classified as either *fat-soluble* or *water-soluble*. The water-soluble vitamins are components of many **coenzymes.** Recall that coenzymes act with enzymes to carry out specialized chemical functions. Once ingested, vitamins undergo chemical modifications that convert them into coenzymes. However, it serves no purpose to consume enormous doses of vitamin tablets because water-soluble vitamins are not stored in the body. Since they are soluble in water, the excess is simply excreted in the urine. Table 17.3 lists the coenzymes that are derived from the water-soluble vitamins and their chemical functions. Table 17.4 provides the major nutritional sources of these common vitamins.

Recently, a great deal of emphasis has been placed on vitamin supplements to combat stress, prevent the common cold, protect against various kinds of cancer and heart disease, offset the symptoms of premenstrual syndrome, delay the aging process, and improve one's sex life! Most nutritionists feel that a well-balanced diet provides all the nutrients, including the vitamins, that the body requires. Indeed, when associations such as the American Cancer Society suggest that certain vitamins might help to prevent cancers, they recommend that these vitamins be obtained from the natural food sources rather than from vitamin supplements.

TABLE 17.3 Water-Soluble Vitamins Required by Humans and Coenzymes Derived from Them

Vitamin	Coenzyme	Function
Thiamine (B_1)	Thiamine pyrophosphate	Decarboxylation reactions
Riboflavin (B_2)	Flavin mononucleotide (FMN)	Carrier of H atoms
	Flavin adenine dinucleotide (FAD)	
Niacin (B_3)	Nicotinamide adenine dinucleotide (NAD^+)	Carrier of hydride ions
	Nicotinamide adenine dinucleotide phosphate ($NADP^+$)	
Pyridoxine (B_6)	Pyridoxal phosphate	Carriers of amino and carboxyl groups
	Pyridoxamine phosphate	
Cyanocobalamin (B_{12})	Deoxyadenosyl cobalamin	Coenzyme in amino acid metabolism
Folic acid	Tetrahydrofolic acid	Coenzyme for 1-C transfer
Pantothenic acid	Coenzyme A	Acyl group carrier
Biotin	Biocytin	Coenzyme in CO_2 fixation
Ascorbic acid (C)	Unknown	Hydroxylation of proline and lysine in collagen

TABLE 17.4 Major Nutritional Sources of Water-Soluble Vitamins Required by Humans and Some Physiological Effects of Deficiencies

Vitamin	Source	Symptoms of Deficiency
Thiamine (B_1)	Brain, liver, heart, whole grains	Beriberi, neuritis, mental disturbance
Riboflavin (B_2)	Milk, eggs, liver	Photophobia, dermatitis
Niacin (B_3)	Whole grains, liver	Pellagra, dermatitis, digestive problems
Pyridoxine (B_6)	Whole grains, liver, fish, kidney	Dermatitis, nervous disorders
Cyanocobalamin (B_{12})	Liver, kidney, brain	Pernicious anemia
Folic acid	Liver, leafy vegetables, intestinal bacteria	Anemia
Pantothenic acid	Most foods	Neuromotor and cardiovascular disorders
Biotin	Egg yolk, intestinal bacteria	Scaly dermatitis, muscle pains, weakness
Ascorbic acid (C)	Citrus fruits, green leafy vegetables, tomatoes	Scurvy, failure to form collagen

The Food and Nutrition Board of the National Research Council/National Academy of Sciences publishes information on the quantities of vitamins and minerals that are required in the diet. These are called **Recommended Dietary Allowances (RDAs)** and are defined as ''the levels of intake of essential nutrients considered adequate to meet the known nutritional needs of practically all healthy persons.'' The RDA is determined by obtaining an estimate of the range of normal human needs. The value at the high end of the range is chosen, and an additional safety factor is added. Thus the RDA is by no means a minimum value, but rather a high estimate of daily requirements. Serious physical problems can follow ingestion of megadoses of many of these vitamins, especially those that are fat-soluble.

Question 17.10 What is the difference between vitamins and coenzymes?

Pantothenic acid

Sections 13.7 and 15.2

Pantothenic acid (Figure 17.3) is essential for the normal metabolism of fats and carbohydrates. Like many other vitamins, pantothenic acid is abundant in meat, fish, poultry, whole-grain cereals, and legumes. Fruits and vegetables, on the other hand, contain only small amounts of pantothenic acid. The daily requirement of pantothenic acid is in the range of 4–7 mg per day. Pantothenic acid deficiency, which is rather rare in the United States except among alcoholics, manifests itself as gastrointestinal, neuromotor, and cardiovascular disorders. Pantothenic acid is converted to its biologically functional form, known as **coenzyme A,** in the body. We have seen that derivatives of coenzyme A are important for the transfer of acyl groups in the metabolism of fatty acids and carbohydrates.

FIGURE 17.3
The structure of pantothenic acid. This vitamin is required as a component of coenzyme A, shown at the bottom.

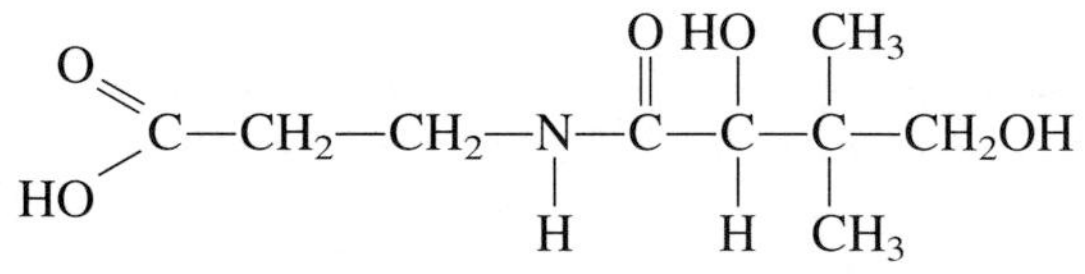

Pantothenic acid

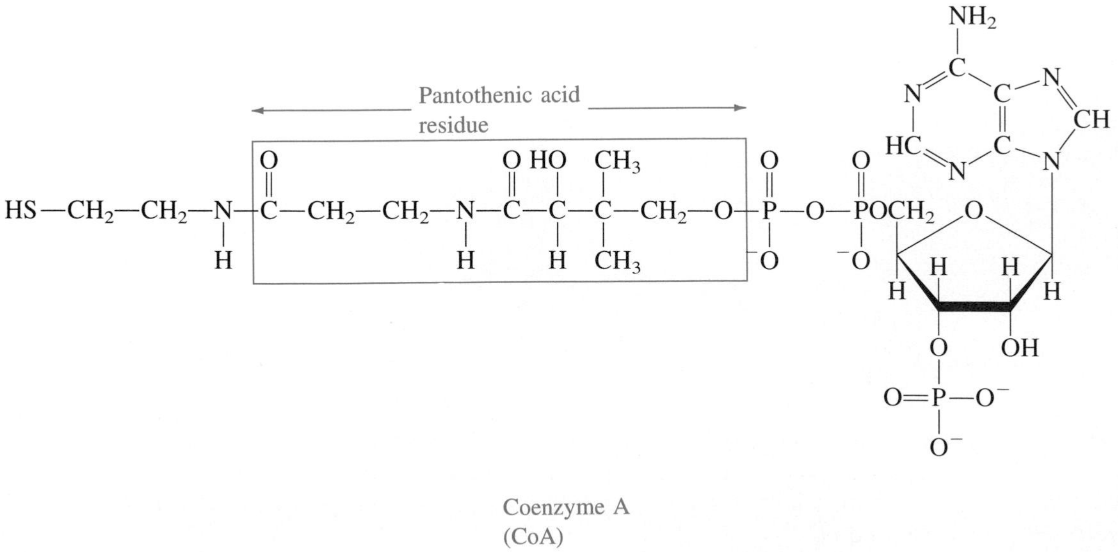

Pantothenic acid residue

Coenzyme A
(CoA)

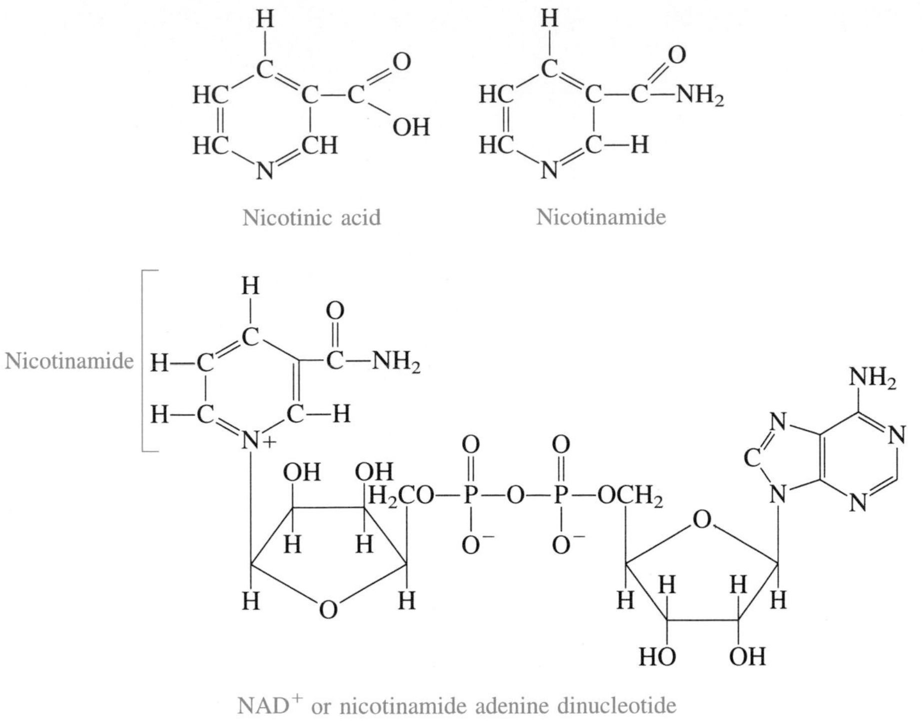

Nicotinic acid

Nicotinamide

Nicotinamide

NAD$^+$ or nicotinamide adenine dinucleotide

FIGURE 17.4
The structures of nicotinic acid, nicotinamide, and NAD$^+$. Nicotinamide is the portion of the coenzymes that serves as the carrier of hydride ions.

Niacin

Niacin (vitamin B_3) refers to both nicotinic acid and nicotinamide. Nicotinamide is an essential precursor for the coenzymes nicotinamide adenine nucleotide (NAD$^+$) and nicotinamide adenine dinucleotide phosphate (NADP$^+$). In addition to the nicotinamide group, these molecules contain an ADP unit (Figure 17.4).

Section 12.7

Niacin is found in fish, lean meat, legumes, milk, and whole-grain and enriched cereals. The RDA of niacin is 20 mg per day. Niacin deficiency leads to dermatitis, diarrhea, dementia, and death. The most common illness that develops from niacin deficiency is **pellagra,** a form of dermatitis. This nutritional disease is found where corn is abundant in the diet and meat is scarce.

Corn actually contains a rather large amount of niacin, but it is present in a form that is not made available to the body simply by cooking. American and South American Indians discovered centuries ago that soaking cornmeal in lime water (dilute calcium hydroxide) proved beneficial. The lime water releases nicotinamide in a form that can then be absorbed through the intestines.

Riboflavin

Riboflavin, or vitamin B_2, is abundant in milk, eggs, and dark green leafy vegetables (Figure 17.5). As a component of the coenzyme flavin adenine dinucleotide (FAD), it is essential for the energy-releasing reactions of the cell. The recommended daily intake of riboflavin is about 1.7 mg. Severe riboflavin deficiency is rare in most parts of the world, but a marginal deficiency of this vitamin is common even in the United States. Mild riboflavin deficiency leads to dry and cracked lips and other mild forms of dermatitis. In severe cases, however, riboflavin deficiency leads to extreme sensitivity to sunlight and retarded growth in children.

Sections 13.3 and 14.3

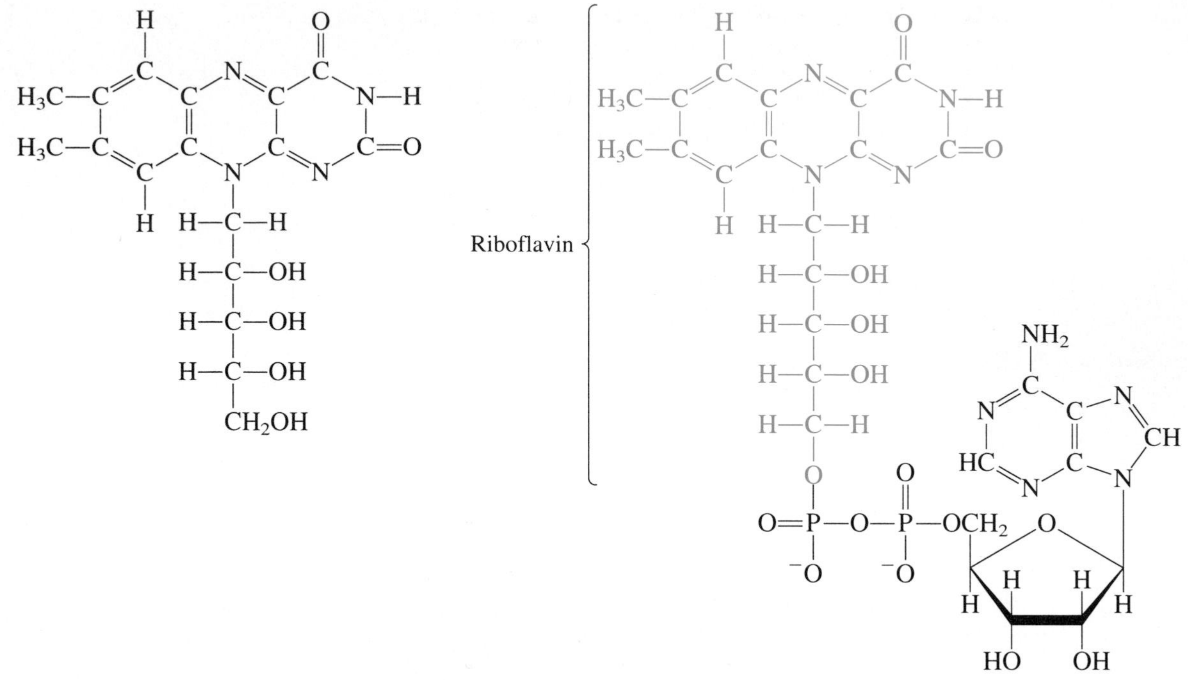

Riboflavin

Riboflavin
(Vitamin B_2)

Flavin adenine dinucleotide
(FAD)

FIGURE 17.5
The structures of riboflavin and flavin adenine dinucleotide (FAD). Riboflavin is the portion of FAD that serves as a carrier of hydrogen atoms in energy-generating reactions.

Thiamine

Thiamine, also known as vitamin B_1, is required in the diet of all animals (Figure 17.6). In its biologically active form it is bonded to a pyrophosphate group. This forms the coenzyme thiamine pyrophosphate, which is required for many decarboxylation reactions, including the decarboxylation of pyruvate to form CO_2 and acetyl CoA in the transition reaction between glycolysis and the citric acid cycle.

Section 13.7

Thiamine is abundant in whole-grain and enriched cereals, meats, legumes, and green leafy vegetables. The RDA of thiamine is about 1.5 mg. Thiamine is lost from whole grains during the refining process. However, thiamine deficiency is largely prevented because many foods, including bread and cereal products, contain thiamine as an additive.

FIGURE 17.6
The structures of thiamine and thiamine pyrophosphate.

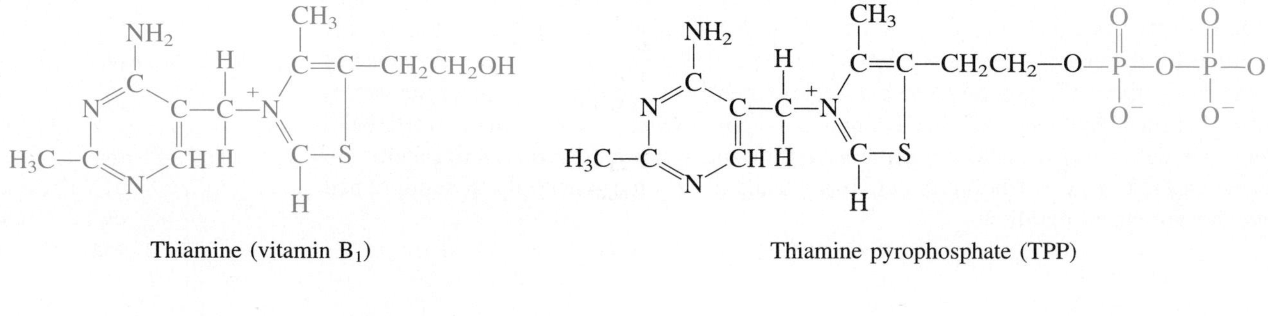

Thiamine (vitamin B_1)

Thiamine pyrophosphate (TPP)

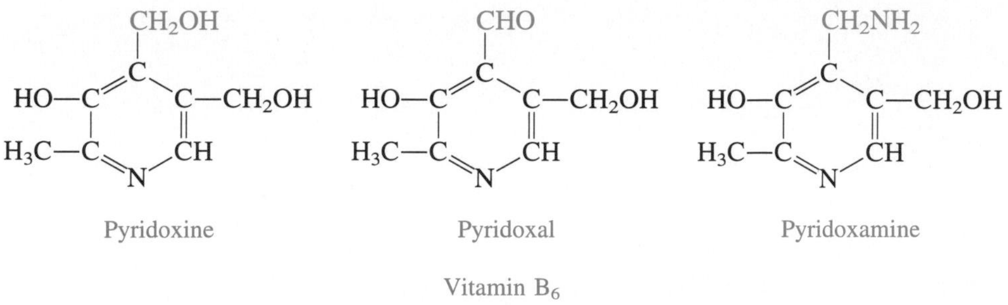

Pyridoxine Pyridoxal Pyridoxamine

Vitamin B_6

FIGURE 17.7
The vitamin B_6 family: pyridoxine, pyridoxal, and pyridoxamine.

Dietary deficiency leads to **beriberi,** a disease that is characterized by muscle weakness and mental instability. Beriberi is most often found in Asia, where rice is a staple of the diet. Although thiamine is abundant in rice and other whole grains, the rice-polishing machinery that was introduced late in the nineteenth century strips off the husks of the grain, and it is the husks that contain the thiamine. Severe malnutrition from such refined rice is the result. Sudden recovery from the effects of beriberi is observed within hours of administration of thiamine.

Pyridoxine

Pyridoxine, also known as vitamin B_6 (Figure 17.7), is required for the synthesis and breakdown of amino acids. Fish, meat, poultry, and leafy green vegetables are excellent sources of vitamin B_6. Because this vitamin is so readily available in a variety of foods, its deficiency is relatively rare. When it does occur, the symptoms include nervousness and muscular weakness. The recommended daily intake of vitamin B_6 is 2.0 mg.

Because vitamin B_6 is a water-soluble vitamin, one would expect that the ingestion of excessive amounts would result simply in the excretion of the excess. However, excess vitamin B_6 (50–100 times the RDA) that is taken to reduce the symptoms of premenstrual syndrome has resulted in peripheral neuropathy in several young women. This is characterized by a numbness in the limbs and a clumsy, stumbling walk.

Folic acid and the sulfanilamide antibiotics

Folic acid is a complicated molecule whose structure consists of three components: a heterocyclic ring system known as *pterin, para-aminobenzoic acid,* and *glutamic acid* (Figure 17.8). It is required for the synthesis of the amino acid methionine and the nucleic acid precursors, the purines and pyrimidines. The RDA of folic acid is only about 0.4 mg. Since such a small amount is required daily, it might be thought that folic acid deficiency would be rare. The opposite is true: Folic acid deficiency is a very common vitamin deficiency. Green vegetables, whole-grain cereals, and meat contain abundant folic acid, but it is destroyed by cooking. Because a deficiency of folic acid results in anemia and growth failure, folic acid is especially necessary for children and pregnant women.

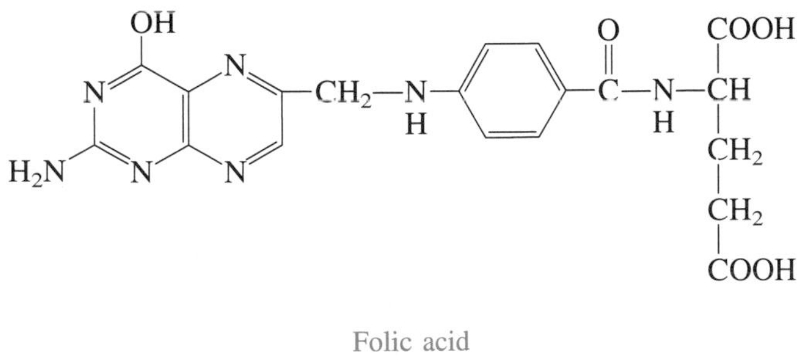

Folic acid

FIGURE 17.8
The structure of folic acid.

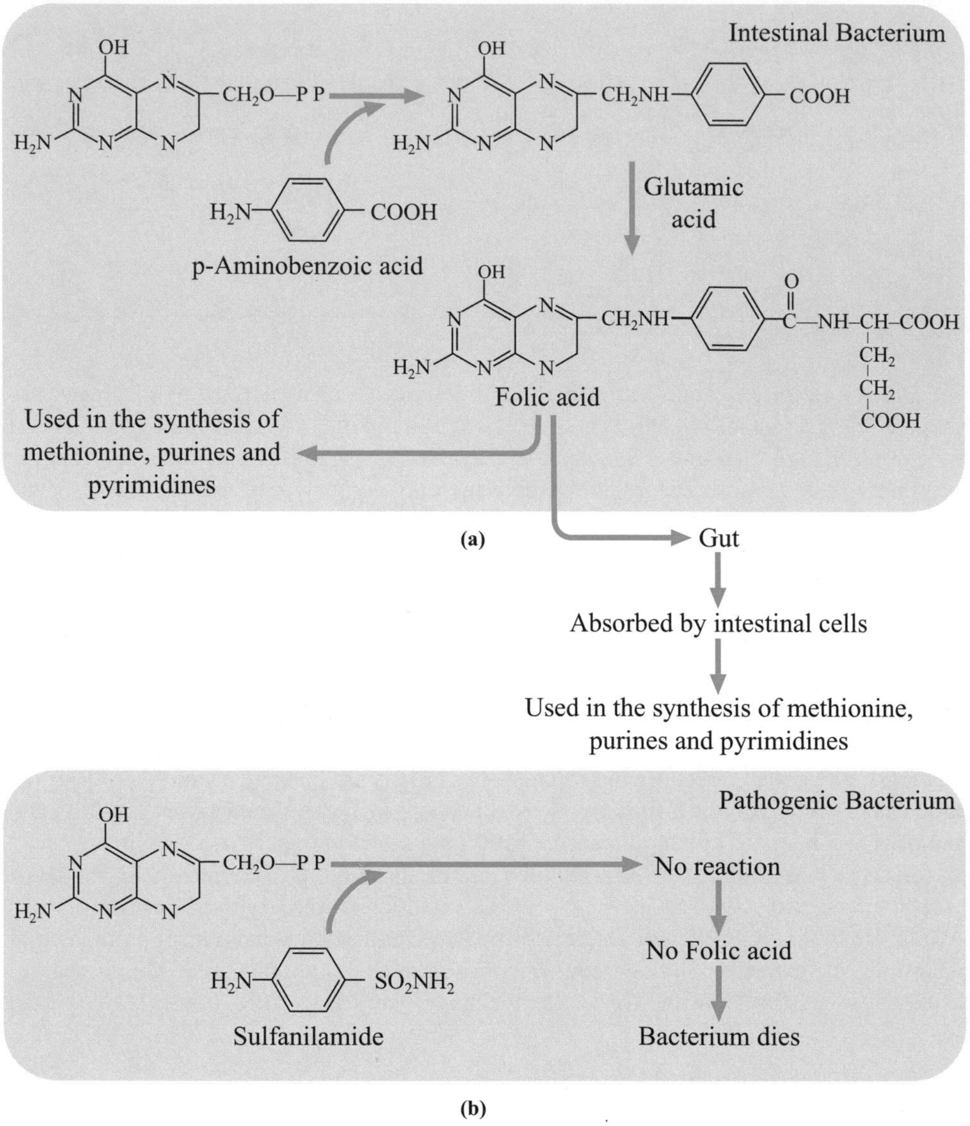

Intestinal Bacterium

p-Aminobenzoic acid

Glutamic acid

Folic acid

Used in the synthesis of methionine, purines and pyrimidines

(a)

Gut

Absorbed by intestinal cells

Used in the synthesis of methionine, purines and pyrimidines

Pathogenic Bacterium

No reaction

No Folic acid

Bacterium dies

Sulfanilamide

(b)

FIGURE 17.9
(a) The pathway of folic acid synthesis in intestinal bacteria. (b) The sulfa drugs, including sulfanilamide, are competitive inhibitors of the enzyme that links PABA to pterin. In the presence of the antibiotic the bacteria are unable to produce folic acid, and they die.

Humans and other mammals are unable to synthesize folic acid and must rely on an exogenous, or outside, source. In addition to the folic acid supplied in the diet, we obtain folic acid from our intestinal bacteria. A portion of the pathway for the synthesis of folic acid is shown in Figure 17.9a. Obviously, one of the very early steps in this pathway requires para-aminobenzoic acid (PABA) as one of the substrates. Without PABA this reaction will not proceed, folic acid will not be made, and the cell will die.

As discussed in Section 12.10, structural analogues are molecules that resemble the normal substrate for an enzyme and inhibit the activity of the enzyme by binding to the active site.

The **sulfanilamide antibiotics,** discovered by Gerhard Domagk in the 1930s, were shown to kill bacteria. Somewhat later, it was demonstrated that this antimetabolite did not work well in the presence of PABA. Careful examination of the structures of PABA and sulfanilamide reveals that they are structural analogues of one another. Since the two structures are so similar, both can bind to the enzyme catalyzing the first step of folic acid synthesis. In other words, sulfanilamide is a **competitive inhibitor** of PABA.

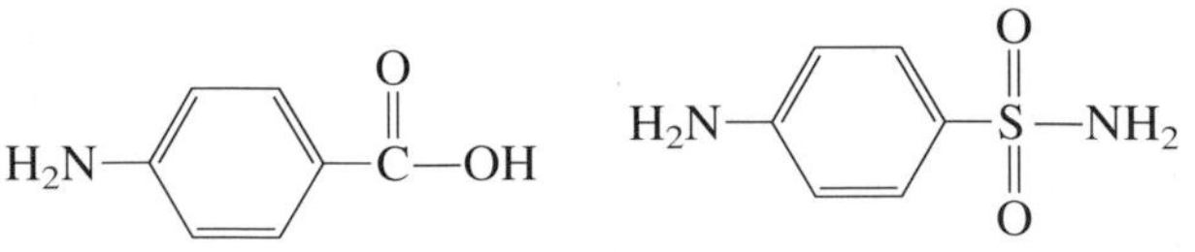

p-Aminobenzoic acid

Sulfanilamide

As shown in Figure 17.9a, if the correct substrate (PABA) is bound by the enzyme, the reaction occurs and the cell lives. However, if sulfanilamide is present in excess over PABA, it will bind more frequently to the active site of the enzyme, no folic acid will be produced, and the bacterial cell will die.

Thus the sulfanilamide antibiotics are called **antimetabolites** because they block a specific bacterial metabolic reaction. Luckily, we can exploit this property for the treatment of bacterial infections. Sulfanilamide and the group of similar compounds that have been developed are powerful weapons against bacterial infection. The use of sulfanilamide and other sulfa drugs has saved countless lives. Although bacterial infection was the major cause of death before the discovery of sulfa drugs and other antibiotics, at present death caused by bacterial infection is relatively rare.

Question 17.11 How do the sulfanilamide drugs selectively kill bacteria while causing no harm to humans?

Biotin

Biotin (Figure 17.10) is involved in carboxylation and decarboxylations in the intermediary metabolism of fats, carbohydrates, and proteins. Liver, egg yolks, cheese, and peanuts are excellent sources of biotin. In addition, it is produced by bacteria in the intestines.

Section 15.2

Although egg yolks are very rich in biotin, by a trick of nature raw egg white is a very poor source of biotin. Indeed, consumption of large quantities of raw eggs (12–24 per day) leads to biotin deficiency. Egg white contains a protein called *avidin,* which binds biotin extremely tightly. (The protein takes its name from its ''avid'' binding to biotin.) Cooking an egg, however, destroys avidin, freeing the biotin; this explains why cooked egg is an excellent nutritional source of biotin. The ESADDI for biotin is 0.3 mg, and in a normal diet, biotin deficiency is almost unknown. However, when it does occur, the symptoms include dermatitis (scaling and hardening of the skin), loss of appetite and nausea, muscle pain, and elevated levels of blood cholesterol.

Vitamin B_{12}

The extraordinary chemical structure of **vitamin B_{12}** is shown in Figure 17.11. B_{12} is a very important vitamin that is needed for the production of red and white blood cells and the normal growth and maintenance of nerve tissue.

A defective mechanism for the uptake of vitamin B_{12} results in **pernicious anemia,** a disease that is characterized by the presence of large immature red blood cells in the blood. Symptoms include a sore tongue, weight loss, and mental and nervous disorders. The damage to the central nervous system can even cause demyelination of the peripheral nerves in the arms and legs. Eventually, this can progress to the spinal cord. The require-

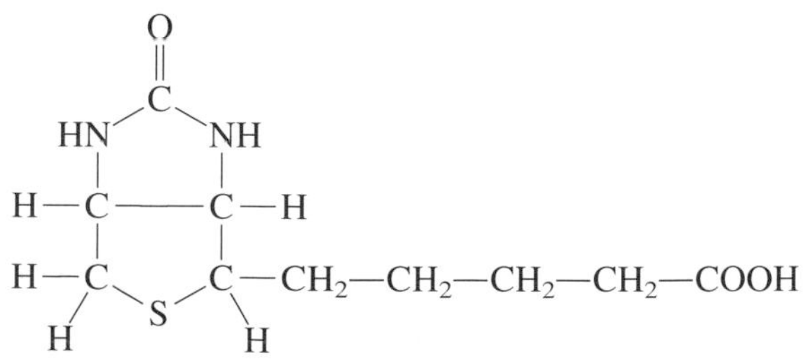

Biotin

FIGURE 17.10
The structure of biotin. This vitamin is involved in carboxylation and decarboxylation reactions in energy metabolism.

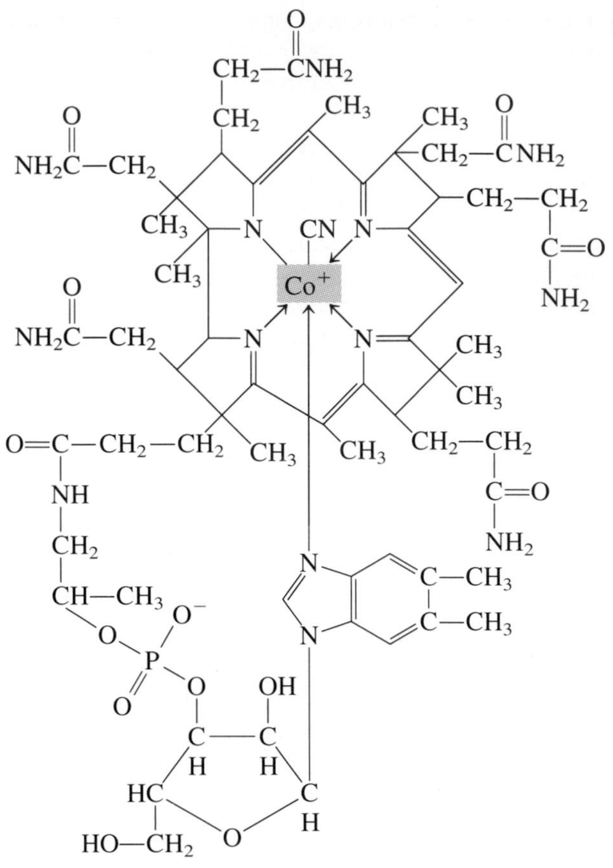

Vitamin B_{12} (cyanocobalamin)

FIGURE 17.11
The structure of vitamin B_{12}. This vitamin is critical for the formation of red and white blood cells and for the normal growth and maintenance of nerve tissue.

ment for vitamin B_{12} is only about 6 μg (0.006 mg) per day. Since many foods, including meats, eggs, and dairy products, contain this vitamin, nearly all diets, except those that are completely devoid of animal products, provide a sufficient amount of vitamin B_{12}. In fact, bacteria in the human intestine produce enough vitamin B_{12} to satisfy the normal daily requirement.

FIGURE 17.12
The structure of vitamin C (ascorbic acid).

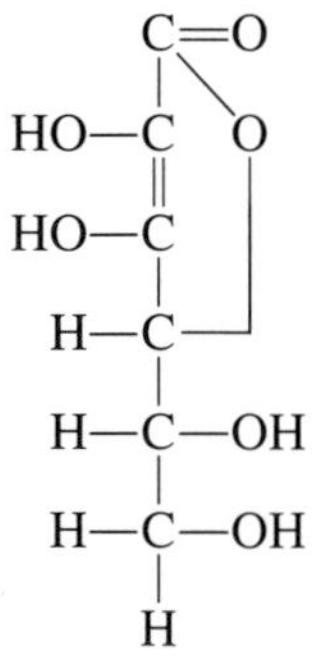

Vitamin C
(ascorbic acid)

Vitamin C

Vitamin C is important in the formation of collagen. It therefore functions in growth and repair of connective tissue, teeth, bones, and cartilage. In addition, it promotes wound healing, enhances absorption of iron, and functions in the biosynthesis of several hormones. Vitamin C also serves as an antioxidant in many biological processes. It is almost a part of folk medicine that large doses of **vitamin C,** or *ascorbic acid* (Figure 17.12), can prevent or cure the common cold and a host of other ailments.

The multiple assertions that have been made for the powers of vitamin C have not, however, been substantiated in extensive clinical testing. Although many individuals recommend megadoses of vitamin C, the RDA is only 60 mg. In fact, the ingestion of large doses, more than 1–2 g daily, has been reported to cause intestinal cramps, nausea, diarrhea, and kidney stones. Fresh fruits, especially citrus fruits, and vegetables, among them potatoes, are rich dietary sources of vitamin C. However, extensive cooking of vegetables or fruits destroys vitamin C.

A deficiency of vitamin C leads to **scurvy,** a disorder that is characterized by bleeding gums, loss of teeth, sore joints, and slow wound healing.

TABLE 17.5 Major Nutritional Sources of Lipid-Soluble Vitamins Required by Humans and Some Physiological Effects of Deficiencies

Vitamin	Source	Symptoms of Deficiency
A, carotene	Egg yolk, green and yellow vegetables, liver, fruits	Night blindness, blindness in children
D_3, calciferol	Milk, action of sunlight on skin	Rickets (malformation of the bones)
E	Vegetable oil	Fragile red blood cells
K	Leafy vegetables, intestinal bacteria	Blood-clotting disorders

TABLE 17.6 Functions of the Lipid-Soluble Vitamins

Vitamin	Function
Vitamin A	Synthesis of visual pigments
Vitamin D	Regulation of calcium metabolism
Vitamin E	Antioxidant, protection of cell membranes
Vitamin K	Required for the carboxylation of prothrombin and other blood-clotting factors

17.4 LIPID-SOLUBLE VITAMINS

Vitamins A, D, E, and K are soluble in lipids and in biological membranes. Many of the functions of lipid-soluble vitamins are intimately involved in metabolic processes that occur in membranes. The common sources and functions of the lipid-soluble vitamins are summarized in Tables 17.5 and 17.6.

Vitamin A

Vitamin A (Figure 17.13) is obtained in the active form called *retinol* from animal sources such as liver and egg yolks. It is also acquired in the precursor form, *provitamin A* or *carotene,* from plant foods. Green and yellow vegetables and fruits are good sources of vitamin A. Carrots are especially rich in this vitamin.

Vitamin A helps to maintain the skin and mucous membranes of the oral cavity and the digestive, respiratory, reproductive, and urinary tracts. Vitamin A is also critical for vision. The aldehyde form of vitamin A, called **retinal,** binds to a protein called **opsin** to form the visual pigment **rhodopsin.** This pigment is found in the **rod cells** of the retina of the eye. These cells are responsible for black-and-white vision. When light is absorbed by rhodopsin, the retinal portion is isomerized to a *trans* isomer, which dissociates from opsin. This dissociation increases the permeability of the rod cell membrane to ions. The influx of ions, in turn, stimulates nerve cells that send signals to the brain. Interpretation of those signals produces the visual image.

See also Chapter 6, "A Human Perspective: The Chemistry of Vision"

Following the initial light stimulus, retinal returns to the *cis* isomer and reassociates with opsin. Thus the system is ready for the next impulse of light. However, some retinal is lost in this process and must be replaced by conversion of dietary vitamin A to retinal. As you might expect, a deficiency of vitamin A can have terrible consequences. In children, lack of vitamin A leads to **xerophthalmia,** an eye disease that results first in night blindness and eventually in total blindness. The disease can be prevented by an

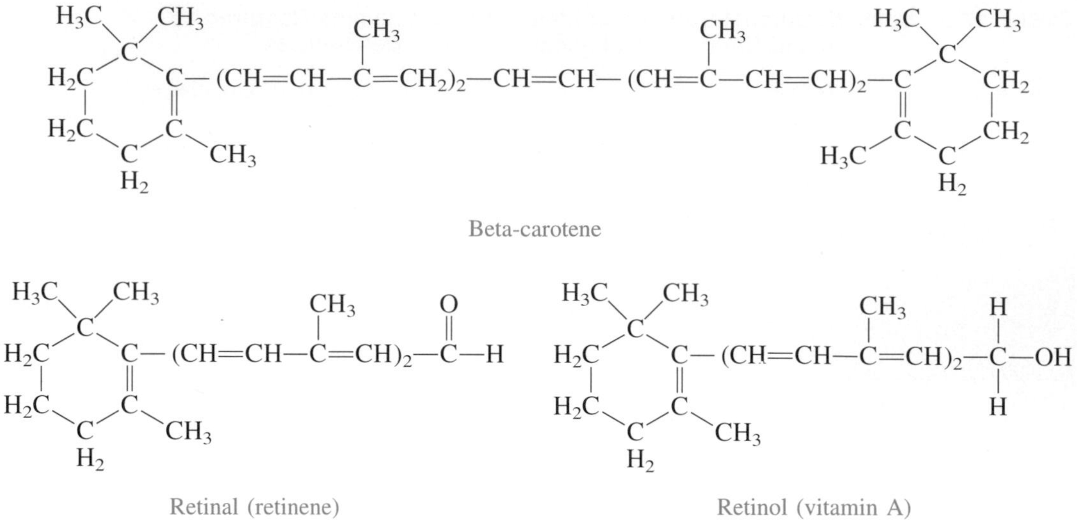

Beta-carotene

Retinal (retinene)

Retinol (vitamin A)

FIGURE 17.13
The structures of β-carotene, retinal, and vitamin A (retinol). Vitamin A is required for the synthesis of the visual pigments in the rods and cones of the eyes.

adequate dietary or supplementary supply of this vitamin. Because vitamin A is stored in the liver, a dose of 0.03 mg will protect a child for six months. Yet in countries that have suffered from cruel famines, even this amount of vitamin A is unavailable, and the burdens of malnutrition and disease lead to total blindness in thousands of children.

The current RDA for vitamin A is expressed in **retinol equivalents (R.E.).** Each R.E. is equal to 1 μg of retinol or 6 μg of β-carotene. For males the RDA is 1000 R.E., and for females it is 800 R.E. Since vitamin A is a fat-soluble vitamin that is stored in the liver, it is also dangerous to ingest quantities larger than the RDA. Manifestations of vitamin A poisoning **(hypervitaminosis)** include elevated pressure of the spinal fluid and around the brain, as well as swelling around the optic nerve. These result in severe headaches. Other symptoms include anorexia, swelling of the spleen and liver (hepatosplenomegaly), irritability, hair loss, and scaly dermatitis. It is interesting to note that early Arctic explorers suffered from hypervitaminosis A, experiencing many of these symptoms. Later it was found that this was the result of eating polar bear liver, which has an unusually high concentration of vitamin A.

Vitamin K

The formation of a blood clot in response to a wound is an intricate process that involves at least a dozen proteins in the blood serum. **Vitamin K** (Figure 17.14) is involved in blood coagulation. In clot formation, molecules of the serum protein **prothrombin** must be proteolytically cleaved to become **thrombin,** which then initiates the final stages of clot formation. This process is aided by Ca^{2+} ions. Prothrombin binds 10–12 Ca^{2+} ions

Sections 12.13 and 12.14

FIGURE 17.14
The structure of vitamin K. This vitamin is a coenzyme involved in the synthesis of prothrombin and several other factors involved in normal blood clotting.

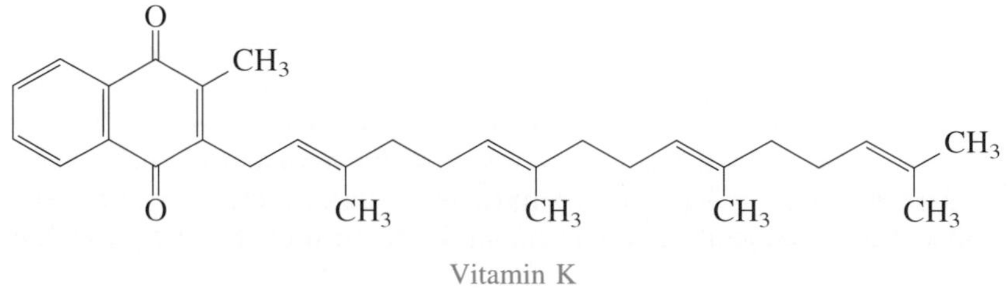

Vitamin K

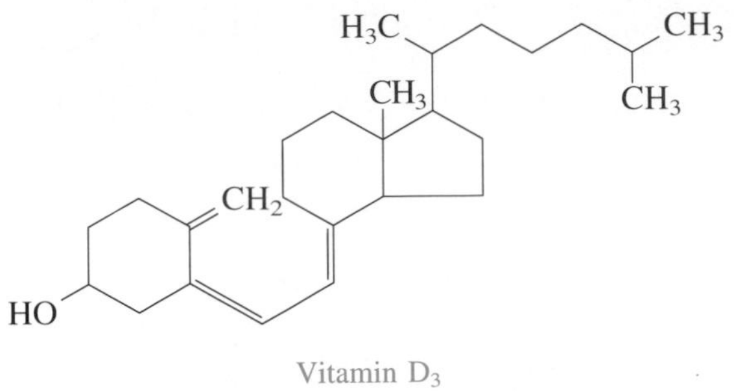

H_3C CH_3 CH_3 CH_3 CH_2 HO

Vitamin D_3

FIGURE 17.15
The structure of vitamin D.

at positions that have the unusual amino acid γ-carboxyglutamate. Since vitamin K is a coenzyme in the carboxylation of the glutamate residues, the lack of vitamin K in the diet leads to a deficiency of blood clotting. Vitamin K is found in leafy vegetables and is conveniently manufactured by intestinal bacteria. It is extremely rare for adults to suffer from vitamin K deficiency, but it is observed in some individuals who are on antibiotic therapy or have fat absorption problems. However, newborns frequently suffer from a vitamin K deficiency because they lack intestinal bacteria. They are often administered injections of vitamin K to prevent excessive bleeding in the early days of their lives.

Since vitamin K is another fat-soluble vitamin, it is possible to suffer from hypervitaminosis K. The symptoms include gastrointestinal disturbances and anemia.

Vitamin D

Vitamin D (Figure 17.15) plays a major role in the regulation of calcium levels and therefore is required for the proper formation of bone and teeth. Liver and fish oils are rich in this vitamin, and the cod liver oil that was traditionally administered to children in New England had great nutritional value.

A deficiency of vitamin D causes **rickets,** a disease that causes children to have soft, deformed, and poorly calcified bones. Vitamin D deficiency is almost totally confined to children, who require a supplement of about 20 mg per day in their diet. The action of sunlight on skin produces adequate amounts of vitamin D in most adults; however, homebound elderly people may suffer from vitamin D deficiency. The vitamin D produced in the skin results from the action of ultraviolet light on 7-dehydrocholesterol (Figure 17.16). This alcohol is then hydroxylated in the liver to produce 25-hydroxy vitamin D_3, also called 25-hydroxycholecalciferol. In the kidney a final hydroxylation produces the hormone, 1,25-dihydroxy vitamin D_3 (1,25-dihydroxycholecalciferol), the active form of the vitamin. 1,25-dihydroxy vitamin D_3 is classified as a hormone because it is synthesized in one part of the body but exerts its effects elsewhere. Only about 1% of the body's calcium exists outside of bone, but regulating the concentration of calcium ions in the blood is critical because these soluble calcium ions are involved in many physiological processes from blood clotting to muscle contraction, including heart function. When the level of calcium in the blood is low, 1,25-dihydroxy vitamin D_3 stimulates the uptake of calcium from the intestines and its transport into the blood. In the kidneys this hormone, along with parathyroid hormone, stimulates the reabsorption of calcium so that it is not lost in the urine. If the blood level of calcium is low enough, 1,25-dihydroxy vitamin D_3 even stimulates the removal of calcium from the bone. This, of course, leads to the weakening of the bone, which can result in *osteomalacia,* very brittle, decalcified bones, later in life.

Section 9.4

People who have chronic kidney failure suffer from painful bone deterioration. Administration of the hormone 1,25-dihydroxy vitamin D_3 in amounts as small as 1 μg per day alleviates this condition. Recent studies suggest that many elderly people in this country may be suffering from vitamin D deficiency because of a decrease in the consumption of milk and limited exposure to sunlight. In fact, elderly people produce only about 30–50% the amount of vitamin D produced by a young adult. This unfortunate fact might not be discovered until an elderly patient is admitted to the hospital with a broken hip or broken back.

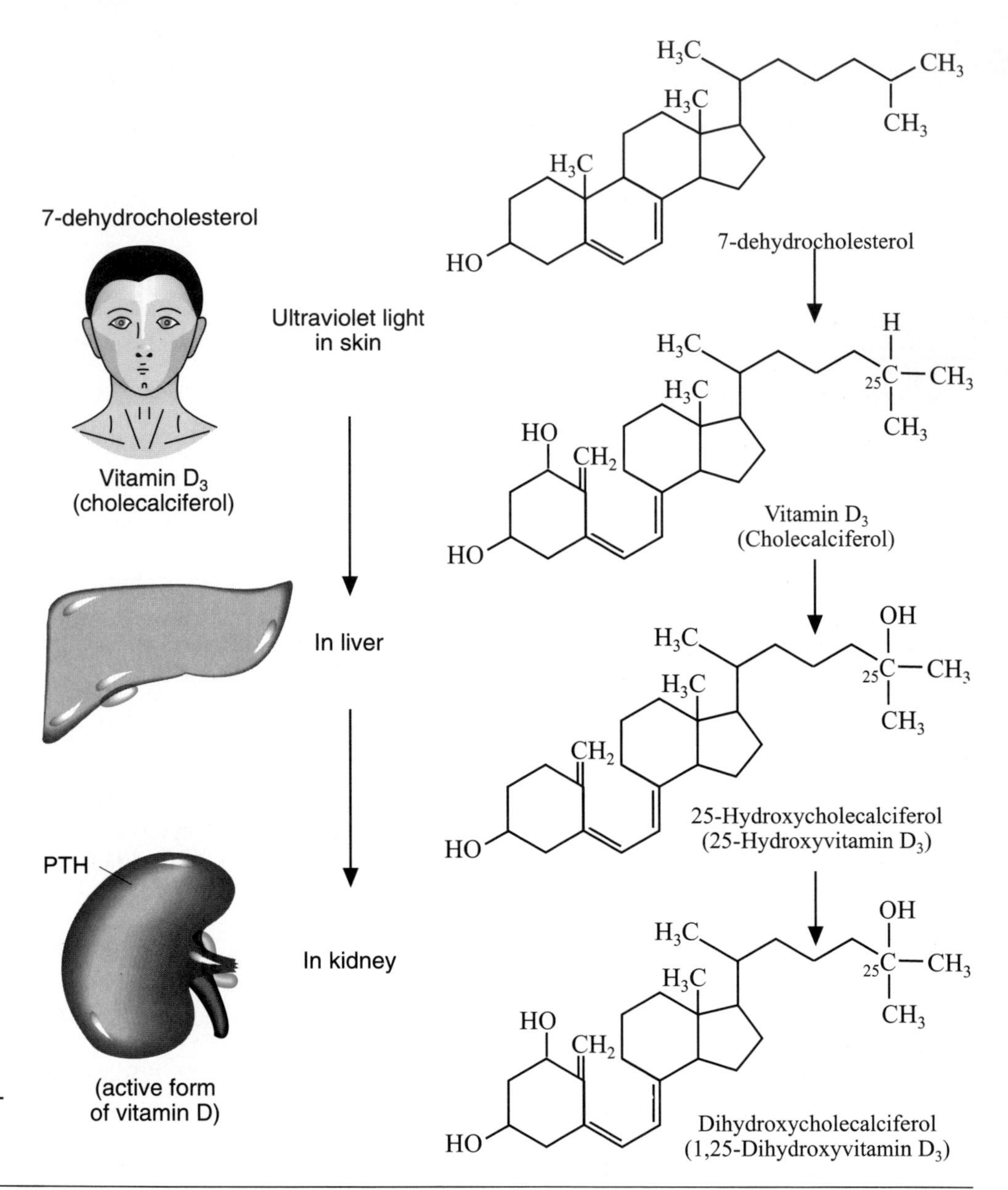

FIGURE 17.16
The pathway of the synthesis of vitamin D_3 in the skin and its conversion to the hormone active in calcium metabolism.

Vitamin D is another of the fat-soluble vitamins. The ingestion of excess vitamin D can result in hypervitaminosis D with the presentation of the following symptoms: renal failure, weight loss, and calcification of soft tissues of the body.

Vitamin E

Vitamin E (Figure 17.17) is the least well understood of the lipid-soluble vitamins. In fact, the term vitamin E actually refers to a family of eight compounds called the *tocopherols*.

Rats that are deprived of vitamin E become infertile, but the reasons for this are unknown. Vitamin E is known to prevent the oxidation of double bonds in the hydrocarbon groups of membrane lipids, and this could be its major function. Since oxidation reactions accelerate with aging, vitamin E might help to retard the aging process. The RDA for vitamin E is expressed in **α-tocopherol equivalents (α-TE)** because this is the most active form of vitamin E. The recommended daily intake is 10 α-TE for males and 8 α-TE for females. This is roughly the amount of vitamin E obtained from a tablespoon of vegetable oil.

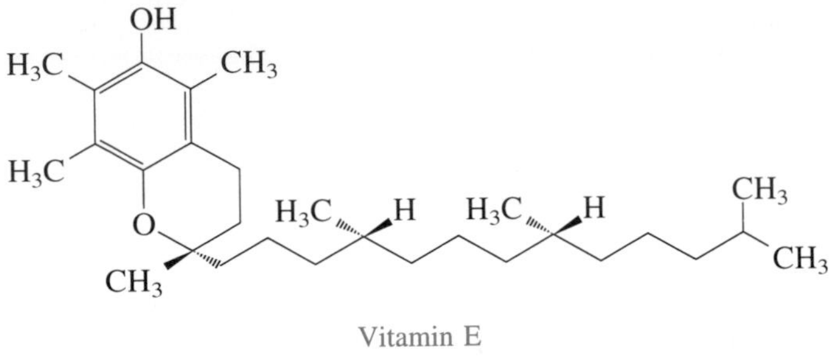

Vitamin E

FIGURE 17.17
The structure of vitamin E. The functions of this mysterious vitamin have not yet been clearly defined; however, it appears to be an antioxidant.

Compared to vitamins A and D, vitamin E is relatively nontoxic at high levels. However, it is unwise to drastically exceed the RDA because high levels of vitamin E could cause diarrhea, nausea, headache, and fatigue.

Question 17.12 Why does hypervitaminosis more commonly result from ingestion of excessive amounts of the fat-soluble vitamins than the water-soluble vitamins?

SUMMARY

A variety of nutrients are required by the body. These serve as sources of metabolic energy, coenzymes or cofactors for enzymes that catalyze the metabolic reactions, structural elements, and critical intracellular and extracellular ions. All of the nutrients required by the body can easily be obtained from a well-balanced diet.

The nutrients in the human diet that provide cells with metabolic energy are carbohydrates, fats, and proteins. Many minerals are required in the diet: calcium and phosphorus are incorporated in bones and teeth; sodium is the major extracellular cation; potassium is the major intracellular cation; magnesium is required for many metabolic processes; iron is required for heme-containing proteins; copper is required for essential enzymes; fluoride prevents dental caries; and iodine is incorporated in certain thyroid hormones.

Vitamins are trace nutrients required in the diet. They are classified as either water-soluble or fat-soluble.

The water-soluble vitamins are essential for many metabolic pathways. They are generally found as components of coenzymes. For instance, riboflavin is a component of FAD; niacin is a component of NAD^+ and $NADP^+$; and pantothenic acid is a component of coenzyme A. All of these coenzymes are required by the cellular energy generating pathways. Vitamin C is the only water-soluble vitamin that is not known to be a component of a coenzyme. It is required for the synthesis of collagen.

Fat-soluble vitamins are involved in many biological processes. Vitamin A is involved in the biochemistry of vision; vitamin K is required for blood clotting; vitamin D participates in calcium metabolism; and vitamin E is implicated in the process of aging.

The absence or deficiency of vitamins results in impaired metabolic activity and in many vitamin-deficiency diseases. Niacin deficiency causes pellagra; thiamine deficiency leads to beriberi; a deficiency in vitamin B_{12} assimilation results in pernicious anemia; and vitamin C deficiency causes scurvy.

Ingestion of large quantities of vitamins can lead to hypervitaminosis. This is particularly true of the fat-soluble vitamins, because the excess is stored in the body rather than being excreted in the urine.

GLOSSARY OF KEY TERMS

acquired pellicle (17.1) a layer of saliva glycoproteins that coats the teeth and allows adherence of oral bacteria onto tooth surfaces.

antimetabolite (17.3) a drug that blocks a specific enzymatic reaction.

beriberi (17.3) a disease that is caused by a dietary deficiency of thiamine. It is characterized by muscle weakness and mental instability.

biotin (17.3) a vitamin that is required for carboxylation and decarboxylation reactions involved in the metabolism of fats, carbohydrates, and protein.

calcitonin (17.4) a hormone produced by the thyroid gland that decreases the level of blood calcium by inhibiting kidney reabsorption and intestinal absorption and causing incorporation of calcium into bone.

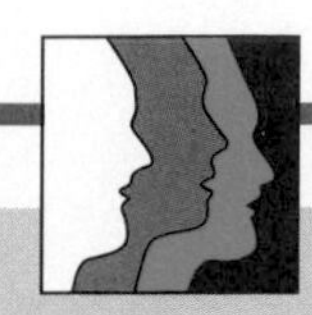

A HUMAN PERSPECTIVE

Osteoporosis and Calcium Metabolism

Osteoporosis, a disease characterized by brittle bones, is responsible for more than a million broken bones—especially in the hips, wrists, and spines—of older Americans. The realization that so many of our elderly people suffer from osteoporosis has spurred great interest in the primary mineral component of bone, calcium. Advertisements in magazines and on television urge us to take calcium supplements to prevent crippling osteoporosis later in life and point out that calcium might help to prevent colon cancer and lower high blood pressure.

But is calcium *alone* enough to protect us from osteoporosis? The answer seems to be a resounding no. Many lines of evidence suggest that calcium supplements by themselves will not even slow down the loss of bone calcium in postmenopausal women.

To understand osteoporosis and the ways in which it can be prevented and treated, we must understand the system of checks and balances that regulates calcium levels in the blood and the bones. Several hormones are involved in this regulation, but the primary controlling hormone is **parathyroid hormone (PTH).** The parathyroid gland responds instantly to fluctuations in the blood concentration of calcium. When the concentration drops, PTH is released. The accompanying figure summarizes the effects of the release of PTH into the blood. In the bone, PTH stimulates resorption, that is, the removal of calcium from the bone, for use in the bloodstream. In the kidneys it stimulates calcium retention, and in the intestines it stimulates the absorption of dietary calcium. The latter effect is not a direct result of the action of PTH but is mediated by the hormone form of vitamin D_3. PTH stimulates the synthesis of 1,25-dihydroxy vitamin D_3 in the kidney. The main target of 1,25-dihydroxy vitamin D_3 is the intestines, where it stimulates the cells of the intestinal wall to absorb dietary calcium. In addition, it enhances the reabsorption of calcium, phosphorus, and sodium in the kidneys. By increasing the PTH sensitivity of some bone cells, 1,25-dihydroxy vitamin D_3 further enhances the resorption of bone calcium.

This complicated cascade is just one part of calcium regulation. When the blood concentration of calcium rises, the thyroid gland secretes another hormone, **calcitonin.**

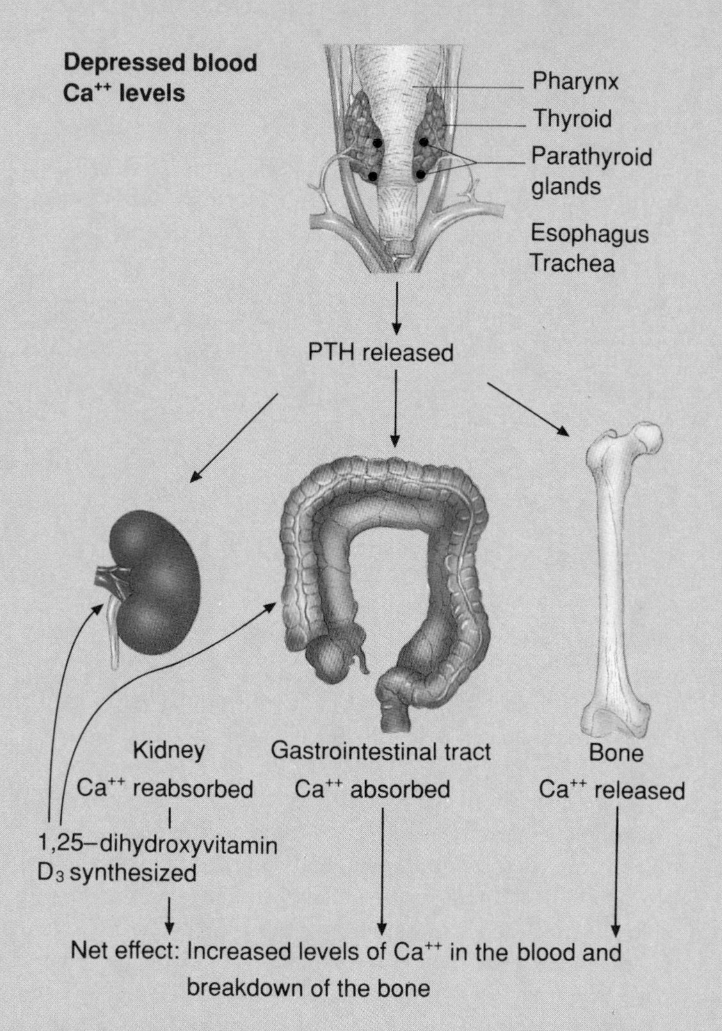

A summary of the hormones that control the amount of calcium in the blood.

carbohydrates (17.1) an organic compound usually characterized by the general formula $(CH_2O)_n$. These are generally sugars and polymers of sugars that are the primary source of energy for the cell.

coenzyme (17.3) an organic group that is required by some enzymes. These generally serve as donors or acceptors of electrons or functional groups in a reaction.

coenzyme A (17.3) a coenzyme that is involved in the transfer of acyl groups in metabolic reactions.

competitive inhibitor (17.3) a chemical that has a structure that is very similar to that of the natural substrate of an enzyme and competes with the natural substrate for binding to the enzyme active site. The result is the inhibition of the reaction catalyzed by the enzyme.

complete protein (17.1) a protein source that contains all the essential and nonessential amino acids.

dental plaque (17.1) an aggregation of dextran and oral bacteria that builds up on teeth and provides an environment in which tooth decay can occur.

dextran (17.1) a polymer of glucose that forms the ground substance of dental plaque.

essential amino acids (17.1) amino acids that cannot be synthesized by the body and therefore must be supplied by the diet.

essential fatty acids (17.1) fatty acids that cannot be synthesized

The effects of calcitonin are opposite those of PTH, that is, calcium is incorporated into bones, and intestinal absorption and kidney reabsorption of calcium are inhibited.

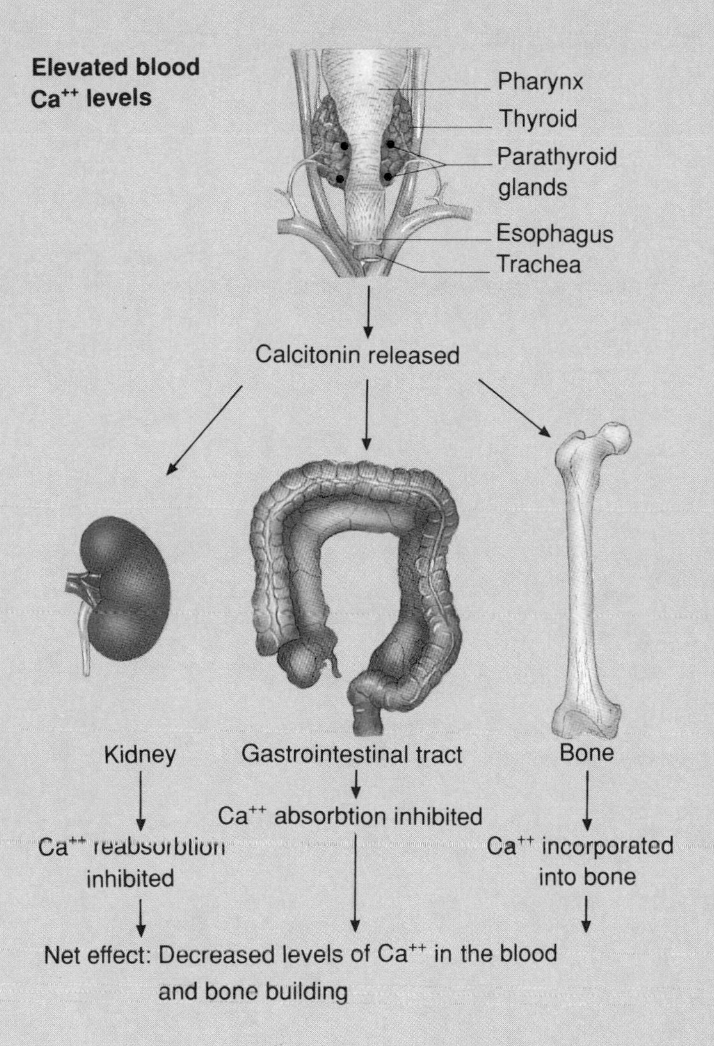

Thus our bones are constantly being broken down and built up in response to the level of calcium in the blood.

Children need extra calcium because their bones are actively growing, but they absorb calcium so efficiently that the RDA of 1000–1200 mg is sufficient. As we get older, we continue to need calcium. Although our bones are no longer growing, they continue to incorporate calcium to become more dense. Eating a well-balanced diet should allow a 35-year-old to have bones that are 10% more dense than those of a 20-year-old. After age 35 there is a tendency to consume less calcium, and it is absorbed less efficiently. In addition, we produce much less vitamin D. These factors cause a drain of bone calcium, and our bones begin to decline in strength. This condition is exacerbated by a number of additional factors, including smoking, drinking large amounts of alcohol or caffeine-containing beverages, and consuming large amounts of protein. In fact, postmenopausal women may lose bone at the rate of 1.5% per year! As a result, they may suffer from osteoporosis in later life.

What can be done to prevent this bone depletion? First, consume foods that are rich in calcium and vitamin D. Skim milk is an excellent source of both. A few minutes a day in the sun also provides the adult RDA of vitamin D. A word of caution: There is no need to spend longer periods in the sun, unprotected from harmful ultraviolet irradiation, to fulfill your vitamin D requirements. Second, get out and exercise. Researchers have determined that load-bearing exercise—walking, running, biking, and weight-lifting—is also necessary to maintain strong bones.

Osteoporosis can be treated with calcium supplements, vitamin D supplements, estrogen therapy, and even injections of the hormone calcitonin. But the bottom line is that this painful and crippling disease can be avoided by eating a diet that is rich in calcium-containing foods, exercising, and avoiding smoking and large amounts of alcohol or caffeine. Following these guidelines in your teens and throughout adulthood will allow you to have strong bones throughout your life.

by the body and therefore must be supplied in the diet.

fatty acids (17.1) long carbon chains with a carboxyl group at one terminus. These are the most concentrated source of energy used by the cell.

folic acid (17.3) a vitamin that is required for the transfer of one-carbon groups for the synthesis of methionine, purines, and pyrimidines.

glucosyl transferase (17.1) an enzyme that cleaves sucrose into glucose and fructose and converts the glucose into dextran.

hydroxyapatite (17.2) a calcium phosphate polymer that crystallizes on collagen and makes up the mineral portion of bone.

hypervitaminosis (17.4) any of several pathological syndromes that result from ingestion of very large quantities of a vitamin.

incomplete protein (17.1) a protein source that does not contain all the essential and nonessential amino acids.

major minerals (17.2) inorganic elements that are required in large amounts (greater than 100 mg/day) by the body.

niacin (17.3) a vitamin that is essential for the energy-generating reactions of the cell. It is a component of two coenzymes: nicotinamide adenine dinucleotide and nicotinamide adenine dinucleotide phosphate. (Also called vitamin B_3.)

nonessential amino acids (17.1) amino acids that can be synthesized by the body.

opsin (17.4) the protein portion of the visual pigment rhodopsin.

osteomalacia (17.4) loss of calcium from bones late in life, as a result of vitamin D deficiency. The result is brittle, decalcified bones.

osteoporosis (17.4) a disease characterized by brittle bones and caused by the gradual release of calcium from the bones.

pantothenic acid (17.3) a vitamin that is essential for the normal metabolism of fats and carbohydrates. It is needed for the synthesis of coenzyme A.

parathyroid hormone (PTH) (17.4) the primary hormone controlling the level of calcium in the blood. It is released when blood calcium levels fall and causes the increase by stimulating kidney reabsorption, intestinal absorption, and bone resorption.

pellagra (17.3) a disease that results from a niacin deficiency. It is characterized by dermatitis, diarrhea, dementia, and death.

pernicious anemia (17.3) a disease that results from a vitamin B_{12} deficiency. The symptoms include a decrease in the amount of hemoglobin in the blood, weight loss, and mental and nervous disorders.

protein (17.1) an organic compound that is primarily composed of carbon, hydrogen, oxygen, nitrogen, and sulfur. Proteins are polymers of amino acids linked to one another by peptide bonds. These represent both an energy source and a source of amino acids for the cell.

prothrombin (17.4) the inactive precursor of thrombin, required for blood clotting.

pyridoxine (17.3) a vitamin that is required for the synthesis and breakdown of amino acids. (Also called vitamin B_6.)

Recommended Dietary Allowance (RDA) (17.3) the daily recommended dose of a vitamin or mineral. The level is determined by considering the range of requirements in humans, using the value at the high end of the range, and adding an additional safety factor.

retinal (17.4) the aldehyde form of vitamin A that binds to the protein opsin to form the visual pigment rhodopsin.

retinol equivalent (R.E.) (17.4) the units that are used to express the RDA for vitamin A.

rhodopsin (17.4) the major visual pigment found in the rod cells of the retina of the eye.

riboflavin (17.3) a vitamin that is required by the energy-releasing reactions of the cell. It is a component of the coenzyme flavin adenine dinucleotide. (Also called vitamin B_2.)

rickets (17.4) a deficiency of vitamin D that results in the formation of soft, deformed bones in children.

rod cells (17.4) the cells of the retina of the eye that are responsible for black-and-white vision.

scurvy (17.3) a disease resulting from a deficiency of vitamin C. It is characterized by bleeding gums, loss of teeth, sore joints, and slow wound healing.

sulfanilamide (17.3) a structural analogue of *para*-aminobenzoic acid (PABA). This antibiotic competes with PABA for binding to the active site of an enzyme required for the synthesis of folic acid.

thiamine (17.3) a vitamin that is required for decarboxylation reactions. It is a component of the coenzyme thiamine pyrophosphate. (Also called vitamin B_1.)

thrombin (17.4) the proteolytic enzyme that converts fibrinogen to fibrin in the last stage of clot formation.

α-tocopherol equivalents (α-TE) (17.4) the units that are used to express the RDA for vitamin E.

tocopherols (17.4) a family of eight compounds collectively referred to as vitamin E.

trace minerals (17.2) inorganic elements that are required in very small amounts for normal cellular function.

vitamin (17.3) an organic substance that is required in the diet in small amounts. Some vitamins are used in the synthesis of coenzymes required for the function of cellular enzymes. Others are involved in calcium metabolism, vision, and blood clotting.

vitamin A (17.4) a vitamin that is essential to vision. (Also called retinol.)

vitamin B_{12} (17.3) a vitamin that is required for the production of red and white blood cells and the growth and maintenance of nerve tissue.

vitamin C (17.3) a vitamin that is required for the growth and repair of connective tissue, teeth, bones, and cartilage. It is involved in the formation of collagen. (Also called ascorbic acid.)

vitamin D (17.4) a vitamin that plays a major role in the regulation of calcium metabolism in the body.

vitamin E (17.4) a vitamin that prevents the oxidation of double bonds in the hydrocarbon chains of membrane lipids.

vitamin K (17.4) a vitamin that is vital to blood clotting. It serves as a coenzyme in the synthesis of prothrombin and other blood-clotting factors in the liver.

xerophthalmia (17.4) the condition caused by a deficiency of vitamin A that results in night blindness and eventually total blindness.

QUESTIONS AND PROBLEMS

Major Dietary Energy Sources

17.13 What are the major nutrients in the following:
a. milk products
b. meats
c. vegetables and fruits
d. grains

17.14 Why is it necessary to mix vegetable proteins to provide an adequate vegetarian diet?

17.15 Why are fats more concentrated energy sources than carbohydrates?

17.16 What class of foods is the best source of dietary protein?

17.17 What is the difference between essential and nonessential amino acids?

17.18 What is the difference between a complete protein and an incomplete protein?

Water-Soluble Vitamins and Coenzymes

17.19 What coenzyme requires pantothenic acid as one of its components?

17.20 What metabolic pathways require coenzyme A?

17.21 What disease results from niacin deficiency?

17.22 What coenzyme contains niacin?

17.23 What are the best dietary sources of riboflavin?

17.24 What condition results from riboflavin deficiency?

17.25 What clinical condition results from thiamine deficiency?

17.26 What is the major physiological function of coenzymes derived from pyridoxine?

17.27 What is the function of *para*-aminobenzoic acid?

17.28 What is the structure of sulfanilamide and how is it related to the structure of folic acid?

17.29 Why are sulfa drugs called antimetabolites?

17.30 What is a major source of biotin?

17.31 Explain why people who consume large amounts of raw eggs can develop biotin deficiency.

17.32 What metal ion is part of the structure of vitamin B_{12}?

17.33 What disease results from a defect in the uptake of vitamin B_{12}?

17.34 What is the physiological function of vitamin C?

17.35 What disease is associated with a deficiency of vitamin C?

Lipid-Soluble Vitamins

17.36 What is the role of vitamin A in the visual cycle?

17.37 What molecule is the precursor of vitamin A?

17.38 What is the function of rhodopsin?

17.39 What is the cause of night blindness?

17.40 What disease results from a deficiency of vitamin D?

17.41 What is the function of vitamin E?

17.42 Why is 1,25-dihydroxycholecalciferol classified as a hormone?

17.43 What process requires vitamin K?

17.44 What is the physiological function of rod cells?

17.45 What vitamin deficiency is associated with xerophthalmia?

Further Problems

17.46 What are the functions of each of the following minerals in human metabolism?

a. chloride
b. magnesium
c. iron
d. fluoride
e. copper
f. iodine

17.47 Why doesn't the body store water-soluble vitamins?

17.48 Why have South American Indians historically soaked corn in lime water?

17.49 Why does polishing rice destroy much of its nutritional value?

17.50 What does avidin do?

17.51 Why does cooking eggs make them a good source of biotin?

17.52 Fill in the blanks:

a. The minerals that are important for bone and teeth are ______ and ______.
b. A deficiency in thiamine causes the condition called ______.
c. The substance that is the precursor to vitamin A is called ______.
d. Amino acids that cannot be synthesized by the body are called ______ amino acids.

17.53 Which dietary supplement is a component of the coenzymes NAD^+ and $NADP^+$?

17.54 List the major biological cations that are important to one's diet.

17.55 List the trace minerals that are important to one's diet.

17.56 List the vitamins that are lipid-soluble.

17.57 What is the major source of magnesium in one's diet?

17.58 List the vitamins and minerals that are toxic if ingested in large quantities.

Appendixes

OUTLINE

APPENDIX I

Metric Units

The metric system was originally developed in France just before the French Revolution in 1789. The modern version of this system is the "Système International" or *S.I. System*. Although the S.I. system has been in existence for over 30 years, it has yet to gain widespread acceptance. To make the S.I. system truly "systematic," it utilizes certain units, especially those for pressure, that many find cumbersome. Applications in this text use the metric system, not the S.I. system. Both the metric and English systems are used to demonstrate conversion of units within a system and conversions from one system to another.

In the metric system there are three basic units. Mass is represented as the *gram,* length as the *meter,* and volume as the *liter*. Any subunit or multiple unit contains one of these unit terms preceded by a prefix indicating the power of ten by which the base unit is to be multiplied. The most common metric prefixes are shown in Table I.1.

$$1 \text{ milliliter (mL)} = \frac{1}{1000} \text{ liter} = 0.001 \text{ liter} = 10^{-3} \text{ liter}$$

A volume unit is indicated by the base unit, liter, and the prefix, milli-, which indicates that the unit is one thousandth of the base unit. In the same way,

$$1 \text{ milligram (mg)} = \frac{1}{1000} \text{ gram} = 0.001 \text{ gram} = 10^{-3} \text{ gram}$$

$$1 \text{ millimeter (mm)} = \frac{1}{1000} \text{ meter} = 0.001 \text{ meter} = 10^{-3} \text{ meter}$$

$$1 \text{ kilogram (kg)} = 1000 \text{ grams} = 10^{3} \text{ grams}$$

$$1 \text{ microgram } (\mu\text{g}) = \frac{1}{1{,}000{,}000} \text{ gram} = 0.000001 \text{ gram} = 10^{-6} \text{ gram}$$

$$1 \text{ decigram (dg)} = \frac{1}{10} \text{ gram} = 0.1 \text{ gram} = 10^{-1} \text{ gram}$$

TABLE I.1 Some Common Prefixes Used in the Metric System

Prefix	Multiple	Decimal Equivalent
mega (M)	10^6	1,000,000.
kilo (k)	10^3	1,000.
deka (da)	10^1	10.
deci (d)	10^{-1}	0.1
centi (c)	10^{-2}	0.01
milli (m)	10^{-3}	0.001
micro (μ)	10^{-6}	0.000001
nano (n)	10^{-9}	0.000000001

UNIT CONVERSION

We must have a *conversion factor,* or series of conversion factors, that relate two units in attempting to convert from one unit to another. The proper use of these conversion factors is referred to as the *factor-label method*.

This method is used in either of the following calculations:

Convert from one unit to another within the *same system* or

Convert units from *one system to another*.

Conversion of units within the same system

In the English system,

$$1 \text{ gallon} = 4 \text{ quarts}$$

Since dividing both sides of the equation by the same term does not change its identity,

$$\frac{1 \text{ gallon}}{1 \text{ gallon}} = \frac{4 \text{ quarts}}{1 \text{ gallon}}$$

The term on the left is equal to unity (1); therefore

$$1 = \frac{4 \text{ quarts}}{1 \text{ gallon}}$$

Now, multiplying any other term by the ratio (4 quarts/1 gallon) will not change the value of the term because multiplication by unity produces the original value. However, there is one important difference: the units will have changed.

Example I.1

Convert 7 gallons to units of quarts.

$$7 \cancel{\text{gal}} \times \frac{4 \text{ qt}}{1 \cancel{\text{gal}}} = 28 \text{ qt}$$

The conversion factor, (4 qt/1 gal), serves as a bridge, or unifying term, between the unit that was given (units of gal) and the unit that was sought (units of qt).

The conversion factor may be written as (4 qt/1 gal) or (1 gal/4 qt), since both are equal to unity. However, only the first factor, (4 qt/1 gal), produces the units that are sought in the problem. If we had set up the problem incorrectly, we would get

$$7 \cancel{\text{gal}} \times \frac{1 \text{ gal}}{4 \text{ qt}} = 1.75 \text{ gal}^2/\text{qt}$$

Incorrect units

Clearly, units of gal^2/qt are not those asked for in the problem, nor are they reasonable units. The factor-label method is therefore a self-indicating system; only if the factor is set up properly will the correct units result.

A variety of commonly used English system conversion factors is included in Table I.2.

TABLE I.2 Some Common Conversion Factors Used in the English System

A. Weight	1 pound (lb) = 16 ounces (oz)	C. Volume	1 gallon (gal) = 4 quarts (qt)
	1 ton (t) = 2000 pounds (lb)		1 quart (qt) = 2 pints (pt)
B. Length	1 foot (ft) = 12 inches (in)		1 quart (qt) = 32 fluid ounces (oz)
	1 yard (yd) = 3 feet (ft)		
	1 mile (mi) = 5280 feet (ft)		

Conversion of units within the metric system may be accomplished by using the factor-label method. Unit prefixes, which dictate the conversion factor, simplify unit conversion.

Example I.2

Convert 10 centimeters to meters.

If we recognize that the prefix *centi-* means 1/100 of the base unit, the meter, the conversion factor is either 1 meter/100 cm or 100 cm/1 meter, each being equal to unity. Only one, however, will result in proper cancellation of units, producing the desired answer to the problem. If we proceed as follows:

$$\underset{\text{Data given}}{10\,\cancel{\text{cm}}} \times \underset{\text{Conversion factor}}{\frac{1\ \text{m}}{100\,\cancel{\text{cm}}}} = \underset{\text{Desired result}}{0.10\ \text{m}}$$

We obtain the desired units, meters (m). Use of the factor 100 cm/1 m would yield an answer that would be numerically incorrect:

$$10\ \text{cm} \times \frac{100\ \text{cm}}{1\ \text{m}} = \underset{\text{Incorrect units}}{1000\ \text{cm}^2/\text{m}}$$

and expressed in meaningless units.

Conversion of units from one system to another

The conversion of a quantity expressed in units of one system to an equivalent quantity in the other system (English to metric or metric to English) requires a *bridging* conversion unit. For example:

Quantity	English		Metric
Mass	1 pound (lb)	=	454 grams (g)
	2.2 pounds (lb)	=	1 kilogram (kg)
Length	1 inch (in)	=	2.54 centimeters (cm)
	1 yard (yd)	=	39.4 meters (m)
Volume	1 quart (qt)	=	0.946 liters (L)
	1 gallon (gal)	=	3.78 liters (L)

The conversion may be represented as a three-step process:

Step 1. Conversion from units stated in the problem to a bridging unit.

Step 2. Conversion to the other system using the bridge.

Step 3. Conversion within the desired system to units required by the problem.

Example I.3

Convert 4 ounces to kilograms.

Step 1. A convenient bridging unit for mass is 1 lb = 454 grams. To use this conversion factor, we relate ounces (given in the problem) to pounds:

$$4\,\cancel{\text{oz}} \times \frac{1\ \text{lb}}{16\,\cancel{\text{oz}}} = 0.250\ \text{lb}$$

Step 2. Employing the bridging unit conversion, we have

$$0.250\,\cancel{\text{lb}} \times \frac{454\ \text{g}}{1\,\cancel{\text{lb}}} = 114\ \text{g}$$

Step 3. Grams may then be directly converted to kilograms, the desired unit:

$$114 \not{g} \times \frac{1 \text{ kg}}{1000 \not{g}} = 0.114 \text{ kg}$$

The calculation may also be done in a single step by arranging the factors in a chain:

$$4 \text{ oz} \times \frac{1 \not{\text{lb}}}{16 \not{\text{oz}}} \times \frac{454 \not{g}}{1 \not{\text{lb}}} \times \frac{1 \text{ kg}}{1000 \not{g}} = 0.114 \text{ kg}$$

APPENDIX II

Significant Figures and Scientific Notation

Data from a scientific experiment convey information about the way in which the experiment was conducted. The degree of uncertainty or doubt associated with measurements is indicated by the number of figures used to represent the information.

SIGNIFICANT FIGURES

Consider the following situation: A student was asked to obtain the length of a section of wire. In the chemistry laboratory the student finds two measuring devices. Not knowing which is more appropriate, the student measured the object using both devices. The following data were obtained:

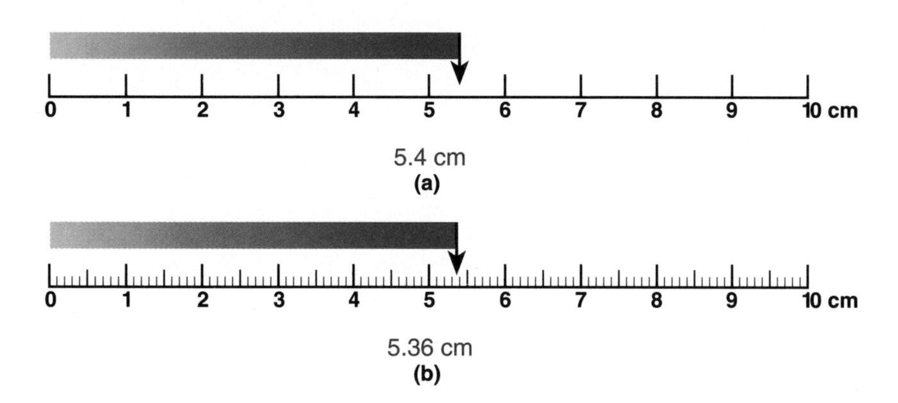

Two questions should immediately come to mind:

1. Are the two answers equivalent?
2. If not, which answer is correct?

In fact, the two answers are *not* equivalent, but *both* are correct. How do we explain this apparent contradiction?

The data are not equivalent because each is known to a different degree of certainty. The answer, 5.36 cm, containing three *significant figures,* specifies the length of the object more exactly than 5.4 cm, which contains only two significant figures. The term significant figures is defined as all digits in a number that are known with certainty *plus the first uncertain digit.*

With ruler I we are certain that the object is at least 5 *cm* long and equally certain that it is *not* 6 cm long because the end of the object falls between the calibration lines 5 and 6. We can only estimate between 5 and 6, since there are no calibration indicators between 5 and 6. The end of the wire appears to be approximately four tenths of the way between 5 and 6, hence 5.4 cm. 5 is known with certainty, 4 is estimated; there are two significant figures.

With ruler II the ruler is calibrated in tenths of centimeters. The end of the wire is at least 5.3 cm and not 5.4 cm. Estimation of the second decimal place between the two closest calibration marks leads to 5.36 cm. In this case, 5.3 is certain and the six is estimated (or uncertain), leading to three significant digits.

Both answers are correct because each is consistent with the measuring device that is used to generate the data. An answer of 5.36 cm obtained from a measurement using ruler I would be *incorrect* because the measuring device is not capable of that exact specification. On the other hand, a value of 5.4 cm obtained from ruler II would be erroneous as well; in that case the measuring device is capable of generating a higher level of certainty (more significant digits) than are actually reported.

In summary, the number of significant figures associated with a measurement is determined by the measuring device. Conversely, the number of significant figures reported indicates the sophistication of the measuring device.

RECOGNITION OF SIGNIFICANT FIGURES

Only *significant* digits should be reported as data or results. However, are all digits, as written, significant digits? Let's look at a few examples.

Example II.1

7.314 has *four* significant digits.

Rule: All nonzero digits are significant. ■

Example II.2

73.14 has *four* significant digits.

Rule: The number of significant digits is independent of the position of the decimal point. ■

Example II.3

60.052 has *five* significant figures.

Rule: Zeros located between nonzero digits are significant. ■

Example II.4

4.70 has *three* significant figures.

Rule: Zeros at the end of a number (often referred to as trailing zeros) are significant if the number contains a decimal point. ■

Example II.5

100 has *one* significant figure.

Rule: Trailing zeros are insignificant because the number does not contain a decimal point. ■

Example II.6

0.0032 has *two* significant figures.

Rule: Zeros to the left of the first nonzero integer are not significant; they serve only to locate the position of the decimal point. ■

SCIENTIFIC NOTATION

It is often difficult to express very large numbers to the proper number of significant figures using conventional notation. Consider the number fifteen thousand, which might have resulted from a measurement capable of generating three significant figures. Writing the number in a conventional fashion as 15,000. (note the presence of a decimal point) shows five digits that, if written, must be assumed to be significant. Alternatively, writing 15000 (no decimal point) indicates only two significant figures. How can we express such a number using only the desired (in this case, three) significant figures? The solution lies in the use of *scientific notation,* also referred to as *exponential notation,* which involves the representation of a number as a power of ten.

If we consider the number 15,000, it is equivalent to 15×1000, which is also 15×10^3. In fact, 15,000 could be represented as:

$15{,}000 \times 10^0$

1500×10^1 (decimal moved *one* place to the left)

150×10^2 (decimal moved *two* places to the left)

15×10^3 (decimal moved *three* places to the left)

1.5×10^4 (decimal moved *four* places to the left)

All appear correct; however, when we superimpose our initial requirement that the data point fifteen thousand has three significant figures, 15.0×10^3 or 1.50×10^4 are logical choices. 1.50×10^4 would be preferred; the usual convention shows the decimal point in standard position—to the right of the leading digit.

Rule: To convert a number greater than 1 to scientific notation, the original decimal point is moved x places to the left, and the resulting number is multiplied by 10^x. The exponent (x) is a *positive* number equal to the number of places that the original decimal point was moved.

Scientific notation is also useful in representing numbers less than 1. For example, the mass of a single helium atom is 0.000000000000000000000006692 gram, a rather cumbersome number as written. Scientific notation would represent the mass of a single helium atom as 6.692×10^{-24} gram. The conversion is illustrated by using a simpler number:

$$0.0062 = 6.2 \times \frac{1}{1000} = 6.2 \times \frac{1}{10^3} = 6.2 \times 10^{-3}$$

or

$$0.0534 = 5.34 \times \frac{1}{100} = 5.34 \times \frac{1}{10^2} = 5.34 \times 10^{-2}$$

Rule: To convert a number less than 1 to scientific notation, the original decimal point is moved x places to the right, and the resulting number is multiplied by 10^{-x}. The exponent $(-x)$ is a *negative* number equal to the number of places that the original decimal point was moved. ■

SIGNIFICANT FIGURES IN CALCULATION OF RESULTS

Addition and subtraction

If we combine the following numbers:

37.68 liters
108.428 liters
6.71862 liters

Our calculator would show a final result of

152.82662 liters

Clearly, the answer, with eight digits, defines the volume of total material much more accurately than *any* of the individual quantities being combined. This cannot be correct; *the answer cannot have greater significance than any of the quantities that produced the answer*. We rewrite the problem

$$\begin{array}{r} 37.68xxx \\ 108.428xx \\ +\quad 6.71862 \\ \hline 152.82662 \end{array} \quad \text{(should be 152.83)}$$

where x = no information; x may be any integer from 0 to 9. Adding 2 to two unknown numbers (in the right column) produces no information. Similar logic prevails for the next two columns. Thus five digits remain, all of which are significant. Conventional rules for rounding off would dictate a final answer of 152.83.

Multiplication and division

In the case of addition or subtraction, the position of the decimal point in the quantities being combined has a bearing on the number of significant figures in the answer. In multiplication and division this is not the case. The decimal point position is irrelevant. It is the number of significant figures in each number being multiplied or divided that is important. Consider:

$$\frac{4.237 \times 1.21 \times 10^{-3} \times 0.00273}{11.125} = 1.26 \times 10^{-6}$$

The answer is limited to three significant figures; the answer can have *only* three significant figures because two numbers in the calculation, 1.21×10^{-3} and 0.00273, have three significant figures and ''limit'' the answer. Remember, *the answer can be no more precise than the* least *precise number from which the answer is derived.*

Exponents

In each case the number of significant figures in the answer is identical to the number contained in the original term. Therefore

$$(8.314 \times 10^2)^3 = 574.7 \times 10^6 = 5.747 \times 10^8$$

and

$$(8.314 \times 10^2)^{1/2} = 2.883 \times 10^1$$

Each answer contains four significant figures.

It is important to note, in operating with significant figures, that defined or counted numbers do *not* determine the number of significant figures.

Example II.7

One pen costs $1.98. What is the cost of four pens?

$$4\ \text{pens} \times \frac{\$1.98}{1\ \text{pen}} = \$7.92 \quad (3\ \text{S.F.})$$

4 is a counted, hence an exact, number. The $1.98 determines the number of significant digits.

Example II.8

How many grams are contained in 0.240 kg?

$$0.240\,\cancel{\text{kg}} \times \frac{1000\ \text{g}}{1\,\cancel{\text{kg}}} = 240\ \text{g}$$

The 1 in the conversion factor is defined, or exact, and does not limit the number of significant digits.

A good rule of thumb is the following: The quantity being converted, not the conversion factor, determines the number of significant figures.

ROUNDING OFF NUMBERS

The use of an electronic calculator generally produces more digits for a result than are justified by the rules of significant figures, based upon the data input. For example, on your calculator, $3.84 \times 6.72 = 25.8048$. The most correct answer would be 25.8, dropping 048.

One generally accepted rule for rounding off states that if the first digit dropped is 5 or greater, we raise the last significant digit to the next higher number. If the first digit dropped is 4 or less, the last significant digit remains unchanged.

Example II.9

Round each of the following to three significant figures:

The symbol $x > y$ implies "x greater than y." The symbol $x < y$ implies "x less than y."

a. $63.6\underline{6}9$ becomes 63.7. Rationale: $6 > 5$

b. $8.77\underline{1}5$ becomes 8.77. Rationale: $1 < 5$

c. $2.22\underline{4}5$ becomes 2.22. Rationale: $4 < 5$

d. $0.000410\underline{9}$ becomes 0.00411. Rationale: $9 > 5$

APPENDIX III

Error, Accuracy, Precision, and Uncertainty

Error may broadly be defined as the difference between the true value and our estimation, or measurement, of the value. *Accuracy* is the absence of error, the agreement between the true value and the measured value.

Only discrete objects—for example, the number of pages in this book or the number of quarters in your pocket—can be measured with perfect accuracy. There are 50 pages in your notebook, *exactly* 50, not 50 1/2 or 49 1/2. In measuring quantities that show continuous variation, such as the weight of this page of the volume of one of your quarters, some uncertainty is present because the answer cannot be expressed with an infinite number of meaningful digits. *Uncertainty* is the degree of doubt in a single measurement. The number of meaningful figures is determined by the measuring device. The presence of some error is a natural consequence of any measurement. This is the reason that replicate measurements of the same quantity should be made whenever practical. *Precision* is a measure of the agreement of replicate measurements.

It is important to recognize that the terms accuracy and precision have very different meanings. As the example in the margin shows, it is possible to have one without the other. However, if scientific measurements are taken with proper attention to experimental detail, the two should go hand in hand.

The first bullseye (A) shows the goal of all experimentation: accuracy *and* precision.

The second bullseye (B) shows the results to be repeatable (good precision); however, some flaw in the experimental procedure has caused the results to center on an incorrect value. This ''flaw'' is consistent, occurring in each replicate measurement.

The third bullseye (C) shows ''accidental'' accuracy. The precision is poor, but the average of these replicates leads to a correct value.

We cannot rely on accidental success; this experiment must be repeated until the precision inspires faith in the accuracy of the method.

Modern measuring devices in chemistry, equipped with powerful computers having immense storage capacity, make literally thousands of individual replicate measurements to enhance the precision of the method.

The last bullseye (D) describes a more common situation. A low level of precision is all too often associated with poor accuracy.

The true value is not ''true'' in the strictest sense of the word. In chemical measurement it is the *best estimate,* based on a large number of replicate measurements of the same quantity. True values, such as those for the speed of light and mass of an atom are periodically revised and refined, based on new information.

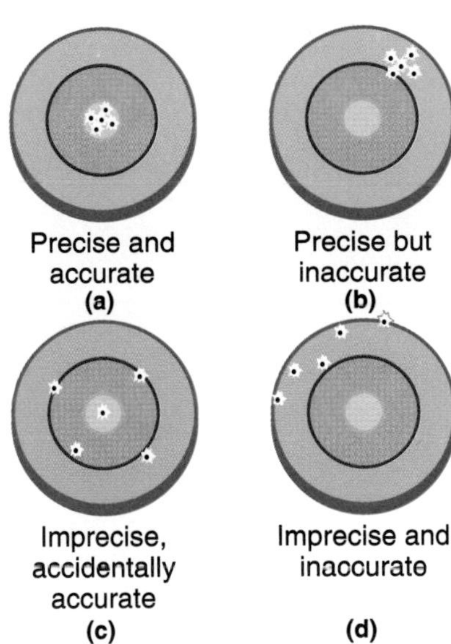

APPENDIX IV

Inorganic Nomenclature

IONIC COMPOUNDS

The "shorthand" symbol for a compound is its *formula,* for example:

$$NaCl, \quad MgBr_2, \quad \text{and} \quad NaIO_3$$

The formula identifies the number and type of the various atoms that comprise the compound. The number of like atoms is denoted by a subscript. The presence of one atom is implied when no subscript is present. For example, the formula NaCl indicates that each *ion pair* consists of one sodium cation, Na^+, and one chloride ion, Cl^-. Similarly, the formula $MgCl_2$ indicates that one magnesium ion and two chloride ions combine to form the compound, and the formula $NaIO_3$ indicates a compound made up of one sodium atom, one iodine atom, and three oxygen atoms (the sodium present as the sodium ion and iodine and oxygen coexisting as the iodate ion, IO_3^-).

Positive ions were formed from elements with the following characteristics:

1. They are located on the left of the periodic table.
2. They are referred to as *metals.*
3. They have low ionization energies and low electron affinities; hence they easily *lose* electrons.

Elements that form negative ions, on the other hand, have the following traits:

1. They are located on the right of the periodic table (but exclude the inert gases).
2. They are referred to as *nonmetals.*
3. They have high ionization energies and high electron affinities; hence they easily *gain* electrons.

In short, metals and nonmetals react to produce ionic compounds resulting from the transfer of one or more electrons from the metal to the nonmetal.

The names given to ionic compounds are based upon their formulas, with the name of the cation before the name of the anion. The cation is simply the name of the element, while the anion is named by using the element's stem name fused to the suffix *-ide.* Some examples follow:

Compound	+ion	and	−ion stem +ide =	compound name
NaCl	sodium		chlor + ide	sodium chloride
Na_2O	sodium		ox + ide	sodium oxide
Li_2S	lithium		sulf + ide	lithium sulfide
$AlBr_3$	aluminum		brom + ide	aluminum bromide
CaO	calcium		ox + ide	calcium oxide

If the cation and anion exist in only one common charged form, there is no ambiguity between formula and name. Sodium chloride *must be* NaCl, and lithium sulfide *must be* Li_2S so that the sum of positive and negative charges is zero. With many elements, such as the transition metals, several ions of different charge may exist. Fe^{2+}, Fe^{3+}, and Cu^+,

TABLE IV.1 Common Monatomic Cations and Anions

Cation	Name	Anion	Name
H^+	Hydrogen	H^-	Hydride
Li^+	Lithium	F^-	Fluoride
Na^+	Sodium	Cl^-	Chloride
K^+	Potassium	Br^-	Bromide
Cs^+	Cesium	I^-	Iodide
Be^{2+}	Beryllium	O^{2-}	Oxide
Mg^{2+}	Magnesium	S^{2-}	Sulfide
Ca^{2+}	Calcium	N^{3-}	Nitride
Ba^{2+}	Barium	P^{3-}	Phosphide
Al^{3+}	Aluminum		
Ag^+	Silver		

Note: The most commonly encountered ions are highlighted.

Cu^{2+} are a few common examples. Clearly, an ambiguity exists if we use the name iron for both Fe^{2+} and Fe^{3+} or copper for both Cu^{1+} and Cu^{2+}. Two systems have developed which avoid this problem: the *stock system* and the *common nomenclature system*.

In the stock system for naming of an ion (the *systematic name*) a Roman numeral indicates the magnitude of charge of the cation. In the older ''common nomenclature'' the suffix *-ous* indicates the lower of the ionic charges, and the suffix *-ic* indicates the higher ionic charge. Consider these examples:

For systematic name:

Formula	+ Ion Charge	Cation Name	Compound Name
$FeCl_2$	2+	iron(II)	iron(II) chloride
$FeCl_3$	3+	iron(III)	iron(III) chloride
Cu_2O	1+	copper(I)	copper(I) oxide
CuO	2+	copper(II)	copper(II) oxide

For common nomenclature:

Formula	+ Ion Charge	Cation	Common ous/ic name
$FeCl_2$	2+	ferr*ous*	ferrous chloride
$FeCl_3$	3+	ferr*ic*	ferric chloride
Cu_2O	1+	cupr*ous*	cuprous oxide
CuO	2+	cupr*ic*	cupric oxide

Systematic names are easier and less ambiguous. Whenever possible, we shall use this system of *nomenclature* or compound naming. The older, less precise -ous/-ic convention is less specific; further, it often relies on the older Latin names of the elements (for example, iron, *ferr-*; L.: *ferrium*).

Ions consisting of only a single atom are said to be *monatomic*. Common monatomic ions are listed in Table IV.1 with the ions of principal importance in this course highlighted.

In contrast, *polyatomic ions,* such as the hydroxide ion, OH^-, are composed of two or more atoms bonded together. These ions, frequently bonded to other ions with ionic bonds, are themselves held together by covalent bonds. The polyatomic ion has an *overall* positive or negative charge. Some common polyatomic ions are listed below in Table IV.2 with the most commonly encountered ions highlighted.

TABLE IV.2 Common Polyatomic Cations and Anions

Ion	Name	Ion	Name
NH_4^+	Ammonium	CO_3^{2-}	Carbonate
NO_2^-	Nitrite	HCO_3^-	Bicarbonate
NO_3^-	Nitrate	ClO^-	Hypochlorite
SO_3^{2-}	Sulfite	ClO_2^-	Chlorite
SO_4^{2-}	Sulfate	ClO_3^-	Chlorate
HSO_4^-	Hydrogen sulfate	ClO_4^-	Perchlorate
OH^-	Hydroxide	$C_2H_3O_2^-$	Acetate
CN^-	Cyanide	MnO_4^-	Permanganate
PO_4^{3-}	Phosphate	$Cr_2O_7^{2-}$	Dichromate
HPO_4^{2-}	Hydrogen phosphate	CrO_4^{2-}	Chromate
$H_2PO_4^-$	Dihydrogen phosphate	O_2^{2-}	Peroxide

Note: The most commonly encountered ions are highlighted.

Examples of formulas of several compounds containing polyatomic ions are shown below:

Formula	Cation	Anion	Name
NH_4Cl	NH_4^+	Cl^+	Ammonium chloride
$Ca(OH)_2$	Ca^{2+}	OH^-	Calcium hydroxide
Na_2SO_4	Na^+	SO_4^{2-}	Sodium sulfate
$NaHCO_3$	Na^+	HCO_3^-	Sodium bicarbonate

Sodium bicarbonate may also be named sodium hydrogen carbonate, a preferred and less ambiguous name. Likewise, Na_2HPO_4 is named sodium hydrogen phosphate, and others are named in a similar fashion.

It is equally important to write the correct formula when given the compound name. It is essential to be able to predict the charge of monatomic ions and the charge and formula of polyatomic ions. Equally important, the relative number of positive and negative ions in the unit must result in a unit (compound) charge of zero.

Example IV.1

Write the formula of sodium sulfate.

Step 1. The sodium ion is Na^+ (group I element). The sulfate ion is SO_4^{2-} (from Table IV.2).

Step 2. Two positive charges are necessary to counterbalance sulfate (two negative charges).

Hence

$$Na_2SO_4$$

Example IV.2

Write the formula of ammonium sulfide.

Step 1. The ammonium ion is NH_4^+ (from Table IV.2). The sulfide ion is S^{2-} (from its position on the periodic table).

Step 2. Two positive charges are necessary to counterbalance sulfide (two negative charges).

Hence

$$(NH_4)_2S$$

Note the parentheses that must be used whenever a subscript is used with a polyatomic ion.

TABLE IV.3 Prefixes Used to Denote Numbers of Atoms in a Compound

Prefix	Number of Atoms	Prefix	Number of Atoms
mono-	1	penta-	5
di-	2	hexa-	6
tri-	3	hepta-	7
tetra-	4	octa-	8

COVALENT COMPOUNDS

Most covalent compounds are formed by the reaction of nonmetals. Compounds characterized by covalent bonding are referred to as *molecules*. Covalent compounds exist as discrete units in the solid state. Each molecule retains its identity in the liquid or vapor state (assuming that no chemical reaction has occurred during the phase change). The existence of this compound unit is a major distinctive feature of covalently bonded substances.

The convention for naming covalent compounds is as follows:

1. The names of the elements are written in the order in which they appear in the formula.
2. A prefix (see Table IV.3) indicating the number of each kind of atom found in the unit is placed before the name of the element.
3. If only one atom of a particular kind is present in the molecule, the prefix mono- is usually omitted.
4. The stem of the name of the last element is used with the suffix -ide (or -ate in the case of a polyatomic anion).

Example IV.3

Name the covalent compound N_2O_4.

Step 1. 2 nitrogen atoms, 4 oxygen atoms

Step 2. di-, tetra-

Step 3. dinitrogen, tetraoxide

Name: dinitrogen tetraoxide

Some other covalent compounds are given below:

Formula	Name
N_2O	Dinitrogen oxide
NO_2	Nitrogen dioxide
SiO_2	Silicon dioxide
CO_2	Carbon dioxide
CO	Carbon monoxide

Many compounds are so familiar to us that their *common names* are used. For example, H_2O is water, NH_3 is ammonia, C_2H_5OH (ethanol) is alcohol, and $C_6H_{12}O_6$ is glucose. It is useful to be able to correlate both systematic and common names with the corresponding molecular formula.

The major disadvantage of common names is that the chemical formula must be memorized; water is H_2O, ammonia is NH_3, and so forth. However, because of the widespread use of common names, they cannot be avoided.

On the other hand, compounds named using Greek prefixes are easily converted to formulas. Consider the following examples.

Example IV.4

Write the formula of nitrogen monoxide.

Answer: Nitrogen has no prefix; *one* is understood. Oxide has the prefix ''mono'': one oxygen. Hence

$$NO$$

Example IV.5

Write the formula of sodium dihydrogen phosphate, which is found in many biochemically important solutions.

Answer: This combines more than one system of nomenclature. Sodium has no prefix; *one* is understood. Hydrogen has the prefix ''di'': two hydrogens. Phosphate is the complex ion PO_4^{3-} (from Table IV.2). Hence

$$NaH_2PO_4$$

APPENDIX V

Concentration of Solutions

The amount of solute dissolved in a given amount of solution is defined as the solution *concentration*. The concentration of a solution has a profound effect on the properties of the solution, both *physical* (melting and boiling points) and *chemical* (solution reactivity). Solution concentration may be expressed in many different units. The most widely used concentration units are considered below.

WEIGHT-VOLUME PERCENT

The concentration of a solution is defined as the amount of solute dissolved in a specified amount of solution, or

$$\text{concentration} = \frac{\text{amount of solute}}{\text{amount of solution}}$$

If we define the amount of solute as the *mass* of solute (grams) and the amount of solution in *volume* units (milliliters), concentration is expressed as the ratio

$$\text{concentration} = \frac{\text{grams of solute}}{\text{milliliters of solution}}$$

We can calculate the *weight-volume percent, % (W/V)*, by multiplying the ratio by the factor, 10^2. This results in

$$\%\ (\text{W/V}) = \%\ \text{concentration} = \frac{\text{grams of solute}}{\text{milliliters of solution}} \times 10^2$$

Example V.1

Calculate the percent composition, or % (W/V), of 3.00×10^2 mL of solution containing 15.0 g of glucose.

There are 15.0 g of glucose, the solute, and 3.00×10^2 mL of total solution. Therefore

$$\%\ (\text{W/V}) = \frac{15.0\ \text{grams glucose}}{3.00 \times 10^2\ \text{mL solution}} \times 10^2$$

$$\%\ (\text{W/V}) = 5.00\%\ \text{glucose}$$

Example V.2

Calculate the number of grams of NaCl in 5.00×10^2 mL of a 10.0% solution.

$$10.0\%\ (\text{W/V}) = \frac{X\ \text{grams NaCl}}{5.00 \times 10^2\ \text{mL solution}} \times 10^2$$

$$X\ 10^2 = [10.0\%\ (\text{W/V})](5.00 \times 10^2\ \text{mL solution})$$

$$X = 50.0\ \text{grams NaCl}$$

If the units of mass are other than grams, or if the solution volume is in units other than milliliters, the proper conversion factor (Appendix I) must be used to arrive at the units used in the equation.

VOLUME/VOLUME PERCENT

If the volume of solute is known, rather than its mass, it is often more convenient to represent solution concentration as *volume/volume percent,* or *%(V/V)*:

$$\%\ (\text{V/V}) = \frac{\text{mL solute}}{\text{mL solution}} \times 10^2$$

The logic and calculations are similar to those demonstrated in the preceding section.

Example V.3

Calculate the % (V/V) of ethyl alcohol if 5.0×10^{-1} mL of the alcohol are dissolved in 10.0 mL of solution.

$$\%\ (\text{V/V}) = \frac{\text{mL solute}}{\text{mL solution}} \times 10^2$$

$$\%\ (\text{V/V}) = \frac{5.0 \times 10^{-1}\ \text{mL}}{10.0\ \text{mL}} \times 10^2$$

$$\%\ (\text{V/V}) = 5.0\%\ \text{ethyl alcohol}$$

WEIGHT/WEIGHT PERCENT

The *weight/weight percent,* or *% (W/W)*, is most useful for mixtures of solids, whose weights (masses) are easily obtainable. The expression used to calculate weight/weight percentages is analogous in form to % (W/V) and % (V/V):

$$\%\ (\text{W/W}) = \frac{\text{grams solute}}{\text{grams solution}} \times 10^2$$

Example V.4

Calculate the % (W/W) of platinum in a gold ring that contains 14.00 grams of gold and 4.500 grams of platinum.

$$\%\ (\text{W/W}) = \frac{\text{grams solute}}{\text{grams solution}} \times 10^2$$

$$\%\ (\text{W/W}) = \frac{4.500\ \text{g platinum}}{4.500\ \text{g platinum} + 14.00\ \text{g gold}} \times 10^2$$

$$\%\ (\text{W/W}) = \frac{4.500\ \text{g}}{18.50\ \text{g}} \times 10^2 = 24.3\%\ \text{platinum}$$

MOLARITY

The chemical equation represents the relative number of *moles* of two or more reactants producing products. When chemical reactions occur in solution, it is most useful to represent their concentrations on a *molar* basis. One concentration unit that accomplishes this is molarity. *Molarity*, symbolized M, is defined as the number of moles of solute per liter of solution, or

$$M = \frac{\text{mol solute}}{\text{L solution}}$$

Example V.5

Calculate the molarity of 2.0 L of solution containing 5.0 moles of NaOH.

$$M_{NaOH} = \frac{5.0 \text{ mol solute}}{2.0 \text{ L solution}}$$

$$M_{NaOH} = 2.5 \text{ mol/L}$$

Remember the need for conversion factors to convert from mass to number of moles.

Example V.6

If 5.00 grams of glucose are dissolved in 1.00×10^2 mL of solution, calculate the M of the glucose solution.

The formula weight of glucose is 180 grams/mole. Therefore

$$5.00 \cancel{\text{grams}} \times \frac{1 \text{ mol}}{180 \cancel{\text{grams}}} = 2.78 \times 10^{-2} \text{ mol glucose}$$

and

$$1.00 \times 10^2 \cancel{\text{mL}} \times \frac{10^{-3} \text{ L}}{1 \cancel{\text{mL}}} = 1.00 \times 10^{-1} \text{ L}$$

and

$$M_{(glucose)} = \frac{2.78 \times 10^{-2} \text{ mol}}{1.00 \times 10^{-1} \text{ L}}$$

$$M_{(glucose)} = 2.78 \times 10^{-1} \text{ mol/L}$$

Example V.7

Calculate the volume of a 0.750 M sulfuric acid (H_2SO_4) solution containing 0.120 moles of solute.

Substituting in our basic expression, we have

$$0.750\ M\ H_2SO_4 = \frac{0.120 \text{ mol } H_2SO_4}{x \text{ L}}$$

$$x = 0.160 \text{ L}$$

DILUTION

Laboratory reagents are often available as concentrated solutions (for example, 12 M HCl, 6 M NaOH) for reasons of safety, economy, and space availability. One must often *dilute* such a solution to a larger volume to prepare a less concentrated solution for the experiment at hand. The approach to such a calculation is outlined below:

If we define

M_1 = molarity of solution *prior to* dilution
M_2 = molarity of solution *after* dilution
V_1 = volume of solution *prior to* dilution
V_2 = volume of solution *after* dilution

and if

$$M = \frac{\text{moles solute}}{\text{V}}$$

and

$$\text{moles solute} = (M)\ (V)$$

The number of moles of solute *prior to* and *after* dilution is unchanged, since dilution involves only addition of extra solvent:

$$\underset{\text{Initial condition}}{\text{moles solute}_1} = \underset{\text{Final condition}}{\text{moles solute}_2}$$

or

$$(M_1)\ (V_1) = (M_2)\ (V_2)$$

Knowing any three of these terms enables one to calculate the fourth.

Example V.8

Calculate the M of 0.050 L of HCl solution, initially 0.10 M, which was diluted to 1.0 L.

$$M_1 = 0.10\ M$$
$$M_2 = x$$
$$V_1 = 0.050\ \text{L}$$
$$V_2 = 1.0\ \text{L}$$

Then

$$(M_1)\ (V_1) = (M_2)\ (V_2)$$

$$x\ M = \frac{(0.10\ M)\ (0.050\ \text{L})}{(1.0\ \text{L})}$$

$$x\ M = 0.0050M \quad \text{or} \quad 5.0 \times 10^{-3}\ M\ \text{HCl}$$

Example V.9

Calculate the volume, in liters, of water that must be added to dilute 20.0 mL of 12.0 M HCl to 0.100 M HCl.

$$M_1 = 12.0\ M$$
$$M_2 = 0.100\ M$$
$$V_1 = 20.0\ \text{mL, which is } 0.0200\ \text{L}$$
$$V_2 = V_{\text{final}},\ \text{solution}$$

Then

$$(M_1)\ (V_1) = (M_2)\ (V_2)$$

$$(12.0\ M)\ (0.0200\ \text{L}) = (0.100\ M)\ (V_{\text{final}})$$

$$V_{\text{final}} = \frac{(12.0\ M)\ (0.0200\ \text{L})}{0.100\ M}$$

$$V_{\text{final}} = 2.40\ \text{L solution}$$

Note that this is the *total final volume*. The amount of water added equals this volume *minus* the original solution volume, or

$$2.40\ \text{L} - 0.0200\ \text{L} = 2.38\ \text{L water}$$

The dilution equation is valid with any concentration units, such as % (W/V) or % (V/V), *as well as* molarity. However, you must be certain to use the same units for both initial *and* final concentration values. Only in this way will proper unit cancellation occur.

REPRESENTATION OF CONCENTRATION OF IONS IN SOLUTION

The concentration of ions in solution may be represented in a variety of ways. The most common include moles per liter (molarity) and equivalents per liter.

Molarity emphasizes the number of individual ions. A one molar solution of Na^+ contains Avogadro's number, 6.02×10^{23} Na^+. In contrast, equivalents per liter emphasizes charge; one equivalent of Na^+ contains Avogadro's number of positive charge.

We define one mole as the number of grams of an atom, molecule, or ion corresponding to Avogadro's number of particles. One *equivalent* of an ion is the number of grams of the ion corresponding to Avogadro's number of electrical charges. Some examples follow:

1 mole Na^+ = 1 equivalent Na^+

1 mole Cl^- = 1 equivalent Cl^-

1 mole Ca^{2+} = 2 equivalents Ca^{2+} (one Ca^{2+} produces 2 units of charge/ion)

1 mole $CO_3{}^{2-}$ = 2 equivalents $CO_3{}^{2-}$ (one $CO_3{}^{2-}$ produces 2 units of charge/ion)

1 mole $PO_4{}^{3-}$ = 3 equivalents $PO_4{}^{3-}$ (one $PO_4{}^{3-}$ produces 3 units of charge/ion)

Changing from moles/L to equivalents/L (or the reverse) can be accomplished by using conversion factors.

Example V.10

Calculate the number of equivalents per liter of phosphate ion, $PO_4{}^{3-}$, in a solution that is 5.0×10^{-3} *M* phosphate.

$$\frac{5.0 \times 10^{-3} \text{ mol } PO_4{}^{3-}}{1 \text{ liter}} \times \frac{3 \text{ mol charge}}{1 \text{ mol } PO_4{}^{3-}} \times \frac{1 \text{ eq}}{1 \text{ mol charge}} = \frac{1.5 \times 10^{-2} \text{ eq } PO_4{}^{3-}}{\text{liter}}$$

APPENDIX VI

Osmosis and Osmotic Pressure

Certain types of thin films, or *membranes,* while appearing impervious to matter, actually contain a network of small holes or pores. These pores may be large enough to allow small *solvent* molecules, such as water, to pass from one side of the membrane to the other. On the other hand, larger *solute* molecules cannot cross the membrane, since their average diameter may exceed that of the membrane pores. Membranes that allow transport of solvent but not solute from one side to the other are termed *semipermeable membranes*. Examples of semipermeable membranes range from synthetics such as cellophane to biological membranes.

Osmosis is the movement of a solvent from a *dilute solution* to a more *concentrated solution* through a *semipermeable membrane*. Pressure must be applied to the more concentrated solution to stop this flow, and the magnitude of the pressure required to just stop the flow is termed the *osmotic pressure*.

The "driving force" for the osmotic process is the need to establish an equilibrium between the solutions on either side of the membrane. Pure solvent enters the more concentrated solution, diluting the more concentrated solution. If this process is successful and concentrations on both sides of the membrane become equal, the "driving force," or concentration differential, disappears, a dynamic equilibrium is established, and the osmotic pressure difference between the two sides becomes equal to zero.

The osmotic pressure, not unlike the pressure exerted by a gas, may be treated quantitatively. Osmotic pressure, symbolized by π, follows the same form as the ideal gas equation:

	Ideal Gas		**Osmotic Pressure**
	$PV = nRT$		$\pi V = nRT$
or		or	
	$P = \frac{nRT}{V}$		$\pi = \frac{nRT}{V}$
and since	$M = \frac{n}{V},$	and since	$M = \frac{n}{V},$
then		then	
	$P = MRT$		$\pi = MRT$

The osmotic pressure may be calculated from the solution concentration at any given temperature. How do we determine "solution concentration"? Recall that osmosis is a *colligative property;* it depends on the concentration of solute particles. Again, it becomes necessary to distinguish between solutions of electrolytes and nonelectrolytes. For example, 1 M glucose solution consists of one mole of particles per liter; glucose is a nonelectrolyte. A solution of 1 M NaCl produces two moles of particles per liter (one mole of Na^+ and one mole of Cl^-). A 1 M $CaCl_2$ solution is 3 M in particles (one mole of Ca^{2+} and two moles of Cl^- per liter).

By convention the molarity of particles in solution is termed *osmolarity,* abbreviated *Osm,* for osmotic pressure calculations.

Example VI.1

Determine the osmolarity of 5.0×10^{-3} M Na_3PO_4.

Na_3PO_4 is an ionic compound and produces an electrolytic solution:

$$Na_3PO_4 \xrightarrow{H_2O} 3Na^+ + PO_4{}^{3-}$$

$$\frac{5.0 \times 10^{-3} \cancel{\text{mol } Na_3PO_4}}{\text{liter solution}} \times \frac{4 \text{ mol particles}}{1 \cancel{\text{mol } Na_3PO_4}} = \frac{2.0 \times 10^{-2} \text{ mol particles}}{\text{liter solution}}$$

and

$$\frac{2.0 \times 10^{-2} \text{ mol particles}}{\text{liter solution}} = 2.0 \times 10^{-2} \text{ Osm}$$

Example VI.2

Calculate the osmotic pressure of a 5.0×10^{-2} M solution of NaCl at 25°C (298 K).

$$\pi = MRT$$

M should be represented as osmolarity:

$$\frac{5.0 \times 10^{-2} \cancel{\text{mol NaCl}}}{\text{liter solution}} \times \frac{2 \text{ mol particles}}{1 \cancel{\text{mol NaCl}}} = \frac{1.0 \times 10^{-1} \text{ mol particles}}{\text{liter solution}}$$

and

$$\pi = \frac{1.0 \times 10^{-1} \text{ mol particles}}{\text{liter solution}} \times \frac{0.0821 \text{ L-atm}}{\text{K-mol}} \times 298 \text{ K}$$

$$= 2.4 \text{ atm}$$

Note that the osmotic pressure is quite high, even in a dilute solution.

APPENDIX VII

Calculation of pH

The *pH scale* correlates the hydronium ion concentration with a number, the pH, which serves as a useful indicator of the degree of acidity or basicity of a solution. The pH scale is somewhat analogous to the temperature scale that is used for assignment of relative levels of "hot" or "cold." The temperature scale was developed to allow us to assess "how cold" or "how hot" an object is. The pH scale specifies "how acidic" or "how basic" a solution is:

1. Addition of an acid (proton donor) to water *increases* the $[H_3O^+]$ and decreases the $[OH^-]$.
2. Addition of a base (proton acceptor) to water *decreases* the $[H_3O^+]$ and increases the $[OH^-]$.
3. $[H_3O^+] = [OH^-]$ when *equal* amounts of acid and base are present.
4. In all of the above cases, $[H_3O^+][OH^-] = 1.0 \times 10^{-14} = K_w$, which is the *autoionization constant* for water.

It is often necessary to calculate the pH of a solution, knowing the hydronium or hydroxide ion concentration. Conversely, we may be required to calculate $[H_3O^+]$ or $[OH^-]$ from the measured pH. Two different approaches, one requiring a working knowledge of logarithms and the other using decimal logic, are reviewed below.

APPROACH I: LOGARITHM-BASED pH CALCULATIONS

The pH of a solution is defined as the negative logarithm of the molar concentration of the hydronium ion:

$$pH = -\log [H_3O^+]$$

Example VII.1

Calculate the pH of a 1.0×10^{-3} *M* solution of HCl.

HCl is a strong acid. If 1 mole of HCl dissociates, it produces 1 mole of H_3O^+. Therefore a 1.0×10^{-3} *M* HCl solution has $[H_3O^+] = 1.0 \times 10^{-3}$ *M*, and

$$pH = -\log [H_3O^+]$$

Consider the logarithm term to be composed of two parts, 1.0 and 10^{-3}. The logarithm of $1.0 = 0$, and the logarithm of 10^{-3} is simply the exponent, -3. Therefore

$$pH = -[0 - 3.00]$$

$$pH = -[-3.00] = 3.00$$

Example VII.2

Calculate the $[H_3O^+]$ of a solution with a $pH = 4$.

$$pH = -\log [H_3O^+]$$

$$4 = -\log [H_3O^+]$$

Multiplying both sides of the equation by -1, we have

$$-4 = \log [H_3O^+]$$

Taking the antilogarithm of both sides (the reverse of a logarithm), we get

$$\text{antilog } -4 = [H_3O^+]$$

The antilog is the exponent of 10; therefore

$$10^{-4} \text{ (or } 1 \times 10^{-4}) = [H_3O^+]$$

Example VII.3

Calculate the pH of a 1.0×10^{-5} *M* solution of NaOH.

NaOH is a strong base. If 1 mole of NaOH dissociates, it produces 1 mole of OH^-. Therefore a 1.0×10^{-5} *M* NaOH solution has $[OH^-] = 1.0 \times 10^{-5}$ *M*. To calculate pH, we need $[H_3O^+]$. Recalling that

$$[H_3O^+][OH^-] = 1.0 \times 10^{-14}$$

$$[H_3O^+] = \frac{1.0 \times 10^{-14}}{[OH^-]}$$

$$[H_3O^+] = \frac{1.0 \times 10^{-14}}{1.0 \times 10^{-5}}$$

$$[H_3O^+] = 1.0 \times 10^{-9}\ M$$

The solution is

$$pH = -\log [H_3O^+]$$

$$pH = -\log [1.0 \times 10^{-9}]$$

$$pH = 9.00$$

Example VII.4

Calculate the $[H_3O^+]$ and $[OH^-]$ of a solution with a pH = 10.00.

$$pH = -\log [H_3O^+]$$

$$10.00 = -\log [H_3O^+]$$

$$-10.00 = \log [H_3O^+]$$

$$\text{antilog } -10 = [H_3O^+]$$

$$1.0 \times 10^{-10}\ M = [H_3O^+]$$

To calculate the $[OH^-]$, we need to solve for $[OH^-]$, using the following expression:

$$[H_3O^+][OH^-] = 1.0 \times 10^{-14}$$

$$[OH^-] = \frac{1.0 \times 10^{-14}}{[H_3O^+]}$$

Substituting the $[H_3O^+]$ from the first part, we have

$$[OH^-] = \frac{1.0 \times 10^{-14}}{1.0 \times 10^{-10}}$$

$$[OH^-] = 1.0 \times 10^{-4}$$

Often, the pH or $[H_3O^+]$ will not be an even number (1.0, 5.0, 1.0×10^{-3}, and so forth). With the advent of inexpensive and versatile calculators, calculations with noninteger numbers pose no great problems. Consider the following examples.

Example VII.5

Calculate the pH of a sample of lakewater that has a $[H_3O^+] = 6.5 \times 10^{-5}$ *M*.

$$pH = -\log [H_3O^+]$$

$$pH = -\log [6.50 \times 10^{-5}]$$

To perform the calculation on your calculator:

1. Enter 6.5×10^{-5} into your calculator. With most calculators the following protocol will work:

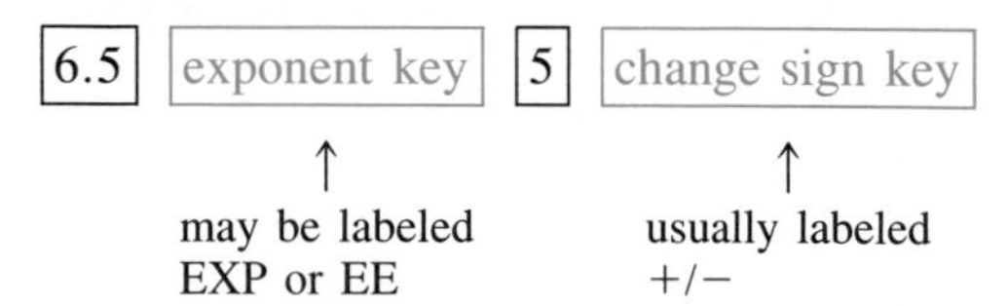

2. Press the LOG key. This should produce a value of −4.1871.
3. Since the pH is negative log, change sign, pH = 4.1871.
4. Round off to 4.19.

With logarithms, in calculations of this type the number before the decimal point is not counted as a significant digit. Therefore, 19, two significant figures, is in agreement with the original data, 6.5×10^{-5}. The pH of this lake, 4.19, is low enough to suspect acid rain.

Example VII.6

The measured pH of a sample of lakewater is 6.40. Calculate $[H_3O^+]$.

An alternative mathematical form of

$$pH = -\log [H_3O^+]$$

is the expression

$$[H_3O^+] = 10^{-pH}$$

which we shall use in situations in which we must solve for $[H_3O^+]$. To perform the calculation on your calculator:

1. Enter 6.40.
2. Press Change sign, +/−.
3. Press 10^x. The result is 3.98×10^{-7} or 4.0×10^{-7} $M = [H_3O^+]$.

Two significant figures agrees with two significant figures in pH = 6.40.

The six examples above illustrate the most frequently used pH calculations. It is important to remember that in the case of a base you must convert the $[OH^-]$ to $[H_3O^+]$, using the expression for the autoionization constant for the solvent water.

APPROACH II: DECIMAL-BASED pH CALCULATION

If you do not have facility with logarithms or a calculator available, it is still possible to approximate the pH of a solution, as well as to approximate acid or base concentration from the pH. To do this, remember:

1. The pH of a 1 *M* solution of any strong acid is 0.
2. The pH of a 1 *M* solution of any strong base is 14.

3. Each tenfold change in concentration changes the pH by one unit. A tenfold change in concentration is equivalent to moving the decimal point one place.
4. A *decrease* in acid concentration *increases* the pH.
5. A *decrease* in base concentration *decreases* the pH.

Consider the following strong acid:

HCl Molarity	pH
1.0×10^{0}	0
1.0×10^{-1}	1
1.0×10^{-2}	2
1.0×10^{-3}	3
1.0×10^{-4}	4
1.0×10^{-5}	5
1.0×10^{-6}	6
1.0×10^{-7}	7

For the case of a strong base,

NaOH Molarity	pH
1.0×10^{0}	14
1.0×10^{-1}	13
1.0×10^{-2}	12
1.0×10^{-3}	11
1.0×10^{-4}	10
1.0×10^{-5}	9
1.0×10^{-6}	8
1.0×10^{-7}	7

Note that for a strong acid the exponent, with the sign changed, is the pH.
Note that for a strong base the exponent, algebraically added to 14, is the pH.

APPENDIX VIII

The Chemical Equation

The *chemical equation* is the shorthand notation for a chemical reaction. It describes all of the substances that react to produce the product(s). The chemical equation also describes the physical state of the reactants and products as solid, liquid, or vapor. It tells us whether the reaction in fact occurs and identifies the solvent and experimental conditions employed, such as heat or electrical energy applied. Most important, the relative number of moles of reactants and products appears in the equation. According to the *law of conservation of mass,* matter cannot be either gained or lost in the process of a chemical reaction. The total mass of the products must be equal to the total mass of reactants. In other words, the law of conservation of mass states that we must have a *balanced chemical equation.*

IMPORTANT FACTORS IN GENERATING A SUITABLE EQUATION

Consider the decomposition of calcium carbonate:

$$\underset{\text{Calcium carbonate}}{CaCO_3(s)} \xrightarrow{\Delta} \underset{\text{Calcium oxide}}{CaO(s)} + \underset{\text{Carbon dioxide}}{CO_2(g)}$$

The factors involved in the construction of this equation are discussed below:

1. *The identity of products and reactants must be specified.* In some cases it is possible to predict the products of a reaction. More often, the fate of a reactant must be verified by chemical analysis. (Generally, you will be given information regarding the identity of the reactants and products.)
2. *Reactants are written to the left of the reaction arrow (→) and products to the right.* The direction in which the arrow points indicates the direction in which the reaction proceeds. In the preceding example, reactants on the left ($CaCO_3$) are converted to products on the right ($CaO + CO_2$) during the course of the reaction.
3. *The physical state of reactants and products is shown in parentheses.* For example:
   - $Cl_2(g)$ or $Cl_2\uparrow$ both mean that chlorine is in the gaseous state.
   - $Mg(s)$ or $Mg\downarrow$ indicates that magnesium is a solid.
   - $Br_2(l)$ indicates that bromine is present as a liquid.
   - $NH_3(aq)$ tells us that ammonia is present as an aqueous solution (dissolved in water).
4. *The symbol Δ over the reaction arrow means that heat energy is necessary for the reaction to occur.* Often, other special conditions are noted above or below the reaction line. For example, ''light'' means that a light source provides energy that is necessary for the reaction. Such reactions are termed photochemical reactions.
5. *The equation must be balanced.* Applying the above factors, the equation for the decomposition of calcium carbonate is written as

$$CaCO_3(s) \xrightarrow{\Delta} CaO(s) + CO_2(g)$$

The equation tells us that solid calcium carbonate, when heated, decomposes to solid calcium oxide and gaseous carbon dioxide. Furthermore, this equation predicts that one mole of calcium carbonate produces one mole of calcium oxide and one mole of carbon dioxide.

THE EXPERIMENTAL BASIS OF A CHEMICAL EQUATION

The chemical equation represents a real chemical transformation. Evidence for the reaction may be based on observations such as:

- the evolution of carbon dioxide when an acid is added to a carbonate,
- the formation of a solid (or precipitate) when a solution of iron ions is made basic,
- the evolution of heat,
- the change in color of a solution upon addition of a second substance.

Many reactions are not so obvious. Sophisticated analytical instrumentation, available to the chemist, allows the detection of subtle changes that would otherwise go unnoticed. Such instruments may measure:

- heat or light absorbed or emitted,
- changes in the way in which the sample behaves in an electric or magnetic field,
- changes in electrical properties.

Whether we use our senses or a one hundred thousand dollar computerized instrument, the ''bottom line'' is the same—we are measuring a change in one or more chemical or physical properties in an effort to assess the changes taking place in a chemical system.

BALANCING CHEMICAL EQUATIONS

A chemical equation tells us both the identity and state of the reactants and products. The chemical equation also shows the *molar quantity* of reactants needed to produce a certain *molar quantity* of products. The chemical equation expresses these quantities in terms of relative numbers of moles.

The number of moles of each product and reactant is indicated by placing a whole-number *coefficient* before the formula of each substance in the chemical equation. A coefficient of 2 (for example, 2NaCl) indicates that two moles of sodium chloride are involved in the reaction. The notation $3NH_3$ signifies three units of ammonia; it also means that three moles of nitrogen atoms and (3×3) or nine moles of hydrogen atoms are involved in the reaction. The coefficient 1 is understood, not written. H_2SO_4 would therefore be interpreted as one mole of sulfuric acid or two moles of hydrogen atoms, one mole of sulfur atoms, and four moles of oxygen atoms.

The equation discussed above,

$$CaCO_3(s) \xrightarrow{\Delta} CaO(s) + CO_2(g)$$

is balanced as written. On the reactant side of the reaction arrow we have

one mole of Ca
one mole of C
three moles of O

On the product side there are

one mole of Ca
one mole of C
three moles of O

Therefore the law of conservation of mass is obeyed.

Now consider the reaction of hydrochloric acid, dissolved in water, with solid calcium metal:

$$HCl(aq) + Ca(s) \longrightarrow CaCl_2(aq) + H_2(g)$$

The equation, as written, is not balanced.

Reactants	**Products**
one mole H atoms	two moles H atoms
one mole Cl atoms	two moles Cl atoms
one mole Ca atoms	one mole Ca atoms

We need two moles of both H and Cl on the left, or reactant, side. An *incorrect* way of balancing the equation is as follows:

$$H_2Cl_2(aq) + Ca(s) \longrightarrow CaCl_2(aq) + H_2(g)$$

Not a correct equation

The equation satisfies the law of conservation of mass; however, we have altered one of the reacting species. Hydrochloric acid is HCl not H_2Cl_2. We must remember that *we cannot alter any chemical species in the process of balancing the equation*. We can introduce *only* coefficients into the equation. To do anything else would alter the chemistry of the reaction. The equation must represent the reaction accurately. The correct equation is

$$2HCl(aq) + Ca(s) \longrightarrow CaCl_2(aq) + H_2(g)$$

Correct equation

Many equations are balanced by trial and error. If the identity of the products and reactants, the physical state, and the reaction conditions are known, the following steps provide a method for correctly balancing a chemical equation:

Step 1. Count the number of atoms of each element on both product side and reactant side.

Step 2. Determine which atoms are not balanced.

Step 3. Balance one atom at a time, using coefficients.

Step 4. After you believe that you have successfully balanced the equation, check, as in step 1, to be certain that mass conservation has been achieved.

Example VIII.1

Apply the steps listed above to the reaction of calcium with hydrochloric acid:

$$HCl(aq) + Ca(s) \longrightarrow CaCl_2(aq) + H_2(g)$$

Step 1. Count the numbers of atoms:

Reactants	**Products**
one mole H atoms	two moles H atoms
one mole Cl atoms	two moles Cl atoms
one mole Ca atoms	one mole Ca atoms

Step 2. The numbers of moles of both H and Cl are not balanced.

Step 3. Insertion of a 2 before HCl on the reactant side should balance the equation:

$$2HCl(aq) + Ca(s) \longrightarrow CaCl_2(aq) + H_2(g)$$

Step 4. Check for mass balance:

Reactants	**Products**
two moles H atoms	two moles H atoms
two moles Cl atoms	two moles Cl atoms
one mole Ca atoms	one mole Ca atoms

Hence the equation is balanced.

Example VIII.2

Hydrogen gas and oxygen gas react explosively to produce water vapor. Recall that hydrogen and oxygen are diatomic molecules. Therefore

$$H_2(g) + O_2(g) \longrightarrow H_2O(g)$$

First, balance the oxygen atoms:

$$H_2(g) + O_2(g) \longrightarrow 2H_2O(g)$$

Balancing oxygen creates an imbalance in hydrogen, so

$$2H_2(g) + O_2(g) \longrightarrow 2H_2O(g)$$

the balanced equation

Example VIII.3

Propane gas, C_3H_8, a fuel, reacts with oxygen to produce carbon dioxide and water vapor. The reaction is

$$C_3H_8(g) + O_2(g) \longrightarrow CO_2(g) + H_2O(g)$$

First, balance carbon; there are three carbon atoms on the left and only one carbon atom on the right:

$$C_3H_8(g) + O_2(g) \longrightarrow 3CO_2(g) + H_2O(g)$$

Next, we balance the hydrogen; there are two hydrogen atoms on the right and eight on the left. We need $4H_2O$ on the right:

$$C_3H_8(g) + O_2(g) \longrightarrow 3CO_2(g) + 4H_2O(g)$$

There are now ten oxygen atoms on the right and two on the left. To balance, we must have five O_2 on the left side of the equation:

$$C_3H_8(g) + 5O_2(g) \longrightarrow 3CO_2(g) + 4H_2O(g)$$

The balanced equation

Remember: In every case, be sure to check the final equation for mass balance.

APPENDIX IX

Chemical Equilibrium

Many chemical reactions do not proceed to "completion." After a period of time, determined by the kinetics of the reaction, the concentration of reactants no longer decreases, and the concentration of products ceases to increase. At this point, a mixture of products and reactants exists, and its composition will remain *constant* unless the experimental conditions are changed. This mixture is in a state of *chemical equilibrium*. The point of equilibrium (how complete the reaction is) may be computed from the first and second laws of thermodynamics. The approach here will be to develop a nonmathematical understanding of equilibrium as it applies to chemical and physical change.

REVERSIBILITY OF CHEMICAL AND PHYSICAL PROCESSES

In our representation of chemical reactions thus far, we have referred to reactants to the left and products to the right of the reaction arrow. However, many reactions may proceed in either direction, left to right *or* right to left, depending upon the experimental conditions. Such reactions are *reversible*. For example, consider the reaction of nitrogen gas and hydrogen gas to produce ammonia:

$$N_2(g) + 3H_2(g) \rightleftharpoons 2NH_3(g)$$

The double arrow ($\rightleftharpoons$) shows that the reaction is reversible. Implicit in this chemical equation is the following: The equilibrium concentration of components is *independent* of the direction of the reaction. If we begin with three moles of hydrogen and one mole of nitrogen in a sealed container at equilibrium, the amount of hydrogen, nitrogen, and ammonia will be *identical to* that formed by starting with two moles of ammonia and allowing the ammonia to decompose to hydrogen and nitrogen. The equilibrium concentration of components is *independent* of direction of the reaction. This is depicted in Figure IX.1.

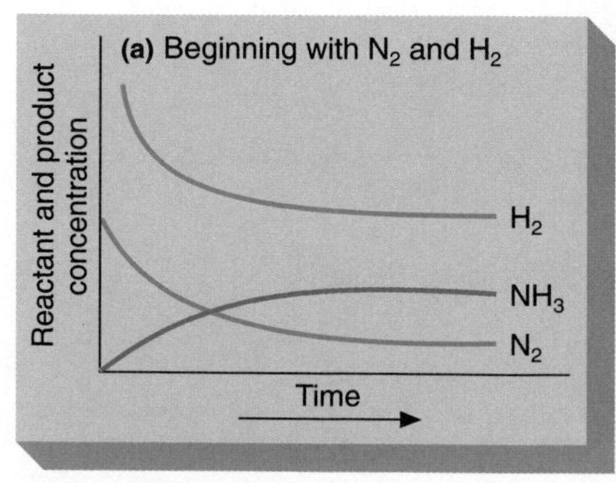

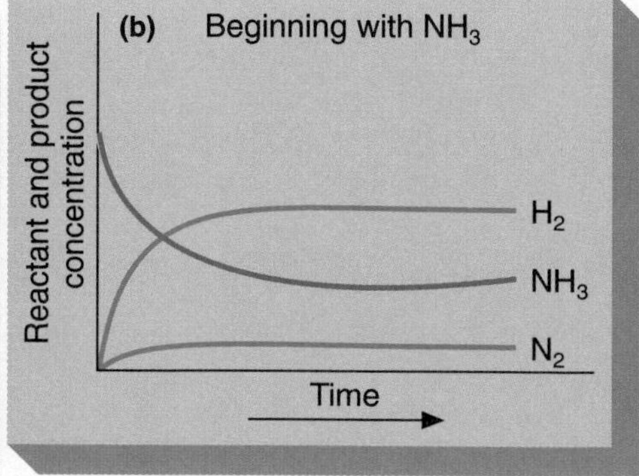

FIGURE IX.1
Variation in concentration of N_2, H_2, and NH_3 as equilibrium is approached. Note that the equilibrium concentrations of the three components are the same in each case. The reaction is written $N_2 + 3H_2 \rightleftharpoons 2NH_3$.

Beginning with a mixture of hydrogen and nitrogen, the rate of the reaction is initially rapid, since the reactant concentration is high; as the reaction proceeds, the concentration of reactants decreases. At the same time, the concentration of the product, ammonia, is increasing. When the equilibrium composition of hydrogen, nitrogen, and ammonia is reached *the rates of the forward and reverse reactions are equal*. Chemists often use the term *dynamic equilibrium* to describe this situation. The concentration of the various species is fixed at equilibrium because product is being *consumed and formed at the same rate*. In other words, the reaction continues indefinitely (dynamic), but the concentrations of products and reactants are fixed (equilibrium). This is a *dynamic equilibrium*.

The change in reaction rates for this reaction as a function of time is depicted in Figure IX.2.

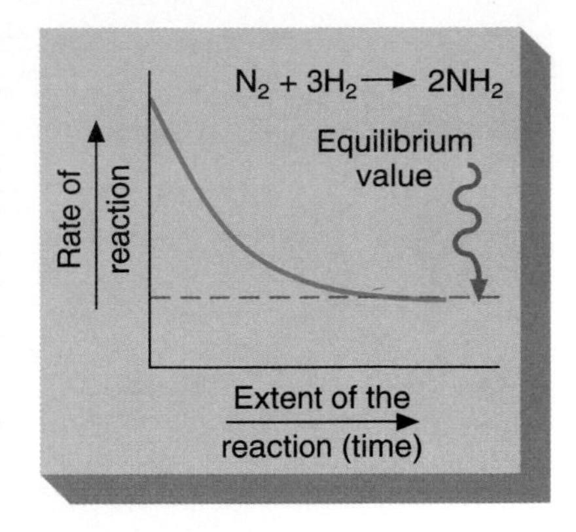

FIGURE IX.2
The change of the rate of reaction of N_2 and H_2 as a function of time. The rate of reaction, initially rapid, decreases as the concentration of reactants decreases and approaches a limiting value at equilibrium. At this point the rates of the forward and reverse reactions are equal.

THE EQUILIBRIUM CONSTANT

The most precise description of an equilibrium process is through the use of the equilibrium constant, K_{eq}.

Consider again the reaction that we have been discussing:

$$N_2(g) + 3H_2(g) \rightleftharpoons 2NH_3(g)$$

We may subdivide this expression into two reactions: a forward reaction

$$N_2(g) + 3H_2(g) \longrightarrow 2NH_3(g)$$

and a reverse reaction

$$2NH_3(g) \longrightarrow N_2(g) + 3H_2(g)$$

If the extent of each reaction is related to the concentration of the reactants remaining, at equilibrium these concentrations will become invariant. The ratio of the concentrations (actually, the concentrations raised to a power equal to the coefficient in the balanced equation) will itself be constant. This ratio is *defined* as the equilibrium constant, K_{eq}.

For the decomposition of ammonia (above),

$$K_{eq} = \frac{[\text{concentration of reactants in the reverse reaction}]^x}{[\text{concentration of reactants in the forward reaction}]^y}$$

$$K_{eq} = \frac{[NH_3]^2}{[N_2][H_2]^3}$$

- *Products* of the overall equilibrium reaction are in the numerator, and *reactants* are in the denominator.
- [ ] represents molar concentration, *M*.
- The exponents correspond to the *coefficients* of the balance equation.

It does not matter what initial amounts (concentrations) of reactants or products we choose. When the system reaches equilibrium, the calculated value of K_{eq} will not change. The magnitude of K_{eq} can be altered only by changing the temperature; thus K_{eq} is temperature dependent. The chemical industry uses this fact to advantage by choosing a reaction temperature that will maximize the yield of a desired product.

LeChatelier's Principle offers some other approaches to alter the yield of an equilibrium reaction.

LeCHATELIER'S PRINCIPLE

In the nineteenth century the French chemist Henri LeChatelier discovered that changes in equilibrium depended upon the "stress" applied to the system. The stress may take the form of an increase or decrease in the temperature of the system at equilibrium or perhaps the concentration of reactant or product.

LeChatelier's Principle states that if a stress is placed on an equilibrium system, the system will respond by altering the equilibrium in such a way as to minimize the stress.

Consider the equilibrium situation discussed earlier:

$$N_2(g) + 3H_2(g) \rightleftharpoons 2NH_3(g)$$

If the reactants and products are present in a fixed volume (such as 1 L) and more NH_3 (the *product*) is introduced into the container, the system will be stressed—the equilibrium will be disturbed. The system will try to alleviate the stress (as we all do) by *removing* as much of the added material as possible. How can it accomplish this? By converting some NH_3 to H_2 and N_2. The equilibrium shifts to the left (←). A new dynamic equilibrium is soon established that is different from the original. The relative amounts of product/reactant in that fixed volume is now different.

Had we added extra H_2 or N_2, the stress would have been applied to the other side of the equilibrium. To minimize the stress, the system would "use up" some of the excess H_2 or N_2 to make product, NH_3. The equilibrium would shift to the right (→).

In summary,

$$N_2(g) + 3H_3(g) \rightleftharpoons 2NH_3(g)$$

$$\text{Product introduced: } \xleftarrow[\text{shifted}]{\text{equilibrium}}$$

$$\text{Reactant introduced: } \xrightarrow[\text{shifted}]{\text{equilibrium}}$$

Remember that the changes discussed above will not change the magnitude of K_{eq}; only a change in temperature will accomplish this.

A P P E N D I X X

Radioactivity

Radioactivity is the process by which atoms emit high-energy particles or rays. These particles or rays are termed *radiation*. Nuclear radiation occurs as a result of an alteration in nuclear composition or structure that occurs because the nucleus is unstable and hence radioactive.

We designate the nucleus by using *nuclear symbols,* which are analogous to the *atomic symbols*. The nuclear symbols consist of the *elemental symbol* (Na, C, Mg, and so forth), the *atomic number* (equivalent to the number of protons in the nucleus) and the *mass number,* which is defined as sum of neutrons and protons in the nucleus.

Be careful not to confuse the mass number (neutrons and protons) with the atomic mass, which includes the contribution of electrons and is a true *mass* figure.

Using nuclear symbolism, the fluorine nucleus is represented:

$$\begin{array}{r} \text{Mass number} \longrightarrow \\ \text{Atomic number} \longrightarrow \end{array} {}^{19}_{9}\text{F} \longleftarrow \text{Atomic symbol}$$

(or nuclear charge)

Not all nuclei are unstable; only unstable nuclei undergo change and produce radioactivity, the process of radioactive decay. Furthermore, not all atoms of a particular element undergo radioactive decay. You will recall that different forms of the same element exist as *isotopes*. One isotope of an element may be radioactive while others of the same element may be quite stable. For this reason, in writing the symbols for a nuclear process, it is important to designate the particular isotope that is involved. This is why the mass number and atomic number are used; these values tell us the number of neutrons in the species and hence the identity of the isotope.

Three types of natural radiation emitted by unstable nuclei are *alpha particles, beta particles,* and *gamma rays*.

Alpha particles (α) contain two protons and two neutrons. An alpha particle is therefore identical to the nucleus of the helium atom (He) or a *helium ion* (He^{2+}). Having no electrons to counterbalance the nuclear charge, the alpha particle may be symbolized as

$${}^{4}_{2}\text{He}^{2+}$$

Because of their relatively large mass, alpha particles emitted by radioisotopes move relatively slowly (approximately 10% of the speed of light) and they are stopped by very little mass—the thickness of a few pages from this book is sufficient.

The *beta particle* (β), in contrast, is a fast-moving electron traveling at approximately 90% of the speed of light as it leaves the nucleus. The beta particle is represented as

$${}^{0}_{-1}\text{e}$$

or

$${}^{0}_{-1}\beta$$

The subscript -1 is written in the same position as the atomic number and, like the atomic number (number of protons), indicates the charge of the particle.

Beta particles are smaller, faster, and are more energetic than alpha particles. As a result, they are more penetrating and are stopped only by more dense materials such as wood or several layers of clothing.

TABLE X.1 A Summary of the Major Properties of Alpha, Beta, and Gamma Radiation

Name and Symbol	Identity	Charge	Mass (amu)	Velocity	Penetration
Alpha (α)	Helium nucleus	+2	4.0026	5–10% the speed of light	Low
Beta (β)	Electron	−1	0.000549	Up to 90% of the speed of light	Medium
Gamma (γ)	Radiant energy similar to X-rays	0	0	Speed of light	High

Gamma rays (γ) are pure energy, in contrast to alpha and beta radiation, which are material (composed of matter). Since pure energy has no mass or charge, the symbol for a gamma ray is simply

$$\gamma$$

Gamma radiation is highly energetic and is the most penetrating form of nuclear radiation. Barriers of lead, concrete, or, more often, a combination of the two are required for protection from this type of radiation.

PROPERTIES OF ALPHA, BETA, AND GAMMA RADIATION

Important properties of alpha, beta, and gamma radiation are summarized in Table X.1. The penetrating power of alpha radiation is very low. Therefore damage to internal organs from this form of radiation is negligible except when an alpha particle emitter is actually ingested. Beta particles are higher in energy; still, they have limited penetrating power. They cause damage principally to the skin but not to internal organs. Shielding is required in working with beta emitters; pregnant women must take special precautions. The great penetrating power and high energy of gamma radiation, in contrast make it particularly damaging to internal organs. One must take precautions in working with any of the forms of radiation. These safety considerations are required, monitored, and enforced in the United States under provisions of the Occupational Safety and Health Act (OSHA).

ARTIFICIAL RADIOACTIVITY

In the examples discussed thus far, all of the radioactive isotopes were naturally occurring. For this reason the radioactivity produced by these unstable isotopes is described as *natural radioactivity*. On the other hand, if a normally stable, nonradioactive nucleus is made radioactive through bombardment with protons, neutrons, or alpha particles, the resulting radioactivity is termed *artificial radioactivity*. The stable nucleus is made unstable by the introduction of ''extra'' protons, neutrons, or both.

The ''bombardment'' process is often accomplished in the core of a *nuclear reactor*, where an abundance of small nuclear particles, particularly neutrons, are available. Alternatively, extremely high-velocity charged particles (such as α and β) may be produced in *accelerators,* such as a cyclotron. Accelerators are extremely large and use magnetic and electric fields to "push and pull" charged particles toward their target at very high speeds.

Many isotopes that are useful in medicine are produced by particle bombardment. Production of medically useful isotopes by neutron bombardment are illustrated in the following examples:

1. Gold-198, *used as a "tracer" in the liver:*

$$^{197}_{79}\mathrm{Au} + {}^{1}_{0}\mathrm{n} \longrightarrow {}^{198}_{79}\mathrm{Au}$$

2. Technetium-99m, *a gamma source for tracer application:*

$$^{98}_{42}\mathrm{Mo} + {}^{1}_{0}\mathrm{n} \longrightarrow {}^{99m}_{43}\mathrm{Tc} + {}^{0}_{-1}\mathrm{e}$$

NUCLEAR STABILITY

We refer to the energy that holds the protons, neutrons, and other particles together in the nucleus as the *binding energy* of the nucleus. This binding energy must be quite large, since identically charged protons in the nucleus exert extreme repulsive forces upon one another. These forces, referred to as "strong forces," must be overcome if the nucleus is to be stable. When an isotope decays, some of this binding energy is released. This energy released is the source of the high-energy radiation emitted and the basis for utilization of atomic power for nuclear bombs and nuclear power plants.

Why are some isotopes more stable than others? The answer to this question is not completely clear. Evidence obtained so far points to several factors being responsible for nuclear stability:

1. Nuclear stability correlates with the ratio of neutrons to protons in the isotope. For example, for light atoms a neutron/proton ratio of 1 characterizes a stable atom.
2. Nuclei with large numbers of protons (84 or more) tend to be unstable.
3. Isotopes containing 2, 8, 20, 50, 82, or 126 protons or neutrons are stable. These *magic numbers* seem to indicate the presence of energy levels in the nucleus, analogous to *s, p, d,* and *f* orbitals in the atom.
4. Isotopes with even numbers of protons or neutrons are generally more stable than those with odd numbers of protons or neutrons.

HALF-LIFE

Not all radioactive isotopes decay at the same rate. The rate of nuclear decay is generally represented in terms of the half-life of the isotope. The *half-life,* $t_{1/2}$, is the time required for one half of a given quantity of a substance to undergo change. Each isotope has its own characteristic half-life, which may be as short as a few millionths of a second or as long as a billion years. Half-lives of some naturally occurring isotopes are given in Table X.2.

TABLE X.2 Half-lives of Selected Radioisotopes

Name	Symbol	Half-Life
Carbon-14	$^{14}_{6}\mathrm{C}$	5720 years
Sodium-24	$^{24}_{11}\mathrm{Na}$	15 hours
Iron-59	$^{59}_{26}\mathrm{Fe}$	45 days
Cobalt-60	$^{60}_{27}\mathrm{Co}$	5.3 years
Uranium-235	$^{235}_{92}\mathrm{U}$	710 million years

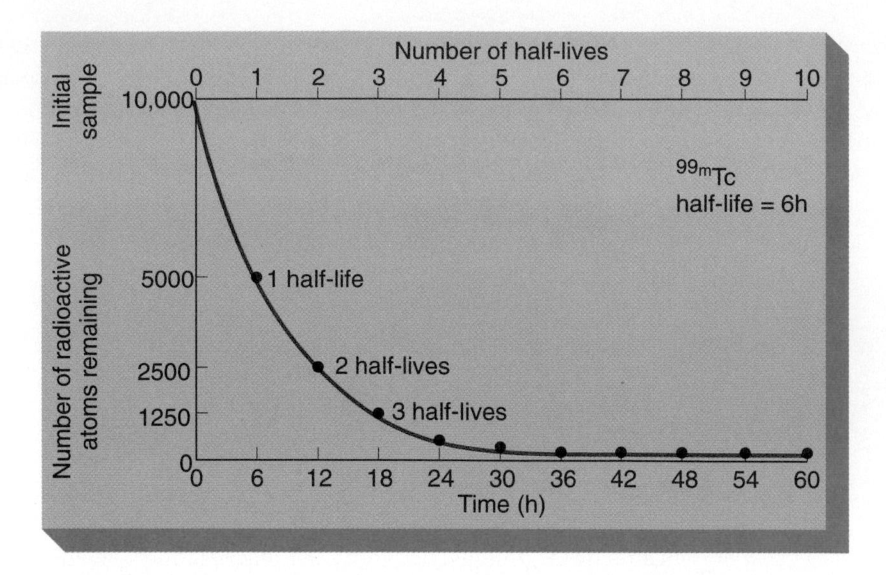

FIGURE X.1
The decay curve for the medically useful radioisotope technetium-99m. Note that the number of radioactive atoms remaining, and hence the radioactivity, asymptotically approaches but never reaches zero. After ten half-lives (60 hours), the level of radiation is indistinguishable from background.

The degree of stability of an isotope is indicated by the isotope's half-life. Isotopes with short half-lives decay rapidly; they are very unstable. This is not meant to imply that substances with long half-lives are less hazardous. Often, just the reverse is true.

Imagine that we begin with 100 mg of a radioactive isotope that has a half-life of 24 hours. After one half-life, or 24 hours, 1/2 of 100 mg has decayed to other products, and 50 mg remain. After two half-lives (48 hours), 1/2 of the remaining material has decayed, leaving 25 mg, and so on.

$$100 \text{ mg} \xrightarrow[\substack{\text{half-}\\\text{life}\\\text{(24 hrs)}}]{\text{one}} 50 \text{ mg} \xrightarrow[\substack{\text{half-}\\\text{life}\\\text{(48 hrs)}}]{\text{two}} 25 \text{ mg} \longrightarrow \text{etc.}$$

Decay of a radioisotope that has a reasonably short $t_{1/2}$ is experimentally determined by following its activity as a function of time. Graphing the results produces a radioactive decay curve (see Figure X.1).

The mass of any radioactive substance remaining after a period of time may be calculated with a knowledge of the initial mass and the half-life of the isotope using the following two expressions:

$$n = \frac{\text{time elapsed}}{t_{1/2}}$$

and

$$\text{mass}_{\text{remaining}} = \frac{\text{mass}_{\text{initial}}}{2^n}$$

where

n = the number of half-lives elapsed
$t_{1/2}$ = half-life of the isotope

Example X.1

A 50-mg sample of iodine-131, which is used in hospitals in the treatment of hyperthyroidism, was saved for 32.4 days. If the half-life of iodine-131 is 8.1 days, how many milligrams remain? First calculate n, the number of half-lives elapsed:

$$n = \frac{\text{time elapsed}}{t_{1/2}} = \frac{32.4 \text{ days}}{8.1 \text{ days}} = 4.0 \text{ half-lives}$$

Then, calculate the amount remaining:

$$\text{mass}_{\text{remaining}} = \frac{\text{mass}_{\text{initial}}}{2^n}$$

$$\text{mass}_{\text{remaining}} = \frac{50 \text{ mg}}{2^{4.0}}$$

$$\text{mass}_{\text{remaining}} = \frac{50 \text{ mg}}{16} = 3.1 \text{ mg}$$

Glossary

absolute specificity (12.5) the property of an enzyme that allows it to bind to, and catalyze the reaction of, only one substrate.

acetal (6.4) the family of organic compounds with the general formula $RCH(OR')_2$; formed via the reaction of two molecules of alcohol with an aldehyde in the presence of an acid catalyst.

acetyl CoA (13.7) a molecule composed of coenzyme A and an acetyl group; the intermediate that provides acetyl groups for complete oxidation by aerobic respiration.

acetylcholine (12.15) a chemical messenger that transmits a message from the nerve cell to the muscle cell.

acetylcholinesterase (12.15) an enzyme that destroys acetylcholine in the neuromuscular junction and thereby stops the nerve impulse.

acid-base reaction (2.1) the transfer of a hydrogen ion from one reactant to another.

acid chlorides (10.6) the family of organic compounds with the general formula

$$\text{(Ar)}\ R-\overset{\overset{\displaystyle O}{\|}}{C}-Cl$$

aconitase (14.2) the enzyme that catalyzes the isomerization of citrate to isocitrate. This is a two-step reaction in which citrate is dehydrated to produce *cis*-aconitate, which is then rehydrated to produce isocitrate.

Acquired Immune Deficiency Syndrome (AIDS) (16.1) a set of symptoms that results from infection with the human immunodeficiency virus. These symptoms include a variety of opportunistic infections and cancers, an impaired immune response, and dementia.

acquired pellicle (17.1) a layer of saliva glycoproteins that coats the teeth and allows adherence of oral bacteria onto tooth surfaces.

activated complex (2.4) the arrangement of atoms at the top of the potential energy barrier as a reaction proceeds.

activation energy (2.4) the threshold energy that must be overcome to produce a chemical reaction.

active site (12.4) the cleft in the surface of an enzyme that is the site of substrate binding and catalysis.

active transport (9.6) the movement of molecules across a membrane against a concentration gradient.

acyl carrier protein (ACP) (15.4) the protein that forms a thioester linkage with fatty acids during fatty acid synthesis.

acyl group (Chapter 6, Introduction; Chapter 7, Introduction) the functional group that contains the carbonyl group attached to one alkyl or aryl group; $\text{(Ar)}R-\underset{|}{C}{=}O$; the functional group found in the derivatives of the carboxylic acids.

addition reaction (4.4, 6.4) a reaction in which two molecules add together to form a new molecule; often involves the addition of one molecule to a double or triple bond in an unsaturated molecule, e.g., the addition of alcohol to an aldehyde or ketone to form a hemiacetal or hemiketal.

adenine (16.1) one of the purine nitrogenous bases in DNA and RNA.

adenosine (16.1) a nucleoside formed from adenine and ribose.

adenosine triphosphate (ATP) (8.5, 13.1) a nucleotide composed of the purine adenine, the sugar ribose, and three phosphoryl groups; the primary energy storage and transport molecule used by the cells in cellular metabolism.

adipocyte (15.1) a fat cell.

adipose tissue (15.1) fatty tissue that stores most of the body lipids.

air oxidation (6.3) the oxidation of a molecule in the presence of air; air is the oxidizing agent.

alanine aminotransferase (14.5) the enzyme that catalyzes the transfer of the α-amino group of alanine to α-ketoglutarate, producing pyruvate and glutamate.

alcohol (5.1) an organic compound that contains an hydroxyl group (—OH) attached to an alkyl group.

alcohol dehydrogenase (13.4) the enzyme that reduces acetaldehyde to produce ethanol in the alcohol fermentation.

aldolase (13.3) the enzyme that cleaves fructose-1,6-diphosphate into dihydroxyacetone phosphate and glyceraldehyde-3-phosphate.

aldose (7.2) a sugar that contains an aldehyde (carbonyl) group.

alkaloids (10.3) a class of naturally occurring compounds that contain one or more nitrogen heterocyclic rings; many of the alkaloids have medicinal and other physiological effects.

alkane (3.1) also called a *saturated hydrocarbon;* a hydrocarbon that contains only carbon and hydrogen, that is bonded together through carbon-hydrogen and carbon-carbon single bonds, and that has the general molecular formula C_nH_{2n+2}.

alkene (4.1) a hydrocarbon that contains one or more carbon-carbon double bonds; an unsaturated hydrocarbon with the general formula C_nH_{2n}.

alkyl group (3.2) a simple hydrocarbon group that results from the removal of one hydrogen from the original hydrocarbon (e.g., methyl, CH_3^-; ethyl, $CH_3CH_2^-$).

alkyl halide (3.7) a substituted hydrocarbon that has the general structure R—X, where R represents any alkyl group and X = (a halogen) F^-, Cl^-, Br^-, or I^-.

alkyne (4.1) a hydrocarbon that contains one or more carbon-carbon triple bonds; an unsaturated hydrocarbon with the general formula C_nH_{2n-2}.

allosteric enzyme (12.9) an enzyme that has an effector binding site as well as an active site; effector binding changes the shape of the active site, rendering it either active or inactive.

amides (10.6) the family of organic compounds with the general formula

$$\text{(Ar)}\ R-\overset{\overset{\displaystyle O}{\|}}{C}-NH_2$$

amines (Chapter 10, Introduction) the family of organic molecules with the general formula $R—NH_2$, R_2NH, or R_3N (R— can equal R— or Ar—); they may be viewed as substituted ammonia molecules in which one or more of the ammonia hydrogens has been substituted by a more complex organic moiety.

α-amino acid (11.2) the subunits of proteins composed of an α-carbon bonded to a carboxylate group, a protonated amino group, a hydrogen atom, and a variable R group.

amino acid residue (11.3) the amino acid unit that remains after a peptide bond has been formed.

aminoacyl synthetase (16.11) an enzyme that recognizes one tRNA and covalently links the appropriate amino acid to it.

aminoacyl tRNA (16.11) a charged tRNA; the transfer RNA covalently linked to the correct amino acid.

aminoacyl tRNA binding site of ribosome (16.11) a pocket on the surface of a ribosome that holds the aminoacyl tRNA during translation.

aminotransferase (14.5) an enzyme that catalyzes the transfer of an amino group from one molecule to another. (Also called transaminase.)

amphetamines (10.6) a family of compounds, many of which are physiologically active and stimulate the central nervous system; they are often called ''uppers'' and are often used in diet pills to help decrease appetite; they are also used to treat psychological disorders such as severe depression.

amphibolic pathway (14.7) a metabolic pathway that functions in both anabolism and catabolism.

amylopectin (7.4) a highly branched form of amylose; the branches are attached to the C-6 hydroxyl by α (1 ⟶ 6) glycosidic linkage; a component of starch.

amylose (7.4) a linear polymer of α-D-glucose in α (1 ⟶ 4) glycosidic linkage that is a major component of starch; a polysaccharide storage form.

anabolism (14.7) energy-requiring biosynthetic pathways.

anaerobic threshold (13.4) the point at which the level of lactate in the exercising muscle inhibits glycolysis and the muscle, deprived of energy, ceases to function.

analgesic (8.5, 10.6) any drug that acts as a pain killer; e.g., aspirin, phenacetin, acetaminophen.

anaplerotic reaction (14.7) a reaction that replenishes a substrate needed for a biochemical pathway.

Andersen's Disease (13.6) a genetic defect in the glycogen branching enzyme resulting in difficulty using glycogen stores efficiently.

anesthetic (4.10, 10.6) any drug that causes lack of sensation in any part of the body (local) or causes unconsciousness (general); an analgesic effect is generally also associated with these compounds.

angular molecule (4.3) a planar molecule with bond angles other than 180°.

anion (1.1) a negatively charged atom.

anomers (7.2) isomers of cyclic sugars that differ in the arrangement of groups around an asymmetric carbon.

antibody (11.8) immunoglobulin; a specific glycoprotein produced by cells of the immune system in response to invasion by infectious agents.

anticodon (16.7) a sequence of three ribonucleotides on a tRNA that is complementary to a codon on the mRNA; codon-anticodon binding results in delivery of the correct amino acid to the site of protein synthesis.

antidiuretic hormone (9.6) a hormone secreted from the anterior pituitary that decreases water excretion by the kidney.

antigen (11.1) any substance able to stimulate the immune system; generally a protein or large carbohydrate.

antimetabolite (17.3) a drug that blocks a specific enzymatic reaction.

antiparallel (16.3) a term describing the polarities of the two strands of the DNA double helix; on one strand the sugar-phosphate backbone advances in the 5′ ⟶ 3′ direction, while on the opposite, complementary strand the sugar phosphate backbone advances in the 3′ ⟶ 5′ direction.

antiseptic (4.10) any compound that has the effect of preventing or inhibiting bacterial infection of the body.

apoenzyme (12.7) the protein portion of an enzyme that requires a cofactor in order to function in catalysis.

arachidonic acid (9.2) a fatty acid that is derived from linolenic acid; the precursor of the prostaglandins.

aromatic compound (4.7) a hydrocarbon that contains a benzene ring or that has properties that are similar to those exhibited by benzene.

Arrhenius Theory (2.2) a theory that defines an acid as a substance that dissociates to produce H^+ and a base as a substance that dissociates to produce OH^-.

aryl halide (4.9) a benzene ring, or other aromatic compound, in which a hydrogen on the ring has been substituted by a halogen atom (F^-, Cl^-, Br^-, or I^-).

aspartate aminotransferase (14.5) an enzyme that mediates the transfer of the α-amino group of aspartate to α-ketoglutarate, producing oxaloacetate and glutamate.

asymmetric carbon (7.2) a chiral carbon; a carbon bonded to four different groups.

atherosclerosis (9.5) deposition of excess plasma cholesterol and other lipids and proteins on the walls of arteries, resulting in a decreased artery diameter and increased blood pressure.

atom (Introduction to Chapter 1) the smallest unit of an element that retains the properties of that element.

atomic mass (1.1) the mass of an atom expressed in atomic mass units.

atomic number (1.1) the number of protons in the nucleus of an atom. It is a characteristic identifier of an element.

ATP synthase (14.3) a multiprotein complex within the inner mitochondrial membrane that uses the energy of the proton gradient to produce ATP. (Also called $F_0\,F_1$ complex.)

autoimmune reaction (11.8) a reaction of the immune system against one's own tissues.

autoionization (2.2) or self-ionization, the reaction of a substance, such as water, with itself to produce a positive and a negative ion.

autoionization constant (2.3) the equilibrium constant that mathematically describes the self-ionization process.

AZT (16.1) 3′-azido-2′,3′-dideoxythymidine, a nucleoside analogue of deoxythymidine that is used to inhibit the growth of the human immunodeficiency virus.

B lymphocytes (11.8) specialized white blood cells of the immune system that produce antibodies.

BAL (5.9) British Anti-Lewisite; used as an antidote to mercury poisoning.

$$\underset{\displaystyle OH}{\underset{|}{CH_2}}—\underset{\displaystyle SH}{\underset{|}{CH}}—\underset{\displaystyle SH}{\underset{|}{CH_2}}$$

BAL

barbiturate (10.6) a class of physiologically active compounds that are derived from amides; the barbiturates are ''downers'' and are used as sedatives; they are also used as anticonvulsants for epileptics and for people suffering from a variety of brain disorders.

base pair (16.3) a hydrogen-bonded pair of bases within the DNA double helix. The standard base pairs always involve a purine and a pyrimidine; in particular, adenine always base pairs with thymine and cytosine with guanine.

Benedict's solution (6.4, 7.2) a buffered solution of Cu^{2+} ions that can be used to test for reducing sugars.

Benedict's Test (6.4) use of a test reagent (two solutions: sodium citrate and aqueous cupric hydroxide) to test for the carbonyl functional group, particularly those found in reducing sugars.

beriberi (17.3) a disease that is caused by a dietary deficiency of thiamine. It is characterized by muscle weakness and mental instability.

bile (15.1) micelles of lecithin, cholesterol, bile salts, protein, inorganic ions, and bile pigments that aid in lipid digestion by emulsifying fat droplets.

binding site (12.4) the chemical groups of the active site of an enzyme that are involved in the specific substrate binding.

biodegradable (4.10) any substance that will break down naturally in the atmosphere.

biological magnification (4.10) the process through which chemicals are concentrated in greater and greater quantities as they are processed through the food chain.

biotin (17.3) a vitamin that is required for carboxylation and decarboxylation reactions involved in the metabolism of fats, carbohydrates, and proteins.

blood-brain barrier (15.5) structures and cells in the brain that select which molecules may enter the brain.

bond angle (1.5) the angle (in degrees) formed by two covalent bonds involving a common atom.

bond length (1.5) the distance between the nuclei of two atoms joined by a chemical bond.

bond strength (1.5) the energy necessary to break a bond between two atoms.

branching enzyme (13.6) the enzyme in glycogenesis that makes the α (1 $\longrightarrow$ 6) glycosidic linkages at the branch points.

Brönsted-Lowry Theory (2.2) a theory that describes an acid as a proton donor and a base as a proton acceptor.

brown fat (14.3) specialized tissue that burns lipids and produces heat.

buffer solution (2.2) a solution that contains a weak acid or base and its salt that is resistant to large changes in pH upon addition of strong acids or bases.

C-terminal residue (11.3) the amino acid residue with a free α-CO_2^- group; the last amino acid in a peptide.

calcitonin (17.4) a hormone produced by the thyroid gland that decreases the level of blood calcium by inhibiting kidney reabsorption and intestinal absorption and causing incorporation of calcium into bone.

carbinol carbon (5.5) that carbon in an alcohol to which the hydroxyl group is attached.

carbohydrate (17.1) an organic compound usually characterized by the general formula $(CH_2O)_n$; generally sugars and polymers of sugars; the primary source of energy for the cell.

carbonyl group (Chapter 6, Introduction) the functional group that contains a carbon-oxygen double bond;

$$\rangle C{=}O;$$

the functional group found in aldehydes and ketones.

carboxyl group (Chapter 6, Introduction; Chapter 15, Introduction) the —COOH functional group; the functional group found in carboxylic acids.

carboxylic acid (Chapter 8, Introduction) the family of organic compounds that contains the —COOH functional group.

carboxylic acid derivative (Chapter 8, Introduction) any of several families of organic compounds that are derived from carboxylic acids and that have the general formula

$$(\mathrm{Ar})\ \ \mathrm{R}-\underset{\displaystyle\mathrm{Z}}{\overset{}{\mathrm{C}}}{=}\mathrm{O};$$

($Z = X$, OR, NH_2, etc.); includes the acid chlorides, esters, amides, and acid anhydrides.

carcinogen (10.6, 16.5) any chemical or physical agent that causes mutations in the DNA that lead to uncontrolled cell growth or cancer.

catabolism (13.1, 14.7) the degradation of fuel molecules and production of energy for cellular functions.

catalase (Chapter 12, Introduction) an enzyme that mediates the conversion of two molecules of hydrogen peroxide into water and oxygen.

catalyst (2.4, 4.6) any substance that increases the rate of a chemical reaction (by lowering the activation energy of the reaction) and that is not destroyed in the course of the reaction.

catalytic cracking (3.4) a process that results in the decomposition of large hydrocarbons to smaller ones under the influence of heat and/or pressure in the presence of a catalyst.

catalytic groups (12.4) the chemical groups of the active site of an enzyme that are involved in catalysis.

catalytic reforming (3.4) a process that results in the rearrangement of one organic molecule into another under the influence of heat and/or pressure in the presence of a catalyst.

cation (1.1) a positively charged atom.

cellulose (7.4) the most abundant organic compound in the world; a polymer of β-D-glucose linked by β (1 $\longrightarrow$ 4) glycosidic bonds.

central dogma (16.6) a statement of the directional transfer of the genetic information in cells: DNA $\longrightarrow$ RNA $\longrightarrow$ protein.

central nervous system (CNS) (10.1) the brain, the spinal cord, and all of the nerves that radiate from the spinal column.

cephalin (9.3) one of the major phospholipids; a complex phosphoglyceride with the amino alcohol ethanolamine attached to the phosphoryl group. (See also phosphatidylethanolamine.)

chain elongation (16.11) the sequential addition of amino acids to the growing peptide chain during translation.

chemical bond (1.3) the attractive electrical force holding two atomic nuclei together in a chemical compound.

chemical energy (2.4) energy stored in substances that can be released during a chemical reaction.

chemiosmotic theory (14.3) the theory that explains the way in which the oxidation of glucose is coupled to the phosphorylation of ADP to generate ATP.

chiral (7.2) molecules that are capable of existing in mirror-image forms.

cholesterol (9.4) a 27-carbon steroid ring structure that serves as the precursor of the steroid hormones.

chylomicron (9.5, 15.1) a plasma lipoprotein (aggregate of protein and triglycerides) that carries triglycerides from the intestines to all body tissues via the bloodstream.

chymotrypsin (12.11) a proteolytic enzyme that is produced in the pancreas and secreted into the small intestines, where it hydrolyzes dietary protein.

citrate synthase (14.2) the enzyme that catalyzes the condensation of the acetyl group of acetyl CoA with oxaloacetate in the first reaction of the citric acid cycle.

citric acid cycle (14.2) a cyclic biochemical pathway that is the final stage of degradation of carbohydrates, fats, and amino acids. It results in the complete oxidation of acetyl groups derived from these dietary fuels. Also called the Krebs Cycle.

closed system (2.4) a system in which the total mass and energy of the system does not change.

coagulation (11.12) the process by which proteins in solution are denatured and aggregate with one another to produce a solid.

codon (16.7) a group of three ribonucleotides on the mRNA that specifies the addition of a specific amino acid onto the growing peptide chain.

coenzyme (5.8, 12.7, 17.3) an organic group that is required by some enzymes, generally serves as a donor or acceptor of electrons or functional groups in a reaction.

coenzyme A (13.7, 17.3) a molecule derived from ATP, the vitamin pantothenic acid, and the amino acid cysteine. Coenzyme A functions in the transfer of acetyl groups in lipid and carbohydrate metabolism.

cofactor (12.7) an inorganic group, usually a metal ion, that must be bound to an apoenzyme to maintain the correct configuration of the active site.

colipase (15.1) a protein that aids in lipid digestion by binding to the surface of lipid droplets and facilitating binding of pancreatic lipase.

collagen (11.6) the most abundant protein in the body. It confers mechanical stability to skin, bone, and tendons.

combination reaction (2.1) the joining of two or more atoms or compounds to produce a product of different composition.

combustion (3.6) the oxidation of hydrocarbons by burning in the presence of air to produce carbon dioxide and water.

common clotting pathway (12.13) the final steps of the formation of a fibrin clot that are the same regardless of whether the intrinsic or extrinsic pathway is used to initiate clot formation.

common system of nomenclature (3.2) the nonsystematic, though older and well-established, system of nomenclature.

competitive inhibitor (17.3) a chemical that has a structure that is very similar to the natural substrate of an enzyme and competes with the natural substrate for binding to the enzyme active site and inhibits the reaction.

complementary strands (16.3) the opposite strands of the double helix are hydrogen-bonded to one another such that adenine and thymine or guanine and cytosine are always paired.

complete protein (17.1) a protein source that contains all the essential and nonessential amino acids.

complex lipid (9.5) a lipid that is bonded to other types of molecules.

condensed structural formula (3.1) a structural formula showing all of the atoms in a molecule and placing them in a sequential arrangement that details which atoms are bonded to each other; the bonds themselves are not shown.

cone (6.3) one of the two types of cells in the retina of the eye responsible for vision; primarily responsible for vision in bright light and for the detection of color.

conformers (3.5) also called *conformational isomers;* discrete, distinct isomeric structures that may be converted, one to the other, by rotation about the bonds in the molecule.

conjugated protein (11.8) a protein that is functional only when it carries other chemical groups attached by covalent linkages or by weak interactions.

Cori Cycle (13.5) a metabolic pathway in which the lactate produced by working muscle is converted back to glucose by gluconeogenesis in the liver.

Cori's Disease (13.6) a genetic defect in the glycogen debranching enzyme that results in inefficient utilization of stored glycogen.

cortisone (9.4) a steroid used to suppress the inflammatory response in the treatment of rheumatoid arthritis, asthma, and many other diseases.

coumarin (12.14) an anticoagulant that reduces blood clotting by reducing the amount of prothrombin and factors VII, IX, and X. It is a competitive inhibitor of vitamin K.

covalent bond (1.3) a pair of electrons shared between two atoms.

creatine phosphate (14.1) a molecule in muscle tissue that serves as a storage reservoir for high-energy phosphate groups.

cristae (14.1) the fingerlike invaginations of the inner membrane of the mitochondria.

cyclic AMP (11.1) a nucleotide chemical messenger that allows cells to respond to certain stimuli, such as binding of enkephalins, to cell surfaces.

cycloalkanes (3.4) cyclic alkanes (saturated hydrocarbons) that have the general formula C_nH_{2n}.

cytidine (16.1) a nucleoside composed of the pyrimidine cytosine and the sugar ribose.

cytochrome *c* (11.4) a protein of 104 amino acid residues that is required for aerobic respiration.

cytosine (16.1) one of the pyrimidine nitrogenous bases found in DNA and RNA.

debranching enzyme (13.6) an α (1 $\longrightarrow$ 6) glycosidase that hydrolyzes the α (1 $\longrightarrow$ 6) glycosidic bonds at the branch points of glycogen in the process of glycogenolysis.

decomposition reaction (2.1) a reaction that involves the breakdown of a substance into two or more substances.

decongestants (10.6) a group of compounds that are useful in the treatment of the symptoms associated with colds and many allergies; they act by causing a shrinking of the membranes that line the nasal passages.

degenerate code (16.10) a term used to describe the fact that several triplet codons may be used to specify a single amino acid in the genetic code.

degradation (4.5) the breakdown of a larger molecule into smaller molecules.

dehydration (5.5) a reaction that involves the loss of a water molecule, e.g., the loss of water from an alcohol and the concomitant formation of an alkene.

denaturation (11.12) the process by which the organized structure of a protein is disrupted, resulting in a completely disorganized, nonfunctional form of the protein.

denatured alcohol (5.3) ethanol (C_2H_5OH) to which a denaturing agent (often methanol) has been added to make it unfit to drink.

dental plaque (17.1) an aggregation of dextran and oral bacteria that builds up on teeth and provides an environment in which tooth decay can occur.

deoxyadenosine (16.1) a nucleoside made up of the purine adenine and the sugar 2′-deoxyribose.

deoxycytidine (16.1) a nucleoside composed of the pyrimidine cytosine and the sugar 2′-deoxyribose.

deoxyguanosine (16.1) a nucleoside made up of the purine guanine and the sugar 2′-deoxyribose.

deoxyribonucleic acid (DNA) (16.1) the nucleic acid molecule that carries all of the genetic information of an organism; the DNA molecule is a double helix composed of two strands, each of which is composed of phosphate groups, deoxyribose, and the nitrogenous bases thymine, cytosine, adenine, and guanine.

deoxyribonucleotide (16.1) a nucleoside phosphate or nucleotide composed of a nitrogenous base in β-N-glycosidic linkage to the 1′-carbon of the sugar 2′-deoxyribose and with one, two, or three phosphoryl groups esterified at the hydroxyl of the 5′-carbon.

deoxythymidine (16.1) a nucleoside composed of the pyrimidine thymine and the sugar 2′-deoxyribose.

destructive distillation (5.3) a distillation, or separation of compounds by differences in boiling point, that also involves the decomposition of the compound(s) in the course of the distillation.

detergent (15.1) a molecule having a highly charged (polar) group and a long hydrocarbon chain. In water the charged end dissolves, but the hydrophobic chain does not. When fats are present, the hydrocarbon chain binds to them, forming micelles.

dextran (17.1) a polymer of glucose that forms the ground substance of dental plaque.

dextrorotatory (7.2) the enantiomer that rotates plane polarized light in a clockwise direction.

diabetes mellitus (15.3, 16.11) a disease caused by the production of insufficient levels of insulin and characterized by the appearance of very high levels of glucose in the blood and urine.

diglyceride (9.3) the product of esterification of glycerol at two positions.

dipeptide (11.3) a molecule formed by condensing two amino acids.

disaccharide (7.3) a sugar composed of two monosaccharides joined through an oxygen atom bridge.

distillation (3.4) the process of separating materials on the basis of differences in boiling points.

disulfide (5.9) an organic compound that contains a disulfide group (—S—S—).

divergent evolution (12.12) the process whereby copies of an identical gene evolve to become dissimilar.

double helix (16.3) the spiral staircaselike structure of the DNA molecule characterized by two sugar-phosphate backbones wound around the outside and nitrogenous bases extending into the center.

double-replacement reaction (2.1) a reaction that involves a chemical change in which cations and anions ''exchange partners.''

eicosanoid (9.2) any of the derivatives of 20-carbon fatty acids, including the prostaglandins, leukotrienes, and thromboxanes.

electron (1.1) a negatively charged particle outside of the nucleus of an atom.

electron affinity (1.2) the energy released when an electron is added to an isolated atom.

electronegativity (1.2) a measure of the tendency of an atom in a molecule to attract shared electrons.

electrophile (3.9) (Gk: *electro*, electron; *phile*, loving) an electron-deficient species such as a hydrogen ion (H^+).

electrophilic substitution (3.9) a substitution reaction in which the incoming substituent that reacts with the parent compound is an electrophile.

electrostatic force (1.3) the attractive force between two oppositely charged particles.

emulsifying agent (9.3) a bipolar molecule that aids in the suspension of fats in water.

enantiomers (7.2) stereoisomers that are mirror images of one another.

end product inhibition (12.9) a means of enzyme regulation in which the product of a reaction inhibits its own synthesis by interacting with the enzyme and forcing the reaction to proceed in the reverse, from product to substrate.

endothermic process (2.4) a process that absorbs energy in a chemical change.

endothermic reaction (13.1) a reaction that requires an input of energy in order to occur.

enediol reaction (7.2) the reaction by which ketoses are converted to aldoses.

enkephalins (11.1) peptide opiates that are involved in the perception of pain.

enolase (13.3) the enzyme that catalyzes the dehydration of 3-phosphoglycerate, which produces phosphoenolpyruvate.

enzyme (11.1, Chapter 12, Introduction) a protein that serves as a biological catalyst.

enzyme assay (12.16) a test to measure the amount of enzyme in a sample, for instance, in the bloodstream.

enzyme-linked immunosorbant assay (ELISA) (12.16) an assay that is used to detect the presence of human immunodeficiency virus (HIV) antibodies in blood serum, based on antigen-antibody binding and using an enzyme reaction as the detection system.

enzyme specificity (Chapter 12, Introduction) the ability of an enzyme to bind to only one, or a very few, substrates and thus catalyze only a single reaction.

enzyme-substrate complex (12.4) a molecular aggregate formed when the substrate binds to the active site of the enzyme.

enzymology (12.16) the study of the function and structure of enzymes.

epinephrine (13.6) a hormone produced by the adrenals in times of stress that stimulates glycogenolysis and thereby increases the concentration of blood glucose. Also called adrenaline.

equilibrium constant (2.2) a number equal to the ratio of the equilibrium concentrations of products to the equilibrium concentrations of reactants, each raised to a power equal to its stoichiometric coefficient.

essential amino acids (17.1) amino acids that cannot be synthesized by the body and must therefore be supplied by the diet.

essential fatty acids (9.2, 17.1) the fatty acids linolenic and linoleic acids, which must be supplied in the diet because they cannot be synthesized by the body.

estrone (9.4) a female sex hormone produced by modification of progesterone.

ether (5.7) an organic compound that contains two alkyl and/or aryl groups attached to an oxygen atom; R—O—R, Ar—O—R, and Ar—O—Ar.

exothermic process (2.4) a process in which energy is released in a chemical change.

exothermic reaction (13.1) a reaction that liberates energy.

extrinsic pathway of blood clotting (12.13) the initial stages of the clotting mechanism that are induced by damage to the tissues surrounding the blood vessel.

facilitated diffusion (9.6) movement of a solute across a membrane from an area of high concentration to an area of low concentration through an integral membrane protein, or permease.

factor VIII (16.11) one of the proteins that is necessary for proper blood clotting.

fast twitch muscle fibers (14.1) muscle fibers that are specialized to provide ATP energy principally by anaerobic glycolysis.

fatty acid (Chapter 8, Introduction, 9.2, 17.1) any member of the family of continuous-chain carboxylic acids that generally contain 4–20 carbon atoms; the most concentrated source of energy used by the cell.

fatty acid synthase (15.4) the multienzyme complex that carries out fatty acid biosynthesis.

feedback inhibition (12.9) excess product of a biosynthetic pathway turns off the entire pathway for its own synthesis.

Fehling's solution (7.2) a buffered solution of Cu^{2+} ions that is used to test for reducing sugars.

fermentation (5.3) the anaerobic (in the absence of oxygen) metabolism or degradation of sugars.

fetal hemoglobin (11.10) the form of hemoglobin produced by the fetus; it has a greater affinity for oxygen than does the mother's adult hemoglobin.

F_0F_1 complex (14.3) the multiprotein complex in the inner mitochondrial membrane that uses the energy of the proton gradient to produce ATP. (Also called ATP synthase.)

F_0F_1 event (14.3) the stage of oxidative phosphorylation in which ATP synthase (F_0F_1 complex) phosphorylates ADP to produce ATP, using the energy of the proton gradient.

fibrin (12.13) the insoluble blood protein that makes up a blood clot.

fibrinogen (12.13) the soluble blood protein that is converted to insoluble fibrin by the enzyme thrombin in the last step of blood clotting.

fibrous protein (11.7) proteins composed of peptides arranged in long sheets or fibers.

first law of thermodynamics (2.4) the law

of conservation of energy; in chemical change, energy cannot be created or destroyed.

Fischer Projection (7.2) a two-dimensional formula used to designate the three-dimensional structure of a molecule. It is drawn as a cross with the chiral carbon in the center.

Fischer-Tropsch Synthesis (6.3) an industrial process used in the preparation of various hydrocarbons and their oxygen-containing derivatives that involves the combination of carbon monoxide and hydrogen under controlled conditions of heat pressure and the use of a catalyst.

fluid mosaic theory (9.6) the model of membrane structure that describes the fluid nature of the lipid bilayer and the presence of numerous proteins embedded within the membrane.

folic acid (17.3) a vitamin that is required for the transfer of one-carbon groups for the synthesis of methionine, purines, and pyrimidines.

formal charge (1.1) the charge on each atom in a molecule.

fructose diphosphatase (13.5) the enzyme that catalyzes the removal of the C-1 phosphoryl group from fructose-1,6-diphosphate in gluconeogenesis.

fumarase (14.2) an enzyme that catalyzes the hydration of fumarate to produce malate.

galactose (7.2) an aldohexose that is a component of lactose, milk sugar.

galactosemia (7.3) a human genetic disease caused by the inability to convert galactose to glucose-1-phosphate.

geometric isomers (4.3) isomers that differ from one another owing to the placement of substituents on a double bond or a ring.

globular protein (11.7) a protein composed of polypeptide chains that are tightly folded into a compact spherical shape.

glucagon (13.6, 15.7) a peptide hormone synthesized by the α-cells of the islets of Langerhans in the pancreas in response to low blood glucose levels; glucagon promotes glycogenolysis and thereby increases the concentration of blood glucose.

glucokinase (13.6) an enzyme that catalyzes the phosphorylation of glucose in the first step of glycogenesis.

gluconeogenesis (13.5) the synthesis of glucose from noncarbohydrate precursors.

glucose (7.2) an aldohexose, the most abundant monosaccharide; it is a component of many disaccharides, such as lactose and sucrose, and polysaccharides, such as cellulose, starch, and glycogen.

glucose-6-phosphatase (13.5) the enzyme that catalyzes the dephosphorylation of glucose-6-phosphate in the last step of gluconeogenesis.

glucosyl transferase (17.1) an enzyme that cleaves sucrose into glucose and fructose and converts the glucose into dextran.

glutamate dehydrogenase (14.5) an enzyme that catalyzes the oxidative deamination of the amino acid glutamate.

glyceraldehyde-3-phosphate dehydrogenase (13.3) the enzyme that catalyzes the oxidation of glyceraldehyde-3-phosphate to 1,3-diphosphoglycerate and the concomitant reduction of NAD^+ in glycolysis.

glyceride (9.3) a lipid that contains glycerol.

glycogen (7.4, 13.6) a highly branched polymer of glucose stored in liver and muscles of animals; a linear backbone of α-D-glucose in α $(1 \longrightarrow 4)$ linkage, with numerous short branches attached to the C-6 hydroxyl group by α $(1 \longrightarrow 6)$ linkage.

glycogen granule (13.6) a core of glycogen surrounded by enzymes that are responsible for glycogen synthesis and degradation.

glycogen phosphorylase (13.6) an enzyme that mediates the first step in glycogenolysis, the displacement of a molecule of glucose from glycogen by the addition of a phosphate group.

glycogen synthetase (13.6) an enzyme that catalyzes the formation of an α $(1 \longrightarrow 4)$ glycosidic bond between the glucose moiety of UDP-glucose and the growing glycogen polymer.

glycogenesis (13.6) the metabolic pathway that results in the addition of glucose to growing glycogen polymers when blood glucose levels are high.

glycogenolysis (13.6) the biochemical pathway that results in the removal of glucose molecules from glycogen polymers when blood glucose levels are low.

glycolysis (13.3) the enzymatic pathway that converts a glucose molecule into two molecules of pyruvate. This anaerobic process generates energy in the form of two ATP and two NADH.

glycoprotein (11.8) proteins conjugated with sugar groups. Often these are receptors on the cell surface.

α $(1 \longrightarrow 6)$ glycosidase (13.6) the enzyme that hydrolyzes the α $(1 \longrightarrow 6)$ glycosidic bond at the branch points of glycogen.

glycosidic bond (7.3) the bond between the hydroxyl group of the anomeric carbon of one sugar and a hydroxyl group of another sugar.

group specificity (12.5) an enzyme that catalyzes reactions involving similar substrate molecules having the same functional groups.

guanine (16.1) one of the purine nitrogenous bases found in DNA and RNA.

guanosine (16.1) a nucleoside composed of the purine guanine and the sugar ribose.

guanosine triphosphate (13.5) a nucleotide composed of the purine guanosine, the sugar ribose, and three phosphoryl groups. It serves as a phosphoryl group donor in the conversion of oxaloacetate to phosphoenolpyruvate during gluconeogenesis.

H^+ transport (14.3) the stage of oxidative phosphorylation in which electrons pass from NADH or $FADH_2$ through a series of electron carriers, causing H^+ to be transported from the mitochondrial matrix to the intermembrane space.

halide exchange (3.9, 5.3) a substitution reaction in which one halogen atom is exchanged for another.

halogenation (3.6) a reaction in which one of the C—H bonds of a hydrocarbon is replaced with C—X bond of a halogen atom (X = Br or Cl, generally).

Haworth Projections (7.2) a means of representing the orientation of substituent groups around a cyclic sugar molecule.

heat energy (2.4) random kinetic energy absorbed or released by a substance or substances.

α-helix (11.5) a right-handed coiled secondary structure maintained by hydrogen bonds between the amide hydrogen of one amino acid and the carbonyl oxygen of an amino acid four residues away.

heme (11.9) the chemical group found in hemoglobin and myoglobin that is responsible for the ability to carry oxygen.

hemiacetal (6.4, 7.2) the family of organic compounds with the general formula RR′C(OR″)(OH); formed via the reaction of one molecule of alcohol with an aldehyde in the presence of an acid catalyst.

hemiketal (6.4, 7.2) the family of organic compounds with the general formula RR″C(OR′) (OH); formed via the reaction of one molecule of alcohol with a ketone in the presence of an acid catalyst.

hemoglobin (11.9) the major protein component of red blood cells. The function of this iron-containing protein is transport of oxygen.

hemophilia (12.13, 16.11) a genetic disorder in the blood-clotting mechanism in which one of the clotting factors is missing or inactive. It is characterized by excessive bleeding that may occur spontaneously or following even minor wounds.

heroin (11.1) a derivative of morphine that is much more addictive than morphine and induces a longer-lived euphoria.

heteroatoms (1.6) atoms other than carbon and hydrogen contained in an organic molecule.

heterocyclic amine (Chapter 10, Introduction) a heterocyclic compound that contains nitrogen in at least one position in the ring skeleton.

heterocyclic compound (5.7) a cyclic compound in which one or more of the carbon atoms in the ring skeleton has been replaced by *hetero* atoms (for example, O, N, P, or S).

hexokinase (13.3) the enzyme that catalyzes the phosphorylation of glucose in the first reaction of glycolysis.

hexose (7.2) a six-carbon monosaccharide.

high-density lipoprotein (HDL) (9.5) a plasma lipoprotein that transports cholesterol from peripheral tissue to the liver.

homeostatic mechanism (14.3) a mechanism that works to maintain a constant internal environment within the body.

homolog (3.1) a member of a homologous series.

homologous series (3.1) a group or series of compounds in which each member of the series differs from the immediately preceding or following member by a fixed number of atoms.

human growth hormone (16.11) a hormone secreted by the anterior pituitary that stimulates the growth of an organism.

human immunodeficiency virus (16.11) the virus that produces the set of symptoms called the Acquired Immune Deficiency Syndrome (AIDS).

hydrate (6.4) any substance that contains water bound to a molecule; often used to define a product that results from the addition to water to a precursor molecule.

hydration (4.4, 5.3) a reaction in which water is added to a molecule, e.g., the addition of water to an alkene to form an alcohol.

hydroforming, platforming, and reforming (4.7) the conversion of aliphatic hydrocarbons to aromatic hydrocarbons under the influence of heat, pressure, and a catalyst.

hydrogenation (4.4, 6.4) a reaction in which hydrogen (H_2) is added (usually) to a double or a triple bond.

hydrohalogenation (4.4) the addition of a hydrohalogen (HCl, HBr, and so on) to an unsaturated bond.

hydrolase (12.1) an enzyme that catalyzes hydrolysis reactions.

hydrolysis (8.5, 11.2) a chemical reaction that involves the reaction of a molecule with water; the process by which molecules are broken into their constituents by addition of water.

hydronium ion (2.2) a hydrated proton.

hydrophilic (11.2) "water loving," polar and ionic amino acids that have a high affinity for water.

hydrophobic (11.2) "water fearing," a nonpolar amino acid that prefers contact with other nonpolar amino acids over contact with water.

hydroxyapatite (11.6, 17.2) a calcium phosphate polymer that crystallizes on collagen and makes up the mineral portion of bone.

hydroxyl group (Chapter 5, Introduction) the —OH functional group.

hyperammonemia (14.6) a genetic defect in one of the enzymes of the urea cycle that results in toxic or even fatal elevations of the concentration of ammonium ions in the body.

hyperglycemia (13.6) blood glucose levels that are higher than normal.

hypertonic solution (9.6) a solution with a greater solute concentration and hence a greater osmotic pressure; the more concentrated solution of two separated by a semipermeable membrane.

hypervitaminosis (17.4) any of several pathological syndromes that result from ingestion of very large quantities of a vitamin.

hypoglycemia (13.6) blood glucose levels that are lower than normal.

hypotonic solution (9.6) a solution with a lesser solute concentration and hence a lesser osmotic pressure; the more dilute solution of two separated by a semipermeable membrane.

IgA (11.8) the type of immunoglobulin that protects body surfaces, such as the gut, oral cavity, and genitourinary tract.

IgD (11.8) an immunoglobulin that is thought to regulate antibody synthesis.

IgE (11.8) an immunoglobulin that is thought to be responsible for allergies.

IgG (11.8) the major immunoglobulin found in blood serum.

IgM (11.8) the first immunoglobulin produced by the B cells in response to an infection.

immune system (11.8) an organized group of cells that defend the body against bacterial, viral, parasitic, and fungal infections.

immunoglobulin (11.1) antibody; a very specific glycoprotein that is formed in response to invasion of the body by infectious agents, such as bacteria and viruses.

incomplete protein (17.1) a protein source that does not contain all the essential and nonessential amino acids.

induced fit model (12.5) the theory of enzyme-substrate binding that assumes that the enzyme is a flexible molecule and that both the substrate and enzyme change their shapes to accommodate one another as the enzyme-substrate complex forms.

initiation codon (16.11) the triplet of ribonucleotides, AUG, on the mRNA that specifies binding of the initiator tRNA, N-formylmethionyl tRNA. This is always the first codon of a gene.

initiation factor (16.11) a protein that is required for formation of the translation initiation complex, which is composed of the large and small ribosomal subunits, the mRNA, and the initiator tRNA, N-formylmethionyl tRNA.

inner mitochondrial membrane (14.1) the highly folded, impermeable membrane within the mitochondrion that is the location of the electron transport system and ATP synthase.

insecticide (4.10) a compound that is used to kill insects; used in the control of insect populations.

insulin (13.6, 16.11) a hormone released from the pancreas in response to high blood glucose levels. Insulin stimulates glycogenesis, fat storage, and cellular uptake and storage of glucose from the blood.

integral protein (9.6) a protein that is embedded within a membrane, traverses the lipid bilayer, and protrudes from the membrane both inside and outside the cell.

intermembrane space (14.1) the region between the outer and inner mitochondrial membranes that is the location of the proton reservoir that drives ATP synthesis.

international unit (12.16) the amount of enzyme needed to catalyze conversion of 1 μmol of substrate to product in one minute at standard conditions of temperature and pH.

intrinsic pathway of blood clotting (12.13) the initial stages of the clotting mechanism that are induced by damage within the blood vessel.

intron (16.9) a noncoding sequence within a eukaryotic gene that must be removed from the primary transcript to produce a functional mRNA.

ion (1.1) a charged atom.

ion pair (1.3) the empirical formula unit for an ionic compound.

ionic bond (1.3) an electrostatic attractive force between ions resulting from electron transfer.

ionization energy (1.2) the energy needed to remove an electron from an atom in the gas phase.

irreversible enzyme inhibitor (12.10) a chemical that binds strongly to the R groups of an amino acid in the active site and eliminates enzyme activity.

ischemia (12.5) interrupted blood flow to an organ.
isocitrate dehydrogenase (14.2) an enzyme that catalyzes the oxidation and decarboxylation of isocitrate to yield α-ketoglutarate.
isoelectric (11.13) a situation in which a protein has an equal number of positive and negative charges and therefore has an overall net charge of zero.
isoelectronic (1.3) atoms and ions containing the same number of electrons.
isoenzymes (12.5) forms of the same enzyme with slightly different amino acid sequences.
isolated system (2.4) a system that cannot interact with its surroundings.
isomerase (12.1) an enzyme that catalyzes the conversion of one isomer to another.
isomers (1.4, 3.3) two or more substances that have the same molecular formula but different structures and hence different properties.
isotonic solution (9.6) a solution that has the same solute concentration (water activity and osmotic pressure) as another solution with which it is being compared; a solution that has the same osmotic pressure as a solution existing within a cell.
isotopes (1.1) atoms of the same element that differ in mass because they contain different numbers of neutrons.
I.U.P.A.C. Nomenclature System (3.2) the International Union of Pure and Applied Chemistry (I.U.P.A.C.) standard, universal system for the nomenclature of organic compounds.
α-keratins (11.5) fibrous proteins that form the covering of most land animals. They are major components of fur, skin, beaks, and nails.
ketal (6.4) the family of organic compounds with the general formula $RR'C(OR'')_2$; formed via the reaction of two molecules of alcohol with a ketone in the presence of an acid catalyst.
ketoacidosis (15.3) a drop in the pH of the blood caused by elevated ketone levels.
α-ketoglutarate dehydrogenase complex (14.2) a multienzyme complex that accomplishes the oxidation and decarboxylation of α-ketoglutarate to produce succinyl CoA.
ketone bodies (15.3) acetone, acetoacetone, and β-hydroxybutyrate produced from fatty acids in the liver via acetyl CoA.
ketose (7.2) a sugar that contains a ketone (carbonyl) group.
ketosis (15.3) an abnormal rise in the level of ketone bodies in the blood.
kinetics (2.4) the study of rates of chemical reactions.
lactate dehydrogenase (13.4) the enzyme that catalyzes the reduction of pyruvate to produce lactate in the lactate fermentation.
lactose (7.3) a disaccharide composed of β-D-galactose and either α- or β-D-glucose in $\beta\ (1 \rightarrow 4)$ glycosidic linkage; milk sugar.
lactose intolerance (7.3) the inability to produce the enzyme lactase, which degrades lactose to galactose and glucose.
LeChatelier's Principle (2.2) a law that states that when a system in equilibrium is disturbed, the equilibrium shifts in the direction that minimizes the disturbance.
lecithin (9.3) a complex phosphoglyceride with the phosphoryl group attached to the hydrophilic amino alcohol choline. (See also phosphatidylcholine.)
leukotriene (9.2) an eicosanoid produced by white blood cells that causes bronchial constriction, such as is associated with asthma.
levorotatory (7.2) the enantiomer that rotates plane polarized light in a counterclockwise direction.
Lewis bases (10.5) electron pair donors, donating a pair of electrons to form a bond with another atom, often a proton.
Lewis symbol (1.3) representation of an atom or ion by using the atomic symbol (for the nucleus and core electrons) and dots to represent valence electrons.
ligase (12.1) an enzyme that catalyzes the joining of two molecules.
linkage specificity (12.5) the property of an enzyme that allows it to catalyze reactions involving only one kind of bond in the substrate molecule.
lipase (15.1) an enzyme that hydrolyzes the ester linkage between glycerol and the fatty acids of triglycerides.
lipid (Chapter 9, Introduction) a member of the group of organic molecules of varying composition that are classified together on the basis of their solubility in nonpolar solvents.
lock-and-key model (12.4) the theory of enzyme-substrate binding that supposes that enzymes are inflexible molecules and that the substrate fits into the rigid active site in the same way a key fits into a lock.
low-density lipoprotein (LDL) (9.5) a plasma lipoprotein that carries cholesterol to peripheral tissues and helps to regulate cholesterol levels in those tissues.
lyase (12.1) an enzyme that catalyzes the cleavage of C—O, C—C, or C—N bonds, thereby producing a product containing a double bond.
lysosome (9.5, 12.8) a membrane-bound vesicle in the cell cytoplasm that contains numerous hydrolytic enzymes that break down macromolecules into simple subunits, which can then be used as an energy source or as substrates in biosynthetic reactions.
major minerals (17.2) inorganic elements that are required in large amounts (greater than 100 mg/day) by the body.
malate dehydrogenase (14.2) an enzyme that catalyzes the oxidation of malate to produce oxaloacetate.
maltose (7.3) a disaccharide composed of two glucose molecules in $\alpha\ (1 \rightarrow 4)$ glycosidic linkage.
Markovnikov's Rule (4.4) the rule that states that a proton, adding to a carbon-carbon double bond, will add to the carbon having the larger number of hydrogens attached to it already.
matrix space (14.1) the region of the mitochondrion within the inner membrane. It is the location of the enzymes that carry out the reactions of the citric acid cycle and β-oxidation of fatty acids.
McArdle's Disease (13.6) a genetic defect in the muscle enzyme glycogen phosphorylase that results in an inability to degrade glycogen in muscle tissue.
messenger RNA (16.7) an RNA species that is produced by transcription and that specifies the amino acid sequence for a protein.
micelle (15.1) an aggregation of molecules having a nonpolar and a polar region. The nonpolar regions of the molecules aggregate, leaving the polar regions facing the surrounding water.
microfibril (11.5) an aggregate of α-keratin protofibrils that possesses great mechanical strength.
mitochondria (14.1) the cellular "power plants" in which the reactions of the citric acid cycle, the electron transport system, and ATP synthase function to produce ATP.
molecular formula (3.1) a formula that provides the atoms and number of each type of atom in a molecule but gives no information regarding the bonding pattern involved in the molecule's structure.
molecule (Introduction to Chapter 1) a unit of two or more atoms held together by chemical bonds.
monoglyceride (9.3) the product of the esterification of glycerol at one position.
monomers (8.5) the individual molecules from which a polymer is formed.
monosaccharide (7.1) the simplest type of carbohydrate, consisting of a single saccharide unit.
morphine (11.1) a narcotic that causes drowsiness, euphoria, mental confusion,

and chronic constipation; it is used as a pain killer.

mutagen (16.5) any chemical or physical agent that causes changes in the nucleotide sequence of a gene.

mutarotation (7.2) the interconversion of α and β anomers.

mutation (16.5) any change in the nucleotide sequence of a gene.

myocardial infarction (12.5) heart attack, damage to the heart muscle caused by an interrupted blood supply.

myoglobin (11.9) the oxygen storage protein found in muscle.

myosin (11.5) one of the major proteins of muscle tissue. It has a rodlike structure of two α-helices coiled around one another.

N-terminal residue (11.3) the amino acid residue with a free α-N^+H_3 group. This is the first amino acid of a peptide.

negative allosterism (12.9) effector binding inactivates the active site of an allosteric enzyme.

neurotransmitter (10.1, 12.6) a chemical substance that acts as a chemical ''bridge'' in nerve impulse transmission; these chemicals are released at the nerve ending of one nerve, travel across the synaptic gap, and react with an adjacent nerve or muscle cell.

neutral glyceride (9.3) the product of the esterification of glycerol at one, two, or three positions.

neutralization reaction (2.2) the reaction between an acid and a base.

neutron (1.1) an uncharged particle, with the same mass as the proton, in the nucleus of an atom.

niacin (17.3) a vitamin that is essential for the energy-generating reactions of the cell. It is a component of two coenzymes: nicotinamide adenine dinucleotide and nicotinamide adenine dinucleotide phosphate. (Also called vitamin B_3.)

nicotinamide adenine dinucleotide (NAD^+) (5.8, 8.5, 13.3) a molecule synthesized from the vitamin niacin and the nucleotide ATP and that serves as a carrier of hydride anions; a coenzyme that is an oxidizing agent used in a variety of metabolic processes.

nonessential amino acid (17.1) any amino acid that can be synthesized by the body.

nonreducing sugar (7.3) a sugar that cannot be oxidized by Fehling's, Benedict's, or Tollens' reagents.

nonshivering thermogenesis (14.3) production of heat by the mitochondria of brown fat tissue.

norlutin (9.4) 17-α-ethynyl-19-nortestosterone, a synthetic steroid hormone administered orally for birth control.

norprogesterone (9.4) one of the first synthetic steroid hormones used as a birth control agent; it had to be administered by injection.

nucleophile (3.9) nucleophile (Gk: *nucleo,* nucleus; *phile,* loving); an electron-rich species, such as the halogen ion (X^-), cyanide ion (CN^-), or hydroxide ion (HO^-).

nucleophilic substitution (3.9) a substitution reaction in which the incoming substituent that reacts with the parent compound is a nucleophile.

nucleoside (16.1) a five-carbon sugar, ribose or 2′-deoxyribose, linked by a β-N-glycosidic linkage to either a purine or a pyrimidine base.

nucleotide (13.1) a molecule composed of a nitrogenous base, a five-carbon sugar, and one, two, or three phosphoryl groups.

nucleotide diphosphokinase (14.2) an enzyme that catalyzes the transfer of a phosphoryl group from GTP to ATP.

nucleus (1.1) the small, dense center of positive charge in the atom.

obligate anaerobe (13.4) an organism that cannot live in an environment containing molecular oxygen (O_2).

octet rule (1.3) a rule predicting that atoms form the most stable molecules or ions when they are surrounded by eight electrons in their highest occupied energy level.

oligosaccharide (7.1) an intermediate-sized carbohydrate composed of from two to ten monosaccharides.

open system (2.4) a system that can exchange matter and energy with its surroundings.

opsin (6.3, 17.4) the protein portion of the visual pigment rhodopsin; it combines with 11-*cis*-retinal to form the protein rhodopsin.

optical activity (7.2) the ability to rotate plane polarized light.

organelles (Chapter 14, Introduction) a membrane-bound structure within the cytoplasm that carries out a specialized, compartmentalized function for the cell.

osmoreceptor (9.6) a receptor in the hypothalamus that detects changes in the osmotic concentration of the blood.

osmosis (9.6) the net flow of a solvent across a semipermeable membrane in response to a water chemical activity gradient.

osmotic pressure (9.6) the net force with which water enters a solution through a semipermeable membrane from a region of pure water; alternatively, the pressure required to stop net transfer of solvent across a semipermeable membrane.

osteomalacia (17.4) loss of calcium from bones late in life as a result of vitamin D deficiency. The result is brittle, decalcified bones.

osteoporosis (17.4) a disease characterized by brittle bones and caused by the gradual release of calcium from the bones.

outer mitochondrial membrane (14.1) the membrane that surrounds the mitochondrion and separates it from the contents of the cytoplasm. It is highly permeable to small ''food'' molecules.

oxidation (2.3) a loss of electrons or increase in oxidation state.

β-oxidation (15.2) the biochemical pathway that results in the oxidation of fatty acids and the production of acetyl CoA.

oxidation (of alcohols) (5.8, 8.3) the conversion of an alcohol to a carboxylic acid via the use of an oxidizing agent.

oxidation (of carbonyl compounds) (6.3) the conversion of an alcohol to an aldehyde or ketone via the use of an oxidizing agent.

oxidation-reduction reaction (2.1) or redox reaction, a reaction involving the transfer of one or more electrons from one reactant to another.

oxidation state (2.3) the apparent charge on an atom in a molecule or the charge of a monatomic ion.

oxidative deamination (14.5) an oxidation-reduction reaction in which NAD^+ is reduced and the amino acid is deaminated.

oxidative phosphorylation (13.3, 14.3) production of ATP using the energy of electrons harvested during biological oxidation-reduction reactions.

oxidizing agent (2.3) a substance that oxidizes, or removes electrons from, another substance; the oxidizing agent is itself reduced in the process.

oxireductase (12.1) an enzyme that catalyzes an oxidation-reduction reaction.

oxygen debt (13.4) the O_2 required after strenuous exercise to restore ATP levels and to convert lactate back to glucose.

pancreatic serine proteases (12.12) a family of proteolytic enzymes, including trypsin, chymotrypsin, and elastase, that arose by divergent evolution.

pantothenic acid (17.3) a vitamin that is essential for the normal metabolism of fats and carbohydrates. It is needed for the synthesis of coenzyme A.

parathyroid hormone (PTH) (17.4) the primary hormone controlling the level of calcium in the blood. It is released when blood calcium levels fall and causes the increase by stimulating kidney reabsorption, intestinal absorption, and bone resorption.

parent chain or parent compound (3.2, 6.2) in the I.U.P.A.C. Nomenclature System the parent chain is the longest continuous carbon chain containing the principal functional group (e.g., carbon-oxygen double bond) in the molecule that is being named.

passive diffusion (9.6) the net movement of a solute from an area of high concentration to an area of low concentration.

pellagra (17.3) a disease that results from a niacin deficiency. It is characterized by dermatitis, diarrhea, dementia, and death.

pentose (7.2) a five-carbon monosaccharide.

pepsin (12.8) a proteolytic enzyme found in the stomach. It catalyzes the breakdown of dietary proteins.

pepsinogen (12.9) the inactive form of pepsin produced in cells lining the stomach that is converted to the active form by proteolytic cleavage in the stomach.

peptide bond (11.3) the covalent linkage between two amino acids in a peptide chain, formed by a condensation reaction.

peptidyl tRNA binding site of ribosome (16.11) a pocket on the surface of the ribosome that holds the tRNA bound to the growing peptide chain.

peripheral protein (9.6) a protein that is bound to either the inner or outer surface of a membrane.

permease (9.6) an integral membrane protein that transports molecules across biological membranes.

pernicious anemia (17.3) a disease that results from a vitamin B_{12} deficiency. The symptoms include a decrease in the amount of hemoglobin in the blood, weight loss, and mental and nervous disorders.

pH optimum (12.8) the pH at which an enzyme catalyzes the reaction at maximum efficiency.

pH scale (2.2) a numerical representation of acidity or basicity of a solution ($pH = -\log [H^+]$).

phenol (5.6) an organic compound that contains an hydroxyl group (−OH) attached to a benzene ring.

phenyl group (5.1) a benzene ring that has had a hydrogen atom removed, C_6H_5—.

pheromone (8.5) any compound involved in chemical communication.

phosphatidate (9.3) a molecule with fatty acids esterified to C-1 and C-2 of glycerol and a free phosphoryl group esterified at C-3.

phosphatidylcholine (9.3) a complex phosphoglyceride, found in cell membranes, in which the phosphoryl group is attached to the hydrophilic amino alcohol choline; one of the major membrane phospholipids. (See also lecithin.)

phosphatidylethanolamine (9.3) a complex phosphoglyceride with the phosphoryl group attached to the hydrophilic amino alcohol; one of the major membrane phospholipids. (See also cephalin.)

phosphoenolpyruvate carboxykinase (13.5) the enzyme that catalyzes the removal of CO_2 from and the addition of a phosphoryl group to oxaloacetate in gluconeogenesis.

phosphofructokinase (13.3) the enzyme that catalyzes the phosphorylation of fructose-6-phosphate to produce fructose-1, 6-diphosphate in glycolysis.

phosphoglucoisomerase (13.3) the enzyme that catalyzes the rearrangement of glucose-6-phosphate to fructose-6-phosphate in glycolysis.

phosphoglucomutase (13.6) the enzyme that catalyzes the isomerization of glucose-1-phosphate to glucose-6-phosphate in glycogen degradation.

phosphoglyceride (9.3) a molecule with fatty acids esterified at the C-1 and C-2 positions of glycerol and a phosphate group esterified at the C-3 position.

phosphoglycerokinase (13.3) the enzyme that catalyzes the first substrate-level phosphorylation in glycolysis, the transfer of a phosphoryl group from 1,3-diphosphoglycerate to ADP to produce ATP.

phosphoglyceromutase (13.3) the enzyme that catalyzes the isomerization of 3-phosphoglycerate to 2-phosphoglycerate in glycolysis.

phospholipid (9.3) a lipid containing a phosphoryl group.

phosphopantetheine (15.4) the portion of coenzyme A and the acyl carrier protein that is derived from the vitamin pantothenic acid. It forms a thioester linkage with fatty acids.

photodimerization (16.5) the photochemical reaction in which the absorption of ultraviolet light causes the production of thymine dimers.

photoreactivation (16.5) a mechanism for repair of thymine dimers. The enzyme that cleaves the dimers, photoreactivating enzyme, uses visible light as an energy source.

pituitary dwarfism (16.11) a genetic disease caused by the inability to produce human growth hormone and characterized by very short stature.

plane polarized light (7.2) light filtered through a polaroid lens.

plasma lipoprotein (9.5) a complex composed of lipid and protein that is responsible for the transport of lipids throughout the body.

β-pleated sheet (11.5) a common secondary structure of a peptide chain that resembles the pleats of an Oriental fan.

point mutation (16.10) the substitution of a single base in a codon. This may or may not alter the genetic code of the mRNA resulting in the substitution of one amino acid in the protein.

polyamide (8.5) a polymer in which the monomeric units result in a series of amide linkages (bonds); nylons.

polyester (8.5) polymer in which the monomeric units result in a series of ester linkages (bonds).

polyhalogenated hydrocarbon (4.10) a compound that contains several halogen substituents.

polymer (8.5) a very large molecule formed by the combination of many very small molecules (called monomers) (e.g., polyamides, nylons).

polymerization (8.5) a reaction that produces a polymer.

polysaccharide (7.1) a large, complex carbohydrate composed of long chains of monosaccharides.

polysome (16.11) complexes of many ribosomes all simultaneously translating a single mRNA.

positive allosterism (12.9) effector binding activates the active site of an allosteric enzyme.

potential energy diagram (2.4) a diagram that represents stored energy as a function of distance or time (as in the progress of a chemical reaction).

precursor (4.9) a molecule that, when reacted with appropriate reagents, gives a desired product; the molecule from which a desired product is immediately derived synthetically.

preproinsulin (15.6, 16.11) the first precursor in insulin biosynthesis, composed of a signal sequence and proinsulin.

primary (1°) alcohol (5.4) an alcohol with the general formula RCH_2OH.

primary (1°) amine (10.1) an amine with the general formula RNH_2.

primary protein structure (11.4) the linear sequence of amino acids in a protein chain, determined by the genetic information of the gene for each protein.

product (Chapter 12, Introduction) the chemical species that results from a chemical reaction and that appears on the right side of a chemical equation.

progesterone (9.4) one of the most important hormones associated with pregnancy; produced by chemical modification of cholesterol.

proinsulin (15.6, 16.11) an intermediate in insulin biosynthesis that has lost the signal peptide and that must be proteolytically cleaved to become the active hormone.

promoter (16.8) the sequence of nucleotides immediately before a gene that is recognized by the RNA polymerase and signals the start point and direction of transcription.

prostaglandins (8.5, 9.2) a family of hormonelike substances derived from the 20-carbon fatty acid arachidonic acid; produced by many cells of the body, they regulate many body functions.

prosthetic group (6.3, 11.8) the nonprotein portion of a conjugated protein that is essential to the biological activity of the protein; often a complex organic compound.

protein (11.3, 17.1) a macromolecule whose primary structure is a linear sequence of α-amino acids and whose final structure results from folding of the chain into a specific three-dimensional structure; proteins serve as catalysts, structural components, and nutritional elements for the cell.

proteolytic enzyme (12.11) an enzyme that hydrolyzes the peptide bonds between amino acids in a protein chain.

prothrombin (12.14, 17.4) the inactive precursor of thrombin, required for blood clotting.

protofibril (11.5) three single α-keratin helices coiled around one another in a bundle.

proton (1.1) a positively charged particle in the nucleus of an atom.

purines (16.1) a family of nitrogenous bases that are components of DNA and RNA and consist of a six-sided ring fused to a five-sided ring. The common purines in nucleic acids are adenine and guanine.

pyridoxal phosphate (14.5) a coenzyme derived from vitamin B_6 that is required for all transamination reactions.

pyridoxine (17.3) a vitamin that is required for the synthesis and breakdown of amino acids. (Also called vitamin B_6.)

pyrimidines (16.1) a family of nitrogenous bases that are components of nucleic acids and consist of a single six-sided ring. The common pyrimidines of DNA are cytosine and thymine; the common pyrimidines of RNA are cytosine and uracil.

pyrophosphorylase (13.6) the enzyme that mediates bond formation between the C-1 phosphoryl group of glucose-1-phosphate and the α phosphoryl group of uridine triphosphate in glycogenesis.

pyruvate carboxylase (13.5) the enzyme that catalyzes the addition of atmospheric CO_2 to pyruvate to produce oxaloacetate in gluconeogenesis.

pyruvate decarboxylase (13.4) the enzyme that catalyzes the removal of CO_2 from pyruvate to produce acetaldehyde in the alcohol fermentation.

pyruvate dehydrogenase complex (13.7) a complex of all the enzymes and coenzymes required for the synthesis of CO_2 and acetyl CoA from pyruvate.

pyruvate kinase (13.3) the enzyme that catalyzes the transfer of a phosphoryl group from phosphoenolpyruvate to ADP with the resultant production of ATP and pyruvate in the last reaction of glycolysis.

quaternary ammonium salt (10.1) an amine salt of the general formula $R_4N^+A^-$ (in which R− can be an alkyl or aryl group or a hydrogen atom and A^- can be any anion.

quaternary protein structure (11.8) aggregation of more than one folded peptide chain to yield a functional protein.

radioactivity (1.1) the process by which atoms emit high-energy particles or rays; the spontaneous decomposition of a nucleus to produce a different nucleus.

rate of a reaction (2.4) the change in concentration of a reactant or product per unit of time.

reaction mechanism (4.5, 5.5) the pictorial, step-by-step process by which reactants are converted to products in a chemical reaction; a series of elementary steps involved in a chemical reaction.

receptor (11.1) a protein on the cell surface that binds to a specific food molecule and facilitates its entry into the cell. Other receptors bind specific chemical signals and direct the cell to respond appropriately.

receptor-mediated endocytosis (9.6) the process by which molecules attach to receptors in the cell surface and are brought into the cell by invagination of the cell membrane.

Recommended Dietary Allowance (RDA) (17.3) the daily recommended dose of a vitamin or mineral, determined by considering the range of requirements in humans, using the value at the high end of the range, and adding an additional safety factor.

reducing agent (2.3) a substance that reduces, or donates electrons to, another substance; the reducing agent is itself oxidized in the process.

reducing sugar (7.2) a sugar that can be oxidized by Fehling's, Benedict's, or Tollens' reagents. This includes all monosaccharides and most disaccharides.

reduction (2.3, 5.8, 6.4) the gain of electrons or decrease in oxidation state; a gain of electrons by a molecule or atom, e.g., the conversion of an aldehyde or ketone to an alcohol via the use of a reducing agent.

regulatory proteins (11.1) proteins that control cell functions such as metabolism and reproduction.

release factor (16.11) a protein that binds to the termination codon in the empty A-site of the ribosome and causes the peptidyl transferase to hydrolyze the bond between the peptide and the peptidyl tRNA.

replication fork (16.4) the point at which new nucleotides are added to the growing daughter DNA strand.

replication origin (16.4) the region of a DNA molecule where DNA replication always begins.

respiratory electron transport system (14.3) the series of electron transport proteins embedded in the inner mitochondrial membrane that accepts high-energy electrons from NADH and $FADH_2$.

retinal (17.4) the aldehyde form of vitamin A that binds to the protein opsin to form the visual pigment rhodopsin.

retinol equivalent (R.E.) (17.4) the units that are used to express the RDA for vitamin A.

retrovirus (16.1) a class of animal viruses with genetic information that is RNA. The RNA is converted to DNA before it can be expressed to produce new virus particles.

reverse transcriptase (16.1) the enzyme produced by retroviruses that synthesizes double-stranded DNA from a single strand of RNA.

reversible, competitive inhibitor (12.10) a chemical that resembles the structure and charge distribution of the natural substrate and competes with it for the active site.

reversible, noncompetitive inhibitor (12.10) a chemical that binds weakly to an amino acid R group or cofactor and inhibits activity. When the inhibitor dissociates, the enzyme is restored to its active form.

rhodopsin (6.3, 17.4) the major visual pigment found in the rod cells of the retina of the eye; a protein formed via a combination of opsin with 11-*cis*-retinal.

riboflavin (17.3) a vitamin that is required by the energy-releasing reactions of the cell; a component of the coenzyme flavin adenine dinucleotide. (Also called vitamin B_2.)

ribonucleic acid (RNA) (16.1) single-stranded nucleic acid molecules that are composed of phosphate groups, ribose, and the nitrogenous bases uracil, cytosine, adenine and guanine.

ribonucleotide (16.1) a ribonucleoside phosphate or nucleotide composed of a nitrogenous base in β-N-glycosidic linkage to the 1′-carbon of the sugar ribose and with one, two, or three phosphoryl groups esterified at the hydroxyl of the 5′-carbon of the ribose.

ribose (7.2) a five-carbon monosaccharide that is a component of RNA and many coenzymes.

ribosomal RNA (16.7) the RNA species that are structural and functional components of the small and large ribosomal subunits.

ribosome (16.7) an organelle composed of a large and a small subunit, each of which is made up of ribosomal RNA and proteins; the platform on which translation occurs and which carries the enzymatic activity that forms peptide bonds.

rickets (17.4) a deficiency of vitamin D that results in the formation of soft, deformed bones in children.

rod cells (6.3, 17.4) the cells of the retina of the eye that are responsible for black-and-white vision and vision in dim light.

saccharide (7.1) a sugar molecule.

saponification (8.5) a reaction in which a soap is produced; more generally, the hydrolysis of an ester by an aqueous base.

saturated compound (4.1) an alkane; a hydrocarbon that contains only carbon and hydrogen bonded together through carbon-hydrogen and carbon-carbon single bonds, and that has the general molecular formula C_nH_{2n+2}; also called saturated hydrocarbon.

saturated fatty acid (9.2) a long-chain monocarboxylic acid in which each carbon of the chain is bonded to the maximum number of hydrogen atoms.

scurvy (11.6, 17.3) a disease of collagen metabolism resulting from a deficiency of vitamin C. The symptoms include bleeding gums, loss of teeth, sore joints, slow wound healing, skin lesions, and fragile blood vessels.

second law of thermodynamics (2.4) a law that states that an increase in entropy (disorder) accompanies any spontaneous process.

secondary (2°) alcohol (5.4) an alcohol with the general formula R_2—CHOH.

secondary (2°) amine (10.1) an amine with the general formula R_2—NH.

secondary protein structure (11.5) folding of the primary structure of a protein into an α-helix or a β-pleated sheet; folding is maintained by hydrogen bonds between the amide hydrogen and the carbonyl oxygen of the peptide bond.

semipermeable membrane (9.6) a membrane that is permeable to the solvent but not the solute; a material that allows the transport of certain substances from one side of the membrane to the other.

sickle cell anemia (11.10) a human genetic disease resulting from inheriting mutant hemoglobin genes from both parents. The disease is fatal because of poor oxygen transport to the tissues.

sickle cell trait (11.10) the condition of having one normal and one mutant hemoglobin gene.

signal peptide (15.6) a sequence of hydrophobic amino acids at the amino terminus of a protein that guides the protein through cellular membranes.

silk fibroin (11.5) fibrous protein that is produced by silkworms and whose structure is an antiparallel β-pleated sheet.

single-replacement reaction (2.1) or substitution reaction, one in which one atom in a molecule is displaced by another.

slow twitch muscle fibers (14.1) muscle fibers that are specialized to produce ATP by oxidative phosphorylation.

soap (8.4) any of a variety of the alkali metal salts of fatty acids.

sphingolipid (9.4) a phospholipid that is derived from the amino alcohol sphingosine, rather than from glycerol.

sphingomyelin (9.4) a sphingolipid that is found in abundance in the myelin sheath that surrounds and insulates cells of the central nervous system.

standard solution (2.2) a solution whose concentration is accurately known.

stereochemical specificity (12.5) the property of an enzyme that allows it to catalyze reactions involving only one stereoisomer of the substrate.

stereochemistry (7.2) the study of the spatial arrangement of atoms in a molecule.

stereoisomers (7.2) a pair of molecules having the same structural formulas and bonding patterns but differing in the arrangement of the atoms in space.

steroid (9.4) a lipid that is derived from cholesterol and composed of one five-sided ring and three six-sided rings. The steroids include sex hormones and anti-inflammatory compounds.

streptokinase (12.5) an enzyme produced by the bacterium *Streptococcus pyogenes* that destroys fibrin clots. It is used to treat heart attack victims.

structural analogue (12.10) a chemical having a structure and charge distribution that are very similar to those of a natural enzyme substrate.

structural formula (1.4, 3.1) a formula showing all of the atoms in a molecule and exhibiting all bonds as lines.

structural protein (11.1) a protein that provides mechanical support for large plants and animals.

substitution reaction (3.9) a reaction that results in the replacement of one group for another.

substrate (Chapter 12, Introduction) the reactant in a chemical reaction that binds to an enzyme active site and is converted to product.

substrate-level phosphorylation (13.3) the production of ATP by the transfer of a phosphoryl group from the substrate of a reaction to ADP.

succinate dehydrogenase (14.2) an enzyme that catalyzes the oxidation of succinate to fumarate.

succinyl CoA synthase (14.2) an enzyme that catalyzes a substrate-level phosphorylation in which the thioester bond of succinyl CoA is cleaved to yield succinate and coenzyme A and GDP is phosphorylated to produce GTP.

sucrose (7.3) a disaccharide composed of α-D-glucose and β-D-fructose in α,β glycosidic linkage; table sugar.

sulfa drugs (10.6) a family of drugs that are derived from the compound sulfanilamide and that have antibacterial properties.

sulfanilamide (17.3) a structural analogue of *para*-aminobenzoic acid. This antibiotic competes with PABA for binding to the active site of an enzyme required for the synthesis of folic acid.

synthesis (4.5) the conversion of one molecule into another molecule.

terminal electron acceptor (14.3) the final electron acceptor in an electron transport system that removes the low-energy electrons from the system. In aerobic organisms the terminal electron acceptor is molecular oxygen.

termination codon (16.11) a triplet of ribonucleotides with no corresponding anticodon on a tRNA. As a result, translation will end, since there is no amino acid to transfer to the peptide chain.

termination reaction (8.5) a reaction that is used to ''terminate'' a polymerization.

tertiary (3°) alcohol (5.4) an alcohol with the general formula R_3C—OH.

tertiary (3°) amine (10.1) an amine with the general formula R_3—N.

tertiary protein structure (11.7) the globular, three-dimensional structure of a protein that results from folding the regions of secondary structure. This folding occurs spontaneously as a result of interactions of the side chains or R groups of the amino acids.

testosterone (9.4) a male sex hormone produced by chemical modification of progesterone.

tetrose (7.2) a four-carbon monosaccharide.

thermodynamics (2.4) the study of energy and its interconversion.

thermogenic hyperphagia (14.3) the ability of brown fat tissue to burn excess calories in the diet.

thermogenin (14.3) a protein in the inner mitochondrial membrane of brown fat that allows the dissipation of the proton gradient and thereby produces heat.

thiamine (17.3) a vitamin that is required for decarboxylation reactions. It is a component of the coenzyme thiamine pyrophosphate. (Also called vitamin B_1.)

thiamine pyrophosphate (13.7) a coenzyme derived from vitamin B_1 (thiamine) that is part of the pyruvate dehydrogenase complex.

thiol (5.9) an organic compound that contains a thiol group (—SH).

thrombin (12.13, 17.4) the proteolytic enzyme that converts fibrinogen to fibrin in the last stage of clot formation.

thromboxane A_2 (9.2) an eiconsanoid produced by blood platelets that stimulates vasoconstriction and platelet aggregation.

thymine (16.1) one of the pyrimidine nitrogenous bases found in DNA but not in RNA.

thymine dimer (16.5) ultraviolet light–induced covalent bonding of two adjacent thymine bases in a strand of DNA.

tissue-type plasminogen activator (TPA) (12.5) an enzyme that is a part of the natural anticlotting mechanism of the human body; used to treat heart attack victims.

titration (2.2) the process of adding a solution from a buret to a sample until a reaction is complete, at which time the volume is accurately measured.

α-tocopherol equivalents (α-TE) (17.4) the units that are used to express the RDA for vitamin E.

tocopherols (17.4) a family of eight compounds collectively referred to as vitamin E.

Tollens' reagent (7.2) an aqueous solution of $Ag(NH_3)_2^+$ that can be used to test for reducing sugars.

Tollens' Test (6.4) use of a test reagent (silver oxide in ammonium hydroxide) to test for the carbonyl functional group; also called the Tollens' Silver Mirror Test.

trace minerals (17.2) inorganic elements that are required in very small amounts for normal cellular function.

transaminase (14.5) an enzyme that catalyzes the transfer of an amino group from one molecule to another. (Also called aminotransferase.)

transamination (14.5) a reaction in which an amino group is transferred from one molecule to another.

transcription (16.6) the synthesis of RNA from a DNA template.

transfer RNA (16.7) small RNAs that bind to a specific amino acid at the 3′-end and mediate its addition at the appropriate site in a growing peptide chain. This is accomplished by recognition of the correct codon on the mRNA by the complementary anticodon on the tRNA.

transferase (12.1) an enzyme that catalyzes the transfer of a functional group from one molecule to another.

transition state (12.6) the unstable intermediate in catalysis in which the enzyme has altered the form of the substrate so that it now shares properties of both the substrate and the product.

translation (16.6) the synthesis of a protein from the genetic code carried on the mRNA.

translocation (16.11) movement of the ribosome along the mRNA during translation.

transport protein (11.1) a protein that transports materials across the cell membrane or from tissue to tissue within the body.

triglyceride (9.3, 15.1) a molecule composed of glycerol esterified to three fatty acids. (Also called triacylglycerol.)

triose (7.2) a three-carbon monosaccharide.

triose isomerase (13.3) the enzyme that catalyzes the isomerization of dihydroxyacetone phosphate to glyceraldehyde-3-phosphate in glycolysis.

tropocollagen (11.6) a triple-stranded helical fiber of collagen.

unsaturated compound (4.1) any hydrocarbon that contains one or more carbon-carbon double or triple bonds.

unsaturated fatty acid (9.2) a long-chain monocarboxylic acid having at least one carbon-to-carbon double bond.

uracil (16.1) the pyrimidine nitrogenous bases found in RNA but not in DNA.

urea cycle (14.6) a cyclic series of reactions that detoxifies ammonium ions by incorporating them into urea, which is excreted from the body.

uridine (16.1) a nucleoside composed of the pyrimidine uracil and the sugar ribose.

uridine triphosphate (13.6) a nucleotide composed of the pyrimidine uracil, the sugar ribose, and three phosphate groups and that serves as a carrier of glucose-1-phosphate in glycogenesis.

vaccination (11.8) the process of immunizing an animal against an infectious agent by injecting small amounts of the antigens of the virus or organism causing the disease.

valence electrons (1.3) electrons in the outermost shell (principal quantum level) of an atom.

valence shell electron pair repulsion (VSEPR) theory (1.5) a model accounting for the geometrical arrangement of shared and unshared electron pairs around a central atom in terms of electron pair repulsion.

very low-density lipoprotein (VLDL) (9.5) a plasma lipoprotein that binds triglycerides synthesized by the liver and carries them to adipose tissue for storage.

vitamin (17.3) an organic substance that is required in the diet in small amounts. Some vitamins are used in the synthesis of coenzymes required for the function of cellular enzymes; others are involved in calcium metabolism, vision, and blood clotting.

vitamin A (17.4) a vitamin that is essential to vision. (Also called retinol.)

vitamin B_{12} (17.3) a vitamin that is required for the production of red and white blood cells and the growth and maintenance of nerve tissue.

vitamin C (11.6, 17.3) a water-soluble vitamin that is required for the growth and repair of connective tissue, teeth, bones, and cartilage. It is involved in the hydroxylation of collagen. (Also called ascorbic acid.)

vitamin D (17.4) a vitamin that plays a major role in the regulation of calcium metabolism in the body.

vitamin E (17.4) a vitamin that prevents the oxidation of double bonds in the hydrocarbon chains of membrane lipids.

vitamin K (12.14, 17.4) a vitamin that is vital to blood clotting; it serves as a coenzyme in the synthesis of prothrombin and factors VII, IX, and X in the liver.

von Gierke's Disease (13.6) a genetic defect in the enzyme glucose-6-phosphatase that results in the inability of the liver to provide a supply of blood glucose.

wax (9.4) a collection of lipids that are generally considered to be esters of long-chain alcohols.

Western blot (12.16) a test that involves the separation of viral proteins by electrophoresis and the detection of those proteins by antigen-antibody binding and enzyme assay.

xeroderma pigmentosum (16.5) a human genetic disease caused by a mutation in a gene for one of the enzymes involved in a DNA repair pathway. Sufferers usually die of skin cancer at an early age.

xerophthalmia (17.4) the condition caused by a deficiency of vitamin A that results in night blindness and eventually total blindness.

zidovudine (16.1) trade name for the drug used to block the replication of human immunodeficiency virus (see AZT).

zymogen (12.9) the inactive form of a proteolytic enzyme.

Answers to Selected Problems

CHAPTER 1

1.1

	Atom	Protons	Neutrons	Electrons
a.	$^{32}_{16}S$	16	16	16
b.	$^{23}_{11}Na$	11	12	11
c.	$^{1}_{1}H$	1	0	1
d.	$^{244}_{94}Pu$	94	150	94
e.	$^{40}_{18}Ar$	18	22	18

1.3 a. *atomic size*—increases from right to left across a period, and down a group

$$F < Cl < Br < I$$

b. *ionization energy*—increases from left to right across a period, and up a group

$$I < Br < Cl < F$$

c. *electron affinity*—increases from left to right across a period, and up a group

$$I < Br < Cl < F$$

d. *electronegativity*—increases from left to right across a period, and up a group

$$I < Br < Cl < F$$

1.5

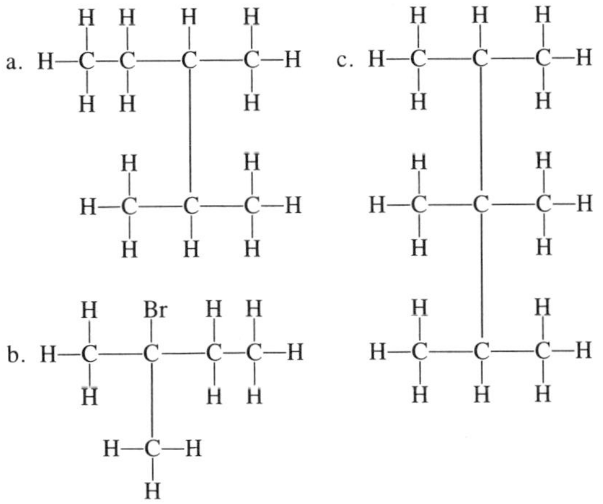

```
      H  H     H     H                 H        H        H
      |  |     |     |                 |        |        |
a. H—C—C—————C—————C—H      c. H—C————————C————————C—H
      |  |     |     |                 |        |        |
      H  H     |     H                 H        |        H
               |                                |
      H        |     H                 H        |        H
      |        |     |                 |        |        |
   H—C—————C—————C—H             H—C————————C————————C—H
      |        |     |                 |        |        |
      H        H     H                 H        |        H
                                                |
      H      Br    H  H                H        |        H
      |      |     |  |                |        |        |
b. H—C—————C—————C—C—H           H—C————————C————————C—H
      |      |     |  |                |        |        |
      H      |     H  H                H        H        H
             |
         H—C—H
             |
             H
```

1.7 a. The H—C—Cl bond angle is greater than the tetrahedral angle of 109.5° due to the size and electronegativity of Cl.
b. The H—C—C bond angle is approximately 109.5°.
c. The C≡C—H bond angle is 180°.
d. The C=C—Br bond angle is greater than the trigonal planar angle of 120° due to the electronegativity of Br.
e. The C—O—H bond angle is slightly less than 109.5°.
f. The C—C≡C bond angle is 180°.

1.9 a. SO_2: $\ddot{O}$=S: ⟷ :O—S: (with the second O single-bonded in the first structure and double-bonded in the second)

SeO_2: :O=Se: ⟷ :O—Se: (with the second O single-bonded in the first structure and double-bonded in the second)

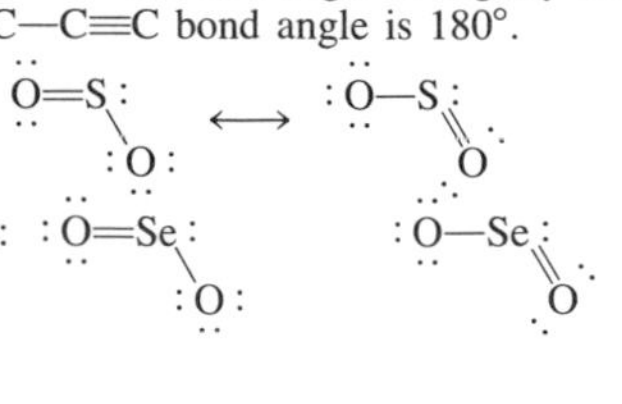

b. Sulfur and selenium are Group VI elements and thus both elements have six valence electrons available for bonding. Since they have the same number of available bonding electrons, the structures they form will have the same geometry.

1.11 a. true b. true c. false—ionization energy *decreases* from top to bottom within a group d. true

1.13 a. Na—sodium b. K—potassium c. Mg—magnesium d. Ca—calcium e. Cu—copper f. Co—cobalt

1.15 The alkali metals are those belonging in Group IA: Li, Na, K, Rb, Cs, Fr

1.17 The halogens are those elements belonging to Group VIIA: F, Cl, Br, I, At

1.19 a. H 1 valence electron b. Na 1 valence electron c. B 3 valence electrons d. F 7 valence electrons e. Ne 8 valence electrons f. He 2 valence electrons

1.21 a. O^{2-}, Ne—isoelectronic, both have 10 electrons
b. S^{2-}, Cl^-—isoelectronic, both have 18 electrons
c. F^-, Cl^-—not isoelectronic, F^- has 10 electrons, Cl^- has 18 electrons
d. K^+, Ar—isoelectronic, both have 18 electrons

1.23 Inert gases are nonreactive because they have a stable octet of electrons. The atom does not need to lose or gain electrons to attain a stable electron configuration.

1.25 a. ·Si· b. :P· c. :Cl· d. He: e. :Se:

1.27 a. $1s^22s^22p^1$ b. $1s^22s^22p^63s^23p^4$ c. $1s^22s^22p^63s^23p^6$
d. $1s^22s^22p^63s^23p^64s^23d^3$ e. $1s^22s^22p^63s^23p^64s^23d^{10}4p^65s^24d^{10}$
f. $1s^22s^22p^63s^23p^64s^23d^{10}4p^65s^24d^{10}5p^4$

1.29 The common feature of the electron configuration of the elements of Group 1A is that they all have one electron in the *s* orbital.

1.31 a. $N < O < F$ b. $Cs < K < Li$ c. $Br < Cl < F$

1.33 a. $F < O < N$ b. $Li < K < Cs$ c. $Cl < Br < I$

1.35 The element, fluorine, has the highest electronegativity, ionization energy, and electron affinity.

1.37 The size of a positive ion is always smaller than its parent ion because the same number of protons is able to pull a smaller number of electrons closer to the nucleus.

1.39 a. $MgCl_2$ ionic b. CO_2 covalent c. NaCl ionic d. H_2S covalent e. NO_2 covalent f. ICl covalent

1.41 Ionic bonding is due to an electrostatic attraction of ions of opposite charge; covalent bonding involves a sharing of electrons. Ionic bonding occurs between metals and nonmetals; covalent bonding occurs between nonmetals. Covalent bonds result in molecules; ionic bonds do not form molecules, but rather crystal arrays (lattices) described by formula units.

1.43

```
a. CO3 2-   [     :O:      ]2-       c. CF4       :F:
            [  :O:C:O:     ]                   :F:C:F:
                                                  :F:

b. NH4 +    [    H    ]+             d. SO3 2-  [    :O:     ]2-
            [  H:N:H  ]                         [  :O:S:O:   ]
            [    H    ]
```

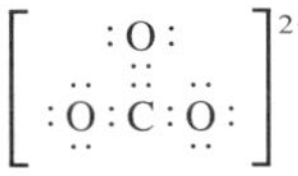

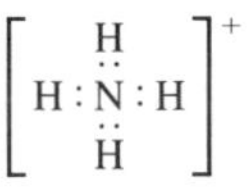

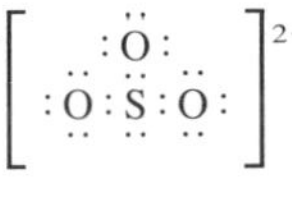

e. HNO_3 :Ö:N:Ö:H (with :Ö: below N) f. Br_2 :B̈r:B̈r:

1.45

Atomic Symbol	# Protons	# Neutrons	# Electrons
a. $^{23}_{11}Na$	11	12	11
b. $^{32}_{16}S^{2-}$	16	16	18
c. $^{16}_{8}O$	8	8	8
d. $^{24}_{12}Mg^{2+}$	12	12	10
e. $^{39}_{19}K^{1+}$	19	20	18

1.47 An ionic compound forms a crystal lattice in which the ions are packed in an orderly array. One cannot discern the formula of the ionic compound from this lattice except by calculation from the empirical formula. When an ionic solid is dissolved in water, the lattice breaks up into separate ions, not the formula unit of the compound. For instance, when sodium chloride dissolves in water, individual sodium and chloride ions are found in solution, not units of sodium chloride. A covalent compound lattice is composed of molecules packed in an orderly array. When a covalent lattice is broken down, individual molecules result. For example, if sugar, $C_{12}H_{22}O_{11}$ is dissolved in water, individual sugar molecules will be found in solution.

1.49 Due to strong attractive forces in the lattice, ionic compounds tend to have higher boiling points than covalent compounds.

1.51 a. Li^+ b. Mg^{2+} c. Be^{2+} d. Al^{3+} e. [H—H—H with H above and below]$^+$

f. Cl^- :C̈l: g. P^{3-} :P̈: h. O^{2-} :Ö: i. S^{2-} :S̈:

j. [H—C(H)(H)—C(=O:)—Ö:]$^-$

1.53 a. Cl and Cl b. H and H c. C and H

:C̈l—C̈l: H—H H—C(H)(H)—H

d., h., i., and j. are ionic compounds. e. O and O

Ö=Ö

f. C and N g. S and O

[:C≡N:]$^-$ Ö=S̈—Ö:

1.55 Resonance can occur when more than one valid Lewis structure can be written for a molecule. Each individual structure that can be drawn is a resonance form. The true nature of the structure for the molecule is the resonance hybrid, which consists of the "average" of the resonance forms.

1.57 For simple, diatomic molecules, a polar bond means the molecule will be polar. For larger molecules, this is not always true. If the molecule is symmetrical (all groups are identical), the molecule may have polar bonds, but be nonpolar overall.

1.59 A polar compound will tend to have a higher melting point because it will have stronger interparticle interactions than a nonpolar compound.

1.61 a. CH_4 H—C(H)(H)—H —nonpolar, all the groups on the central atom are equivalent

b. NH_3 H—N̈(H)—H —polar, the central nitrogen has four nonidentical groups: three H's and one lone pair of electrons

c. CO_3^{2-} [:Ö—C(=O:)—Ö:]$^{2-}$ ⟷ [Ö=C(—Ö:)—Ö:]$^{2-}$ ⟷ [:Ö—C(—Ö:)=Ö]$^{2-}$

—nonpolar, all the groups on the central carbon are equivalent by resonance

d. CH_3OH H—C(H)(H)—Ö—H

—polar, all the groups on the central carbon are not identical.

e. HCOOH H—C(=O:)—Ö—H

—polar, all the groups on the central carbon are not identical.

f. benzene (CH)$_6$ ring ⇌ (CH)$_6$ ring

—nonpolar, all the carbons are equivalent by resonance, the molecule is perfectly symmetrical.

g. NH_4^+ [H—N(H)(H)—H]$^+$

—nonpolar, all groups on the central nitrogen are identical.

1.63 Carbon has four valence electrons and needs four more to complete its octet. There are four ways to complete the octet, as shown below:

$\dot{\underset{x}{C}}$ structures (four electron-dot forms of C with x and · electrons)

CHAPTER 2

2.1 a. DR (double replacement) b. SR (single replacement) c. DR (double replacement) d. C (combination) e. D (decomposition) f. D (decomposition) g. DR (double replacement) h. C (combination)

2.3 $CO_2 + H_2O \rightleftharpoons H_2CO_3 \rightleftharpoons H_3O^+ + HCO_3^-$

a. Since the $[H_2CO_3]$ increases, the equilibrium will continue to shift to the right, and an increase in $[H_3O^+]$ is observed.

b. Since the $[H_2CO_3]$ decreases, the equilibrium will continue to shift to the left, and a decrease in $[H_3O^+]$ is observed.

2.5 *Decomposition reactions:*

$CuCO_3 \longrightarrow CuO + CO_2$
$2KClO_3 \longrightarrow 2KCl + 3O_2$
$Pb(NO_3)_2 \longrightarrow PbO + NO_2 + NO + O_2$
$Ba(OH)_2 \longrightarrow BaO + H_2O$

2.7 *Replacement reactions:*

$Mg + 2HCl \longrightarrow MgCl_2 + H_2$
$Cu + 2AgNO_3 \longrightarrow Cu(NO_3)_2 + 2Ag$
$BaCl_2 + K_2SO_4 \longrightarrow BaSO_4 + 2KCl$
$HBr + KOH \longrightarrow KBr + H_2O$

2.9 a. $CH_4(g) + 2O_2(g) \longrightarrow 2H_2O(l) + CO_2(g)$
b. $C_3H_8(g) + 5O_2(g) \longrightarrow 4H_2O(l) + 3CO_2(g)$
c. $4NH_3(g) + 3O_2(g) \longrightarrow 2N_2(g) + 6H_2O(l)$

2.11 Strong electrolytes dissociate *completely* in aqueous solution. A few examples are shown below:

$NaCl(aq) \longrightarrow Na^+(aq) + Cl^-(aq)$
$Ca(NO_3)_2(aq) \longrightarrow Ca^{2+}(aq) + 2NO_3^-(aq)$
$CuSO_4(aq) \longrightarrow Cu^{2+}(aq) + SO_4^{2-}(aq)$
$H_2SO_4(aq) \longrightarrow H^+(aq) + HSO_4^-(aq)$
$NH_4Cl(aq) \longrightarrow NH_4^+(aq) + Cl^-(aq)$

2.13 Metal carbonates will decompose into a metal oxide and give off carbon dioxide gas when heated:

$$CuCO_3(s) \longrightarrow CuO(s) + CO_2(g)$$

Metal hydroxides will decompose into a metal oxide and give off gaseous water when heated:

$$Ba(OH)_2(s) \longrightarrow BaO(s) + H_2O(g)$$

Active metals will produce a salt and hydrogen gas when combined with an acid:

$$Mg(s) + 2HCl(aq) \longrightarrow MgCl_2(aq) + 2H_2(g)$$

Metal carbonates will produce a salt, carbon dioxide, and water when combined with an acid:

$$K_2CO_3(s) + 2HCl(aq) \longrightarrow 2KCl(aq) + CO_2(g) + H_2O(l)$$

2.15 a. An Arrhenius acid is a substance which will produce H^+ in water.
b. An Arrhenius base is a substance which will produce OH^- in water.

2.17 $CH_3NH_2(aq) + H_2O(l) \longrightarrow CH_3NH_3^+(aq) + OH^-(aq)$
a. Methylamine is acting as a Bronsted base.
b. A Bronsted base is a proton acceptor.

2.19 A neutralization involves the complete reaction of an acid and a base to form a salt and water.

2.21 *1* A known amount of analyte is placed into an Erlenmeyer flask, and a few drops of indicator solution are added.

2 The buret is filled with the titrant, and an initial reading is taken.

3 The titrant is delivered into the Erlenmeyer flask until an indicator color change shows the endpoint has been reached.

4 A final volume reading is taken from the buret.

5 The quantity of the analyte is calculated based on the known stoichiometry of the reaction taking place, and the data recorded during the titration.

2.23 a. True. Since appreciable quantities of both weak acid and its conjugate base are present, the equilibrium can easily shift both ways, depending on how the equilibrium is stressed.
b. True. An acidic buffer consists of a weak acid and its salt.
c. False. An alkaline buffer consists of a weak base and its salt.
d. True. Blood pH cannot vary by more than 0.1 pH unit without adversely affecting the body's metabolism. Many biochemical processes in the body are very pH sensitive.

2.25 a. S = 0 (the oxidation number of any element is 0)
b. H = +1, S = +4, O = −2
c. H = +1, S = +6, O = −2
d. S = +4, O = −2
e. S = +6, O = −2

2.27 a. The change in oxidation number of nitrogen is from +3 to +5. This indicates a loss of electrons; therefore, oxidation has occurred.
b. The change in oxidation number of manganese is from +7 to +2. This indicates a gain of electrons; therefore, reduction has occurred.
c. The change in oxidation number of chlorine is from −1 to 0. This indicates a loss of electrons; therefore, oxidation has occurred.
d. The change in the oxidation number of chromium is from +3 to +6. This indicates a loss of electrons; therefore, oxidation has occurred.
e. The change in oxidation number of oxygen is from −1 to −2. This indicates a gain of electrons; therefore, reduction has occurred.

2.29

a.	Zn oxidized reducing agent	HNO_3 reduced oxidizing agent
b.	Cl_2 reduced oxidizing agent	KI oxidized reducing agent
c.	$Cr_2O_7^{2-}$ reduced oxidizing agent	Fe^{2+} oxidized reducing agent

2.31 a. +6 b. +4 c. 0 d. 0 e. −2

2.33 The first law of thermodynamics states that energy cannot be created or destroyed, only converted from one form to another.

2.35 a. An exothermic reaction liberates heat energy.
b. An endothermic reaction consumes heat energy.
c. A closed system is one in which the total mass and energy of the system does not change.
d. An open system can exchange matter and energy with its surroundings.
e. An isolated system cannot interact with its surroundings.

2.37 a. *melting of a solid metal*—entropy increases; liquids have greater disorder than solids.
b. *boiling of water*—entropy increases; gases have greater disorder than liquids.
c. *burning a log in a fireplace*—entropy increases; the major products are gaseous (CO_2 and H_2O vapor); gases have much greater disorder than solids, such as wood.
d. *combustion of gasoline*—entropy increases; the products are gaseous; gases have greater disorder than liquids.
e. *condensation of water vapor on a cold surface*—entropy decreases; gases have greater disorder than liquids.

2.39 Entropy increases when a dinner plate falls to the floor and shatters. Entropy decreases if one glues the pieces together to reform the plate; this requires some expenditure of energy.

2.41 Energy is released when new bonds form.

2.43 A glass of iced tea feels cold because your hand is warmer than the glass of tea. The heat is traveling from your hand to the glass of tea. Since your hand is losing heat, it feels colder. A cup of hot chocolate feels hot because your hand is cooler than the cup of hot chocolate. The heat is traveling from the cup of hot chocolate to your hand. Since your hand is gaining heat, it feels warmer.

2.45 a. During the course of a reaction, the highest energy state reached is called the *activated complex*.
b. The state at which the forward rate of reaction equals the reverse rate of reaction is called *dynamic equilibrium*.
c. The steps which show the path a reaction takes is called a *mechanism*.
d. The rate determining step in a mechanism is the *slow* step of the mechanism.

2.47 An increase in temperature will increase the rate of a reaction.

2.49 A catalyst will increase the rate of reaction by providing an easier path for the reaction to occur.

CHAPTER 3

3.1 a. carbon tetrachloride b. 2,2-dimethylpropane
c. 2,2-dimethylpentane d. 1,2,3-tribromopropane
e. 3-methylheptane f. 1-bromo-5-iodo-2-methylhexane
g. 1,3-dibromo-2,2-dimethylpropane
h. 1-chloro-4-ethyl-5-methylhexane

3.3

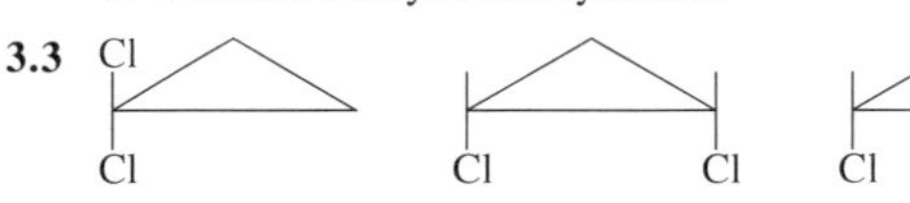

3.5 The four monobrominated products of 2-methylbutane are: 1-bromo-2-methylbutane, 1-bromo-3-methylbutane, 2-bromo-2-methylbutane, and 2-bromo-3-methylbutane

3.7 a. octane b. 1-bromopentane c. 1,3-dichlorohexane

3.9 a. H_3O^+ electrophilic b. H^- nucleophilic c. H^+ electrophilic
d. $^+CH_3$ electrophilic e. HO^- nucleophilic
f. $^+NO_2$ electrophilic g. Br^+ electrophilic

3.11 a. $CH_3CH(Br)CH_2CH_3$ (Br on C-2)

b. $CH_3—C(Cl)(CH_3)—CH_3$

c. $CH_3—C(CH_3)_2—CH_2CH_2CH_2CH_3$

d. $Cl—C(I)(I)—Cl$

e. CH_3CH_2–cyclohexane–CH_2CH_3

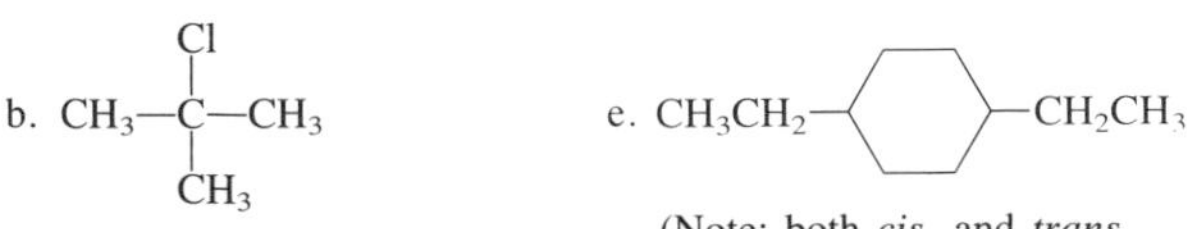

(Note: both *cis*- and *trans*-forms are possible)

3.13 a. structural isomers: have the same molecular formula but differ in the way in which the atoms are joined together:

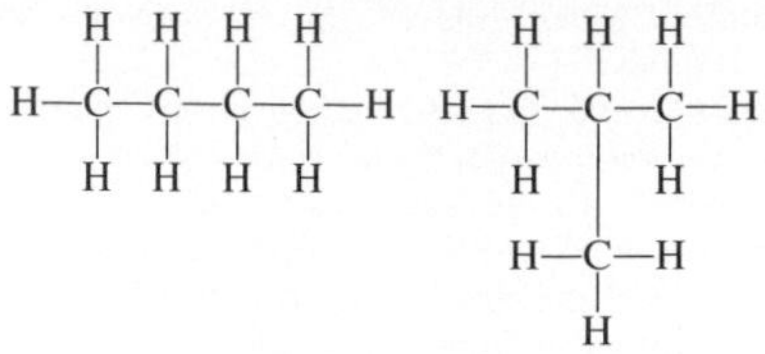

b. hydrocarbons: organic compounds containing only carbon and hydrogen atoms:

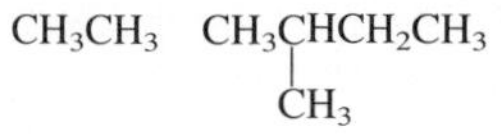

CH_3CH_3 $CH_3CHCH_2CH_3$ (with CH_3 on the second carbon)

c. alkyl groups: simple hydrocarbon groups that result from the removal of one hydrogen atom from the original hydrocarbon (e.g., methyl, CH_3—; ethyl, CH_3CH_2—).

d. cracking (catalytic): process in which large hydrocarbons are broken down into smaller hydrocarbons by the use of heat, pressure, and a catalyst, for example:

$$C_{20}H_{42} \xrightarrow[\text{catalyst}]{\text{Heat, pressure,}} C_1 \text{ through } C_{19} \text{ hydrocarbons}$$

e. reforming (catalytic): process in which an organic molecule is rearranged to a different molecule by use of heat, pressure, and a catalyst, for example:

$$CH_3CH_3 \xrightarrow[\text{catalyst}]{\text{Heat, pressure,}} CH_2{=}CH_2$$

f. combustion: the burning of a substance in the presence of oxygen:

$$CH_4 + 2O_2 \longrightarrow CO_2 + 2H_2O$$

3.15 a. There are two possible isomers of C_3H_7I:

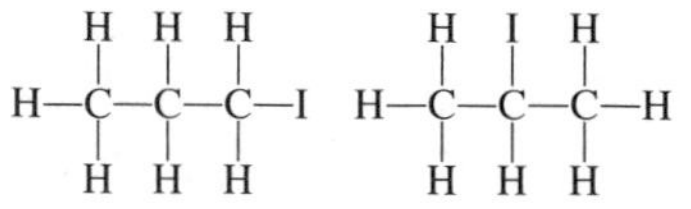

1-iodopropane 2-iodopropane

b. There are four possible isomers of C_4H_9F:

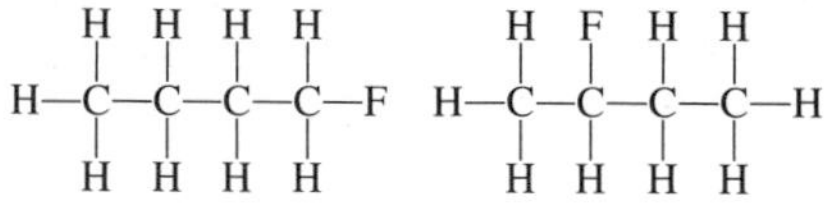

1-fluorobutane 2-fluorobutane

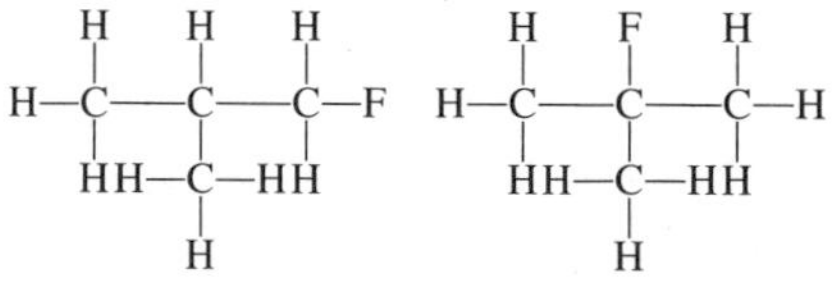

1-fluoro-2-methylpropane 2-fluoro-2-methylpropane

c. There are eight possible isomers of alcohols with the molecular formula $C_5H_{11}OH$:

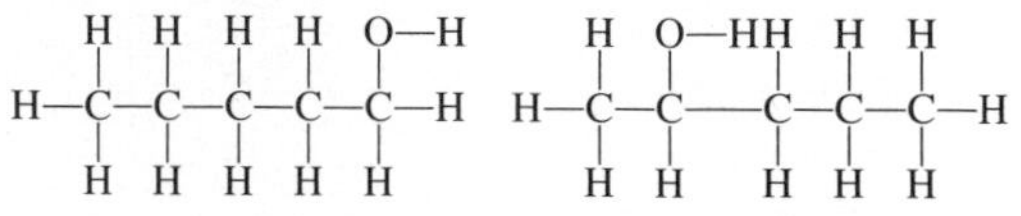

1-pentanol 2-pentanol

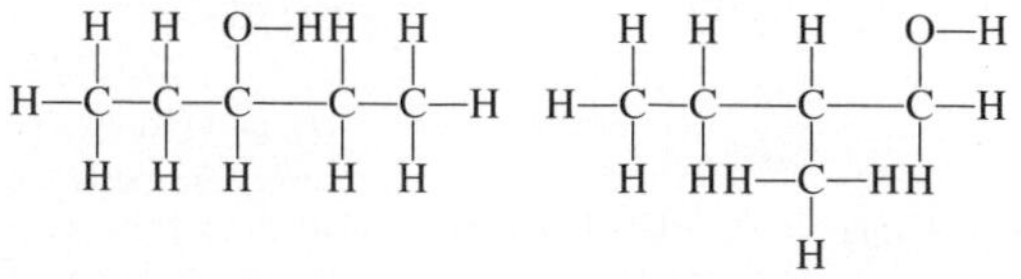

3-pentanol 2-methylbutanol

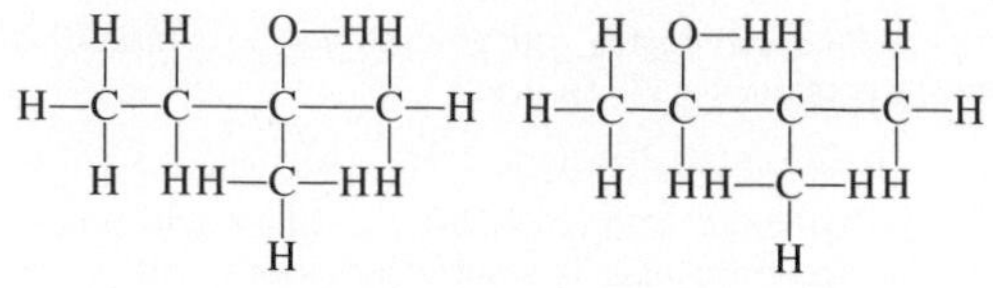

2-methyl-2-butanol 3-methyl-2-butanol

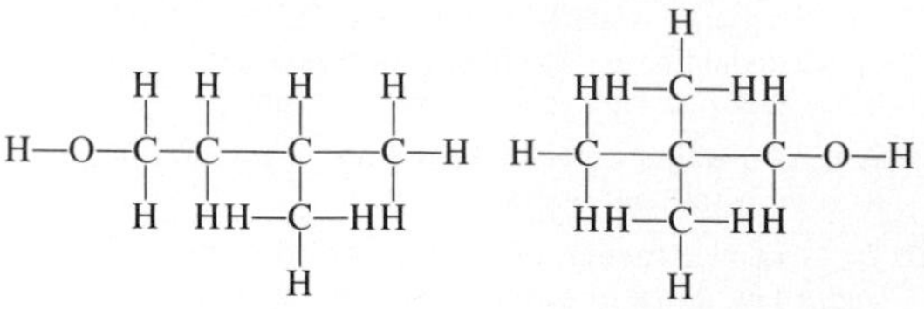

3-methylbutanol 2,2-dimethylpropanol

3.17 There are 14 isomers with the molecular formula $C_4H_6Cl_2$, all of which are cyclic:

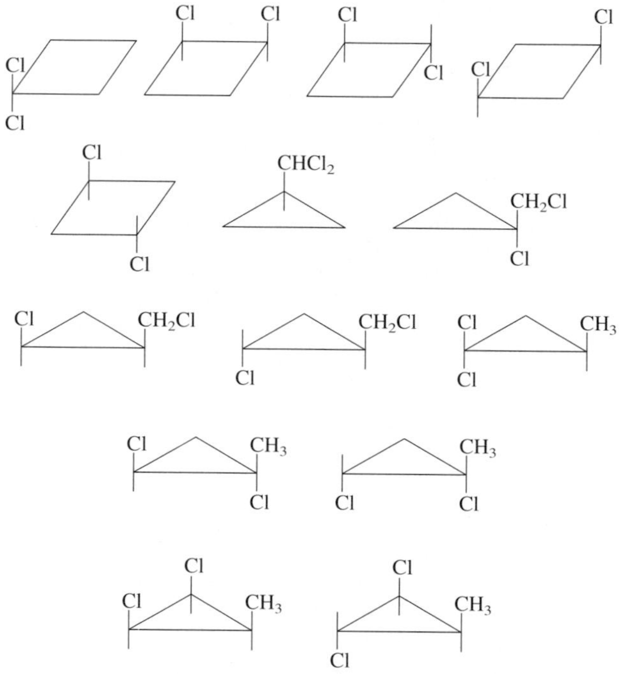

There are 27 linear isomers of the molecular formula, $C_4H_6Cl_2$:

$Cl_2C{=}CHCH_2CH_3$ $ClCH{=}C(Cl)CH_2CH_3$ (cis and trans) $ClCH{=}CHCH(Cl)CH_3$ (cis and trans)

$ClCH_2CH{=}CHCH_2Cl$ (cis and trans) $CH_2{=}CCl_2CHCH_3$ $CH_2{=}C(Cl)CH_2CH_2Cl$

$CH_2{=}CHC(Cl)_2CH_3$ $CH_2{=}CHCH(Cl)CH_3$ $CH_2{=}CHCH_2CHCl_2$

$Cl_2CHCH{=}CHCH_3$ (cis and trans) $ClCH_2C(Cl){=}CHCH_3$ (cis and trans) $ClCH_2CH{=}C(Cl)CH_3$ (cis and trans)

$ClCH_2CH{=}CHCH_2Cl$ (cis and trans) $CH_3C(Cl){=}C(Cl)CH_3$ (cis and trans) $Cl_2C{=}C(Cl)CH_3$

$ClCH{=}C(CH_3)CH_2Cl$ (cis and trans) $CH_2{=}C(CH_3)CHCl_2$ $CH_2{=}C(CH_2Cl)CH_2Cl$

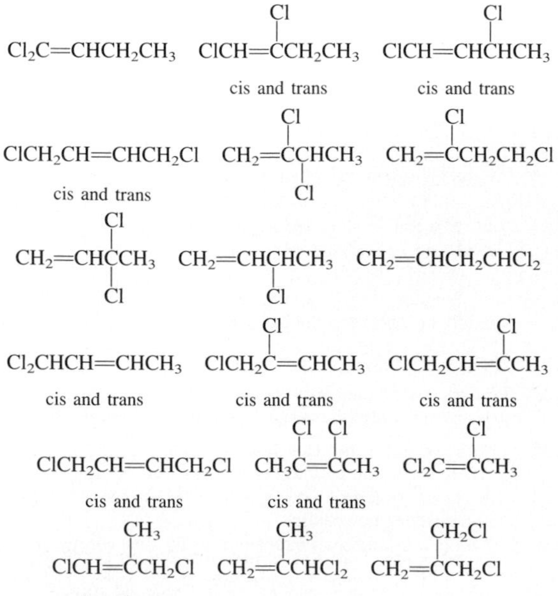

3.19 a. alkanes b. alkenes or cycloalkanes

3.21 a. nonpolar, polar b. benzene c. raise d. isooctane

3.23 There are seven possible ways of arranging the —CH_3 and —Br groups on a cyclohexane ring:

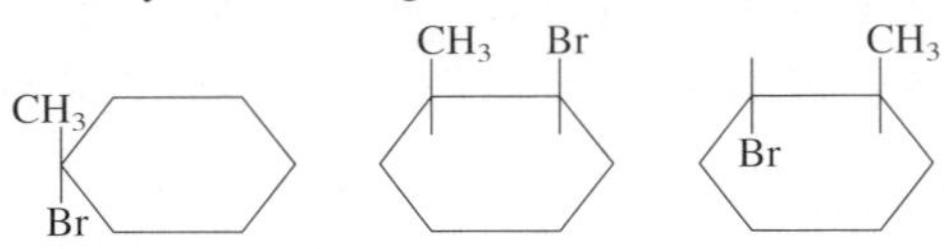

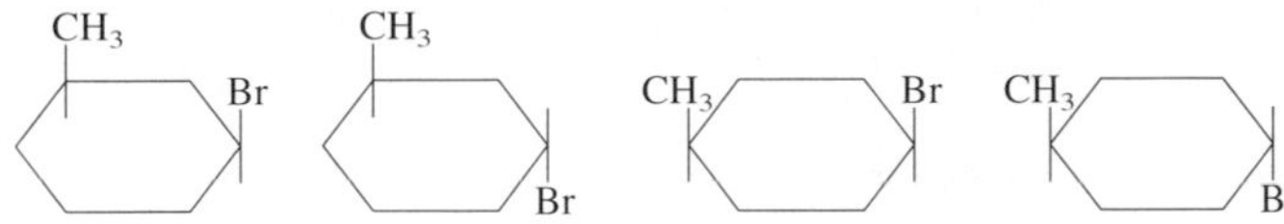

3.25 a. methyl alkyl halide:

H
Br—C—H
H

b. primary alkyl halide:

Br
CH_3—C—H
H

c. secondary alkyl halide:

CH_3
CH_3—C—Br
H

d. tertiary alkyl halide:

CH_3
CH_3—C—Br
CH_3

3.27 There are nine isomers with the molecular formula $C_4H_8Br_2$:

Br H H H
Br—C—C—C—C—H
H H H H

H Br H H
Br—C—C—C—C—H
H H H H

H H Br H
Br—C—C—C—C—H
H H H H

H H H H
Br—C—C—C—C—Br
H H H H

H Br H H
H—C—C—C—C—H
H Br H H

H Br Br H
H—C—C—C—C—H
H H H H

Br H H
Br—C——C——C—H
HH—C—HH
H

H Br H
Br—C——C——C—H
HH—C—HH
H

H H H
Br—C——C——C—Br
HH—C—HH
H

3.29 Catalytic cracking is a process that results in the decomposition of large hydrocarbons into smaller fragments under the influence of heat and/or pressure in the presence of a catalyst. For example:

$$C_{40}H_{90} \xrightarrow[\text{catalyst}]{\text{Heat,}} 5C_8H_{18}$$

Catalytic reforming is a process that results in the rearrangement of one organic molecule into another under the influence of heat and/or pressure in the presence of a catalyst. For example:

$$CH_3CH_2CH_2CH_2CH_3 \xrightarrow[\text{catalyst}]{\text{heat}} \text{(cyclopentane)}$$

3.31 The steps used to name a compound using the I.U.P.A.C. nomenclature system are:

1. Find the longest continuous carbon chain.
2. Number the chain in the direction that gives substituents the lowest possible numbers; the length of the carbon chain gives the parent name of the alkane.
3. Identify the substituents. For each alkyl substituent, take the parent name, drop the -e and add a -yl ending. For halides, take the first syllable in the element name and add an -o suffix.
4. Each substituent should have a number corresponding to the carbon number of the parent chain to which it is bonded. Separate the number and the substituent name with a dash (-). If the same substituent occurs more than once, cite the carbon numbers involved, separated by commas (,).
5. Arrange the substituents in alphabetical order; halogens, however, precede alkyl groups.
6. Place the parent name at the end of the name.

3.33 Both a and b are identical (both are *n*-hexane); c is a structural isomer of a and b (c is 2-methylpentane).

3.35 a. incorrect; the correct name is 3-methylhexane b. incorrect; the correct name is 2-methylbutane c. incorrect; the correct name is 3-methylheptane d. correct e. incorrect; the correct name is 1,2-dibromocyclohexane

3.37 a. degradation b. synthesis c. synthesis

3.39 Cl—$CH_2CH(CH_3)CH_2CH_3$ + CH_3—C(CH_3)(Cl)—CH_2CH_3 + $CH_3CH(CH_3)CH(Cl)CH_3$ + $CH_3CH(CH_3)CH_2CH_2Cl$

1-Chloro-2-methylbutane, 2-Chloro-2-methylbutane, 2-Chloro-3-methylbutane, 1-Chloro-3-methylbutane

3.41 9 moles of fluorine

3.43 a. $CH_3(CH_2)_5CH_2I$ + NaBr

b. (cyclopentyl)—Cl + HCl

c. $CH_3(CH_2)_5CH_2OH$ + NaI

3.45 For halogenated butanes:

1. Boiling point increases as the mass of the halogen increases; boiling point increases as the number of halogens increases.
2. Melting point increases as the mass of the halogen increases; melting point increases as the number of halogens increases.
3. Alkyl halides are generally water insoluble independent of the mass or number of halogens on the molecule.
4. Density increases as the mass of the halogen increases; density increases as the number of halogens increases.

3.47 The two chair conformations of chlorocyclohexane are

Cl (axial) and Cl (equatorial)

axial Cl equitorial Cl

The conformation showing the chlorine in the equatorial position is preferred. Placement of the bulkier group in the equatorial position is more stable, since less crowding is present.

3.49 a. $CH_3CH_2CH_2$—Br and $CH_3CH(Br)CH_3$

1-Bromopropane 75% 2-Bromopropane 25%

b. Br—$CH_2CH(CH_3)CH_3$ and CH_3—C(CH_3)(Br)—CH_3

1-Bromo-2-methylpropane 90% 2-Bromo-2-methylpropane 10%

3.51 The relative unreactive properties of hydrocarbons may be accounted for by their nonpolarity and by the strength and lack of reactivity of C—C and C—H bonds.

3.53 a. Oxidation is defined as either a gain of oxygen, a loss of hydrogen, or a loss of electrons. Reduction is defined as a loss of oxygen, a gain of hydrogen, or a gain of electrons.
b. The combustion of methane is both an oxidation and a reduction. One cannot have only oxidation take place without a reduction. In the combustion of methane, methane is oxidized—a gain of oxygen has occurred. Oxygen has been reduced—a gain of hydrogen has occurred.

3.55 The balanced chemical equation is

$$C_5H_{10} + Br_2 \xrightarrow{\text{Heat}} C_5H_9Br + HBr$$

0.0500 mol C_5H_{10} will react 3.51 g C_5H_{10} were consumed

3.57

Vitamin A

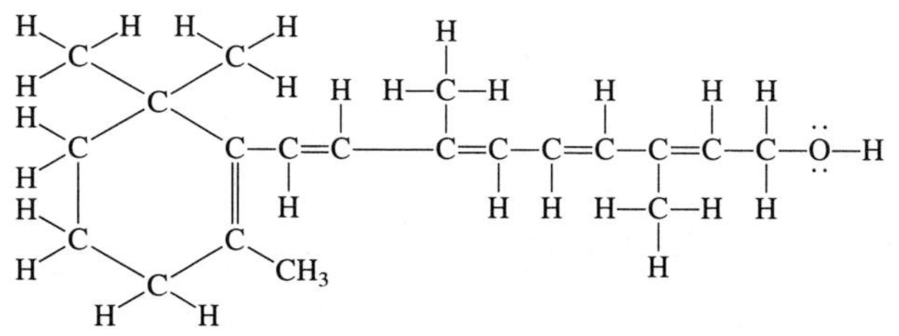

Demerol

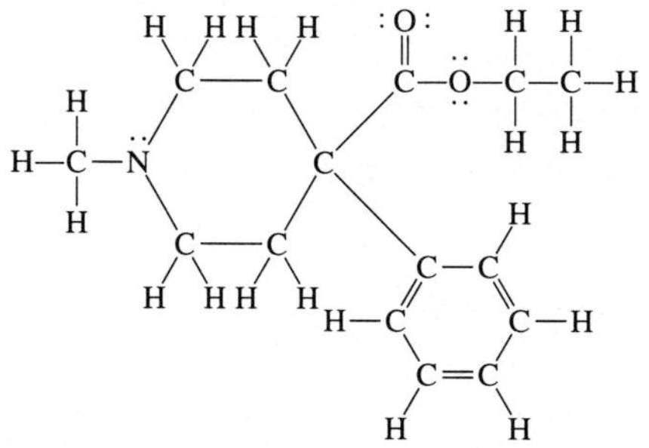

Benzocaine

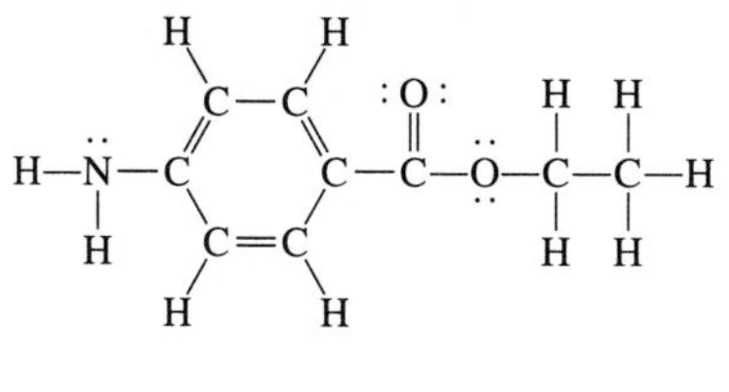

Testosterone

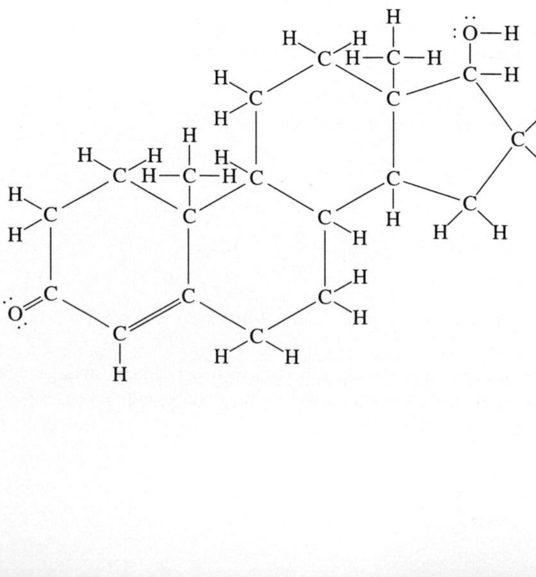

CHAPTER 4

4.1 a. $BrCH_2CH_2—C≡C—CH_2CH_3$ b. $CH_3—C≡C—CH_3$
c. $Cl—C≡C—Cl$ d. $H_2C=CH—Cl$
e. $HC≡C—CH_2CH_2CH_2CH_2CH_2CH_2CH_2—I$
f. $CH_3CH_2—C≡C—CH_2CH_3$

4.3 a. $CH_3CH_2CH_2CH_3$ b. $CH_3CH_2CH_2CH_3$ c. 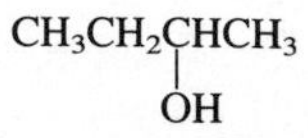 $CH_3CH_2CH(OH)CH_3$
d. 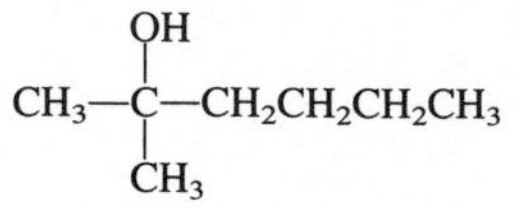 $CH_3—C(OH)(CH_3)—CH_2CH_2CH_2CH_3$

4.5 a. 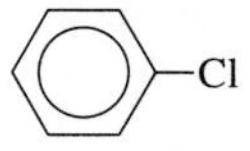 (benzene ring)—Cl
b. 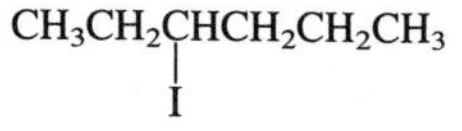 $CH_3CH_2CH(I)CH_2CH_2CH_3$
c. 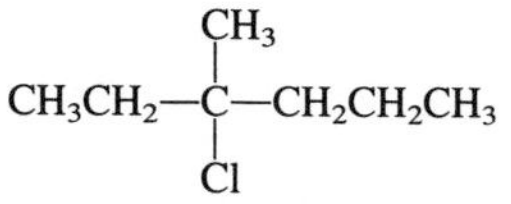 $CH_3CH_2—C(CH_3)(Cl)—CH_2CH_2CH_3$
d. 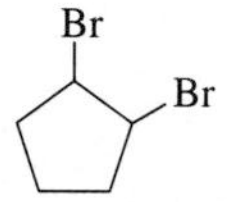 (cyclopentane ring with Br, Br)
e. 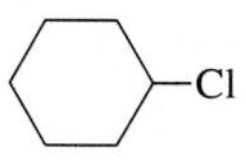 (cyclohexane ring)—Cl
f. 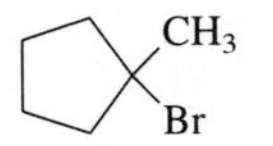 (cyclopentane ring with CH_3 and Br)

4.7 a. $CH_3CH_2—C≡C—CH_2CH_3$
b. 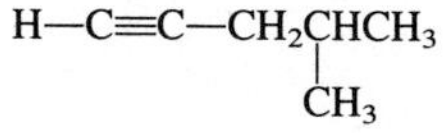 $H—C≡C—CH_2CH(CH_3)CH_3$
c. $ClCH_2—C≡C—C(CH_3)_2—CH(CH_3)CH_2CH_3$
d. 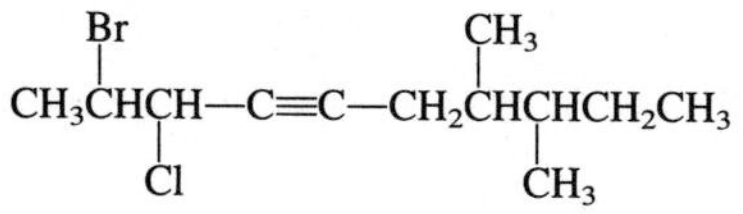 $CH_3CH(Br)CH(Cl)—C≡C—CH_2CH(CH_3)CH(CH_3)CH_2CH_3$
e. 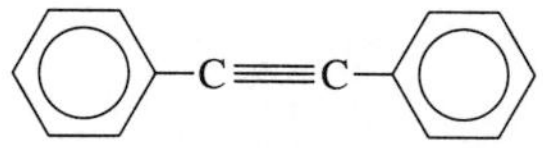 (benzene ring)—C≡C—(benzene ring)

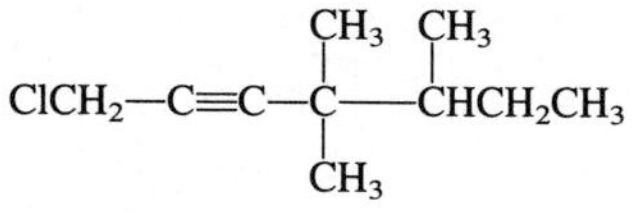

4.9 a. 2-methyl-2-butene b. *cis*-3-methyl-3-hexene
c. *trans*-5,5-dimethyl-2-hexene d. 1,3-dimethylcyclopentene
e. *trans*-4-ethyl-3,6-dimethyl-3-octene

4.11 a. 2-methyltoluene b. 2-nitrotoluene
c. 3-chloro-6-ethylcyclohexene d. 3-bromo-5-chlorotoluene
e. 4-bromo-2,6-dinitrotoluene

4.13 a. 3-methyl-1-pentene b. 2,5-dimethyl-2-hexene
c. 7-bromo-1-heptene d. 4-chloro-3-methylbutyne
e. 5-ethyl-3,4-diiodoheptane f. 6-chloro-1-heptyne
g. 4-bromo-*t*-butylcyclohexene
h. *cis*-1-bromo-2-chlorocyclopentane i. 4-chlorotoluene
j. *cis*-3-hexene k. 4-nitroaniline
l. 3-amino-1-bromocyclopentane

4.15 a. 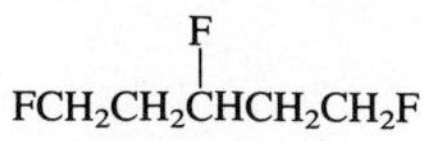 $FCH_2CH_2CH(F)CH_2CH_2F$
b. 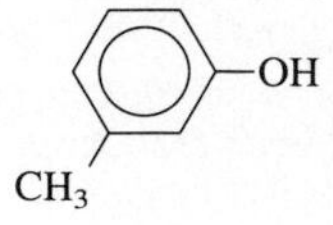 (benzene ring with OH and CH_3)

c. CH_3 / $CH_2CH_2CH_2CH_2CH_3$ on C=C, H and H (cis)

d. $H—C{\equiv}C—CH_2CH_2CH_2CH_2CH_2CH_3$

e. $CH_3CH_2CH_2—C{\equiv}C—CH_2CH_2CH_3$

f. $C_6H_5—CH(CH_3)CH_3$ (benzene ring with $\overset{CH_3}{|}CHCH_3$)

g. $H_2C{=}CH—C(CH_3)_2CH_2CH(CH_3)CH_2CH_3$

h. benzene ring with O_2N, O_2N, NO_2

i. $Br—C{\equiv}C—CH(Cl)CH_2CH_2CH_2CH_3$ j. Cl, Cl / C=C / F, H

4.17 Polynuclear aromatic hydrocarbons bind to the DNA of cells. This causes errors during DNA replication. If the cell's DNA is inaccurately replicated, mutations can arise that result in uncontrolled growth of the cell.

4.19 a. cyclopentane ring—OH
cyclopentanol

c. cyclopentane ring
cyclopentane

b. cyclopentane ring—Cl
chlorocyclopentane

d. cyclopentane ring—I
iodocyclopentane

4.21 a. $CH_3CH_2CH_2CH_2CH{=}CH_2$
b. $BrCH_2CH{=}CHCH_2CH_2CH_3$
c. $CH_3C{\equiv}CCH_2CH_2CH_2CH_3$
d. $CH_2{=}CHCH{=}CHCH{=}CHCH{=}CH_2$

4.23 *alkanes:* general formula: C_nH_{2n+2}, nonpolar, saturated, single bonds only, all carbons are tetrahedral
alkenes: general formula: C_nH_{2n}, relatively nonpolar, contain one or more carbon-carbon double bonds, double-bonded carbons are planar, unsaturated
alkynes: general formula: C_nH_{2n-2}, nonpolar, unsaturated, contain one or more carbon-carbon triple bonds, triple-bonded carbons are linear
aromatic: must have a benzene ring or have properties similar to benzene

4.25 a. 1-iodobutane b. water c. 1-iodobutane
d. chlorobenzene e. 1-iodooctane

4.27 DDT has been banned from most applications in the United States. Due to its stability, it accumulates in biological tissue and has been implicated in the deterioration of the central nervous system, cancer, and even death in humans. Additionally, its concentration increases continually throughout the food chain, posing a special threat to species at the top of the food chain.

4.29 A synthesis is the formation of a compound; a degradation is the breakdown of a compound.

For example, an ester may be synthesized from a carboxylic acid and an alcohol:

$$CH_3CH_2\underset{\|}{C}(O)—OH + HO—CH_3 \longrightarrow CH_3CH_2\underset{\|}{C}(O)—OCH_3 + H_2O$$

As an example of a degradation, the same ester may be degraded back to its parent carboxylic acid and alcohol:

$$CH_3CH_2\underset{\|}{C}(O)—OCH_3 \xrightarrow[H^+]{H_2O,} CH_3CH_2\underset{\|}{C}(O)—OH + HO—CH_3$$

4.31 The "curly arrows" used in a reaction mechanism show the movement of electrons (and thereby bonds) in each step of the reaction.

4.33 a. H_2, Pt b. H_2O, H^+ c. HBr d. $CO_2 + H_2O$ e. Cl_2, heat f. cyclopentene

4.35 Add bromine in CCl_4.

4.37 a. *geometric isomers:* compounds with the same molecular formula but different structure by virtue of hindered rotation due to a ring or a double bond (*cis* or *trans* isomers)
conformational isomers: the same compound in a different configuration due to bond rotation
b. Geometric isomers can be separated from one another; conformational isomers cannot.
c. Geometric isomers can be separated because they are two different compounds and will have different physical properties. Conformational isomers generally cannot be separated because they are three-dimensional forms of the same compound and differ only because of bond rotation.

4.39 $CH_3CH_2—CH{=}CH_2 \xrightarrow{H^+} CH_3CH_2—\overset{+}{C}H—CH_2(H)$ + $:\ddot{Br}:^-$ $\longrightarrow CH_3CH_2—CH(Br)—CH_2(H)$

4.41 a. 2 moles Br_2 b. 1 mole Br_2 c. 3 moles Br_2 d. 3 moles Br_2

4.43 $CH_2{=}CHCH_2CH_2CH_3$ $CH_3CH{=}CHCH_2CH_3$ (*cis-* and *trans-*) $CH_2{=}CHCH(CH_3)CH_3$

$CH_3—C(CH_3){=}CHCH_3$ $CH_2{=}C(CH_3)—CH_2CH_3$

4.45 a. *trans*-1,2-dimethylcyclohexane

CH₃, CH₃ on chair cyclohexane

b. *trans*-1,3-dimethylcyclohexane

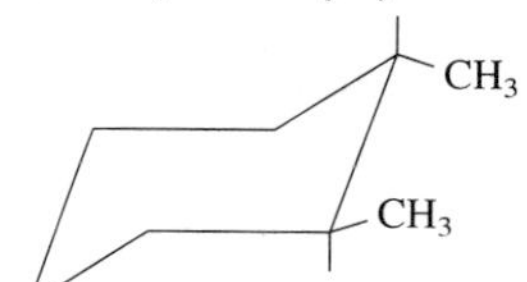

c. *trans*-1,4-dimethylcyclohexane

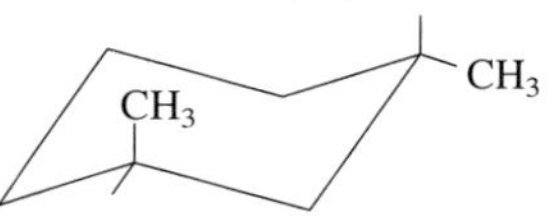

4.47 cyclopropane with Br, Br (cis)

cyclopropane with Br, Br; cyclopropane with Br, Br — mirror images

Each isomer has different physical properties.

4.49 There are multiple listings for information about the hazards of benzene usage available in the library. A few are referenced below:
G.D. Muir, ed., *Hazards in the Chemical Laboratory,* The Chemical Society, London, 1977
E. Meyer, *The Chemistry of Hazardous Materials,* Englewood Cliffs, N.J., Prentice Hall, Inc., 1977

4.51 The LD_{50} of a given substance is determined on laboratory animals and listed as a given amount per kilogram of body weight. Since the dosage is known per kilogram of body weight, the toxic dose can be scaled to humans.

4.53 The compound in vial A decolorized Br_2, hence it is an alkene. The compound in vial B did not decolorize bromine, hence it is a cycloalkane.

CHAPTER 5

5.1 a. methanol b. ethanol c. 1-butanol d. diethyl ether
e. 2,3-butanediol

5.3 a. CH_3CH_2OH
b. $CH_3CH(OH)CH_3$
c. $CH_3CH_2CH(OH)CH_3$
d. $CH_3CH_2CH(OH)CH_3$
e. $CH_3C(OH)(CH_3)CH_3$
f. $CH_3CH(OH)$–cyclohexyl

5.5 a. $CH_3CH{=}CH_2$
b. $CH_3CH{=}CHCH_3 + CH_3CH_2CH{=}CH_2$
c. $CH_2{=}C(CH_3){-}CH_2CH_3 + CH_3{-}C(CH_3){=}CHCH_3$
d. $CH_2{=}C(CH_3){-}CH_3$

5.7 penthrane: 2,2-dichloro-1,1-difluoro-1-methoxyethane
ethrane: 2-chloro-1-(difluoromethoxy)-1,1,2-trifluoroethane

5.9 a. $CH_3CH_2CH_2{-}OH < CH_2(OH){-}CH_2{-}CH_2(OH) < CH_2(OH){-}CH(OH){-}CH_2(OH)$
Least soluble — Most soluble
b. 1-hexanol < *n*-pentylalcohol < ethylene glycol
Least soluble — Most soluble

5.11 Ethyl alcohol has a higher boiling point because it is able to form intermolecular hydrogen bonds; ethanethiol cannot form intermolecular hydrogen bonds.

5.13 Ethyl alcohol has the higher boiling point because it is able to form intermolecular hydrogen bonds; diethyl ether cannot form intermolecular hydrogen bonds.

5.15 a. 1-heptanol b. 2-propanol c. 2,2-dimethyl-1-propanol
d. 2-methoxybutane e. 4-bromo-1-hexanol f. cyclopentanol
g. 3,3-dimethyl-2-hexanol h. 3-ethyl-3-heptanol i. cycloheptanol
j. 3-methylphenol

5.17 a. 1-propanethiol b. 2-butanethiol c. 2-methyl-2-butanethiol
d. 1,4-cyclohexanedithiol e. 1,2-propanedithiol
f. benzenethiol g. 2-pentanethiol h. 1-heptanethiol

5.19 The carbinol carbon is the carbon atom to which the hydroxyl group is attached.

5.21 $CH_3CH_2CH_2CH_3 < CH_3CH_2CH_2{-}O{-}CH_2CH_3$
Lowest b.p.
$< CH_3C({=}O)CH_2CH_2CH_3 < CH_3CH(OH)CH_2CH_3$
Highest b.p.

5.23 methyl alcohol: gasoline additive (''dry gas''); ethyl alcohol: alcoholic beverages; isopropyl alcohol: window cleaners, rubbing alcohol; dodecyl alcohol (lauryl alcohol): shampoos; menthol: flavoring in candies, pharmaceuticals; ethylene glycol: antifreeze
(This is just a partial list; you may have found others.)

5.25 a. primary b. secondary c. tertiary d. tertiary e. primary
f. tertiary g. secondary

5.27 Ethers do not have polar hydroxyl groups that can hydrogen bond with water; hence they are less water soluble. Alcohols have polar hydroxyl groups that can hydrogen bond with water; hence they are more water soluble.

5.29 [Structures: benzene ring with OH and O_2N, NO_2, NO_2 substituents; benzene ring with CH_3 and O_2N, NO_2, NO_2 substituents]

Picric acid — Trinitrotoluene (TNT)

Picric acid has a polar hydroxyl group, whereas TNT does not. This polar hydroxyl group will make picric acid more water soluble because the hydroxyl group can hydrogen bond with water.

5.31 [Structures: benzene ring with a bond; benzene ring with –OH]

Phenyl — Phenol

5.33 a. 2-nitrophenol b. 4-isopropylphenol
c. 3-bromo-5-chlorophenol d. 4-bromo-2-methylphenol

5.35 The I.U.P.A.C. rules for naming alcohols are:
1. Find the longest continuous chain containing the alcohol group. Assign the chain the parent alkane name according to the number of carbons in the longest chain.
2. Number the chain in the direction that gives the alcohol group the lowest possible number.
3. Name and place the substituents in alphabetical order. Separate numbers with commas and letters and numbers with a dash (-).
4. Take the parent alkane name, drop the -e, and add the suffix -ol. The number location of the alcohol group should precede the parent name.

5.37 a. $CH_3{-}C({=}O){-}CH_2CH_3$ b. no reaction c. cyclohexanone ($C{=}O$ in a six-membered ring) d. no reaction e. no reaction

5.39 a. $CH_3CH_2{-}O{-}CH_2CH_3 + H_2O$
b. $CH_3{-}O{-}CH_2CH_3 + CH_3{-}O{-}CH_3 + CH_3CH_2{-}O{-}CH_2CH_3 + H_2O$
c. $CH_3CH(CH_3){-}O{-}CH_3 + CH_3CH(CH_3){-}O{-}CH(CH_3)CH_3 + CH_3{-}O{-}CH_3 + H_2O$
d. cyclopentyl$-CH_2{-}O{-}CH_2-$cyclopentyl $+ H_2O$

5.41
$$CH_3{-}C(CH_3)(\ddot{O}H){-}CH_2CH_3 \xrightarrow{H^+} CH_3{-}C(CH_3)(\overset{+}{O}H_2){-}CH_2CH_3$$
$$\downarrow -H_2O$$
$H_2\ddot{O}$: removes H from carbocation $CH_2(H){-}\overset{+}{C}(CH_3){-}CH(H)CH_3$:
$$CH_2{=}C(CH_3){-}CH_2CH_3 \longleftarrow CH_2{-}\overset{+}{C}(CH_3){-}CHCH_3 \Longrightarrow CH_3{-}C(CH_3){=}CHCH_3$$
$$+\ H{:}\ddot{O}{:}H^+ \text{ (} H_3O^+ \text{)}$$

5.43
$$CH_3CH{=}CH_2 \xrightarrow{H_2O,\ H^+} CH_3{-}CH(OH){-}CH_3 \xrightarrow{[O]} CH_3{-}C({=}O){-}CH_3$$

5.45 C_6H_5–C(=O)–CH_3 $\xleftarrow{K_2Cr_2O_7,\ H^+}$ C_6H_5–CH(OH)–CH_3

$\uparrow$ $H_2O,\ H^+$

C_6H_5–CHBr–CH_2Br $\xleftarrow{Br_2}$ C_6H_5–CH=CH_2

C_6H_5–CHBr–CH_3 $\xleftarrow{HBr}$ C_6H_5–CH=CH_2

5.47 a. CH_3CH_2—S—CH_2CH_3 Diethyl sulfide c. $CH_3CH_2CH_2CH_2$—S—$CH_2CH_2CH_2CH_3$ Di-*n*-butyl sulfide

b. CH_3—S—$CH_2CH_2CH_3$ Methyl propyl sulfide d. CH_3CH_2—S—C_6H_5 Ethyl phenyl sulfide

5.49 Methanal is more toxic (LD_{50} of 0.07 g/kg) than ethanal (LD_{50} of 0.9 g/kg). Fewer grams of methanal per kilogram produce a toxic effect.

5.51 a. 2-chlorotetrahydrofuran: (tetrahydrofuran ring with Cl on carbon 2) c. 2,5-dibromofuran: (furan ring with Br on carbons 2 and 5)

b. furan: (furan ring)

5.53 HO—C(=O)—CH(NH_2)—CH_2—S—S—CH_2—CH(NH_2)—C(=O)—OH

5.55 CH_3—C(CH_3)(OH)—CH_3 CH_2=C(CH_3)—CH_3

Compound A Compound B

CHAPTER 6

6.1 Good solvents should dissolve a wide range of compounds. Simple ketones are considered "universal solvents" because they contain both a polar group (carbonyl group) and nonpolar side chains. These groups allow interaction with both polar and nonpolar solutes.

6.3 a. 3-iodobutanone b. 1,3-difluorobutanone c. 3-methylheptanal d. 3-methyl-2-hexanone e. 1-fluoro-3-methyl-2-pentanone f. 3,5-dibromopentanal

6.5 a. CH_3CH_2—C(=O)—CH_3 Butanone d. no reaction

b. CH_3CH(CH_3)—C(=O)—H Methylpropanal e. $CH_3(CH_2)_6$—C(=O)—CH_3 2-Nonanone

c. (cyclopentane ring)=O Cyclopentanone f. $CH_3(CH_2)_8$—C(=O)—H Decanal

6.7 a. (1) reduction (2) reduction (3) reduction (4) oxidation (5) reduction

b. For reductions, a reducing agent such as hydrogen with a platinum catalyst would be used. For oxidations, an oxidizing agent such as potassium permanganate would be used.

6.9 a. methanal: H—C(=O)—H

b. 7,8-dibromooctanal: H—CH(Br)—CH(Br)—CH$_2$—CH$_2$—CH$_2$—CH$_2$—CH$_2$—C(=O)—H

c. acetone: H—C(H)(H)—C(=O)—C(H)(H)—H

d. *o*-bromobenzophenone: (2-bromophenyl)—C(=O)—C_6H_5

e. hydroxyethanal: H—C(H)(OH)—C(=O)—H

f. 3-chloro-2-pentanone: H—C(H)(H)—C(=O)—C(Cl)(H)—C(H)(H)—C(H)(H)—H

g. benzaldehyde: C_6H_5—C(=O)—H

h. triiodoacetone: I—C(I)(I)—C(=O)—C(H)(H)—H

i. 2-chloropropionaldehyde: H—C(H)(H)—C(H)(Cl)—C(=O)—H

j. butyraldehyde: H—C(H)(H)—C(H)(H)—C(H)(H)—C(=O)—H

6.11 a. acetone b. methyl ethyl ketone c. acetaldehyde d. propionaldehyde e. methyl isopropyl ketone f. diethyl ketone

6.13 The rules for naming aldehydes using the I.U.P.A.C. Nomenclature system are as follows:

1. Find the longest continuous chain containing the aldehyde group.
2. Take the parent name of the longest chain, drop the -e, and add the -al suffix.
3. Number the chain, beginning at the aldehyde carbon.
4. Identify the substituents on the parent chain.
5. Place the name and number of the substituents in alphabetical order and follow with the parent name of the aldehyde.

6.15 There are 7 possible isomeric aldehydes and ketones with the molecular formula, $C_5H_{10}O$:

$CH_3CH_2CH_2CH_2$—C(=O)—H pentanal

CH_3—CH(CH_3)—CH_2—C(=O)—H 3-methylbutanal

CH_3CH_2—CH(CH_3)—C(=O)—H 2-methylbutanal

CH_3—C(CH_3)(CH_3)—C(=O)—H 2,2-dimethylpropanal

CH_3—C(=O)—$CH_2CH_2CH_3$ 2-pentanone

CH_3—C(=O)—CH(CH_3)—CH_3 3-methylbutanone

CH_3CH_2—C(=O)—CH_2CH_3 3-pentanone

6.17 A = 120° B = 120° C = 109.5° D = 109.5°

6.19 a. CH_3CH_2—C(OH)(OCH_2CH_3)—CH_3

b. CH_3—C(OH)(OCH_2CH_3)—C_6H_5

c. (cyclopentane ring carbon bearing OH and OCH_2CH_3)

6.21 a. acetal b. hemiketal c. hemiketal d. hemiacetal e. ketal f. hemiketal

6.23 H—C(H)(H)—C(=O)—C(H)(H)—C(H)(H)—C(H)(H)—C(H)(H)—C(H)(H)—H

2-heptanone

H—C(H)=C(H)—C(=O)—C(H)(H)—C(H)(H)—C(H)(H)—C(H)(H)—C(H)(H)—H

1-octene-3-one

6.25 The solubility of aldehydes and ketones in water is due to the ability of the polar carbonyl group to hydrogen bond to water. Simple aldehydes and ketones have short, nonpolar alkyl chains which will not greatly affect the overall polarity of the molecule. For larger aldehydes and ketones, the nonpolar alkyl chain eventually overcomes the polarity of the carbonyl group and reduces the solubility in water.

6.27 $CH_3-\overset{O}{\overset{\|}{C}}-OH \xrightarrow{Ag(NH_3)_2^+}$ NO REACTION

$\downarrow H_2, Pt$

$CH_3-\overset{O}{\overset{\|}{C}}-H \xrightarrow{H_2,\ Pt} CH_3-\overset{OH}{\overset{|}{C}H_2}$

6.29 $CH_3CH_2CH_2CH_2-OH \xrightarrow[K_2CrO_4]{(1)}$

$CH_3CH_2CH_2-\overset{O}{\overset{\|}{C}}-H \xrightarrow[KMnO_4]{(2)} CH_3CH_2CH_2-\overset{O}{\overset{\|}{C}}-OH$

$\downarrow CH_3OH,\ H^+$

$CH_3CH_2CH_2-\overset{OCH_3}{\overset{|}{\underset{OCH_3}{\underset{|}{C}}}}-H \xleftarrow[H^+]{CH_3OH} CH_3CH_2CH_2-\overset{OH}{\overset{|}{\underset{OCH_3}{\underset{|}{C}}}}-H$

6.31 $CH_3CH_2-\overset{O}{\overset{\|}{C}}-CH_2CH_3 \xrightarrow{H_2,\ Pt} CH_3CH_2\overset{OH}{\overset{|}{C}}HCH_2CH_3 \xrightarrow[heat]{H_2SO_4}$

$CH_3CH{=}CHCH_2CH_3$

6.33 a. $CH_3-\overset{OH}{\overset{|}{C}H}-CH_2\overset{Br}{\overset{|}{C}}HCH_3 \xrightarrow{[O]} CH_3-\overset{O}{\overset{\|}{C}}-CH_2\overset{Br}{\overset{|}{C}}HCH_3$

b. 3-methylcyclopentanone $\xrightarrow{[R]}$ 3-methylcyclopentanol (ring with O / OH and CH_3)

c. cyclopentanone $+ Br_2 \xrightarrow{heat}$ 2-bromocyclopentanone + 3-bromocyclopentanone

d. $CH_3-\overset{O}{\overset{\|}{C}}-CH_2CH_3 \xrightarrow[H^+]{CH_3CH_2OH} CH_3-\overset{OH}{\overset{|}{\underset{OCH_2CH_3}{\underset{|}{C}}}}-CH_2CH_3$

e. $CH_3-\overset{O}{\overset{\|}{C}}-CH_2CH_3 \xrightarrow[H^+]{\text{2 moles } CH_3CH_2OH} CH_3-\overset{OCH_2CH_3}{\overset{|}{\underset{OCH_2CH_3}{\underset{|}{C}}}}-CH_2CH_3$

6.35 a. $CH_3\overset{CH_3}{\overset{|}{C}}HCH_2\overset{OH}{\overset{|}{C}}HCH_3 \xrightarrow{[O]} CH_3\overset{CH_3}{\overset{|}{C}}HCH_2-\overset{O}{\overset{\|}{C}}-CH_3$

b. $HO-CH_2CH_2CH_2CH_2-OH \xrightarrow{[O]} H-\overset{O}{\overset{\|}{C}}-CH_2CH_2-\overset{O}{\overset{\|}{C}}-H$

c. $C_6H_5-CH_2CH_2OH \xrightarrow{[O]} C_6H_5-CH_2-\overset{O}{\overset{\|}{C}}-H$

d. $HO-CH_2CH_2\overset{OH}{\overset{|}{C}}HCH_3 \xrightarrow{[O]} H-\overset{O}{\overset{\|}{C}}-CH_2-\overset{O}{\overset{\|}{C}}-CH_3$

e. $CH_3-\overset{CH_3}{\overset{|}{\underset{CH_3}{\underset{|}{C}}}}-CH_2CH_2CH_2-OH \xrightarrow{[O]} CH_3-\overset{CH_3}{\overset{|}{\underset{CH_3}{\underset{|}{C}}}}-CH_2CH_2-\overset{O}{\overset{\|}{C}}-H$

f. $HO-C_6H_4-OH \xrightarrow{[O]} O{=}C_6H_4{=}O$

6.37 a. $C_6H_5-CH_3 \xrightarrow[FeBr_3]{Br_2}$ o-bromotoluene (CH_3, Br) + $Br-C_6H_4-CH_3$ (para)

$\downarrow KMnO_4,\ \Delta$

o-bromobenzoic acid ($-\overset{O}{\overset{\|}{C}}-OH$, Br)

b. o-xylene (CH_3, CH_3) $\xrightarrow[light,\ \Delta]{Br_2}$ (CH_3, CH_2Br)

$\downarrow KMnO_4$

($-\overset{O}{\overset{\|}{C}}-OH$, CH_2Br)

6.39 *Oxidation*—benzaldehyde is oxidized to benzoic acid present as the benzoate anion in basic solution:

$$2\ OH^- + C_6H_5-\overset{O}{\overset{\|}{C}}-H \longrightarrow C_6H_5-\overset{O}{\overset{\|}{C}}-O^- + 2\ H_2O + e^-$$

Reduction—silver(I) reduced to silver metal

$$Ag(NH_3)_2^+ + e^- \longrightarrow Ag + 2\ NH_3$$

6.41 (2,4-dinitrophenylhydrazine structure: $O_2N-C_6H_3(NO_2)-NH-NH_2$)

amino group

nitro group

6.43 a. $O{=}N(-O)-C_6H_3(N{=}O\text{–}O)-\overset{H}{N}-N{=}\overset{H}{C}-C_6H_5$

b. 2,4-dinitrophenylhydrazone: $O=N(O)-C_6H_3(N(=O)O)-NH-N=C(CH_3)-CH_2CH_2CH_3$

c. $O=N(O)-C_6H_3(N(=O)O)-NH-N=C$ (cyclopentylidene, C_5H_8)

d. $O=N(O)-C_6H_3(N(=O)O)-NH-N=C(CH_3)-CH_2-C(CH_3)=N-NH-C_6H_3(N(=O)O)-N(=O)O$

6.45 a. $CH_2{=}CHCH_2CH_3 \xrightarrow{O_3} H{-}\underset{H}{C}{=}O + O{=}\underset{H}{C}{-}CH_2CH_3$

b. $CH_3CH{=}CHCH_2CH_2CH_3 \xrightarrow{O_3} CH_3{-}\underset{H}{C}{=}O + O{=}\underset{H}{C}{-}CH_2CH_2CH_3$

c. $(CH_3CH_2)(CH_3)C{=}CH(CH_2CHCH_3(CH_3)) \xrightarrow{O_3} CH_3CH_2{-}\underset{CH_3}{C}{=}O + O{=}\underset{H}{C}{-}CH_2\underset{CH_3}{CH}CH_3$

6.47 $H_2C{=}CH_2 \xrightarrow{O_3}$ [Unstable intermediate] $\longrightarrow$ [Ozonide] $\xrightarrow[Zn]{H_2O}$ $H_2C{=}O$

Ethene — Unstable intermediate — Ozonide

6.49 $CH_3CH_2CH_2{-}\overset{O}{\overset{\|}{C}}{-}CH_3$ (Compound A) $\quad CH_3CH_2CH_2\overset{OH}{CH}CH_3$ (Compound B)

$CH_3CH_2CH_2CH{=}CH_2$ (Compound C) $\quad CH_3CH_2CH{=}CHCH_3$ (Compound D)

CHAPTER 7

7.1 An aldose sugar contains an aldehyde group; a ketose contains a ketone group.

7.3 The aldose or ketose moiety is shown in bold:

- CH_3–**C=O**–H–C–OH–CH_2–OH — Ketose
- **H**–**C=O**–H–C–OH–H–C–OH–HO–C–H–CH_2–OH — Aldose
- CH_2–OH–**C=O**–HO–C–H–H–C–OH–H–C–OH–CH_2–OH — Ketose
- **H**–**C=O**–H–C–OH–CH_2–OH — Aldose
- CH_3–**C=O**–H–C–OH–H–C–OH–H–C–OH–CH_2–OH — Ketose
- **H**–**C=O**–HO–C–H–H–C–OH–HO–C–H–CH_2–OH — Aldose

7.5 a. H–C=O–H–C–OH–CH_2–OH — Aldose

b. CH_2OH–C=O–H–C–OH–CH_2–OH — Ketose

c. H–C=O–H–C–OH–HO–C–H–H–C–OH–CH_2OH; CH_2OH–C=O–H–C–OH–H–C–OH–CH_2–OH

A pentose may be either an aldose or a ketose.

d. H–C=O–H–C–OH–H–C–OH–CH_2OH; CH_2OH–C=O–H–C–OH–CH_2–OH

A tetrose may be either an aldose or a ketose.

e. CH_2OH–C=O–H–C–OH–CH_2–OH

A ketotetrose

f. H–C=O–H–C–OH–H–C–OH–HO–C–H–CH_2OH — An aldopentose

g. CH_2OH–C=O–HO–C–H–H–C–OH–H–C–OH–CH_2OH — A ketohexose

7.7

H
C=O
H—C—OH
HO—C—H
HO—C—H
H—C—OH
CH_2—OH

D-Galactose → β-D-Galactose ⇌ α-D-Galactose

(β-D-Galactose ring: CH_2OH, HO, H, OH, H, OH, H, H, OH; α-D-Galactose ring: CH_2OH, HO, H, OH, H, H, OH, H, OH)

7.9 A saccharide is a sugar molecule having the general formula $C_x(H_2O)_y$.

7.11 a. β-D-glucose b. β-D-fructose c. α-D-galactose

7.13 a. hemiacetal b. hemiketal c. hemiacetal

7.15 The linkage joining two sugars of a disaccharide is called a glycosidic bond. There are two types of glycosidic bonds. If the linkage is formed between the anomeric hydroxyl group in the beta position of one sugar and a nonanomeric hydroxyl group on the other sugar, the linkage is called a β-glycosidic bond. If the linkage is formed between the anomeric hydroxyl group in the alpha position on one sugar and a nonanomeric hydroxyl group on the other sugar, the linkage is called an α-glycosidic linkage.

7.17 mammalian milk

7.19 the inability to convert galactose to glucose-1-phosphate

7.21 dulcitol

7.23 People with type O blood are univeral donors because type O blood does not contain type A or type B antigens. Type O blood does contain both anti-A and anti-B antibodies; hence people with type O blood are not universal recipients.

7.25 To receive blood transfusions successfully, one must transfuse the same blood type or be of blood type AB (universal recipient). For the Incas to successfully perform blood transfusions without knowledge of blood types, the population must have been predominately of one blood type, possibly mostly of type AB.

7.27 Cellulose is a structural component of rigid cell walls.

7.29 Amylose consists of glucose molecules linked via an α(1 ⟶ 4) glycosidic bond. Amylopectin is similar to amylose in that the main chain of the polysaccharide is also linked via an α(1 ⟶ 4) glycosidic bond; however, amylopectin also has glucose sugars branching off the main chain of the polysaccharide by an α(1 ⟶ 6) glycosidic bond. Glycogen is almost identical to amylopectin, differing only by the fact that there are more branches and they are shorter.

7.31 Lysozyme hydrolyzes the glycosidic bonds linking the peptidoglycan chains that compose the cell walls. If the cell wall of a bacterium is ruptured, the cell dies.

7.33

H	H
C=O	C=O
H—C—OH	HO—C—H
CH_2—OH	CH_2—OH

7.35

H	H	H	H
C=O	C=O	C=O	C=O
H—C—OH	HO—C—H	HO—C—H	H—C—OH
H—C—OH	HO—C—H	H—C—OH	HO—C—H
H—C—OH	HO—C—H	H—C—OH	HO—C—H
H—C—OH	HO—C—H	H—C—OH	HO—C—H
CH_2—OH	CH_2—OH	CH_2—OH	CH_2OH

H	H	H	H
C=O	C=O	C=O	C=O
H—C—OH	HO—C—H	H—C—OH	HO—C—H
HO—C—H	H—C—OH	H—C—OH	HO—C—H
H—C—OH	HO—C—H	HO—C—H	H—C—OH
H—C—OH	HO—C—H	H—C—OH	HO—C—H
CH_2—OH	CH_2—OH	CH_2—OH	CH_2OH
H	H	H	H
C=O	C=O	C=O	C=O
H—C—OH	HO—C—H	H—C—OH	HO—C—H
H—C—OH	HO—C—H	H—C—OH	HO—C—H
H—C—OH	HO—C—H	HO—C—H	H—C—OH
HO—C—H	H—C—OH	HO—C—H	H—C—OH
CH_2—OH	CH_2—OH	CH_2—OH	CH_2OH
H	H	H	H
C=O	C=O	C=O	C=O
H—C—OH	HO—C—H	H—C—OH	HO—C—H
HO—C—H	H—C—OH	HO—C—H	H—C—OH
HO—C—H	H—C—OH	H—C—OH	HO—C—H
H—C—OH	HO—C—H	HO—C—H	H—C—OH
CH_2—OH	CH_2—OH	CH_2—OH	CH_2OH

7.37 A starch is composed of many sugars linked by glycosidic bonds. Think of a starch as a natural polymer of sugars.

7.39 $C_3H_6O_3 + 2Ag(NH_3)_2^+ + 2H_2O \longrightarrow C_3H_6O_4^- + 2Ag + 4NH_4^+ + OH^-$

7.41 Sucrose is a nonreducing sugar because it does not contain an oxidizable hemiacetal group.

7.43 The term D- designates the specific arrangement of atoms in a molecule with one or more asymmetric carbons. It *does not* designate the direction of rotation of plane polarized light. A D-sugar has the priority group (the hydroxyl group) on the penultimate carbon positioned on the right.

7.45 Both L-alanine and L-glyceraldehyde have their priority groups on the chiral carbon on the *left*.

7.47 The structures of β-D-glucose and β-D-galactose are similar in that both have the hydroxyl group on the anomeric carbon over the ring (on the same side as the terminal —CH_2—OH group).

7.49 The D- representation of a structure places the priority group on a chiral center on the right. The L- representation of a structure places the priority group on a chiral center on the left. The D- or L- configuration does not specify the direction of the rotation of plane polarized light.

7.51 The chiral centers are indicated with an asterisk (*):

a. $CH_3CH(CH_3)CH(Br)CH_2CH_3$ (* on the carbon bearing CH_3)

b. $CH_3CH(OH)CH_2CH(Br)—C(=O)—H$ (* on the carbon bearing OH and * on the carbon bearing Br)

c. CH_2=$CHCH(OH)CH_3$ (* on the carbon bearing OH)

d. (ring with OH and CH_3; * on the carbon bearing CH_3)

e. (ring with CH_2OH, O, OH, HO, HO, OH; * on each ring carbon)

7.53 There are two chiral centers in aspartame; each is marked with an asterisk (*):

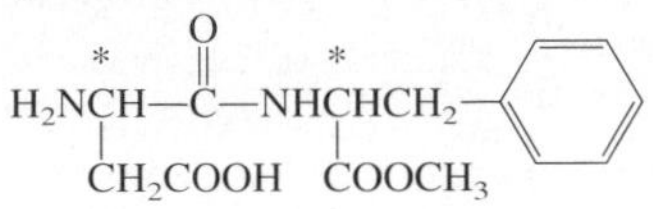

$H_2N\overset{*}{C}H(CH_2COOH)-\overset{O}{\overset{\|}{C}}-NH\overset{*}{C}H(COOCH_3)CH_2-C_6H_5$

7.55

a. CH_3 / H—C—OH / Br and CH_3 / HO—C—H / Br

b. CH_3 / H—C—OH / H—C—OH / CH_2Br and CH_3 / HO—C—H / HO—C—H / CH_2Br

c. $CH{=}CH_2$ / CH_3CH_2—C—H / OH and $CH{=}CH_2$ / H—C—CH_2CH_3 / OH

d. HC=O / H—C—OH / HO—C—H / CH_2OH and HC=O / HO—C—H / H—C—OH / CH_2OH

e. CH_3 / H—C—Br / HO—C—H / CH_3 and CH_3 / Br—C—H / H—C—OH / CH_3

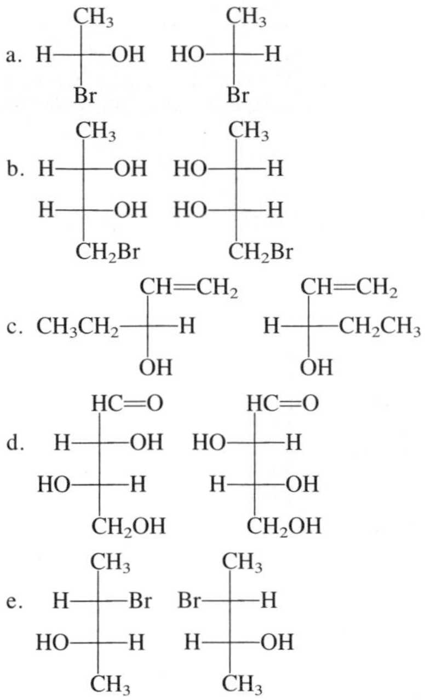

7.57 Corn syrup and maple syrup contain sucrose, a disaccharide of glucose and fructose; milk contains lactose, a disaccharide of glucose and galactose.

Candy bars contain sucrose: a disaccharide of glucose and fructose; some contain maltose: a glucose sugar.

Corn starch contains amylose and amylopectin, polymers of glucose.

CHAPTER 8

8.1

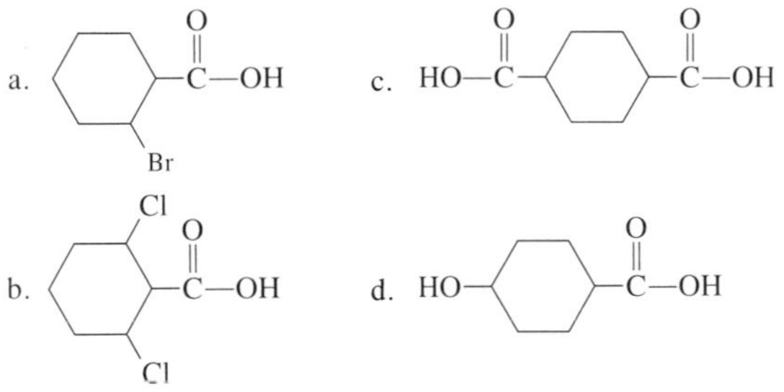

a. 2-bromocyclohexane—$\overset{O}{\overset{\|}{C}}$—OH (ring with Br)

b. cyclohexane ring with two Cl—$\overset{O}{\overset{\|}{C}}$—OH

c. HO—$\overset{O}{\overset{\|}{C}}$—(cyclohexane)—$\overset{O}{\overset{\|}{C}}$—OH

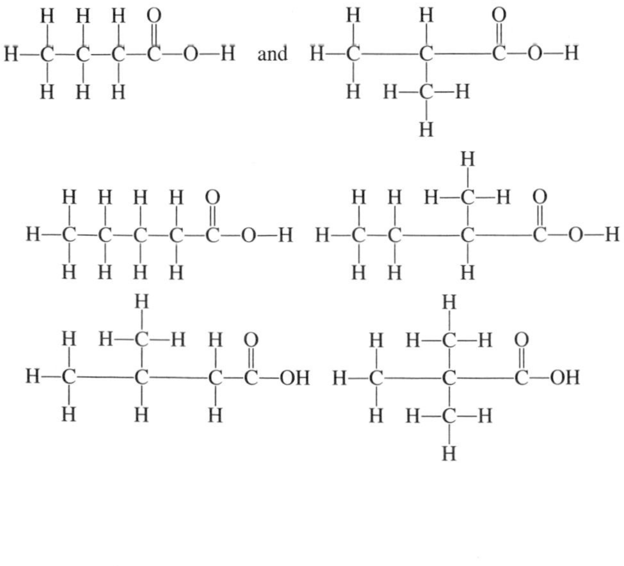

d. HO—(cyclohexane)—$\overset{O}{\overset{\|}{C}}$—OH

8.3 $H-CH_2-CH_2-CH_2-\overset{O}{\overset{\|}{C}}-O-H$ and $H-CH_2-CH(CH_3)-\overset{O}{\overset{\|}{C}}-O-H$

$H-CH_2-CH_2-CH_2-CH_2-\overset{O}{\overset{\|}{C}}-O-H$ $\quad$ $H-CH_2-CH_2-CH(CH_3)-\overset{O}{\overset{\|}{C}}-O-H$

$H-CH_2-CH(CH_3)-CH_2-\overset{O}{\overset{\|}{C}}-OH$ $\quad$ $H-CH_2-C(CH_3)_2-\overset{O}{\overset{\|}{C}}-OH$

8.5 $Br-CH_2-\overset{O}{\overset{\|}{C}}-Cl$ $\quad$ $Cl-C_6H_4-\overset{O}{\overset{\|}{C}}-Cl$

2-Bromoethanoyl chloride $\quad$ *p*-Chlorobenzoylchloride

$CH_3-\overset{O}{\overset{\|}{C}}-OCH_2CH_3$ $\quad$ $CH_3(CH_2)_2-\overset{O}{\overset{\|}{C}}-O-\overset{O}{\overset{\|}{C}}-(CH_2)_2CH_3$

Ethyl acetate $\quad$ Butanoic anhydride

$CH_3(CH_2)_4-\overset{O}{\overset{\|}{C}}-O-\overset{O}{\overset{\|}{C}}-(CH_2)_4CH_3$

Hexanoic anhydride

$CH_3CH_2CH_2CH(Br)-\overset{O}{\overset{\|}{C}}-O-\overset{O}{\overset{\|}{C}}-CH(Br)CH_2CH_2CH_3$

2-Bromopentanoic anhydride

$CH_3CH(CH_3)-O-\overset{O}{\overset{\|}{C}}-C_6H_5$

Isopropyl benzoate

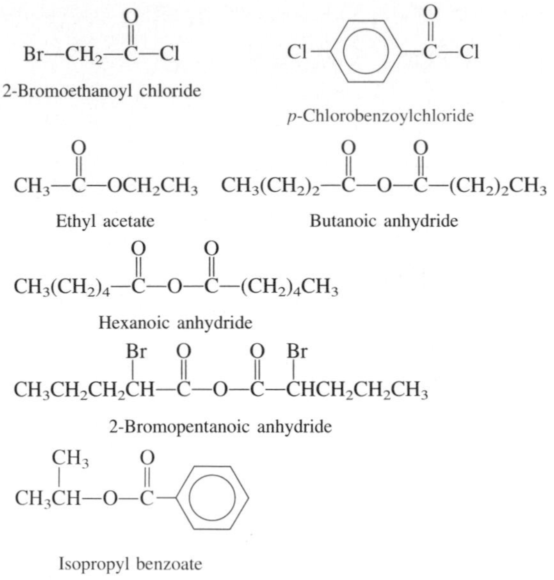

8.7 Carboxylic acids have significantly higher boiling points than alcohols of comparable molecular weight because of the strength of the hydrogen bonding interactions between the carboxyl groups.

8.9 a. $CH_3CH_2COO^-K^+ + H_2O$
b. $(CH_3CH_2CH_2COO^-)_2Ba^{2+} + H_2O$
c. 2 (cyclopentane)—COO^-Na^+ $+ CO_2 + H_2O$
d. $C_6H_5-COO^-Na^+ + H_2O$

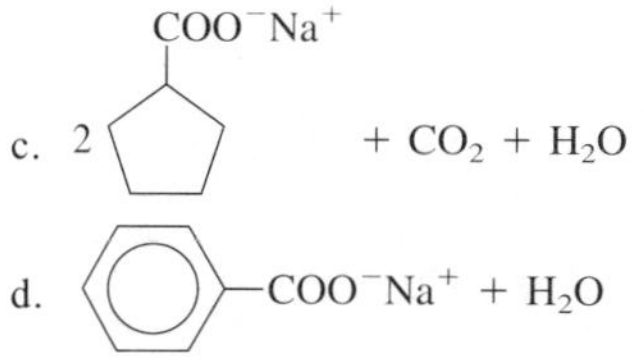

8.11 a. $KMnO_4$, followed by $SOCl_2$
b. $KMnO_4$, followed by $CH_3(CH_2)_3OH$
c. $KMnO_4$, followed by $SOCl_2$, followed by $CH_3COO^-Na^+$

8.13 Treatment with NaOH gives

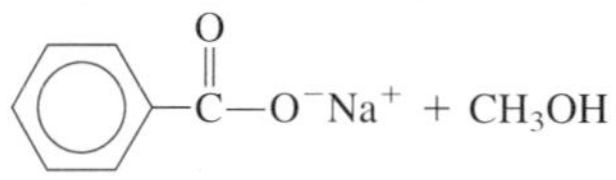

$C_6H_5-\overset{O}{\overset{\|}{C}}-O^-Na^+ + CH_3OH$

Subsequent treatment with HCl gives

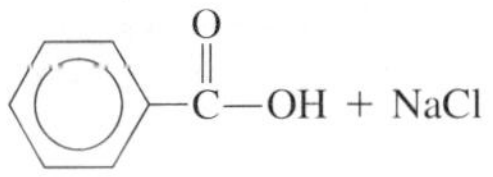

$C_6H_5-\overset{O}{\overset{\|}{C}}-OH + NaCl$

8.15

$C-\overset{O}{\overset{\|}{C}}-\underline{H}$	$C-\overset{O}{\overset{\|}{C}}-\underline{C}$	$C-\overset{O}{\overset{\|}{C}}-\underline{OH}$	$C-\overset{O}{\overset{\|}{C}}-\underline{OC}$
Aldehyde	Ketone	Carboxylic acid	Ester

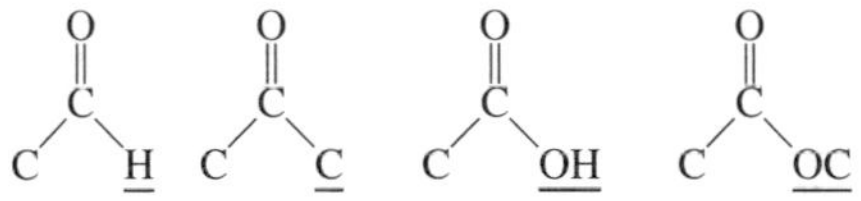

All carbonyl compounds will have the carbonyl group and a carbon on one side of the C=O in common. It is what is on the other side of the C=O that determines the functional group (shown underlined above). The geometry around the carbonyl group will always be trigonal planar.

8.17 a. 3-methylbutanoic acid b. 3-nitrobenzoic acid
c. 4-bromo-3-methylhexanoic acid d. 4-ethylbenzoic acid
e. 4-ethylhexanoic acid f. cyclopentanecarboxylic acid

8.19 a. $CH_3CH_2-C(CH_3)_2-CH_2CH_2-\overset{O}{\overset{\|}{C}}-OH$ $\quad$ b. $CH_3CH(Br)CH(CH_3)CH_2-\overset{O}{\overset{\|}{C}}-OH$

c. (benzene ring with O_2N and NO_2)—$\overset{O}{\overset{\|}{C}}$—OH $\quad$ d. (cyclohexane ring with CH_3)—$\overset{O}{\overset{\|}{C}}$—OH

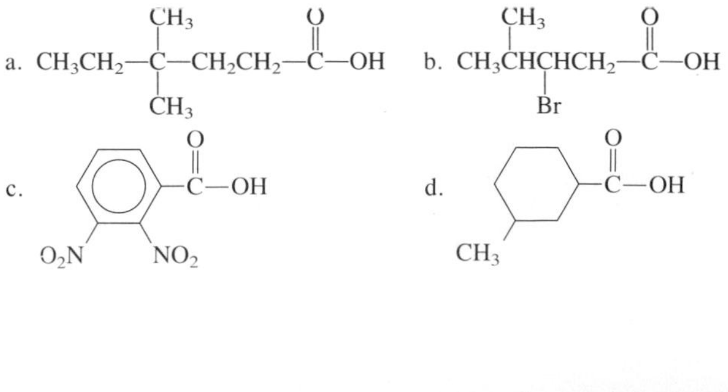

8.21 $-\left[-O-CH_2CH_2-\overset{O}{\overset{\|}{C}}-O-CH_2CH_2-\overset{O}{\overset{\|}{C}}-\right]_n-$

A polyester

$-\left[-O-CH_2CH_2CH_2-\overset{O}{\overset{\|}{C}}-NH-CH_2CH_2CH_2CH_2CH_2CH_2-\right]_n-$

A nylon

8.23 a. $CH_3CH_2CH_2CH_2CH_2-\overset{O}{\overset{\|}{C}}-O^-Na^+$

b. $CH_3CH_2CH_2CH_2CH_2CH_2CH_2CH_2-OH$

c. $CH_3CH_2-\overset{O}{\overset{\|}{C}}-OCH_3$

d. $CH_3CH_2-O-CH_2CH_3$

e. $CH_3-\overset{O}{\overset{\|}{C}}-OH$

f. $CH_3CH_2CH_2-\overset{O}{\overset{\|}{C}}-OH$

8.25 a. $CH_3CH_2CH_2-\overset{O}{\overset{\|}{C}}-OCH_2CH_3 + HCl$

b. 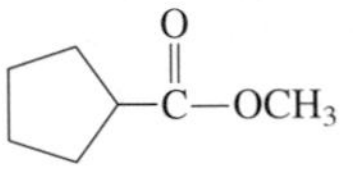 cyclopentyl$-\overset{O}{\overset{\|}{C}}-OCH_3$

c. $CH_3CH_2-\overset{O}{\overset{\|}{C}}-OH + CH_3CH_2OH$

d. $CH_3-\overset{O}{\overset{\|}{C}}-O\overset{CH_3}{\overset{|}{C}}HCH_2CH_3 + H_2O$

8.27 $CH_3CH_2CH_2CH_2CH_2-\overset{O}{\overset{\|}{C}}-OH$

8.29 $C_6H_4(-\overset{O}{\overset{\|}{C}}-OH)(-O-\underset{O}{\underset{\|}{C}}-CH_3)$ (ortho)

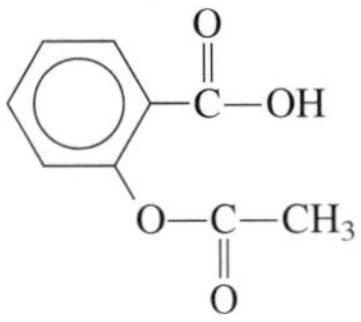

8.31 Most pheromone devices work by attracting one sex of the insect, usually the male, into a trap. The attractant typically consists of the female's sex pheromone. Since only males are trapped, they cannot mate, which lowers the population of the insect in the vicinity of the trap. Pheromones are used in "roach motels," Japanese beetle traps, silverfish traps, etc.

8.33 $\left[-\overset{O}{\overset{\|}{C}}-C_6H_4-\overset{O}{\overset{\|}{C}}-OCH_2CH_2-O-\right]_n$

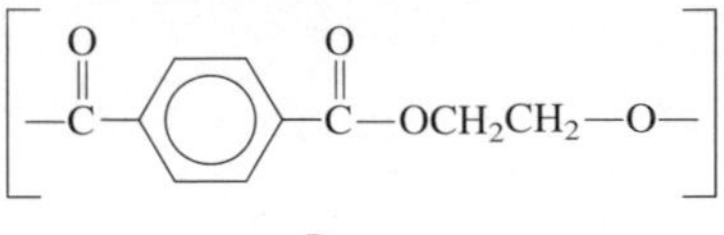

Dacron

8.35 Butanoic acid reacted with H_2 and Pt to form butanol. Butanol is then reacted with butanoic acid under acidic conditions to form the ester.

8.37 ATP is the designation for *a*denosine *tri*phosphate and consists of the base, adenosine, linked with three phosphate groups via an ester-type linkage. The energy stored in these phosphate bonds is released when these bonds are broken. This energy can be used by the body for its metabolic processes.

8.39 a. $CH_3CH_2CH_2CH_2CH_2-\overset{O}{\overset{\|}{C}}-OH + HCl$

b. $CH_3CH_2CH_2CH_2CH_2-\overset{O}{\overset{\|}{C}}-NH-CH_3 + HCl$

c. 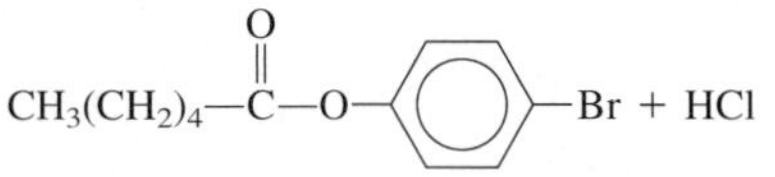 $CH_3(CH_2)_4-\overset{O}{\overset{\|}{C}}-O-C_6H_4-Br + HCl$

d. $CH_3(CH_2)_4-\overset{O}{\overset{\|}{C}}-OCH_2CH_2-C_6H_5 + HCl$

e. $CH_3(CH_2)_4-\overset{O}{\overset{\|}{C}}-O^-Na^+ + Na^+ + Cl^- + H_2O$

f. $CH_3(CH_2)_4-\overset{O}{\overset{\|}{C}}-NH_2 + NH_4Cl$

8.41 a. [R] b. [R] c. $SOCl_2$ d. $NH(CH_3)_2$ e. [R]

8.43 $H_2C(-O-NO_2)-CH(-O-NO_2)-CH_2(-O-NO_2)$

H—C(H)—O—N(O)=O
H—C—O—N(O)=O
H—C(H)—O—N(O)=O

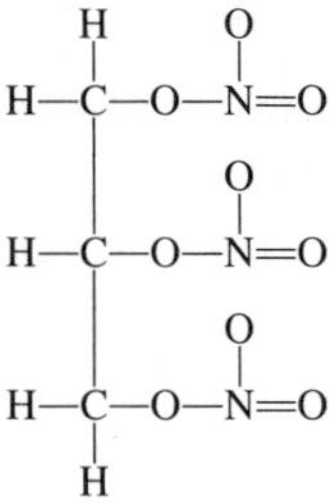

CHAPTER 9

9.1 a. 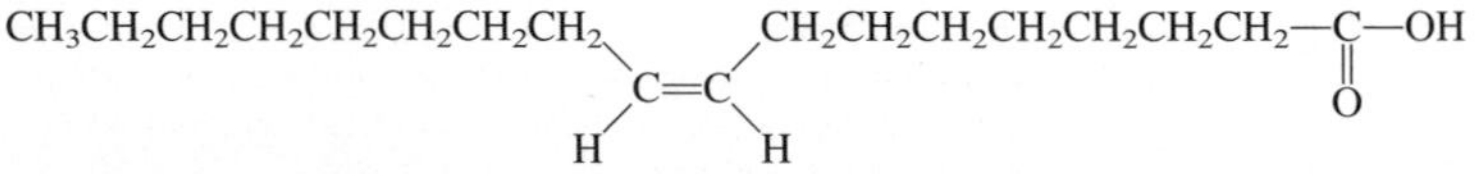 $CH_3CH_2CH_2CH_2CH_2CH_2CH_2CH_2-CH=CH-CH_2CH_2CH_2CH_2CH_2CH_2CH_2-\overset{O}{\overset{\|}{C}}-OH$ (cis; both H on same side)

b. $CH_3CH_2CH_2CH_2CH_2CH_2CH_2CH_2CH_2CH_2CH_2-\underset{O}{\underset{\|}{C}}-OH$

c. 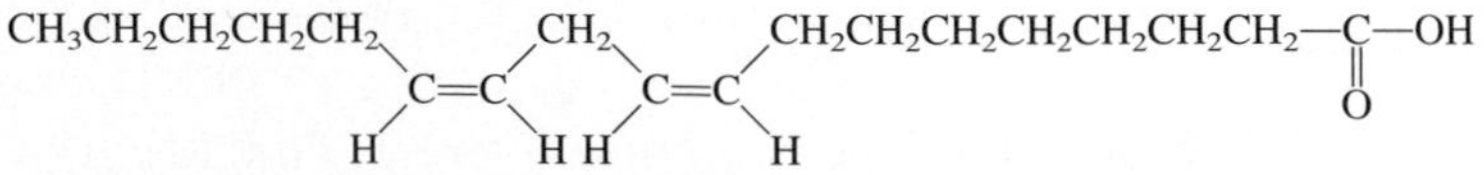 $CH_3CH_2CH_2CH_2CH_2-CH=CH-CH_2-CH=CH-CH_2CH_2CH_2CH_2CH_2CH_2CH_2-\underset{O}{\underset{\|}{C}}-OH$ (cis, cis)

d. $CH_3CH_2CH_2CH_2CH_2CH_2CH_2CH_2CH_2CH_2CH_2CH_2CH_2CH_2CH_2CH_2CH_2CH_2-\underset{O}{\underset{\|}{C}}-OH$

9.3 The melting points of unsaturated fatty acids are typically lower than those of saturated fatty acids. An unsaturated fatty acid is usually found as an oil at room temperature, whereas a saturated fatty acid is solid.

9.5 a. Esterification of oleic acid

$CH_3CH_2CH_2CH_2CH_2CH_2CH_2CH_2$ $\quad$ $CH_2CH_2CH_2CH_2CH_2CH_2CH_2{-}C{-}OH$
$C{=}C$ $\qquad$ ($C{=}O$)
$H \quad H$

$\downarrow$ ROH

$CH_3CH_2CH_2CH_2CH_2CH_2CH_2CH_2$ $\quad$ $CH_2CH_2CH_2CH_2CH_2CH_2CH_2{-}C{-}OR + H_2O$
$C{=}C$ $\qquad$ ($C{=}O$)
$H \quad H$

b. Acid-base reaction of oleic acid

$CH_3CH_2CH_2CH_2CH_2CH_2CH_2CH_2$ $\quad$ $CH_2CH_2CH_2CH_2CH_2CH_2CH_2{-}C{-}OH$
$C{=}C$ $\qquad$ ($C{=}O$)
$H \quad H$

$\downarrow$ NaOH

$CH_3CH_2CH_2CH_2CH_2CH_2CH_2CH_2$ $\quad$ $CH_2CH_2CH_2CH_2CH_2CH_2CH_2{-}C{-}O^-Na^+ + H_2O$
$C{=}C$ $\qquad$ ($C{=}O$)
$H \quad H$

a. Esterification of lauric acid

$CH_3CH_2CH_2CH_2CH_2CH_2CH_2CH_2CH_2CH_2CH_2{-}C{-}OH$ ($C{=}O$)

$\downarrow$ ROH

$CH_3CH_2CH_2CH_2CH_2CH_2CH_2CH_2CH_2CH_2CH_2{-}C{-}OR + H_2O$ ($C{=}O$)

b. Acid-base reaction of lauric acid

$CH_3CH_2CH_2CH_2CH_2CH_2CH_2CH_2CH_2CH_2CH_2{-}C{-}OH$ ($C{=}O$)

$\downarrow$ NaOH

$CH_3CH_2CH_2CH_2CH_2CH_2CH_2CH_2CH_2CH_2CH_2{-}C{-}O^-Na^+ + H_2O$ ($C{=}O$)

a. Esterification of linoleic acid

$CH_3CH_2CH_2CH_2CH_2$ $\quad$ CH_2 $\quad$ $CH_2CH_2CH_2CH_2CH_2CH_2CH_2{-}C{-}OH$
$C{=}C$ $\quad$ $C{=}C$ $\qquad$ ($C{=}O$)
$H \quad H\,H \quad H$

$\downarrow$ ROH

$CH_3CH_2CH_2CH_2CH_2$ $\quad$ CH_2 $\quad$ $CH_2CH_2CH_2CH_2CH_2CH_2CH_2{-}C{-}OR + H_2O$
$C{=}C$ $\quad$ $C{=}C$ $\qquad$ ($C{=}O$)
$H \quad H\,H \quad H$

b. Acid-base reaction of linoleic acid

$CH_3CH_2CH_2CH_2CH_2$ $\quad$ CH_2 $\quad$ $CH_2CH_2CH_2CH_2CH_2CH_2CH_2{-}C{-}OH$
$C{=}C$ $\quad$ $C{=}C$ $\qquad$ ($C{=}O$)
$H \quad H\,H \quad H$

$\downarrow$ NaOH

$CH_3CH_2CH_2CH_2CH_2$ $\quad$ CH_2 $\quad$ $CH_2CH_2CH_2CH_2CH_2CH_2CH_2{-}C{-}O^-Na^+ + H_2O$
$C{=}C$ $\quad$ $C{=}C$ $\qquad$ ($C{=}O$)
$H \quad H\,H \quad H$

a. Esterification of stearic acid

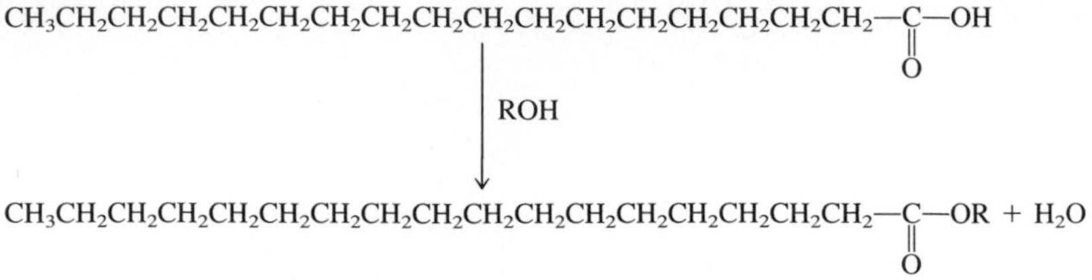

$$CH_3CH_2CH_2CH_2CH_2CH_2CH_2CH_2CH_2CH_2CH_2CH_2CH_2CH_2CH_2CH_2CH_2{-}\underset{\underset{O}{\|}}{C}{-}OH$$

$$\Big\downarrow \text{ROH}$$

$$CH_3CH_2CH_2CH_2CH_2CH_2CH_2CH_2CH_2CH_2CH_2CH_2CH_2CH_2CH_2CH_2CH_2{-}\underset{\underset{O}{\|}}{C}{-}OR + H_2O$$

b. Acid-base reaction of stearic acid

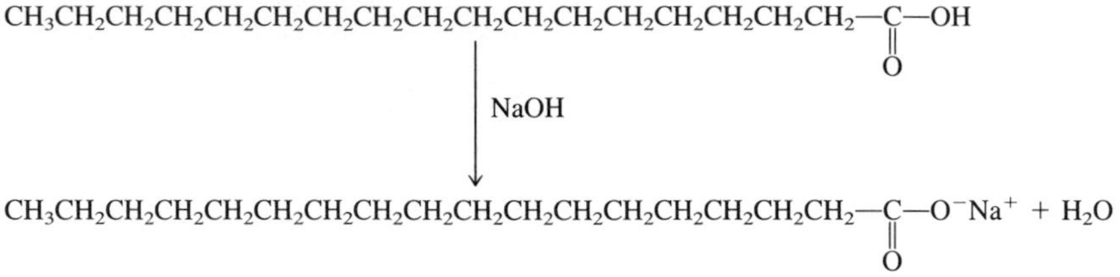

$$CH_3CH_2CH_2CH_2CH_2CH_2CH_2CH_2CH_2CH_2CH_2CH_2CH_2CH_2CH_2CH_2CH_2{-}\underset{\underset{O}{\|}}{C}{-}OH$$

$$\Big\downarrow \text{NaOH}$$

$$CH_3CH_2CH_2CH_2CH_2CH_2CH_2CH_2CH_2CH_2CH_2CH_2CH_2CH_2CH_2CH_2CH_2{-}\underset{\underset{O}{\|}}{C}{-}O^-Na^+ + H_2O$$

9.7 The monoglycerides will contain only one of the glycerol alcohols forming an ester linkage with the fatty acid:

a. 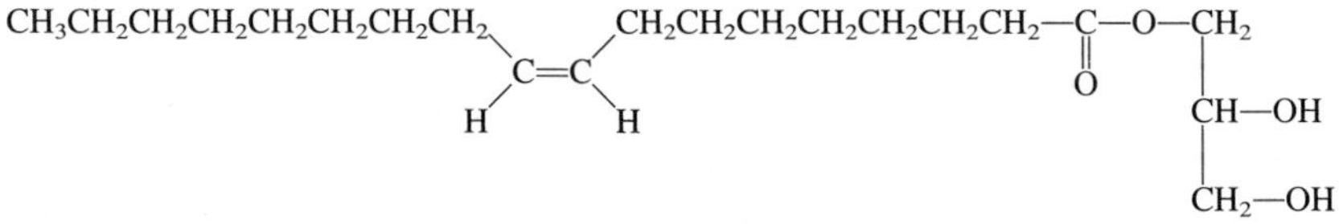 $CH_3CH_2CH_2CH_2CH_2CH_2CH_2CH_2(H)C{=}C(H)CH_2CH_2CH_2CH_2CH_2CH_2CH_2{-}\underset{\underset{O}{\|}}{C}{-}O{-}CH_2$
$\qquad\qquad\qquad\qquad\qquad\qquad\qquad\qquad\qquad\qquad\qquad\qquad\qquad |$
$\qquad\qquad\qquad\qquad\qquad\qquad\qquad\qquad\qquad\qquad\qquad\qquad\qquad CH{-}OH$
$\qquad\qquad\qquad\qquad\qquad\qquad\qquad\qquad\qquad\qquad\qquad\qquad\qquad |$
$\qquad\qquad\qquad\qquad\qquad\qquad\qquad\qquad\qquad\qquad\qquad\qquad\qquad CH_2{-}OH$

b. $CH_3CH_2CH_2CH_2CH_2CH_2CH_2CH_2CH_2{-}\underset{\underset{O}{\|}}{C}{-}O{-}CH_2$ (glycerol: $CH_2{-}O{-}$ / $CH{-}OH$ / $CH_2{-}OH$)

c. $CH_3CH_2CH_2CH_2CH_2CH_2CH_2CH_2CH_2CH_2CH_2CH_2CH_2CH_2CH_2{-}\underset{\underset{O}{\|}}{C}{-}O{-}CH_2$ (glycerol: $CH_2{-}O{-}$ / $CH{-}OH$ / $CH_2{-}OH$)

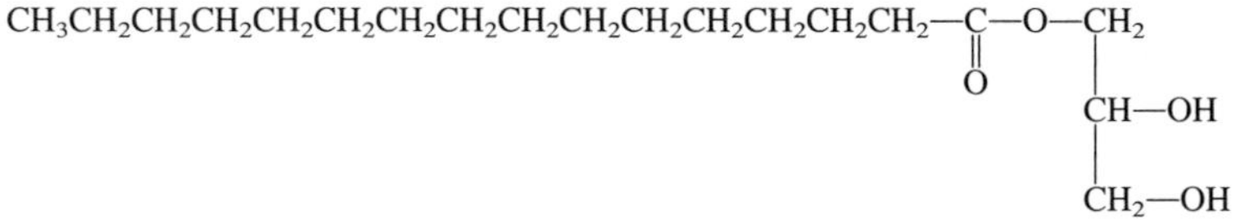

d. 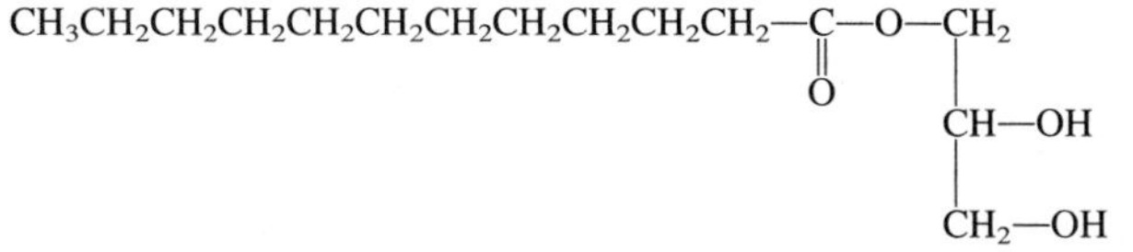 $CH_3CH_2CH_2CH_2CH_2CH_2CH_2CH_2CH_2CH_2CH_2{-}\underset{\underset{O}{\|}}{C}{-}O{-}CH_2$ (glycerol: $CH_2{-}O{-}$ / $CH{-}OH$ / $CH_2{-}OH$)

The diglycerides will have two of the glycerol alcohols forming ester linkages with two fatty acids:

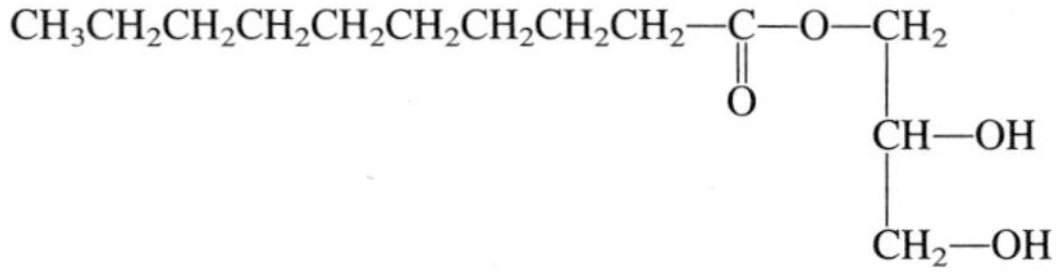

a. 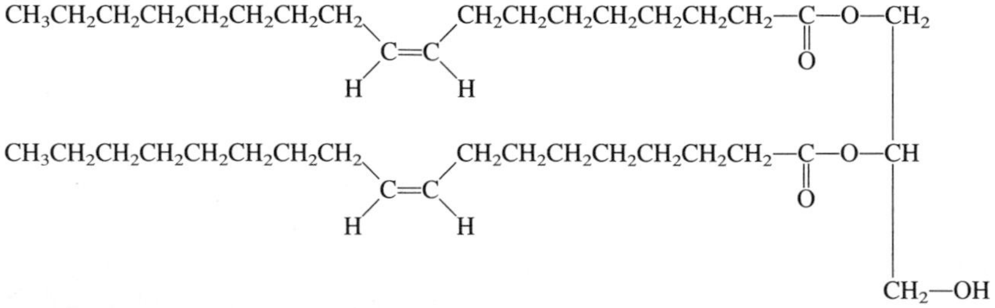 $CH_3CH_2CH_2CH_2CH_2CH_2CH_2CH_2(H)C{=}C(H)CH_2CH_2CH_2CH_2CH_2CH_2CH_2{-}\underset{\underset{O}{\|}}{C}{-}O{-}CH_2$

$CH_3CH_2CH_2CH_2CH_2CH_2CH_2CH_2(H)C{=}C(H)CH_2CH_2CH_2CH_2CH_2CH_2CH_2{-}\underset{\underset{O}{\|}}{C}{-}O{-}CH$

$CH_2{-}OH$

b. $CH_3CH_2CH_2CH_2CH_2CH_2CH_2CH_2CH_2{-}\underset{\underset{O}{\|}}{C}{-}O{-}CH_2$

$CH_3CH_2CH_2CH_2CH_2CH_2CH_2CH_2CH_2{-}\underset{\underset{O}{\|}}{C}{-}O{-}CH$

$CH_2{-}OH$

c. $CH_3CH_2CH_2CH_2CH_2CH_2CH_2CH_2CH_2CH_2CH_2CH_2CH_2CH_2CH_2{-}\underset{\underset{O}{\|}}{C}{-}O{-}CH_2$

$CH_3CH_2CH_2CH_2CH_2CH_2CH_2CH_2CH_2CH_2CH_2CH_2CH_2CH_2CH_2{-}\underset{\underset{O}{\|}}{C}{-}O{-}CH$

$CH_2{-}OH$

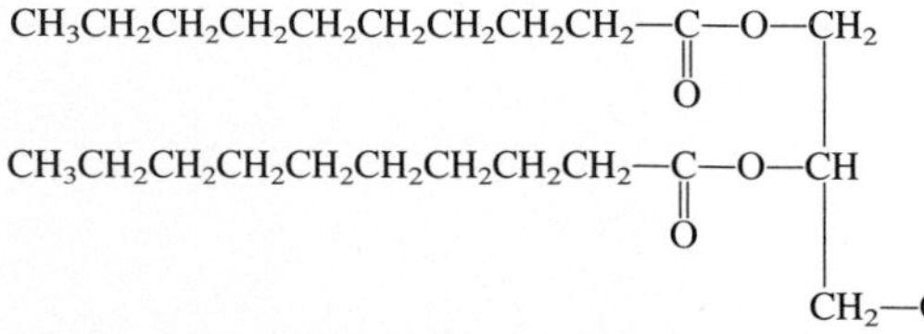

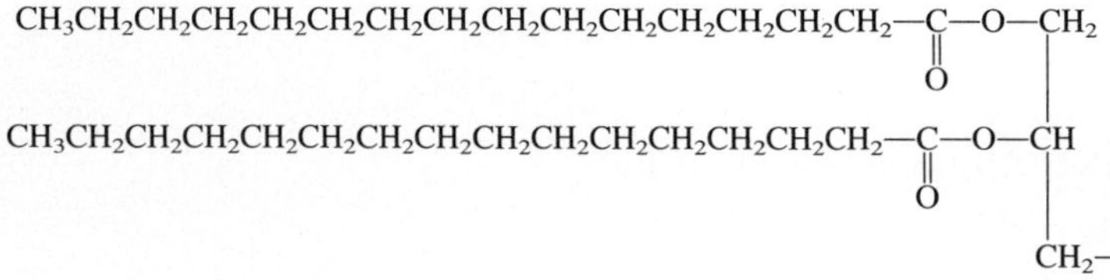

d. 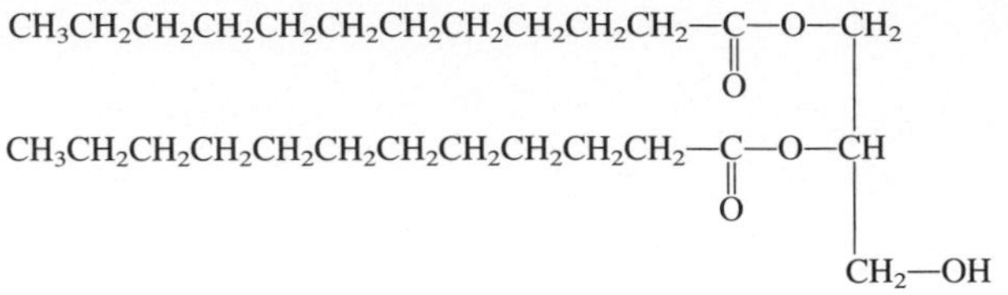 $CH_3CH_2CH_2CH_2CH_2CH_2CH_2CH_2CH_2CH_2CH_2-\underset{\|}{C}(=O)-O-CH_2$

$CH_3CH_2CH_2CH_2CH_2CH_2CH_2CH_2CH_2CH_2CH_2-C(=O)-O-CH$

CH_2-OH

The triglycerides would have all three of the glycerol alcohol groups forming ester linkages with three fatty acids:

a. 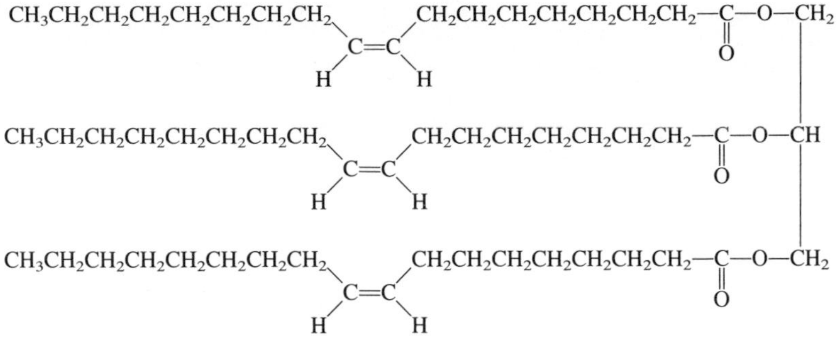 $CH_3CH_2CH_2CH_2CH_2CH_2CH_2CH_2(H)C=C(H)CH_2CH_2CH_2CH_2CH_2CH_2CH_2-C(=O)-O-CH_2$

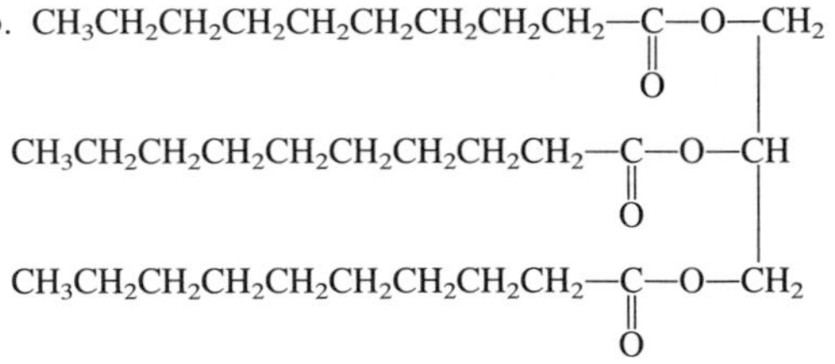

$CH_3CH_2CH_2CH_2CH_2CH_2CH_2CH_2(H)C=C(H)CH_2CH_2CH_2CH_2CH_2CH_2CH_2-C(=O)-O-CH$

$CH_3CH_2CH_2CH_2CH_2CH_2CH_2CH_2(H)C=C(H)CH_2CH_2CH_2CH_2CH_2CH_2CH_2-C(=O)-O-CH_2$

b. $CH_3CH_2CH_2CH_2CH_2CH_2CH_2CH_2CH_2-C(=O)-O-CH_2$

$CH_3CH_2CH_2CH_2CH_2CH_2CH_2CH_2CH_2-C(=O)-O-CH$

$CH_3CH_2CH_2CH_2CH_2CH_2CH_2CH_2CH_2-C(=O)-O-CH_2$

c. $CH_3CH_2CH_2CH_2CH_2CH_2CH_2CH_2CH_2CH_2CH_2CH_2CH_2CH_2CH_2-C(=O)-O-CH_2$

$CH_3CH_2CH_2CH_2CH_2CH_2CH_2CH_2CH_2CH_2CH_2CH_2CH_2CH_2CH_2-C(=O)-O-CH$

$CH_3CH_2CH_2CH_2CH_2CH_2CH_2CH_2CH_2CH_2CH_2CH_2CH_2CH_2CH_2-C(=O)-O-CH$

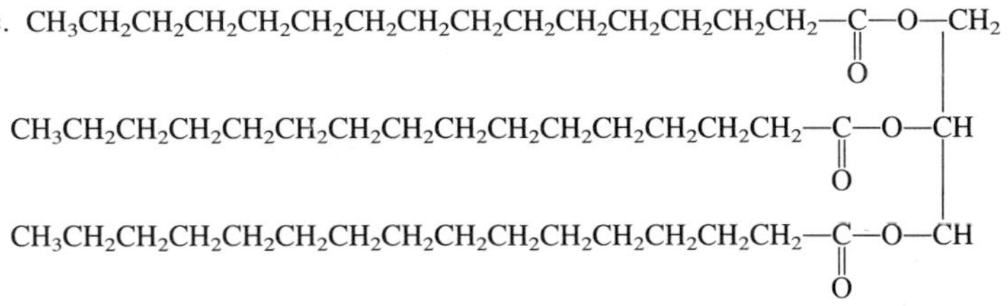

d. 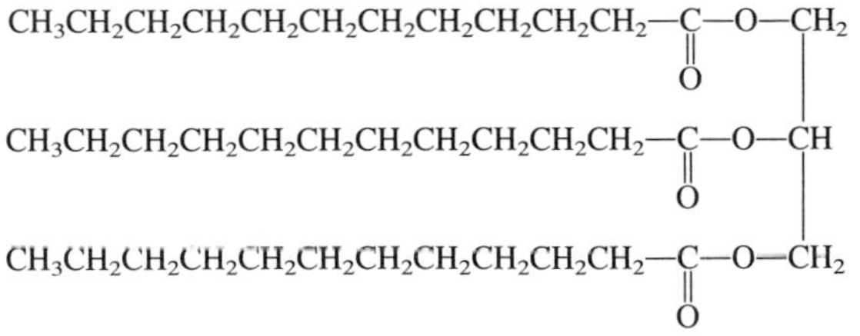 $CH_3CH_2CH_2CH_2CH_2CH_2CH_2CH_2CH_2CH_2CH_2-C(=O)-O-CH_2$

$CH_3CH_2CH_2CH_2CH_2CH_2CH_2CH_2CH_2CH_2CH_2-C(=O)-O-CH$

$CH_3CH_2CH_2CH_2CH_2CH_2CH_2CH_2CH_2CH_2CH_2-C(=O)-O-CH_2$

9.9 Fused rings have connecting sides.

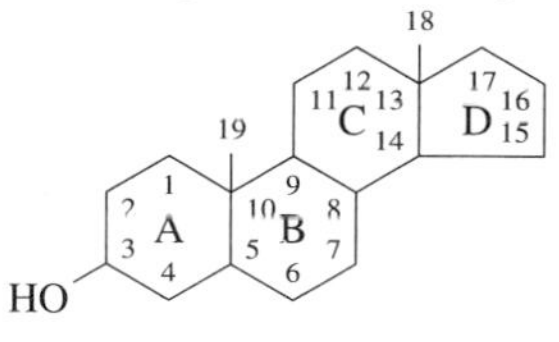

HO; rings A, B, C, D; positions 1–19

Steroid nucleus

9.11 Myricyl palmitate is the ester of the fatty acid palmitic acid, $C_{15}H_{31}COOH$, and the fatty alcohol myricol, $C_{30}H_{61}OH$.

9.13 The most important physical property differentiating decanoic acid, *trans*-5-decenoic acid, and *cis*-5-decenoic acid is the melting points of these compounds. Saturated fatty acids have higher melting points than unsaturated fatty acids; hence decanoic acid would have the highest melting point of these three fatty acids. For unsaturated fatty acids, the *cis* isomer would have a lower melting point than the *trans* isomer because it causes the hydrocarbon chain to bend. The bend in the hydrocarbon chain makes the formation of the solid more difficult.

9.15

```
CH3CH2CH2CH2CH2CH2CH2CH2       H
                        \     /                              O
                         C=C                                 ||
                        /     \                              
                       H       CH2CH2CH2CH2CH2CH2CH2—C—O—CH2
                                                              |
CH3CH2CH2CH2CH2CH2CH2CH2        CH2CH2CH2CH2CH2CH2CH2—C—O—CH
                        \      /                      ||    |
                         C=C                          O     |       O
                        /     \                             |       ||
                       H       H                            CH2—O—P—O⁻
                                                                    |
                                                                    O⁻
```

9.17 The function of essential fatty acids is to form arachidonic acid, a compound that is used by the body to make eicosanoids. The essential fatty acids cannot be manufactured by the body and must be supplied by one's diet.

9.19 Inflammation is a response to tissue damage and results in swelling, redness, fever, and pain. Prostaglandins promote fever and pain of the inflammatory response. Aspirin reduces inflammation by blocking the formation of prostaglandins.

9.21 Effects of prostaglandins are:

1. promotion of fever and pain in inflammatory response
2. promotion of the onset of labor during childbirth
3. inhibition of platelet aggregation (prevent clotting of blood)
4. inhibition of the secretion of stomach acid
5. dilation of blood vessels
6. promotion of bronchodilation

9.23

```
                        O
                        ||
          CH3(CH2)16—C—O—CH2
                            |
           O                |          Lecithin
           ||               |
CH3(CH2)16—C—O—CH
                  |
                  |     O              CH3
                  |     ||             | +
                  CH2—O—P—O—CH2CH2—N—CH3
                        |              |
                        O⁻             CH3
```

9.25

```
                                                                       O
                                                                       ||
CH3CH2CH2CH2CH2CH2CH2CH2CH2CH2CH2CH2CH2CH2CH2CH2CH2—C—O—CH2
  Stearic acid acyl tail                                               O      |
                                                                       ||     |
      CH3CH2CH2CH2CH2CH2CH2CH2CH2CH2CH2CH2CH2CH2CH2—C—O—CH
      Palmitic acid acyl tail                                          O      |
                                                                       ||     |
CH3CH2CH2CH2CH2CH2CH2CH2        CH2CH2CH2CH2CH2CH2CH2—C—O—CH2
                        \      /
  Oleic acid             C=C
  acyl tail             /     \
                       H       H
```

9.27 Cholesterol is found in the hydrophobic regions of the cell membrane and, as a result, controls the fluidity of the cell membrane.

9.29 The basic structure of a cell membrane is a bilayer of phospholipids with the hydrophobic fatty acyl groups facing the center of the bilayer and the polar ''heads'' of the ester portion of the phosholipids lining the exterior and interior of the cell membrane.

9.31 The lipid bilayer structure allows lipids to rapidly diffuse within the bilayer. The nonpolar hydrocarbon tails of the lipid are ''sandwiched'' between the polar heads of the lipid. This constructs both a polar and a nonpolar barrier as part of the cell membrane structure. Since lipids are mostly nonpolar, they will diffuse within the nonpolar portion of the lipid bilayer.

9.33 Triglycerides form three ester linkages at each of the three glycerol alcohol moieties. Phosopholipids form two ester linkages at two of the glycerol alcohol sites and a phosphate ester linkage at the third glycerol alcohol. Triglycerides are found in fat cells and serve as a source of reserve energy for the body. Phospholipids are found mainly in cell membranes.

9.35 The lipid bilayer is very structured. The nonpolar hydrocarbon tails are packed side by side in the middle of the bilayer, and the polar heads form a ''rigid'' shell on the surfaces of the cell membrane. This prevents proteins from ''flip-flopping'' or revolving within the membrane.

9.37 The four types of plasma lipoproteins are: chylomicrons, very low-density lipoproteins, low-density lipoproteins, and high-density lipoproteins.

9.39 Lysosomes are cell organelles that contain hydrolytic enzymes. Lysosomes fuse with endosomes carrying the low density lipoprotein complex. The hydrolytic enzymes digest the complex and cholesterol is released into the cell cytoplasm.

9.41 The inability of a person to synthesize receptors for plasma lipoproteins is a genetic defect that results in a greatly increased tendency toward atherosclerosis.

9.43 Simple diffusion involves the net movement of a solute across a membrane from an area of high concentration to an area of low concentration. In facilitated diffusion, a protein carrier, known as a permease, provides a channel through which the molecules diffuse.

9.45 Facilitated diffusion involves a protein carrier (permease) that provides a channel for diffusion of metabolites from an area of higher concentration to an area of lower concentration. No energy expenditure is necessary during this diffusion process. Active transport involves the movement of metabolites from an area of lower concentration to an area of higher concentration. Since this movement is against a concentration gradient, energy must be expended during the process.

9.47 One Na^+-K^+ ATPase molecule hydrolyzes one ATP molecule, allowing two K^+ ions to enter the cell for every three Na^+ ions that leave the cell.

9.49 Each permease has a charge and a conformation specific to the metabolite that it will transport (like a ''custom fit'').

9.51 The physiological effects of digitalis are an increase in the stroke volume, an increase in the strength of the contraction of the heart muscle, and a reduction of the heart rate.

9.53 Digitalis is called a cardiotonic steroid because it is a mixture of steroids that stimulate the heart to beat more efficiently.

9.55 Insulin increases the rate of glucose transport into cells by a factor of three to four.

9.57 The permeases actively transport metabolites such as glucose across a cell membrane by binding the metabolite to a stereospecific site on the protein. The three-dimensional structure of D-glucose will fit into the permease site; since L-glucose has a different three-dimensional arrangement, it will not fit into the permease site, nor will any molecule other than D-glucose.

9.59 O ... COOH

OH OH

Prostaglandin E_1

The letter designation, E, indicates the prostaglandin group and the number, 1, indicates the number of carbon-carbon double bonds.

9.61 Arachidonic acid

COOH + $4H_2$

↓ Ni catalyst

$CH_3(CH_2)_{18}$—C(=O)—OH

Arachidic acid

9.63 Cortisone is used in treatment of rheumatoid arthritis, asthma, and a variety of other diseases. Some side effects include fluid retention, congestive heart failure, gastric ulcers, and neurological symptoms.

CHAPTER 10

10.1 a. methanol b. ethylamine c. ethanol d. *n*-propylamine e. ethylene glycol f. ethylene glycol

10.3 The boiling points of methylamine, dimethylamine, and trimethylamine are −6.3°C, 7°C, and 3.5°C, respectively. As expected, the boiling point of dimethylamine is higher than that of methylamine because of its higher molecular weight. The boiling point of trimethylamine is *lower* than that of dimethylamine, even though it has a higher molecular weight. The lower boiling point is due to its lack of hydrogen bonding ability since there are no hydrogens attached to nitrogen.

10.5 diphenylamine: $(C_6H_5)_2$N—H triphenylamine: $(C_6H_5)_3$N

10.7 a. CH_3CH_2—N(CH_3)—CH_2CH_3

b. $CH_3CH_2CH(CH_3)$—N—$CH(CH_3)CH_2CH_3$ with N bearing $CH_3CHCH_2CH_3$ ($CH_3\overset{|}{C}HCH_2CH_3$)

c. $CH_3CH(NH_2)CH_2CH_2$—C(=O)—H

d. $CH_3CH(CH_3)CH_2CH_2$—NH_2

e. C_6H_5—N(H)—CH_3

f. CH_3CH_2—N(H)—$CH_2CH_2CH_2CH_2CH_2CH_3$

10.9 a. CH_3—C(=O)—NH_2 d. H—C(=O)—N(CH_3)—CH_2CH_3 or CH_3—N(CH_3)—C(=O)—CH_3

b. $CH_3CH_2CH(CH_3)$—NO_2 e. cyclopentyl—NO_2

c. 1-bromo-2-nitrobenzene (NO_2, Br on benzene ring) f. $CH_3CH_2CH_2CH_2CH_2$—NO_2 or $CH_3CH_2CH_2CH_2$—C(=O)—NH_2

10.11 a. NH_3 b. CH_3NH_2 c. $(CH_3)_2NH$

10.13

	I.U.P.A.C. Name	Common Name
a.	1-aminooctane	*n*-octylamine
b.	4-chloroaniline	*p*-chloroaniline
c.	2-aminobutane	*sec*-butylamine
d.	3-amino-3-methylhexane	

10.15 a. propanamide b. N,N-dimethylethanamide c. pentanamide d. 5-bromo-2,5-dimethylheptanamide e. 3-bromo-pentanamide f. 3-bromobenzamide

10.17 $CH_3CH_2CH_2CH_2$—NH_2 1-aminobutane is a 1° amine.

$CH_3CH_2CH(NH_2)CH_3$ 2-aminobutane is a 1° amine.

$CH_3CH_2CH_2NHCH_3$ N-methyl 1-aminopropane is a 2° amine.

$CH_3CH_2NHCH_2CH_3$ N-ethyl aminoethane is a 2° amine.

CH_3—CH($NHCH_3$)—CH_3 N-methyl 2-aminopropane is a 2° amine.

$CH_3CH(CH_3)CH_2$—NH_2 2-methyl-1-aminopropane is a 1° amine.

CH_3—C(CH_3)$_2$—NH_2 2-methyl-2-aminopropane is a 1° amine.

CH_3CH_2—N(CH_3)—CH_3 N,N-dimethyl aminoethane is a 3° amine.

10.19

Putrescine	Cadaverine
H_2N—$CH_2CH_2CH_2CH_2$—NH_2	H_2N—$CH_2CH_2CH_2CH_2CH_2$—NH_2
More polar, therefore more water soluble	Higher molecular weight, therefore higher boiling point

10.21 a. di-*n*-pentylamine:

$CH_3CH_2CH_2CH_2CH_2$—NH—$CH_2CH_2CH_2CH_2CH_3$

b. 3,4-dinitroaniline:

O_2N—$C_6H_3(NO_2)$—NH_2

c. 3,4-dimethyl-4-aminoheptane:

$$CH_3CH_2CH_2-\overset{NH_2}{\underset{CH_3}{C}}-\overset{CH_3}{C}HCH_2CH_3$$

d. *t*-butyl-*n*-pentylamine:

$$CH_3-\overset{CH_3}{\underset{CH_3}{C}}-NH-CH_2CH_2CH_2CH_2CH_3$$

e. 2,3-diaminohexane:

$$CH_3-\overset{NH_2}{C}H-\underset{NH_2}{C}H-CH_2CH_2CH_3$$

f. triethylammonium iodide:

$$\left[CH_3CH_2-\overset{CH_3CH_2}{\underset{CH_3CH_2}{\overset{+}{N}}}-H\right][I^-]$$

10.23 There are 17 isomeric amines with the molecular formula $C_5H_{13}N$. Each isomer has been named by using the I.U.P.A.C. system. Where appropriate, the common name is given in parentheses.

$CH_3CH_2CH_2CH_2CH_2-NH_2$ — 1-Aminopentane (*n*-pentylamine)

$CH_3CH_2CH_2\underset{NH_2}{C}HCH_3$ — 2-Aminopentane

$CH_3CH_2\underset{NH_2}{C}HCH_2CH_3$ — 3-Aminopentane

$CH_3CH_2\overset{CH_3}{C}HCH_2-NH_2$ — 2-Methyl-1-aminobutane

$CH_3\overset{CH_3}{C}HCH_2CH_2-NH_2$ — 3-Methyl-1-aminobutane

$CH_3\overset{CH_3}{C}H-\overset{CH_3}{C}H-NH_2$ — 2,3-Dimethyl-2-aminobutane

$CH_3CH_2-\overset{CH_3}{\underset{CH_3}{C}}-NH_2$ — 2-Methyl-2-aminobutane

$CH_3-\overset{CH_3}{\underset{CH_3}{C}}-CH_2-NH_2$ — 2,2-Dimethyl-1-aminopropane

$CH_3CH_2CH_2CH_2-NHCH_3$ — N-Methyl-1-aminobutane (*n*-butyl methylamine)

$CH_3CH_2\overset{CH_3}{C}H-NHCH_3$ — N-Methyl-2-aminobutane (*sec*-butyl methylamine)

$CH_3\overset{CH_3}{C}HCH_2-NHCH_3$ — N-Methyl-2-methyl-1-aminopropane (isobutyl methylamine)

$CH_3-\overset{CH_3}{\underset{CH_3}{C}}-NHCH_3$ — N-Methyl-2-methyl-2-aminopropane

$CH_3\overset{CH_3}{C}H-NH-CH_2CH_3$ — N-Ethyl-2-aminopropane (ethyl isopropylamine)

$CH_3CH_2CH_2-NH-CH_2CH_3$ — N-Ethyl-1-aminopropane (ethyl *n*-propylamine)

$CH_3CH_2-\underset{CH_3}{N}-CH_2CH_3$ — N-Ethyl-N-methyl-aminoethane (diethylmethylamine)

$CH_3\overset{CH_3}{C}H-\underset{CH_3}{N}-CH_3$ — N,N-Dimethyl-2-aminopropane (isopropyl dimethylamine)

$CH_3CH_2CH_2-\overset{CH_3}{N}-CH_3$ — N,N-Dimethyl-1-aminopropane (dimethyl *n*-propylamine)

10.25 The monomer unit for the condensation polymer is

$$HO-\overset{O}{\overset{\|}{C}}-CH_2CH_2CH_2-\overset{H}{N}-H$$

This allows water to be ''condensed'' between monomer units to form the polymer:

$$HO-\overset{O}{\overset{\|}{C}}-CH_2CH_2CH_2-\overset{H}{N}-\boxed{H\quad HO}-\overset{O}{\overset{\|}{C}}-CH_2CH_2CH_2-\overset{H}{N}-H$$

$$\downarrow$$

$$-\overset{O}{\overset{\|}{C}}-CH_2CH_2CH_2-\overset{H}{N}-\overset{O}{\overset{\|}{C}}-CH_2CH_2CH_2-\overset{H}{N}-$$

10.27

amide

Saccharin

10.29 The Lewis acid/base reaction between trimethylamine and hydrochloric acid is

$$(CH_3)_3N:+H^+Cl^- \rightleftharpoons (CH_3)_3N:H^+ + Cl^-$$

and, the Brönsted-Lowry acid/base reaction between trimethylamine and hydrochloric acid is

$$(CH_3)_3N + HCl \rightleftharpoons (CH_3)_3NH^+ + Cl^-$$

Proton acceptor (base) — Proton donor (acid) — Conjugate acid — Conjugate base

10.31 The stronger base is italicized:

a. $CH_3CH_2CH_2NH_2$ or *$(CH_3CH_2CH)_3N$*

3° amines are stronger bases than 2° or 1° amines.

b. *$CH_3CH_2CH_2NH_2$* or $NH_2-C_6H_5$

Aliphatic amines are stronger bases than aromatic amines.

c. *NH_3* or NH_4^+

Able to accept a proton — Unable to accept a proton

d. *$CH_3CH_2CH_2CH_2NH_2$* or $CH_3CH_2-O-CH_2CH_3$

More water soluble, therefore easier to accept a proton

10.33 1-butanol

10.35 A buffer consists of appreciable quantities of a weak acid and its salt or a weak base and its salt. Buffers have the capability of resisting pH change when quantities of acid or base are added.

10.37 a. 1° amine b. 1° amine c. 1° amine d. 1° amine e. 1° amine f. 1° amine g. 2° amine h. 3° amine i. 1° amine j. 2° amine

10.39 The increased water solubility of amine salts is due to their ionic character. Water is able to solvate the alkyl ammonium ion and its counter ion.

10.41 a. HCl

b. $CH_3CH_2-\overset{O}{\overset{\|}{C}}-NH_2 + HCl$

c. $C_6H_5-NH_2$

d. $CH_3CH_2CH_2CH_2CH_2CH_2C{\equiv}N$
e. $(CH_3)_3C{-}NH_3^+$, HSO_4^-
f. (1) NH_3 (2) CH_3NH_2

10.43 a. $SOCl_2$, followed by NH_3, followed by a reducing agent, [R]
b. reducing agent, [R] c. Br_2, $FeBr_3$ d. Br_2, $FeBr_3$, followed by $SOCl_2$, followed by NH_3, followed by a reducing agent, [R]
e. H_2SO_4 f. HCl

10.45 The strength of a base is proportional to the concentration of OH^- it can establish in water. This is controlled by the relative stability of its conjugate acid (the protonated amine). For aromatic amines, the protonated amine stability is less than that of the unprotonated amine. Aromatic compounds gain stability through conjugation present in the ring and side groups, such as the amine. The unprotonated amine has an electron pair that can be delocalized and conjugated with the aromatic ring. The protonated amine no longer has this electron pair available for conjugation. The lowered stability of the protonated amine results from decreased conjugation; thus it is a very weak base. The aliphatic amines are stronger bases because they behave purely as Lewis bases, accepting a proton from water.

10.47

$$\left[-HNCH_2CH_2CH_2CH_2CH_2{-}\overset{O}{\overset{\|}{C}}{-}NHCH_2CH_2CH_2CH_2CH_2{-}\overset{O}{\overset{\|}{C}}{-}\right]_n$$

10.49 The formation of barbituric acid from urea and diethyl malonate is a condensation reaction. The compound that is ''condensed'' is ethanol. The condensation results in the formation of amide bonds.

10.51 Aspirin (acetylsalicylic acid) contains a carboxylic acid group, which makes it acidic. Tylenol, acetominophen, is an amine that is alkaline. The acidity of aspirin causes irritation of the stomach lining, which results in an upset stomach.

10.53 $CH_3\overset{O}{\overset{\|}{C}}{-}NH{-}C_6H_4{-}OH$

CHAPTER 11

11.1

a. $H_3^+N{-}C(COO^-)(H){-}H$ — Glycine (Gly)

b. $H_2^+N{-}C(COO^-){-}H$ with ring $H_2C{-}CH_2{-}CH_2$ — Proline (Pro)

c. $H_3^+N{-}C(COO^-){-}H$, CH_2–phenyl — Phenylalanine (Phe)

d. $H_3^+N{-}C(COO^-){-}H$, CH_3 — Alanine (Ala)

e. $H_3^+N{-}C(COO^-){-}H$, $CH_2{-}SH$ — Cysteine (Cys)

f. $H_3^+N{-}C(COO^-){-}H$, $CH_2{-}S{-}S{-}CH_2$, $H_3^+N{-}C{-}H$, COO^- — Cystine (cys-cys)

11.3 The peptide bond consists of an amide group. An amide does not have free rotation around the carbon-nitrogen bond because of resonance.

11.5 The L prefix in the name of an amino acid indicates that the priority group (the amine) is on the left of a Fisher projection. The L comes from the Latin *levo*, meaning left.

11.7 The α-amino acids are chiral because the α-carbon is bonded to four different groups.

11.9 The ten amino acids that have hydrophobic side chains are

Glycine: $+NH_3{-}C(H)(H){-}COO^-$

Alanine: $+NH_3{-}C(H)(CH_3){-}COO^-$

Valine: $+NH_3{-}C(H){-}COO^-$, $CH(H_3C)(CH_3)$

Isoleucine: $+NH_3{-}C(H){-}COO^-$, $CH_3{-}CH{-}CH_2CH_3$

Leucine: $+NH_3{-}C(H){-}COO^-$, $CH_2{-}CH(H_3C)(CH_3)$

Methionine: $+NH_3{-}C(H){-}COO^-$, $CH_2{-}CH_2{-}SCH_3$

Phenylalanine: $+NH_3{-}C(H){-}COO^-$, CH_2–phenyl

Tryptophan: $+NH_3{-}C(H){-}COO^-$, CH_2–indole (N–H)

11.11 The oxidation of two cysteine residues produces cystine:

$$+NH_3{-}C(H)(CH_2SH){-}COO^- + +NH_3{-}C(H)(CH_2SH){-}COO^- \xrightarrow{[O]} +NH_3{-}C(H){-}COO^-,\ CH_2{-}S{-}S{-}CH_2,\ +NH_3{-}C(H){-}COO^-$$

11.13 a. his-trp-cys:

$$+NH_3{-}\underset{CH_2}{\overset{H}{C}}{-}\overset{O}{\overset{\|}{C}}{-}\overset{H}{N}{-}\underset{CH_2}{\overset{H}{C}}{-}\overset{O}{\overset{\|}{C}}{-}\overset{H}{N}{-}\underset{CH_2SH}{\overset{H}{C}}{-}COO^-$$

(side chains: imidazole HN/+HN; indole NH; SH)

b. gly-leu-ser:

$$+NH_3{-}\underset{H}{\overset{H}{C}}{-}\overset{O}{\overset{\|}{C}}{-}\overset{H}{N}{-}\underset{CH_2}{\overset{H}{C}}{-}\overset{O}{\overset{\|}{C}}{-}\overset{H}{N}{-}\underset{CH_2}{\overset{H}{C}}{-}COO^-$$

(side chains: $CH_2{-}CH(H_3C)(CH_3)$; $CH_2{-}OH$)

c. arg-ile-val

$$^{+}NH_3-CH(CH_2CH_2CH_2NH{-}C(=NH_2^{+})NH_2)-C(=O)-NH-CH(CH(CH_3)CH_2CH_3)-C(=O)-NH-CH(CH(CH_3)_2)-COO^{-}$$

11.15 $^{+}HN_3-CH(CH_3)-C(=O)-NH-CH(CH_2OH)-COO^{-}$ The peptide bond is the amide linkage shown in bold.

11.17 a. *wool fibers:* the most prominent structural feature is the α-helical secondary structure of the peptides.
b. *silk fibroin:* the most prominent structural feature is an antiparallel β-pleated sheet.

11.19 In an α-helix, each carbonyl oxygen is hydrogen bonded to an amide hydrogen four residues away in the chain. The hydrogen bonds are parallel to the long axis of the helix. The helix spirals in a right-handed fashion. These hydrogen-bonding interactions hold the helix in place. In a β-pleated sheet, the peptide chain is almost completely extended. Each carbonyl oxygen and amide hydrogen are participating in hydrogen-bonding interactions. The overall appearance resembles the folds of a pleated drapery.

11.21 A conjugated protein is a peptide that can function only if it is bound to another chemical group.

11.23 Tropocollagen consists of a triple-stranded helix.

11.25 $^{+}H_3N-CH(CH_2CH_2CH(OH)CH_2NH_3^{+})-COO^{-}$ Hydroxylysine; HO-substituted pyrrolidinium ring ($N^{+}H_2$) bearing COO^{-}

Hydroxylysine

11.27 The oxidation of myoglobin would involve the loss of an electron from the iron(II) present to form iron(III). Oxygenation of myoglobin would involve the binding of oxygen (O_2) to myoglobin (transfered from hemoglobin).

11.29 Sickle cell anemia trait is carried in the genes of an individual. If the individual has inherited the sickle cell gene from both parents, the disease will be manifest. If the gene is inherited from only one parent, that individual will carry the sickle cell trait but will not be severely affected by the disease because he or she also has a copy of the gene for normal β-hemoglobin.

11.31 Immunoglobulin G contains four peptide chains linked by disulfide bonds. It has a Y shape with an antigen binding site at each tip of the Y.

11.33 The major function of IgA is the protection of mucous membranes of the gut, oral cavity, and genitourinary tract.

11.35 IgM is the first antibody produced by the immune system when an organism has invaded the body. IgD regulates the synthesis of antibodies. IgE is produced as an allergic response to dust and pollen.

11.37 The B cell is a specialized white blood cell that produces the antibodies needed to fight an infection.

11.39 A single antigen can bind to many different immunoglobulins because it has many different antigenic sites.

11.41 The primary sequence of a protein is encoded in the genes. This code is translated during protein synthesis. The R groups of the amino acid in the primary structure dictate the way the protein will fold.

11.43 The coiled structure of fibrous proteins accounts for their water insolubility and their mechanical strength (elasticity).

11.45 a. *α-keratins:* the major structural feature is the three-stranded protofibril.
b. *silk fibroin:* the major structural feature is the antiparallel β-pleated sheet.
c. *collagen:* the major structural feature consists of three left-handed helices wrapped in a right-handed fashion.
d. *myosin:* the major structural feature consists of two α-helices coiled around one another.

11.47 The resonance forms of the amide bond result in resonance stability which accounts for the rigidity of the peptide bond:

$$-C(=O)-NH- \longleftrightarrow -C(-O^{-})=NH^{+}-$$

11.49 Valine is a neutral, nonpolar residue. Glutamate is an acidic, polar residue. Replacement of valine by glutamate radically alters the hydrophobic interaction of that portion of the protein. As a result, the folding of the protein is altered leading to polymerization of hemoglobin molecules when oxygen is released. The long polymers of hemoglobin cause sickling of the cells. This defect leads to the disorder called sickle cell anemia.

11.51 Hydrogen bonding interactions are responsible for the secondary and, in part, for the tertiary folding of proteins.

11.53 a. A protein is active only when folded in its proper configuration. One factor influencing protein folding is pH. Pepsin is properly folded at pH 2, while trypsin is properly folded at pH 8.5.
b. The pH of one's stomach is acidic. An enzyme such as pepsin must be active in this environment. The pH of one's intestines is alkaline. An enzyme such as trypsin must be active in this environment.

CHAPTER 12

12.1 a. sucrose b. pyruvate c. succinate

12.3 a. transferase b. ligase c. isomerase d. oxireductase e. transferase f. hydrolase

12.5 The induced fit model assumes that the enzyme is flexible. Both the enzyme and the substrate are able to change shape to form the enzyme-substrate complex. The lock-and-key model assumes that the enzyme is inflexible (the lock) and the substrate (the key) fits into a specific rigid site on the enzyme to form the enzyme-substrate complex.

12.7 Coenzymes that function in the body's metabolism are often modified water-soluble vitamins.

12.9 A decrease in pH tends to denature a protein. Since a change in the conformation of the protein will drastically alter its active site, it will no longer be able to bind the substrate for which it was specifically designed, and activity decreases.

12.11 Irreversible inhibitors block the active site of an enzyme and eliminate catalysis at that site. This disrupts the metabolic pathway, and the cell cannot function. The binding of the inhibitor is often tighter than the normal substrate.

12.13 Structural analogues can mimic an enzyme substrate because of their similarity in structure to the substrate and thus act as an inhibitor.

12.15 Chymotrypsin cleaves peptide bonds on the carbonyl side of aromatic residues.
a. ala-phe-ala: the bond between phenylalanine (carbonyl side) and the second alanine will be cleaved.
b. tyr-ala-tyr: the bond between the first tyrosine (carbonyl side) and alanine will be cleaved.
c. trp-val-gly: the bond between tryptophan (carbonyl side) and valine will be cleaved.
d. phe-ala-pro: the bond between phenylalanine (carbonyl side) and alanine will be cleaved.

12.17 There are three major types of hemophilia. Hemophilia A is the most common and results from the production of abnormal clotting factor VIII. Hemophilia B results from a lack of clotting factor IX. Hemophilia C results from a lack of clotting factor XI. Both hemophilia A and B are sex linked (genes are on the X-chromosome). The gene for hemophilia C is on another chromosome. All three forms of hemophilia are passed on genetically.

12.19 green, leafy vegetables and liver, intestinal bacteria

12.21

Substrate	**Enzyme**
1. urea	e. urease
2. hydrogen peroxide	c. peroxidase
3. lipid	a. lipase
4. aspartic acid	f. aspartase
5. glucose-6-phosphate	b. glucose-6-phosphatase
6. sucrose	d. sucrase

12.23 a. cleaving a carboxylic acid group from citrate b. adds a phosphate group to tyrosine c. reduction of oxalate d. oxidation of nitrite e. interconversion of *cis* and *trans* isomers

12.25 A general enzyme-mediated equilibrium could be described as

$$\text{enzyme} + \text{substrate} \rightleftharpoons \text{enzyme} + \text{product}$$

$$K_{eq} = \frac{[\text{product}]}{[\text{substrate}]}$$

An enzyme catalyzes a reaction by lowering the activation energy for the reaction. The equilibrium constant, K_{eq}, is not altered because both the rate of the forward reaction and the rate of the reverse reaction are increased. The enzyme allows equilibrium to be reached faster. The enzyme is always regenerated during the reaction and is not consumed.

12.27 Doubling the substrate concentration will double the reaction rate.

12.29 a.

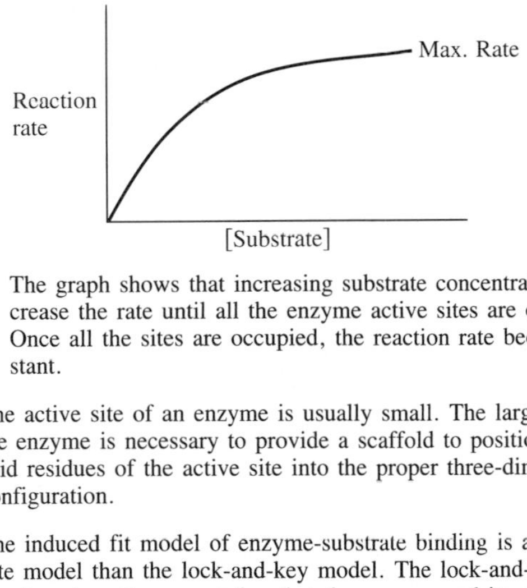

Max. Rate

Reaction rate

[Substrate]

b. The graph shows that increasing substrate concentration will increase the rate until all the enzyme active sites are occupied. Once all the sites are occupied, the reaction rate becomes constant.

12.31 The active site of an enzyme is usually small. The large structure of the enzyme is necessary to provide a scaffold to position the amino acid residues of the active site into the proper three-dimensional configuration.

12.33 The induced fit model of enzyme-substrate binding is a more accurate model than the lock-and-key model. The lock-and-key model was based on the assumption that the enzyme and its active site were rigid. Enzymes have been shown to be flexible molecules as opposed to rigid ones. The induced fit model assumes the flexible enzyme has a specific combination of charged polar and nonpolar groups. The substrate will have complementary chemical groups to fit the active site. The active site will conform to the substrate in a flexible manner.

12.35

$$E + S \xrightleftharpoons{\text{Step I}} ES \xrightleftharpoons{\text{Step II}} ES^{*} \xrightleftharpoons{\text{Step III}} EP \xrightleftharpoons{\text{Step IV}} E + P$$

Step I involves the formation of the enzyme-substrate complex. Step II involves the formation of the transition state from the enzyme-substrate complex. Step III involves the formation of the enzyme-product complex from the transition state. Step IV involves the formation and release of the product and regeneration of the enzyme.

12.37 maintaining the proper configuration of the active site

12.39

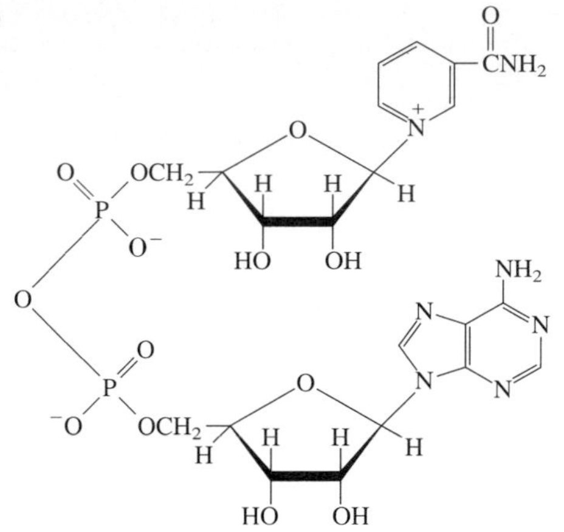

O, CNH$_2$, N$^+$, O, OCH$_2$, H, H, H, H, P, O$^-$, HO, OH, NH$_2$, O, N, N, O, N, N, P, O, $^-$O, OCH$_2$, H, H, H, H, HO, OH

NAD$^+$ is a carrier of hydride anions; oxireductase.

12.41 a. true b. false c. true d. true e. true

12.43 The optimum pH for enzyme activity corresponds to the pH at which the rate of the reaction is maximized. For most enzymes, this pH would correspond to a physiological pH of 7. A few enzymes must operate in different environments, such as the stomach, where pH $<<$ 7.

12.45 In general, an increase in temperature will increase a reaction rate. For enzymes, this holds true until higher temperatures are reached. At high temperatures, a protein will denature (unfold), upon which the reaction rate will decrease or the reaction will cease altogether. The folding of a protein is responsible for the shape of the active site. If the protein unfolds, the active sites are no longer available for catalysis.

12.47 Lysosomal enzymes are found within the lysosome, which maintains an internal pH of 4.8. Thus lysosomal enzymes operate maximally under these pH conditions. These enzymes, if released into the cytoplasm, could destroy cellular components needed for the cell's survival. The cell protects itself from accidental release of lysosomal enzymes by maintaining a cytoplasmic pH of 7.0–7.3. If any lysosomal enzyme is released into the cytoplasm, it becomes inactive owing to the drastic change from its optimal pH.

12.49 Many cell reactions are equilibrium reactions. If an excess of product is made, the reaction proceeds in reverse to relieve stress on the equilibrium (LeChatelier's Principle).

12.51 Negative allosterism involves the conversion of an active site to an inactive configuration. Positive allosterism involves the conversion of an active site to an active configuration.

12.53 A zymogen is an enzyme that is initially inactive but is eventually converted to an active form by proteolytic cleavage.

12.55 The compound will behave as a competitive, reversible inhibitor. Competitive, reversible inhibitors are structural analogues that resemble a substrate and are able to fit into the active site but do not initiate a reaction.

12.57 Competitive, reversible enzyme inhibitors are structural analogues that resemble a substrate and are able to fit into the active site but do not initiate a reaction.

12.59 Chymotrypsin, trypsin, and elastase all hydrolyze peptide bonds.

12.61 Chymotrypsin cleaves on the carbonyl side of aromatic residues. Specifically, in this example chymotrypsin cleaves on the carbonyl side of tryptophan.

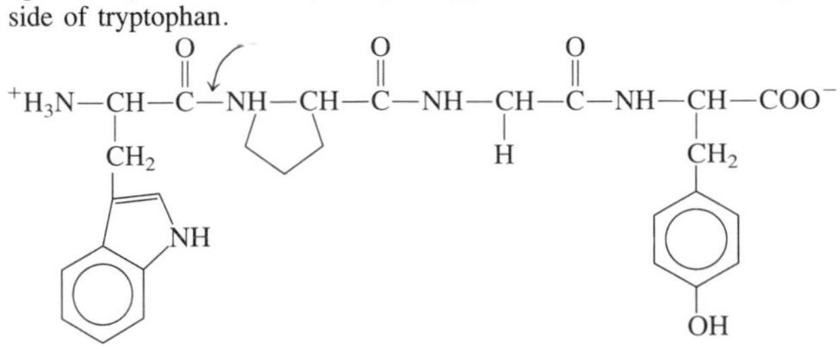

^+H_3N—CH—C(=O)—NH—CH—C(=O)—NH—CH—C(=O)—NH—CH—COO$^-$

CH$_2$, NH, H, CH$_2$, OH

12.63 Chymotrypsin, trypsin, and elastase have different pockets for the amino acid side chains of their substrates. This accounts for the difference in specificity. That results in cleavage of peptide chains at different amino acid residues.

12.65 The structure of γ-carboxyglutamic acid in a complex with Ca^{2+} is

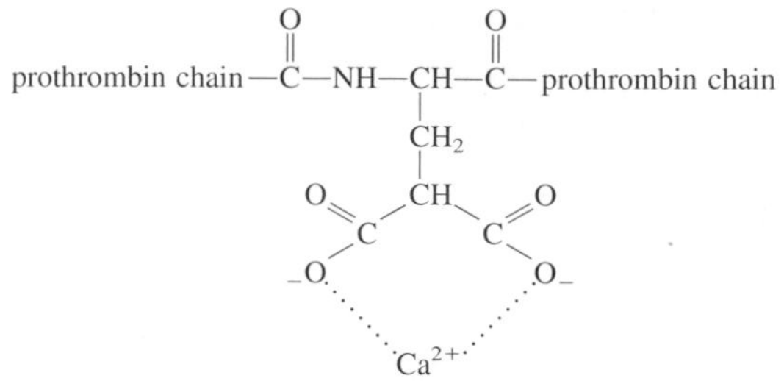

```
                    O            O
                    ‖            ‖
prothrombin chain—C—NH—CH—C—prothrombin chain
                         |
                        CH2
                         |
                O       CH       O
                 ⸌C⸍        ⸌C⸍
                 |            |
               ⁻O             O⁻
                  ·.        .·
                     Ca2+
```

12.67 A compound in spoiled sweet clover acts as an anticoagulant by behaving as a competitive inhibitor of vitamin K.

12.69 creatine phosphokinase (CPK), lactate dehydrogenase (LDH), and aspartate aminotransferase (AST/SGOT)

12.71 A catalyst does not alter the equilibrium constant because it affects both the forward and reverse reaction rates in the same way.

12.73 Refrigeration lowers their activity and thus preserves their shelf life.

12.75 Succinylcholine acts as a structural analogue of acetylcholine and competitively inhibits receptor binding. This inhibits muscle contraction and thus acts as a muscle relaxer.

12.77 Organofluorophosphates are nerve poisons that act as irreversible, noncompetitive inhibitors. They bind covalently to the active site of acetylcholinesterase and inactivate this enzyme.

12.79 The ELISA test for AIDS involves an enzyme-immunoassay that makes use of the specificity of antigen-antibody binding and an enzyme to detect the presence of the AIDS virus. First, viral antigens are produced by growing the virus in tissue culture and purifying the viral proteins. A series of diluted samples of the patient's serum are introduced to the antigen. If HIV antibodies are present, they will bind the viral antigens. To make the complex formed visible, an enzyme bound to a second antibody is introduced into the reaction mixture, followed by a substrate that would produce a colored product. The enzyme that is often used is horseradish peroxidase, and its substrate is ortho-phenylenediamine.

CHAPTER 13

13.1 The generation of ATP by glycolysis is unique in that it occurs by substrate-level phosphorylation.

13.3 Since all organisms use glucose as an energy source for the glycolysis pathway, it must have been one of the first pathways to evolve.

13.5 Both are anaerobic processes. Both processes reoxidize the NADH produced in glycolysis by donating the electrons to pyruvate or to a product produced from pyruvate. Because pyruvate is utilized and NADH is reoxidized, the glycolytic pathway can continue to function and produce ATP.

13.7 Glucagon is a hormone released by the pancreas when blood sugar levels are depleted. It functions by altering the activity of glycogen phosphorylase and glycogen synthetase. Glycogen phosphorylase is activated to produce more glucose, and glycogen synthetase is inactivated.

13.9 Insulin is produced by the pancreas in response to a rise in blood glucose levels. It causes an increase in the uptake of glucose in the liver, heart, skeletal muscle, and adipose tissue. It is in the liver where insulin induces glycogen synthesis and storage by regulating the enzymes involved in glycogen metabolism.

13.11 Acetyl CoA production from pyruvate requires four vitamins: vitamin B_1, vitamin B_2, niacin, and pantothenic acid.

13.13 Five cellular functions that require ATP energy are: 1. movement of ions into and out of the cell 2. movement of ions into and out of cellular organelles 3. biosynthesis 4. contraction of muscles 5. movement of cilia

13.15 cause pyruvate to build up, which would, in turn, inhibit energy production by glycolysis

13.17 2 ATP/glucose

13.19 aerobic conditions

13.21 The balanced equation for glycolysis is

$$C_6H_{12}O_6 + 2\ ADP + 2\ P_i + 2\ NAD^+ \longrightarrow 2C_3H_3O_3 + 2\ ATP + 2\ NADH + 2H_2O$$

13.23 Since seals must spend a considerable amount of time submerged and are thus unable to replenish their oxygen, much of their metabolism of glucose will be under anaerobic conditions. Anaerobic conditions lead to lactate formation.

13.25 NADH is formed during oxidation of PGAL, and NADH is reoxidized to NAD^+ when lactate is formed from pyruvate.

13.27 fructose-6-phosphate + ATP → fructose-1,6-diphosphate + ADP + H^+

13.29 produce glucose from noncarbohydrate precursors in order to replenish liver and muscle glycogen stores or to raise blood glucose levels.

13.31 pyruvate

13.33 The conversion of pyruvate to phosphoenolpyruvate in gluconeogenesis is achieved by two enzymes: pyruvate carboxylase and phosphoenolpyruvate carboxykinase. The former carboxylates pyruvate using atmospheric CO_2, and the latter removes the CO_2 and adds a phosphoryl group.

13.35 pyrophosphorylase or glycogen synthetase

13.37 If an individual is unable to produce the enzymes or hormones that regulate glycogenolysis, hypoglycemia results.

13.39 Insulin stimulates the enzymes glucokinase and glycogen synthetase. As a result, glycogen production and storage in the liver are increased. An increase in glycogen production will lower blood glucose levels.

13.41
a. synthesis of glucose from noncarbohydrate precursors
b. any metabolic reactions in the cell that break down molecules for the production of energy
c. the metabolic pathway that breaks down glucose into pyruvate and generates energy for the cell via ATP and NADH production
d. metabolic pathway that requires oxygen
e. metabolic pathway that requires no oxygen
f. any metabolic reaction or pathway with pyruvate as the initial substrate and that results in reoxidation of NADH and production of a stable fermentation end product

13.43 liver

13.45 Steps 1, 3, and 9 in glycolysis are irreversible. Step 1 involves the production of glucose-6-phosphate from glucose. Step 3 involves the production of fructose-1,6-diphosphate from fructose-6-phosphate. Step 9 involves the production of pyruvate and ATP from phosphoenolpyruvate.

13.47
a. phosphorylation of glucose to glucose-6-phosphate in the first step of glycolysis
b. transfer of a phosphoryl group from phosphoenolpyruvate to ADP in the final step of glycolysis
c. transfer of a phosphoryl group on the third carbon of 3-phosphoglycerate to the second carbon to produce 2-phosphoglycerate in Step 7 of glycolysis
d. oxidation of the aldehyde group of glyceraldehyde-3-phosphate to a carboxylic acid in Step 5 of glycolysis

13.49 Von Gierke's disease is a genetic defect in the enzyme glucose-6-phosphatase. If a person lacks this enzyme, the liver cannot produce sufficient blood glucose. Excess glucose-6-phosphate builds up in the liver, causing enlargement of the liver.

13.51 During fermentation, CO_2 is produced. If fermentation takes place in a sealed bottle, enough pressure will eventually build up to cause the bottle to explode.

CHAPTER 14

14.1 Mitochondria serve as the "powerhouse" of a cell. Most of the cell's energy production occurs in the mitochondria.

14.3

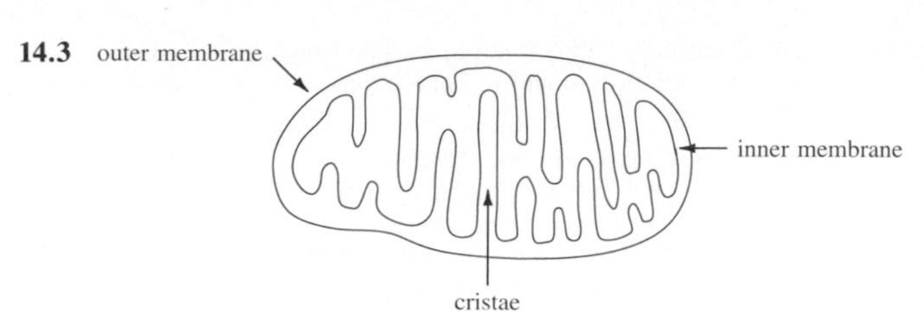

14.5 Compartmentalization allows for diverse and independent metabolic processes to occur simultaneously in the cell without interference.

14.7 $NAD^+ + H^- \longrightarrow NADH$

14.9 The citric acid cycle is regulated to provide efficient energy production in the cell. Energy production should increase when required by the cell. Conversely, energy production should decrease when demands by the cell are not as great in order to conserve energy.

14.11 The first step in the degradation of aspartate is the removal of the α-amino group using the enzyme aspartate aminotransferase. In the process aspartate is converted into oxaloacetate and α-ketoglutarate accepts the amino group to become glutamate. Oxaloacetate enters the citric acid cycle directly.

14.13 A decrease in the amount of oxaloacetate in the citric acid cycle will not critically affect energy production in a cell because a secondary source is available from the urea cycle. Fumarate from the urea cycle is capable of producing oxaloacetate in the citric acid cycle.

14.15 replenish a substrate needed for a pathway

14.17 area between the outer and inner mitochondrial membranes that serves as a proton reservoir that drives ATP synthesis

14.19 The inner membrane of the mitochondria is highly folded and is the site of electron transport and ATP synthesis. The outer mitochondrial membrane separates the mitochondria from cellular cytoplasm and is highly permeable to ''food'' molecules.

14.21 transport proteins, respiratory electron transport proteins, and ATP synthetase.

14.23 a. false b. false c. true d. true

14.25 Three NAD^+ ions are reduced to NADH in one turn of the citric acid cycle.

14.27 CO_2

14.29 Nucleotide diphosphokinase catalyzes the transfer of a phosphoryl group from GTP to ADP. This generates ATP and GDP.

14.31 36 molecules of ATP

14.33 Oxaloacetate is the ''starting point'' for the citric acid cycle. It accepts an acetyl group from acetyl CoA to form citrate.

14.35 Aminotransferases remove the α-amino group from an amino acid during degradation and transfer it to an α-keto acid

14.37 The glutamate family of aminotransferases is important because the α-keto acid corresponding to glutamate is α-ketoglutarate. α-Ketoglutarate is a citric acid cycle intermediate.

14.39 a. pyruvate b. α-ketoglutarate c. oxaloacetate

14.41 Two ATP are produced from each $FADH_2$.

14.43 ATP synthase

14.45 Respiratory electron transport chain complexes are found in the inner mitochondrial membrane.

14.47 Alanine enters into the citric acid cycle via pyruvate and acetyl CoA and thus 15 ATP molecules will be produced.

14.49 The urea cycle occurs only in the liver because the enzyme arginase, which catalyzes the last step, is only found in the liver.

14.51 Fumarate couples the urea cycle and the citric acid cycle.

14.53 a. The source of the amino groups is NH_4^+ from the blood and the amino acid aspartate.
b. The source of the carbonyl group is CO_2.

14.55 Glycolysis produces 2 ATP molecules per molecule of glucose. The citric acid cycle, along with oxidative phosphorylation, produces 36 ATP molecules upon complete oxidation. The citric acid cycle is an aerobic process that occurs in mitochondria. Glycolysis takes place in the cytoplasm of a cell and is an anaerobic process.

14.57 fast twitch; slow twitch

14.59 The pyruvate dehydrogenase complex is inhibited by NADH, ATP, and acetyl CoA. Citrate synthase is inhibited by high concentrations of ATP. Isocitrate dehydrogenase is inhibited by ATP and NADH and stimulated by ADP. The α-ketoglutarate dehydrogenase complex is inhibited by succinyl CoA, NADH, and ATP.

CHAPTER 15

15.1 Fatty acids are mainly stored in the form of triglycerides in adipose tissue.

15.3 The ω (omega) of ω-phenyl-labeled fatty acids indicates that the terminal methyl group of the fatty acid has been replaced by a phenyl group.

15.5 Ketone bodies result when oxaloacetate concentrations are too low to allow acetyl CoA to enter the citric acid cycle. This condition often occurs during starvation, with a low-carbohydrate diet, or with the disease diabetes mellitus.

15.7 An adipocyte is a fat cell. It contains a large fat droplet that occupies almost the entire volume of the cell.

15.9 Lipases hydrolyze triglycerides.

15.11 A chylomicron is an aggregate of protein and triglycerides and carries triglycerides from the intestines to all body tissues by means of the bloodstream.

15.13 Colipase is a protein that binds to the surface of lipid droplets in the intestines and facilitates binding of pancreatic lipases. The lipases can then hydrolyze the ester linkages between the glycerol and fatty acids. This releases fatty acids and monoglycerides that can be absorbed by intestinal cells.

15.15 The β-oxidation of 5-phenylpentanoic acid produces benzoate ion and 2 acetyl CoA molecules.

15.17

$$C_3H_7CH_2CH_2\overset{O}{\overset{\|}{C}}{-}S{-}CoA \xrightarrow{FAD} C_3H_7CH{=}CH\overset{O}{\overset{\|}{C}}{-}S{-}CoA + FADH_2$$

$$\xrightarrow{H_2O} C_3H_7\overset{OH}{\overset{|}{C}H}{-}CH_2\overset{O}{\overset{\|}{C}}{-}S{-}CoA \xrightarrow[\text{coenzyme A}]{NAD^+,}$$

$$NADH + C_3H_7\overset{O}{\overset{\|}{C}}{-}S{-}CoA + CH_3\overset{O}{\overset{\|}{C}}{-}S{-}CoA$$

15.19 3 ATP molecules

15.21 Ketone bodies consist of β-hydroxybutyrate, acetoacetate, and acetone:

$$CH_3\overset{OH}{\overset{|}{C}H}CH_2\overset{O}{\overset{\|}{C}}{-}O^- \qquad CH_3{-}\overset{O}{\overset{\|}{C}}{-}CH_3 \qquad CH_3{-}\overset{O}{\overset{\|}{C}}{-}CH_2\overset{O}{\overset{\|}{C}}{-}O^-$$

β-Hydroxybutyrate   Acetone   Acetoacetone

15.23 When aerobic respiration of fatty acids is inhibited, ketogenesis takes place instead.

15.25 The body is able to take ketone bodies and reconvert them into acetyl CoA. The heart derives most of its metabolic energy from oxidation of ketone bodies. Other tissues, including the brain, increasingly rely on ketone bodies when glucose is unavailable. This occurs during starvation.

15.27 NADPH acts as a reducing agent during fatty acid synthesis. NADH is produced during fatty acid oxidation.

15.29 Fatty acid biosynthesis occurs in the cytoplasm; β-oxidation occurs in the mitochondria.

15.31 regulation of blood glucose concentration

15.33 The major fuel for the heart is ketone bodies. The brain relies on glucose for fuel. The liver's main source of fuel is the carbon skeletons of amino acids such as alanine.

15.35 Fatty acid biosynthesis occurs.

15.37 Preproinsulin contains an N-terminal sequence of 23 amino acids that directs movement of the protein during the early stages of synthesis. In proinsulin this signal sequence has been cleaved off. Proinsulin is the immediate precursor to insulin.

15.39 The hydrophobic amino acid residues of the signal sequence allow movement of preproinsulin through cellular membranes during the early stages of synthesis. This ensures that insulin will be properly secreted from the cell when it is called for.

15.41 Insulin helps regulate blood sugar levels. When the blood sugar level increases substantially from 10 mM, insulin production begins to lower blood glucose levels. Insulin stimulates glycogen production while concurrently inhibiting glycogenolysis and gluconeogenesis.

15.43 the inability to produce insulin

15.45 Ketoacidosis is a result of excessive ketone body production. Elevated ketone body concentrations overwhelm the buffering capacity of the blood, causing the blood pH to drop. Dehydration also results, as ketone bodies cause the kidneys to excrete huge amounts of water. Such conditions exist during starvation or with diabetics.

15.47 increases lipolysis

15.49 a. triglycerides b. lipase c. 12 d. muscle

15.51 Bile salts are able to form micelles that have both hydrophilic and hydrophobic moieties. The hydrophilic portion of the micelle is on the outer surface; the hydrophobic portion is in the center of the micelle. Cholesterol, being hydrophobic, will associate with the central hydrophobic portion of the micelle and thus be emulsified in water.

CHAPTER 16

16.1

Purine Pyrimidine

16.3 A *nucleoside* consists of a sugar (either ribose or deoxyribose) and a purine or pyrimidine. A *nucleotide* consists of a nucleoside with a phosphate ester bonded to one of the sugar hydroxyl groups.

16.5

16.7 a. *double helix:* The double helix is the structural feature comprising the "backbone" of DNA; the two antiparallel strands of nucleotides form the spiral molecule of DNA.

b. *template:* A template is a "master copy" from which new copies are made. In this case, DNA itself is the template used to replicate DNA. The double helix is pulled apart during DNA replication, much like opening a zipper. Each individual strand becomes the template for the complementary strand.

c. *the experiment of Meselson and Stahl:* This work gave experimental evidence that a parent DNA molecule replicates by opening the double helix into single strands, each strand acting as the template for synthesizing the new complementary strand. Thus each new DNA molecule will contain one strand from the parent DNA molecule, as proved by isotopic labeling. Cells containing "heavy" DNA (completely substituted with a "heavy" isotope of nitrogen) were allowed to grow one generation in a medium containing only "light" nitrogen. The isolated DNA of the daughter cells was found to contain one "heavy" strand and one "light" strand.

16.9 There are three major classes of RNA:

1. *messenger RNA (mRNA):* carries genetic information for a protein from DNA to the ribosomes
2. *ribosomal RNA (rRNA):* structural component of a ribosome that also plays a role in protein synthesis
3. *transfer RNA (tRNA):* translates the genetic code carried by mRNA into the primary sequence of amino acids in a protein

16.11 The genetic code is said to be degenerate because more than one set of triplet codons can code for a given amino acid.

16.13 Each ribosome is separated by approximately 80 nucleotide residues. For a 7200-nucleotide mRNA, one would find 7200/80, or 90, ribosomes.

16.15 The P-site and A-site are the two locations for binding tRNA molecules to the ribosome. The function of the P-site, called the peptidyl tRNA binding site, is to hold tRNA covalently bound to the growing protein chain. The function of the A-site, called the aminoacyl-tRNA binding site, is to hold the tRNA carrying the next amino acid to be added to the protein.

16.17 The ring structure of a pyrimidine is

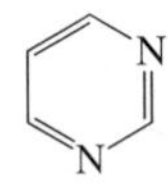

16.19 The adenine-thymine base pair is linked by two hydrogen bonds.

16.21 The adenine-thymine base pair is held together by only two hydrogen bonds, whereas the guanine-cytosine base pair is held together by three hydrogen bonds. Breaking two hydrogen bonds takes less energy than breaking three hydrogen bonds; thus it is easier to disrupt the adenine-thymine base pair than a guanine-cytosine base pair.

16.23

16.25 DNA replication is semiconservative because one strand of DNA serves as the template for the second strand (daughter strand). The bases along the parent strand define which base will be opposite during replication:

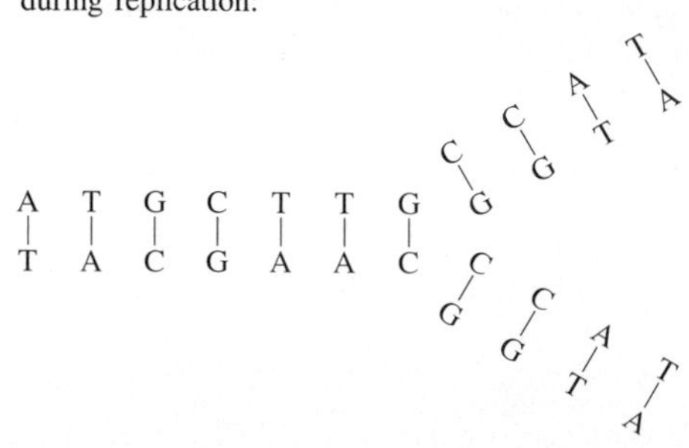

16.27 A replication fork is the site where deoxyribonucleotides are added to the growing daughter DNA strand.

16.29 Parent strand

ATGCGGCTAGAATATTCCATATGGCCGGTATTCG

| | | | | | | | | | | | | | | | | | | | | | | | | | | | | | | | | |

TACGCCGATCTTATAAGGTATACCGGCCATAAGC

Daughter strand

16.31 Ultraviolet (UV) light induces photodimerization of DNA. Sometimes DNA repair mechanisms make an error when trying to correct a pyrimidine dimer. Such an error results in a change in the nucleotide sequence of the DNA. By definition, this is a mutation.

16.33 Ultraviolet (UV) light causes photodimerization of DNA.

16.35 The disease xeroderma pigmentosum results from a genetic defect in the repair endonuclease gene or another DNA repair gene. Individuals with this disease are very susceptible to skin cancers.

16.37 Using sickle cell anemia as an example, the change of one amino acid in the primary sequence of a protein can vastly affect its function. The sequence change in sickle cell anemia consists of a valine substituting for a glutamic acid residue in the primary sequence. Glutamic acid has a side chain that can form negatively charged ionic bridges with oppositely charged residues. These interactions must be important in the tertiary (folding) structure of normal hemoglobin. Since valine has a nonpolar, hydrophobic side chain and therefore cannot form the same kind of interactions, the folding of the protein will be affected in the region of the valine residue.

16.39 The central dogma of molecular biology states that the direction in which genetic information is transferred is:
DNA ⟶ RNA ⟶ protein.

16.41
a. carries genetic information for a protein from DNA to the ribosomes
b. a structural and functional component of ribosomes
c. translates the genetic code carried by mRNA into the primary sequence of amino acids in a protein

16.43 There are 64 codons constituting the genetic code. Most amino acids are specified by more than one codon.

16.45

$$\begin{array}{c} \overset{+}{NH_3} \\ | \\ HC{-}CH_3 \\ | \\ O{=}C \\ | \\ O \\ | \\ \text{adenosine} \end{array}$$

16.47 Gene sequence

ATGCCTAGAGGCTGATATTGCTAGCTCTGGTCAT

mRNA sequence

UACGGAUCUCCGACUAUAACGAUCGAGACCAGUA

16.49 An intervening sequence is a noncoding sequence in a gene and is often referred to as an intron.

16.51 Properties of the genetic code are:
1. The genetic code is highly degenerate. The term "degenerate" means that there is more than one codon for most amino acids. Only methionine and tryptophan have only one codon.
2. The genetic code is resistant to mutation. For most amino acids, the first two nucleotides of the codon determine the identity of the amino acid; the third nucleotide of the codon is variable. Thus a point mutation of the third nucleotide will not affect which amino acid is incorporated in the primary sequence of the protein.

16.53 The aminoacyl-tRNA synthetase recognizes the correct amino acid for a specific tRNA and covalently links this amino acid to the 3′ end of that tRNA.

16.55 In bacteria, the first amino acid in a protein is N-formylmethionine. The N-formylmethionyl-tRNA carries this amino acid and is often called the initiator tRNA.

16.57 Elongation of the peptide chain occurs in three steps:
1. An aminoacyl-tRNA enters the empty A-site.
2. The peptide chain is shifted to the tRNA occupying the A-site as a result of the formation of a peptide bond between the amino acid bound to the aminoacyl-tRNA and the nascent peptide chain. This is mediated by the catalytic action of peptidyl transferase.
3. The uncharged tRNA leaves, and the ribosome moves to the next codon.

16.59 The two tRNA binding sites on the ribosome are the aminoacyl-tRNA binding site (A-site) and the peptidyl-tRNA site (P-site).

16.61 Break the genetic code into codons, and identify the amino acid encoded by each codon in the sequence.

AUG UGU AGU GAC CAA CCG AUU UCA CUG UGA

met --- cys --- ser --- asp --- gln --- pro --- ile --- ser --- leu   Stop codon

(Initiator)

└──────── Protein ────────┘

16.63 Reverse transcriptase is able to produce a double stranded DNA copy of the original single stranded RNA. It does so by reading a single strand of RNA to produce a single strand of DNA. The RNA strand is then destroyed and the newly synthesized DNA is read to produce the complementary strand of DNA.

16.65 Mutagens are chemicals that cause changes in the nucleotide sequence of cellular DNA (mutations). If the mutation occurs in a gene involved in the control of normal cell division, the result will be uncontrolled cell division, cancer. Thus, a mutagen is often also a carcinogen.

16.67 The genes for the insulin A and B chains were synthesized with an AUG initiator codon and a UGA terminator codon added to each gene. Each synthesized gene was then cloned into a bacterial plasmid near a promoter that could be turned on or off, at will, by the addition or removal of an inducer compound. Finally the recombinant plasmids were introduced into bacteria. When the bacteria were grown in the presence of the inducer, they produced large amounts of the insulin A or B peptides. The purified A and B peptides were mixed and allowed to self-assemble into the functional insulin protein.

CHAPTER 17

17.1 Complex sugars, such as starch and cellulose, not only provide the body with a source of energy, but also act as a source of fiber.

17.3 Saturated fatty acid intake should be low in one's diet. Studies have indicated a strong correlation between high levels of saturated fatty acids and heart disease. Recent studies also have linked saturated fatty acids to increased incidence of various cancers such as colon, esophageal, stomach, and breast cancers.

17.5 Vegetables will vary in amino acid composition. By eating a variety of vegetables, one can ensure that all amino acid requirements are met.

17.7 meats, eggs, legumes, and other vegetables

17.9
a. major component of teeth and bones; factor in the clotting of blood and muscle function
b. along with calcium, a major component of bone; also found in nucleic acids and other biologically important molecules
c. one of the electrolytes found in blood; helps maintain the proper fluid level inside and outside of the cell
d. along with sodium and chloride, the third common component of electrolytes found in blood

17.11 Sulfanilimide drugs bind to the active site of an enzyme in the pathway for folic acid biosynthesis, inhibiting the production of this vitamin. The bacteria must produce their own folic acid; humans obtain it from the diet. Thus, the bacteria are selectively killed.

17.13
a. calcium, vitamins A and D (as supplements), phosphorus, protein, and riboflavin
b. protein, thiamine, niacin, vitamin B_{12}, iron, and other minerals
c. carotene (Vitamin A precursor), vitamin C, water-soluble vitamins, protein, fiber
d. carbohydrates, fiber, and protein

17.15 Oxidation of fats produces more energy than oxidation of carbohydrates.

17.17 Nonessential amino acids can be manufactured by one's body. The eight essential amino acids cannot be made by the body and must be supplied by one's diet.

17.19 coenzyme A

17.21 pellagra

17.23 milk, eggs, and liver

17.25 beriberi, neuritis, and mental disturbances

17.27 important substrate in the production of folic acid

17.29 Sulfa drugs are antimetabolites because they function as structural analogues of para-aminobenzoic acid (PABA). Since they are structurally similar to PABA, they enter the enzyme active site, but no bond can be formed between pterin and the sulfa drug. Thus the synthesis of folic acid is inhibited when sulfa drugs are present.

17.31 Although biotin is abundant in egg yolks, raw egg whites contain a protein called avidin, which binds tightly to biotin, making it unavailable to the body. This protein functions when the egg is raw; cooking the egg destroys avidin and releases biotin.

17.33 pernicious anemia

17.35 scurvy

17.37 carotene

17.39 vitamin A deficiency

17.41 anti-oxidant for cell membrane lipids

17.43 clotting of blood

17.45 vitamin A

17.47 They are excreted in one's urine.

17.49 Much of the thiamine present in rice is found in the seed coat. Polishing rice removes the seed coat and thereby removes most of the thiamine.

17.51 By cooking an egg, the avidin binding the biotin is destroyed, and biotin is released and made available as a nutrient.

17.53 niacin

17.55 cobalt, zinc, copper, iron, iodine, fluoride, nickel, and selenium

17.57 The major source of magnesium in one's diet is green, leafy vegetables. These green plants contain large amounts of chlorophyll. Each chlorophyll molecule contains a magnesium ion.

Credits

Table and Figure Credits

Page 218, Figure 9.11: From Stuart Ira Fox, *Human Physiology,* 3d ed. Copyright © 1990 Wm. C. Brown Communications, Inc., Dubuque, Iowa. All Rights Reserved. Reprinted by permission.

Page 222, Figure 9.15: From Stuart Ira Fox, *Human Physiology,* 3d ed. Copyright © 1990 Wm. C. Brown Communications, Inc., Dubuque, Iowa. All Rights Reserved. Reprinted by permission.

Page 228, Figure 9.23: From David C. Nieman, Diane E. Butterworth, and Catherine N. Nieman, *Nutrition,* Revised First Edition. Copyright © 1992 Wm. C. Brown Communications, Inc., Dubuque, Iowa. All Rights Reserved. Reprinted by permission.

Page 266, Figure 11.10: From Lubert Stryer, *Biochemistry,* 3d ed. Copyright © 1988 Lubert Stryer. W.H. Freeman and Co., New York.

Page 267, Figure 11.12: From Eugene Morkin, ''Contractile Proteins of the Heart'' in *Hospital Practice,* Vol. 18, issue 6, pages 97–112. Copyright © HP Publishing Company, New York.

Page 269, Figure 11.16: From Sylvia S. Mader, *Biology,* 3d ed. Copyright © 1990 Wm. C. Brown Communications, Inc., Dubuque, Iowa. All Rights Reserved. Reprinted by permission.

Page 270, Figure 11.19: From Robert F. Weaver and Philip W. Hedrick, *Genetics*. Copyright © 1989 Wm. C. Brown Communications, Inc., Dubuque, Iowa. All Rights Reserved. Reprinted by permission.

Page 271, Figure 11.21: From Stuart Ira Fox, *Human Physiology,* 3d ed. Copyright © 1990 Wm. C. Brown Communications, Inc., Dubuque, Iowa. All Rights Reserved. Reprinted by permission.

Page 273, top left and right: From Robert M. Coleman, Mary F. Lombard, Raymond E. Sicard, and Nicholas J. Rencricca, *Fundamental Immunology*. Copyright © 1989 Wm. C. Brown Communications, Inc., Dubuque, Iowa. All Rights Reserved. Reprinted by permission.

Page 273, bottom: From Robert F. Weaver and Philip W. Hedrick, *Genetics*. Copyright © 1989 Wm. C. Brown Communications, Inc., Dubuque, Iowa. All Rights Reserved. Reprinted by permission.

Page 305, Figure 12.16: From A. Marchard, ''Case of the Month, Circulating Anticoagulants: Chasing the Diagnosis,'' *Diagnostic Medicine,* June 1983, p. 14. Adapted by permission.

Page 308, Figure 12.20: From John W. Hole, Jr., *Human Anatomy and Physiology,* 5th ed. Copyright © 1990 Wm. C. Brown Communications, Inc., Dubuque, Iowa. All Rights Reserved. Reprinted by permission.

Page 323, Figure 13.6: From Kent M. Van De Graaff, *Human Anatomy,* 3d ed. Copyright © 1992 Wm. C. Brown Communications, Inc., Dubuque, Iowa. All Rights Reserved. Reprinted by permission.

Page 380, box table: Courtesy of Metropolitan Life Insurance Company.

Page 381, box table: From American Health Foundation, *The Book of Health*. Copyright © 1981 American Health Foundation, Franklin Watts, Inc. Reprinted by permission of Franklin Watts, Inc.

Page 382, Figure 15.3: From Stuart Ira Fox, *Human Physiology,* 3d ed. Copyright © 1990 Wm. C. Brown Communications, Inc., Dubuque, Iowa. All Rights Reserved. Reprinted by permission.

Page 383, Figure 15.5: From Stuart Ira Fox, *Human Physiology,* 3d ed. Copyright © 1990 Wm. C. Brown Communications, Inc., Dubuque, Iowa. All Rights Reserved. Reprinted by permission.

Page 385, Figure 15.7: From Stuart Ira Fox, *Human Physiology,* 3d ed. Copyright © 1990 Wm. C. Brown Communications, Inc., Dubuque, Iowa. All Rights Reserved. Reprinted by permission.

Page 420, Figure 16.14: From Robert F. Weaver and Philip W. Hedrick, *Genetics*. Copyright © 1989 Wm. C. Brown Communications, Inc., Dubuque, Iowa. All Rights Reserved. Reprinted by permission.

Page 428, Figure 16.20: From Sylvia S. Mader, *Biology,* 3d ed. Copyright © 1990 Wm. C. Brown Communications, Inc., Dubuque, Iowa. All Rights Reserved. Reprinted by permission.

Photo Credits

Table of Contents

Introduction: Tom McHugh/National Museum of Natural History; Chapter 2: Eric Kamp/Phototake; Chapter 5: Martin Rotker/ Phototake; Chapter 8: Daniel Brody/Stock Boston; Chapter 9: Hans Pfletschinger/Peter Arnold; Chapter 10: Jackie Lewin, Royal Free Hospital/Science Library/Photo Researchers; Chapter 11: David Scharf/Peter Arnold; Chapter 12: R. Feldmann/D. McCoy/ Rainbow; Chapter 14: CNRI/Phototake; Chapter 17: Manfred Kage/ Peter Arnold

Text

Page 5: Richard Feldmann/Phototake; **Page 62:** Ellis Herwig/Stock Boston; **Page 123:** Elizabeth Wilcox/Phototake; **Page 173:** R. Feldmann/D. McCoy/Rainbow; **Page 214:** Martin Rotker/ Phototake; **Page 219:** R. H. Albertin; **Page 220:** Hans Pfletschinger/Peter Arnold; **Page 228:** Robert Becker, Ph. D./ Custom Medical Stock Photo; **Page 274:** SIU/Peter Arnold; David Scharf/Peter Arnold; Jackie Lewin, Royal Free Hospital/Science Photo Library/Photo Researchers; **Page 311:** Hank Morgan/Science Source/Photo Researchers; Hank Morgan/Photo Reserchers; **Page 352:** C. CNRI/Phototake; **Page 362:** top, CNRI/Phototake; bottom, Dr. Luz Mangurian/Dept. of Biological Sciences, Towson State University

Index

PRINCIPAL FUNCTIONAL GROUPS IN ORGANIC COMPOUNDS

Type of Compound	Structural Formula	Condensed Formula	Chapter Reference	Example		
				Structural Formula	IUPAC Name	Common Name
Alcohol	$R—O—H$	ROH	12	$CH_3CH_2—O—H$	Ethanol	Ethyl alcohol
Aldehyde	$R—\overset{O}{\overset{\|}{C}}—H$	$RCHO$	13, 14	$CH_3\overset{O}{\overset{\|}{C}}—H$	Ethanal	Acetaldehyde
Amide	$R—\overset{O}{\overset{\|}{C}}—\overset{H}{\overset{\|}{N}}—H$	$RCONH_2$	17	$CH_3\overset{O}{\overset{\|}{C}}—\overset{H}{\overset{\|}{N}}—H$	Ethanamide	Acetamide
Amine	$R—\overset{H}{\overset{\|}{N}}—H$	RNH_2	17, 18	$CH_3CH_2\overset{H}{\overset{\|}{N}}—H$	Aminoethane	Ethyl amine
Carboxylic acid	$R—\overset{O}{\overset{\|}{C}}—O—H$	$RCOOH$	15, 16, 18	$CH_3\overset{O}{\overset{\|}{C}}—O—H$	Ethanoic acid	Acetic acid
Ester	$R—\overset{O}{\overset{\|}{C}}—O—R'$	$RCOOR'$	15, 16, 23	$CH_3\overset{O}{\overset{\|}{C}}—OCH_3$	Methyl ethanoate	Methyl acetate
Ester	$R—O—R'$	ROR'	12	CH_3OCH_3	Methoxymethane	Dimethyl ether
Halide	$—Cl$ (or $—Br$, $—F$, $—I$	RCl	10	CH_3CH_2Cl	Chloroethane	Ethyl chloride
Ketone	$R—\overset{O}{\overset{\|}{C}}—R'$	$RCOR'$	13, 14	$CH_3\overset{O}{\overset{\|}{C}}CH_3$	Propanone	Acetone